15th International Conference on X-ray Absorption Fine Structure

(XAFS15)

Journal of Physics: Conference Series Volume 430

Beijing, China
22-28 July 2012

ISBN: 978-1-62748-307-0
ISSN: 1742-6588

Printed from e-media with permission by:

Curran Associates, Inc.
57 Morehouse Lane
Red Hook, NY 12571

Some format issues inherent in the e-media version may also appear in this print version.

Printed by Curran Associates, Inc. (2013)

For permission requests, please contact the Institute of Physics
at the address below.

Institute of Physics
Dirac House, Temple Back
Bristol BS1 6BE UK

Phone: 44 1 17 929 7481
Fax: 44 1 17 920 0979

techtracking@iop.org

Additional copies of this publication are available from:

Curran Associates, Inc.
57 Morehouse Lane
Red Hook, NY 12571 USA
Phone: 845-758-0400
Fax: 845-758-2634
Email: curran@proceedings.com
Web: www.proceedings.com

TABLE OF CONTENTS

THEORY

DATA ANALYSIS

NEW TECHNOLOGY AND DEVICES

APPLICATIONS IN NANOSCIENCE AND TECHNOLOGY

APPLICATIONS IN LIFE SCIENCE

APPLICATIONS IN CHEMISTRY

APPLICATIONS IN CATALYTIC SCIENCE

APPLICATIONS IN SURFACE AND INTERFACE SCIENCE

APPLICATIONS IN MATERIALS SCIENCE

APPLICATIONS IN ENERGY AND ENVIRONMENTAL SCIENCE

APPLICATIONS IN MAGNETIC AND RELATED MATERIALS SCIENCE

APPLICATIONS IN NUCLEAR SCIENCE

APPLICATIONS IN DISORDERED SYSTEMS

APPLICATIONS IN EXTREME CONDITIONS

APPLICATIONS FOR TIME-RESOLVED EXPERIMENTS

XMCD TECHNOLOGY AND ITS APPLICATIONS

ADVANCED METHODS

XAFS COMBINED WITH OTHER EXPERIMENTAL METHODS

OTHER RELATED STUDIES

PREFACE

The 15th International Conference on X-ray Absorption Fine Structure (XAFS15) was held on 22-28 July 2012 in Beijing, P. R. China. About 340 scientists from 34 countries attended this important international event.

Figure 1. Main hall of XAFS15.

The rapidly increasing application of XAFS to the study of a large variety of materials and the operation of the new SR source led to the first meeting of XAFS users in 1981 in England. Following that a further 14 International Conferences have been held. Comparing a breakdown of attendees according to their national origin, it is clear that participation is spreading to include attendees from more and more countries every year.

The strategy of development in China of science and education is increasing quickly thanks to the large investment in scientific and technological research and infrastructure. There are three Synchrotron Radiation facilities in mainland China, Hefei Light Source (HLS) in the National Natural Science Foundation of China (NSRL), Beijing Synchrotron Radiation Facility (BSRF) in the Institute of High Energy Physics, and Shanghai Synchrotron Radiation Facility (SSRF) in the Shanghai Institute of Applied Physics. More than 10000 users and over 5000 proposals run at these facilities. Among them, many teams from the USA, Japan, German, Italy, Russia, and other countries. More than 3000 manuscript were published in SCI journals, including (incomplete) Science (7), Nature (10), Nature Series (7), PNAS (3), JACS (12), Angew. Chem. Int. Ed. (15), Nano Lett. (2), etc.

In XAFS15, the participants contributed 18 plenary invited talks, 16 parallel invited talks, 136 oral presentations, 12 special talks, and 219 poster presentations. Wide communication was promoted in the conference halls, the classical banquet restaurant, and the Great Wall.

Figure 2. Parallel hall **Figure 3. Communication** **Figure 4. Poster room**

This volume contains 136 invited and contributed papers, accepted after a rigorous peer review procedure. A group of about 90 outstanding scientists in the field reviewed and suggested revisions of the manuscripts to improve scientific presentation. As a result, we believe the entire volume has reached a high standard. The 19 topics covered are listed as follows:

- Theory
- Data analysis
- New technology and devices of XAFS
- Applications in Nano science and technology
- Applications in Life Science
- Applications in Chemistry
- Applications in Catalytic Science
- Applications in Surface and Interface Science
- Applications in Material Science
- Applications in Energy and Environmental Science
- Applications in Magnetic and Related Material Science
- Applications in Nuclear Science
- Applications in Disordered Systems
- Applications in Extreme Conditions
- Applications for Time-resolved experiments
- XMCD technology and its applications
- Advanced methods (e.g., new coherent sources and spectroscopic imaging techniques)
- XAFS combined with other experimental methods
- Other related studies

We hope this volume will be a useful reference for the ongoing scientific activity in XAFS.

We would also like to express our sincere appreciation to the sponsors for their generous support: Chinese Academy of Science, National Natural Science Foundation of China, China Center of Advanced Science and Technology World Laboratory, University of Science and Technology of China, National Synchrotron Radiation Laboratory, Institute of High Energy Physics Chinese Academy of Sciences, and our commercial sponsors (AREVA, Xi'an Action Power Electric Co., Ltd).

Finally, we would like to acknowledge the entire local organizing staff (names are given below) and particularly the collaborators and members of the XAS group at the National Natural Science Foundation of China and Institute of High Energy Physics Chinese Academy of Sciences for their efforts to make the XAFS15 conference a success.

Ziyu Wu
Chair of the Conference and Proceedings Editor
Hefei, P. R. China, 28 September 2012

Committees and Staff

Chair of the Conference
Ziyu Wu

International Advisory Committee
Adam Hitchcock, Canada
Adriano Filipponi, Italy
Alain Manceau, France
Alexander Soldatov, Russia
Andrea Di Cicco, Italy
Britt Hedman, USA
Bruce Bunker, USA
Calogero R. Natoli, Italy
Christopher T. Chantler, Australia
Frank M. F. De Groot, Netherlands
Hiroyuki Oyanagi, Japan
Ingolf Lindau, USA
J. Mustre de Leon, México
James E Penner-Hahn, USA
Joaquin Garcia Ruiz, Spain
John Evans, UK
John J. Rehr, USA
Kiyotaka Asakura, Japan
Majed Chergui, Switzerland
Mark Newton, UK
Shiqiang Wei, P. R. China
Tsun-Kong Sham, Canada
Ziyu Wu, P. R. China

International Program Committee
Antonio Bianconi, Italy
Augusto Marcelli, Italy
Emad Flear Aziz, Germany
Jinghua Guo, USA
Joly Yves, France
Masaharu Nomura, Japan
Maurizio Benfatto, Italy
Pieter Glatzel, France
Shiqiang Wei, China
Tiandou Hu, China
Toshihiko Yokoyama, Japan

Way-Faung Pong, Taiwan
Xinyi Zhang, China
Yi Xie, China
Yuying Huang, China
Zhonghua Wu, China
Ziyu Wu, China

Local Organizing Committee
Bo He
Fengchun Hu
Haifeng Zhao
Jing Zhang
Meijuan Yu
Qin Yu
Shuo Zhang
Wangsheng Chu
Wei He
Wei Xu
Wensheng Yan
Xiaomei Gong
Xing Chen
Yang Zou
Yi Xia
Zheng Jiang
Zhi Xie
Zhihu Sun
Zhiyun Pan

Additional Staff
Chengxun Liu

Peer review statement

All papers published in this volume of *Journal of Physics: Conference Series* have been peer reviewed through processes administered by the proceedings Editors. Reviews were conducted by expert referees to the professional and scientific standards expected of a proceedings journal published by IOP Publishing.

New Developments in FEFF: FEFF9 and JFEFF

Kevin Jorissen and John J. Rehr

Department of Physics, University of Washington, Seattle, WA 98195, USA

kevinjorissen.pdx@gmail.com

Abstract. The ab initio core-level spectroscopy code FEFF9 has seen many new developments in recent years. We describe the addition of new physics and new features designed to calculate more accurate spectra. We also present the user-friendly Java-based GUI JFEFF that simplifies running FEFF on platforms ranging from personal computers to high-performance parallel systems and virtual cloud platforms.

1. New developments in FEFF9

FEFF is an implementation of Real-Space Multiple Scattering theory (RSMS) and is capable of calculating a wide range of x-ray and electron spectra. While it was originally a single-purpose EXAFS code [1], the current version FEFF9 [2,3] calculates self-consistent electronic structure over a wide energy range and produces XANES, XES, EELS, NRIXS and Compton scattering, as well as the local Density of States. FEFF was designed from the ground up for excited states properties within a relativistic, all electron framework that includes inelastic losses, self-energy effects, and vibrational damping.

Recently the treatment of the self-energy has been improved by adding a "many-pole self-energy" model (MPSE) [4] that tends to correct the width and energy position of features in the spectrum. Debye-Waller factors can now be obtained from ab initio codes to take vibrational damping of the spectrum into account [5], leading to improved quantification of EXAFS data.

FEFF has traditionally been a real-space code and can therefore treat molecules and aperiodic materials, unlike band-structure codes. FEFF9 can also treat crystals in reciprocal space, leading to faster and more accurate results for such systems.[6] Moreover a core hole can be added for Final State Rule (FSR) calculations without the cumbersome introduction of a supercell. The k-space input has been much improved over previous versions and now accepts standardized CIF files, which are available in large online crystal structure databases. Previous bugs were eliminated.The code has been updated to the FORTRAN90 standard, leading to more efficient memory management. MPI functionality allows the most time-consuming algorithms to be distributed over a parallel environment providing excellent speedup characteristics.

2. The JFEFF GUI

Many state-of-the-art materials science codes are difficult to use: they require that a user learn the finesses of computational work (e.g. working on a Linux command line interface (CLI), or using a Fortran compiler), as well as the peculiarities of a specific code. This is a barrier that deters occasional users. Steps have been taken to reduce this barrier for the FEFF9 code.

To address this difficulty we have developed a Java Graphical User Interface (GUI), JFEFF. (Fig. 1) This GUI can read all FEFF9 input files and allows the user to set all FEFF9 options. It is capable of running the FEFF9 code in a variety of ways, e.g. in serial or parallel mode, either on the local computer or on a remote computer using SSH access over a network. Finally, JFEFF can display the results of the calculation. All this functionality is accessible without a CLI and it makes FEFF9 a useful tool even for researchers who are not specialized computational scientists. JFEFF works across all major platforms (Linux, MS Windows, Mac OS X).

Documentation for the FEFF9 code is at http://www.feffproject.org/feff/html/documentation.html. The JFEFF installer, which is available from the FEFF project website (http://www.feffproject.org), installs not only the Java GUI and compiled FEFF9 binaries, but also a Users Guide in PDF format and a collection of 20 example calculations with all input and output files. Separate installers are available for Windows, OS X, and Linux. JFEFF also offers live documentation: when hovering the mouse over an input option, the appropriate documentation is shown in a help window.

3. Cloud Computing Platform

Cloud Computing refers to the use of virtualized online resources (e.g., the Amazon Web Services (AWS) Elastic Compute Cloud (EC2) [7]) for computational work. This is a very attractive paradigm for users who don't own high-performance computing (HPC) hardware and do not have access to traditional computing facilities. For users whose computational needs fluctuate (e.g. researchers who occasionally need model calculations for their measurements), it can be far more attractive to rent resources per hour than to invest in ownership, especially when the burden of maintenance, configuration and optimization is born by the provider.

We have developed a Scientific Cloud Computing (SCC) platform [8], which is integrated in JFEFF. The user can simply select to run a calculation in the cloud. The GUI takes care of everything else: virtual machines are launched in the EC2 cloud, preconfigured with the FEFF9 code and a fully functional parallel environment. JFEFF then executes and monitors the calculation in the cloud and brings the results back to the local machine, where they are stored and displayed to the user. Meanwhile the user never has to leave the GUI environment or perform any kind of configuration. JFEFF has a "demo mode" that allows a new user to try a few SCC calculations. For ongoing access the user must set up an AWS account with modest per-hour charges.

4. Conclusion

The FEFF9 code is a significantly improved version of the ab initio spectroscopy code FEFF. In particular, FEFF9 includes JFEFF, a graphical user interface that greatly simplifies the calculation of x-ray and other spectra. JFEFF is well documented and comes with many examples. The built-in, automated Scientific Cloud Computing platform offers convenient and cheap HPC access. As a result, FEFF9 is not only a more powerful package for spectroscopy simulation, but is also much easier to use.

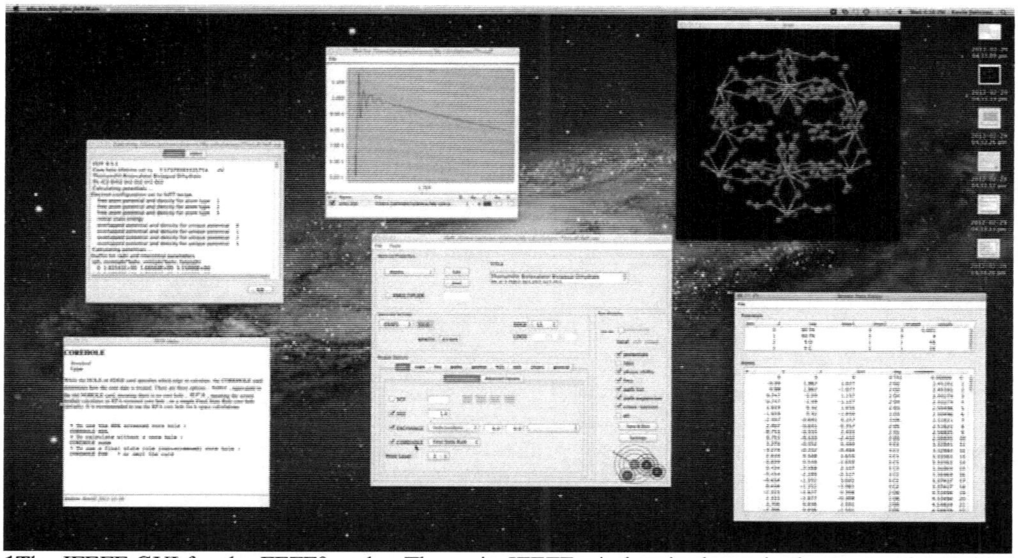

Figure 1 The JFEFF GUI for the FEFF9 code. The main JFEFF window is shown in the centre. Around it from left to right are the help window (showing documentation on the COREHOLE card, which the mouse is hovering over in the main window) ; the runtime window (showing runtime information on the FEFF9 calculation); the plot window (showing the EXAFS spectrum calculated by FEFF9); the JMol window (showing the material visually); and the atoms window (showing xyz Cartesian coordinates for the structure). For calculations using periodic boundary conditions in k-space, the atoms window is unavailable. The atom positions would be specified through a cif file instead.

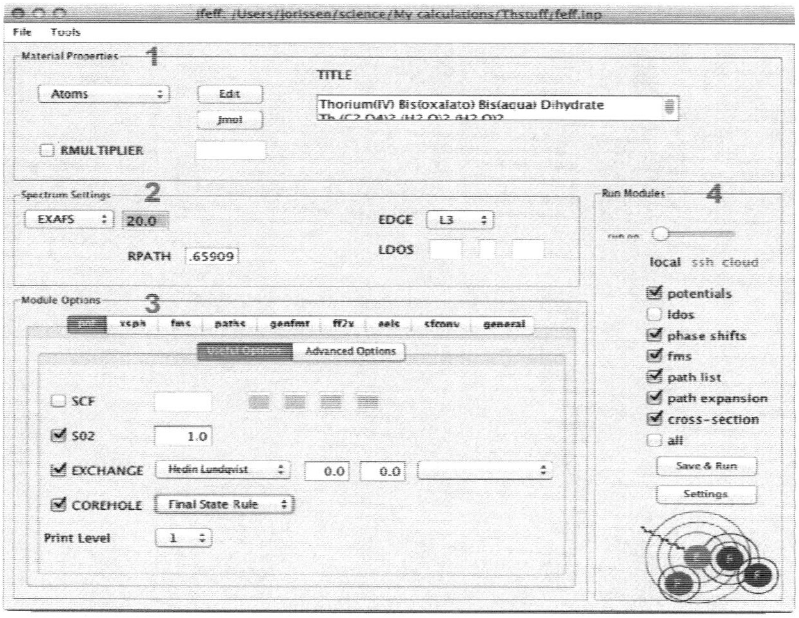

Figure 2 In JFEFF, the input for a FEFF9 calculation is organized in four panels. 1- the material ; 2- the spectrum ; 3- technical options for FEFF9 ; 4- the runtype. An existing calculation can be loaded from the menu (File/Open). The slider in panel 4 controls where the actual calculation takes place: on the local computer, on a remote computer, or on the EC2 cloud. Other new JFEFF functionality includes EELS calculations, NRIXS calculations, and reciprocal space calculations for crystals. Once all options are configured correctly, clicking the "Save & Run" button starts the calculation.

Acknowledgments

Primary support for the FEFF project comes from DOE-BES grant DE-FG03-97ER45623. Primary support for the UW Scientific Cloud Computingproject comes from NSF grant OCI-1048052. Additional EC2 cloud computer time is provided by an Amazon AWS in Education research grantPC3VBYVHQ3TASL8.

References

[1] Stern, E. A., D. E. Sayers, and F. W. Lytle, 1975, Phys. Rev. B **11**, 4836.

[2] Rehr, J. J., Kas, J. J., Vila, F. D., Prange, M. P., and Jorissen, K., Parameter-free calculations of X-ray spectra with FEFF9, Physical Chemistry Chemical Physics 12 (2010) 5503-5513.

[3] Rehr, John J., Kas, Joshua J., Prange, Micah P., Sorini, Adam P., Takimoto, Yoshinari, and Vila, Fernando, Ab initio theory and calculations of X-ray spectra, Comptes Rendus Physique 10 (2009) 548-559.

[4] Kas, J., Sorini, A., Prange, M., Cambell, L., Soininen, J., and Rehr, J., Many-pole model of inelastic losses in x-ray absorption spectra, Physical Review B 76 195116 (2007).

[5] Vila, F., Lindahl, V., and Rehr, J., X-ray absorption Debye-Waller factors from ab initio molecular dynamics, Physical Review B 85 024303 (2012).

[6] Jorissen, K. and Rehr, J. J., Calculations of electron energy loss and x-ray absorption spectra in periodic systems without a supercell, Physical Review B 81 245124 (2010).

[7] Amazon Web Services, http://aws.amazon.com/ (2011).

[8] Jorissen, K., Vila, F., and Rehr, J. J., A high performance scientific cloud computing environment for materials simulations, Computer Physics Communications 183 1911-19 (2012).

15th International Conference on X-ray Absorption Fine Structure (XAFS15) IOP Publishing
Journal of Physics: Conference Series **430** (2013) 012002 doi:10.1088/1742-6596/430/1/012002

Theory of Electric Quadrupole Transition in the XMCD Spectra at the Yb L_3 Edge of Mixed Valence Yb Compounds in High Magnetic Fields

Akio Kotani[1,2]

[1] Photon Factory, High Energy Accelerator Research Organization, 1-1 Oho, Tsukuba, Ibaraki 305-0801, Japan

[2] RIKEN SPring-8 Center, 1-1-1 Kouto, Sayo, Hyogo 679-5148, Japan

E-mail: kotani@post.kek.jp

Abstract. The spectra of X-ray magnetic circular dichroism by the electric quadrupole transition (EQ-XMCD) are theoretically studied at the Yb L_3 edge of mixed valence compounds, $YbInCu_4$ and $YbAgCu_4$, around the field-induced valence transition in high magnetic fields. Using a new formulation with a single impurity Anderson model, which directly connects the EQ-XMCD spectra with the weights of the Yb $4f$ angular momentum j_z in the ground state, we show that the EQ-XMCD intensity of $YbAgCu_4$ is more than twice as large as that of $YbInCu_4$. This explains the reason for that the EQ-XMCD spectra was very recently measured successfully for $YbAgCu_4$, whereas no experimental evidence of EQ-XMCD has so far been obtained for $YbInCu_4$. Different EQ-XMCD intensities for the two materials and a peculiar field dependence of EQ-XMCD intensities are discussed on the basis of the weights of j_z in the ground state.

1. Introduction

X-ray magnetic circular dichroism (XMCD) gives us the bulk-sensitive, element-specific and site-selective information on electronic states in ferromagnetic and ferrimagnetic materials [1]. Most of mixed valence rare earth compounds are paramagnetic, but by applying high magnetic fields they are magnetized so that the measurements of XMCD become possible [2].

Recently experimental observations of X-ray absorption spectroscopy (XAS) and its XMCD were successfully made at the Yb L_3 edge of mixed valence Yb compounds, $YbInCu_4$ and $YbAgCu_4$, by utilizing pulsed high magnetic fields up to about 40T [3, 4, 5, 6]. By applying magnetic fields, the Yb valence number of $YbInCu_4$ and $YbAgCu_4$ changes rapidly around 30 T, which is denoted by the field-induced valence transition. Investigations of XMCD spectra for these materials have so far been made for the electric dipole (ED) transition measured around the field-induced valence transition, but another interesting subject is those due to the electric quadrupole (EQ) transition, which is stimulated by new experimental findings [6].

For $YbInCu_4$, the present author analyzed theoretically the experimental XMCD spectra [4] due to the ED transition with a single impurity Anderson model (SIAM) [7], but no clear signature of the XMCD due to the EQ transition (denoted by EQ-XMCD) was recognized in the experimental data. For $YbAgCu_4$, on the other hand, the EQ-XMCD spectra have been measured very recently by Nakamura *et al.*, and the confirmation of the EQ-XMCD spectra

Journal of Physics: Conference Series **430** (2013) 012002 doi:10.1088/1742-6596/430/1/012002

has also been made by theoretical calculations (by the present author) [6]. It is necessary to calculate the EQ-XMCD spectra for both $YbInCu_4$ and $YbAgCu_4$ and compare the results, in order to understand why the EQ-XMCD spectra could be measured for $YbAgCu_4$ but could not for $YbInCu_4$. For this purpose, we present in this paper a new formulation of the EQ-XMCD spectra with SIAM, where the EQ-XMCD spectra are directly connected with the weights of the Yb $4f$ angular momentum j_z in the ground state, by extending the theory of XMCD spectra due to the ED [2] and EQ [6] transitions. A different field dependence in the calculated EQ-XMCD spectra for $YbInCu_4$ and $YbAgCu_4$ will be explained on the basis of the difference in the field-induced valence transition of the two materials, which is clearly seen from the field dependence of the weights of j_z in the ground state.

2. Model and Formulation

As a model to describe the XMCD spectra by the EQ transition at the Yb L_3 edge of $YbInCu_4$ and $YbAgCu_4$ in high magnetic fields, we use SIAM [8, 9]. The Hamiltonian of the system is written, in the hole picture, as [7]

$$\begin{aligned} H &= \sum_{\nu}(\epsilon_{f,\nu} + g_J \mu_B j_z h) f_\nu^\dagger f_\nu - \sum_{\mu} \epsilon_{p,\mu} p_\mu^\dagger p_\mu + \sum_{k,\nu} \epsilon_k a_{k,\nu}^\dagger a_{k,\nu} \\ &+ \sum_{k,\nu}(V_k a_{k,\nu}^\dagger f_\nu - V_k^* f_\nu^\dagger a_{k,\nu}) + U_{ff} \sum_{\nu>\nu'} f_\nu^\dagger f_\nu f_{\nu'}^\dagger f_{\nu'}, \end{aligned} \tag{1}$$

where the first, second and third terms represent, respectively, the single hole energies of the Yb $4f$ states, the Yb $2p$ states and the ligand band, the fourth term is the hybridization between Yb $4f$ and ligand band states, and the last term represents the Coulomb interaction between Yb $4f$ holes. Here, g_J is the Landé g-facter, μ_B is the Bohr magneton, h is the external magnetic field, and the indices ν and μ denote a set of quantum numbers (j, j_z) and (j', j'_z) of the $4f$ (and ligand band) and $2p$ states, respectively, with the JJ coupling scheme (j and j' being fixed at 7/2 and 3/2). For more details on the model, see ref. [7].

The effect of the hybridization between the $4f$ and ligand-band states is taken into account within the lowest order approximation of the $1/N_f$ expansion method [8], where N_f (= 8 for j =7/2) is the number of degeneracy of the $4f$ state in zero magnetic field. We disregard the effect of the $4f^2$ hole configuration by putting $U_{ff} = \infty$. The ground state $|g\rangle$ (with energy E_g) is obtained by diagonalizing the Hamiltonian H in the form:

$$|g\rangle = c_0^{(g)}|0\rangle + \sum_{k,j_z} c_{k,j_z}^{(g)} |k, j, j_z\rangle. \tag{2}$$

Here, $|0\rangle$ is the state where all the holes (except for $4f$ holes) below the Fermi level ϵ_F (= 0) are occupied, and $|k, j, j_z\rangle = f_{j,j_z}^\dagger a_{k,j,j_z}|0\rangle$.

By the EQ transition, a $4f$ hole with $\nu = (7/2, j_z)$ makes a transition to the $2p_{3/2}$ state with $\mu = (3/2, j'_z)$, so that the final state is expressed as

$$|f_{k,\nu,\mu}\rangle = p_\mu^\dagger a_{k,\nu}|0\rangle, \tag{3}$$

where the hole with (k, ν) in the ligand band corresponds to a hole included in the state $|k, j, j_z\rangle$ in the ground state $|g\rangle$. The energy $E_{k,\nu,\mu}$ of this final state depends on k but does not depend on ν and μ because of the degeneracy. The XAS spectrum by the EQ transition for $\pm$ helicity light is given by

$$F^\pm(\omega) = \sum_{k,\nu,\mu} |\langle f_{k,\nu,\mu}|M_{\rm EQ}^\pm(\nu;\mu)|g\rangle|^2 \frac{\Gamma/\pi}{(\omega - E_{k,\nu,\mu} + E_g)^2 + \Gamma^2}, \tag{4}$$

where $M^{\pm}_{\rm EQ}(\nu;\mu)$ is the matrix element of the atomic EQ transition between ν and μ states. We take a geometry where the magnetic field is antiparallel to the direction of the X-ray propagation, then $M^{\pm}_{\rm EQ}(\nu;\mu)$ is expressed as

$$M^{\pm}_{\rm EQ}(\frac{7}{2},j_z;\frac{3}{2},j'_z) = \langle 4f\frac{7}{2}j_z|r^2C^{(2)}_{\pm 1}|2p\frac{3}{2}j'_z\rangle = \sum_{m_p,m_f,s_f}(-1)^{3-m_f}\langle\frac{7}{2}j_z|3m_f\frac{1}{2}s_f\rangle$$
$$\times \langle 1m_p\frac{1}{2}s_f|\frac{3}{2}j'_z\rangle\begin{pmatrix} 3 & 2 & 1 \\ -m_f & \pm 1 & m_p \end{pmatrix}\langle 4f||r^2C^{(2)}||2p\rangle, \quad (5)$$

by using the Clebsch-Gordan coefficients, the 3-j symbol and the reduced matrix element.

Now we introduce a very convenient expression of the EQ-XMCD spectrum to study the difference in EQ-XMCD spectra for $YbInCu_4$ and $YbAgCu_4$. We put approximately

$$E_{k,\nu,\mu} - E_g = E_0\delta_{k,k_0}, \quad (6)$$

where δ is the Kronecker δ-function and $k = k_0$ corresponds to the Fermi level. Then the XMCD spectrum is expressed as

$$F_{\rm XMCD}(\omega) = F^+(\omega) - F^-(\omega) = \sum_{j_z} A^{(g)}_{j_z} B_{j_z}(\omega), \quad (7)$$

where

$$A^{(g)}_{j_z} = \sum_k |c^{(g)}_{k,-j_z}|^2, \quad (8)$$

$$B_{j_z}(\omega) = \sum_{j'_z}(|M^+_{-j_z j'_z}|^2 - |M^-_{-j_z j'_z}|^2)\frac{\Gamma/\pi}{(\omega - E_0)^2 + \Gamma^2}. \quad (9)$$

We note that $A^{(g)}_{j_z}$ is the weight of the Yb $4f$ angular momentum j_z in the ground state and $B_{j_z}(\omega)$ is the contribution from each j_z to the EQ-XMCD spectrum. It is checked that $E_{k,\nu,\mu}$ has a prominent peak for ϵ_k just on the Fermi level for $YbInCu_4$ and $YbAgCu_4$, so that the approximate relation (6) is well acceptable.

3. Calculated results

For $YbInCu_4$, the value of $A^{(g)}_{j_z}$ for each j_z was already calculated as a function of the magnetic field in ref. [7], and the result is reproduced in Fig. 1. For details on the calculation of $A^{(g)}_{j_z}$, see ref. [7]. The function $B_{j_z}(\omega)$, which is common for $YbInCu_4$ and $YbAgCu_4$, can easily be calculated, and the result is shown in Fig. 2, where the maximum amplitude of $B_{j_z}(\omega)$ is normalized to be unity and we put $E_0 = 0$ and $\Gamma = 2.5$ eV. By using these results, the XMCD spectrum of $YbInCu_4$ is calculated for various magnetic field strengths, and the result is shown in Fig. 3. It is seen that the XMCD spectrum is negative and its amplitude first increases with the magnetic field, but it takes a maximum around $30 \sim 40$ T and decreases with further increase in the magnetic field. The peak value of the XMCD spectrum is plotted as a function of the magnetic field in Fig. 4. In order to understand the field-dependence of the XMCD spectra, it is important to note from Fig. 2 that the states of $j_z = 5/2$ and $3/2$ cause the negative XMCD, those of -5/2 and -3/2 cause the positive one with the same amplitudes, but those of 7/2 and -7/2 give no XMCD because of the selection rule. At zero magnetic field, all j_z states have the same weight, resulting in no XMCD signal, but by applying the magnetic field the weights of $j_z = 5/2$ and $3/2$ become larger than -5/2 and -3/2, so that the amplitude of the negative XMCD increases. For magnetic fields higher than the field-induced valence transition around 30 T, the

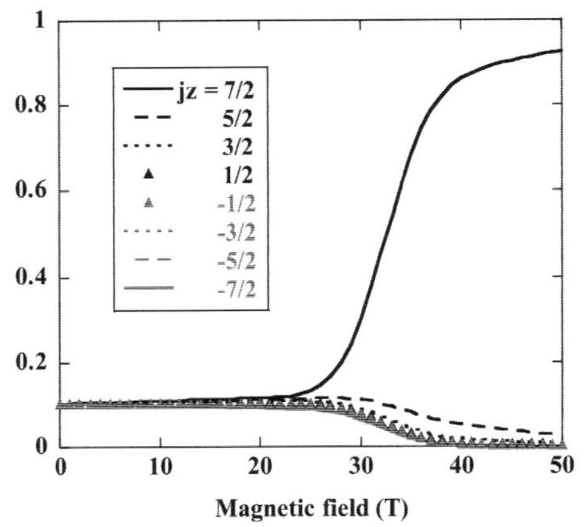

Figure 1. The weights $A^{(g)}_{j_z}$ of the states j_z in the ground state of $YbInCu_4$ as a function of the magnetic field.

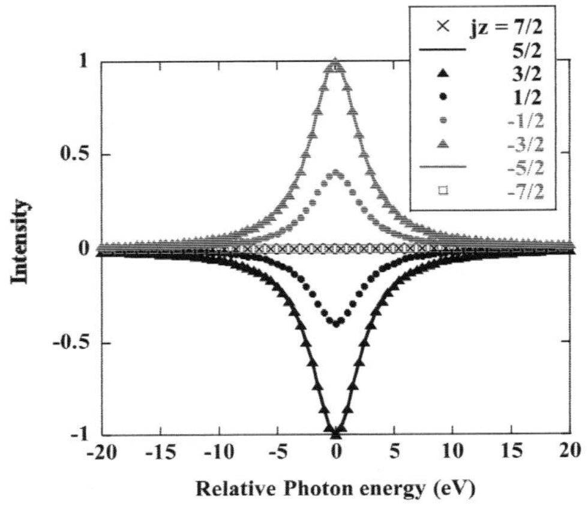

Figure 2. The XMCD-EQ spectral function $B_{j_z}(\omega)$ for various values of j_z as a function of the relative photon energy.

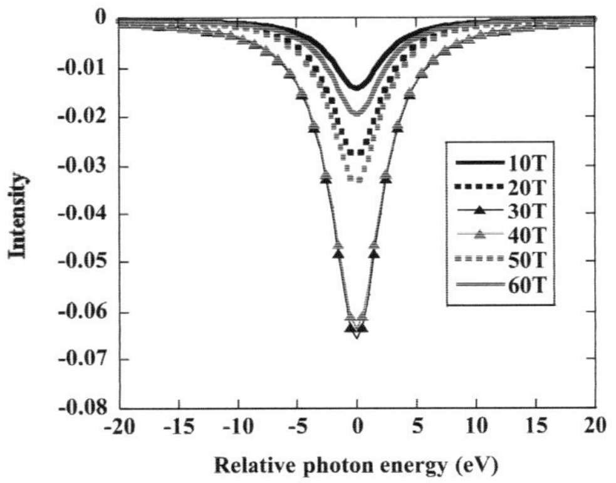

Figure 3. The XMCD-EQ spectra of $YbInCu_4$ for various values of magnetic field.

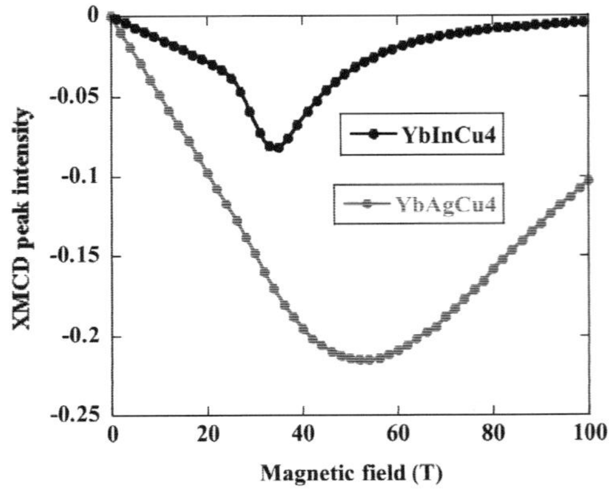

Figure 4. The peak intensities of XMCD-EQ spectra of $YbInCu_4$ and $YbAgCu_4$ as a function of magnetic field.

weight of $j_z = 7/2$ increases rapidly, so that the XMCD amplitude decreases, and it vanishes in the limit of high magnetic field because the ground state tends to a pure $j_z = 7/2$ state.

Similar calculations are also performed for $YbAgCu_4$. The parameters of SIAM are determined so as to reproduce the experimental magnetization curve and valence numbers [6], as in the case of $YbInCu_4$. The results of $A^{(g)}_{j_z}$, $F_{\mathrm{XMCD}}(\omega)$ and the XMCD peak intensity are shown in Figs. 5, 6, and 4, respectively. As seen from Figs. 1 and 5, the field-induced valence transition in $YbAgCu_4$ is much broader and more gradual than that in $YbInCu_4$, so that the induced magnetization below 30 T is larger and the weight of the $j_z = 7/2$ state above 30 T is smaller than those in $YbInCu_4$. This is the reason for that the intensity of the XMCD signal is larger for $YbAgCu_4$ than for $YbInCu_4$. Very recently the EQ-XMCD spectra for $YbAgCu_4$ were measured by Nakamura *et al.* [6] and the behavior of its peak intensity shown in Fig. 4 was confirmed for the magnetic field up to 35 T, but no clear evidence of EQ-XMCD was observed

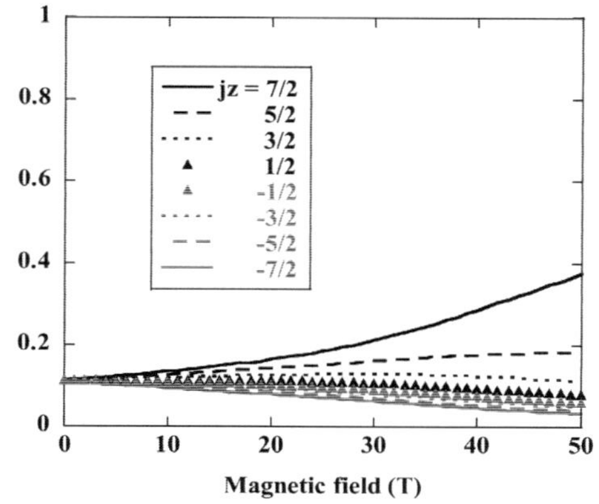

Figure 5. The weights $A_{j_z}^{(g)}$ of the states j_z in the ground state of $YbAgCu_4$ as a function of the magnetic field.

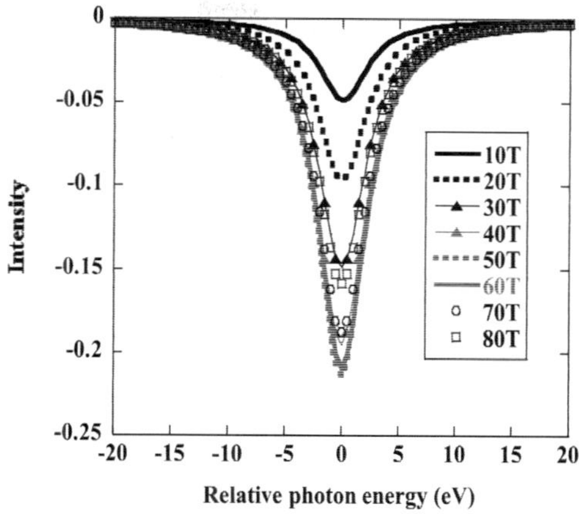

Figure 6. The XMCD-EQ spectra of $YbAgCu_4$ for various values of magnetic field.

experimentally for $YbInCu_4$ because of its smaller intensity.

In conclusion, we can say that the EQ-XMCD intensity for $YbInCu_4$ is so small as to be comparable with or smaller than the experimental noise level, but that for $YbAgCu_4$ is larger than the noise level. Experimental observations of the EQ-XMCD spectra with higher resolution for $YbInCu_4$ and those for magnetic fields higher than 35 T for both $YbInCu_4$ and $YbAgCu_4$ are desirable in the future investigations.

Acknowledgments
The author would like to thank Prof. Y. H. Matsuda for providing experimental data and for useful discussions. He was partially supported by a Grant-in-Aid for Scientific Research C (No. 90029504) from the Japan Society for the Promotion of Science.

References
[1] See, for instance, de Groot F M F and Kotani A 2008 *Core Level Spectroscopy of Solids* (Boca Raton: CRC/London: Taylor & Francis)
[2] Kotani A, Matsuda Y H and Nojiri H 2009 *J. Phys.: Conf. Series* **190** 012013
[3] Matsuda Y H, Inami T, Ohwada K, Murata Y, Nojiri H, Murakami Y, Ohta H, Zhang W and Yoshimura K 2007 *J. Phys. Soc. Jpn.* **75** 34702
[4] Matsuda Y H, Her J L, Inami T, Ohwada K, Ouyang K, Okada K, Nojiri H, Mitsuda A, Wada H, Yoshimura K, Amitsuka H, Kawamura N, Suzuki M and Kotani A 2009 *J. Phys.: Conf. Series* **190** 012019
[5] Matsuda Y H, Nakamura T, Her J-L, Kindo K, Michimura S, Inami T, Mizumaki M, Kawamura N, Suzuki M, Chen B, Ohta H and Yoshimura K 2012 *J. Phys. Soc. Jpn.* **81** 015002
[6] Nakamura T, Matsuda Y H, Her J-L, Kindo K, Michimura S, Inami T, Mizumaki M, Kawamura N, Suzuki M, Chen B, Ohta H, Yoshimura K and Kotani A 2012 *J. Phys. Soc. Jpn.* **81** 114702
[7] Kotani A 2012 *Eur. Phys. J. B* 85: 31
[8] Gunnarsson O and Schönhammer K 1983 *Phys. Rev. B* **28** 4315
[9] Kotani A, Jo T and Parlebas J C 1988 *Adv. Phys.* **37** 37

Real Space Green's Function Approach to Resonant Inelastic X-Ray Scattering and HERFD XAS

Joshua Kas[1,*], John Rehr[1], Moniek Tromp[2], Rowena J. Thomas[3], and Pieter Glatzel[4]

[1]Physics, University of Washington, Seattle, 98195, USA
[2]Strukturanalytik in der Katalyse, Chemie, Catalysis Research Center, Technische Universitt Mnchen, 85748 Garching, Germany
[3]University of Southampton, Chemistry, Southampton, SO17 1BJ, United Kingdom
[4]3European Synchrotron Radiation Facility, B.P. 220, F-38043 Grenoble, France

E-mail: *joshua.j.kas@gmail.com

Abstract. We present a theory of resonant inelastic x-ray scattering (RIXS) and high energy resolution fluorescence detected (HERFD) x-ray absorption spectra (XAS) based on the real space multiple scattering Green's function formalism. The Kramers-Heisenberg formula is transformed into an expression involving Single particle Green's functions. Simplifying assumptions give the RIXS cross section in terms of a convolution of an effective absorption spectrum with the x-ray emission spectrum. Given the RIXS cross section, calculations of the HERFD XAS are also possible. Self-energy effects are calculated using an efficient many-pole model. Further many-body effects can be treated via a convolution with an effective spectral function. The approach is implemented as an extension of the FEFF9 code. Calculated results are compared with experimental RIXS of WO_2.

1. Introduction

Resonant inelastic x-ray scattering is an important probe of electronic structure of both ground and excited states.[1, 2, 3, 4] Recent experimental advanced have allowed a large number of materials studies. Several methods are available to treat RIXS theoretically. The most prolific approach of treating RXES is the charge transfer multiplet (CTM) model, which has been very successful in explaining transitions to localized d- and f-states.[5] However, the multiplet approach depends on a number of parameters which are generally varied to fit experimental spectra. There are a few ab-initio methods available, including configuration interaction based theory, ab-initio CTM, DFT based models, and a Bethe-Salpeter based approach.[6, 7] Here we present an ab-initio theory of RXES and high energy resolution fluorescence detected (HERFD) x-ray absorption spectra (XAS) based on the real space multiple scattering Green's function formalism.[8] The Kramers-Heisenberg formula is transformed into an expression involving Single particle Green's functions. Simplifying assumptions give the RIXS cross section in terms of a convolution of an effective absorption spectrum with the x-ray emission spectrum. The method has been implemented via an extension of the FEFF9 code, which has been used extensively for other core level spectroscopies.[9, 10] The benefit of this approach is that it is relatively efficient, requires no assumption on symmetry, and has the capability of calculating spectra over a wide range of energies. In addition, extensions of the theory to include multiple configurations could allow the unified treatment of atomic multiplet and solid-state effects.

2. Theory

We begin with the Kramer-Heisenberg equation for the RIXS cross section,

$$\frac{d^2\sigma}{d\Omega d\omega} = \frac{\omega}{\Omega}\sum_F \left| \frac{\langle F|\Delta_2^\dagger|M\rangle\langle M|\Delta_1|I\rangle}{E_M - \Omega - E_I + i\Gamma_M} \right| \delta(\Omega - \omega + E_I - E_F), \tag{1}$$

where Ω is the frequency of the incident x-ray and ω is the frequency of the outgoing x-ray. After some manipulation and several approximations, the cross section can be expressed in terms of a convolution of an effective x-ray absorption spectrum (XAS) with the x-ray emission spectrum (XES), i.e.,

$$\frac{d^2\sigma}{d\Omega d\omega} = \frac{\omega}{\Omega}\int d\omega_1 \frac{\mu_{XES}(\omega_1)\mu_{XAS}(\Omega, \Omega - \omega - \omega_1 + E_b)}{|\omega - \omega_1 + i\Gamma_b|^2}, \tag{2}$$

where E_b is the deep core-level energy, and Γ_b the inverse core-hole lifetime. The effective XAS can be written in terms of single particle Green's functions,

$$\mu_{XAS}(\omega_1, \omega_2) = -\frac{1}{\pi}\text{Im}\left[\langle b|T^\dagger(\omega_1)g^c(\omega_2 + E_c)T(\omega_1)|b\rangle\right], \tag{3}$$

where $T(\omega)$ replaces the usual dipole transition operator which is related to the single particle Green's function calculated in the presence of the deep hole, and is Green's function calculated in the presence of the shallow hole. If the core-hole potentials are essentially identical in the intermediate and final states, the effective XAS in the above expression becomes equivalent to the normal absorption spectrum, giving

$$\frac{d^2\sigma}{d\Omega d\omega} = \int d\omega_1 \frac{\mu_{XES}(\omega_1)\mu_{XAS}(\Omega - \omega - \omega_1 + E_b)}{|\omega - \omega_1 + i\Gamma_b|^2}. \tag{4}$$

If we also ignore matrix elements in the XES and XAS spectra, the expression reduces to one involving a convolution of the occupied and unoccupied densities of states.[11, 12] We have implemented the above formula using an extension of the FEFF9.0 code, which has the capability of calculating a large number of other spectroscopies in addition to RXES, including XANES/EXAFS, EELS, NRIXS, DAFS/DANES, and Compton scattering.

3. Results and Discussion

Figure 1 shows a comparison of the calculated W L3-valence RXES of WO_2. Overall, the results are in good qualitative agreement with experiment, reproducing all peaks in the RIXS map, with the exception of the elastic line, which was not calculated. The experimental spectrum is slightly more asymmetric along the incident energy direction. This could be due to the Lorentzian approximation used for the core-hole spectral function, which should be replaced by an asymmetric Fano line shape. Two intense peaks can be seen at approximately 10207 eV incident energy and 10210 eV incident energy. The peak positions along this axis correspond to dipole transitions from the deep core level, and are thus related to the unoccupied density of d-states calculated in the presence of the deep core-hole. These two peaks are reproduced with lower intensity at approximately 7 eV higher energy transfer. The structure along this (energy transfer) axis is related to the dipole transition from the occupied valence band to the deep core-hole and thus related to the occupied density of d-states. In addition to the absense of the elastic peak in the calculations, there are other discrepancies between the theoretical and experimental results, the most obvious being the difference in relative peak positions as well as amplitudes, especially along the energy transfer axis. There are a variety of approximations either in the derivation of the theory, or in implementation which may contribute to these differences. Several major approximations in implementation are; 1) the atomic sphere approximation on the potentials, 2) as currently implemented, the transition matrix elements $T(\omega)$ are assumed to be site diagonal, 3) the Green's functions are also taken to be diagonal in angular momentum, 4) the deep core state is described as a single atomic state. These approximations lead to a variety of errors in the spectrum, for example, the use of the spherical approximation can change the electronic structure (DOS) and thus the peak positions and heights of the specrum. In addition, for some systems it has been shown that crystal momentum is approximately conserved in valence RIXS experiments, however, our implementation does not allow for this conservation due to the site diagonal approximation, as well as the desciption of the core state as a single site quantity. It should be noted however, that the degree to which the crystal momentum is truly conserved depends on the inelastic mean free path of the photo-electron, which is effected primarily by the deep core-hole lifetime, but also by vibrational and static disorder in the system, as well as quasi-particle lifetime effects. The conservation of crystal momentum has been seen in systems such as diamond, where the core-hole width is approximately 0.1 eV, compared to 4.7 eV for the W L_3 edge. This points to a physical picture in which the RIXS signal in this case can be described in terms of local excitations around an atomic core state, which is supported by the agreement between theory and experiment. In addition to the above mentioned approximations, we have neglected a variety of many-body excitation effects. Strong many-body effects can be expected to drastically alter the features of the spectrum, causing large shifts in peaks, and possibly causing additional peaks, for example in the case of strong charge transfer excitations, however, we see no evidence of this in these results. We also note that previouse calculations using a convolution of angular momentum projected densities of states produced similar results to ours in the case of WO_2.[13]

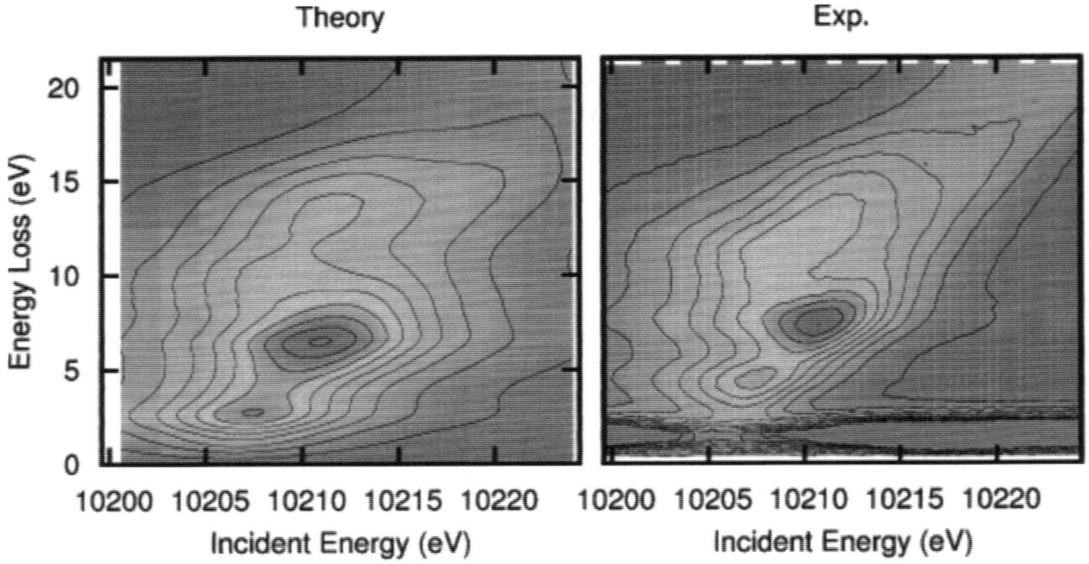

Figure 1. Calculated (left) W L_3−Valence RIXS of WO_2 compared to experiment (right).

4. Conclusion

In summary, we have developed and implemented an ab initio real-space multiple scattering Green's function approach to RIXS, which includes core-hole and matrix-element effects on the spectrum. We have applied the approach to the W L_3-valence RIXS of WO_2 and find reasonable agreement with experiment. Possible explanations of the differences between theory and experiment include the spherical muffin-tin approximation and the use of site and angular momentum diagonal approximations. In the future we plan to improve our implementation to include the full site and angular momentum dependence, which will allow us to investigate the effect of our approximations on the spectrum.

This work was supported in part by US DOE Grant DE-FG03-97ER45623 and facilitated by the DOE CMCSN (JK, JR), and EPSRC (EP/E060404/1) (MT). The ESRF is acknowledged for providing beam time and technical support (P.G.).

References

[1] Glatzel P, de Groot F M F and Bergmann U 2009 *Synchrotron Radiation News* **22** 12
[2] de Groot F and Kotani A 2008 *Core Level Spectroscopy of Solids* (Taylor and Francis CRC press)
[3] Kotani A and Shin S 2001 *Rev. Mod. Phys.* **73** 203–246
[4] Schülke W 2007 *Electron Dynamics by Inelastic X-Ray Scattering* (Oxford University Press)
[5] Kotani A 2005 *Eur. Phys. J. B* **47** 3–27
[6] Ikeno H, Mizoguchi T and Tanaka I 2011 *Phys. Rev. B* **83**(15) 155107
[7] Soininen J A and Shirley E L 2001 *Phys. Rev. B* **64** 165112
[8] Kas J J, Rehr J J, Soininen J A and Glatzel P 2011 *Phys. Rev. B* **83**(23) 235114
[9] Rehr J J, Kas J J, Vila F D, Prange M P and Jorissen K 2010 *Phys. Chem. Chem. Phys.* **12** 5503 – 5513
[10] Rehr J J and Albers R C 2000 *Rev. Mod. Phys.* **72** 621
[11] Glatzel P, Singh J, Kvashnina K O and van Bokhoven J A 2010 *J. Am. Chem. Soc.* **132** 2555
[12] Jiménez-Mier J, van Ek J, Ederer D L, Callcott T A, Jia J J, Carlisle J, Terminello L, Asfaw A and Perera R C 1999 *Phys. Rev. B* **59** 2649–2658
[13] Smolentsev N, Sikora M, Soldatov A V, Kvashnina K O and Glatzel P 2011 *Phys. Rev. B* **84** 235113

EXAFS parameters and VDOS in zincblende structures

P Fornasini[1], R Grisenti[1] and N Abd el All[2]

[1] Dipartimento di Fisica, Università di Trento, Via Sommarive 14, I-38123 Povo (Trento), Italy
[2] Physics Department, Assiut University, Assiut, Egypt

E-mail: paolo.fornasini@unitn.it

Abstract. The local dynamical behaviour of the tetrahedrally coordinated semiconductors Ge, CdTe and CuCl, affected by negative thermal expansion at low temperatures, has been recently investigated by EXAFS. Parallel and perpendicular first-shell mean square relative displacements (MSRDs) are here discussed in terms of correlation and anisotropy. The Einstein frequencies are compared with the vibrational densities of states (VDOS). For all considered systems, the perpendicular Einstein frequency is situated between the VDOS structures due to transverse and longitudinal acoustic modes, while the parallel frequency is situated between the VDOS structures due to longitudinal acoustic and optic modes.

1. Introduction

The parallel mean square relative displacement (MSRD) of nearest-neighbour atoms can be directly obtained from the analysis of the EXAFS signal [1]. The comparison of the nearest-neighbours bond thermal expansion measured by EXAFS with the lattice thermal expansion measured by Bragg diffraction gives information on the perpendicular MSRD [2]. The knowledge of both parallel and perpendicular MSRDs allows a deeper understanding of the local vibrational dynamics, in terms of correlation of vibrational motion and of anisotropy of relative thermal ellipsoids.

Three crystals with the diamond or zincblende structure, characterized by different ionicity values and different strength of low-temperature negative thermal expansion (NTE), have been recently investigated by EXAFS: Ge [3], CdTe [4] and CuCl [5]. The comparison shows that a correlation exists between the bond ionicity, the negative thermal expansion strength and a number of quantities measured by EXAFS: the stronger is the NTE, the larger are the positive bond expansion, the relative atomic vibrations perpendicular and parallel to the bond, as well as their anisotropy [4].

In this paper, the local dynamic properties of the three systems are further discussed and the Einstein frequencies of the parallel and perpendicular MSRDs are phenomenologically compared with the vibrational densities of states (VDOS).

Journal of Physics: Conference Series **430** (2013) 012004 doi:10.1088/1742-6596/430/1/012004

2. MSRDs correlation and anisotropy

The main properties of the parallel and perpendicular thermal MSRDs $\langle \Delta u_\parallel^2 \rangle$ and $\langle \Delta u_\perp^2 \rangle$, respectively, can be synthesised by the frequency ν of the best-fitting correlated Einstein models

$$\langle \Delta u_\parallel^2 \rangle = \frac{h}{8\pi^2 \mu \nu_\parallel} \coth\left(\frac{h\nu_\parallel}{2k_B T}\right), \quad \langle \Delta u_\perp^2 \rangle = \frac{h}{4\pi^2 \mu \nu_\perp} \coth\left(\frac{h\nu_\perp}{2k_B T}\right). \tag{1}$$

The two models differ by a factor 2 because the perpendicular MSRD is defined as a 2-dimensional projection of the total MSRD on the plane perpendicular to the bond [6].

For all crystals here considered, the perpendicular MSRDs are significantly larger than twice the parallel MSRDs. Correspondingly, the perpendicular Einstein frequencies $\nu_\perp$ are smaller than the parallel frequencies $\nu_\parallel$ (Table 1, where the data for copper are shown for comparison). This difference corresponds to a stronger parallel than perpendicular correlation and gives rise to a non-negligible anisotropy of relative vibrations.

The extent of correlation can be measured by the dimensionless correlation parameters [7, 8]

$$\phi_\parallel = \frac{U_\parallel^{(b)} + U_\parallel^{(a)} - \langle \Delta u_\parallel^2 \rangle}{2\left[U_\parallel^{(b)} + U_\parallel^{(a)}\right]^{1/2}}, \quad \phi_\perp = \frac{U_\perp^{(b)} + U_\perp^{(a)} - \langle \Delta u_\perp^2 \rangle/2}{2\left[U_\perp^{(b)} + U_\perp^{(a)}\right]^{1/2}}, \tag{2}$$

where $U_\parallel$ and $U_\perp$ are the uncorrelated MSDs of absorber and backscatterer atoms parallel and perpendicular to the bond direction, respectively, which can be calculated from the anisotropic displacement parameters (ADP) U_{ij} measured by diffraction. The correlation parameters for copper and for the three diamond-zincblende structures here considered are listed in Table 1. The parallel and perpendicular correlations are very similar for copper, strongly different for Ge and still more different for CdTe and CuCl.

The parallel Einstein frequency $\nu_\parallel$ is connected to an effective bond-stretching force constant $k_\parallel = \mu(2\pi\nu_\parallel)^2$, where μ is the reduced mass. Similarly, the perpendicular Einstein frequency can be connected to an effective force constant $k_\perp = \mu(2\pi\nu_\perp)^2$. One should not confuse these effective force constants with the force constants of the Born-von Karman models of lattice dynamics [4]. The effective force constants (Table 1), being independent of masses, can be more suitable than the frequencies to compare different isostructural compounds. Actually, the frequency values $\nu = (k/\mu)^{1/2}/2\pi$ represent the response of the pair of atoms to the interaction force, and depend on the inertia, which is measured by the reduced mass. As is evident in Table 1, the effective force constants show a neat decreasing trend when passing from Ge to CdTe to CuCl, while the Einstein frequencies for CdTe and CuCl are nearly equal.

Table 1. Einstein frequencies ν, correlation parameters ϕ, effective force constants k for parallel and perpendicular MSRDs, and anisotropy parameters $\xi = k_\parallel/k_\perp$. The data of Cu are shown for comparison.

		Cu	Ge	CdTe	CuCl
Einstein frequencies	$\nu_\parallel$ (THz)	4.9	7.7	3.89	3.93
	$\nu_\perp$ (THz)	4.6	4.6	1.83	1.66
Correlation	$\phi_\parallel$	0.44	0.76	0.83	0.8
	$\phi_\perp$	0.42	0.31	0.19	-0.2
Force constants	$k_\parallel$ (eV/Å^2)	3.2	8.5	3.71	1.4
	$k_\perp$ (eV/Å^2)	2.6	2.9	0.81	0.26
Anisotropy	$\xi = k_\parallel/k_\perp$	1.2	2.9	4.6	5.4

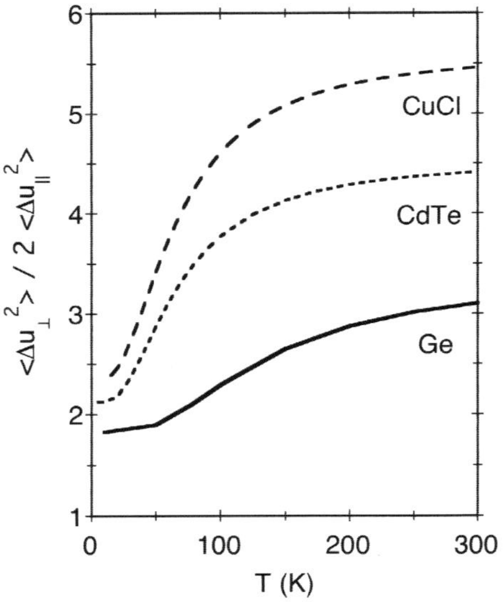

(a)

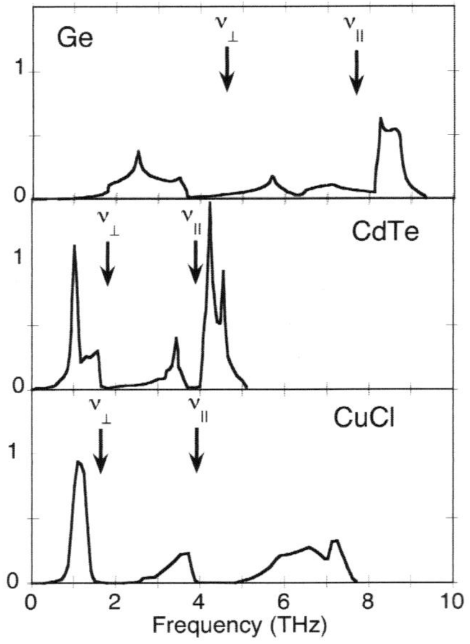

(b)

Figure 1. (a) Perpendicular to parallel MSRD ratios $\langle\Delta u_\perp^2\rangle/2\langle\Delta u_\parallel^2\rangle$ for Ge, CdTe and CuCl. (b) VDOS for Ge [9], CdTe [10] and CuCl [11]. The vertical arrows correspond to the Einstein frequencies best fitting the parallel and perpendicular MSRDs.

The perpendicular to parallel anisotropy of the nearest-neighbours relative vibrations can be measured by the ratio $\langle\Delta u_\perp^2\rangle/2\langle\Delta u_\parallel^2\rangle$, which is generally temperature dependent. Fig. 1 (a) shows the ratios of the best-fitting perpendicular and parallel Einstein models for Ge, CdTe and CuCl. A temperature independent measure of anisotropy can be obtained by considering the high-temperature asymptotic behaviour of the Einstein models

$$\xi = \lim_{T\to\infty} \frac{\langle\Delta u_\perp^2\rangle}{2\langle\Delta u_\parallel^2\rangle} = \frac{2k_BT/k_\perp}{2k_BT/k_\parallel} = \frac{k_\parallel}{k_\perp}\,. \tag{3}$$

The ratio $\xi = k_\parallel/k_\perp$ is equal to one for perfectly isotropic relative motion. For copper, ξ is nearly one; for the three diamond-zincblende crystals here considered, ξ is significantly larger than one: the ellipsoids of relative thermal displacements are disk-shaped. The value of ξ increases in going from Ge to CdTe to CuCl (Table 1), say increasing the ionicity. It is worth noting that the atomic thermal displacements measured by Bragg diffraction are isotropic for symmetry reasons.

The extent of the parallel and perpendicular force constants as well as the anisotropy of relative vibrations can be correlated to the strength of negative thermal expansion [4].

3. Comparison with the vibrational densities of states

The MSRDs depend on the contributions of all normal modes; for the parallel MSRDs [1]

$$\langle\Delta u_\parallel^2\rangle = \frac{1}{\mathbf{N}}\sum_{\vec q,s} \langle|Q(\vec q,s,t)|^2\rangle \left|\left(\frac{\vec w_b(\vec q,s)e^{i\vec q\cdot\vec R}}{\sqrt{m_b}} - \frac{\vec w_a(\vec q,s)}{\sqrt{m_a}}\right)\cdot\hat R\right|^2 \tag{4}$$

where $\vec q$ and s are mode wavevector and branch index, respectively.

A reliable reproduction of parallel and perpendicular MSRDs requires a homogenous sampling of the first Brillouin Zone, taking into account the inter-cell phaseshifts of each mode as well as the mode eigenvectors, their phase relationships and their projection along the bond direction and in the perpendicular plane. The knowledge only of the VDOS is clearly insufficient to this purpose; nevertheless, interesting qualitative clues can be obtained from the comparison of the MSRDs Einstein frequencies and the VDOS.

The VDOS for Ge [9], CdTe [10] and CuCl [11] are shown in Fig. 1 (b). Three structures can be identified in the VDOS of the three crystals: a low-frequency structure due to the transverse acoustic (TA) modes, an intermediate structure due to the longitudinal acoustic (LA) modes and a high-frequency structure due to the optic modes.

The terms "longitudinal" and "tranverse" of lattice dynamics are relative to the direction of the normal mode wavevectors $\vec{q}$, which densely and isotropically fill up the first Brillouin zone, while the attributes "parallel" and "perpendicular" of MSRDs refer to the nearest-neighbour bonds, limited to the $\langle 111 \rangle$ directions for the diamond-zincblende structures.

In spite of this non trivial difference, a neat correlation can be found between the Einstein frequencies of the MSRDs and the VDOS structures. For all crystals here considered: *a)* the perpendicular Einstein frequency $\nu_\perp$ is situated between the VDOS structures due to the TA modes and the LA modes, respectively; *b)* the parallel Einstein frequency $\nu_{||}$ is situated between the VDOS structures due to the LA modes and the optic modes, respectively.

The contribution of normal modes to the MSRDs is inversely proportional to the square of the frequency, since $\langle |Q(\vec{q}, s, t)|^2 \rangle = \langle E(\vec{q}, s) \rangle / \omega^2(\vec{q}, s)$. As a consequence, optic modes are relatively less important than acoustic modes.

4. Conclusions

The first-shell parallel and perpendicular MSRDs of Ge, CdTe and CuCl have been comparatively analyzed. When the ionicity of the system increases (from Ge to CdTe to CuCl), both parallel and perpendicular effective force constants decrease, the parallel correlation increases, the perpendicular correlation decreases and the perpendicular to parallel anisotropy increases.

A phenomenological correlation has been established between the Einstein frequencies of the parallel and perpendicular MSRDs and the VDOS structures due to TA, LA and optic modes for Ge, CdTe and CuCl.

Accurate dynamical calculations sampling the eigenfrequencies and the eigenvectors within the first Brillouin zone would help to quantitatively evaluate the contribution of the different vibrational modes to the parallel and perpendicular MSRDs.

References

[1] Beni G and Platzman P M 1976 *Phys. Rev. B* **14** 1514–1518
[2] Fornasini P, a Beccara S, Dalba G, Grisenti R, Sanson A, Vaccari M and Rocca F 2004 *Phys. Rev. B* **70** 174301
[3] Dalba G, Fornasini P, Grisenti R and Purans J 1999 *Phys. Rev. Lett.* **82** 4240–4243
[4] Abd el All N, Dalba G, Diop D, Fornasini P, Grisenti R, Mathon O, Rocca F, Thiodjio Sendja B and Vaccari M 2012 *J. Phys.: Condens. Matter* **24** 115403
[5] Vaccari M, Grisenti R, Fornasini P, Rocca F and Sanson A 2007 *Phys. Rev. B* **75** 184307
[6] Vaccari M and Fornasini P 2006 *J. Synchrotron Rad.* **13** 321–325
[7] Booth C H, Bridges F, Bauer E D, Li G G, Boyce J B, Claeson T, Chu C W and Xiong Q 1995 *Phys. Rev. B* **52** R15745–R15748
[8] Sanson A, Rocca F, Dalba G, Fornasini P, Grisenti R, Dapiaggi M and Artioli G 2006 *Phys. Rev. B* **73** 214305
[9] Nilsson G and Nelin G 1972 *Phys. Rev. B* **6** 3777–3786
[10] Talwar D N and Holliday K 1999 *Physica B* **263-264** 540–545
[11] Ma Y, Tse J S and Klug D D 1986 *Phys. Rev. B* **34** 3970–3979

Positive or Negative Thermal Expansion in Anharmonic Vibrational Systems in EXAFS study

Dong-Seok Yang[1] and Jay Min Lee[2]

[1] Department of Physics Education,Chungbuk National University, Cheongju, 361-763, Korea
[2] Beamline Research Division, Pohang Accelerator Laboratory, POSTECH, 790-784, Korea

E-mail: `dsyang@chungbuk.ac.kr`

Abstract. An EXAFS expression was introduced for analysis of positive or negative thermal expansion. The EXAFS parameters were directly related to the interatomic potential. Also, it was shown that the sign of anharmonic potential parameter k_3 depends on the sign of thermal expansion coefficient. Simulation works showed that the positive or negative thermal expansion can be sensitively measured by EXAFS study.

1. Introduction

Bond stretching, causing the local thermal expansion of chemical bond, occurs frequently in molecules or lattice relaxed with the thermal disturbance beyond the harmonic vibrational limit. Recently the negative thermal expansion has been extensively studied for the practical application to the non-expandable alloys [1, 2, 3, 4, 5]. The positive/negative thermal expansion is relevant to the bond stretching/contraction configuration, severely subjected to the interatomic potential of atoms in solid [6, 7, 8, 9]. Although these local thermal expansion phenomena have been investigated in EXAFS study, considerable progress has not been made yet because of experimental limitation and theoretical difficulties [10, 20, 12, 13]. It has been shown that the local thermal expansion occurs in the anharmonic vibrational systems [8, 9] which can be related to a asymmetric pair distribution [14]

In this study, we demonstrated a simple criterion to identify the bond stretching/contraction in a theoretical manner, using a pair distribution function applicable to the positive/negative thermal expansion [9].

2. EXAFS expression for thermal variational systems

The volume of solid materials can be expanded or contracted according to the characteristics of bonding configuration and effective potential between constituent atoms. The structures of solids are more or less deformed by changing lattice parameters as temperature increases. This means that the origin of positive or negative thermal expansion may cause the apparent deformation of amplitude and phase of EXAFS spectra. The deformed spectra require a proper EXAFS expression, explicitly indicating the positive/negative thermal expansion.

In anharmonic system, the EXAFS equation can be expressed for both positive and negative thermal expansion by

$$\chi(k) = \frac{NF(k)S_0^2(k)e^{-2R/\lambda}}{kR^2}A_c(k)sin(2kR + \varphi(k) + \phi_c(k)) \quad (1)$$

where

$$A_c = (1 + (2k\delta)^2)^{-\frac{\sigma^2}{2\delta^2}} \quad (2)$$

and

$$\phi_c(k) = \frac{\sigma^2}{\delta^2}(tan^{-1}(2k\delta) - (2k\delta)) \approx (-\frac{8}{3}\sigma^2\delta k^3 + \cdots) \quad (3)$$

where N is the number of neighbor atoms, R is the average interatomic distance, σ^2 is the Debye-Waller parameter, and δ is the parameter of thermal expansion. The backscattering amplitude $F(k)$ and the phase $\varphi(k)$ are available from FEFF [15, 16, 17]. Yet, asymmetric pair distribution function is implicit, even though it can be demonstrated in terms of anharmonic potential. Here, δ is valued with polarity of thermal expansion, either positive or negative. Look how Debye-Waller parameter (σ^2) and thermal expansion parameter (δ) are related to the parameters of anharmonic potential

$$V(r) = \frac{1}{2}k_0(r - r_0)^2 - k_3(r - r_0)^3 + k_4(r - r_0)^4 \quad (4)$$

where r_0 is the equilibrium distance, k_0 is the force constant in attraction, k_3 is the skew parameter ascribing the local environments of solid atoms, and $k_4(= \frac{9k_3^2}{4k_0})$ is the extra constraint for the anharmonic vibrational model. As well as δ-parameter, k_3 indicates the positive or negative thermal expansion. By setting $\delta = R - r_0$, we have the following relationship between σ^2, δ, k_0, and k_3 [8, 18]

$$\sigma^2 = \frac{k_BT}{k_0}. \quad (5)$$

and

$$\delta = \frac{3k_3}{k_0^2}k_BT. \quad (6)$$

where k_B is Boltzmann constant and T is absolute temperature.

The relevant thermal expansion coefficient (α) can be expressed by

$$\alpha = \frac{\Delta\delta}{r_0\Delta T}. \quad (7)$$

The anhamonic potential gives a pair distribution function as follows

$$g(r) = \frac{e^{-\frac{V(r)}{k_BT}}}{\int_0^\infty e^{-\frac{V(r)}{k_BT}}dr}. \quad (8)$$

The general EXAFS spectrum is convoluted with the pair distribution $g(r)$ by

$$\chi(k) = \int_0^\infty \frac{NF(k)S_0^2(k)e^{-2r/\lambda}}{kr^2}g(r)sin(2kr + \varphi(k))dr. \quad (9)$$

From the anharmonic pair distribution function (Eq.8), we can see the explicit variation of the EXAFS spectrum. Without such explicit representation of g(r) function, we may use the EXAFS equation (Eq.1) instead in case for slightly anharmonic system (e.g. $\delta < 0.03\mathring{A}$).

3. EXAFS simulation for positive and negative thermal expansion

Let us consider the case of positive thermal expansion with the potential parameters of $k_0 = 47N/m$ and $k_3 = +2.56 \times 10^{11}N/m^2$, bond distance is $R = 2.54\mathring{A}$. Fig.1 shows the temperature dependence of pair distributions for positive thermal expansion. Fig.2 shows the EXAFS spectra of a copper calculated by Eqs.4, 8, and 9, respectively, for 300 K, 500 K and 700K. As apparently shown, the amplitude is decreasing and the phase of EXAFS spectra is relatively expanded as temperature increases. This is a dilemma about the polarity of thermal expansion, less indicating that the bond is positively expanded as temperature increases.

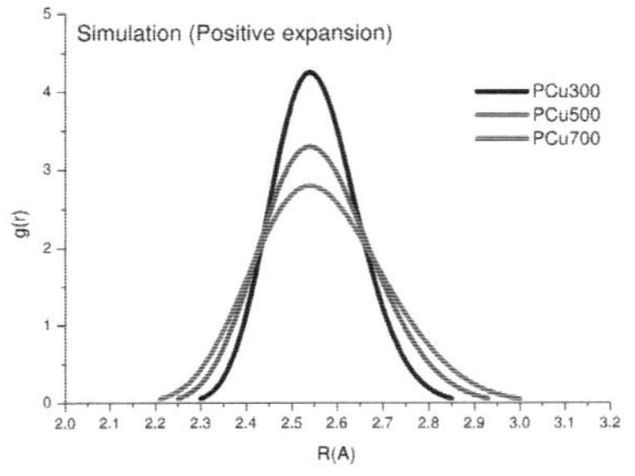

Figure 1. Pair distribution function for positive thermal expansion.

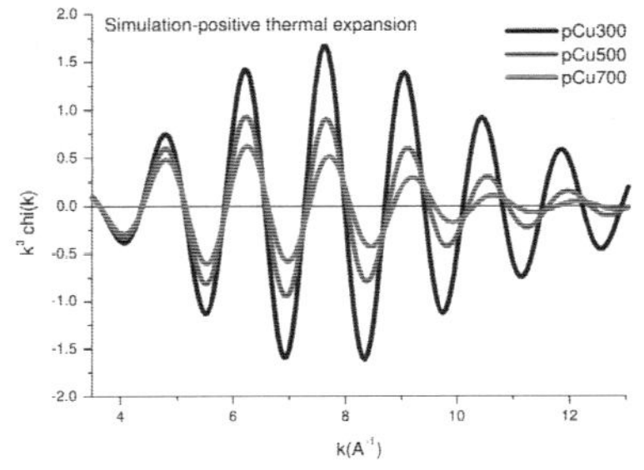

Figure 2. Simulated EXAFS spectra for positive thermal expansion.

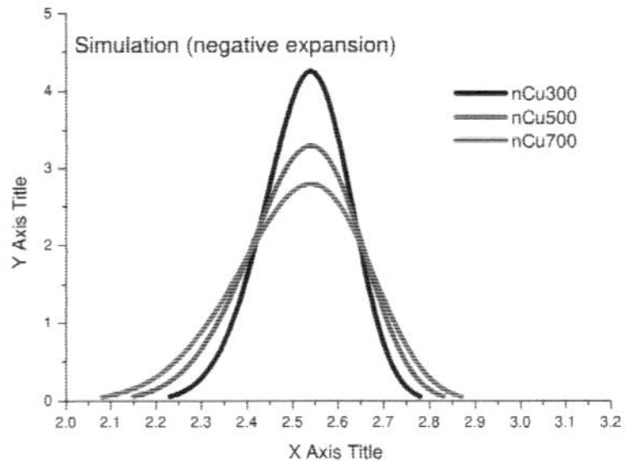

Figure 3. Pair distribution function for negative thermal expansion.

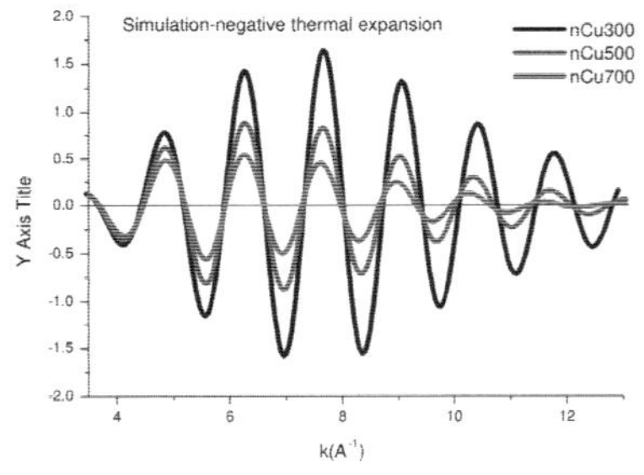

Figure 4. Simulated EXAFS spectra for negative thermal expansion.

When the simulated spectra were fitted to Eq.1, we have $\delta = 0.0148\mathring{A}$ for the spectrum at 300K, and $\delta = 0.0252\mathring{A}$ at 500K. As expected, δs are all positive indicating that this material is positively expanded as temperature increases. The estimated value of the thermal expansion coefficient is $\alpha = 20.47 \times 10^{-6}/K$. It is obvious that δ is positive for the positive thermal expansion. On the contrary, let us consider negative simulation with negative thermal expansion. EXAFS spectra for NTE was simulated with the potential parameters of $k_0 = 47N/m$ and $k_3 = -2.56 \times 10^{11}N/m^2$, bond distance is $R = 2.54\mathring{A}$. Fig.3 shows the pair distributions for negative thermal expansion. Fig.4 shows the simulated spectra, respectively, for 300 K, 500 K and 700K. As apparently shown, the decreasing amplitude is the same but the phase is reversed compared to the case of PTE. Simply, this reverse phenomenon in phase feature indicates the

the negative thermal expansion. When the simulated spectra in Fig. 4 were fitted to Eq. 1, we obtained $\delta = -0.0155\mathring{A}$ for the spectrum at 300 K and $\delta = -0.0273\mathring{A}$ at 500 K. The values of δ are all negative. Also, the estimated thermal expansion coefficient is $\alpha = -23.9 \times 10^{-6}/K$. Analytically, because negative thermal expansion is the reverse direction of positive thermal expansion, the modified EXAFS equation (Eq(1)) works for the practical analysis.

As discussed with our computation simulation, the EXAFS spectra are very sensitive for the local structural variation. The origin of negative thermal expansion was not interpreted in our study. We only elucidated the numerical validity associated with the unusual case of negative thermal expansion. Without doubt, EXAFS is a powerful tool to quantitatively characterize the positive or negative thermal expansions. In a theoretical aspect, although the modified EXAFS expression (Eq.1) is practically useful to analyze the anharmonic vibrational system, the correction phase term ($\phi_c(k)$ in Eq.3) is the artificial strangeness, lacking interpretation of interatomic potential. To avoid such an ambiguity, we need to perform the inverse computation with establishing the measurement error bar. Our study of STIC(Synthesized Topological Inverse Computation) [19] is promising to make a progress for general cases, even highly disordered configurations such as amorphous or non-crystalline materials - complex multicomponent systems.

In summary, quantitatively we mentioned that the modified EXAFS expression is valid only for slightly anharmonic system (i.e. $\delta < 0.03\mathring{A}$). Also, we elucidated the structural insight of positive or negative thermal expansion under establish anharmonic potential. It was clearly simplified that the sign of anharmonic potential parameter k_3 determines the trend of the positive or negative thermal expansion.

References

[1] T H K Barron, J A Birch and G K White 1977 *J.Phys.C* **10** 1617
[2] T. A. Mary, J. S. Evans, T. Vogt, A. W. Sleight 1996 *Science* **272** 90
[3] Brendan J. Kennedy, Yoshiki Kubota, Kenichi Kato 2005 *SolidStateCommunications* **136** 177-180
[4] Chen W. Li, Xiaoli Tang, J. A. Munoz, J. B. Keith, S. J. Tracy, D. L. Abernathy, and B. Fultz 2011 *Phys.Rev.Lett.* **107** 195504
[5] Zi-Kui Liu, Yi Wang and Shun-Li Shang 2011 *ScriptaMaterialia* **65** 664-667
[6] P. Eisenberger, George S. Brown 1979 *SolidStateCommunications* **29** 481-484
[7] E. D. Crozier, A. J. Seary 1980 *CanadianJournalofPhysics* **58** 1388-1399.
[8] Frenkel A I and Rehr J J 1993 *Phys.Rev.B* **48** 585-588.
[9] D.S. Yang, S.K. Joo 1998 *SolidStateCommunications* **105** 595-599
[10] D. E. Sayers, E. A. Stern, and F. W. Lytle 1971 *Phys.Rev.Lett.* **27** 1204
[11] G. Dalba, P. Fornasini, R. Grisenti, and J. Purans 1999 *Phys.Rev.Lett.* **82** 4240-4243
[12] A. Sanson and F. Rocca, G. Dalba, P. Fornasini, and R. Grisenti, M. Dapiaggi and G. Artioli 2006 *Phys.Rev.B* **73** 214305.
[13] M. Vaccari, R. Grisenti, and P. Fornasini, F. Rocca and A. Sanson 2007 *Phys.Rev.B* **75** 184307
[14] Dong S. Yang, David R. Fazzini, Timothy I. Morrison, Larc Troger, Grant Bunker 1997 *JournalofNon − CrystallineSolids* **210** 275-286
[15] B. Ravel and M. Newville 2005 *J.Synchrotron.Rad.* **12** 537
[16] M. Newville 2001 *J.SynchrotronRad.* **8** 322
[17] S. I. Zabinsky, J. J. Rehr, A. Ankudinov, R. C. Albers and M. J. Eller 1995 *Phys.Rev.B* **52** 2995
[18] C. Kittel 1996 *IntroductiontoSolidStatePhysics* (7th edition) (New York: John Wiley and Sons, Inc.) chapter 6
[19] J. M. Lee, D. S. Yang and G. Bunker *EXAFSPhaseRetrievalSolutionTrackingforComplexMulti − ComponentSystem : SynthesizedTopologicalInverseComputation* to be presented in XAFS15
[20] G. Dalba, P. Fornasini, R. Grisenti, and J. Purans 1999 *Phys.Rev.Lett.* **82** 4240-4243

Journal of Physics: Conference Series **430** (2013) 012006 doi:10.1088/1742-6596/430/1/012006

Path aggregation techniques for EXAFS visualization and analysis

Bruce Ravel

National Institute of Standards and Technology, Gaithersburg, MD 20899, USA

E-mail: `bravel@bnl.gov`

Abstract. The path expansion model used by FEFF considers the EXAFS spectrum as the sum of contributions from scattering geometries of two or more atoms in a cluster surrounding the absorbing atom. This model and its representation as a collection of files is the basis for a number of popular EXAFS analysis programs, including one co-developed by me. Here I present several tools which build upon the scattering path model to articulate analysis and visualization capabilities not normally included in FEFF-based analysis software. These include the generation of arbitrary single-scattering paths not represented in FEFFs input structure, consideration of non-Gaussian and non-cumulant distribution functions, easy visualization of arbitrary path summations, and others. Each of these tools is tightly integrated into the ARTEMIS EXAFS analysis program, making these analysis and visualization tools easily and widely available.

The widely used FEFF program provides a real-space, multiple-scattering approach to the calculation [1] of the X-ray absorption spectrum. As applied to the quantitative analysis of the extended XAFS region, FEFF provides a solution based on path enumeration. The input to the calculation includes a list of Cartesian coordinates of atoms in a cluster with one atom selected as the absorber. First, atomic potentials are computed using a muffin tin approximation and the list of Cartesian coordinates. Later versions of FEFF offer improvements to the potential and energy loss models, however the details of the potential calculation are not important to the context of this paper. FEFF then enumerates all scattering geometries [2] within the cluster up to some order of scattering. This enumeration is exhaustive (no scattering geometries are missed) and recognizes degeneracy (geometries contributing identically to the spectrum are flagged). FEFF also uses a simple algorithm to exclude geometries that contribute scant spectral weight. Finally, FEFF computes the spectral contribution from each scattering geometry.

This *path expansion* is the basis of the EXAFS data analysis program IFEFFIT [3], as well as several other popular analysis packages. This path-based representation of the EXAFS analysis problem has proven to be powerful and flexible. IFEFFIT and other analysis program using this representation have been applied successfully to complex analysis problems from all of the many scientific disciplines which use absorption spectroscopy.

The reason that the path expansion works so well is that it maps naturally onto the problem of EXAFS data analysis. In the simple case of a single-shell analysis, the first-shell single-scattering path is readily understood as the theoretical representation of the single shell data. Core complex analyses – multiple coordination shells, shells with more than one scattering species, multiple scattering contributions – the value of the path expansion is even more evident. In general, in a fit, there is value to parametrizing things such as ΔR and σ^2 separately for different shells and different scattering species. A path-by-path parametrization of these terms in the EXAFS equation is inherent to IFEFFIT.

As valuable as the path expansion has proven to be in the nearly two decades that FEFF has been available in its current form (since the release of FEFF6 in 1995), it has its limitations, mostly related to its input structure. In normal use, FEFF's output is strictly related to the input data. Contributions from scattering geometries not represented in the input cluster cannot be computed. While small changes to inter-atomic spacings can be accommodated by the parametrization of ΔR, this approach is unreliable for large changes. This is because the effect of ΔR in IFEFFIT is only to change terms in the EXAFS equation which depend explicitly on R. The scattering amplitude and phase shift also have an R dependence which is not considered by IFEFFIT, resulting in large inaccuracies when a ΔR parameter becomes large.

In this paper, I present several features newly available in the ARTEMIS [4] EXAFS data analysis program. Each of these uses FEFF in a creative way to enable a data visualization or analysis capability not readily available when used in its conventional manner.

1. Virtual path

A *virtual path* is a grouping of two or more normal paths for the sake of visualization. Each path in the group is kept up to date with respect to its own parametrization of the EXAFS equation. This group is added together and the sum is Fourier transformed and plotted as a unit. While this serves no direct role in the evaluation of a fit, it is a powerful visualization tool.

Consider an EXAFS analysis on a mixed phase system, for instance a mixture of a metal oxide and a metal sulfide. FEFF calculations will be run on the structures of both phases. Paths will be selected from each calculation for use in the fitting model. One virtual path – called a *VPath* in ARTEMIS – could be made using the paths from the oxide calculation. Another VPath could then be made using the paths from the sulfide calculation. Plotting VPaths along with the data, as in Fig. 1, makes it clear how each component of the sample contributes to the spectrum.

2. Quick first shell path

In the case of data requiring only a first shell analysis, the problem with FEFF is not that the input data limits the scope of the problem, but that input data must be provided at all for such a simple analysis problem. An common example of this situation is a measurement of a glassy material. With no long range order, the metal site in the glass has a well defined first shell, but little or no structure beyond that.

To expedite a simple first shell analysis, ARTEMIS offers a *quick first shell*, or *QFS*, path. The user is prompted for the atomic species of the absorber and the scatterer, the edge symbol, and a nominal distance between the two atoms. From this, ARTEMIS generates an input file for FEFF for a rock-salt structured crystal containing the two specified elements. This results in a sixfold coordinated structure about the absorber from which FEFF will compute muffin tin potentials for absorber and scatterer which are reasonable for the specified distance. After the calculation, only the first shell path is retained and the rest of the calculation is discarded. As a user option, ARTEMIS can generate a set of fitting parameters to complete a simple, single-shell fitting model. This feature of ARTEMIS was inspired by similar functionality in the SIXPACK [5] program.

3. Single scattering path

The limitation of the QFS path is that it is only applicable to single scattering paths in the first coordination shell. The user-supplied QFS distance parameter *is* the near neighbor distance. If that number is unphysically large, the muffin tin potentials for the absorber and scatterer will be inaccurate. Still, the concept of generating a single scattering path at an arbitrary distance is useful.

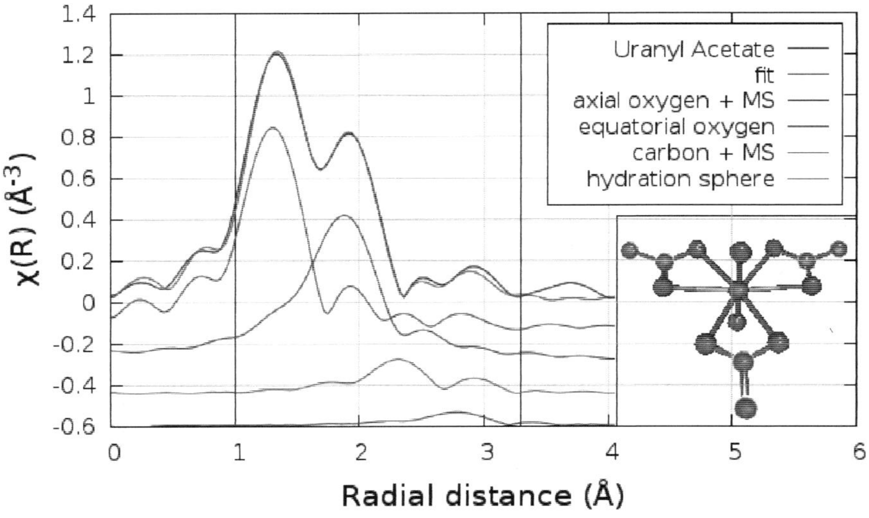

Figure 1. This fit to data collected on a solution of uranyl acetate demonstrates two virtual paths and a single scattering path. The inset shows the structure of the uranyl complex, which includes carbon atoms (gray) bound to the uranium (blue) by a bidentate bridge of equatorial oxygens (red). Two very short axial oxygens (red) extend out of the plane formed by the other atoms. The signal from the axial oxygens is shown by the first offset trace. This signal is a VPath composed of the single scattering from the axial oxygens and three significant multiple scattering paths. Note that the spectral weight from this VPath extends throughout the entire fitting range. The second offset trace is a normal path representing the equatorial oxygen atoms. The third offset trace is a VPath containing single scattering from the carbon and multiple scattering from carbon and the bridging oxygen. The final trace is an SSPath, discussed below.

In ARTEMIS, an *SSPath* uses the potentials from a successful FEFF calculation to compute a single scattering path for a scatterer at an arbitrary distance not represented in the input cluster. The user chooses the scatterer species from among those present in the FEFF calculation and the distance between the two atoms. ARTEMIS then instructs the GENFMT part of FEFF to use the previously computed potentials with the specified single scattering path.

Consider a mineral sample which contains an unknown secondary phase with the same elemental composition as the primary phase. An SSPath can be used to investigate the possibility of the second phase contributing a single scattering path at a distance not represented in the structure of the primary phase without requiring that further FEFF calculations be made on candidate structures.

Figure 1 shows a fit to uranyl acetate in solution. The structure used in the FEFF calculation was crystalline sodium uranyl triacetate. Paths for the surrounding oxygen and carbon atoms were included in the fit. An SSPath representing the hydration sphere was computed using the potential of the equatorial oxygen and a position of 3 Å. This provides a plausible explanation for part of the spectral weight around 3 Å in the data.

4. Histogram path

Another shortcoming of FEFF is that its input structure represents a static configuration of atoms. In a case where variation about that static configuration is well represented by Gaussian disorder – perhaps with small skewness and kurtosis represented by third and fourth cumulants

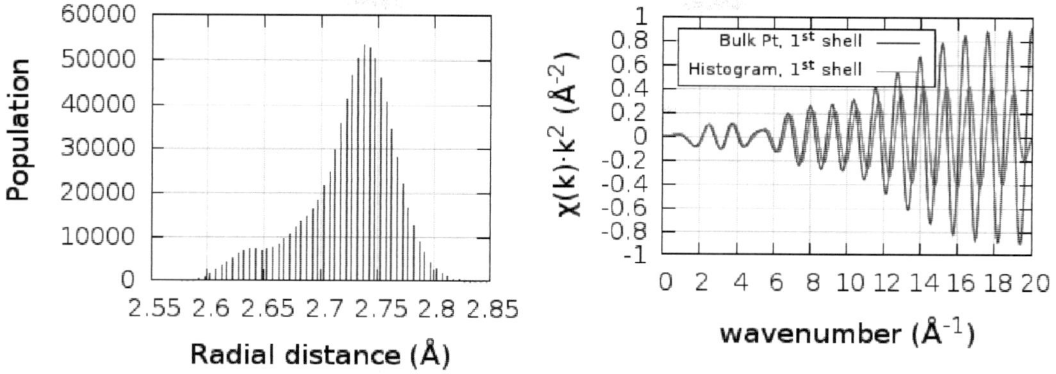

Figure 2. (Left) The distribution of first shell neighbors in a Pt nanoparticle as determined from a molecular dynamics calculation. (Right) $\chi(k)$ for the first shell in bulk Pt compared with the $\chi(k)$ for a Pt atom distributed over the histogram. Note the differences in both amplitude and phase. Both are shown with $N = 1$, $S_0^2 = 1$ and $\sigma^2 = 0$.

– the conventional path expansion works very well. For more complex distributions, for example a multi-modal distribution, a cumulant expansion is a poor approximation.

For materials with substantial disorder, one approach to understanding the nature of that structure is to employ a molecular dynamics (MD) simulation. Typically an MD simulation is made on a cluster of some size and over some number of time steps. The distribution of atoms in a coordination shell can be found by extracting all absorber-scatterer pairs within some band of distances. This ensemble of pairs can be binned into a histogram. Shown on the left of Figure 2 is an example of the first shell histogram extracted [7] from an MD simulation on a small platinum nanoparticle. Contraction at the surface of the particle results in the peak at short distance.

Given a histogram representing a distance distribution, ARTEMIS computes an SSPath at the distance of each bin. Each such SSPath is weighted by the population of its bin. These SSPaths are added and normalized by the sum of the bin populations. The resulting object, then, is the contribution to the EXAFS from a single atom distributed over the histogram.

5. Empirical path

The final path-like object implemented in ARTEMIS is a small refinement of the concept of an empirical standard, which is useful for the rare situation when FEFF may not adequately compute a fitting standard. One such situation is the case of a near-neighbor atom with an intervening hydrogen atom. While the scattering from the hydrogen atom is typically quite weak and may contribute scant spectral weight to the EXAFS, its presence can alter the scattering potential of the non-hydrogen neighbor in ways that are poorly approximated by FEFF's muffin tin model (particularly in FEFF6 or FEFF7).

The use of empirical standards [8] goes back to the early days of the EXAFS technique. The algorithm in ARTEMIS is much the same. Reference data are Fourier filtered to isolate the spectral contribution from the scatterer in question. In many implementations this Fourier filtered standards is fitted directly to the data, which are also filtered using identical transform parameters. The twist here is that the filtered reference data are exported in the format FEFF uses for the contribution from a scattering path. This export format saves the $\chi(k)$ function as a complex number, the imaginary part of which is extracted by IFEFFIT for use in a fit. Although FEFF exports this function on a rather sparse grid, because these amplitude and phase functions are relatively slowly varying functions of k, empirical paths for scatterers at any distance can be reliably extracted from a reference material. The advantage of extracting the empirical standard to the format of a FEFF path is that it is incorporated into a fit with IFEFFIT and ARTEMIS like any other path.

6. Implementation

The recurring theme among all these path-like objects is that each is expressed in the form of a conventional FEFF path. In ARTEMIS, each data set included in a fitting model accumulates a list of paths as the theory used to describe that data. That list can contain any combination of conventional paths and the path-like objects described here. For each path or path-like object, the EXAFS equation must be parametrized and evaluated. Each type of path requires specification of the same set of parameters. These include an amplitude term (incorporating both degeneracy and S_0^2), an energy shift, a ΔR, and a σ^2.

Each of these path-like objects is implemented in ARTEMIS and thus is available for any analysis problem. The histogram path requires a specific implementation based upon the source of the molecular dynamics output. To date, implementations for DL_POLY [9], LAMMPS [10], and VASP∗ [11] have been written and new sources for structural data can be included with limited effort.

The application of FEFF to EXAFS analysis has proven to be a great success over the past two decades. That FEFF is both robust and flexible is evident from its deployment as the theoretical back end to dozens of data analysis software packages. That said, there are still novel and creative ways that this solid framework can be used to solve interesting data analysis problems. This short paper documents some of these features of my own software package. The analysis tools discussed here are broadly useful and could easily be implemented by other FEFF-using analysis packages.

Acknowledgments

Without FEFF and IFEFFIT, my work would not be possible. John Rehr along with his many students and Matt Newville are responsible for those fine programs. The many users of my software have been a joy and an inspiration for me. A few deserve special mention for exceptional contributions over the years, including Shelly Kelly, Scott Calvin, Eric Breynaert, and Stephen Price.

References

[1] Rehr J J and Albers R C 2000 *Rev. Mod. Phys.* **72**(3) 621–654
[2] Zabinsky S I, Rehr J J, Ankudinov A, Albers R C and Eller M J 1995 *Phys. Rev. B* **52**(4) 2995–3009
[3] Newville M 2001 *Journal of Synchrotron Radiation* **8** 322–324
[4] Ravel B and Newville M 2005 *Journal of Synchrotron Radiation* **12** 537–541
[5] Webb S 2012 Sam's Interface for XAS Package software website URL `http://sixpack.sams-xrays.com/`
[6] Lee P A and Pendry J B 1975 *Phys. Rev. B* **11**(8) 2795–2811
[7] Price S W T, Zonias N, Skylaris C K, Hyde T I, Ravel B and Russell A E 2012 *Phys. Rev. B* **85**(7) 075439
[8] Bunker G 1983 *Nuclear Instruments and Methods in Physics Research* **207** 437 – 444 ISSN 0167-5087
[9] Smith W, Yong C and Rodger P 2002 *Molecular Simulation* **28** 385–471
[10] Plimpton S 1995 *Journal of Computational Physics* **117** 1 – 19 ISSN 0021-9991
[11] Kresse G and Furthmüller J 1996 *Phys. Rev. B* **54** 169

∗Any mention of commercial products is for information only; it does not imply recommendation or endorsement by NIST.

Journal of Physics: Conference Series **430** (2013) 012007 doi:10.1088/1742-6596/430/1/012007

Larch: An Analysis Package for XAFS and Related Spectroscopies

Matthew Newville

Center for Advanced Radiation Studies, University of Chicago, Building 434A, Argonne National Laboratory, Argonne. IL 60439, USA

E-mail: `newville@cars.uchicago.edu`

Abstract. LARCH, a package of analysis tools for XAFS and related spectroscopies is presented. A complete rewrite of the IFEFFIT package, the initial release of LARCH preserves the core XAFS analysis procedures such as normalization, background subtraction, Fourier transforms, fitting of XANES spectra, and fitting of experimental spectra to a sum of FEFF Paths, with few algorithmic changes made in comparison to IFEFFIT. LARCH is written using Python and its packages for scientific programming, which gives significant improvements over IFEFFIT in the ability to handle multi-dimensional and large data sets, write complex analysis scripts, visualize data, add new functionality, and customize existing capabilities. Like the earlier version, LARCH can run from an interactive command line or in batch-mode, but LARCH can also be run as a server and accessed from clients using standard inter-process communication techniques available in a variety of computer languages. LARCH is freely available under an open source license. Examples of using LARCH are shown, future directions for development are discussed, and collaborations for adding new capabilities are actively sought.

1. Introduction

The basic algorithms for XAFS analysis of extracting $\chi(k)$ from measured absorption spectra and determining structural information from $\chi(k)$ are well-established and mature. While there may be room for improvement in the algorithms, most existing analysis packages[1, 2, 3, 4, 5, 6, 7] vary more in details and in emphasis on particular features or user interface than on the core algorithms used. As the field matures and synchrotron beamlines become more sophisticated, XAFS data is coming at much higher rates, either as traditional spectra measured at higher rates[8], or by using fast area detectors [9], or collected in fluorescence mode using energy dispersion onto area detectors[10]. In addition, XAFS is increasingly being embedded in complimentary measurements, such 3-dimensional, full field imaging datasets[11], x-ray fluorescence mapping, elastic x-ray scattering, and inelastic resonant x-ray scattering. The existing software tools for XAFS analysis are generally found in specialized, single-purpose packages designed to work with individual $\mu(E)$ spectra, and are unable to seamlessly deal with many of the forms of XAFS data now being collected.

The IFEFFIT package[12] is a widely used XAFS analysis package, with its success due to a combination of being freely available, having high quality algorithms and a flexible interface separating core analysis engine from graphical user interface, the existence and support for easy-to-use GUI applications such as ATHENA and DEMETER[13], SIXPACK[14], and PRESTOPRONTO[15], an active mailing list, and frequent training courses. Now over a decade old,

Journal of Physics: Conference Series **430** (2013) 012007 doi:10.1088/1742-6596/430/1/012007

IFEFFIT has many limitations that make it unable to meet the growing demands for quantity and type of data detailed above that a modern XAFS analysis package should have. The most severe of these limitation – that only a fixed and rather small amount of computer memory could be used – resulted from IFEFFIT (like its predecessors FEFFIT, AUTOBK, and FEFF) being implemented in standard FORTRAN-77. Other limitations, such as a weak built-in scripting language and poor quality graphics, also resulted from using FORTRAN-77. Additionally, the code base IFEFFIT was complex, making maintenance, minor modifications to existing algorithms, and the addition new features difficult. Thus, to meet the dynamic needs of modern XAFS data, including the ability to analyze hundreds to thousands of spectra or to analyze spectra embedded in larger data sets, a complete reimplementation of IFEFFIT was undertaken. In this report, LARCH, a replacement for IFEFFIT is described. The fundamental change is that LARCH is written entirely in Python, a popular, cross-platform, and open source interpreted language known for its readability and clarity, and with a very large set of built-in libraries and efficient, well-tested libraries for scientific computing.

2. LARCH: A replacement for IFEFFIT

LARCH maintains the basic model of an interactive command-line session that was an important feature of IFEFFIT. This approach preserves data within a session, but allows for dynamic and open-ended analysis routes, so that the user can re-visit analysis steps at will and mix analysis of multiple spectra while also maintaining a history of processing steps. This interactivity does not mean one is restricted to typing at a command-line shell interface, however. Scripts can be used for batch processing, or more complex modules and programs be written.

Though IFEFFIT also featured an interactive session, weaknesses in its implementation made it difficult to work well with many sets of data. The pre-defined limits on memory mentioned above were part of this. Though allowing for thousands of separate arrays, analyzing more than a few hundred data sets in a single session was not possible. In addition, IFEFFIT did not support features necessary for moderately complex scripts such as conditionals and loops, supported only one dimensional arrays, and had no multi-dimensional or flexible data structures. Finally, IFEFFIT used a single, flat "name space" so that important quantities, such as `e0` (E_0), could have only value at a time. Together, these limitations made interactive sessions and scripting more challenging than they should be. Wrappers in languages such as Perl[16] or Python[17], which created and ran a single IFEFFIT session by sending commands constructed in the wrapping language, were needed for any sustained analysis session or Graphical User interface, such as ATHENA[13] or SIXPACK[14].

LARCH overcomes the limitations described above and provides a much richer scripting language. By using Python, limits on memory are not set by the program, but by the host computer and operating system. The well-tested and supported `numpy`[18] and `scipy`[19] modules provide efficient and high-quality numerical arrays and interfaces to basic numerical libraries for procedures such as Fourier transforms, spline interpolation, linear algebra, and non-linear least squares optimization to Python, and are used throughout LARCH.

The text of one or more lines of LARCH code goes through a minimal translation step to convert it into Python code which is then compiled with Python's own compiler tools into an abstract syntax tree that can be evaluated in a straightforward, if recursive, way. This gives LARCH a complete language, including conditional statements (`if-then-else`), loops (`for` and `while`), exception handling, function definitions, multi-dimensional arrays with complex slicing operations, while sparing the need to parse and support a fully-featured language. In the translation and interpretation steps, LARCH is able to modify some of the assumptions made in Python, such as how program symbols are retrieved and how code blocks are specified, making LARCH a dialect of Python that is specially tuned for scientific computing. Compared to Python, the LARCH dialect makes these changes:

Table 1. Core XAFS analysis functions in LARCH.

function	**description**
`pre_edge()`	Find E_0, Pre-edge subtraction, normalization
`autobk()`	post-edge background subtraction
`ftwindow()`	calculate a Fourier transform window function
`xafsft()`	forward ($k \to R$) Fourier transform
`xafsift()`	reverse ($R \to k$) Fourier transform
`feff_path()`	Read and define a FEFF Path, with Parameters
`feffit_transform()`	Define Fourier transform for `feffit()`
`feffit_dataset()`	Define data set, Paths for `feffit()`
`feffit()`	Perform fit of a set of Paths to a set of data sets
`minimize()`	General purpose least-squares fit

(i) LARCH uses code blocks for `if`, `for`, etc that end with `endif`, `endfor`, etc instead of using significant indentation level.

(ii) LARCH uses a more complex namespace for program variables and built-in functions, so that names for variables and functions are organized in a hierarchy, but can be easily accessed.

(iii) one cannot create new classes of objects inside LARCH.

Being a dialect of Python means that all variables and functions in LARCH are Python objects. Thus LARCH has basic data types such as integers, floats, complex numbers, and strings, and richer objects including multi-dimensional arrays of floats, heterogeneous lists of other object, and associative arrays. This is an important improvement over IFEFFIT. In addition, two important LARCH-specific data types are also added:

Groups: An empty container to hold other objects. Many LARCH functions work with Groups, either returning them or using them to put multiple, related outputs. In this way, an entire data set (raw intensities, $\mu(E)$, $\chi(k)$, $\chi(R)$, values for E_0, edge jump etc) can be contained in a single Group.

Parameters Because non-linear least-squares fits are an integral part of XAFS analysis, used for XANES fitting, background subtraction `autobk` and fitting of FEFF paths to $\chi(k)$ spectra, fitting parameters in LARCH are defined as Parameter. These can either be variable or fixed in a fit, or *constrained* to a mathematical expression of other Parameters.

Extending LARCH functionality is very easy. Since LARCH is so closely based on Python, any Python module can be imported and used directly from LARCH. For functionality that interacts more closely with the LARCH infrastructure (for example, creating Groups), a *plug-in* mechanism exists. Finally, like Python, LARCH is cross-platform. Once the required numpy and scipy libraries are installed, LARCH runs without any compilation step or changes on Windows, Mac OS X, and Linux.

3. Using LARCH

LARCH has built-in many numerical functions from `numpy` and `scipy` (`sqrt`, `log`, and a host of others) that work efficiently on entire arrays at once, without the need for explicit looping. In addition, LARCH provides functions for high-quality graphical plots for X-Y line plots and for 2-dimensional intensity maps, using the `matplotlib`[20] library. Of course, LARCH also adds several XAFS-specific functions for working with XAFS data. A partial list of these functions is given in Table 1.

```
dat = read_ascii('raw_xafs.dat', labels='en time i0 i1')
dat.mu = -ln(dat.i1/ dat.i0)

pre_edge(dat.en, dat.mu, group=dat)

autobk(dat.en, dat.mu, rbkg=1.2, kweight=2,group=dat)
xafsft(dat.k, dat.chi, kmin=3, kmax=16, dk=5,
      kweight=2, window='kaiser', group=dat)

newplot(dat.k, dat.chi*dat.k**2, win=1,
    xlabel='$k \, (\mathrm{\AA}^{-1})$',
    ylabel='$k^2\chi(k) \, (\mathrm{\AA^{-2}})$')
plot(dat.k, dat.kwin, color='black', style='dashed')

newplot(dat.r, dat.chir_mag, win=2,
     xlabel='$R \, (\mathrm{\AA}) $ ',
     ylabel='$|\chi(R)| \, (\mathrm{\AA^{-3}})$')
plot(dat.r, dat.chir_re, win=2)
```

Figure 1. Example LARCH session, in which data is read from an ASCII file, $\mu(E)$ is computed, E_0 and edge jump are found with `pre_edge()`, $\mu_0(E)$ and $\chi(k)$ are found with `autobk()`, $\chi(k)$ is transform to $\tilde{\chi}(R)$ with `xafsft()`, and data arrays are plotted.

With many built-in functions, LARCH can be used as a high-level command-line shell. An example session to read raw XAFS data from an ASCII file, compute $\mu(E)$, $\chi(k)$, and $\chi(R)$, and plot some of the results is shown in Figure 1, with the graphical outputs shown in Figure 2. Note that a subset of the TEX typesetting system is understood by the `matplotlib` plotting library, and used here for axis labels. Though not shown, interactive zooming, printing, copying to system clipboard, and customization of plot properties such as line colors, symbols, legends, and labels are readily available to the user from the plot window.

3.1. Fitting XAFS Data

Of course, a primary need for XAFS analysis is to compare experimental to theoretical standards from first principles calculations such as those from FEFF[21]. Continuing the use of a "Sum-of-Paths" approach, LARCH provides functions for defining **FEFF Paths** with Path Parameters such as E_0, S_0^2, δR, and σ^2, and refining these to match data[22], usually in R-space, but with some flexibility in "fitting space". Using Parameters mentioned in section 2, there is great flexibility in how Path Parameters can be varied, and a full range of algebraic constraints can be applied, to build elaborate structural models for the refinements. In addition, simultaneous refinement of multiple data sets can be performed without much effort.

As with IFEFFIT, fit statistics, estimates of uncertainties in fitted variables, and correlations between variables are automatically determined. Interestingly, the basic fitting algorithm is the same[23], and very little systematic difference is seen between fits using IFEFFIT and LARCH. LARCH's ability to put a set of Parameters into a Group allows sets of fitting variables for different fitting models to be isolated and more readily compared.

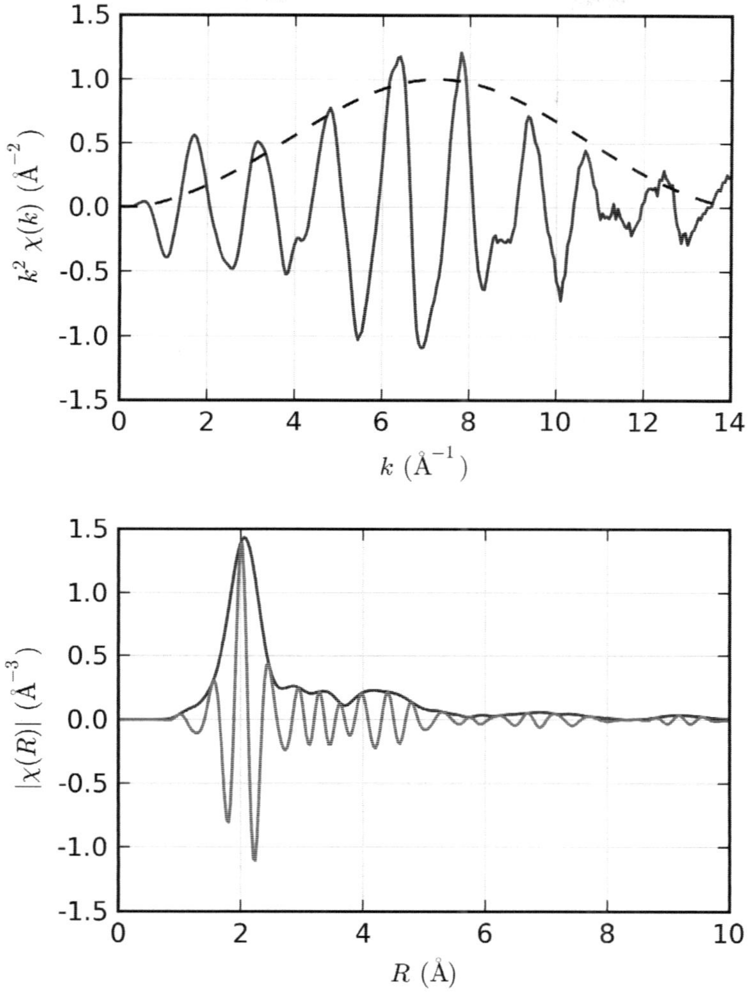

Figure 2. Plots of $k^2\chi(k)$ (top) and $\chi(R)$ (bottom) from the LARCH session shown in Figure 1 (with data from Se K edge of CoSe. The images above are extracted directly from LARCH without further processing.

3.2. Efficiency Tests

While using Python and its numerical libraries makes LARCH flexible, easy to use, and easy to implement and maintain, Python is not generally considered fast for numerical processing, especially in comparison to FORTRAN. However, because IFEFFIT contained a poor scripting language within it and tried to manage a memory pool, much of its expected efficiency was lost. Thus, when strictly limited to core, non-graphical processes, LARCH can be faster than IFEFFIT. As a benchmark appropriate for QEXAFS work, a loop that goes through a set of raw data files, reading the file, running `autobk()` and `xafsft()` to extract $\chi(R)$, is shown in Figure 3. A similar script using IFEFFIT and its Python wrapper is more complex, limited in how many spectra can be preserved, and about $\sim 2.5\times$ slower than the LARCH script.

```
filelist = ['xafsscan.001', 'xafsscan.002', 'xafsscan.003' ]
save_groups = {}
for fname in filelist:
    dat = read_ascii(fname, labels='en time i0 i1')
    dat.mu = -ln(dat.i1/ dat.i0)
    autobk(dat.en, dat.mu, rbkg=1.2, kweight=2, group=dat)

    xafsft(dat.k, dat.chi, kmin=3, kmax=16, dk=5,
           kweight=2, window='kaiser', group=dat)
    save_groups[fname] = dat
endfor
```

Figure 3. Example of looping over a set of files, and running `autobk()`, and `xafsft()` for each set. The use of `save_groups` preserves the group used for each data set, so that it can be retrieved after the loop.

3.3. Remote Server Mode
Being able to access the functionality of LARCH as a calculation engine from other applications, such as GUIs, is an important priority. To facilitate this need, LARCH can be run as a server, with a unique session per server instance. This allows client applications to connect to one or more separate server processes either on the local machine or remote machines. Since each server is a separate process, they can easily make use of multiple cores on all platforms. Server-mode also allows multiple clients to share a single LARCH session if desired. Communication between client application and LARCH server uses the standard XML-RPC protocol, which is widely supported in a variety of languages (Perl, Python, C++, Java, .Net, PHP, javascript) without the need for a LARCH-specific wrapper per language.

3.4. Additional Features
Though not discussed in detail for lack of space, LARCH is not intended to be restricted to XAFS analysis. Because LARCH can use any Python module and is already using its basic scientific libraries, adding more functionality is straightforward, as will be discussed further below. Though early in its development, LARCH already has available functionality for the following tasks:

- accessing tabulated x-ray data for emission and absorption energies[24], elastic x-ray scattering, anomalous scattering factors[25] and other data.
- preliminary analysis and visualization of x-ray fluorescence spectra and x-ray fluorescence maps from x-ray microprobe measurements.
- reading, manipulating, and displaying TIFF images, used by many diffraction cameras.
- reading data from HDF5 data files, used for data storage at many synchrotrons.
- interacting with the EPICS experimental control system, used at many synchrotrons.

Some of these require additional python modules to be installed.

4. Conclusion and Future Directions
The initial release of LARCH described here should not be considered to be a final description of its capabilities. Indeed, the main effort so far has been to reproduce the functionality of IFEFFIT, so as to be a starting place for future developments in analysis algorithms for XAFS, XANES,

and related techniques, and to provide a proof-of-concept for additional functionality outside of XAFS analysis, including more complete x-ray fluorescence analysis and experimental control. Though any list of development details will become out-dated, a few possible improvements for XAFS analysis can be mentioned. Some of these exist in other analysis programs, but may not be widely supported and available. The most obvious development would be to add a full set of linear algebra techniques, including principle component analysis and target transformation.[15] Alternative background subtraction and normalization methods[26] could also be added. Another development would be to include a complete statistical treatment of $\mu(E)$ that is carried fully through the analysis, as shown recently.[27] Finally the application of wavelet transforms[28] or other transformations beyond the simple XAFS conventions of applying a k-weight and Fourier window function, may be used during refinement to better isolate spectral and structural signatures. Other improvements are certainly possible, and the LARCH infrastructure is better designed to make such developments possible.

Acknowledgments

I would like to thank Bruce Ravel, Tom Trainor, Peter Eng, Joanne Stubbs, Tony Lanzirotti, Ken McIvor, and Darren Dale for discussions, and to the IFEFFIT mailing list for encouragement and continued motivation. This work was funded by through grants to GeoSoilEnviroCARS from the National Science Foundation – Earth Sciences (EAR-1128799) and Department of Energy - Geosciences (DE-FG02-94ER14466).

[1] Filipponi A, Cicco A D and Natoli C R 1995 *Physical Rev. B* **52** 15122–15134
[2] Tomic S, Searle B G, Wander A, Harrison N M, Dent A J, Mosselmans J F W and Ingelsfield J E 2005 *CCLRC Technical Report DL-TR* **001** 1
[3] Ressler T 1998 *Journal of Synchrotron Radiation* **5** 118–122
[4] George G N and Pickering I J 1995 EXAFS-PAK Tech. rep. Stanford Synchrotron Radiation Laboratory Stanford, CA
[5] Klementev K 2001 *Journal of Physics D-Applied Physics* **34** 209–217
[6] Taguchi T, Ozawa T and Yashiro H 2005 *Physica Scripta* **T115** 205–206
[7] Michalowicz A, Moscovici J, Muller-Bouvet D and Provost K 2009 *Journal of Physics: Conference Series* **190** 12034–12035
[8] Stötzel J, Luetzenkirchen-Hecht D and Frahm R 2011 *Journal of Synchrotron Radiation* **18** 165–175
[9] Tanida H, Yamashige H, Orikasa Y, Oishi M, Takanashi Y, Fujimoto T, Sato K, Takamatsu D, Murayama H, Arai H, Matsubara E, Uchimoto Y and Ogumi Z 2011 *Journal of Synchrotron Radiation* **18** 919–922
[10] Hayashi H, Kawata M, Takeda R, Udagawa Y, Watanabe Y, Takano T, Nanao S and Kawamura N 2004 *Journal of Electron Spectroscopy and Related Phenomena* **136** 191–197
[11] Meirer F, Cabana J, Liu Y, Mehta A, Andrews J C and Pianetta P 2011 *Journal of Synchrotron Radiation* **18** 773–781
[12] Newville M 2001 *Journal of Synchrotron Radiation* **8** 322–324
[13] Ravel B and Newville M 2005 *Journal of Synchrotron Radiation* **12** 537–541
[14] Webb S M 2006 SixPack `http://sixpack.sams-xrays.com/`
[15] Prestipino C 2011 PrestoPronto `http://code.google.com/p/prestopronto/`
[16] Wall L 1997 Perl Programming Langauge `http://perl.org/`
[17] von Rossum G and Drake F L 2003 Python Reference Manual `http://python.org/`
[18] Oliphant T E 2006 Numerical Python `http://numpy.scipy.org/`
[19] Jones E, Oliphant T, Peterson P *et al.* 2001 SciPy: Open source scientific tools for Python `http://scipy.org/`
[20] Hunter J D 2007 matplotlib `http://matplotlib.sourceforge.net/`
[21] Zabinsky S I, Rehr J J, Ankudinov A, Albers R C and Eller M J 1995 *Physical Rev. B* **52** 2995–3009
[22] Newville M 2001 *Journal of Synchrotron Radiation* **8** 96–100
[23] Garbow B S, Hillstrom K E and Móre J J 1980 MINPACK-1 `http://www.netlib.org/minpack`
[24] Elam W T, Ravel B D and Sieber J 2002 *Radiation Physics and Chemistry* **63** 121–128
[25] Chantler C T 2000 *Journal of Physical and Chemical Reference Data* **29** 597–1048
[26] Weng T C, Waldo G S and Penner-Hahn J E 2005 *Journal of Synchrotron Radiation* **12** 506–510
[27] Chantler C T 2012 *Journal of Synchrotron Radiation* **19** 145–158
[28] Munõz M, Argoul P and Farges F 2003 *American Mineralogist* **88** 694–700

GPU Computing in XAFS

K Pedersen and G Bunker
Department of Physics, Illinois Institute of Technology, Chicago, IL

E-Mail: bunker@iit.edu

Abstract: Modern *graphics processing units* (GPUs) are powerful parallel computers. Porting the mathematically intense, parallelizable algorithms of XAFS to run on GPUs should expand the scale of tractable XAFS calculations. Part of a time-limiting subroutine in FEFF 8.4 (*fms*) was converted to run on a GPU, showing significant performance gains. With parallel CPUs sharing GPU resources, the runtime of the modules calling *fms* should accelerate by one or two orders of magnitude versus a single-threaded CPU process (at least for large atomic clusters,). New or existing XAFS algorithms and programs (e.g. FEFF 9) should show similar acceleration when ported to GPUs, as long as they meet certain criteria (outlined in detail).

1. Introduction to GPU Computing

A GPU is essentially a cluster computer condensed onto a single board. Like a cluster, a GPU (or *device*) has a number of parallel processing nodes, called *streaming multiprocessors* (SMs) [1]. Since the SMs all reside on the same board and share the same on-chip global memory space (RAM), GPUs sidestep much of a cluster's overhead. GPUs are controlled by their computer's CPU (or *host*) [2].

The transistors of a GPU are optimized for FLOPS (i.e. math), severely limiting a device's ability to process control logic (like *if-then-else*) [2]. Each branch in a GPU control tree executes *in series*, and *every branch* (even those bypassed by a given control instance) gets a pass through the SM (this is called *divergence*). Bypassed branches, while computationally impotent, still eat up SM clock cycles. In order to benefit from GPU conversion, (***rule 1***) algorithms or subroutines must operate in an almost rote manner on many different inputs, with minimal inline control [2]. Hence the GPU niche is large *for/DO* loops which fill or map matrices (e.g. matrix operations) [1]. Sections of an algorithm that rely heavily on control statements should be handled by the host, which can then outsource niche GPU sections to the device. When the device is busy the host is not idle; after issuing instructions to the device, the host can immediately return to its own tasks (this is called asynchronous operation).

1.1. Compute Unified Device Architecture (CUDA) and CUDA C

The transistors of Nvidia CUDA devices have the versatility to tackle mathematical tasks not usually required in graphics rendering [1]. Using CUDA C, an extension of C, users can harness the power of their CUDA devices with a simple API. CUDA C is free, well supported, and well documented, but it only works on Nvidia GPUs. An open-source alternative, OpenCL, is supported on GPUs from a number of manufacturers, including Nvidia. This paper focuses solely on CUDA and CUDA C.

To translate an algorithm into CUDA C, it is broken into irreducible sets of instructions called *kernels*, which the host calls sequentially. Each parallel element of a kernel's task is carried through its code by a *thread*. Threads are organized into small, 2D arrays called *blocks* (1024 threads max). Each block is executed by a single SM, with a device's multiple SMs executing many blocks concurrently. Blocks are arranged into a 3D array called a *grid*, which is a kernel's work order [1]. Based on its location in its 2D block, and its block's location in the 3D grid, a thread can identify which element in a data structure to handle [1]. To complete a kernel call, blocks are taken from the grid and distributed to the SMs. Since devices do not schedule block execution deterministically, (***rule 2***) blocks must be able to execute in *any order*; algorithms with data interdependencies must be broken into autonomous kernels [1]. Also, since neither the device nor the host can directly access each other's RAM; (***rule 3***) host memory needed by the device, and device results needed by the host, must be transferred [1].

Published under licence by IOP Publishing Ltd

Device RAM (global memory) is external to SMs, so SM read latency is much slower than SM clock speed [2]. To prevent SM idleness (waiting for global memory calls to return), threads in the same block are bundled into *warps* (cadres of 32 threads). Via SIMD, all 32 threads simultaneously complete the same kernel instruction until they collectively read from global memory; if they access a contiguous region, this *coalesced* read forms a single transaction [2]. The SM requests the memory, sets the warp aside, and picks up an idle warp whose memory has arrived and been cached, processing the new warp until *it* reads global memory. Since each SM can currently house only 8 blocks at a time, (***rule 4***) blocks should contain as many threads as possible so there are always idle, ready warps [2]. Nonetheless, SM memory bus bandwidth is finite, so (***rule 5***) global memory traffic should be kept to a minimum. Just like carpooling, coalescing reduces global memory traffic by reducing redundancy; but utilizing several other types of GPU memory (shared, constant, textures, registers) is crucial [1]. Learning how to use these memories effectively is the main challenge of GPU computing [1, 2].

1.2. CUBLAS and CULA

Maintaining their original API, BLAS and LAPACK have been translated into CUDA C (as CUBLAS and CULA, respectively). CUBLAS is distributed free by Nvidia. CULA uses CUBLAS under the hood, and is licensed by *EM Photonics*, with yearly subscriptions for commercial and academic use. Both libraries can be called on Nvidia devices via CUDA C, or directly from FORTRAN.

2. FEFF 8.4

FEFF 8.4 is software, written in FORTRAN 77 and licensed by the University of Washington, which simulates XAFS spectrum ab initio. To calculate self-consistent atomic potentials, the SCF module is activated, creating a long iterative loop [3]. Activating the FMS module calculates the full multiple scattering, which is also extremely time-consuming. However, using these modules helps generate the best simulated spectra, which are most useful for fitting experimental data [4].

2.1. Subroutine fms

The beating heart of both the SCF and FMS modules is the subroutine *fms*. For a cluster of *n* atoms, *fms* uses a *DO* loop to fill the atomic scattering () and electron propagator () matrices for a given energy. It then calculates the Green's function, , where *I* is the identity matrix [4]. By default, is inverted via LU **f**actorization, and is found via LU **s**olving (with back-substitution) using the LAPACK routines *cgetr**f*** and *cgetr**s*** [3]. Since *DO* loops are a GPU niche and CULA allows LAPACK to be called on a device, *fms* is a good candidate for GPU conversion.

2.2. Tiling : A CUDA algorithm

To fill and on the device rather than the host, the symmetric *n*x*n* matrix storing a cluster's weighted inter-atomic distances must exist in device global memory; the easiest way to accomplish this is to create on the device. Each element , where *k* is the complex wavenumber. First, a cluster's list is downloaded into device global memory. Then each is assigned to a thread, which will need 48B of global memory (2 s with 3 *floats* each). Reading each $2n$ times, once for every element in its corresponding row and column, would overtax the global memory bus. Using *shared memory* and *tiling* can mitigate the redundancy and congestion [1].

On a GPU, matrices can be divided into a regular lattice and assembled from a number of small, identical *tiles*. Each tile is a thread block with threads (with for warp efficiency) [2]. The largest square tile is a *square warp* (32x32). When tiling with square warps, each block needs 64 s; 32 for the rows it spans, and 32 for its columns. Each block can reduce its calls to global memory 32-fold by caching these 192 *floats* in shared memory, which is internal to its SM, and is quickly accessed by the entire thread block (though inter-block sharing is prohibited by *rule 2*) [1]. The caching of each float is handled by a thread. As only 192 threads per block (6 warps) need to participate, the caching is delegated to the first 6 warps via an *if* statement. Such inline control usually leads to GPU inefficiency, but extra SM passes occur only when threads *within the same warp* diverge [2]. After the cache is filled, each thread calculates its . Finally, each warp performs a coalesced write to global memory (adjacent threads filling adjacent elements in). Each column of tiles is a 1D grid (see *fig. 1*), with its own kernel call, because (***rule 6***) grids – a kernel's work order – must have a regular shape.

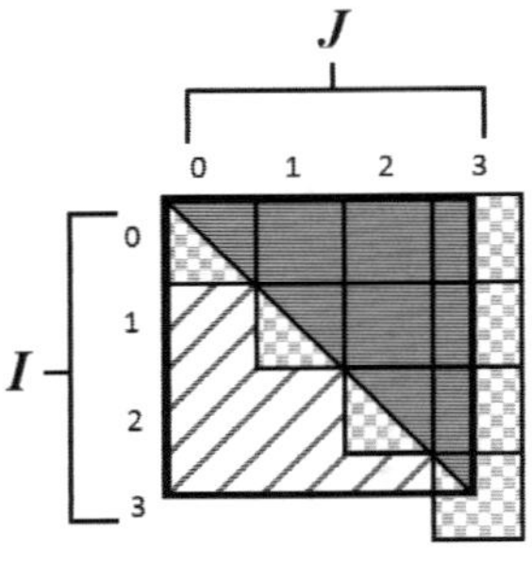

Figure 1. Tiling with lattice locations , where . GPUs have two classes of redundancy. *Small* redundancies (checkered) are a necessary evil, since they usually require inline control to avoid. Cluster size *n* is usually not a multiple of tile size, so tiled blocks often spill the boundaries of to fully cover it. Writing a control statement to avoid out-of-bounds memory access creates a divergence in at least *n* warps, so it is actually faster to allocate enough global memory to handle OOB indices and fill this extra space with garbage [2]. In contrast, *large* redundancies (striped), like filling *both* halves of a symmetric matrix, can be avoided (even though the additional small redundancy of sub-diagonal filling cannot). The ratio of the small redundancies to target data (solid) decreases substantially as *n* grows much larger than tile width.

2.3. Gfms (GPU fms)

The CUDA C kernels for filling and were incomplete at the time of publishing; the matrices were filled by the host into "pinned" host RAM, via modified FEFF 8.4 CPU code, then were copied into device global memory. Copying is fastest to/from "pinned" host RAM, which cannot be sent to a page file [1]. A simple CUDA C kernel calculated , and was created using CULA to replace the original LAPACK calls. The GPU used was an Nvidia Tesla C2050 with error correcting code.

3. *fms* (CPU) vs. *Gfms* (GPU)

The runtimes of *fms* and *Gfms* were evaluated on crystalline MnO with varying cluster size *n*. The matrix is , where (is the largest orbital mapped). The device's internal clock was used for device benchmarks and the function *clock* in "time.h" was used for CPU/host benchmarks.

Comparing *table 1* with *figure 2* shows the acceleration (a smaller exponent) that each section of *fms* attained on the device, especially the time-limiting LU factorization (*b*). The amount of time the host spent controlling the device (*fig. 2d*) reveals that CULA calls are not asynchronous, using many host clock cycles. Calling *Gfms* asynchronously would require the host to use a CPU worker-thread for CULA control; otherwise the host will get relegated to CULA control until each CULA call returns.

Table 1. Runtime of Single-threaded CPU *fms* in milliseconds

	Filling &	(*a*)	*cgetrf* (*b*)	*cgetrs* (*c*)
CPU (ms)				

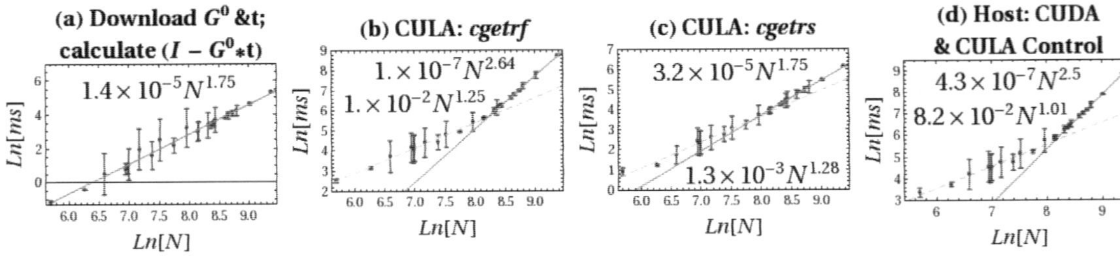

Figure 2. *Gfms* runtimes on an Nvidia Tesla C2050. Several FEFF 8.4 processes used the same device simultaneously. When two processes attempted simultaneous use of the device, one was forced to wait, creating inconsistent timings (large error bars). A device queue should alleviate this problem. The early fit lines (dashed) are very fast, but the asymptotic fit lines (solid) are the true device complexity.

3.1 Anticipated Gains with On-device and Filling

In *fig.* 2, (*a*) and *cgetrs* (*c*) accelerated from to . It is reasonable to assume that the unfinished and filling kernel will also boost from to , yet retain its CPU coefficient, thus allowing the runtime of a completed *Gfms* to be estimated.

The SCF module can call *fms* hundreds of times. The host spends a lot of time preparing each energy level (*k)* for its *Gfms* call, much longer than *Gfms* runtime. If *Gfms* is called asynchronously, then as long as the device uploads its result before the host calls *Gfms* on the *next* energy level, *Gfms* runtime will be completely hidden by host prep. To prevent device downtime, multiple host threads could prep energy levels in parallel on multiple CPU cores. With *x* host threads, as long as the device time needed by each *Gfms* call is as long as the *Gfms*-prep time, there should be no conflicts in device usage. Both host and device will be constantly busy, and *Gfms* will realize the full GPU benefit.

The FMS module calls *fms* continuously (no prep). *Gfms* leaves nothing for the host to do but device control. Retaining *some* of the and filling on the host might balance the workload better.

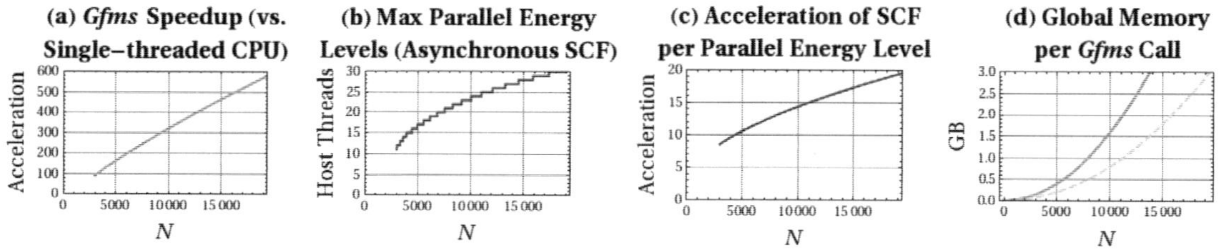

Figure 3. Estimates of *Gfms* (*a*) and SCF (*b, c*) performance. A C2050's 3GB of global memory sets a hard limit on *N*. In (*d*): actual use by *Gfms* (solid) and theoretical gains from an efficiency upgrade (dotted).

3.2 Error Considerations

Floating point arithmetic is non-associative [1]. Since blocks execute in random order, *unit-in-the-last-place* errors occur in GPU calculations, when compared to CPU results. For and , *eq. 1* was used to compare a matrix uploaded from the device to one generated by the original FEFF 8.4 code (*fig. 3 b,c*).

(1)

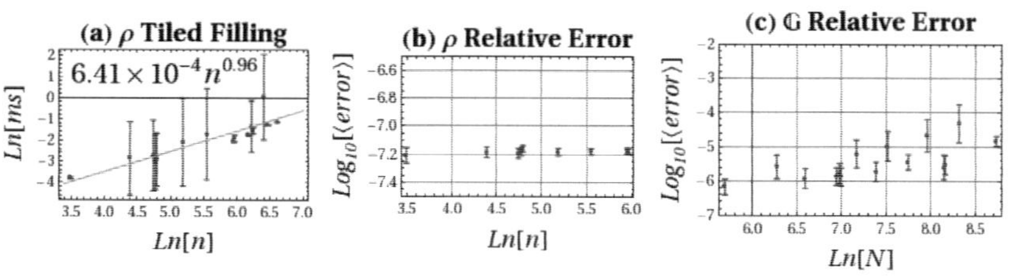

Figure 4. fills quickly on the device (*a*). Direct CPU runtime was unavailable, as filled faster than the resolution (10 ms) of the CPU timing function *clock*. The anticipated CPU complexity is .

4. Discussion

FEFF 8.4 achieved significant performance gains on the GPU; a completed, asynchronous *Gfms* should expand them. Many XAFS algorithms are candidates for GPU acceleration (e.g. FEFF 9). Even cluster computers stand to benefit from GPU computing; a distributed host can control *many* devices.

5. References

[1] Sanders J and Kandrot E 2011 *CUDA by Example* Addison-Wesley
[2] Kirk DB and Wen-mei WH 2010 *Programming Massively Parallel Processors* Elsevier
[3] Ankudinov A L, Ravel B and Rehr JJ *et al* 2005 FEFF 8.4 Source Code
[4] Ravel B 2005 *Journal of Alloys and Compounds* **401** 1-2 pp 118-126

Journal of Physics: Conference Series **430** (2013) 012009 doi:10.1088/1742-6596/430/1/012009

The application of molecular dynamics to fitting EXAFS data

Stephen W.T. Price[1*], Nicholas Zonias[1], Chris-Kriton Skylaris[1], Andrea E. Russell[1*] & Bruce Ravel[2*]

[1] Chemistry, University of Southampton, Highfield, Southampton, SO17 1BJ, UK

[2] National Institute of Standards and Technology, 100 Bureau Drive MS 8520, Gaithersburg MD 20899, USA

Email: s.w.t.price@soton.ac.uk; a.e.russell@soton.ac.uk; bravel@bnl.gov

Abstract. In recent work, a new method for using the outputs of molecular dynamics simulations to generate enhanced inputs for EXAFS analysis of nanoparticles was reported. Although currently limited to the first coordination shell, the method was demonstrated to improve both the quality of the fit, and the cross-correlation of the EXAFS data with that from TEM and XRD. Here we discuss further the justifications behind this assertion.

1. Introduction

Measuring the coordination number of highly disordered systems using EXAFS can be a complex task. When cross-correlating the data with XRD and TEM, the average size determined by EXAFS is smaller [1-3], most often a result of the failure of the fitting model to accurately account for the high degree of disorder present. One of the most common approaches to analysing EXAFS involves quantifying the atom-atom pair distribution, or radial distribution function (RDF), extracted from the data by approximating a Gaussian or near-Gaussian distribution, allowing for the determination of an average coordination number relative to the position of the absorber to be determined, along with the disorder of the absorbing atoms at that distance to be measured. Previously, the addition of higher order cumulants [4] to the analysis has been used to account for small deviations from a Gaussian distribution of bond lengths, whether through skewedness or kurtosis of the distribution. This is effective for small degrees of disorder, however variations on bond length within a sample can give rise to a distribution that is highlynon-Gaussian and therefore cannot be modelled using the cumulant method. Much theoretical work has been reported on the accurate determination of coordination number from EXAFS [5-8]however this has not been directly applied to fitting real data. In these studies the assumption was still on a Gaussian distribution of bond lengths for each coordination shell, with no account for uneven bond length contraction such as that brought about at the surface of a nanoparticle due to termination of the structure.

The effect of surface disorder on the RDF of nanoparticles has recently been investigated by Frenkel [9] and highlights how the increased bond length contraction present at the surface results in a deviation from a Gaussian bond length distribution. This contraction in surface bond length has been observed in coherent diffraction studies of nanoparticles [10]. In recent work we illustrated how the output of a molecular dynamics simulation could be used to provide an enhanced structural input for EXAFS fitting [11]. The nanoparticle structure was simulated [12] using DL_POLY 2.0; the histograms generated from the output and scattering paths generated using Artemis [13]. Full details

Journal of Physics: Conference Series **430** (2013) 012009 doi:10.1088/1742-6596/430/1/012009

of the nanoparticle simulations and characterisation have been published previously [11]. Briefly, four approximately spherical Pt nanoparticles consisting of 55, 177, 381 and 767 atoms, representing sizes of 1.39, 1.94, 2.50 and 3.05 nm respectively were simulated, and the outputs used to generate the histograms that provided the EXAFS fitting inputs.

2. Interpretation of the fits using the molecular dynamics input and histogram approach

Computation of the "true" coordination number of a nanoparticle from MD simulations over a range of temperatures has been shown to differ from those determined from the standard EXAFS analysis [1, 14, 15]. The difference between the real and apparent coordination number is exacerbated by both increases in temperature of the measurement and decreases in the size of the particles under investigation.

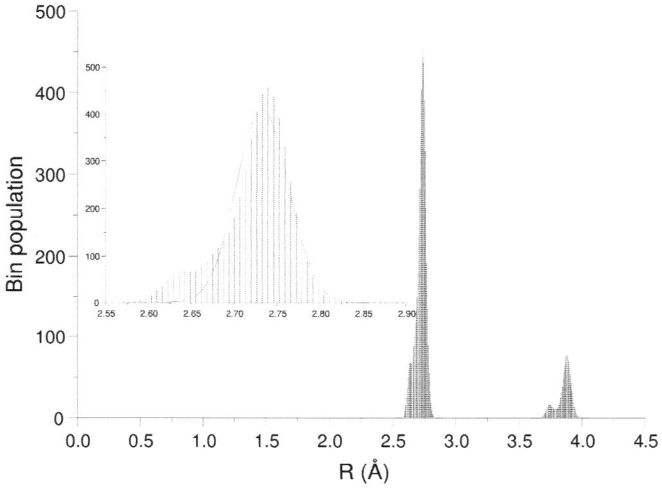

Figure 1: Sample histogram generated from MD output, with Gaussian fit over first shell (red line - inset). Vertical lines are the sum of nearest neighbour bond distances forming a radial distribution function[11].

Table 1: Cross correlation of average particle sizes determined by XRD, TEM and EXAFS analyses (XRD is Rietveld determined average crystallite size). XRD size determination was not possible for the 10 wt.% sample due to very poor crystallinity. The TEM results are volume weighted and the error reported is the standard deviation of the particle size distribution. EXAFS particle size was determined based on the formulae of Jentys [16] and are the average of fits at 20, 150 and 300 K; the reported error is the variance. Results previously published by Price *et al.* [11].

Sample wt.% Pt/C	Average particle size / nm					
	XRD	TEM		EXAFS		
	D_{Vol}	Particles observed	D_{Vol}	Standard	Cumulant	Sutton-Chen
10	---	354	1.6 ± 0.5	1.5 ± 0.1	1.8 ± 0.4	1.6 ± 0.1
20	1.5	445	2.4 ± 0.7	1.9 ± 0.3	2.1 ± 0.4	2.0 ± 0.4
40	1.8	600	2.6 ± 0.7	2.3 ± 0.3	1.9 ± 0.3	2.5 ± 0.3
60	2.3	433	3.0 ± 0.8	2.7 ± 0.3	2.0 ± 0.4	2.9 ± 0.3

For larger nanoparticle systems the contribution of the anharmonicity towards the overall RDF is small as a result of the vast majority of atoms residing in the core of the nanoparticle. In these cases it may be possible for the addition of higher cumulants to adequately model this disorder and therefore provide an improved fit. With smaller nanoparticles, as is the case with those studied here, there is a significant contribution to the RDF from the asymmetry of surface atom contractions (Figure 1) that cannot be adequately described by a Gaussian distribution. As a result the addition of higher cumulants fails to give any physically realistic improvement to the fit. This is most simply observed when comparing the trend in the average particle size (Table 1). The standard EXAFS analysis, XRD and TEM determined particle sizes all follow the same trend with particle loading (as do the results from the MD analysis) however there is no clear trend when adding higher cumulants to the analysis.

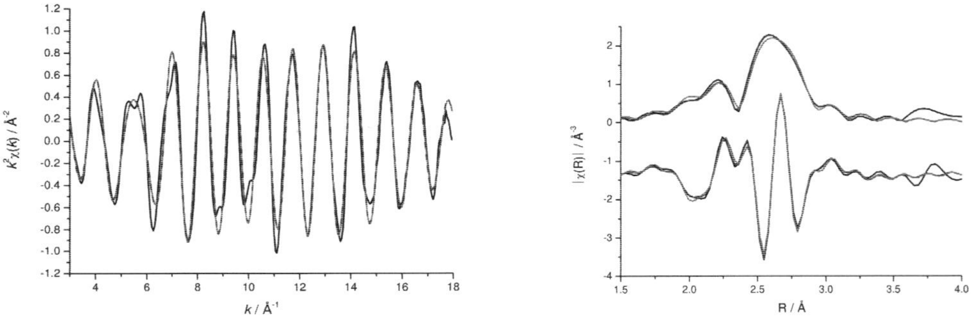

Figure 2: k^2 weighted experimental data (black) and fit to standard model (red), and Sutton-Chen (blue dashed) potentials, along with k^2 weighted magnitude and real Fourier transform for 40 wt.% Pt/C at 150 K[11].

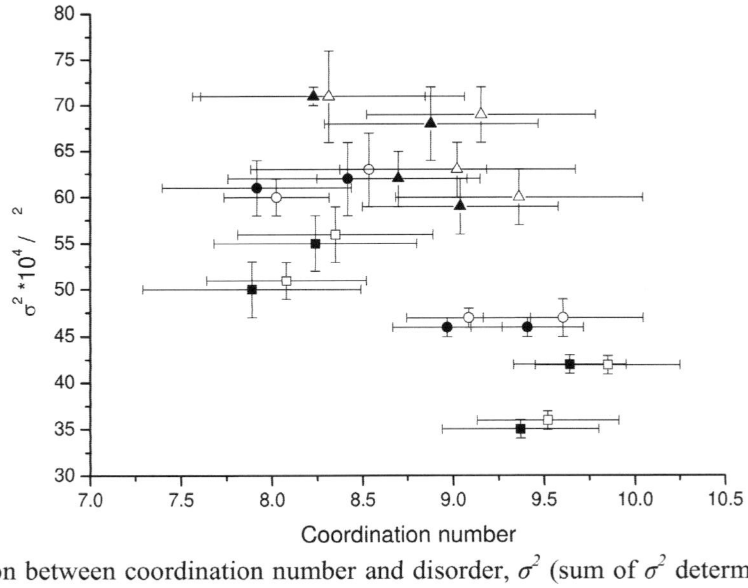

Figure 3: Correlation between coordination number and disorder, σ^2 (sum of σ^2 determined from fit and σ^2 contained in MD output). Closed points are those from the standard method, open points are those from the MD method. The 20 K MD simulations were used as the inputs for the MD approach; experimental data at 20 K is represented by a square marker (□), at 150 K by a circular marker (○) and at 300 K by a triangular marker (Δ).

As previously discussed [11] there is little visual difference in the fits to the data between the standard approach and that using the MD approach, as illustrated in Figure 2. The improvement in the fit of the MD approach over the standard approach is asserted to arise from the reduction of errors for each determined parameter, and the improved goodness of fit. Nominally, a necessary condition for asserting an improved fit is that the parameters differ outside of the error bars. Whilst the coordination numbers and disorder determined from the fits do not necessarily differ outside of the error bars (Figure 3), the trend for each parameter is such that in all cases the average coordination number determined from the fit is larger when using the MD approach. When converted into an average particle size [16] this brings the EXAFS into better correlation with TEM and XRD. The relationship between coordination number and σ^2 in both models and the improvement that the MD approach offers can be better understood by comparing the relationship between the two parameters(Figure 3).

The σ^2 values obtained using the RDF histogram are less than those obtained using the standard model. The disorder of the first shell was calculated directly from the simulation,i.e. the mean square relative displacement of all nearest neighbour atom-atom pair bond lengths, (dotted red arrows, Figure 4) and combined with the σ^2determined during the fit (red open points, Figure 4) to provide a more direct comparison (open black points Figure 4) with the disorder determined using the standard method (closed black points Figure 4). The smaller nanoparticle simulations contained a higher degree of disorder than the larger simulations. For example, the σ^2 value for the 177 atom simulation (providing the best input to fit the 10 wt.% Pt/C sample – three furthest left open points in Figure 3 and 4) was 0.0016 Å^2 whereas the σ^2 value for the 381 atom simulation (providing the best input to fit all other samples) was only 0.0011 Å^2. Within error, the combinedσ^2 values fall in the same range as that determined by the standard method however there is no clear trend; some values are marginally larger, some smaller, and some the same.

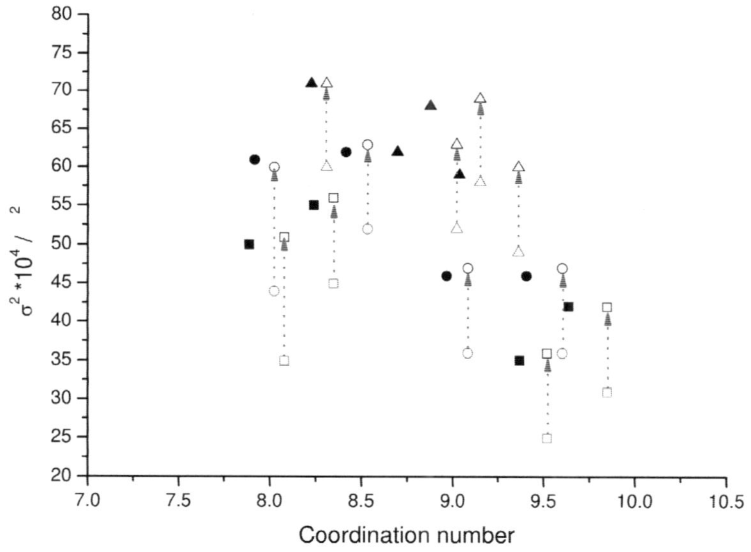

Figure 4: Correlation between coordination number and disorder, σ^2. The black closed points are those determined from the standard fitting approach. The red open points are the values determined from the fit using the histogram without correction. The black open points are the values once the σ^2 computed by the MD has been added to that determined in Demeter. The red arrows are there to link each before and after pair of points. The 20 K MD simulations were used as the inputs for the MD approach; experimental data at 20 K is represented by a square marker (□), at 150 K by a circular marker (○) and at 300 K by a triangular marker (Δ).Error bars are not included for ease of interpretation.

The first shell coordination numbers increase across all of the fits performed resulting in a much clearer trend than with the disorder. Although this is not a large increase in coordination number, and falls within the generally accepted error limits [17] for coordination number determination, the improved qualities of fit indicate that the larger size given by the MD approach is physically reasonable. Indeed when combined with TEM and XRD measurements (Table 1), the MD approach gives a better interpretation of the data than the standard approach and overcomes some of the limitations in applying bulk theory to the analysis of nanoparticles.

Using a more structurally disordered input from the MD gives an improved model to fit the nanoparticles. The MD simulations clearly do not completely model the disorder of the nanoparticle as a σ^2 term is still required in the fit [11, 18]; however it is unlikely that any of the nanoparticles are the perfect "spheres" modelled as is often assumed when fitting nanoparticles.

3. Conclusions

It is worth emphasising that although all of the coordination numbers determined from the fits fall within the error bars, every single time the results consistently shift in the same direction. This shift is as expected, both in light of the cross-correlation with other measurements, and with previous studies of the determination of nanoparticle coordination number [1, 14, 15].For any one individual measurement a comparison of two different fits where the fitted parameters fall within the error bars would hold little to no significance. However, given the same trend occurs for four different samples, measured over a range of temperatures, is independent of the MD simulation used as the input, and that it brings the EXAFS into better agreement with other measurements, we assert that the use of the MD simulations and the histogram approach provides an improved method for fitting nanoparticle EXAFS data. Even with the best experimental set-up there will always be some error associated with the raw data that has an effect on the parameters determined and the quality of fit; however, as the same data set is used to compare each fitting method, this will not be the cause of any differences in the parameters determined from each fitting method. This improved fitting method can also double as a measurable metric for evaluating the effectiveness of molecular dynamics simulations, whether classical or non-classical, for modelling nanoparticle behaviour. Currently, the histogram approach can be applied to single scattering paths generated in Artemis from the molecular dynamics outputs of DL_POLY [12], LAMMPS [19] and VASP [20], with limited effort required to include other sources.

Acknowledgements

The authors acknowledge the EPSRC and Johnson Matthey for funding and would like to thank Dr B. Theobald, Dr T. Hyde and Dr D. Ozkaya of Johnson Matthey Technology Centre, Sonning Common for supplying the catalyst samples, and for XRD and TEM analysis. N.Z. would like to thank the Southampton Nanoforum for a PhD studentship. C.-K.S would like to thank the Royal Society for a University Research Fellowship. Use of the National Synchrotron Light Source, Brookhaven National Laboratory, was supported by the U.S. Department of Energy, Office of Science, Office of Basic Energy Sciences, under Contract No. DE-AC02-98*C*H10886.

References

[1] B. S. Clausen, H. Topsøe, L. B. Hansen, P. Stoltze, and J. K. Nørskov 1994, *Catal. Today* **21**, 49.
[2] E. Curis and S. Benazeth 2005, *J. Synchrotron Rad.* **12**, 361.
[3] T. Shido and R. Prins 1998, *J.Phys. Chem. B* **102**, 8426.
[4] G. Bunker 1983, *Nuclear Instruments & Methods* **207**, 437.
[5] R. B. Greegor and F. W. Lytle 1980, *J. Catal.* **63**, 476.
[6] A. I. Frenkel 2007, *Z. Fur Kristallographie* **222**, 605.
[7] J. M. Montejano-Carrizales, F. Aguilera-Granja, and J. L. Moran-Lopez 1997, *NanoStructured Mat.* **8**, 269.
[8] A. M. Beale and B. M. Weckhuysen 2010, *Phys. Chem. Chem. Phys.* **12**, 5562.

[9] A. Yevick and A. I. Frenkel 2010, *Phys. Rev B*81, 115451.
[10] W. J. Huang, R. Sun, J. Tao, L. D. Menard, R. G. Nuzzo, and J. M. Zuo 2008, *NatureMat.*7, 308.
[11] S. W. T. Price, N. Zonias, C.-K. Skylaris, T. I. Hyde, B. Ravel, and A. E. Russell 2012, *Phys. Rev. B*85, 075439.
[12] W. Smith and T. R. Forester 1996, *J. Mol. Graphics*14, 136 (1996).
[13] B. Ravel and M. Newville 2005, *J. Synchrotron Rad.*12, 537.
[14] B. S. Clausen, H. Topsoe, L. B. Hansen, P. Stoltze, and J. K. Norskov 1993, *Japanese Journal of Applied Physics Part 1-Regular Papers Short Notes & Review Papers*32, 95.
[15] B. S. Clausen, L. Grabaek, H. Topsoe, L. B. Hansen, P. Stoltze, J. K. Norskov, and O. H. Nielsen 1993, *J. Catal.*141, 368.
[16] A. Jentys 1999, *Phys. Chem. Chem. Phys.*1, 4059.
[17] M. Vaarkamp 1998, *Catal. Today*39, 271.
[18] O. M. Roscioni, N. Zonias, S. W. T. Price, A. E. Russell, T. Comaschi, and C.-K. Skylaris 2011, *Phys. Rev. B*83, 115409.
[19] S. Plimpton 1995, *J. Comp. Phys.*117, 1.
[20] G. Kresse and J. Furthmuller 1996*Phys. Rev. B*54, 11169.

Journal of Physics: Conference Series **430** (2013) 012010 doi:10.1088/1742-6596/430/1/012010

Copper *L*-edge spectra: multiplet vs. multiple scattering theory

W Xu[1*], X-L Zhang[1], Z-Y Guo[1], C Si[1], Y-D Zhao[1], A Marcelli[2,3*], D-L Chen[1], and Z-Y Wu[3*]

[1]*Institute of High Energy Physics, Chinese Academy of Sciences, Beijing, 100049, China*

[2]*Laboratori Nazionali di Frascati - INFN, Frascati, 00044, Italy*

[3]*NSRL, University of Science and Technology of China, Hefei 230026, China*

E-mail: xuw@mail.ihep.ac.cn; marcelli@lnf.infn.it; wuzy@ustc.edu.cn

Abstract. Cu $L_{2,3}$-edge X-ray absorption spectra of the orthorhombic La_2CuO_4 have been investigated using the multiplet method and the multiple scattering theory. We found that *d* states of Cu^{2+} are not localized and both long- and medium- range atoms contribute to the $L_{2,3}$ edge spectrum, i.e., the transition from 2p electrons to partial and local empty 3d states. The behaviour is different from that observed in many other transition metals in which localized *d* states are merely influenced by the crystal field with a strong spin-orbit interaction.

Introduction

The transition metal $L_{2,3}$ edge X-ray absorption spectroscopy (XAS) has been widely used to investigate *3d* empty states in many materials and probes both the spin-orbit interaction and the crystal-field contribution. The La_2CuO_4 is the simplest cuprate, which is characterized by a LaO-CuO_2-LaO unit cell also known as a single-layer cuprate. [1] Cu L_3-edge XAS spectra of cuprates were investigated to extract information regarding their electronic structure [2,3], the valence state of copper, [4] site symmetry [5], while complementary or verified results can be achieved from interpreting both edges. [6,7] The L_2 and L_3 edge spectra probe transitions from $2p^{1/2}$ and $2p^{3/2}$ orbital levels, respectively, to 3d empty states. These soft X-ray absorption edges have been widely utilized to investigate charge-transfer effects, Coulomb interactions and other electronic effects in many transition metal oxides.[2-7] The theoretical framework to interpret $L_{2,3}$ edge is based either on the atomic-like multiplet method or on the single-particle approximation full multiple scattering theory.

The arrangement of CuO_6 octahedrons with respect to the lanthanum atoms is affected by the different oxygen atoms. As shown in Figure.1, the tilting of the distorted CuO_6 octahedrons is characteristic of the orthorhombic symmetry. Consequently, it is extremely interesting to reconstruct the local structure around Cu atoms and its electronic interaction with neighbouring atoms, i.e., lanthanum and oxygen.

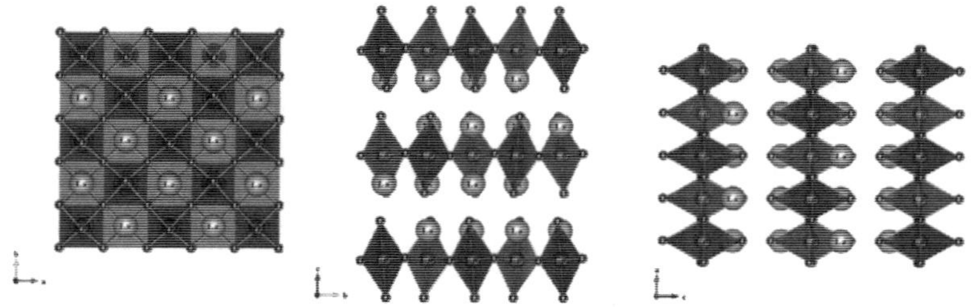

Figure 1 The crystal structure of the La_2CuO_4 observed from different axis (see text).

Experiment

The La_2CuO_4 polycrystalline ceramic was synthesized by Spark Plasma Sintering method X-ray absorption. Details about the synthesis and characterization are reported elsewhere [8]. The dense ceramic pellets were grounded into fine powders using a mortar. The powder was evenly brushed on the surface of a carbon tape that was stick to a copper sample holder for the X-ray absorption spectroscopy (XAS) measurements. XAS spectra were collected at the 4B7B endstation of the Beijing Synchrotron Radiation Facility.

Calculation details

Cu $L_{2,3}$ XANES spectra have been simulated using two different approaches: the multiplet theory which due to localized nature of L_{23} edges of transition metals emphasizes crystal field effects and spin-orbit interactions; and the multiple scattering theory which describes the interactions between the central photo absorber and the coordination scatterers in the real space. For the multiplet we employed the strategy implemented in the CTM4XAS package. Details about parameters are presented and discussed in the next section. For the multiple scattering calculations we used the FDMNES code (2012). [9] The muffin-tin potential and Hedin-Lundqvist exchange potential were chosen in the Green mode, while the non-muffin-tin Finite Difference Method was also used for comparison.

Results and discussions

Multiplet

The multiplet method [10] emphasizes the atomic spin-orbit interactions of the photo absorber and the crystal field effects imposed by the coordinated ligands. Although it is easy to reproduce the L edge for other metal edges as shown in Ref. [10], it is not possible to reproduce Cu L_{23} XANES spectra of the La_2CuO_4 system by playing the orbital-charge related parameters. In Figure 2, we present here only the evolution of both shape and position of L_2 and L_3 edges with respect to different spin-orbital couplings of 2p core states and 3d valence states. The energy difference between L_2 and L_3 edges are linearly dependent by both 2p and 3d spin-orbital couplings. However, the multiplet approach is not sufficient to reproduce Cu the L_{23} XANES spectra of the La_2CuO_4 system.

Full Multiple Scattering Theory (FMST)

The measured Cu $L_{2,3}$ edge XANES are shown in Figure 3(a). They are pronounced $2p^{1/2}$-3d and $2p^{3/2}$-3d transitions whose spectral shape can be qualitatevely fitted using two lorentzian profiles and one arctangent step function. The central energy for L_2 and L_3 edges is 950.5 eV and 930.5 eV, respectively; while the weak bump feature appearing in between edges has a maximum at 938.2 eV. Besides the peak positions, the peak intensity of L_2 and L_3

edges as obtained by fitting is 6.1 and 10.7, respectively, giving rise to a branch ratio of approximately 3:5.

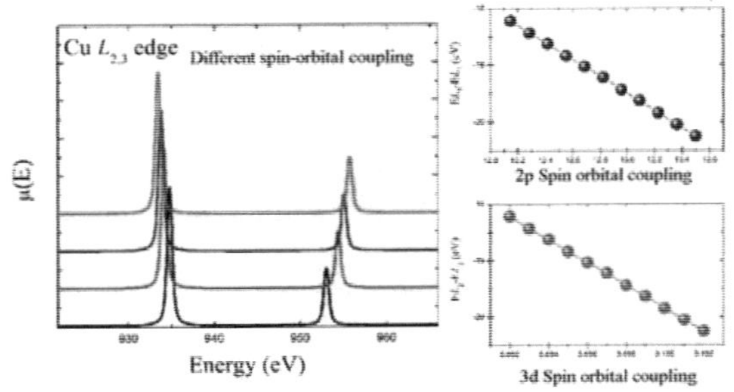

Figure 2 Comparison of Cu $L_{2,3}$ edge XAS spectra calculated using the multiplet approach with different atomic spin-orbit coupling parameters.

As shown in Figure 3(b)-(e), we compare spectra simulated using different clusters in the framework of both the full multiple scattering theory with muffin-tin potentials, as well as the finite difference method (FDM). The calculation reproduces the L_2 and L_3 peak features well, indicating that the calculations converges at 8 Å. Nevertheless, new bump features appear in between L_2 and L_3 peak for clusters larger than 8 Å. This discrepancy is due to overestimation of multiple scattering from long-distance atoms. This is reasonable due to the localization of *d* states. The small energy shift at the L_2 edge for a cluster of 8 Å can be assigned to both distortion of the local structure around copper atoms or defects. The finite difference method (FDM) overcome the limitation of the muffin-tin potential but is limited by computation power. By using a small cluster, it is demonstrated the calculation can be slightly improved using full potential method, as shown in Figure 3(e).

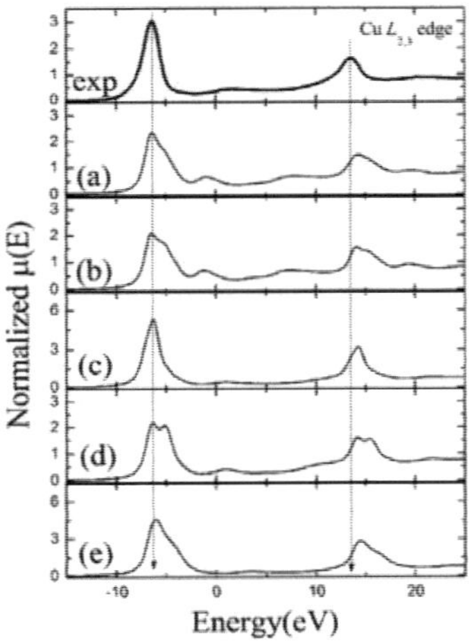

Figure 3 Comparison of experimental and theoretical spectra at the Cu $L_{2,3}$ edge of the La_2CuO_4 system. Theoretical spectra have been calculated with the Green mode using a cluster of 10 Å (a); 9 Å (b); 8 Å (c); 6 Å (d), and (e) the FDM mode using a cluster of 6 Å.

Conclusions

We compared theoretical XAS spectra obtained with the atomic multiplet theory and the full multiple scattering approach. The multiplet theory cannot yet reproduce the experimental spectra while the full multiple scattering theory well reproduce both peaks although the calculation is limited by the localization of *d* states and to a cluster size of ~8 Å. Simulations point out that the full potential method have space for improvement in the description of these systems. Moreover, for this particular system, an integration of the full multiple scattering single-particle picture with the atomic screening field method appears mandatory.

Acknowledgments
The research has been supported by the National Natural Science Foundation of China (Grant No. 11105172). We thank Y. Joly for many fruitful discussions.

References

[1] N. Poccia, A. Ricci, G. Campi, M. Fratini, A. Puri, D. Di Gioacchino, A. Marcelli, M. Reynolds, M. Burghammer, N.L. Saini, G. Aeppli and A. Bianconi, Optimum inhomogeneity of local lattice distortions in La_2CuO_{4+y}, PNAS 109, 15685-15690 (2012)
[2] A. Bianconi, J. Budnick, G. Demazeau, A. M. Flank, A. Fontaine, P. Lagarde, J. Jegoudez, A. Revcolevski, A. Marcelli, and M. Verdaguer, Cu L_3 x-ray absorption of formally trivalent Cu compounds Physica. C **153-155** (1988) 117.
[3] A. Bianconi, P. Castrucci, M. de Simone, A. Fabrizi, M. Pompa, A. Marcelli, A.M. Flank and P. Lagarde, Nuovo Cimento **25** (1990) 737.
[4] M. Grioni, J. B. Goedkoop, R. Schoorl, F. M. F. de Groot, J. C. Fuggle, F. Schäfers, E. E. Koch, G. Rossi, J. M. Esteva, and R. C. Karnatak, Studies of copper valence states with Cu L_3 x-ray-absorption spectroscopy, Phys. Rev. B **39** (1989) 1541.
[5] M. Pompa, C. Li, A. Bianconi, A. Congiu Castellano, S. Della Longa, A. M. Flank, P. Lagarde, and D. Udron, Full multiple scattering analysis of linearly polarized Cu L_3-edge XANES of La_2CuO_4 Physica. C **184** (1991) 51.
[6] L. Braicovich, M. Moretti Sala, L. J. P. Ament, V. Bisogni, M. Minola, G. Balestrino, D. Di Castro, G. M. De Luca, M. Salluzzo, G. Ghiringhelli, and J. van den Brink, Momentum and polarization dependence of single-magnon spectral weight for Cu L_3 -edge resonant inelastic x-ray scattering from layered cuprates, Phys. Rev. B **81** (2010) 174533
[7] M. M. Sala, V. Bisogni, C. Aruta, G. Balestrino, H. Berger, N. B. Brookes, G. M. d. Luca, D. D. Castro, M. Grioni, M. Guarise, P. G. Medaglia, F. M. Granozio, M. Minola, P. Perna, M. Radovic, M. S. T. Schmitt, K. J. Zhou, L. Braicovich and G. Ghiringhelli, Energy and symmetry of dd excitations in undoped layered cuprates measured by Cu L_3 resonant inelastic x-ray scattering, New J. Phys. **13** (2011) 043026.
[8] Yuan-Hua Lin, Yong Liu, Bo-Ping Zhang, Ce-Wen Nan, Jing-Feng Li, and Zhijian Shen, Electrical Transport Properties of La_2CuO_4 Ceramics Processed by the Spark Plasma Sintering J. Am. Ceram. Soc. **90** (2007) 4005.
[9] Y. Joly, X-ray absorption near-edge structure calculations beyond the muffin-tin approximation, Phys. Rev. B **63** (2001) 125120.
[10] F. de Groot, Multiplet effects in X-ray spectroscopy, Coord. Chem. Rev. **249** (2005) 31.

Journal of Physics: Conference Series **430** (2013) 012011 doi:10.1088/1742-6596/430/1/012011

Coordination and Oxidation State Analysis of Cobalt in Nanocrystalline $LiGa_5O_8$ by X-ray Absorption Spectroscopy

B Yildirim and H Riesen

School of Physical, Environmental and Mathematical Sciences

The University of New South Wales, UNSW Canberra (ADFA), Canberra, ACT 2600, Australia

E-mail: h.riesen@adfa.edu.au

Abstract The luminescence of nanocrystalline $LiGa_5O_8:Co^{2+}$ has been assigned to Co^{2+} in tetrahedral sites. However, x-ray absorption spectroscopy (XAS) reveals that there is a significant fraction of cobalt ions that substitute into the lattice in the octahedral sites and, in this case, predominantly in the 3+ oxidation state.

1. Introduction

We have recently reported some persistent spectral hole-burning properties of nanocrystalline $LiGa_5O_8:Co^{2+}$ (LGO:Co) [1]. We have concluded that the hole-burning mechanism is most likely based on the photoionisation of Co^{2+} ions in the tetrahedral sites and the subsequent trapping of photo-electrons by Co^{3+} ions that may exist in the structure.

The Co^{2+} ions replace some of the Ga^{3+} ions when doped into the lithium gallate host which includes tetrahedral and octahedral sites. Macfarlane et al. [2] and Donegan et al. [3] suggested that Co^{2+} ions can enter the crystal lattice in both of these sites. It has been assumed that the percentage of cobalt ions occupying octahedral sites is lower than for tetrahedral sites [2, 3].

To verify these assumptions and in order to gain a better understanding of valence and coordination state distributions of cobalt ions in LGO, XAS experiments were conducted at the Australian Synchrotron. XAS experiments were conducted in the x-ray absorption near-edge structure (XANES) and extended x-ray absorption fine structure (EXAFS) regions. XANES, in particular pre-edge transitions from 1s to unoccupied 3d states, give the most valuable information on charge distribution. In addition to the comparison of the intensities and the shifts of the XANES first derivatives with reference samples, correlating the experimental results with numerical simulations of XANES and EXAFS provide quantitative information about the coordination and oxidation states of the cobalt ions.

2. Experimental details

The XAS experiment was performed at the Australian Synchrotron. A silicon 111 monochromator with a focusing ability from 0.1x0.5 mm to 1x2 mm was used. The spectrum was measured using a well-powdered sample grinded with cellulose, the mixture was pressed into a pellet and placed in an aluminium sample holders using Kapton tape, and the calibration was performed by measuring a cobalt foil in each run.

IFEFFIT ver. 1.2.11c [4] was used for the XAS analysis and the lattice host was modeled using starting parameters from the work of Ahman *et al.* [5]. Two different configurations of the Co atoms, in the octahedral and tetrahedral sites in place of the Ga atoms, were considered separately and corresponding scattering patterns were generated in ARTEMIS [6]. The same method was used for simulations of the XANES spectra in JFEFF 9.0 [7].

As Co was not only in the tetrahedral but also in the octahedral sites in this host, the fitting parameters were different for the two configurations. The best values were obtained separately and then the final EXAFS fit was obtained by merging of both the octahedral and tetrahedral peak fits. Satisfactory results were obtained when the first shell scatterings and low wave number (k) values were taken while the different valence states were omitted in the peak fitting.

3. Results and discussion

3.1 XANES

As a general qualitative trend, it is well established that the higher the oxidation state, the higher the energy positions for all features in XANES spectra, with a rule of thumb of 3eV per unit of oxidation state. Also, the larger the edge width, the more pronounced is the mixed valence state. In XANES studies of LGO:Co, the Co_3O_4 normal spinel has been used as a reference [8] to examine the oxidation and coordination states of cobalt in LGO since it is well known to contain both tetrahedrally coordinated Co^{2+} and octahedrally coordinated Co^{3+} ions in a 1:2 ratio [9].

Instead of a direct comparison of the XANES peak, we analysed the oxidation states of cobalt by a comparison of the peaks and intensities in the first derivative of the XANES spectra. The first LGO:Co peak was at 7709 eV (pre-edge) and those of its first-order derivatives at 7717 eV, 7721 eV and 7725 eV. Similarly, the first peak of Co_3O_4 was at 7709 eV (pre-edge) and those of its first-order derivatives at 7717 eV, 7722 eV and 7727 eV which were very close to the LGO:Co peaks. In Figure 1, XANES and their first derivative spectra are presented.

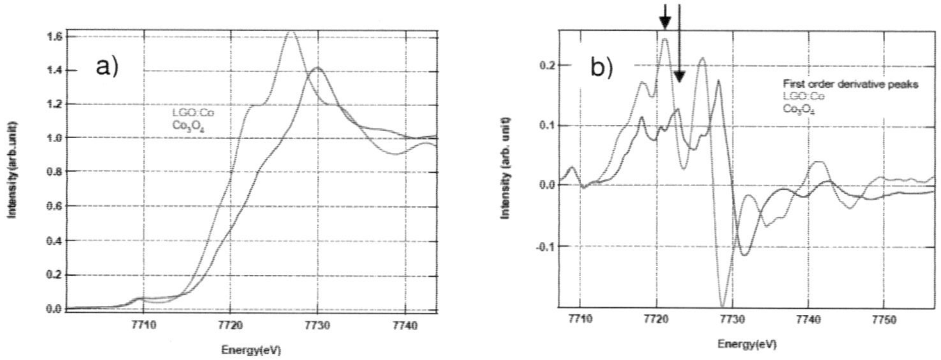

Figure 1: a) Normalised Co K-edge spectrum of LGO:Co (red), and the reference sample Co_3O_4(blue) b) First derivative of the spectra (same colours)

The derivative peaks at 7717 eV and 7721 eV were assigned to Co^{2+} and Co^{3+} respectively and these peaks were also observed in $[Co(NH_3)_5Cl]Cl_2$ and $[Co(NH_3)_6](SbCl_6)$ reference samples with edge energies at 7717 eV and 7722 eV respectively [10]. The shoulder on the 7717 eV peak in the first derivative spectrum of the LGO, around 7715 eV, is most probably due to the 1s to 4s transition.

The same percentage (50:50) of Co^{2+} was used for both the octahedral and tetrahedral sites of the host structure for the XANES simulations by JFEFF 9.0. The Co^{2+} peak was shifted around 1.6 eV on derivative peaks in simulated spectra (The simulations are not illustrated here). In the LGO:Co sample, the shift is around 0.8 eV for the prominent peak of Co^{3+} with respect to Co^{3+} peak position of $[Co(NH_3)_6](SbCl_6)$ [10] and Co_3O_4 (marked with arrows in Figure 1b) .The shift is due to existence of some Co^{3+} ions in the tetrahedral sites. The Co^{2+} ions in octahedral sites should be less than tetrahedrally coordinated ones since no shift was observed on the Co^{2+} peak.

The Co^{3+}/Co^{2+} peak intensity ratios are around 3:2 for LGO:Co and 6:5 for Co_3O_4 but the latter should be around 2:1. For the latter, there appears to be another peak in between the ones of Co^{2+} and Co^{3+} which could be resolved. This feature between the 2+ and 3+ peaks was assumed to be scattering from distorted octahedral sites. Taking into account the integrated intensity of this peak would increase the Co^{3+}/Co^{2+} ratio to around 2:1 for Co_3O_4. The distorted octahedral site was also observed in the pre-edge of Co_3O_4. In particular, the pre-edge of Co_3O_4 broadens and shifts towards the edge which indicates that the cause of the broadening is more likely due to distorted Co ions in the octahedral sites.

The pre-edge peak was also observed for LGO:Co confirming the existence of Co ions in the tetrahedral sites (Figure 1). The pre-edge is most intense when the first coordination shell lacks inversion symmetry, such as in a tetrahedral symmetry, while they are also pronounced for octahedral coordination in distorted symmetries [10, 11].

3.2 EXAFS

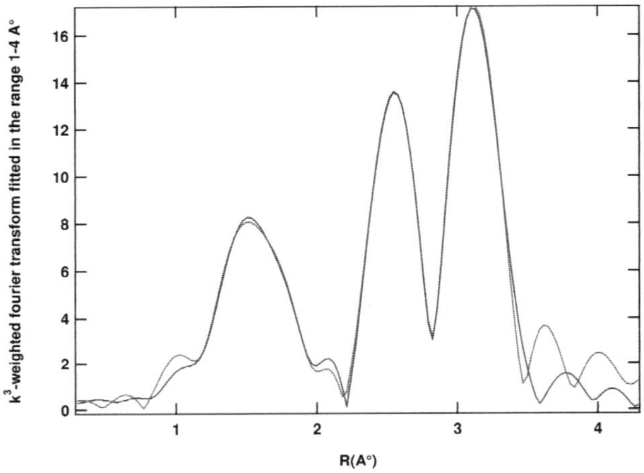

Figure 2: Weighted Fourier transform (k^3 (k)) of Co K-edge EXAFS (red) and its fit obtained with ARTEMIS (blue) (as we are interested in first shell scattering, R is taken in the range of 1-4 Å)

Different amplitude factors were used for the first shell fit to compensate for the percentage difference of octahedral Co ions which was not expected to be exactly 50:50 (in ARTEMIS, one set of scatterings from the first shell of the tetrahedrally coordinated Co and one from the octahedrally coordinated Co ions were taken for the final fit). The scatterings from the first shell lay in between 1-2 Å with Co-O bond lengths of 2.05 (±0.02) Å for the octahedral and 1.7760 (±0.005) Å for the tetrahedral configurations. The peaks from the second shell, which were mainly from the Ga atoms (Li atoms have a low scattering factor) showed two distinct peaks, the first formed by the Ga atoms closest to the octahedrally coordinated Co ion, at 2.9 Å, and the second by those further away, at 3.4 Å. All the other scatterings from the Ga atoms of the tetrahedrally coordinated Co lay between 3.4 and 3.5 Å.

4. Conclusion

From XAS measurements it follows that a significant percentage of Co ions enter the LGO lattice, indeed, also in the octahedral sites, and that both the +2 and +3 oxidation states are present. According to a XANES derivative peak analysis the Co^{3+}/Co^{2+} ratio is around 3:2 and both XANES and EXAFS spectra show that Co^{3+} ions exist predominantly in octahedral sites of the host structure. This result is commensurate with conclusions reached by Mössbauer spectroscopy of $LiGa_5O_8:Fe^{3+}$ in which Fe^{3+} enters the lattice predominantly in the Ga octahedral sites without any charge compensation requirement and is highly stable in these sites [12].

Acknowledgements

We would like to thank the Australian Synchrotron for access to the XAS beam line and for the provision of a travel allowance (AS093/XAS1957). BY acknowledges UNSW for a postgraduate research scholarship and for postgraduate research student support (PRSS) to attend XAFS-15.

References

[1] Yildirim B and Riesen H 2010 *Phys. Proc.* **3** 1547-1551
[2] Macfarlane R M and Vial J C 1986 *Phys. Rev.* B **34** 1
[3] Donegan J F Anderson F G Bergin F J Glynn T J and Imbusch G F 1992 Phys. Rev. B **45** 563
[4] M. Newville 2001 *J. Synchrotron Rad.* **8** 322-324
[5] Ahman J Goran Svensson G and Albertsson J 1996 *Acta Chem. Scand.* **50** 391
[6] Ravel B and Newville M 2005 *J. Synchrotron Rad.* **12** 537-541
[7] Rehr J J Kas J J Vila F D Prange M P Jorissen K 2010 *Phys. Chem. Chem. Phys.* **12** 5503-5513
[8] Co_3O_4 data was downloaded from International X-ray Absorption Society (IXAS) database , http://ixs.iit.edu/database/data/Farrel_Lytle_data/RAW/Co/index.html
[9] Murakami M Yatsumoto and Y Hasegawa T 2004 *J. Appl. Phys.* **95** 5330
[10] Moen A Nicholson D G Ronnnig M Lamble G M Lee J F and Emerich H 1997 *J. Chem. Soc.* **93** 4071-4077
[11] Groot F De Vanko G and Glatzel P 2009 *J. Phys: Condens. Matter* **21** 104207
[12] Yildirim B Stewart G and Riesen H 2012 *Proceedings of the 36th Annual Condensed Matter and Materials Meeting* 37

Journal of Physics: Conference Series **430** (2013) 012012 doi:10.1088/1742-6596/430/1/012012

An efficient implementation of the reverse Monte Carlo method for EXAFS analysis in crystalline materials

Janis Timoshenko, Alexei Kuzmin and Juris Purans

Institute of Solid State Physics, University of Latvia, Kengaraga street 8, Riga, LV-1063, Latvia

E-mail: janis.timoshenko@gmail.com

Abstract. Lattice dynamics of crystalline materials can be efficiently studied using reverse Monte Carlo (RMC) simulations of EXAFS spectra. In this work we demonstrate potentiality of the method on an example of the thermal disorder study in rhenium trioxide ReO_3.

1. Introduction

Reverse Monte Carlo (RMC) method is an atomistic simulation technique that allows one to reconstruct the 3D atomic structure of the material by minimizing the difference between its structure-related experimental and calculated properties [1]. In our recent work [2] we have shown that the RMC is a useful tool for the analysis of the extended x-ray absorption fine structure (EXAFS) in crystalline materials. The efficient implementation of the RMC algorithm allows us to obtain reliable results by simulating configurations with relatively small number of atoms (several hundreds in contrast to several thousands, usually required for the analysis of disordered materials). Our approach [2] significantly reduces the computational costs and accounts for the multiple-scattering (MS) contributions thus allowing one to extract from EXAFS spectrum the information not only on the interatomic distances, but also on the bond angles, anisotropy of atomic vibrations and even bond strength. Considering the MS effects is essential for crystals with cubic symmetry as rhenium trioxide (ReO_3), where the MS contributions are strongly enhanced due to the "focusing effects" within the collinear atomic chains [3, 4].

Here we present the results of the RMC-EXAFS study of lattice dynamics in ReO_3 over a broad range of temperatures from 10 K to 573 K.

2. RMC simulations for the model data

Before applying the proposed method to experimental data, we have validated it using the model Re L_3-edge EXAFS signal for ReO_3, obtained from molecular dynamics (MD) simulations [5]. The RMC calculations have been carried out for $4\times4\times4$ supercell (256 atoms) and 40000 iterations. A difference between the model and RMC EXAFS spectra has been minimized by the best fit of their wavelet transforms, i.e., simultaneously in k-space (from 3 to 18 Å^{-1}) and in R-space (from 0 to $R_{\max}$=4.5 Å) [2, 6]. The EXAFS spectra for the MD model and during the RMC simulations have been calculated using ab initio self-consistent real space multiple-scattering FEFF8 code [7]. As one can see in Fig. 1, due to the cubic perovskite-type structure

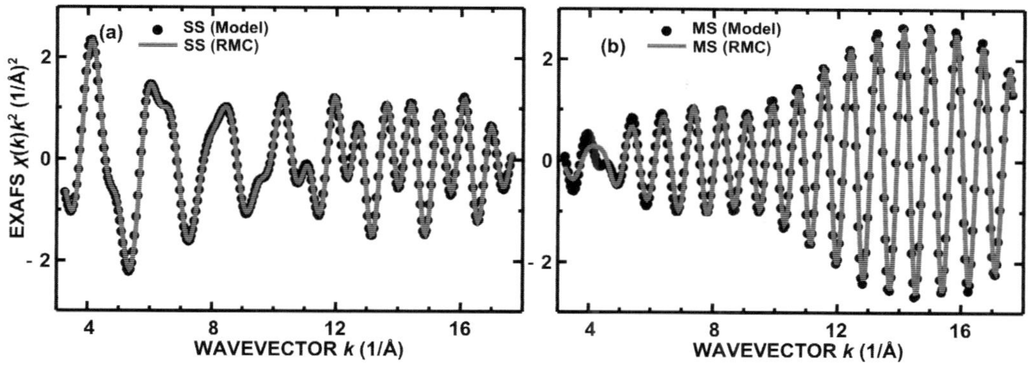

Figure 1. Model (dots) and reconstructed by the RMC simulation (lines) Re L_3-edge EXAFS spectra $\chi(k)k^2$ for ReO_3. The SS and MS contributions are shown separately.

of ReO_3, the contribution of MS effects to the total EXAFS signal is comparable to the one from SS paths. Moreover, our RMC scheme allows us to reconstruct and analyze simultaneously both SS and MS parts of the EXAFS signal.

The evaluated values of the mean-square radial displacements (MSRD) for the first two coordination shells (Re_0–O_1 and Re_0–Re_2) differ from that in the MD model by only about 0.0001 $Å^2$, and the average value of the Re_0–O_1–Re_2 angle is determined with noteworthy precision better than 0.01°. The MSRD value for the third coordination shell (Re_0–O_3) shows slightly worse agreement (within about 10%) due to the fact that this MSRD value is large and therefore the contribution of the third shell to the total EXAFS signal is relatively small.

3. RMC analysis of experimental data

Next we applied our method to the experimental temperature-dependent Re L_3-edge EXAFS spectra for ReO_3 [8]. The experimental EXAFS spectra for T=10 K and 573 K and the corresponding theoretical EXAFS spectra, calculated for the final atomic configurations obtained in the RMC simulations, are shown in Fig. 2. The agreement between the experimental and simulated EXAFS spectra is very good and implies that the obtained atomic configurations are representative sample of ReO_3 structure. One can consider separately the SS and MS contributions and analyze for the first-time the temperature dependence of the MS effects. Note that the MS effects give the maximum contribution to the total EXAFS signal in ReO_3 at about k=16 $Å^{-1}$, where it is about two times larger than the contribution of the SS effects, and this ratio is nearly independent on temperature.

The conventional EXAFS analysis can provide in ReO_3 only very limited accuracy beyond the first coordination shell due to the pronounced MS effects. Using RMC simulations, in turn, the structural information can be obtained straightforwardly. The radial distribution functions (RDFs) for Re_0–O_1, Re_0–Re_2, and Re_0–O_3 pairs are shown in Fig. 3(a), and the temperature dependences of the MSRDs, extracted from the RDFs, are given in Fig. 3(b) for the first three coordination shells. It should be emphasized that the proposed method gives us reliable information even at low temperatures - in the region where, for example, the classical molecular dynamics cannot be applied. For the first coordination shell Re_0–O_1, the RMC results are close to that obtained using the FEFFIT code [9].

Temperature dependences of the MSRDs in Fig. 3(b) can be approximated using the correlated Einstein model [10]. The Einstein frequency for the first coordination shell ω_E=94.9 THz is in good agreement with previously published result (ω_E=94.1 THz [11]). The Einstein frequencies for the second (36.2 THz) and third (60.8 THz) coordination shells are

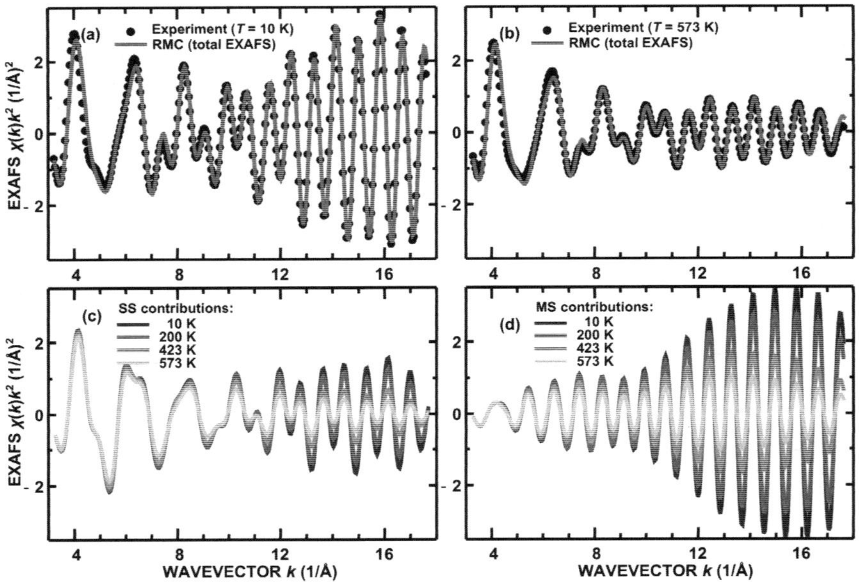

Figure 2. Experimental (dots) and RMC (lines) Re L_3-edge EXAFS spectra $\chi(k)k^2$ for ReO_3 at temperatures T=10 K (a) and 573 K (b). Temperature dependences of the SS (c) and MS (d) contributions are also shown.

obtained for the first time. It is important to note that while the Einstein frequency for the second shell (Re_0–Re_2 pairs) is significantly smaller than for the first one, due to the larger masses of involved rhenium atoms, the effective bond-strength coefficient $\kappa = \omega_E^2\mu$ (μ is the effective mass) for the Re_0–Re_2 pairs (203 N/m) is almost as large as for the Re_0–O_1 pairs (222 N/m) and is about twice larger than the effective bond-strength coefficient (91.2 N/m) for the third coordination shell.

This means that a strong correlation is present between the motion of the neighbouring Re_0 and O_1 atoms (the fact, usually emphasized in the explanation of negative thermal expansion (NTE) of ReO_3 lattice [12, 13]), but also between the motion of the next-neighbouring Re_0 and Re_2 atoms.

The NTE phenomenon in ReO_3 is usually explained in terms of the rigid-unit model (RUM) [12]. The strong correlation in the Re_0–O_1 atomic motion leads to the nearly temperature-independent Re_0–O_1 bond length, thus ensuring the rigidity of the ReO_6 octahedra. At the same time, an increase of temperature raises the vibration amplitude of oxygen atoms in the direction, orthogonal to the Re_0–Re_2 bonds. The two effects together result in a reduction of the Re_0–Re_2 interatomic distances. According to this model, the average value of the Re_0–O_1–Re_2 angle should decrease with the increasing temperature, and this effect can be accurately determined from the results of the RMC simulations. The described trend is clearly seen in Figs. 3(c) and (d). The average value of the Re_0–O_1–Re_2 angle decreases almost linearly from 174.7° at 10 K to 173.3° at 573 K. In turn, the statistical dispersion of the Re_0–O_1–Re_2 angle values increases from 3.1° at 10 K to 3.7° at 573 K (we used the so-called median absolute deviation as a robust estimator of dispersion). Thus, the temperature dependence of the Re_0–O_1–Re_2 angle agrees well with the structural model of the NTE in ReO_3.

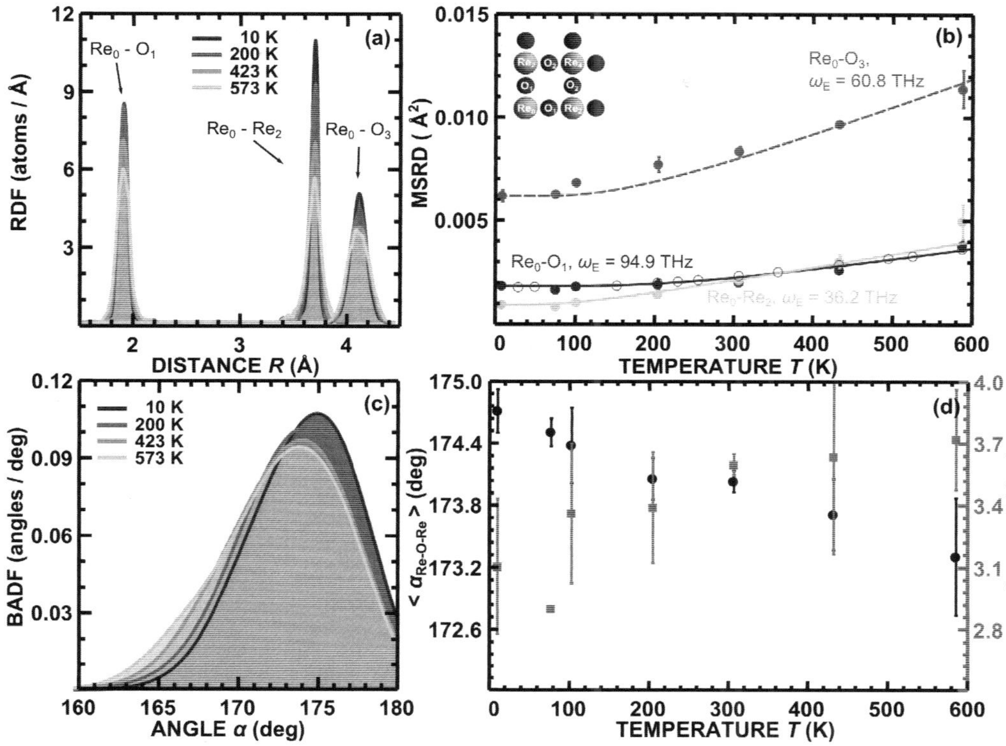

Figure 3. (a) Temperature dependent Re–O and Re–Re RDFs obtained from the RMC simulations. (b) Temperature dependences of the MSRDs for the Re–O and Re–Re pairs: filled circles - the RMC data; solid lines - approximations by Einstein model; open circles - the FEFFIT results, (c) Temperature dependence of the Re_0–O_1–Re_2 bond angle distribution function (BADF). (d) The average value of the Re_0–O_1–Re_2 angle (black circles) and its statistical dispersion σ (red squares).

4. Conclusions

We have demonstrated that reverse Monte Carlo (RMC) method can be successfully used to interpret EXAFS spectra of crystalline materials even in case, when the multiple-scattering effects are very pronounced. The analysis of the Re L_3-edge EXAFS data from the second and third coordination shells of rhenium in ReO_3 has been carried out for the first time taking into account both thermal disorder and multiple-scattering effects. The obtained results are in agreement with the rigid unit model of lattice dynamics in ReO_3. We affirm the strong correlation (i) between displacements of oxygen and nearest rhenium atoms and (ii) between oxygen motion in the direction orthogonal to the Re_0–Re_2 bond and the variation of the average Re_0–Re_2 distance. Our results reveal also the strong correlation between the motion of two nearest rhenium atoms.

Acknowledgments

This work was supported by ESF Projects Nos. 2009/0202/1DP/1.1.1.2.0/09/APIA/VIAA/141, 2009/0216/1DP/1.1.1.2.0/09/APIA/VIAA/044 and Latvian Government Research Grant No. 09.1518.

References

[1] McGreevy R and Pusztai L 1988 *Mol. Simul.* **1** 359

[2] Timoshenko J, Kuzmin A and Purans J 2012 *Comp. Phys. Commun.* **183** 1237
[3] Alberding N, Crozier E, Ingalls R and Houser B 1986 *J. Physique IV (France)* **47** C8–681
[4] Kuzmin A, Purans J, Benfatto M and Natoli C 1993 *Phys. Rev. B* **47** 2480
[5] Kalinko A, Evarestov R, Kuzmin A and Purans J 2009 *J. Phys.: Conf. Ser.* **190** 012080
[6] Timoshenko J and Kuzmin A 2009 *Comp. Phys. Commun.* **180** 920
[7] Ankudinov A, Ravel B, Rehr J and Conradson S 1998 *Phys. Rev. B* **58** 7565
[8] Purans J, Dalba G, Fornasini P, Kuzmin A, de Panfilis S and Rocca F 2007 *AIP Phys. Conf. Ser.* **882** 422
[9] Newville M, Ravel B, Haskel D, Rehr J, Stern E and Yacoby Y 1995 *Physica B* **208** 154
[10] Rehr J and Albers R 2000 *Rev. Mod. Phys.* **72** 621
[11] Dalba G, Fornasini P, Kuzmin A, Purans J and Rocca F 1995 *J. Phys.: Condens. Matter* **7** 1199
[12] Tao J and Sleight A 2003 *J. Solid State Chem.* **173** 442
[13] Chatterji T, Henry P, Mittal R and Chaplot S 2008 *Phys. Rev. B* **78** 134105

EXAFS Phase Retrieval Solution Tracking for Complex Multi-Component System: *Synthesized Topological Inverse Computation*

Jay Min Lee[1], Dong-Seok Yang[2] and Grant B. Bunker[3]

[1]Beamline Division, Pohang Accelerator Laboratory, POSTECH, 790-784, Korea,
[2]Department of Physics Education, Chungbuk National University, 361-763, Korea,
[3]Department of Physics, Illinois Institute of Technology, Chicago, IL 60616, USA

Correspondence e-mail: jaymin@postech.ac.kr

Abstract. Using the FEFF kernel **A(k,r)**, we describe the inverse computation from χ(k)-data to g(r)-solution in terms of a singularity regularization method based on complete Bayesian statistics process. In this work, we topologically decompose the system-matched invariant projection operators into two distinct types, ($\mathbf{A^+AA^+A}$) and ($\mathbf{AA^+AA^+}$), and achieved *Synthesized Topological Inversion Computation (STIC)*, by employing a 12-operator-closed-loop emulator of the symplectic transformation. This leads to a numerically self-consistent solution as the optimal near-singular regularization parameters are sought, dramatically suppressing instability problems connected with finite precision arithmetic in ill-posed systems. By statistically correlating a pair of measured data, it was feasible to compute an optimal EXAFS phase retrieval solution expressed in terms of the complex-valued χ(k), and this approach was successfully used to determine the optimal g(r) for a complex multi-component system.

1. Introduction

The numerical inverse computation method, which is fundamentally similar to a feedback stabilization of complex systems, has been investigated with substantial effort in a wide range of applications but there has been limited progress relative to its potential. Common numerical difficulties associated with the ill-posed nature of the problem stem from the statistical treatment of noise, the system configuration complexity in its topological aspects. Recent studies have shown the common applicability of concepts from topological mathematics, leading to computational feasibility of the phase retrieval solution [1]. Among these inverse-problem solving recipes, the singularity regularization method [2,3,4] has been received with significant interest as a unique way to seek the most reasonable solution under the so-called Bayesian statistics process [5,6], preserving maximum entropy under the time-sequential probability.

In EXAFS the essential goal is to construct the radial distribution function g(r) from χ(k) data in the presence of noise. Although this is a linear transformation, it is unstable under inversion, and therefore stabilization procedures (regularization) and regularization parameters are typically used. [2,4] There is an inherent tradeoff between smoothness of the solution and its accuracy. Commonly, the goal is to determine the optimal tradeoff, but yet insufficient to reveal a *topology-invariant* trajectory through the regularization parameter space. Emphasized, nature exists in a complex dynamic system, and therefore Bayesian treatment – based on causality of prior and posterior – should substitute Fourier filtering (originator of signal deformation), upon which EXAFS analysis was historically based [3,7]. Despite widespread interest by EXAFS investigators, optimal inverse computation code has not yet been developed in the framework of a linear matrix system. This paper describes an effort to address that need.

2. Numerical Procedures

Due to availability of complex-number kernel FEFF [8], $\mathbf{A}_{\mathbf{real}}(k,r)+i\mathbf{A}_{\mathbf{imaginary}}(k,r)$, the general EXAFS equation can be expressed as a linear matrix relationship: $[\chi_{complex}(k)]=[\mathbf{A}_{complex}(k,r)][g(r)]$ or $[\chi_{imag}(k)]=[\mathbf{A}_{imag}(k,r)][g(r)]$. [9] We defined the complex-number representation, $[\mathbf{A}_{\mathbf{imag}}; \mathbf{A}_{\mathbf{real}}]$-arrangement, as shown in figure(1). In topological aspect, the complementary pair $[\chi_{complex}(k), g(r)]$ describes the dynamic phase space participating in self-consistent solution domain, but real EXAFS measurement can see only the imaginary part, $\chi_{imag}(k)$. In a topological analogy with the dynamical phase spaces, (x-position, p-momentum), we take the conjugate observables, $(g(r), \chi(k))$. Simply, we can express $\mathbf{A}|g\rangle = \mathbf{A}[(\mathbf{A}^+\mathbf{A})\ldots(\mathbf{A}^+\mathbf{A})]|g\rangle$ where $\mathbf{A}^+$ represents the Hermitian conjugate matrix, i.e. the transposed complex conjugate. Consequently, $\mathbf{A}|g\rangle = \mathbf{Q}^2..\mathbf{A}..\mathbf{P}^2|g\rangle$ illustrates two distinct projection blocks, $(\mathbf{P}^2=\mathbf{A}^+\mathbf{A}\mathbf{A}^+\mathbf{A})$ and $(\mathbf{Q}^2=\mathbf{A}\mathbf{A}^+\mathbf{A}\mathbf{A}^+)$. The symplectic topological essence built into these phase-space projection filters is the system-matched (chemical-specific) character, ensuring the full performance of both "denoising" and "feature preservation".

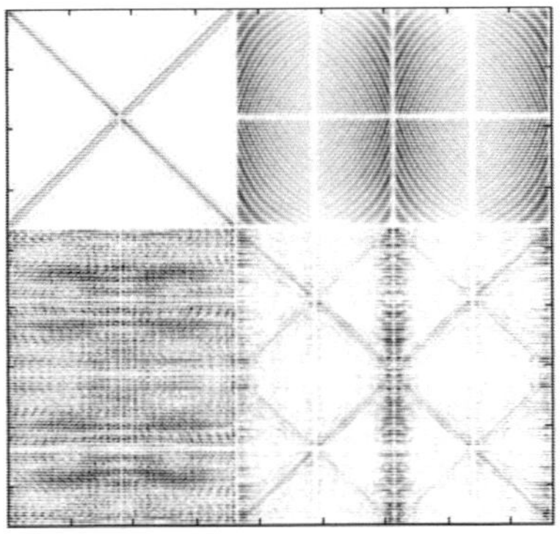

Figure (1)
Color-map matrix $[\mathbf{Q},\mathbf{A}; \mathbf{A}^{-1},\mathbf{P}]$ for complete Hilbert four-space set with $\mathbf{A}$ (forward transformation), $\mathbf{A}^{-1}$ (inverse transformation), $\mathbf{Q}=\mathbf{A}\mathbf{A}^{-1}$ (k-space projection) and $\mathbf{P}=\mathbf{A}^{-1}\mathbf{A}$ (r-space projection). Red/blue colors indicate positively/negatively valued. From the Zn K-edge FEFF $\mathbf{A}(k,r)$ matrix, for two different backscattering species of Zn-O and Zn-S. $\mathbf{A}$-matrix consists of eight Hanning-windowed segments in k[3, 12]-range and r[1.5, 4]-range, respectively, with intervals 0.05 and 0.02. This is corresponding to k^3-weighted EXAFS equation, $[\chi_{imag}(-k); \chi_{real}(k)] = [\mathbf{A}_{imag};\mathbf{A}_{real}]$ $[g_2(r);g_1(-r);g_1(r);g_2(-r)]$.

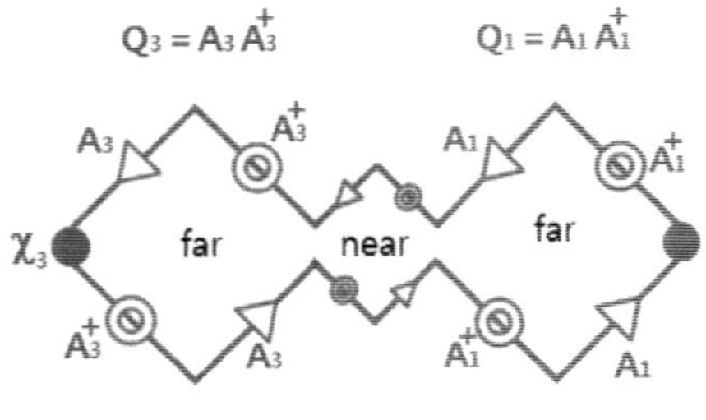

Figure (2)
Illustration of STIC(*synthesized topological inversion computation*) methodology. The symplectic transformation linkage is applied to eliminate the numerical discontinuity arising at near field zone. The equivalent inverse matrix $(\mathbf{A}^{-1})$ is related to:

$$\mathbf{J}_1 = \mathbf{A}_1\mathbf{A}_1^{-1} = Q_{1f}Q_{1n}(Q_{3f}Q_{3f}Q_{3n})Q_{1f}$$
$$\mathbf{J}_3 = \mathbf{A}_3\mathbf{A}_3^{-1} = Q_{3f}Q_{3n}(Q_{1f}Q_{1f}Q_{1n})Q_{3f}$$

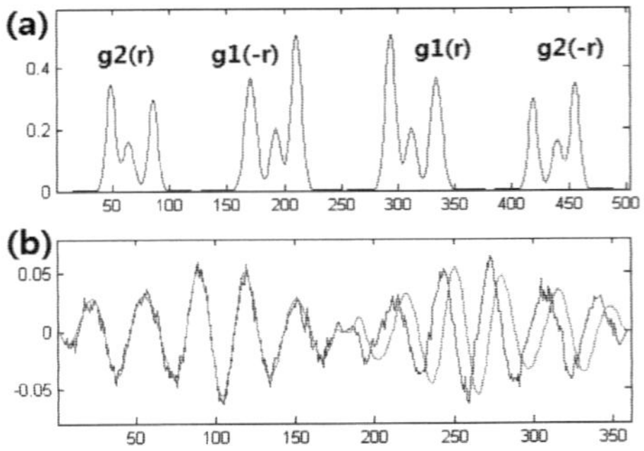

Figure (3)
(a) The simulated RDF-g(r) including two different backscattering atoms. (b) The phase retrieved $\chi(k)$ from imaginary space (blue-colored, simulated data with noise) to complex-number space (red-colored). Denoising efficiency is observed significantly.

To start our numerical procedure, let us introduce a novel singularity regularization method using dual parameters, fulfilling the complete feedback cycle requirement, as the following equations:

$$\boldsymbol{Reg}\,(\mathbf{K},\alpha,\beta) = (\mathbf{K}^T\mathbf{K} + \alpha\,\mathbf{I})^{-1}\,\mathbf{K}^T\,[\,\mathbf{I} + \alpha(1+\beta)\;(\mathbf{K}\mathbf{K}^T + \beta\,\mathbf{I})^{-1}\,]$$

$$\mathbf{W}(\mathbf{A},\alpha) = \boldsymbol{Reg}\,(\mathbf{A},\alpha,\alpha) \quad \text{when } \|\mathbf{A}^T\mathbf{A}\|=1$$

$$\mathbf{Z}\,(\mathbf{W},\alpha) = \left(\frac{\mathbf{W}}{\ }\right)\boldsymbol{Reg}\,(\|\mathbf{W}^T\mathbf{W}\|,\alpha,\alpha)$$

where **W** and **Z** pertain to, respectively, $\mathbf{A}^+$ and **A** matrices during progressive computation process. The parameters, α and β, are chosen for the projection domain, respectively, in r-space and k-space. As close as regularization parameters are approaching to zero, the optimal tracking status - between smoothing and distortion - gives rise to serious numerical instability relevant to ill-posed problem. (Refer to Tikhonov's regularization [10]) Nevertheless, our numerical approach is attainable to force an isotropic selection (i.e. $\alpha=\beta$) into the near-singularity regime with extraordinary stability under the topological feature momentum preservation. To do this, we chose $\mathbf{W_1=W+\lambda WQ}$ (where $\mathbf{Q=ZW}$ initially) with λ–parameter, the optimal Lagrangian path under the constraint of $\mathbf{W_1Q_1A}$ feature preservation. It is a way that a primitive set ($\mathbf{A}$, $\mathbf{A}^+$) is consistent with ($\mathbf{AA^+A}$, $\mathbf{A^+AA^+}$). Straightforwardly, by successive selection of regularization parameters, [(α_3, β_3), α_2] for far-field and [α_1,α_0] for near-field distant ($\alpha_3>\alpha_2>\alpha_1>\alpha_0$) from the inherent numerical singularity, we can enter a numerical gateway to the inverse solution. Subsequently, we need to suppress a numerical discontinuity at the vicinity of singularity. As shown in figure(2), we employed a 12-operator-closed-loop emulator of the symplectic transformation, **S**. ($\mathbf{A_3=SA_1}$ with $\mathbf{S^2=-I}$) [11] This STIC (*Synthesized Topological Inverse Computation*) architecture is the singularity inhibitor algorithm. In quantum physics, the symplectic transformation implies that $\psi_{complex}$ and $i\Psi_{complex}$ are indistinguishable. Equivalently, in our numerical recipe, a conjugated pair of ($\chi_{imaginary}$; χ_{real}) and (χ_{real}; $-\chi_{imaginary}$) was taken to simplify complex-number matrix manipulation. At every cycle of STIC-loop, the polarity of $\chi_{complex}(k)$ is just reversed. As a consequence, the ill-posed instability was dramatically suppressed under this STIC architecture.

Dissimilar to simple mathematical solution formula, a real world requires formulating the $\mathbf{A}^{-1}$ equivalent operator, in terms of successive matrix multiplications. By satisfying a numerical self-consistency in conjugate observables, we can assert that a hypothesis solution of the priori-probability g(r) can be computed from the posteriori-probability χ(k). This statistical approach satisfies the Bayesian requirement. According to the STIC methodology in figure(2), become a genuine inverse kernel. To increase the precision of inverse computation, primarily required is the numerical presetting of $\mathbf{A_r}$-matrix refinement ($\mathbf{A_r=A_1+\lambda[J_1A_1-J_3A_3]}$) in advance to other consecutive computations. In our experience, the numerical truncation error (a type of piled-up residuals producing r-domain glitches) can be removed with a better numerical presetting of A_r-matrix refinement. Therefore, instead of raw FEFF kernel generating such numerical glitches, we prefer to upgrade FEFF kernel (normally, norm($\mathbf{A_r}$-$\mathbf{A_1}$)/norm($\mathbf{A_1}$) ~ 0.2%) by own iterative algorithm in progress. [13]

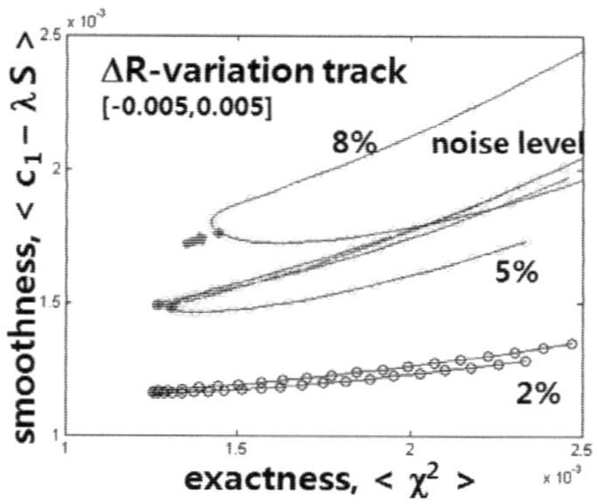

Figure (4)
Examples of STIC fine-tip probe for EXAFS applications. The best MEM pinning –red arrow - for g(r) solution is slightly deviated from the parabolic apex. Even at excessive noise statistics, the solution tracking is independently stable and accurate. The simulated ΔR-variation track was chosen in [-0.005, 0.005]Å ranges for two counterpart configurations; [R_1, R_2, R_3]=

[2.33+ΔR, 2.69, 3.13-ΔR] and [2.45+ΔR, 2.76, 3.19-ΔR], respectively, for $g_1(r)$ and $g_2(r)$, as shown in figure 3(a).

Throughout our study, we established the following procedures: (i). The g(r)-model vector, $[g_2(r);g_1(-r);g_1(r);g_2(-r)]$ as shown in figure(3-a), was chosen with six shells from two different chemical species, i.e. multi-component complex system. (ii). A pair of measurement data, $\chi_{imag}(k)$, was plugged into one input set for individual simulation. (iii). In figure(3-b), $[\chi_{imag}(-k); \chi_{real}(k)]$ was computed by an iteration method, multi-component filtering and Kramers-Kronig transformation. [12] (ix). A paired set of $\chi_{complex}(k)$ and trial g(r) were plugged into the cross-correlation analysis, evaluating the numerical fidelity of inverse solution in terms of MEM (Maximum Entropy Method) estimator, minimizing χ^2-λS where χ^2–fitting and S-entropy.

The patterned g(r)-function representation can be probed within a selected region of variation. The quality of S-entropy can be traced by evaluating the covariance matrix between symplectic observables (using the MATLAB function corrcoef(x1, x2)). A principal result is that this inverse computation approach dramatically enhances the numerical accuracy. Our numerical computation results indicate good performance even at relatively high noise level (e.g. 5%, noise-to-signal ratio). In a case of arbitrary track shown in figure(4), the tracking curve edge is highly sharpened at the pinning position for g(r)-solution. (This ΔR-variation track is different from conventional L-curve [4].) Also, our simulation results indicate that the optimal λ-parameter in (χ^2-fitting, S-entropy) domain is empirically proportional to the noise level. Surprisingly, even at excessive noise level, our simulation results showed numerical feasibility of system-matched "de-noising" filtering and the ultra-fine probing despite the complexity of the multi-component system. Without our STIC method, such precise tracking is not achieved due to the hindrance of noise problem. When a trial g(r) is approaching closely to the true solution, it is possible to keep tracking while preserving solution stability. In case of huge complexity, it is still promising to track the derivative trends approaching the ultimate solution. Advanced procedures will be described later in another publication.[13]

Beyond this specific application to EXAFS data analysis, this work demonstrates that the completeness of numerical inverse computation was possible using topological methods. Mathematics is universal and this method can be applied elsewhere. It is hoped that this method will open a numerical gateway to understanding a broad class of numerically ill-posed problems in inverse computation.

References

[1] E.J.Candes, Y. Eldar, T. Strohmer, and V. Voroninski. "Phase retrieval via matrix completion", Technical report, Stanford University, 2011.
[2] Gocha Khelashvili and Grant Bunker, J. *Synchrotron Rad.* 6, 271-273 (1999)
[3] Jay Min Lee and Dong-Seok Yang, "X-ray Absorption Fine Structure, XAFS13", by B. Hedman and P.Pianetta (ed.), American Institute of Physics, p138-140 (2007)
[4] A. Rossberg and H. Funke, *J.Synchrotron Rad.* 17, p280-288 (2010)
[5] H.J.Krappe and H.H.Rossner, *Phys. Rev.* B 66, 184303 (2002)
[6] E.T.Jaynes, "Bayesian Methods: General Background." In Maximum-Entropy and Bayesian Methods in Applied Statistics, by J. H. Justice (ed.). Cambridge: Cambridge Univ. Press, (1986)
[7] D.E.Sayers, E.D.Stern, and F.W.Lytle, *Phys. Rev. Lett.*, 27(18), 1204 (1971)
[8] J.J.Rehr, FEFF project, http://leonardo.phys.washington.edu/feff/
[9] Grant Bunker, "Introduction to XAFS: A Practical Guide to X-ray Absorption Fine Structure Spectroscopy", Cambridge University Press, 2010
[10] A.N.Tikhonov and V.Ya.Arsenin, "Solution of Ill-Posed Problems", John Wiley and Sons Inc., New York, 1981
[11] Dusa McDuff and D. Salamon, "Introduction to Symplectic Topology", Oxford University Press, 1998. ISBN 0-19-850451-9
[12] J.O.Cross et al, *Phys. Rev.* B 58, 11215-11225 (1998)
[13] to be published.

Ab-initio Calculation of the XANES of Lithium Phosphates and $LiFePO_4$

Y M Yiu*[1], Songlan Yang[1,2], Dongniu Wang[1,2], Xueliang Sun[2], and T K Sham [1]

[1] *Department of Chemistry, University of Western Ontario, London, N6A 5B7, Canada*
[2] *Department of Mechanical and Materials Engineering, University of Western Ontario, London N6A 5B7 Canada*
*Corresponding author, email: yyiu@uwo.ca

Abstract
Lithium iron phosphate has been regarded as a promising cathode material for the next generation lithium ion batteries due to its high specific capacity, superior thermal and cyclic stability [1]. In this study, the XANES (X-ray Absorption Near Edge Structure) spectra of lithium iron phosphate and lithium phosphates of various compositions at the Li K, P $L_{3,2}$, Fe $M_{3,2}$ and O K-edges have been simulated self-consistently using ab-initio calculations based on multiple scattering theory (the FEFF9 code) and DFT (Density Functional Theory, the Wien2k code). The lithium phosphates under investigation include $LiFePO_4$, γ-Li_3PO_4, $Li_4P_2O_7$ and $LiPO_3$. The calculated spectra are compared to the experimental XANES recorded in total electron yield (TEY) and fluorescence yield (FLY). This work was carried out to assess the XANES of possible phases presented in $LiFePO_4$ based Li ion battery applications [2].

1. Introduction

Currently, lithium-ion batteries employ $LiCoO_2$ as cathode material and carbon as anode material. This limits applications to small-scale batteries due to the high cost and toxicity of Co [3]. In recent years, lithium iron phosphate ($LiFePO_4$) has been considered to be a potential cathode material for rechargeable lithium ion batteries due to its high specific capacity (170 mAh/g) and a discharge voltage of 3.4 V between loading of Li^+ versus Li. [4]. $LiFePO_4$ becomes desirable for the next generation of Li-ion cells over $LiCoO_2$ because of its low raw material cost, environmental friendliness and excellent cycling capacity. The key limitation of this material has been extremely low electronic conductivity (10-9 S/cm). It has been reported that either particle size reduction or coating with carbon or metal ion doping greatly enhances the electrochemical behavior [5].

2.Theory

2.1. Multiple Scattering Theory - Calculation by FEFF9

The XANES (X-ray Absorption Near Edge Structure) spectra of these Li compounds with the above-mentioned crystal models have been calculated by the self-consistent multiple scattering theory using the FEFF9 code [6]. This theory is based on the real-space Green`s function approach [7]. The FEFF code is arranged in primary subroutines that correspond to the calculation of XAFS [8]. First it calculates the atomic potentials and densities. Then it generates a muffin-tin scattering potential that includes the excited-state self-energy and calculated energy reference. It also calculates the complex partial wave phase shifts. Finally, it computes the cuved-wave scattering amplitudes and phase shifts and the exact single-scattering curved -wave XAFS spectra.

2.2 DFT and GGA Theory - Calculation by WIEN2k

The theoretical XANES spectra are also calculated self-consistently by the Density Functional Theory (DFT) with Generalized Gradient Approximation (GGA) using the WIEN2k program [9]. Full potential augmented plane wave method has been used in the numerical calculation. The code has been built based on the Density Functional Theory [10] which provides an ab-initio method of calculating matters with a relatively homogenous distribution of electrons around the atomic sites. It further uses the Generalized Gradient Approximation [11] to contribute a first derivative correction to DFT. Crystals are described by the unit cell which have atoms occupying the symmetric atomic sites with interstitial area. Atomic sites are spheres of bases, which are generated by a set of non-overlapping spherical harmonics Y_{lm}. The interstitial regions are expressed by set of plane waves. The set of plane waves, e^{ikr}, has been extended to 20000 plane waves in the calculation of the systems in the present work. The boundary condition between the atomic sites and the interstitial area satisfies the Dirichlet condition. Scalar relativistic density function was used with GGA correction. Self-consistent minimization of energy was achieved with 1000 k points to keep a better accuracy in each calculation. The theoretical XANES spectra have been calculated using the electrical dipole transition given by the selection rule of $\Delta l = \pm 1$ (l is the angular momentum quantum number) and convoluted by instrumental Guassian and core-hole lifetime Lorentzian broadenings [12].

2.3 Structures of the Li phosphate crystals

The Li compounds, $LiFePO_4$, γ-Li_3PO_4, $Li_4P_2O_7$, and $LiPO_3$ have been modelled as crystals with structural symmetries of Pnma [13], Pmnb [14], P-1 [15], and P2/n [16] respectively. The crystal constants and geometries of these Li compounds are listed in table 1. $LiFePO_4$ has an olivine structure, in which Li, Fe, and P atoms occupy 4a, 4c, and 4c sites, respectively.

Table 1. Crystal constants of $LiFePO_4$, γ-Li_3PO_4, and $Li_4P_2O_7$, and $LiPO_3$.

	a (Å)	b (Å)	c (Å)	α (^{0})	β (^{0})	γ (^{0})
$LiFePO_4$	10.332	6.01	4.692	90	90	90
γ-Li_3PO_4	6.1113	10.4612	4.9208	90	90	90
$Li_4P_2O_7$	8.5613	7.11	5.1851	111.441	89.986	103.065
$LiPO_3$	13.074	5.4068	16.452	90	99	90

3. Experiment

All the samples are very fine powders from Phostech Lithium Inc. The XANES measurements were performed at the Canadian Light Source (CLS) on the Variable Line Spacing Plane Grating Monochromator (VLS PGM) beam-line for the Li K-edge and P $L_{2,3}$-edge spectra, and the high resolution Spherical Grating Monochromator (SGM) beam-line for the O K-edge and Fe $L_{2,3}$-edge spectra. Spectra were recorded in fluorescence yield (FLY) using a micro-channel plate detector and total electron yield (TEY) by measuring the sample current with a current amplifier.

4. Results and Discussions

XANES spectra are characterized by intense resonance features, arising from excitations of core-level electrons to unoccupied orbitals that are bound, quasi-bound and continuum levels following mainly electrical dipole transition. The XANES spectra of Li K edges, P L_{32} edges, O K edges, and Fe $M_{3,2}$, edges of the Li compounds as described by the above-mentioned crystal structures have been calculated numerically by FEFF9 and WIEN2k self-consistently. The calculated results together with the experimental spectra are plotted in figure 1 to 3. All experimental spectra reported were normalized to one after divided by the intensities of the incident beam (I_0), which have been measured simultaneously with a refreshed gold mesh located after the last optical elements of the beam-line.

The selection rule for the dipole transition follows $\Delta l = \pm 1$, ▮ $j = 0, \pm 1$ where l and j are the orbital and spin-orbit quantum numbers, respectively. Thus Fe $M_{3,2}$ (3p) and $L_{3,2}$-edge (2p) probe the unoccupied states of Fe 3d and 4s character. The Li and O K-edges (1s) probe the unoccupied states of the p character of Li and O respectively. The P $L_{3,2}$ edge probes states of s and d characters of P. The

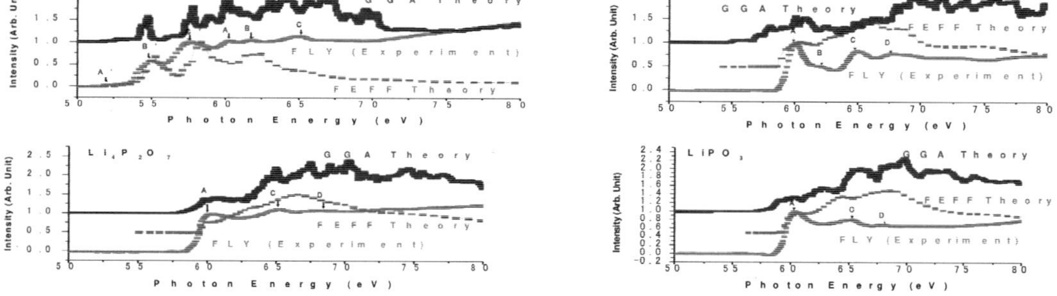

Figure 1. The experimental and theoretical XANES at Li K-edge of $LiFePO_4$, γ-Li_3PO4, $Li_4P_2O_7$, and $LiPO_3$.

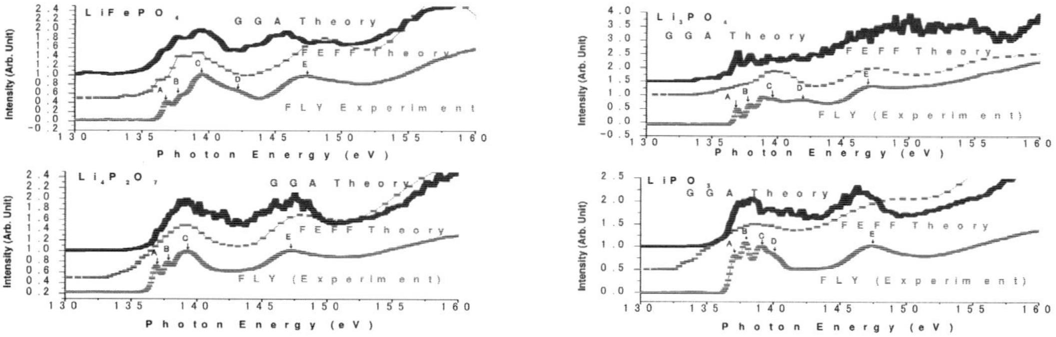

Figure 2. The experimental and theoretical XANES at P L_{32}-edge of $LiFePO_4$, γ-Li_3PO_4, $Li_4P_2O_7$, and $LiPO_3$.

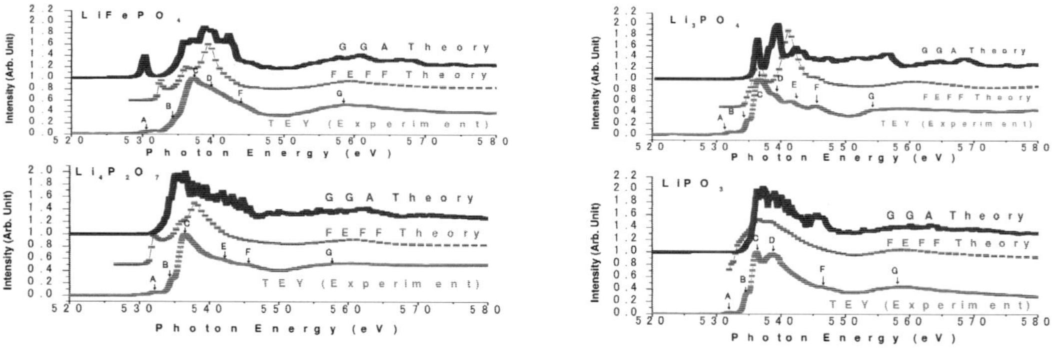

Figure 3. The experimental and theoretical XANES at O K-edge of $LiFePO_4$, γ-Li_3PO_4, $Li_4P_2O_7$, and $LiPO_3$.

partial densities of state of Li and Fe of $LiFePO_4$ have also been evaluated by WIEN2k. The peak positions of the XANES of the Li compounds are shown by the arrows in figure 1 to 3. In figure 1, the peaks A', B' and C' of $LiFePO_4$ are due to the Fe $M_{3,2}$ edges XANES of $LiFePO_4$. The calculated spectra from FEFF has been shifted 2 eV to fit the experimental Fe $M_{3,2}$ edge of $LiFePO_4$. The onsets of the peak A in the calculated Li K-edge of γ-Li_3PO_4, $Li_4P_2O_7$, and $LiPO_3$ are not as acute as that of the experimental FLY of these samples due to smoother rising of the calculated un-occupied densities of states of Li p character above the Fermi levels. In figure 2, there are moderate agreement between the experimental and calculated spectra: the C and E peaks follow the same trends. The A, B and D peaks of the experimental P $L_{3,2}$ XANES smooth out in the FEFF 9 calculation due to the finite sizes of the crystal clusters of about 50 atoms constrained by the full multiple scattering. All the theoretical XANES spectra show there are peaks around 160 eV for these model Li phosphates. In figure 3, the theoretical O K XANES of the Li compounds compose the average O K edges from different symmetric sites of O in the crystals. There are 3, 3, 7, and 15 O symmetric sites in $LiFePO_4$, γ-Li_3PO_4, and $Li_4P_2O_7$, and $LiPO_3$, respectively. The A peaks of the GGA calculated O K XANES spectra are higher than that of the experimental TEY spectra and show that the effect of the local environment of the O sites contributes substantially to the theoretical spectra. TEY measurements are surface sensitive, since low energy electrons have a short escape depth, while the x-ray fluorescent yield is bulk sensitive unless the sample is thick and self absorption occurs. The discrepancy between the TEY of the O K-edge and the calculated spectra by FEFF9 is due to the small cluster sizes for the FEFF calculation and the surface oxides of the samples which deviate from the crystal models.

5. Conclusion

XANES probes the local structure and bonding of the absorbing atom in a chemical environment and is element specific. Thus it can provide information of the local environment of Li and other elements of these Li phosphate compounds, in particular $LiFePO_4$, to help the design of future batteries. XANES simulation provides a useful tool to investigate the subtle variation of the local environment.

Acknowledgements
This research was supported by Discovery and Engage grants from Natural Sciences and Engineering Research Council of Canada (NSERC), Canada Research Chair (CRC) Program, Canada Foundation for Innovation (CFI), Ontario Innovation Trust (OIT), Ontario Research Fund (ORF), Ontario Early Researcher Award (ERA) and the University of Western Ontario. CLS is supported by CFI, NRC, NSERC, CIHR and the University of Saskatchewan. Technical support from the CLS Staffs is gratefully acknowledged.

References

[1] Padhi A K, Nanjundaswamy K S and Goodenough J B 1997 *J Electrochem Soc.* **144** 1188
[2] Yang S, Wang D, Liang G, Yiu Y M, Wang J, Liu L, Sun X and Sham T K 2012 *Energy & Environmental Science* **5** 7007
[3] Tarascon J M and Armand M 2001 *Nature* **414** 359; Whittingham M S 2004 *Chem. Rev.* **104** 4271
[4] Chung S Y, Bloking J T and Chiang Y M 2002 *Nat. Mater.* **1** 123; Huang H, Yin S C and Nazar L F 2001 *Electrochem. Solid-State Lett.* **4** A170; Barker J, Saidi M Y and Swoyer J L 2003 *Electrochem. Solid-State Lett.* **6** A53
[5] Shi S, Liu L, Ouyang C, Wang D, Wang Z, Chen L, and Huang X 2003 *Phys. Rev. B* **68** 195108
[6] Rehr J J, Kas J J, Prange M P, Sorini A P, Takimoto Y and Vila F 2009 *C. R. Physique* **10** 548
[7] Rehr J J and Albers R C 2000 *Rev. Mod. Phys.* **72** 621
[8] Mustre de Leon J, Rehr J J, Zabinsky S I and Albers R C 1991 *Phys. Rev.* B **44** 4146
[9] Blaha P, Schwarz K, and Sorantin P, and Trickey S B 1990 *Computer Phys. Comm.* **59** 399
[10] Hohenberg P and Kohn W 1964 *Phys. Rev.* **136** B864 ; Kohn W and Sham L J 1965 *Phys. Rev.* **140** A1133
[11] Perdew J P and Wang Y 1992 *Phys. Rev.* **B 45** 13244
[12] Fuggle J C and Inglesfield J E 1992 *Unoccupied Electronic States: Fundamentals for XANES, EELS, IPS, and BIS* (Springer-Verlag, Berlin Heidelberg)
[13] Streltsov V A, Belokoneva E L, Tsirelson V G, and Hansen N K 1993 *Acta Cryst.* **B49** 147
[14] Wang B, Chakoumakos B C, Sales B C, Kwak B S, and Bates J B 1995 *J. of Solid State Chem.* **115** pp 313-323
[15] Daidouh A, Veiga M L, Pico C, and Martinez-Ripoll M 1997 *Acta Cryst.* **C53** pp 167-169
[16] Murashova E V and Chudinova N N 2001 *Crystallography Reports* **46** pp 942-947

Journal of Physics: Conference Series **430** (2013) 012015 doi:10.1088/1742-6596/430/1/012015

Impact of the number of fitted Debye-Waller factors on EXAFS fitting

K Provost[1], E C Beret[2], D Muller[1], E Sanchez Marcos[2] and A Michalowicz[1]

[1] Equipe SAX, ICMPE UMR7182 CNRS-UPEC, Thiais, 94320, France
[2] Departamento de Quimica Fisica, Universidad de Sevilla, Sevilla, 41012, Spain

E-mail: provost@u-pec.fr

Abstract. EXAFS fit applied to asymmetric systems may imply the fit of parameters for many scattering paths. We illustrate on some examples how fitting too many independent DWs may lead to incorrect distances and mask some structural details. We question the physical meaning of the fitted DW values and we propose some ideas to avoid this problem.

1. Introduction

EXAFS applied to systems with asymmetric coordination spheres may imply the fit of parameters for many scattering paths. In such a situation raises the question of how to deal with the Debye-Waller factors (DWs) in the fit procedure. A usual approach consists in fitting a first DW for single scattering in the first coordination sphere and a second one for all the other scattering paths. Less usual is the fit of a unique DW for all scattering paths, which leads to a potential overestimation of amplitudes beyond the first coordination sphere [1]. In the hope to improve the fit, and in absence of external information, it stands tempting to fit as many DWs as many shells introduced in the fit procedure.

It is well-known that, in EXAFS fit procedure, amplitude parameters are highly correlated. It can be a source of unrealistic error bars or fitted values. When working on a Nitrile Hydratase model complex (FL1O2) and on platinum based anticancer drugs (figure 1), we observed that, when we fitted too many DWs, we had troubles in the structure determination [2]. The aim of this paper is to illustrate some drawbacks of a large number of fitted DWs, and to question the physical meaning of such fitted DWs in a case where the actual distance ditributions are known *a priori*: the fit of average spectra issued from a molecular dynamics trajectory [3,4].

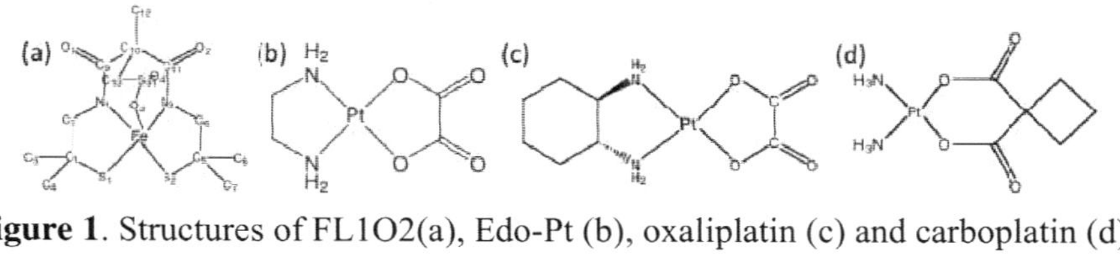

Figure 1. Structures of FL1O2(a), Edo-Pt (b), oxaliplatin (c) and carboplatin (d).

2. Experiments

2.1. Nitrile hydratase model complex FL1O2.

Pellets of FL1O2 [5] were prepared as a homogeneous mixture of pure product and cellulose. XAS spectra were recorded in transmission mode on the EXAFS13 spectrometer, at LURE-DCI, at the Fe K-edge, using a Si(111) channel-cut monochromator. EXAFS spectra were extracted using standard

Journal of Physics: Conference Series **430** (2013) 012015 doi:10.1088/1742-6596/430/1/012015

procedure available in MAX [6], and fitted using phases and amplitudes calculated with FEFF [7] on the basis of the crystal structure [5] with RoundMidnight. Because of difficulties to obtain the 1st coordination sphere distances, we tested 108 fits conditions. The criteria for fit acceptance were: a fitted distance differing from the crystallographic one by less than 0.03Å, a 0.02-0.15Å range for DWs, and error bars lower or equal to the value. The detailed study will be published elsewhere.

2.2. Platinum complexes.

Experimental XAS spectra were obtained for a 10mg/ml solution of oxaliplatin, at the Pt L_{III} edge, at ELETTRA, as described in ref. [4]. They were extracted using MAX [6]. EdoPt simulated spectra were calculated on the basis of a Car-Parrinello trajectory (CP-MD) of Edo-Pt in water [3,4]. Both experimental and simulated spectra were fitted in the same conditions using RoundMidnight [6], on the basis of a 7-shell model including: 1st coordination sphere single scattering (SS) (4PtN/O), 2nd coordination sphere SS ($2PtC_{oxalate}$, $2PtC_{ethylenediamine}$), single and multiple scattering (MS) of oxalate (2PtO, 4PtCO, 2PtCOC) and MS through platinum (4PtNPtO). We tested different amplitude parameters and different constraints on distances and DWs. In all cases, a unique distance and a unique DW are used for oxalate. We reduced the number of fitted distances from 5 to 4 or 3 by using distance shifts constraints (cf. §3.2). We tested three empirical choices for DWs fitting: (i) a unique DW for all scattering paths (1DW); a first DW for 1st coordination sphere SS and a second one for all other signals (2DW); a first DW for the 1st coordination sphere SS, a second one for 2nd coordination sphere SS, a 3rd one for oxalate, using a ratio of 2 between the 1st DW and DW for MS in the 1st coordination sphere [8] (3DW); two more choices are based on the use of CP-MD DWs (MDV: no DW fitted, MDR: one DW fitted). The detailed study will be published elsewhere.

For reaction of carboplatin and oxaliplatin with methionine, experimental XAS spectra were obtained and treated as described in ref. [1].

3. Results and discussion

3.1. Nitrile hydratase model complex

The theoretical signal calculated on the basis of the crystallographic structure [5] fits well the experimental signal (figure 2a): the fit of a global DW and an E_0 shift, without any structural rearrangement, leads to a QF=1.66 (σ^2=56(3).10^{-4}Å^2, ΔE=0.7(2)eV). However, when the distances were fitted, among the 108 tested fit conditions, only few gave fitted distances in agreement with crystallography: 33 fits presented incorrect distances and incorrect amplitude parameters, 45 incorrect distances and correct amplitude parameters, 1 correct distances and incorrect amplitude parameters, and only 29 presented all parameters correct. When fitting a unique DW for first coordination sphere, 44% of fits were correct (100% with fitting conditions reducing the signal weight at low k), while for three DWs, 0% were correct.

Table1. Example of fit parameters for 1DW or 3DW, all other fit conditions identical

	1DW	3DW	Cryst.
R_S (Å)	2.23(1)	2.23(1)	2.22(1)
σ_S^2 (Å^2)	$22(6)10^{-4}$	$26(8)10^{-4}$	
R_N (Å)	1.94(1)	2.00(2)	1.95(1)
σ_N^2 (Å^2)	$22(6)10^{-4}$	$0(324)10^{-4}$	
R_O (Å)	1.99(2)	1.87(2)	2.00(1)
σ_O^2 (Å^2)	$22(6)10^{-4}$	$11(45)10^{-4}$	
QF	1.66	1.55	

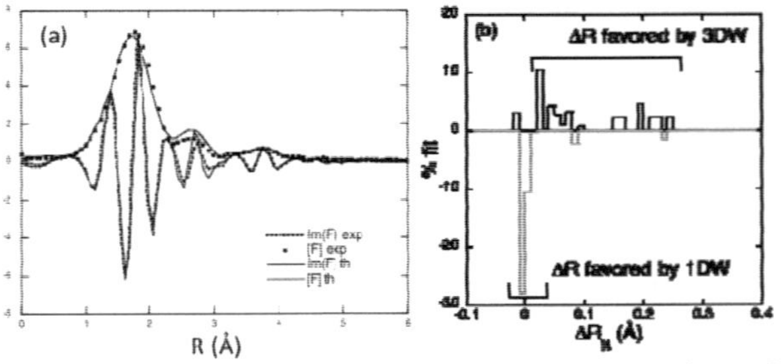

Figure 2. (a) Global fit of FL1O2 spectra: Fourier transform of the experimental signal and of the theoretical signal based on the crystal structure, with a global DW and a global ΔE.
(b) Histogram of distance difference $\Delta R=R_{fit}-R_{crystal}$ for FeN scattering path: $\%fits_{3DW}-\%fits_{1DW}$ as a function of ΔR.

Table 1 presents an example of fitted parameters when going from 1 to 3 DWs fitted: fitting 3DWs induces a 0.06Å increase of FeN distance and a 0.12Å decrease of FeO distance. When considering all fitting conditions, it is possible to calculate, for each shell (2 FeS, 2 FeN or 1 FeO), and each distance difference $\Delta R=R_{fit}-R_{crystal}$ value, the corresponding percentage of fits, for 1 DW and for 3DW, and to calculate the difference: $\%fit_{3DW}-\%fit_{1DW}$. Figure 2b presents the results obtained for FeN distance. Fitting 1 DW favors perfect agreement with crystallography ($\Delta R\approx 0$Å) for all shells, while fitting 3 DW favors $|\Delta R|$ larger than 0.1Å especially for FeN and FeO distances. Thus, for FL1O2 compound, fitting three independent DWs impairs the structure determination.

3.2. Oxaliplatin and EdoPt: systematic study

Figure 3 presents the percentage of good and physically inacceptable fits for all distances constraints (a), and the percentage of fits presenting at least one incorrect distance when all distances are fitted (b), as function of the DW constraints. As long as we fit no more than two independent DWs, all percentages stand stable. But fitting three independent DWs induces a huge decrease for good fits, and an increase for inacceptable ones and fits presenting at least one incorrect distance.

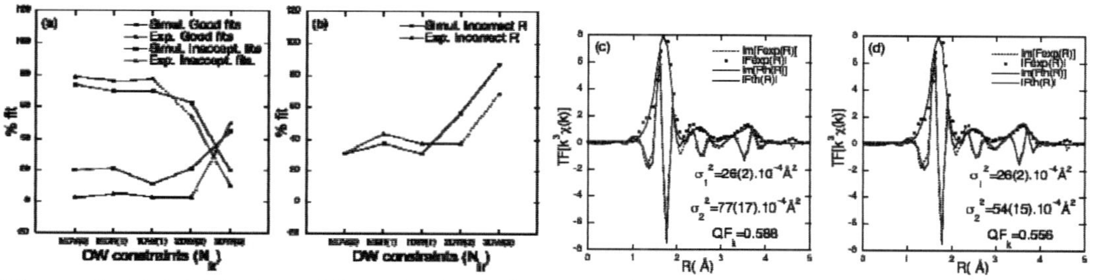

Figure 3. (a)-(b) % of good fits, physically inacceptable ones (a) and fits with incorrect distances (b) for a simulated spectrum and an experimental one. N_{fit} is the number of fitted DWs. (c)-(d): Compensation of interference effect by a 2nd DW fitted: (c) 0Å distance shifts; (d) 0.1Å distance shift.

A way to decrease the risk of obtaining wrong distances beyond the 1st shell is to apply ΔR distance shifts constraints: it is possible to fit $PtC_{oxalate}$ distance, and to set $PtC_{ethylenediamine}$ one to $PtC_{oxalate}+\Delta R$, where ΔR is fixed. A similar distance shift can be applied between the whole oxalate signal and MS through platinum. We tested two distance shifts: $\Delta R=0$Å corresponding to distance equality, and $\Delta R=0.1$Å close to the distance shifts observed in oxaliplatin crystal structure [9] and EdoPt optimized one [3]. Significant fit improvements are tested by the QF ratio QF_{max}/QF_{min} which should be much greater than 1. When the systematic errors are minimized, the 0.1Å distance shift always presents a lower QF than 0Å distance shift with significant QF ratios (1.43-2.08), provided than no more than a unique DW is fitted. When 2 independent DWs are fitted, the QF ratios are much less significant (1.16-1.22). Figure 3c and 3d present the experimental and theoretical Fourier transform for a simulated spectrum, fitted with 2 DWs and distance shifts of 0Å (figure 3c) or 0.1Å (figure 3d). Changes in interference effects, between 0Å and 0.1Å distance shifts, are compensated by the 2nd DW fitted: it becomes impossible to distinguish the two structural models on the basis of EXAFS fit.

3.3. Platinum anticancer drugs degradation products by methionine.

Reaction of carboplatin and oxaliplatin with an excess of methionine leads to degradation products with PtN_2S_2 or PtN_3S first coordination sphere [1]. From the complete modeling, fitting a unique DW, we concluded that carboplatin and oxaliplatin degradation products were *cis*-$Pt(Met)_2$ and Pt(Dach)(Met), respectively, in agreement with other experimental and theoretical studies [1]. Table 2 presents the quality factors obtained for all the models tested. When a second DW is fitted, most fits present inacceptable distances. The QF ratios are decreased for carboplatin, and, for oxaliplatin, the fit would bring to conclude on the most chemically unlikely model: $Pt(Dach)(S\text{-}Met)_2$.

Table 2. QF values obtained for 1DW and 2DW. (1) S-monodentate fixation of methionine; in all other cases S,N-bidentate fixation. (2) Fits with inacceptable distances.

		$Pt(NH_3)_2(Met)_2$	*cis*-$Pt(Met)_2$	*trans*-$Pt(Met)_2$	Pt(Dach)(Met)	$Pt(Dach)(Met)_2$
CarboPt	1DW	1.82(1)	0.70	0.93		
	2DW	0.61(1)	0.50(2)	0.66(2)		
OxaliPt	1DW		1.32	1.34	1.03	1.38(1)
	2DW		0.71(2)	0.70(2)	0.80(2)	0.59(1,2)

3.4. Validity of the fitted DW values
When working on CP-MD based spectra, the actual distance distributions are known and it is possible to check whether the fitting procedure restitute the right DW values or not. Using EdoPt simulated spectra we applied this strategy. Table 3 presents the CP-MD DW values and the fitted ones. DW error bars are large for scattering paths beyond the first coordination sphere. For 3DW, they are larger than the values difference. Comparison CP-MD and fit results shows that neither 2DW nor 3DW succeed in restituting the right values. Lower QF values can be explained by the compensation of systematic errors by the introduction of other systematic errors.

Table 3. CP-MD and fitted DW values for an EdoPt simulated spectrum.

	CP-MD		2DW		3DW	
	σ_i^2 (Å^2)	σ_i^2/σ_1^2	σ_i^2 (Å^2)	σ_i^2/σ_1^2	σ_i^2 (Å^2)	σ_i^2/σ_1^2
1: PtN/O	0.0024	1	0.0027(5)	1	0.0026(4)	1
2-3: PtC	0.0035	1.46	0.0068(23)	2.5(6)	0.0056(16)	2.1(6)
4-5-6: oxalate signal	0.0033	1.37	0.0068(23)	2.5(6)	0.0063(17)	2.4(4)
7: MS through Pt	0.0044	1.83	0.0068(23)	2.5(6)	0.0052	2

4. Conclusion
In EXAFS fit of complexes with asymmetric coordination spheres, correlations between distances and DWs are not negligible. They may cause instabilities in the fit process resulting in physically unreasonable distances. It may happen as well in the first coordination sphere as well as beyond. In all cases tested, 3DW led to bad results. Structural interference effects related to the presence of many similar scattering paths with different distances may be compensated when a second DW is fitted. Thus fitting many DW may mask some structural details. At least, the fit procedure applied to CP-MD based spectra proves that the fit of many DW does not give a more accurate description of the system physical reality. Thus the number of independent DWs should be reduced as much as possible. Introducing theoretical external information should be a solution to reach this aim.

References
[1] Provost K, Bouvet-Muller D, Crauste-Manciet S, Olivi L, Vlaic G and Michalowicz A 2009 *J. Conf. Series* 190 012206
[2] Papers in preparation.
[3] Beret E, Pappalardo RR, Marx D and Sanchez Marcos E 2009 *ChemPhysChem* 10 1044
[4] Beret E, Provost K, Muller D and Sanchez Marcos E 2009 *J. Phys. Chem. B* 113 12343
[5] Heinrich L, Li Y, Vaissermann J, Chottard G and Chottard J.C. *Angew. Chem. Int. Ed.* 1999 38 3526
[6] Michalowicz A, Moscovici J, Bouvet-Muller D and Provost J 2009 *J. Conf. Series* 190 012034
[7] Ankudinov A L, Ravel B, Rehr J J and Conradson S D Phys. 1998 *Phys. Rev. B* 58 7565
Ankudinov AL, Bouldin C E, Rehr J J, Sims J and Hung H 2002 *Phys. Rev. B* 65 104107
[8] Yokoyama T, Ohta T and Sato H 1997 *Phys. Rev. B* 55 11320
[9] Bruck M A, Bau R, Noji M, Inagaki K and Kidani Y *Inorg. Chim. Acta* 1984 92 279

Journal of Physics: Conference Series **430** (2013) 012016 doi:10.1088/1742-6596/430/1/012016

MAX (Multiplatform Applications for XAFS) New Features

A Michalowicz*, J Moscovici, D Muller-Bouvet and K Provost

Institut de Chimie et des Matériaux, CNRS-UPEC, 2 rue H. Dunant, F-94320 Thiais, France

*Corresponding author, email: michalov@u-pec.fr

Abstract. MAX is a complete suite of XAFS data analysis computer programs, written with the cross-platform object scripting langage Livecode [1] working on Windows, MacOSX and Linux systems, freely available on our web site since 2007 [2]. The 2009 version presentation is available in the XAFS14 conference proceedings [3] and regularly updated on the www.xafs.org web site [4].

1. Introduction

MAX is composed of 5 modules (figure 1):

1) **Absorbix**, 2) **Cherokee**, 3) **Crystalffrev**, 4) **Roundmidnight** (divided in RoundMidnightRev, RoundMidnightFit and RoundFitAuto), which will be described in detail only for their original functions, new features and improvements included since XAFS14 conference.

5) **Straight no Chaser**, which is totally new and deserves to be described in more details.

Figure 1 MAX modules

2. Modules description

2.1. Absorbix

Absorbix is a utility for sample absorbance calculations. It can calculate and optimize total absorbance and edge jump of pure samples, mixtures of solid powdered sample with solid solvents (recently improved), and liquid solutions. It includes also fluorescence self-absorption corrections, for EXAFS and for XANES

2.2. Cherokee

Cherokee includes raw data conversion to absorption curves (with various input data files formats), predge and atomic absorption baselines removal, XANES and EXAFS normalization, Fourier Transform and Filtering, experimental EXAFS data statistical error estimation. The main new feature of this program is the Multiple Data modules, which add to the standard single spectrum data treatment, the possibility to open, calculate and display quickly spectra issued from multiple data files. This is especially useful for time-resolved experiments with many data files to treat in almost automatic sequences.

2.3. Crystalffrev

Like Bruce Ravel's ATOMS included in ARTHEMIS code [5], **CRYSTALFFREV** is a crystallographic tool for the conversion of crystal structures clusters centered on absorbing atoms, in order to created FEFF input files.

The main differences are :

a) FEFF7 and FEFF 8 entries (FEFF9 available soon)

b) Import/Export file formats. In addition to a specific Crystalffrev input/output file format, CIF **and PDB** import/export file formats are also available.

c) Crystal Space group search. A more complete compilation of the most recent version of the IUCr International Tables of Crystallography space groups, with a space groups synonym dictionary and search tools for alternate definitions, and non-conventional crystallographic symmetry operations editing.

d) Cluster rotation. The possibility to rotate the local cluster axis in order to choose the orientation of the z axis (very useful for axial ligand analysis of macrocycle metalloenzyme active sites).

e) Crystal disorder tools. Partial substitution and vacancies random distribution generator for multiple partial occupancy sites in disordered materials.

New: A normally distributed random atomic position generator in order to simulated translational disorder and random radial distribution function of selected atoms in the cluster .

In development:

1) FDMNES [6] input files editing.

2) Non gaussian RDF generator for disordered crystal structures.

2.4. Roundmidnight

Roundmidnight is our EXAFS fitting program, composed of 3 modules: RoundMidnightRev, RoundMidnightFit and RoundFitAuto. The fit of theory versus experimental EXAFS is based on the standard multiple scattering and spherical waves EXAFS formula as documented in FEFF *ab initio* code, and the general function minimization software MINUIT from the CERN computer library [6].

Roundmidnight is totally compatible with FEFF, both in the fitting module, and in the multiple scattering paths analysis module. However, any other sources of phase and amplitude functions are also admitted. The empirical mean free path curve can be fitted on model standards and used in place to, or in combinaison with, the FEFF reduction factor and calculated photoelectron mean free path in order to improve the EXAFS fit in the low k ($<3Å^{-1}$) domain.

The fitting procedure follows the IXS Standard and Criteria error subcommittee definitions and recommendations [7] on statistical error estimation (both students statistics on several exafs recordings and fourier noise filtering methods), reduced Δc^2 (quality factor QF, calculated either in k or R spaces), parameter correlations and fitting convergence criteria. In addition, we have implemented à F-test calculator [8]. A "batch" module is available, in order to edit, run and analyze the results of multiple fits of input files.

In the previous versions RoundmidnightRev was only able to prepare input files for the fitting routine RoundmidnightFit (which was opened by hand by the user) and analyse the result. The most important new feature since 2009 is the possibility to run automatic fits directly from the main program RoundMidnightRev in one button click, since the new Roundfitauto external routine is launched and run automatically. RoundMidnightRev contains also a FEFF scattering paths analysis module.

2.5. Straight no Chaser.
This new member of the MAX family is a linear algebra fit program in order to decompose mixed spectra in their model components. The linear decomposition of mixed xafs spectra on the basis of a priori known model spectra was developed first.

The program provides fitted molar or mass ratio (if the user provides the model compounds molar mass), and their standard error bars. It works on normalized XANES, complete normalized absorption or EXAFS spectra.

Linear Least Squares Fit is performed by Gauss-Jordan matrix inversion or by Single Value Decomposition (SVD, preferable). It provides statistical Quality Factor (QF), error bars and correlation matrix, tools to check and minimize the correlations and improve the fit accuracy, when possible.

Designed to treat multiple set of spectra, Straight no Chaser can quickly open multiple model compounds and sample spectra in a few clicks.

We propose two special features :

1) A Progressive Linear Least Squares method in order to find the best composition using three constrains: i) fitted coefficients must be positive, ii) errors on the fitted coefficients must be less than x% of the values, x set by the user, iii) when both i and ii criteria are fulfilled, choose the composition with the lowest QF. This procedure was found to be very useful when the actual number of components among the possible model compounds spectra is unknown and some of the model spectra are highly correlated.

2) Fit automatically a series of N spectra (N only limited by the computer memory, and the computing time). This module can be used to fit automatically series of spectra issued from a Quick EXAFS experiment and plot the composition evolution versus time or any physical parameter.

When the *a priori* knowledge of the model compounds in the unknown samples is incomplete, other linear algebra statistical tools, based on SVD, must be available: Principal Component Analysis and Target Transformation [9] modules are under development and will be available soon in Straight no Chaser.

References

[1] http://www.runrev.com/products/livecode/livecode-platform-overview/
[2] ICMPE SAX group web page. http://www.icmpe.cnrs.fr/SAX/
[3] Michalowicz A, Moscovici J, Muller-Bouvet D and Provost K, Journal of Physics, Conference Series 190 (2009) 012034-012035
[4] http://xafs.org/Software/
[5] Ravel B, J. Synchrotron Radiat. (2005) vol. 12, pp. 537-541
[6] Joly Y, Phys. Rev. B 63, 125120-125129 (2001)
[7] IXS Standards and Criteria Subcommittee Reports. http://ixs.iit.edu/subcommittee_reports/sc/
[8] Michalowicz A, Provost K, Laruelle S, Mimouni A and Vlaic G, *J. Synchrotron Rad.* (1999). 6, 233
[9] a) Wasserman S R, *J. Phys. IV France* 7 (1997) C2-203
b) Malinowski E R, Anal. Chem. 49 (1977) 606
c) Ressler T, et al., *Env. Sci. & Technol.* 34 (2000)

Journal of Physics: Conference Series **430** (2013) 012017 doi:10.1088/1742-6596/430/1/012017

A hard X-ray nanospectroscopy station at SPring-8 BL39XU

M. Suzuki,[1,2] N. Kawamura,[1] M. Mizumaki,[1] Y. Terada,[1,2] T. Uruga,[1] A. Fujiwara,[1] H. Yamazaki,[1,2] H. Yumoto,[1,2] T. Koyama,[1,2] Y. Senba,[1,2] T. Takeuchi,[1,2] H. Ohashi,[1,2] N. Nariyama,[1,2] K. Takeshita,[1,2] H. Kimura,[1,2] T. Matsushita,[1,2] Y. Furukawa,[1,2] T. Ohata,[1,2] Y. Kondo,[3] J. Ariake,[3] J. Richter,[4] P. Fons,[4] O. Sekizawa,[5] N. Ishiguro,[6] M. Tada,[6] S. Goto,[1,2] M. Yamamoto,[1,2] M. Takata,[1,2] and T. Ishikawa[1,2]

[1] JASRI/SPring-8, Sayo, Hyogo 679-5198, Japan
[2] RIKEN SPring-8 Center, Sayo, Hyogo 679-5148, Japan
[3] Akita Industrial Technology Center, Akita 010-1623, Japan
[4] AIST, Tsukuba, Ibaraki 305-8568, Japan
[5] The University of Electro-Communications, Chofu, Tokyo 182-8585, Japan
[6] Institute for Molecular Science, Okazaki, Aichi 444-8585, Japan

E-mail: m-suzuki@spring8.or.jp

Abstract. We upgraded the hard X-ray spectroscopy beamline BL39XU at SPring-8 by installing Kirkpatrick-Baez mirrors to equip it with submicron-focusing capability. At the new experimental end-station located 74 m from the undulator source, an X-ray nanoprobe with a size of 120×100 nm^2 has become available at 5–16 keV. This upgrade is fully compatible with the X-ray polarization tunability and helicity-switching between right- and left-circular polarizations; these are unique features of the beamline, which is equipped with a diamond X-ray phase retarder. A spatial resolution of $\approx$100 nm for X-ray absorption fine structure and X-ray magnetic circular dichroism (XMCD) spectroscopy has been achieved. We show a magnetization reversal process of an individual magnetic dot in bit-patterned perpendicular recording media using element-specific XMCD magnetization measurements to demonstrate the performance of the new station.

1. Introduction

X-ray spectroscopic techniques, X-ray absorption fine structure (XAFS), and X-ray magnetic circular dichroism (XMCD) are useful tools for characterizing functional materials. These techniques allow chemical, structural, and magnetic characterization in both element-specific and site-specific ways. The techniques use hard X-rays and offer bulk sensitivity, a variety of sample environments, and easy combination with an external field or optical excitations. Our goal is to combine a micro/nanoprobe capability with hard X-ray spectroscopy using a focused X-ray beam. We are aiming at XAFS and XMCD at 100-nm spatial resolution. With this capability, one can perform two-dimensional XAFS mapping to obtain a spatial distribution of chemical and bonding states and local symmetries of a sample with nanostructures. Local XANES/EXAFS measurements allow us to investigate particular parts of nanostructures or very small samples, such as single catalyst particles [1]. An XMCD technique using a nanofocusing beam also allows scanning element-specific magnetic domain imaging and XMCD spectroscopy and element-specific magnetization (ESM) curve measurements in a local area of samples with magnetic domains or magnetic nanostructures [2]. Single magnetic device elements or small magnetic particles of submicron dimensions can be studied in this manner. In this paper, we describe the upgrade of an existing beamline BL39XU at SPring-8 to include features for hard X-ray nanospectroscopy. Early results of XMCD magnetometry in CoPt perpendicular magnetic dots are presented to demonstrate the performance of the new experimental station.

2. Upgrade of BL39XU with a new experimental station

We upgraded BL39XU at SPring-8, which has been used for XMCD spectroscopy in the hard X-ray region. The beamline is equipped with a Si 111 double-crystal monochromator and a diamond X-ray phase plate to generate circularly polarized X-rays. The existing experimental hutch is located 45 m from the source and equipped with several apparatus for XMCD experiments, including an electromagnet, a 10-T superconducting magnet for high-field experiments, and diamond anvil cells for XMCD studies at high pressures of up to 170 GPa [3]. In 2011, we extended the beamline and constructed a new experimental hutch dedicated to nanospectroscopy applications, located 74 m from the source, as shown in Fig. 1. We installed Kirkpatrick-Baez mirrors in the hutch to focus the X-ray beam down to 100 nm [4].

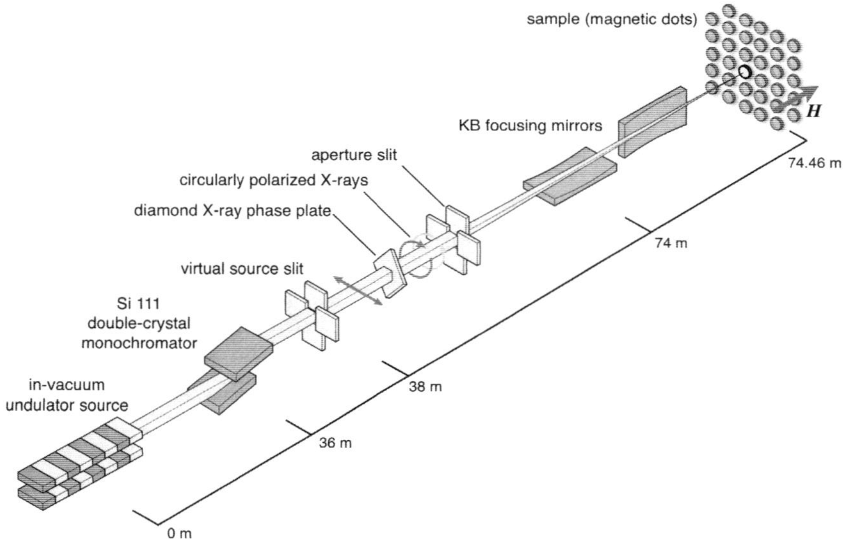

Figure 1. Layout of the nanospectroscopy station at the SPring-8 BL39XU beamline.

Table 1 summarizes the parameters of the focused X-ray beams. The available X-ray energy ranges from 5 to 15 keV, which covers the absorption edges of most $3d$ transition metals, rare earth elements, and $5d$ noble metals, such as Pt, a vital element in some catalysis reactions and magnetic materials for data storage. One can choose two focusing modes—"high-resolution" and "high-flux" modes—by changing the opening of the virtual source slit 36 m upstream of the focusing mirror. In the high-resolution mode with the small opening of the virtual source slit [8 (V) $\times$ 18 (H) μm^2], the typical beam size is 120 (V) $\times$ 100 (H) nm^2. The available photon flux is on the order of 10^9 photons/s in the spot. When higher flux is needed, a flux higher than 10^{12} photons/s in a spot of 300 $\times$ 250 nm^2 is obtained with a virtual source slit of 2000

(V) × 54 (H) μm^2 in size, i.e., fully open in the vertical direction. This photon flux value is incredibly high compared with the total flux at a typical undulator beamline of SPring-8, which is on the order of 10^{13} photons/s and that at a standard bending magnet beamline, which is with 10^{10}–10^{11} photons/s.

One of the most important features of the beamline is the control of the X-ray polarization state using the diamond phase plate. We have confirmed that the phase plate has no influence on the focused beam size by comparing focused beam profiles of ≈100 nm obtained under three different conditions: without a phase plate and with a 0.73-mm-thick phase plate in the 220 Laue geometry that was tuned for left- and right-circular polarizations. No significant increase in the beam size was observed within the reproducibility of the measurement (≈±10 nm) by a knife-edge scan. The current upgrade was shown to be fully compatible with polarization control using a diamond phase plate, and it would directly lead to XMCD applications using a focused beam.

Table 1. Parameters of the focused X-ray beams.

	High-resolution mode	High-flux mode
X-ray energy (keV)	5–15	5–15
beam size (nm^2)	120 (V) × 100 (H)	300 (V) × 250 (H)
photon flux (photons/s)	6.2×10^9	1.7×10^{12}

3. XMCD magnetometry in a single magnetic dot

To demonstrate the performance of our hard X-ray nanoprobe, we performed XMCD magnetometry in a single magnetic dot. We studied a prototype of bit-patterned media, which is a candidate for ultra-high-density recording media in the near future. The sample consisted of CoPt magnetic dots of 200-nm diameter arranged on a matrix with a 1-μm spacing. A magnetic thin film with a structure of $Co_{80}Pt_{20}$(15 nm)/Ru(15 nm)/NiW(5 nm)/C(3 nm)/Si substrate was deposited by dc magnetron sputtering, and dots were patterned by electron beam lithography and Ar beam milling [2].

We focused an X-ray beam of size 320 (V) × 290 (H) nm^2 onto a single magnetic dot and successfully recorded XAS and XMCD spectra at the Pt L_3 edge for the single dot (see Fig. 2). In the XAS spectrum, the W L_2 (11.542 keV) absorption from the NiW underlayer overlaps that of the Pt L_3 (11.565 keV), which demonstrates the bulk sensitivity of our hard X-ray nanoprobe. The XMCD signal was observed only at the Pt L_3 edge because Pt has a net magnetic moment due to hybridization with Co, but W does not. We recorded ESM curves of a single dot by monitoring the XMCD amplitude at the Pt L_3 edge as a function of the external magnetic field. In Fig. 3, the solid circles show a hysteresis loop measured for a single magnetic dot. The magnetization of the dot switches very rapidly at the coercive field, demonstrating that such a small dot has a small number of (less than a few) magnetic domains and that the magnetization reversal process is nearly coherent. For comparison, a hysteresis loop recorded using our previous focusing setup with a 2-μm beam is shown [5], as denoted by the line in the figure. With a larger X-ray beam spot, about 100 dots were included in the footprint of the focused beam, and the measured signal was an average for these dots. The resulting hysteresis loop, therefore, changes more slowly compared with that for a single dot. Coercivity fields are not exactly the same for many dots, and there is some distribution in the switching field. Using the 300-nm focused beam, we were able to characterize the switching field for 100 individual dots by recording each ESM curve for these dots [6].

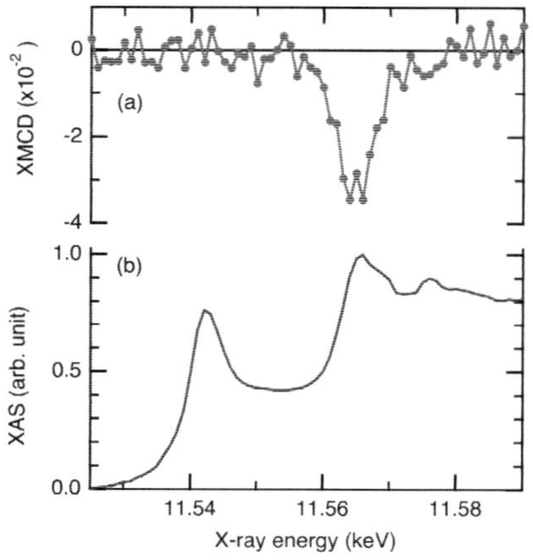

Figure 2. (a) X-ray magnetic circular dichroism and (b) X-ray absorption spectra of a $Co_{80}Pt_{20}$ magnetic dot of 200-nm diameter, measured at the Pt L_3 edge.

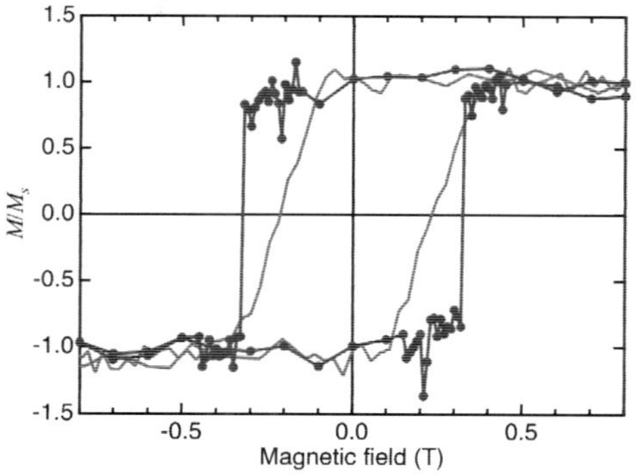

Figure 3. Element-specific magnetization curves of $Co_{80}Pt_{20}$ magnetic dots, recorded with a 300-nm focused X-ray beam (solid circles) and 2-μm beam (line).

4. Summary

In summary, we upgraded the SPring-8 BL39XU beamline by equipping it with submicron-focusing capability and achieved a focused X-ray beam of extremely high flux for a beam size of 300 nm and a reasonable flux for a beam size of 100 nm. Ongoing studies using the submicron-focused X-ray beam include (i) XAFS analysis of single particles of micro/nanocatalysts [1], (ii) *in situ* and time-resolved XAFS in phase-change memory devices, and (iii) two-dimensional mapping of chemical, structural, and magnetic states in $Nd_2Fe_{14}B$ sintered permanent magnets. We have a number of instrumentation issues that need to be addressed: we hope to achieve higher stability of the focused beam in the short and long term and expand the variety of sample environments; we are also considering the use of a fast and large-area X-ray detector to accept strong X-ray fluorescence signals from the high input photon flux on the sample that has been achieved by this upgrade.

Acknowledgments

This work was performed with the approval of SPring-8 (Nos. 2011A1808, 2011A2058, 2011A2065, and 2011A2066) and was supported by the Research-Network Building Program for Reduction of Carbon-dioxide Emission and a Grant-in-Aid for Scientific Research (No. 23360016) from the Ministry of Education, Culture, Sports, Science and Technology (MEXT), Japan.

References

[1] Tada M, Ishiguro N, Uruga T, Tanida H, Terada Y, Nagamatsu S, Iwasawa Y and Ohkoshi S 2011 *Phys. Chem. Chem. Phys.* **13** 14910
[2] Kondo Y, Ariake J, Chiba T, Taguchi K, Suzuki M, Kawamura N and Honda N 2011 *Physics Procedia* **16** 48
[3] Kawamura N, Ishimatsu N and Maruyama H 2009 *J.Synchrotron Rad.* **16** 730
[4] Koyama T and *et al* 2011 *Proc. SPIE* **8139** 81390I
[5] Suzuki M, Takagaki M, Kondo Y, Kawamura N, Ariake J, Chiba T, Mimura H and Ishikawa T 2007 *AIP Conf. Proc.* **879** 1699
[6] Kondo Y, Suzuki M and Ariake J 2012 *International Conference of the Asian Union of Magnetics Society (ICAUMS2012), Oct 2–5, 2012, Nara*

Journal of Physics: Conference Series **430** (2013) 012018 doi:10.1088/1742-6596/430/1/012018

Pt $L_{3,2}$-edge whiteline anomaly and its implications for the chemical behaviour of Pt $5d_{5/2}$ and $5d_{3/2}$ electronic states – a study of Pt-Au nanowires and nanoparticles

T K Sham[1*], M J Ward[1], M W Murphy[1], L J Liu[1] and W Q Han[2]

[1]Department of Chemistry, University of Western Ontario London N6A 5B7 Canada and [2]Center for Functional Nanomaterials, Brookhaven National Laboratory, Upton, New York 11973, United States

tsham@uwo.ca

Abstract. We report the $L_{3,2}$-edge whiteline anomaly observed in PtAu nanowire, PtAu and Pt nanoparticles deposited on Si nanowire, and their comparison with that of Pt metal. It is found that charge redistribution upon the formation of these materials can indeed be tracked with the $L_{3,2}$ whiteline intensity. The implications of these findings are discussed.

1. Introduction

The whiteline (WL) anomaly is often referred to as the non-statistical distribution of the intensity of the sharp resonance just above the threshold of the $L_{3,2}$-edge absorption edge. The most well known case is the Pt $L_{3,2}$-edge WL anomaly in Pt metal, of which the L_3-edge exhibits an intense WL while the L_2-edge does not. This is a classic example of non-statistical distribution of $L_{3,2}$ WL intensity, hence the distribution of the densities of states of the $5d_{5/2}$ and $5d_{3/2}$ electrons since dipole transition allows for the probing of unoccupied states of both $5d_{5/2}$ and $5d_{3/2}$, and $5d_{3/2}$ only, at the Pt L_3 and L_2-edge, respectively. The interpretation of this phenomenon as the result of relativistic effect and the disparity in the filling of the 5d electrons in Pt ($5d_{3/2}$ gets filled first) was first provided by Mott.[1] He pointed out that this was because the empty 5d states in Pt metal are mainly with $j = 5/2$ character.

This anomaly has since been revisited on several occasions.[2-5] For example, Brown, Peierls and Stern reported that the difference in the area under the curve of the L_3, and L_2 edge WL can be correlated quantitatively to the 5d hole counts in Pt metal; Matthesis and Dietz extended the discussion to include $d_{5/2}$ and $d_{3/2}$ hole counts and correlated them to the area under the curve derived from both edges as the difference between the XANES and an arctan-like function (represented by the Au $L_{3,2}$-edge XANES of the Au metal). Mansour, Cook and Sayers applied this approach to derive relativistic d-hole counts in Pt metal and related systems. [4] This anomaly was also found at the $M_{3,2}$ edge of Pt metal, confirming the non-statistical distribution of the unoccupied $5d_{5/2}$ and $5d_{3/2}$ states in Pt metal [5]. In Pt compounds, however, such as Pt(II) and Pt(IV) species, both edges exhibit WL whose intensities are closer to the statistical expectation. Thus this anomaly is associated with a nearly filled 5d band in Pt metal or related metallic Pt systems.

The advent of nanoscience and nanotechnology has prompted us to revisit this phenomenon, since many Pt and Au nanostructure systems exhibit potential for the enhancing catalytic performance. We ask whether or not this anomalous behavior is seen in PtAu nanoalloys, and if the chemistry of the $5d_{5/2}$ and $5d_{3/2}$ electrons in metallic Pt nanostructures can be tracked using Pt $L_{3,2}$-edge WL. We have

Journal of Physics: Conference Series **430** (2013) 012018 doi:10.1088/1742-6596/430/1/012018

examined the chemical behavior of the Pt $L_{3,2}$-edge WL intensity in Pt-Au nanowire, and Pt-Au nanoparticles and Pt nanoparticles dispersed on silicon nanowires. It is found that there is a subtle but noticeable redistribution of $5d_{5/2}$ and $5d_{3/2}$ charges (holes) relative to Pt metal upon nano Pt-Au alloy formation. The preliminary result of this study is reported herein.

2. Experiment

The Au-Pt nanowire (NW) was prepared using a solution method as describe previously.[6] The specimens had been characterized using XRD and TEM, showing a hybrid structure of crystalline nanowires with nodules of fcc crystal structure. The dimension is on the order of 5 – 10 nm. The Pt nanoparticle (NP) and Pt-Au nanoparticle on Si specimens were prepared by an electrode-less technique.[7] A PtAu bulk alloy sample, $PtAu_6$ were also run as references. The Pt $L_{3,2}$-edge XAFS measurements were conducted at the PNC-XSD (sector 20) BM beam line of the Advanced Photon Source (APS) in transmission mode. APS operates at 100 mA in the top-up mode.

3. Results and Discussions

Figure 1 shows the XAFS of the PtAu NW at the Pt $L_{3,2}$ and Au $L_{3,2}$-edge together with the TEM image. It can be seen that the PtAu NW shows a larger Au edge jump indicating that it is Au rich. The edge jump analysis using tabulated values from McMaster et. al.[8] indicates that the atomic ratio of Au/Pt is 1.92, and more interestingly, the Pt L_2-edge exhibits no WL while the Pt L_3-edge exhibits an intense WL. In the following, the anomaly is tracked via (i) comparison of PtAu nw with Pt metal, (ii) comparison of the AuPt NW L_3 intensity with that of its L_2-edge and (iii) comparison of the area under the curve (WL intensity) using Au $L_{3,2}$-edge as the background.

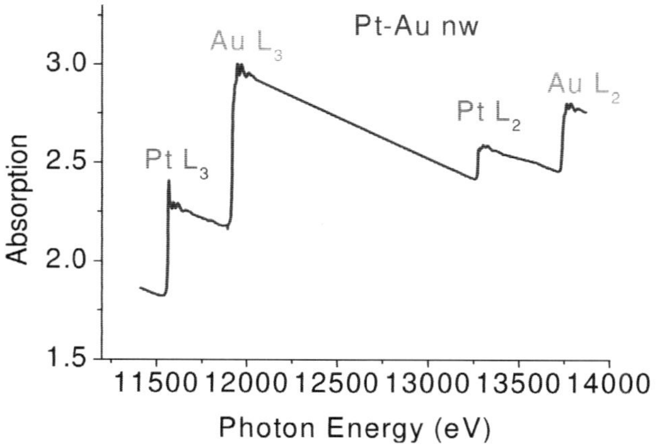

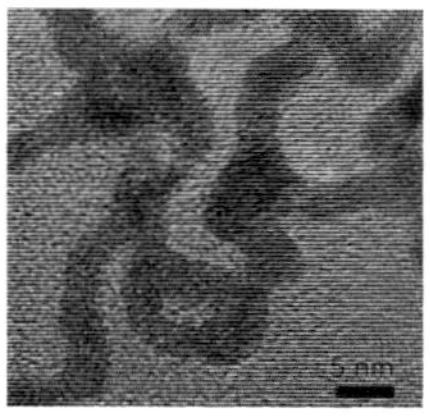

Figure 1. XAFS of PtAu NW at the Pt $L_{3,2}$ and Au $L_{3,2}$-edges; the inset displays the TEM of the sample showing chains and nodule of crystalline structures

3.1 Comparison with Pt metal whitlines

Figure 2 shows the Pt $L_{3,2}$ WL of PtAu NW and that of Pt metal. It appears at first glance that the PtAu NW L_3-edge WL is less intense than that of the Pt metal although it is blue-shifted (+0.9 eV) and broadened (FWHM by1.4 eV). The L_2-edge WL of PtAu NW appears more intense than that of Pt metal and exhibits the same shift to higher energy. This observation indicates that charge redistribution takes place upon the formation of PtAu NW. The difference curve between the WL of PtAu NW and Pt metal (not displayed) however shows comparable Pt L_3 WL intensity in PtAu NW relative to Pt and a small net gain in the intensity of the Pt L_2 WL indicating that $d_{3/2}$ charge is slightly depleted at the Pt site in PtAu NW. The considerable mismatches in the oscillation maxima in energy (*k* space) show that alloying takes place in this system (it is not core-shell). This will be discussed further below.

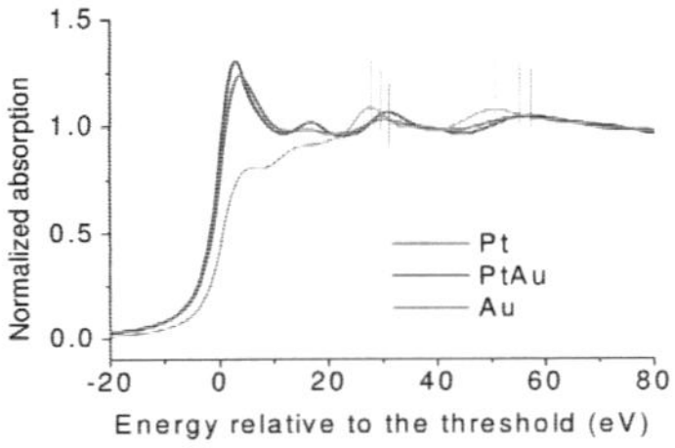

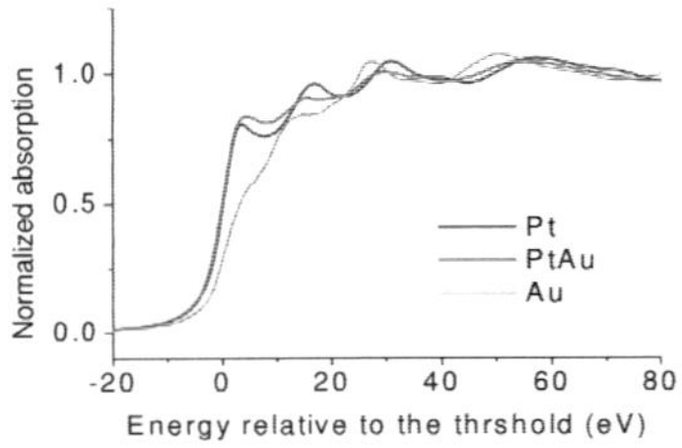

Figure 2. XANES of PtAu NW at the Pt L_3 (left), and L_2-edge (right), the Au foil XANES are also shown; photon energy is relative to the threshold; vertical bars mark the maxima of the oscillations

3.2 Difference curve between Pt L_3 and L_2 XANES

We use the difference curve between the Pt L_3 and L_2 edge XANES to obtain the difference in the area under the WL, $\Delta A = A_3(exp) - cA_2(exp)$ where c is a scaling factor, the ratio of the edge jump between Pt L_3 and L_2 edge, A_3 and A_2 are the experimentally determined area under the WL. If $5d_{5/2}$ is the dominant component of the unoccupied DOS, ΔA scales with the $5d_{5/2}$ hole counts (e.g. 0.3 from theory).[3] Figure 3 shows such a comparison. ΔA thus obtained is ~1.4×10^4 $cm^{-1}eV$ in both cases.

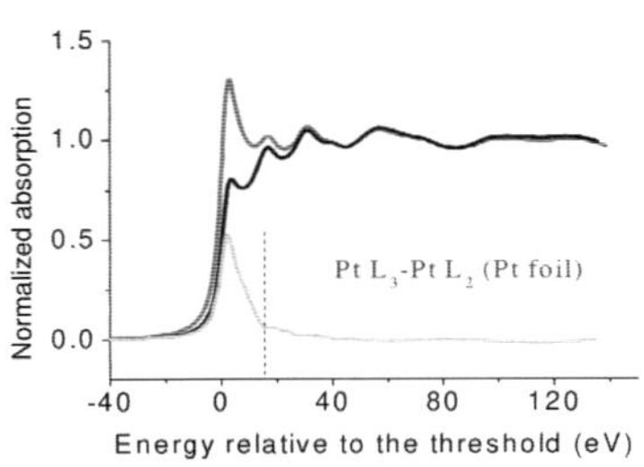

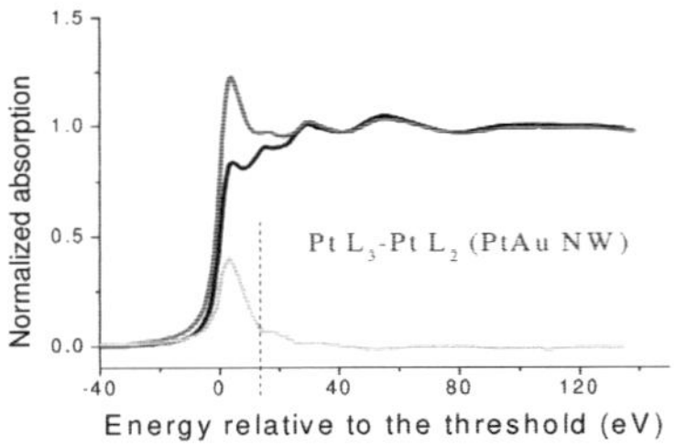

Figure 3. Difference spectrum between area under the curve for Pt (left) and PtAu NW (right); the vertical line shows the upper bound of the energy to which the area under the curve is integrated.

3.3 WL intensity

The Pt WL intensity can be obtained by removing an arctan-like back ground represented by the Au $L_{3,2}$-edge XANES since Au has the same fcc structure as Pt and exhibits no WL. Analysis using tabulated edge jumps (2.5×10^3 cm^{-1} for L_3 and 1.16×10^3 cm^{-1} for L_2)[8] yields nearly identical area under the curve of A_3 of 1.43×10^4 and 1.44×10^4 cm^{-1} eV and A_2 of 0.35×10^4 and 0.37×10^4 cm^{-1} eV for Pt and PtAu NW, respectively, yielding $h_{5/2}/h_{3/2}$ ratio of 0.29 and 0.27. This result shows that aside from for $5d_{3/2}$ to $5d_{5/2}$ charge redistribution (seen in WL shift and broadening), there is little detectable net charge count change on and off the Pt site upon PtAu NW formation compared to Pt metal.

3.4 The Au site

Since metallic gold has a nominally full d band, its WL in PtAu alloy is less sensitive unless a significant amount of d charge is depleted. However, the position of the oscillations in the XANES is informative since the progressive mismatching of the oscillations reflects whether or not Pt has an Au nearest neighbor since the *k* dependence of the backscattering amplitude and phase of Au is distinctly different from that of Pt even if they are both fcc with similar lattice constants. Close examination of Figure 2 and Figure 4 reveals that the oscillation maxima in the alloy occurs in between those of Au and Pt foils indicating that the nearest neighbour of Pt contains Au atom in the PtAu NW.

3.5 Pt NP and PtAu NP on SiNW

Figure 5 shows the Pt L_3-edge XANES of Pt/SiNW, PtAu/SiNW, alloy $PtAu_6$, and Pt metal. Together with the oscillation patterns shown in Figure 2 for PtAu NW compared with Pt, the characteristic

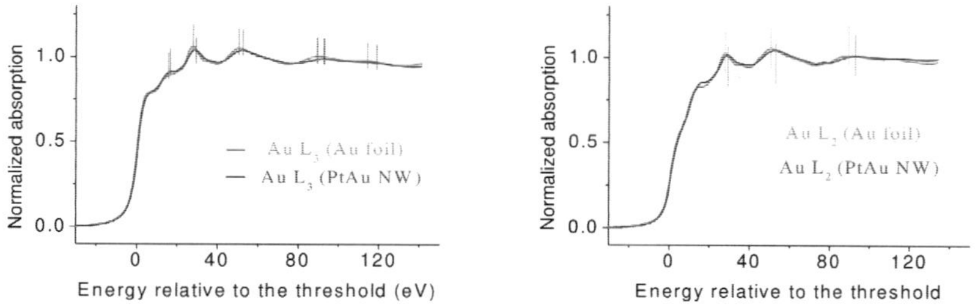

Figure 4. XANES of PtAu NW and Au at the Au L_3, and L_2-edges (left and right), the Au XANES are also shown; photon energy is relative to the threshold; vertical bars mark the maxima of the oscillations

oscillations beyond the WL reveal the following (i) all structures are fcc-like, (ii) The oscillations for Au metal appear at significantly lower *k* than that of Pt metal, (iii) PtNP/SiNW shifts slightly to high *k* whereas PtAu/SiNW shifts moderately to lower *k* and (iv) PtAu NW and $PtAu_6$ both shift to lower *k* in accordance with their stoichiometry (more dilute the sample, more significant the shift). These results indicate that Pt's nearest neighbor contains Au, likewise Au neighbors containing Pt; more quantitative information awaits further analysis.

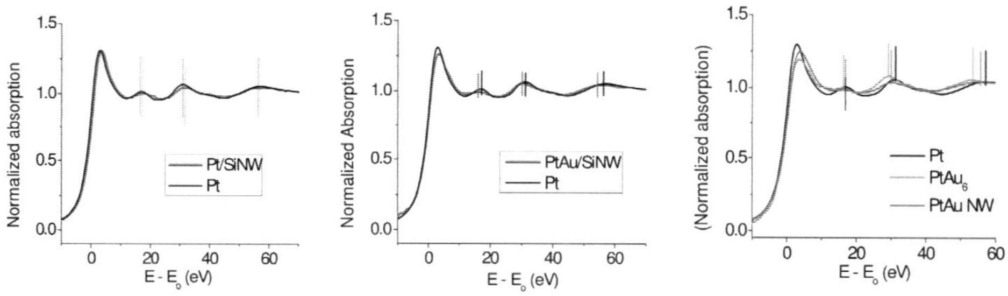

Figure 5. Pt L_3–edge XANES (left to right) of Pt NP/SiNW, PtAu NP/SiNW and PtAu NW and $PtAu_6$ alloy, vertical bars track the movement of the oscillation maxima.

4. **Conclusion**

The results reports show that the 5d charge redistribution in metallic Pt systems can be tracked using the Pt $L_{3,2}$-edge WL anomaly. Thus it is a great tool to study the performance of catalyst.

Acknowledgement

Research at Western is supported by NSERC, CRC, and CFI of Canada and OIT (Ontario); PNC/XSD at APS is supported by the US Department of Energy (DOE), Basic Energy Sciences, the Major Resource Support (NSERC), the University of Washington, and Simon Fraser University. Technical assistance of Robert Gordon is greatly appreciated.

References

[1] Mott N F, *Proc. Phys. Soc. (London), A* **62**, 416 (1949).
[2] Brown M, Peierls R E, Stern E A, *Phys. Rev. B* **15**, 738 (1977).
[3] Matthesis L M, Deitz R E, Phys. Rev. B 22 1663 (1980)
[4] Mansour A N, Cook Jr J W, Sayers D E, *J. Phys. Chem.* **88**, 2330 (1984).
[5] Sham T K, Naftel S J, Coulthard I, *J. Appl. Phys.* **79**, 7134 (1996).
[6] Teng X, et al., *J. Phys. Chem. C Lett.* 112, 14696 (2008)
[7] Coulthard I, Sham T K, *Solid State Comm.* **105,** 751-754 (1998)
[8] McMaster W H et al. *Compilation of X-ray Cross Sections* (NTI services, Springfield, VA1960)

Development of Fast Scanning Microscopic XAFS Measurement System

T. Tsuji[1], T. Uruga[1,2], K. Nitta[1], N. Kawamura[1], M. Mizumaki[1], M. Suzuki[1,4], O. Sekizawa[2], N. Ishiguro[3], M. Tada[3], H. Ohashi[1], H. Yamazaki[1,4], H. Yumoto[1,4], T. Koyama[1,4], Y. Senba[1,4], T. Takeuchi[1,4], Y. Terada[1,4], N. Nariyama[1,4], K. Takeshita[1,4], A. Fujiwara[1], S. Goto[1,4], M. Yamamoto[1,4], M. Takata[1,4], and T. Ishikawa[1,4]

1 JASRI/Spring-8, Sayo, Hyogo 679-5198, Japan

2 The Univ. of Electro-Communications, Chofu, Tokyo 182-8585, Japan

3 Institute for Molecular Science, Myodaiji, Okazaki, Aichi 444-8585, Japan

4 RIKEN Harima Institute, Sayo, Hyogo 679-5148, Japan

E-mail:t.tsuji@spring8.or.jp

Abstract. We have developed a fast scanning microscopic X-ray absorption fine-structure (XAFS) measurement system using a 100–300 nm focused X-ray beam at the BL39XU nano-scale analysis station at SPring-8. This system provides two-dimensional XAFS images constructed from X-ray fluorescence (XRF) images measured by fast continuous 2-dimensional scanning at all XAFS measurement energies. High-quality XAFS spectra at each position were obtained by correcting shift in the XRF images. We applied our proposed method to measuring single catalyst particles, and succeeded in measuring the microscopic x-ray absorption near-edge structure (XANES) images that were 2–4 μm square with a spatial resolution of 300 nm.

1. Introduction

It is essential to measure single particles or single cells of devices and not their averages to investigate the dynamic aspect of structures and electronic states of nano/meso-scale particle-based functional materials and devices. We need to carry out measurements with a spatial resolution of less than the size of the subject to do this. We previously conducted μ-XAFS measurements of single catalyst particles using an X-ray μ-beam of 800–1000 nm at SPring-8 [1]. Nano-scale analysis stations were recently constructed at beamline BL37XU and BL39XU at SPring-8 in 2011 [2]. These stations supply high flux 100-nm focused X-ray beams, which enable spatially resolved measurements of the internal electric state and structure of meso-scale particles and device cells to be conducted.

Full-field measurements are generally more effective for site-dependent studies than point-by-point measurements, since the former can be conducted with almost no time lag when different sites are measured, and in many cases there is no information about the sites of specific points in the samples. It is also not easy to conduct μ-XAFS measurements at fixed points because of shift in the position irradiated by X-ray beams due to vibration and/or thermal drift in the focusing devices and sample stages during XAFS measurements. This is more serious when carrying out measurements with smaller X-ray beams.

Therefore, we developed a system for fast two-dimensional scanning microscopic XAFS measurements and its design and some preliminary experimental results are presented in this report.

2. Experimental and data correction method

The measurements were conducted at the BL39XU nano-scale analysis station at SPring-8 (8 GeV, 100 mA) [2]. The X-rays were monochromatized by using an Si (111) double crystal monochromator. Incident X-rays were focused with an Rh-coated Kirkpatrick-Baez mirror (KB mirror). The focused beam was 300 (H) ×260 nm (V). The beam flux incident on the sample was 8×10^9 photons/sec at 5.8 keV.

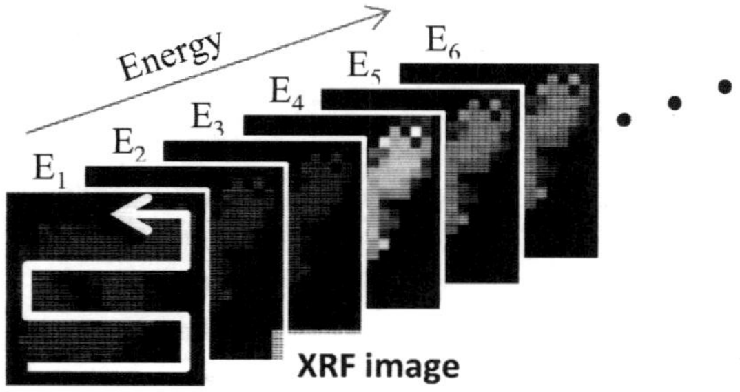

Figure 1. Schematic of method of fast scanning microscopic XAFS.

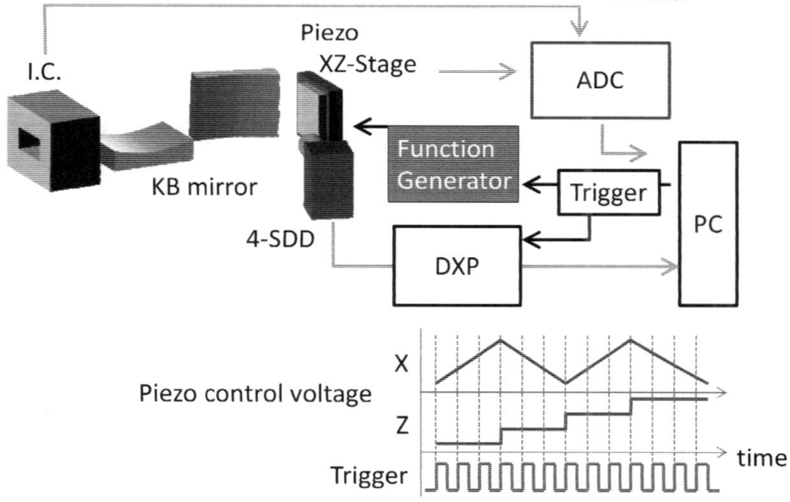

Figure 2. (a) Schematic of fast system for scanning microscopic XAFS measurements and (b) voltage wave form to piezo stage.

Figures 1 and 2 have schematics of the scanning method and proposed system for fast scanning microscopic XAFS measurements. The sample holder is mounted on a 2-axis piezo positioning stage (PI, P-541.2DD) to continuously scan the sample position. The maximum scanning speed of this piezo stage is about 500 Hz with the load of the sample holder (about 1 kg in weight). Piezo stage scanning is controlled with an analog voltage wave generated by a function generator (Fig. 2 (b)). The voltage waveforms applied to the X and Z piezo stages are programmed with a PC for scanning along the route in Fig. 1. Fluorescence X-rays from the sample are measured with a four-element silicon drift detector (SDD) (Vortex, ME4) and the X-ray intensity incident on the sample is measured with an ionization chamber (IC) (Oken, S-1194A1). The output voltages of the IC and the piezo stage's 2-axis sensor are stored in ADC memory. The output signals of the SDD are stored in the memory of a digital X-ray processor (DXP) (XIA, DXP-XMAP). Data storage in the ADC and DXP is synchronized with a trigger signal from the PC. Storage data sets are transferred to the PC's hard disk after each XRF measurement, thus enabling us to obtain fast 2-dimensional XRF images. Spatially resolved XAFS images can be obtained by carrying out these measurements at each energy point of the XAFS measurements.

The data points of 2D XRF images obtained by continuously scanning with the piezo stage are on an unequally-spaced grid. Consequently, all measured 2D XRF data are linearly interpolated at equally spaced positions. The position of the focused X-ray beam on the sample also slightly shifts at the same position of the piezo stage during the XAFS measurements due to the displacement of the X-ray optics and/or sample stages. The shift in interpolated XRF image data are corrected by calculating the shift of the center of gravity coordinates for particle-like samples or by finding a peak for the cross-correlation function of XRF data and that of the reference.

3. Results

Figure 3 shows a linearly interpolated 2D XRF image of the Ce-Lα fluorescence of a single particle of catalyst mounted on an SiC membrane that was 1000 nm thick. The sample was a $Pt/CeZrO_x$ catalyst, and its particles were 1.0 (H) × 1.3 μm (V) in size. The scanning area was 3.5 (H) × 3.4 μm (V), and there were 36 × 33 scanning points. The total time for the measurements was 30 sec. High speed continuous 2D scanning was achieved by using the piezo stage while continuously storing the output of the high accuracy position encoder. This measurement system combined with a high flux focused X-ray beam at the BL39XU nano-scale analysis station rapidly provided high S/N ratio 2D XRF images with a spatial resolution of 100–300 nm.

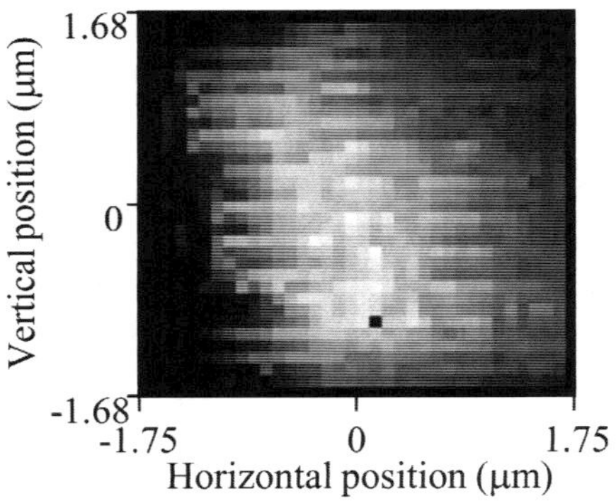

Figure 3. Linearly interpolated Ce-Lα XRF image of single particle catalyst. Incident X-ray energy: 5.8 keV.

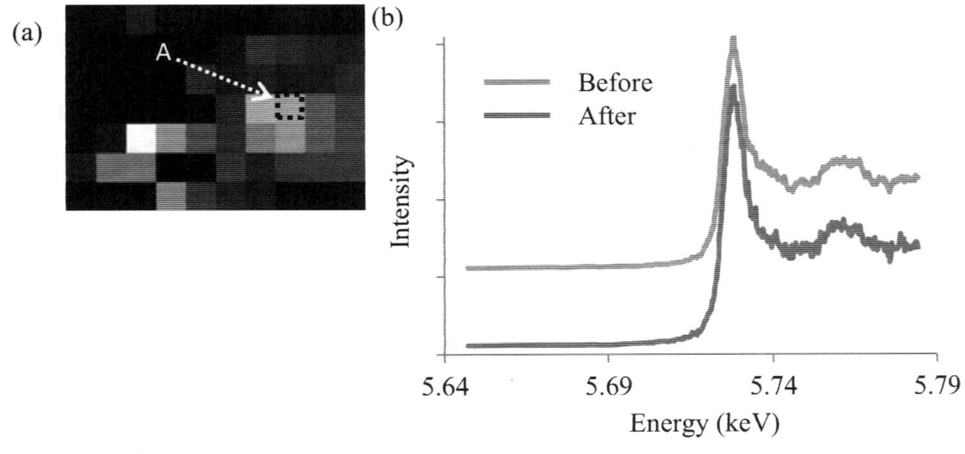

Figure 4. Ce Lα XRF images (a) (incident X-ray energy: 5.73 keV), and Ce L_3-XANES spectra at point A before and after correction of shift in center of gravity.

Figure 4 (a) shows Ce Lα XRF image. Figure 4 (b) shows the Ce L_3-edge XANES spectra at point A in Fig. 4 (a) before and after correction of shift in center of gravity. The scanning area of the XRF image was 2.9 (H) × 1.8 μm (V) and its pitch was 300 (H) × 260 nm (V). There were 10 × 7 scanning points and the XRF images were measured at 280 energy points. The time to measure one XRF image was 30 sec and that to measure the XANES spectrum was 140 min. The shift in the field of view of the XRF images was corrected by estimating the shift in the center of gravity of the particles in the selected area. Figure 5 shows the shift in position of the center of gravity of the particles in the XRF images. The maximum deviations were about 200 nm (H) and 110 nm (V), and the standard deviations were 33 nm (H) and 22 nm (V). Here, the quality of the XANES spectra was only slightly improved by this correction (Fig. 4 (c)) as the shift in the XRF images was relatively small. The time to measure the XANES of one area of 300× 260 nm (V) was 2 min. A reduced spatial resolution of 1/3 and measurement time of 1/20 were achieved using this measurement system in comparison with our previous studies on similar samples [1]. In addition, it was possible to rapidly obtain spatially resolved information about local structures and chemical states in areas that were several micrometers square of meso-scale samples.

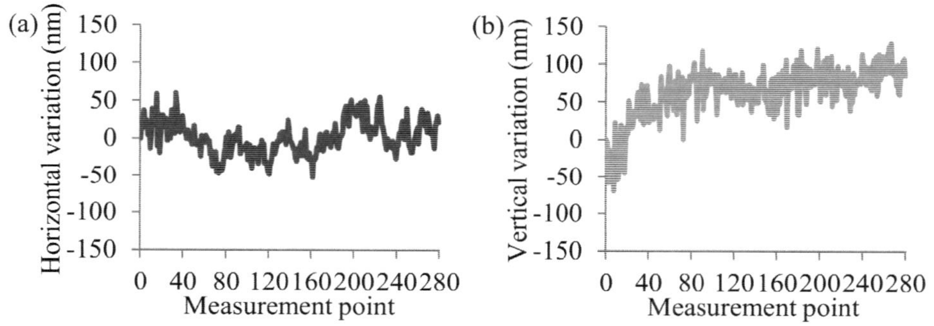

Figure 5. Shift in center of gravity of particle (a) horizontally and (b) vertically around point A in Fig. 4 (a).

4. Summary

We developed a fast system for scanning microscopic XAFS measurements with a spatial resolution of 100–300 nm at the BL39XU nano-scale analysis station. This system of measurement rapidly provided 2D XAFS spectra images. We expect to be able to apply this method to in-situ temporally and spatially resolved XAFS measurements of meso-scale samples in the near future.

5. Acknowledgments

This research was carried out with the approval of SPring-8 (Nos. 2011A2060, 2011B1889 and 2011B2100). It was also supported in part by the Polymer Electrolyte Fuel Cell program by the New Energy and Industrial Technology Development Organization (NEDO) of Japan and by the "Low-Carbon Research Network (LCnet)" funded by the Ministry of Education, Culture, Sports, Science and Technology (MEXT), Japan.

References

[1] Tada M, Ishiguro N, Uruga T, Tanida H, Terada Y, Nagamatsu S, Iwasawa Y, and Ohkoshi S 2011 *Phys. Chem. Chem. Phys.* **13** 14910–14913

[2] Koyama T, Yumoto H, Terada Y, Suzuki M, Kawamura N, Mizumaki M, Nariyama N, Matsushita T, Ishizawa Y, Furukawa Y, Ohata T, Yamazaki H, Takeuchi T, Senba Y, Matsuzaki Y, Tanaka M, Shimizu Y, Kishimoto H, Miura T, Kimura H, Takeshita K, Ohashi H, Yamamoto M, Goto S, Takata M, and Ishikawa T 2011 *Proc. of SPIE* **8139** 81390I

Journal of Physics: Conference Series **430** (2013) 012020 doi:10.1088/1742-6596/430/1/012020

New XAFS beamline for structural and electronic dynamics of nanoparticle catalysts in fuel cells under operating conditions

O Sekizawa[1], T Uruga[1, 2], M Tada[3], K Nitta[2], K Kato[2], H Tanida[2], K Takeshita[2,4], S Takahashi[2,4], M Sano[2], H Aoyagi[2], A Watanabe[2], N Nariyama[2,4], H Ohashi[2,4], H Yumoto[2], T Koyama[2], Y Senba[2], T Takeuchi[2], Y Furukawa[2,4], T Ohata[2,4], T Matsushita[2,4], Y Ishizawa[2], T Kudo[2,4], H Kimura[2,4], H Yamazaki[2], T Tanaka[2, 4], T Bizen[2,4], T Seike[2,4], S Goto[2,4], H Ohno[2], M Takata[2, 4], H Kitamura[2,4], T Ishikawa[2,4], T Yokoyama[3] and Y Iwasawa[1]

[1] The University of Electro-Communications, Chofu, Tokyo 182-8585, Japan
[2] JASRI/SPring-8, Sayo-gun, Hyogo 679-5198 Japan
[3] Institute for Molecular Science, Okazaki, Aichi 444-8585, Japan
[4] Riken Harima Institute, Sayo-gun, Hyogo, 679-5148 Japan

E-mail: sekizawa@pc.uec.ac.jp

Abstract. We are currently constructing a new X-ray absorption fine structure (XAFS) beamline BL36XU at SPring-8 dedicated for the structural and electronic analysis of the dynamic events on polymer electrolyte fuel cell (PEFC) cathode catalysts for the development of next-generation PEFCs. To investigate the cathode catalyst nanoparticles in PEFCs under the operating conditions, the beamline is designed to provide time- and spatially resolved XAFS techniques having a time resolution of 100 μs, spatial resolution of 200 nm, and depth resolution of 1 μm. We report the outline and design of the new beamline.

1. Introduction

Recently, the development of a low carbon society has become a countermeasure against problems of global warming or exhaustion of resources. Fuel cells that generate electricity from hydrogen and oxygen are one of the most promising technologies for green energy creation. Crucial issues for developing next generation polymer electrolyte fuel cells (PEFCs) are improvement of activity, high durability, and reduction in the cost of cathode catalysts. To make PEFCs reality particularly in automobiles, it is necessary to clarify the dynamic aspects of structures and electronic states, and their kinetics/dynamics of cathode catalysts in membrane electrode assemblies (MEAs), and chemical reaction mechanisms at cathode catalyst surfaces under PEFC operating conditions and in the deterioration processes.

Time- and spatially resolved X-ray absorption fine structure (XAFS) techniques are very powerful for investigation of the structure and chemical states of nanoparticle catalysts under in-situ working conditions. To clarify the elementary steps in the practical catalytic reaction mechanism and deterioration process of PEFCs in real time, XAFS measurements with time resolutions of 100 μs to 1 s is required. Also, the catalyst layer of MEAs is a micro-heterogeneous dispersed system consisting of a spatially complex mixture of catalyst, support, ionomer, and water, which requires spatial resolutions of 100 nm to 10 μm for XAFS measurements. However, there is no beamline capable of conducting XAFS measurements with such high time and spatial resolutions.

Therefore, we started to construct a new XAFS beamline BL36XU at SPring-8, which provides high time- and spatially resolved XAFS methods specialized for the structural and electronic analysis of the dynamic events on the cathode catalysts in PEFC MEAs.

2. Light source and optics

The target specifications of BL36XU are as follows: time resolutions of 800 μs by quick XAFS and 100 μs by energy dispersive XAFS (DXAFS), a 2-dimensional in-plane spatial resolution of 200 nm, and 3-dimensional resolution of 1 μm for membrane electrode assembly (MEA) samples. The energy range is from 4.5 to 35 keV, which covers absorption edges of almost all metal elements used for the PEFC catalysts.

Figure 1 is a schematic arrangement of the beamline. The design of the synchrotron light source and optics is a SPring-8 standard, which results in both quick XAFS measurements with 10-ms time resolution and spatially resolved measurements using 100-nm-order X-ray beams [1]. The light source is an in-vacuum type tapered undulator to adjust the energy width of an X-ray suited for quick XAFS and DXAFS measurements. The hutches consist of one optics hutch and one experimental hutch. The experimental hutch is located 77 m far from the light source to obtain a high reduction ratio of the focusing mirror, which was designed for spatially resolved XAFS using a 100-nm focused beam.

The main X-ray optics consists of four mirrors (M1, M2, M3, and M4) and two monochromators. The M1 is a liquid-nitrogen cooled horizontal deflection flat mirror, and M2 is a water-cooled spherically bent flat mirror for horizontal focusing. The M1 and M2 are placed apart at a distance of 7.5 m with a glancing angle of 2 mrad to separate reflected X-rays by 30 mm in the horizontal direction from high energy X-rays and gamma-rays emitted from the light source. This configuration enables the rejection of the high energy components by a gamma-ray stopper and downsizing of the monochromator. The downstream mirrors, M3 and M4, are water-cooled spherically bent flat mirrors for vertical focusing and higher harmonics rejection. The M1 and M2 are 1000 mm long, and M3 and M4 are 700 mm long. The substrates of the mirrors are Si. The coatings are stripes of Pt and Rh, which are switched depending on the energy region of the XAFS measurement.

Servomotor-driven compact monochromators located downstream of the M2 mirror results in10-ms time-resolved quick XAFS measurement [2]. To cover a wide energy range, two monochromators having a channel-cut Si (111) crystal for 4.5-28 keV and a Si (220) crystal for 7-35 keV are tandemly arranged. The crystals are cooled with liquid nitrogen. The monochromators can be switched using vertical translation stages while maintaining a vacuum.

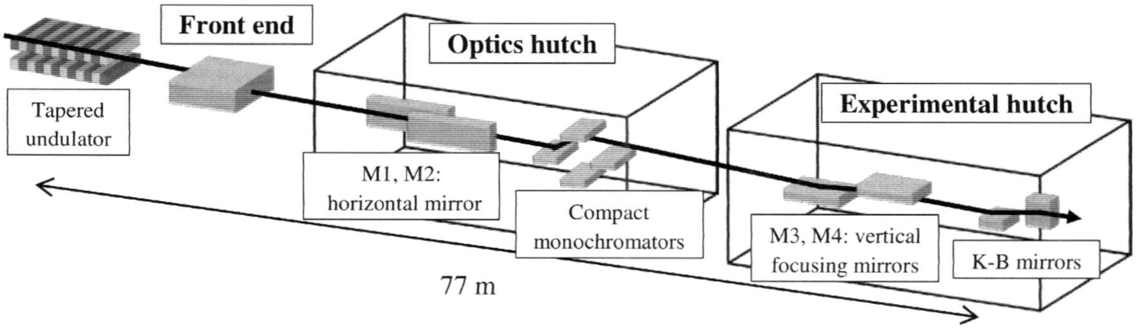

Figure 1. Schematic of the new beamline BL36XU.

3. XAFS measurement system

3.1. *Time-resolved XAFS*

3.1.1. Quick XAFS. Figure 2 is a schematic drawing of a 10-ms time-resolved QXAFS measurement system using servomotor-driven monochromators. The X-ray detectors are fast response ionization chambers having narrow gap electrodes for transmission-mode measurements and a Si PIN photodiode (Hamamatsu photonics, S3584-09) for fluorescence-mode measurements. This system is a base system for XAFS measurements of this beamline. For long measurements, the step-scan XAFS method is applied, and the undulator gap is not tapered and is adjusted to maximize X-ray flux at each measurement point.

For 800-μs time-resolved QXAFS measurement, the servomotor-driven monochromator is removed from the beam axis, and instead a newly developed galvano scanner motor driven monochromator is set [3]. The monochromator has a compact Si crystal (less than 30(L)×15(H)×13(W) mm) installed in a He chamber and has no active cooling devices to lower the rotational inertia for fast angle scanning. Calculation by the synchrotron radiation calculation code SPECTRA [4] showed that the heat load of the undulator radiation on the monochromator crystal can be reduced to less than 10 watts by rejecting higher harmonics and low energy parts using the M3 and M4 mirrors and an absorber, respectively, and by limiting incident beam size using the front end slit. This heat load is acceptable for maintaining the Bragg condition of the crystal. This system results in fast quick XAFS measurements, but causes a slight loss of beam flux. To cover a wide energy range, three monochromators, each having Si (111) for 4.5-8 keV, Si (111) for 7-14 keV, and Si (220) for 12-35 keV, will be installed in the experimental hutch and switched using a vertical translation stage.

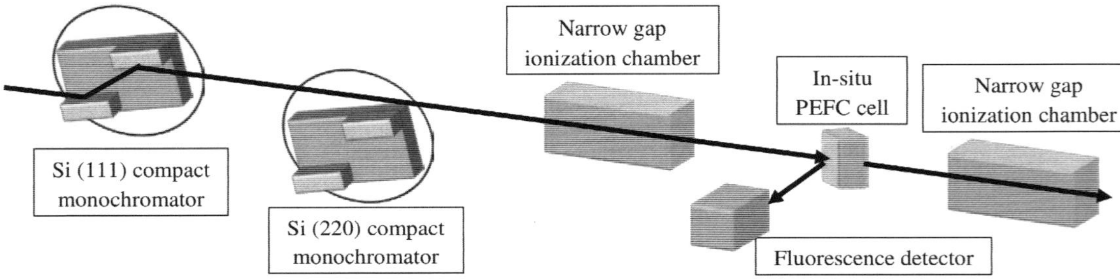

Figure 2. Schematic of the 10-ms time-resolved QXAFS optics and measurement system.

3.1.2. DXAFS. Figure 3 is a schematic drawing of a 100-μs time-resolved DXAFS optics and measurement system. DXAFS method will be especially used for model samples of PEFC with high catalyst concentration to obtain information on the faster elementary steps in a cathode catalytic reaction if anything. The X-ray beam incident on the bent crystal polychromator placed in the experimental hutch is expanded in the horizontal direction by the M2 mirror to achieve the energy band required for DXAFS measurement. To cover a wide energy range, two types of polychromator configurations, *i.e.* Bragg configuration for 4.5-12 keV and Laue configuration for 12-35 keV, will be installed. The polychromator consists of a common elliptical bender and two exchangeable crystalline holders for each configuration (Toyama). The crystalline holders are cooled with water. The position-sensitive X-ray detector is a fast 2-dimensional X-ray imaging detector that is composed of a visible light conversion unit called an imaging unit (Hamamatsu Photonics, AA40), which converts the incident X-rays to visible light by a crystalline scintillator (YAP: Ce), and a high-speed CMOS camera (Phantom, V711, maximum frame rate: > 10^4 fps, full well capacity: 23000 e).

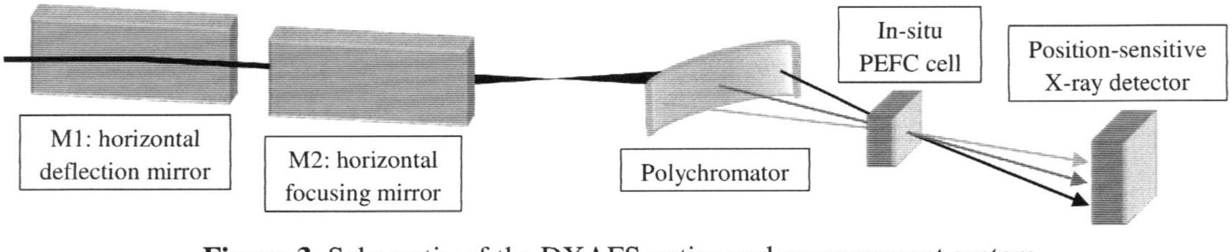

Figure 3. Schematic of the DXAFS optics and measurement system.

3.2. Spatially resolved XAFS

3.2.1. Two-dimensional imaging XAFS. Two types of 2-dimensional imaging XAFS methods will be available in BL36XU depending on the catalyst concentration in MEAs. For high concentration samples, the transmission microscopic XAFS method will be used. X-ray images transmitted through the sample are measured using a 2-dimensional X-ray imaging detector consisting of an imaging unit having a magnification ratio of 50 (Hamamatsu Photonics, AA50) and a low noise CMOS camera (Hamamatsu photonics, Orca flash 4.0). The expected field of view and spatial resolution are about 300×300 μm and 1×1 μm, respectively.

For low concentration samples, a newly developed fast scanning microscopic XAFS method will be used [5]. In this method, 2-dimensional X-ray fluorescence (XRF) images are measured by fast scanning of a sample using a high-speed piezo positioning stage (PI, P-541.2DD) at each energy point of the XAFS measurement; as a result, a 2-dimensional fluorescence XAFS image is obtained. The designed field of view and spatial resolution is 40×40 μm and 200×200 nm, respectively. The X-ray beam is focused at the sample using a Kirkpatrick-Baez (KB) mirror (JTEC). To cover a wide energy range, two sets of KB mirror systems will be installed, one is Rh coated with a glancing angle of 4 mrad for 4.5-15 keV and the other is Pt coated with a glancing angle of 2 mrad for 15-30 keV. The two KB mirror systems are switched over using horizontal translation stages. Fluorescence X-rays from the sample are detected using a 4-element silicon drift detector (Vortex, ME4) or 21-element Ge detector (Canberra).

3.2.2. Depth-resolved XAFS. An exit angle-resolved fluorescence-yield XAFS method [6] will be used as a depth-resolved XAFS method. Fluorescence X-rays are detected using a 2-dimensional pixel detector (DECTRIS, PILATUS 300 K-W). The depth-resolution of this method highly depends on the flatness of the sample surface and thickness of the cap layer. Therefore, it may be suitable for flat model samples rather than real PEFC samples. We are currently developing a depth-resolved XAFS method applicable for real PEFC samples.

4. Conclusion

We described the design of the new XAFS beamline BL36XU at SPring-8 to investigate dynamic reactions on cathode catalyst nanoparticles distributed in MEAs of PEFC using newly-developed high time- and spatially resolved XAFS techniques. BL36XU will be commissioned from Oct. 2012, and opened for users in Jan. 2013.

5. Acknowledgments

We thank Drs. T. Nonaka and K. Dohmae (Toyota Central R&D Labs) for their helpful discussion. Part of this work was conducted with the approval of SPring-8 Nos. 2011A2060 and 2011B2100. This work was supported by New Energy and Industrial Technology Development Organization (NEDO), Japan.

References

[1] Hirose Y 2010 *SPring-8 Reserch Frontiers 2009* 170
[2] Nonaka T, Dohmae K, Araki T, Hayashi Y, Hirose Y, Uruga T, Yamazaki H, Mochizuki T, Tanida H and Goto S 2012 *Rev. Sci. Instrum.* **83** 083112
[3] Uruga T, Tanida H, Inoue K, Yamazaki H and Irie T 2007 *AIP Conf. Proc.* **882** 914
[4] Tsuji T, et al. 2012 this proceedings
[5] Tanaka T and Kitamura H 2001 *J. Synchrotron Rad.* **8** 1221
[6] Shinoda K, Sato S, Suzuki S, Uruga T, Tanida H, Toyokawa H, Terada Y and Takagaki Y 2009 *J. Surface Analysis* **15** 295

In situ two-dimensional micro-imaging XAFS with CCD detector

H Tanida[1,*], H Yamashige[1], Y Orikasa[2], T Fujimoto[1], M Oishi[1], H Murayama[1], H Arai[1], M Katayama[3], Y Inada[3], T Ohta[3], Y Uchimoto[2] and Z Ogumi[1]

[1]Office of Society-Academia Collaboration for Innovation, Kyoto University, Gokasho, Uji, Kyoto 611-0011, Japan
[2]Graduate School of Human and Environment Studies, Kyoto University, Yoshida Nihon-matsu, Sakyo, Kyoto 606-8501, Japan
[3]SR center, Ritsumeikan University, 1-1-1 Noji-Higashi, Kusatsu, Shiga 525-8577, Japan

E-mail: tanida@saci.kyoto-u.ac.jp

Abstract. In situ two-dimensional (2D) micro-imaging X-ray absorption fine structure (XAFS) measurements were performed in transmission mode using a charge coupled device (CCD) detector, phosphor screen, and magnifying lens. This method makes it possible to display a 2D image with a spatial resolution of around 2 μm at each energy point in a XAFS spectrum. The method was applied to in situ transmission micro-imaging XAFS measurement with a quick scanning technique.

1. Introduction

X-ray absorption fine structure (XAFS) is a powerful technique used to investigate the local structure of amorphous and solution samples such as battery materials [1]. It is important to fully understand the concentration profile and the chemical state of the element in a given material especially in terms of chemical reactions. We recently developed a two-dimensional (2D) x-ray imaging method with a pixel array detector for in situ analysis of a large sample area [2]. In the present study, we developed a 2D x-ray micro-imaging method with a charge coupled device (CCD) detector for in situ analysis. The imaging data obtained with this method show the distribution of particles in the material, and XAFS reveals the kind of element and the chemical state. X-ray microscopy and tomography combined with an X-ray absorption method is an effective technique that has been previously studied [3]. In addition, a 2D X-ray imaging technique with a micro-beam and translation stage has been reported for battery samples [4]. Recently, a 3D computed tomographic X-ray near edge fine structure (XANES) technique was proposed [5]. A CCD detector with a high space resolution of a few micrometers is usually used for imaging measurements. The high spatial resolution reveals in detail the distribution of transition metals and their chemical states in the materials. In this study, we performed in situ 2D micro-imaging XAFS with a CCD detector for a micro-scale spatial resolution analysis in a lithium ion battery composite electrode. It has been well known that electrode reactions in batteries proceed inhomogeniously and the electronic states (caused by electrochemical charging/discharging) are different from part to part in micron to millimeter ranges. It is important to elucidate such non-uniform distribution of the reaction process in the electrode to understand the battery performances.

2. Experimental

The data were taken at BL37XU in SPring-8, Hyogo, Japan [6]. The X-ray beam was from an in-vacuum type undulator. The monochromator was a double crystal fixed-exit type, and Si (111) was used as the monochromator crystal. Two Rh-coated mirrors were used to remove higher harmonics. The X-ray beam was around 1 mm high by 2 mm wide on the sample.

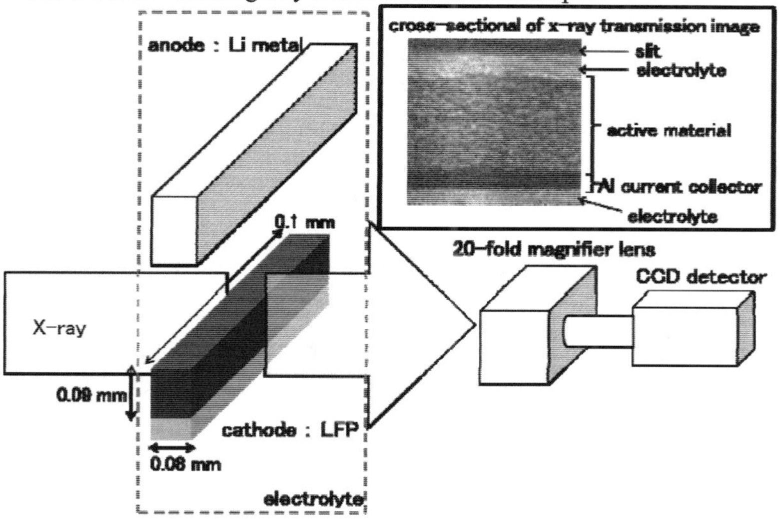

Figure 1. Schematic diagram of 2D XAFS measurement of electrode sample with CCD and magnifier lens.

The intensity of the incident X-rays was measured in an ionization chamber that was 17 cm long and filled with nitrogen gas. A current signal from the ionization chamber was measured as a voltage value by a current amplifier (Keithley 428, Keithley, USA), and the voltage value was converted to a frequency value by a voltage-to-frequency (VF) converter (DS-VFC2, Seiko EG&G, USA). The pulse value from the VF converter was counted by a scaler, (Ortec 974, SEIKO EG&G Ortec, USA), which generated a timing trigger signal. A schematic of the system is shown in figure 1. The CCD detector system was from Hamamatsu Photonics, Japan. The system consisted of a C9100-14 CCD, a 20-fold magnification lens, and a Lu2SiO5 (LSO) phosphor screen holder. The LSO was high-resolution phosphor, and the thickness of the quartz glass holder was 0.01 mm. The phosphor screen holder was set in front of the CCD detector. The CCD acquired images in synchronization with the trigger signal of the scaler. The specifications of the CCD with the 20-fold magnification lens were 0.65 μm per pixel for the spatial resolution and 0.665×0.665 mm for the effective field of view.

Electrodes were fabricated by mixing 80% cathode active material: $LiFePO_4$ (LFP), 10% acetylene black, and a 10% polyvinylidene fluoride (PVDF) binder; they were then pasted on an aluminum current collector. The grain size was 100 nm. We observed a cross section of the composite electrode by cutting the electrode with a microtome. The cathode between the electrodes was 90 μm in size: 80 μm thick for the X-ray, and 10 mm long. The electrodes were in a laminated aluminum bag with an electrolyte solution and Li metal as an anode. The sample electrodes in the bag worked as a battery in the X-ray absorption spectrum measurement. A cross-sectional x-ray transmission image of the sample is also shown in figure 1, which was trimmed from the field of view image. A tantalum blade slit was used to shape the incident X-ray beam so that it would hit only the cathode electrode. The representing of the upper blade of the slit appears between the cathode and anode in the image in figure 1. The electrolyte, the active material, and a Al current collector were also observed. The X-ray did not transmit the slit or the Al foil, and thus, a dark shade was observed. The electrolyte did not absorb the X-ray, resulting in a bright shade. The active material included LFP as well as conductive and binder materials, and they appear as a gray shade in the image.

3. Results and Discussion

Figure 2 shows images acquired under the Fe K absorption edge, at the midpoint of the edge, and at the top of the edge. It took 300 ms to obtain the images at each energy point. The absorption spectrum of the three points is shown at the bottom of figure 2. The contrast in these images is different from that in the incident X-ray photon energy images because the absorption by the Fe ions in the sample depends on the energy. Above the edge the image is darker than that under the edge. However, each image appears to be uniform because of the homogenous distribution of Fe ions in the sample. The particle size of LFP is much smaller than the special resolution of the CCD. If the Fe ions in the sample were heterogeneously distributed in a few tenths of μm, a large bright spot would be observed.

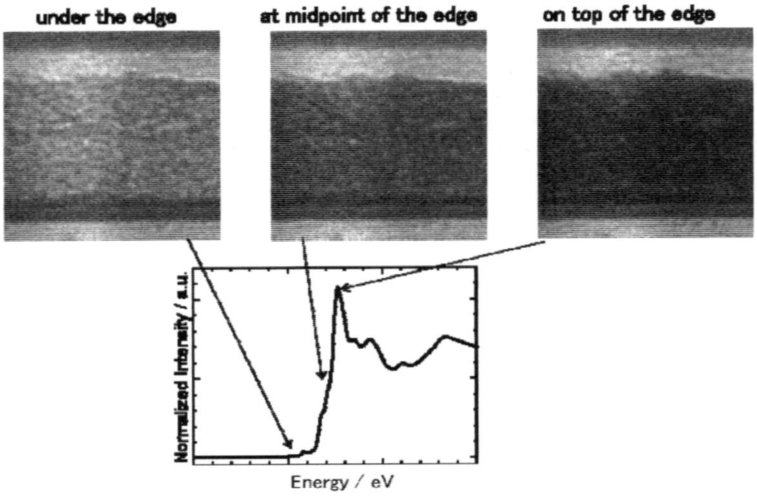

Figure 2. CCD images of the sample at different energy of the Fe K edge XANES spectrum.

Figure 3 shows a cross-sectional image and XANES spectra at different positions and electric potentials. The spectra were obtained at the positions (I1), (I2), and (I3) in figure 3 (a), which were at the near end against the anode, inside the cathode, and at the near end against the Al current collector, respectively. The spectra in figures (b) and (c) are plotted against the sum of 4×4 pixels, which corresponds to 2.6×2.6 μm as the spatial resolution. In this method the spatial resolution was checked by a measurement of a micro-image pattern chart for X-ray, and the spatial resolution was measured at around 2 μm. Although the specification of the CCD with the 20-fold magnification lens was 0.65 μm, the spreading of the fluorescent screen and the diffusion of the X-ray beam would reduce the spatial resolution of this method. Therefore, 4×4 pixels of the CCD image were averaged, which corresponded to the actual spatial resolution. The XANES spectra of the CCD image are normalized by blank images obtained without a sample. The most accurate normalization method would use a blank image at each energy point simultaneously. However, such a method is hard to combine with the in situ method used in this study. The top-off mode operation of the storage ring minimizes normalization effects.

The edge jump of the spectra in figure 3 depends on the concentration profile of Fe ions at various positions in the electrode. The Fe ions near the anode were smaller than those inside the cathode. However, no difference in the noise level was observed after charging the sample battery. In contrast, no differences in the spectral features were observed between figures (b) and (c). With the electrochemical reaction, the valence state of Fe ions would change, and the features of the spectra would be affected. The spectra in figures (b) and (c) were obtained before and while charging the sample battery, respectively. If the progress of the reaction depends on the distance from the anode, the spectra would be different in figure 3. In this study, the distribution of the electro-chemical reaction between the electrodes was not clearly observed. In future, in situ measurement at several positions of the electrode, several potentials, and several reaction rates will be attempted using a quick scanning XAFS technique.

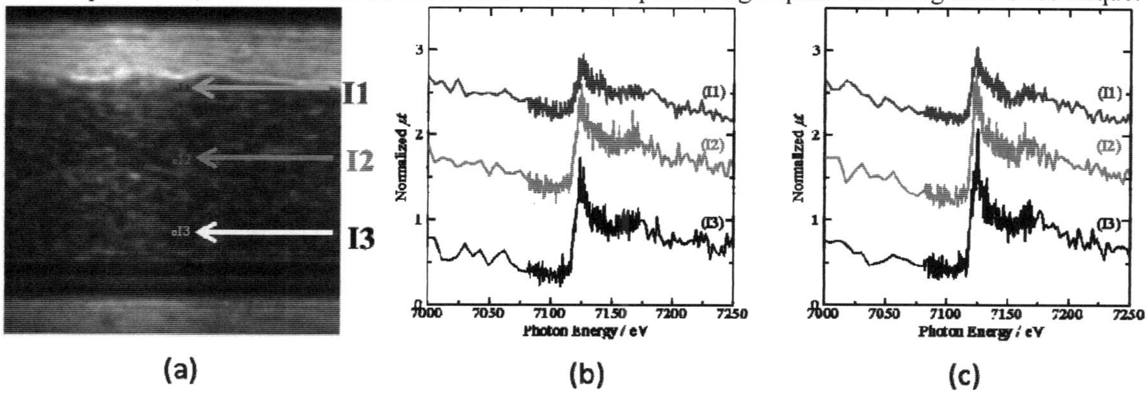

Figure 3. Fe-K edge XANES spectra at different positions of the sample. XANES spectra were obtained at three positions: I1 (near anode), I2 (inside), and I3 (far anode) in transmission image (a). The spectra were at the potential before charging (b) and during charging (c) of the sample battery.

4. Conclusion

In situ transmission 2D micro-imaging XAFS by a CCD with magnification lens was successfully demonstrated at the spatial resolution of around 2 μm. The method is widely applicable to the field of materials science for solutions, amorphous materials, and batteries to observe the distribution of the chemical reaction in an area of a few micrometers. This method is suitable for in situ and time resolved measurement of 2D micro-imaging by the transmission mode, especially when observing the reaction distribution and inhomogeneity of the electrode at quick charge and discharge of lithium ion batteries.

Acknowledgments

This work has been performed under the approval of SPring-8 (Proposal Nos. 2010B1031, 2011A1016 and 2011B1026). This work was supported by the Research and Development Initiative for Scientific Innovation of New Generation Battery (RISING) project from the New Energy and Industrial Technology Department Organization (NEDO) in Japan.

References

[1] McBreen J 2009 *J Solid State Electrochem.* **13** 1051
[2] Tanida H, Yamashige H, Orikasa Y, Oishi M, Takanashi Y, Fujimoto T, Sato K, Takamatsu D, Murayama H, Arai H, Matsubara E, Uchimoto Y and Ogumi Z 2011 *J Synch. Rad.* **18** 919
[3] Grunwaldt J-D and Schroer C G 2010 *Chem. Soc. Rev.* **39** 4656. *Ibid.* 474
[4] Nakai I 1998 *Denki Kagaku* **66** 968 (Japanese)
[5] Meirer F, Cabana J, Liu Y, Mehta A, Andrews JC and Pianetta P 2011 *J Synch. Rad.* **18** 773
[6] Terada Y, Goto S, Takimoto N, Takashita K, Yamazaki H, Shimizu Y, Takahashi S, Ohashi H, Furukawa Y, Matsushita T, Ohata T, Ishizawa Y, Uruga T, Kitamura H, Ishikawa T and Hayakawa S, *AIP Conf. Proc.* 2004 **705** 376

High throughput data acquisition at the XAS and SUL-X beamline at ANKA

Stefan Mangold[1]*, Ralph Steininger[1], Thomas Spangenberg[1]

1 Karlsruhe Institute of Technology, Institute for Synchrotron Radiation (KIT-ISS), P.O. Box 3640, D-76021 Karlsruhe, Germany

stefan.mangold@kit.edu

Abstract. The ANKA beamlines XAS and SUL were upgraded with a faster data acquisition and higher degree of automation, because of the increased number of in-situ measurements and higher demand on measurement on a number of different edges during one beamtime. We optimized the data acquisition by using the multi element semiconductor detector together with the XMAP-electronic (XIA LLC) in the mapping mode. By using the hardware-triggered mode with buffered read out the fluorescence detector data collection is done without no further delay.

1. Introduction

Over the last nearly 10 years of user operation at ANKA, the demand of users changed quite significantly. There are an increased number of in-situ measurements and also most beamtimes includes spectra taken at different absorption-edges. To enhance the performance of the XAS and SUL-X beamine further, we optimized the data acquisition and also reduced the time needed for the set-up of the beamline for measurements at a given absorption-edge.

2. Upgrades

2.1. Data acquisition upgrade

The wasted time for data read out at the beamlines is diminished quite dramatically. We use the multi element semiconductor detector together with the XMAP-electronic (XIA LLC) in the mapping mode, which allows a read-out of the full fluorescence spectra of all channels with a full data rate of 50 Hz. Because the struck counter card produces the initial trigger pulse, any in spec implemented stepwise or continuous scan mode can be used with this type of data acquisition. This enables the change of acquisition time and step size during Q-XAFS scans. Conversion boxes are allowing to transfer the trigger signal via a cat5-cable over large distances and to split the signal to up to 4 devices per trigger box (see figure 1). A nearly unlimited number of boxes can be stringed together to trigger various detectors. Special care was taken during installation of the trigger line to induce no further noise into the XMAP-electronic due to the trigger signal. The data acquisition with the xmap-detector electronic is done via a tango-server, which was developed in cooperation with Soleil [1]. A second tango-server is needed to combine the data from the counter card with the data from the fluorescence detector, which than produce again spec-like data files. Two further tango-servers act as the gateway to the personal and machine safety systems and the visualization software IgorPro [2].

Figure 1. Detector electronic including triggering box

Table 1: Tango server needed for the data acquisition with the X-Map detector electronic and their functionality.

Server name	Main Task and Usage
ds_MCADxpXmap	Needed for loading of configs, Arm for mapping mode, getting the raw data in chunks of up to eighty spectra per channel from the digital electronic
ds_xmap_data	Creates all data, which is not produced by the server above during Mapping mode (e.g. Software-Roi's, MCA-Sum)
ds_likeSpecdata	Combines data of point detectors (over counter-card), e.g. transmission data with the fluorescence data
rato_Abs	Serves as an interface to get data from spec and to control spec remote – used for the interface with the visualisation based on IgorPro [2]

With the above described set-up it was possible to shorten the dark time (time without measurements during a continuous scan) to below 12 (30) ms at the XAS (SULX) beamline. This enables the measurement of Q-XAFS scans with full fluorescence spectra at the Cu-K-edge with 2000 energy points per scan and 0.5 eV resolution in 50 seconds (measurement time 13 ms per point). Additionally slow Q-XAFS scans in 3-6 min with nearly no dark time compared to the acquisition time will most probably replace the standard step scans on the mid term, because they enable to check for beam damage and any time dependent change of the photon beam or sample consistence or sample environment. If we compare step-by-step EXAFS scans with reasonable time (40 min) to Q-XAFS with the same accumulated time, we do see better quality of the accumulated Q-XAFS data in transmission mode (figure 2 & 3) and the same or slightly inferior quality of the fluorescence mode (figure 4). We hope to enhance the quality of the Q-XAFS scan even further with larger energy step sizes and resulting longer acquisition times in the pre-edge and EXAFS region for this type of scans.

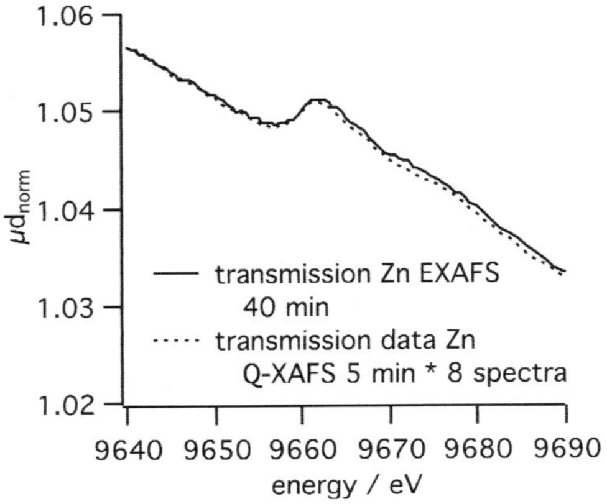

Figure 2. These two figures have been placed side-by-side to save space.

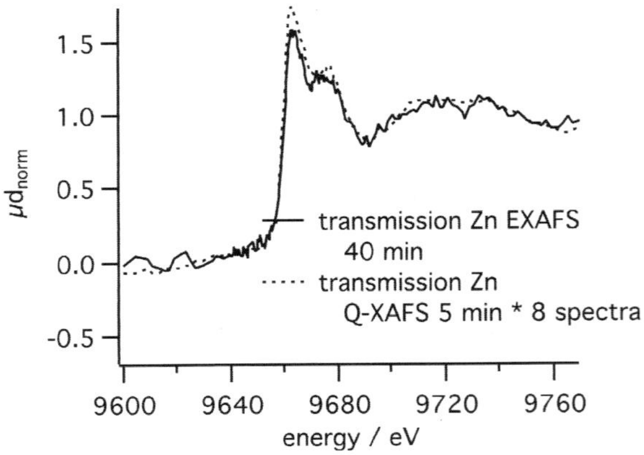

Figure 3. These two figures have been placed side-by-side to save space.

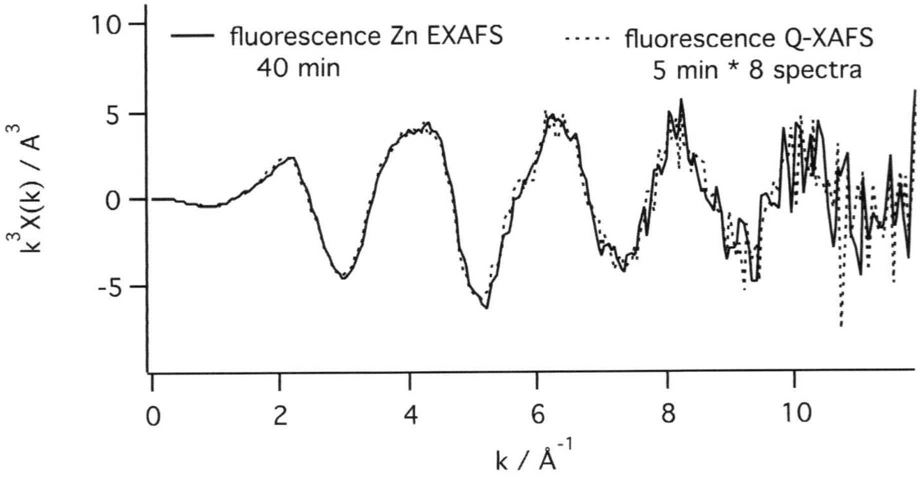

Figure 4. Fluorescence EXAFS data

2.2. Automation

We installed on both beamlines an automatic filling system for the ionization chambers (based on code on a Siemens S7 system), which adjusts the absorption of the chambers to reasonable values for a selected photon energy. The GUI can be seen in figure 5, which also shows the scheme of the system. The refilling systems do no only simplifies the handling of the beamline, but also speeds up the process. In cooperation with a logging mechanism any filling during the measurement time is recorded. The system can be controlled with scripts via spec and/or tango.

The time for beamline alignment is reduced to an absolute minimum at the XAS beamline by automatic vertical alignment of the components, stand-alone filling of the ionization chambers and scripts for activation of low energy mode (high harmonic reduction mirror in) and high resolution mode with Si<111> crystal. Any interruption of the storage ring operation (injection, beam dump) is detected and the beamline reactivates the scans after restore of the storage ring operation. The automatic change from 3.5 KeV to 21 KeV in less than 15 min with just one command and no further user input allows to measuring plenty of different absorption edges in one beamtime with minimum of beamtime loss.

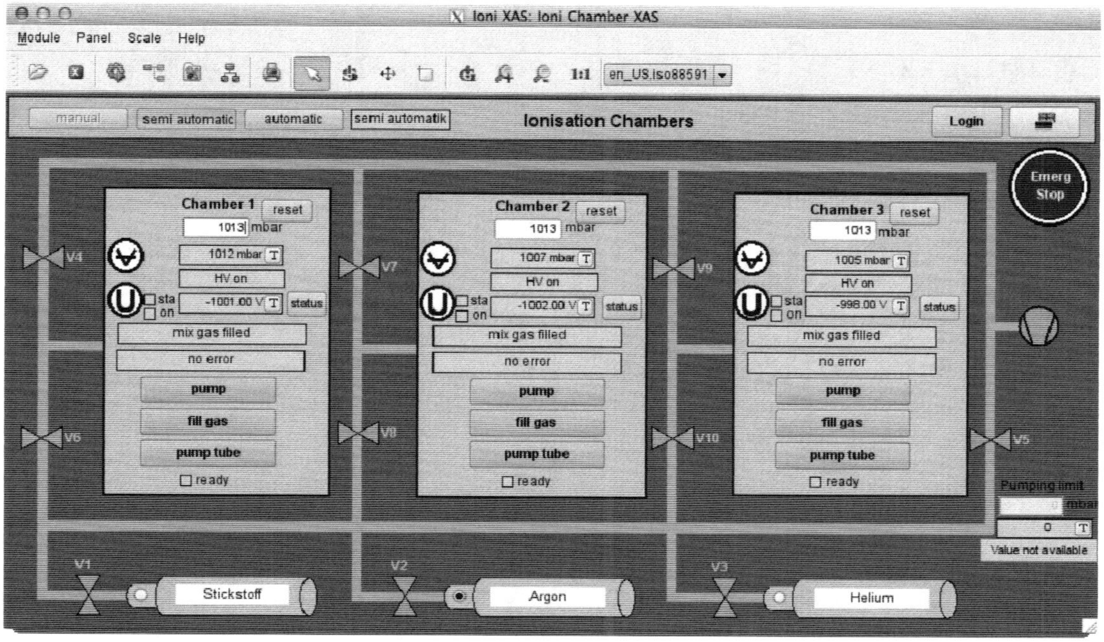

Figure 5. Automatic refilling system for ionisation chambers

3. Conclusion and outlook

The implemented upgrade on the data acquisition system offers high-speed data acquisition in fluorescence mode with complete fluorescence spectra for each energy step combined with a high degree of flexibility. Without any change of the hardware any motor movement can used for fast scanning options (on basis of a simple implementation). In user operation is already the Q-XAFS scan option; the mapping option is up till now in testing. With the next upgrades we believe to be able to achieve even better data-quality for the Q-XAFS scan, which will lead to a replacement of the standard Q-XAFS scans.

Due to the high automation of the beamline the user can focus on their experiments and not on the preparation of the beamline. The misalignment due to wrong optimization of the beamline is thereby drastically reduced.

References

[1] Synchrotron SOLEIL - L'Orme des Merisiers Saint-Aubin - BP 48 91192 GIF-sur-YVETTE CEDEX
[2] IgorPro - http://www.wavemetrics.com/

Journal of Physics: Conference Series **430** (2013) 012023 doi:10.1088/1742-6596/430/1/012023

Performance of B18, the Core EXAFS Bending Magnet beamline at Diamond

A J Dent[1], G Cibin[1,3], S Ramos[1], S A Parry[1], D Gianolio[1], A D Smith[2], S M Scott[1], L Varandas[1], S Patel[1], M R Pearson[1], L Hudson[1], N A Krumpa[2], A S Marsch[1] and P E Robbins[1]

[1]Diamond Light Source Ltd. Diamond House Harwell Science and Innovation Campus, Didcot Oxfordshire OX11 0DE United Kingdom
[2]STFC, Daresbury Laboratory Keckwick Lane Daresbury Warrington WA4 4AD United Kingdom

Abstract. B18 has been operational since April 2010 and has hosted a total of 63 EXAFS user experiments up to Jan 2012. B18 is contributing to research programs across a wide range of scientific disciplines, e.g. solid state physics and materials, catalysis, chemistry, soft matter, surfaces and biomaterials. We present a review of its present performance and capabilities.

1. Introduction

B18, Core EXAFS is the Bending Magnet beamline within the Diamond Spectroscopy Village.B18 complements I18, the micro-focus beamline, and I20, I20-1 the twin-branch, multi-pole wiggler-based beamlines, proving a scanning and a dispersive section, and in the future I08-STXM and I14, the Diamond Phase III nano-focus spectroscopy and SAXS beamline under construction.

Considering this ample portfolio, the B18 design has focused on reaching a high flexibility on a stable beamline, to provide the UK user community an effective and reliable tool, able to deliver high quality data and allowing fast experimental turnover.

2. Beamline design and optical layout

B18 covers a wide energy range, from 2 to 35 keV, covering K-edges from phosphorus to iodine, and LIII-edges from yttrium to uranium. This requires using mirrors with at least two different metal coatings. The choice for B18 has been to use Pt, allowing to reach high energies with a mirror of reasonable length (1.2m, at 2.3mrad grazing angle), and Cr. The latter is used for the low energies, to avoid the Pt M absorption edges and provide a cutoff energy at 22 keV useful for harmonic rejection in the commonly used, intermediate energy range. However, the availability of two coating branches usually requires horizontal translation of mirrors to swap between configurations. Instead, the B18 design profits from the wide fan available from a Diamond bending magnet front-end (3 mrad) and adopts a configuration with two beamline branches, fixed in position, 1mrad wide. This layout almost eliminates the need to realign the beamline after every mirror configuration change. The first mirror (Si, coated with two stripes, Pt and Cr as mentioned) is always fully exposed to the 3 mrad-wide white

[3] Corresponding author. E-mail: giannantonio.cibin@diamond.ac.uk

beam, reducing instabilities due to thermal drifts likely after each branch change with a conventional configuration. The operational beamline branch is selected just downstream of the primary mirror, using a pair of secondary white beam slits, which translate horizontally to allow an 1 mrad-wide section of the beam to reach the monochromator.

The first mirror is flat, cylindrically bent to provide optimal collimation in the vertical direction. This setup provides consistently the best energy resolution and high flux, removing the need to adjust the gap of vertical selection slits to match the beam divergence with the desired monochromatic beam resolution.

The beamline routinely provides focused beam, to allow the study of small samples with high flux density. The beam is focused using a double toroidal mirror placed after the monochromator. The mirror is a single bent, 1.2m long, silicon unit, with two cylindrical grooves, slightly divergent (1.4 mrad) so that the cylinder axes are aligned with the two beam branches. The mirror is placed at 25m from the bending magnet source, and focuses the beam on the sample placed a further 12.5m downstream. This focusing geometry (2:1) in combination with the primary collimating mirror minimizes optical aberrations and provides a regular, elliptical focal shape. The beam can be quickly focused down to 150 x 200 μm2 (HxV), or defocused to about 700 x700 μm by changing the mirror pitch to move the position of the horizontal focus and then adjusting the mirror bending radius to further expand the vertical footprint on the sample.

Running at low energies requires removal of the higher energy harmonics. To this purpose, we use short mirrors placed in the beam path at high grazing angles. When working at energies lower than 7 and 11 keV on the two branches, we insert a pair of pre-aligned and parallel Si mirrors, Pt- and Ni-coated, with a 7 to 17 mrad incidence angle. The mirrors are separated vertically by 2 mm, so the beam height change of approximately 4 mm is compensated by lifting the experimental table, which also translates horizontally (by *circa* 60 mm) according to the focal position of the selected branch.

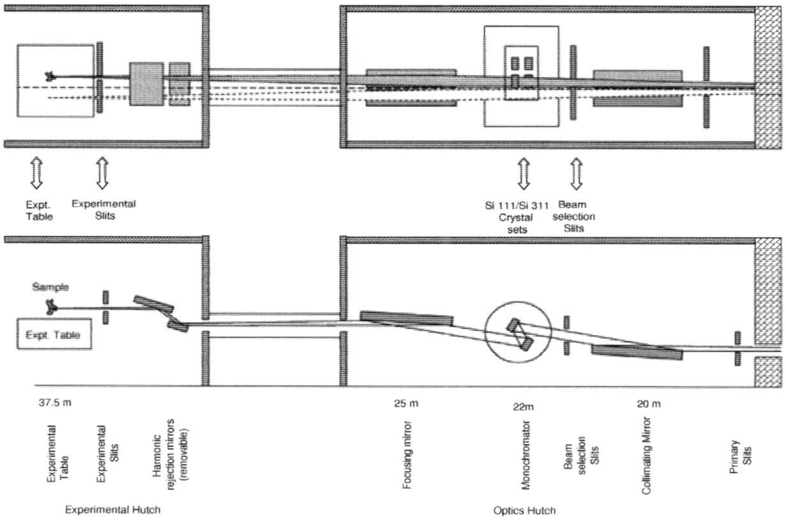

Figure 1. B18 optical layout.

2.1. *Monochromator*

The double crystal monochromator (from IDT) hosts two pairs of Si(111) and Si(311) crystals. The whole mechanics is in vacuum. This allows rapid crystal change without breaking the vacuum, reduces leak risks due to the absence of rotating mechanical feedthroughs and minimizes interferences to the motion from the torque exerted on the primary axis. The monochromator mechanics translates horizontally to four predetermined positions, allowing use of both crystals on each beamline branch. The first crystals are directly cooled with water. To guarantee a good thermal control and the absence of vibrations, we developed a Peltier-based water chiller system, including a pressurized water vibration damping system. An intermediate vacuum path surrounds the water circuit, protecting the beamline vacuum by avoiding direct water-to vacuum joints.

The monochromator is designed to carry out both conventional and continuous-scan (QEXAFS) measurements. The Bragg axis is fitted with an in-vacuum, DC brushless motor that guarantees a continuous motion with a wide angular speed range, from 0.4 degrees/s to 0.1 mdeg/s. This means we have the possibility to acquire full QEXAFS spectra in the full energy range of the monochromator (including the low energy range, from 2.05 kev at 70 degrees), with fixed exit, with acquisition times variable from a few seconds to several minutes. In addition from the possibility to study time-dependent phenomena, continuous scan operations have demonstrated a decisive advantage over the step - scan acquisition strategy traditionally used on EXAFS beamlines even for long acquisition times. Indeed, by minimizing the acquisition dead times, slow fluorescence experiments on diluted samples profit sensibly from the adoption of the QEXAFS setup. This is therefore the default mode of operation for the beamline.

3. The continuous scan (QEXAFS) acquisition system

The acquisition system in Quickexafs mode is based on a dedicated hardware, developed at Diamond. The system acquires directly the quadrature signals from the monochromator encoders, providing a 250 ns triggering signal used to synchronize all the acquisition electronics, through the beamline time frame generator. The system is partially independent of the monochromator motion control system, which takes care only of the angular motion with the desired speed, and provides the overall gating signal to the triggering system, used to calculate the absolute angular position. This design implies that the gating signals are hardware-based, and do not rely on slow software readouts of the encoder positions. This guarantees a very high angular positioning accuracy (limited by the encoder interpolated step size only) and fast response, with a maximum pulse output frequency of 1.4 MHz. It has to be noted that the amount of data generated is the same as per a single traditional step scan, and does not require dedicated preprocessing, such as binning and averaging, as required for fast-scanning Qexafs mechanisms.

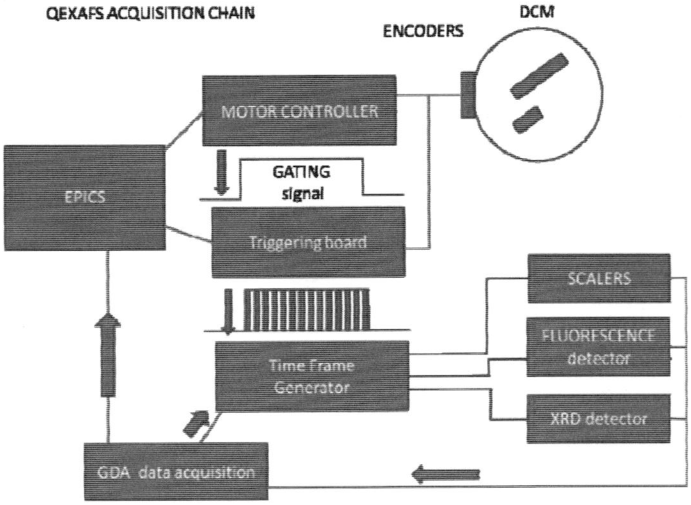

Figure 2. B18 Quick Exafs acquisition system schematics.

To evaluate the quality of the QEXAFS data obtainable on B18, we ran a series of comparative measurements with the classic step-scan modality , on the same samples representing very different data acquisition conditions. The measurements consisted in the acquisition of step-scan and qexafs acquisitions on a Cu foil, in transmission mode, and on a copper nitrate solution, 10 mMolar, in fluorescence mode, using B18 9-element Ge detector with XSPRESS-II acquisition electronics [2].

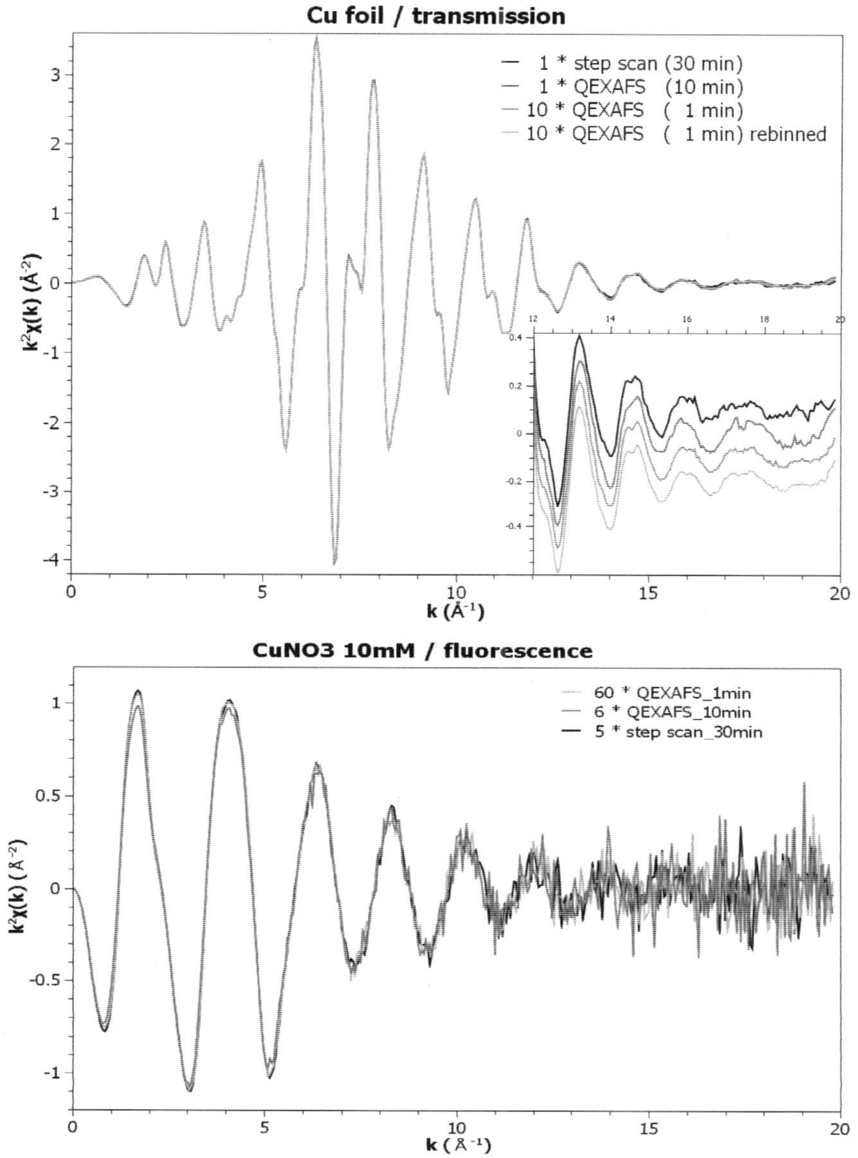

Figure 1. Comparison of step-scan and QEXAFS acquisition on a Cu foil and a 10 mM copper nitrate solution in fluorescence using B18 9-element Ge detector with XSPRESS-II [2] electronics. Top: Cu foil, RT. 1 x step-scan (30min), 1 x 10-min QEXAFS, 10 x 1-min QEXAFS, 10 x 1-min QEXAFS, rebinned as per step-scan. Bottom: CuNO3, 10mM ,5 x 30 min step scan (blue) , 60 x 1-min and 6 x 10-min QEXAFS.

We extracted then the experimental noise by subtraction of a smoothed signal calculated using a simple running-average on 5 points after normalization. The spectra were normalized adopting a linear fit for the pre-edge region and a second order polynomial fit for the post-edge region. The noise on the normalized spectra is evaluated then by calculating the standard deviation of the residual in a range far away from the absorption edge, in a region from 500 eV after the edge to the end of the spectrum.

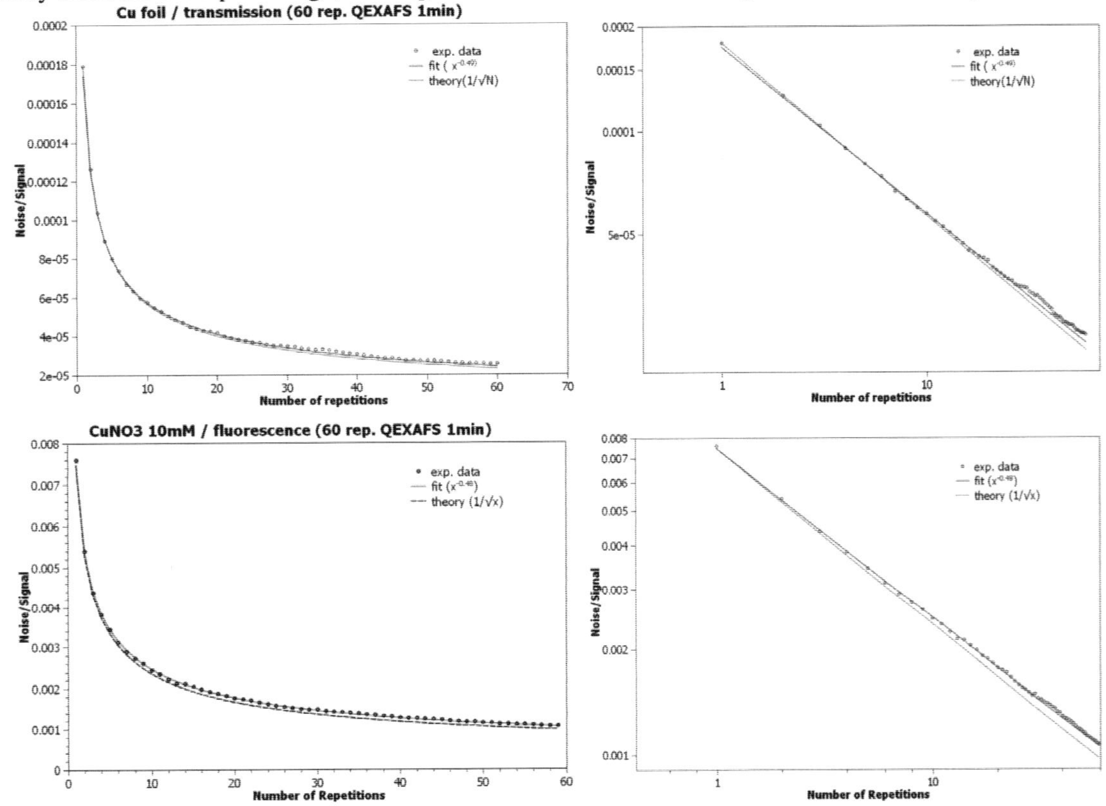

Figure 4. Noise (N/S, normalized) behaviour for QEXAFS acquisitions on Cu foil (top) and 10 mM fluorescence (bottom panel). Average over up to 60 scans– 1 minute each. In red, $N^{-1/2}$ theoretical curve.

In Figure 3 and following we report the result of this analysis. We can anticipate that the results show a decisive advantage in terms of total acquisition time for the continuous scan mode. The N/S estimates (standard deviation of the residual) are plotted as a function of the number of averaged spectra for Qexafs repeated scans. Experimental data (dots) trend is plotted along with the result of a power function curve fit and the ideal $1/N^2$ trend. The noise, both in transmission and fluorescence modes, follows closely the 1/sqrt(N) curve, down to a N/S level (signal is the absorbance normalized to the absorption jump in both cases) in the range of 2 10^{-5} for transmission for a total 1h acquisition time, and is remarkably consistent with photon-limited statistics as B18 estimated flux is in the range 10^{10}-5 10^{11} ph/s (with 5 10^{11} ph/s one can expect, on a foil, N/S to be ~ 1.10^{-5}, for 1s acquisition /point). It has to be noted in addition that the noise values shown are calculated for constant step, constant monochromator speed acquisition. This means that binning the data on a constant photoelectron *k*-step mesh as usually done during EXAFS analysis will result in increasing acquisition times, proportional to k^2, and correspondingly increased acquisition statistics. For fluorescence data, the data quality is naturally dominated by the fluorescence photon counting statistics. In the case of transmission measurements on copper, the experimental curves tend to deviate slightly from the theoretical behaviour (the fitted power exponent is 0.49), showing a weak tendency to reach a plateau.

A close inspection of the averaged data shows some indication of little dispersive-like localized structures coming from a residual signal from non-fully compensated monochromator glitches (whose amplitude can reach 10% of the monochromatic beam intensity in the region explored) , not visible above the noise in the non-averaged, single spectra. Those structures are usually removed by the user during the preprocessing steps of the EXAFS analysis, indicating that, also for realistic experiment conditions, a beamline 'detection limit' had still to be reached in the present measurements. In addition, because the signal has been acquired using the conventional voltage-to-frequency converter system (with 100 kHz/V conversion ratio) feeding pulses to the beamline scalers, we are actually reaching the resolution limit of the acquisition system, as the measurement on both I0 and It channels resulted in 7000/4500 pulses respectively per step in a 1 minute scan. A simple error propagation calculation on the absorbance expression shows that this resolution is matching the N/S value we get from the experimental data on a single scan. This indicates that for faster acquisition times, a higher resolution system as the 24 bit simultaneous ADC system under commissioning at the moment on B18 will be needed to reach photon-counting statistics in this case.

4. Sample environments and detectors

As mentioned before, B18 default setup consists in a set of 30cm-long ionization chambers for intensity monitoring and transmission measurements, and a 9-element Ge detector with XSPRESS-II [2] electronics. In addition a 4 element SDD and a gas microstrip are available on the beamline for fluorescence experiments at low energies, and a high resolution Si-microstrip detector (Dectris, Mythen [3]) covering 60 degrees.

The B18 sample space has been designed to be very flexible to accommodate complex user experimental setups. Experiments requiring stop-flow systems, electrochemical cells, low temperature XEOL detection have been successfully installed and are commonly used on the beamline. The beamline provides to users several default sample environments. All operate in transmission and fluorescence geometry for XAS, and additionally, simultaneous XRD measurements are possible in the majority of cases. Environment hardware and software controls have been integrated into the B18 data acquisition software.

For low temperature studies, three cryostats are available operating down to 77 K. Two systems are used for loading single samples with a rapid cool down, and a large cryostat for loading and automated translation for up to 50 samples is available for high-throughput experiments. In the Soft X-ray experimental section (in vacuum, 10-5 mbar), up to 10 samples can be loaded on a cold finger sample holder. This automated motion stage operates down to 77 K. Where samples are not suitable for use with a conventional cryostat, a variable temperature cryojet operating down to 77 K, is available. In this case, a RT N2 gas flow shield limits condensation and sample icing. Finally, a low-vibration pulse tube refrigerator operates down to 1.5 K and holds 3 samples. After the first cryostat cooling, the cool down time to base temperature for a new sample is ~60 minutes.

High temperature experiments, typically for catalysis studies are quite common on B18. We can provide as default a purpose-built, furnace operating up to 1000˚C. The atmosphere can be continuously purged with bespoke gas mixtures. Samples (one only is allowed in the heated chamber) are mounted on a rail system allowing rapid sample exchange. For liquids, a heating liquid sample system (250°C) incorporates a variable path length liquid sample holder, sample being held between 2 windows (typically mica, Mylar or Kapton, others feasible). If the experiment requires precise control of the sample atmosphere, a heated catalysis reaction system (developed by Diamond, Industrial Group) allows a controlled flow of User defined gas mixture over a sample held in a capillary (typically quartz). A rapid gas switching manifold allows the gas mixture being flowed over the sample to be quickly changed (switching <1s) to change the reaction conditions. An integrated controllable hot gas blower can rapidly heat the sample up to 700˚C. Finally, a High throughput system (Southampton University) [4] is available to the Users: This system was designed for catalyst research but is ideal for any experiment that requires spectroscopic analysis of samples, use of a controlled gas atmosphere, or use of the other analysis techniques available. The system is fitted with an automated motion sample manipulator to move between several samples. The sample can be heated up to 900°C. A confocal Raman system with adjustable optics and CCD camera can position the Raman sampling point coincident with X-ray beam position on the sample. A turbo pump allows the sample chamber to be evacuated and operated under vacuum; and a mass flow controller allows the chamber to be purged and filled with bespoke gas mixtures. An internal capillary adjacent to the sample is connected to a mass spectrometer to monitor gas/sample evolution in real time. The data is collected simultaneously with the XAS/XRD data acquisition.

5. Conclusions

B18, after the initial periods of commissioning with users started on April 2010, is now fully operational. The beamline can provide reliably high quality data in a wide energy range, and the continuous scan acquisition mode is becoming increasingly adopted due to its clear advantage in terms of data collection times respect to the usually adopted step scan mode. The beamline design allows quick turnaround times, allowing experiments requiring different optical configurations to be scheduled without interruptions, improving the overall efficiency. With the availability of a wide portfolio of detection systems and sample environments, B18 is providing the user community an all-round, reliable and effective instrument for XAS spectroscopy.

References

[1] Dent A J, Cibin G, Ramos S, Smith A D, Scott S M, Varandas L , Pearson M, Krumpa N, Jones C 2009 *Journal of Physics: Conference Series* **190** 012039

[2] Farrow, R., Derbyshire G E , Dobson B R , Dent A J , Bogg D, Headspith J, Lawton R , Martini M, Buxton M 1995 *Nucl. Instr. Meth. Phys. Res. B* **97** 567

[3] Schmitt B, Brönnimann C, Eikenberry E F ,Gozzo F, Hörmann C,Horisberger R, Patterson B 2003 *Nucl. Instr. Meth. Phys. Res. A* **501**, 267-272

[4] M. Tromp, Russu S, Dent A J, Mosselmans J F W , Harvey I, Hayama S, Russell A E, Guerin S, Hayden B E ,Suchsland J-P, Meacham K, Surridge M, Frey J G, Tsapatsaris N, Beesley A M,Schroeder S L M,Newton M A, Fiddy S, Safonova O V , Glatzel P, Binsted N, Evans J 2007 AIP, **CP882** 858

Journal of Physics: Conference Series **430** (2013) 012024 doi:10.1088/1742-6596/430/1/012024

200 Hz observation of hydrogenation reaction of Pd nanoparticles by using dispersive XAFS optics

D Matsumura[1], Y Okajima[1,2], Y Nishihata[1] and J Mizuki[1,3]

[1]Quantum Beam Science Directorate, Japan Atomic Energy Agency, 1-1-1 Koto, Sayo, Hyogo 679-5148, Japan

[2]SPring-8 Service Co., Ltd., 1-20-5 Shingu-cho Koto, Tatsuno, Hyogo 679-5165, Japan

[3]School of Science and Technology, Kwansei Gakuin University, 2-1 Gakuen, Sanda, Hyogo 669-1337, Japan

E-mail: daiju@spring8.or.jp

Abstract. Observation of X-ray absorption fine structure (XAFS) spectroscopy with dispersive optics is attractive application for the study of chemical reaction from a viewpoint of real-time-resolved structural change. We have constructed dispersive optics for XAFS spectroscopy measurement at SPring-8 BL14B1 and succeeded to detect real-time-resolved structural transformation of Pd nanoparticles under hydrogenation reaction with high reaction rate at room temperature. 200 Hz observation of the hydrogenation reaction of Pd nanoparticles reveals the expansion of Pd-Pd interatomic distance with a time step of 5 ms by analysing extended X-ray absorption fine structure (EXAFS) region.

1. Introduction

Palladium is one of the classical materials for the hydrogen storage system and well known to show high performance for hydrogenation reaction because of the small activation barrier for the surface adsorption and the exothermal reaction for the inner absorption [1]. Hydrogen atoms are known to be absorbed into the octahedral interstitials of Pd face-centered-cubic (fcc) structure and hydrogenation reaction is realized in the wide ranges of the hydrogen pressure and the sample temperature. The low-concentrate phase of Pd-H system creates the solid-solution with high solubility (α phase). Some types of Pd-H chemical bondings are generated in the high-concentrate phase and metal hydride phase with largely expanded lattice is established (β phase).

Although there is a significant phase boundary between the α phase and the β phase in the bulk Pd, the Pd metal fine particles show smooth change between two phases [2]. Surface effect [3] and size effect [4] were suggested for the explanation of the peculiar properties of hydrogenation in the Pd metal fine particles. It is necessary to observe the structure of the Pd nanoparticles during hydrogen absorption reaction in order to judge the formation of hydride phase over the whole of particles.

There are also open questions about the dynamic absorption process because most of dynamic studies were carried out from the viewpoint of pressure change of surrounding hydrogen gases. Direct observation of the dynamical structural and electronic changes of Pd metal nanoparticles during hydrogenation reaction, which consists of surface adsorption and following inner penetration processes, has been achieved by our group using X-ray absorption fine structure (XAFS) with dispersive optics system [5,6]. The continuous structural change of Pd nanoparticles under hydrogenation reaction in millisecond order was revealed by the real-time-resolved XAFS spectra. Surface dissociative adsorption process of hydrogen molecular gases was found to be the limiting-rate process of hydrogenation reaction of Pd nanoparticles.

In this study, we report 200 Hz observation of hydrogenation reaction of Pd nanoparticles by using dispersive XAFS system. As far as we know, 200 Hz observation is the fastest operation in the meaningful real-time-resolved XAFS observation including extended X-ray absorption fine structure (EXAFS) region. We will here show the details of the observed spectra and structural parameters.

2. Experiments

A powdered γ-Al_2O_3 was used for the impregnation method with dilute aqueous palladium nitric acid, $Pd(NO_3)_2$. Following drying and calcination at 500 °C, Pd(4 wt%)/Al_2O_3 sample was prepared. Evaluated mean diameter of Pd nanoparticles by the CO pulse adsorption method is 5.1 nm.

XAFS spectra were taken at the beamlines of BL14B1 and BL28B2 at SPring-8 with transmission and dispersive modes [7,8]. Both beamlines are used for white X-ray study and have similar optics for dispersive XAFS system [5,6,9-12]. Main results were obtained at BL14B1. Dispersed X rays were generated by a curved polychromator with Si(422) reflection plane and Laue configuration. Samples were set at the position of 2000 mm distance from the polychromator. Transmitted X rays were converted to visible light by a P43 phosphor screen and detected by a charge coupled device (CCD) camera (640 × 480 pixels, 12 bits, 120 Hz for full region detection). 200 Hz detection is realized in the CCD camera by using 2 × 2 binning mode. Neither vertical focusing nor cutting higher-order harmonics by a mirror was adopted.

The samples were reduced by hydrogen gases at 200-400 °C in the XAFS sample cell just before the XAFS study. The sample cell was evacuated about 15 minutes at room temperature in order to ensure the complete pure metal phase before the Pd *K*-edge X-ray observation. Hydrogen gas was stored in the inlet line after closing the valve connected to the sample cell. After closing the vacuum

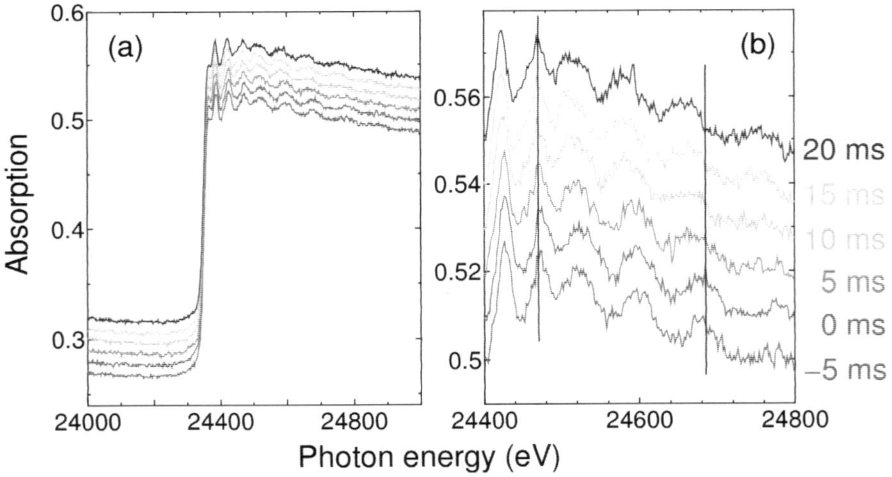

Figure 1. Pd *K*-edge XAFS spectra for the Pd(4 wt%)/Al_2O_3 during hydrogenation reaction with 660 kPa hydrogen gas at room temperature. XAFS spectra were observed by dispersive mode at a rate of 200 Hz. Hydrogen dosing signal is set to 0 ms. Wide energy region (a) and focusing on oscillation region (b) are displayed. Vertical lines are eye guides for the contraction of oscillation wavelength after hydrogen dosing.

line, an opening signal from the valve connected to the inlet line was used to start the XAFS measurement. The XAFS measurements of Pd nanoparticles on Al_2O_3 during H_2 dosing were operated at room temperature by 200 Hz rate (sampling time for one spectrum is only 5 ms) with the real-time-resolved mode. No data accumulation by the repetition of the reaction was adopted.

3. Results

Figure 1 shows the results of the 200 Hz observation of XAFS spectra for the Pd(4 wt%)/Al_2O_3 taken with dispersive mode just after the 660 kPa hydrogen dosing at room temperature. It is recognized that the oscillations in the extended region from Pd absorption edge are clearly observed even in the case of 200 Hz observation mode. The oscillation wavelength contracts just after the hydrogen dosing, indicating that the Pd-Pd interatomic distance is elongated by the hydrogen absorption. Figure 2 shows EXAFS functions and Fourier transform intensities of the spectra depicted in figure 1. We are also convinced that the Pd-Pd interatomic distance is elongated as the hydrogenation reaction proceeds.

EXAFS curves were fitted for getting the structural parameters [13]. Free parameters were coordination number, interatomic distance, and Debye–Waller factor. Figure 3 shows the time-resolved observation of the expansion of the Pd-Pd interatomic distance for the Pd nanoparticles with the hydrogen pressure dependence at room temperature. This figure clearly demonstrates that the hydrogen absorption rate strongly depends on the surrounding hydrogen pressure. The reaction rate increases as the hydrogen pressure increases. This indicates the non-zeroth order of the chemical reaction rate. The half-life reaction time is only 5 ms at the highest hydrogen pressure of 660 kPa in the figure 3. It is noted that high-rate observation itself is easy in dispersive optics because it only depends on the frame rate of one- or two-dimensional detector. However, meaningful high-rate observation is not easy. In this study, it is revealed that the reaction time between hydrogen gases and Pd nanoparticles reaches about 5 ms. The reaction time of 5 ms can be determined by 200 Hz and above real-time-resolved observation only. We have first succeeded to observe meaningful XAFS spectra including EXAFS region with a high rate of 200 Hz in real-time-resolved mode.

The monotonic slope of the value of the interatomic distance implies that Pd undergoes smooth change from α to β phase in nanoparticles. The Pd nanoparticles directly change to β phase without

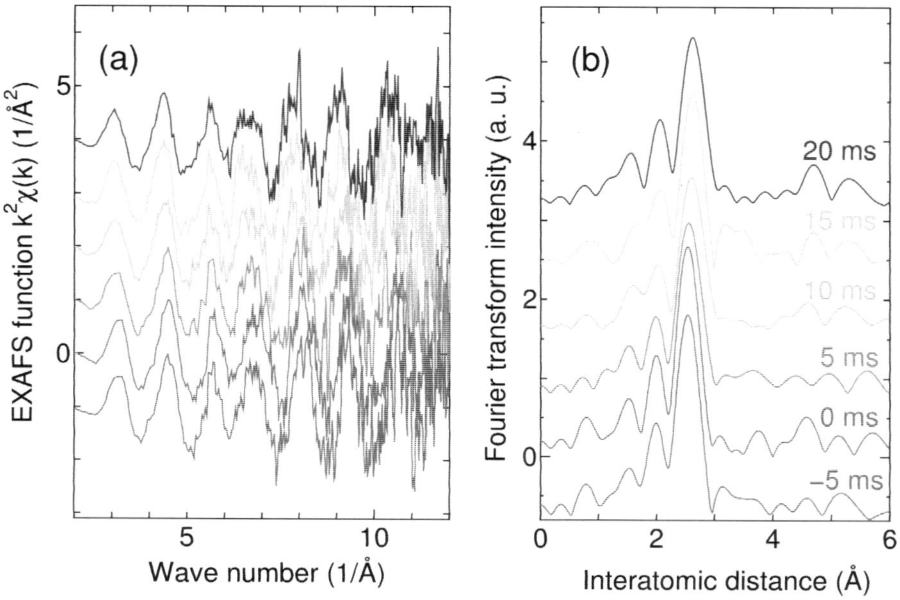

Figure 2. k^2-weighted EXAFS functions (a) and Fourier transform intensities (b) for the spectra depicted in figure 1.

duration time for coexistence of α and β phases. Figure 3 also indicates that the saturated value of the interatomic distance also depends on the surrounding hydrogen pressure. The saturated value of the interatomic distance increases as the hydrogen pressure increases. Nanoparticle Pd shows a smooth change to β phase with a change of hydrogen pressure although a sudden change to β phase occurs in the bulk Pd. The observed two hydrogen pressure dependences which are the reaction rate and the saturated interatomic distance clearly suggest that the interatomic distance directly reflects the hydrogen absorption property of the Pd nanoparticle.

The chemical reaction order was estimated from the inverse of the half-life reaction time under the assumption that the kinetics of the absorption reaction is constructed by the one chemical reaction step [6]. The value of the reaction order was determined to 0.84 ± 0.08. The hydrogen absorption process normally consists of the two reaction steps: the surface dissociative adsorption and the inner diffusive penetration. If the reaction rate of hydrogenation is limited by the surface adsorption step, the chemical reaction order must be evaluated to 1 because the hydrogen in the gas phase is evaluated as H_2 molecules. On the other hand, if the hydrogenation reaction is limited by the inner diffusion step, the chemical reaction order must be evaluated to 2 because the diffusion is operated by H atoms. The determined chemical reaction order in this study is almost one. This indicates that, even in the case of high hydrogen pressure and short reaction time, the hydrogenation reaction of the Pd nanoparticle is limited by the surface dissociative adsorption step, which is a same result with the previous work which was operated lower hydrogen pressure than this study [6].

4. Discussion

Dispersive optics is suitable for very high-rate real-time-resolved XAFS observation. Quick motion of monochromator is also used for the high-rate real-time-resolved XAFS observation, which is now approaching 100 Hz observation [14,15]. This technique is the exact opposite of dispersive optics from the viewpoint of the motion of optics. Dispersive optics shows no mechanical motion during XAFS observation, which brings about high data stability and low sample generality. Successive XAFS spectra taken by dispersive optics are very stable without any energy correction in one dataset. However, dispersive optics is almost only used under transmission mode. It is also noted that quick motion of monochromator is difficult in high-energy region because of the narrow bandwidth of diffracted X rays.

The hydrogen absorption reaction for the metallic system is largely different from the other normal gas-solid chemical interaction from the viewpoints of the time-dependence. The hydrogen atom sometimes fast diffuses to the inner side of metal without large structural rearrangement because of the small atomic radius of the hydrogen atom. This character of the hydrogen absorption is quite unique

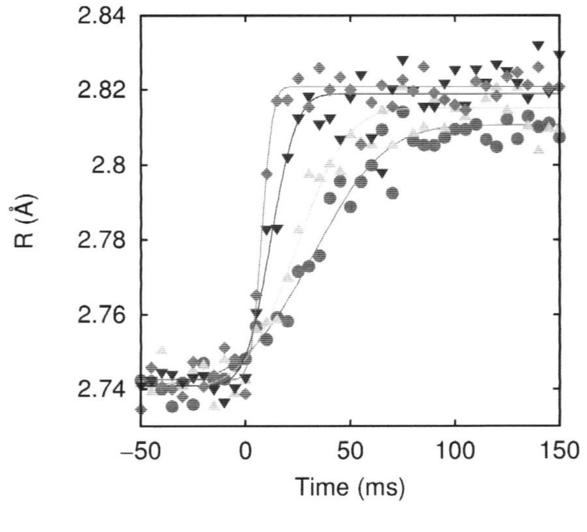

Figure 3. Real-time-resolved change of the interatomic distance of the Pd(4 wt%)/Al_2O_3 during the hydrogenation reaction at room temperature. The spectra were continuously observed at 200 Hz and the sampling time of each point is 5 ms. Partial hydrogen pressures at the sample position are 160 kPa (●), 230 kPa (▲), 410 kPa (▼) and 660 kPa (◆).

when comparing with the other light-element addition reactions in the gas-solid reaction systems such as the oxidation, nitrogenation, and carbonization. It is important to adopt the method with high frame rate and high sensitivity for making a reliable analysis in order to observe fast and subtle reaction. XAFS technique is an element selective method and can be used as a time-resolved analysis by adopting dispersive optics, which is very useful for the research of the real-time-resolved atomic and electronic structure during hydrogenation reaction.

The high-pressure experiments around the atmospheric pressure such as this study are seldom conducted in the case of the bulk Pd surface. The hydrogen absorption kinetics under low pressure has been studied [16]. It was pointed out that the inner diffusion step limits the rate of the hydrogen absorption reaction for the clean surface of Pd. It is deduced that the hydrogen diffusion step is more critical than the case of the low pressure study because the absorbed hydrogen largely elongates the interatomic distance of Pd in the hydride phase. Therefore, the hydrogen diffusion step should limit the reaction rate for the hydrogen absorption reaction in the high-pressure region. However, the surface dissociative adsorption step is the key process for the hydrogen absorption reaction in this study. Two reasons are considered for the change of the limiting-rate step. One is the difference of the surface in nanoparticle and single crystal. There are many steps and kinks in the surface of nanoparticle, which may change the surface dissociative adsorption property of Pd nanoparticles [17]. Another is the change of the diffusive penetration rate of the hydrogen atoms in the nanoparticles. The penetration mechanism can be influenced by the size of the particles because the size of the particles is as small as nanometer range [18].

Although the limiting-rate process for the hydrogen absorption reaction of Pd nanoparticles was revealed by the kinetic analysis, there are open and important problems to be resolved from the viewpoint of the time-resolved observation. Pd nanoparticles are not energetically stabilized during the middle stage of the hydrogenation reaction proceeds. It is a problem whether the hydrogen atoms are thinly diffused over the whole particle or densely reside at the surface of the particle in this non-equilibrium state. The hydrogen permeation for Pd membranes with a thickness of micrometer order was discussed from the competition of inner diffusion, surface adsorption, and desorption processes [19]. The diffusion process is still unclear in the case of nanometer size and high hydrogen concentration phase. Width of the domain walls was discussed from the conception named coherent and incoherent models as for the domain boundary in the bulk Pd-H system [20]. There are stress-free surface and stressful interface in the case of nanoparticles. It is interesting to determine the model of domain walls in nanoparticles because the interface stress can be controlled by changing support material and/or particle size. The support and/or size dependent study can enable us to understand the hydrogen absorption property of nanoparticle with more deeply interesting hydrogenation mechanism.

5. Conclusion

Hydrogenation reaction of Pd nanoparticles at room temperature was studied by using real-time-resolved XAFS spectra including EXAFS region. Dispersive XAFS optics at SPring-8 beamline enables us to observe the dynamic structural change of Pd nanoparticles with 5 ms of reaction time by 200 Hz detection mode. Surface dissociative adsorption process is the limiting-rate process even in such short time reaction.

Acknowledgements

We are grateful to Mr. K. Kato for constructing the dispersive XAFS system and stimulating discussion. We thank Mr. S. Higuchi for the experimental support. We also thank to Mr. M. Taniguchi, Mr. S. Araki and Mr. S. Mitachi for the sample preparation. This work was supported by the New Energy and Industrial Technology Development Organization (NEDO) under "Advanced Fundamental Research Project on Hydrogen Storage Materials (Hydro-Star)".

References

[1] Frieske H and Wicke E 1973 *Ber. Bunseges. Physik. Chem.* **77** 50

[2] Yamauchi M, Ikeda R, Kitagawa H and Takata M 2008 *J. Phys. Chem.* C **112** 3294
[3] Sachs C, Pundt A, Kirchheim R, Winter M, Reets M T and Fritsh D 2001 *Phys. Rev.* B **64** 075408
[4] Eastman J A, Thompson L J and Kestel B J 1993 *Phys. Rev.* B **48** 84
[5] Matsumura D, Okajima Y, Nishihata Y and Mizuki J 2010 *Mater. Res. Soc. Symp. Proc.* **1262** W06-10
[6] Matsumura D, Okajima Y, Nishihata Y and Mizuki J 2011 *J. Alloy. Compd.* **509S** S849
[7] Matsushita T and Phizackerley R P 1981 *Jpn. J. Appl. Phys.* **20** 2223
[8] Kaminaga U, Matsushita T and Kohra K 1981 *Jpn. J. Appl. Phys.* **20** 355
[9] Okajima Y, Matsumura D, Nishihata Y, Konishi H and Mizuk J 2007 *AIP Conf. Proc.* **879** 1234
[10] Matsumura D, Okajima Y, Nishihata Y, Mizuki J, Taniguchi M, Uenishi M and Tanaka H 2009 *J. Phys.: Conf. Ser.* **190** 012154
[11] Matsumura D, Nishihata Y, Mizuki J, Taniguchi M, Uenishi M and Tanaka H 2010 *J. Appl. Phys.* **107** 124319
[12] Matsumura D, Ohyama T, Okajima Y, Nishihata Y, Li H W and Orimo S 2011 *Mater. Trans.* **52** 635
[13] Yokoyama T, Hamamatsu H and Ohta T 1998 *EXAFSH 2.4* (The University of Tokyo)
[14] Stötzel J, Lützenkirchen-Hecht D and Frahm R 2011 *J. Synchrotron Rad.* **18** 165
[15] Stötzel J, Lützenkirchen-Hecht D and Frahm R 2010 *Rev. Sci. Instrum.* **81** 073109
[16] Kay B D, Peden C H F and Goodman D W 1986 *Phys. Rev.* B **34** 817
[17] Doyle A M, Shaikhutdinov S K, Jackson S D and Freund H J 2003 *Angew. Chem. Int. Ed.* **42** 5240
[18] Pundt A, Suleiman M, Bähtz C, Reetz M T, Kirchheim R and Jisrawi N M 2004 *Mater. Sci. Eng.* B **108** 19
[19] Ward T L and Dao T 1999 *J. Membr. Sci.* **153** 211
[20] Zabel H and Peisl H 1979 *Phys. Rev. Lett.* **42** 511

Journal of Physics: Conference Series **430** (2013) 012025 doi:10.1088/1742-6596/430/1/012025

Hydrogenated carbon nanotubes: x-ray absorption spectroscopy and *ab initio* simulation analysis

M Brzhezinskaya[1], G Yalovega[2], V Shmatko[2], A Klyushin[3], E Bogoslavskaya[4], A Krestinin[5], I Bashkin[6]

[1]Institut für Nanometeroptik und Technologie, Helmholtz-Zentrum Berlin für Materialien und Energie GmbH, Albert-Einstein-Str. 15, 12489 Berlin, Germany

[2]Southern Federal University, Zorge Str. 5, 344090 Rostov-on-Don, Russia

[3]Fritz-Haber-Institut der Max-Planck-Gesellschaft, Faradayweg 4-6, 14195 Berlin, Germany

[4]Don State Technical University, Gagarina Sq. 1, 344010 Rostov-on-Don, Russia

[5]Institute of Problems of Chemical Physics RAS, Academician Semenov Avenue 1, 142432 Chernogolovka (Moscow Region), Russia

[6]Institute of Solid State Physics RAS, Academician Ossipyan Str. 2, 142432 Chernogolovka (Moscow Region), Russia

E-mail: maria.brzhezinskaya@helmholtz-berlin.de

Abstract. The paper presents the results of investigation of the chemical bond formation in hydrogenated single-walled carbon nanotubes using X-ray absorption spectroscopy. All measurements were performed with the use of synchrotron radiation at BESSY II. The C1*s* absorption spectra point to formation of covalent chemical bonding between the hydrogen and carbon atoms in H-SWNTs. The investigation of the H-SWNT local structure has been carried out on the basis of the semi-empirical PM6 method (MOPAC 2009) and Finite Difference method. The possibility of H-SWNTs dehydrogenation was also investigated.

1. Introduction

Since oil resources are impermanent, hydrogen is gaining more and more importance as highly efficient and ecological energy carrier that is very promising for wide use in the power industry, especially as vehicle fuel. Therefore, during recent decades researchers continue seeking for techniques for hydrogen storage and transportation. None of the current methods for hydrogen storage (at high pressure, in the absorbed state at low temperatures, in the liquid state, in the form of metal hydrides or intermetallics) ensures safe reversible storage and compact transportation of gaseous hydrogen. It was found out that carbon nanotubes (CNTs) may be used as hydrogen reservoirs [1]. Therefore, a lot of investigations are now aimed at studying peculiar features of hydrogen sorption by carbon nanotubes that exhibit extremely high porosity.

In this work, high-resolution near edge X-ray absorption fine structure spectroscopy (NEXAFS) technique was used to elucidate the nature of chemical bonding between carbon and hydrogen atoms on the surface and inside hydrogenated single-walled carbon nanotubes (H-SWNTs). The possibility of H-SWNTs dehydrogenation was also investigated.

2. Methods

SWNTs were synthesized by the electroarc method using nickel-yttrium catalyst. The purified nanotubes had a narrow diameter distribution with the average value of ~1.5 nm. High-purity SWNTs (~ 98 wt.%) were prepared in the form of paper (SWNT paper). The SWNTs were hydrogenated at P~5.0 GPa and T~500°C and contained 5.335 wt.% of hydrogen [2]. Measurements were performed at the Russian-German beamline BESSY II (Berlin, Germany) by using experimental station Mustang. NEXAFS spectra of all the samples were recorded at the C 1*s* absorption edge in the total electron yield mode with the photon energy resolution of 75 meV. In all the dehydrogenation experiments, temperature was controlled by a thermocouple and pyrometer. Variations in the sample electron structure were monitored by NEXAFS.

Investigation of the local structure of hydrogenated single-walled carbon nanotubes was performed based on advanced theoretical analysis of experimental NEXAFS data. In order to get detailed understanding of the origin of the chemical bond between C and H atoms in H-SWNTs, semiempirical calculations were carried out using the MOPAC2009 code with PM6 parameterization.

Models of C-H chemical bonds where H atoms are connected to the SWNT outer surface were investigated. Fig. 1 shows the model structure of C-H chemical bonds that is a result of geometry optimization. In the case when hydrogen was located on the nanotube surface, formation of a tetrahedral structure with C-H bonds was observed.

Theoretical C K-edge XANES was analyzed using the non-muffin-tin approach, the finite difference method (FDMNES 2012) [3]. The C1*s* absorption spectra of H-SWNTs were calculated for two cases of absorbing atoms. In the first case, the atom unbound with hydrogen was assumed to be absorbing; in the other case, that bound with hydrogen was chosen.

3. Results and discussion

In the C1*s* absorption spectra of carbon nanotubes, the main structures *A–a* and *B–F* reflect transitions of the carbon atom 1*s* electrons into the conductivity band free states that are formed from the 2*p* electron states of carbon atoms and have π and σ symmetry, respectively [4]. Significant differences in the structure of C1*s* spectra of the pristine SWNTs and H-SWNTs are observed. There is a drastic decrease in the *A* peak intensity, emergence of new features B_1* (288.8 eV), B_2* (287.3 eV), *B**, *C**, *D** instead of *B–D* structures, and formation of isolated band *E*–F** (Fig. 2). Note that, in addition to new absorption bands in H-SWNT spectra, *A* band observed in the pristine nanotube spectrum still persists thus proving incompleteness of nanotube hydrogenation.

It would be natural to associate the changes observed in the H-SWNT C1*s* spectra with chemical interaction between the H atoms and nanotube wall. What is significant is that the SWNT C1*s* spectrum variations taking place after SWNT hydrogenation are similar to those after SWNT fluorination (F-SWNTs). The F-SWNT C1*s* spectra that we have measured under the same conditions exhibit similar main features. Among them, the most considerable ones were reduction of π- and σ-resonances (peaks *A* and *B*, respectively), rise of new spectral features in the 286 to 292 eV range, and band *E*–F** at the energies of 303 to 310 eV. Similarity of H-SWNT and F-SWNT C1*s* spectra allows us to assume that the chemical binding of fluorine and hydrogen atoms with carbon ones occurs in both cases mainly by binding the fluorine and hydrogen atoms to carbon atoms on nanotube sidewalls with forming σ(C–F) and σ(C–H) bonds due to covalent mixing of F2*p*- and C2p_z-, and H1*s*- and C2p_z- valence electron states, respectively [5]. Obviously, this binding of fluorine and hydrogen atoms modifies the coordination of carbon atoms from triangular in original nanotubes to nearly tetrahedral in fluorinated and hydrogenated nanotubes; this is possible only when sp^2 valence state hybridization of carbon atoms changes for sp^3 hybridization in F-SWNTs and H-SWNTs. At the same time, the H-

Journal of Physics: Conference Series **430** (2013) 012025 doi:10.1088/1742-6596/430/1/012025

SWNT spectrum has one more new peak at 287.3 eV as compared with the F-SWNT spectrum. This peak is commonly attributed to C-H bonds in the C1s NEXAFS of hydrocarbons [4]. The two new peaks in the C1s spectrum of H-SWNTs could arise due to existence of two different manners in which H atoms are connected to the SWNT graphene surface with forming σ(C–H) bonds with different value of charge transfer between C and H atoms.

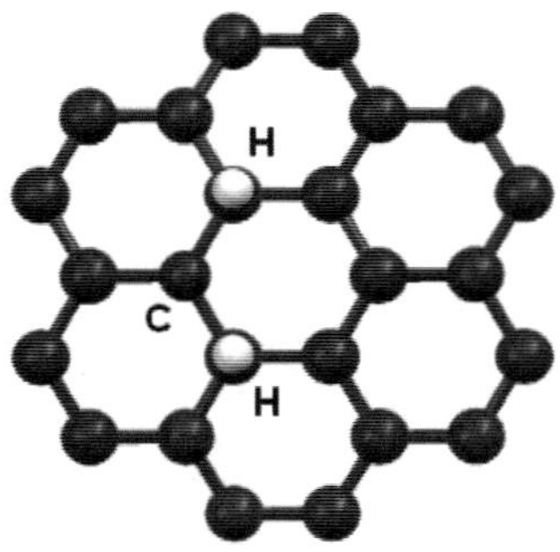

Figure 1. The model structure of H-SWNTs used for calculations.

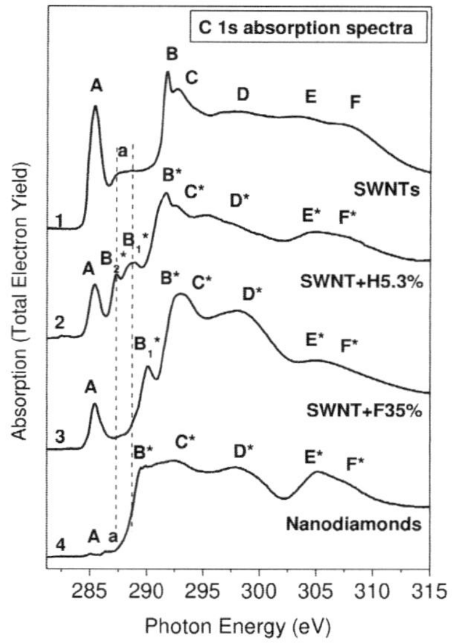

Figure 2. C1*s* absorption spectra of the SWNTs, SWNTs+H5.3%, SWNTs+F35% and nanodiamonds.

The model of H-SWNTs with 1.6 nm diameter was generated, and nanotube fragments were used in calculations. As a result of geometry optimization, formation of tetrahedron with C-H and C-C bond lengths of 1.2 and 1.5 Å, respectively, took place. The C K-edge spectra of the optimized H-SWNT model were calculated. Results of comparison between the experimental and theoretical spectra are presented in Fig. 3. The H-SWNT theoretical spectra (Fig. 3, curves 3 and 4) reproduce all the main details of the experimental spectrum. This reproducibility confirms our conclusion about formation of C-H chemical bond at the SWNT surface. Fig. 3 shows that, when a carbon atom unbound with hydrogen was chosen to be absorbing, the theoretical spectrum (Fig. 3, curve 3) exhibits both features B_1* and B_2* in the 286 to 292 eV energy range. On the other hand, when the carbon atom bound with H was chosen an absorbing (Fig. 3, curve 4), only one feature B_2* is observed. This can be regarded as an evidence of the fact that the measured C1s absorption spectrum of H-SWNTs is the summed spectrum of carbon atoms of two types with different C-H bond forces and different value of electron transfer between the carbon and hydrogen atoms.

Investigation of CNT dehydrogenation process makes it possible to estimate the type and force of carbon-hydrogen bonds and to find out whether the graphene mesh is distorted. Thermal annealing is one of the possible methods of H-CNT dehydrogenation. Fig. 4 presents the results of H-SWNT dehydrogenation process controlled by NEXAFS. The figure demonstrates that even after annealing at the temperature equal to that of H-SWNT hydrogenation (T_H~500°C) dehydrogenisation of H-SWNTs was impossible. This means that the results of hydrogenation are more complicated for H-SWNTs (as well as for F-SWNTs) than for graphene and graphite. It is natural to assume that H-SWNT dehydrogenation temperature is higher than T_H and will be equal to T_H of H-MWNTs. Whereas, our data obtained earlier for F-CNT annealing showed that annealing of F-SWNTs and F-MWNTS started at the temperatures equal to the fluorination temperature (T_F) of F-MWNTs whose fluorination temperature of ~200°C is higher than F-SWNTs.

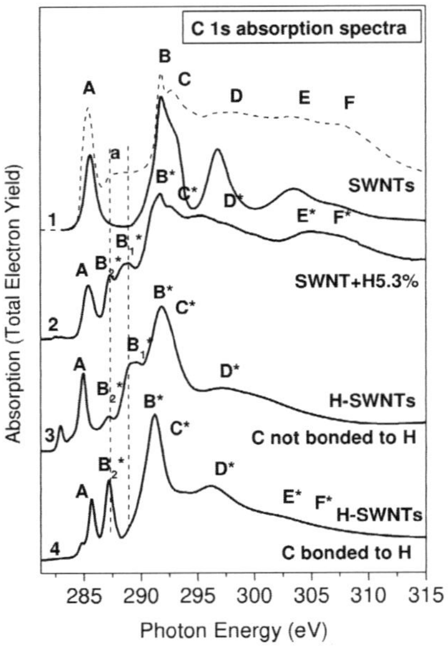

Figure 3. Measured and calculated *CK*-edge spectra of SWNTs and H-SWNTs.

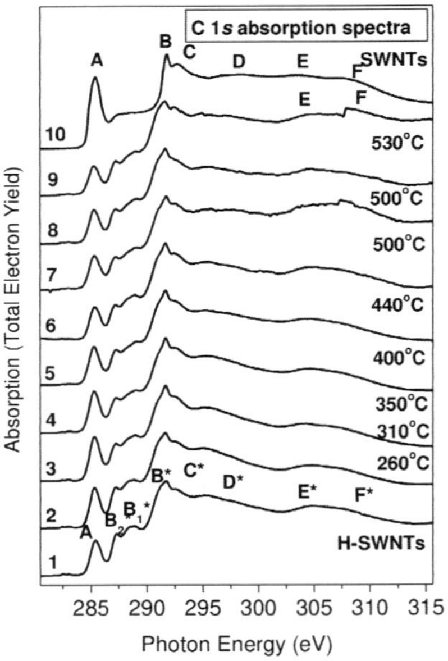

Figure 4. *CK*-edge spectra of SWNTs+H5.3% before and during annealing.

4. Conclusions

Investigation of chemical bonds formation in hydrogenated single-walled carbon nanotubes by using X-ray absorption spectroscopy has been performed. It was found out that SWNT hydrogenation is accompanied by chemical binding of hydrogen and carbon atoms on the tube side walls. Hydrogen atoms do not substitute carbon atoms in graphene layers of SWNTs but they join perpendicularly to them due to covalent mixing between C $2p_z$ and H $1s$ states. The coordination of carbon atoms changes from sp^2-triangular to sp^3-tetrahedral.

Acknowledgements

This work was financially supported by the Russian Foundation for Basic Research (project no. 09-02-01278), the bilateral Program "Russian-German Laboratory at BESSY" and a Marie Curie FP7-Reintegration-Grants within the 7th European Community Framework Program (Project No. PCIG10-GA-2011-297905).

References

[1] Dillon A C, Jones K M, Bekkedahl T A, Kiang C H, Bethune D S and Heben M J 1997 *Nature (London)* **386** 377
[2] Meletov K P, Maksimov A A, Tartakovskii I I, Bashkin I O, Shestakov V V, Krestinin A V, Shulga Yu M, Andrikopoulos K S, Arvanitidis J, Christofilos D and Kourouklis G A 2007 *Chem. Phys. Lett.* **433** 335
[3] Joly Y J *Synchr. Rad.* 2003 **10** 58
[4] Stöhr J 1992 *NEXAFS Spectroscopy* in *Springer Series in Surface Science* (Berlin: Springer-Verlag)
[5] Brzhezinskaya M M, Muradyan V E, Vinogradov N A, Preobrajenski A B, Gudat W and Vinogradov A S 2009 *Phys. Rev. B* **79** 155439

OD-XAS and EXAFS: Structure and Luminescence in Ge Quantum Dots

A Karatutlu[1,*], W R Little[1], A V Sapelkin[1], A Dent[2], F Mosselmans[2], G Cibin[2] and R Taylor[2]

[1] *Centre for Condensed Matter and Materials, School of Physics and Astronomy, Queen Mary, University of London, London, E1 4NS, United Kingdom*

[2] *DiamondLight Source Ltd, Rutherford Appleton Laboratory, Didcot, OX11 0DE, Oxon, United Kingdom*

Ali Karatutlu, email: a.karatutlu@qmul.ac.uk

Abstract:
The source of visible luminescence from Ge nanocrystalline materials has been attributed to various phenomena such as the quantum confinement effect, surface contributions, and so on. In order to understand the origin of this light emission we have performed extended x-ray absorption fine structure (EXAFS) and optically-detected x-ray absorption spectroscopy (OD-XAS) using x-ray excited optical luminescence (XEOL) measurements on a range of Ge quantum dots prepared by various routes including stain etching, sol-gel method, and laser ablation. Such XEOL and OD-XAS studies on Ge and Ge-O rich regions allowed us to determine specificlocal sites responsible for PL.

1. Introduction

Reducing the size down to a certain value, known as the Bohr exciton radius for asemiconductor (for example silicon [1], or germanium [2]) causessignificant changes in the light emitting properties due to the well known quantum confinement effect. Shortly after these discoveries it became obvious that the photoluminescence in Ge nanocrystals is not only size dependent but also may be due to surface species such as oxides and hydrides, or defects [3]. It has been over two decades since the effect was first observed but the topic is still contentious as direct observation of the source of the photoluminescence (PL) is challenging despite intense research[4].

Extended x-ray absorption fine structure (EXAFS) by transmission mode is accepted as a conventional local structural measurement technique[5].This technique enables one to determine chemical composition, interatomic distances, and number of neighbours around the absorbent atom [6]. Here we used a technique known as optically detected x-ray absorption spectroscopy (OD-XAS) using x-ray excited optical luminescence (XEOL) to probe thesubstructure responsible for the light emission. To this end simultaneous OD-XAS and EXAFS measurements were taken, including XEOL from Ge nanoparticles. The nanoparticles were prepared by various routes such as stain etching, liquid-phase pulsed laser ablation (LP-PLA), and sol-gel method. Characterisation of the Ge nanoparticles has also been performed using transmission electron microscopy, and Raman and PL spectroscopy techniques.

2. Experimental Procedure

Ge powder, supplied by Sigma-Aldrich, of over 99.999 % purity was used for both stain (chemical) etching andLP-PLA. Stain etching was performed with the solution of $HF:H_3PO_4:H_2O_2$ having the chemical volume ratio of 15 ml:15 ml:0,075 ml respectively which was found to be optimal in [7]. The procedure took place over 3 hours under indoorfluorescent illumination.For LP-PLA a Nd:Yag (Surelite I-10) was used. Using a combination of dichroic mirrors the beam was focused to an area of 4x1 mm^2 inside a quartz cuvette filled with n-hexane. The output of the laser was kept at 108 mJ. The pulse frequency and the pulse duration were 1 Hz and 13 ns respectively. The cuvette contained 17 mg of Ge powder and n-hexane to a depth of a 1.5 mm. Laser Ablation was performed for 7 hours. The hexane was allowed to fully evaporate before suspending the residue in ethanol and centrifuging to separate the germanium nanoparticles from the remaining Ge powder. The method for sol-gel synthesis [8] of Ge nanoparticles embedded in silica (SiO_2) is three successive processes including the sol-gel method , and then 48 hours of heating treatment in air at 600 °C, followed by a heating treatment in H_2(5%)/Ar(95%) medium at 700 °C.

Raman spectroscopy, PL spectroscopy (Renishaw 1000) and TEM measurements (JEOL JEM 2010 Microscope) of the resultant product have been analyzed in order to find the size of Ge nanoparticles prepared by the each route.OD-XAS and EXAFS including simultaneous XEOL measurements of the same area of each sample have been carried out at beamline B18 at Diamond Light Source UK. The equipment for OD-XAS includes a TRIAX180 spectrometer, a photomultiplier tube,Hamamatsu R3809U-50 MCP in addition to an X-ray detection system (details given in the appendix) in order to record the X-ray emitted luminescence. For OD-XAS and EXAFS, the Ge K-edge was recorded as a function of energy over the range of 11.05-12 KeV. XEOL measurements were collected from the same area as a function of wavelength centered at 700 nm for the each set of the samples. Demeter 0.9.8 [9] was used for EXAFS analysis.

3. Results and Discussions

We present the details of the size analyses and how to distinguish various subsets of Ge nanoparticles by OD-XAS and EXAFS measurements. Raman spectra of Ge nanoparticles formed by (a) stain etching, (b) sol-gel synthesis and (c) LP-PLA have been studied for the size analyses and shown in figure 1. The analyses have been performed under the assumption of the phonon confinement model [10].

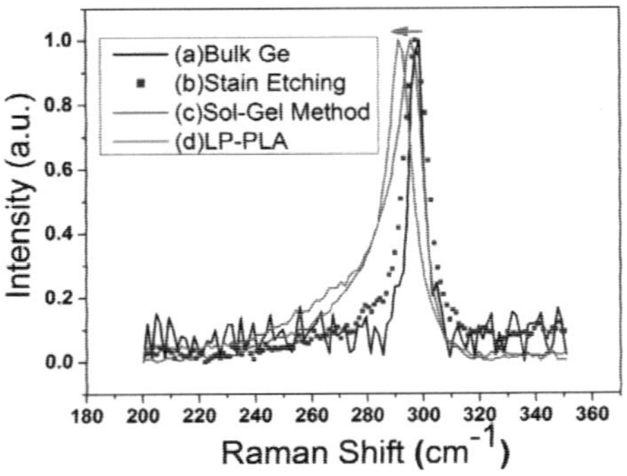

Figure 1 Normalised Raman Shift of Ge nanocrystals from right to left (a) bulk Ge and by using (b) stain etching, (c) sol-gel method and (d) LP-PLA.

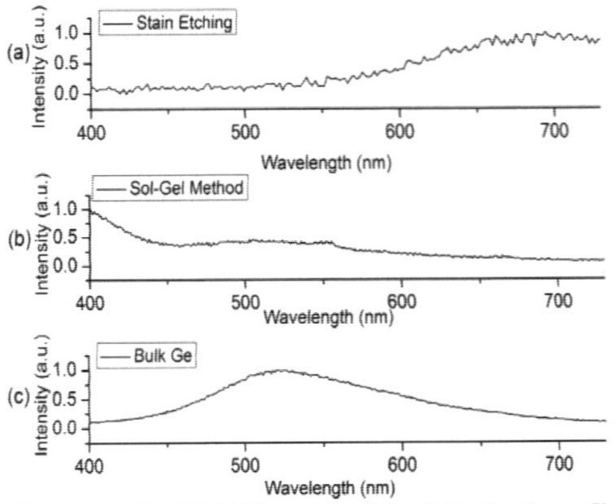

Figure 2 XEOL at 100 °Kof the Ge nanoparticlesprepared by (a) stain etching, (b) Sol-gel method, (c) of the bulk Ge.

The size of the Ge nanoparticles fabricated by stain etching, sol-gel synthesis and LP-PLA were found to be approximately 6 nm, 5 nm, and 6 nm respectively both from TEM and Raman measurements.Raman measurements in figure 1 showing that the Ge nanocrystals [11] synthesizedby these routes are of a similar size, . XEOL signals in figure 2 gives some indication of differences in the origins of sites responsible for light emission. Further, OD-XAS reveals the different subset of the structures which are optically active at around 700nm.

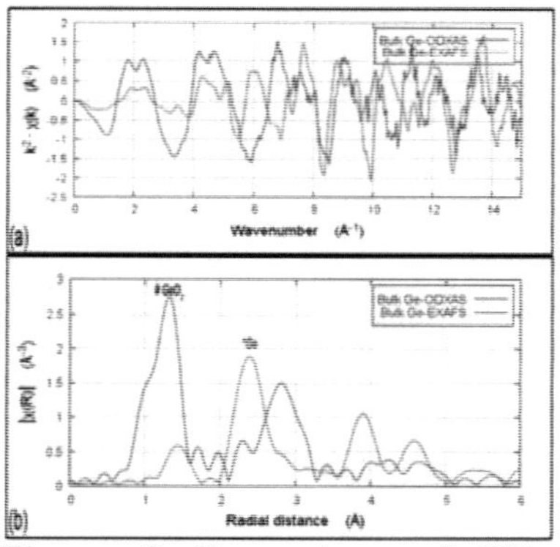

Figure 3 Comparison of OD-XAS and EXAFS of Ge K edge of bulk Ge (a) in k-space (b) in R-space.

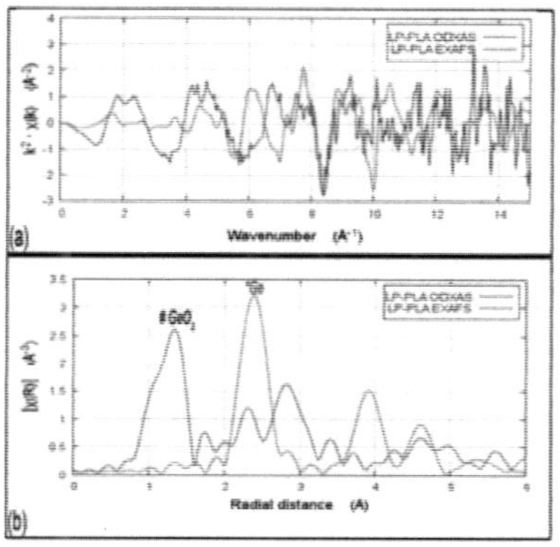

Figure 4 Comparison of OD-XAS and EXAFS of Ge K edge of LP-PLA (a) in k-space (b) in R-space.

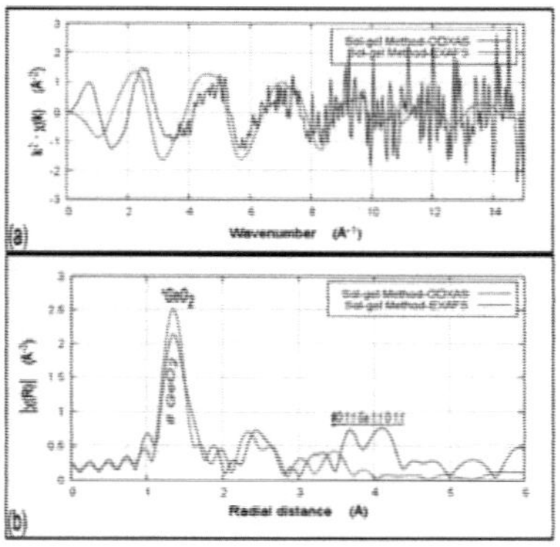

Figure 5 Comparison of ODXAS and EXAFS of Ge K edge of Ge nanoparticles embedded in SiO_2 formed by sol-gel synthesis (a) in k- space (b) in R-space.

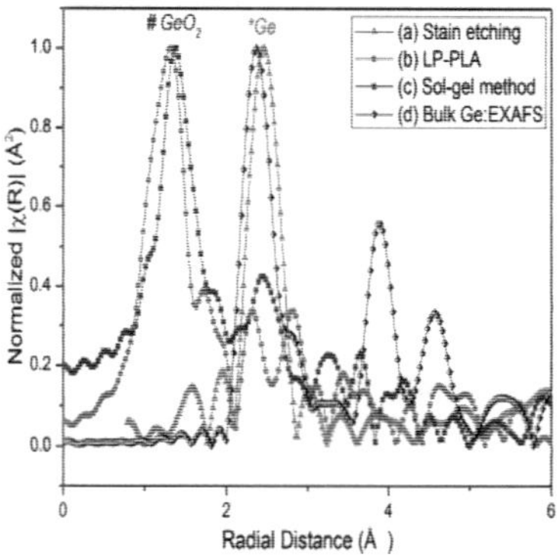

Figure 6 Comparison of OD-XAS of Ge K edge of Ge nanoparticles synthesized by (a) stain etching, (b) LP-PLA, (c) sol-gel method and (d) EXAFS of Bulk Ge.

Here we identify the origin of the PL fromthe structuraldata by using OD-XAS. Aclear difference was observed by comparing OD-XAS and EXAFS in R-space over the Ge K edge shown in figures3(b), 4(b),and 5(b). In figure 3(b), OD-XAS probes native oxide sites in bulk Ge. Furthermore, in figure 4(b) the Ge nanoparticles formed by LP-PLA reveals that theoptical contribution is primarily from the GeO_2 region. However, in figure 5(b), EXAFS and ODXAS of Ge nanocrystals embedded in $SiO_2$are almost identical until 3 Å. There are limits to sensitivity in a multicomponent system with various emitting sites [12]. Figure 6 indicates that OD-XAS is sufficiently sensitive to contributions from Ge nanoparticles synthesized by stain etching, LP-PLA and sol-gel method.In the case of stain etching, the main contribution to the PL is due to Ge atoms with an inter-atomic spacing of 2.44±0.01 Å which only differs by0.09 Å from EXAFS of Ge K edge of bulk Ge, shown in figure 6(d).

4. Conclusion

OD-XAS is found to be a very promising technique for achieving the specificity of the sub-nanoparticle resolution when investigating Ge nanoparticles synthesized by various routes. We show that by using OD-XAS and EXAFS on the Ge K edge the subsets of the structure responsible for the emission process have been determined. The main difference between OD-XAS and EXAFS stems from the fact that OD-XAS probes only the sites of the sample that contribute to light emission, as in the case of sol-gel method and LP-PLA or structurally disordered sites of nano-crystalline germaniumas in the case of stain etching. Despite having shown how the subsets of the structure contribute to the final decay process, the exact process for excitation and emission is not yet completely understood.Particular attention is drawn to the roles played bythe surface, the interfacial region, and the core.

5. Acknowledgement

The following are greatly acknowledged by the authors: the financial support by the Queen Mary, University of London Research Fund and SEPnet and, V. T. Karavanski (for stain etching sample preparation).

6. Appendix

The details and supplementary results are given in this part of the paper.

7.References

[1] CanhamL T 1990 Silicon quantum wire array fabrication by electrochemical and chemicaldissolution of wafers *Applied Physics Letter* **57** 1046-1048

[2] Maeda Y, Tsukamoto N, Yazawa Y, Kanemitsu Y and Masumoto Y 1991 Visible photoluminescenceof Ge microcrystals embedded in SiO_2 glassy matrices *Applied Physics Letters***59** 3168

[3] DelerueC 1998 Optical band gap of Si nanoclusters *Journal of Luminescence***80** 65–73

[4] Daldosso N, Luppi M, Ossicini S, Degoli E, Magri R, Dalba G, Fornasini P, Grisenti R, RoccaF, Pavesi L, Boninelli S, Priolo F, Spinella C and Iacona F 2003 Role of the interface region optoelectronic properties of silicon nanocrystals embedded in SiO_2*Physical Review B***68** 1–8

[5] DowsettM G, Adriaens A, Jones G K C, Poolton N, Fiddy S and Nikitenko S Opticallydetected X-ray absorption spectroscopy as a means of monitoring corrosion layers on copper *Analytical Chemistry***80**8717-24

[6] Erenburg S B, Trubina S V, Bausk N V, Dvurechenskii a V, Nikiforova I, Mansurov V G, Zhuravlev K S and Nikitenko S G 2009 Microstructure of quantum dots ensembles by EXAFS spectroscopy *Journal of Physics: Conference Series***190** 012131

[7] Karavanskii V A, Lomov A A, Sutyrin A G, Bushuev V A, Loikho N N, Melnik N N, Zavaritskaya T N and Bayliss S 2003 Raman and X-ray studies of nanocrystals in porous stain-etched germanium *Thin Solid Films***437** 290–6

[8] Henderson E J, Seino M, Puzzo D P and Ozin G A 2010 Colloidally Stable GermaniumNanocrystals for Photonic Applications *ACS Nano***4** 7683–91

[9] Ravel B, Newville M 2005 *ATHENA*, *ARTEMIS*, *HEPHAESTUS*: data analysis for X-ray absorption spectroscopy using *IFEFFITJournal of Synchrotron Radiation***12**5375–41

[10] Campbell H I and Fauchet P M 1986 The effects of microcrystal size and shape on the one phonon Raman spectra of crystalline semiconductors *Solid State Communications***58** 739–41

[11] Sapelkin A V, Karavanskii V A, Kartopu G, Es-Souni M and Luklinska Z 2007 Raman study of nano-crystalline Ge under high pressure *Physica Status Solidi (B)***244** 1376–80

[12] Emura S, Moriga T, Takizawa J, Nomura M,Bauchspiess K R; Murata T; Harada K; Maeda H 1993 Optical-luminesce yield spectra produced by x-ray excitation *Physical Review B***47** 6918–30

Appendix

The experimental details in addition to the complementary results are given in this part. Figure 7 shows PL signal from Ge nanoparticles produced by (a) stain etching, (b) LP-PLA, (c) sol gel synthesis.

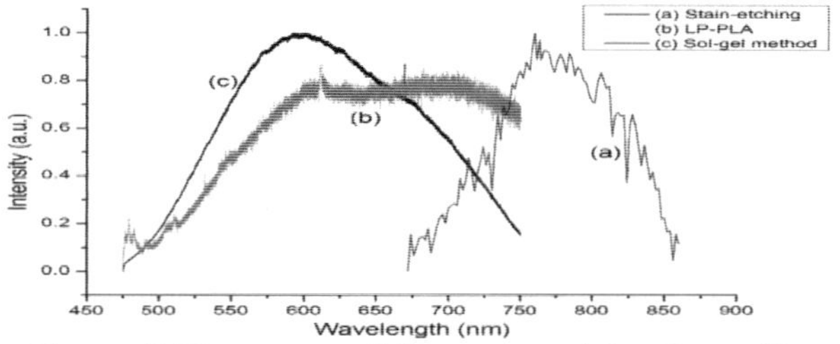

Figure 7 PL spectra of Ge nanoparticles formed by (a) stain etching (b) LP-PLA, (c) sol gel synthesis.Photoluminescence (PL) spectrum has been recorded from the each of the samples with excitation by HeCd laser at 473 nm.

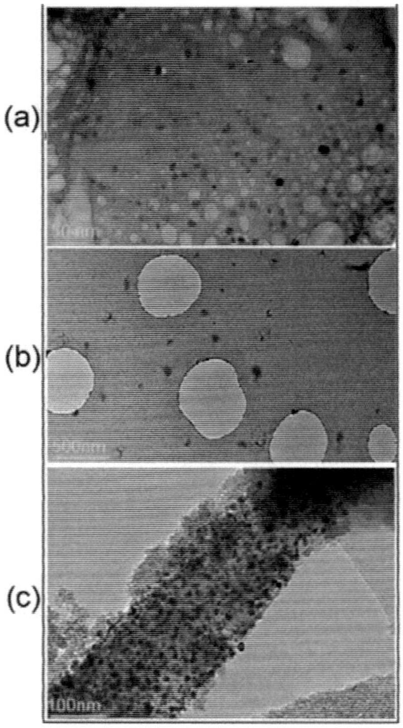

Figure 8 TEM micrograph of Ge nanoparticles from top to down prepared by (a) stain etching (b) LP-PLA and (c) Sol-Gel Method.

TEM micrograph in figure 8 shows Ge nanoparticles formed by (a) stain etching, (b) LP- PLA and (c) sol-Gel Method. Size analyses in table 1 reveals that the size of the Ge nanoparticles prepared by stain etching and the sol-gel method are in agreement but those formed by LP-PLA are not, due to the agglomeration issue which results in particle sizes well out of the quantum confinement region (about 24 nm for Ge nanoparticles given in [2].

Journal of Physics: Conference Series **430** (2013) 012026 doi:10.1088/1742-6596/430/1/012026

Table 1 Size analyses of the Ge nanoparticles by Transmission electron microscopy, Raman spectroscopy and PL spectroscopy.

Method Technique	Transmission Electron Microscopy (nm)	Raman Spectroscopy Size (nm)	Raman Spectroscopy FWHM (cm^{-1})
(a) Stain Etching	10 ± 4	6.9	10
(b) LP-PLA	41 ± 22	6.2	12
(c) Sol-Gel Method	10 ± 6	5.3	14

The equipment for OD-XAS includes also a photomultiplier tube,Hamamatsu R3809U-50 MCP in order to record the X-ray emitted luminescenceusing an optical fibre in addition to the detection system consisting of a Triax 190 spectrometer equipped with Synapse CCD and a Newport VIS Femtowattphotoreceiver.All measurements have been performed at a temperature of 100 K using a cryojetsystem.

For the fits of OD-XAS and EXAFS spectra of bulk Ge, the k-weight value such as 1, 2 and 3 is shown in figure 9. The best fit was obtained at k-weight 2 and the rest of the fittings were performed at k-weight 2 for the fits of each sample.

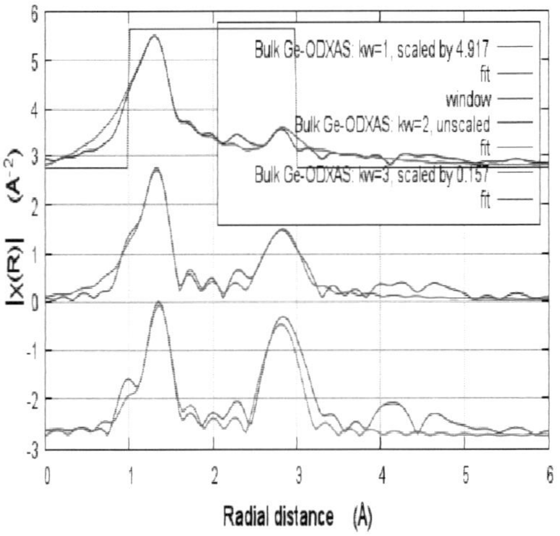

Figure 9 OD-XAS of Ge K edge of bulk Gein R-space at k-values 1, 2 and 3.

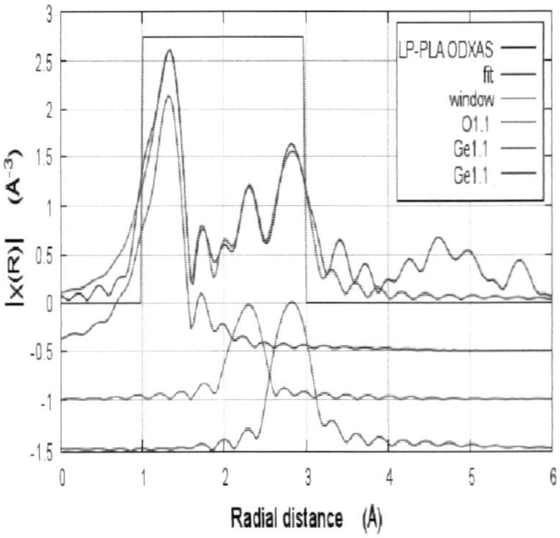

Figure 10 OD-XAS of Ge K edge of Ge nanocrystals formed by LP-PLA and the fit in R-space including the first three scattering paths.

As an example of the fits performed for each sample with a k-value of 2, the fit for theGe K edge of Ge nanoparticles formed by LP-PLA in R-space with the contributions from the each scattering path is given in figure 10.

Probing vacancies in NiO nanoparticles by EXAFS and molecular dynamics simulations

A Anspoks, A Kalinko, R Kalendarev and A Kuzmin
Institute of Solid State Physics, University of Latvia, Riga, LV-1063, Latvia
E-mail: andris.anspoks@cfi.lu.lv

Abstract. In this work we apply a combination of classical molecular dynamics (MD) and *ab initio* multiple-scattering EXAFS approach (MD-EXAFS methodology) to study the influence of point defects (Ni vacancies) on the structure and lattice dynamics of NiO nanoparticles. A NiO nanoparticles model, able to reproduce the experimental Ni K-edge EXAFS spectra, has been developed and allowed us to obtain original information on the particle size, the amount of Ni vacancies, thermal disorder and atomic structure relaxation.

1. Introduction

Nanosized nickel oxide (NiO) shows an increase of the lattice parameter upon decreasing of the particle size, as is evidenced by x-ray diffraction [1]. EXAFS spectroscopy reveals a decrease of the first shell Ni–O_1 average distance by ~0.02-0.04 Å and an increase of the mean second shell Ni–Ni_2 distance by ~0.01-0.02 Å for pure $Ni_{1-x}O$ and vanadium doped $Ni_{1-x}V_xO_y$ thin films [2, 3]. At the same time, an elongation of the first shell Ni–O_1 and the second shell Ni–Ni_2 distances has been found in ultra fine NiO particles dispersed on activated carbon fibers [4]. Such structure relaxation is closely connected with the nanoparticles size and surface as well as the presence of defects in the bulk. Recently, the presence and the role of the Ni vacancies have been studied by EXAFS in sputtered NiO films [5, 6] and nanoparticles [7] based on the analysis of the first two shells. It has been proposed that the distribution of the Ni vacancies in NiO nanoparticles, having a size of 10–18 nm, can be considered within a core-shell model [7].

In this work we evaluate the effects of the NiO nanoparticles size and the presence of Ni vacancies employing a recently developed simulation method [8], based on the use of classical molecular dynamics with *ab initio* multiple-scattering EXAFS calculations (MD-EXAFS methodology). Such approach allowed us to go beyond the limitations of the conventional EXAFS analysis procedure.

2. Experimental and data analysis

Two different NiO nanocrystalline samples, nanoparticles (nano-NiO) and thin film (tf-NiO), were studied in comparison with commercial microcrystalline NiO (c-NiO, Aldrich, 99%), having green color. Black color of nano-NiO and dark brown color of tf-NiO indicate the presence of the Ni vacancies [9]. Sample preparation details were reported previously in [10, 11]. According to the Scherrer's method and assuming the cubic crystallites shape, the average size of nanocrystallites was around 6 nm. The Ni K-edge x-ray absorption spectra were measured in

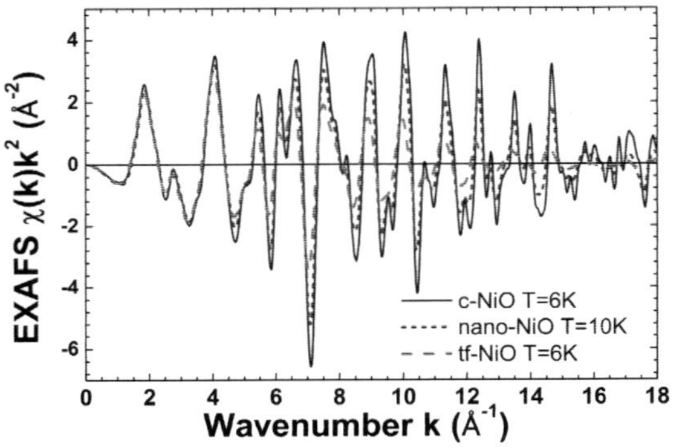

Figure 1. Low temperature Ni K-edge EXAFS spectra $\chi(k)k^2$ for polycrystalline NiO, NiO nanoparticles and thin film.

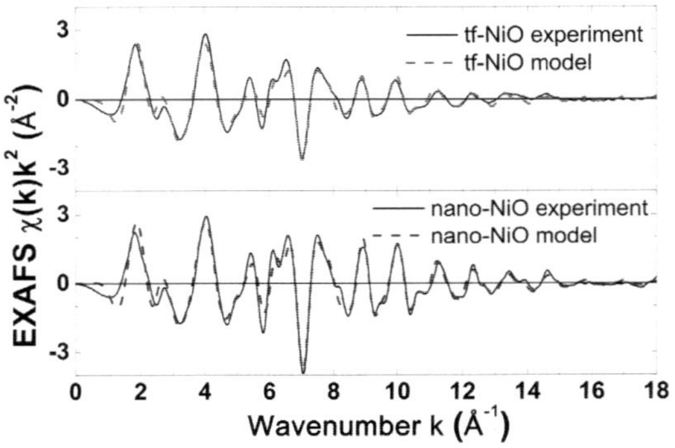

Figure 2. Comparison of the experimental and configuration-averaged Ni K-edge EXAFS spectra $\chi(k)k^2$ for nano-NiO and tf-NiO at T=300 K.

transmission mode at the HASYLAB/DESY C1 bending-magnet beamline in the temperature range from 6 K to 300 K. Details of experiments can be found in [10, 11].

The EXAFS oscillations $\chi(k)$ were extracted and analysed following the conventional procedure [12] using the EDA software package [13]. Noticeable difference between the EXAFS spectra of microcrystalline and nanocrystalline samples is clearly visible due to the static disorder caused by the atomic structure relaxation near the nanocrystallite surface and around defects (Fig. 1). The contribution from the first two coordination shells was analysed in the single-scattering approximation [11]. The values of the interatomic distances for the 1st (Ni–O_1, ±0.002 Å) and 2nd (Ni–Ni_2, ±0.003 Å) coordination shells of Ni at low temperature are, respectively, 2.088 Å and 2.952 Å for c-NiO, 2.082 Å and 2.961 Å for nano-NiO, 2.084 Å and 2.969 Å for tf-NiO. This result is in agreement with previous EXAFS works [2, 3].

3. MD-EXAFS simulations

The scheme of our MD-EXAFS calculations is shown in Fig. 3. First, we selected the cubic model for the nanoparticles with the size $L \times L \times L$. It was generated starting from the unit cell containing 4 Ni and 4 O atoms and having the lattice parameter a_0=4.173 Å as for c-NiO [1]. The required concentration of Ni vacancies (C_{vac}) was introduced by removing randomly selected Ni atoms in order to achieve uniform distribution of the vacancies in the particle.

Our force-field (FF) potential model includes two-body central force interactions between atoms and is described by a sum of the Buckingham and Coulomb potentials with the values reported in [11, 14]. The charge of nickel atoms $Z_{\rm Ni}$ was selected as the optimization parameter, and the charge of the oxygen atoms $Z_{\rm O}$ was calculated to fulfill electroneutrality condition. In our model all Ni ions have the same charge and so do all oxygen ions. All other Buckingham potential parameters have been left unchanged as for c-NiO [11, 14].

MD simulations were performed using the DLPOLY4.02 code [15], which is suitable for both crystalline and nanosized materials with huge number of the particles. Crystalline c-NiO was modeled in the isothermal-isobaric NPT ensemble using the supercell $6a_0 \times 6a_0 \times 6a_0$ and 3D periodic boundary conditions. The nanosized NiO particles were simulated in the canonical NVT ensemble with the nanoparticle placed in the middle of a large empty box. All simulations were performed at 300 K and zero pressure corresponding to the conditions of the EXAFS experiment. Technical details of calculations have been reported previously in [10, 11].

For each nanoparticle with the given size L and Ni vacancy concentration C_{vac} we performed

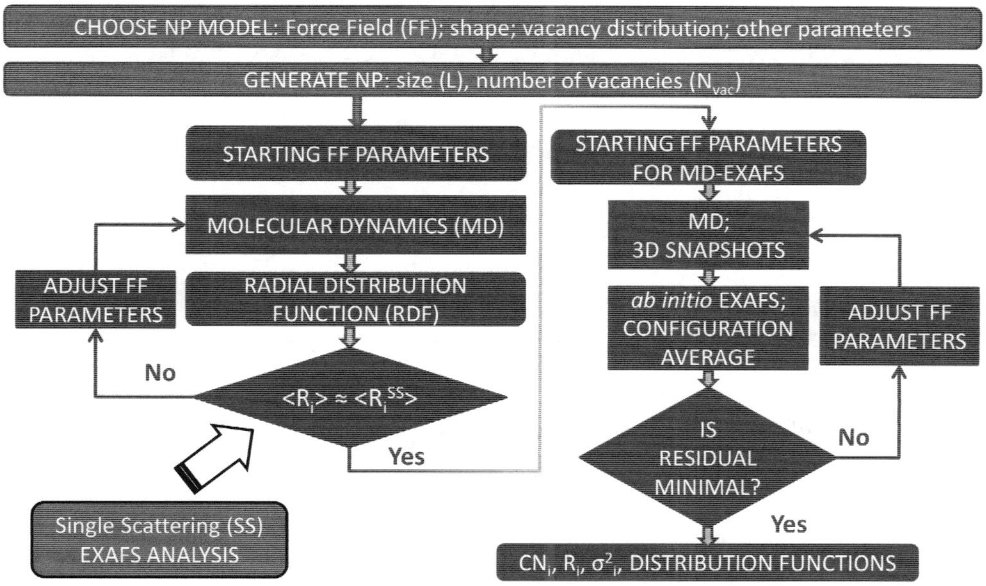

Figure 3. Scheme of the MD-EXAFS calculations for nanoparticles (NP).

MD calculations with two different FF parameter values (in our case two different $Z_{\rm Ni}$ values: +2.0 and +1.9). Then we found the corresponding interatomic distance R(Ni–Ni$_2$) and compared it with the value obtained from the conventional EXAFS analysis using the single-scattering approximation R^{SS}(Ni–Ni$_2$). If two values do not coincide within the desired precision, then we adjust FF parameters for the next iteration. This process is repeated until the required precision is achieved. At the end, we found for each nanoparticle with the size L and the vacancy concentration C_{vac} the FF parameters, which give the same value of the average Ni–Ni$_2$ distance as the one obtained from conventional EXAFS analysis.

At the next step (see the right part of the scheme in Fig. 3), the residual between the experimental and configuration-averaged (CA) EXAFS spectra for the model nanoparticle is used as a criterion for the FF model optimization and to evaluate the quality of the nanoparticle model. At each iteration step, we calculate a set of atomic configuration snapshots using classical MD. The number of snapshots depends on the nanoparticle material, size and shape as well as on the experimental conditions, i.e. pressure and temperature, and should ensure the convergence of the CA EXAFS spectrum. We found that EXAFS spectra from at least 4000 absorbers are required to achieve convergence. For EXAFS calculations we used *ab initio* real-space multiple-scattering FEFF8 code [16], and the technical details of calculations were reported in [10, 11].

Further we will describe the algorithm of the FF optimization. We start from the results of the previous step and calculate the CA EXAFS spectra for two MD simulations, obtained using slightly different (usually ±0.005) $Z_{\rm Ni}$ values. Then we compare residuals between the experimental and CA EXAFS spectra and guess new value of $Z_{\rm Ni}$ in order to minimize the residual value. Here we change $Z_{\rm Ni}$ value in steps within the desired precision (in our case ±0.005). When the residual minimum is reached, we stop the MD-EXAFS calculations and evaluate all other required properties from the MD data, for example, values of coordination numbers CN_i, interatomic distances R_i, and MSRDs σ_i^2 for each i coordination shell.

As the first application of the described procedure we optimized the ions charges for c-NiO. The best fit to its EXAFS spectrum was achieved with $Z_{\rm Ni}$=+2.015 and $Z_{\rm O}$=−2.015.

Next we have studied the reliability of our FF model in a description of the local atomic structure relaxation around the Ni vacancy in the bulk of c-NiO. Our results show that the six nearest oxygen atoms move away from the center of the vacancy by ≈0.2 Å, thus significantly shortening the distances to nearest Ni atoms. At the same time, twelve nearest Ni atoms move slightly towards the center of the vacancy by ≈0.08 Å, thus increasing the distance with their neighbouring Ni atoms. Such relaxation of atomic structure, induced by the presence of single Ni vacancy, plays significant role in the average structure relaxation, especially for the first coordination shell of Ni, as was found recently [5]. The results of our simulations are in excellent agreement with those from the recent *ab initio* calculations [17, 18], which predict the relaxation of O atoms by about 0.17 Å outwards from the Ni vacancy site. This result confirms that our simple FF potential model reproduces well structure relaxation around the Ni vacancies.

Finally we have simulated NiO nanoparticles, starting from the initially cubic shape. The size of the cube L and the concentration of the nickel vacancies C_{vac} were chosen as model parameters, which were optimized by minimizing the residual between the experimental and CA Ni K-edge EXAFS spectra (Fig. 2). A well defined residual minimum has been found indicating the reliability of the model. The best agreement was achieved for nanoparticles with non-zero concentration of the Ni vacancies: $C_{\text{vac}}\approx$0.4-1.2% for NiO nanoparticles with the size of $L\approx$3.6-4.2 nm and $C_{\text{vac}}\approx$1.6-2.0% for NiO thin film with the size of $L\approx$1.3-2.1 nm.

4. Conclusions

We report on the recently developed complex simulation approach, combining classical MD and *ab initio* multiple-scattering EXAFS theory, which allows us to reliably simulate both bulk and nanosized materials using small number of adjustable model parameters (L, C_{vac}, Z). The method leads to better use of the rich information hiding in the EXAFS spectra, which is particulary interesting for nanosized objects. The application of the method to nickel oxide allowed us to evaluate the size of nanoparticles and the concentration of nickel vacancies.

Acknowledgments

The authors are grateful to Dr. R. Chernikov for assistance during EXAFS measurements at DESY. This work was supported by ESF Project 2009/0202/1DP/1.1.1.2.0/09/APIA/VIAA/141 and Latvian Government Research Grant No. 09.1518. The EXAFS experiments at HASYLAB/DESY have been supported from the European Community's Seventh Framework Programme under grant agreement No. 226716 (Project I-20100110 EC).

References

[1] Li L, Chen L, Qihe R and Lia G 2006 *Appl. Phys. Lett.* **89** 134102
[2] Kuzmin A, Purans J and Rodionov A 1997 *J. Phys.: Condens. Matter* **9** 6979
[3] Avendaño E, Kuzmin A, Purans J, Azens A, Niklasson G A and Granqvist C G 2005 *Phys. Scripta* **T115** 464
[4] Hattori Y, Konishi T and Kaneko K 2002 *Chem. Phys. Lett.* **355** 37
[5] Jang W L, Lu Y M, Hwang W S, Hsiung T L and Wang H P 2009 *Appl. Phys. Lett.* **94** 062103
[6] Jang W L, Lu Y M, Hwang W S, Dong C L, Hsieh P H, Chen C L, Chan T S and Lee J F 2011 *Europhys. Lett.* **96** 37009
[7] Mandal S, Banerjee S and Menon S R 2009 *Phys. Rev.* B **80** 214420
[8] Kuzmin A and Evarestov R A 2009 *J. Phys.: Condens. Matter* **21** 055401
[9] Sato H, Minami T, Takata S and Yamada T 1993 *Thin Solid Films* **236** 27
[10] Anspoks A, Kuzmin A, Kalinko A and Timoshenko J 2010 *Solid State Commun.* **150** 2270
[11] Anspoks A and Kuzmin A 2011 *J. Non-Cryst. Solids* **357** 2604
[12] Aksenov V L, Kuzmin A Yu, Purans J and Tyutyunnikov S I 2001 *Phys. Part. Nucl.* **32** 675
[13] Kuzmin A 1995 *Physica* B **208-209** 175
[14] Fisher C A J 2004 *Scr. Mater.* **50** 1045
[15] Todorov I T, Smith W, Trachenko K and Dove M T 2006 *J. Mater. Chem.* **16** 1611
[16] Ankudinov A L, Ravel B, Rehr J J and Conradson S D 1998 *Phys. Rev.* B **58** 7565
[17] Ferrari A M and Pisani C 2007 *J. Chem. Phys.* **127** 174711
[18] Park S, Ahn H S, Lee C K, Kim H, Jin H, Lee H S, Seo S, Yu J and Han S 2008 *Phys. Rev.* B **77** 134103

Journal of Physics: Conference Series **430** (2013) 012028 doi:10.1088/1742-6596/430/1/012028

The Role of Hydrogen in Formation of Pd XANES in Pd-Nanoparticles

A L Bugaev[1]*, V V Srabionyan[1], A V Soldatov[1], L A Bugaev[1] and **J A van Bokhoven[2,3]**

[1] Research center for nanoscale structure of matter, Southern Federal University, Rostov-on-Don, Russia

[2] Paul Scherrer Institute, Villigen, Switzerland

[3] ETH Zurich, Zurich, Switzerland

*E-mail: arambugaev@gmail.com

Abstract. The effects of hydrogen in Pd nanoparticles on the features of Pd K-edge X-ray absorption near edge structure (XANES) are studied. Simulations indicate a linear dependence of the intensities of XANES maxima upon the concentration of hydrogen in the bulk of Pd clusters at low H-concentrations and stepwise increase of the first near edge peak at high concentrations. The effects of the diffusion of H atoms over the interstitial sites are considered. The application of the obtained results to the experimental spectra measured in PdH nanoparticles of sizes from 1.3nm to 10.5nm indicates low H-concentrations and high mobility of H-atoms in the bulk of the nanoparticles.

1. Introduction

Palladium is well known for its hydrogen absorption ability [1] and has been extensively studied as a hydrogen storage material for many decades [2-4]. Palladium nanoparticles are of great interest because of their large active surface area. The lattice parameter for fcc palladium increases after hydride absorption [5]. For bulk palladium hydride different phases are presented depending on the hydrogen concentration and these phases correspond to different lattice parameters [6, 7]. However, one should use the term "phase" very carefully when speaking about nanoparticles. Numerous studies observe lattice expansion for PdH_x nanoparticles [8-13], but numerical values of the lattice parameter changes differ from that of bulk samples.

X-ray absorption spectroscopy (XAS) [14] is a powerful technique for obtaining structural information of nanostructured samples and it was applied to Pd clusters in numerous works [8, 10, 12, 13]. Lattice expansion was commonly observed from Pd K-edge extended X-ray absorption fine structure (EXAFS) data analysis, however no direct information on hydrogen presence, its concentration and positions in fcc of metallic Pd nanoparticles was obtained. Significant changes in Pd L3-edge XANES for bulk and nanostructured PdH were analyzed in [13, 15].

In this work we found hydrogen effects on Pd K-edge X-ray absorption near edge structure (XANES). The performed calculations show a very strong sensitivity of the first near edge peaks relative intensity on hydrogen presence, total H concentration and positions occupied by H atoms in fcc. of palladium, which must be taken into consideration for a correct theoretical description of experimental data. A significant increase of the first peak in the theoretical spectrum calculated for a simple model of a PdH cluster is observed. Calculations show a linear dependence of that XANES peak on the hydrogen concentration in the Pd cluster. Changes in Pd K-edge XANES caused by H migration from octahedral to tetrahedral interstitial sites were analyzed. Results of analysis indicated a very high probability of hydrogen diffusion through the bulk of Pd nanoparticles of different sizes in the experiment made by our coauthors [13], and also explain changes in XANES for previous experimental works [8, 10].

2. Pd K-edge XANES calculation

2.1. General description

Theoretical Pd K-edge XANES spectra were calculated for small (135 Pd atoms) nanoclusters of bare Pd and Pd-hydride. The Pd local structure used for spectra modeling was taken as in the fcc lattice of Pd-foil. The fcc-like local structure of Pd-nanoparticles was confirmed by the Fourier Transform of experimental [13] Pd K-edge EXAFS for nanoparticles of the sizes from 1.3 nm to 10.5 nm [16]. Small clusters of 19 Pd atoms were used for hydrogen migration effects calculation.

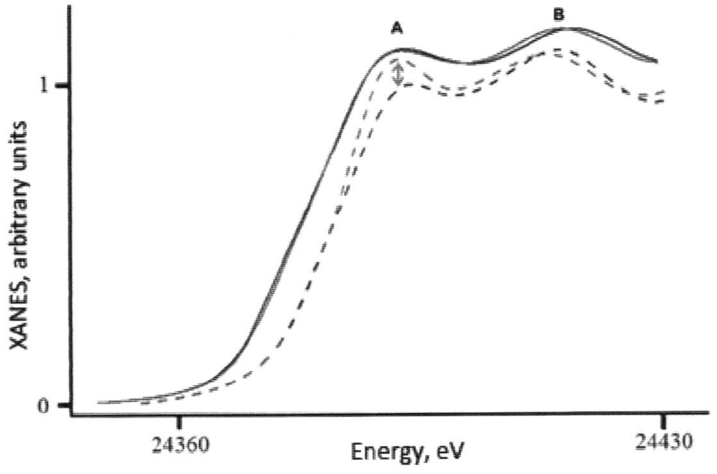

Figure 1. Experimental (solid) and theoretical (dash) Pd K-edge XANES for bare (black) and hydride (red) palladium nanoparticles.

XANES spectra were calculated using two different approaches: the multiple scattering technique (FEFF9) within muffin-tin approach [17, 18] and the finite difference method (FDMNES) with full potential [19]. Analysis of calculated spectra showed that first

Journal of Physics: Conference Series **430** (2013) 012028 doi:10.1088/1742-6596/430/1/012028

near edge peaks are very sensitive to hydrogen presence and its location in the lattice. A model of Pd cluster with H atoms located in octahedral interstitial site gives an increase of peak A (fig.1). These results are in contradiction with experimental data [13] where these changes are negligible, which should indicate low hydrogen concentration in the bulk of experimentally observed nanoparticles. As both muffin-tin and full potential approaches indicate this effect, less time-consuming FEFF calculations were chosen for further investigation.

Relative intensity of peak A (I_A/I_B) was taken as a numerical description of observed changes. This value showed stability to the changes in cluster size, interatomic distances and FEFF parameters: region of self-consistent potential (SCF), full multiple scattering area (FMS), type of exchange-correlation potential. The following structural changes were then studied: total hydrogen concentration in Pd cluster, hydrogen migration from octahedral to tetrahedral interstitial site near the absorbing Pd atom. Calculations showed that the value of I_A/I_B is significantly affected only by H atoms presence in the nearest coordination shell. Dependence on the further hydrogen is negligible and I_A/I_B tends to its value in the bare Pd cluster. Therefore only nearest shell hydrogen atoms modifications were involved in our study.

2.2. Hydrogen concentration effects

A strong dependence of the relative intensity I_A/I_B on the H-atoms' concentration was observed. The number of hydrogen atoms in the octahedral interstitial sites of the nearest to absorbing Pd atom coordination shell was varied from 0 to 6. Further shells were not affected for the reasons mentioned above. For each configuration Pd K-edge XANES spectra were calculated. Different configurations of hydrogen atoms in 6 octahedral interstitial sites possible for models containing 2-4 H atoms were taken into consideration: XANES spectra for non-equivalent configurations were calculated and summed using corresponding weight coefficients. Calculated relative intensities I_A/I_B are shown in fig.2.

A linear dependence of I_A/I_B on the number of H atoms is observed for small H concentration. When the number of H atoms corresponds to a β-phase concentration (>60%) a non-linear increase occurs. Linear dependence of intensity increase is observed due to the fact that at low concentrations H-atoms do not interact with each other and contribute to Pd K-edge XANES independently. We suppose that a non-linear increase for high concentrations indicates the presence of inter-hydrogen correlations which is consistent with strain field effects responsible for the well-known difficulty in obtaining high loaded PdH samples [20].

According to the obtained results hydrogen concentration in the bulk of a palladium clusters must be considered for theoretical XANES calculation. However, a determination of hydrogen concentration from an experimental XANES without any complementary knowledge appears to be a difficult problem as concentration is not the only parameter responsible for peak A intensity changes.

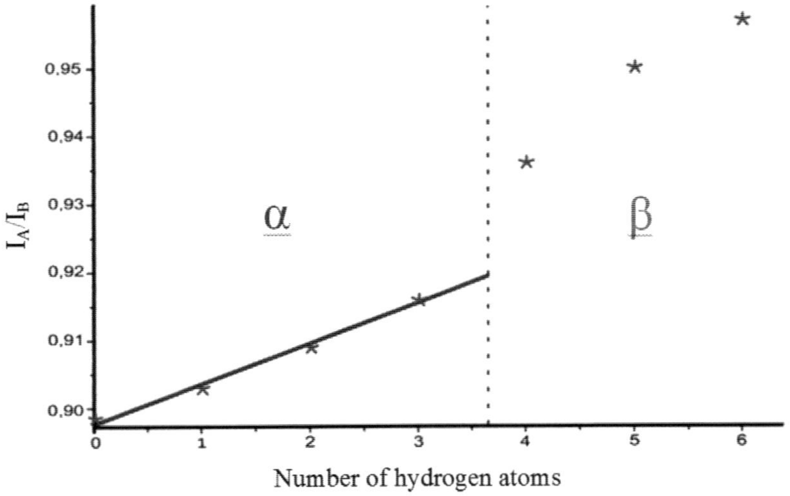

Figure 2. Increase of the relative intensity of the first peak of calculated Pd K-edge XANES with the increase of number of H-atoms in the nearest octahedral interstitial sites (stars) and linear approximation for intensity behavior at low concentrations.

2.3. Hydrogen migration

Numerous works have studied a quantum delocalization and diffusion of hydrogen in bulk metal samples [21-23]. One should expect that light hydrogen atoms are in permanent motion in palladium nanoclusters and are able to diffuse through different interstitial sites in fcc of palladium. We studied an effect of hydrogen diffusion on Pd K-edge XANES in a simple model of a 19 atom Pd cluster with one H atom migrating from one octahedral interstitial site to another via neighboring tetrahedral one.

Fig.3. illustrates peak A intensity dependence on the hydrogen position during H atom migration from octahedral to tetrahedral sites. Calculation of the relative intensity for the bare 19 atom Pd cluster was also performed and is plotted as a red dashed line. The maximum increase of peak A was observed when H atom was located in octahedral position; tetrahedral position also corresponds to a local maximum but with much weaker intensity changes. According to ab initio calculations for hydrogen migration in Pd lattice [24] potential energy has minima at both of these interstitial sites. When the H atom was located on a triangle face connecting an octahedron and a tetrahedron, which is a saddle position, opposing changes of peak A intensity were achieved.

Such dependence means that intensity changes should become more negligible with the higher mobility of hydrogen. According to the results reported by our coauthors [13] interatomic distances in supported Pd hydride nanoparticles were bigger than distances in bulk palladium hydride which should decrease energy barriers between the neighboring interstitial sites and should favor more intense diffusion.

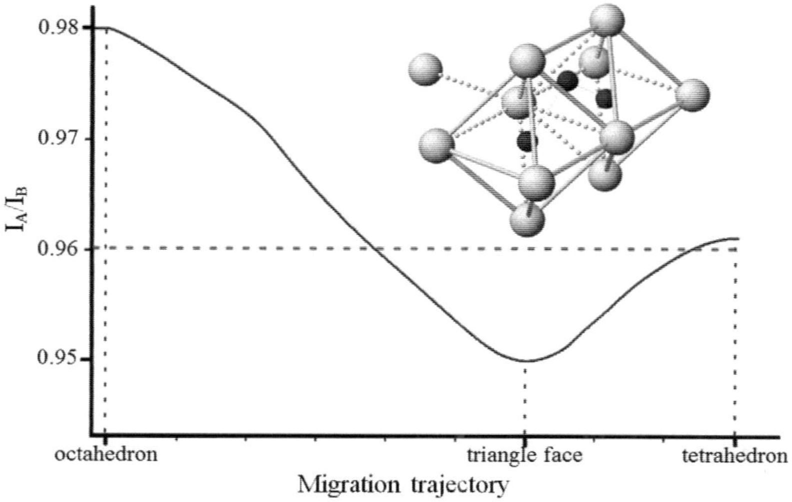

Figure 3. Dependence of the relative intensity of peak A upon the position occupied by the hydrogen atom. On the horizontal axis the generalized coordinate for hydrogen migration from octahedron to tetrahedron is plotted. The red dashed line is a relative intensity value for the model with no hydrogen. Plotted cluster geometry illustrates two neighboring octahedral interstates connected via a tetrahedron.

3. Summary and conclusion

Calculation showed that the hydrogen concentration and the effect of hydrogen diffusion must be reflected in Pd K-edge XANES. These results compared with experimental data [13] indicate both low hydrogen concentration in the bulk of Pd nanoparticles and high mobility of hydrogen atoms. It should be mentioned that in works [8, 10] an increase of peak A is observed.

However these parameters should correlate with each other. A probability of existence of an unoccupied interstate near the existing H atom decreases with an increase of hydrogen concentration and should prevent diffusion. On the other hand diffusion magnitude increases with an increase of interatomic distances which also depend on hydrogen concentration. All these relations should make the process of determination of described parameters more complicated. Thereby Pd K-edge XANES data for PdH_x nanoparticles may be used to give complementary information on EXAFS analysis, Pd L_3-edge XANES, X-ray diffraction data and other techniques.

4. Acknowledgements

Authors would like to thank Russian Ministry of Education and Science for the financial support (project #11.519.11.3005).

5. References

[1] Alefeld G and Volkl J 1978 *Hydrogen in Metals I: Basic Properties* (Springer).
[2] Kishore S, Nelson J A, Adair J H, and Eklun P C 2005 *Journal of Alloys and Compounds* **389** 234.
[3] Yamauchi M, Ikeda R, Kitagawa H, and Takata M 2008 *Journal of Physical Chemistry C* **112** 3294.
[4] Zacharia R, Kim K Y, Fazle Kibria A K M, and Nahm K S 2005 *Chemical Physics Letters* **412** 369.
[5] Griessen R and Feenstra R 1985 *Journal of Physics F: Metal Physics* **15** 1013.
[6] Fukai Y 2005 *The metal-hydrogen system : basic bulk properties* (Springer Berlin ; New York).
[7] Worsham J E, Wilkinson M K, and Shull C G 1957 *Journal of Physics and Chemistry of Solids* **3** 303.
[8] Davis R, Landry S, Horsley J, and Boudart M 1989 *Physical Review B* **39** 10580.
[9] Kochubey D I, Feodorov V K, Williams C, Nogin Y N, Stenin M V, Ryndin Y A, Thomas J M, and Zamaraev K I 1990 *Catalysis Letters* **5** 349.
[10] McCaulley J A 1993 *The Journal of Physical Chemistry* **97** 10372.
[11] Nosova L V, Stenin M V, Nogin Y N, and Ryndin Y A 1992 *Applied Surface Science* **55** 43.
[12] Stakheev A Y, Mashkovskii I C, Tkachenko O P, Klementiev K V, Grünert W, Baeva G N, and Kustov L M 2010 *Russian Chemical Bulletin* **58** 280.
[13] Tew M W, Miller J T, and van Bokhoven J A 2009 *The Journal of Physical Chemistry C* **113** 15140.
[14] Koningsberger D C and Prins R 1987 *X-ray absorption : principles, applications, techniques of EXAFS, SEXAFS, and XANES* (Wiley New York).
[15] Soldatov A V, Della Longa S, and Bianconi A 1993 *Solid State Communications* **85** 863.
[16] Srabionyan V V, Bugaev A L, Pryadchenko V V, Avakyan L A, Van Bokhoven J A, and Bugaev L A *(to be published)*.
[17] Rehr J J, Kas J J, Prange M P, Sorini A P, Takimoto Y, and Vila F 2009 *Comptes Rendus Physique* **10** 548.
[18] Rehr J J, Kas J J, Vila F D, Prange M P, and Jorissen K 2010 *Physical chemistry chemical physics : PCCP* **12** 5503.
[19] Joly Y 2001 *Physical Review B* **63**.
[20] De Ninno A, Violante V, and La Barbera A 1997 *Physical Review B* **56** 2417.
[21] Flynn C and Stoneham A 1970 *Physical Review B* **1** 3966.
[22] Nishijima M, Okuyama H, Takagi N, Aruga T, and Brenig W 2005 *Surface Science Reports* **57** 113.
[23] Seymour E, Cotts R, and Williams W 1975 *Physical Review Letters* **35** 165.
[24] Koroteev Y M, Gimranova O V, and Chernov I P 2011 *Physics of the Solid State* **53** 896.

Journal of Physics: Conference Series **430** (2013) 012029 doi:10.1088/1742-6596/430/1/012029

A Comparative XAFS Study of Gold-thiolate Nanoparticles and Nanoclusters

D M Chevrier[1], A Chatt[1], T K Sham[2] and P Zhang[1]*

[1] Department of Chemistry, Dalhousie University, Halifax, NS, Canada, B3H 4R2
[2] Department of Chemistry, University of Western Ontario, London, ON, Canada, N6A 6B7

E-mail: peng.zhang@dal.ca

Abstract. Tiopronin-capped gold nanoparticles and gold nanoclusters of sizes 3.0 and 1.5 nm, respectively, were investigated with XAFS at the gold L_3-edge. The specific EXAFS fitting procedure is discussed for obtaining reliable fit parameters for each system. The difficulties and challenges faced when analysing EXAFS data for gold nanoparticles and nanoclusters are also mentioned. Fitting results for gold nanoparticles reveal a small amount of surface Au-thiolate interactions with a large Au-Au metal core. For gold nanoclusters, only a one-shell fit was obtainable. Instead of Au-Au metal core, long-range interactions are expected for gold nanoclusters. Tiopronin-capped gold nanoclusters are proposed to be polymeric in nature, which helps explain the observed red luminescence.

1. Introduction

X-ray absorption of fine structure (XAFS) has been a technique used to aid and supplement the characterization of thiol-capped gold nanoparticles (Au NPs) [1–3]. More recently, XAFS has also become an important tool for understanding the structural environment of novel gold nanoclusters (Au NCs) where other characterization techniques are unable to provide information on the gold local structure [4,5]. Utilizing XAFS for studying Au NCs has been successful in detecting changes in structure due to cluster size and ligand environment [6,7]. Although the XAFS technique is very sensitive for identifying small structural changes, there are still many challenges researchers face when working with Au NP and NC XAFS data.

This study will focus on the XAFS analysis (near-edge and extended x-ray absorption of fine structure (EXAFS)) of tiopronin-capped (thiol-capped) Au NPs and NCs with nominal sizes of 3.0 nm and 1.5 nm, respectively. Comparing these two samples will illustrate the dramatic transition in gold structure from nanoparticles to nanoclusters as seen from XAFS results. The smaller tiopronin-capped Au NC studied here has been shown to have luminescent properties, therefore a short discussion will be included to comment on tiopronin-capped Au NC optical properties.

2. Experimental Methods

Tiopronin-capped Au NPs and NCs were synthesized following a standard colloidal preparation method [8] with additional characterization published earlier [9]. XAFS data for tiopronin-capped Au NPs and Au NCs were collected at the Advanced Photon Source (APS), operating at 6.9 GeV, in Argonne National Laboratory (ANL). Dried powder of each sample was packed and sealed in a kapton pouch for solid-state measurements. The Au L_3-edge XAFS signal was measured in transmission mode at room temperature and atmospheric pressure. EXAFS fitting was performed with the WinXAS 3.1 software package along with theoretical simulations from the FEFF8.2 computational program. The crystal structure of thiol-capped Au_{25} nanocluster was constructed to perform FEFF calculations for obtaining theoretical scattering paths used in the EXAFS fitting. Specific fitting procedures are discussed in the following section as they pertain to each sample.

3. Results and Discussion

3.1 Technical EXAFS fitting results and discussion

The raw data are presented in k-space for Au foil, Au NPs, and Au NCs in figure 1. A k^1-weighting was used to compare the EXAFS oscillations between samples. The frequency of oscillations is similar for Au foil and Au NPs, indicating the Au NPs measured retain the Au fcc structure. Au NCs display an EXAFS oscillation pattern considerably different from that of Au foil and Au NPs. XAFS data quality was good for Au foil and Au NPs with a k-range up to 13.5 Å^{-1}. The data quality for Au NCs was slightly poorer with a useable k-range only up to 11.5 Å^{-1}. This was due to monochromator glitches in the high k-region of the EXAFS oscillation. Before beginning EXAFS refinements, the reduction factor (S_o^2) was determined by fixing a CN of 12 for Au foil while running all other parameters free. This was done for both k^1- and k^3-weighting, obtaining values of 0.81 and 0.90, respectively. Error bars in the EXAFS fits (table 1) were determined using the double residual method [1].

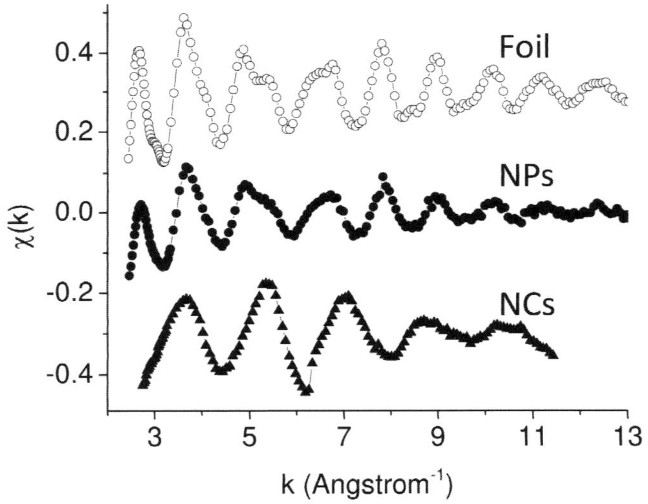

Figure 1. k^1-weighted k-space of Au Foil, Au NPs, and Au NCs.

3.1.1 Tiopronin-capped Au NPs

Fourier-transformed (FT) EXAFS for 3.0 nm Au NPs was transformed using a k-range of 3.3 – 11.8 Å^{-1} and a k^1-weighting. As seen in figure 2, there is a large contribution from Au – Au bonding with a maximum at approximately 2.5 Å, not phase corrected. With a qualitative observation, Au – S contribution appears to be weak. To begin the fitting procedure for Au NPs, a preliminary fit was first done for the larger shell (Au – Au), obtaining one-shell refinement results shown in table 1. The bond length is the most reliable parameter to carry over for two-shell fitting, reducing the number of free running parameters ((# free running parameters)/2 $\leqslant$ # independent parameters is desirable). A k^1-

weighting was chosen for Au NPs to enhance low k-range features and fit the Au – S contribution. For an optimal two-shell fit, a Au – Au distance of 2.85 Å and a reduction factor of 0.81 were fixed with E_0 values correlated for both shells. By adjusting these parameters for the fitting procedure, we were able to reduce the number of free running parameters to obtain a reliable fit for both shells, seen in figure 2a.

From the two-shell fitting, the Au – Au CN changes from 9.7 to 9.1 once the Au – S is included in the fitting. A similar Debye-Waller value (σ^2) is found for both one-shell and two-shells. The Au – S bond length for these Au NPs is 2.46 Å. As seen from the Au – S fit in figure 2b, the amount of Au – S is very small due to the large size of the Au core. A CN value of 0.12 was determined from our fitting, which is reaching the lower limit for a reliable CN result. If the Au core were any larger for this system, Au – S contributions would most likely be unattainable. This can be a problem when using EXAFS to determine the nature of the interactions between capping ligand and surface Au when the averaged EXAFS signal for Au is heavily weighted for the metallic core.

Table 1. EXAFS fitting results for Au Foil, Au NPs, and Au NCs.

	Bond	CN	R (Å)	σ^2 ($Å^2$)	E_0 (eV)
Au Foil	Au-Au	12(f)	2.86(1)	0.0070(5)	0.6(5)
Au NPs (one-shell)	Au-Au	9.7(2)	2.85(1)	0.0095(3)	0.9(3)
Au NPs (two-shell)	Au-S	0.1(1)	2.46(9)	0.007(5)	0.3(9)
	Au-Au	9.1(6)	2.85(1)	0.0091(6)	0.3(9)
Au NCs	Au-S	1.7(2)	2.30(1)	0.0060(8)	-3(4)

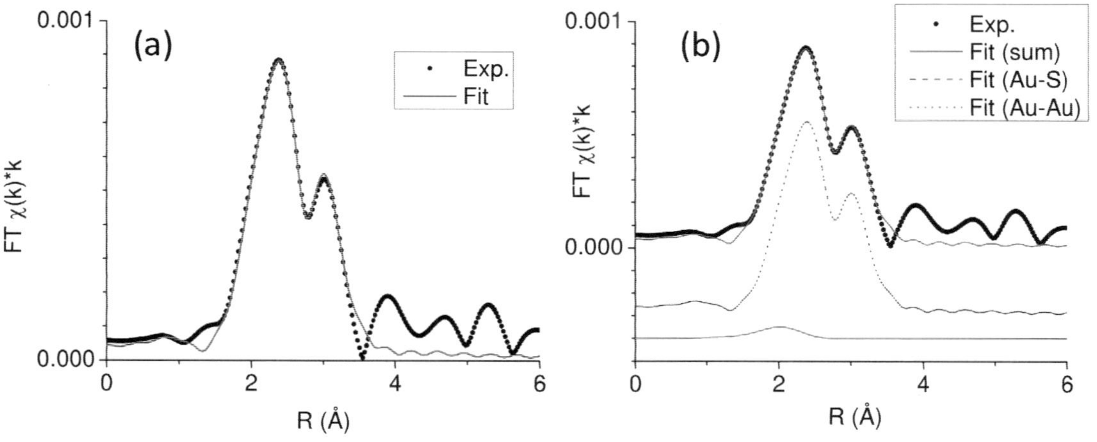

Figure 2. FT-EXAFS of Au NPs with experimental and fitted spectra (a). The contribution of Au-S and Au-Au shells are shown (b) underneath the fitted FT-EXAFS of Au NPs.

3.1.2 Tiopronin-capped Au NCs

FT-EXAFS for Au NCs is shown in figure 3 with a k^1-weighting and a k-range of 3.3-11.5 $Å^{-1}$ used for the transformation. Contrary to the 3.0 nm Au NP, the 1.5 nm Au NC has a very weak Au – Au shell and a much larger Au – S shell with a maximum around 2.0 Å, not phase corrected. Because of the weak Au – Au contribution, refinements with k^3-weighting were also attempted with identical k-range to k^1-weighting. A similar two-step, two-shell fitting method was attempted for Au NCs as for Au NPs.

First step included a one-shell fit of Au – S with a fixed reduction factor. A bond length of 2.30 Å was carried over to two-shell fitting. Both k^1- and k^3- weightings were used for fitting Au NCs. For the two-shell fitting, Au – S bond length and reduction factor were fixed with correlated E_0 values for both shells. Unlike Au NPs, there is a large amount of disorder in the Au – Au shell, resulting in high Debye-Waller values ($\sigma^2 > 0.05$ Å^2). When σ^2 values were correlated for both shells, the CN values for Au – Au were unreasonably high. Correlating both E_0 and σ^2 obtained CN < 0.5 for Au – Au. All fitting attempts (with both k-weightings) to incorporate the Au – Au shell produced varying bond lengths ranging from 2.9-3.2 Å. From our results, we concluded that a two-shell fitting of the prepared Au NCs was inconclusive. For this reason, only a one-shell fit was reported in table 1 with k^1-weighting for comparison to Au NPs. The difference in Au – S bond lengths obtained from Au NPs (2.46 Å) and Au NCs (2.30 Å) could be explained by the difference in surface capping interactions. The longer Au – S bond length for Au NPs also has a significantly larger margin of error (table 1) than Au NCs due to the small contribution in the EXAFS fit (figure 2b).

A number of factors affect the refinement results for Au – Au fitting for Au NCs. Since the shell is weak, it is sensitive to the chosen k-range and k-weighting. The quality of the data hindered our fitting result because we could not use a higher k-range to confirm the nature of Au – Au bonding. This is a challenge for fitting other Au NC systems where the Au – Au environment can have multiple shells, as seen with thiol-capped Au_{25} [6,7]. In the case of small cluster sizes, purity is crucial for obtaining high quality data for multi-shell Au – Au fitting.

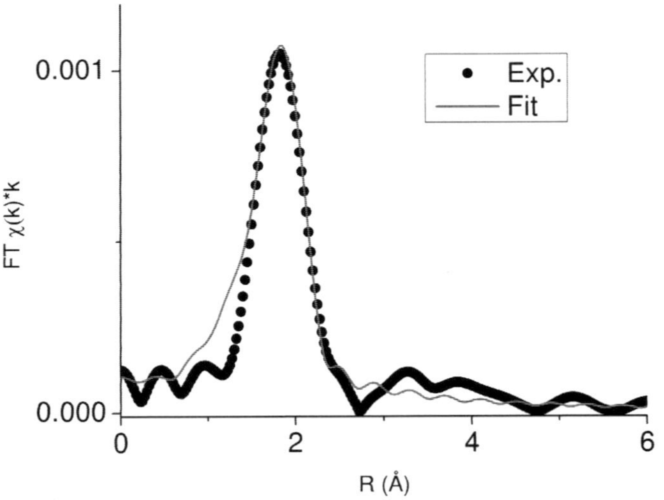

Figure 3. FT-EXAFS of Au NCs with experimental and fitted spectra.

3.2 Tiopronin-capped Au NC optical properties

From our refinement results for Au NCs, we believe the absence of a Au core and low CN values for Au – Au interactions indicates possible polymeric character of S(R)-Au-S(R) units with long-range Au – Au bonding. We believe the lack of Au core and presence of longer Au-Au contacts is responsible for the absence of a SPR band and luminescence behaviour, as seen in the previous work [9], based on the availability of discrete electronic transitions in smaller cluster sizes [10].

The near-edge spectra for Au foil, Au NPs, and Au NCs, shown in figure 4, illustrates the difference in d-electron occupancy for each sample. Au NPs and Au Foil have similar near-edge features, in agreement with our EXAFS results. There is a shift in higher energy for both Au NPs and Au NCs relative to the foil, with Au NCs having higher edge energy. The white line intensity compliments this trend, as Au NCs appear to have more Au^+ character than Au NPs and Au Foil. The difference in valence state and near-edge features supports the proposed polymeric character of these Au NCs. With a higher Au^+ composition, Au – Au interactions are expected to be long-range and varied [11,12]. These sorts of interactions are seen in numerous Au^+-thiol complexes where luminescence is observed [13,14].

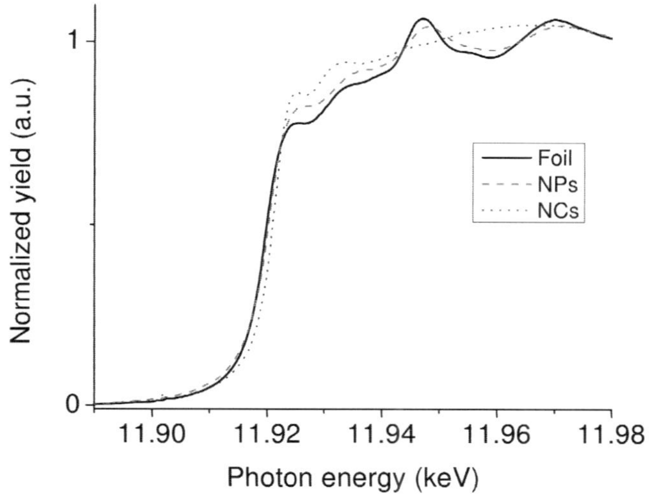

Figure 4. Near-edge spectra comparison of Au Foil, Au NPs, and Au NCs.

4. Conclusion

XAFS analysis was used to compare the structural environment of tiopronin-capped Au NPs and NCs. Au NPs were easily fit for both Au – Au and Au – S parameters where Au NCs only had reliable fitting for Au – S bonding. Even though Au – S contribution was very little for Au NPs, using a k^1-weighting produced reliable fitting results. Due to limited k-range and data quality, Au – Au bonding for Au NCs was unsuccessful, even with k^3-weighting. From our XAFS measurements, Au NCs appear to mainly consist of polymeric character with little to no metallic core. Fitting long-range Au-Au bonding was inconclusive due to the varied bond length results when EXAFS fitting conditions were changed. We conclude from this work that fitting procedures for nanoparticles and nanoclusters require certain conditions and procedures for reliable EXAFS fitting. We also report that the higher composition of Au^+ and evidence of polymeric -S(R)-Au-S(R)- structure is important for understanding the luminescent properties of the smaller tiopronin-capped Au NCs.

Acknowledgements

The authors from Dalhousie University would like to thank the Natural Sciences and Engineering Research Council (NSERC) of Canada for financial support in the form of Discovery Grants. Work done at the University of Western Ontario was supported by NSERC, CFI, CRC, and OIT.

References

[1] Zanchet D, Tolentino H, Martins Alves M C, Alves O L and Ugarte D 2000 *Chem. Phys. Lett.* **323** 167
[2] Zhang P and Sham T K 2003 *Phys. Rev. Lett.* **90** 3
[3] Shibata T, Tostmann H, Bunker B, Henglein A, Meisel D, Cheong S, and Boyanov M 2001 *J. Synchrotron Rad.* **8** 545
[4] Wu Z, MacDonald M A, Chen J, Zhang P and Jin R 2011 *J. Am. Chem. Soc.* **133** 9670
[5] MacDonald M A, Zhang P, Chen N, Qian H and Jin R 2011 *J. Phys. Chem. C* **38** 65
[6] MacDonald M A, Zhang P, Qian H and Jin R 2010 *J. Phys. Chem. Lett.* **1** 1821
[7] MacDonald M A, Chevrier D M, Zhang P, Qian H and Jin R 2011 *J. Phys. Chem. C* **115** 15282
[8] Templeton A C, Chen S, Gross S M and Murray R W 1999 *Langmuir* **15** 66
[9] Sham T K, Kim P-S G and Zhang P 2006 *Solid State Communications* **138** 553
[10] Dickson R M, Nicovich P R and Zheng J 2007 *Annu Rev Phys Chem.* **58** 409
[11] de la Riva H, Pintado-Alba A, Nieuwenhuyzen M, Hardacre C and Lagunas M C 2005 *Chem. Commun. (Cambridge, England)* 4970
[12] Simpson C A, Farrow C L, Tian P, Billinge S J L, Huffman B J, Harkness K M and Cliffel D E 2010 *Inorg. Chem.* **49** 10858
[13] Jones W B, Yuan J, Narayanaswamy R, Young M A, Elder R C, Bruce A E and Bruce M R M 1995 *Inorg. Chem.* **34** 1996
[14] Forward J M, Bohmann D, Fackler Jr. J P and Staples R J 1996 *Inorg. Chem.* **34** 6330

Journal of Physics: Conference Series **430** (2013) 012030 doi:10.1088/1742-6596/430/1/012030

XANES: observation of quantum confinement in the conduction band of colloidal PbS quantum dots

IN Demchenko[1*], M Chernyshova[2], X He[3], R Minikayev[1], Y Syryanyy[1], A Derkachova[1], G Derkachov[1], WC Stolte[4,5], E Piskorska-Hommel[6], A. Reszka[1], H Liang[3]

[1]Polish Academy of Sciences, Institute of Physics, Warsaw, 02-668, Poland
[2]Institute of Plasma Physics and Laser Microfusion, Warsaw, 01-497, Poland
[3]Department of Mechanical Engineering, Texas A&M University, College Station, TX 77843-3123, USA
[4]Department of Chemistry, University of Nevada, Las Vegas, NV 89154-4003, USA
[5]Lawrence Berkeley National Laboratory, ALS, Berkeley, CA 94720-8235, USA
[6]Institute of Solid State Physics, University of Bremen, Otto-Hahn-Allee NW1, 28359 Bremen, Germany
E-mail: demch@ifpan.edu.pl

Abstract. The presented investigations aimed at development of inexpensive method for synthesized materials suitable for utilization of solar energy. This important issue was addressed by focusing, mainly, on electronic local structure studies with supporting x-ray diffraction (XRD) and transmission electron microscopy (TEM) analysis of colloidal galena nano-particles (NPs) and quantum dots (QDs) synthesized using wet chemistry under microwave irradiation. Performed x-ray absorption near edge structure (XANES) analysis revealed an evidence of quantum confinement for the sample with QDs, where the bottom of the conduction band was shifted to higher energy. The QDs were found to be passivated with oxides at the surface. Existence of sulfate/sulfite and thiosulfate species in pure PbS and QDs, respectively, was identified.

1. Introduction

Development of inexpensive methods for synthesized materials that demonstrate both visible light activity and high solar energy utilization along with stability is still a challenge. Here we propose to explore structures based on lead sulfide (PbS) as one of the promising materials for photovoltaic (PV) applications considering its tuneable broad band absorption in the quantum regime, equal confinement of electrons and holes and a long lifetime of the excited states. Such low-dimensional semiconductor structures have recently provoked an increasing interest due to their peculiar optical and electronic properties as compared to bulk materials [1]. The size of QDs limits the electron energies available within them due to quantum mechanics considerations. The latter, in turn, define the position of energy bands and, consequently, material bandgaps and optical properties depending on QDs shape and composition [2]. One of the advantages of QDs is an ability to grow them easily up to a desired size. This means they can be tuned across a wide variety of bandgaps *without* changing an underlying material or synthesizing techniques. Such easy bandgap tuning makes them desirable for solar cell use. In this respect QDs are similar to the existing, but expensive, GaAs tandem cells having the calculated efficiency of the same order. The colloidal QDs, particularly lead sulfide, are expected to further improve this. As opposed to CdS and ZnS, the bandgap width of PbS is very sensitive to finite size effects manifesting the incredibly high rise of optical absorption energy of PbS QDs versus bulk PbS. Variation of the size of PbS QDs could provide the optical bandgap, E_g, to be increased up to 2.8

* To whom any correspondence should be addressed.

against the "bulk" value of ~0.4 eV at room temperature [3]. That means the mentioned QDs have bandgaps which can be tuned into the far infrared (IR) energy region that normally remains untapped via traditional materials. Let us note that half of the whole solar energy reaching the Earth is in the IR, mainly in the near IR, region. Embedded into QDs solar cell IR-sensitive materials are just as easy to use as any others, enabling of cost-effective capturing of much more energy. Sensitivity of lead sulfide QDs to the visible and near-IR radiation in 400-3000 nm wavelength range makes them very promising for PV applications. Their additional advantage is the ability to be sprayed onto flexible surfaces, including plastics. It enables the production of less expensive and more durable solar cells compared to the common silicon-based solar cells. This paper presents the preliminary research of electronic local structure of unoccupied states under QDs size modification with supporting investigations of XRD and TEM. To our knowledge this is the first x-ray absorption investigations on these materials, which reveal the detailed local electronic structure of QDs in comparison to the bulk one. That kind of in-depth knowledge of the properties of these materials would be invaluable for their prompt application in solar energy conversion devices.

2. Experimental

Investigated colloidal PbS NPs and QDs were grown by wet chemical and radiation method. Particular details of synthesis routine could be found in [4].

XANES measurements at the sulfur *K* edge were performed at room temperature on the 9.3.1 beam line [5] at the Advanced Light Source (Berkeley, USA). X-ray absorption spectra were gathered in the total fluorescence yield (TFY) and the total electron yield (TEY) detection modes with a total energy resolution of ~0.6 eV. Bulk sensitive TFY data were collected using a Hamamatsu Si photodiode while rather surface sensitive TEY - by measuring the sample drain current. The samples for XANES measurements were prepared by placing 2 ml of colloidal QDs solution on quartz substrates which were subsequently dried on a hot plate (at level of 100 °C) for 15 min.

3. Results and discussion

3.1. ***Experimental observations***

The results of synchrotron XRD measurements for colloidal PbS QDs confirmed their single-phase character with NaCl structure. The details of XRD and TEM observations and analysis are not presented here due to the article size restriction and could be found elsewhere [6].

According to the synthesis procedure described in [4], nano-particles of different sizes, ~50 nm (NPs) and ~5 nm (QDs), were grown as was manifested by TEM (data are not presented here). A narrow size distribution of QDs is observed. The high resolution (HR)TEM investigations confirmed the monocrystalline PbS QDs formation. In accordance with synchrotron XRD result, three main *fcc* diffraction patterns, i.e. 200, 111, and 220, were identified.

XANES measurements were devoted to directly evaluate the eventual modification of conductive band position/profile of bulk PbS versus PbS QDs. **Fig. 1** shows the bonding symmetry for the bulk and QDs of PbS along with $PbS(VI)O_4$, $Na_2S(IV)O_3$ standard samples near the sulfur *K* edge spectra recorded in TFY, TEY, and transmission (TR) detection modes. The transitions from an atomic-like S $1s$ state to the unoccupied S $3p$ σ^* antibonding orbitals are probed.

For bulk PbS target the self-absorption (SA) effect [7, 8] was visible in experimental data (intensity damping of the first resonance, see **Fig. 1 A)**). The differences in the spectral profiles for bulk PbS XANES spectra collected in TFY and TEY detection modes in the higher energy region, 2.476-2.486 keV, definitely indicate on the surface oxidation. From comparison of investigated here XANES spectra to $PbSO_4$ and Na_2SO_3 standards and the literature data [9], the feature at ~2.478 keV can be attributed to the absorption of sulfur in the form of sulfite ($S(IV)O_3$), whereas the second resonance at ~2.482 keV mainly in the form of sulfate ($S(VI)O_4$), **Fig. 1 A**.

A *blue shift,* visible for the absorption threshold (inflection point), is about 1.67 eV for PbS QDs with respect to the bulk one in consequence of decreasing the conductive band minimum *p*-S states contribution, antibonding orbital of t_{1u} symmetry, at about 2.470 keV (see the shoulder marked as 1), **Fig. 1 B**. Whereas the absorption onset positions for bulk and QDs PbS obtained under the technique

Journal of Physics: Conference Series **430** (2013) 012030 doi:10.1088/1742-6596/430/1/012030

mentioned in [10], so called extrapolation and arctangent subtraction data analysis methods, make possible to conclude that this shift is less than 1 eV.

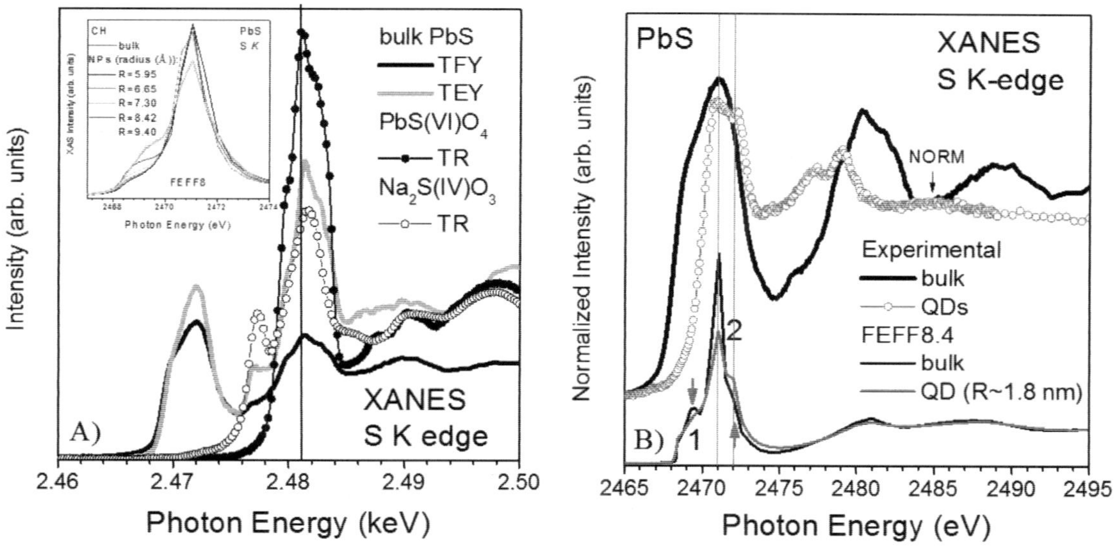

Fig. 1. A) S *K* edge TFY (black) and TEY (gray) XANES of bulk PbS compared to standard samples gathered in TR detection mode: $PbSO_4$ (solid circles line), Na_2SO_3 (open circles line). B) Experimental S *K* edge TFY XANES spectra of bulk PbS (thick black) and sample with QDs (open red circles) compared to calculated by FEFF8 PbS XANES spectra corresponding to bulk (thin black) and QD (thin red) models.

As it is well known, the energy gap of semiconductor separating the conduction and valence energy bands is its defining feature. In semiconductors of macroscopic sizes the gap width is a fixed parameter determined by the material's peculiarity. It is changed in the case of nano-scale semiconductor particles smaller than ~10 nm. This size range already corresponds to the regime of quantum confinement, when the spatial extent of the electronic wave function is comparable with the quantum dot size. As a result of such "geometrical" constraints, electrons "feel" the presence of the particle boundaries and respond to the particle size changes by adjusting their energy. This phenomenon is known as the quantum-size effect and it plays an important role in QDs. Here we assume that a *blue shift* of the absorption threshold (**Fig. 1 B**) occurs due to quantum confinement. Our suggestion is bolstered up by the fact that contrary to Si, Cu_2S, $CuInS_2$, $CuInGaSe_2$ and $CuZnInS_2$ having small Bohr exciton radii and generally preserving their bulk properties, PbS has large Bohr exciton radius, ~ 20 nm [11]. Since the QDs radius of PbS investigated here is much smaller than the Bohr radius of the exciton in the initial bulk material, optical and electronic properties of PbS QDs may exhibit their great enhancement allowing the wide tunability, even producing bandgaps that can reach the red edge of the visible light. It was shown that the optical absorption onset of PbS QDs can be tuned from 3000 nm to < 600 nm as a function of QD size [12]. Precise examination of the TEM data shown in [4] reveals the dimension of the smaller PbS QDs to be of order of 5 nm. If these QDs are the main contributor to the aggregate, they will account for the *blue shift* in the XANES spectrum of PbS QDs.

Other noticeable differences in XANES spectra were found comparing the above edge regions, around ~2.480 keV, **Fig. 1 B**. The sharp features around ~2.480 keV in colloidal PbS QDs can presumably be attributed to the absorption of sulfur in the form of thiosulfate [8] versus sulfate/sulfite revealed at the surface of bulk PbS.

As observed there are no differences in TFY and TEY XANES of QDs spectra profiles, except a small discrepancy in the resonances intensity (data are not presented here). As particles become smaller, a bigger fraction of the atoms is located on the surface, resulting in the increased surface sensitivity (we are considering here the influence of "surface atoms" on the final XANES spectral profile) often visible in TEY measurements compared to TFY. As it is shown the investigated QDs have the average size of about 5 nm. For such small objects the contribution of surface component from each QD is hardly distinguishable within the differences between TEY and TFY spectra, since even for the soft X-rays the probing depths for TEY and TFY are at about 5 and 100 nm, respectively [13]. These values are larger for semi-hard X-rays. It makes clear why TFY and TEY XANES spectra for the sample with QDs provide the similar profiles.

Finally, the XRD analysis of PbS QDs does not expose the existence of other, than cubic, phases of PbS with the rock-salt structure [6], whereas all standards, discussed under XANES qualitative analysis, have low-symmetry crystallographic structure, like orthorhombic or monoclinic, which was not detected by the XRD. It is likely that the surface of PbS QDs is oxidized, forming a thin structurally disordered in macro-scale, thus invisible to XRD, passivated layer of SO_x. We cannot exclude the possibility that the thiosulfate comes from the oxidation of unreacted S^{2-} in the system. Qualitative analysis of PbS QDs and thiosulfate [8] XANES spectra allows us to preliminarily conclude, that the surface sulfur atoms in the state of thiosulfate have a minor contribution into the spectrum profile close to conduction band minimum. To conclusively determine a measure of the surface oxidation and to estimate precisely its eventual influence on the observed blue shift in comparison to the quantum-size effect further photoelectron spectroscopy and high resolution TEM measurements with the following advanced analysis are necessary and are currently ongoing. These results will be published elsewhere.

3.2. ***Size effect. FEFF8***

The calculations of PbS cluster of increasing size were done by FEFF8 by adding the successive coordination shells to the central S atom until radius (R) of 9.4 Å. It was found for a cluster of increasing size that the S *p* unoccupied states form a shoulder at the absorption edge enhancing its intensity (inset in **Fig. 1 A**) until cluster radius 9.4 Å, where no further modification of near-structure profile was found. Next, the calculations for cluster of fixed diameter of 3.6 nm were repeated taking into account a different number of inequivalent sites of S close to the surface region, defined as sites with different coordination numbers. Such an approach was previously described in details in [14]. It concerned the relation between the total number of atoms in full shell cluster and the percentage of surface atoms. Our result is presented in **Fig. 1 B** at the bottom. A significant effect of decreasing the shoulder intensity (marked as 1) and increasing the states marked as 2 was observed for the sample with QDs compared to unoccupied states of bulk PbS. Similar dependence is clearly visible in the experimental XANES spectra, establishing, thus, that electronic structure of unoccupied states close to the absorption threshold is very sensitive to size effect and surface atoms contribution in the final XANES profile.

4. Conclusions

Presented here preliminary XANES studies seem to be promising tool for the determination of the local structure of our samples. In case of PbS QDs with size of about 5 nm, the sulfur *p* states at the bottom of the conduction band are blue shifted as compared to bulk PbS. This observation suggests a widening of the energy gap caused by the quantum confinement effect. The QDs were found to be passivated with layer of SO_x at the surface. Existence of sulfate/sulfite and thiosulfate species in pure PbS and QDs, respectively, was identified. To decide on the surface oxidation further photoelectron spectroscopy and high resolution TEM measurements are necessary.

Acknowledgements

The synthesis work was partially supported by the Texas Engineering Experiments Station and the Texas A&M University. X-ray absorption measurements were performed at the Advanced Light Source, which is supported by DOE (DE-AC03-76SF00098).

References

[1] Tang J Kemp KW Hoogland S Jeong KS Liu H Levina L Furukawa M Wang X Debnath R Cha D Chou KW Fischer A Amassian A Asbury JB Sargent EH 2011 *Nature Materials* **10** 765
[2] Demchenko IN Lawniczak-Jablonska K Kret S Novikov AV Laval J-Y Zak M Szczepanska A Yablonskiy AN Krasilnik ZF 2007 *Nanotechnology* **18** 115711
[3] Hoffmann E Entel P 1999 From solids to clusters World Scientific
[4] He X Demchenko IN Stolte W C van Buuren A Liang H 2012 *J. Phys. Chem. C* DOI: 10.1021/jp304728u
[5] Ng W Jones G Perera RCC Hansen D Daniels J Hemmers O Glans P Whitfield S Wang H Lindle DW 1996 *Rev. Sci. Instrum.* **67** 3374
[6] Demchenko IN Chernyshova M He X Minikayev R Syryanyy Y Derkachova A Derkachov G Stolte WC Liang H 2012 *X-Ray Spectrometry* submitted
[7] Pfalzer P Urbach JP Klemm M Horn S denBoer ML, Frenkel A I Kirkland JP 1999 *Phys. Rev. B* **60** 9335
[8] Tröger L Arvanitis D Baberschke K Michaelis H Grimm U Zschech E1992 *Phys. Rev. B* **46** 3283
[9] Fleet ME 2005 *The Canad. Mineral.* **43** 1811
[10] Lee JR Meulenberg RW Hanif KM Mattoussi H Klepeis J E Terminello LJ van Buuren T 2007 *Phys. Rev. Let.* **98** 146803
[11] Wise FW 2000 *Acc. Chem. Res.* **33** 773
[12] Hines MA Scholes GD 2003 *Adv. Mater.* **15** 1844
[13] Troger L Arvanitis D Rabus H Wenzel L Baberschke K 1990 *Phys. Rev. B* **41** 7297
[14] Bazin D Sayers D Rehr JJ Mottet C 1997 *J. Phys. Chem. B* **101** 5332

Thermal effects on Rhodium nanoparticles supported on carbon

G Agostini[1], R Grisenti[2], C Lamberti[1], A Piovano[3] and P Fornasini[2]

[1] Dipartimento di Chimica, Università di Torino, I-10125 Torino, Italy

[2] Dipartimento di Fisica, Università di Trento, Via Sommarive 14, I-38123 Povo (Trento), Italy

[3] Institut Laue Langevin, F-38042 Grenoble, France

E-mail: giovanni.agostini@unito.it

Abstract. EXAFS measurements have been made in the temperature range 5 - 300 K on rhodium nano-clusters of average diameters 15.9 and 11.5 Å (rms dispersion 7.2 and 4.7 Å, respectively) supported on carbon, as well as on a Rh reference foil. The preliminary results of the first shell analysis are presented. The Debye temperature is slightly smaller in n-Rh with respect to bulk and decreases when the cluster size decreases. The results of amplitude analysis (coordination number and static DW) are sensitive to the inclusion of the 4th cumulant. In going from bulk Rh to n-Rh and decreasing the nanocluster size the average coordination number decreases and the static disorder increases. A contraction of the average nearest-neighbour distance is observed at 5 K, -0.004 Å and -0.009 Å for the larger and smaller clusters, respectively, accompanied by a very slight thermal expansion.

1. Introduction

In the last years, a big effort has been made to identify the correlation between the structure of metal nano-particles and their catalytic activity [1, 2]; the relevance not only of the ratio between surface and bulk atoms but also of the type of surface sites and of the interaction with the support has been evidenced. Standard laboratory techniques are generally insufficient to lead to a complete characterisation of nano-particles and EXAFS can give relevant complementary information on their structural and vibrational properties. Many EXAFS studies have been performed on supported and un-supported metallic clusters, but no definitive general assessments can still be made, in view of the large variability of production methods, sizes and size distributions, shapes, as well as interactions with the support and with the environment [3].

The lack of definitive conclusions make sensible to perform EXAFS measurements on a relatively new class of systems, rhodium clusters, which are actually employed for hydrogenation reactions [4, 5]. EXAFS measurements have been performed as a function of temperature (5 to 300 K) on two n-Rh samples supported on carbon, with different average diameters, determined by HAADF-STEM: 14.9 Å (rms dispersion 7.2 Å) and 10.7 Å (rms dispersion 4.7 Å), hereafter samples Rh48 and Rh72, respectively. A 12 μm reference metal foil (b-Rh) was measured too. The preliminary results of the first-shell analysis are here presented.

2. Experiment and data analysis

The n-Rh samples have been prepared following the deposition-precipitation method on activated carbon from wood with 5 weigth % Rh loading and characterized by laboratory techniques such

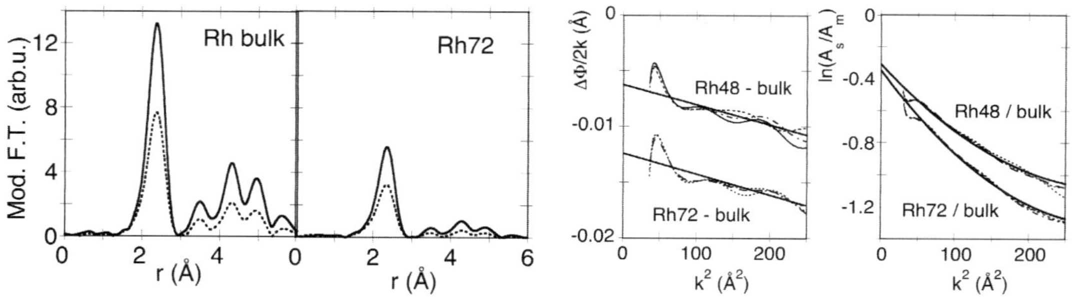

Figure 1. Left: Fourier Transforms of b-Rh and Rh72 at 5 K (continuous lines) and 300 K (dotted lines). Right: Phase differences $\Delta\Phi/2k$ and logarithm of amplitude ratios of Rh48 and Rh72 spectra measured at 5 K with respect to b-Rh at 5 K; continuous lines are the best fits.

as HAADF-STEM and chemisorption. EXAFS measurements at the Rh K edge have been performed in standard transmission configuration at the BM23 beamline of ESRF. Electron energy and average current were 6 GeV and 190 mA, respectively. The X-ray beam was monochromatized by two silicon crystals with flat (311) parallel reflecting faces; the harmonics influence was reduced by reflection from two Pt-coated mirrors. The photon fluxes were measured by ionization chambers filled with krypton gas. A Pd-Rh alloy foil maintained at room temperature and placed among the second and the third ionisation chambers gave the reference for the edge position. The samples were measured in vacuum. The temperature was varied from 5 to 300 K in a liquid helium cryostat. Three spectra were measured at each temperature. No significant edge shift was detected between b-Rh and n-Rh samples, suggesting that the interaction of n-Rh with the carbon support is negligible.

The EXAFS signals $\chi(k)$ were extracted according to a well established procedure [6]. The Fourier Transform moduli for b-Rh and for Rh72 at 5 and 300 K are shown in Fig. 1, left. The first-shell quantitative analysis has been performed by the ratio method [7, 8], taking the 5 K average spectrum of b-Rh as reference for the separate analysis of phases and amplitudes. The right panels of Fig. 1 show the phase differences $(\Phi_s - \Phi_m)/k$ and and the logarithms of amplitude ratios $\ln(A_s/A_m)$ for all the spectra (s) of Rh48 and Rh72 measured at 5 K compared with b-Rh (m) at the same temperature. Phase differences are fitted by a straight line whose intercept and slope give the difference of the first and third cumulants, respectively, of the effective distance distributions of nano-clusters with respect to bulk. Logarithms of amplitude ratios are best-fitted by a polynomial in k^2, whose intercept is $\ln(N_s/N_m)$ and the linear and quadratic terms give the difference of the second and fourth cumulants, respectively, of nano-clusters with respect to bulk. The results from the ratio method have been confirmed by non-linear best fit of theoretical spectra generated by the FEFF6 code [9, 10].

3. First-shell results

3.1. Thermal and static disorder, coordination number

To quantify the effects of thermal disorder, a Debye correlated model has been fitted to the temperature dependence of the second cumulant of the three samples. For b-Rh, a Debye temperature θ_D^{b}=359.3±0.7 was found. The results for nanoclusters are summarised in Table 1. The Debye temperature decreases in going from b-Rh to Rh48 to Rh72.

The comparison between n-Rh samples and b-Rh allowed us to evaluate the ratio of coordination numbers and the static contribution to the width of the distribution of distances, σ^2_{st} (Table 1). The average coordination number decreases when going from b-Rh to Rh48 to Rh72. The contribution of static disorder is larger for Rh72 than for Rh48 (the structural disorder

Journal of Physics: Conference Series **430** (2013) 012031 doi:10.1088/1742-6596/430/1/012031

Table 1. First-shell Debye temperatures, coordination numbers, static disorder and bond-length contraction at 5 K. The apices "n" and "b" stay for nano, and bulk, respectively.

		Without 4th cumulant			With 4th cumulant		
	θ_D^n/θ_D^b	N^n/N^b	σ^2_{stat} (Å^2)	ΔC_1^* (Å)	N^n/N^b	σ^2_{stat} (Å^2)	ΔC_1^* (Å)
bulk	1	1	0	0	1	0	0
Rh48	0.959±0.013	0.691	0.00172	-0.0046	0.735	0.00241	-0.0039
Rh72	0.955±0.008	0.649	0.00207	-0.0104	0.707	0.00302	-0.0095

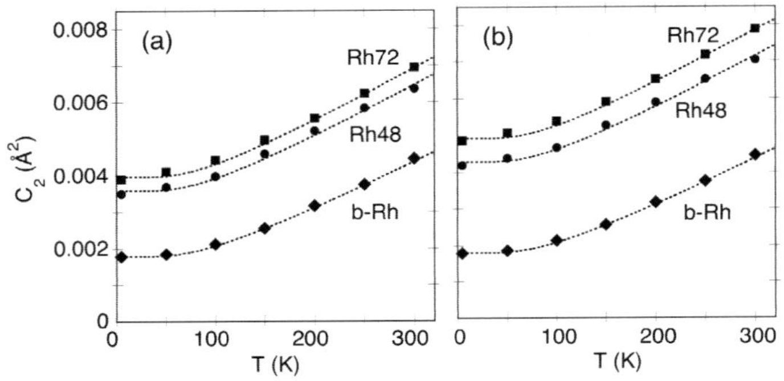

Figure 2. Second cumulant of the first coordination shell. The static contribution is determined by comparing the amplitudes of the EXAFS signals of nano-Rh and bulk Rh taking into account only the difference of the second cumulant (a) or also the difference of the 4th cumulant (b).

in bulk Rh was considered negligible). Both coordination numbers and static contributions to disordered have been found independent of temperature. The absolute values of the second cumulants of the three samples (static plus thermal disorder) are shown in Fig. 2.

It is noteworthy that the inclusion, in the amplitude analysis, of the fourth cumulant, which measures the symmetric deviations of the distribution of distances from a normal shape (Fig. 1, right) gives rise to significantly higher coordination numbers and static disorder contributions (Table 1 and Fig. 2) with respect to a standard analysis including only the second cumulant.

3.2. Nearest neighbour distance and thermal expansion

The relation between the first cumulant C_1 of the effective distribution, directly determined by the ratio method, and the first cumulant C_1^* of the real distribution depends on the width of the distribution, measured by the second cumulant, as well as on the photo-electron mean free path. As previously observed, different values of the second cumulant were obtained depending on the exclusion or inclusion of the fourth cumulant in the data analysis. As a consequence, the differences between the average distance in n-Rh samples and in b-Rh are different for the two procedures of data analysis, including or not the fourth cumulant (Table 1 and Fig. 3).

The comparison of the average nearest-neighbour distances $\langle r \rangle = C_1^*$ of the three samples is made in Fig. 3, where the nearest-neighbour distance in bulk Rh at 5 K has been taken as reference. For b-Rh, no static disorder is assumed and the average $\langle r \rangle$ is assumed to be only over the thermal configurations. For the n-Rh samples, the average is over the convolution of both thermal and structural configurations. There is an evident contraction of the average nearest-neighbour distance in going from bulk to Rh48 to Rh72. The thermal expansion is slightly weaker for n-Rh than for bulk Rh.

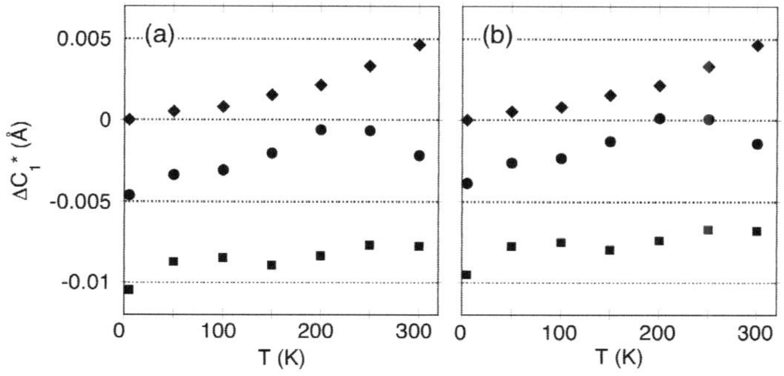

Figure 3. Average nearest-neighbours distances in bulk Rh (diamonds), in Rh48 (circles) and Rh72 (squares) relative to the distance in bulk Rh at 5 K. The real distances were calculated from the effective distances (as obtained by the ratio method) assuming a mean free path $\lambda = 9$ Å and a distribution width evaluated taking into account only the second cumulant (a) or also the 4th cumulant (b).

4. Conclusions

The absence of edge shift suggests that n-Rh clusters have negligible interaction with the carbon support. The Debye temperature has been found slightly smaller in n-Rh with respect to b-Rh and decreasing when the cluster size decreases. The results of amplitude analysis (coordination number and static DW) are non-neglibly sensitive to the inclusion of the 4th cumulant. In going from b-Rh to n-Rh and decreasing the nanocluster size, the average coordination number decreases and the static disorder increases. A contraction of the average NN distance is observed at 5K, -0.004 Å (Rh48) and -0.009 Å (Rh72), accompanied by a very slight thermal expansion.

Further work is in progress *a)* to extend the data analysis to the outer shells, *b)* to take into account the effects of the dependence of the photoelectron mean free path on the cluster size, *c)* to check the reliability of the cumulant approach to account for structural disorder in nano-particles, and, in case, to understand the physical meaning of the fourth cumulant.

Acknowledgments

We are grateful to Chimet SpA, Arezzo (Italy), for sample prepation and to G. Salviati and G. Bertoni of CNR-IMEM for assistance in HAADF-STEM measurements. We acknowledge ESRF for provision of synchrotron radiation facilities under project HS-3309 and are grateful to S. Pascarelli, O. Mathon for scientific support and technical assistance.

References

[1] Bell A T 2003 *Science* **299** 1688–1691
[2] van der Eerden A M J, Visser T, Nijhuis T A, Ikeda Y, Lepage M, Koningsberger D C and Weckhuysen B M 2005 *J. Am. Chem. Soc.* **127** 3272–3273
[3] Roldan Cuenya B, Alcántara Ortigoza M, Ono L K, Behafarid F, Mostafa S, Croy J R, Paredis K, Shafai G, Rahman T S, Li L, Zhang Z and Yang J C 2011 *Phys. Rev. B* **84**(24) 245438
[4] Ros T G, Keller D E, van Dillen A J, Geus J W and Koningsberger D C 2002 *J. Catal.* **211** 85102
[5] Tauchert M E, Incarvito C D, Rheingold A L, Bergman R G and Ellman J A 2012 *J. Am. Chem. Soc.* **134** 1482–1485
[6] Vaccari M, Grisenti R, Fornasini P, Rocca F and Sanson A 2007 *Phys. Rev. B* **75** 184307
[7] Bunker G 1983 *Nucl. Instrum. Methods Phys. Res.* **207** 437–444
[8] Dalba G, Fornasini P and Rocca F 1993 *Phys. Rev. B* **47** 8502–8514
[9] Rehr J J, Zabinsky S I and Albers R C 1992 *Phys. Rev. Lett.* **69** 3397
[10] Ravel B and Newville M 2005 *J. Synchrotron Rad.* **12** 537–541

Journal of Physics: Conference Series **430** (2013) 012032 doi:10.1088/1742-6596/430/1/012032

XAFS studies of Au nanocrystals passivated by different surfactants

Yuanyuan Li, Hao Cheng, Lina Yang, Zhihu Sun*, and Shiqiang Wei*

National Synchrotron Radiation Laboratory, University of Science and Technology of China, Hefei, Anhui 230029, P.R. China

*Corresponding author. E-mail: zhsun@ustc.edu.cn, sqwei@ustc.edu.cn

Abstract. The electronic and atomic structures of Au nanocrystals affected by different surfactants (PPh_3, PVP, and dodecanethiol) capping were studied by X-ray absorption fine structure (XAFS) at Au L3-edge. For PPh_3, PVP, and dodecanethiol-capped Au nanocrystals, the surface Au atoms interact with P, O, and S atoms respectively; however, only for dodecanethiol-capped Au nanocrystals, a significant electron transfer between Au and S atoms is observed. The extend-XAFS results reveal that the atomic structure disorder of Au-Au shell gradually increases from 0.0095 to 0.0152 $Å^2$ when the surfactants goes from PPh_3 to dodecanethiol. These results suggest that the surfactants would change the electronic and atomic structures of Au nanocrystals rather than merely passivates on the surface of Au nanocrystals.

1. Introduction

Gold nanocrystals possess unique electronic, optic and surface properties, and have been widely used in fields such as catalysis, electronics, optics, biology and medicine [1-4]. For nanomaterials, surfactants are very important because they not only make the nanocrystals stabilized but also link nanocrystals with the outsides, enabling the particles to be applied in different environment. For example, in the field of biology and medicine, gold nanocrystals are usually covered with a modified thiolate ligands, the end of which are initially connected with nucleotides, proteins or antibodies [5,6]. Gold nanocrystals show high catalytic activities by covering with well-defined polymers known as dendrimers which does not fully passivate particles so that substrates can access the particle surface [7]. Besides, surfactants have great influence on the electronic and atomic structures of nanocrystals which can be used to modify the chemical and physical properties of gold nanocrystals [8,9]. Therefore, it is important to study the surfactant-effects on gold nanocrystals for fundamental research and for scientific applications.

Because different surfactant-effects on gold nanocrystals result in different properties of gold nanocrystals, understanding the interaction between surfactants and particles is essential and has been widely studied for decades. Since the interaction between surfactants and particles can directly cause the variation of electronic and atomic structures of gold nanocrystals, investigation on the electronic and atomic structures of gold nanocrystals passivated by different surfactants can give some hints to the surfactant-nanoparticle interaction.

In this work, x-ray absorption spectroscopy (XAFS) was employed to study the local atomic and electronic structures of gold nanocrystals covering with different surfactants (riphenylphosphine

Journal of Physics: Conference Series **430** (2013) 012032 doi:10.1088/1742-6596/430/1/012032

(PPh_3), polyvinylpyrrolidone (PVP), and dodecanethiol ($C_{12}H_{26}S$)). The samples were prepared by wet chemical method. The aim is to determine the effects of surfactants on the structures of gold nanocrystals.

2. Experimental

PPh_3 covered gold nanocrystals [10]: 0.050g of $AuClPPh_3$ and 0.088g of tert-butylamine-borane powder were mixed in 40 ml of benzene to form a clear solution. The mixture was kept stirring at room temperature for one day to complete the reaction.

PVP covered gold nanocrystals [11]: To the aqueous solution of $HAuCl_4$ (1mM, 50ml) was added 555.5mg (0.0139mmol) PVP. The mixture was further stirred for 30 minutes under a bath of 0℃. Then, the aqueous solution of $NaBH_4$ (0.1M, 5ml) was rapidly added into the mixture under vigorous stirring. The mixture was kept stirring at room temperature for one day to complete the reaction.

Dodecanethiol capped gold nanocrystals: 0.050g of $AuClPPh_3$ and 0.088g of tert-butylamine-borane powder were mixed in 40 ml of benzene to form a clear solution. Then 0.4mmol dodecanethiol was added into the solution. The mixture was kept stirring at room temperature for one day to complete the reaction.

XAFS measurement: The Au L_3-edge XAFS spectra were measured at the U7C beamline of the National Synchrotron Radiation Laboratory (NSRL), China. For XAFS measurements, the solutions were transferred to a 10-mm-thick Teflon cell to obtain optimized edge-jumps.

3. Results and discussion

The TEM images of Au nanocrystals covered by different kinds of surfactants are shown in fig.1. It can be observed that PPh_3-Au nanocrystals tend to be polydisperse, however it turns much better when Au nanocrystals covered with PVP. For dodecanethiol-Au nanocrystals, the particles are evenly spaced and orderly arranged. The insets in fig.1 display the average sizes and corresponding size distributions of Au nanocrystals. By changing the passivated surfactants to PPh_3, PVP, and dodecanethiol, the sizes of Au nanocrystals gradually vary to 7.2±0.4, 3.5±0.4, and 3.1±0.1 nm, respectively. From the above, it can be suggested that different surfactants have different effects on the size of Au particles and among these three surfactants, dodecanethiol shows the strongest ability in narrowing the size distribution of gold nanocrystals.

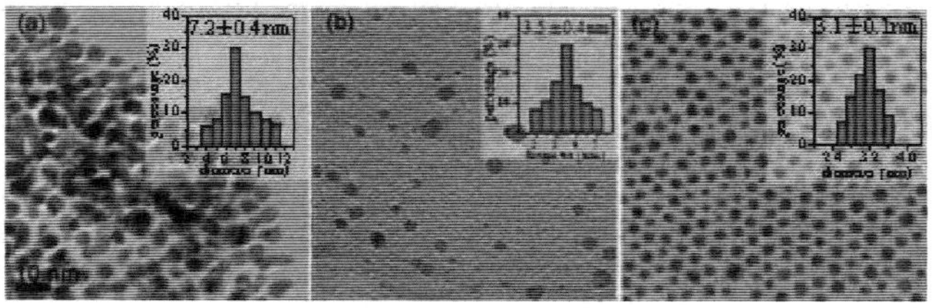

Figure 1. TEM images of Au nanocrystals covered by (a) PPh_3, (b) PVP, (c) dodecanethiol; the insets display corresponding size distributions

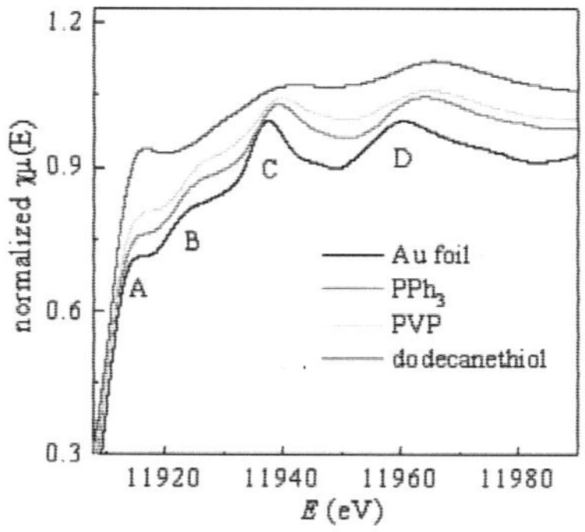

Figure 2. The normalized XANES spectra of Au nanocrystals capped by PPh_3, PVP and dodecanethiol. The spectrum of Au foil is displayed for comparison

Surfactant-induced electronic and atomic structure changes of Au nanocrystals could be obtained by means of x-ray absorption fine structure (XAFS, including XANES and EXAFS). The normalized Au L_3 edge XANES spectra of Au nanocrystals capped by PPh_3, PVP, and dodecanethiol are shown in fig.2. The XANES spectrum of Au foil is also displayed for comparison. The XANES spectra exhibit four characteristic peaks A-D locating at around 11915, 11924, 11937 and 11960eV, respectively. When covered with dodecanethiol, Au nanocrystals exhibit much more intense white-line peak A (corresponding to the electronic transition from the 2p3/2 to the 5d5/2, 3/2 states), indicating an increase in d-hole population just above the Fermi level, in good agreement with electro-negativity considerations (S is slightly more electronegative than Au) [12]. As compared with Au foil, peak B, C and D of Au nanocrystals are shifted to higher energy and reduced in intensity. The peak B that is apparent for Au foil is absent for Au nanocrystals capped by dodecanethiol. The weakness of the intensity could be attributed to the increase of the structure disorder of Au nanocrystals.

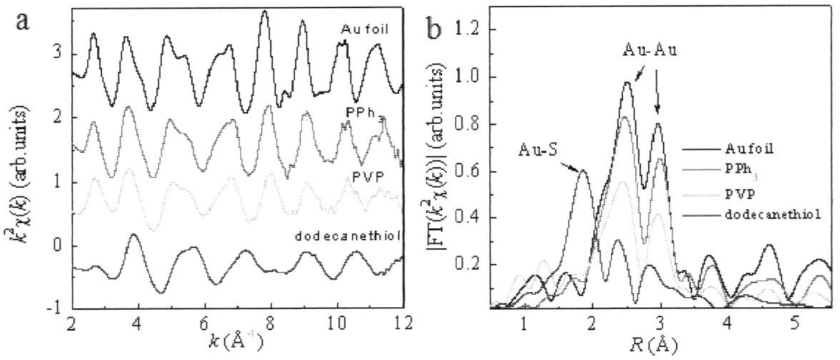

Figure 3. (a) $k^2\chi(k)$ and (b) Fourier transformed $k^2\chi(k)$ spectra of gold nanocrystals passivated by different surfactants.

Table 1. The fitting parameters and corresponding results

Sample	Bond	R (Å)	N	$\sigma^2(10^{-3}$Å)	ΔE (eV)
Au-foil	Au-Au	2.88±0.01	12	8.4±0.4	3.6±0.4
PPh_3	Au-Au	2.82±0.01	10.4±0.4	9.5±0.4	4.1±0.4
PVP	Au-Au	2.81±0.01	10.2±0.4	12.9±0.6	4.1±1.0
dodecanethiol	Au-S	2.30±0.03	1.2±0.2	5.0±0.6	2.7±0.5
	Au-Au	2.79±0.02	10.5±0.2	15.2±1.8	0.5±0.3

The $k^2\chi(k)$ and Fourier transformed (FT) $k^2\chi(k)$ spectra of gold nanocrystals passivated by different surfactants are displayed in fig.3 (a) and (b). The spectra of Au foil are plotted for comparison. As shown in fig. 3 (a), the shape of the $k^2\chi(k)$ spectra of samples capped with PPh_3 and PVP are similar with that of Au foil. When Au nanocrystals are covered with dodecanethiol, the profile of the oscillation curve is quite different with that of Au foil especially in the range of 6-12 Å^{-1}. The local atomic structure of Au nanocrystals covered with different surfactants can be more explicitly manifested by Fourier transformed $k^2\chi(k)$ spectra. As shown in fig. 3 (b), the peaks in the range of 2.5-3 Å are the nearest Au-Au coordinations. When covered with PPh_3 and PVP molecules, the FT spectra of Au nanocrystals keep basic features of Au foil except for the reduced intensity. However, when covered with dodecanethiol, Au-S coordination appears in the low R-region and the intensities of Au-Au coordination are sharply decreased. For all samples, Au-Au coordinations move towards low R range comparing with that of Au foil. Quantitative least-squares curve-fitting (Tab.1) shows Au-Au bond length of thiol-capped particles is 2.79 Å while that of Au foil is 2.88 Å. From PPh_3, PVP, dodecanamine to dodecanethiol, structural disorder degree of Au-Au increases from 0.0095, 0.0129, to 0.0152 $Å^2$. These results indicate that surfactants are not merely passivates the surface of Au nanocrystals and by forming chemical bonds like Au-S, the atomic structure of gold nanocrystals could also be greatly distorted [13].

Based on the above results, it can be suggested that different surfactants has different effects on Au nanocrystals. PPh_3 and PVP weakly cover the particle surface, and they stabilize the particles by physical adsorption. However, dodecanethiol stabilizes the nanocrystals by forming Au-S bond which may result in great structure distortion of the particles.

4. Conclusions

In summary, we prepared gold nanocrystals covered by different surfactants with wet chemical methods. By using XANES and EXAFS, we compared their local electronic and atomic structures. The XANES results show that the electron-transfer occurs when gold nanocrystals is passivated by dodecanethiol ligands, which results in the change of the electronic structure of gold nanocrystals. And based on the EXAFS results, the passivated surfactants cause distortion of atomic structure of gold nanocrystals. For PPh_3, PVP, and dodecanethiol covered Au nanocrystals, the atomic structure disorder of Au-Au shell was gradually increased from 0.0095 to 0.0152 Å. And dodecanethiol interacted with Au nanocrystals by forming strong Au-S bond, which may contribute to the greatly distorted structure.

Acknowledgments

This work is supported by the National Natural Science Foundation of China (Grant Nos. 11175184, 11135008, 10979044, 11079032, 11079004 and 10905058). The authors would like to thank NSRL for the synchrotron radiation beamtime.

References

[1] Giljohann D A, Seferos D S, Daniel W L, Massich M D, Patel P C and Mirkin C A 2010 *Angew. Chem. Int. Ed.* **49** 3280
[2] Daniel M C and Astruc D 2003 *Chem. Rev.* 104 293
[3] Rosi N L and Mirkin C A 2005 *Chem. Rev.* 105 1547
[4] Talapin D V, Lee J S, Kovalenko M V and Shevchenko E V 2009 *Chem. Rev.* **110** 389
[5] Sperling R A, Rivera Gil P, Zhang F, Zanella M and Parak W J 2008 *Chem. Soc. Rev.* **37** 1896
[6] Wilson R 2008 *Chem. Soc. Rev.* **37** 2028
[7] Crooks R M, Zhao M, Sun L, Chechik V and Yeung L K 2000 *Acc. Chem. Res.* 34 181
[8] Zhao D, Timmons D J, Yuan D Q and Zhou H C 2011 *Acc. Chem. Res.* **44** 123
[9] Love J C, Estroff L A, Kriebel J K, Nuzzo R G and Whitesides G M 2005 *Chem. Rev.* **105** 1103
[10] Zheng N, Fan J and Stucky G D 2006 *J. Am. Chem. Soc.* **128** 6550
[11] Tsunoyama H, Sakurai H, Negishi Y and Tsukuda T. 2005 *J. Am. Chem. Soc.* **127** 9374
[12] Zhang P and Sham T K 2003 *Phys. Rev. Lett.* **90** 245502
[13] Garzón I L, Rovira C, Michaelian K, Beltrán M R, Ordejón P, Junquera J, Sánchez-Portal D, Artacho E and Soler J M 2000 *Phys. Rev. Lett.* **85** 5250

Journal of Physics: Conference Series **430** (2013) 012033 doi:10.1088/1742-6596/430/1/012033

X-ray absorption fine structure spectroscopy studies of thiol-capped copper nanoparticles

Y Y Huang, T Yao, Z H Sun, and S Q Wei*

National Synchrotron Radiation Laboratory, University of Science and Technology of China, Hefei 230029, People's Republic of China

*Corresponding author. E-mail: sqwei@ustc.edu.cn

Abstract. Monodisperse Copper nanoparticles with controllable sizes of 0.6 and 1.8nm were synthesized via modifying Brust-Schiffrin two-phase method. This route involves a slow reduction of $[TOA]_2[CuBr_4]$ precursor in toluene, with particles sizes rationally controlled by the S/Cu ratio. Size-dependent inter-atomic distance contraction in the thiol-capped copper nanoparticles was investigated by Extended X-ray Absorption Fine Structure (EXAFS). The analysis revealed a shortened Cu-Cu inter-atomic distance, indicating strong surface interaction, in accord with the larger effect expected for ultrasmall particals (<2nm).

1. Introduction

Synthesizing nanoparticles with desired sizes is important in exploring their either unusual or different properties from corresponding bulk materials [1]. Nanostructured copper particles have been found to possess fascinating properties in catalysis [2], optics [3-5] and sensors [6,7]. Thus, the synthesis of copper nanoparticles with controllable sizes and electronic properties is vital to both fundamental science and technical applications.

Although there are a number of approaches available to synthesize copper NP, few methods have been established to obtain stable and size controllable copper nanoparticles as they undergo rapid oxidation in air and aqueous media. In previous reports, copper nanoparticles have been synthesized with molecularly capped surface to keep them stable. Most of the strategies are performed by the manipulation of reaction temperature or concentration of precursor and reducing agent or other extreme condition for the synthesis of copper nanoparticles in two-phase liquid/liquid system or in a specific single-phase solvent in the presence of capping agents [8,9]. However, few reports provide even smaller (<2 nm) Cu NPs that behave more like a molecule. It is expectable that ultrasmall nanoparticles possess special characters such as strong surface interaction and lattice contraction [10].

This work describes an investigation of synthesizing thiol capped copper nanoparticles with controlled sizes. X-ray absorption fine structure (XAFS) was used to investigate the atomic structures. With the advantage of the sensitivity of the XAFS spectroscopy to the local structure around Cu atoms, combined with the theoretical calculations, we aim to reveal the local structure of the Cu atoms in ultrasmall nanoparticles(0.5-2 nm).

2. Experimental

The synthesis of DT-capped copper nanoparticles was achieved by the procedure modified Brust-Schiffrin two-phase method as illustrated schematically in figure 1 [11]. $CuCl_2 \cdot 2H_2O$ was dissolved in water in the presence of potassium bromide to convert Cu^{2+} to $CuBr_4^{2-}$. This solution was then mixed with a solution of TOABr in toluene under vigorous stirring to induce a phase transfer of the $CuBr_4^{2-}$ from the aqueous phase to the organic phase. After being stirred for 45 min, the aqueous solution was removed. The copper toluene solution was stirred under argon for 30 min purge to eliminate all oxygen from the system. Then, different amount of decanethiol was added. Next, a solution of sodium borohydride in water was added dropwise. The solution became a deep cloudy orange, indicating the formation of copper nanoparticles. The solution was stirred under argon for 5 h before further analysis. The Cu k-edge XANES and EXAFS measurements were conducted at U7C stations of the National Synchrotron Radiation Laboratory, China. High Resolution Transmission Electron Microscopy (HRTEM) measurements were carried out using a JEOL 2010 system. The calculations were performed with the Feff 8.2 code [12].

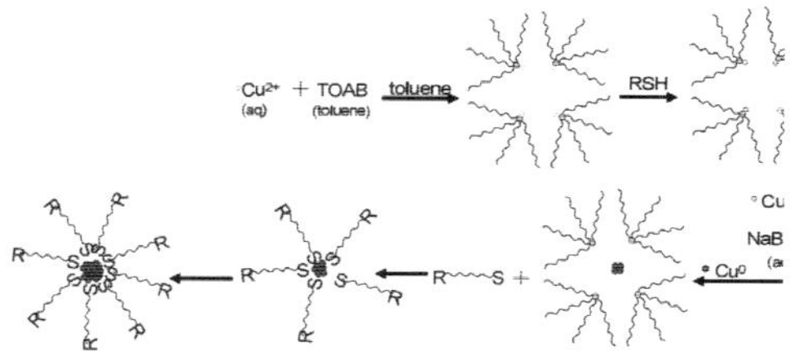

Figure 1. Schematic presentation of the Cu NPs formation process

3. Results and Discussion

The TEM images of the Cu nanoparticles with tailored sizes are shown in Figure 2(a)-(b), where the insets display their corresponding size distributions. Ultrasmall Cu nanoparticles of ~0.6 nm in size are obtained when S/Cu ratio is 5. By reducing the amount of dodecanethiol, the size of Cu nanoparticles increases to ~1.8 nm. As illustrated in Figure 1, before the addition of the reductant $NaBH_4$, the copper atom is in the state of $[TOA]_2[CuBr_4]$ with the absence of Cu-S bond [13]. Upon the addition of $NaBH_4$, the complex was reduced slowly, due to the stronger steric hindrance. Thus, the initial naked, micelle-encapsulated NP distribution is narrow. Then the thiolate generated from the reduction of thiol diffuses through the TOA shell and forms the Cu-S bonds, and the ligand-protected copper NPs were formed. The increase of the concentration of thiol is expected to increase the probability of collision between thiolate and the initial NPs. With more Cu-S bonds formed, further growth of first formed Cu core inside the micelles would be impeded, as indicated in Figure 2.

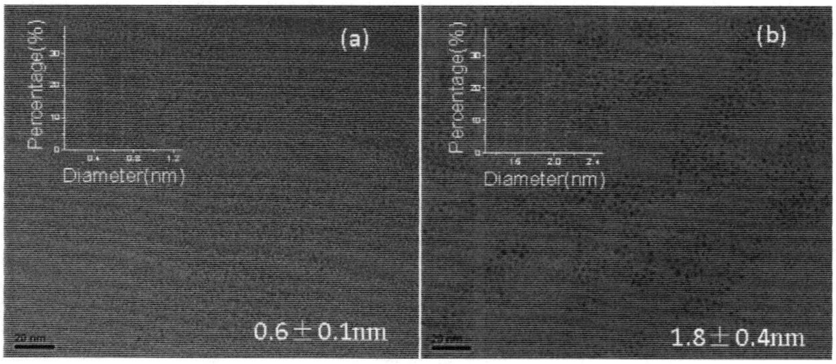

Figure 2. The TEM images (a)-(c) of copper nanoparticles with tailor sizes as 0.6 and 1.8 nm by adding dodecanethiol with varied S/Cu ratio (5 and 2.5) and corresponding size distributions. The scale bar is the same.

Figures 3(a) and (b) displays the Cu *K*-edge extended XAFS $k^3\chi(k)$ functions and their Fourier transform (FT) for as-prepared Cu NPs and Cu foil. We have also performed quantitative curve-fitting for the XAFS spectra, and the obtained structural parameters are summarized in Table 2. First of all, the oscillatory functions of prepared Cu nanoparticles shown in Fig. 3(a) are significantly different with that of Cu foil. Owing to the reduced particle size, an overall attenuation and broadening is observed over the whole XAS spectra. It is mainly due to the decrease of the mean coordination number and the absence of higher order coordination shells as indicated in Table 2. The presence of the ligand on the surface may also contribute to the oscillation damping at high *k*-values. Fig. 3(b) presents two peaks at about 1.2 and 1.9 Å corresponding to the Cu-S shell and Cu-Cu shell. With high S/Cu ratio, the smaller Cu NP show stronger oscillation in Cu-S shell, in accordance with the increase of the Cu-S coordination numbers. The obvious shift of the second peak for Cu NP compared to bulk Cu is also probably resulting from the size effect of lattice contraction, as significantly shortened Cu-Cu distance (R_{Cu-Cu}) is observed in Table 2. The Cu-Cu Debye-Waller factor (σ^2) indicates that the system disorder increases for smaller particles, as expected due to the larger surface/volume ratio. Generally, the prepared nanoparticles process obvious lattice contraction effect and strong surface interaction.

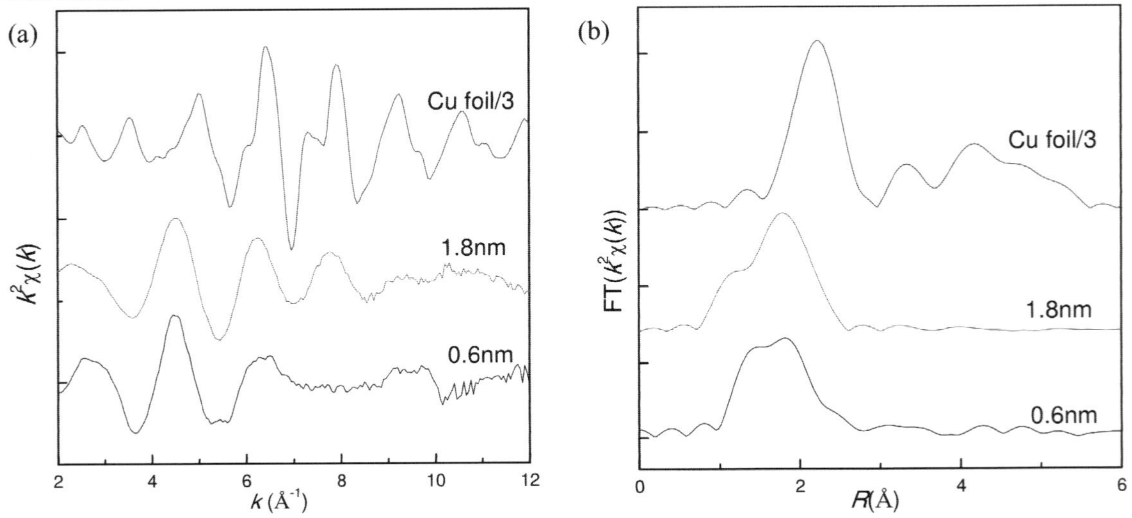

Figure 3. (a) Experimental EXAFS $k^2\chi(k)$ functions of prepared Cu NPs and Cu foil. (b) Experimental Fourier transforms (FT) curves of prepared Cu NPs and Cu foil.

Table 2. Structure parameters around Cu atoms for prepared Cu NPs and Cu foil.

Sample	Bond type	N	R(Å)	$\sigma^2(10^{-3}Å^2)$
Bulk	Cu-Cu	12	2.56	-
1.8 nm	Cu-Cu	5.2±0.2	2.25±0.01	11.7±0.1
	Cu-S	1.1±0.2	1.91±0.01	12.3±0.3
0.6 nm	Cu-Cu	4.9±0.3	2.24±0.01	13.2±0.2
	Cu-S	1.4±0.2	1.92±0.01	8.9±0.2

The XANES spectra of the two Cu nanoparticles samples reveal similar features to the bulk one in general (Figure 4). Nevertheless, their main oscillations show obvious attenuation and broadening, especially for the white line (arising from p→d dipole transition). The intensity of white line is very strong for elements with unoccupied d orbital. Although the d band is nominally fully filled in Cu atoms, a small white line can still be detected in the XANES of bulk Cu due to s-p-d hybridization

[14,15]. The size effect of small nanoparticles leads to an increase of d electron density. Meanwhile, in the Cu-SR system, the electronegativity of Cu- and SR- is different leading to an electrical potential, which induces the d-charge transfer from copper to sulfur atoms. Then the d-hole counts at the copper site of the nanoparticles would increase [16]. In despite of the strong surface interaction, there is still a significant decrease in the XANES spectra of the Cu K-edge. It is indicated a strong size effect in the samples, as reflected by the much shorter Cu-Cu distance.

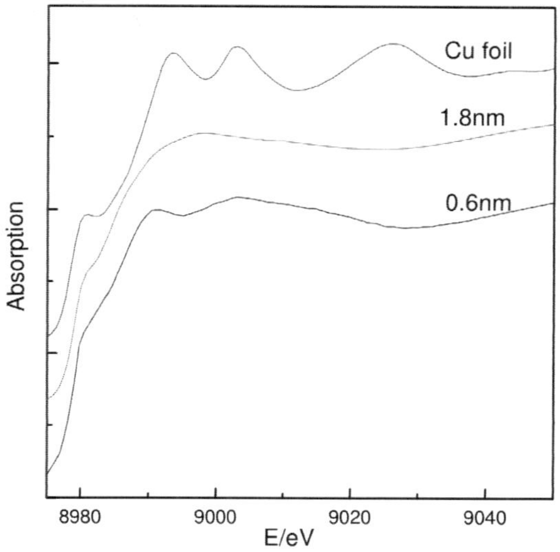

Figure 4. The XANES spectra of Cu k-edge for samples and foil.

4. Conclusions

Stable Cu NPs have been prepared with controllable size in the range of 0.6 and 1.8nm by varying the S/Cu molar ratio in the two-phase system, and the XAS spectra (XANES and EXAFS regions) have been analyzed in detail. The thiol-capped Cu NPs show rather strong surface interaction. Furthermore, we have demonstrated that copper nanoparticles smaller than 2.0 nm present a much strong size-dependent inter-atomic distance contraction. These results reinforce importance of even smaller size effect on nanoparticles, which should not be neglected for research of nanometer science.

Acknowledgments

This work was supported by the National Basic Research Program of China (No. 2009CB939901), the National Natural Science Foundation of China (11135008, 10725522, 11079004, 10605024, 11175184, 20701036, and 10734070). The authors would like to thank NSRL for the beam time.

References

[1] Alivisatos, A P 1996 *J. Phys. Chem.* 100 13226
[2] Kidwai M, Bansal A, Saxena S, Aerry and Mozumdar S 2006 *Tetrahedron Lett* 47 8049
[3] Chan G H, Zhao J, Hicks E M, Schatz G C and Van Duyne R P, 2007 Nano Lett. 7 1947
[4] Huang H H, Yan F Q, Kek Y M, Chew C H, Xu G Q, Ji W, Oh P S and Tang SH 1997 *Langmuir* 13 172
[5] Zhang Y X, Aslan K, Previte M J R and Geddes C D 2007 *Appl. Phys. Lett.* 90 173116
[6] Athanassiou E K, Grass R N and Stark W J 2006 *Nanotechnology* 17 1668
[7] Muraviev D N, Macanás J, Farre M, Muñoz M and Alegret S 2006 *Sens and Actuators B* 118 408
[8] Mott D, Galkowski J, Wang L Y, Luo J and Zhong C J 2007 *Langmuir* 23 5740
[9] Dong T Y, Wu H H and Lin M C 2006 *Langmuir* 22 6754
[10] Zhang P, Sham T K 2002 *Appl. Phys. Lett.* 81 736
[11] Brust M, Walker M, Bethell D, Schiffrin D J and Whyman R 1994 *J. Chem. Soc., Chem. Commun.* 801
[12] Ankudinov A L, Ravel B, Rehr J J, and Conradson S D 1998 *Phys. Rev. B* 58 7565
[13] Li Y and Zaluzhna O 2011 *J. Am. Chem. Soc.* 133 2092
[14] Lytle F, Bunsen-Ges B 1987 *Phys. Chem.* 91 1251
[15] Mattheiss L F, Dietz R E 1980 *Phys. Rev. B* 22 1663
[16] López-Cartes C, Rojas T C, Litrán R, Martínez-Martínez D, de la Fuente J M, Penadés S and Fernández A 2005 *J. Phys. Chem.* B 109 8761

Journal of Physics: Conference Series **430** (2013) 012034 doi:10.1088/1742-6596/430/1/012034

XANES Study of the Radiation Damage on Alkanethiolates-Capped Au Nanoparticles

J M Ramallo-López[1], L J Giovanetti[1], F C Vicentin[2], F G Requejo[1]

[1] Instituto de Investigaciones Fisicoquímicas Teóricas y Aplicadas (INIFTA, Facultad de Ciencias Exactas, Universidad Nacional de La Plata – CONICET), Suc. 4 CC 16, (1900) La Plata, Argentina

[2] Laboratório Nacional de Luz Síncrotron, CNPEM - Caixa Postal 6192, 13083-970 Campinas-SP, Brazil

E-mail: ramallo@inifta.unlp.edu.ar

Abstract. The radiation damage during XANES experiments on alkanethiols capped gold nanoparticles has been investigated. Different carbon length chains and nanoparticle sizes have been studied. Changes in the spectra after 45 minutes of irradiation, using a bend magnet, were observed for chains with more than 6 carbon atoms and are associated with the cleavage of S-C bonds and formation of atomic sulphur on the nanoparticle surface.

1. Introduction

Self-assembled monolayers (SAM) of *n*-alkanethiolates on Au(111) have been extensively studied [1]. However the nature of the metal-sulfur bond and the spatial arrangements of sulfur groups on the gold surface are still controversial. It is now accepted the formation of an ordered sulfur lattice based on a $(\sqrt{3}x\sqrt{3})R30°$ overlayer [2-6]. This structure corresponds to a molecule-molecule spacing of ~ 5 Å and an area per molecule of ~ 21.6 $Å^2$. In this full-coverage phase, which corresponds to the highest possible packing of the molecules, S:Au ratio is 3:1. In contrast, the structure of *n*-alkanethiolates on the surface of Au nanoparticles (NPs) is not so well understood, although the thiol-capping strategy is widely used to fabricate highly stable NPs [2]. In the last years many works have dedicated to understand the SAM of alkanethiols on the surface of gold NPs [8]. Hostetler et al. suggested from thermal gravimetric analysis (TGA) results that small Au NPs (<4.4 nm) have a higher density of alkanethiols per gold atom that an ideal, perfectly flat single crystal Au(111) surface [9]. Authors have attributed this higher coverage to the occupancy of alternate binding sites (edges and corners). We have found similar results and attributed this higher coverage to higher accessibility of S atoms due to the high curvature of the NPs [10]. Jadzinsky et al. have also found a higher sulfur coverage on thiol-capped gold NPs [11].

X-ray based techniques as X-ray Photoemission Spectroscopy (XPS) is nowadays a common technique used to the study of thiol-capped gold NPs and lots of works have been made to establish the radiation damage produced during experiments on thiols although discrepancies on which is the effect of irradiation on thiols and which are the sulfur species produced because of the irradiation still exist [8, 12]. On the other hand, X-ray absorption spectroscopies, like XANES, have been only used more recently to study the state of the sulfur on the surface of the NPs [13, 14] and because of this, very little information exist on how the X-ray beam affect the structure of the thiols. As X-ray absorption spectroscopies studies can be very time consuming, exposing the sample to high doses of X-rays, and changes on the state of the absorber can be produced during the experiments. This exposure has the consequence that the final spectrum represents an average over the different states of the absorbing atoms during the time of the experiment. In this work we study the radiation damage produced by X-rays coming from synchrotron light during XANES experiments on S-Au and S-C bonds in alkanethiolates-capped gold NPs.

2. Experimental

The gold NPs used in this study have average core diameters of 1.5 nm, and 2.2 nm, and are stabilized by butanethiol, hexanethiol and nonanethiol. The particle core size is related to the synthesis conditions [9] and it was already determined in all samples by HRTEM [15]. The synthesis of gold NPs with different core dimensions in toluene/*H2O* from alkanethiols was analogous to the Brust reaction [7] and was reported previously [10].

X-ray absorption fine structure measurements with soft and hard X-rays were obtained at the SXS and XAS beamlines respectively at the Laboratorio Nacional de Luz Sincrotron (LNLS), Campinas, Brazil. XANES spectra were obtained at the *S K*-edge (2472 eV) and EXAFS and XANES spectra were obtained at the *Au* L_3- edge (11919 eV). For XANES experiments at the *S K*-edge, all samples were carefully washed to evaporate all free thiols from the solution. NPs diluted in dichloromethane and the solvent was completely evaporated in vacuum in order to lower the boiling point of the solvent. This cycle was repeated until all free thiols were completely removed for each sample. In order to obtain the desired high resolution of about 0.5 eV a *Si* (111) double-crystal monochromator with a slit aperture of 1.0 mm was used. Details of the experimental setup of the SXS beam line have been published elsewhere [16]. The X-ray absorption spectra were recorded in total electron yield (TEY) mode, collecting the emitted current for each photon-energy with an electrometer connected to the sample. Experiments were performed in a vacuum of 10^{-8} mbar at room temperature. The energy scale was calibrated setting the *Mo* L_3-edge, defined by the first inflection point of the X-ray absorption spectrum of *Mo* metallic foil sample to 2520 eV. The final TEY XANES spectra were obtained after background subtraction and normalization to the post-edge intensity. XANES spectra were taken in quick scans to avoid as much as possible radiation damage during each scan. This resulted is statistically poor spectra, but important features were strong enough as to be evaluated and compared. In all cases two spectra were taken, being the second one taken after 45 minutes of constant irradiation. XANES spectra at the Au L_3 edge were taken in transmission mode.

3. Results

Figure 1 shows the XANES spectra at the sulfur K-edge of NPs with two different sizes capped with hexanethiols after washing in dichloromethane. Figure 2 shows the XANES spectra for a set of three NPs with the same size (1.5 nm) and different thiol lengths: (a) butanethiol, (b) hexanethiol and (c) nonanethiol. Two spectra are shown for each sample: the first one obtained after putting the samples in the chamber and a final one taken 45 minutes after first acquisition with continuous exposure of the sample to the X-ray beam.

The two peaks at 2470 eV and 2478 eV present in the spectra of all samples corresponds to the S-Au and S-C bonds as has been shown previously [13, 14]. Changes in the relative intensity of the two peaks between samples with different size (Figure 1) are evident and have been already discussed [10]. We interpreted the different intensity of the peaks as an effect of molecular distortions that change the orientation of the final state orbital, particularly those deriving from S-C bonds. We proposed that the effect of distortions is a more efficient space filling of the more open ligand shells of small particles and are driven by optimization of the van der Waals energy. A second effect that would be modifying these peaks intensities is the existence of a sulfur-enrichment at the NPs surface as the nanoparticle is smaller. Differences between the intensities of the peaks for the samples with the same size and different capping lengths are not so clear. Now, apart from these sources of changes in the NPs spectra, differences in the relative intensities of the two peaks are observed for almost all samples after 45 minutes of exposure to the X-rays. These changes are evidencing modifications in the nature of the S-C and S-Au bond and thus changes in the sulfur state during the experiments which are produced by radiation damage of the thiols due to the X-ray beam.

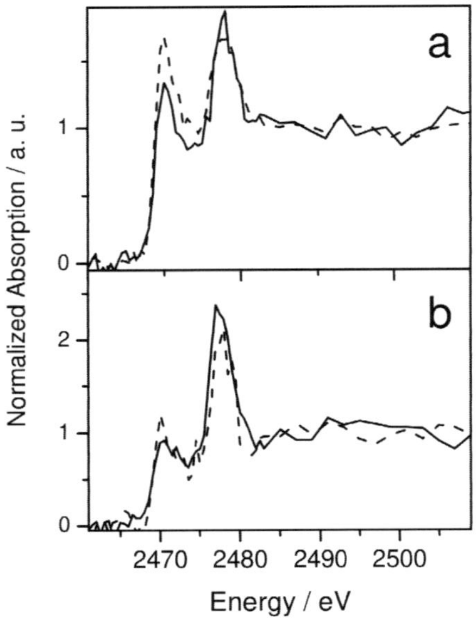

Figure 1. XANES spectra at the sulfur K-edge of 1.5 nm (a) and 2.2 nm (b) hexanethiol capped gold nanoparticles. Solid lines show the first spectra and dashed line a spectra taken after 45 minutes of irradiation.

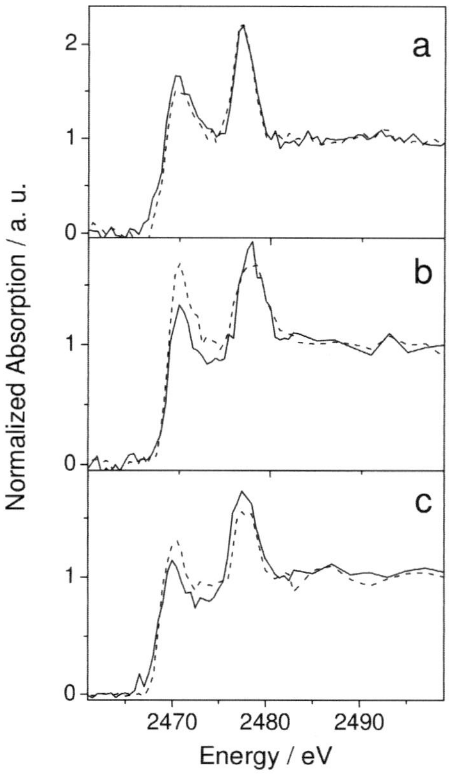

Figure 2. XANES spectra at the sulfur K-edge of 1.5 nm gold nanoparticles capped with (a) butanethiol, (b) hexanethiol and (c) nonanethiol. Solid lines show the first spectra and dashed line a spectra taken after 45 minutes of irradiation.

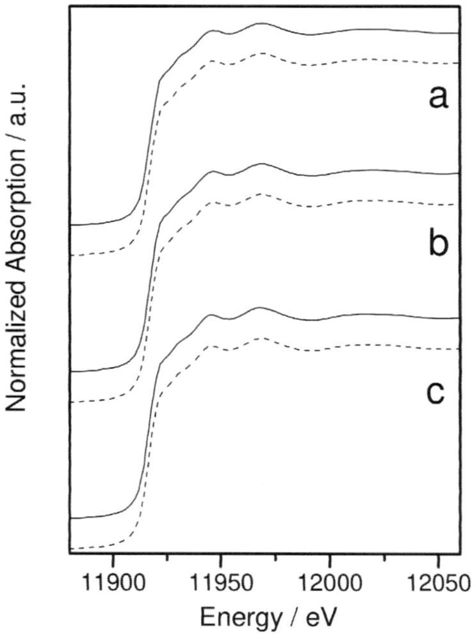

Figure 3. XANES spectra at the Au L_3 edge of 1.5 nm gold nanoparticles capped with (a) butanethiol, (b) hexanethiol and (c) nonanethiol. Solid lines show the first spectra and dashed line a spectra taken after 45 minutes of irradiation. The spectra are vertically displaced for comparison purposes.

Journal of Physics: Conference Series **430** (2013) 012034 doi:10.1088/1742-6596/430/1/012034

4. Discussion

The effect of X-ray irradiation on thiols self assembled monolayers have been extensively studied in the last years [17-19]. However, discrepancies on which is the effect of irradiation on thiols and which are the sulfur species produced because of the irradiation still exist [12]. Lamont et al. suggested that the new sulfur structures observed by XPS are due to atomic sulfur on gold, resulting from the fission of the C–S bond and desorption of the hydrocarbon chain fragment [20]. Kondoh and Nozoye studied the effect of electron irradiation of (CH_3S) self-assembled monolayers on Au(111) and found that the methyl moiety desorbs via S-C bond cleavage, leaving a sulfur adatom on the surface [21]. They observed the existence of sulfur atoms or sulfur island on Au(111) with STM. Zharnikov et al. proposed that there is an incorporation of sulfur into the alkyl chains via bonding to irradiation-induced carbon radicals in the adjacent aliphatic chains [22]. Zubrägel et al. [23] proposed the existence of dialkyl disulfide species, where one S atom interacts with the Au substrate and the other one is slightly higher. The desorption of the alkanethiol chain has also been observed [24].

The trend observed in the XANES spectra at the sulfur K-edge for the two NPs after 45 minutes of exposure to the X-ray beam is very similar both samples (see figure 1): an apparent increase of the intensity of the S-Au peak and a decrease of the intensity of the S-C peak. In addition, a decrease in the minimum between both peaks is observed which could be related to a change in the edge position. The effect on each peak and in the edge position will be discussed later. This tendency is more evident for smaller NPs while is weak for the biggest ones. When the length of the carbon chain is changed, similar results are obtained (figure 2). However for the NPs capped with butanethiols, differences between the two spectra are not so clear and considering the noise level of the spectra they can be considered as identical. It is possible that, because a lower stability of the short carbon chain of butanethiol, damages are so fast that cannot be followed by our experiments. Theoretical calculations indicate that the Au-S binding energy does not depend on the carbon chain length [25]. However, our results may be indicating that different radiation induced reactions may be taken place for short and long thiol chains. In addition, it has been reported that butanethiols SAMS on Au(111) can desorb just because of ultra high vacuum conditions [26] so this possibility should not be ruled out.

No radiation damage effect is observed in the Au L_3 XANES spectra of the 1.5 nm nanoparticles capped with different alkanethiols. As it can be seen from figure 3, the initial spectrum and the last one taken after 45 minutes of irradiation show no differences. Although the spectra represent the average over all Au atoms present in the NPs, almost one half of all Au atoms are placed at the nanoparticle surface so, if electronic changes are due to the different species produced by irradiation on the NPs on the surface, we expected to be able to observe these changes. However, it has already been shown that changes in the Au 5d electron density because of the alkanethiol adsorption is difficult to quantify from XANES spectroscopy at the Au L_3 edge. The absence of any change in the spectra shows that the Au L_3 edge is not as sensible as the sulfur K-edge to the radiation damage.

To evaluate quantitatively these changes in the K-edge XANES profiles, a simple curve fitting analysis was carried out by using a set of two Gaussians and an arctangent function as shown in figure 3. The simulated curves fit the experimental spectra very well in the whole region. Note that this simulation method was employed not for a strict determination of the electronic transitions but simply for evaluation of the peaks intensity and position in each spectrum. Changes in both peaks intensities are observed, however, the S-Au peak at 2470 eV decreases its intensity after irradiation, contrary to what was supposed previously from the visual inspection of the spectra. This is a consequence of the shift in the edge position to lower energy, indicating the reduction of the average oxidation state of sulfur. A decrease of the S-C peak at 2478 eV is also observed. Considering the proposed models for the sulfur species formed after irradiation with X-rays, our results are consistent with the cleavage of some C-S bonds, and the consequent appearance of atomic sulfur on the nanoparticle surface as has already been reported for XPS radiation damage [27]. Being XANES an average technique, the cleavage of some S-C bonds would result in a decrease of this interaction in the total XANES spectrum as it is observed. In addition, the appearance of atomic sulfur on the surface would result in a reduction of the average oxidation state of sulfur, shifting the edge position to lower energies. Also, the S-Au interaction would be reduced and thus the intensity of the peak assigned to this bond will decrease as observed. In the same direction, as the van der Waals forces that limit the sulfur density would be reduced because of the smaller density of carbon chains in this new configuration produced by the S-C cleavage, a less compact SAM arrangement could be formed on the nanoparticle surface which could also lead to a decrease of the S-Au interaction.

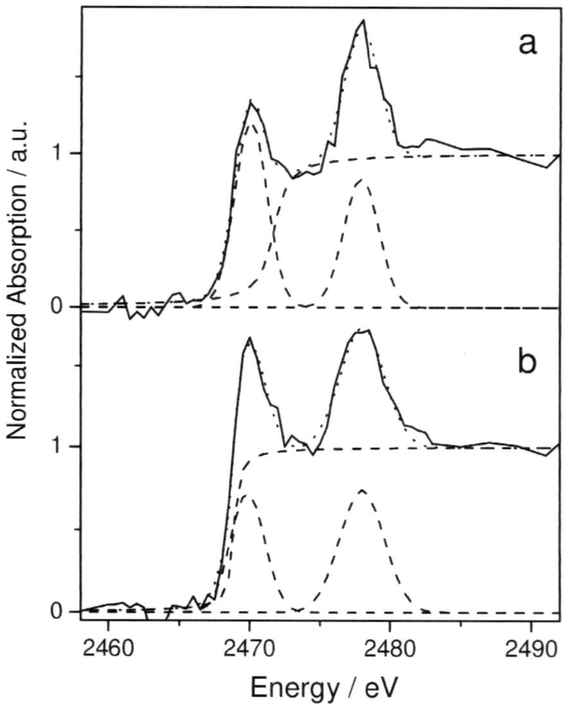

Figure 4. Linear combination fit of the initial (a) and final (b) XANES spectra at the sulfur K-edge of 1.5 nm gold nanoparticles capped with hexanethiol using 2 Gaussians and one arctangent functions. Solid line: experimental data, dashed line: individual functions, pointed line: total fitted function.

An important remark is that, even though, the irradiation with X-rays leads to a change in the intensity of both characteristic peaks of the XANES spectrum at the sulphur K-edge, this effect is small compared with the intrinsic effect due to the particle size. As it can be seen from figure 1, the difference of the relative intensities of these peaks in each sample is smaller than the differences from on sample to the other. As a consequence, the use of XANES to study the S-Au interaction in thiol-capped nanoparticles could still valid, but short time acquisition would be desired in order to minimize the radiation damage during the experiment. However, some concerns have to be made when trying to evaluate in a quantitatively manner the intensities of the peaks appearing in the XANES spectrum, as changes in their intensities are expected during the acquisition due to radiation damage as has been shown. It would be more convenient to perform a bigger number of fast scans even if the statistics is poor than to perform smoother scans with long acquisition time at each energy. Because of this, only qualitative analysis of the XANES spectra should be made when the samples are exposed for long periods of time to the X-ray beam or even in cases where fast spectra are taken as in dispersive XAS but there is no control of the radiation damage.

5. Conclusions

Radiation damage on alkanethiol-capped Au NPs was observed during the acquisition of XANES spectra at the sulfur K-edge in ultra high vacuum conditions. Changes in the S-Au and S-C peak intensities of the XANES spectra are observed, indicating changes in the sulfur species adsorbed on the nanoparticle surface probably due to cleavage of the carbon chains and the appearance of atomic sulfur on the nanoparticle surface. Because of this, fast scan experiments should contribute to minimize radiation damage and improve quantitative analysis of XANES spectra of alkanethiol capped Au nanoparticles at the sulfur K edge.

Acknowledgments

This work was supported by LNLS, Brazil (Project SXS-5729 and XAS-3492), ANPCyT (PICT-2008-00038) and CONICET (PIP 112-200801-03079), Argentina.

References

[1] Love J C, Estroff L A, Kriebel J K, Nuzzo, R G and Whitesides, G M 2005 *Chem. Rev.* **105** 1103 and references therein
[2] Nuzzo R G, Zegarski B R and Dubois L H 1987 *J. Am. Chem. Soc.* **109** 733
[3] Strong L and Whitesides G M 1988 *Langmuir* **4** 546
[4] Chidsey C E D, Liu G -Y, Rowntree P and Scoles G 1989 *J. Chem. Phys.* **91** 4421
[5] Chidsey C E D and Loiacono D N 1990 *Langmuir* **6** 709
[6] Dubois L H, Zegarski B R, and Nuzzo R G 1993 *J. Chem. Phys.* **98** 678
[7] Brust M, Walker M, Bethell D, Schriffin D and Whyman R 1994 *J. Chem. Soc. Chem. Commun.* 801
[8] Vericat C, Benitez G A, Grumelli D E, Vela M E and Salvarezza R C 2008 *J. Phys.: Condens. Matter* **20** 184004
[9] Hostetler M J, Wingate J E, Zhong C J, Harris J E, Vachet R W, Clark M R, Londono J D, Green S J, Stokes J J, Wignall G D, Glish G L, Porter M D, Evans N D and Murray R W 1998 *Langmuir* **14** 17
[10] Ramallo-López J M, Giovanetti L J, Requejo F G, Isaacs S R, Shon Y S and Salmeron M 2006 *Physical Review* B **74** 073410
[11] Jadzinsky P D, Calero G, Ackerson C J, Bushnell D A and Kornberg R D 2007 *Science* **318** 430
[12] Duwez A –S 2004 *J. Elect. Spectrosc. Relat. Phenom.* **134** 97
[13] Zhang P, Sham T K 2003 *Appl. Phys. Lett.* **82** 1778
[14] Zhang P, Sham T K 2003 *Phys. Rev. Lett.* **90** 245502
[15] Shon Y-S and Cutler E 2004 *Langmuir* **20** 6626
[16] Abbate M, Vicentin F C, Compagnon-Cailhol V, Rocha M C and Tolentino 1999 *J. Synchr. Rad.* **6** 964
[17] Hesiter K Zharnikov M Grunze M Johansson L S O and Ulman A 2001 *Langmuir* **17** 8
[18] Ishida W T, Choi N, Mizutani W, Tokumoto H, Kojima I, Azehara H, Hokari H, Akiba U and Fujihira M 1999 *Langmuir* **15** 6799
[19] Dannenberger O, Weiss K, Himmel H -J, Jäger B, Frey M S, Heister K, Zharnikov M and Grunze M 2000 *Phys. Chem. Chem. Phys.* **2** 1979
[20] Lamont C L A and Wilkes J 1999 *Langmuir* **15** 2037
[21] Kondoh H and Nozoye H 1998 *J. Phys. Chem.* B **102** 2367
[22] Zharnikov M, Geyer W, Gölzhäuser A, Frey S and Grunze M 1999 *Phys. Chem. Chem. Phys.* **1** 3163
[23] Zubrägel Ch, Deuper C, Schneider F, Neumann M, Grunze M, Schertel A and Wöll Ch 1995 *Chem. Phys. Lett.* **238** 308
[24] Büttner M, Belser T and Oelhafen P 2005 *J. Phys. Chem. B* **109** 5464
[25] Krüger D, Fuchs H, Rousseau R, Marx D and Parrinello M J 2001 *Chem. Phys.* **115** 4776
[26] Hutt D A and Leggett G J 1997 *Langmuir* **13** 3055

[27] Gonella G, Cavalleri O, Terreni S, Cvetko D, Floreano L, Morgante A, Canepa M and Rolandi R, 2004 *Surface Science*, **566–568** 638

Journal of Physics: Conference Series **430** (2013) 012035 doi:10.1088/1742-6596/430/1/012035

Metal induced crystallization of amorphous silicon thin films studied by x-ray absorption fine structure spectroscopy

K Lakshun Naidu[1,2], Md Ahamad Mohiddon[1,2], M Ghanashyam Krishna[2] , G Dalba[1] and F Rocca[3,*]

[1] Department of Physics, University of Trento, 38123 Povo (Trento) Italy
[2] School of Physics, University of Hyderabad, Hyderabad-500 046, India
[3] IFN-CNR, Institute for Photonics and Nanotechnologies, Unit "FBK-Photonics" of Trento, 38123 Povo (Trento) Italy

*Corresponding author, email: rocca@science.unitn.it

Abstract. The role of thin metallic layer (Chromium or Nickel) in the crystallization of a-Si film has been studied using X-ray absorption fine structure spectroscopy (XAFS). The films were grown at different substrate temperatures in two different geometrical structures : (a) a 200 nm metal layer (Cr or Ni) was deposited on fused silica (FS) followed by 400 nm of a-Si and (b) the 400 nm a-Si layer was deposited on FS followed by 200 nm of metal layer. XAFS measurements at Cr K-edge and Ni K-edge were done at BM08 - GILDA beamline of the European Synchrotron Research Facility (ESRF, Grenoble, F) in fluorescence mode. To understand the evolution of the local structure of Cr/Ni diffusing from bottom to top and from top to bottom, total reflection and higher incidence angles were employed. The relative content of metal, metal oxide and metal silicides compounds on the upper surface and/or in the bulk of different films has been evaluated as a function of thermal treatment.

1. Introduction

Amorphous silicon crystallizes at 750 C, or higher, temperature [1]. The crystallization temperature can be strongly reduced (in the range 165-300 °C) by low-doping with metals or by depositing a thin metal layer onto a-Si film [2]. The mechanism is known as Metal Induced Crystallization (MIC). Various conflicting explanations for the MIC phenomenon have been discussed in the literature [3].
Aim of the present work is to investigate the evolution of the local environment of two metal dopants (Cr or Ni) deposited at different substrate temperatures below or above a layer of a-Si using X-ray absorption fine structure (XAFS) technique. The result is a detailed study of changes with temperature around Cr or Ni, due to different metal diffusion and to the formation of new silicide phases.

2. Experimental and data analysis methods.

Films composed of two nominal layers were deposited onto fused silica (FS) substrates using electron beam evaporation at 4 x10^{-6} torr in both possible combinations: Cr or Ni (nominal 200 nm) metals were deposited before (i.e. on bottom) or after (i.e. on top) the deposition of a 400 nm thick a-Si layer. The starting materials were granular pure Silicon powders (99.999%), Chromium granules (99.99% pure) and Nickel granules (99.99% pure). The films were annealed for 30 min inside the vacuum chamber at the different substrate temperatures selected for the deposition. The deposition conditions were decided in order to discriminate: a) the possible different diffusion rate from bottom up to the surface of metals; b) the influence of temperature on chemical reactions of metals with a-Si. In parallel studies, we already showed that the Cr films deposited at 200 °C do not contain significant modifications in the metal chemistry, because changes appear only at higher deposition or annealing

temperature [4]. We present here results on films deposited at 200 °C or 500 °C below a-Si, while 400 °C has been selected for deposition above a-Si, to reduce oxidation effects.
XAFS measurements of the films were performed at Cr or Ni K-edges (5989 eV and 8400 eV respectively) in fluorescence mode with a Ge multidetector at BM08-GILDA (ESRF, France) [5]. Total Reflection geometry was used, when possible, to study the metal environment near the surface, while different higher incidence angles were selected to investigate the whole sample [6]. Reference samples for Cr edge were measured in transmission mode at the XAFS beamline of ELETTRA (Trieste, Italy). Later the energy axis of the spectra was carefully aligned with respect to the energy axis of XAS spectra of a Cr foil measured at GILDA. Reference materials for Nickel where measured in transmission mode at GILDA.
Quantitative analysis of EXAFS has been carried out using FEFF8 and FEFFIT codes [7,8]. Details of theoretical calculations are deeply discussed and commented in parallel papers [4,9]. In summary, the FT-EXAFS spectra of films were best-fitted using a set of fixed distances (R) and coordination numbers (N) as pre-determined by studying the crystal structure of reference materials: thus the best fit was a simple linear combination of the minimal number of shells of the reference materials, each of them having a free DW factor. The fit results give us the relative content of metal, oxide and silicides present in the investigated sample region.

3. Results

3.1 EXAFS analysis of Reference materials

The reference compounds were metallic foils of Cr and Ni, and Cr_2O_3, $CrSi_2$, NiO, Ni_2Si, $NiSi_2$ powders, measured in transmission in optimal standard conditions. The E_0 and S_0^2 values of each reference compound were carefully evaluated through FEFFFIT analysis and used later as fixed parameters for the best fit of films, whose EXAFS was extracted using E_0 of metals. In particular, we determined E_0 as 5992 eV for metallic Cr, 5998 eV for Cr_2O_3 , 5996 eV for $CrSi_2$, 8337.5 eV for metallic Ni foil , 8342 eV for NiO , 8341 eV for Ni_2Si and 8337.8 eV for $NiSi_2$.
We considered the following shells to fit the spectra of the references at the corresponding distances R (and coordination numbers N): to fit metallic Cr six Cr-Cr shells; to fit Cr_2O_3 three Cr-O shells and four Cr-Cr shells; to fit $CrSi_2$ three Cr-Si shells and two Cr-Cr shells.
Similarly, to fit metal Ni spectra we considered seven Ni-Ni shells; to fit Ni_2Si three Ni-Si shells and six Ni-Ni shells; to fit $NiSi_2$ one Ni-Ni shell and five Ni-Si paths; for NiO we considered three Ni-O shells, three Ni-Ni shells and five multiple scattering (MS) paths.
The obtained S_0^2 values were used to evaluate later the relative content of these reference compounds in the investigated films.

3.2 EXAFS analysis of films with Cr

Figure 1 shows near edge x-ray absorption spectra (NEXAFS) of films deposited at 200 °C and 500 °C with Cr as bottom layer and deposited at 400 °C with Cr as top layer along with the reference samples. The EXAFS Fourier transforms (FT) with the best fits in dotted line are shown in Figure 2. The film deposited at 200 °C has only partial resemblance with metallic Cr spectrum, suggesting the presence of other Cr compounds along with metallic Cr. On the film deposited below a-Si at 500 °C, we did not detect any fluorescence signal when probed in TR geometry, indicating that Cr was not able to diffuse up to the surface of the film. The two EXAFS spectra of film deposited at 500 °C probed at 10° and 45° have similar features. Both the spectra can be reproduced by a linear combination of shells from Cr_2O_3 and $CrSi_2$. The relative percentage of Cr compounds obtained from the EXAFS quantitative analysis are presented in Table 1, with error bars estimated on the basis of different extractions and best fit procedures. From Table 1 we may note that, for the film deposited above a-Si at 200 °C, only a small fraction of Cr metal reacted with a-Si during deposition to form $CrSi_2$, but we have quantified a significant fraction of Cr in oxidized state. When the substrate temperature is raised to 500 °C, the

Journal of Physics: Conference Series **430** (2013) 012035 doi:10.1088/1742-6596/430/1/012035

bottom metal layer contribution disappears and a larger fraction of Cr is transformed into $CrSi_2$ and Cr_2O_3. Since Oxygen was totally absent in the deposition chamber, we explain the Cr_2O_3 formation as due to a successive oxidation of metal atoms or nanoparticles diffused inside the a-Si layer, but unreacted and highly unstable. The TR spectrum of film deposited on top at 400 °C shows on the surface the typical features of both Cr_2O_3 and metallic Cr, while the corresponding one at 10° shows a complete absence of Cr_2O_3, (note the absence of the peak of Cr-O at 2 Å in the FT), and the presence of the features of metallic Cr and $CrSi_2$. From Table 1 it may be observed that on the surface about 50 % of unreacted Cr was oxidized, while a large part of Cr atoms diffused into the a–Si layer were stabilized in met-Cr nanoclusters, or able to react forming a more stable $CrSi_2$ compound.

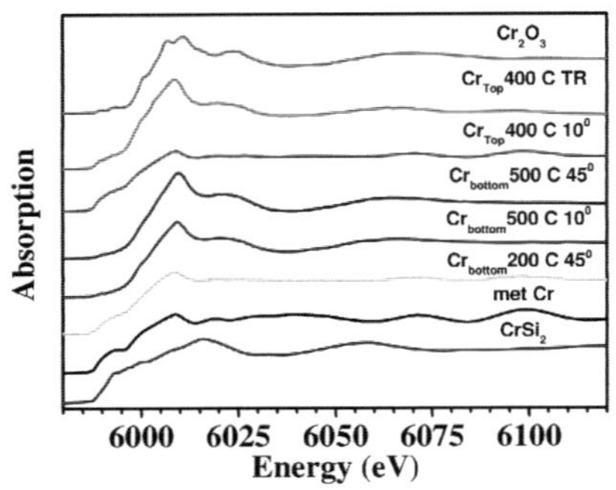

Figure 1 Cr K edge XANES of FS/Cr (200 nm)/a-Si (400 nm) stack deposited at 200 °C and 500 °C, and of FS/Si (400 nm)/Cr (200nm)stack deposited at 400 °C, along with those of metal Cr, Cr_2O_3 and $CrSi_2$.

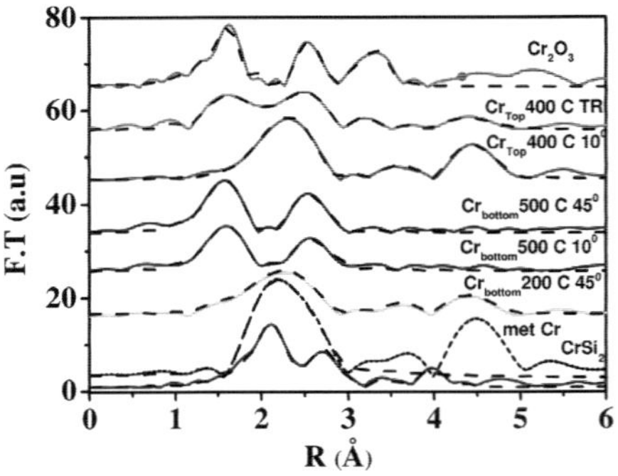

Figure 2 FT EXAFS signal of FS/ Cr (200 nm)/ a-Si (400 nm) stacks at 200 °C and 500 °C, and of FS/Cr (200 nm)/ a-Si (400 nm) stack at 400 °C, along with those of the three references. The best fits of FT signals are shown by dotted lines.

Table 1 The percentage of different Cr compounds present in films with Cr as bottom and top layer in a-Si films.

Sample	**Cr (%)**	**Cr_2O_3 (%)**	**$CrSi_2$ (%)**
FS/Cr/*a*-Si 200°C 45°	64 ± 10	21 ± 10	15 ± 10
FS/Cr/*a*-Si 500°C 10°	---	48 ± 10	52 ± 10
FS/Cr/*a*-Si 500°C 45°	---	62 ± 10	37 ± 10
FS/*a*-Si/Cr 400°C TR	55 ± 10	45 ± 10	---
FS/*a*-Si/Cr 400°C 10°	61 ± 10	---	39 ± 10

3.3 EXAFS analysis of film with Ni

The XAS and FT of films with Ni as bottom layer deposited at 200 °C and 500 °C, and with Ni as top layer deposited at 400 °C are shown in figure 3 and figure 4. The films were probed at 2° and TR mode. The quantitative results are shown in Table 2. The film deposited at 200 °C appears to contain almost metal Ni. Contrary to the Cr case, on the surface investigated by TR spectrum, we could detect Ni atoms diffused from the bottom layer, in met-Ni clusters or partially reacted with a-Si to form Ni_2Si. The film deposited at 500 °C contains combined features of $NiSi_2$ and NiO in the whole layer. As substrate temperature is raised to 500 °C, metal Ni not only diffuses into a-Si layer towards the upper surface of the film, but there is an increasing formation of $NiSi_2$. From the TR spectrum it is shown that about 75 % of metal Ni at the surface of the film evolves as $NiSi_2$. The remaining 25 % Ni is embedded in the $NiSi_2$ layers.

The films with Ni top layer at 400 °C had resemblance with NiO and metal Ni. It is clear from the analysis that Ni is unable to react with Si to form $NiSi_2$ .The amount of Ni oxidation is more on the surface, with NiO about 77 % on surface and about 56 % inside the film.

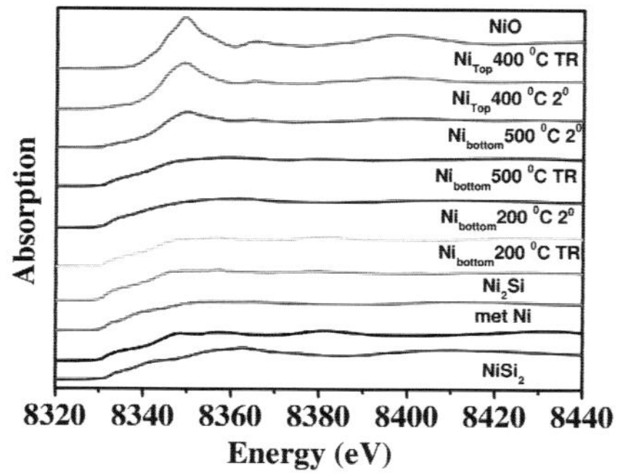

Figure 3 Ni K edge XAS of FS/Ni (50 nm)/a-Si (400 nm) stacks deposited at 200 °C and 500 °C, and FS/Si (400 nm)/Ni (50 nm)stack deposited at 400 °C, along with those of metal Cr, Cr_2O_3 and $CrSi_2$.

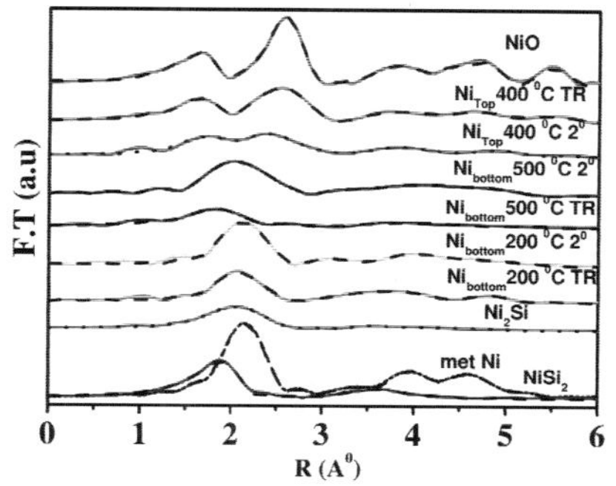

Figure 4 FT EXAFS signal of FS/ Ni (50 nm)/ a-Si (400 nm) stacks deposited at 200 °C and 500 °C, and of FS/Ni (50 nm)/ a-Si (400 nm) stack deposited at 400 °C, along with those of three references. The best fits of FT signals are shown by dotted lines

Table 2 The percentage of different Ni compounds present in films with Ni as bottom and top layer in a-Si films.

Sample	Ni (%)	NiO (%)	$NiSi_2$	Ni_2Si
FS/Ni/*a*-Si 200°C TR	57 ± 10	----	---	43 ± 10
FS/Ni/*a*-Si 200°C 2°	100	----	---	----
FS/Ni/*a*-Si 500°C TR	25 ± 10	----	75 ± 10	----
FS/Ni/*a*-Si 500°C 2°	51± 5	----	49± 5	----
FS/*a*-Si/Ni 400°C TR	23 ± 10	77 ± 10	----	----
FS/*a*-Si/Ni 400°C 2°	44 ± 5	56 ± 5	----	------

4. Conclusions:

Films containing metal (Cr / Ni) layer and a-Si are deposited using electron beam evaporation at higher substrate temperatures. X-ray absorption spectra measured at Total Reflection and higher angles to study the local environment around metal at surface and inside the film. The EXAFS analysis is able to explain the diffusion of metal and formation of silicide at different temperatures. The metal silicides and their diffusion across the film promote the crystallizing of a-Si.

5. References:

[1] Fan J C C and Anderson H, *J. Appl. Phys.*52, 4003 (1981)
[2] Wang ZM, Wang JY, Jeurgens LPH and Mittemeijer EJ *Phys Rev Lett* **100**:125503(2008)
[3] Wang JY, He D, Zhao YH and Mittemeijer EJ, *Appl. Phys. Lett.***88**, 061910–11 (2006)
[4] Mohiddon MA, Naidu KL, Krishna MG, Dalba G and Rocca F, *Phys. Status Solidi C* **9**, 1493-1995 (2012)
[5] Pascarelli S, Boscherini F, D'Acapito F, Hrdy J, Meneghini C, and S. Mobilio, *J.Synchrotron Radiat.* **3**, 147 (1996)
[6] D'Acapito F, Milita S., Satta A. and Colombo L., *J. of Appl Physics* **102**, 043524 (2007)
[7] Newville M, *J. Synchrotron Radiat.* **8**, 322 (2001)
[8] Raveland B. M. Newville, *J. Synchrotron Radiat.* **12**, 537 (2005)
[9] Mohiddon M A, K Lakshun Naidu, Ghanashyam Krishna M, Dalba G and Rocca F, *in preparation* (2012)

Journal of Physics: Conference Series **430** (2013) 012036 doi:10.1088/1742-6596/430/1/012036

Theoretical study of the local atomic and electronic structure of dimetacyano azobenzene molecules on Bi (111) substrate

V Mazalova[1], A Soldatov[1], A Krüger[2], W Kuch[2]

[1]Department of Physics, Southern Federal University, Zorge 5, 344090, Rostov-on-Don, Russia

[2]Institut für Experimentalphysik, Freie Universität Berlin, Arnimallee 14, 14195, Berlin, Germany

E-mail: mazalova@sfedu.ru

Abstract. The present work is devoted to the local atomic and electronic structure of dimetacyano (DMC) molecules deposited on a Bi (111) substrate before and after irradiation by X-rays and UV light using density functional theory (DFT) and a theoretical analysis of the X-ray absorption near edge structure (XANES) spectroscopy. As a result of the calculations the low-energy structure for DMC azobenzene molecules on a Bi (111) substrate was obtained.

1. Introduction

Using controlled reversible processes in molecules deposited on metal and non-metal surfaces develops new ways for the miniaturization of electronic devices. In particular the family of switchable molecules offers the possibility to reversibly control their properties by external parameters like photons, electrons, protons, temperature, or electrical currents – a so-called molecular switch. In this sense the direct adsorption of molecular switches on metallic and non-metallic surfaces is an interesting field in the ongoing development of bottom-up technologies. There are several classes of photoactive chemical compounds which are able to undergo reversible transformations between two or more stable molecular states. These states usually differ, e.g., in their chemical and/or geometric structure. Molecules for which the reversible process is given by a change between different stable geometries while the bond order of each atom is mainly preserved are referred to as conformational switches. For instance, the azobenzene molecules in gas phase and solution are subject to photoisomerization, during which changes are accompanied by corresponding changes in chemical structure. This property makes them suitable for using as building blocks in molecular motors and optical storage devices. On the other hand, deposition of molecules on surfaces leads to a restriction of the molecular switch, which occurs due to the fact that the isomerization process becomes influenced not only by an atomic bonding in the molecule but also by the electronic interactions between molecules and surroundings [1-3].

There are several scanning tunnel microscopy studies (STM) devoted to azobenzene molecules on Cu(001) and Au(111) surfaces [4,5,6]. The presence of an interaction between adsorbate and surface has been shown during the switching an azobenzene single molecule on Cu (001) surface at 5 K [4]. After switching, the resulting *cis*-isomer was found to be stabilized by chemisorption, preventing a reversible switching. It has been shown, that during the adsorption of an azobenzene monolayer on Cu (001) at the room temperature structures similar to the *cis*-form are formed, whereas, on Au(111) the

Journal of Physics: Conference Series **430** (2013) 012036 doi:10.1088/1742-6596/430/1/012036

trans-form is more preferable [5]. It means that it is possible to use different types of substrates to control the switching process and provide reversibility. Also, it was revealed that the self-assembly and switching behavior is highly dependent on the chemistry and corrugation of the surface [6]. The atomic and electronic properties of DMC azobenzene molecules on Cu(001) were studied by angle-dependent near edge X-ray absorption fine structure (NEXAFS), X-ray photo electron spectroscopy (XPS) and DFT calculations [7]. It has been shown that the molecule coverage and chemical character of the substrate and its temperature have an influence on the adsorption state of DMC azobenzene molecules.

The present study provides additional structural information which leads to a more detailed picture of the mechanisms of adsorption. We used the density functional theory (DFT) and molecular dynamic simulations (MOPAC) to provide information on the atomic structure of adsorbed molecules.

2. Methods of calculation

The atomic geometry of DMC azobenzene molecules was analysed on the basis of density functional theory (DFT) and molecular dynamic simulations (MOPAC method) implemented in the ADF program code [8,9]. The electronic configuration of the cluster was described by a core double zeta, valence triple zeta, doubly polarized (TZ2P) basis set of Slater-type orbitals. The bismuth substrate was frozen during the geometry optimization and a double zeta (DZ) basis set was used.

The DFT-D3 exchange-correlation correction parameterized for B3LYP functional was applied during DFT calculations. The DFT-D3 dispersion correction is less empirical, i.e., the most important parameters are computed from first principles by standard Kohn-Sham DFT. The DFT-D3 implementation based on the paper by Grimme [10] was used to take into account weak van der Waals substrate – adsorbate interactions. This approach provides high accuracy for both "light" and "heavier" atoms.

3. Results and discussion

It is known that DMC azobenzene molecule change their geometric structure after irradiation during the photoisomerization of *trans*- and *cis*- isomers. The two isomers can be switched with particular wavelengths of light - ultraviolet or blue light. The *cis* isomer is less stable than the *trans* (for instance, it has a distorted configuration and is less delocalized than the trans configuration). Thus, *cis*-azobenzene will thermally relax back to the *trans* via *cis*-to-*trans* isomerization.

The present research was concentrated on the theoretical prediction of the atomic and electronic structure of DMC azobenzene molecules deposited on a Bi(111) substrate before and after irradiation by UV light. In order to study the geometric structure the relaxation of DMC azobenzene molecules deposited on the bismuth surface density functional theory was performed. At first, we studied the structural changes of the free DMC molecules depending on the spin state. Figure 1 shows the proposed changing the *trans*- DMC azobenzene molecule geometry as a result of changing its spin state. The initial structure corresponds to the *trans*-isomer of the molecule in the ground state S_0 (1). After photoexcitation the molecule passes into the excited electronic state S_2, which corresponds to the *cis*-isomer of DMC azobenzene molecules (2). The excited state is a triplet. In the process of relaxation of the molecule it can take the geometry 3 or 4 with equal probability (the bonding energy for these two configurations has almost the same value). In the DFT approach used there, the bonding energy is calculated as the energy of the cluster minus the energy of the constituent atoms.

In the next step the most probable geometry of the DMC azobenzene molecule on a Bi(111) substrate in the ground and excited state (corresponding to the states before and after irradiation) was found. The simulation results are presented in table 1. The geometrical optimization was performed using ab initio DFT. For each of the resulting structures the values of a bonding energy in order to select energetically the most favorable configuration were calculated. For *trans* isomer of DMC azobenzene molecule only the model with a highest bond energy is presented. The DMC azobenzene molecule in this model is arranged horizontally with respect to the substrate and the azo nitrogen atoms are located directly above the bismuth atoms.

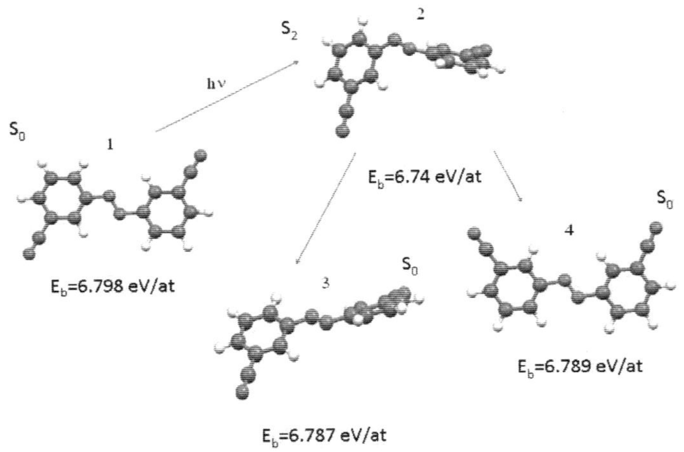

Figure 1. Schematic representation of the *trans-cis* photoisomerization pathways of DMC azobenzene molecules. E_b is a bond energy per atom.

In table 1 the two models for the excited DMC azobenzene molecule on the Bi(111) substrate are shown. They are quite close in bonding energies and in order to choose the more reasonable model in a follow-up study the analysis of angle-depended NEXAFS spectra should be done.

Table 1. The relaxed geometry structures of a single DMC azobenzene molecule on top of a Bi cluster from a Bi(111) substrate, resulting from DFT geometry optimization. The substrate atoms were fixed during optimization.

Spin state	Optimized DMC molecule on the Bi(111) substrate	Bonding energy per atom, eV
S_0 *trans*-DMC	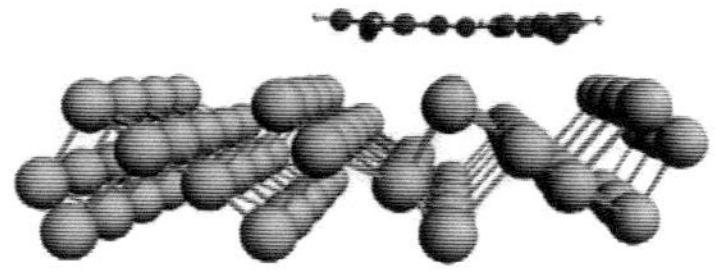	4.242
S_2 *cis*-DMC	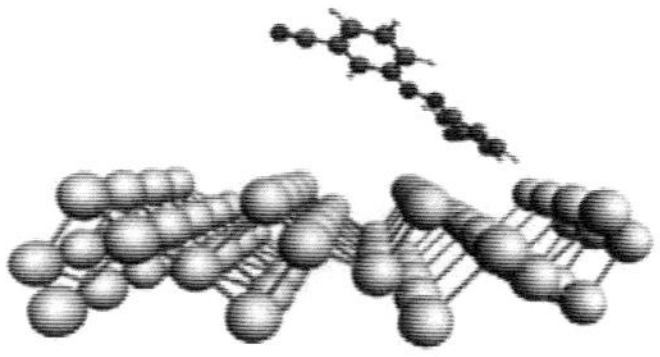	4.042
S_2 *cis*-DMC	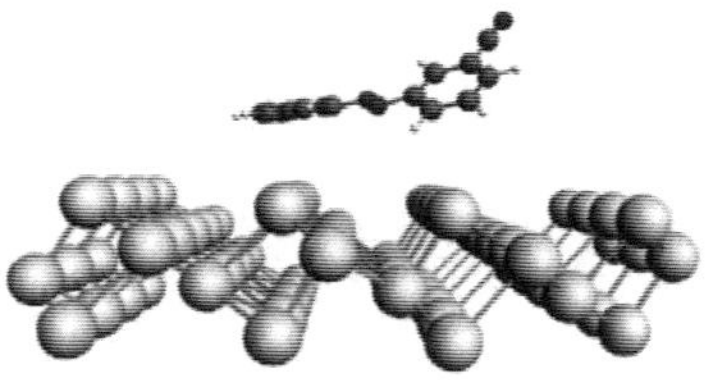	4.054

Journal of Physics: Conference Series **430** (2013) 012036 doi:10.1088/1742-6596/430/1/012036

We calculated angle-dependent N K-edge NEXAFS spectra for *trans*- and *cis*- isomers of the DMC azobenzene molecule in order to understand the geometry of the molecules related to a Bi(111) surface. On the figure 2 the theoretical spectra for two isomers and experimental spectra for *trans*-DMC azobenzene molecule deposited on a Bi(111) are compared. The inserted pictures show the highest occupied molecular orbitals (HOMOs) and the lowest unoccupied molecular orbitals (LUMOs) at the DMC azobenzene molecule. The energy position and relative intensity of the features on the simulated N K-edge NEXAFS spectra for trans- isomer well describe the experimental spectrum which let us to assume the flat settle of DMC molecules on the Bi(111) substrate. As one can see from the bottom two pictures of the figure 2 the simulated spectra for *cis*- isomer are characterized by the high intensity resonance from cyano nitrogen group which lies perpendicular to a surface for both normal and grazing X-ray incidence.

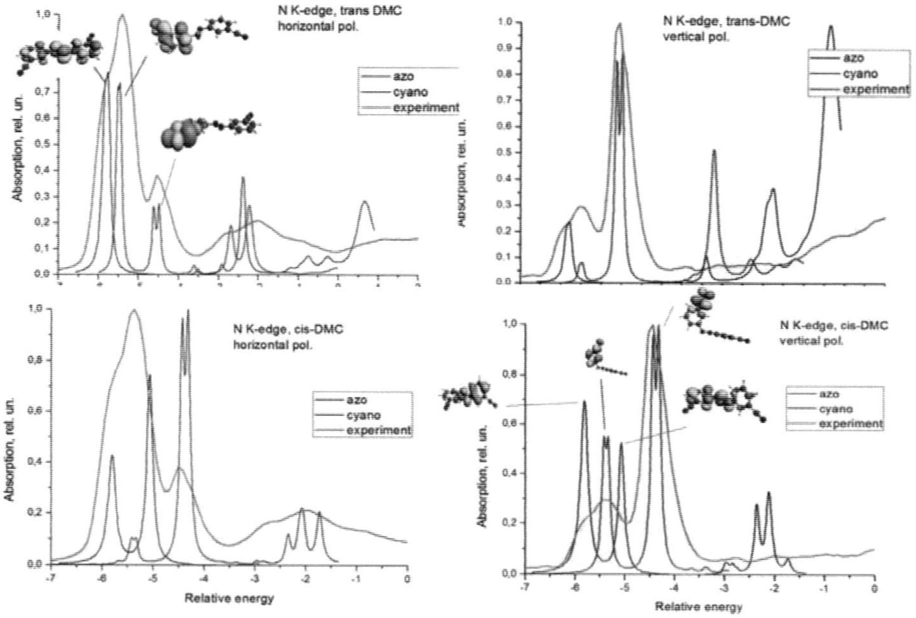

Figure 2. Angle-dependent N K-edge absorption spectra of 1 multilayer DMC molecules evaporated on Bi(111) substrate at 110 K temperature compared with simulated NEXAFS spectra of free trans- and cis- DMC molecules. The black and blue curves correspond to the azo and the cyano excitation centers.The red lines correspond to experimental spectra at 90° (normal) and 20° (grazing) X-ray incidence.

In order to predict a possible arrangement of DMC azobenzene molecules on a Bi(111) substrate the geometrical optimization by a semiempirical MOPAC method implemented in the code ADF2012 was performed. We optimized several models in which two DMC azobenzene molecules were placed on the top of a bismuth cluster. Some of the models are presented in table 2. The energetically most preferable for a non-excited DMS molecule is to symmetrically settle parallel to the bismuth surface.

Table 2. The relaxed geometric structure of two DMC azobenzene molecules on top of a Bi cluster from Bi(111) substrate resulting from MOPAC geometry optimization. The substrate atoms were fixed in space during optimization.

Optimized DMC molecules on the Bi(111) substrate	Total energy, eV
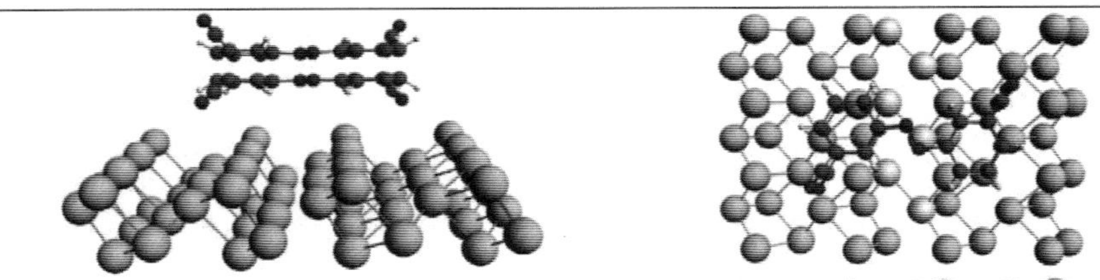	-12 204.153
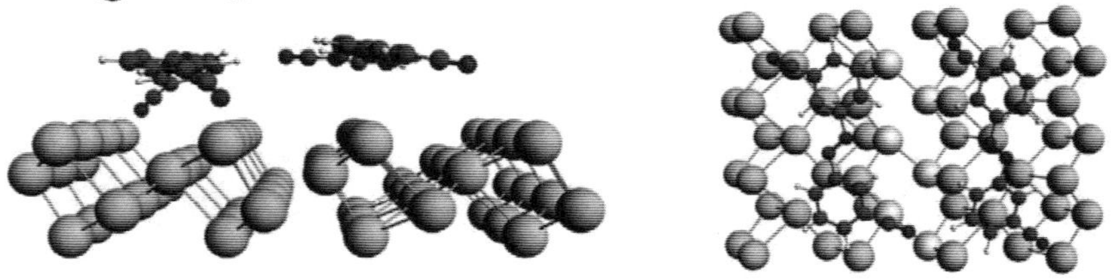	-12 214.250
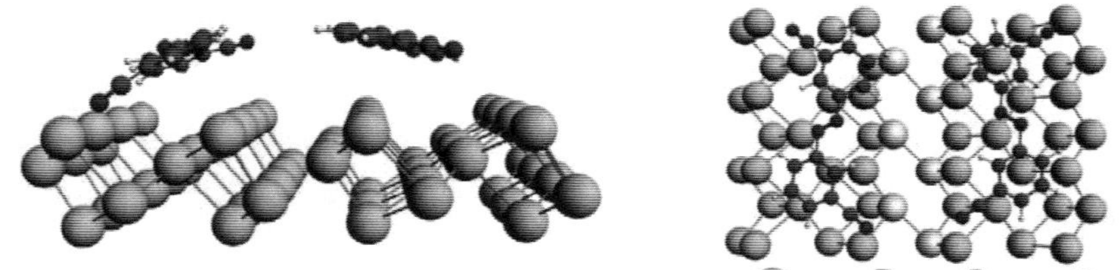	-12 213.277
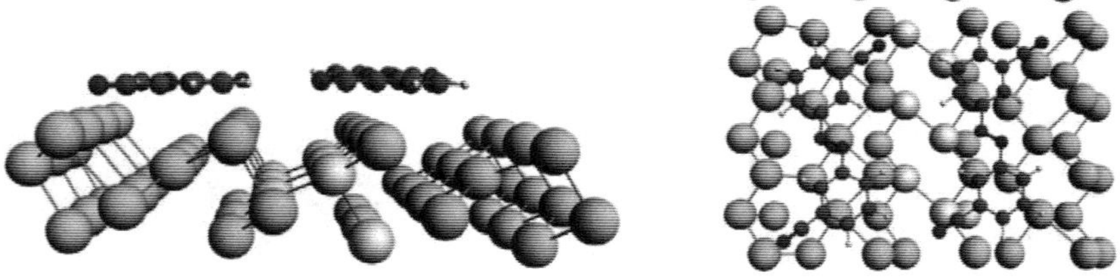	-12 214.617

4. Conclusions

The research has involved the theoretical modelling of the atomic and electronic structures of DMC azobenzene switchable molecules on a Bi(111) substrate using the density functional theory (DFT) and molecular dynamics (MOPAC) techniques.

The most probable structure of trans- and cis- isomers of the DMC azobenzene molecule on a Bi(111) substrate was found by means of DFT calculations. In order to proof our theoretical predictions, the combination of experiment and theory of angle-dependant X-ray absorption near fine structure (NEXAFS) spectroscopy for the DMC azobenzene molecule on a Bi(111) substrate was applied. We found the planar geometry of the *trans*- DMC azobenzene molecule relative to a Bi(111) substrate.

Acknowledgements

V.M. acknowledges the German Academic Exchange Service (DAAD) for support of this research during the Mikhail Lomonosov and G-RISC scholarships.

References

[1] Petek H, Nagano H, Weida M J and Ogawa S 2001 Surface Femtochemistry: Frustrated Desorption of Alkali Atoms from Noble Metals. *J. Phys. Chem.* B **105** 6767

[2] Dutton G, Quinn D P, Lindstrom C D and Zhu X-Y 2005 Exciton dynamics at molecule-metal interfaces: C_{60}/Au(111) *Phys. Rev.* B **72** 045441
[3] Yang A, Shipman S T, Garrett-Roe S, Johns J, Strader M, Szymanski P, Muller E and Harris C 2008 Two-Photon Photoemission of Ultrathin Film PTCDA Morphologies on Ag(111) *J. Phys. Chem.* C **112** 2506
[4] Henningsen N, Rurali R, Franke K J, Fernandez-Torrente I and Pascual J I 2008 Trans to cis isomerization of an azobenzene derivative on a Cu(100) surface *Appl. Phys.* A **93** 241
[5] Piantek M, Miguel J, Bernien M, Navío C, Krüger A, Priewisch B, Rück-Braun K and Kuch W 2008 Adsorption of carboxymethylester-azobenzene on copper and gold single crystal surfaces *Appl. Phys.* A **93** 261
[6] Alemani M, Selvanathan S, Ample F, Peters M V, Rieder K-H, Moresco F, Joachim C, Hecht S and Grill L 2008 Adsorption and Switching Properties of Azobenzene Derivatives on Different Noble Metal Surfaces: Au(111), Cu(111), and Au(100) *J. Phys. Chem.* C **112** 10509
[7] Piantek M, Miguel J, Krüger A, Navıo C, Bernien M, Ball D K, Hermann K and Kuch W 2009 Temperature, Surface, and Coverage-Induced Conformational Changes of Azobenzene Derivatives on Cu(001) *J. Phys. Chem.* C **113** 20307
[8] te Velde G, Bickelhaupt F M, van Gisbergen S J A, Fonseca Guerra C, Baerends E J, Snijders J G and Ziegler T 2001 Chemistry with ADF *J. Comput. Chem.* **22** 931
[9] *ADF2012* Available at: http://www.scm.com.
[10] Grimme S, Anthony J, Ehrlich S and Krieg H 2010 A consistent and accurate ab initio parametrization of density functional dispersion correction (DFT-D) for the 94 elements H-Pu *J. Chem. Phys.* **132** 154104

Journal of Physics: Conference Series **430** (2013) 012037 doi:10.1088/1742-6596/430/1/012037

Synthesis and Characterization of Cu-Pt Bimetallic Nanoparticles

Xusheng Zheng[1,2], Shoujie Liu[1], Xing Chen[1], Jie Cheng[1,2], Cheng Si[2], Zhiyun Pan[1], Augosto Marcelli[3], Wangsheng Chu[1,2]*, Ziyu Wu[1,2] *

[1]National Synchrotron Radiation Laboratory, University of Science and Technology of China, Hefei, 230029, China
[2]Beijing Synchrotron Radiation Facility, Institute of High Energy Physics, Beijing, 100049, China
[3]Instituto Nazionale di Fisica Nucleare, Laboratori Nazionali di Frascati, P.O. Box 13, 00044, Frascati, Italy

*E-mail: chuws@ustc.edu.cn; wuzy@ustc.edu.cn

Abstracted: Pt-based alloys have recently triggered a lot of attentions due to their important potential industrial applications. They provide great opportunities for the development of low-cost and high-performance fuel-cell catalysts. Many studies have already pointed out the excellent physico-chemical properties of Pt-based alloys, intimately related to their internal structure. Great efforts have been spent to characterize shape, homogeneity, dispersion, alloying extent and kinetic growth of Pt-based nano-particles. Here, we present Cu-Pt bimetallic nano-particles synthesized by the thermal decomposition method under oleylamine and OE coordination. HRTEM images show that Cu-Pt nanostructures having size of about 1.2 nm includes about 35 atoms capped by the surfactant with OA. Accurate structural information of this system has been obtained by XRD and XAFS. A charge transfer mechanism has been observed and Pt occupied Cu sites in these Cu-Pt nanoparticles.

Introduction

An increasing attention has been devoted to Pt-based alloys due to their important potential industrial applications[1,2], which provide a great opportunity for the development of low-cost and high-performance industrial fuel-cell catalysts[3,4]. A variety of Pt-Me (Me: Metal) alloy nano-particles (NPs) such as Pt-Cu, Pt-Fe, Pt-Co, Pt-Mn have been synthesized with a precise control of size, shape and chemical composition[5-7]. These Pt-based alloys exhibit perfect properties hardly achievable by Pt metal Pt[8-11], especially in catalytic reactors [12,13,14].

Excellent physico-chemical properties of Pt-Me alloy nano-particles depend on several factors, such as the extent of the alloying or the atomic distribution and the surface segregation of the particles [15]; the particle homogeneity and structure and shape of particles [16]; the ensemble effect, ligand and geometrical effects in Pt-Me alloys [17]. Therefore, in order to get the ultimate goal, i.e., the control of material properties, a lot of efforts have been devoted to characterize shape, homogeneity, dispersion, alloying extent, and kinetic growth of alloy nano-particles. Although it is well known that the structure of a material strongly influences its properties, it remains a challenge to understand both structure and dynamics of alloy nano-materials. It is then mandatory identify and set up accurate methods to characterize the internal structure of a bimetallic nano-particle [18] at the atomic level.

In this work, we focus on the structure of Cu-Pt nano-clusters with size of ~1 nm by using advanced X-ray absorption spectroscopy (XAS), combined with transmission electron microscopy (TEM) and X-ray diffraction (XRD). Since the catalytic activity of the bimetallic cluster is strongly affected by the structure, but also by the atomic distribution, we will try to provide precise details of the atomic distribution of Cu-Pt bimetallic clusters.

Experimental

Cu-Pt nano-crystals were synthesized in agreement with a previous reported [19] with some modifications. In brief, a ratio of molar $Pt(acac)_2$ and $Cu(acac)_2$, 210 mg of the 1,2-hexadecanediol, 0.6 ml of oleic acid and 0.4 ml of oleylamine were mixed together with 5 ml of 1-octadecene in a 100 ml three-neck round-bottom flask. The flask was purged with Ar, heated to 120 °C for 20 min and then heated to 225 °C at a rate of 20 °C/min. After 30 min, the black-colored solution was cooled to room temperature. The resultant products were isolated by centrifugation, washed with ethanol for several times and resolved in the hexane for the other measurements.

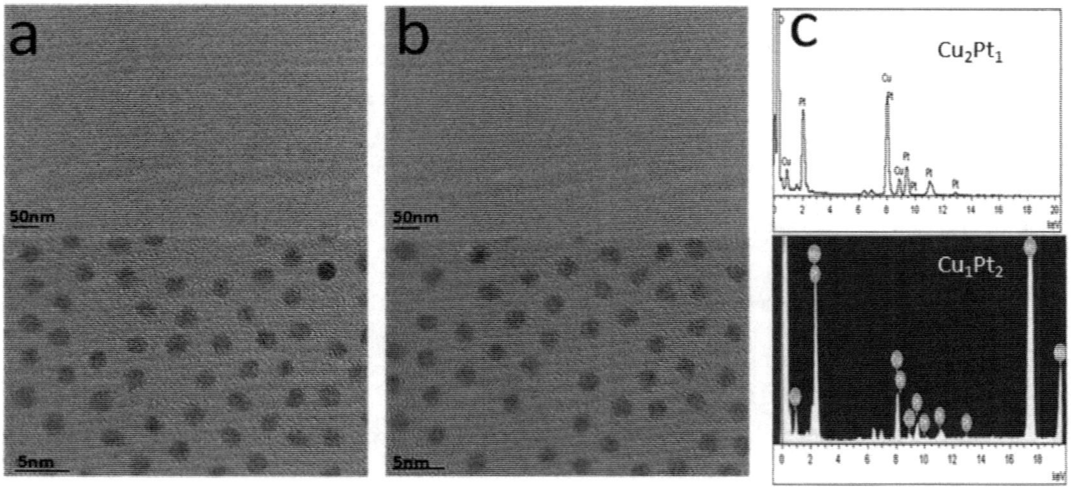

Fig.1 TEM images of Cu_2Pt_1 (a) and Cu_1Pt_2 (b) nano-particles; (c) EDS of Cu-Pt nano-particles.

The Cu-Pt alloy nano-crystals with different molar ratios have been characterized by XRD, TEM and XAS, i.e., both X-ray absorption near-edge structure (XANES) and extended X-ray absorption fine-structure (EXAFS). XRD measurements were performed using a PANalytical X'pert pro diffractometer using a graphite monochromatized Cu-Kα radiation source. Samples were also characterized with a JEOL JEM2100 transmission electron microscope operating at 200 keV. The samples for TEM were prepared by placing one drop of the hexane solution onto a carbon-coated Ni grid and allowing it to dry in air. XAS spectra at the Pt L_3- and Cu K-edge were collected in the transmission mode at the beamline U7C of the National Synchrotron Radiation Laboratory (NSRL, Hefei). The storage ring was working at the energy of 800 MeV with an average electron current of 150 mA. Hard X-rays were monochromatized with a Si(111) double-crystals monochromator. The incident and output beam intensities were recorded using ionization chambers filled by a 25% argon

Journal of Physics: Conference Series **430** (2013) 012037 doi:10.1088/1742-6596/430/1/012037

doped nitrogen mixture and argon gas, respectively. XAS data were analyzed by UWXAFS3.034 and USTCXAFS3.035 software packages.

Results and discussion

1. XRD and TEM characterizations

TEM images (Fig.1a and 1b) show highly homogenous Cu-Pt nano-clusters. The magnified TEM images show nano-clusters with size of ~1.2 nm, which contains approximately 40 atoms. The Cu/Pt ratio of these structures can be obtained with the Energy Dispersive Spectrometry (EDS) (Fig1.c). It points out that the ratio of Cu_2Pt_1 and Cu_1Pt_2 samples are consistent with that of $Cu(acac)_2$ and $Pt(acac)_2$ raw materials. The XRD patterns of the Cu-Pt bimetallic nano-clusters with different molar ratios are shown in Fig.2 and rules out the presence of Cu, Pt metal and their oxides in the prepared samples. Due to the small size of the nano-cluster, XRD patterns of CuPt samples present only the peak of (111). We may recognize that these clusters have very similar XRD patterns with only a small shift of the (111) peak (Fig.2a). On the basis of the Scherrer formula, the sizes of both Cu_2Pt_1 and Cu_1Pt_2 were evaluated to be about 1.2 nm from the half width of the diffraction peaks, in good agreement with TEM data.

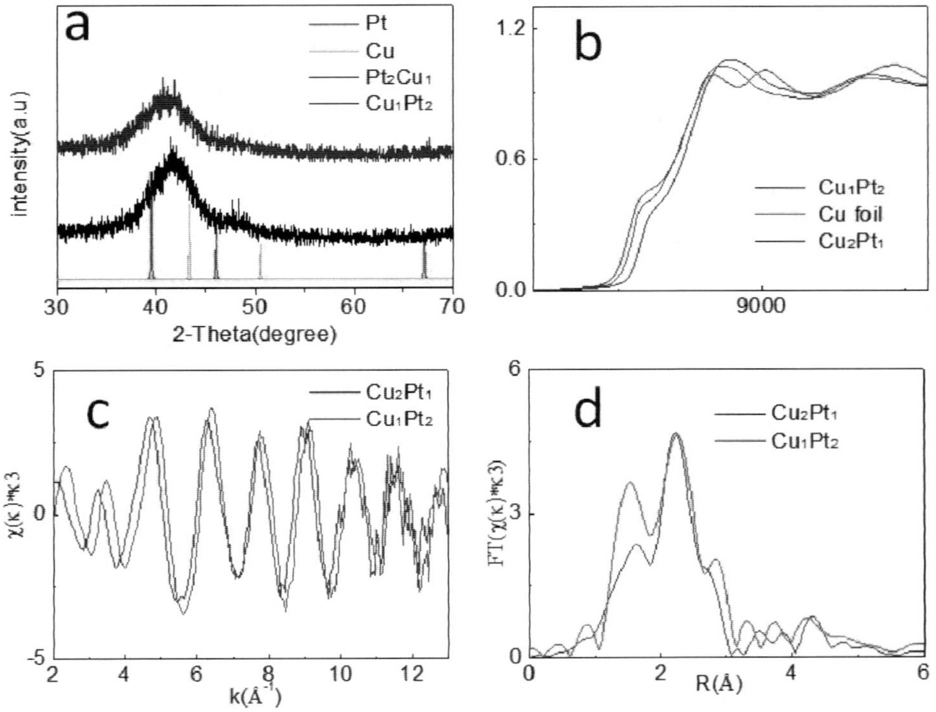

Fig.2 XRD patterns (a) and Cu K-edge XANES spectra (b) of Cu-Pt bimetallic nano-clusters with different molar ratios; Cu K-edge EXAFS of Cu_2Pt_1 and Cu_1Pt_2 bimetallic nano-clusters (c) and their FTs (d).

2. XANES spectroscopy at Cu K-edge and Pt L_3-edge

Fig. 2b shows Cu K-edge XANES spectra of Cu-Pt clusters with different molar ratios. The main peak at 8995 eV, which corresponds to the transition from 1s to 4p, reflects the character of the 4p unoccupied states[8]. The increasing intensity of the main feature means that the number of unoccupied states of the Cu 4p electrons in the Cu_1Pt_2 cluster increases compared to the Cu_2Pt_1 cluster, with a decreasing Cu composition. The threshold peak which occurs at 8980 eV is due to the hybridization between p- and d-states and probes the 3d character of the unoccupied states. The increasing intensity of this feature probes also the enhancement of the p-d hybridization at Cu sites upon alloying. Fig.3a shows Pt L_3-edge XANES spectra of Cu-Pt nanoclusters. The Pt L_3-edge XANES is a dipole allowed transition from $2p_{3/2}$ to $5d_{3/2}$ and $5d_{5/2}$ states. However, it only involves unoccupied states of $5d_{5/2}$ due to the lack of $5d_{3/2}$ holes in Cu-Pt nanoparticles. Indeed, it is very clear that the intensity of the white line (WL) decreases, increasing the Cu concentration. The behavior can be explained because in these nanostructures $5d_{5/2}$ Pt unoccupied states were partially occupied by electrons transferred from the Cu site. From the theoretical point of view, the loss of Cu electrons induces an increase of the Cu chemical value. The hypothesis is confirmed by the shift of the Cu edge towards high energy (Fig. 2b). Therefore, an effective charge transfer is detected in Cu-Pt nanoparticles, similar to what has been observed in Cu-Pt and Cu-Pd alloys by Lee and Cho with XANES experiments [20]. Increasing the Cu/noble metal molar ratio they also observed a clear reduction of the Pt L_3-edge WL intensity.

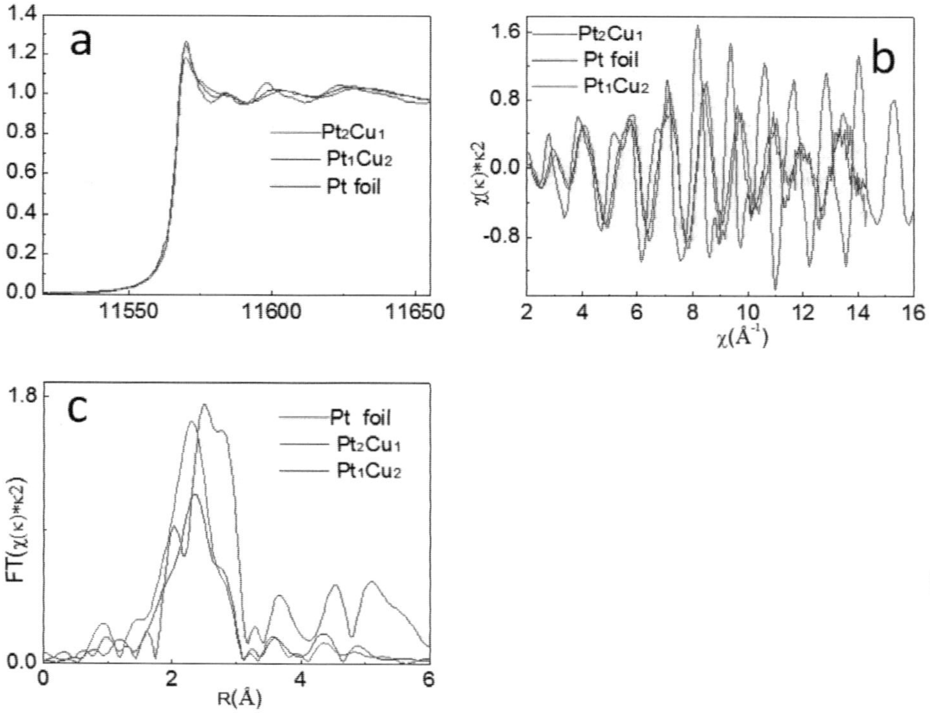

Fig.3 (a) Pt L_3-edge XANES spectra of Cu-Pt clusters with different molar ratios; (b) Pt L_3-edge EXAFS oscillation functions and (c) the corresponding FTs of Pt foil, Cu_2Pt_1 and Cu_1Pt_2 bimetallic nano-clusters.

3. EXAFS spectroscopy for Cu K- and Pt L_3-edge.

Fig.2d and 3c show the k^3-weighted Fourier transforms (FT) of the EXAFS of the two Cu-Pt samples at the Cu K-edge and at the Pt L_3-edge, respectively. Figures clearly show that the FTs of the Cu K-edge EXAFS of the CuPt NPs have a double peak in the region 1.7-3.3 Å (Fig. 2d) with a lower-R (1.8-2.0 Å) and a higher-R (2.6-2.9 Å) contribution. Both can be associated to first shell (the nearest) metal-metal bonds. The splitting of the first shell peak is due to the interference between the EXAFS oscillations of Cu-Cu and Cu-Pt bonds, whose phase shift and amplitude are different. However, as the Cu concentration increases, the intensity of the peak at lower R decreases. Moreover, the EXAFS oscillations at the Pt L_3-edge (Fig. 3b) of as-prepared samples show significant differences in contrast to those of Pt foil, not only in the phase but also in the intensity. From the EXAFS analysis, we may also conclude that due to the formation of Cu-Pt bond the bond distances in our samples decrease, as clearly shown from the FTs in Fig. 3c. The FTs of the Cu-Pt NPs also show that the Pt-Cu distance is ~2.7 Å, which is just the length of the Cu-Cu bond in agreement with the claiming that in these nano-particles Pt occupies Cu sites.

Conclusions

In summary, synthesized Cu-Pt alloy nano-particles with different molar ratios were homogenous systems with an average size of ~1 nm as characterized by XRD, TEM, EDS and XAS. Moreover, from EXAFS results Pt atoms occupy Cu sites and Pt-Cu bonds are formed. An evident charge transfer from Cu to Pt is also detected by XANES in Cu-Pt nano-clusters. All together, the structural and electronic characterizations provide useful information to understand the properties of such small nano-clusters

Acknowledgements

This work was partly supported by the Knowledge Innovation Program of the Chinese Academy of Sciences (KJCX2-YW-N42), the Key Important Project of the National Natural Science Foundation of China (10734070), the National Basic Research Program of China (2009CB930804) and the National Natural Science Foundation of China (NSFC 11179023 and 11179001).

References

[1] Y.-J. Zhao, S.-S. Chng, T.-P. Loh, 2007 J Am Chem Soc 129 492 – 493

[2] R. Narayanan, M. A. El-Sayed 2004 Nano Lett 4 1343 – 1348

[3] Stamenkovic, V. R.; Mun, B. S.; Mayrhofer, K. J. J.; Ross, P. N.; Markovic, N. M., 2006 J Am Chem Soc 128(27) 8813-8819

[4] Son, S. U.; Jang, Y.; Park, J.; Na, H. B.; Park, H. M.; Yun, H. J.; Lee, J.; Hyeon, T., 2004 J Am Chem Soc 126 (16) 5026-5027

[5] Koh, S.; Strasser, P., 2007 J Am Chem Soc 129(42) 126

[6] Sun, S. H.; Murray, C. B.; Weller, D.; Folks, L.; Moser, A., 2000 Science 287(5460) 1989-1992

[7] Kang, Y. J.; Murray, C. B., 2010 J Am Chem Soc 132(22) 7568

[8] Stamenkovic, V. R.; Mun, B. S.; Arenz, M.; Mayrhofer, K. J. J.; Lucas, C. A.; Wang, G. F.; Ross, P. N.; Markovic, N. M., 2007 Nat Mater 6 (3) 241-247

[9] I.-S. Park, K.-S. Lee, J.-H. Choi, H.-Y. Park, Y.-E. Sung, 2007 J Phys Chem C 111 19126-19133

[10] H. Ye, R. M. Crooks, 2007 J Am Chem Soc 129 3627-3633
[11] W. Chen, J. Kim, S. Sun, S. Chen, 2008 J Phys Chem C 112 3891-3898
[12] Howard, L. E. M.; Nguyen, H. L.; Giblin, S. R.; Tanner, B. K.; Terry, I.; Hughes, A. K.; Evans, J. S. O., 2005 J Am Chem Soc 127(29) 10140-10141
[13] W. Chen, J. Kim, S. Sun, 2007 S. Chen, Langmuir 23 11303 –11310
[14] Y. Li, X. L. Zhang, R. Qiu, R. Qiao, Y. S. Kang, 2007 J Phys Chem C 111 10747-10750
[15] Shibata, T.; Bunker, B. A.; Zhang, Z. Y.; Meisel, D.; Vardeman, C. F.; Gezelter, J. D., 2002 J Am Chem Soc 124(40) 11989-11996
[16] Zhang, J. T.; Tang, Y.; Weng, L.; Ouyang, M., 2009 Nano Lett 9(12) 4061-4065
[17] Jiang, H. L.; Akita, T.; Ishida, T.; Haruta, M.; Xu, Q., 2011 J Am Chem Soc 133(5) 1304-1306
[18] Stamenkovic, V. R.; Fowler, B.; Mun, B. S.; Wang, G. F.; Ross, P. N.; Lucas, C. A., 2007 Science 315(5811) 493-497
[19] Xu, D.; Bliznakov, S.; Liu, Z. P.; Fang, J. Y.; Dimitrov, N., 2010 Angew Chem Int Edit 49(7) 1282-1285
[20] Chen, X.; Chu, W. S.; Cai, Q.; Xia, D. G.; Wu, Z. H.; Wu, Z. Y., 2006 Radiat Phys Chem 75(11) 1622-1625

In-situ observation of Cu-Pt core-shell nanoparticles in the atomic scale by XAFS

Xusheng Zheng[1,2], Shoujie Liu[1], Xing Chen[1], Jie Cheng[1,2], Qing Ye[1], Zhiyun Pan[1], Augosto Marcelli[3], Wangsheng Chu[1,2]* , Ziyu Wu[1,2]*

[1]National Synchrotron Radiation Laboratory, University of Science and Technology of China, Hefei, 230029, China
[2]Beijing Synchrotron Radiation Facility, Institute of High Energy Physics, Beijing, 100049, China
[3]Instituto Nazionale di Fisica Nucleare, Laboratori Nazionali di Frascati, P.O. Box 13, 00044, Frascati, Italy

*E-mail: chuws@ustc.edu.cn; wuzy@ustc.edu.cn

Abstract: Bimetallic nanoparticles play an important role in potential industrial applications, such as catalysis, optoelectronics, information storage and biological labeling. Herein, homogeneous Cu-Pt core-shell nanoparticles with the averaged size of 8 nm have been synthesized by chemical methods. Cu atoms diffusion process, which motivated by heating, was observed in-situ by using temperature-dependent x-ray absorption fine-structure (XAFS) spectroscopy. Results show that Cu diffuse gradually from Cu core to Pt shell in these nanoparticles with increasing temperature. We also found the surface ligand (O) bonded Pt at the room temperature and were removed gradually by heating the sample. The analysis of the diffusion process in bimetallic nanoparticles will provide important guideline for their designing and tuning.

Introduction

It is well known that bimetallic nanoparticles exhibit perfect physical and chemical properties that differ from corresponding bulk materials because the addition of a second atomic species not only changes the structure of nanoclusters but also controls their electronic structure. Bimetallic nanoparticle has spawned an enormous research effort aimed at application in catalysis[1,2], optoelectronics[3,4], information storage[5,6], biological labeling[7,8]. Recently, controlling the external morphologies (e.g., shape, size and uniformity) has become focus of many studies in order to exploit the relationship between the structure and properties, which is necessary to apply nano-materials in nanotechnology[9]. However, internal structure of nanoparticles which influencing the physical property directly are very important to be paid attention as well[10,11]. Although synthesis and characterization of bimetallic clusters have motivated numerous experimental and theoretical studies, the atomic distribution, structural dynamic and thermodynamic behavior of bimetallic nanoparticles, which directly affects their properties, still remain less explored.

It is common sense that bimetallic clusters exhibit high performance usually by annealing process. It may be due to that the surfactants were removed by high temperature and the internal structure of bimetallic nanoparticles formed prior coordination. A few researches reported experimental evidence of temperature effects on the bimetallic internal structure[12,13]. For instance, the process of thermodynamic behavior of Pt-Cu nanoparticles aggregating in various temperatures was investigated by XRD to exhibit the diffusion and nucleation process. However, XRD technique only provided long-range order but not short-range order[14,15]. Pt-Cu metallic nanoalloys were usually studied at room temperature in order to avoid removal of stable ligand and increasing of the size of nanoparticles. Herein, we focus on the structure and the diffusion of Cu-Pt nanoparticles in different temperatures by in-situ XAFS technique.

Experimental methods

Core-shell Cu-Pt nanoparticles were synthesized with a similar scheme employed in the Fe-Pt synthesis, washed by ethanol and segregated by centrifuging. Then 22 mg dried core-shell Cu-Pt nanoparticles were dissolved with 100 mg graphite powder as well as plenty of hexane. The solution was mixed homogeneously by Ultrasonic. Finally the mixed solution was dried in vacuum drying oven and pressed into a pellet in a stainless steel holder and used to collect the XAFS data.

The same apparatus for temperature-dependant in-situ XAFS study on atomic kinetics across the MIT in crystalline VO_2[16] was employed to probe the thermodynamic process of core-shell nanoparticles, which can preserve the precision of XAFS data enough to explain the real process. Powder X-ray diffraction (XRD) data were collected at U7B stations of the National Synchrotron Radiation Laboratory (NSRL, Hefei). Energy-dispersive spectroscopy (EDS) analysis and high-resolution transmission electron microscopy (TEM) image were obtained by a JEM-2010 TEM.

Results and discussion

Fig.1a shows TEM image of the as-prepared Cu-Pt core-shell nanoparticles, exhibiting a homogenous core-shell shape with the averaged size of 8 nm. And the corresponding size distribution, which will benefit extraction of precise structure parameter, was presented in the inset of Fig.1a. In order to check if the nanoparticles were core-shell structure, we performed the Cu K-edge and Pt L_3-edge XAFS measurements at room temperature. Fig.1b shows the Fourier transforms (FTs), looking at the data in the R-space (real space), of EXAFS spectra at Pt L_3-edge. Two intense peaks emerge in the region 1.5-3.1 Å for the as-made sample (green solid line), consisting of a lower-R peak (1.5-1.8 Å) and a higher-R peak (2.5-2.2 Å), much different from the signal of the Pt foil (red solid line). It can be easily found out that these two peaks correspond to the Pt-ligand bond and the Pt-Pt bond, respectively. In contrast, Cu K-edge FT-EXAFS spectra shown in the Fig.1c was similar to the Cu foil except for the amplitude. The missing of the signals of Cu-O and Cu-Pt peak indicates that the Cu cluster in our sample presents in the form of elemental crystal. It was well-known that Cu clusters were extremely oxidized to form CuO or CuO_2 if they were exposed to air[17]. Therefore, Cu clusters in the as-made sample should be protected by Pt shells to avoid oxidizing. Based on all above analyses, we can get a simple and clear view about the structure of the as-prepare Pt-Cu nanoparticles: Pt coordinated with ligand of PVP in the shell and protected Cu core from oxidization, and Cu formed core in the nanoparticles without oxidization and alloying.

In order to obtain quantitative information of the local structure, FT-EXAFS of the first shell (0.17-0.33 nm) were fitted (see circle lines in Fig1b and c) using theoretical phase-shift and amplitude

Journal of Physics: Conference Series **430** (2013) 012038 doi:10.1088/1742-6596/430/1/012038

generated by FEFF 8 program. The coordination number *N* of Pt can give some useful information to prove core-shell structure. In Table 1, N_{Pt-Pt} + N_{Pt-Cu} is 12 in total, indicating that the NP surface is covered entirely with Pt atoms, in agreement with the report by Zhang[18]. It can also be concluded that the Cu-Pt nanoparticles present the core-shell structure.

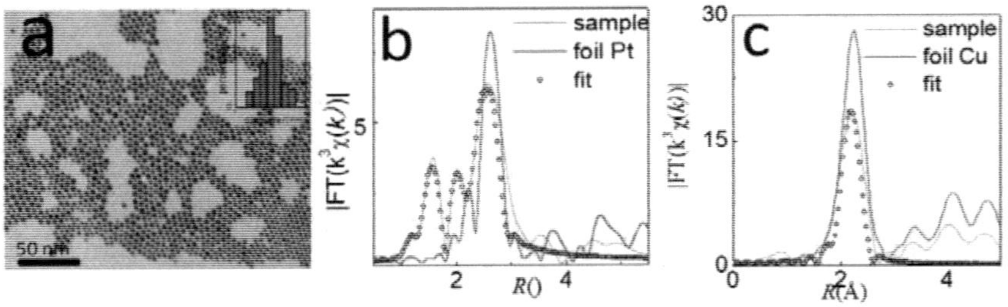

Fig.1 (a) TEM images of Cu-Pt core-shell nanoparticles; (b) and (c) the FT-EXAFS spectra of Cu-Pt core-shell nanoparticles at Pt L_3-edge and Cu K-edge.

Pt L_3-edge XAFS spectra of diffusion process in the temperature range 25-600℃ were shown in Fig 2. The extended XAFS (EXAFS) functions χ (k) (Fig. 2a) and the corresponding FTs (Fig. 2b) exhibit remarkable and systematic temperature evolution. As the temperature increases, it was found that the EXAFS oscillations of the Cu-Pt nanoparticle changed significantly. This change is originated from the fact that, as the diffusion ocuurs between Cu and Pt atoms, the average number of Cu neighbor atoms surrounding the Pt absorbers increases, resulting in the formation of the Cu-Pt alloy. This observation can be also indentified with the Fourier transform profiles in the real space. The FT curve at room temperature shows a intense peak in the range of 1.5-2 Å, which was typical characterization of the Pt-ligand (O). It was interesting to find that the intensity of the peak of Pt-ligand about 1.5 Å decreased and disappeared at 230℃, suggesting that the surface ligand can protect Cu-Pt nanoparticles surface at high temperature.

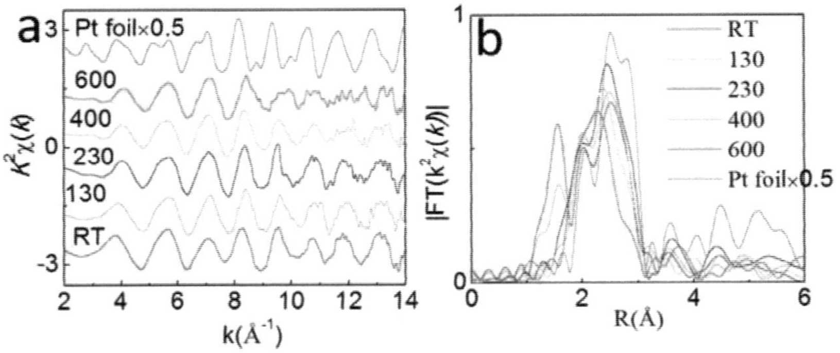

Fig. 2 (a) $k^2\chi(k)$ of Cu-Pt nanoparticles at Pt L_3-edge and (b) Corresponding FTs vs. temperature.

According to the parameters from the EXAFS fits in the table 1, the coordination number of the Pt-ligand, exactly the Pt-O bond, drops from 3.1 at room temperature to 1.6 at 130 ℃ and disappear at 230 ℃. With temperature increasing, the peak shifts to lower R and gradually becomes more intense. This may arise from the formation of the Pt-Cu alloy when Cu atoms gradually diffuse to the Pt shell. The structure parameters in the table 1 can perfectly explain the change. At the interface of the core-shell at room temperature, only a small amount of Pt-Cu bonds existed. When heating the sample, the coordination number of the Pt-Cu bond increased from 1 at room temperature to 4 at 600 ℃, suggesting the diffusion was finished. Meanwhile, that of the Pt-Pt bond decreased gradually from 6.2 to 0, suggesting the total Cu-Pt alloy formed at the temperature of 600 ℃. Therefore, from the change of the coordination number of Pt-Pt and Pt-Cu, it can be estimated that the percentage of Cu-Pt alloy at the interface was 2%, 10%, 30%, 60% and 100% at 25℃, 130℃, 230℃, 400℃ and 600℃, respectively.

Table 1 Structural parameters of bond distance R, coordination number N, and Debye factor from EXAFS fits of the as-prepared Cu-Pt nanoparticle at different temperature.

Sample	Bond	R (Å)	N	σ^2 (10^{-3}Å^2)
Rt- Pt	Pt-O	2.00±0.01	3.1±0.8	8.0±0.8
	Pt-Cu	2.64±0.01	1.0±0.8	6.0±0.8
	Pt-Pt	2.87±0.01	6.2±0.8	3.0±0.8
130- Pt	Pt-O	2.03±0.01	1.6±0.8	12.0±0.8
	Pt-Cu	2.65±0.01	1.8±0.8	16.1±0.8
	Pt-Pt	2.85±0.01	5.3±0.8	4.7±0.8
230- Pt	Pt-O	2.66±0.01	1.4±0.8	6.0±0.8
	Pt-Cu	2.34±0.01	1.3±0.8	13.6±0.8
	Pt-Pt	2.83±0.01	3.3±0.8	5.0±0.8
400- Pt	Pt -Cu	2.68±0.04	2.3±0.8	7.2±0.8
	Pt- Pt	2.80±0.05	2.8±0.8	3.5±0.8
600- Pt	Pt -Cu	2.65±0.03	4±0.8	8.1±1.2

Conclusions

In summary, we prepared homogeneous Cu-Pt core-shell nanoparitcles with average size of 8 nm by chemical methods. By performing temperature-dependent x-ray absorption fine structure measurements, we studied in situ the local structures of core-shell nanoparticles at different temperatures. The result shows the surface ligands bonded Pt at the room temperature and were removed gradually by heating the sample. The temperature evolution of the coordination number of the Pt-Cu and Pt-Pt bonds also provides insight into the extent of the alloying at the interface between the Pt shell and the Cu core of the Cu-Pt nanoparticles. This detailed information about the diffusion process should be useful for designing and controlling bimetallic nanoparticles for high implication.

Acknowledgements

This work was partly supported by the Knowledge Innovation Program of the Chinese Academy of Sciences (KJCX2-YW-N42), the Key Important Project of the National Natural Science Foundation of China (10734070), the National Basic Research Program of China (2009CB930804) and the National

Natural Science Foundation of China (NSFC 11179023 and 11179001).

References

1. Son, S. U.; Jang, Y.; Park, J.; Na, H. B.; Park, H. M.; Yun, H. J.; Lee, J.; Hyeon, T., 2004 *J Am Chem Soc* **126(16)** 5026-5027
2. Stamenkovic, V. R.; Mun, B. S.; Arenz, M.; Mayrhofer, K. J. J.; Lucas, C. A.; Wang, G. F.; Ross, P. N.; Markovic, 2007 *N. M., Nat Mater* **6(3)** 241-247
3. Doering, W. E.; Nie, S. M., 2003 *Anal Chem* **75 (22)** 6171-6176
4. Schrinner, M.; Ballauff, M.; Talmon, Y.; Kauffmann, Y.; Thun, J.; Moller, M.; Breu, J., 2009 *Science* **323(5914)** 617-620
5. Nguyen, H. L.; Howard, L. E. M.; Giblin, S. R.; Tanner, B. K.; Terry, I.; Hughes, A. K.; Ross, I. M.; Serres, A.; Burckstummer, H.; Evans, J. S. O., 2005 *J Mater Chem* **15(48)** 5136-5143
6. Wang, D. S.; Peng, Q.; Li, Y. D., 2010 *Nano Res* **3(8)** 574-580
7. Lee, J. H.; Huh, Y. M.; Jun, Y.; Seo, J.; Jang, J.; Song, H. T.; Kim, S.; Cho, E. J.; Yoon, H. G.; Suh, J. S.; Cheon, 2007 J. *Nat Med* **13(1)** 95-99
8. Pabi, S. K.; Murty, B. S., 1996 *B Mater Sci* **19(6)** 939-9
9. Zhang, J. T.; Tang, Y.; Weng, L.; Ouyang, M., 2009 *Nano Lett* **9(12)** 4061-4065
10. Lim, B.; Yu, T. Y.; Xia, Y. N., 2010 *Angew Chem Int Edit* **49(51)** 9819-9820
11. Zhang, J. L.; Vukmirovic, M. B.; Xu, Y.; Mavrikakis, M.; Adzic, R. R., 2005 *Angew Chem Int Edit* **44(14)** 2132-2135
12. Chu, W. S.; Zhang, S.; Yu, M. J.; Zheng, L. R.; Hu, T. D.; Zhao, H. F.; Wu, Z. Y., 2009 *J Synchrotron Radiat* **16** 30-37
13. Chu, W. S. ; Marcelli, A. ; Hu, T. D., 2009 *New J Phys* **11** 083005
14. Shibata, T.; Bunker, B. A.; Zhang, Z. Y.; Meisel, D.; Vardeman, C. F.; Gezelter, J. D., 2002 *J Am Chem Soc* **124 (40)** 11989-11996
15. Wu, J. B.; Qi, L.; You, H. J.; Gross, A.; Li, J.; Yang, H., 2012 *J Am Chem Soc* **134(29)** 11880-11883
16. Yao, T.; Zhang, X. D.; Sun, Z. H.; Liu, S. J.; Huang, Y. Y.; Xie, Y.; Wu, C. Z.; Yuan, X.; Zhang, W. Q.; Wu, Z. Y.; Pan, G. Q.; Hu, F. C.; Wu, L. H.; Liu, Q. H.; Wei, S. Q., **2010** *Phys Rev Lett* **105 (22)** 226405
17. Yin, M.; Wu, C. K.; Lou, Y. B.; Burda, C.; Koberstein, J. T.; Zhu, Y. M.; O'Brien, S., 2005 *J Am Chem Soc* **127 (26)** 9506-9511
18. Liu, F.; Wechsler, D.; Zhang, P., 2008 *Chem. Phys. Lett.* **461 (4-6)** 254-259

Journal of Physics: Conference Series **430** (2013) 012039 doi:10.1088/1742-6596/430/1/012039

Hg and Se Speciation in Liver Tissue of Marine Birds

Cahit Karanfil[1], Karyn Bischoff[2], Grant Bunker[3]

1) Muğla University, Muğla, Turkey

2) Cornell University, Ithaca, NY, USA

3) Illinois Institute of Technology, Chicago, IL, USA

E-mail: bunker@iit.edu

Abstract. X-ray fluorescence microprobe maps of Hg and Se distribution were made on liver tissue slices from marine birds (Cormorants) from Florida's Everglades. The birds bio-concentrate these toxic pollutants that are ingested from their diet of fish. The fluorescence maps show strong localization of Mercury and Selenium into "hot spots" less than 100 microns in size. XAFS measurements were made on hot spots to demonstrate that the Hg and Se are indeed are chemically bound, supporting the idea that moderate levels of Se may have a protective effect against Hg through its complexation.

1. Introduction

Hg pollution and toxicity is a serious health concern. Environmental Hg is generated from both natural and anthropogenic sources, commonly as an airborne pollutant from the burning of coal and municipal waste, volcanic eruptions, and as a water pollutant from natural and industrial emissions. Recent reviews can be found in [1,2]. It has been observed that elevated Hg concentrations in tissues are often accompanied by elevated Se concentrations. This work is an investigation into the speciation of Hg and Se in tissues using x-ray fluorescence mapping and XAFS. Previous electron microscopic/x-ray microprobe studies by Martoja and Viale [3] on liver tissue from marine mammals has shown that Hg and Se are present in small particles which appeared to have the approximate stoichiometry HgSe; these were proposed to be the mineral Tiemannite. Subsequent XAFS studies by Arai et al [4] were done on liver tissue *extracts* from fur seals and albatrosses that indicated coordination of Se by Hg in a manner consistent with HgSe deposition. Together these studies suggest that Hg is complexed by Se (and perhaps S) into an inorganic form that renders the Hg immobile and less toxic, as proposed by Martoja and Viale [3]. However, those measurements were made on different samples, from different species, with different techniques, and the only successful XAFS measurements were on tissue *extracts*, not whole tissues.

Here we report spatially resolved x-ray fluorescence and XAFS measurements on the *same* sample consisting of *whole* liver tissue sections (rather than extracts) from a cormorant (a fish-eating marine bird) obtained from Florida's Everglades.

Journal of Physics: Conference Series **430** (2013) 012039 doi:10.1088/1742-6596/430/1/012039

2. Materials and Methods

Initial investigations were made on the speciation of Hg in liver tissues from cormorants known to have high Hg exposure. Focussed x-ray beams were used on the BioCAT undulator beamline at the APS, using a multilayer array analyzer. Initially these measurements were done on bulk tissue specimens, but the average concentration was too low to obtain good XAFS data. However, in one measurement, at a particular location, a surprisingly large signal was obtained, suggesting that there were "hot spots" of high Hg concentration. Subsequently, spatial maps of x-ray fluorescence were made on 20 micrometer thick liver tissues slices using the BioCAT undulator beamline at the APS with a Vortex silicon drift detector. The step size is 0.14 mm in the short direction (x) and 0.10 in the long direction (y). Because statistical correlations between the bulk Hg and Se contents were known from the literature, Hg and Se maps were measured, as were Fe and elastic scatter as controls. Hg and Se XAFS measurements were made on selected regions of high concentration. More than 300 one-minute long continuous scans were summed together to obtain sufficient signal to noise. Here we report Se K-edge results.

3. Results

Fluorescence maps are shown in Figure 1. The Hg and Se evidently are mostly concentrated in "hot spots" that are smaller than the spatial resolution of approximately 100 micrometers.

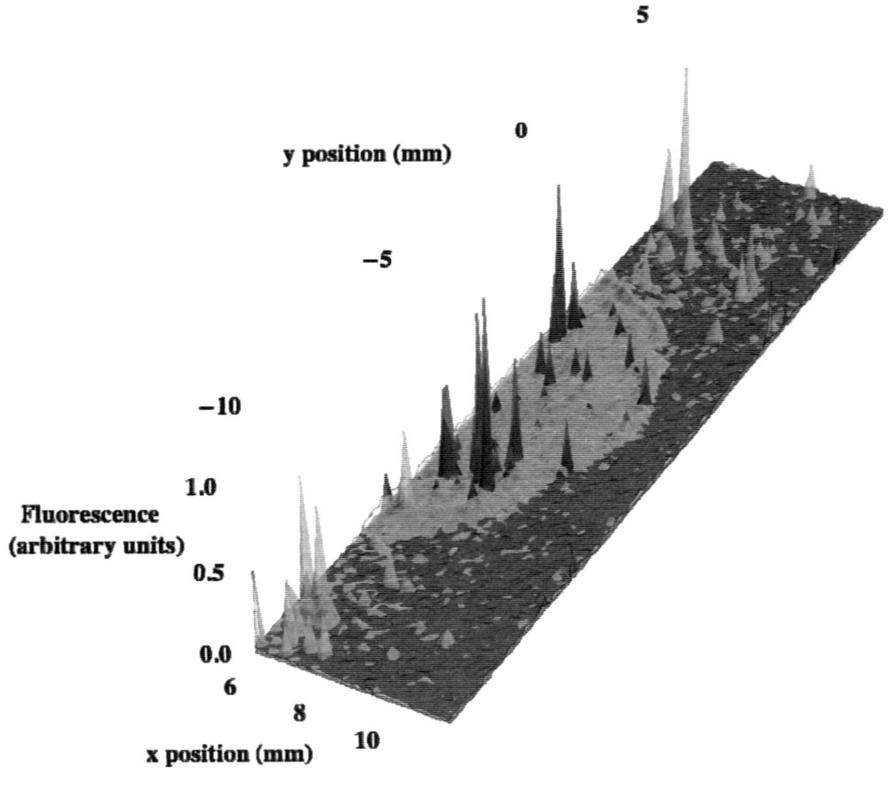

Figure 1: Spatial maps of fluorescence of Hg (red), Se (blue), and Fe (yellow). This shows the existence of "hot spots" of high concentration of both Hg and Se. The peaks corresponding to Hg and Se overlap so that they appear purple. The yellow Fe peaks are not spatially well-correlated with Hg and Se. The spatial resolution is approximately 100 micrometer.

XAFS measurements were subsequently made on one of the hot spots. Several hundred quick-XAFS scans of one-minute duration were summed together to obtain sufficient signal to noise. We report results on first shell Se edge EXAFS analysis. The Hg L_I edge is several hundred eV below the Se K-edge, which complicates the EXAFS analysis for the Se edge, because of the limited data range. Although the Hg L_I edge EXAFS overlaps the Se EXAFS range, it varies more slowly there because of the quadratic dependence of E on k. It varies sufficiently slowly that it appears as a background that is spectrally distinct from the Se edge EXAFS. Therefore conventional methods of analysis are sufficient for this purpose. Only the first shell was analyzed because the limited information content in the data restricts the number of parameters.

The data were reduced using Mathematica 8 and Athena [5], and then fit using Artemis [5], using the mineral HgSe (Tiemannite) as a theoretical standard (see Figure 2). Tiemannite is of the Zincblende structure, with each Hg and Se tetrahedrally surrounded by atoms of the other type. Fourier transforms of the Se K-edge data from k=2 to 10.5 $Å^{-1}$ are shown in figure 3; no phase correction was done, so the true distances are approximately 0.4 Angstrom longer than it appears. The background peak at R~1.0 Å is primarily due to the Hg L_I edge EXAFS extending into the Se transform range, but it is of no consequence because it is spectrally isolated at low-R.

The first shell Fourier transform peak is split because of a Ramsauer-Townsend resonance in Hg's backscattering amplitude, which has a minimum at about k=6 Å, giving the appearance of a beat. A value of $S_0^2 = 0.85$ was used, and delta E_0 was 4.10 ± 1.43 eV. The first shell distribution of atoms is well fit as 4 ± 1 Hg atoms at 2.61 ± 0.02 Å and $\sigma^2 = 0.010 \pm .003$ $Å^2$. This large value of σ^2 is presumably due to structural disorder. The number of independent points was 7, variables 4, reduced chi square 15.1, R-Factor .056. Error estimates presented are fitting errors estimated by Artemis.

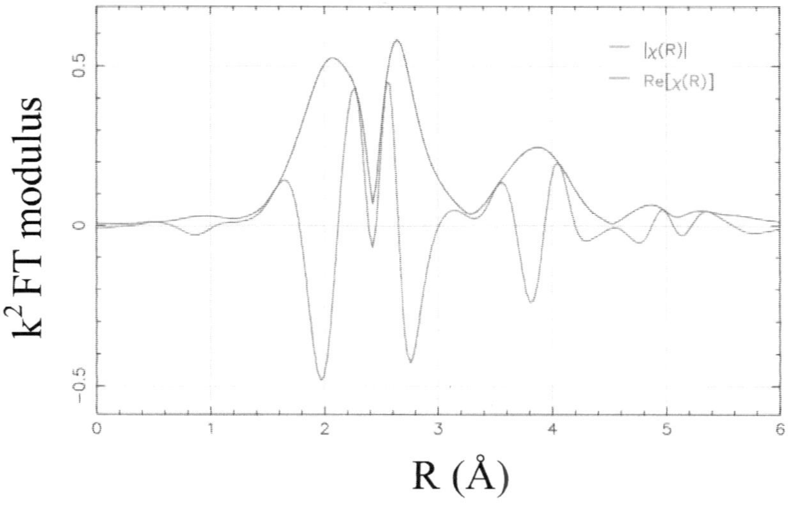

Figure 2: Fourier Transform of theoretical spectrum of Tiemannite, with DWFs calculated by FEFF8 in Debye approximation at sample temp 300K and Debye temperature 200K. The nearest neighbor peaks are due to 4 Hg atoms at 2.63 Å, and the peak at 3.8 Å is Se. The first shell peak splitting is due to a Ramsauer-Townsend-type resonance in the backscattering amplitude of Hg.

Journal of Physics: Conference Series **430** (2013) 012039 doi:10.1088/1742-6596/430/1/012039

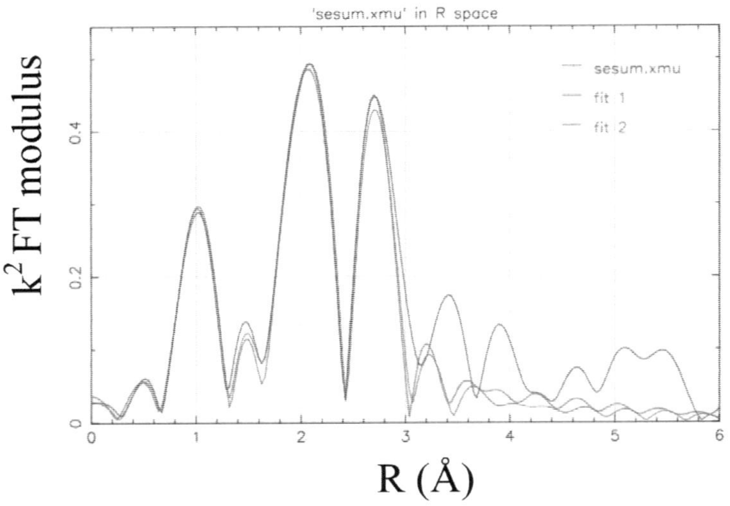

Figure 3: Magnitude of k^2 Fourier transform of summed Se data and two R-space fits to data fit between R=1.6-3.2 Å. The (background-fitted) peak at R=1.0 Å is due to Hg EXAFS extending into the Se transform range, and is of no consequence. The data within the fit range correspond to a single shell of 4 Hg at 2.61 Å; the peak splitting is due to the shape of the Hg scattering amplitude function.

4. Discussion and Conclusions

Electron microscopic studies in marine mammals previously have shown evidence for nanoparticles containing Hg and Se which were assumed to be HgSe [3]. Although the nearest neighbor environment of the Se is indeed similar to that in the mineral Tiemannite, the second and higher shells appear to differ from this simple picture. In Tiemannite there is also a well ordered set of Se atoms in the second coordination shell, and such a shell is not evident in the sample Fourier transforms. The observed suppressed second shell signal could be due to several factors: large structural disorder; reduced number of neighbors, if the particles are nano-scale; or mixed Se and S coordination, which would tend to cause cancellations of the EXAFS contributions through interference.

Direct evidence is presented here for chemical bonding between the Hg and Se that is contained in small particles in liver tissue of cormorants. This is consistent with the proposal [1] that Se may offer a protective effect against Hg through its complexation into inorganic particles.

References

[1] "Environmental Chemistry and Toxicology of Hg", Guangiang Liu, Yong Cai, Nelson O'Driscoll editors, John Wiley and Sons, Inc., Hoboken New Jersey, (2012)
[2] "Se and Hg in organisms: Interactions and Mechanisms", Dan-Yi Yang, Yu-Wei Chen, John M. Gunn, Nelson Belzile, Environ. Rev. 16, 71-92, (2008)
[3] "Accumulation de granules de séléniure mercurique dans le foie d'odontocètes: un mécanisme possible de détoxication du methyl-mercure par le sélénium.", Martoja, R., D. Viale, 1977, C. R. Acad. Sc. Paris, Série D, 285: 109-112
[4] "Chemical Forms of Hg and Cadmium Accumulated in Marine Mammals and Seabirds as Determined by XAFS Analysis", Teruko Arai, Tokutaka Ikemoto, Akiko Hokura,Yasuko Terada,Takashi Kunito, Shinsuke Tanabe, Izumi Nakai, Environ. Sci. Technol. 2004, 38, 6468-6474
[5] "ATHENA, ARTEMIS, HEPHAESTUS: data analysis for X-ray absorption spectroscopy using IFEFFIT", B. Ravel and M. Newville, J. Synchrotron Rad. 12, pp 537--541 (2005)

Journal of Physics: Conference Series **430** (2013) 012040 doi:10.1088/1742-6596/430/1/012040

EXAFS studies of prostate cancer cell lines

J. Czapla[1], W. M. Kwiatek[1,*], J. Lekki[1], A. Kisiel[2], R. Steininger[3], J. Goettlicher[3]

[1] The Henryk Niewodniczanski Institute of Nuclear Physics, Polish Academy of Sciences, Krakow, Poland

[2] Institute of Physics, Jagiellonian University, ul. Reymonta 4, 30-059 Kraków, Poland

[3] Karlsruhe Institute of Technology, Institute for Synchrotron Radiation, Hermann-von-Helmholtz-Platz 1, D-76344 Eggenstein-Leopoldshafen, Germany

* Corresponding author, e-mail: Wojciech.Kwiatek@ifj.edu.pl

Abstract. Sulphur plays a vital role in every human organism. It is known, that sulphur-bearing compounds, such as for example cysteine and glutathione, play critical roles in development and progression of many diseases. Any alteration in sulphur's biochemistry could become a precursor of serious pathological conditions. One of such condition is prostate cancer, the most frequently diagnosed malignancy in the western world and the second leading cause of cancer related death in men. The purpose of presented studies was to examine what changes occur in the nearest chemical environment of sulphur in prostate cancer cell lines in comparison to healthy cells. The Extended X-ray Absorption Fine Structure (EXAFS) spectroscopy was used, followed by theoretical calculations. The results of preliminary analysis is presented.

1. Introduction

Prostate cancer is recognized as one of the major medical problems facing the male population. Nevertheless, the factors that determine the risk of developing clinical symptoms are not well known [1]. The role of sulphur in prostate cancer progression may be significant for understanding the process of carcinogenesis. In human body sulphur is found mostly as a component of protein-building amino acids (methionine, cysteine and cystine), and thus it participates in a wide range of essential biological functions. J. Mates *et al.* presented function of different compounds containing sulphur in cancer progression. They discussed that cellular methylation and antioxidant metabolism are linked by the transsulfuration pathway, which converts the methionine cycle intermediate, homocysteine, to cysteine, the limiting reagent in glutathione synthesis [2]. Glutathione, on the other hand, has many physiological functions, including its involvement in the defence against reactive oxygen species. Disturbance in this biochemical balance could lead to serious pathological conditions. One of the methods to get information about these transformations is to examine the changes occurring in the nearest chemical environment of sulphur existing in sulphur-bearing compounds in biological systems. In this study the Extended X-ray Absorption Fine Structure (EXAFS) spectroscopy was used in order to establish whether and what changes take place around the sulphur atom in chosen compounds in prostate cancer cells.

EXAFS spectra contain information about the average local atomic environment around sulfur. The spectra can be modeled to determine bonding distances and neighboring atom types and changes that occur to sulfur under different conditions. It has been used since more specific analysis of the local neighbourhood of sulphur atoms can be obtained from such studies. EXAFS is now often used in biology and medicine in the determination of very complex sequences of processes occurring in living organisms. XANES studies of the same material, evaluated before, showed the differences in the ratio of sulphur chemical species between various cell types. Therefore the EXAFS analysis was performed in order to improve the information gained from XANES.

2. Experimental

EXAFS measurements of S K edge were performed at the SUL-X beamline of the synchrotron radiation source ANKA, Karlsruhe (Germany), in fluorescence mode under high vacuum to avoid absorption of the radiation in air. A 7–element Si(Li) fluorescence detector (Gresham, now SGX Sensortech) was used. The experiments on SUL-X beamline can be done at the energy range 2,14 - 20 keV, with a minimum beam size of 30 μm x 25 μm (horizontal x vertical), obtainable by focusing with Kirkpatrick-Baez mirrors and the downsizing horizontal and vertical slit gaps. The dimension of the beam spot used for this experiment was 350 μm x 150 μm and 50 μm x 50 μm for reference compounds and for cell samples, respectively. The beam size of 50 μm x 50 μm was chosen for cell samples in order to obtain the best signal to noise ratio and to be close to the cellular size. The energy was varied with the double crystal monochromator (Si(111) crystal pair) in steps of 5 eV, 2 eV and 0,2 eV in the region of -100 to -50 eV before the edge, -50 to -20 eV before the edge and in the edge region, respectively. The post-edge region was collected to k=9,43 $Å^{-1}$ with k increment of 0,05 $Å^{-1}$. X-rays with higher harmonic energies were suppressed by detuning the monochromator second crystal to about 50% of its maximum intensity. Energy has been calibrated to 2481.4 eV at the maximum of the sulphate white line obtained from the scotch tape.

The experimental material consisted of four commercially available cell lines: three from metastasized prostate cancer (PC3, LNCaP, DU145) and one from the peripheral zone of the prostate, used as a control (PZ-HPV-7). The organic reference compounds were analysed in parallel to provide spectral information of interactions between sulphur and ligands expected, such as oxygen, carbon, nitrogen and other sulphur atoms.

The data were evaluated using the two of the IFEFFIT package programs: ATHENA and ARTEMIS [3]. The spectra were averaged from data collected at six different points on each sample. Firstly, the baseline correction and normalization was performed. Then, the spectra were changed from energy dependence to photoelectron wave number (k). The EXAFS signal $\chi(k)$ was extracted from experimental data, and Fourier transformed from k to R space using the data range from k=2.2 to 7.5 $Å^{-1}$ using a Hanning window. The theoretical amplitudes and phases for the chosen scattering paths were calculated using FEFF8.4 [4].

3. Results and discussion

In case of biological samples, the XANES and EXAFS signal is the average of the contribution from all sulphur atoms present in different cell structures. In this EXAFS studies we focused only on the nearest sulphur neighbours since further coordination shells requires more detailed analysis.

Base on the preliminary analysis of EXAFS data for different cell lines, as it can be noticed on Fig. 1, we can assume that first coordination shells around sulphur will be similar in all studied samples. Main differences are in the shape and magnitude of first and second feature of Fourier transform.

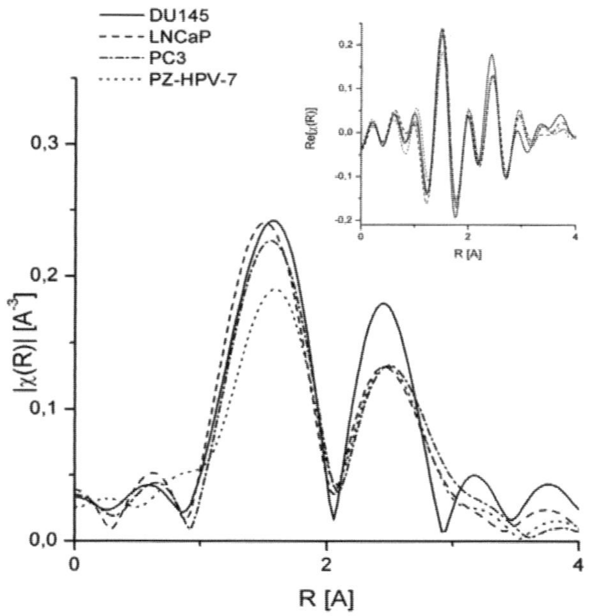

Fig. 1. Fourier transform and its Re part of the sulphur K-edge EXAFS signal for prostate cancer cell lines and control cell line.

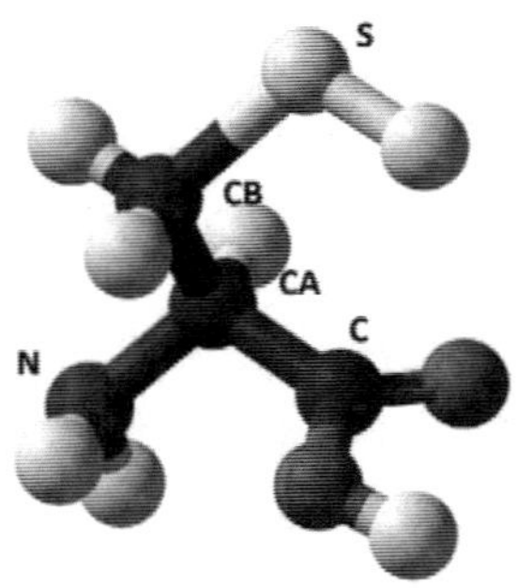

Fig. 2. The structure of cysteine [5].

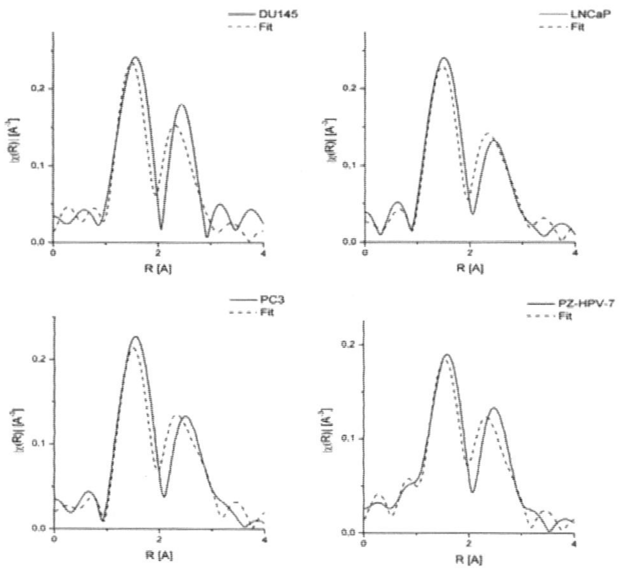

Concerning the results of XANES experiment, performed on the same samples, we determined that the most abundant of sulphur species in cells are amino acids and their content is 60%-75%. For that reason, in order to perform calculations, the crystallographic data for cysteine were used. In cysteine, sulphur is bind directly to the carbon (CB) as can be seen on Fig.2, there are also another two carbon atoms (CA and C) and one nitrogen atom (N) in the nearest neighbourhood. The positions of these atoms were used in theoretical calculations, as the assumption was done that this kind of nearest environment is most likely in studied cells. The coordination number was determined from crystal structure and not included into the fit.

Fig. 3. Comparison of experimental and fitted spectra of different cell lines.

The other tested crystallographic structure was methionine, that is the second common sulphur-bearing amino acid and has two carbon atoms directly bind to sulphur. Nevertheless this calculations didn't give physically meaningful results.

The results of the fits made in ARTEMIS are presented on Fig. 3. The agreement between fitted and experimental data for first two peaks in the case of each sample is fairly good. Additionally the distances between sulphur central atom and its nearest neighbours for different cell lines were determined and the results are presented in Table 1.

Table 1. The distances between sulphur and neighbouring atoms determined during ARTEMIS fits along with parameters of the fit.

Cell line	Calculated parameters			
	S-CB	S-CA	S-N	S-C
DU145	**R [Å] 1,59 ± 0,02** σ^2 [Å^2] 0,029 ± 0,00 ΔE_0 [eV] -7,9 ± 4,8	**R [Å] 2,58 ± 0,02** σ^2 [Å^2] 0,029 ± 0,007 ΔE_0 [eV] -12,8 ± 2,8	**R [Å] 2,98 ± 0,02** σ^2 [Å^2] 0,019 ± 0,018 ΔE_0 [eV] -1,4 ± 1,9	**R [Å] 3,05 ± 0,02** σ^2 [Å^2] 0,029 ± 0,007 ΔE_0 [eV] -7,9 ± 4,8
	S_0^2 0,45 ± 0,31			
LNCaP	**R [Å] 1,61 ± 0,01** σ^2 [Å^2] 0,021 ± 0,007 ΔE_0 [eV] -7,0 ± 2,9	**R [Å] 2,60 ± 0,01** σ^2 [Å^2] 0,018 ± 0,008 ΔE_0 [eV] -13,3 ± 2,0	**R [Å] 2,99 ± 0,01** σ^2 [Å^2] 0,018 ± 0,012 ΔE_0 [eV] -1,2 ± 1,4	**R [Å] 3,06 ± 0,01** σ^2 [Å^2] 0,018 ± 0,008 ΔE_0 [eV] -13,3 ± 2,0
	S_0^2 0,49 ± 0,19			
PC3	**R [Å] 1,62 ± 0,02** σ^2 [Å^2] 0,014 ± 0,005 ΔE_0 [eV] -5,7 ± 3,3	**R [Å] 2,61 ± 0,02** σ^2 [Å^2] 0,010 ± 0,006 ΔE_0 [eV] -14,1 ± 2,9	**R [Å] 3,01 ± 0,02** σ^2 [Å^2] 0,005 ± 0,008 ΔE_0 [eV] -0,6 ± 1,8	**R [Å] 3,07 ± 0,02** σ^2 [Å^2] 0,005 ± 0,008 ΔE_0 [eV] -0,6 ± 1,8
	S_0^2 0,32 ± 0,16			
PZ-HPV-7	**R [Å] 1,62 ± 0,02** σ^2 [Å^2] 0,019 ± 0,006 ΔE_0 [eV] -9,3 ± 3,6	**R [Å] 2,61 ± 0,02** σ^2 [Å^2] 0,017 ± 0,008 ΔE_0 [eV] -13,6 ± 2,7	**R [Å] 3,01 ± 0,02** σ^2 [Å^2] 0,015 ± 0,011 ΔE_0 [eV] -1,4 ± 1,9	**R [Å] 3,07 ± 0,02** σ^2 [Å^2] 0,017 ± 0,008 ΔE_0 [eV] -13,6 ± 2,7
	S_0^2 0,20 ± 0,11			
Theoretical model	1,80	2,80	3,20	3,26

As it is shown, the calculated distances vary from the theoretical values obtained from crystallographic model, but as it was mentioned before this may be related with the EXAFS signal averaging from different sulphur structures. The results for cell lines are similar, the most different values were obtained for DU145 cells and PZ-HPV-7 (control) cells nevertheless this differences are not statistically significant.

4. Conclusions

The use of EXAFS technique combined with theoretical calculations showed that the complex biological systems, such as cells, can be analysed with this method. The model presented, which works fairly well for the measured spectra and contains 3 C atoms and 1 N atom in the nearest neighborhood of sulphur occurring in cells, corresponds to the structure of amino acid cysteine, although the distances between individual atoms are slightly different than in theoretical model of cysteine. For this fitted structure no significant differences in atom distances were found between various cell lines nevertheless, as in the case of XANES analysis, the most diversified results were obtained for DU145 and control cells. That could indicate the fact that the structure of sulphur compounds in DU145 cells is somehow distorted.

Acknowledgements

The research leading to these results has received funding from the European Community's Seventh Framework Programme (FP7/2007-2013) under grant agreement nº 226716. We acknowledge ANKA for granting the beamtime in the proposal BIO-12.

References

1. Aus G, Abbou CC, Bolla M 2005 *Eur Urol* **48** 546
2. Mates JM, Segura JA, Alonso FJ, Marquez J 2012 *Frontiers in Bioscience* **S4** 722
3. Ravel B, Newville M 2005 *J. Synchrotr. Radiat.* **12** 537
4. Ankudinov AL, Ravel B, Rehr JJ, Conradson SD 1998 *Phys. Rev. B* **58** 7565
5. www.wikipedia.com

Journal of Physics: Conference Series **430** (2013) 012041 doi:10.1088/1742-6596/430/1/012041

Characterization of the Cu (II) and Zn (II) binding to the Amyloid-β short peptides by both the Extended X-ray Absorption Fine Structure and the Synchrotron Radiation Circular Dichroism spectroscopy

Zhiyin Zhang*, Shuaishuai Sun, Jianhua Xu, Jing Zhang, Yan Huang, Bingbing Zhang and Ye Tao

Beijing Synchrotron Radiation Facility, Institute of High Energy Physics, Chinese Academy of Sciences, Beijing, 100049

*Corresponding autor. E-mail: zhangzhiyin@ihep.ac.cn

Abstract. Alzheimer's disease (AD) is a progressive and devastating neurodegenerative pathology, clinically characterized by dementia, cognitive impairment, personality disorders and memory loss. It is generally accepted that, misfolding of Aβ peptides is the key element in pathogenesis and the secondary structure of Aβ can be changed to major β-strand with reasons unknown yet. Many studies have shown that the misfolding may be linked with some biometals, mainly copper and zinc ions. To characterize interactions of Aβ and metal ions, we utilized both the extended X-ray fine structure spectroscopy (EXAFS) and the synchrotron radiation circular dichroism spectroscopy (SRCD). Aβ (13-22), Aβ (13-21), Aβ (E22G) and Aβ(HH-AA) were selected to study the mechanism of copper and zinc binding to Aβ. We found that Cu interaction with H13 and H14 residues led to the disappearance of the PPΠ, while the Cu binding E22 residue caused a remarkable conformation change to β-sheet enrichment. The Zn ion, in contrast, made little effect on the conformation and it coordinated to only one histidine (H residue) or not.

1. Introduction

It is generally believed that polypeptide amyloid-β, the main component of plaque deposits in Alzheimer's disease (AD), is probably a causative of AD [1, 2]. In our bodies, Aβ produced by sequential proteolytic cleavage of the amyloid precursor protein (APP) is normal and soluble with little content of β-strand [2]. While, it is yet unknown that how soluble Aβ is transformed into insoluble, fibrous and eventually aggregated one which can possibly lead to AD [2]. X-ray diffractions on the amyloid fibrils of Aβ indicate a cross-β conformation with its β-sheet and backbone hydrogen bonds directing perpendicular and parallel to the fibril axis respectively [1]. The conformation changes of native Aβ may be linked with some biometal ions, such as copper and zinc. There is evidence that copper and zinc are elevated in amyloid plaque deposits with concentrations of 0.4 mM and 1 mM, respectively [1]. In vitro, Aβ deposits induced by copper can be dissolved backwards after the additions of metal chelators [3]. Mean while, the amyloid precipitate in vivo extracted from the postmortem tissue of AD patient behaves the same above after being doped by metal chelators [4]. Recently, researches attract much interest in the short-range structure of metal ions coordinating to Aβ. Histidines (H6, H13 and H14) are considered to be the most attracting binding sites for metal ions.

The conformation structure around metal ions sensitively depends on the solution condition, such as buffer and pH [5 6 7]. Furthermore, Electron Paramagnetic Resonance (EPR) spectroscopy on copper after binding to Aβ demonstrates that in low pH, copper can coordinate to two histidines [8].

In our studies, four short sequences of human Aβ, Aβ (13-22), Aβ (13-21), Aβ (E22G) and Aβ (HH-AA) were selected as short peptide models for detection of conformation changes triggered by copper and zinc in vitro. Synchrotron radiation circular dichroism spectroscopy (SRCD), as one of the most sensitive tool for measurement of protein's secondary structure, was used to detect long-range structure of Aβ after being coordinated to metal ions. While short-range structure around metal ions was probed by the extended X-ray fine structure spectroscopy (EXAFS) fitting analysis. By both SRCD and EXAFS analysis, we found that metal ions binding to at least two histidines led to the disappearance of PPΠ [9], a secondary structure of protein.

2. Materials and Methods

Aβ (13-22), Aβ (13-21), Aβ (E22G) and Aβ (HH-AA), as four short peptides, were chosen with purity exceeding 98%. The sequence of Aβ (13-22) is HHQKLVFFAE. Ac- and -NH_2 represent acetylation of the N-terminal amino group and amidation of the C-terminal amino group, respectively. The stock solution of Aβ short peptides (~2mM) has been prepared by dissolving them in MES (2-(N-Morpholino) ethanesulfonic acid). The pH of the solution was adjusted to pH 5.6 by adding NaOH stock solution. The stock solution were stirred, sonicated for 10 minutes and centrifuged for 3 minutes at 12000 g. The supernatant was mixed by copper or zinc at mole ratio of 1:0.5. The mixture was centrifuged (3minutes at 12000 g) to get supernatant for measurements of SRCD and EXAFS.

SRCD spectra about measurements of Aβ were performed at the beamline 4B8 of the Beijing Synchrotron Radiation Facility (BSRF) [10]. CaF_2 sample cell with optic length of 0.05 mm was used for holding solution Aβ. CD spectrum of each sample was recorded at least three times, with baseline to be extracted later. Online analysis [11, 12] for secondary structures of Aβ was followed.

The Cu and Zn K-edge EXAFS spectra were collected at the 1W1B beamline of BSRF. The monochromator is double-crystal Si (111). Data was collected in fluorescence mode using Lytle ionization chamber filled by Ar. Ni and Cu filter were inserted before the detector to inhibit scattering signal for Cu and Zn EXAFS data collecting, respectively. Athena was used for subtraction and normalization. While, Artemis was used for fitting analysis of k or R space spectra with a predicted model

3. Results and Discussion

Figure 1 displays the SRCD spectra of the four short peptides of Aβ (13-22), Aβ (13-21), Aβ (E22G) and Aβ (HH-AA) after being reacted with Cu^{2+}. First of all, the SRCD shapes (solid curves) of Aβ short peptides in their native state are almost the same and no obvious conformation distinction about their secondary structure can be found, except for the Aβ (HH-AA). The spectra of the four Aβ sequences similarly show the CD positive peak at about 220nm, which manifests possibly a considerable PPΠ helix contribution. However, the CD patterns (dotted lines) in Fig. 1 display obvious differences for the four Aβ short peptides after reaction with Cu^{2+}. In contrast, their spectra show almost the same CD negative peak at about 220nm, except for the Aβ (HH-AA). The changes of CD peak at ~220nm after addition of copper demonstrate that reactions of copper and Aβ can lead to the disappearance of PPΠ and two histidines play significantly important role in that process.

Figure 2 exhibits the Cu K-edge EXAFS spectra in k space for the four Aβ short peptides. All the EXAFS experimental spectra display the similar split peak at almost 4$Å^{-1}$, except for the Aβ (HH-AA), in which histidine has been substituted by A. The split peak at ~4$Å^{-1}$ is generally considered as a fingerprint that metals coordinate to the histidines in BioXAS. Therefore copper binds to H residue, except for the Aβ (HH-AA). Data has been best fitted by the 2-histidine geometry model (data not shown). So, combined with the results of SRCD spectra in Fig. 1, EXAFS patterns in Fig. 2 elucidates that copper coordinating to two histidines can lead to the disappearance of PPΠ.

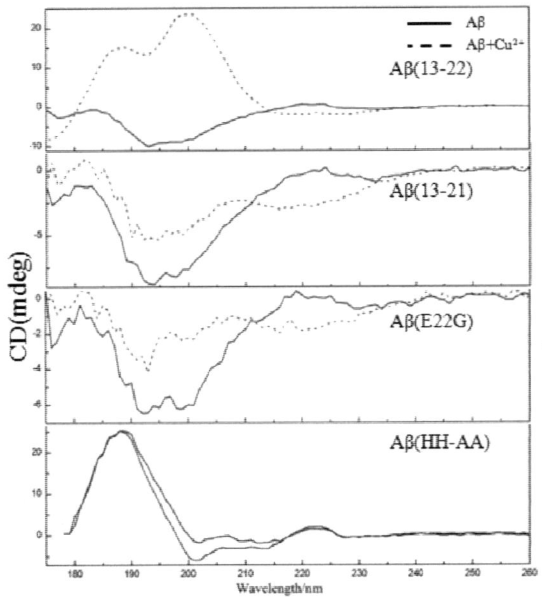

Figure 1. The SRCD spectra of Aβ (13-22), Aβ (13-21), Aβ (E22G) and Aβ (HH-AA) after additions of Cu^{2+} at molar ratio of 1:0.5.

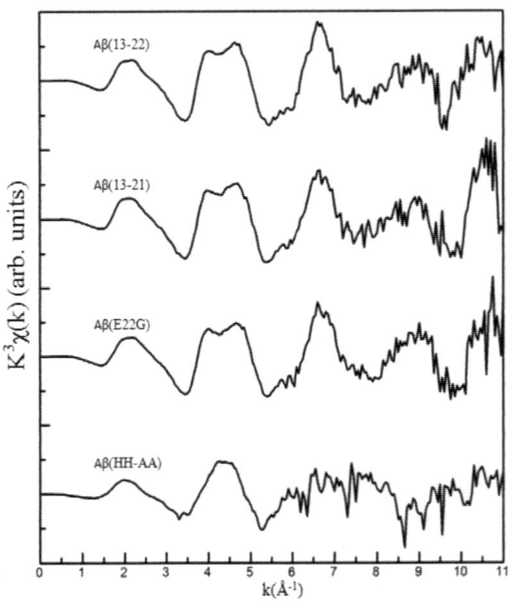

Figure 2. Cu K-edge EXAFS spectra of Aβ (13-22), Aβ (13-21), Aβ (E22G) and Aβ (HH-AA) after additions of Cu^{2+} at molar ratio of 1:0.5.

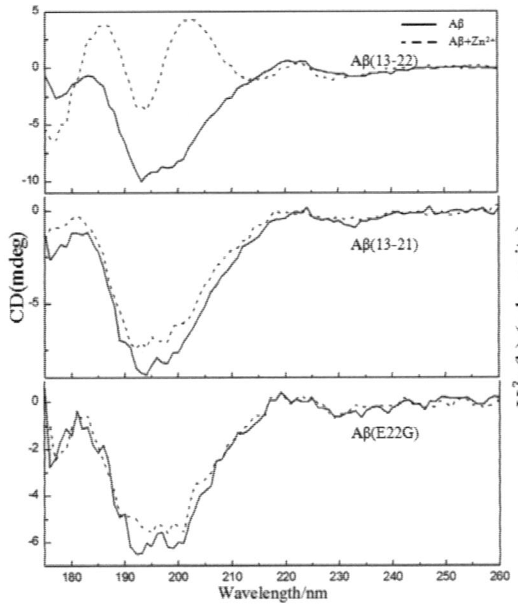

Figure 3. The SRCD spectra of Aβ (13-22), Aβ (13-21) and Aβ (E22G) after additions of Zn^{2+} at molar ratio of 1:0.5.

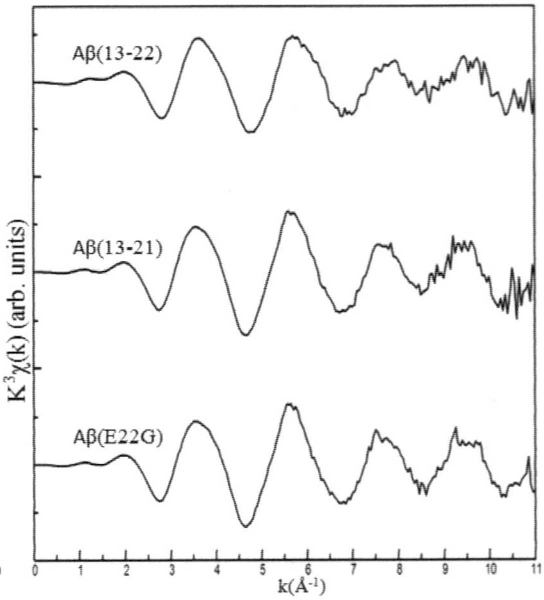

Figure 4. Zn K-edge EXAFS spectra of Aβ (13-22), Aβ (13-21) and Aβ (E22G) after additions of Zn^{2+} at molar ratio of 1:0.5.

Figure 3 displays the SRCD spectra of Aβ short peptide models of Aβ (13-22), Aβ (13-21) and Aβ (E22G) after reaction with Zn^{2+}. For CD positive peak at about 220 nm, there are no obvious changes for all the Aβ short peptides after interaction with zinc. That is to say, the content of ΠPP remains nearly the same after the additions of Zn^{2+}. So, reactions of zinc and Aβ fail to trigger the disappearance of PPΠ secondary structure.

Figure 4 shows the Zn K-edge EXAFS spectra in k space for Aβ short peptides of Aβ (13-22), Aβ (13-21) and Aβ (E22G). Compared to Cu K-edge EXAFS spectra in Fig. 2, there are no split peak at ~4 $Å^{-1}$ for Zn EXAFS spectra in Fig. 4, which manifests that zinc may not coordinate to histidine. Data has been best fitted by the 1-histidine geometry model (data not shown). So, combined with our results of SRCD spectra in Fig. 3, the Zn EXAFS spectra elucidate that zinc binding to at most one histidine can't lead to the disappearance of PPΠ.

4. Conclusion

In summary, using SRCD and EXAFS spectroscopy, we present long-range and short-range structure change of Aβ short peptide models after reaction with metal ions, respectively. SRCD demonstrates that PPΠ disappears after reaction with copper, while EXAFS elucidates that copper has coordinated to two histidines. So, it shows that Cu^{2+} binding to two histidines leads to the disappearance of PPΠ. However, PPΠ remains the same after reaction with zinc by SRCD and Zn^{2+} binds to at most one histidine by EXAFS. Compared to copper, Zn^{2+} binding to at most one histidine can't trigger the disappearance of PPΠ.

References

[1] Faller P, Hureau C 2009 Dalton. Trans. 1080
[2] Roberson E D, Mucke L 2006 Science 314 781
[3] Atwood C S, Moir R D, Huang X D, Scarpa R C, Michael N, Bacarra E, Romano D M, Hartshorn M A, Tanzi R E, Bush A I 1998 The Journal of Biological Chemistry 273 12817
[4] Cherny R A, Legg J T, Mclean C A, Fairlie D P, Huang X D, Atwood C S, Beyreuther K, Tanzi R E, Masters C L, Bush A I 1998 The Journal of Biological Chemistry 274 23223
[5] Faller P 2009 ChemBioChem 10 2837
[6] Roychaudhuri R, Yang M F, Hoshi M M, Teplow D B 2009 J. Biol. Chem. 284 4749
[7] Budimir A 2011 Acta. Pharm. 61 1
[8] Alies B, Eury H, Bijani C, Rechignat L, Faller P, Hureau C 2011 Inorg. Chem. 50 11192
[9] Woody R W 2009 J. Am. Chem. Soc. 131 8234
[10] Tao Y, Huang Y, Gao Z H, Zhuang H, Zhou A Y, Tan Y L, Li D W, Sun S S 2009 J. Synchrotron Rad. 16 857
[11] Whitmore L, Wallace B A 2004 Nucleic Acids Research 32 W668
[12] Lees J G, Miles A J, Wien F, Wallace B A 2006 Bioinformatics 22 1955

Journal of Physics: Conference Series **430** (2013) 012042 doi:10.1088/1742-6596/430/1/012042

Amyloid-β peptide active site: theoretical Cu K-edge XANES study

A. P. Chaynikov[1], M. A. Soldatov[1], V. Streltsov[2] and A. V. Soldatov[1]

[1] Research center for nanoscale structure of matter, Department of Physics, Southern Federal University, Rostov-on-Don, Russia
[2] Commonwealth Scientific and Industrial Research Organization Materials Science and Engineering, and Preventative Health Flagship, Parkville, Victoria, Australia

E-mail: chaynap@rambler.ru

Abstract. This article is dedicated to the local atomic structure analysis of the copper binding site in amyloid-β peptide. Here we considered two possible structural models that were previously obtained by means of EXAFS analysis and density functional theory simulations. We present the calculations of Cu K-edge XANES spectra for both models and make comparison of these spectra with experiment.

1. Introduction

Alzheimer's disease (AD) is a progressive, irreversible brain disease that destroys memory and thinking skills. AD is the most common form of dementia. Most often, AD is diagnosed in people older than 65 years [1]. However, the disease is not associated with the age only. A small percentage of people have early-onset AD in their 40 or 50 years. An estimated 5.4 million Americans of all ages have AD in 2012 and of these 5.2 million people aged 65 and older [2] and 200 thousands individuals are under the age of 65 years [3].

Currently, AD is incurable, and treatments for symptoms are available only. Today, many scientists are searching for new and more effective treatments that can help delay the onset of the disease, alleviate the symptoms and prevent its appearance.

At the cellular level, the AD is related to the death of nerve cells and loss of synapses in the cerebral cortex. The disease is also characterized by the presence of misfolded protein depositions or amyloid plaques in the brain. The main component of these plaques is the amyloid-β peptide (Aβ). Its predominant length is 42 amino acids: DAEFRH$_6$DSGYEVH$_{13}$H$_{14}$QKLVFFAEDVGSNKGAIIGLM-VGGVVIA. The Aβ peptide is cleaved from the membrane-bound amyloid precursor protein (APP) by the β/γ-secretase pathway. In vitro Aβ can bind to Cu, Zn and Fe with relatively high affinity and increased content of these metals are found in AD amyloid plaques [4,5]. There is evidence that the neurodegeneration observed in AD is related to the toxicity from reactive oxygen species (ROS) produced in the brain by the Aβ peptide bound to copper ions [5,6]. The ROS oxidize essential cellular components such as proteins, lipids and DNA. It leads to cell damage and oxidative stress. According to studies [5,7] the Aβ-Cu^{2+} complexes can participate in extensive redox chemical reactions that produce H_2O_2 and other ROS from molecular oxygen. As a result of these redox reactions the copper ion Cu^{2+} is reduced to Cu^{1+}. Knowledge of the electronic and atomic structures of the nearest environment of Cu^{2+} is required for deeper understanding this process. Recent researches show particular interest to the role amyloid-β redox for oxidative stress [8-11]. It may be critical to the etiology of AD.

2. Models and method

This article focuses on the structure of the Aβ-Cu^{2+} active site. We consider two model (Model 1 and Model 2) which structures were described in [8]. These models were proposed by means EXAFS analysis of the Aβ peptides complexed with Cu^{2+} in solution under a range of buffer conditions and density functional theory (DFT) simulations.

In both models Cu^{2+} is six-coordinated. In the Model 1 the copper ion is surrounded by three nitrogen atoms belonging to the imidazole rings of histidines (His-6, His-13, His-14), two oxygen atoms belonging to the carboxylate side chain of glutamic acid (Glu-11) and water molecule (W1). The Cu^{2+} coordination environment is similar octahedral, the three nitrogens and one oxygen of glutamic acid (O^1_{Glu-11}) are located in the equatorial plane and other oxygen of glutamic acid (O^2_{Glu-11}) and water molecule are located on the axis. This model includes two additional solvent oxygen (O_{sol}) atoms, which may belong to the carboxylate group from the N-terminal amino acids such as asparagine. They may be involved in a hydrogen bonding with water molecule W1 to stabilize the copper binding site.

The Model 2 also includes the three imidazole equatorial rings (His-6, His-13, His-14), axial water molecule W1, and solvent oxygen atoms. But carboxylate side chain of Glu-11 is replaced by phenol ring of tyrosine (Tyr-10) and second axial water molecule is added. The distances between the copper ion and the neighboring atoms are listed in 'table 1'.Both models with the notation adopted above are shown in 'figure 1'.

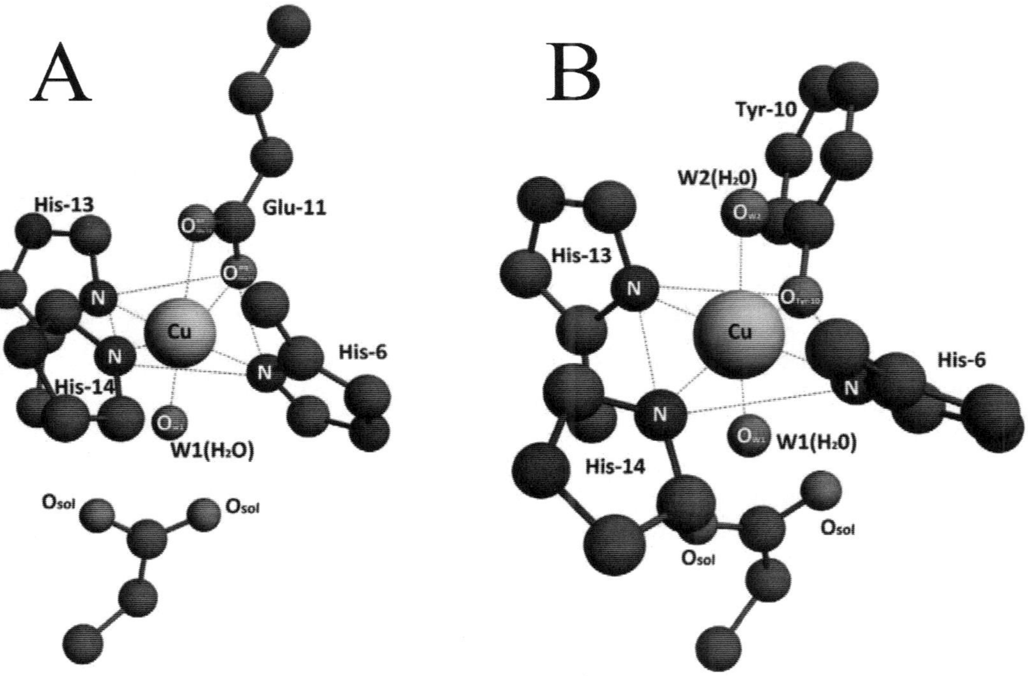

Figure 1. A ball-and-stick representation of the Model 1 – A and Model 2 – B.

Table 1. The distances between the copper ion and the neighboring atoms (in Å) for two model structures of the copper binding site in Aβ [9].

	Model 1	**Model 2**
$d(Cu^{2+}\text{-}N_{His\text{-}6})$	1.91	1.91
$d(Cu^{2+}\text{-}N_{His\text{-}13})$	1.99	1.97
$d(Cu^{2+}\text{-}N_{His\text{-}14})$	2.09	2.11
$d(Cu^{2+}\text{-}O_{W1})$	2.03	2.32
$d(Cu^{2+}\text{-}O^{eq}_{Glu\text{-}11})$	1.94	
$d(Cu^{2+}\text{-}O^{ax}_{Glu\text{-}11})$	2.27	
$d(Cu^{2+}\text{-}O_{Tyr\text{-}10})$		1.97
$d(Cu^{2+}\text{-}O_{W2})$		1.96

In order to study these models the XANES spectroscopy was used. The theoretical XANES spectra were calculated using two one-electron approaches. The first one is based on the real space Green's function formalism in considering the full multiple scattering a photoelectron wave on a self-consistent potential of muffin tin (MT) type and is implemented in the FEFF9.03 code [12]. The second one uses the finite difference method to solve the Schrodinger equation and is carried out in the FDMNES program [13].

3. Results and discussions

The 'figure 2' shows the comparison of the experimental Cu K-edge XANES spectrum of the truncated $A\beta_{1\text{-}16}\text{-}Cu^{2+}$ complexes dissolved in phosphate buffer [8] with the theoretical spectra for both models calculated with FEFF9.03 program. The radius of the cluster in which the multiple scattering procedure is performed, was equal to 8 Å and the cluster included all atoms of each model (30 for Model 1 and 32 for Model 2). The exchange part of the scattering potential was Hedin-Lundqvist type [14]. The position of the absorption edge is determined by the self-consistent calculations, but in the 'figure 2' both theoretical spectra are shifted by 5.1 eV to lower energies. This is done in order to make the comparison of theory with experiment easier. Most likely the cause of this shift is the incompleteness of models, which don't have hydrogen atoms. Hydrogen atoms can influence the self-consistent procedure and lead to a shift of the Fermi level.

The experimental spectrum has a number of features which are identified in 'figure 2' as A, B, C, D, E and F. The low-intensity peak A (at ≈8979 eV) is caused by the 1s→3d quadrupole-allowed transitions in Cu^{2+} [15,16]. The small shoulder B (at ≈8987-8988 eV), which occurs from either a vibronic allowed 1s→4s transition or 1s→4p transitions with a metal-ligand charge-transfer shakedown, has been explained mainly to scattering by the axial ligands [16]. The main peaks C and D (at ≈9897 eV and ≈9001 eV, respectively) is assigned to the 1s→4p (or 1s→continuum) transition. There are also features E and F, which are located at ≈9010 eV and 9045 eV. They are related to multiple scattering in first shell ligands [15,17], and its form and position can depend, for example, on the orientation of the imidazole rings [16].

The shoulder B is observed in both the theoretical spectra, but for Model 1 is more pronounced. The shape of peaks C and D is very different between the models. The intensity ratio of the peaks C and D for Model 2 more agrees with the experimental curve. But the feature E, which is observed in spectrum of Model 1 is absent or greatly reduced for Model 2. The peak F in both cases is shifted to lower energies (by 11 eV for Model 1 and by 5 eV for Model 2).

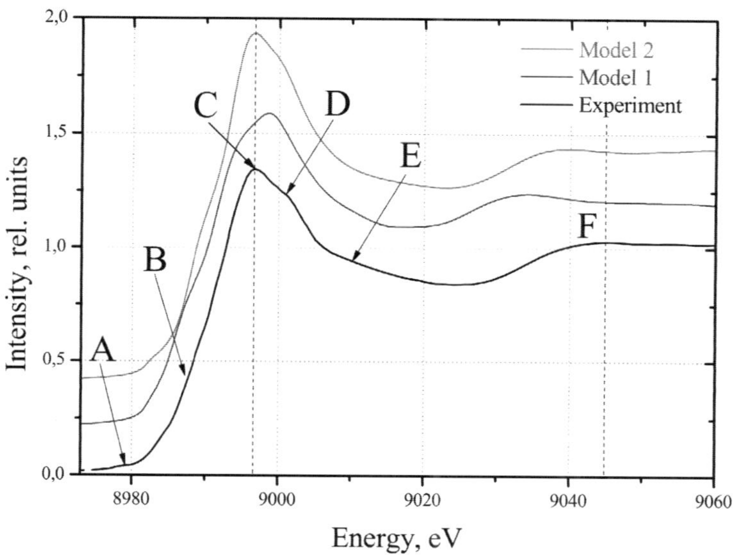

Figure 2. The comparison of simulated Cu K-edge XANES spectra for both models of the copper binding site in Aβ (FEFF9.03) with experimental spectrum of the truncated $A\beta_{1\text{-}16}$-Cu^{2+} complexes in phosphate buffer [9]. From top to bottom: Model 2 (red line), Model 1 (blue line), experiment (black line). The spectral features are marked as A, B, C, D, E and F.

These deviations of the theoretical spectra from the experiment may be associated with features of the considered models and/or the calculation method. In order to answer the question, we calculated the spectra using a method, which is based on the finite difference method to solve the Schrodinger equation (code FEMNES). This approach is more time-consuming, but at the same it has the opportunity to perform calculations beyond the MT approximation, which in this case is not entirely justified. The calculation results are presented in the 'figure 3'. The Schrodinger equation was solved on a grid of points in real space, which is a sphere centered on the Cu^{2+} and the radius of 5.5 Å. Thus the calculation included 27 atoms for Model 1 and 29 atoms for Model 2. It should be noted that the Model 1 curve has a near-edge feature A and this peak is shifted to higher energies on ≈1 eV. The spectrum of Model 2 doesn't have the peak A, but a new peak A' is observed there (at ≈8985-8986 eV). In the experimental spectrum it is not explicitly pronounced. Perhaps the peak A' is smoothed by experimental resolution. The shoulder B on the theoretical spectra is located closer to the main maximum C than on the experimental spectrum. Its position and shape differ between the two calculated spectra and this is in agreement with the nature of this feature [16]. The feature D is absent on the theoretical spectra and there is a small dip in its place for Model 2. The shoulder E is also absent on the spectrum for Model 1. The maximum F for both models is shifted to lower energies by 6 eV relative to corresponding experimental peak.

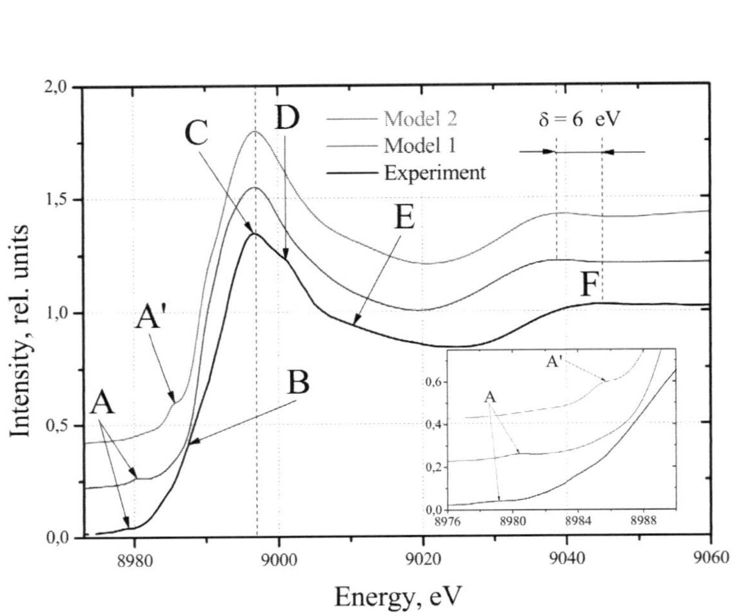

Figure 3. The comparison of non-MT theoretical Cu K-edge XANES spectra for both models of the copper binding site in Aβ (FDMNES) with experimental spectrum of the truncated $A\beta_{1\text{-}16}$-Cu^{2+} complexes in phosphate buffer [9]. From top to bottom: Model 2 (red line), Model 1 (blue line), experiment (black line).

4. Conclusions

We calculated the Cu K-edge XANES spectra for the two models of the copper binding site in Aβ-Cu^{2+} complex using the two approaches: full multiple scattering with MT approximation (FEFF9.03) and finite difference method to solve the Schrodinger equation (FDMNES). It was found that spectra simulated by FEFF9.03 program show better overall agreement with experimental XANES spectrum with some preference to Model 2. The intensity ratio of the peaks C and D for Model 2 more agrees with the experimental curve ('figure 2'). However, the statistical analysis of EXAFS fittings of three two models gave preference to Model 1 [8] which is in agreement with other biophysical studies. For example, recent spectroscopic measurements have confirmed that Tyr-10 and Met-35 are not key residues in metal binding [18-20]. This inability to precisely define the exact metal coordination sites in Aβ suggests that metal coordination may be pleomorphic [19] and we should model a superposition of different binding modes. Also, the use of MT approximation in this case is not entirely justified, since the structure of the Aβ-Cu^{2+} active site has cavities in which the potential is not constant would be more appropriate. However, these calculations are very useful for preliminary analysis and identification of the most significant effects. Next step will be analysis of the spectra calculated beyond the MT approximation in FEFF9.03.

5. Funding

The study was supported by The Ministry of education and science of Russia, project 14.A18.21.1070.

References

[1] R. Brookmeyer, S. Gray and C. Kawas 1998 *American Journal of Public Health* **88** 1337–1342.
[2] L.E. Hebert, P.A. Scherr, J.L. Bienias, D.A. Bennet, D.A. Evans 2003 *Archives of Neurology* **60** 1119–1141.

[3] Alzheimer's Association 2006 Early-Onset Dementia: A National Challenge, A Future Crisis *Alzheimer's Association*.
[4] M.A. Smith, P.L. Harris, L.M. Sayre and G. Perry 1998 *Proc. Natl. Acad. Sci. USA* **94** 9866–9868.

[5] K.J. Barnham, C.L. Masters and A.I. Bush 2004 *Nat. Rev. Drug Discovery* **3** 205–214.
[6] A.I. Bush 2003 *Trends Neurosci.* **26** 207–214.
[7] D.G. Smith, R. Cappai and K.J. Barnham 2007 *Biochim. Biophys. Acta* **1768** 1976-1990.
[8] V.A. Streltsov, S.J. Titmuss, V.C. Epa, K.J. Barnham, C.L. Masters, and J.N. Varghese 2008 *Biophysical Journal* **95** 3447–3456.
[9] V.A. Streltsov and J.N. Varghese 2008 *Chem. Commun.* **2008** 3169-3171.
[10] V. A. Streltsov 2008 *Eur. Biophys* **37** 257-263.
[11] V.A. Streltsov, J.N. Varghese, C.L. Masters and S.D. Nuttall 2011 *Journal of Neuroscience* **31** 1419–1426.
[12] J.J. Rehr and R.C. Albers 2000 *Rev. Mod. Phys.* **72** 621.
[13] Y. Joly 2001 *Phys. Rev. B* **63** 125120.
[14] L. Hedin and S. Lundqvist 1971 *J. Phys. C* **4** 2064.
[15] L.S. Kau, D.J. Spira-Solomon, J.E. Penner-Hahn, K.O. Hodgson and E.I. Solomon 1987 *J. Am. Chem. Soc.* **109** 6433–6442.
[16] R.W. Strange, L. Alagna, P. Durham and S.S. Hasnain 1990 *J. Am. Chem. Soc.* **112** 4265-4268.
[17] J.L. Fulton, M.M. Hoffman, J.G. Darab, B.J. Palmer and E.A. Stern 2000 *J. Phys. Chem.* **104** 11651-11663.
[18] S.C. Drew, C.L. Masters and K.J. Barnham 2009 *J. Am. Chem. Soc.* **131** 8760-8761.
[19] S.C. Drew, C.J. Noble, C.L. Masters, G.R. Hanson and K.J. Barnham 2009 *J. Am. Chem. Soc.* **131** 1195-1207.
[20] F. Bousejra-ElGarah, C. Bijani, Y. Coppel, P. Faller, and C. Hureau 2011 *Inorg Chem* **50** 9024-9030.

Journal of Physics: Conference Series **430** (2013) 012043 doi:10.1088/1742-6596/430/1/012043

Quantitative investigation of the local structure around cobalt ion in two different peptide deformylase by XANES spectroscopy

Yu Wang[1,2], Peixin Cui[1,2], Xiaoyun Guo[3], Haifeng Zhao[2], Feifei Yang[1], Meijuan Yu[2], Yuhui Dong[2], Yaning Xie[2], Weimin Gong[3], Wangsheng Chu[1,2*], Augusto Marcelli[4] and Ziyu Wu[1,2*]

[1]National Synchrotron Radiation Laboratory, University of Science and Technology of China, Hefei, 230029, China

[2]Institute of High Energy Physics, Chinese Academy of Sciences, Beijing, 100049, China

[3]National Laboratory of Biomacromolecules, Institute of Biophysics, Chinese Academy of Sciences, 100101, Beijing, China

[4]Instituto Nazionale di Fisica Nucleare, Laboratori Nazionali di Frascati, P.O. Box 13, 00044, Frascati, Italy

Email: wuzy@ustc.edu.cn, chuws@ustc.edu.cn

Abstract. The local atomic environment of a cobalt ion bound to the foldings of the active site of *Li*PDF (Co-*Li*PDF) and *Ec*PDF (Co-*Ec*PDF) in a pH 8.0 buffer solution has been determined using X-ray absorption near-edge spectroscopy (XANES) combined with *ab initio* calculation. Both active metal sites have been perfectly reconstructed using suitable initial structure modes. Results show that the nearest water (Wat1), which can be deprotonated and turned into hydroxyl, locates at 1.91 Å from the metal centre in Co-*Li*PDF while a high enzymatic activity occurs at 2.15 Å in the Co-*Ec*PDF. The stronger water-bonding of cobalt makes the distance between water and Glu144 (wat1-Glu144) longer. Such lengthening weakens the polarization capability of Glu144 to wat1 being responsible of the lower enzymatic activity of Co-*Li*PDF respect to Co-*Ec*PDF.

1. Introduction

Peptide deformylase (PDF, EC 3.5.1.27) is an important metalloprotein in particular because of its role in drug design target researches since the deformyling apparently does not occur in eucaryotic cells [1-4]. In eubacterium, during the elongation, the formyl group of the polypeptide chain is removed hydrolytically by the PDFs, apparently universally conserved and essential for bacterial survival [5].

A previous work has shown that the activity of *Escherichia coli* PDF (*Ec*PDF) is strongly affected by changes of the metal ion. On the basis of the presence of a classical HEXXH zinc-binding motif, PDF was originally thought to be a zinc enzyme. Later studies demonstrated that the active site metal

* The corresponding author Email: wuzy@ustc.edu.cn, chuws@ustc.edu.cn

Journal of Physics: Conference Series **430** (2013) 012043 doi:10.1088/1742-6596/430/1/012043

in wild-type PDF is iron. Actually, Fe^{2+} ion in *Ec*PDF is easily oxidized to Fe^{3+} so that *Ec*PDF loses its activity. *Ec*PDF with Co^{2+} replacing Fe^{2+} shows a similar high catalytic activity with the advantage of an improved stability while Zn-*Ec*PDF shows a decreased activity [6]. In contrast to *Ec*PDF, Co^{2+}-bound *Leptospira interrogans* PDF (Co-*Li*PDF) is much less active than Zn-*Li*PDF [7]. Little is known about chemical and physical mechanisms responsible for the difference in the enzymatic activity of Co-*Li*PDF and Co-*Ec*PDF in particular because of the lack of structural information around the metal sites of these proteins, largely due to the difficulty to obtain crystallized PDF sample.

X-ray absorption spectroscopy on biological systems (*Bio*XAS) is a short-range probe sensitive to variations in the local structure and its use has brought considerable contributions to the reconstruction of the local environment of specific atoms in many biological systems [8-12]. With XAS biological samples can be studied in a variety of forms, including oriented single protein crystals, disoriented micro-crystals, frozen or room-temperature solutions and proteins inserted in membranes. Here, we present Co *K*-edge XAS spectra of Co-*Li*PDF and Co-*Ec*PDF in a pH 8.0 buffer solution. Combining *ab initio* multiple scattering calculations, we applied the MXAN package to analyze data and extracted subtle local structure information around metal sites for *Li*PDF and *Ec*PDF, especially the position of bonded water molecules.

2. Materials and methods

E.coli BL21(DE3) cells carrying plasmid pET-22b-def [6] and plasmid pET22b-*Li*PDF [7] were expressed, respectively. Cells were harvested by centrifugation and resuspended in 25 ml of buffer A (50 mM Tris-HCl7.5, 10 mM NaCl). After sonication, cell debris was removed by centrifugation. The supernatant (25 ml) was applied onto a HiTrapTM 5 ml Q HP (Amersham Pharmacia Biotech) column equilibrated with buffer B (50 mM Tris-HCl pH 8.5, 10 mM NaCl). The column was eluted with 250 ml of buffer B plus a linear gradient of 10mM - 1M NaCl (AKTA purifier 900, Amersham Pharmacia Biotech). The fractions with *Li*PDF and *Ec*PDF activity were concentrated and applied to a SuperdexTM 75 column (16/60, Amersham Pharmacia Biotech) pre-equilibrated with buffer C (50 mM Tris-HCl pH 8.5, 50 mM NaCl). The sample homogeneity was monitored by SDS-PAGE and the protein concentration was determined by Braford assay with BSA as the standard.

To get samples of Co-PDF, we first obtained 30 mg of metal-free PDF protein by adding 15 ml of 0.1 M EDTA to 0.5 ml of 60 mg/ml protein for 2 hours at 277 K. The additional EDTA was then removed by direct exchange dialysis and concentrated centrifugal tube. The procedure was repeated twice by direct exchange dialysis (dilute 100*100 times) and concentrated centrifugal tube (dilute 15*15 times). Metal-free PDF (0.5 ml, 20 mg/ml) was incubated with 1.2 mM $CoCl_2$ 60 ml for 2 hours at 277 K and then concentrated to 20 mg/ml. The purified Co-*Li*PDF and Co-EcPDF were dialyzed against 50 mM Tris-HCl pH 8.0. The sample homogeneity was monitored by SDS-PAGE. Finally, all PDF concentration was about 60 mg/ml and the Co^{2+} concentration ~200 ppm. The protein concentration was determined by Braford assay with BSA as the standard [6].

Measurements at the Co *K*-edge were carried out at the beamline 1W1B of the Beijing Synchrotron Radiation Facility (BSRF) in the fluorescence yield (FY) mode at room temperature. The typical energy of the storage ring was 2.5 GeV and experiments were performed with a decreasing electron current from 250 mA to 160 mA. The incident beam intensity was monitored using an ionization chamber flowed by a 25% argon-doped nitrogen mixture while the fluorescence signal was collected by means of a Lytle detector flowed by argon gas. After XAFS measurement, we recycled the samples and measured again the activity of the samples. Data point out that proteins did not lose their activity. The behavior is probably associated to the high radiation resistance of *Li*PDF.

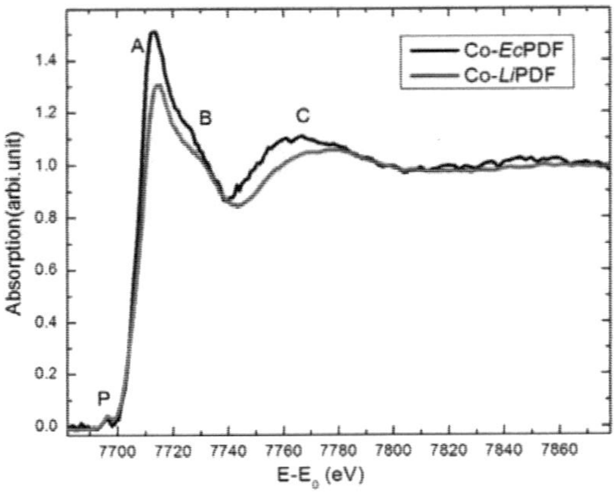

Figure 1. Co *K*-edge XANES spectrum of Co-*Li*PDF (red curve) and Co-*Ec*PDF in a buffer solution at pH 8.0

In order to extract structural/geometrical information around the metal site, we performed a quantitative analysis of the XANES spectrum using the MXAN code. This package is capable of performing a quantitative analysis of a XANES spectrum from the absorption edge up to 200 eV via comparison between experimental data and many theoretical calculations obtained by changing relevant geometrical parameters of the atomic site [13-14]. The X-ray absorption cross-sections have been calculated using the full multiple scattering (MS) scheme in the framework of the muffin-tin (MT) approximation for the shape of the potential [15-16]. In particular, the exchange and correlation part of the potential have been determined on the basis of the local density approximation of the self energy. Inelastic processes have been taken into account by a convolution with a broadening Lorentzian function having an energy-dependent width of the form $\Gamma(E) = \Gamma_c + \Gamma_{mfp}(E)$. The constant part Γ_c includes the core-hole lifetime (1.8 eV) [17] and the experimental resolution (1.5 eV), while the energy-dependent term represents all intrinsic and extrinsic inelastic processes [18-19]. The muffin-tin radii were chosen according to the Norman criterion with a 15% overlap [20-21]. This method takes into account for MS events in a rigorous way through the evaluation of the scattering path operator [15, 22]. Its reliability has been successfully tested over the years in a number of different applications [13, 18, 23]. The spectrum is convergent when the cluster includes 39 atoms within a radius of 5.5 Å. Therefore the local structure of both Co-*Li*PDF and Co-*Ec*PDF active centers can be properly represented by a cluster that contains 50 atoms within 6 Å from the metal ion.

3. Results and Discussion

In Fig.1, we reported the Co *K*-edge XANES spectrum of Co-*Ec*PDF (black curve) and Co-*Li*PDF (red curve) in a buffer solution at pH 8.0. Spectra are characterized by four main features: P, A, B and C. As we know, XAS features in this region are mainly dominated by scattering events of neighbor atoms around the photoabsorber. An obvious pre-edge peak can be also observed in the spectrum due to unoccupied 3d states. The main peak A is due to the scattering among the nearest atoms. The prsenece of this structure confirms the atomic coordination of Co. The valley B is related to higher order MS pathways of the excited photoelectrons involving nearest neighboring atoms while the position of peak C is correlated to the distance of first-shell atoms. According to Natoli's rule [reference] it can be seen that the subtle difference of peak C between the two XANES spectra indicates that the average distance of the first-shell atoms around Co in Co-*Ec*PDF is longer than in Co-*Li*PDF.

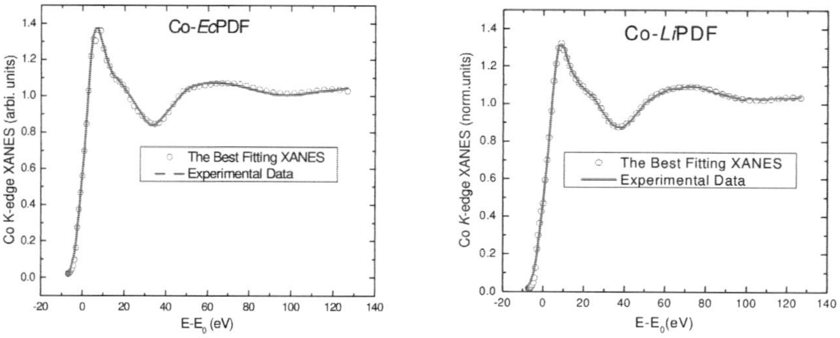

Figure 2. Best fit of the XANES spectra (blue dots) of Co-*Ec*PDF (left) and Co-*Li*PDF (right) in buffer solution at pH 8.0, compared with experimental spectra (red line).

In order to obtain accurate structural information around the metal site, based on the available crystallographic structure of *Li*PDF (PDB code 1Y6H(B)) [24] and *Ec*PDF (PDB code 1XEM) [25] we performed also a quantitative analysis of XANES spectra using the MXAN code. Fig. 2 shows the best fits (blue dots) of experimental (red curve) XANES spectra of Co-*Ec*PDF (left) and Co-*Li*PDF (right) in a buffer solution at pH 8.0. As we can see, all features in the near-edge region of the Co *K*-edge have been successfully reproduced. The square residue values for these calculations are 1.362 and 0.453, respectively. Some refined critical distances of the best fits are summarized in Tables 1. Comparisons of the local structure obtained by best fits of the Co-*Ec*PDF and the Co-*Li*PDF are plotted in Fig.3.

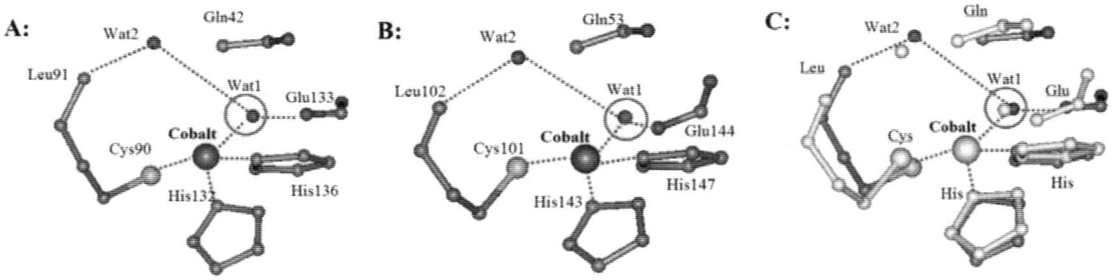

Figure 3. A: the local structure of the best fit of Co-*Ec*PDF; B: the local structure of the best fit of Co-*Li*PDF; C: comparison between the local structure of the best fit of Co-*Ec*PDF (colored stick) and Co-*Li*PDF (thick yellow sticks) at pH 8.0.

In the catalytic process of Co-*Ec*PDF, the most important step is associated to the proton of the catalytic water molecule (wat1) that is transferred to the amide at the N-terminus of the peptide with the help of Glu133 (*Li*PDF: Glu144). The added positive charge makes possible to nitrogen leaving the group [25-26]. Our result shows a fundamental difference between the local structures around metal site of Co-*Ec*PDF and Co-*Li*PDF: the position of the wat1. The Co-wat1 distance in Co-*Li*PDF is 1.91 Å, much shorter that of Co-*Ec*PDF that is 2.15 Å. The stronger water-bonding of cobalt affects the distance between water and Glu144 (wat1-Glu144) that become longer (3.25 Å). The mechanism weakens also the polarization of Glu144/133 to wat1 and prevents the transfer of the proton of wat1 to the amide at the N-terminus. This is probably the main reason at the base of the lower enzymatic activity of Co-*Li*PDF respect to Co-*Ec*PDF.

Table 1: Refined critical distances and error of the best fits of Co-*Li*PDF and Co-*Ec*PDF

Atom pairs	1XEM Model structure (Å)	Co-*Ec*PDF pH 8.0 distance (Å)	1Y6H(B) Model structure (Å)	Co-*Li*PDF pH 8.0 distance (Å)
S^2 (square residue)	-	1.362	-	0.453
Cobalt-His147 (His 136)	2.00	1.90±0.02	2.15	1.95±0.02
Cobalt-His143 (His 132)	2.00	2.04±0.02	2.07	1.81±0.02
Cobalt-Cys101 (Cys 90)	2.31	2.29±0.03	2.40	2.31±0.03
Cobalt-Wat1	2.32	2.15±0.03	2.31	1.91±0.03
Wat1-Glu144 (Glu 133)	2.85	2.78±0.06	2.67	3.25±0.07

4. Conclusion

X-ray absorption spectroscopy (XAS) is a powerful tool to investigate the local coordination environment of specific atoms with sub-atomic resolution. In this work we investigated the active site of *Ec*PDF and *Li*PDF by using XANES combined with *ab initio* multiple scattering calculations. All features in the near-edge region of the Co *K*-edge have been successfully reproduced and detailed structural information around the metal ion has been obtained. From the analysis, we show that the bond between the Co ion and the catalytic water molecule in Co-*Li*PDF is much shorter than in Co-*Ec*PDF, explaining the lower enzymatic activity of the former. Further researches on PDF containing different metal ions and with different pHs solution can be considered to improve our understanding of pH-dependent and metal-dependent enzymatic activity providing relevant structural information for any further applications of these proteins, in particular for the specific design of new drugs.

Acknowledgements

This work was partly supported by the Knowledge Innovation Program of the Chinese Academy of Sciences (KJCX2-YW-N42), the Key Important Project of the National Natural Science Foundation of China (10734070), the National Natural Science Foundation of China (NSFC 10805055 and 11275227) and the National Basic Research Program of China (2009CB930804).

References

[1] Meinnel T, Mechulam Y and Blanquet S 1993 *Biochimie.* **75** 1061-1075.
[2] Adams J M 1968 *J. Mol. Bio.* **33** 571-589.
[3] Ball L A and Kaesberg P 1973 *J. Mol. Biol.* **79** 531-537.
[4] Mazel D, Pocher S and Marliere P 1994 *EMBO J.* **13** 914-923.
[5] Rajagopalan P T R, Yu X C and Pei D 1997 *J. Am. Chem. Soc.* **119** 12418–12419.
[6] Groche D, Becker A, Schlichting I, Kabsch W, Schultz S and Wagner A F 1998 *Biochem. Biophys. Res. Commun.* **246** 342-346.
[7] Li Y, Ren S and Gong W 2002 *Acta Cryst. D.* **58** 846-848.
[8] Rehr J J and Albers R C 2000 *Rev. Mod. Phys.* **72** 621-654.
[9] Hayakawa K, Hatada K, D'Angelo P, Della Longa S, Natoli C R and Benfatto M 2004 *J Am Chem Soc.* **126** 15618–15623
[10] Wu Z Y, Tao Y, Benfatto M, Xian D C and Jiang J Z 2005 *J. Synchrotron Rad.* **12** 98-101.
[11] Nordlander E, Lee S C, Cen W, Wu Z Y, Natoli C R, Di Cicco A, Filipponi A, Hedman B, Hodgson K O and Holm R H 1993 *J. Am. Chem. Soc.* **115** 5549–5558.
[12] Conradson S D, Burgess B K, Newton W E, Di Cicco A Filipponi A, Wu Z Y, Natoli C R, Hedman B and Hodgson K O 1994 *Proc. Natl. Acad. Sci.* **91** 1290-1293.
[13] Longa S D, Arcovito A, Girasole M, Hazemann J L and Benfatto M, 2001 *Phys. Rev. Lett.* **87** 155501.
[14] Benfatto M and Wu Z Y 2003 *Nuclear Science and Techniques.* **14** 9-19.
[15] Natoli C R and Benfatto M 1986 *J. Phys. (Paris) Colloq.* **47** 11-23.

[16] Wu Z Y, Ouvrard G, Lemaux S, Moreau P, Gressier P, Lemoigno F and Rouxel J 1996 *Phys. Rev. Lett.* **77** 2101-2104.
[17] Fuggle J C and Inglesfield J E 1992 *Unoccupied Electronic States, Topics in Applied Physics, (Berlin: Springer) appendix B* p347.
[18] Benfatto M, Della Longa S and Natoli C R 2003 *J. Synchrotron Rad.* **10** 51-57.
[19] Roscioni O M, D'Angelo P, Chillemi G, Longa S D and Benfatto M 2005 *J.Synchrotron Rad.* **12** 75-79.
[20] Norman J G 1974 *Mol. Phys.* **81** 1191-1198.
[21] Joly Y 2001 *Phys. Rev. B.* **63** 125120.
[22] Wu Z Y, Xian D C, Natoli C R, Marcelli A, Mottana A and Paris E 2001 *Appl. Phys. Lett.* **79** 1918–1920.
[23] Zhang X W, Yan X J, Zhou Z R, et al. 2010 *Science.* **328** 240–243.
[24] Zhou Z, Song X, Li Y and Gong W 2004 *J. Mol. Biol.* **339** 207-215.
[25] Jain R, Hao B, Liu P R and Chan M K 2005 *J. Am. Chem. Soc.* **127** 4558-4559.
[26] Becker A, Schlichtin L, Kabsch W, Groche D, Schultz S and Volker Wagner A F 1998 *Nature Structural Biology* **5** 1053-1058.

X-ray absorption studies of copper(II) mixed ligand complexes with tetramethylethylenediamine as primary ligands

S K Joshi[1], V K Hinge[2] and B D Shrivastava[3]

[1]Physics Department, Government P G Arts and Science College, Ratlam-457001, India
[2]Physics Department, Ujjain Engineering College, Ujjain 456010, India
[3] School of Studies in Physics, Vikram University, Ujjain 456010, India

E-mail: joshisantoshk@yahoo.com

Abstract. X-ray absorption near edge structure (XANES) measurements have been made on the K-edge of copper in some of its biologically important complexes involving tetramethylethylenediamine (tmen) as primary ligand. Other inorganic anionic ligands are oxalate anion (ox), acetylacetonate anion(acac), glycinate(gly) anion and ethylenediamine(en). Various X-ray absorption parameters in the near edge region, e.g., chemical shift, i.e., shift of K-edge of copper in complex with respect to K-edge of copper metal, energy position of the principal absorption maximum and edge-width have been determined. The K-edge has been found to split into two components, i.e., K_1 and K_2 in the complexes. This feature suggests square planar geometry for the complexes. The chemical shift data suggests that in the complexes copper is in oxidation state +2 and that the relative ionic character of the complexes is in the following sequence: $Cu(tmen)(ox).4H_2O$ < $Cu(tmen)(acac)(ClO_4)$ < $Cu(tmen)(gly)ClO_4$ < $Cu(tmen)(en)SO_4.4H_2O$. For these four complexes, we observe that the sequence of variation of chemical shifts follows the same trend as obtained in the available spectrochemical series, i.e., oxalate < glycinate < ethylenediamine. The effective nuclear charge has been estimated from the chemical shift data.

1. Introduction

It is well known that the fine structure in the vicinity of an X-ray absorption discontinuity is influenced by the immediate surroundings of the absorbing atom. Studies of the absorption edge and the associated fine structure help in understanding the environment of a metal ion in its complexes and biologically important molecules. Chemical shifts as obtained by X-ray absorption spectroscopic studies have yielded useful information in various kinds of samples having biological relevance and utility in pharmacology. In the present paper, we have studied X-ray absorption near edge structure (XANES) at the K-edge of copper and have also measured various X-ray absorption parameters in some copper complexes involving tetramethylethylenediamine (tmen) as primary ligands and the other ligands being oxalate anion(ox), acetylacetonate anion(acac), glycinate anion(gly) and ethylenediamine(en). The ligands used are common ones (en, gly, ox and acac) with no bulky groups around the donor atoms, little steric hindrance is expected when one of them combines with copper (II) together with a tmen molecule. It was found that all these ligands form quite stable mixed chelates with tmen [1].

2. Experimental

All the complexes were prepared according to the standard methods reported in literature [1] and their purity was checked. All the mixed chelates obtained are fine crystals or apparently crystalline powders. A sealed Machlett tube with a tungsten target was employed as X-ray radiation source and was operated at 17 kV and 15 mA. The X-ray absorption spectra at the K-edge of copper were recorded photographically on a bent crystal 0.4 m Cauchois-type transmission spectrograph employing a mica crystal, oriented to reflect from (100) planes, for dispersion. For calibrating the spectra, the tungsten emission lines $WL\alpha_2$, $WL\beta_4$ and $WL\beta_6$ were used as reference lines. The dispersion of the spectrograph was about 0.8 eV/10^{-2} mm of the X-ray film. The resolution of the spectrograph was tested by recording the Mo $K\beta_{1,3}$ doublet as two separate lines, which was adequate for this study. The measurements at the K absorption edge were made on digital records obtained using a Carl-Zeiss G II microphotometer. Computer programs *Origin and Athena* [2] were used for data analysis. Derivative spectra were used for obtaining the K-edge position. The error in the measurement of K-edge was ± 0.5 eV.

3. Results and Discussion

Figure 1 shows the copper K-absorption edges for all the four complexes studied. The curves in this figure represent the normalized K absorption spectra. The copper K-edges in all of the complexes have been found to split into two components, i.e., K_1 and K_2. The energies of the copper K-edge (E_{K1} and E_{K2}) and the principal absorption maximum (E_A) along with the values of the edge-width (E_A-E_{K1}) and the chemical shift ΔE_{K1} are given in Table 1. It can be readily seen from this table that the K-absorption edge of copper is shifted towards the high energy side in all the complexes studied as compared to the K-absorption edge in the metal.

Table 1. Copper K-edge XANES data for the copper complexes

S. NO.	Complex	E_{K1} (eV)	E_{K2} (eV)	E_A (eV)	Chemical Shift (eV)	Z_{eff}	Shift in principal absorption maxima (eV)
	Copper metal	8980.5	8988.2	8992.5	-	-	-
1	Cu(tmen)(ox).$4H_2O$	8985.9	8990.7	8999.0	5.4	0.67	18.5
2	Cu(tmen)(acac) (ClO_4)	8988.3	8993.6	9002.9	7.8	0.92	22.4
3	Cu(tmen)(en)SO_4. $4H_2O$	8988.7	8994.0	8998.9	8.2	0.96	18.4
4	Cu(tmen)(gly)ClO_4	8988.5	8995.0	9001.1	8.0	0.94	20.6

3.1 *Chemical shift*

The shift of the X-ray absorption edge i (i = K, L, M....) of an element in a complex with respect to that of the pure element is written as: $\Delta E_i = E_i(\text{complex}) - E_i(\text{metal})$. It is well known that the position of the K-edge depends on the valence state of the absorbing ion. But the valence state is not the only factor that governs the magnitude of the chemical shift. At least two other factors that must be considered to explain the values of the chemical shifts are stereochemistry and covalent bonding [3]. For the present complexes, the chemical shift values lie in the range 5.4 - 8.2 eV. Earlier workers [4, 5, 6] have reported the chemical shift values of various copper (II) complexes between 5.7-12.9 eV. Hence, on the basis of values of the chemical shifts, all our complexes are found to have copper in oxidation state +2.

The chemical shift values for the four complexes studied in the present work follow the sequence: Cu(tmen)(ox).$4H_2O$<Cu(tmen)(acac)(ClO_4) < Cu(tmen)(gly)ClO_4 < Cu(tmen)(en)SO_4.$4H_2O$. For the three complexes, we observe that the sequence of variation of chemical shifts follows the same trend as obtained in the spectrochemical series, i.e., oxalate < glycinate < en. The (acac) complex finds a place between oxalate and glycinate complex. Though we know that the major contribution towards the magnitude of chemical shift depends on the ligands surrounding the central metal ion copper in the first coordination shell, but the effect of distant neighbours in the second coordination shell is not denied. Such a observation of contribution of second and higher shell ligands has been observed earlier by Gupta *et al.* [7]. As is well known, an ionic bonding enhances the chemical shift, whereas a covalent bonding suppresses it [3]. Hence, the above order may also be taken as representative of the relative ionic character of the bonding in these complexes.

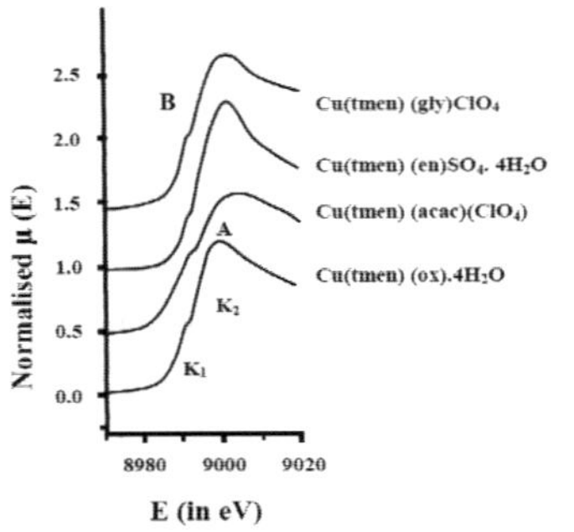

Figure 1. Copper K-edge XANES spectra for the copper complexes. The different spectra have been shifted vertically for better presentation.

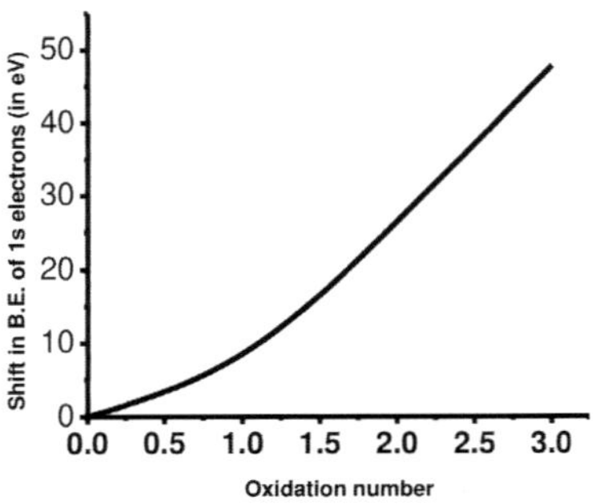

Figure 2. Plot of oxidation number vs. shift in B.E. of 1s electrons.

3.2 *Effective nuclear charge* Z_{eff}

Z_{eff} (electrons/atom) has been obtained from the measured chemical shift by using the semi-experimental method by employing the procedure suggested by Nigam and Gupta [8].According to this method the shift in the 1s electron energy under different oxidation state of copper atom is determined from Hatree Fock (HRF) calculations of 1s atomic function. A graph was plotted between the theoretical shift in the binding energy and the oxidation number [9] for copper (Figure 2). The effective nuclear charge Z_{eff} on the copper atom in the complexes studied was then determined from this plot corresponding to the measured value of the edge shifts. The data given in Table 1 indicate that Z_{eff} varies between 0.67 – 0.96 electrons/atom in the copper complexes studied.

3.3 *Splitting of the main edge*

The copper K-edges, as shown in Figure 1, have been found to split into two components, i.e., K_1 and K_2 in all of the complexes. This feature suggests square planar structure for the complexes investigated. In an extensive XANES study of Cu(I) and Cu(II) compounds Kau *et al.* [10] have found that a peak about 8984 eV is diagnostic of Cu(I) and has been assigned to the Cu(I) $1s \longrightarrow 4p_{x,y}$ transition. Further, no Cu(II) complex exhibits a significant peak below 8985 eV. Berry *et al.*[11] have reported that in Cu(II) compounds a shoulder B in the absorption edge has been observed which is strongly z-polarized and has been assigned Cu(II) $1s \longrightarrow 4p^*_{x,y}$ transition simultaneous with ligand to metal charge transfer. Another characteristic feature of Cu (II) is the maximum A at about 9000 eV which also corresponds to $1s \longrightarrow 4p_z$ transition as well as to the transitions to continuum states. In the spectra of presently studied complexes, the peak like feature at 8984eV, characteristic of Cu (I), is absent. Further, in the absorption edge the shoulder B is present and the feature A at about 9000eV is also present, which are characteristic of Cu(II). Thus, copper has been found to have oxidation state +2 in the complexes, on this basis also.

3.4 *Edge-width*

In Table 2, we have reported the values of the edge-width (E_A-E_{K1}). The edge-widths are related to the elctronegativity difference between metal atom (X_M) and its neighbouring atom (X_L) in the coordination sphere in a manner proposed by Nigam and Srivastava [12] as: $[\, E_W.\, \sum (X_M \sim X_L)]^{1/2}$ = Constant , where X_M and X_L are the Pauling's electronegativities of the central metal atom and the nearest neighbours respectively, and the summation on the right hand side extends over all the atoms in the coordination sphere. In Table 2 we have reported the stoichiometry of the complexes. In general the edge-width values depend not only on stoichiometry but also on the nature of bonding, i.e., ionic/ covalent. As seen from Table 2 the constant has a variation from 10.1 to 12.6 the slightly different values for the edge-widths are not unusual as the stoichiometry is different.

Table 2. Values of X_L, $\sum(X_M - X_L)$ and constant

S. No.	Complex	Stoichiometry Cu: N: O	Donor environment	Edge - width E_W (eV)	$\sum X_L$	$\sum(X_M - X_L)$	Constant
1	$Cu(tmen)(ox).4H_2O$	1: 2: 2	2N, 2O	13.1	13.0	11.1	12.0
2	$Cu(tmen)(acac)(ClO_4)$	1: 2: 2	2N, 2O	14.6	13.0	11.1	12.6
3	$Cu(tmen)(en)SO_4.4H_2O$	1: 4: –	4N	10.2	12.0	10.1	10.1
4	$Cu(tmen)(gly)ClO_4$	1: 3: 1	3N, O	12.6	12.5	10.6	11.5

$\sum X_M$ (for copper) = 1.9

3.5 *Principal absorption maximum*

For copper metal, the value of the energy of the principal absorption maximum (E_A) is 8992.5 eV and for all the complexes it is shifted towards the higher energy side (Table 1). Since, the principal absorption maximum occurs due to the transitions from 1s orbital to the unoccupied antibonding molecular orbitals [1s⟶4p (t_{1u}*) in octahedral], the principal absorption maximum, therefore, shifts to the higher energy side. In some cases the strength of the bond has been decided on the basis of the shift values of the principal absorption maximum. In general, the greater the overlap of metal and ligand orbitals, the greater the strength of the bond. Since strengthening of M-L bond would lead to a greater stabilisation of bonding MOs, a shift in the principal absorption maximum towards the high-energy side with increasing M-L bond strength would be expected.

4. Conclusions

The present study on the XANES at the K-edge of copper in all the complexes suggests that copper has valence state +2 in all the complexes. For all the four complexes we observe that the sequence of variation of chemical shifts follows the same trend as obtained in the spectrochemical series. The relative ionic character of the complexes has been found to have the following sequence: $Cu(tmen)(ox).4H_2O$ < $Cu(tmen)(acac)(ClO_4)$ < $Cu(tmen)(gly)ClO_4$ < $Cu(tmen)(en)SO_4.4H_2O$. The estimated effective nuclear charge (Z_{eff}) has been found to vary between 0.67 – 0.96 electrons/atom in the copper complexes studied.

References

[1] Fukuda Y and Sone K, 1972 *Bulletin of the Chemical Society of Japan*, **45** 465
[2] Ravel B and Newville M, 2005 *J. Synchrotron Rad.* **12** 537
[3] Agarwal B K and Verma L P, 1970 *J. Phys. C: Solid State Phys*. **39** 535
[4] Vyas R K, Joshi S K, Shrivastava B D, Sapre V B, Palikundwar U A and Patel R N, 2009 *Proc. Nat. Acad. Sci. India* **79 A** 125
[5] Katare R K, Joshi S K, Shrivastava B D, Patel R N, Pandeya K B and Mishra A, 2002 *X- ray Spectrom*. **31** 327
[6] Ballal M M and Mande C, 1978 *J. Phys.C: Solid State Phys.* **11** 837
[7] Gupta R D, Joshi S K, Shrivastava B D, Shah M C and Prasad J, 2005 *Natl. Acad. Sci. Lett.* **28** 39
[8] Nigam A K, Gupta M K, 1974 *J. Phys. F : Metal Physics* **4** 1084
[9] Clementi E and Roetti C, 1974 *At. Data Nucl. Data Tables* **14** 177
[10] Kau L S, Spira-Solomon D J, Penner-Hahn J E, Hodgson K O and Solomon E I, 1987 *J. Am. Chem. Soc.* **109** 6433
[11] Berry A J, Hack A C Mavrogenes J A, Newville M and Sutton S R, 2006 *Am. Mineral.* **91** 1773.
[12] Nigam H L and Srivastava U C, 1971 *Chem. Comm.* **14** 761

Journal of Physics: Conference Series **430** (2013) 012045 doi:10.1088/1742-6596/430/1/012045

Study of XAFS spectroscopic methods of speciation using mixtures of Cu(I) and Cu(II) chlorides

A Gaur[1*], B D Shrivastava[1] and S Khalid[2]

[1]School of Studies in Physics, Vikram University, Ujjain 456010, India
[2]Brookhaven National Laboratory NSLS, Bldg. 725D Upton, NY 11973

[*]Email: abhijeetgaur9@gmail.com

Abstract. Speciation concerns determination of the chemical forms along with the relative quantities of the different species in a given sample. The aim of the present work is to make a comparative study of the different methods of speciation using X-ray absorption fine structure (XAFS) spectroscopy. For this purpose, mixtures have been prepared by mixing $CuCl_2$ and CuCl in different proportions. The X-ray absorption spectra have been recorded at the copper K-edge in the mixtures and the two chlorides separately. The different characteristic features of the XANES spectra of the two chlorides, useful for speciation, have been identified. Firstly, Principal component analysis (PCA) and target transformation (TT) methods have been used to check the number and identity of components in the mixtures. After the identification of the components, the percentages of the species in the mixtures have been quantitatively determined by linear combination fitting (LCF) of XANES, derivative XANES, and EXAFS (k^3chi and k^2chi) spectra. The other methods of speciation which have been employed are normalized difference absorption edge spectra analysis (NDAES), methods based on derivative XANES spectra of species and the method based on the relative position of the absorption edge. Results obtained from these methods have been compared and their relative merits discussed. It is probably for the first time that such a study has been done. The first section in your paper

1. Introduction

X-ray absorption spectroscopy (XAS) is one of the best known structural techniques for speciation, i.e., for determination of the chemical forms along with the relative quantity of the different species in a given sample. For this purpose one usually models X-ray absorption fine structure (XAFS) spectra of the sample by using well-defined chemical species as standards. This can be done both in the X-ray absorption near edge structure (XANES) and extended X-ray absorption fine structure (EXAFS) regions. For speciation, several approaches for the analysis of XAFS data have been developed, the commonly used being: principal component analysis (PCA) along with target transformation method (TT) [1] and linear combination fitting (LCF) [2-4]. Other methods which can also be used are: derivative XANES spectra methods [5], normalized difference absorption edge analysis (NDAES) [6] and a method based on position of the edge. In earlier studies, we have applied these methods of speciation to only a single mixture of copper oxides and the LCF has been performed only in XANES spectra [7, 8]. In the present work, by taking five mixtures having specific percentages of Cu(I) and Cu(II) chlorides we have studied all of these methods of speciation and the LCF has been performed not only in XANES spectra but also in derivative XANES and EXAFS (k^3chi and k^2chi) spectra. The aim of the present work is to make a comparative study of these methods of speciation and to discuss their relative merits.

2. Experimental

Commercially available 99.99% pure CuCl and $CuCl_2$ powders were mixed in the specific ratios by weight to prepare five mixtures (numbered as M1, M2, M3, M4 and M5). The 'as prepared' percentages of CuCl and $CuCl_2$ in these mixtures are given in Table 1. Absorption screens were prepared of the finely powdered pure CuCl, $CuCl_2$ and their mixtures. The X-ray absorption spectra for these absorption screens were recorded at room temperature at the EXAFS beamline X-19A at National Synchrotron Light Source (NSLS), Brookhaven National Laboratory, Upton, New York, USA. The values of edge position, as determined from first inflection point, have been found to be to be 8979.0, 8981.9 and 8985.9 eV for Cu foil, CuCl and $CuCl_2$, respectively, representing the expected trend, i.e., increase in edge energy as the oxidation number increases. The positions of the K-absorption edge in the mixtures have been found to be in between those of CuCl and $CuCl_2$.

3. XANES features

Figure 1 shows the normalized K-edge XANES spectra of copper metal, cuprous chloride, cupric chloride and their mixtures. For the Cu(I) compound, a feature P at 8986 eV and in case of Cu(II) compound, a weak feature W at 8978 eV, absorption-edge shoulder S and an intense white line C at 8995 eV have been observed which have been taken as diagnostic features for speciation.

4. Data analysis, Results and Discussions:

In the present study, for PCA and target transformation, *SIXPack* data analysis software has been used. EXAFS data analysis software *Athena* has been used for performing LCF, for constructing the difference spectra and the derivative spectra and determining the position of the edge corresponding to the point having $\mu x = 0.5$ in the normalized spectrum. After confirming the number and identity of components in the five mixtures by using PCA and Target transformation (details of which are not given here), we have applied the following methods for determining the proportion of the $CuCl_2$ and CuCl in the mixtures and the results obtained are given in table 1.

Table 1. Results of speciation as obtained from different methods. 'I' is the percentage of CuCl and 'II' is the percentage of $CuCl_2$ in the mixture M1 to M5.

	Mixture →		M1		M2		M3		M4		M5	
S. No	Method of speciation ↓	max. % error	I	II	I	II	I	II	I	II	I	II
1.	XANES LCF	± 3	13.6	86.4	34.7	65.3	47.3	52.7	52.9	47.1	76.0	24.0
2.	Derivative XANES LCF	± 5	12.5	87.5	33.6	66.4	44.3	55.7	51.3	48.7	75.2	24.8
3.	EXAFS LCF (k^2 chi)	± 7	12.7	87.3	41.2	58.8	52.1	47.9	58.3	41.7	75.8	24.2
4.	EXAFS LCF (k^3 chi)	± 6	11.7	88.3	39.8	60.2	51.8	48.2	58.6	41.4	74.3	25.7
5.	NDAES	± 5	13.8	86.2	34.2	65.8	46.7	53.3	51.3	48.7	75.2	24.8
6.	Derivative Cu(I) spectra	± 3	18.0	82.0	36.5	63.5	47.0	53.0	53.0	47.0	77.0	23.0
7.	Derivative Cu(II) spectra	± 5	15.2	84.8	39.4	60.6	54.5	45.5	60.6	39.4	75.6	24.4
8.	Position of the edge	± 10	9.0	91.0	42.8	57.2	51.3	48.7	57.0	43.0	84.9	15.1
	As prepared →		15.0	85.0	35.0	65.0	50.0	50.0	55.0	45.0	75.0	25.0

4.1 Methods based on derivative spectra: The derivative spectra of the five mixtures and the standards CuCl and $CuCl_2$ are shown in figure 2. The derivative spectra of a mixture has two distinct features: 'A' corresponding to the Cu (I) component and 'B' corresponding to the Cu(II) component. These two features have been used to determine the percentage of a species in the mixture in two ways one utilizing feature 'A' and another feature 'B'. The percentages of CuCl and $CuCl_2$ in the mixtures, as determined from these two ways, i.e., from derivative Cu(I) spectra and derivative Cu(II) spectra, are given in table 1.

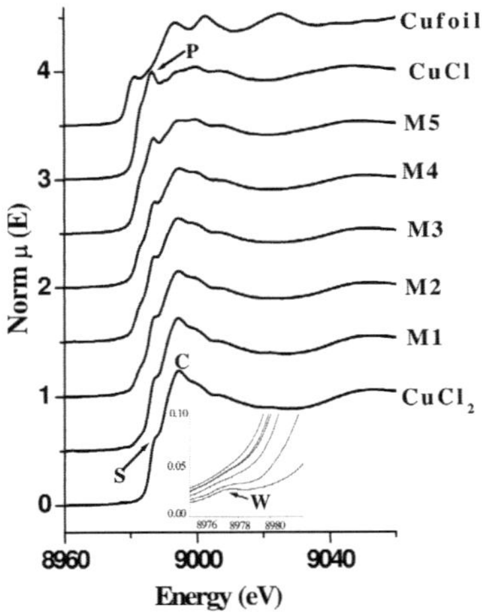

Figure 1. XANES spectra at the copper K-edge in metal, cuprous chloride, cupric chloride and their five mixtures M1 to M5.

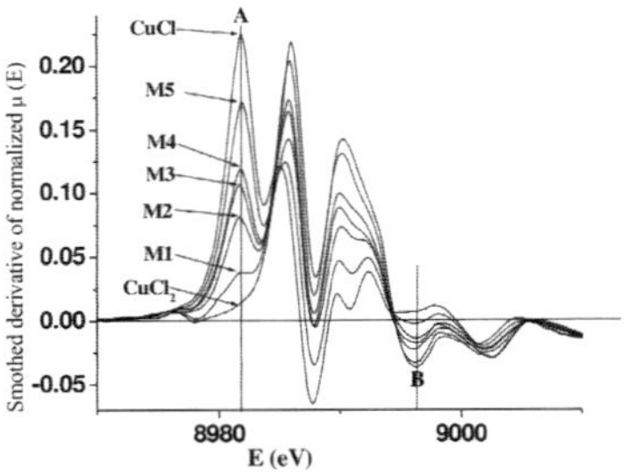

Figure 3. First derivatives of XANES region for CuCl, $CuCl_2$ and their mixtures M1 to M5. Feature A corresponds to the Cu (I) component and B corresponds to the Cu(II) component.

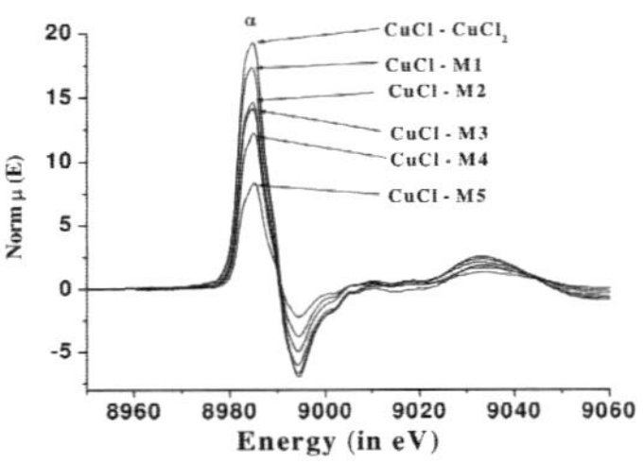

Figure 4. Normalized difference absorption edge spectra (NDAES) for XANES at copper K-edge.

4.2 Normalized Difference Absorption Edge Analysis (NDAES): Figure 3 shows the normalized difference edge absorption spectra for all the mixtures. The difference spectra show a positive peak α at 8983-8984 eV and a broad negative feature at 8990- 9000 eV. The amount of cuprous ion present can be determined from the height of this positive peak α. The height of this peak has been measured precisely in all the mixtures and the percentages of CuCl and $CuCl_2$ in the mixtures, as determined from this method are given in table 1.

4.3 Linear Combination Fitting (LCF): In the present work, LCF has been performed in the XANES spectra (over the range 8960-9040 eV), derivative XANES spectra (over the range 8960-9040 eV) and EXAFS spectra (k^2chi and k^3chi) (over the range 2-12 Å^{-1}). The percentages of CuCl and $CuCl_2$ in the mixtures, as determined from LCF are given in table 1. The figures 4(a-d) show the results for mixture M3 (taken as a representative case). The results of XANES, derivative XANES LC fitting and EXAFS LC fitting (k^3 chi and k^2 chi) are shown in figures 4(a), 4(b), 4(c) and 4(d) respectively. In case of

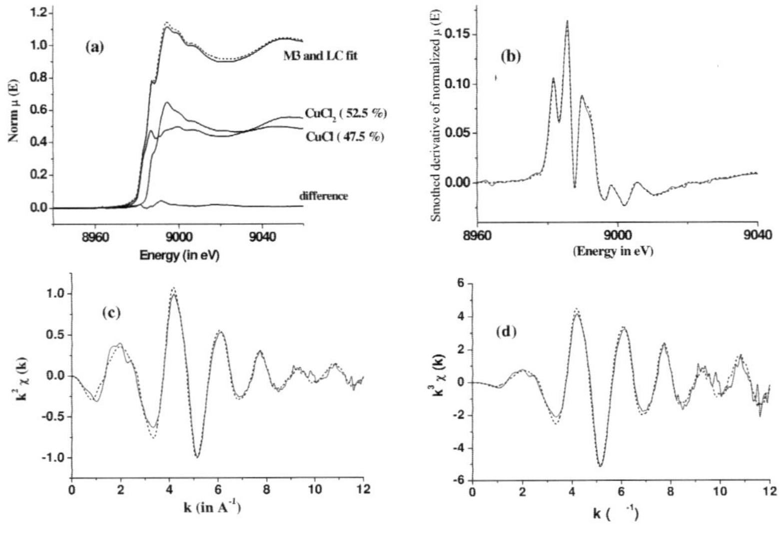

Figure 4. Results of XANES, derivative XANES LC fitting, EXAFS LC fitting (k^3 chi) and EXAFS LC fitting (k^2 chi) for mixture M3 (taken as a representative example) are shown in fig. 4(a), 4(b), 4(c) and 4(d) respectively. Dotted and solid lines denote experimental and LC fit data respectively. In fig. 4(a) fractions of CuCl and $CuCl_2$ making up the fitted spectra are also shown. The goodness-of-fit between experimental data and the LC fit data can be seen from the little difference between the two data, plotted as difference curve.

XANES LC fitting, the figure 4(a) also shows the fractions of the CuCl and $CuCl_2$ making up the fitted spectra. In the case of M3, the R-factor and chi-square values are 0.00017 and 0.01537, 0.003003 and 0.00066, 0.020219 and 0.53792 and 0.045977 and 30.04687, for the LCF performed in XANES, derivative XANES and EXAFS (k^3chi and k^2chi) spectra, respectively.

From table 1, it can be seen that the XANES methods give more accurate results as compared to the EXAFS (k^nchi) methods for estimating the percentages of components in the mixtures. This may be due to the fact that difference between the XANES spectra of the two copper chlorides is more significant than between their corresponding EXAFS spectra. Also, the absorption edge, which is the dominant feature in XANES spectra, occurs at significantly offset absorption edge energies for both the chlorides making XANES methods more reliable.

4.4 Method based on position of the edge: In this method, position of the edge is arbitrarily defined as the energy corresponding to normalized $\mu x = 0.5$ [5]. The fraction y_m of Cu(II) species in the mixture can be determined from relation $y_m = (E_m - E_{CuCl}) / (E_{CuCl2} - E_{CuCl})$, where E_{CuCl}, E_{CuCl2} and E_m are the energies of the position of the edge in the CuCl and $CuCl_2$ and the mixture respectively. The percentages of CuCl and $CuCl_2$ in the mixtures as determined from this method are given in table 1.

5. Conclusions:

The percentages of CuCl and $CuCl_2$ in the mixtures were estimated by different methods, namely, linear combination fitting (LCF) (in XANES, derivative XANES, and EXAFS (k^3chi and k^2chi) spectra), normalized difference absorption edge analysis (NDAES), two methods based on derivative spectra of Cu(I) and Cu(II) species and a method based on the relative position of the edge. A comparison of the results obtained from the different methods with the as prepared composition of the mixtures shows that LCF (in XANES spectra) appears to be the best method for speciation.

The methods of speciation described in the present work can be used for biologically or environmentally important systems, for example, which contain XAS-active metals in several oxidation states, each of which generates quite different fingerprint features in their XAFS spectra. The methods can also be utilized to solve the phase speciation problem of multi-component mixtures and to investigate the speciation of metal in the electrokinetic remediation (EKR) process [9]. Also these methods can be useful for molecular characterization of copper in soils [3]. The methods can also be used for the study of copper-based catalysts under in situ or in operando conditions [5].

Acknowledgement
Thanks are due to Madhya Pradesh Council of Science & Technology (MPCST), Bhopal (India) for a research grant.

References

[1] Beauchemin S, Hesterberg D and Beauchemin M 2002 *Soil Sci. Soc. Am.* **66** 83.
[2] Kelly S D, Hesterberg D and Ravel B 2008 *Chapter 14 in Methods of Soil Analysis. Part 5. Mineralogical Methods* (Madison, Soil Science Society of America)
[3] Legros S, Chaurand P, Rose J, Masion A, Briois V, Ferrasse J H, Macary H S, Bottero J Y, Doelsch E 2010 *Environ. Sci. Technol.* **44** 6926.
[4] Huggins F E, Kim D J, Dunn B C, Eyring E M, Huffman G P 2009 Spectrochim. Ac. A 72 975.
[5] Lamberti C, Bordiga S, Bonino F, Prestipino C, Berlier G, Capello L, D'Acapito F, Xamena F X and Zecchina A 2003 *Phys. Chem. Chem. Phys.* **5** 4502.
[6] Kau L S, Spira-Solomon D J, Penner-Hahn J E, Hodgson K O and Solomon E I 1987 *J. Am. Chem. Soc.* **109** 6433.
[7] Gaur A, Shrivastava B D, Joshi S K, 2009 *J. Phys.: Conf. Series* **190** 012084.
[8] Gaur A And Shrivastava B D, 2012 *Acta Phys. Polon. A* **121** 1111.
[9] Farquar G R, Alderman S L, Poliakoff E D, Dellinger B, 2003 *Environ. Sci. Technol.* **37** 931.

XAFS and XEOL of tetramesityldigermene – An electronic structure study of a heavy group 14 ethylene analogue

Matthew J Ward, Paul A Rupar, Michael W Murphy, Yun-Mui Yiu, Kim M Baines and Tsun-Kong Sham*

Department of Chemistry, University of Western Ontario, London, N6A5B7, CANADA
* Corresponding author, email: tsham@uwo.ca

Abstract. Digermene, the germanium analogue of ethylene, has a multiple bonding motif that differs greatly from that of alkenes and exhibits no pure σ or π type bonds. The electronic structure of digermenes is difficult to study experimentally due to their reactivity, and is computationally challenging because of their shallow potential energy surfaces. Using X-ray absorption near edge structures at both the germanium K and L edges we have been able to directly probe the unoccupied electronic states, or the lowest unoccupied molecular orbital (LUMO), and LUMO+ etc. in the Ge=Ge bond of tetramesityldigermene. We have demonstrated that the LUMO, LUMO+, etc. are composed of hybrid Ge 4s and 4p orbitals. Additionally, our data suggest that the LUMO exhibits relatively more Ge 4s character, whereas the LUMO+ and LUMO+2 exhibit relatively more Ge 4p character. An X-ray excited optical luminescence study of Ge_2Mes_4 revealed one broad optical emission band at 620 nm, which is significantly red shifted compared to the known energy gap of this molecular germanium compound.

1. Introduction

Multiple bonding in heavier group 14 elements (E = Si, Ge, and Sn) is of great interest due to the structural, and thus electronic structure (bonding) changes about the central element that take place upon descending the group. The multiple bonding motif in ethylene has been well described by both valence bond (VB) and molecular orbital (MO) theory and involves one C-C σ and one C-C π bond, which is consistent with the experimental planar geometry of the molecule. In contrast, most digermenes ($R_2Ge{=}GeR_2$) are pyramidal at germanium and assume a trans-bent geometry. Digermenes have been investigated intensely since the first reports of their isolation [1-4]. The simplest bonding model consists of two polar dative bonds between two germylene fragments [5]. Digermenes are difficult to study due to their reactivity with many different substrates [6-9]. Computational studies of digermenes are also made challenging because shallow potential energy surfaces result in problematic geometry optimizations and difficulties in studying electronic transitions [7, 9, 10].

To directly study the electronic structure of tetramesityldigermene [Ge_2Mes_4 (henceforth denoted as **1**), Mes=2,4,6-trimethylphenyl], a prototypical digermene, we have measured the germanium K and L-edge X-ray absorption near edge structures (XANES). The Ge K and $L_{3,2}$ edges of **1** were modeled using the *ab initio* self-consistent real-space-multiple-scattering (RSMS) code of FEFF9. Additionally **1** was observed to visibly luminesce under soft X-ray excitation and we measured X-ray excited optical luminescence (XEOL) from the sample. Herein, we report the preliminary findings of this study.

Journal of Physics: Conference Series **430** (2013) 012046 doi:10.1088/1742-6596/430/1/012046

2. Experimental

Germanium K-edge XANES for **1** was collected on the PNC/XSD BM-20 beam-line at the Advanced Photon Source (APS) at Argonne National Laboratory (see M. Ward *et al Chem. Commun.* **2010,** *46,* 7016 and supporting information for more detail). Ge $L_{3,2}$-edge XANES for **1** was collected on the SGM beam-line at the Canadian Light Source (CLS) in total electron yield (TEY) and total fluorescence yield (FY) detection modes. Briefly, a powdered sample of **1** was prepared on carbon tape in a glove bag attached to the vacuum sample load-lock chamber of the SGM beam-line under nitrogen purge and then immediately put under high vacuum. XEOL spectra of **1** were collected on the SGM beam-line using an Ocean Optics QE 65000 spectrophotometer in the UV-vis-NIR (200-900 nm) spectral region. The Ge K and $L_{3,2}$-edge XANES were modeled for **1** using the *ab initio* self-consistent real-space-multiple-scattering (RSMS) code of FEFF9 [11], and included all electrons of molecular clusters based on the molecular structures as experimentally determined by X-ray crystallography, which are then described by a real space relativistic Green's function formalism [12].

3. Results and Discussion

The change from planar to trans-pyramidal geometry at the metal center, and thus, from D_{2h} to C_{2h} molecular symmetry, upon descending the group has been thoroughly studied for heavier homonuclear group 14 alkene analogues [5,13]. Second order Jahn-Teller mixing of σ^* with π and σ with π^* molecular orbitals leads to a simplified MO picture for heavier group 14 alkenes in which both the highest occupied molecular orbital (HOMO) and the lowest unoccupied molecular orbital (LUMO) are non-bonding, n_- (b_u) and n_+ (a_g) respectively [13]. The experimental and theoretical Ge K and $L_{3,2}$-edge XANES are shown in figure 1 (a) and (b), respectively.

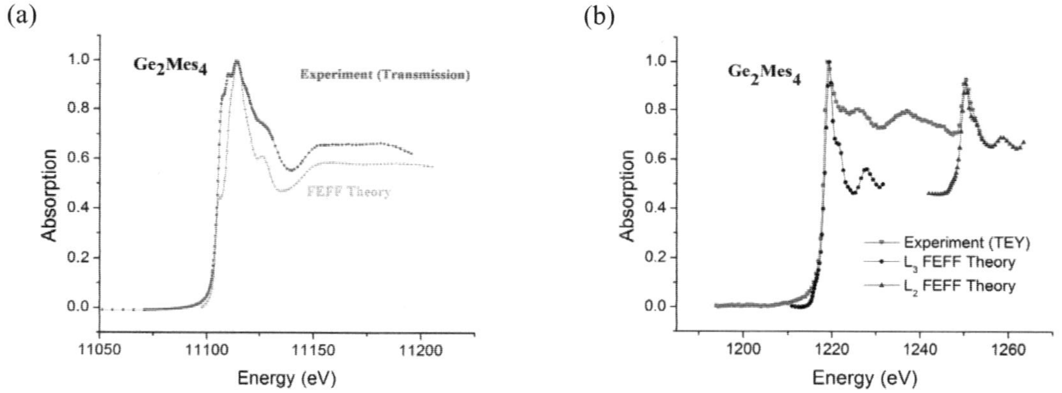

Figure 1 – Experimental and calculated XANES of **1** at the (a) Ge K-edge and (b) Ge $L_{3,2}$-edge

Compound **1** has trans-bent (θ) and twist (δ) angles of 33.4° and 2.9° respectively, (figure 2) and a Ge–Ge bond length of 2.2856(8) Å [14]. The theoretical spectrum of the Ge K-edge is in general agreement with the experimental data for **1**; however, the intensity of the first resonance (LUMO) is diminished and the second and fourth resonances observed in the experimental data appear as shoulders on the third resonance in the calculated spectra. The experimental and theoretical Ge $L_{3,2}$-edge XANES spectra of **1** are shown in figure 1b. The calculated spectra are in general agreement with the experimental TEY spectrum in the near-edge region; however, FEFF overestimates the white-line intensities at both edges and the L_3-edge multiple-scattering peak at ca. 1226 eV is blue-shifted ~ 2 eV to 1228 eV. FY results (not shown) are consistent with the TEY spectra suggesting a uniform composition with no surface oxidation. Figure 2(a) depicts a simple orbital diagram of **1** showing the HOMO-, HOMO, LUMO, and LUMO+. On the basis of the simplified bonding model depicted in figure 2(a), the bonding between Ge centers in **1** is very different from that of alkenes which would render a planar structure. Regardless of the bonding schemes used to describe this situation, Ge 4s and

4p orbitals will participate. Thus the unsaturation at the Ge site can be probed using Ge XANES via dipole transitions at both the K and $L_{3,2}$ edges, probing unoccupied electronic states, or the Ge 4p and 4s character respectively in the LUMO and LUMO+ orbitals. Figure 2(b) shows the Ge K-edge and $L_{3,2}$-edge XANES displayed in energy relative to the threshold of the Ge K-edge and the Ge L_3-edge ($E_o = 0$ eV).

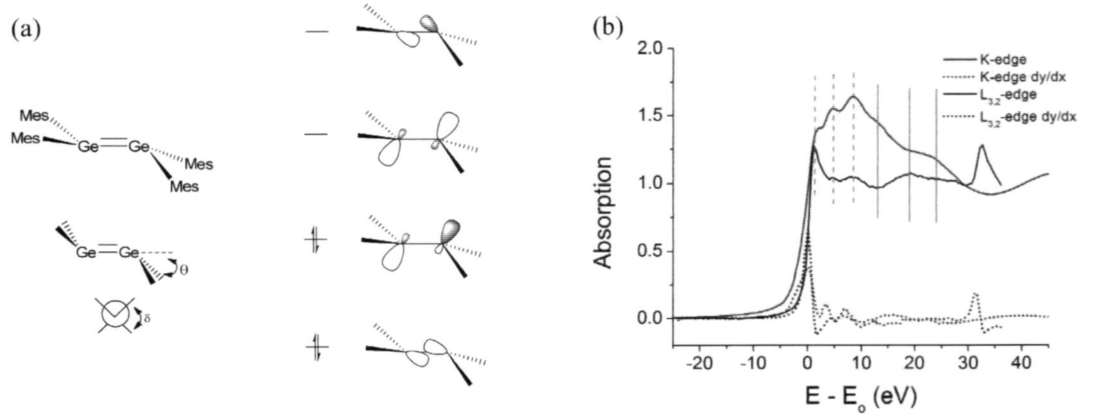

Figure 2 – (a) Simplified orbital diagram of Ge_2Mes_4. (b) XANES at the Ge K and $L_{3,2}$ edges calibrated with respect to the threshold energy (E_o); the first derivative (dy/dx) is also shown.

The first three resonances (figure 2(b), vertical dashed lines) are in phase and the relative intensity of these resonances suggests that the LUMO (first peak) has more Ge 4s character compared to the LUMO+, which is in contrast to alkenes where the LUMO (π^*) has considerable 2p character in comparison to the LUMO+ (σ^*). The next three resonances (figure 2(b), vertical solid lines) are more than 10 eV above the threshold, and are out of phase. We tentatively attribute them to EXAFS with multiple scattering, of which the continuum of the p wave and s/d waves are out of phase. The observation of relatively more Ge 4p character in the LUMO+ than the LUMO from analysis of the Ge K-edge XANES of **1** is, at first appearance, at odds with previous reports of bonding in digermenes. From previous bonding models one would expect to see more p character in the LUMO since the LUMO is proposed to be a distortion of the π^* orbital via second order Jahn-Teller (SOJT) mixing of π^* with σ whereas the LUMO+ is due to SOJT mixing of σ^* with π and subsequent distortion of a σ orbital. Natural bond orbital (NBO) calculations for **1** at the TPSSTPSS/6-31G level, however, show a LUMO with 90.34% s and 9.66% p character, which is in agreement with our observations. The Ge L_3-edge XANES of **1** is also consistent with the NBO analysis as more Ge 4s character is observed in the LUMO than the LUMO+, however, the white-line (LUMO) is very intense suggesting that there is the possibility of some d-orbital participation. A more thorough theoretical treatment of the Ge K and L edges using a density functional based model, such as WIEN 2K, may be able to provide more detail on the degree of orbital participation and mixing in the LUMO and LUMO+ peaks observed in the experimental spectra.

Ge_2Mes_4 (compound **1**) also emits visible light upon X-ray excitation. The XEOL spectra for **1**, collected with 1200 eV excitation energy, is shown in figure 3. The XEOL spectrum of **1** exhibits a single peak at ~ 2 eV (620 nm). Its energy is considerably red-shifted compared with the known HOMO-LUMO separation of **1**. Thus, the observed emission could be due to one of the following effects: (1) solid state effects which influence the energy levels of the HOMO and LUMO bands, narrowing the band gap, (2) formation of excimers involving neighboring molecules, and (3) decomposition of the molecules into Mes_2Ge: germylene type moiety. More work needs to be carried out to pin point its origin.

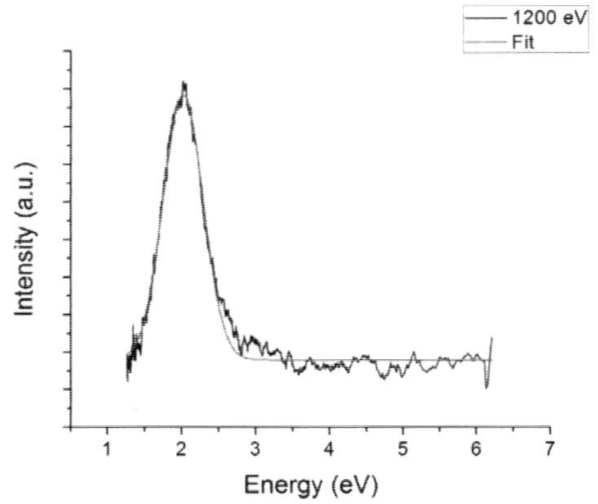

Figure 3 – XEOL of Ge_2Mes_4 measured below the Ge L-edge.

4. Conclusion

By directly probing the unoccupied electronic states in Ge_2Mes_4 *via* germanium K and L edge XANES we have demonstrated that the LUMO is composed of hybridized Ge 4s and 4p atomic orbitals, of which there is more 4s character. The LUMO+1 and LUMO+2 are also composed of hybridized Ge 4s and 4p atomic orbitals, however, they exhibit more 4p character than the LUMO. Ge_2Mes_4 exhibits an optical luminescence emission band at 620 nm which is significantly red shifted when compared to the HOMO-LUMO separation. This shift in the optical emission band may be a result of solid state effects, excimer formation, or decomposition into germylene fragments.

Acknowledgements

PNC/XSD facilities at the Advanced Photon Source (APS), and research at these facilities, are supported by the U.S. Department of Energy (U.S. DOE) - Basic Energy Sciences, a Major Resources Support grant from NSERC, the University of Washington, the Canadian Light Source (CLS), and the APS. Use of the APS at Argonne National Laboratory was supported by the U.S. DOE under Contract No. DE-AC02-06CH11357. Research performed at the CLS, was supported by the NSERC, NRC, CIHR, the Province of Saskatchewan, Western Economic Diversification Canada, and the University of Saskatchewan. Research at The University of Western Ontario (UWO) was supported by NSERC, CRC, CFI, and OIT. M.J.W. acknowledges the financial support of the QEIIGSST and OGS scholarship programs and a GTRA award jointly funded by Research and Development Services and the Faculty of Science at UWO. We thank Margaret Hanson for performing NBO calculations.

References

[1] Masamune S, Hanzawa Y and Williams D J 1982 *J. Am. Chem. Soc.* **104** 6136
[2] Snow J T, Murakami S, Masamune S and Williams D J 1984 *Tetrahedron Letters* **25** 4191
[3] Park J, Batcheller, S A, Masamune, S 1989 J. Organomet. Chem. **367** 39.
[4] Schäfer H, Saak W, and Weidenbruch M 1999 *Organometallics* **18** 3159
[5] Power P P 1999 *Chem. Rev.* **99** 3463
[6] Leigh W J, Harrington C R, and Vargas-Baca I 2004 *J. Am. Chem. Soc.* **126** 16105
[7] Leigh W J and Harrington C R 2005 *J. Am. Chem. Soc.* **127** 5084
[8] Tokitoh N, Okazaki R, in The Chemistry of Organic Germanium, Tin, and Lead Compounds, ed. Z. Rappoport, Wiley and Sons Ltd., Chichester, 2002, vol. 2, pp. 843 and references therein.
[9] Kira M 2011 *Organometallics* **30**, 4459.
[10] Jacobsen H and Ziegler T 1994 *J. Am. Chem. Soc.* **116** 3667
[11] Ankudinov A L, Ravel B, Rehr J J and Conradson S D 1998 *Phys. Rev.* B **58** 7565
[12] Ankudinow A L, Conradson S D, deLeon J M and Rehr J J 1998 *Phys. Rev.* B 1**57** 7518
[13] Fischer R C and Power P P 2010 *Chem. Rev.* **110** 3877
[14] Hurni K L, Rupar P A, Payne N C and Baines K M 2007 *Organometallics* **26** 5569

EXAFS study of mixed ligand copper complexes having nicotinic acid as one of the ligands

S K Joshi[1], A Gaur[2],*, B. D. Shrivastava[2], D. C. Gaur[3], J. Prasad[4], K. Srivastava[4], S. N. Jha[5], D. Bhattacharyya[5] and A. Poswal[5]

[1]Physics Department, Government P G Arts and Science College, Ratlam-457001, India
[2]School of Studies in Physics, Vikram University, Ujjain-452001, India
[3]Physics Department, Government P. G. College, Rajgarh (Biaora)-465661, India
[4]Department of Chemistry, University of Allahabad, Allahabad-211002, India
[5]Applied Spectroscopy Division, Bhabha Atomic Research Centre, Mumbai-400085, India

email: abhijeetgaur9@gmail.com

Abstract. Copper K-edge EXAFS spectra of the copper (II) complexes $[Cu_3(NA)_4(dca)_2(H_2O)_8].2H_2O$ (**1**) and $[Cu_2(2,2'\text{-bipy})(NA)_2](ClO_4)2.H_2O$ (**2**) (where NA = Nicotinic acid; dca = dicyanamide anion; 2,2'-bipy = 2-2' bipyridine) having different coordination environments and geometries have been investigated. The complex **1** is supposed to have six coordinated geometry and complex **2** five coordinated geometry. EXAFS has been recorded at the K-edge of copper in the complex at the dispersive EXAFS beamline at 2 GeV Indus-2 synchrotron source at RRCAT, Indore, India. Theoretical models have been generated for both the complexes separately using their reported crystal structures. These theoretical models have been fitted to the respective experimental EXAFS data and the structural parameters, which include bond lengths, coordination numbers and thermal disorders, have been determined. The results obtained have been found to be comparable with those reported using crystallography. The coordination geometry about the copper (II) ion has been depicted for both the complexes. Further, copper has been found to be in +2 oxidation state in both the complexes. The study shows that the dispersive EXAFS beamline can be used to obtain information about the coordination geometry of a sample, though the beamline provides limited range of EXAFS data.

1. Intoduction

The nicotinate ion, NA^-, is a quite interesting ligand: first of all, its exodentate nature makes it a valuable tecton for designing high-dimensional coordination polymers. The nicotinato ligand combines the coordination ability of the pyridyl moiety with the various coordination modes of the carboxylato group. The nicotinic acid itself, acting as a ligand, can generate extended structures: it coordinates with the nitrogen atom to the metal ion, the structure being further expanded on the base of the classical synthon [1, 2].

In the present work, the extended X-ray absorption fine structure (EXAFS) has been studied in the two coordination complexes involving nicotinate ion: $Cu_3(NA)_4(dca)_2(H_2O)_8].2H_2O$ (**1**) and $[Cu_2(2,2'\text{-bipy})(NA)_2](ClO_4)2.H_2O$ (**2**) (where NA = Nicotinic acid; dca = dicyanamide anion; 2,2'-bipy = 2-2' bipyridine). Complex **1** is supposed to have six coordinated geometry and the complex **2** five

Journal of Physics: Conference Series **430** (2013) 012047 doi:10.1088/1742-6596/430/1/012047

coordinated geometry. The aim of the present work is to investigate the copper K-edge EXAFS spectra of the copper (II) complexes having different coordination environments and geometries and to show that the dispersive EXAFS beamline can be used to obtain information about the coordination geometry of a sample, though the beamline provides limited range of EXAFS data. To our knowledge, the EXAFS study of these mixed ligand complexes have not been made earlier. Madalan *et al.* [3] have only reported the crystal structure (from XRD studies) and magnetic properties of the complexes.

2. Experimental

The complexes have been prepared and characterized by following the standard methods [3]. The complexes were finely powdered and then absorption screens were prepared by carefully spreading calculated amount of the powder on 1sq.cm. square area of kapton tape. The copper K-edge EXAFS spectra of these absorption screens have been recorded at the BL-8 dispersive EXAFS beamline at 2 GeV Indus-2 synchrotron source at Raja Ramanna Center for Advanced Technology (RRCAT), Indore, India described earlier [4-6]. It is worth mentioning here that for the setting of the bent crystal to record the XAFS spectra at the K-edge of copper, the spectra can be recorded only upto ~ 300 eV (i.e., k = 9 $Å^{-1}$) above the edge.

3. Data analysis

In the present work, EXAFS data of the complexes have been analyzed using the available computer software packages *Athena* and *Artemis* [7]. Figures 1(a, b) shows the normalized EXAFS spectra at the K-edge of copper in the complexes. $\chi(k)$ vs. k spectra have been shown in fig. 2(a, b). Using the available crystallographic data [3], the theoretical models for these complexes have been generated which are then fitted to the experimental EXAFS data to obtain the structural parameters. Two theoretical models have been generated, one for each complex by using its own crystal structure. The theoretically calculated contribution to $|\chi(R)|$ by the different scattering paths in the complexes **1** and **2** are shown in figures 3(a) and 3(b) respectively. The experimental $|\chi(R)|$ is also shown in these figures. For generating the theoretical models the procedure outlined by Kelly *et al.* [8] has been followed. Accordingly, a single value of S_0^2 and ΔE_0 is used for all the paths in the fitting but different values of ΔR can be defined for different paths. Unique Debye-Waller factor (σ^2) values are given to

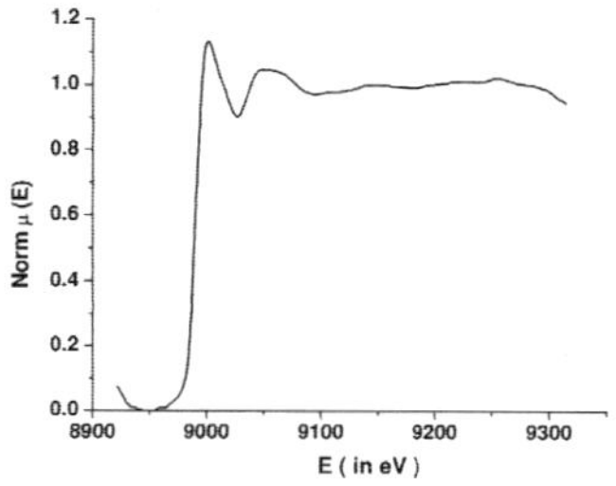

Fig 1 (a) μ (E) vs. E of **1**

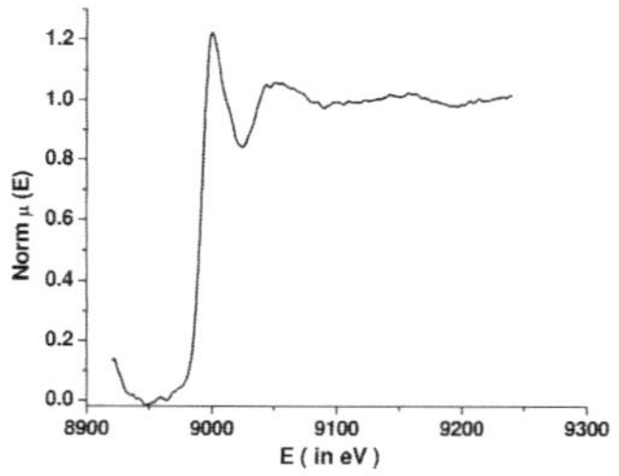

Fig 1(b) μ (E) vs. E of **2**

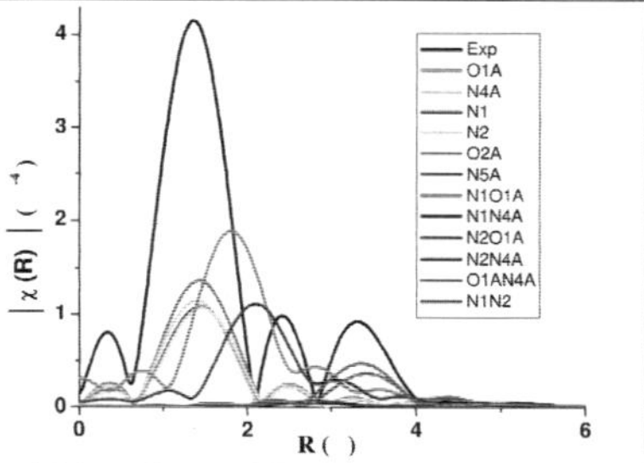

Fig 3(a) Paths used in theoretical model in **1**

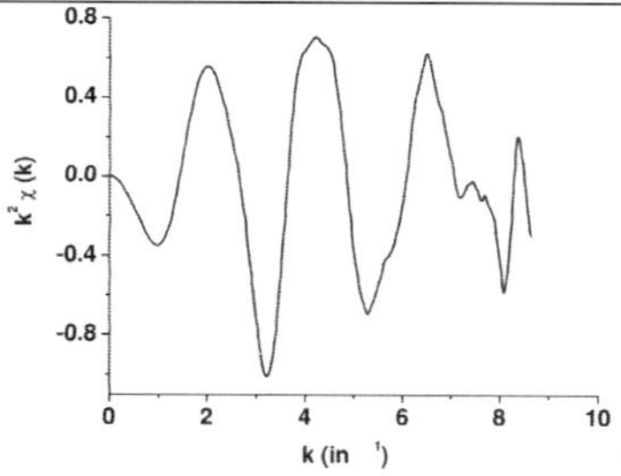

Fig 2(a) χ (k) vs. k of **1**

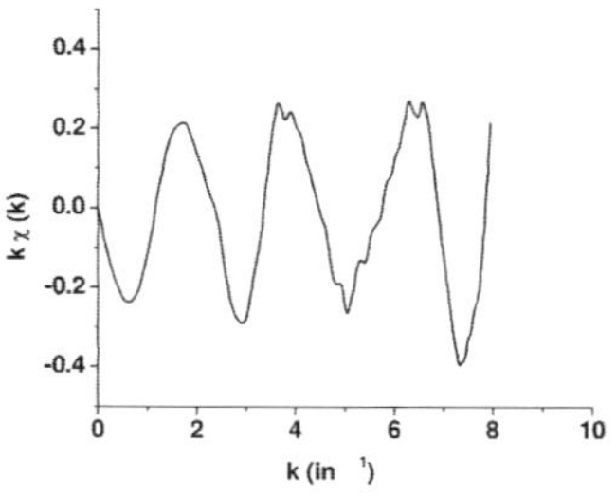

Fig 2(b) χ (k) vs. k of **1**

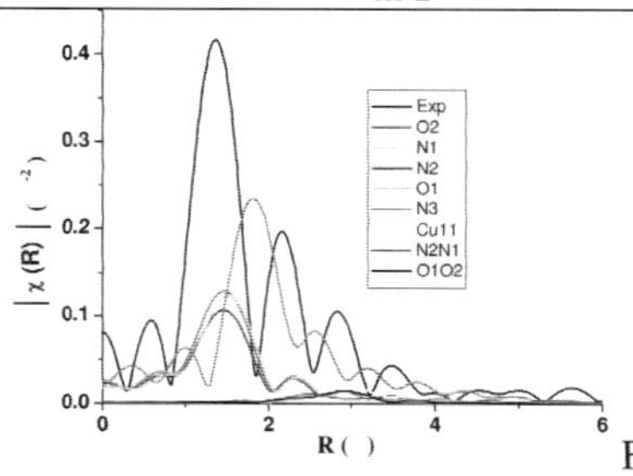

Fig 3(b)Path used in theoretical model in **2**

each of the scattering paths in the models. One ΔR has been defined for the axial O and N paths and another ΔR has been defined for the apical O and N path. In case of both the complexes, there is a N apical path, while in case of **1** there is additional O apical path. The resulting fitted curves for **1** and **2** are shown in figures 4(a) and 4 (b), respectively. The parameters so obtained for **1** and **2** are given in tables 1(a) and 1(b), respectively.

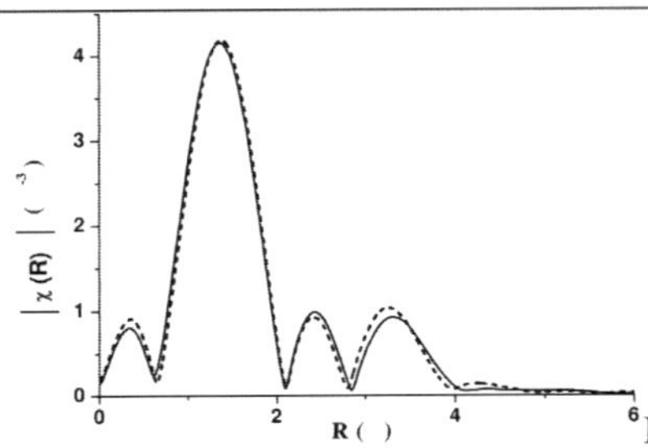

Fig 4(a) Fitting of theoretical model in **1**

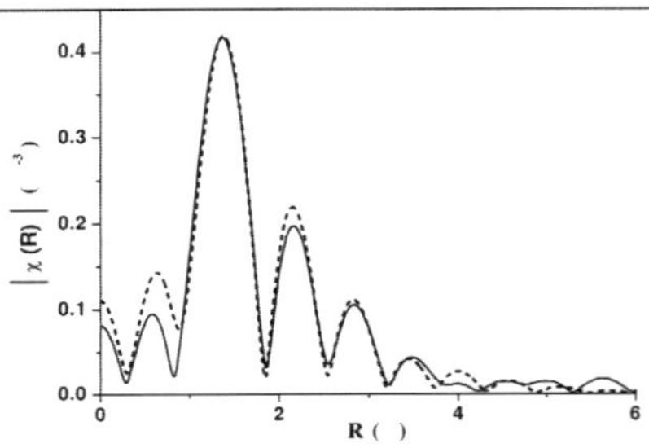

Fig 4(b) Fitting of theoretical model in **2**

Table 1(a-b) The EXAFS fitting results for complexes 1 and 2. The XRD results for the complexes (from Madalan *et al.* 2005) are also given.

Atomic pair	**EXAFS results**				**XRD results**
	N	R(Å)	delr(Å)	σ^2(Å^{-2})	R
Cu1-O1A	1	1.94	-0.028	0.0033 ± 0.0012	1.97
Cu1-N4A	1	1.95	-0.028	0.0033 ± 0.0012	1.98
Cu1-N1	1	1.99	-0.028	0.0033 ± 0.0012	2.02
Cu1-N2	1	2.00	-0.028	0.0033 ± 0.0012	2.02
Cu1-O2A	1	2.31	0.012	0.0096 ± 0.0054	2.30
Cu1-N5B	1	2.65	0.025	0.0096 ± 0.0054	2.62

Atomic pair	**EXAFS results**				**XRD results**
	N	R(Å)	delr(Å)	σ^2(Å^{-2})	R
Cu1-O2	1	1.98	0.018	0.0069 ± .0014	1.96
Cu1-N1	1	2.00	0.018	0.0069 ± .0014	1.98
Cu1-N2	1	2.00	0.018	0.0069 ± .0014	1.98
Cu1-O1	1	2.02	0.018	0.0069 ± .0014	2.00
Cu1-N3	1	2.33	0.134	0.0256 ± .0020	2.20

4. Results and discussions

The energies of the K-edge of copper are found to be 8988.42 eV and 8987.68 eV for complexes **1** and **2,** respectively. Taking the energy of copper metal K-edge as 8980.5 eV, the chemical shifts are found to be 7.92 eV and 7.18 eV for complexes **1** and **2** which suggest that copper is in +2 oxidation state in these complexes. [9, 10]. The results of the EXAFS analysis and the discussions are as follows:

4.1. Complex 1: For the analysis of the EXAFS data of **1**, the input parameters are R_{bkg}=1.02 Å, k-range: k_{min}=2.64 Å^{-1}, k_{max}=7.06 Å^{-1}, k_w= 3 and R range is 1.0 - 5.0 Å. In the fitting procedure, 12 paths obtained from Feff calculations have been used. The results obtained from fitting are given in table 1(a), which gives the local structure parameters for complex **1**. The value of reduced chi-square (χ_v^2) is 188.34, S_0^2 value obtained is 0.75 with an error of ± 0.13 and ΔE_0 value is 1.25 eV with an error of ± 1.11. From the present EXAFS analysis, the axial Cu-O bond distance is found to be 1.94 Å. Three axial Cu-N distances are observed in the range of 1.95 - 2.00 Å. The equatorial distance Cu-O bond, has a value of 2.32 Å and the equatorial distance Cu-N bond, has a value of 2.64 Å.

Complex 2: For the For the analysis of the EXAFS data of **2**, the input parameters are R_{bkg}=1.0 Å, k-range: k_{min} = 2.46 Å^{-1}, k_{max} = 7.89 Å^{-1}, k_w= 1 and R ranges of 1.2 - 5.0 Å. In the fitting procedure, the 8 paths obtained from Feff calculations have been used. The results obtained from fitting are given in table 1(b), which gives the local structure parameters for **2** obtained using the theoretical model. The value of reduced chi-square (χ_v^2) is 146.49, S_0^2 value obtained is 0.80 with an error of ± 0.10 and ΔE_0 value is 1.87 eV with an error of ± 0.0.92. From the present EXAFS analysis, two nitrogen atoms have been found at a distance of 1.99 and 2.00 Å and two oxygen atoms have been found at a distance of 2.02 and 1.98 Å. The apical nitrogen atom has been found at a distance of 2.33Å.

As can be seen from table 1(a) and (b), the results obtained for the bond distances are in agreement with the crystallographic bond distances. Thus, the theoretical models generated by us appear to be quite reasonable. Using the present results, the inner coordination sphere around the copper (II) centers in **1** and **2** has been depicted in fig. 5(a) and (b), respectively. These figures are based on the figures given by Madalan *et al.* [3] but show the distances obtained by us in the present study.

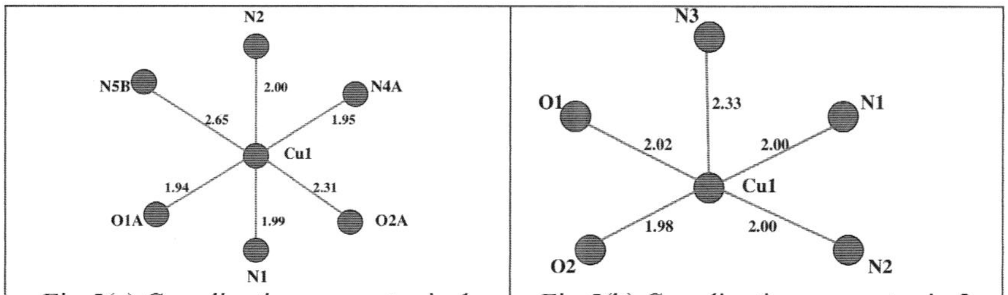

Fig 5(a) Coordination geometry in **1** | Fig 5(b) Coordination geometry in **2**

5. Conclusions

The EXAFS spectra have been recorded at the Cu K-edge in copper(II) complexes $[Cu_3(NA)_4(dca)_2(H_2O)_8].2H_2O$ **1** and $[Cu_2(2,2'\text{-bipy})(NA)_2]$ $(ClO_4)_2.H_2O$ **2** using the dispersive EXAFS beamline at 2.5 GeV Indus-2 synchrotron source at RRCAT, Indore, India. Theoretical models have been generated for the complexes using their available crystallographic data and then fitted to the respective experimental data to obtain the structural parameters, which include bond-lengths, coordination numbers and thermal disorders. The results obtained have been found to be comparable with the crystallographic results. Based on the present EXAFS data analysis the coordination geometry around the Cu(II) atom in the complexes has been depicted.

As already mentioned, the aim of the present work was to investigate the copper K-edge EXAFS spectra of the copper (II) complexes having different coordination environments and geometries. From the EXAFS analysis, it has been shown that complex **1** has six coordinated geometry and complex **2** five coordinated geometry. The study shows that the dispersive EXAFS beamline available at Indus-2 synchrotron source can be used satisfactorily to obtain information about the samples having different coordination geometries, though the EXAFS data is obtained over a limited range on this beamline.

Acknowledgement

Thanks are due to Madhya Pradesh Council of Science & Technology (MPCST), Bhopal (India) for a research grant.

References

[1] Chapman M E, Ayyappan P, Foxman B M, Yee G and Lin W, 2001 *Cryst. Growth Des.* **1** 159.
[2] Qin Z, Jennings M C and Puddephat R J, 2002 *Inorg. Chem.* **41** 5174.
[3] Madalan A M, Paraschiv C, Sutter J P, Schmidtmann M, Muller A and Andruh M, 2005 *Crys. Growth Des.* **5** 707.
[4] Bhattacharyya D, Poswal A K, Jha S N, Sangeeta, Sabharwal S C, 2009 *Bull. Mater. Sci.* **32** 103
[5] Bhattacharyya D, Poswal A K, Jha S N, Sangeeta, Sabharwal S C, 2009 *Nucl. Instrum. Methods A* **609** 286,
[6] Gaur A, Johari S, Shrivastava B D, Gaur D C, Jha S N, Bhattacharyya D, Poswal A and Deb S K, 2011 *Sadhana* **36** 339
[7] Ravel B and Newville M, 2005 *J. Synchrotron Radiat.* **12** 537.
[8] Kelly S D, Hesterberg D and Ravel B 2008 *Chapter 14 in Methods of Soil Analysis. Part 5. Mineralogical Methods* (Madison, Soil Science Society of America)
[9] Kau L S, Spira-Solomon D J, Penner-Hahn J E, Hodgson K O and Solomon E I 1987 *J. Am. Chem. Soc.* **109** 6433.
[10] Fulton J L, Hoffman M M, Darab J G, Palmer B J, Stern E A, 2000 *J. Phys. Chem. A*, **104** 11651

EXAFS study of tetramethylethylenediamine copper(II) complexes with Cl and Br as secondary ligands

V K Hinge[1], A Gaur[2], Davood Ah. Dar[2], S K Joshi[3], B. D. Shrivastava[2]* J. Prasad[4], K. Srivastava[4], S. N. Jha[5], D. Bhattacharyya[5] and A. Poswal[5]

[1]Physics Department, Ujjain Engineering College, Ujjain-456010, India.
[2]School of Studies in Physics, Vikram University, Ujjain-452001, India
[3]Physics Department, Government P G Arts and Science College, Ratlam-457001, India
[4]Department of Chemistry, University of Allahabad, Allahabad 211002, India.
[5]Applied Spectroscopy Division, Bhabha Atomic Research Centre, Mumbai-400085, India.
email: rashmibasant@gmail.com

Abstract. Extended X-ray absorption fine structure (EXAFS) have been recorded at the K-edge of copper in binuclear copper(II) complex [Cu(tmen)Cl_2] (**1**) and its analogous complex [Cu(tmen)Br_2] (**2**) (where tmen = tetramethylethylenediamine), using the dispersive EXAFS beamline at 2 GeV Indus-2 synchrotron source at RRCAT, Indore, India. Theoretical model has been generated for **1** using its available crystallographic data [1] and then fitted to the experimental EXAFS data of **1** to obtain its structural parameters, which include bond-lengths, coordination numbers and thermal disorders. The results obtained have been found to be comparable with the crystallographic results. As complex **2** is analogous to complex **1**, the same theoretical model, which was generated for **1**, has been fitted to the experimental EXAFS data of **2**. The structural parameters, thus determined for **2** have been reported. The coordination geometry around the copper center in both the complexes has been depicted showing the bond-lengths obtained in the present study. It has been demonstrated that if crystal structure is available for a complex from X-ray crystallographic studies, the structural parameters can be determined for an analogous complex from EXAFS studies. Further, the values of the chemical shifts suggest that copper is in +2 oxidation state in these complexes.

1. Introduction

In the present work EXAFS of the two copper complexes, [Cu(tmen)Cl_2] (**1**) and [Cu(tmen)Br_2] (**2**), (tmen = tetramethylethylenediamine), at the K–edge of copper has been recorded using the dispersive EXAFS beamline at the synchrotron source. The theoretical model for the complex **1** has been generated from the available crystallographic data [1] and then fitted to its experimental EXAFS data to obtain the structural parameters, which include bond-lengths, coordination numbers and thermal disorders. The structural parameters thus obtained are found to be in good agreement with XRD results. As complex **2** is analogous to the complex **1**, the same theoretical model which was generated for complex **1**, is fitted to the EXAFS data of complex **2**. The structural parameters thus obtained for **2** have been reported and the coordination geometry around the copper site has been depicted.

2. Experimental details

The complexes have been prepared and characterized by standard methods [1]. The K- absorption spectra at the K-edge of copper in the complexes have been recorded at the BL-8 dispersive EXAFS beamline at 2 GeV Indus-2 synchrotron source at RRCAT, Indore, India [2-4]. The beamline has 460mm long Si(111) crystal having 2d value equal to 6.2709Å mounted on an elliptical bender. The radiation transmitted through the sample is detected by a position sensitive CCD detector having 2048×2048 pixels. For energy calibration of that particular setting, absorption spectra at Cu K-edge of Cu metal foil and at Lu L_3-edge of Lu_2O_3 powder have been recorded under the same setting of crystal bender and the goniometer. Taking the values of the energies of Cu K-edge in metal and Lu L_3-edge in Lu_2O_3 as 8980.50 and 9249.00 eV, respectively, the CCD channels have been calibrated [5].

3. Data analysis

The EXAFS data have been analyzed by using the computer programs Athena and Artemis [6]. The analysis procedure has two stages: reduction of the measured absorption spectra to EXAFS and analysis of the EXAFS data to obtain the structural parameters. Normalized μ(E) versus E (energy) curve for the two complexes under investigation are shown in figure 1, χ(k) versus k curves are shown in figure 2 and the Fourier transformed curves are shown in figure 3 and figure 4. The Fourier transformed data is then fitted with the theoretical model in R-space to obtain the fitting parameters.

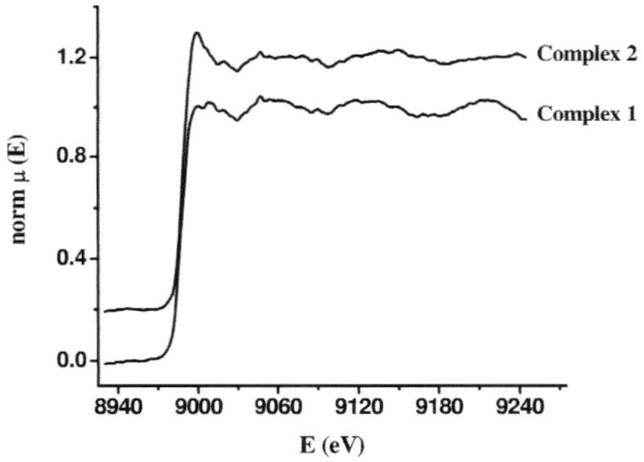

Fig. 1. Normalized μ(E) versus Energy (eV)

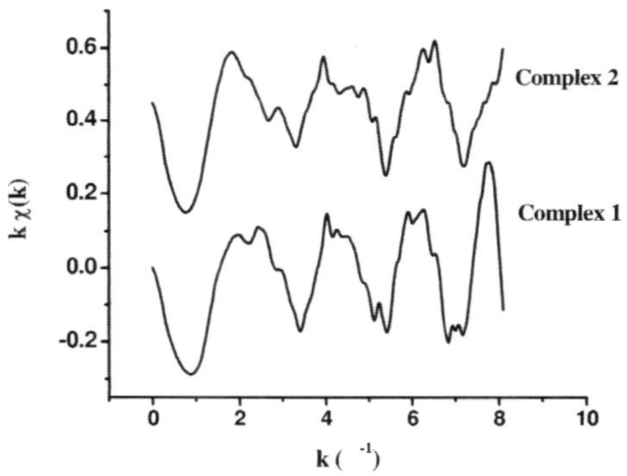

Fig. 2. χ (k) spectrum obtained from Fig. 1.

The crystallographic data is available for **1**, from which the theoretical model for this complex has been generated using the following parameters as input in Artemis: Space group: P -1, Cell constants: a = 7.713Å, b = 8.560Å, c = 9.249Å; α = 98.44, β = 94.39, γ = 118.63 and Cluster size = 8 Å. The remaining input positional parameters have been taken from reference [1]. For generating the theoretical model, we have followed the procedure outlined by Kelly et al. [7]. Accordingly, a single value of S_0^2 and ΔE_0 is used for all the paths in the fitting but different values of ΔR and σ_i^2 can be used for different paths. The resulted fitting curve for complex **1** is shown in figure 3. The structural parameters thus obtained for **1** are given in table 1 and are comparable with the crystallographic results.

The crystallographic data for **2** is not available in the literature. As the theoretical model generated by us for **1** fits well to its experimental EXAFS data and because the complex **2** is analogous to **1**, we have fitted the same theoretical model, which has been generated for **1**, to the experimental EXAFS data of **2**, to determine the structural parameters of **2**. The resulting fitted curve is shown in figure 4 and the structural parameters so obtained are given in table 1.

4. Results and discussion

The energies of the K-edge of copper are found to be 8986.73 and 8988.97eV for complex **1** and complex **2** respectively. The chemical shifts are thus found to be 6.23 and 8.47 eV for **1** and **2** respectively, which indicates that copper is in +2 oxidation state in the two complexes [8]. The

EXAFS analysis procedure and the structural parameters obtained by fitting the theoretical model to the two complexes are given in table 1 and are discussed below:

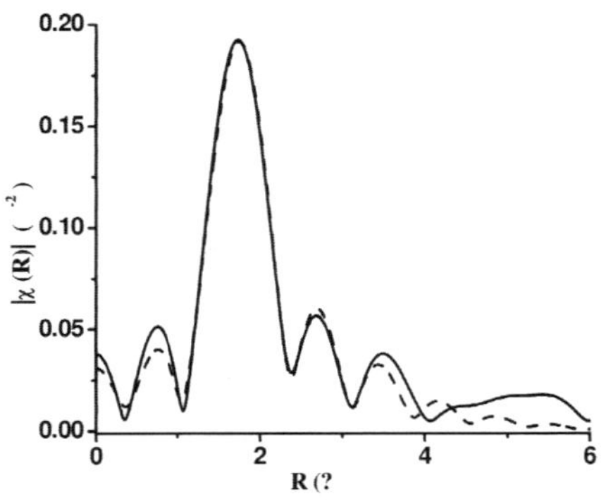

Fig. 3. Fitting of theoretical model in **1**. Solid line is experimental data and dashed line is modelled fit.

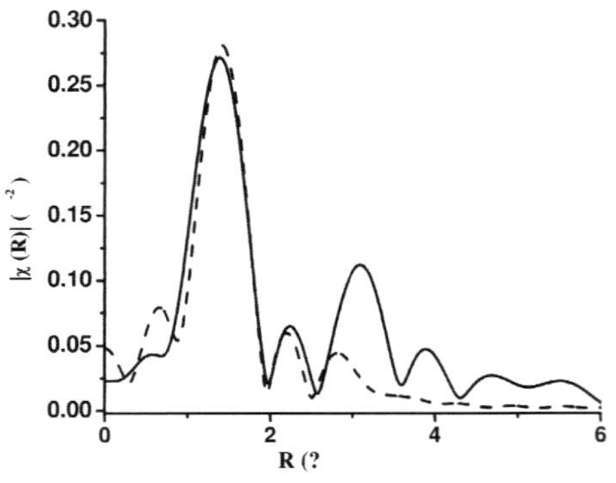

Fig. 4. Fitting of theoretical model in **2**. Solid line is experimental data and dashed line is modelled fit.

Table1. The EXAFS fitting results for complex 1 and 2.

EXAFS results for **1**					XRD [1]
Atomic pair	N	R(Å)	ΔR(Å)	σ^2 (Å^{-2})	R(Å)
Cu-N(1)	1	2.05	0.001	0.016 ± 0.003	2.05
Cu-N(2)	1	2.08	0.001	0.016 ± 0.003	2.08
Cu-Cl(1)	1	2.10	-0.160	0.004 ± 0.009	2.26
Cu-Cl(2)	1	2.10	-0.160	0.004 ± 0.009	2.26
Cu-Cl(1′)	1	2.89	-0.255	0.005 ± 0.009	3.15

EXAFS results for **2**			
Atomic pair	N	R(Å)	σ^2 (Å^{-2})
Cu-N(1)	1	1.91	0.003 ± 0.014
Cu-N(2)	1	1.94	0.003 ± 0.014
Cu-Br(1)	1	2.17	0.003 ± 0.014
Cu-Br(2)	1	2.17	0.003 ± 0.014
Cu-Br(1′)	1	3.19	0.006 ± 0.021

4.1. Complex **1**: For the analysis of EXAFS data of complex **1**, the input parameter, R_{bkg} , was set to 1.05Å and the Fourier transform was performed over the k-range: k_{min}= 2.31Å^{-1}, k_{max}= 7.38Å^{-1}. Theoretically modeled data were fitted in the R-space to the experimental data using kw=1. Fitting was performed for coordination shells in the R range of 1-5 Å using seven single scattering and five multiple scattering paths obtained from the FEFF calculations and the results obtained are given in table 1. The S_0^2 value obtained is 0.80, ΔE_0 is also reasonable, i.e., 3.81 eV. Our results for the bond lengths, given in table1, are in good agreement with the crystallographic bond distances reported by Estes *et al.* [1]. In complex **1**, the coordination about copper is a distorted tetragonal pyramid, with two chlorine (Cl(1) and Cl(2)) and two nitrogen (N(1) and N(2)) atoms in the base and the out of plane chlorine (Cl(1')) atom at the apex [1]. The distances of the four basal atoms Cu-N(1), Cu-N(2), Cu-Cl(1) and Cu-Cl(2), and the apical chlorine Cl(1') atom have been found from the present EXAFS data analysis as 2.05, 2.08, 2.10, 2.10 and 2.89 Å, respectively. Based on the structure reported in reference [1], the coordinate geometry around copper atom has been depicted in figure 5. The distances shown in this figure are those which have been obtained by us in the present study.

4.2. Complex **2**: For analyzing the EXAFS data of complex **2**, the input parameter, R_{bkg}, was set to 1.0Å and the Fourier transform was performed over the k-range: k_{min}= 2.45Å^{-1}, k_{max}= 7.82Å^{-1}. As already pointed out above, because complex **2** is analogous to **1**, we have fitted the same theoretical model, which has been generated for 1, to the experimental EXAFS data of 2 using kw=1. Fitting was performed in the R range of 1-5 Å using six single scattering and seven multiple scattering paths. The results obtained are given in table 1. The obtained values of S_0^2 and ΔE_0 are 0.80 and -1.35 eV, which are reasonable. As complex **2** is analogous to complex **1**, the coordination about copper is expected to be distorted tetragonal pyramid, with two bromine (Br(1) and Br(2)) and two nitrogen (N(1) and N(2)) atoms in the base and the out of plane bromine (Br(1')) atom at the apex. From the present EXAFS data analysis, the distances of the four basal atoms Cu-N(1), Cu-N(2), Cu-Br(1) and Cu-Br(2), and the apical bromine Br(1') atom have been found to be 1.91, 1.94, 2.17, 2.17 and 3.19 Å respectively. Based on these results, the coordinate geometry around the copper atom has been shown in figure 6.

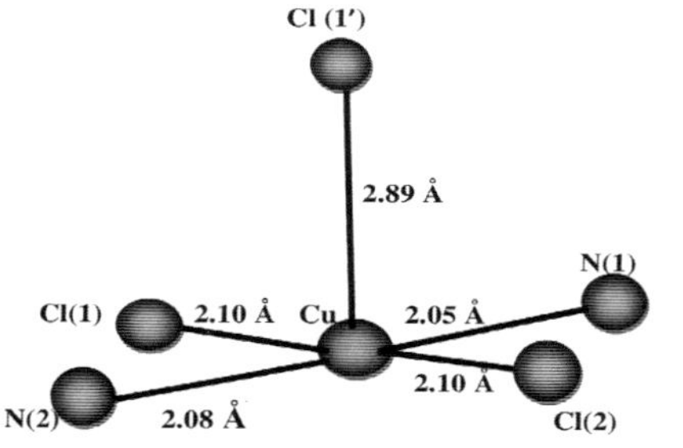

Fig.5. Coordination geometry in complex **1**

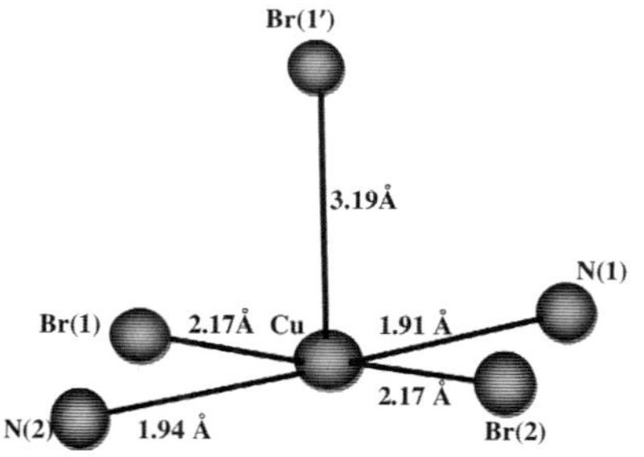

Fig.6. Coordination geometry in complex **2**

5. Conclusions

The EXAFS spectra at the K-edge of copper have been recorded for two copper complexes: [Cu(tmen)Cl2] (**1**) and [Cu(tmen)Br2] (**2**), (tmen = tetramethylethylenediamine). Using the available crystallographic data of complex **1**, theoretical model has been generated which is then fitted to the experimental EXAFS data of **1** to obtain the bond-lengths, coordination numbers and thermal disorders of complex **1**. The structural parameters, thus obtained, are found to be comparable with the crystallographic results. The crystal structure of complex **2** is not available. As the two complexes are analogous to each other, the theoretical model generated for **1** was fitted to the experimental EXAFS data of **2** to determine the structural parameters of **2**. The coordination geometry around the copper atom in the complexes has been depicted showing the present results. Thus, the present study has demonstrated that, if crystallographic data is available for a complex, structural parameters can be determined for an analogous complex from EXAFS studies. Further, from the values of chemical shifts, the copper metal in both the complexes has been found to be in +2 oxidation state.

Acknowledgement

Thanks are due to Madhya Pradesh Council of Science & Technology (MPCST), Bhopal (India) for a research grant.

References

[1] Estes E D, Estes W E, Hatfield W E and Hodgson D J 1975 *Inorganic Chem.* 1**4** 106.
[2] Das N C, Jha S N, Bhattacharyya D, Poswal A K, Sinha A K and Mishra V K 2004 *Sadhana* **29** 545.
[3] Bhattacharyya D, Poswal A K, Jha S N, Sangeeta and Sabharwal S C 2009 *Bull. Mater. Sci.* **32** 103.
[4] Bhattacharyya D, Poswal A K, Jha S N, Sangeeta and Sabharwal S C 2009 *Nucl. Instrum. Methods* A **609** 286.
[5] Gaur A, Johari A, Shrivastava B D, Gaur D C, Jha S N, Bhattacharyya D, Poswal A and Deb S K 2011 *Sadhana* **36** 339.
[6] Ravel B and Newville M 2005 *J. Synchrotron Rad.* **12** 537.
[7] Kelly S D, Hesterberg D, Ravel B 2008 *Chapter 14 in Methods of Soil Analysis, Part 5,MineralogicalMethods,* (Medison, Soil Science Society of America) pp 387-463.
[8] Kau L S, Spira-Solomon D J, Penner-Hahn J E, Hodgson K O and Solomon E I, 1987 *J. Am. Chem. Soc.* **109** 6433.

Journal of Physics: Conference Series **430** (2013) 012049 doi:10.1088/1742-6596/430/1/012049

Structural study of the Cu^{2+}-loaded copper hexacyanoferrate electrode deposited on indium tin oxide substrate

M Giorgetti[1], G Aquilanti[2] and M Minicucci[3]

[1] Department of Physical and Inorganic Chemistry, University of Bologna, Bologna, 40137, Italy

[2] Elettra - Sincrotrone Trieste S.C.p.A., s.s. 14, km 163.5, 34149 Basovizza, Trieste, Italy

[3] CNISM, School of Science and Technology, Physics Division, Università di Camerino, 62032 Camerino (MC), Italy

E-mail: giuliana.aquilanti@elettra.trieste.it

Abstract. We report the structural characterization by XAFS spectroscopy at the Cu K-edge of three different CuHCF-based electrodes deposited on ITO (indium tin oxide) substrate. The electrochemical characterization confirms the occurrence of typical CuHCF films on the ITO surface. The EXAFS analysis evidences the strong multiple scattering effects in these compounds which makes the EXAFS probe effective for distances up to 5 Å from the photo-absorber. The presence of Cu-N-C-Fe linear chains with a high multiplicity has allowed us to determine accurately the local environment of Cu which also includes the Fe atom. The XAFS characterization of the Cu^{2+}-loaded CuHCF shows the best agreement with the structural model for copper hexacyanoferrate, which also features $[Fe(CN)_6]^{3-}$ ion vacancies.

1. Introduction

Prussian Blue (PB) and related transition metal hexacyanoferrates belong to a class of polymeric inorganic compounds which have been extensively studied for many years [1]. The copper analogous CuHCF, has been successfully used for its electrocatalytic properties, demonstrating good capabilities in the determination of the hydrogen peroxide [2]. Recently, we have described a new experimental procedure, involving the insertion of Cu^{2+} ions into previously deposited CuHCF films on graphite substrate, which leads to a Cu^{2+}-loaded CuHCF displaying an increased response sensitivity, with respect to the pristine electrode [3]. A structural study of several electrodeposited CuHCF based mainly on X-ray absorption technique [4] has clarified the key role of the subsequent Cu^{2+} insertion step. In particular it has been observed that it is not possible to obtain a pure electrosynthesized CuHCF by one step because the film deposition is accompanied by a second component, the Prussian Blue (PB) and that a successive second step consisting of Cu^{2+} ions intercalation enhances the amount of the pure CuHCF phase. The paper has also evidenced the strengths and limitations of the EXAFS approach in these systems, by presenting the comparison between the single metal edge data analysis (Cu or Fe) and the multiple-edge one, concluding that the best fit can be obtained by using the single-edge approach at the Cu K-edge. In the present paper, taking advantage of the former study [4], we present an EXAFS analysis of three CuHCF films deposited on an ITO substrate. As suggested from previous work, we considered the Cu K-edge data only, because the PB component spoils the EXAFS signal at the Fe K-edge making the structural refinement not reliable. The CuHCF films on ITO substrates have been recently proposed as precursors for hydroxide films which have catalytic ability towards alcohol oxidation, giving anodic current density of about 10 fold higher than that of any metal hydroxide modified electrodes reported [5].

2. Experimental

The modification of an ITO electrode surface with copper hexacyanoferrate (CuHCF) was performed with a CHI 660C (CH Instruments, Inc.) potentiostat, *via* potentiodynamic procedure[4]. A three-electrode conventional cell was employed, with an ITO electrode as the working, a saturated calomel electrode (SCE) as the reference, and a Pt wire as the counter electrode. Three electrodes were obtained: the *oxidised* CuHCF (**1**), the *reduced* CuHCF (**2**) and the *Cu^{2+}-inserted* CuHCF (**3**). X-ray absorption spectroscopy experiments were performed at the XAFS beamline of Elettra Synchrotron Light Laboratory, in Basovizza (Italy)[6]. The storage ring operated at 2.0 GeV in top up mode with a typical current of 300 mA. The beam was monochromatized using a fixed exit monochromator equipped with a pair of Si(111) crystals. The data were recorded Cu K-edges in fluorescence mode using a large area Si drift diode detector. Four to five spectra were collected and averaged to improve statistics. The energies were defined by assigning the first inflection point of the spectra of the metallic copper to 8979 eV. A spectrum of metallic Cu for calibration purpose was recorded simultaneously in each scan. The EXAFS analysis has been performed by using the GNXAS package[7,8] that takes into account the multiple scattering (MS) theory. The theoretical signal is calculated *ab-initio* and contains the relevant two-body $\gamma^{(2)}$, the three-body $\gamma^{(3)}$ and four body $\gamma^{(4)}$ MS terms. The model includes the relevant set of MS signals obtained from typical structure of hexacyanoferrates[9].

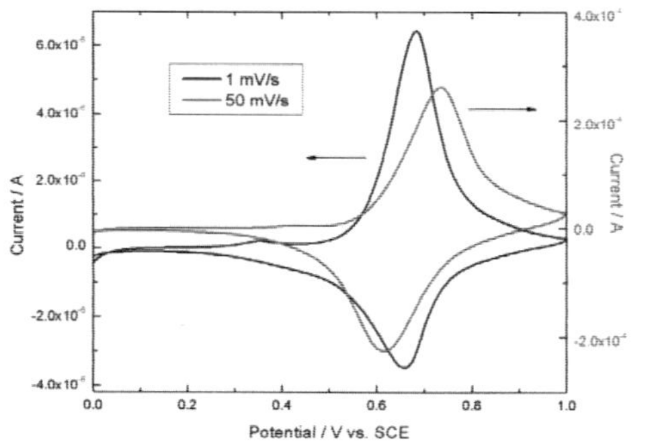

Figure 1. Cyclic voltammetry curves at two different scan rate of electrode **3** at 1 and 50 mV/s. Reference: SCE, Counter: Pt.

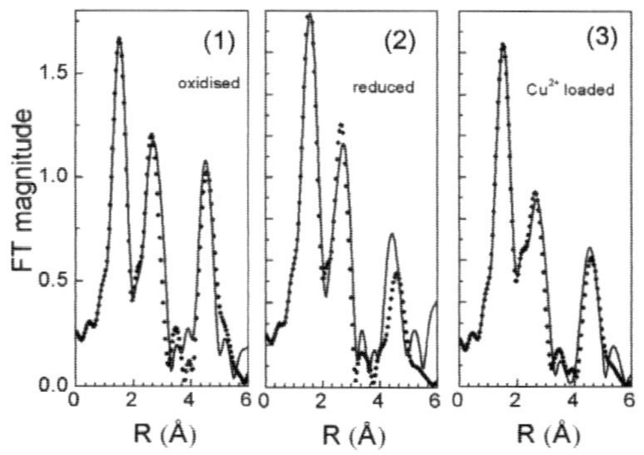

Figure 2. FT curves (theoretical: continuous line and experimental: dotted line) of the EXAFS signals for electrodes **1**, **2**, and **3**.

3. Results and Discussions

The voltammetric characterization of CuHCF deposited on ITO shows, as displayed in figure 1 for the electrode **3** at two different scan rate, a well defined redox system centred at about +0.65 V vs. SCE which is related to the redox reaction involving the ferric and ferrous states of the hexacyanoferrate [4]. The plot at 50 mV/s (red curve) displays a separation of the oxidation and reduction waves of about 100 mV, which is slightly more than the theoretical value of 59 mV per electron for a reversible system in a semi-infinite diffusion approximation (bulk). Interestingly, this peaks separation reduces to few mV when the scan rate is as low as 1 mV/s (blue curve), which is a typical behaviour of a *thin layer* (some hundreds of monolayers) of electroactive material. In addition since these curves have been recorded on the same electrodes after the XAFS measurements, they provide direct evidence for the absence of any beam damage on the electrodeposited material.

The EXAFS spectra of the investigated electrodes were fitted to a structural model based on the *Pm3m* structure of metal HCF in order to retrieve structural information of the electrodeposited film. The applied fitting procedure is similar to that presented in our previous publication [4] about electrodeposited films on a different substrate. The details on the structural model as well as on the choice of the single individual signals which mostly contribute to the overall EXAFS oscillations are available in reference 4.

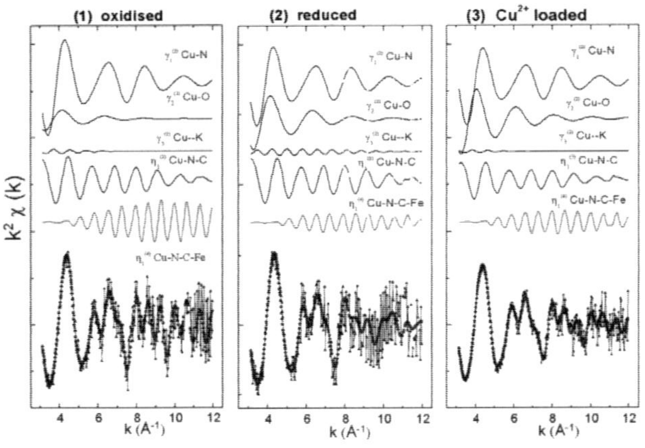

Figure 3. Details of the EXAFS analysis for the investigated electrodes. Each panel of the figure displays the various single contribution to the EXAFS signals and the comparison of the experimental (-) and the theoretical signals (•). This figure underlines the peculiarity of the present EXAFS data analysis where the signal intensity of the four body contributions (signal marked in red) is comparable to the first shell ones.

The best-fit results at the Cu K-edge concerning the fitting procedure for the investigated electrode are illustrated in figures 2 and 3. Figure 3 reports the details of the EXAFS analysis, in terms of single contributions to the theoretical signals and the comparison of the total theoretical signal with the experimental one. The theoretical curves match well with the experimental ones, even though the small signal to noise ratio (S/N) of samples **1** and **2** makes the fitting procedure less reliable. The single theoretical contribution displayed in figure 3 confirm that all these atomic contribution above specified are important in the determination of the experimental signals and that the four body contribution due to the Fe atom (by the Cu-N-C-Fe linear chain) cannot be excluded because it modulates the entire EXAFS signal. Figure 2 shows the comparison between the calculated and the experimental Fourier transform muduli.

Table 1. Structural parameters from EXAFS fitting results of electrodes **1**, **2** and **3**. The estimated parameter errors are indicated in parentheses.

	sample **1**	sample **2**	**sample 3**
d(Cu-N)/ Å	1.98(3)	1.99(4)	**1.97(2)**
σ^2(Cu-N)/ $Å^2$	0.006(3)	0.004(3)	**0.004(2)**
CN1 (Cu-N)	3.5(5)	3.1(5)	**2.1(2)**
d(Cu-O)/ Å	2.00(5)	2.03(5)	**2.09(2)**
σ^2(Cu-O)/ $Å^2$	0.022(8)	0.02(1)	**0.019(4)**
CN2 (Cu-O)	1.0(5)	2.8(8)	**2.6(2)**
Fe-C / Å	1.90(3)	1.95(8)	**1.90(2)**
σ^2 Fe-C / $Å^2$	0.0015(9)	0.009(5)	**0.009(3)**
C≡N / Å	1.15(2)	1.13(4)	**1.16(2)**
σ^2 C≡N / $Å^2$	0.002(2)	0.002(2)	**0.007(3)**
CN Cu-K (FIX)	0.2	2.9	**1.4**
E_0 Cu	8988(2)	8990(2)	**8991.1(7)**
S_0^2 Cu	0.82(8)	0.7(1)	**0.96(4)**
GNXAS residual / (10^{-6})	28.1	62.1	**6.0**

Structurally speaking, considering that the FT curve is related to the radial atomic distribution around the photoabsorbing atom, the FT curves of figure 2 evidences three main peaks, which are related to three different main atomic shells around the copper atom. Specifically, the first shell is due to the N and O atoms, the second shell to the C atoms, and the third shell to the Fe atoms. The different intensities of the FT curves, especially those related to the –Cu-N-C-Fe chains, i.e: the peaks at about 5 Å, are related to the different extent of the "soluble" structure of the copper hexacyanoferrate, which can be related to the concentration of the $[Fe(CN)_6]^{3-}$ ion vacancies [10].

The inter-atomic distances and the corresponding EXAFS Debye-Waller factor with their associated errors are shown in Table 1. They were determined by correlation maps (contour plots) for highly correlated parameters. Because of the noise of samples **1** and **2** the corresponding structural data obtained from the EXAFS analysis are associated not only to large errors with respect to those of electrode **3**, but also to the higher values of the residual.

The atomic distances related to the Fu-C-N-Cu linear chain, i.e the Fe-C, C≡N, and Cu-N have found to be in line with the values of the bulk CuHCF compound [11] and in good agreement to those concerning Cu^{2+}-inserted electrodeposited on graphite [4]. The coordination number Cu-K has been used as fixed parameter on the basis of the elemental analysis. The Cu-O distance, due to the 'insoluble' structure of metal hexacyanoferrate, has been quoted 2.09(2) Å with a coordination number (CN2) of 2.6(2). The coordination number for the Cu-N interaction (CN1) is found to be 2.1(6) which is lower than the ideal value of 4.5 for the insoluble structure. CN1 is related to the mean number of -Fe-C-N-Cu- linear chains, and CN2 indicates the mean number of water molecules linked to Cu. It is worth to recall that both Fe-C, C≡N distances are determined directly from the fitting of the four body $\gamma^{(4)}$ MS term. Those values are very close to that of the Cu^{2+}-inserted electrodeposited on graphite [4] and indicates a number of $[Fe(CN)_6]^{3-}$ ion vacancies close to 55%, in accordance to the recently observed samples of porous copper-based PB analogues [12].

4. Conclusions

The analysis of the -Cu-N-C-Fe- four-body contribution of the EXAFS signal in a set of three CuHCF electrodes synthesized on ITO substrates has been presented. Regardless the presence of two metal sites (Cu and Fe) which could be used for the data analysis, the presence of a two phases in different proportions constitutes a strong limitation for the EXAFS analysis of Fe K-edge data, and hence only the Cu K-edge data have been presented. Here, the most reliable data concerns the Cu^{2+}-inserted CuHCF, where the bond lengths associated to the -Cu-N-C-Fe- chain are similar to those of the bulk copper hexacyanoferrate. CuHCF on ITO also features a large presence of $[Fe(CN)_6]^{3-}$ ion vacancies with respect to bulk samples.

Acknowledgments

XAS measurements at ELETTRA were supported by Sincrotrone Trieste S.C.p.A. (proposal # 20095163). Thanks are also due to RFO funding from the University of Bologna. L. Guadagnini is acknowledged for the help in the electrodes preparation.

References

[1] De Tacconi N R, Rajeshwar K and Lezna R O 2005 *Chem. Mater.* **15** 3046.
[2] Guadagnini L, Giorgetti M, Tarterini F and Tonelli D 2010 *Electroanalysis* **22** 1695.
[3] Guadagnini L, Tonelli D and Giorgetti M 2010 *Electrochimica Acta* **55** 5036.
[4] Giorgetti M, Guadagnini L, Tonelli D, Minicucci M and Aquilanti G 2012 *Phys. Chem. Chem. Phys.* **14** 5527.
[5] Ganesh V, Maheswari D L and Berchmans S 2011 *Electrochimica Acta* **56** 1197.
[6] Di Cicco A, Aquilanti G, Minicucci M, Principi E, Novello N, Cognigni A and Olivi L 2009 *J. Phys.: Conf. Ser.* **190**, 012043.
[7] Filipponi A, Di Cicco A and Natoli C R 1995 *Phys. Rev. B: Condens. Matter* **52** 15122.
[8] Filipponi A and Di Cicco A 1995 *Phys. Rev. B: Condens. Matter* **52** 15135.

[9] Giorgetti M and Berrettoni M 2008 *Inorg. Chem.* **47** 6001.
[10] Giorgetti M, Della Longa S and Benfatto M 2009 *J. Phys. Confer. Series* **190** 012145.
[11] Avila M, Reguera L and Rodriguez-Hernandez J. *et al.* 2008 *J. Solid State Chem.* **181** 2899.
[12] Jiménez-Gallegos J *et al* 2010 *J. Phys. Chem. C* **114** 5043.

Journal of Physics: Conference Series **430** (2013) 012050 doi:10.1088/1742-6596/430/1/012050

In Situ XAS of the Solvothermal Decomposition of Dithiocarbamate Complexes

Husn-Ubayda Islam[1,3], Anna Roffey[1], Nathan Hollingsworth[1], Richard Catlow[1], Mariette Wolthers [2], Nora De Leeuw[1], Wim Bras[3], Gopinathan Sankar[1], Graeme Hogarth[1]

[1] Department of Chemistry, University College London (UCL), London, WC1H 0AJ, UK
[2] Faculty of Geosciences, Utrecht University, 3584 CD Utrecht, The Netherlands
[3] DUBBLE, European Synchrotron Radiation Facility (ESRF), Grenoble, 38000, France

Abstract. An *in situ* XAS study of the solvothermal decomposition of iron and nickel dithiocarbamate complexes was performed in order to gain understanding of the decomposition mechanisms. This work has given insight into the steps involved in the decomposition, showing variation in reaction pathways between the iron and nickel dithiocarbamates, and the non-innocent role of oleylamine as the solvent and capping agent in the reaction.

1. Introduction

Solvothermal synthesis of nanoparticles from single source precursors (SSPs) is a widely researched area.[1,2] There is a long standing interest in metal dithiocarbamate complexes as SSPs for the synthesis of sulfides due to their versatility in stabilizing transition metals in a range of oxidation states.[3,4,5] Our group is interested in the synthesis of iron and nickel sulfide nanoparticles which are important materials in several fields, including the naturally occurring conversion of CO_2 to small organic molecules which led to the suggestion that they played a role in pre-biotic catalysis.[6]

The solvothermal synthesis method generally involves heating a molecular complex in a solvent which may also act as a capping agent. It has been noted that the solvent has an effect on the size and morphology of the resulting nanoparticles,[2,7] however, it is generally believed that the solvent is innocent, acting as a heat sink and coordinating to the nanoparticles as they are forming thereby directing and capping their growth.

This paper reports an *in situ* XAS investigation that has enabled the proposition of possible mechanisms of sulfide particle formation in the solvothermal decomposition of nickel(II) bis-, and iron(III) tris-isobutyl dithiocarbamates. *In situ* characterization is a valuable tool for understanding fundamental mechanistic detail of the structural changes that occur during a reaction. Through this study, it has been found that the decomposition pathways vary significantly between the two dithiocarbamate complexes, and the solvent oleylamine plays a non-innocent role in both mechanisms.

2. Experimental

XAS data was acquired on beamline BM26A at the ESRF, Grenoble[8]. The beamline is equipped with a double crystal Si(111) monochromator and a 9 element Germamium solid state detector for fluorescence measurements. The *in situ* solvothermal cell, developed by the catalysis group at UCL, was used to conduct *in situ* measurements. The reaction chamber at the centre of the cell has a capacity of about 400μL and is sealed between two 127μm Kapton windows. The cell can reach 200° C. A calibration of the temperature inside the reaction chamber was performed prior to on-line experiments to account for heat loss during *in situ* measurements. In a typical on-line experiment, 400μL of a stock organic solution containing 0.1mmol dithiocarbamate complex and 20mL oleylamine, was loaded into the solvothermal cell and sealed. The temperature was set to 240° C at a ramp rate of 1° C/min: on account of heat loss, the maximum temperature within the cell reached 170°

C. XAS measurements lasted approximately 12 minutes and were acquired up to $10A^{-1}$ in fluorescence on iron and nickel K-edges. Temperature was recorded at the beginning of each scan. After careful analysis of the XANES data, further measurements were performed to acquire XAFS data at temperatures of interest. Data was acquired typically up to $14A^{-1}$ using the set up above. Solid metal dithiocarbamate data was also acquired in transmission for comparison. Data was processed using the Horae Athena software, and XAFS data was analyzed using Excurv version 9.273.

3. Results and Discussion

3.1. Dithiocarbamate Complexes in Oleylamine Environment

A direct comparison between the metal dithiocarbamate complexes in the solid state and when dissolved in oleylamine reveals unexpected results. For both the nickel and the iron dithiocarbamate species, a significant variation in the XAFS data is observed which strongly indicates the presence of metal-oleylamine interactions.

3.1.1. Nickel(II) bis-isobutyl dithiocarbamate. Modelling of XAFS data for nickel(II) bis-dithiocarbamate produces four Ni-S bond distances at 2.20Å as expected according to XRD data indicating that the ligand bidentate geometry remains intact in the solid state. As a complex dissolved in oleylamine, the Ni-S bond distances increase to between 2.37 and 2.51Å and an additional two Ni-N distances are observed at 2.05Å. A summary of the parameters are listed in Table 1 and XAFS fits are shown in Figure 1. The latter result reveals that oleylamine forms chemical bonds with the nickel centre of the dithiocarbamate complex– most likely at axial positions of the square planar molecule. The new bonding causes an increase in bond length of the Ni-S distances.

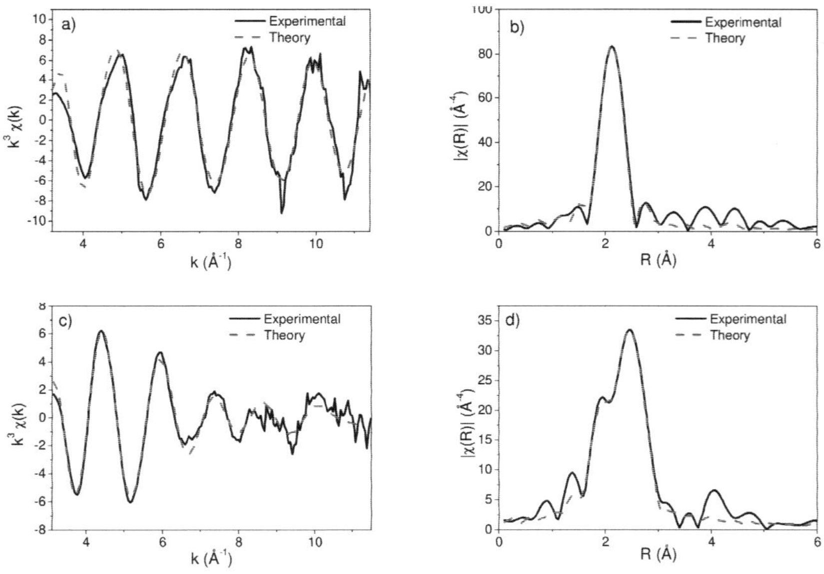

Figure 1: Experimental and theoretical a) XAFS and b) Fourier transform of solid nickel(II) bis dithiocarbamate, and c), and d) Fourier transform of oleylamine dissolved nickel(II) bis dithiocarbamate.

3.1.2. Iron(III) tris-isobutyl dithiocarbamate. The extensive changes seen in the XAFS of iron(III) tris-dithiocarbamate upon dissolution in oleylamine is initially an unexpected result as this complex is a fully coordinated species and therefore oleylamine interaction would require the breakdown of existing Fe-S bonding. Data of the solid species is in agreement with XRD data, indicating a six coordinated iron centre of 2.30Å Fe-S distances. XAFS modeling of the iron dithiocarbamate in oleylamine indicates three Fe-S distances at 2.21Å, and a further three sulfur atoms at a longer distance of 3.08Å thus indicating that the bidentate ligand becomes monodentate, with weak interactions of the furthest sulfurs with the iron(III) centre. In addition, modeling reveals the presence of two Fe-N distances at 1.90Å arising from direct coordination of oleylamine to the iron centre. The parameters are listed in Table 1 and XAFS fits are shown in Figure 2.

Table 1: List of fitting parameters derived from XAFS analysis of nickel(II) bis- and iron(III) tris- isobutyl dithiocarbamate complexes as solids and as dissolved in oleylamine.

Sample	Scatter	*N*	*R* (Å)	σ^2 (Å^2)	*F*
Ni(^{i}Bu$_2$NCS$_2$)$_2$	S	4	2.20	0.007	24
	C	2	2.70	0.006	
	N	2	4.33	0.010	
Ni(^{i}Bu$_2$NCS$_2$)$_2$ in oleylamine	N	2	2.05	0.006	22
	S	2	2.37	0.008	
	S	2	2.51	0.009	
	C	2	2.76	0.007	

Sample	Scatter	*N*	*R* (Å)	σ^2 (Å^2)	*F*
Fe(^{i}Bu$_2$NCS$_2$)$_3$	S	6	2.30	0.011	16
	C	3	2.79	0.011	
	N	3	4.09	0.012	
Fe(^{i}Bu$_2$NCS$_2$)$_3$ in oleylamine	N	2	1.90	0.006	36
	S	3	2.21	0.009	
	S	3	3.08	0.012	
	C	3	2.69	0.006	

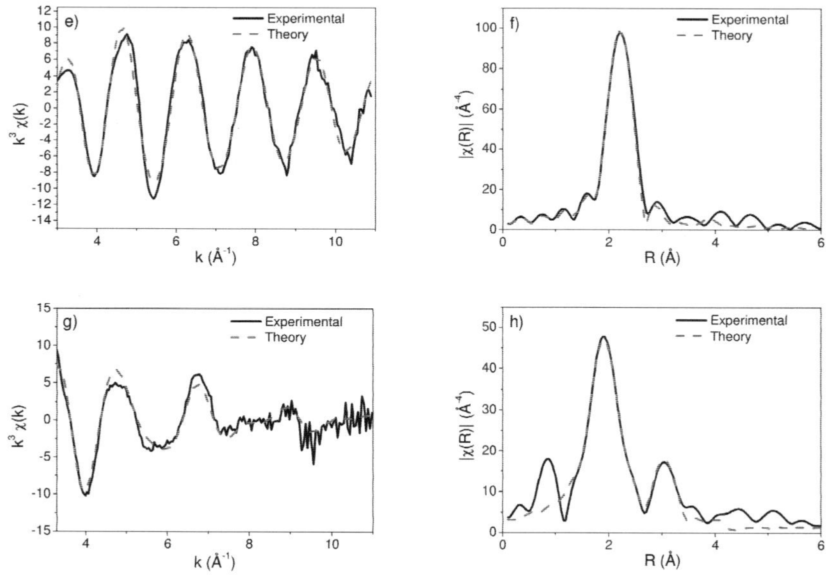

Figure 2: Experimental and theoretical a) XAFS and b) Fourier transform of solid iron dithiocarbamate, and c), and d) Fourier transform of oleylamine dissolved nickel dithiocarbamate.

3.2. Nickel(II) bis-isobutyl dithiocarbamate decomposition

In situ XANES data of the nickel(II) bis dithiocarbamate decomposition, shown in Figure 3a, reveals that between room temperature and 118° C, a continuous transformation occurs to the nickel environment. At 118° C, XAFS can be fit to the original square planar conformation of nickel(II) bis-dithiocarbamate. This confirms that oleylamine dissociates from the nickel centre between the two temperatures, reforming the square planar complex. At 123° C, an abrupt change in the XANES and XAFS indicates the formation of the NiS structure. In summary, *in situ* data shows that a slow dissociation of oleylamine from the nickel centers occurs. When this has reached completion, a facile process converts the dithiocarbamate complex into a sulfide. It is unclear at this stage whether the mechanism leading to the formation of NiS is initiated by temperature, vacant coordination, or both.

3.3. *Iron(III) tri- isobutyl dithiocarbamate decomposition*

In situ XAS data of iron(III) tris dithiocarbamate decomposition is significantly diffrent from that of the nickel(II) complex (Figure 3b). The iron(III) dithiocarbamate-oleylamine structure is stable up to 60° C according to XANES. At 60° C, an edge shift indicates reduction from iron(III) to iron(II). XAFS analysis reveals that the modified complex consists of two bidentate dithiocarbamate ligands and two chemically bound oleylamine molecules, the XAFS is fit equally within statistical error to both cis and trans conformations. By 70° C, oleylamine no longer forms chemical bonds to the iron(II) centre. It is possible that the oleylamine ligands migrate to the carbon backbone, similar to a mechanism reported by Jung *et al.*[9] The dithiocarbamate ligands bind with two long and two short distances at 2.23 and 2.46Å. Between 70° C and 120° C a four coordinated sulfide structure is formed which remains up to the 170° C temperature limit. The structure is not the same as the six coordinated pyrrhotite formed under laboratory conditions at higher temperatures of up to 240° C leading to the conclusion that phase change occurs between 170° C and 240° C.

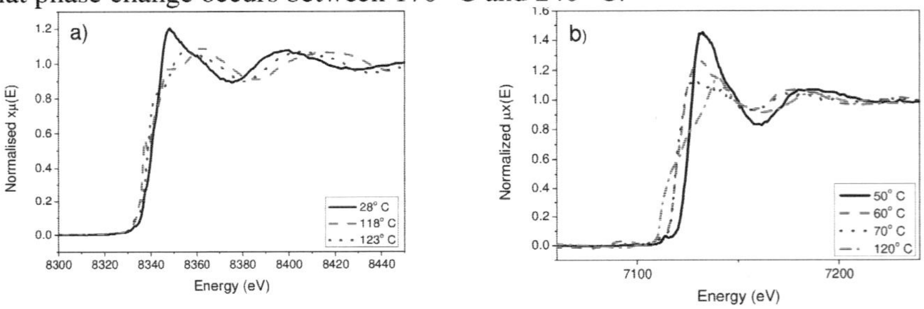

Figure 3: a) In situ XAS of nickel(II) bis isobutyl dithiocarbamate decomposition, and b) in situ XAS of iron(III) tris isobutyl dithiocarbamate decomposition.

4. Conclusions

XAFS data reveals significant differences between the well-studied solid state structures of the dithiocarbamate complexes, and the structures dissolved in oleylamine prior to decomposition. These severe changes to the metal dithiocarbamate coordination environment influence the decomposition pathway. The mechanistic pathway is unique to each system with the nickel complex decomposing in a two-step process and the iron complex in a three-step process including a bulk change in oxidation state.

References

[1] Fan D *et al* 2007 *Coordination Chemistry Reviews* **251** 1878
[2] Gao M R *et al* 2012 *Small* **8** 13
[3] Akhtar M et *al* 2012 *Journal of Crystal Growth* **346** 106
[4] Han W and Gao M 2008 *Crystal Growth & Design* **8** 1023
[5] Hogarth G 2005 *Progress in Inorganic Chemistry* **53** 71
[6] Cody G D 2004 *Annual Review of Earth and Planetary Sciences* **32** 569
[7] Vanitha P V and O'Brien P 2008 *Journal of the American Chemical Society* **130** 17256
[8] Nikitenko S *et al* 2008 *Journal of Synchrotron Radiation* **15** 632
[9] Jung Y K *et al* 2012 Journal of the American Chemical Society **132** 178

Journal of Physics: Conference Series **430** (2013) 012051 doi:10.1088/1742-6596/430/1/012051

Reduction Kinetics of Nickel Species Supported on Silica

S Yamashita, M Katayama and Y Inada*

Graduate School of Science and Engineering, Ritsumeikan University, Kusatsu, Shiga 525-8577, Japan.

E-mail: yinada@fc.ritsumei.ac.jp

Abstract. The reaction mechanism of reduction processes of NiO supported on silica has been clarified by means of the time-resolved dispersive XAFS (DXAFS) technique. The reduction of NiO shows two-phase changes. The first reaction corresponds to the reduction of NiO to the metallic Ni particle, and the sintering of the formed Ni particles occurs in the second step. The reduction process is composed of the surface reduction of the NiO particle and the successive oxygen diffusion from the particle core to the surface. The first-order kinetics suggests that the latter is the rate-determining step. The conditional rate constants indicate that the diffusion of oxygen atoms is relatively faster for the lower Ni loading.

1. Introduction

The nickel catalysts supported on silica are used to generate hydrogen by the CO_2 and steam reforming reactions of alkane, and the reactions are important to solve energy problems of modern society [1]. But the efficiency and lifetime are largely dependent on not only the preparation method but also the loading of active Ni species. In this study, the electronic state and local structure of Ni species supported on silica have been studied by varying the loading of Ni. The Ni species have been characterized by *in situ* XAFS experiments, and the kinetic studies of the reduction of NiO have been carried out by the time-resolved dispersive XAFS (DXAFS) technique.

2. Experimental

The supported Ni catalysts were prepared by the impregnation method using SiO_2 distributed by the Catalysis Society of Japan (JRC-SIO-10). The loading of Ni was changed in the range of 5–20 wt.%.

The DXAFS measurements were carried out at NW2A station of PF-AR (KEK, Japan) [2]. Polychromatic X-rays were obtained with a Si(111) bent crystal (Bragg type) and were detected by a linear photodiode array detector equipped with a scintillation (CsI:Tl). Sample powder was placed in an *in situ* cell for the DXAFS measurements. The H_2 gas was rapidly introduced into the evacuated cell at 873 K to follow the reduction process of Ni species. DXAFS spectra were recorded every 30 ms.

The *in situ* XAFS measurements were performed at BL-12C station of PF (KEK, Japan). The Si(111) double-crystal monochromator was detuned to remove higher order harmonics. The quick-scanning mode was applied for all measurements. Sample powder was first treated in the *in situ* cell by the diluted H_2 gas flow at 873 K, and was cooled to room temperature by keeping the gas flow. The gas was changed to the diluted O_2 (10 vol.% balanced by He, 200 cm^3/min), and the sample was heated to 873 K at 10 K/min (temperature-programmed oxidation: TPO). Thereafter, the gas was changed to the diluted H_2, and the sample was heated again to 873 K at 10 K/min (temperature-programmed reduction: TPR). The quick-XAFS measurements were carried out during these processes with the interval of 2 min.

3. Results and discussion

An example of the XANES spectral change is given in Figure 1 for the TPR process for Ni/SiO_2 with the Ni loading of 15 wt.%. The XANES spectrum at the beginning is consistent with that of NiO, clearly indicating that the Ni species are quantitatively oxidized at the previous TPO process. The final XANES spectrum is almost in accordance with that of Ni foil, suggesting that the treatment of the Ni catalyst at 873 K under H_2 quantitatively reduces NiO to the metallic Ni(0) particle. The similar XANES spectral changes were observed for other samples with different Ni loadings.

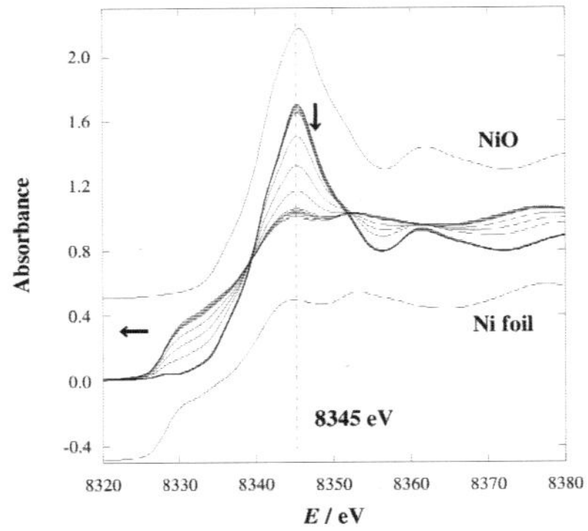

Figure 1. XANES spectral change during the TPR process from room temperature to 873 K under diluted H_2 gas flow for Ni/SiO_2 with the Ni loading of 15 wt.%.

The X-ray absorbance changes at 8345 eV, which corresponds to the white line peak of NiO, are shown in Figure 2 as a function of temperature during the TPR process. The X-ray absorbance is almost unchanged until *ca.* 553 K and is drastically decreased between 553 K and 673 K for all samples. The XANES spectral change reveals that the reduction of the metallic Ni(0) particle proceeds at the temperature range of 553–673 K. The surface reduction of the NiO particle and the progress into the core of the particle are thus considered to occur in that temperature range without any correlations to the Ni loading. This finding is supported by the almost constant $N_{Ni\text{-}Ni}$ values determined by the

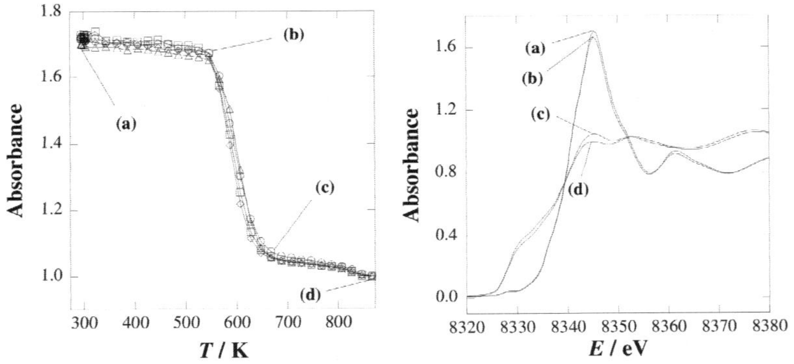

Figure 2. The changes of X-ray absorbance (left panel) at 8345 eV (broken line in Figure 1) are plotted vs. temperature during the TPR processes for the supported Ni catalysts with the Ni loading of 5 wt.% (circles), 7 wt.% (squares), 10 wt.% (diamonds), and 15 wt.% (triangles) and comparison of XANES spectra (right panel) of 15 wt.% Ni/SiO at the points of (a)–(d).

curve-fitting procedure of the EXAFS data (see Table 1). After the reduction is almost completed at *ca.* 673 K, the X-ray absorbance is further decreased with increasing the temperature (see Figure 2(c) and (d)). The spectral change at this step is very similar to the corresponding difference between the Ni(0) nanoparticle and the bulk metal. The treatment of Ni(0) particles under the H_2 atmosphere causes the sintering of Ni(0) particles at temperature over *ca.* 700 K.

Table 1. Structure parameters of reduced Ni species on SiO_2.

Sample	T / K	Interaction	N	R / pm	σ^2 / $10^{-5}nm^2$
Ni foil	rt	Ni-Ni	12	248±1	6.2±0.2
5 wt.% Ni/SiO_2	670	Ni-Ni	10.9±1.2	250±2	12.2±1.4
	873	Ni-Ni	11.8±0.6	250±1	15.2±0.5
7 wt.% Ni/SiO_2	670	Ni-Ni	11.4±1.2	250±1	12.5±0.8
	873	Ni-Ni	11.5±0.6	250±1	14.8±0.5
10 wt.% Ni/SiO_2	670	Ni-Ni	11.2±1.2	250±1	12.2±0.9
	873	Ni-Ni	11.2±0.5	250±1	14.5±0.4
15 wt.% Ni/SiO_2	670	Ni-Ni	10.0±2.7	249±1	11.3±2.2
	873	Ni-Ni	10.9±0.7	250±1	14.4±0.6

The X-ray absorbance changes at 8345 eV obtained by the DXAFS measurements are shown in Figure 3 as a function of time during the reduction process. The reduction of NiO shows two-phase changes. The first reaction, which is completed within *ca.* 5 s, corresponds to the reduction of NiO to the metallic Ni(0) particle, supported by the XANES spectral change. The sintering of the formed Ni(0) particles proceeds at the slow second step in the order of minutes. These results are in agreement with the observations of *in situ* XAFS experiments (*vide supra*). The X-ray absorbance change during the reduction of NiO is represented by a single exponential function. Because the reduction process must be composed of the surface reduction of NiO particles to generate surface Ni(0) state and the successive oxygen diffusion from the NiO core to the surface, the first-order kinetics on Ni suggests that the latter process is a rate-determining step.

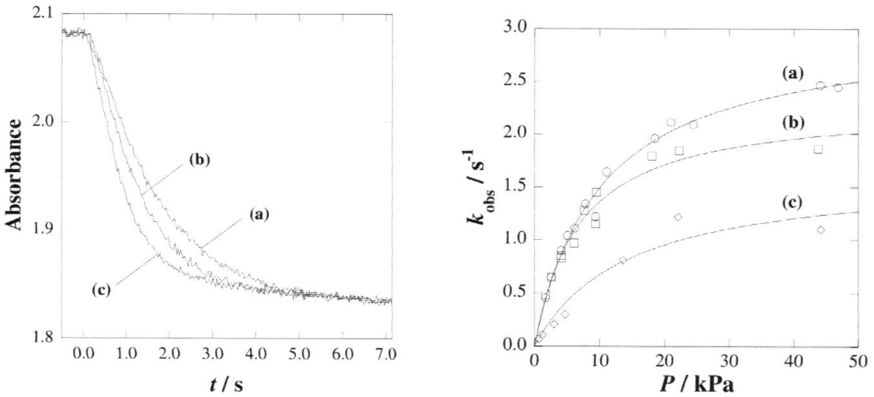

Figure 3. The change of X-ray absorbance at 8345 eV is plotted vs. time during reduction processes at 873 K for 5 wt.% NiO/SiO_2 in A. The H_2 pressure was 2.6 kPa (a), 4.1 kPa (b), and 9.4 kPa (c). In B, the conditional first-order rate constants for the fast reduction process are plotted against H_2 pressure for 5 wt.% (a), 10 wt.% (b), and 15 wt.% (c) NiO/SiO_2.

As schematically shown in Figure 4, the first reduction process is initiated by an adsorption of H_2, and the adsorption equilibrium constant is represented by K_A. The surface NiO species (NiO*) is reduced by an adsorbed H_2 with the rate constant of k_1 to form the surface Ni(0) species (Ni*), which is reversely oxidized with the rate constant of k_{-1}. The internal O atoms are diffused with the rate

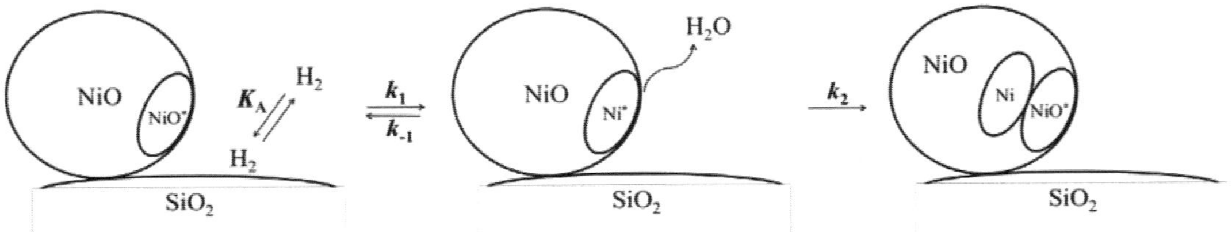

Figure 4. The reaction mechanism for reduction process of the supported NiO species on SiO_2.

constant of k_2 from the NiO core to the surface Ni* species, leading to the regeneration of NiO*. The change of the number of Ni* (N_{Ni*}) is then given in Eq. (1),

$$\frac{d\,N_{Ni^*}}{d\,t} = k_1\theta_{H_2}N_{NiO^*} - k_{-1}N_{Ni^*} - k_2N_{Ni^*}N_{NiO} \quad (1)$$

where θ_{H2} and N_X denotes the surface coverage of H_2 and the number of species X, respectively. Assuming the steady state to the Ni* species, the N_{Ni*} term is represented by Eq. (2) under the approximation that the k_2 step is rate determining,

$$N_{Ni^*} = \frac{K_1\theta_{H_2}N^S}{1+K_1\theta_{H_2}} \quad (2)$$

where K_1 represents the equilibrium constant of the surface redox reaction ($K_1 = k_1/k_{-1}$) and N^S denotes the total number of surface nickel species (NiO* and Ni*). The conditional first-order rate constant (k_{obs}) is thus derived as Eq. (3) by assuming that N^S is constant vs. time.

$$-\frac{dN_{NiO}}{dt} = \frac{k_2K_1\theta_{H_2}N^S}{1+K_1\theta_{H_2}}N_{NiO} = k_{obs}N_{NiO} \quad (3)$$

The hydrogen pressure dependence of k_{obs} is expected to show a saturating behavior according to Eq. (3). The limiting value of k_{obs} under high H_2 pressure is then approximated to Eq. (4),

$$k_{obs} \approx k_2K_1N^S \quad (4)$$

where the K_1 value is treated to be much smaller than 1, since the reduction of NiO is observed with clear isosbestic points. The k_{obs} values are plotted vs. P_{H2} in Figure 3B, in which the saturating dependence is clearly demonstrated. It has clarified that the limiting k_{obs} value under the low Ni loading is larger than that under the high Ni loading, indicating that the migration of internal O atoms is relatively faster for small Ni(0) particles. This mechanism has been clarified for the first time by applying the kinetic analysis to the solid-phase redox reactions for the supported Ni species.

Acknowledgment
The XAFS measurements at Photon Factory (Tsukuba) have been performed under the approval of the Photon Factory Program Advisory Committee (Proposal No. 2011G531).

References

[1] J. Sun, X. Qiu, F. Wu, W. Zhu, W. Wang, and S. Hao, *Int. J. Hydrogen Energy*, 29, 1075 (2004).
[2] Y. Inada, A. Suzuki, Y. Niwa, and M. Nomura, *AIP Conf. Proc.*, 879, 1230 (2006).

Journal of Physics: Conference Series **430** (2013) 012052 doi:10.1088/1742-6596/430/1/012052

EXAFS and XANES structural characterization of bimetallic AuPd vapor derived catalysts.

A Balerna[1], C Evangelisti[2], E Schiavi[3], G Vitulli[4], L Bertinetti[5], G Martra[5] and S Mobilio [1,6]

[1] INFN, Frascati National Laboratory, Frascati, 00044, Italy
[2] CNR, Institute of Molecular Science and Technologies (ISTM), Milano, 20138 Milano, Italy
[3] Department of Chemistry and Industrial Chemistry, University of Pisa, Pisa, 56126, Italy
[4] Advanced Catalysts S.r.l., Pisa, 56126, Italy
[5] Department of Chemistry IFM & NIS Centre of Excellence, Torino, 10125, Italy
[6] Department of Physics, University of Roma TRE, Roma, 00146, Italy

E-mail: antonella.balerna@lnf.infn.it

Abstract. Using an innovative procedure known as metal vapor synthesis (MVS) to prepare bimetallic catalysts, starting from Au and Pd vapors, [AuPd] co-evaporated and [Au][Pd] separately evaporated bimetallic catalysts were achieved. After being tested, the catalytic activity and selectivity of the [AuPd] catalyst turned out to be higher than the [Au][Pd] ones. Using EXAFS spectroscopy it was shown that, in the [AuPd] samples, small bimetallic AuPd nanoparticles were present, having an Au rich core surrounded by an AuPd alloyed shell while in the [Au][Pd] sample there was the presence of monometallic Au and Pd nanoparticles showing some alloying only in the boundary regions. The EXAFS results were also qualitatively confirmed by the XANES spectra.

1. Introduction

Bimetallic catalysts are of great interest due to their catalytic, electronic and optical properties that are different from the ones of their pure constituent metals. These catalysts are not only characterized by particles size and shape but also by the atomic ratios of the metals and their different electronic and chemical properties [1,2]. Among many metallic combinations, gold-palladium systems have been extensively studied due to their peculiar catalytic properties in many catalytic processes [3]. Au and Pd can form solid solutions at any ratio [4] and depending on the preparation method, AuPd bimetallic systems can contain different structural features including core-shell structure [5], segregated Au and Pd monometallic domains [6], uniform Au-Pd alloys [7] or a mixture of them [8]. In this paper, we report the XAFS (X-ray Absorption Fine Structure) structural characterization of AuPd bimetallic systems prepared by using Au and Pd vapors as reagents (metal vapor synthesis, MVS) [9] following two different synthetic procedures. In the first procedure, Au and Pd vapors are simultaneously co-condensed with acetone vapor in order to obtain a close interaction between the two metals, leading to a bimetallic AuPd/acetone solvated metal atoms (SMA) solution. Then, AuPd/acetone SMA is used as starting material to synthesize bimetallic [AuPd] catalysts supported on γ-Al_2O_3. In the second procedure, Au and Pd are evaporated separately in two different evaporation reactions; therefore, Pd/acetone and Au/acetone SMA are achieved. Afterwards, Pd and Au

SMA are mixed together and used as starting materials to obtain supported [Au][Pd] bimetallic systems. Since the bimetallic [AuPd] system shows higher catalytic activity and selectivity than the analogous [Au][Pd] system [9] their XAFS structural characterization becomes quite important. In order to have a further insight into their morphology and dimensions, all catalysts were also investigated by high resolution transmission electron microscopy (HRTEM)[9].

2. Results and Discussion

X-ray absorption spectra at the Au L_3 edge and at the Pd K-edge were taken at the GILDA beamline [10] of the European Synchrotron Radiation Facility (ESRF, Grenoble). Measurements were performed on two reference samples (Au and Pd foils) and on a complete set of two monometallic Au (1%) and Pd (0.5%) and two bimetallic [Au][Pd], [AuPd] samples supported on γ-Al_2O_3 [9]. Only the catalysts at the Pd K edge were measured in fluorescence mode while transmission mode was used for all the other XAFS spectra. To perform an accurate data analysis [11,12] data fitting was performed in both R-space and k-space with k^2 (Pd) and k^3 (Au) weights; also an estimation of the accuracy of the obtained structural parameters, compatible with the data quality and k range used [13], was achieved. In Fig. 1 differences between the EXAFS spectra of the [Au][Pd] and the [AuPd] samples are clearly visible. The values of coordination numbers (N), nearest-neighbors distances (R) and DebyeWaller factors (σ^2) are reported in Table 1. Results achieved show a different environment for Au and Pd. In the monometallic Au/γ-Al_2O_3 case, the reduction in the coordination numbers, N, was used to estimate the average nano-particle size [11]. The case of the monometallic Pd is more complicated because EXAFS data analysis indicates the presence of a Pd-O and a Pd-Pd coordination. As clearly visible in Table 1, small and comparable amounts of oxidized Pd are present in all the catalysts and this can probably exclude any significant effect on the different catalytic activities.

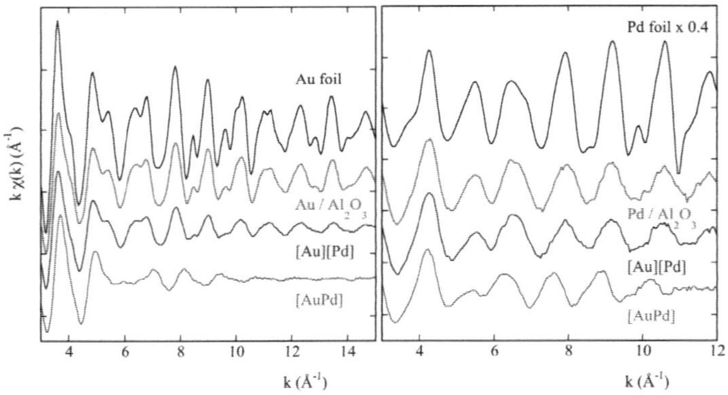

Figure 1. Experimental EXAFS spectra at the Au L_3 edge and at the Pd K edge.

In the bimetallic samples the presence of Au-Pd bonds was clearly observed (Tables 1). In the [Au][Pd] case, the number of Pd atoms among the Au neighbors (and the number of Au atoms among the Pd neighbors) is rather small (about10% and about 20% respectively); on the other hand, in the [AuPd] case, the presence of Au-Pd bonds is much higher. The TEM analysis [9] showed that in the [Au][Pd] sample there is the presence of at least two families of nanoparticles: 70% of the nanoparticles exhibits a size distribution between 2 nm and 3 nm and a $D_m = 2.6$ nm but there is also a minor population of larger particles with sizes between 4 nm and 5.5 nm. In the [AuPd] case TEM analysis shows nanoparticles homogeneously and symmetrically distributed around a mean size of about 2.4 nm. Taking into account the TEM analysis and the

Table 1. Results of the EXAFS fitting procedure.

Sample	R-factor	Shell	N	R($\AA$)	σ^2
$Au\ foil$	0.002	Au-Au	12(fixed)	2.875	0.0020
$Au(1\%)$	0.006	Au-Au	10.9	2.867	0.0031
$[Au][Pd]$	0.006	Au-Au	10.2	2.854	0.0062
		Au-Pd	0.9	2.793	0.0063
$[AuPd]$	0.005	Au-Au	5.5	2.815	0.0087
		Au-Pd	3.6	2.794	0.0071
$Pd\ foil$	0.003	Pd-Pd	12(fixed)	2.745	0.0022
$Pd(0.5\%)$	0.01	Pd-O	0.9	2.00	0.002
		Pd-Pd	5.2	2.74	0.007
$[Pd][Au]$	0.01	Pd-O	1.1	2.00	0.002
		Pd-Pd	4.3	2.74	0.008
		Pd-Au	1.2	2.79	0.012
$[PdAu]$	0.009	Pd-O	0.7	2.01	0.003
		Pd-Pd	2.5	2.80	0.005
		Pd-Au	3.6	2.79	0.006

small amount of heteroatomic Au-Pd bonds present in the [Au][Pd] sample a possible explanation of the achieved results could be that probably some smaller Pd nanoparticles sit on the surface of the larger Au ones giving rise to alloying only in the boundary regions and to the two families of nanoparticles sizes observed. In the case of the [AuPd] sample, at the Pd K-edge the Pd-Pd coordination number is smaller than the Pd-Au one while the interatomic distances are comparable. The presence of a Pd-Pd distance comparable to the Pd-Au one suggests the presence of a well-defined Pd-Au phase. Since the Pd:Au molar ratio is equal to 1 in the presence of a homogenous Pd-Au alloy, the Pd-Pd and Pd-Au coordination numbers should be the same but this is not the case because the number of Pd and Au atoms around the Pd absorber is different. At the Au L_3 edge the Au-Au distance of about 2.82 $\AA$ is different from the Au-Pd and the Pd-Pd ones and is contracted with respect to the Au bulk value. This, together with the value of the Au-Au coordination number rather higher than the Au-Pd one, indicates the presence of small Au-rich cores. From the value of the Au-Au distance [14], we quote a dimension of about 11 $\AA$ corresponding to a nanoparticle of about 55 atoms. From all these considerations, we conclude that in our [AuPd] sample, bimetallic Au-Pd nanoparticles, composed of an Au rich core surrounded by an AuPd alloyed shell, are present. A physical reason of such a structure could be attributed to kinetically more favorable nucleation steps of Au atoms in acetone solution during the melting of the acetone solid matrix, followed from a slower aggregation of Au and Pd atoms on preformed Au seeds but also by the high thermodynamic stability at low temperatures ($<$500 K) of Au-Pd bimetallic particles having a Au rich shell [15,16]. This can be related to the different stabilization properties of acetone towards Pd and Au atoms, as observed in the [Au][Pd] sample. The different structural results achieved by the EXAFS data analysis for the [AuPd] and [Au][Pd] samples can be also qualitatively observed in the normalized experimental X-ray Absorption Near Edge Structure (XANES) spectra reported in Fig. 2. In Fig. 2 , the peak indicated by the arrow at the Au L_3 edge is known as white line and its intensity is related to the presence of unoccupied 5d states (d-holes) [17]. In the nanoparticles, the reduced number of Au-Au bonds results in an increase in the 5d level occupancy and hence in a decrease in its intensity. As shown by Liu et al. [16] in Au-Pd clusters, a further reduction in the intensity of the white line can be due to a charge transfer from Pd to Au. Therefore, the strong electronic effect observed in Fig. 2 comparing the [Au][Pd] sample to the [AuPd] one can be related to the increased Au-Pd coordination, which causes a further filling of the Au d band.

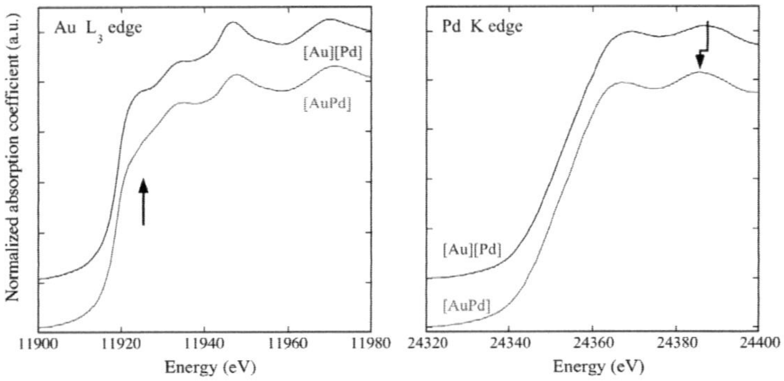

Figure 2. XANES spectra at the Au L_3 edge and at the Pd K edge.

In the Pd XANES the small shifts to higher energies observed in the [Au][Pd] spectra come from lattice contraction. The opposite shift observed in the [AuPd] sample is associated to the large amount of Au atoms present around Pd , which, as given by EXAFS data analysis, results in an increase in the interatomic distances, in agreement with Ref. [5].

3. Conclusions

Investigations on bimetallic Au-Pd catalysts prepared using different procedures have shown different structural features. Simultaneous evaporation of the two metals produces nanoparticles having an Au rich core surrounded by an Au-Pd alloyed shell. Separate evaporation produces monometallic Au and Pd nanoparticles and a small presence of hetero-atomic Au-Pd bonds could be given by the sitting of the smaller Pd nanoparticles on the surface of the larger Au ones giving rise to some alloying only in the boundary regions.

4. References

[1] Toshima N and Yonezawa T 1998 *New J. Chem.* **22** 1179
[2] Guzci L 2005 *Catal. Today* **101** 53
[3] Hutchings G J 2008 *Chem. Commun.* **10** 1148
[4] Okamoto H and Massalki T B 1985 *Bull. Alloy Phase Diagrams* **6** 7
[5] Marx S and Baiker A 2009 *J. Phys. Chem.* C **113** 6191
[6] Harada M, Asakura K and Toshima N 1993 *J. Phys. Chem.* **97** 5103
[7] Ding Y, Fan F, Tian Z, and Lin Wang Z 2010 *J. Am. Chem. Soc.* C **132** 12480.
[8] Beck A, Horvath A, Schay Z, Stefler G, Koppany Z, Sajo I, Geszti O and Guczi L 2007 *Top. Catal.* **44** 115
[9] Evangelisti C, Schiavi E, Aronica L A, Caporusso A M, Vitulli G, Bertinetti L, Martra G, Balerna A and Mobilio S 2012 *J. of Catalysis* **286** 224
[10] Pascarelli S *et al* 1995 *ESRF Newslett.* **23** 17
[11] Comaschi T, Balerna A and Mobilio S 2008 *Phys. Rev.* B **77** 075432.
[12] Rehr J J and Albers R C 2000 *Rev. Mod. Phys.* **72** 621.
[13] Li G G, Bridges F and Booth C H 1995, *Phys. Rev.* B **52** 6332
[14] Balerna A, Bernieri E, Picozzi P, Reale A, Santucci S, Burattini E and, Mobilio S 1985 *Phys. Rev.* B **31** 5058.
[15] Herzing A A, Carley A F, Edwards J K, Hutchings G J and Kiely C J 2008 *Chem. Mater.* **20** 1492.
[16] Liu H B, Pal U, Medina A, Maldonado C and Ascencio J A 2005, *Phys. Rev.* B **71** 075403.
[17] Lin C M, Hung T L, Huang Y H, Wu K T, Tang M T, Lee C H, Chen C T and Chen Y Y 2007 *Phys. Rev.* B **75** 125426

Journal of Physics: Conference Series **430** (2013) 012053 doi:10.1088/1742-6596/430/1/012053

Kinetic Study of Reduction Reaction for Supported PdO Species by Means of Dispersive XAFS Method

M Katayama[1], Y Niwa[2], K Doi[1], S Yamashita[1], and Y Inada[1,*]

[1] Department of Applied Chemistry, Ritsumeikan University, Kusatsu 525-8577, Japan
[2] Photon Factory, High Energy Accelerator Research Organization, Tsukuba 305-0801, Japan

E-mail: yinada@fc.ritsumei.ac.jp

Abstract. Time-resolved dispersive XAFS experiments have been performed to study the reduction reaction of supported PdO species on γ-Al_2O_3.It has been clarified that the reduction of PdO by H_2 proceeds with an unique kinetic behavior, *i.e.*, the initial part of the reduction is expressed by the first-order kinetics on Pd and the reaction order is changed to be zero as the progress of the reduction. The observation is different from that for the other supported metal oxides, for which the reduction is first order through the reaction. The reduction process of PdO is divided into the surface reduction to form the surface Pd(0) species and the succeeding oxygen exchange between inner PdO and surface Pd(0). When the latter step is faster than the former, the reduction rate of PdO is explained in terms of the zero-order kinetics. The pressure dependences of the conditional first- and zero-order rate constants suggest that the Langmuir-type molecular adsorption of H_2 precedes the surface reduction of PdO particles.

1. Introduction

The supported Pd species are well known as useful active catalysts for various chemical reactions, such as hydrogenation, reforming, detoxification of toxic exhausts, etc [1]. The catalytic activity is largely affected by the chemical state, the particle size, and the interaction with the supporting material for the Pd species. It has been also reported that the Pd catalyst has higher activity at a partially reduced state of PdO [2]. The microscopic speciation and the mechanistic investigation in the atomic scale are thus important to understand the catalytic performance. The X-ray absorption fine structure (XAFS) is the most powerful analysis tool for such purposes, and its time-resolved spectroscopy can reveal the dynamic changeover of the active species in response to the reaction environments. Kinetic studies for the supported Pd species are essential to reveal the redox mechanisms and to understand the reactivity of the catalysis reactions. The clarified reaction mechanism may become a useful guide for catalysis design with the improved performance. In this study, we have focused on the reduction mechanism of PdO supported on γ-Al_2O_3, which is closely related to the preparation and activation of Pd catalysts.

2. Experimental

The 10 wt% Pd catalysts supported on γ-Al_2O_3 were prepared by the impregnation method. A mixture of acetone solution of palladium acetate and γ-Al_2O_3 was dried at 383 K for 12 h, and the obtained powder was calcined in air at 673 K for 3 h. The particle size of Pd with the radius distribution in the range of 4-11 nm was estimated by the transmitted electron microscopy after the reduction treatment.

The Dispersive XAFS (DXAFS) measurements were carried out at the NW2A beamline of PF-AR (KEK, Japan) to study the kinetics for the reduction of PdO with H_2[3]. A Si(511) bent crystal (bend radius of 900 mm) was used as the Laue-type polychromator for the DXAFS instrument. The powder of Pd/ γ-Al_2O_3 was set in an *in situ* DXAFS chamber. The chamber was first evacuated and the reaction gas (H_2 for reduction) was rapidly injected into the chamber. The DXAFS measurements were started just before the injection and the spectra were recorded with the time resolution of at most 10 ms.The quantity of injected H_2 molecules were in excess to the amount of Pd species, and thus the pseudo-first-order condition was ensured for PdO. Because the reduction by H_2 and oxidation by O_2 occurred reversibly, the initial PdO state was reproduced by the treatment with O_2 after the time-resolved monitoring for the reduction process. All DXAFS measurements were performed at 823 K.

3. Results and discussion

A representative spectral change and the extracted absorbance changes at 24.37 keV, which corresponds to the white line peak top of PdO, are shown in Fig. 1.The initial and final spectrum is perfectly in agreement with that of PdO and Pd, respectively, indicating that the redox reaction proceeds quantitatively and reversibly.The existence of some isosbestic points indicates that there are no observable metastable intermediates.Therefore, the relative X-ray absorbance shown in Fig. 1(b) is proportional to the content of the PdO species.Interestingly, it is found that the X-ray absorbance varies exponentially at the beginning of the reduction and then changes linearly.The reproducibility has been observed for this behavior, which is independent on the injection pressure of H_2, the amount of sample powder, the loading of Pd, and the measurement temperature in the range from 400 to 600 °C.Such unique time-course changes are clearly different from the standard first-order kinetics which is represented by a single exponential function.Only a single-exponential or a linear function does not reproduce the observations, which are only explained in terms of a combination of first- and zero-order reaction kinetics as described below.

The reduction of PdO by H_2 is divided into three steps, *i.e.*, the adsorption of H_2 (eq. (1)), the reduction of surface PdO (PdO^*) to generate surface Pd(0) (Pd^*) as shown by eq. (2), and the O atom

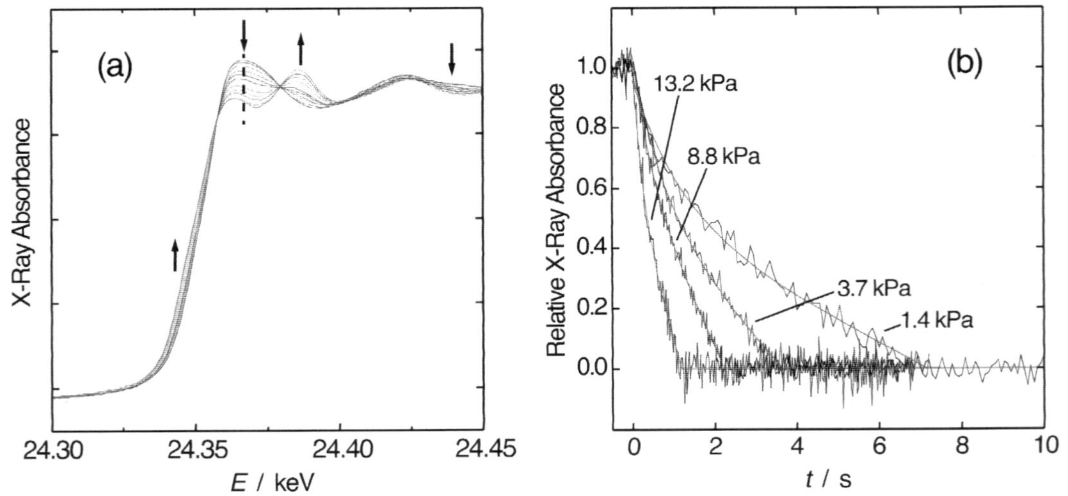

Figure 1. Time-resolved XANES spectra (a) for the reduction of PdO with H_2 and the absorbance change at 24.37 keV (b) under various H_2 pressures. The calculated curves are included in (b).

exchange between Pd^* and internal PdO (eq. (3)).

$$H_2 \overset{K_A}{\rightleftharpoons} H_2^{(a)} \tag{1}$$

$$PdO^* + H_2^{(a)} \xrightarrow{k_1} Pd^* + H_2O \tag{2}$$

$$Pd^* + PdO \xrightarrow{k_2} PdO^* + Pd \tag{3}$$

Because the PdO particles are quantitatively reduced on γ-Al_2O_3 under the current experimental conditions, the sequential extraction (eq. (2)) and supply (eq. (3)) of O atoms at the particle surface is necessary to reduce the internal part of the PdO particle.

At the early stage of the overall reduction process, the Pd^* species is not considered to be sufficiently accumulated to proceed the oxygen exchange represented by eq. (3). The reaction rate is thus given by eq. (4) by neglecting eq. (3),

$$-\frac{dN_{PdO}}{dt} = k_1 N_{PdO^*}\theta_{H_2} = k_1\theta_{H_2}\alpha N_{PdO} \tag{4}$$

whereN_x denotes the number of species x, θ_{H2} is the coverage of adsorbed H_2, and α means the ratio of surface atoms for a PdO particle.The equation (4) indicates that the PdO reduction shows the first-order kinetics on Pd, *i.e.*, the absorbance is changed exponentially vs. time. By accumulating the Pd^* species on the PdO particles, it has been found that the reaction changes to follow the zero-order kinetics on Pd.The diffusion of O atoms must play a key role in this stage.First, we suppose the steady-state assumption for the Pd* species.

$$\frac{dN_{Pd^*}}{dt} = k_1 N_{PdO^*}\theta_{H_2} - k_2 N_{Pd^*} N_{PdO} = 0 \tag{5}$$

The total number (N_{Pd}^S) of surface Pd species is defined by eq.(6).

$$N_{Pd}^S = N_{PdO^*} + N_{Pd^*} = \frac{k_1\theta_{H_2} + k_2 N_{PdO}}{k_2 N_{PdO}} N_{PdO^*} \tag{6}$$

The reduction rate of PdO is then written as eq. (7) by using eqs. (5) and (6).

$$-\frac{dN_{PdO}}{dt} = k_2 N_{Pd^*} N_{PdO} = \frac{k_1 k_2 \theta_{H_2} N_{PdO}}{k_1\theta_{H_2} + k_2 N_{PdO}} N_{Pd}^S \tag{7}$$

When the O atom exchange (eq. (3)) is faster than the surface reduction step (eq. (2)), eq. (7) is approximated to be eq. (8).

$$-\frac{dN_{PdO}}{dt} \approx k_1\theta_{H_2} N_{Pd}^S \tag{8}$$

By assuming N_{Pd}^S as a constant vs. time, the reduction rate of PdO becomes independent on the number of Pd species, leading to the zero-order rate law for the Pd species. According to eqs. (4) and (8), the conditional rate constants in the early and later stage are then expressed by $k_1^{obs} = k_1\theta_{H2}\alpha_{Pd}$ and $k_2^{obs} = k_1\theta_{H2}N_{Pd}^S$, respectively.

The faster oxygen migration (eq. (3)) than the preceding surface reduction (eq. (2)) keeps N_{PdO^*} to be constant during the progress of the reduction.It is concluded that the apparently unchanged N_{PdO^*} causes the zero-order kinetics on PdO.This reduction mechanism is contrast to the other supported metal oxide, such as NiO, for which the reduction is first-order on the NiO species.The linearly aligned O atoms in the crystal structure of PdO[4] may contribute to the faster oxygen migration rather than NiO with the rock salt structure [5], in which the oxygen migration is inhibited by the alternatively arranged Ni atoms.

The conditional rate constants, k_1^{obs} and k_2^{obs}, have been determined under various H_2 pressures (P_{H2}).The H_2 pressure dependence of θ_{H2} is given by eq. (9) and (10) for the Langmuir-type molecular adsorption and the dissociative adsorption in terms of the corresponding equilibrium constant, $K_A^{(m)}$ and $K_A^{(d)}$, respectively.

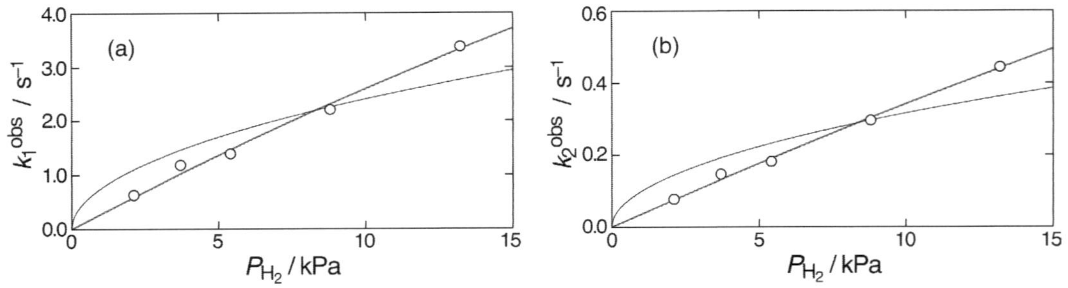

Figure 2. Pressure dependence of conditional rate constants, k_1^{obs} (a) and k_2^{obs} (b). The calculated lines are depicted for molecular adsorption (red line) and dissociative adsorption model (black curve).

$$\theta_{H_2} = \frac{K_A^{(m)} P_{H_2}}{1 + K_A^{(m)} P_{H_2}} \tag{9}$$

$$\theta_{H_2} = \frac{\left(K_A^{(d)} P_{H_2}\right)^{1/2}}{1 + \left(K_A^{(d)} P_{H_2}\right)^{1/2}} \tag{10}$$

The curve-fitting results are compared in Fig. 2 for the two adsorption models.It is found that the molecular adsorption model reproduces well the observed rate constants for both k_1^{obs} and k_2^{obs}.In addition, the almost linear dependence shown in Fig. 2 indicates that the molecular adsorption equilibrium of H_2 is inclined to the left hand side under the high measurement temperature.

4. Conclusions

The solid-phase reduction of PdO particles supported on γ-Al_2O_3 has been clarified by means of the time-resolved DXAFS method.The unique kinetic behavior has been explained by the combined first- and zero-order rate law.The latter is interpreted in terms of a faster oxygen migration in the PdO particle than the surface reduction step. The H_2 pressure dependence of the rate constants reveals that the Langmuir-type molecular adsorption of H_2 precedes the surface reduction of PdO particles.

Acknowledgment

DXAFS measurements have been performed under the approval of the Photon Factory Program Advisory Committee (Nos. 2009G603 and 2011G531).

References

[1] Thomas J M and Thomas W J 1997 *Principles and Practice of Heterogeneous Catalysis* (Weinheim: VCH).
[2] Yazawa Y, Yoshida H, Takagi N, Komai S-I, Satsuma A and Hattori T 1999 *J. Catal.* **187** 15.
[3] Inada Y, Suzuki A, Niwa Y and Nomura M 2007*AIP Conf. Proc.* **879** 1230.
[4] Glemser O and Peuschel G 1955 *Z. Anor. Chem.* **281** 44.
[5] Sasaki S, Fujino K and Takeuchi Y 1979 *Proc. Jpn. Acad. B* **55** 43.

Journal of Physics: Conference Series **430** (2013) 012054 doi:10.1088/1742-6596/430/1/012054

Structure and reducibility of a Fe/Al_2O_3 catalyst for selective catalytic reduction studied by Fe K-edge XAFS spectroscopy

A Boubnov[1], H Lichtenberg[1], S Mangold[2] and J-D Grunwaldt[1,*]

[1]Karlsruhe Institute of Technology (KIT), Institute for Chemical Technology and Polymer Chemistry (ITCP), Kaiserstr. 12, 76131 Karlsruhe, Germany and Institute for Catalysis Research and Technology (IKFT), Hermann-von-Helmholtz-Platz 1, 76344 Eggenstein-Leopoldshafen, Germany
[2]Karlsruhe Institute of Technology (KIT), Institute for Synchrotron Radiation (ISS), Hermann-von-Helmholtz-Platz 1, 76344 Eggenstein-Leopoldshafen, Germany

E-mail: grunwaldt@kit.edu

Abstract. EXAFS and pre-edge information from the Fe K-edge absorption spectra is used in this study to characterise the local environment and geometry of Fe-centres in a 1% Fe/Al_2O_3 model catalyst. The EXAFS results reveal clusters of 2-3 Fe oxo-moieties dispersed on the Al_2O_3-support. The Fe^{3+} centres are coordinated by 6 O-atoms in a strongly distorted octahedral geometry. This is supported by the pre-edge peak, which is far more intense than in α-Fe_2O_3 absorption data acquired for comparison. For preliminary investigations, catalytic tests for selective catalytic reduction (SCR) of NO_x by ammonia and *in situ* XANES studies of Fe/Al_2O_3 are compared with previously published data for Fe/zeolite systems. The low SCR activity of Fe/Al_2O_3 (compared to Fe/zeolite catalysts) in which the Fe^{3+} species do not change their electronic state under SCR-relevant conditions (reference temperature 250°C in our case) correlates well to the significantly higher temperatures required to reduce these species to Fe^{2+}. The difference in reducibility (and consequently in the SCR-activity) between the two systems probably results from differences in the structure and the electronic interaction of the Fe oxo-moieties and the particular catalyst support.

1. Background

Iron oxide based catalysts show a high efficiency in the selective catalytic reduction (SCR) of NO_x by ammonia and can potentially replace the environmentally harmful vanadium oxide-based catalysts [1, 2]. In order to further evaluate the potential of iron oxide-based catalysts in environmental catalysis, structure-activity relationships, i. e. correlations between structural properties and catalytic performance, have to be studied. These investigations are crucial for a fundamental understanding of the origin of catalytic activity.

This X-ray absorption spectroscopy study focuses on a model Fe/Al_2O_3 catalyst with 1% Fe loading, which was recently studied for SCR applications [3]. In Fe/Al_2O_3 systems the highly dispersed Fe^{3+} oxo-species have shown to be similar to those in previously studied Fe/zeolite catalysts [2, 4]. However, the specific catalytic activity was significantly lower [3]. First studies have also indicated that in Fe/Al_2O_3 materials the oxidation state does not change under SCR conditions, whereas Fe/zeolite catalysts show a partial reduction of Fe^{3+} to Fe^{2+} [4, 5]. Here, this issue is further addressed by identifying the structure of the Fe^{3+} clusters in the Fe/Al_2O_3 system and by studying their reducibility.

2. Experimental

The Fe/Al_2O_3 catalyst was prepared by incipient wetness impregnation of γ-Al_2O_3 with aqueous $Fe(NO_3)_3$ followed by drying and calcination at 600°C. XAFS measurements at the Fe K-edge were carried out at the ANKA synchrotron light source (Karlsruhe, Germany) at the XAS beamline using a double crystal monochromator equipped with Si(111) crystals. The monochromator was detuned to 70% of the maximum intensity for rejection of higher harmonics. The samples (catalyst and α-Fe_2O_3 as comparison reference) were mixed with cellulose, pressed into pellets, and EXAFS spectra were measured in transmission mode. For *in situ* temperature-programmed reduction (TPR) measurements, the catalyst with a sieve fraction of 100-200 μm was exposed to a flow of 5% H_2 in balance He in a capillary reactor as described previously [4-6]. Here, the XANES spectra were acquired in continuous quick-scanning XAFS mode (QXAFS, ca. 4 min/scan). XANES and EXAFS data were analysed using the Athena and Artemis extensions of IFEFFIT software [7]. The spectra were energy-calibrated and normalized. The pre-edge peak was extracted from the absorption edge by interpolating points several eV before and after the pre-edge region, similar to the method in [8]. For EXAFS data fitting the k^3-weighted EXAFS functions were Fourier-transformed in the k-range of 2.0 - 12.1 Å^{-1}, Fourier-filtered (R-range 1.0-2.1 Å for α-Fe_2O_3 and 1.0-3.8 Å for Fe/Al_2O_3) and fitted with single-scattering paths generated by FEFF6. In all fits an electronic amplitude reduction factor S_0^2 = 0.86 was used, as determined by constraining the number of O-atoms in the first shell of α-Fe_2O_3 to 6.

3. Results and Discussion

The similarity of the overall shapes of the XANES spectra of α-Fe_2O_3 and the 1% Fe/Al_2O_3 catalyst (figure 1A) indicate similar electronic structures. The coinciding positions of the absorption edges reveal that the catalyst contains trivalent iron. The pre-edge peak in the spectra corresponds to the dipole-forbidden $1s \rightarrow 3d$ transition. It gains intensity through the deviation from centrosymmetric geometry and from $3d$-$4p$ orbital mixing; its position indicates the oxidation state of iron in the sample [9]. The inset of figure 1A shows this feature in greater detail. In this case, the low intensity of the pre-edge peak for α-Fe_2O_3 matches well the slightly distorted octahedral coordination (the high-energy tail corresponds to other transitions which will not be considered [8]). The main pre-edge peak position (7114 eV) is consistent with trivalent Fe in both α-Fe_2O_3 and the catalyst. The high intensity of the pre-edge peak for the catalyst indicates either a very high degree of distortion of octahedral FeO_6-moieties or tetrahedrally-coordinated Fe^{3+} [8].

Figure 1B shows Fourier-transformed EXAFS spectra of α-Fe_2O_3 and the 1% Fe/Al_2O_3 catalyst. Both samples show Fe-O shells in the region of 1-2 Å. Further contributions appear in the interval 2-4 Å. For α-Fe_2O_3 they primarily arise from the strong backscattering of the Fe neighbours in the crystal, whereas the catalyst only shows weak features in this range: the expected neighbours at these distances would be Al atoms, which are significantly weaker backscatterers than Fe atoms. Possibly there is a smaller number of Fe atoms in this coordination shell compared to α-Fe_2O_3. Note that Fe was introduced into the system by impregnation, i.e. the Fe atoms should be mostly located on the surface of the Al_2O_3 crystallites and thus mainly have metal neighbours on one side. The formation of aluminates cannot be excluded but is usually expected at higher temperatures [10].

A derivation of the atomic structure is provided by the EXAFS data analysis presented in Table 1. Concerning the structure of α-Fe_2O_3, the focus is on the distorted octahedral FeO_6-moiety and therefore the analysis of further shells was omitted. The two calculated Fe-O bond distances at 1.92 Å and 2.07Å both with a coordination number of 3, are close to the expected values based on crystallographic data. The two Fe-O shells refined from the spectrum of the 1% Fe/Al_2O_3 catalyst summed up to a coordination number of 4.7 O, suggesting a mixture of tetrahedral and octahedral coordination. The second Fe-O bond distance (2.45 Å) differs significantly from that for α-Fe_2O_3, indicating a highly distorted geometry of the FeO_6-moieties. This is in accordance with the information provided by the pre-edge peak discussed above. Further EXAFS contributions in the 1% Fe/Al_2O_3 are assigned to Fe-Al and Fe-Fe shells (table 1). The Fe-Al contributions in the catalyst data indicate that Fe_xO_y clusters are well dispersed on or partly incorporated into the Al_2O_3 surface (Fe-Al coordination numbers of 1.9, note the rather weak backscattering). A Fe-Fe coordination number of 1-2 suggests that the clusters are likely a mixture of binuclear and oligonuclear clusters. This is also supported by UV-vis measurements [3]. A binuclear cluster which might be present among other structures derived from the EXAFS analysis is shown in figure 1C. The large variation of the Fe-O bond distances may originate from the fact that some of the O-atoms form bonds between the metal ions (Fe, Al), while others point away from the Al_2O_3 surface as oxo- or hydroxyl-groups.

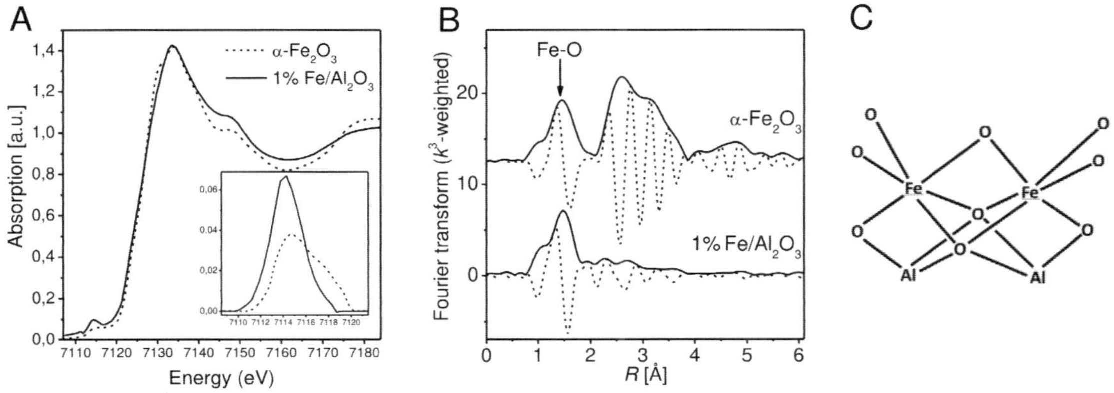

Figure 1. (A) XANES spectra of α-Fe_2O_3 and 1% Fe/Al_2O_3 catalyst with extracted pre-edge feature as inset. (B) Corresponding Fourier transformed EXAFS spectra. Solid line: magnitude; dashed line: imaginary part. (C) Suggested structure of a binuclear iron oxo-cluster based on the EXAFS analysis of 1% Fe/Al_2O_3.

Table 1. EXAFS analysis results α-Fe_2O_3 (only first shell analysis for α-Fe_2O_3) and of 1%Fe/Al_2O_3 (Debye-Waller factor of Fe-Al was constrained to 0.008)

Sample	Path	d [Å]	CN	σ^2 [Å^2]	ΔE_0, R-factor
α-Fe_2O_3	Fe-O	1.92	3	0.005	-5.1 eV, 0.029
	Fe-O	2.08	3	0.008	
1% Fe/Al_2O_3	Fe-O	1.92	2.8	0.004	-9.5 eV, 0.17
	Fe-O	2.45	1.9	0.004	
	Fe-Al	3.28	1.9	0.008	
	Fe-Fe	3.42	1.1	0.010	

The *in situ* XANES spectra during the TPR of the 1% Fe/Al_2O_3 catalyst are shown in figure 2A. The noise level in these spectra is higher since they were acquired in the QXAFS mode. Nevertheless they clearly reveal the trends of interest. The shift of the pre-edge peak (inset of figure 2A) from 7114 eV to 7112 eV and the absorption edge shift by 5 eV to lower energy are evidence of the reduction of Fe^{3+} to Fe^{2+}. The presence of isosbestic points at *e.g.* 7130 eV and 7155 eV indicates that only two types of species are involved in this transformation. The metallic state of Fe is not observed in this experiment (absorption edge of metallic Fe is at 7112 [11]), supporting the observation of the Fe oxo-clusters being well-dispersed and stabilised against sintering and therefore full reduction.

The results of linear combination fitting (figure 2B; reference spectra: first and last spectrum of the series) illustrate the reduction profile, with the onset of significant structural changes taking place above ca. 350°C. Fe^{3+} species are mainly reduced to Fe^{2+}-species. This temperature range is much higher than the reduction temperatures reported for Fe/zeolite systems [4, 5], which can be correlated to the significantly lower SCR-activity of the Fe/Al_2O_3 system at 250°C compared to Fe/zeolite. The reduction profile in terms of the Fe^{3+}-fraction of a Fe/BEA zeolite studied in [4] is also shown in figure 2B to support this hypothesis. The turnover frequency of selective catalytic reduction of NO at ca. 250°C was more than an order of magnitude higher over Fe/BEA (compared to Fe/Al_2O_3), accompanied by half of the Fe in a reduced state (Fe^{2+}) at this temperature.

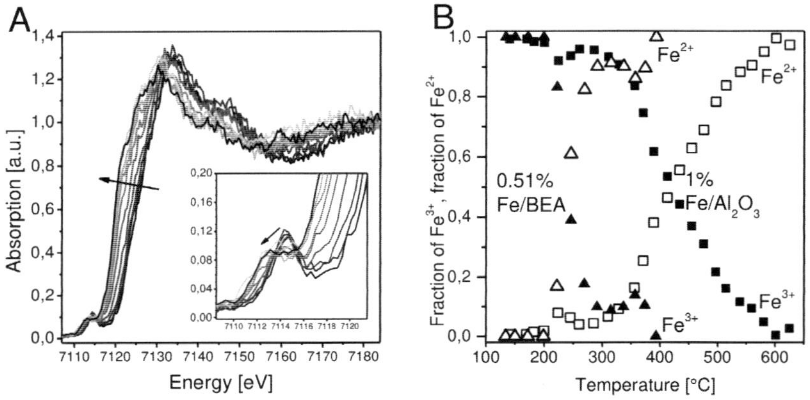

Figure 2. (A) Rapid *in situ* XANES spectra of the 1% Fe/Al_2O_3 catalyst during hydrogen-TPR: selected spectra collected in the temperature range 130°C-625°C. Inset shows details of the pre-edge feature from the same spectrum series (not extracted). (B) Results of linear combination fitting (1% Fe/Al_2O_3 and 0.51% Fe/BEA zeolite for comparison) of the initial and final XANES spectra in the series (cf. text).

4. Conclusion

This study shows that the reducibility of Fe in the present Fe/Al_2O_3 system is significantly lower than that in known Fe/zeolite systems, possibly due to the tight attachment to the Al_2O_3 matrix. The structure of iron on alumina was analysed in detail using XANES and EXAFS spectroscopy. The SCR-activity seems to be strongly influenced by the attachment of iron (and the support itself).

5. Acknowledgements

The authors thank ANKA (Karlsruhe, Germany) for providing beamtime. KIT and the BMBF project "MatAkt" are gratefully acknowledged for funding, Kevin Utermann and the workshop at KIT for designing the new *in situ* setups, and Sven Kureti, Andreas Pacher (Bergakademie Freiberg), Maria Casapu (KIT) and Matthias Bauer (TU Kaiserslautern) for valuable discussions and support.

References

[1] Kumar M S, Schwidder M, Grünert W and Brückner A 2004 *J. Catal.* **227** 384
[2] Balle P, Geiger B and Kureti S 2009 *Appl. Catal.* B **85** 109
[3] Pacher A, Boubnov A, Kureti S and Grunwaldt J-D **in preparation**
[4] Høj M, Beier M J, Grunwaldt J-D and Dahl S 2009 *Appl. Catal.* B **93** 166
[5] Klukowski D, Balle P, Geiger B, Wagloehner S, Kureti S, Kimmerle B, Baiker A, Grunwaldt J-D 2009 *Appl. Catal.* B **93** 185
[6] Grunwaldt J-D, Hannemann S, Göttlicher J, Mangold S, Denecke M A and Baiker A 2005 *Phys. Scr.* **T115** 769
[7] Newville M 2001 *J. Synchr. Rad.* **8** 322
[8] Petit P-E, Farges F, Wilke M and Solé V A 2001 *J. Synchr. Rad.* **8** 952
[9] Westre T E, Kennepohl P, DeWitt J G, Hedman B, Hodgson K O and Solomon E I 1997 *J. Am. Chem. Soc.* **119** 6297
[10] Bolt P H, Habraken F H P M and Geus J W 1998 *J. Solid State Chem.* **135** 59
[11] Kraft S, Stumpel J, Becker P and Kuetgens U 1996 *Rev. Sci. Instrum.* **67** 681

A new metal exchanged zeolite for a present environmental problem. An in-situ XAS study.

C Alonso-Escobar[1,2], C Franch-Martí[1], A E Palomares[1], F Rey[1] and G Guilera[2,*]

[1] Instituto de Tecnología Química (CSIC-UPV), Av. De los Naranjos s/n, 46022 Valencia, Spain.

[2] ALBA Synchrotron Light Facility, Crta. BP 1413 de Cerdanyola a Sant Cugat del Vallès, Km. 3,3 08290 Cerdanyola del Vallès, Barcelona, Spain.

gemma.guilera@cells.es

Abstract. The medium pore zeolite, TNU-9, is prepared and studied for the selective catalytic reduction (SCR) of NO using C_3H_8 as the reducing agent. The catalytic activity of TNU-9 zeolites for the SCR is comparable to other known highly active zeolites but with the advantage of TNU-9 of having almost the same catalytic performance in the presence of H_2O during reaction. The nature and behaviour of Cu and Co active sites contained in the TNU-9 catalysts have been studied under operation conditions using X-ray Absorption Spectroscopy (XAS) to understand the key parameters controlling the performance of this reaction.[1] It was found that the well dispersed Cu and Co centres need to be in a mixed valence state to obtain good catalytic results for the SCR and that the catalytic performance is related to the topology of the TNU-9 itself.

1. Introduction

The current tendency of modern petrol–fuelled and diesel cars is to run under high oxygen to fuel ratios to maximize fuel efficiency. This operation mode makes impossible the use of the traditional three-way catalysts in future mobile sources which are not able to fulfill the emission regulations of NO_x. An alternative to that is the selective catalytic reduction (SCR) of NO_x with hydrocarbons in the presence of O_2 [1,2]. Some medium and large metal exchanged zeolites have shown to catalyze the reduction of NO_x with hydrocarbons such as propane, propylene and ethylene [1-3]. Nevertheless, the catalytic performance strongly diminishes in the presence of water, SO_2 or after a hydrothermal treatment of the original catalyst, probably due to a modification of the environment of the copper active sites and to the low hydrothermal stability of the material [4].

Our group has found that a new medium pore size zeolite, TNU-9, when exchanged with Cu or Co ions present a high activity towards the reduction of NOx with propane under realistic working conditions. Additionally, the zeolite presents high resistance to water decomposition, especially when exchanged Co is present as balancing cation, which makes the catalyst a potential candidate for on-board applications.

[1] The full description of this work has been recently accepted for publication in Journal of Catalysis.

To the best of our knowledge, the nature and behaviour of the metallic active sites of Cu and Co containing TNU-9 catalysts have been studied for the first time under operation conditions using X-Ray Absorption Spectroscopy at the Cu and Co K-edges. The catalytic and structure-electronic results of the metal containing TNU-9 are compared with the best reported zeolites for NO_x SCR.

2. Experimental

TNU-9 was synthesized following the procedure described in [5].

The metal exchange was performed in an aqueous solution containing appropriate concentrations of $Cu(CH_3COO)_2{\bullet}4H_2O$ or $Co(CH_3COO)_2{\bullet}4H_2O$ to achieve the desired amount of metal in the zeolite (2-3 wt%) and a degree of interchange above 70%.

The surface area and micropore volume of the catalysts were determined from the adsorption branch of the isotherms by applying the BET equation and the t-plot formalism, respectively. The catalysts retain most of their micropore volume and surface area upon cationic exchange indicating that no pore blocking, or zeolite destruction, takes place during catalyst preparation.

X-ray absorption spectra (XANES and EXAFS) at the Co K-edge (7709 eV) were collected on the Super-XAS beamline at the Swiss Light Source (SLS). XANES spectra at the Cu K-edge (8978 eV) were taken on BM25 (SpLine beamline) at the ESRF. In both cases, a double crystal Si(111) monochromator, detuned by 50% to reject higher energy harmonics, was used to select the X-ray radiation energy. The measurements were performed in transmission mode and ionization chambers recorded the intensity signal of the incident and transmitted X-ray beam. Co or Cu reference foils were measured in parallel for energy calibration. Other Co and Cu reference compounds were also measured by XAS during the experiment; *i.e.* CoO, Co_3O_4, $CoMoO_4$, $CoAl_2O_4$, $Co(CH_3COO)_2{\bullet}4H_2O$, Cu_2O and CuO.

The NOx reduction was carried out in an in-situ cell built as a result of collaboration between the Instituto Químico de Valencia (ITQ; CSIC-UPV) and ALBA Synchrotron Light Facility. All parameters related to the experimental environment of the sample within the cell (temperature, pressure, cooling circuit, gas flows and mixtures, gas evacuation system and safety logic) were controlled by an integrated control unit built by the same groups [6]. The sample was presented as a pressed disk of 13 mm diameter. XAS measurements were carried out at ambient pressure and reproducing the same gas and temperature conditions as in the catalytic measurements. The sample was activated by heating it at 550 ^{o}C under He. The catalytic reaction was carried out at 450 ^{o}C with 700 ppm of NO_x and 390 ppm of C_3H_8 balanced in He. Different amounts of O_2 (0-14.5%) were introduced during the reaction. The total flow rate in all experiments was 120 $mL{\bullet}min^{-1}$. A 10 min stabilization period was left between changes in gas composition and before collecting the absorption data.

EXAFS data analysis followed standard procedures using the software package Viper v. 10.1 for Windows [7].

3. Results and Discussion

In order to understand why certain medium pore zeolites, when exchanged with Cu or Co, are active towards the Selective Catalytic Reduction (SCR) of NOx with propane in the presence of O_2, is important to discern the electronic and structural changes occurring in the course of the activation and reaction on these catalysts. This in-situ approach allows correlating the catalytic performance with the structuro-electronic changes under the optimal reaction conditions.

For the case of the Cu exchange zeolites, in-situ XANES measurements at the Cu K-edge were performed. Inspection of the XANES spectrum of Cu-TNU-9 taken at RT in He (Figure 1) showed that on average the Cu centers in this zeolite are present as Cu(II). The absence of any pre-peak in the spectrum suggested that the local average structure is very symmetric around Cu, similar to the monomeric copper acetate structure in aqueous solution.

Following the experimental procedure for NO SCR catalytic testing of the Cu-catalysts, the Cu-TNU-9 sample was activated at 550°C in He prior to XAS measurement. It was observed that, after heating at 550°C, the Cu(II) sites were completely reduced to Cu(I) as shown by the characteristic red shift in Figure 2. Subsequently, the Cu-TNU-9 sample was brought to the reaction temperature (450°C), with no change in the XANES region (Figure 2). The fact that the Cu-catalysts studied herein are not active under reaction conditions in the absence of O_2 (see Figure 3) suggests that the active sites in Cu-TNU-9 are not solely Cu(I) species. At the maximum conversion rate, when 0.33 % O_2 was added into the reaction feed, partial oxidation of the Cu(I) sites was observed, demonstrating that a mixed –or intermediate- oxidation state is necessary to attain high catalytic performance. XANES spectra of Cu-TNU-9 under reaction conditions (450 °C, 700 ppm of NO_x and 390 ppm of C_3H_8) and different contents of O_2 are shown in Figure 4.

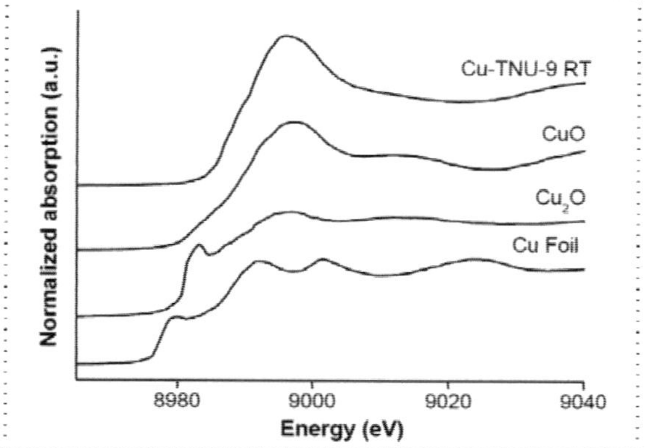

Figure 1. Cu K-edge XANES spectra of reference compounds (Cu foil, Cu_2O and CuO) and Cu-TNU-9 at RT.

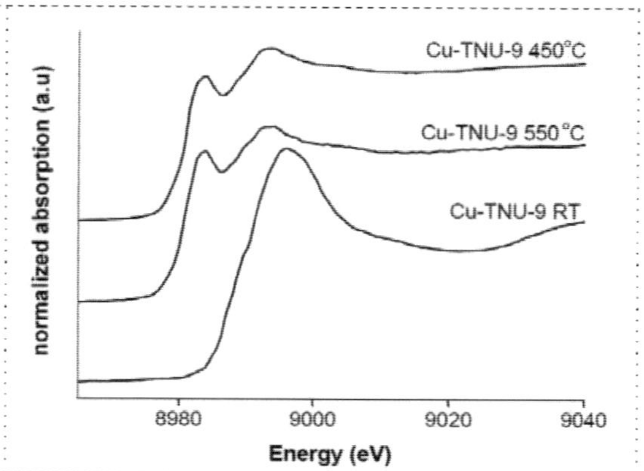

Figure 2. Cu K-edge XANES spectra of Cu-TNU-9 in He at different temperatures of catalyst treatment: RT, 550°C and 450°C.

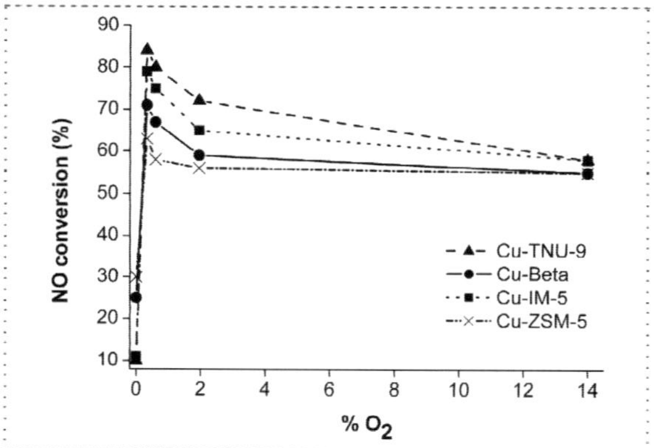

Figure 3. Influence of oxygen concentration on the SCR of NO with propane. (▲) Cu-TNU-9; (●) Cu-Beta; (■) Cu-IM-5; ( ) Cu-ZSM-5. Experimental conditions: 700 ppm NO, 390 ppm C_3H_8, N_2 as carrier, T= 450 °C, total flow: 900 ml.min^{-1}, 0.6 gr. of catalyst.

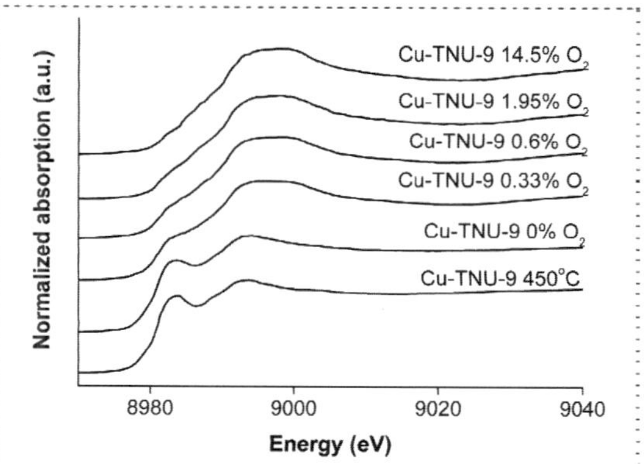

Figure 4. Cu K-edge XANES spectra of Cu-TNU-9 obtained under catalytic conditions (T= 450 °C; 700 ppm NO and 390 ppm C_3H_8 balanced in He) and at different O_2 concentrations (0-14.5 %).

Figure 5 shows the Co K-edge XANES spectra of reference compounds with different oxidation states (Co^0, Co^{2+} and $Co^{2+/3+}$) and of the Co-TNU-9 at RT and after activation at 450°C. The energy threshold and the XANES features show that at RT the Co-TNU-9 contains mostly Co(II) sites and that the average environment around Co is octahedral. After activation the threshold shift towards lower energies indicates that the Co sites are present in a reduced Co^{2-x} form. No Co metal particles were observed. The presence of a pre-edge peak suggests that at 450°C the Co atoms occupy a 4-fold site, probably in a distorted tetrahedral environment.

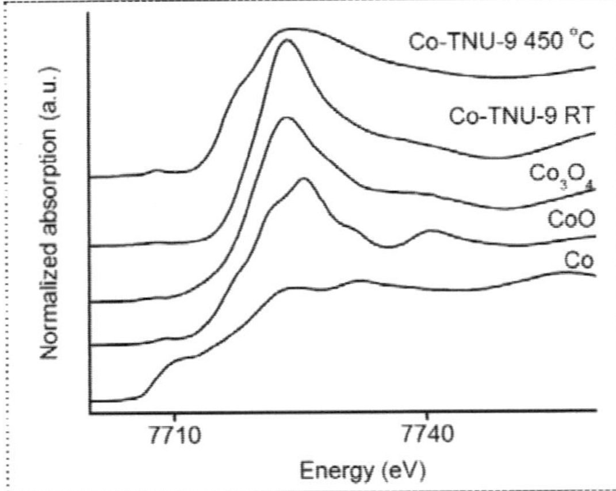

Figure 5. Co K-edge XANES spectra of reference samples (Co, CoO and Co_3O_4); and Co-TNU-9 at RT and at 450 °C in He.

When the zeolite was submitted to operation conditions under the required gas atmospheres of NO and propane, and different amounts of O_2 (0-14.5%) were flown, no apparent changes with respect to the sample at 450°C were observed. Additional EXAFS spectra were acquired to see if there were perceptible changes at a structural level. The analysis of the EXAFS spectra obtained under working conditions were in agreement with the XANES observations, but no structural change was distinguished during the reaction. This analysis also showed that the zeolite contained isolated, or *quasi*-isolated, Co sites, highly dispersed across the channels of the zeolites, with no extraframework phase present.

4. Conclusions

In-situ XAS measurements of Cu and Co exchanged TNU-9 catalyst under real catalytic conditions were performed to understand the electronic and structural behavior of metal sites and to identify which are the key parameters on the catalyst activity for the SCR of NOx.

From these studies we could conclude that:

(a) the metal ions are well dispersed in the pores of the TNU-9 zeolite structure and are present as isolated or *quasi*-isolated small particles; as deduced from both XAS and XRD studies.

(b) during activation at 550°C in He the Cu and Co sites get reduced completely to Cu(I) and mostly to a reduced Co^{2-x} form, respectively. However, it is necessary the presence of metal centers having mixed valence state to obtain good catalytic results in this reaction.

(c) the catalytic reaction rate per unit of surface area is related to the topology of the TNU-9 itself.

Acknowledgements

We acknowledge the MICINN (project no.: ICTS-2008-18, MAT2009-14528-C02-01 and MAT2009-14528-C02-02) for financial support and for a FPI fellowship. Authors thank Super-XAS (SLS) and SpLine (ESRF) beamlines for the beamtime allocated and the scientific support received. We thank J. Hernández and J. J. Cortés for his work on the construction of the in situ cells.

References

[1] Burch R, Breen J P and Meunier F C, 2002 *Appl. Catal. B* **246** 283
[2] Corma A, Palomares A and Fornes V Res. *Chem. Intermed.* 1998 **24** 613
[3] Shibata J, Takada Y, Shichi A, Satakawa S, Satasuma A and Hatttori T *Appl. Catal. B: Environ.* 2004 **54** 137
[4] Gómez S A, Campero A, Martínez-Hernández A and Fuentes G A *Appl. Catal. A* 2000 **197** 157
[5] Palomares A E, Márquez F, Valencia S and Corma A, *J. Mol. Catal. A: Chemical* 2000 **162** 175
[6] see: http://www.cells.es/Beamlines/CLAESS/sample.html, July 2012
[7] Klementiev K V *J. Phys. D: Appl. Phys.* 2001 **34** 209

Journal of Physics: Conference Series **430** (2013) 012056 doi:10.1088/1742-6596/430/1/012056

Structure and formation process of the water-pretreatedPt/FeOx-Al_2O_3 catalyst capable of CO oxidation below room temperature

A Tomita[1], K Shimizu[2], K Kato[3] and Y Tai[1]

[1] Materials Research Institute for Sustainable Development, National Institute of Advanced Industrial Science and Technology (AIST), Nagoya 463-8560, Japan
[2] Catalysis Research Center, Hokkaido University, Sapporo 001-0021, Japan
[3] Japan Synchrotron Radiation Research Institute, Sayo-gun, Hyogo 679-5198, Japan

Email: tai.y@aist.go.jp

Abstract. Structure and formation process of water-pretreatedPt/FeOx-Al_2O_3 catalyst capable of CO oxidation below room temperaturehas been investigated usingPt L_{III}X-ray absorption fine structure (XAFS) measurementand high-angle annular dark-field scanning transmission electron microscopy (HAADF-STEM).Before the water-pretreatment,Pt monomers and small clusters were detected while formation of Ptnanoparticles (PtNPs) having an average diameter of ≈1.3 nm and Pt-O-Fe interaction were confirmed after the pretreatment, implying that Pt monomers and small clusters aggregated at FeOx on alumina upon the pretreatment.The present results seem to indicate that the formation of Pt monomersand small clusters prior to the water-pretreatment is important for the high mobility of the Pt species and the contact formation between the Pt particles and FeO_xon the resultant catalyst which is a key to the high CO oxidation activity at low temperatures.

1. Introduction

The use of additives to promote catalytic reactions over platinum group metals (PGMs) is employed in various applications such as the three-way catalyst (TWC) for purification of automotive exhaust. To optimize interactions between PGMs and promoters, it is necessary for uniform contacts to exist between them. We have recently reported that in Pt/Fe-containing alumina catalysts prepared and treated with water under moderate conditions, Pt nanoparticles (PtNPs) and iron oxides form intimate contacts with each other and as a result, the catalyst exhibited CO oxidation activity even at temperatures below 0 °C [1].

Obviously, the key process for the catalyst preparation is the hydrogen treatment with water at ≈200 °C prior to measurement of CO conversions (water-pretreatment). From transmission electron microscopy (TEM), it was confirmed that on the water-pretreated catalysts, mobility of Pt species was enhanced and larger NPs were formed. However, the growth of the PtNPs was suppressed when the catalyst support contained small amount of FeOx. These results suggest that PtNPs were formed at FeOx sites upon the water-pretreatment.

Although, the effect of the water-pretreatment on the catalyst structure has been clarified to a certain extent, the role of each step in the catalyst preparation is not clarified. In this paper, we will analyze the structure of the catalyst at each step of the catalyst preparation in detail using Pt L_{III}X-ray absorption fine structure (XAFS) measurement andhigh-angle annular dark-field scanning TEM (HAADF-STEM).

2. Experimental

2.1. Catalyst Preparation

The method of catalyst preparation has been described elsewhere; thus, is described here only briefly. γ-alumina sample containing small amount of Fe [ALO-5 (Fe content: 0.48 wt%, Japan Reference Catalyst (JRC))] was used as a catalyst support. A HNO_3 solution of $Pt(NH_3)_2(NO_3)_2$ (Tanaka precious metals, Pt content: 50 g/L) was added to a slurry containing the γ-alumina (1 g). The mixture was then heated on a steam bath, with stirring, to evaporate water. The resultant solid was (I) calcined at 200 °C for 2 h in a flow of dry air (300 mL/min) and then (II) it was reduced at 200 °C in a flow of H_2/Ar (1:1, 50 mL/min). The catalysts (100mg) were wetted with deionized water at room temperature and (III) heat-treated under a flow of H_2 (33.3 mL/min) in a Pyrex tube at 200 °C for 30 min prior to measurement of CO conversions. We denote the final treatment, the water addition and the following H_2 reduction, as "water-pretreatment". In addition to this preparation procedure, the water addition was examined at the steps (I) and (II). For the measurement of catalytic activity, the reactant gas [CO (1%), O2 (0.5%), N2 (98.5%)] was passed through the catalytic bed at a flow rate of 33.3 mL/min [hourly gas space velocity (HGSV) = 20,000 mL/(h·g-cat.)].The specific surface area, pore diameter, and chemical compositions of the ALO-5 alumina have also been described previously [2].

2.2. Catalyst Characterization

High-angle annular dark-field scanning transmission electron microscopy (HAADF-STEM) images were recorded using a Hitachi HD-2700 electron microscope with a Cs corrector operated at an accelerating voltage of 200 kV. For the STEM observations, an ethanol dispersion of the ground catalysts was dried on a carbon-coated microgrid.

Pt L_{III} XAFS measurements for the catalysts (2.5wt% Pt/ALO-5) prior to the water-pretreatment have been performed in air at the BL11 in the SAGA-LS (Saga, Japan, Proposal No.100657N). The catalystsafter the water-pretreatment have been measured at the BL01B1 in SPring 8 (Hyogo, Japan, Proposal No.2011B1337) in the *in situ* transmission QXAFS mode. A pelletized catalyst was water-pretreated in a flow cell with kapton windows, and then measured under He atmosphere at room temperature.The $k^3\chi(k)$ functions in the range of values of $k$typically from 3.0 to 16.0 Å were Fourier-transformed using Hanning windows to obtain FTs.

3. Results and discussion

We examined the addition of water at each step of the catalyst preparation (I, II, and III). Table 1 compares the CO conversions at 25 °C for the catalysts (5wt% Pt) added ≈0.15 mL of deionized water prior to step I, II, and III together with that without water addition. Apparently, only the water addition prior to step III was effective; the conversion for the catalysts prepared with the water additions at other steps were the same as that of the catalyst prepared without the water addition.Accordingly, we investigated the structure of the catalyst at each step in the preparation procedure in which water is added prior to step III using Pt L_{III} XAFS spectroscopy and HAADF-STEM observation. We compare Fourier transforms (FTs) of PtL_{III} EXAFS oscillations at each step of the catalyst preparation before the water-pretreatment together with that of the HNO_3 solution of $Pt(NH_3)_2(NO_3)_2$ in Figs. 1 and 2. Interestingly, a few peaks were detected at an R range from 2.4 to 3.8 Å. This result is indicative

Table 1. CO conversions over the catalysts (5wt% Pt)treated with water at each preparation step

Timing of water addition	STEP I	Step II	Step III	No addition
CO conversion at 25 °C (%)	14	27	100	20

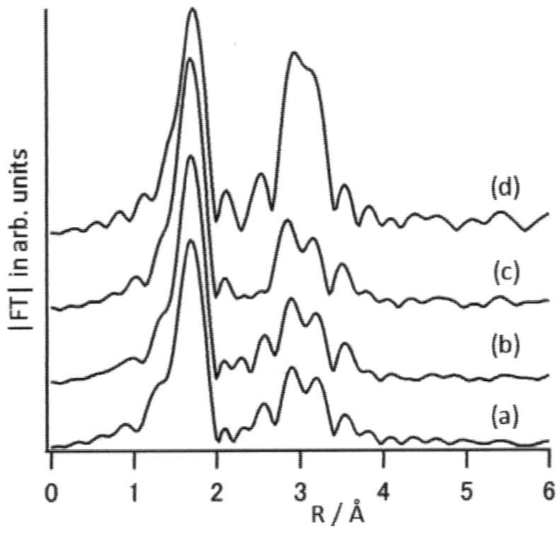

Figure 1.FTs of EXAFS oscillations of (a) HNO_3 solution of $Pt(NH_3)_2(NO_3)_2$, (b) the catalyst after impregnation, (c) the catalyst after calcination, and (d) PtO_2.

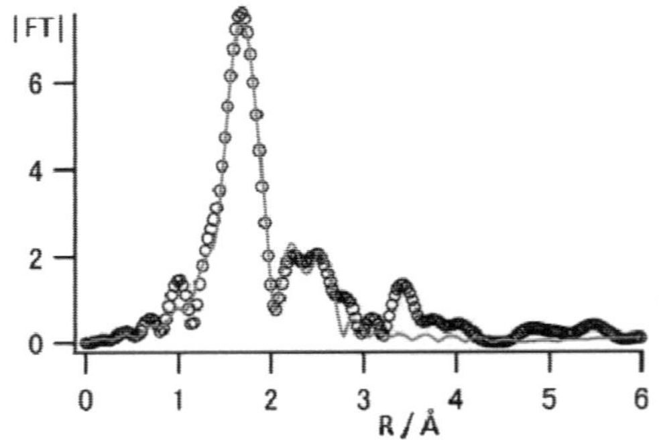

BS	N	R/Å	ΔE/eV	DW/Å
O	4.2	2.01	17.3	0.067
Pt	0.8	2.53	0.95	0.063

Figure 2.FT of EXAFS oscillations of the catalyst after H_2 reduction and the fitting parameters. Open circles represent the experimental data and the solid line indicates the fitting curve.

of the formation of second or higher coordination shells in the starting Pt solution. Similar higher shell formation due to oligomerization have been reported for Pt nitrate [3] and $Pt(NH_3)_2(NO_3)_2$ solutions [4]. Thus, oligomerization of $Pt(NH_3)_2(NO_3)_2$ might had been occurred in our starting Pt solution during storage. The impregnated catalyst before calcination exhibited identical FT to that of the starting Pt solution. The FT of the calcined catalyst was also similar to that of thestarting solution except for the disappearance of a small peak at ≈2.5 Å. However, after reduction [step (II)], the FT changed significantly as can be seen in Fig. 2. The fitting parameters, $N_{Pt\text{-}O}$=4.2, $R_{Pt\text{-}O}$=2.01Å, $N_{Pt\text{-}Pt}$=0.8, and $R_{Pt\text{-}Pt}$=2.53 Å, were derived. These values can be explained by assuming an ensemble of monomers and very small clusters coordinated with oxygen [5, 6]. Asakura et al. reported that Pt monomers on the (111) surface of α-alumina exhibited $N_{Pt\text{-}O}3\pm0.5$, which indicates that Pt monomers occupy threefold sites on the surface [5]. The present value, $N_{Pt\text{-}O}$=4.2, is slightly larger than their value. However, in the present experiment, Pt species are supported on γ-aluminaparticles of which the surfaces may well have some undulations and some of the Pt species are expected to exist at the

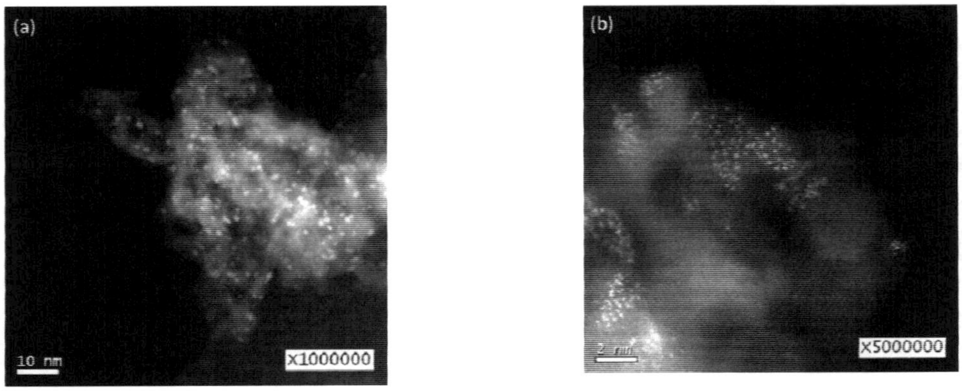

Figure 3. HAADF-STEM images of the catalyst after H_2 reduction taken at different magnifications.

Journal of Physics: Conference Series **430** (2013) 012056 doi:10.1088/1742-6596/430/1/012056

boundaries of the γ-aluminaparticles. These factors possibly increase the oxygen coordination number. Short Pt-Pt distances, less than 2.6 Å, have been often reported for small Pt cluster species [6].

To clarify further the state of Pt dispersions, we have observed the catalyst after reduction with HAADF-STEM. Figure 3 displays the results. At low magnification, Pt aggregates with an average diameter of 1~2 nm was observed [Fig.3(a)]. However, the image taken at high magnification [Fig.3(b)] showsa lot of Pt atoms separated from others. Although we cannot identify dimer or other small cluster species unambiguously, these observations seem to be consistent with the EXAFS results. It is known that Pt particles with diameters less than ≈1nm tend to decay in air [7]. Thus, the

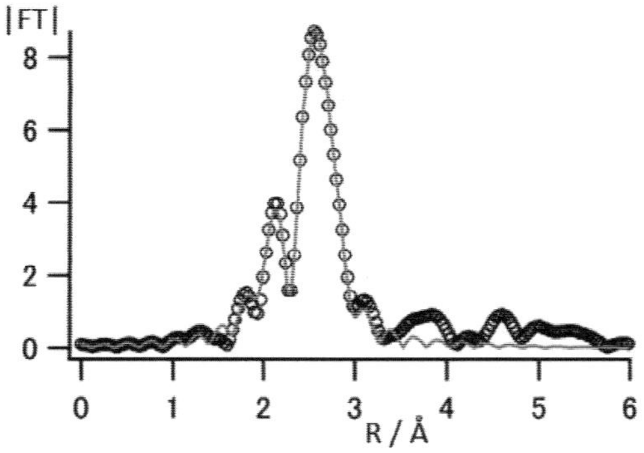

BS	N	R/Å	ΔE/eV	DW/Å
Pt	6.6	2.72	6.6	0.083
Fe	0.7	3.05	-7.6	0.059

Figure 4.FT of EXAFS oscillations of the catalyst after the water-pretreatement and the fitting parameters. Open circles represent the experimental data and the solid line indicates the fitting curve.

aggregated structures seen in Fig. 3(a) probably resulted from the decay of small Pt particles formed in the reduction step [step (II)].

After the water-pretreatment, the formation of Pt particles was evidenced by the fitting parameters due to Pt backscatter, $N_{Pt\text{-}Pt}$=6.6, $R_{Pt\text{-}Pt}$=2.72 Å (Fig.4). In addition, the best-fit curve was obtained assuming Fe backscatter with $N_{Pt\text{-}Fe}$=0.7 and $R_{Pt\text{-}Fe}$=3.05 Å. These results seem to indicate that Pt particles having an average diameter of ≈1.3 nm[8] were formed and pinned atFeO_x sites,possibly forming thin-layered islands or small clusters on the alumina surface [9], through Pt-O-Fe interaction during the water-pretreatment [1].

As indicated in table 1, the addition of water was effective only when it was done after reduction [step III]. The present results seem to indicate that the formation of atomically dispersed Pt species prior to the water-pretreatment is important for their high mobility and the contact formation between the resultant PtNPs and FeO_x.

References

[1]Tomita A, Shimizu K,Kato K,Tai Y, Catal. Commum. **17**, 194 (2012).
[2] Okazaki N, Shiina Y, Itoh H, Tada A, Iwamoto M, Catal. Lett. **49**169 (1997).
[3] Dou D, Liu D -J, Williamson W B, Kharas K C, Robota H J, Appl. Catal. B-Env. **30**, 11 (2001).
[4] Rochon F D, Buculei V, Can. J. Chem. 82, 1606 (2004).
[5] Asakura K, Chun W -J, Shirai M, Tomishige K, Iwasawa Y, J. Phys. Chem. **101**, 5549 (1997).
[6] see for example Chun W -J, Tomishige K, Hamakado M, Iwasawa Y, Asakura K, JCS Faraday Trans. **91**, 4161 (1995).
[7] Joyner R, J. Chem. Soc. Faraday Trans. I, **76**, 357 (1980).
[8] De Graaf J, Van Dillen J, De Jong K P and Konigsberger D C, J. Catal. **203**, 307 (2001).
[9] Tomita A, Shimizu K, Kato K, Akita T, Tai Y, submitted to J. Phys. Chem. C.

One body, many heads; the *Cerberus* of catalysis. A new multi-purpose in-situ cell for XAS at ALBA.

G Guilera[1], F Rey[2], J Hernández-Fenollosa[2] and J J Cortés-Vergaz[2]

[1] ALBA Synchrotron Light Facility, Crta. BP 1413 de Cerdanyola a Sant Cugat del Vallès, Km. 3,3 08290 Cerdanyola del Vallès, Barcelona, Spain.

[2] Instituto de Tecnología Química (CSIC-UPV), Av. De los Naranjos s/n, 46022 Valencia, Spain.

gemma.guilera@cells.es

Abstract. A new multi-purpose in-situ cell and its control system have been developed for synchrotron-based techniques as are X-Ray Absorption spectroscopy (XAS) and X-Ray Diffraction (XRD). The cell is made of a stainless steel 'body' and three different exchangeable 'heads' to tackle different scientific areas that include solid-gas catalysis, solid-liquid catalysis and electrocatalysis. The different versions of the cell are herein described and their functionality is exemplified by some case studies.

1. Introduction

Nowadays, nobody questions the importance that catalysis has in our developed economies. It is estimated that catalysis controls more than 90% of the world's chemical manufacturing processes to produce 60% of the chemical products our society uses for different purposes, such as plastics, synthetic fibers and pharmaceutical and agricultural products. This broad usage range leads us to understand that catalysis is an interdisciplinary area that combines both fundamental and applied science with major contributions from chemistry, physics and material science.

While homogeneous catalysis has been studied in depth in common chemical laboratories, there are still formidable problems in the unambiguous identification of the nature of the active species in heterogeneous catalysis, in the elucidation of catalytic reaction pathways/mechanisms and in the understanding of the role that the surface structure and nanostructure play. In practice, the performance of a catalyst is governed by several factors including its nanostructure, chemistry, surface structure and electronic properties on a fine (atomic) scale.

As a result, during the last few decades the community of catalysis has created its own scientific and technological space in the synchrotron world to fill this gap. The effect of this intrusion is a better understanding of many industrially relevant reaction mechanisms from an electronic and structural point of view and the consequent development of novel, more efficient and environmentally friendlier catalysts.

In-situ investigations of these catalytic reactions under real operation conditions, using synchrotron techniques such as X-Ray Absorption Spectroscopy or X-Ray Diffraction appear to be determinant for the understanding of these processes at an atomic level. Only in this way one can extract catalytically relevant information that can be of real value to produce novel materials or improve present catalytic

Journal of Physics: Conference Series **430** (2013) 012057 doi:10.1088/1742-6596/430/1/012057

processes. These in-situ experiments can be carried out under static or dynamic conditions, depending on the reaction and the information that needs to be extracted from it. At present, it starts to be more and more frequent the use of secondary techniques such as Raman or UV-Visible, to obtain synchronous complementary information in order to be able to unambiguously correlate electronic, structural and functionality data; thus obtaining a more complete scenario of the process. The current tendency of expert research groups is to develop tailored catalytic cells, where the reactions take place, that are compatible with synchrotron measurements, as well as other ancillary infrastructure that can be adjusted to specific beamlines at synchrotron sources. It exists nowadays a broad myriad of cells developed by academic and industrial research groups for synchrotron measurement purposes that permit the study of a very diverse type of systems ranging from solid-gas catalysis, liquid base catalysis, solid-liquid catalysis, electrocatalysis, biocatalysis, etc. Each and every one of these cells has its own peculiarity because they are generally developed for specific needs (*i.e.* for time-resolved studies, for plug-flow systems, to obtain good spectroscopic data, for high pressure studies, etc.), and it is practically impossible to design a single cell that performs equally well for all desired requirements [1-4].

CLÆSS (Core Level Absorption&Emission SpectroscopieS) beamline at ALBA Synchrotron Light Facility, which will open its doors to users in Fall 2012, has devoted part of its time to developing new in-situ cells and infrastructure for catalysis and chemistry, with the aim of creating a dedicated facility for catalysis at ALBA. This facility will be available to all the user community and can be especially valuable for those groups that either do not have the resources to build such specialized infrastructure or do not have extended experience in synchrotron-based catalytic experiments.

The developments presented herein have been done in collaboration with the Instituto de Tecnología Química (CSIC-UPV) of Valencia that is a world's recognized institution in catalysis and synthesis of novel materials –especially zeolites–, and it has a strong proven experience in the development of catalytic reactors.

Our developments have been conceived with the idea of offering robust, flexible and easy-to-use equipment –so that any experienced or inexperienced user is able to use them in a very short time– whilst maintaining the scientific and technical rigour and requirements needed to perform the different types of experiments. For that we have designed a multi-purpose cell based on a stainless steel 'body' and three different exchangeable 'heads' to tackle different scientific areas that include solid-gas catalysis, solid-liquid catalysis and electrocatalysis; reason why the cell has been named as the *Cerberus* of catalysis.

2. Multi-purpose cell for solid gas catalysis

The first and more general cell developed of this series is the multi-purpose cell for solid-gas catalysis (Figure 1). This in-situ cell was created with the aim to cover most of the standard experiments involving a reaction between a solid catalyst and a gas or mixture of gases, empathizing on the quality of the spectroscopic data.

The cell is built to be compatible with X-ray absorption measurements performed in absorption and /or fluorescence mode and with X-ray diffraction measurements at CLÆSS and MSPD beamlines of the ALBA synchrotron, respectively. This is because in catalysis crystalline and amorphous high-surface-area materials often occur beside each other. Then, it is important to see how these distinct phases evolve or transform during the course of the reaction while keeping the same sample environment and reaction conditions. Consequently, correlations between long and short-range structural order, electronic changes and catalytic performance can be accomplished.

The body of the cell is 90 mm wide x 140 mm height, and has an internal volume of 100 mL. It has four easily exchangeable windows (Kapton, vitreous C or other material) of diameter 25 mm and an angular aperture of 44°. It has also four other small ports to grant access to fiber probes when other secondary techniques are required. The head of the cell, which can be rotated at different angles around 180°, contains the feedthroughs for all electrical connections and supports the sample holder which is surrounded by a Thermocoax heater. The sample holder can accommodate a pellet of 13 mm of diameter that can be easily mounted and exchanged. The device can work at temperatures between RT - 700°C and at pressures up to 20 bars.

A movable and very reliable control system has been built to control the temperature and pressure on the reactor, the water stabilization circuit and the flow rate of up to six different gases. All safety components are in-built in this equipment.

Figure 1. External view of the multi-purpose in-situ cell for solid-gas catalysis.

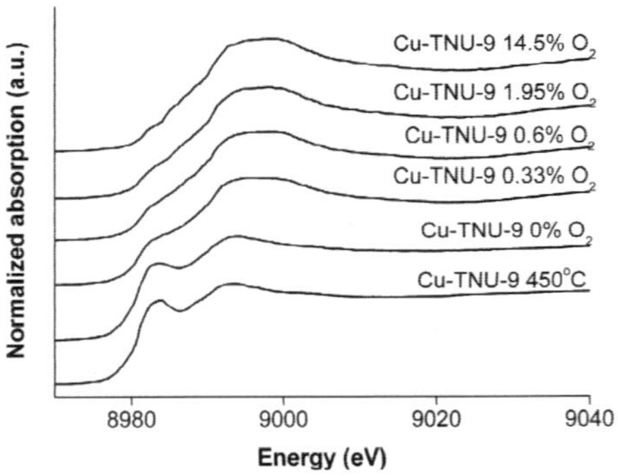

Figure 2. Cu K-edge XANES spectra of Cu-TNU-9 obtained under catalytic conditions (T= 450 °C; 700 ppm NO and 390 ppm C_3H_8 balanced in He) and at different O_2 concentrations (0-14.5 %).

This cell has been successfully tested at SLS (Super-XAS beamline) and at the ESRF(SpLine beamline, CRG) to study the Selective Catalytic Reduction of NO_x with propene on Co/Cu exchange TNU-9 zeolite at 450 °C after adding different amounts of O_2. The in-situ XAS experiment was performed in transmission mode on the Co and Cu K-edges. Figure 2 shows an extract of the absorption data obtained with this cell [5].

3. Electrochemical cell

The second type of head of this cell contains a special holder to accommodate thin solid samples that are used as half electrochemical cell instead of using the simpler pellet holder of the multi-purpose

Figure 3. External view of the in-situ cell while heating the sample at 700°C.

Figure 4. Internal view (upside down) of a prototype of half electrochemical cell

solid-gas cell. Several Pt wires are needed for *(i)* applying current to the sample (two wires at the front side of the sample) and *(ii)* performing electrical measurements of voltage-reference-current (three wires on the back of the sample). This cell can be heated up to 900°C, reason why the ceramic parts are made of alumina.

This cell was tested for the in-situ study of a set of perovskites based solid oxide fuel cell (SOFC). EXAFS measurements in fluorescence mode were performed at the Fe and Co K-edges on the Super-XAS beamline at SLS in order to understand the changes occurring on the perovskite electrode from an electronic and structural point of view. The experiment was done under real working conditions; *i.e.*, high temperature (600 – 900°C) and applied current densities (0 – 0.5 mA/cm^2). The sample was indeed the cathode of an electrochemical cell and was prepared by serigraphy. The characteristics of the chemical constituents were: electrolyte material (0.2-0.8 mm-thick yttria-stabilized zirconia); top layer of gadolinia-doped ceria to prevent the reaction of Zr with Sr and La (thickness layer less than 1 μ m); working electrode ($La_{0.6}Sr_{0.4}MO_{3-\delta}$ (M = Fe or Co)); and gold porous current collector net (thickness ~ 1-2 μ m); counter electrode (porous platinum or gold of thickness ~ 5 μ m). The results are under study.

4. Solid-liquid cell

Finally, the third type of head of the cell was designed such that a liquid could be introduced inside to perform solid-liquid catalysis. This cell is depicted in Figure 5. Again, the head contains the electrical connections that are directly linked to the same control system mentioned previously. In this case the system has an additional HPLC pump that allows introducing liquid in the reactor at the desired flow. The sample holder is very similar to the multi-purpose solid-gas cell but the pellet inside an inclusion (0.3 mm gap at each side of the pellet) where the liquid reactant flows through in al laminar way.

The cell can be heated up to 450 °C in temperature.

This cell has been preliminary tested on the XAFS beamline at ELETTRA to study in-situ the selective hydrogenation of cinnamaldehyde by 2 wt% Pd/Sn-Beta zeolite catalyst. EXAFS spectra in transmission mode were measured at the Pd K-edge (24350 eV). The reaction was followed at 60°C using cinnamaldehyde and H_2 dissolved in toluene as reactants with a 1mL/min flow. Figure 6 shows the EXAFS signal obtained with this cell before and during the selective hydrogenation reaction.

Figure 5. External view of the multi-purpose in-situ cell for solid-liquid catalysis.

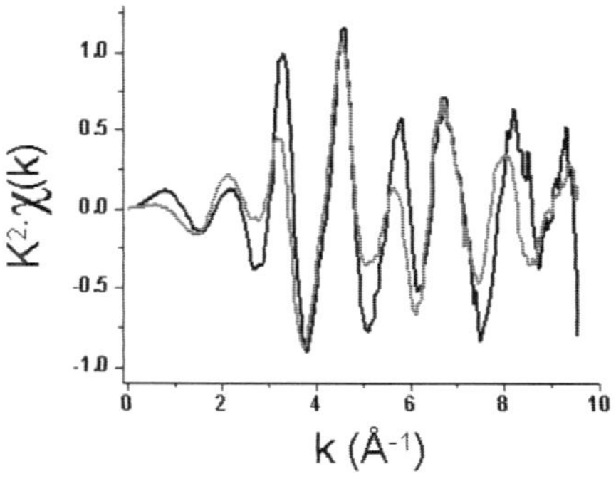

Figure 6. k^2 weighted EXAFS signal obtained in transmission at the Pd K-edge before (black) and during (red) the selective hydrogenation of the cinnamaldehyde by 2 wt% Pd/Sn-Beta zeolite catalyst.

Acknowledgements

We acknowledge the MICINN (project no.: ICTS-2008-18, MAT2009-14528-C02-01 and MAT2009-14528-C02-02) for financial support and for a FPI fellowship. Authors thank Super-XAS (SLS), SpLine (ESRF-CRG) and XAFS (ELETTRA) beamlines for the beamtime allocated and the scientific support received.

References

[1] Grunwaldt J-D and Baiker A, 2005 *Phys. Chem. Chem. Phys.* **7** 3526 and references therein.

[2] Tinnemans S J, Mesu J G, Kervinen K, Visser T, Nijhuis T A, Beale A M, Keller D E, van der Eerden A M J and Weckhuysen B M, 2006 *Catalysis Today* **113** 6 and references therein.

[3] Hannemann S, Casapu M, Grunwaldt J-D, Haider P, Trüssel P, Baiker A and Welter E, 2007 *J. Synchrotron Rad.* **14** 345.

[4] Rosciano F, Holzapfel M, Kaiser H, Scheifele H W, Ruch P, Hahn M, Kotz R and Novak P, 2007 *J. Synchrotron Rad.* **14** 487.

[5] Franch-Martí C, Alonso-Escobar C, Jorda J L, Peral I, Hernández-Fenollosa J, Corma A, Palomares A E, Rey F and Guilera G, accepted for 2012 publication in *Journal of Catalysis*.

Colloidally prepared platinum nanoparticles deposited on iron oxide studied by XAFS

E Piskorska-Hommel[1], D Arndt[2], T Wilkens[1], J Falta[1], and M Bäumer[2]

[1]Institute of Solid State Physics, University of Bremen, Otto-Hahn-Allee NW1, 28359 Bremen, Germany
[2]Institute of Applied and Physical Chemistry, University of Bremen, Leobener Str. NW2, 28359 Bremen, Germany

E-mail: e.piskorska@ifp.uni-bremen.de

Abstract. Colloidal chemistry is an attractive method for the preparation of metal nanoparticles with well-defined shape and size rendering. This preparation technique is particularly interesting for the fabrication of heterogeneous catalysts, where size and shape play a key role for the catalytic properties. Nevertheless, the metal nanoparticles synthesized by colloidal approaches are often surrounded by a shell of organic ligands stabilizing the particles during the growth process. The influence of the ligands on the interaction with the support and the catalytic reaction was studied by x-ray absorption spectroscopy (XAS). It was found that the activity of the metallic Pt nanoparticles previously protected by organic ligands is higher in CO reaction than for unprotected nanoparticles deposited on iron oxide.

1. Introduction

The catalytic properties of the metal nanoparticles depend on their size, shape and – if present – on the nature of a surrounding ligand or polymer shell [1]. Heterogeneous catalysts in the form of supported nanoparticles are traditionally prepared by wet-impregnation or chemical precipitation methods. These methods often offer a limited control over the size of the noble metal catalyst which additionally depends on the nature of the supporting material. Generally, the nanoparticles are deposited on supports, such as metal oxides, zeolites or carbon. However, due to undesired metal-support interactions during deposition the support can change or diminish catalytic activity of the noble metal catalyst. Here, the use of preparation methods allowing to prepare uniform nanoparticles and to control the metal-support interactions is advantageous. This is the case for the colloidal synthesis of the metal catalysts [2]. However, a problem could arise from the fact that the prepared nanoparticles are often stabilized by organic ligands added during the synthesis. On the one hand they can e.g. stabilize the particles against aggregation and agglomeration [3]; on the other hand they could have detrimental effects on the catalytic activity in the case of heterogeneous gas-phase catalysis. Therefore, the usual approach is to remove the ligands prior to the catalytic reaction.

[1] To whom any correspondence should be addressed.

On the contrary, our group and others have shown that the performance of an iron supported Pt catalyst can be enhanced with respect to CO-oxidation by the use of such organic stabilizers [2, 4]. The enhanced effect can be attributed to the ligand-induced blocking of a first, strong chemical interaction that occurs during the deposition of the colloidal platinum nanoparticles onto the iron oxide support. In order to achieve comprehensive knowledge about this class of materials, X-ray absorption spectroscopy (XAS) was applied. Specifically, a ligand stabilized platinum iron oxide catalyst prepared by a one pot method in diethylene glycol was studied.

2. Experimental

For the preparation of the catalyst, first nanocrystalline iron oxide was prepared by the so-called polyol method in diethylene glycol, i.e., the decomposition of iron acetylacetonate at 200°C for 2 hours [5]. After a cool down period, platinum acetylacetonate was added to the solution and stirred for 30 minutes. For the formation of the platinum nanoparticles in the presence of iron oxide, the solution was heated up for 2 hours at 180°C again. At the end of the synthesis, the catalyst was precipitated by acetone and dried under air. Under these conditions, residual diethylene glycol (DEG) will stay adsorbed on the surface of the particles thus acting as ligands. These ligand-capped Pt nanoparticles will be called P1 in the following. After using them for CO oxidation (called activated as the ligands desorb during the process, as discussed in the following), the particles, called P2, were studied by x-ray absorption spectroscopy (XAS). The x-ray absorption fine structure (XAFS) measurements were performed at beamline C at the Hasylab, DESY in Hamburg, Germany. The beam was monochromatized by a Si(311) monochromator. The samples were pressed into pellets and fixed onto the sample holder. The XAFS spectra were recorded in fluorescence mode using the 7 cell SDD detector.

XAFS spectra were processed according standard steps by applying Athena software; the structural parameters were obtained using the program Artemis [6]. The scattering paths Pt-Pt and Pt-Fe for metal Pt used for fitting of the EXAFS data were generated by the FEFF6 code [7]. The EXAFS oscillation in the k range 4–11 $Å^{-1}$ were Fourier transformed (FT's), using a Kaiser- Bessel window. The fitting of one coordination shell was performed in R space in the range 1.3- 3.3 Å.

The size of the metal Pt nanoparticles was determined by transmission electron microscopy (TEM) for sample P1 and P2, respectively. For the CO oxidation experiments, a suitable laboratory reactor (fixed-bed type reactor with a heating system) was used. Precautions were taken to avoid mass-transport limitations and electrostatic charging. The powder catalysts were pressed and sieved into 0.31-0.45 mm grains, from which 10 mg were then mixed with quartz (800 mg, diameter 0.4-0.8 mm, Roth, calcinated at 1000 °C for 6 h) and placed between quartz wool into the reactor. The measurements were performed under continuous flow conditions with 3 vol.-% CO in synthetic air (total flow: 50 mL/min) unless otherwise noted. The components in the product stream were analysed with a photometric detector for CO/CO2 (URAS 3G, Hartmann & Braun).

3. Results and Discussion

The XANES spectra collected at the L_3 edge for metal Pt nanoparticles supported on Fe_3O_4 and Pt foil are depicted in Figure 1. The spectra acquired for the nanoparticles show features characteristic for metallic Pt. A double peak structure at the 11.578 keV and 11.590 keV after the white line are present in both studied samples. The intensity of the white line of the sample P2 is higher than for metallic Pt and sample P1, indicating the existence some oxidized Pt.

On the contrary, the EXAFS data for sample P2 could be well fitted assuming the presence of only metallic Pt. The determined Pt-Pt bond lengths are shorter by about 0.03 Å as compared to bulk Pt, 2.74 Å and 2.77 Å respectively, Table 1. The best fit for sample P1 was obtained assuming the presence Pt-Pt and Pt-Fe distances in the model. The small inclusion of Fe was found to be in the range of ~ 8% and could be the result of an interaction with the support. The Pt-Pt distance was found to be 2.72 Å in comparison to 2.77 Å for bulk Pt. The Pt-Fe bond lengths were also shorter than for bulk fcc PtFe, namely 2.68 Å as compared to 2.72 Å for bulk FePt. The coordination numbers 12 found for both samples are in agreement with coordination number for metallic Pt nanoparticles and compatible with an average nanoparticles size of 5 nm [8]. The shorter Pt-Pt distances observed for sample P1 in comparison to bulk Pt could be the result of the particle size. The adsorption of residual carbon monoxide on the sample P2 after catalysis might result in bond relaxation thereby expanding the particles [9].

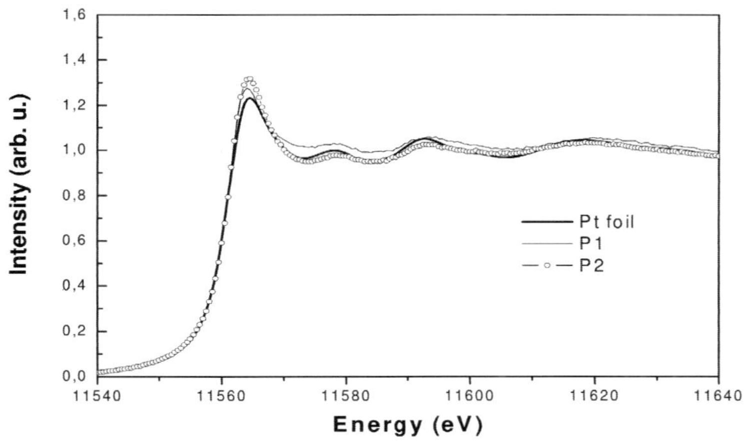

Figure 1. Pt L_3 edge XANES spectra for Pt/Fe_3O_4 ligand-capped (DEG) (thin line) (P1) and activated (circle) (P2) in comparison with reference Pt foil.

Table 4. The numerical results of the fitting EXAFS oscillation recorded for investigated Pt/Fe_3O_4 ligand-capped ((P1) and activated (P2), respectively.

	Pair	*R (Å)*	σ^2 (Å^2)	*N*
P1	Pt-Pt	2.72(2)	0.010	11
	Pt-Fe	2.68(1)	0.003	1
P2	Pt-Pt	2.74(1)	0.009	12
Pt foil	Pt-Pt	2.77(2)	0.005	12

In order to study the influence of organic ligands on the catalytic properties, CO oxidation was performed for all catalysts. Figure 2 shows the catalytic performance of unprotected 2 nm Pt nanoparticles (preparation: see ref. [2]), 2 nm particles protected with dodecylamine (DDA) (another type of ligand: for details see ref. 2) and the new DEG covered platinum particles 3 and 5 nm size described in this paper. The activity of the unprotected Pt nanoparticles is comparable to the activity of nanoparticles on irreducible supports, such as γ–Al_2O_3 [10]. On the other hand, if DDA capped platinum nanoparticles are deposited on the iron oxide support, the catalysts shows high catalytic activity after desorption/decomposition of the ligands above 160°C. If the catalytic reaction is repeated in a second cycle, activity is also observed below 160°C. Even at 100°C, appreciable conversion rates are detected. A comparable situation, however, with even higher activities in the second cycle of the catalytic CO-oxidation, can be reached by the use of the DEG covered platinum-iron oxide catalysts. After the activation of the catalyst in the first cycle at a temperature of 170°C - which can be assigned to desorption/decomposition temperature of DEG -, the catalysts show a 20% higher conversion rate in the low temperature region (<160°C), compared to the DDA-capped platinum nanoparticles on iron oxide. As concluded from the infrared spectra acquired for the DEG covered particles, the adsorption of linear CO on platinum (not shown here) lead to a band at 2044 cm^{-1}, which indicates metallic state at the surface of the platinum nanoparticles. However, by applying reaction conditions, a shift of this band to lower wavelength (2035 cm^{-1}) was observed which might be attributable to an incorporation of iron atoms into the platinum lattice [11].

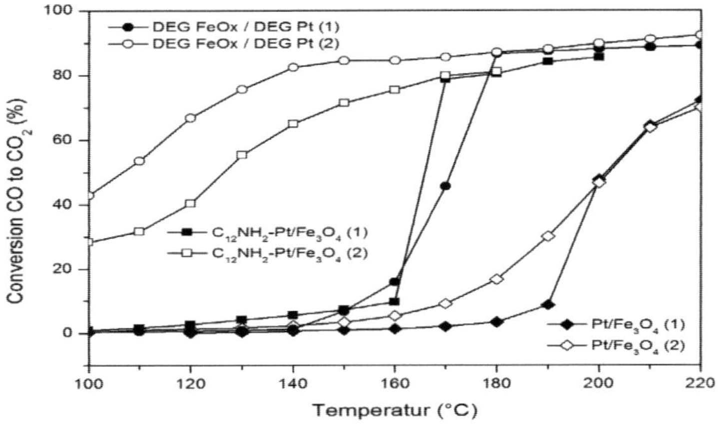

Figure 2. Catalytic activity for CO oxidation of the unprotected, DDA- and DEG-capped platinum nanoparticles supported on iron oxide.

The XANES data after catalysis show a small indication of oxidized Pt, while the EXAFS data show no differences to the situation before catalysis (metallic Pt for sample P2). This behaviour is in marked contrast to smaller unprotected and DDA capped particles. As detailed in ref. 2, they get largely (unprotected) or partially (ligand-capped) oxidized. Nevertheless, the small (3 nm) and large (5 nm) particles show the same behaviour under catalytic conditions. As was concluded for the small particles, an SMSI effect with the iron oxide, coming into action after desorption of the ligands in the first catalytic run, gives rise for the remarkable low temperature catalytic activity in the second and following runs. This SMSI effect can – independent of the particle size – only be achieved by using a ligand assisted deposition technique. Further experiments are under way to elucidate the exact mechanism.

4. Conclusion

The XAFS data prove that Pt nanoparticles colloidally prepared by an one pot method in DEG and in the presence of iron oxide, have a strong metallic character. The DEG solvent and simultaneously ligand solution protect the formed metal Pt nanoparticles as well as the iron oxide nanoparticles against a strong primer interaction, previously observed for smaller particles. Nevertheless, the ligands can be removed by activation at temperatures of 170°C. In the following the particles are active for CO oxidation at temperatures significantly lower than observed for conventional Pt catalysts. This activity can be ascribed to an SMSI effect between Pt and the iron oxide being only effective if the particles are protected by the ligands while being deposited or created in the presence on the support.

Acknowledgments

This work was performed within the Func-band project supported by Senator of science and education of the free hanseatic city of Bremen. The authors would like to thank Hasylab, DESY, Hamburg for beamtime at the beamline C and Roman Chernikov for support during the experiment.

References

[1] Burch R and Loader P K 1994 *Appl. Catal. B* **5** 149
[2] P. Sonström, D. Arndt, X. Wang, V. Zielasek and M. Bäumer 2011 *Angewandte Chemie International Edition* 50 3888
[3] Sun Y-N, Giordano L, Goniakowski J, Lewandowski M, Qin Z-H, Noguera C, Shaikhutdinov S, Pacchioni G and Freund H-J 2010 *Angewandte Chemie International Edition*, 49 4418
[4] Liu L, et al., 2010 Journal of Catalysis **274**(1) 1
[5] Wang X, Sonström P, Arndt D, Stöver J, Zielasek V, Borchert H, Thiel K, Al-Shamery K and Bäumer M 2011 *Journal of Catalysis* 278 143
[6] Ravel B and Newville M 2005 J. Synch. Rad. 12 537
[7] Zabinsky S I, Rehr J J, Ankudinov A, Albers R C, and Eller M J 1995 Phys.Rev. B 52 2995
[8] Frenkel A I, Hills C W and Nuzzo R G 2001 J. Phys. Chem. B 105 12689
[9] Alayon E M C, *et al* 2009 *J. Phys.: Conf. Ser.* **190** 012152
[10] Fenske D, et al., 2010 Chemcatchem 198
[11] Wadayama T, et al. 2008 Journal of Physics: Conference Series **100**(1) 012007

Journal of Physics: Conference Series **430** (2013) 012059 doi:10.1088/1742-6596/430/1/012059

Exploring the local structure of Fe in $Co_{3-x}Fe_xO_4$ electrode by XAFS

W R Cheng†, J F He†, Y Wu, Y H Peng, Q H Liu*, J H Huang and S Q Wei*

National Synchrotron Radiation Laboratory, University of Science and Technology of China, Hefei, Anhui 230029, P.R. China

*Corresponding authors. E-mail: qhliu@ustc.edu.cn, sqwei@ustc.edu.cn
†These authors contributed equally to this work.

Abstract. Determining the local structure of dopant atoms is critical for understanding the mechanism of enhanced photoelectrochemical activity in doping-modified semiconductor photoelectrodes. Here, we have synthesized Fe-doped Co_3O_4 nanowire arrays photoanodes on Ti foil and detected the local structure and role of Fe atoms via X-ray absorption fine structure (XAFS) spectra and electrochemical impedance measurements. The XAFS analysis shows that Fe dopants are doped into Co_3O_4 and occupy the octahedral sites. Coupling with Co atoms at octahedral sites, Fe dopants significantly increases the carrier density by an order of magnitude and results in a five-fold increase of the photoconversion efficiency. These results provide some guidance for optimization of Co_3O_4 as a practical photoanode material.

1. Introduction

Efficient production of hydrogen from water and sunlight is a strategy to meet the energy demand for all over the world [1]. To reach high conversion efficiency in the water splitting process, photoanode materials need to satisfy the essential requirements: ample solar light absorption, facile photo-carrier transport and catalytic oxygen evolution reaction (OER) [2]. Co_3O_4 is an ideal candidate photoanode material due to a small band-gap of ~1.6 eV [3] and its catalytic effect for water oxidation [4]. However, the photoconversion efficiency of Co_3O_4 is limited by the poor conductivity.

For a long time, there is hardly any report about photo response property of Co_3O_4. Recently, Woodhouse et al reported that Fe-doped Co_3O_4 is deemed to be useful for photoelectrolysis of water due to well photo-response property [5]. To clarify the role of Fe dopants, the structural and electronic properties of the Co-Fe-O ternary spinel oxides system were studied using density-functional theory calculation [6]. However, since there is no report about occupation sites of Fe atoms, the calculated results via assuming Fe atoms occupation of tetrahedral or/and octahedral sites are disorderly and shallow. It is necessary to clarify the sites of Fe dopants in Fe-doepd Co_3O_4 in order to figure out the reason of enhanced photocurrent.

In this work, vertical orient Fe-doped nanowire arrays were synthesized on Ti foil by hydrothermal method. Co_3O_4 nanowire arrays were used as the matrix due to the large surface area of nanowire which was benefit for Fe deposition and doping [7]. It is found that Fe dopants successfully doped into lattice of Co_3O_4 by substitution for Co of octahedral sites. After Fe doping, the conductivity of Fe-doped sample significantly increase.

2. Experimental

Co_3O_4 nanowire arrays were prepared according to the recently reported hydrothermal method [8]. For Fe doping, pristine sample was dipped into the same reaction solution as prior work [9], maintaining at 120 °C for 30 min. The precursor was marked as Fe-deposited sample and it was annealed at 600 °C in air for 4 hours before becoming Fe-doped sample. The Co and Fe k-edges X-ray absorption fine structure (XAFS) spectroscopy were recorded in the fluorescence mode at U7 C beamline in the National Synchrotron Radiation Laboratory (NSRL).

3. Results and discussion

To study the performance of Fe-doped sample, IPCE and photocurrent measures were carried out. The light absorption of pristine sample obviously decreases at around 800 nm presented in Fig 1a, corresponding to the band-gap of Co_3O_4 of ~1.6 eV [10]. It is of high interest that uniformly high IPCE values is observed for Fe-doped sample though the slight blue shift of absorption edge, with the value of ~10% until 450 nm and remaining ~1 % at 650 nm at 0.1 V vs Ag/AgCl as shown in Fig 1c. In contrast, the IPCEs of pristine and Fe-deposited samples in the whole region are all under 3% and sharply decreases as the irradiation wavelength increases, tailing off at around 600 nm. The IPCEs of Fe-doped sample is much higher than that of Fe-deposited sample, indicating there is some structural change after second annealing. As a matter of course, the increased IPCEs of Fe-doped sample result in a strong raise of photocurrent to 400 A/cm^2 at 0.1 V vs Ag/AgCl (see Fig 1b) , corresponding to a photocurrent conversion efficiency of 0.37% presented in Fig 1d, which is near five-fold higher than that of pristine and Fe-deposited samples.

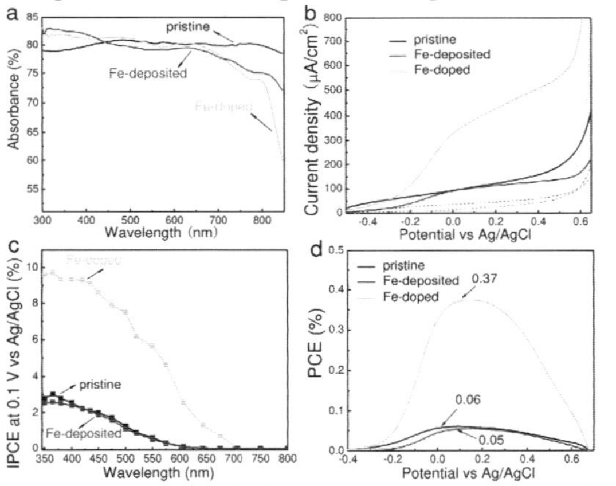

Figure 1. (a) UV-vis absorption spectra, (b) Current density, (c) IPCE, and (d) PCE of pristine, Fe-deposited, and Fe-doped samples.

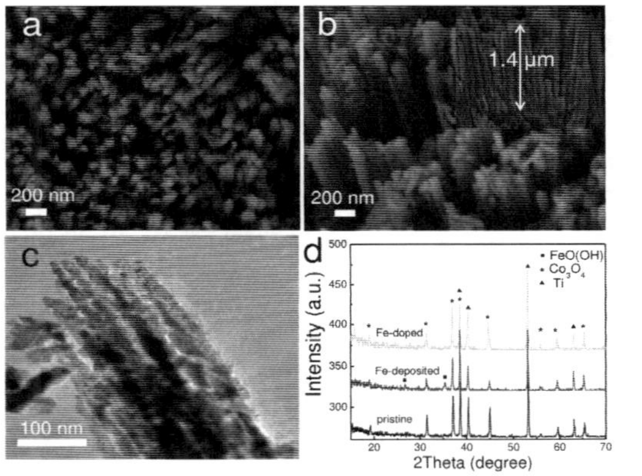

Figure 2. Top view FESEM imagine of Fe-deposited (a) and Fe-doped (b) sample (inset of Fig b exhibits corresponding cross-section imagine). (c) shows TEM imagine of a bunch of nanowires. (d) displays the XRD pattern of pristine, Fe-deposited and Fe-doped sample.

The structural details of the samples are exhibited to find out the origin of the enhanced photocurrent by Fe doping. FESEM and TEM imagines demonstrate standing bunches of nanowires with diameter of ~160 nm, length of ~1.4 μm and each bunch consists of about ten nanowires. As shown in Fig 2a, there are lots of flower-like clusters of ferric compound with size of ~200 nm homogeneously paving on the top of Fe-deposited sample and these ferric compounds disappear after second annealing (see Fig 2b). The XRD patterns of samples presented in Fig 2c suggest all diffraction peaks of pristine sample could be well indexed to crystalline Co_3O_4 (JCPDS 73-1701). But for Fe-deposited sample, there are two new peaks at 2θ of 26.6° and 35.5° , corresponding to (310) and (211) faces of FeO(OH) [11]. However, no peak corresponding to Fe_2O_3 or FeO(OH) is observed in XRD pattern of Fe-doped sample, indicating Fe atoms have doped into the lattice of Co_3O_4 as well.

To clarify the atomic structure of Fe dopants, X-ray absorption fine structure (XAFS) technique was used as a sensitive local structure probe [12]. The extended-XAFS (EXAFS) $k^3\chi(k)$ functions and their Fourier transforms (FTs) at Co and Fe K-edge of Fe-doped and/or pristine samples are shown in Fig 3a,b along with that of Fe_2O_3 and Fe_3O_4 as references. Obviously, the existence of Fe_2O_3 and Fe3O4 can be safely excluded since theirs EXAFS spectral of Fe K-edge in region of 5-11 $Å^{-1}$ are significantly different from that of Fe-doped sample as well as the corresponding FT curves. Similarly, Fe atoms are likely to dope into the lattice of Co_3O_4 by substitution of Co sites, since their EXAFS spectral and corresponding FTs curves are similar to each other. In spinel structure of Co_3O_4, there are two different Co sites: tetrahedral (tet) and octahedral (oct) and the ratio of tet:oct is 1:2. For Co of tetrahedral sites, there are 4 nearest Co-O bonds and 12 next-nearest Co-Co bonds at distances around 1.9 Å and 3.3 Å, respectively. On the contrary, for Co of octahedral sites, the coordination numbers for the nearest Co-O bonds, next-nearest and third-nearest Co-Co bonds is all 6 with corresponding distances around 1.9Å, 2.8Å and 3.3Å, respectively. Considering the theory ratio of tet:oct of 1:2, the theory calculated coordination numbers of the nearest Co-O bond, next-nearest and third-nearest Co-Co bonds are 5.3, 6 and 8 respectively. The fitting results summarized in table 1 show that at the central atoms of Co, the bond lengths of nearest Co-O bond, next-nearest and third-nearest Co-Co bonds are 1.93 Å, 2.85 Å and 3.34 Å with corresponding coordination numbers of 5.3, 6.0 and 7.8 respectively. These fitted parameters are closed to the theory values mentioned above, suggesting the ratio of tet:oct in Fe-doped sample is near 1:2. For the central atoms of Fe, the bond lengths corresponding to nearest Fe-O bond, next-nearest Fe-Co and third-nearest Fe-Co bonds are fitted to be 1.97 Å, 2.87 Å and 3.41 Å with coordination numbers of 5.5, 6 and 5.6 respectively, all a little larger than that of the central Co atom [6]. The fitted coordination numbers indicate that Fe atoms favor the substitution of octahedral sites with proportion of at least 90 percent as shown in Fig 3b (empty circle). According to walsh's theory calculation work, the octahedral Fe sites favor coupling with Co of octahedral sites, resulting in boarder conduction band which increase the conductivity of Fe-doped Co_3O_4 [6]. The enhanced conductivity is useful for carriers and photo-carriers transfer.

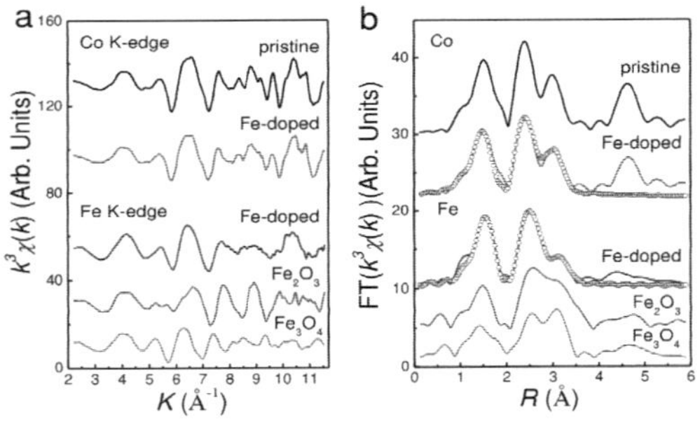

Figure 3. (a) oscillation function $k^3\chi$ (k) at the Co K-edge and Fe K-edge of pristine, Fe-doped, Fe_2O_3, Fe_3O_4 and (b) their FTs curves (the empty circles show the fitting results).

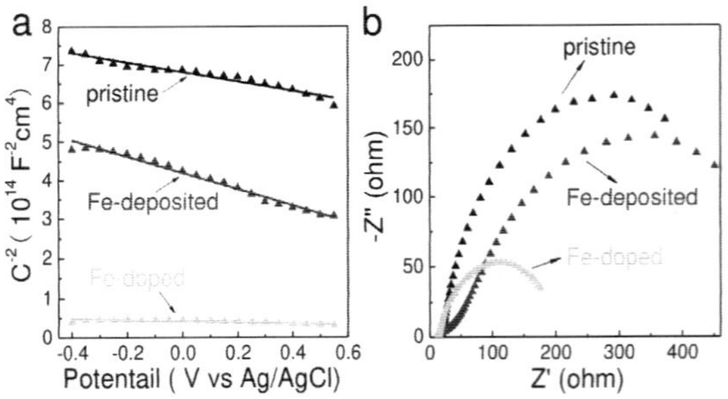

Figure 4. (a) Mott-Schottky plots and (b) Nyquist plots of pristine, Fe-deposited, and Fe-doped samples.

The conjecture of increased conductivity of Fe-doped sample can be confirmed by the results of electrochemical impedance measurement (Mott-Schottky) shown in Figure 4a,b, which was used to determine the capacitance and impedance of the samples. Carrier density can be calculated from the slope of Mott–Schottky plots using the equation: $N_d = (2/e_0\varepsilon\varepsilon_0)\ [d(1/C^2)/dV]^{-1}$ Where ε of Co_3O_4 is 12.9 [10]. Three samples all show p-type behavior according to the negative slopes shown in Fig 4a. The calculated carrier density of pristine sample is 8.8×10^{16} cm^{-3}, consistent with other reported results [13], which is closed to that of Fe-deposited sample (5.2×10^{16} cm^{-3}). Whereas, that of Fe-doped sample is calculated as 8.6×10^{17} cm^{-3}, an order of magnitude larger than that of pristine one. The enhanced carrier density, which is linearly proportional to conductance according to the formula: $\sigma = ne_0\mu$, leads to lower impedance for Fe-doped sample as shown in Nyquist plots (see Figure 3b), in which circle radius is positive correlation with impedance. Thus, the coupling between the octahedral sites Co and Fe is benefit for enhanced photocurrent. These results may provide some guidance for the future design and optimization of Co_3O_4 as photoelectrode.

Table 1. The fitted structure parameters around Co and Fe atoms in Fe-doped sample

Central atoms	Path	R (Å)	N	σ^2 (Å^2)
Co	Co-O	1.93 ± 0.01	5.3 ± 0.5	0.005 ± 0.0002
	Co-Co	2.85 ± 0.01	6.0 ± 0.6	0.007 ± 0.0008
	Co-Co	3.34 ± 0.02	7.8 ± 0.8	0.008 ± 0.0009
Fe	Fe-O	1.97 ± 0.01	5.5 ± 0.3	0.006 ± 0.0007
	Fe-Co	2.87 ± 0.02	6.0 ± 0.5	0.008 ± 0.0007
	Fe-Co	3.41 ± 0.02	5.6 ± 0.6	0.009 ± 0.0008

4. Conclusions

In summary, we have synthesized the vertical orient Fe-doped Co_3O_4 nanowire arrays on Ti foil by hydrothermal. SEM and XRD results show clusters of FeO(OH) disappear after second annealing, indicating Fe atoms dope into the matrix. The XAFS analysis further confirm the Fe atoms have substituted for Co of octahedral sites in the lattice of Co_3O_4, resulting in a five-fold increase of photoelectric conversion efficiency of 0.37%. By electrochemical impedance analysis, the enhanced photocurrent is attributed to increased carrier density and decreased transfer barrier by Fe doping. These results may provide some guides for improved Co_3O_4 as a photoanode.

Acknowledgments

This work was supported by the National Natural Science Foundation of China (Grant Nos. 11135008, 11105151, 10979044, and 10979041) and Knowledge Innovative Program of The Chinese Academy of Sciences (KJCX2-YW-N40).

References

[1] Michael W and Parkinson B A 2009 *Chem. Soc. Rev.* **38** 197
[2] Michael G 2001 *Nature* **414** 338
[3] Varghese B, Mukherjee B, Karthik K R G, Jinesh K B, Mhaisalkar S G, Tok E S and Sow C H 2012 *J. Appl. Phys.* **111** 104306
[4] Liang Y Y, Li Y G, Wang H L, Zhou J G, Wang J, Regier T and Dai H J 2011 *Nat. Mater.* **10** 780
[5] Woodhouse M and Parkinson B A 2008 *Chem. Mater.* **20** 2495
[6] Walsh A, Wei S H, Yan Y F, Al-Jassim M M, Turner J A, Woodhouse M and Parkinson B A 2007 *Phys. Rev. B* **76**,165119
[7] Li Y G, Hasin P and Wu Y Y 2010 *Adv. Mater.* **22** 1926
[8] Xia X H, Tu J P, Zhang Y Q, Mai Y J, Wang X L, Gu C D and Zhao X B 2012 *RSC Adv.* **2** 1835
[9] Vayssieres L, Beermann N, Lindquist S E and Hagfeldt A 2001 *Chem. Mater.* **13** 233
[10] Cheng C S, Serizawa M, Sakata H and Hirayama T 1998 *Mater. Chem. Phys.* **53** 225
[11] Deliyanni E A, Bakoyannakis D N, Zouboulis A I, Matis K A and Nalbandian L 2001 *Microporous Mesoporous Mater.* **42** 49
[12] Yao T, Zhang X D, Sun Z H, Liu S J, Huang Y Y, Xie Y, Wu C Z, Yuan X, Zhang WQ, Wu Z Y, Pan G Q, Hu F C, Wu L H, Liu Q H and Wei S Q 2010 *Phys. Rev. Lett.* **105** 226405
[13] Thota S, Kumar A and Kumar J 2009 *Mater. Sci. Eng. B* **164** 30

Study of formation process of metal nanoparticles on metal oxides by *in-situ* XAFS technique

T Shishido[1,2], H Nasu[1], LDeng[1], K Teramrua[1,2,3] and T Tanaka[1,2]

[1]Department of Molecular Engineering, Graduate School of Engineering, Kyoto University, Kyoto, Japan

[2]Elements Strategy Initiative for Catalysts and Batteries, Kyoto University, Kyoto, Japan

[3]Precursory Research for Embryonic Science and Technology (PRESTO), Japan Science and Technology Agency (JST), Saitama, Japan

Email: shishido@moleng.kyoto-u.ac.jp , tanakat@moleng.kyoto-u.ac.jp

Abstract. The formation process of platinum nanoparticles (PtNPs) on metal oxide (SiO_2) was investigated by various characterization techniques, especially, *in situ* time-resolved Pt-L_3 edge XAFS (QXAFS and DXAFS) combined with Q-Mass spectroscopy. A series of XAFS spectra was analyzed to evaluate the change in local structure, electronic state, and dispersion of PtNPs during their formation and growth process.

1. Introduction

Metal nanoparticles (NPs) are getting much more attention than ever in both fields of fundamental science and engineering. Metal NPs have several unique properties or applications, such as surface plasmon resonance, magnetism, imaging, or catalysis [1]. Various researchers focus their efforts on controlling NP size and/or shape. Their formation mechanismsin solution are also vigorously investigated, but still under discussion. Recently, we proposed the formation process of AuNPs in the presence of the equimolar of dodecanethiol to $HAuCl_4$ using *in situ* millisecond time-resolved QXAFS measurement [2]. We also investigate the formation process of RhNPs from rhodium trichloridetrihydrate ($RhCl_33H_2O$) in ethylene glycol with polyvinylpyrrolidone (PVP) at elevated temperature [3].A combination of *in situ* time-resolved XAFS and ICP(Inductively Coupled Plasma)-MS(Mass) techniques reveals that the formation process is a first-order reaction, which indicates that RhNPswith a uniform sizeappear consecutively and these Rh NPs do not aggregate with each other.

。。On the other hand, it is well known that metal nanoparticles (NPs)supported on metal oxide (solid surface) are effectivein various catalytic reactions. The interaction betweenmetal oxide surface and metal NPs is considered to have played an important effecton controlling shape (morphology), dispersion (size), and catalytic activityof metal NP. Although the states(local environment and electronic structure) of metal NPs on various metal oxides have been studied by using XAFS technique,the detailed mechanism about the formation of metalNPs on metal oxides surface remains unclear.In the present study, we have tried to investigate the dynamic formation process of PtNPson metal oxide surface (SiO_2).For this purpose, we applied *in-situ*time resolved XAFS (Quick XAFS and Energy-dispersive XAFS (DXAFS)) combined with Q-mass.

2. Experimental

○○ Platinum was loaded on SiO_2 (home made, 622 m^2 g^{-1})by an impregnation with an aqueous solution of $Pt(NH_3)_4(NO_3)_2$at 353 K. The impregnated sample was dried at 353 K in the air. The content of platinum was 2 wt.%.

Pt L_3-edge XAFS data (QXAFS) were collected at a facility of the BL01B1 station of SPring-8 of Japan Synchrotron Radiation Research Institute (8 GeV, 100 mA), Hyogo, Japan. The spectrum was recorded in a transmittance mode at room temperature, using a Si(111) two-crystal monochromator. The photon energy was calibrated by Pt foil.The wafer of the sample diluted with BN was mounted in an *in-situ* flow cell. DXAFS measurements were recorded at BL28B2 beam line and the spectra were collected using a bent Si(111) polychromator and a position sensitive CCD detector at room temperature.The data acquisition consisted from 100 spectra with a time resolution of 55ms (5.5 s/spectrum). The exhaust gases during XAFS measurements were analyzed by Q-mass spectrometer (Pfeiffer Omni star). The data reduction was performed by the REX2000 Ver.2.5.9 program (Rigaku) and FEFF8.40.

3. Results and Discussion

3.1. Formation process of Pt NPs on SiO_2in He

Figure 1 showsa series of *in situ* time-resolved Pt L_3-edge XANES spectra (QXAFS) of 2.0 wt% Pt/SiO_2in He flowwith increasing temperature and the reference spectra ofPt foil. Figure 2 shows the change in the white line intensity around 11565 eV and Q-mass spectra against the temperature.The intensity of white line slightly increased from 460 to 500 K, then remarkably decreased.The intensity of white line at 700 K is almost similar to that of Ptfoil, indicating thereduction of Ptcationspecies to Pt^0during the He treatment.The shift of X-ray absorption edge energy (E_0) to lower energy regionalso indicates a reduction of Pt^{2+}cation to Pt^0 species. Q-mass spectra clearly indicate that the formation of H_2O (18), N_2 (28) around 520 K as shown in Fig.2. NH_3 (m/e=16), NO (30), and N_2O (44) were also observed (data not shown). The formation of these species was observed up to 560 K. These results indicate that the decomposition of ligands in $Pt(NH_3)_4(NO_3)_2$ took place above 460 K and the ligands was completely decomposed up to 560 K.

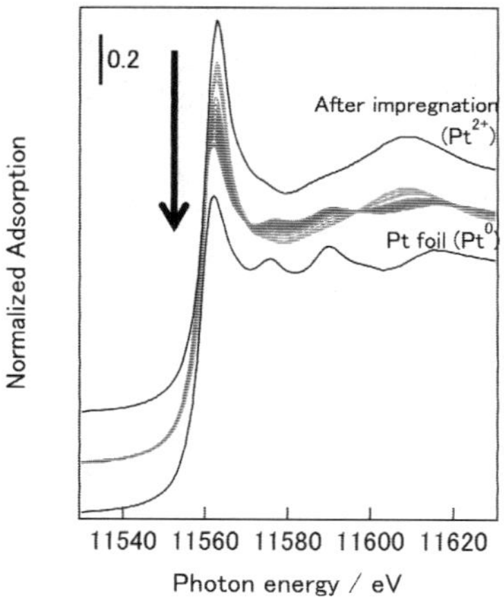

Figure 1: Pt L_3-edge XANES spectra of 2.0 wt% Pt/SiO_2in He (RT to 773 K)

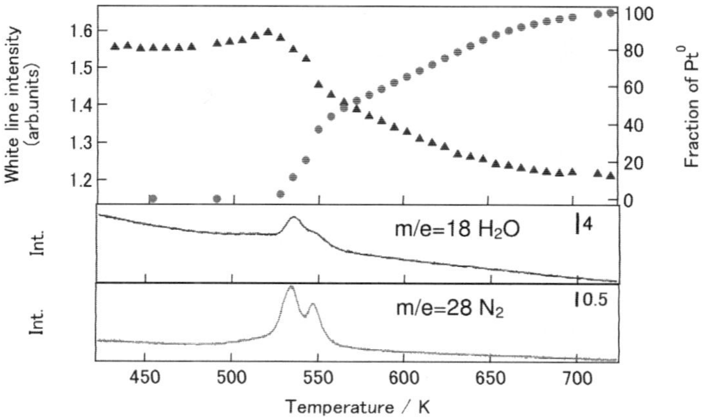

Figure 2: Change in the white line intensity of Pt L_3-edge XANES spectra and Q-mass spectra against the temperature

Figure 3showsk^3-weighted EXAFS spectra and their Fourier transforms (FTs) ofofPt/SiO_2 sample treated with He (phase shift was not corrected). The oscillation of EXAFS spectrum of Pt/SiO_2remarkably changed above 520K. The peak at 1.8Å can be assigned to the contribution of the

Pt-N (or Pt-O, *vide infra*) shell. The second shell due to Pt-Pt was observed above 542 K,although the intensity of Pt-Pt shell (2.5-3.0 Å) was weak due to thermal vibration of a lattice of Pt. This suggests that highly dispersed Ptcationspecies as amonomer is present on Pt/SiO_2, and that a part of Ptcationspecies was reduced to Pt^0 to form PtNPs. The change in the coordination number (CN) of Pt-N (or Pt-O) and Pt-Pt shell was shown in Fig. 4.

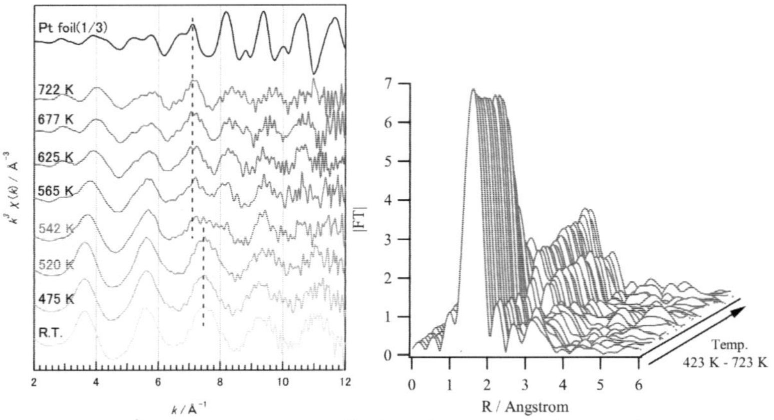

Figure 3: k^3-weighted EXAFS oscillation of 2.0 wt% Pt/SiO_2 and their Fourier transforms. (in He, RT to 773 K)

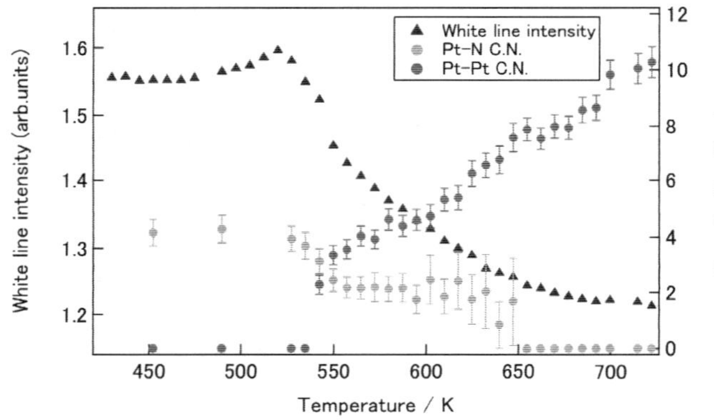

Figure 4: Change in the coordination number (CN) of Pt-N (or Pt-O) and Pt-Pt shell of 2.0 wt% Pt/SiO_2 (in He, RT to 773 K)

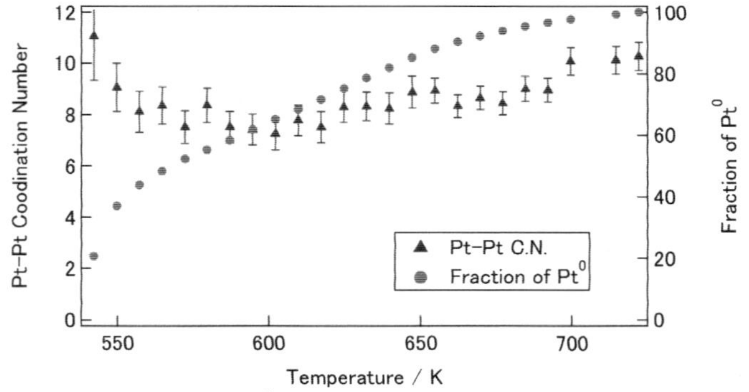

Figure 5: Fractions of $Pt^0$and the average coordination number (CN) of Pt-Pt bond in Pt NPs. Fraction of Pt0 was estimated from the linear combination fitting of XANES spectra of the initialstate (Pt^{2+}; before heating) and Pt foil. The average CN of Pt-Pt bond: the CN of Pt-Pt bond x fraction of Pt^0/100.

Above 520 K, the CN of Pt-N decreased with increase in temperature and reached at ca.2 even after the complete decomposition of ligands in $Pt(NH_3)_4(NO_3)_2$ (> 560 K). Based on Q-mass spectra, above 560 K, nonitrogen was present on Pt/SiO_2. This suggests that the formation of Pt-O bond (oxygen was lattice oxygen of SiO_2 support; Pt-O-Si bond) proceeds simultaneously with the decomposition of ligands in $Pt(NH_3)_4(NO_3)_2$. On the other hand, the CN of Pt-Pt bond rapidly increased at 540 K and then linearly increased with the temperature. This increasing in CN of Pt-Ptbonds suggests the growth of Pt NPs during He treatment at a high temperature. However, based on TEM images, the average particle size of PtNPs on Pt/SiO_2 treated with He at 623 K was almost similar to that treated at 723 K (1.7-0.4 nm at 623 K and 1.7ₒ-0.6 nm at 723 K), indicating that the growth of Pt NPs did not proceed in He (at least, up to 723 K).

All Pt L_3-edge XANES spectra were reproduced with the linear combination fitting of XANES spectra of the initial state (Pt^{2+}; before heating) and Pt foil. The fraction of Pt^0estimated from the coefficient of the liner combination fitting increased with temperature (Fig. 5). Moreover, the average CN of Pt-Pt of PtNPs, estimated from the degree of fraction of Pt^0 (= the degree of reduction of platinum species), was almost constantup to 773 K, suggesting that the particle size of PtNPswas independent on temperature, in other words, PtNPswith a uniform size appear consecutively and these PtNPs do not aggregate with each other with increase in temperature.

。。Based on these results, the formation process of PtNPs on SiO_2under He flow was proposed as follows. First, platinum precursorwas decomposed. Then, the formation of intermediate species, which have Pt-O bond (CN of Pt-O is 2; Si-O-Pt-O-Si), takes place. Next, the intermediate species were reduced to form Pt^0 nuclear. Finally, the Pt^0 nuclear aggregated to form PtNPs having uniform particle size.

3.2. Formation process of Pt NPs on SiO_2in H_2

The change in the white line intensity of Pt L_3-edge XANES spectra ofPt/SiO_2 and Q-mass spectra against the temperature in H_2 flow was measured (data not shown).The intensity of white line rapidly decreasedfrom 420 K, which is about 40 K lower than in He. The formation of H_2O was also observed around 420 K. In order to reveal the initial formation process of PtNPson SiO_2in H_2, DXAFS technique was applied. Figure 6 shows the time course of the intensity of white line, the CNs of Pt-Pt and Pt-N (or Pt-O) shells, and Q-Mass spectra at 473 K in H_2 flow. After the introduction of H_2, the intensity of the white line immediately decreased within 40 s. In response to changes in the intensity of white line, the CN of Pt-Pt bond increased while the CN of Pt-N decreased. In this stage, no formation of NH_3 was observed, although a quite small amount of N_2 and N_2O were formed. These results suggest that the following processes take place at the initial stage of the formation of PtNPs on SiO_2 in H_2; 1) $Pt(NH_3)_4(NO_3)_2$ react with silanols groups ($Pt(NH_3)_4(NO_3)_2 + 2SiOH \rightarrow Pt(NH_3)_4(OSi)_2 + 2HNO_3$), 2) a part of precursor ($Pt(NH_3)_4(OSi)_2$) was decomposed to form PtNPs, and the formed NH_3 react with HNO_3 ($Pt(NH_3)_4(OSi)_2 + 2HNO_3 + H_2 \rightarrow Pt^0 + 4NH_4NO_3 + 2SiOH$), 3) a rest of precursor ($Pt(NH_3)_4(OSi)_2$) and NH_4NO_3were decomposed to form PtO and N_2 (O_2 and H_2O), respectively ($Pt(NH_3)_4(OSi)_2 \rightarrow PtO + 4NH_3 +$ Si-O-Si, $NH_4NO_3 \rightarrow 2N_2 + 4H_2O + O_2$).

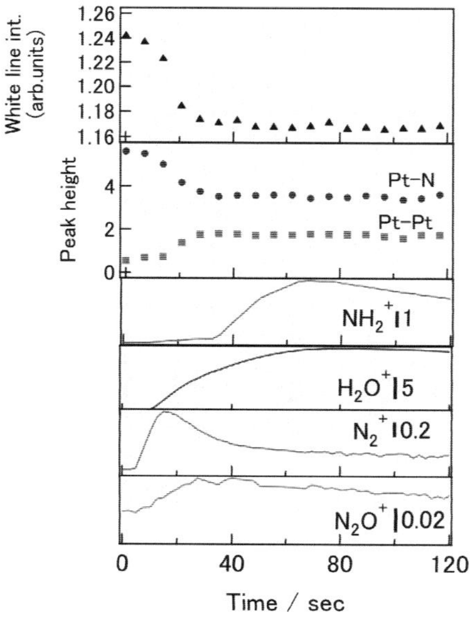

Figure 6: Intensity of white line of Pt L_3-edge XANES (DXAFS), the CN of Pt-PtandPt-N shells, and Q-Mass spectra. (at 473 K, in 20 vol.% H_2/He)

4. Conclusion

A combination of *in situ* time-resolved XAFS and Q-Mass techniques reveals the following formation process of PtNPs on SiO_2 in He and H_2. The reduction much more easily proceeds in the case of H_2. Under the He atmosphere, PtNPs with a uniform particle size formed through the reduction of intermediate species (Si-O-Pt-O-Si). While in the case of H_2, a part of PtNPs formed by the direct decomposition of $Pt(NH_3)_4(OSi)_2$, and other part of PtNPs formed form the reduction of PtO.

Acknowledgement

The X-ray absorption experiments have been performed under the approval of the JASRI (Proposal No. 2011A1173, 2011B1403 and 2011B1393).

References

[1] YanN,Xiao,C,KouY,*Coord. Chem. Rev.*, 2010, **254**, 1179.
[2] Ohyama J, Teramura K, Higuchi Y, Shishido T, Hitomi Y, Kato K, Tanida H, Uruga T, Tanaka T,*ChemPhysChem*, 2011, **12**, 127.
[3] Asakura H, Teramura K,Shishido T, Tanaka T, Yan N, Xiao C, Yao S,Kou Y, *Phys. Chem. Chem. Phys.*, 2012, **14**, 2983.

Journal of Physics: Conference Series **430** (2013) 012061 doi:10.1088/1742-6596/430/1/012061

EXAFS Model of 2-Dimensional Platinum Clusters

S D Kelly, M E Charochak, N Blackwell and S R Bare

UOP LLC, a Honeywell Company, Des Plaines, IL 60017, USA

E-mail: Shelly.Kelly@Honeywell.com

Abstract. Highly dispersed supported platinum clusters have been extensively studied because of their many uses as the active metallic component in heterogeneous catalysts, including catalytic reforming. EXAFS modelling has been used to determine the average coordination number for the first through fifth Pt shells and thereby determine the average cluster size and shape. Several studies have shown that Pt clusters on γ-Al_2O_3 are truncated to form plate-like clusters in the hexagonal (111) plane as evidenced by an overly reduced second shell coordination number, which would be zero for a truly 2-dimensional cluster. In our system of Pt on γ-Al_2O_3 the average Pt morphology is determined, on the average, to be small ~13 atom clusters that are 2-diminsional in the hexagonal (111) plane morphology as determined by first and, nearly zero, second nearest neighbour coordination numbers of 4.4 ± 0.6 and 0.5 ± 0.3, respectively.

1. Introduction

The diverse applications of the catalytic properties of supported Pt clusters have resulted in extensive studies to elucidate the relationship between Pt cluster structure and catalytic performance. Several studies have reported Pt EXAFS modelling up to the 5th shell from catalysts with Pt loading of 1-2wt% [1-3]. Munoz-Paez et al. [1] found that depending on the details of oxidation/reduction that fully reduced Pt clusters have a 1st and 2nd CN of 7.2-10.1 ± 0.3 and 2.0-2.8 ± 0.3, respectively, and characterized these clusters as spherical. In the study by Giovanetti et al. [3] the increased flatness of Pt clusters were identified by a decrease in the first nearest-neighbor (NN) coordination number (CN) and an increase in average cluster diameter determined using SAXS. In this study the first and second CN of the spherical to flattened clusters were reported to change from 9.3 ± 0.5 to 7.2 ± 0.7 and from 4 ± 1 to 4 ± 2, respectively, illustrating the clusters are quite large with significant number of layers in the <110> direction (2nd CN > 0). Cuenya et al. [2] described the EXAFS results for a ~22 atom cluster of two atomic layers prepared by the inverse micelle method. Cuenya et al. [2] showed Pt clusters can be truncated to form plate-like clusters in the hexagonal (111) plane as evidenced by an overly reduced second shell coordination number, which would be zero for a truly 2-dimensional cluster. In this paper, we describe the EXAFS modelling of 2-dimensional ~13 atom (<1nm diameter) hydrogen-saturated Pt clusters supported on γ-Al_2O_3.

2. Methods

Chloroplatinic acid was impregnated onto γ-Al_2O_3 with a surface area of 162 m^2/g. The Pt loading was determined using ICP to be 0.4wt%. The catalyst was calcined at 525°C prior to the *in situ* XAFS experiments and then reduced *in situ* to a temperature of 300°C at a ramp rate of 10°C/min in flow of 100% H_2 using a custom designed XAFS cell [4]. After in situ reduction, Pt EXAFS spectra were collected at room temperature in a flow of 100% H_2 and are assumed to represent Pt clusters saturated with hydrogen. The sample is referred to as 300°C-reduced. EXAFS spectra were collected in fluorescence mode at the MR-CAT beamline 10ID at Argonne National Laboratory.

Table 1. EXAFS paths and parameterization

	Bulk Values			EXAFS Parameterization[a]		
Path	Description	CN	Reff (Å)	CN	ΔR	σ^2
Pt0-O	Support O	6	2.5	**NO**	**delRO**	**sigO**
Pt0-Pt1	1st Shell	12	2.77	**NPt1**	α·Reff	Debye(**Td**, Temp)
Pt0-Pt2	2nd Shell	6	3.92	**NPt2**	α·Reff	Debye(Td, Temp)
Pt0-Pt1b-Pt1a	Triangle	48	4.26	NPt1·4	α·Reff	Debye(Td, Temp)
Pt0-Pt3	3rd Shell	24	4.80	**NPt3**	α·Reff	Debye(Td, Temp)
Pt0-Pt1a-Pt1b	Triangle	48	5.17	NPt1·4	α·Reff	Debye(Td, Temp)
Pt0-Pt3-Pt1	Triangle	96	5.17	NPt3·4	α·Reff	Debye(Td, Temp)
Pt0-Pt4	4th Shell	12	5.54	**NPt4**	**α4**·Reff	Debye(**Td4**, Temp)
Pt0-Pt4-Pt1	Collinear MS	24	5.54	NPt4·2	α4·Reff	Debye(Td4, Temp)
Pt0-Pt1-Pt0-Pt1	Collinear MS	12	5.54	NPt1	α·Reff	Debye(Td, Temp)
Pt0-Pt1-Pt4-Pt1	Collinear MS	12	5.54	NPt4	α4·Reff	Debye(Td4, Temp)
Pt0-Pt1-Pt3-Pt1	Triangle	48	5.54	NPt3·2	α·Reff	Debye(Td, Temp)
Pt0-Pt5	5th Shell	24	6.09	**NPt5**	α·Reff	Debye(Td, Temp)

[a]One energy shift parameter for all paths was also determined.

The EXAFS spectra were modeled using FEFFIT[5] and FEFF7.0[6] for the theoretical calculations based on structure of bulk fcc Pt. The EXAFS models include single and multiple scattering paths[7,8] from Pt and also a long Pt-support O signal[9]. Each of the 5 shells of Pt atoms about the absorbing (Pt0) atom is denoted with the nomenclature of Pt1, Pt2, Pt3, Pt4 and Pt5. These paths and their parameterization are listed in Table 1. There are 13 parameters used to describe the model: 6 coordination numbers (CN), 3 change in path length (ΔR), 2 Debye Temperatures (Td), one σ^2, and an energy shift parameter (ΔE). The ΔR values are described in terms of an expansion/contraction term α which is a constant factor multiplied by the reference path length (Reff). The mean square displacement values (σ^2) are described in terms of a Debye model with characteristic Debye temperature Td. The value for S_0^2 (0.93 ± 0.05) was determined from Pt foil. The data range from 3.5 to 10 $Å^{-1}$ was used in the Fourier transform with k-weight of 1, 2 and 3. The initial model was applied to the Fourier transformed (FT) range of 1.2 to 6 Å, containing 24 independent points and 13 parameters. Further tests were performed to determine if each of the Pt shells required independent α and Td values. Comparisons of the reduced-$\chi 2$ values were made over the data range dominated by the higher shells from 3.2 to 6.0 Å containing 15 independent points and 7 parameters.

3. Results and Discussion

The Pt L_3-edge magnitude and real part of the EXAFS spectra are shown in Figure 1 for the 300°C-reduced and Pt foil. The EXAFS spectrum from the *in situ* reduced catalyst has significantly smaller amplitude compared to Pt foil. The real part of the FT shows that the signal at 2Å differs from Pt foil and the modeling confirms that it is dominated by Pt0-O from the Al_2O_3 support. The dramatic difference in the catalyst spectrum as compared to the Pt foil spectrum is a result of the nanometer-sized Pt clusters in the former.

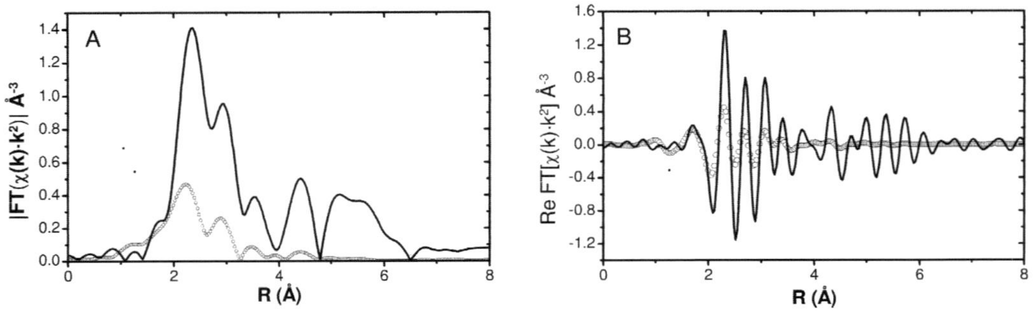

Figure 1. Magnitude (A) and real part of Fourier transform of 300C-reduced (symbols) and Pt foil (solid)

Five test models allowed each of the 5 shells in turn to have independent α and Td parameters but only independent values for the 4th shell were found to have a significant difference in the ability of the model to reproduce the measured spectrum. The reduced- χ^2values are compared over the data range where the 3^{rth} through 5th shells contribute to the data between 3.2 and 6.0 Å. The measure of the quality of the model, the reduced- $\chi^2$for the best-fit model with unique values for the 4th Pt shell is 2 times smaller as compared to the test model with only one α and Td, indicating that independent 4th shell parameters are supported by the measured data. The magnitude and real part of the 300°C-reduced spectrum and the best-fit model are shown in Figure 2. The EXAFS results are listed in Table 2 for CN, ΔR, and σ^2. The best-fit values for α, α4, Td, and Td4 are -0.018 ± 0.001, -0.026 ± 0.004, and 198 ± 13K, and 174 ± 55K, respectively.

Table 1. EXAFS modelling results. Only values for single scattering paths are reported.[a]

Path	CN	R (Å)	σ^2 (x10^{-3} Å^2)
Pt0-O	1.5 ± 0.3	2.497 ± 0.015	8 ± 2
Pt0-Pt1	4.4 ± 0.6	2.721 ± 0.003	7 ± 1
Pt0-Pt2	0.5 ± 0.3	3.848 ± 0.005	9 ± 1
Pt0-Pt3	1.3 ± 0.8	4.713 ± 0.006	9 ± 1
Pt0-Pt4	1.1 ± 0.8	5.395 ± 0.023	12 ± 8
Pt0-Pt5	1.6 ± 1.4	6.085 ± 0.007	10 ± 1

[a]Energy shift parameter was determined to be -0.4 ± 0.5 eV

The EXAFS CN can be used to determine the average cluster geometry [Frenkel 1999]. The nearly zero value for NPt2, and the very small value for NPt3 suggests that the cluster conforms to a single layer in the hexagonal Pt (111) plane. A schematic of such a cluster shown in the inset of Figure 2 and is consistent with the EXAFS determined CNs. This cluster has NPt1, NPt2, NPt3, NPt4, and NPt5 values of 3.7, 0, 2.8, 2.3, and 1.9 respectively. The EXAFS model indicates that on the average each Pt atom in the cluster is bound to at least one oxygen atom of the support (NO =1.5). This is also consistent with a single layer of Pt atoms bound to the alumina surface.

In this paper we have describe the details of the EXAFS model used to accurately reproduce the measured Pt spectrum from reduced hydrogen saturated 0.4wt% Pt on γ-Al_2O_3. The small first shell coordination number of 4.4 ± 0.6 and the nearly zero 2nd shell CN are consistent with a 2-dimensional 13 atom cluster on the γ-Al_2O_3 support.

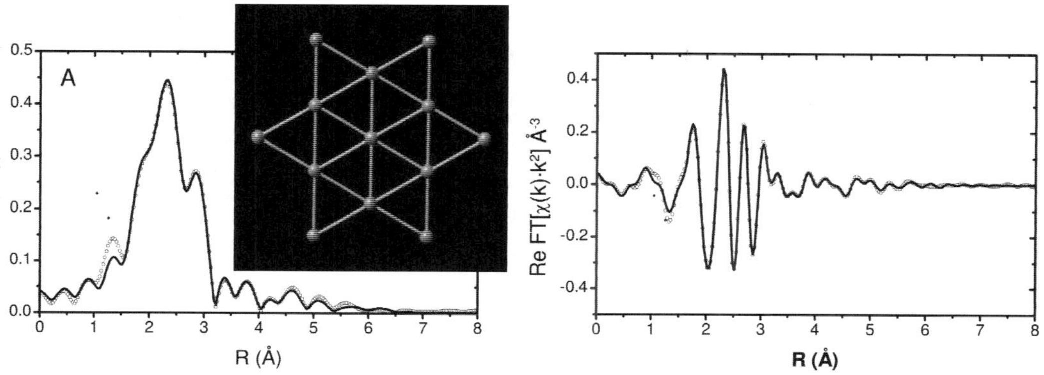

Figure 2. Magnitude (A) and real part of Fourier transform of 300C-reduced (symbols) and model (solid). The inset shows a 13 atom Pt cluster that is consistent with the EXAFS data.

References

[1] Munozpaez A and Koningsberger D C 1995 Decomposition of the Precursor [Pt(Nh3)(3)](Oh)(2), Genesis and Structure of the Metal-Support Interface of Alumina-Supported Platinum Particles - a Structural Study Using Tpr, Ms, and Xafs Spectroscopy *J Phys Chem-Us* **99** 4193-4204

[2] Cuenya B R, Ortigoza M A, Ono L K, Behafarid F, Mostafa S, Croy J R, Paredis K, Shafai G, Rahman T S, Li L, Zhang Z and Yang J C 2011 Thermodynamic properties of Pt nanoparticles: Size, shape, support, and adsorbate effects *Phys Rev B* **84** 245438

[3] Giovanetti L J, Ramallo-Lopez J M, Foxe M, Jones L C, Koebel M M, Somorjai G A, Craievich a F, Salmeron M B and Requejo F G 2012 Shape Changes of Pt Nanoparticles Induced by Deposition on Mesoporous Silica *Small* **8** 468-473

[4] Bare S R, Mickelson G E, Modica F S, Ringwelski a Z and Yang N 2006 Simple flow through reaction cells for in situ transmission and fluorescence x-ray-absorption spectroscopy of heterogeneous catalysts *Rev Sci Instrum* **77** 023105

[5] Newville M 2001 IFEFFIT: Interactive EXAFS analysis and FEFF fitting *J. Synchrot. Radiat.* **8** 322-324

[6] Rehr J J, Deleon J M, Zabinsky S I and Albers R C 1991 Theoretical X-Ray Absorption Fine-Structure Standards *J. Am. Chem. Soc.* **113** 5135-5140

[7] Frenkel A I 1999 Solving the structure of nanoparticles by multiple-scattering EXAFS analysis *J Synchrotron Radiat* **6** 293-295

[8] Witkowska A, Di Cicco A and Principi E 2007 Local ordering of nanostructured Pt probed by multiple-scattering XAFS *Phys Rev B* **76** 104110

[9] Vaarkamp M, Miller J T, Modica F S and Koningsberger D C 1996 On the relation between particle morphology, structure of the metal-support interface, and catalytic properties of Pt/gamma-Al2O3 *J Catal* **163** 294-305

Journal of Physics: Conference Series **430** (2013) 012062 doi:10.1088/1742-6596/430/1/012062

Application Ce L_1 HERFD XAS to determine the atomic structure of CeO_2 based nano-catalysts under working conditions

N Smolentsev[1], A A Guda[1,*], O V Safonova[2*], M A Soldatov[1], C Paun[3], G Smolentsev[2], J A van Bokhoven[2,3] and A V Soldatov[1]
[1] Research center for nanoscale structure of matter, Southern Federal University, Rostov-on-Don, 344090, Russia
[2] Paul Scherrer Institute, Villigen, 5232, Switzerland
[3] ETH Zurich, Zurich, 8093, Switzerland
*Corresponding authors, email: guda_sasha@mail.ru, olga.safonova@psi.ch

Abstract. Ce L_1-edge high-energy-resolution fluorescence detected (HERFD) X-ray absorption spectroscopy (XAS) was applied to unravel the electronic structure of Pt-promoted CeO_2 nanoparticles under different redox conditions. The Ce $L\gamma_3$ emission line was chosen for partial fluorescence yield detection to improve the XAS spectral resolution. Oxygen vacancies were formed on the surface of ceria nanoparticles upon reducing in 5% CO in He. Theoretical Ce L_1-edge simulations showed good agreement with the experimental data.

1. Introduction

Cerium dioxide is a unique material that can reversibly store oxygen in its structure and provide it for catalytic reactions under reducing conditions. The catalytic properties of CeO_2-based materials are strongly related to the surface morphology and the presence of Ce-substituting dopants [1,2]. Nano-sized CeO_2 is of special interest because of the large surface area and the high concentration of unsaturated bonds. However its relation between structure of Ce on the atomic scale and reactivity still remains unclear. Here, we report the oxidation state and local environment of Ce atoms in CeO_2 nanoparticles measured under different redox conditions (in CO and O_2 atmospheres) using *in situ* Ce L_1 HERFD XAS and compare these to the theoretical simulations. We chose the Ce L_1 edge instead of the more commonly investigated Ce L_3 edge because it is not affected by multi-electronic transitions and thus can be simulated within the conventional one-electron transition approximation. To improve the energy reolsution, we measured HERFD XAS [3,4].

2. Methods

The polyhedral CeO_2 nanoparticles (NPs) of averaged size of 9 nm were prepared by a hydrothermal method that is described in [5]. Platinum nanoparticles (1.5 wt%) with the size varying between 1 and 2 nm were deposited using incipient wetness impregnation method using platinum nitrite solution, calcined in air at 400°C and reduced in 5% H_2 at 350°C.

Transmission XAS experiments were performed at a laboratory Rigaku R-XAS absorption spectrometer at Southern Federal University using a Ge 311 monochromator. Ce L_1 HERFD XAS measurements were also performed at the SuperXAS beamline (Swiss Light Source, PSI, Switzerland) using a double-crystal Si (111) incident beam monochromator and multi-crystal Johann-type emission spectrometer [6]. The spectrometer was equipped with three Si(111) crystals with a bending radius of 1000 mm, and which operated at the 333 reflection tuned to the maximum of the Ce $L\gamma_3$ emission line (N_3 to L_1 transition).

Journal of Physics: Conference Series **430** (2013) 012062 doi:10.1088/1742-6596/430/1/012062

Ce L_I-edge spectra were simulated using the full-potential linear augmented plane-wave approximation (FLAPW) implemented in the Wien2k program package [7]. The generalized gradient approximation in the scheme of Perdew–Burke–Ernzerhof (GGA-PBE) [8] was used for the treatment of exchange-correlation effects. The structures of CeO_2 and oxygen deficient Ce_7O_{12} phases were used for simulations.

3. Results and discussion

The natural width of Ce L_I and N_3 levels are 4.2 and 1.6 eV respectively [9,10]. Therefore, HERFD XAS at the Ce $L\gamma_3$ (L_I-N_3) emission line results in a better energy resolution compared to measuring XAS in transmission mode. The estimated total HERFD broadening [11], assuming a crystal analyzer resolution 2 eV and monochromator resolution 0.6 eV, is 2.7 eV at Ce $L\gamma_3$. Figure 1 (left panel) compares Ce L_I HERFD with transmission XAS. Features B and C become sharper using $L\gamma_3$ fluorescence detection. Since Ce L_I HERFD XAS spectra are rich in features they provide enhanced sensitivity to the local atomic and electronic structure of Ce atoms in the presence of oxygen vacancies. The right panel shows the simulated Ce L_I spectrum in CeO_2 using the FLAPW approximation. The simulations were performed both in the ground state and by constructing a supercell and a core hole on Ce 2s level. The ground state approximation reproduced well the experimental data; introducing a core hole did not change the calculated spectrum significantly.

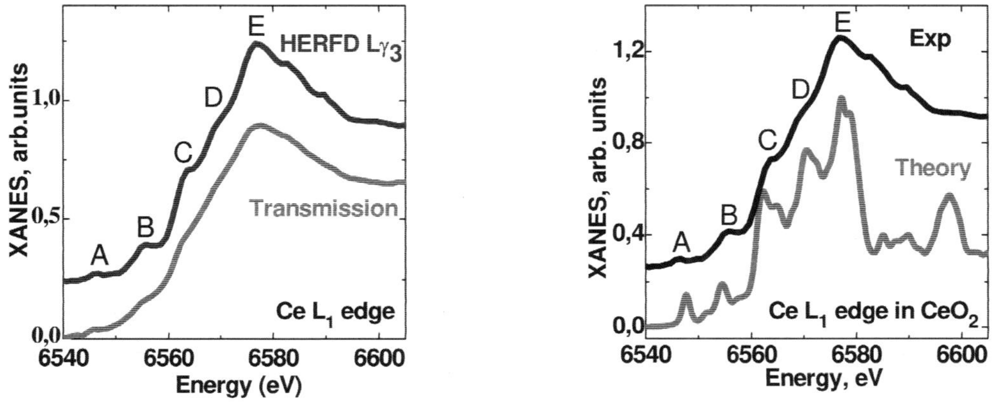

Figure 1. (Left) Ce L_I XAS of bulk CeO_2 detected using transmission mode (red curve), HERFD using Ce $L\gamma_3$ emission line (blue curve). (Right) Comparison of experimental (black curve) and simulated (red curve) spectra of CeO_2.

Figure 2 shows the effect of oxygen vacancies formation on the Ce L_I HERFD XAS spectra of Pt-CeO_2 nano-catalysts during interaction with CO at 150°C. During this process both the oxidation state and the local atomic structure of Ce atoms changed. A Ce_7O_{12} structure [12] was used as model a CeO_2 structure with high concentration of oxygen vacancies. Two non-equivalent crystallographic positions of Ce in Ce_7O_{12} can be considered as Ce atoms with respectively one and two oxygen vacancies in the first coordination sphere. The experimentally observed variations of the spectral shape, mainly visible in peak C, were reproduced. The feature at the lower energy is reminiscent of reduced cerium.

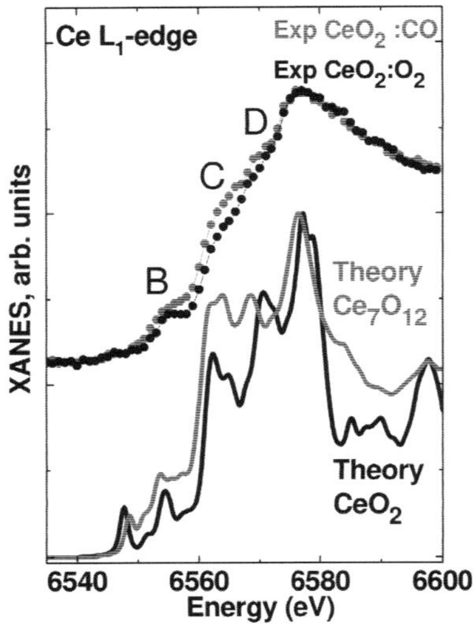

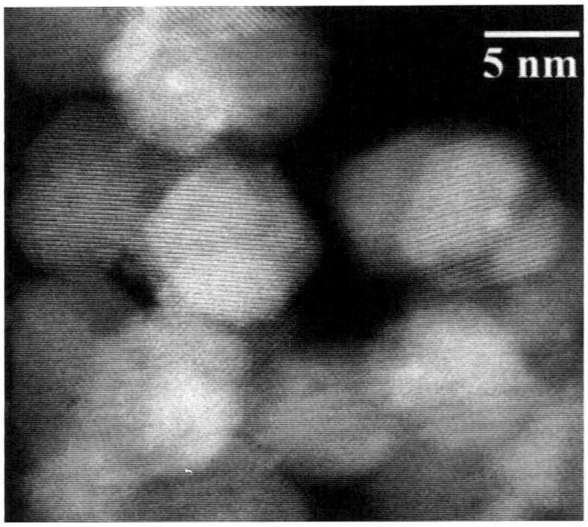

Figure 2. (Left): Experimental Ce L_1 HERFD XAS spectra of 1.5%Pt/CeO_2 polyhedral nanoparticles under 4% O_2 (black circles) respectively 5% CO at 150°C (red circles). Bottom: calculated X-ray XAS spectra of CeO_2 (black line) respectively Ce_7O_{12} (red line). (Right) STEM image of 1.5%Pt-CeO_2 catalyst.

4. Conclusions

Ce L_1 HERFD XAS enables distinguishing the oxidation state of cerium. The Ce local atomic structure in well-defined 1.5%Pt-CeO_2 ceria catalysts was determined under different redox conditions. A clear change in the Ce L_1 near-edge was observed while cycling the atmosphere between 4% O_2 and 5% CO. Unlike the Ce L_3 edge, the L_1 HERFD XAS can be easily reproduced using readily available theoretical codes within the one-electron approximation. The spectral features observed experimentally were successfully compared to theoretical simulations of stoicheometric CeO_2 and an oxygen-deficient Ce_7O_{12} structure. Ce L_1 HERFD XAS provides a unique opportunity to determine the structure of oxygen deficient ceria.

Acknowledgements

We thank the UGINFO computer center of the Southern Federal University, Russia for providing computing time. The work was partially supported by the Ministry of Education and Science of the Russian Federation, state contract № 11.519.11.2039. We thank the SLS for the beam time at the SuperXAS beamline.

References

[1] Trovarelli A 2002 *Catalysis by ceria and related materials* ed G J Hutchings, Catalytic Science Series, Imperial College Press: London.
[2] Si R and Flytzani-Stephanopoulos M, 2008 *Angew. Chem. Int. Ed.* **47** 2884.
[3] Hämäläinen K, Siddons D P, Hastings J B, Berman L E 1991 *Phys. Rev. Lett.* **67** 2850.
[4] Safonova O V, Tromp M, van Bokhoven J A, de Groot F M F, Evans F, Glatzel P 2006 J. Phys. Chem B. **110** 16162-16164
[5] Paun C, Safonova O V, Szlachetko J, Abdala P, Nachtegaal M, Sa J, Kleymenov E, Cervellino A, Krumeich F, van Bokhoven J A 2012 *J. Phys. Chem. C* **116** 7312-7317

[6] Kleymenov E, van Bokhoven JA, David C, Glatzel P, Janousch M, Alonso-Mori R, Studer M, Willimann M, Bergamaschi A, Henrich B, Nachtegaal M 2011 *Rev. Sci. Instrum.* **82** 065107.
[7] Schwarz K and Blaha P 2003 *Comput. Mater. Sci.* **28** 259.
[8] Perdew J P, Burke S, Ernzerhof M 1996 *Phys. Rev. Lett.* **77** 3865.
[9] Fuggle J C and Inglesfield J E 1992 *Topics in Applied Physics* Vol. **69** ("Unoccupied Electronic States") edited by J.C. Fuggle, J.E. Inglesfield (Springer, Berlin, 1992)
[10] Campbell J L, Rapp T 2001 *Atomic Data and Nuclear Data Tables* **77** 1.
[11] Glatzel P, Weng T, Kvashnina K, Swarbrick J, Sikora M, Gallo E, Smolentsev N, Kvashnin Y, Kavcic M, Alonso Mori R 2012 *Journal Electronic Spectroscopy and Related Phenomena*, submitted.
[12] Kuemmerle E A, Heger G 1999 *Journal of Solid State Chemistry* **147** 485.

Journal of Physics: Conference Series **430** (2013) 012063 doi:10.1088/1742-6596/430/1/012063

XSW-XAFS characterization of ion-irradiated *Pt/Ni/C* multilayer

Nitya Ramanan[1], Debdutta Lahiri [1,*], Sumalay Roy[2,†], Surinder M Sharma[1] and **B N Dev[2]**
[1]High Pressure and Synchrotron Radiation Physics Division, Bhabha Atomic Research Centre, Mumbai 400085, India
[2] Department of Materials Science, Indian Association for the Cultivation of Science, Jadavpur, Kolkata-700032, India
*E-mail: debduttalahiri@yahoo.com

Abstract. We employ XSW assisted XAFS at *Ni K* edge to characterize ion-irradiated *Pt/Ni/C* multilayer, particularly the nanoclusters formed within *C* layer, revealed by X-TEM and angle-resolved fluorescence studies. Retrieving the structural model from XAFS coordination results involved intriguing steps such as accounting for the intensity variation across the layers, determination of extra pre-diffused *Ni* into *C* layer (beyond the reflectivity determined roughness) and decoupling interfacial, layer and cluster coordination. The clusters are determined to be *Ni* centered *Ni-Pt* bimetallic nanoclusters (*Ni:Pt* = 60:40), formed due to irradiation induced diffusion of *Pt* atoms from the disrupted *Pt/C* interface towards pre-diffused *Ni* atoms present in *C* layer. They are highly disordered beyond nearest neighbor and resemble glassy structure which could find wide-scale applications in magnetic devices.

1. Introduction

In this work, we employ XAFS in conjunction with X-ray Standing Wave (XSW)[1-2] to characterize structural changes in *Pt/Ni/C* multilayer (ML), following irradiation with 2MeV Au^{2+} of fluence 2×10^{15} ions/cm^2[3]. The purpose of ion irradiation was simulation of possible high energy x-ray induced defects. XTEM[3] and angle resolved fluorescence studies at 14 keV (Figure 1(c-d)) on the ion-irradiated ML revealed completely mixed *Ni-Pt* layers and nano-cluster (of diameter ~ 1.7 nm) formation *C* layer). Such nanoclusters on C matrix could find application as patterned ultra-high density recording media [4] if their composition is ferromagnetic, conductive ink for fabrication of electronic circuit elements [5], if bimetallic etc. Ion irradiation method, being dry, is particularly useful in fabricating clusters for electronic devices where wetting is an issue.

XAFS study is objected at addressing questions pertaining to nature, composition and degree of order of these nanoclusters. Our results would establish whether *Ni* and *Pt* would form isolated clusters or core-shell particles or alloy in *C* matrix. These implications cannot be drawn simply from bulk *Ni-Pt* phase diagrams. Factors such as surface energetics [6], *Ni/C* and *Pt/C* interactions would play an important role.

XAFS was employed with XSW–the latter results from the superposition of incident and Bragg diffracted x-rays from the ML. The antinode of XSW is centered successfully in each layer of the ML by scanning across the Bragg angle [2]. Hence, XSW selectively filters out information from different

† Present Address: Max Planck Institut fuer Microstructurephysik, Weinberg 2, 06120 Halle, Germany.

locations of the ML. The grazing angle of incidence is set such that antinodes and nodes of XSW are respectively located at the centre of the *C* layer and *Ni/Pt* interface (Figure 1(e)) and rising edge of the antinodes coincides with *Ni/C* interfaces. Therefore, fluorescence mode *Ni K* edge XAFS yields structural information of *Ni/C* interface and around *Ni* diffused deeper into *C* layer. In irradiated ML, antinodes and nodes of XSW are respectively located at centre of C layer and centre of *Ni+Pt* mixed layer (Figure 1(f)). Therefore, fluorescence mode *Ni K* edge XAFS yileds structural information of *Ni/C* interface, *C* layer and *Ni+Pt* mixed layer.

2. Experimental Details

Fluorescence mode XAFS at *Ni K* edge (8.333 keV), in conjunction with XSW, was recorded on (i) Pristine *Pt/Ni/C* ML ($d_{Pt+Ni+C}$ ~70Å ×15 layers) on float glass substrate (deposited by ion beam sputtering), and (ii) irradiated *Pt/Ni/C* ML (with 2MeV Au^{2+} ions at a fluence of 2×10^{15} ions/cm^2, at ROEMO (E2) beamline (HASYLAB, Germany). XSW was used to selectively filter out information from *Ni/C* interface, as is also evident from Figure 1(b). To maintain the antinodal-nodal positions during energy scan of XAFS experiments, the angle of incidence was appropriately adjusted continuously. The fluorescence photons were collected using a *Si (Li)* detector.

XAFS data were processed using ATHENA and the structural parameters (coordination number (N), Debye Waller Factor (DWF) and bond length (R)) were retrieved using FEFF6 and FEFFIT.

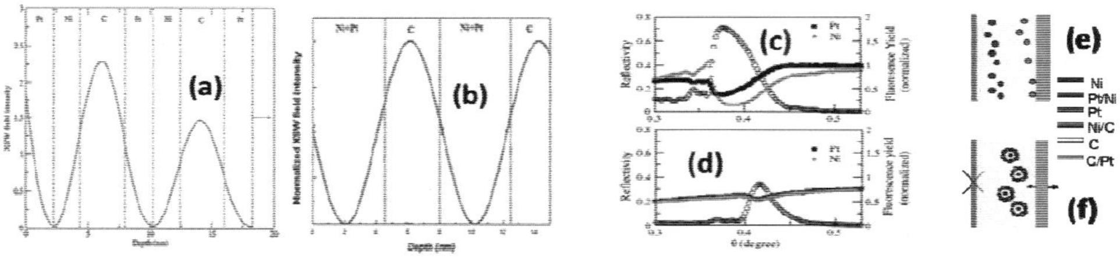

Figure 1. Positions of antinodes and nodes of the XSW (a) before and (b) after irradiation; Fluorescence results (c) before and (d) after irradiation; Structural model for C layer and its interfaces (e) before and (f) after irradiation.

3. Results and discussion

EXAFS data for the first coordination shell (R = 1.2-3.4Å in Figure 2. (b)) around *Ni* were analyzed. k^1-k^2 weighted Fourier Transforms for each individual dataset (transformed over k=2.2-8Å^{-1}) were simultaneously fit to increase the degrees of freedom and reduce uncertainties in N, DWF and R. Various structural models and several fitting strategies were considered to derive unique conclusion. We considered several strategies for fitting, viz. including (or excluding) third cumulant in disorder, correcting for background contribution, constraining the structure around *Ni* to be fcc, or the coordination number for pristine and irradiated MLs to be same. Figure 2. (c) shows an example of fit quality (*R*-factor = 0.001-0.014).

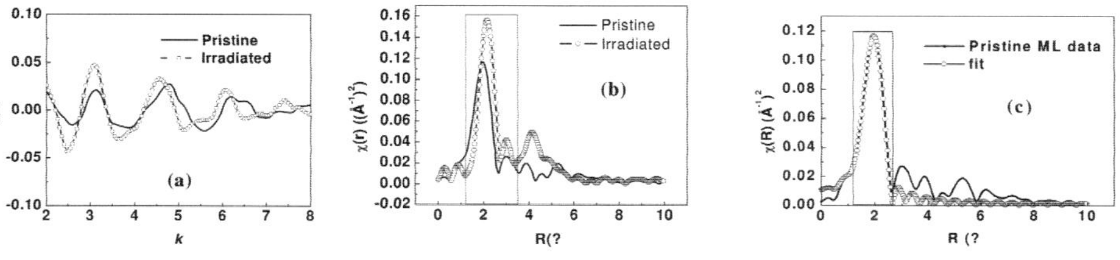

Figure 2. (a) XAFS data at *Ni K* edge; (b) Fourier transform of XAFS data; (c) Example of fit quality.

3.1. Pristine ML: XAFS data for this ML was fit with *Ni* and *C* nearest neighbors. Attempts to include *Pt* nearest neighbors resulted in bad fit. This is consistent with the pristine ML structure from reflectivity. Part of the *Ni* layer, *Ni/C* interface and sporadically diffused *Ni* atoms in *C* layer contribute to the XAFS data (Figure 1(e)). The fit results obtained were: (i) $N_{Ni\text{-}Ni} = 5.46(12)$, $N_{Ni\text{-}C} = 0.88(18)$; (ii) DWF (*Ni-Ni*) =0.0180(4) Å^2, DWF (*Ni-C*) = 0.016(6) Å; $R_{Ni\text{-}Ni}$ = 2.445(3) Å, $R_{Ni\text{-}C}$ = 1.964(7) Å. Extracting the structural model from these fit results involved rigorous considerations and simulations:

a. From reflectivity measurements, *Ni/C* interfacial roughness $\sigma_{Ni/C}$ = 1.7Å [7] and thickness of *Ni* layer = 22Å. We preliminarily consider the roughness to represent the interface thickness and the distribution of *Ni* atoms within the *Ni* layer to be uniform. From these assumptions, we derive the fraction of *Ni* atoms at *Ni/C* interface= 7%.
b. Not all of these *Ni* atoms are x-ray excited. The number of x-ray excited *Ni* atoms ($(N_{Ni})_{ex}$) (that contribute to XAFS signal), at different locations of the ML, follow the normalized intensity variation plot. The location-specific fractions of $(N_{Ni})_{ex}$ are thus extracted.
c. The structure of *Ni* -layer is *fcc* and the *Ni/C* interfacial plane consist of {111}:{110}:{100} = 1:1:1. The weight-averaged surface coordination for *Ni* atom at this interface would be 3.67.
d. We calculate the weighted average $N_{Ni\text{-}Ni}$ and maximum $N_{Ni\text{-}C}$ as:

$$N_{Ni-Ni} = \frac{[(12 \times (\mathrm{N_{Ni}})_{ex} \text{ in Ni layer}) + (8.33 \times (\mathrm{N_{Ni}})_{ex} \text{ at Ni/C}]}{\text{Total } (\mathrm{N_{Ni}})_{ex}} \quad -(1) \qquad (N_{Ni-C})_{\max} = \frac{3.67 \times (\mathrm{N_{Ni}})_{ex} \text{ at Ni/C}}{\text{Total } (\mathrm{N_{Ni}})_{ex}} \quad -(2).$$

A sharp interface, would have yielded $N_{Ni\text{-}Ni}$=11.63 [$(N_{Ni\text{-}Ni})_{XAFS}$=5.46] and a maximum average interfacial $N_{Ni\text{-}C}$ = 0.37 [$(N_{Ni\text{-}C})_{XAFS}$ = 0.879]. These being different from our XAFS results suggest diffusion of *Ni* atoms into *C* layer. $N_{Ni\text{-}C}$ now has contributions from *Ni/C* interface and *Ni-C* bonds within *C* layer. To estimate the % of the latter, we included *Ni/C* contribution from *C* layer in eqn. (2) and further decoupled the interfacial and layer contributions to $N_{Ni\text{-}C}$. $(N_{Ni\text{-}C})_{C\ layer}$= $(N_{Ni\text{-}C})_{XAFS}$ -$(N_{Ni\text{-}C})_{interfacial}$=0.51. *C* layer being amorphous *C*, the coordination within the layer = 2. Therefore, the % of *Ni* atoms diffused deeper into the *C* layer =0.51/2 = 26% (beyond reflectivity determined roughness [7]). The final model for pristine ML includes a significant proportion of sporadically diffused *Ni* in the *C* layers (Figure 1(e)).

3.2. Irradiated ML: At the very outset, it may be clarified that Au^{2+} beam undergoes elastic collisions with atoms in the ML and gets deposited onto the substrate [3]. XAFS data for this ML has contributions from *Ni+Pt* mixed layer, *Ni/C* interface and *C* layer. Besides, X-TEM images the irradiated ML revealed the presence of nanoclusters within *C* layers. Hence, the data was fit by considering *Ni,C* and *Pt* atoms as nearest neighbors. The fit results obtained were : (i) $N_{Ni\text{-}Ni}$ = 6.24(17), $N_{Ni\text{-}C}$ = 0.33(18), $N_{Ni\text{-}Pt}$ = 4.14(14); (ii) DWF (*Ni-Ni*) =0.0210(5) Å^2 , DWF (*Ni-C*) = 0.016(6) Å^2, DWF (*Ni-Pt*) = 0.024(2) Å^2 ; $R_{Ni\text{-}Ni}$ = 2.612(1) Å, $R_{Ni\text{-}C}$ = 1.964(7) Å, $R_{Ni\text{-}Pt}$ = 2.727(7)Å. Disruption of *Pt/C* interface and subsequent diffusion of *Pt* atoms into *C* layer[8-10] (consistent with corresponding 400% increase in $\sigma_{Pt/C}$ [3]) leads to the emergence of $N_{Ni\text{-}Pt}$. To extract the structural model from XAFS results, the following were involved:

a. Steps a-d from Section 3.1 were repeated for irradiated ML (Note that *Ni* layer is now replaced by *Ni+Pt* mixed layer). Clearly from eqn (2), contribution to $N_{Ni\text{-}C}$ is from *Ni/C* interface alone, hence *Ni* atoms in *C* layer have no *C* nearest neighbors.

b. Deduction of the structure of nanoclusters within *C* layer involved decoupling the contributions to $N_{Ni\text{-}Ni}$ from *Ni/C* interface, mixed *Ni-Pt* layer and *C* layer. The % of *Ni* atoms in *C* layer, *Ni+Pt* mixed layer and *Ni/C* interface are respectively 26% (from (i)), 64% and 10%. Considering coordination around *Ni* in *Ni-Pt* mixed layer, *C* layer and *Ni/C* interface to be 6 (since the layers are completely mixed), n and 8.33 respectively, n was calculated using:

$$(N_{Ni-Ni})_{XAFS} = \frac{n \times (\mathrm{N_{Ni}})_{ex} \text{ in C layer } + 6 \times (\mathrm{N_{Ni}})_{ex} \text{ in Ni + Pt layer } + 8.33 \times (\mathrm{N_{Ni}})_{ex} \text{ at Ni/C}}{\text{Total } (\mathrm{N_{Ni}})_{ex}}$$

Similarly the *Ni-Pt* coordination in *C* layer was deduced (note that this does not have any contribution from *Ni/C* interface). The calculated values of N_{Ni-Ni} and N_{Ni-Pt} are listed in Table 1. Various models for nanoclusters considered are listed in Table 1, and most of them are ruled out.

The final model for *Ni-Pt* clusters in *C* layer is *Ni* centered *Ni-Pt* bimetallic clusters, highly disordered beyond first neighbor, and having *Ni:Pt* = 60:40 (Figure 1(f)). This composition is reported to be ferromagnetic [11] (using FEFFIT, we simulated XAFS data (Fourier transform) for different values of N_{Ni-Ni} and N_{Ni-Pt} and observed that for both N_{Ni-Ni} and N_{Ni-Pt} , there is a resolution limit of $\Delta N=2$).These clusters resemble the short range order of multi component glasses, with the *C* matrix (probably) helping in the amorphous structure[12]. The first shell radius of these clusters is estimated to be 0.25nm (corresponding to XAFS derived R_{Ni-Pt}).

Table 1. Various models for Nanoclusters.

	XAFS result	Isolated Ni fcc clusters	Isolated Pt fcc clusters	Ni fcc core Pt shell	Pt fcc core Ni shell	Ni centered icosahedral bimetallic	Pt centered icosahedral bimetallic	Ni centered Ni-Pt bimetallic
N_{Ni-Ni}	6.35	12	0	9(minimum)		$N_{Ni-Ni}+N_{Ni-Pt}=7$	$N_{Ni-Ni}+N_{Ni-Pt}+N_{Pt-Pt}=7$	$9\leq N_{Ni-Ni}+N_{Ni-Pt}\leq 12$
N_{Ni-Pt}	3.64	0	0	3(maximum)	0.3			
		Ruled out	Ruled out	Ruled out	Ruled out	Ruled out	Ruled out	Final model

4. Conclusion

We have characterized ion-irradiated *Pt/Ni/C* ML using a combination of XSW and *Ni K* edge XAFS. The XTEM-XRF identified nanoclusters in *C* layer have been determined to be *Ni* centered *Ni-Pt* bimetallic (*Ni:Pt* = 60:40). They are formed due to diffusion of *Pt* atoms from the disrupted *Pt/C* interface towards pre-diffused *Ni* atoms in *C* layer, following irradiation. These clusters have ferromagnetic composition, highly disordered beyond first neighbor and resemble short range ordered metallic glasses. They show good promise as magnetic devices.

5. References

[1] Heald S M, Tranquada J M 1989 *J. Appl. Phys.* **65** 290.
[2] Ghose S K, Dev B N 2001 *Phys. Rev. B* **63** 245409.
[3] Roy S, Ghatak J, Dev B N 2012 *Appl. Surf. Sci.* **258** 3967.
[4] Bera S, Satpati B, Goswami D K, Bhattacharjee K, Satyam P V, Dev B N 2006 *J. Appl. Phys.* **99** 074301.
[5] Chopra N, Wu Y, Kazmaeir P M 2009 *US Patent App. Publication* US 2009/0274834 A1.
[6] Lahiri D, Chattopadhyay S, Bunker B A, Doudna C M, Bertino M F, Blum F, Tokuhiro A, Terry J 2005 *Physica Scripta*. **T115** 776–780.
[7] Roy S, Dev B N 2011 *Appl. Surf. Sci.* **257** 7566.
[8] Bera S, Goswami D K, Bhattacharjee K, Dev K, Kuri G, Nomoto K, Yamashita K 2003 *Nuc. Inst. and Meth. in Phy. Res. B* **212** 530–534.
[9] Lodha G S, Yamashita K, Suzuki T, Hatsukade, Tamura K, Ishigami T , Takahama S, Namb Y 1994 *Appl. Optics* **33** 5869.
[10] Evans B L, B J Ken 1987 *Appl. Optics* **26** 4491.
[11] Kumar U, Mukhopadhyay P K, Sanyal B., Eriksson O, Nordblad P, Paudyal D, Tarafder K, Mookerjee A 2006*Phys. Rev. B* **74** 064401; Cadeville M, Dahmani C E, Kern F 1986 *J. Mag. Mag. Mat.* **54-57** 1055.
[12] Wang W-H, Wei Q, Bai H Y 1997 *Appl. Phys. Lett.* **71** 58.

Journal of Physics: Conference Series **430** (2013) 012064 doi:10.1088/1742-6596/430/1/012064

Interaction of Nanostructured Calcium Silicate Hydrate with Ibuprofen Drug Molecules: X-ray Absorption Near Edge Structure (XANES) Study at the Ca, Si and O K-edge

X X Guo[1], T K Sham[1,*], Y J Zhu[2] and Y F Hu[3]

[1] Department of Chemistry, University of Western Ontario, London, Ontario, N6A 5B7, Canada

[2] Shanghai Institute of Ceramics, Chinese Academy of Sciences, Shanghai, 200050, China

[3]Canadian Light Source, Saskatoon, Saskatchewan, S7N 0X4, Canada

[*]Corresponding author, email: tsham@uwo.ca

Abstract. Mesoporous calcium silicate hydrate (CSH) nanostructure has been proven to be bioactive and biocompatible, and has a bright future in the application of bone treatment among other applications. X-ray absorption near edge structure (XANES) is a powerful tool for the study of the interactions of calcium silicate hydrates with drug molecules because it is element specific and it probes the unoccupied electronic states. Herein, we report the use of the calcium, silicon and oxygen K-edge XANES spectroscopy to identify how drug molecules interact with different groups in calcium silicate hydrate mesoporous nano-carriers with different morphologies. Significant changes are observed in XANES spectra after drug loading into the calcium silicate hydrate system, especially at the Si and O K-edge. The implications of these findings are discussed.

1. Introduction:

Calcium silicate hydrates are used in the application of bone regeneration and drug delivery due to their good bioactivity: the transformation to hydroxyapatite in simulated body fluid (SBF) [1]. Although CSH which is served as drug carrier has been studied extensively, little is known about the interaction between the drug molecules and the drug delivery carriers. X-ray absorption near edge structure (XANES) is a powerful tool for identifying the molecule carrier interaction. It is elemental specific, and probes the unoccupied electronic states above the HOMO in the molecules and the Fermi level in metals and the semiconductors [2]. Herein, we report the XANES study of the interaction between a model drug, ibuprofen (IBU) and calcium silicate hydrate nanostructure with different morphologies at the Ca, Si and O K-edge in order to find out how drug molecules interact with different functional groups in drug carriers.

2. Experimental:

The synthetic methods of CSH mesoporous microspheres have been described previously [3]. Briefly, $Ca(NO_3)_2$ and tetraethyl orthosilicate (TEOS) mixed solution were irradiated ultrasonically for 1h under ambient conditions. The CSH nanosheets were synthesized using a controlled precipitation reaction between calcium salt and silicate salt. The as-synthesized CSH nano-carriers were added to

IBU-hexane solution and oscillated at 37 °C for 24h, and then IBU loaded CSH nano-carriers were separated via centrifugation.

XANES measurements were conducted at the Canadian Light Source located at the University of Saskatchewan. The Soft X-ray Microcharacterization Beamline (SXRMB) with an energy range of 1.7 to 10 keV is used for the Ca and Si K-edge XANES measurement. The High Resolution Spherical Grating Monochromator (SGM) beamline with an energy range of 250-2000 eV, is used for the O K-edge test. The detection modes are total electron yield (TEY) and X-ray fluorescence yield (FLY). In this paper, we only use TEY XANES for the discussions and illustrations simply because both TEY and FLY have the same results since the specimens are homogeneous and optically thin and show no detectable thickness effects.

3. Results and Discussions:

3.1 Ca K-edge

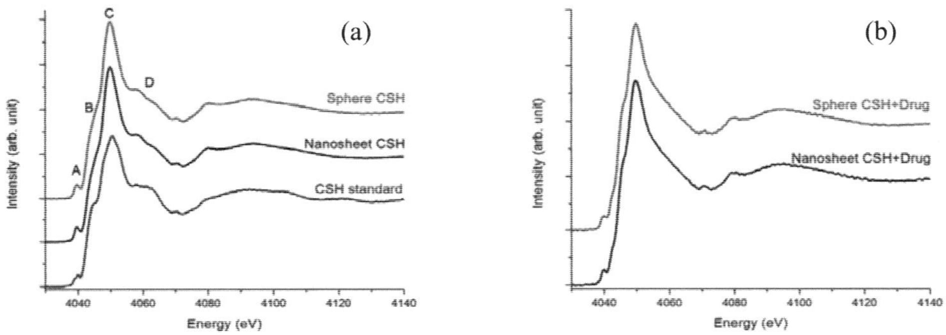

Figure 1 Ca K-edge XANES total electron yield (TEY) spectra of calcium silicate hydrate with (b) and without (a) drug molecules

Comparison with CSH commercial sample (noted as CSH standard), the nanostrutured CSH samples (noted as Nanosheet CSH and Sphere CSH) share the same features although there are subtle differences. Pre-edge peak A is due to the 1s to 3d quadruple transition and shoulder B is assigned to the 1s to 4s transition [4]. Although both of them are transition forbidden, the reason why we can still observe the resonance is that they must be hybridized with ligand p character final states. Main resonance C is due to the 1s to np transitions, allowed by the dipole transition selection rules. The shoulder after the main resonance (D) is mainly from multiple scattering processes.

Discernible change of CSH nanosheets and microspheres before and after drug loading is shoulder D, which smears out after the loading. We attribute this observation to drug loading which changes the local order and modifies the multiple scattering pathways. Furthermore, a subtle change in shoulder B becomes apparent from the first derivative spectra (Figure 2a). It can be associated with the interaction between Ca ions and carboxyl groups of IBU. This assignment is supported by the observation in which drug loaded CSH microspheres exhibit very similar derivative spectra as that of calcium acetate monohydrate (figure 2b dot line).

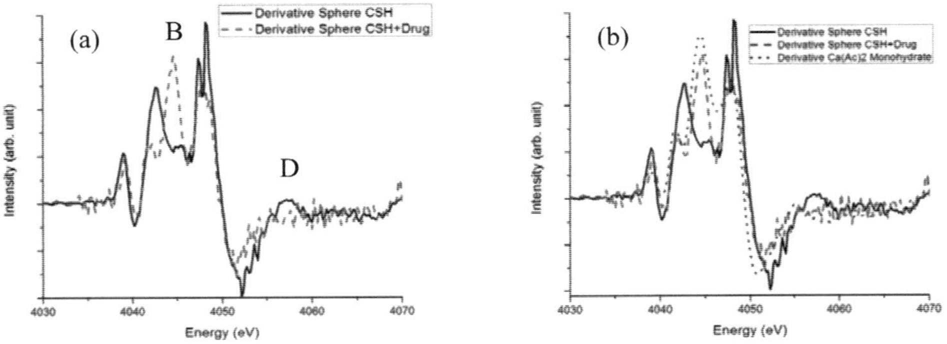

Figure 2 First derivative spectra of CSH mesoporous microspheres with and without drug molecules (a) and comparison of calcium acetate monohydrate (b)

3.2 Si K-edge

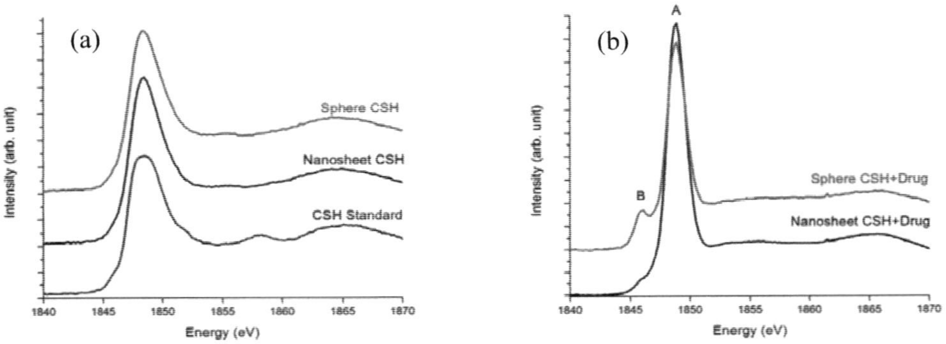

Figure 3 Si K-edge XANES total electron yield (TEY) spectra of calcium silicate hydrate with (b) and without (a) drug molecules

Figure 3 shows all the Si K-edge data. From figure 3(a), we see that all the spectra of the CSH system without drug are almost identical. There is only one prominent peak which is identified as the Si 1s to t_2 transition for Si (IV) in a tetrahedral oxygen ligand environment [5]. The most amazing observation is that after the drug is loaded (figure 3(b)), small feature (labelled as B) emerges at ~2 eV below the main resonance (A) which sharpens considerably as if the loading induces "crystallization" (ordering). The dramatic changes at the Si K-edge can be ascribed to the strong interaction between silanol groups (≡SiOH) on the CSH surface and carboxyl groups (-COOH) of IBU. After the interaction, because of the presence of the alkyl groups, which are known to be excellent electron density donors, in the structure (Si-O-C bonds), we can observe a small peak at the lower energy level.

3.3 O K-edge

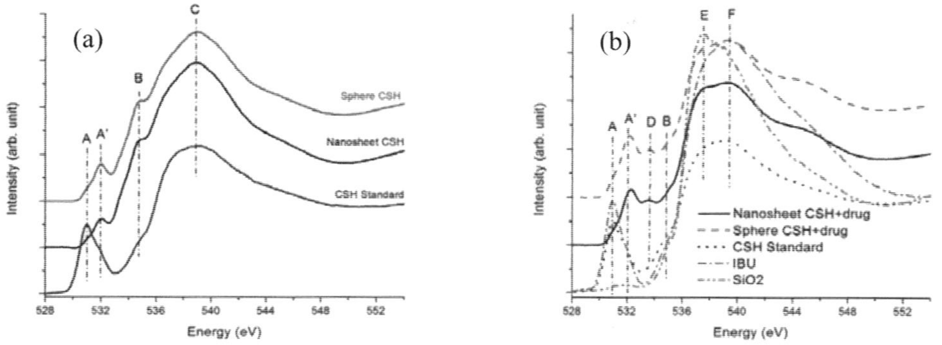

Figure 4 O K-edge XANES total electron yield (TEY) spectra of calcium silicate hydrate without (a) and with (b) drug molecules, comparison with IBU and SiO_2

The O K-edge data are shown in figure 4. Before drug loading, the pre-edge peaks (A, A') are induced by radiation damage from the incident beam [6, 7]. Shoulder (B) is due to 1s to 2p transition mixed with s orbital of cations which oxygen is bound to, and non-bridging oxygens (NBOs) are the main contributors of this shoulder. The main resonance (C) is ascribed to the 1s to 2p anti-bounding states transition [7].

Significant changes are observed after IBU is loaded. First, another small pre-edge peak (D) appears, which is assigned to interaction due to the presence of carboxyl groups [8]. This observation indicates that drug molecules can either adsorb physically or be loaded via hydrogen bonding on the CSH surface. Second, the intensity of B decreases significantly due to the interaction of NBOs with drug molecules. Meanwhile, the main resonance is resolved into two features (E and F) which can be ascribed to the combination of SiO_2 (dash dot dot line) and IBU drug molecules (dash dot line).

4. Conclusion:

Electronic structures of CSH nanosheets and CSH mesoporous microspheres assembled by nanosheets were measured by XANES. Besides the hydrogen bonding, chemical interaction of the drug and the carriers CSH was observed at Ca, Si and O K-edges. Calcium ions interact with carboxyl groups, exhibiting calcium acetate like features. The most amazing observation is the stronger interaction between silicates and carboxylic groups, forming Si-O-C bonds. CSH mesoporous microsphere nano-carriers exhibit stronger interactions with drug molecules than nanosheets.

Acknowledgement:

The assistances from the SGM beamline scientist Tom Regier are gratefully acknowledged. Research at the University of Western Ontario is supported by NSERC, CRC, CFI, and OIT. The Canadian Light Source is supported by CFI, NSERC, NRC, CHIR, and the University of Saskatchewan. Research at Shanghai Institute of Ceramics is supported by the National Natural Science Foundation of China (51172260) and Science and Technology Commission of Shanghai (11nm0506600).

References:

[1] Siriphannon P *et al* 2002 *J. Biomed. Mater. Res.* **60**(1) 175-185
[2] Sham T K 2008 *Int. J. Nanotechnol.* **5**(9-12) 1194-1246
[3] Wu J, Zhu Y J, Cao S W, Chen F 2010 *Adv. Mater.* **22**(6) 749-753
[4] Eichert D *et al* 2005 *Spectrochim. Acta B* **60**(6) 850-858
[5] Li D *et al* 1993 *Solid State Commun.* **87**(7) 613-617
[6] Jiang N, Spence J C H 2006 *Ultramicroscopy* **106**(3) 215-219
[7] Henderson G S, Neuville D R, Cormier L 2009 *Chem. Geol.* **259**(1-2) 54-62
[8] Kim K. *et al* 2011 *Carbon* **49**(5) 1745-1751

Electronic Structures and Optical Properties of α-Al_2O_3Nanowires

Zhiqiang Wang[1], Chunlei Li[2], Lijia Liu[1] and Tsun-Kong Sham[1]

[1] Department of Chemistry, University of Western Ontario, London, N6A5B7, Canada

[2]Faculty of Science, North China University of Technology, Beijing 100041, China

E-mail: zwang443@uwo.ca; tsham@uwo.ca

Abstract. The electronic structure and optical properties of α-Al_2O_3 nanowires (NWs) have been investigated using X-ray absorption near-edge structures (XANES) and X-ray excited optical luminescence (XEOL). The XANES were recorded in total electron yield (TEY) and total fluorescence yield (TFY) across the K- and $L_{3,2}$-edges of aluminium and the K-edge of oxygen. The results indicate that the NWs are of a core/shell structure with a single-crystalline core and an amorphous shell. The XEOL spectra of the NWs show an intense peak at 404 nm, which comes from the F centre located in the amorphous shell of the NWs. The implication of these findings and the sensitivity of XEOL for defect detection are discussed.

1. Introduction

Aluminum oxide (Al_2O_3), an insulator with a wide band gap (9 eV), has many promising properties such as high radiation resistance, high thermal conductivity, high optical transparency, low dielectric loss, and low permeability of alkali ions and other impurities [1]. Recently, α-Al_2O_3 (corundum) nanostructures have drawn much attention due to their excellent optical and mechanical properties and thermal stability. Various α-Al_2O_3 nanostructures, such as nanowires [2], nanobelts [3], nanotubes [4], and nanosheets [3], have been synthesized and have shown excellent dielectric and unique optical properties. The optical properties of α-Al_2O_3 nanostructures depend on the structure and properties of color centers (such as F, F^+, F_2, F_2^+, and F_2^{2+} centers) associated with oxygen vacancies in them, which are strongly related to their morphology, crystallinity and electronic structure. In this work, we studied the electronic structure and optical properties of α-Al_2O_3 nanowires (NWs) using X-ray absorption near-edge structures (XANES) in combination with X-ray excited optical luminescence (XEOL). A detailed XANES analysis of the NWs as well as the standard samples (metallic Aland α-Al_2O_3 powders) indicates that the NWs are of a core/shell structure with a crystalline core and an amorphous shell, which is consistent with TEM observation. The XEOL spectra of the NWs show a strong emission at 404 nm, which is attributed to the F center located in the shell of the NWs and the core-shell interface.

2. Experimental

2.1. Sample preparation and characterization

Standard α-Al_2O_3 (0.8-1 μm, 99.98%) and Al (17-30 μm, 99%) powders were purchased from Alfa Aesar.The synthesis of α-Al_2O_3 NWs was carried out in a horizontal tube furnace, which has been reported elsewhere [5]. The morphologies of the product were characterized by transmission electron microscopy (TEM; JOEL JEM 3000F).

2.2. Synchrotron measurements
Synchrotron measurements were carried out at the variable line spacing plane grating monochromator (VLS PGM) and high-resolution spherical grating monochromator (SGM) beamline at the Canadian Light Source (CLS), the University of Saskatchewan. The samples were mounted on conventional carbon tapes with an angle of 45° facing toward the photon beam. XANES were recorded in total electron yield (TEY), total fluorescence yield (TFY), and photoluminescence yield (PLY) where applicable. The TEY was detected with the specimen current and the TFY was measured by detecting the X-ray fluorescence photons using a micro channel plate. XEOL spectra were collected using a dispersive optical spectrometer (QE65000, Ocean Optics). All XANES and XEOL spectra were normalized to the incident photon flux collected on a refreshed Au grid.

3. Results and Discussion

3.1. Morphological and microstructural characterization
A TEM image of α-Al_2O_3 NWs is shown in Figure 1a. Tens of the NWs with the diameter of about 10-15 nm were grown on the aluminium particle (the upper-left corner of Figure 1a) and spread out from that like a bunch of grass. Figure 1b shows a typical high-resolution TEM (HRTEM) image of a single NW. It is interesting to point out that the NW is of core/shell structure. The core with a diameter of ~ 10 nm is single crystalline with an inter-planar spacing of 0.43 nm, in agreement with the *d* value (d = 0.433 nm) of the (003) planes of the α-Al_2O_3 crystal, which indicates that the NW grew along the <001> direction. While, a shell with a thickness of ~ 3 nm is found on the surface of the core showing poor crystallinity (or amorphous-like).

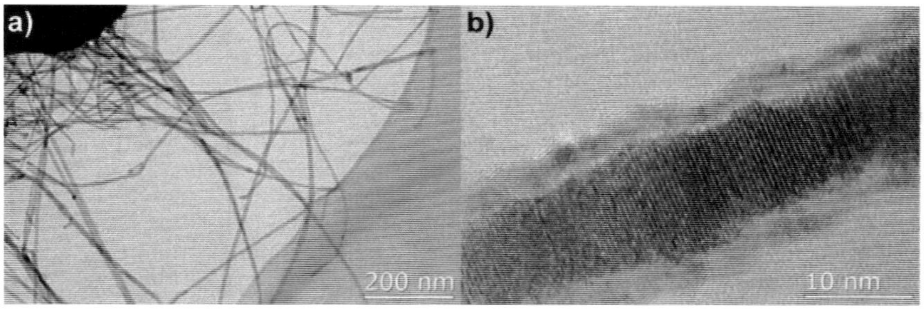

Figure 1. TEM (a) and HRTEM (b) images of the as-prepared α-Al_2O_3 NWs.

3.2. XANES of α-Al_2O_3NWs: Al $L_{3,2}$-edge, K-edge, and O K-edge
Al $L_{3,2}$-edge XANES ofα-Al_2O_3NWsand standard α-Al_2O_3 and metallic Alpowders are shown in Figure 2a.Let us firstlook at the α-Al_2O_3 powders. The TEY of α-Al_2O_3 powdersexhibits two sharp peaks at 79.5 and 80.0eV, and four broad peaks with the centre at 81.2, 83.9, 86.4, and 88.7eV, which are the characteristic features of α-Al_2O_3. The two sharp peaks with higher intensity are attributed to the excitation from Al $2p_{3/2}$ and $2p_{1/2}$ to 3s (3d), respectively. The four broad peaks at higher energies correspond to the unoccupied density of states (DOSs) of s (d) character due to multiple scattering. The TFYshows similar features as the TEY but the first two peaks are much shaper, indicating that the bulk of the NW is more order than that of the surface.
Next, wemove to the XANES of the as-prepared NWs. At first glance the TEY of the NWs does not show any characteristic features of α-Al_2O_3. At the Al $L_{3,2}$-edge, the X-ray attenuation lengthis 30-40 nm. However, the TEY, whichis very surface sensitive, can only detect the chemical information of the shell in the NWs.For standard metallic Al powders, their surfaces wereeasily oxidized in air and

formed a thin layer of amorphous Al_2O_3. The surface sensitive TEY of the NWs exhibits similar features to that of metallic Al powders. It indicates that the shell of the NWs is amorphous Al_2O_3, which is in agreement with the observation by TEM.The bulk sensitive TFY of the NWs shows the features from both the α-Al_2O_3 core (feature *b* and *c*) and the amorphous shell (feature *a* and *d*) but the spectrum is partially inverted due to self-absorption (thickness effect) [6]. For thick samples, as the excitation energy increases abruptly at the edge, the absorption of the element increases but the effective penetration depth of the X-ray in the sample decreases, which tends to compensate for the increase in a total absorption situation. This results in a nonlinear distortion of the TFY spectra.Therefore, there is an energy shift on feature *b* and *c* in the TFY of the NWs.

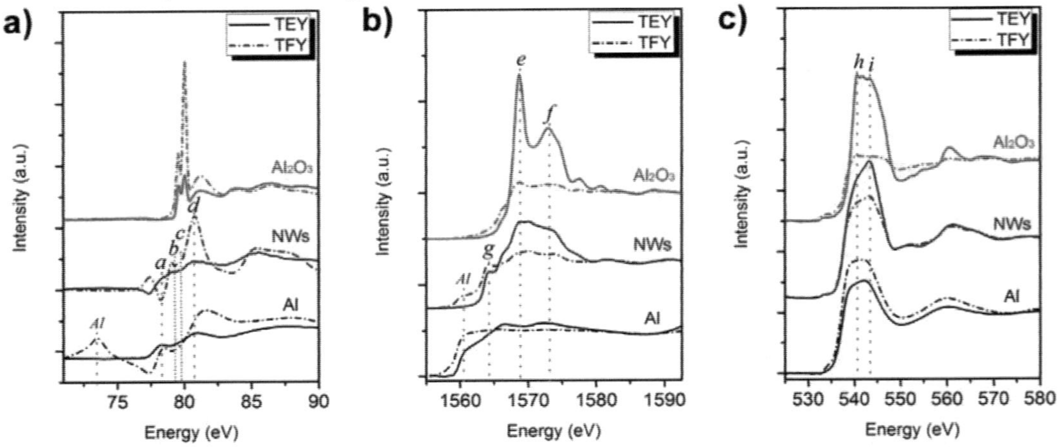

Figure 2. Al $L_{3,2}$-edge (a), K-edge (b), and O K-edge (c) XANES of α-Al_2O_3MPs, NWs, and Al powders.The solid and dash-dot lines are TEY and TFY, respectively.

With the excitation energies increasing to Al K-edge, the XANES explores more "bulk" information of the samples. The Al K-edge XANES ofα-Al_2O_3NWs and standard α-Al_2O_3 and Alpowderare shown in Figure 2b. The TEY ofα-Al_2O_3powders shows a shoulder at 1566.4 eV and six peaks at 1568.8, 1572.8, 1577.6, 1580.8, 1583.1, and 1588.5 eV. The intense absorption at 1568.8 eV (white line) is attributed to the Al 1s - 3p transition. Here, it should be noted that the TEY of the core/shell NWs displays some differences to that ofα-Al_2O_3powders. In the TEY of the NWs, the intensities of the two main peaks at 1568.8 and 1572.8 eV (feature *e* and *f*) drop off a lot, and a new shoulder at 1564.3 eV (feature *g*) appears. This shoulder is associated with some Al^{3+} in the shell and the interface between the core and shell being reduced to Al^{2+} or Al^{1+}, which results in that the main peak intensity decreases, and at the same time a shoulder at lower energy appears. The TFY of α-Al_2O_3NWs and standard α-Al_2O_3 and Alpowdersshow similar features as TEY but the peaks are broadened and strongly attenuated due to self-absorption (thickness effect) [6]. Metallic Al signal was found in the TFY of the NWs, which comes from the residual Al particles that were used as starting materials for the synthesis of NWs (e.g. the large particle in the upper-left corner of Figure 1a).

The O K-edge XANES of α-Al_2O_3 NWs, powder and Al powder are shown in Figure 2c. Both TEY and TFY of α-Al_2O_3NWs andpowder exhibit a pre-edge peak at 533.4 eV which corresponds to adsorbed O species[7], a main peak with the center at 542.6 eV which is attributed to O 1s-2p transition, and a few weak peaks at higher energies due to multiple scattering. Inα-Al_2O_3 powder, feature *h*(540.5 eV) has the maximum intensity. However, in NWs, feature *i*(543.5 eV) is the strongest. Compared to Al $L_{3,2}$-edge and K-edge, O K-edge has moderate energies which still can give us some information of the surface (e.g. the shell of NWs and surface of Al powders). In the case of NWs, O K-edge XANES shows the information of both the amorphous shell and single-crystalline core. The amorphous shell contributes more on feature *i* (see the XANES of Al powders), therefore, feature *i* is more intense than feature *h* in NWs.

3.3. XEOL of α-Al_2O_3NWs
The XEOL spectra of the NWs with excitation energy tuned across Al $L_{3,2}$-edge are shown in Figure 3. With different excitation energies, NWs emit bright luminescence at 404 nm, which is attributed to F center (oxygen vacancy with two electrons). The intensity of luminescence decreases when approaching the edge jump and increases above the edge jump.As shown in the inset of Figure 3, the zero order PLY of the NWswas recorded to monitor the absorption across Al $L_{3,2}$-edge. The TEY of standard α-$Al_2O_3$and Al powders are also displayed for comparison. We can clearly see that the PLY of the NWs is nearly identical to the TEY of Al powders (although inverted) but very dissimilar to that of α-Al_2O_3powders, indicating that the photoluminescence in the NWs originates from the oxygen site of the amorphous shell and the interface, not the single-crystalline core. It is understandable that large amounts of oxygen vacancies were generated on the surface of the NWs and the core-shell interface in an oxygen-deficient environment, which leads to the strong luminescence in XEOL. The inversion of PLY indicates that the energy transfer from Al $L_{3,2}$-edge absorption to the optical channel is less efficient when the edge turns on, but it may also be due to the residual Al particles with the size of 0.5-1 μm.

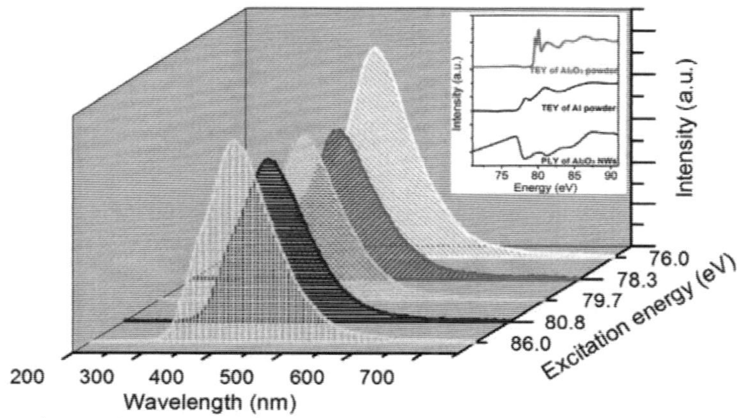

Figure 3. Al $L_{3,2}$-edgeXEOL spectra of α-Al_2O_3 NWs with different excitation energies. Inset is the PLY of the NWs and TEY of standard Al_2O_3 and Al powders.

4. Conclusion
Electronic structures and optical properties of α-Al_2O_3 NWs were examined using XANES and XEOL. The XANES results indicate that the NWs are of a core/shell structure with a single-crystalline core and an amorphous shell. The XEOL spectra of the NWs show an intense emission at 404 nm, which comes from the F center located in the amorphous shell of the NWs.

Acknowledgements
Research at the University of Western Ontario is supported by NSERC, CFI, OIT, and CRC (TKS). CLS is supported by CFI, NSERC, CHIR, NRC, and the University of Saskatchewan.

References
[1] Itou M, Fujiwara A and Uchino T 2009 *J. Phys. Chem. C***113** 20949.
[2] Fang X, Ye C, Xu X, Xie T, Wu Y and Zhang L 2004 *J. Phys.: Condens. Matter***16** 4157.
[3] Fang X, Ye C, Peng X, Wang Y, Wu Y and Zhang L 2003 *J. Mater. Chem.***13** 3040.
[4] Li Y, Bando Y and Golberg D 2005*Adv. Mater.***17**1401.
[5] Li C, Huang H, Yang S, Zheng R. Yang W, Liu Z and Ringer S 2009 *Mater. Lett.***63** 1016.
[6] Zschech E, Troger L, Arvanitis D, Michaelis H, Grimm U and Baberschke K1992*Solid State Commun.***82**1.
[7] Kovalev V, Krupnov A, Pogosbekyan M and Sukhanov L 2010 *Mosc. Univ. Mech. Bull.***4** 85.

Journal of Physics: Conference Series **430** (2013) 012066 doi:10.1088/1742-6596/430/1/012066

New insights into the role of Mn and Fe in coloring origin of blue decorations of blue-and-white porcelains by XANES spectroscopy

Jian Zhu [1,2], Wugan Luo [1,2], Dongliang Chen [4], Wei Xu [4], Chaofang Ming [1,2], Changsui Wang[1,2], Lihua Wang [3]*

[1]Graduate University of Chinese academy of Science, Beijing, 100049, People's Republic of China,

[2]The Joint Laboratory of Human Evolution and Archaeometry, Beijing, 100086, People's Republic of China,

[3] Shanghai Synchrotron Radiation Facility, Shanghai Institute of Applied Physics, Chinese Academy of Sciences, Shanghai,201204, People's Republic of China

[4] Institute of High Energy Physics, Beijing, 100049, People's Republic of China

*Correspondence email: wanglihua.ssrf@gmail.com

Abstract. Blue and white porcelain is one of the most valuable ancient ceramics varieties in ancient China. It is well known for its beautiful blue decorations. However, the origin of its blue color has not been very clear till now. In this research, two blue and white porcelains from Jingdezhen, Jiangxi province were selected and Mn and Fe K-edge XANES spectra were recorded from blue decorations with or without transparent glaze. Results showed that Mn K-edge XANES features were almost identical between different samples while that of iron changed. The above findings indicated the positive role of iron in the variation of blue decorations. As for manganese, although more system researches were need, its negative role on the variations of the tone of blue decorations was obtained. On the other hand, the paper also revealed the XAFS results will be affect by the glaze layer above the pigment. These findings provided us more information to understand the coloring origin of blue decorations of blue-and-white porcelain by means of XANES spectroscopy.

1. Introduction

Blue-and-white porcelain is well-known for its skillful techniques and great artistic merits all around the world. Especially in Ming and Qing dynasty, blue-and-white porcelain reached the height of its technical excellence and played an important role in the cultural exchange between the east and the west[1]. So far, blue-and-white porcelain have been attracting more and more researchers to investigate its manufacturing techniques, the provenance of cobalt blue pigment ore, the dating and authentication, and so on. Coloring origin of the blue color in blue-and-white porcelain is one of these hot topics. Previous studies pointed out that transition elements such as cobalt, iron and manganese made the contribution to blue color And among these elements, cobalt took up the main role in blue color while iron and manganese were responsible for the variation of the tone of blue color. [2]. Although coloring variations are usually obtained by modulating the oxidation states of transition elements, for a long time, the science study on the blue and white porcelain remained in the level of the chemical composition. And little is known about the elements' oxidation and chemical environment which is directly correlated to its coloring origin.

X-ray absorption fine structure (XAFS) is a powerful technique, based on the development of synchrotron radiation source, to investigate the local electron and geometrical structure around an absorbed atom by analyzing the fine structure above the absorption edge. Up to date, XAFS has been applied to solve a lot of important research issues in a wide range of scientific fields, including material science, catalysts research, environment science, biology, and so on. In archaeometry and art conversation, XAFS also has its own advantages, especially in these researches on coloring mechanism of glaze and glass with amorphous phase. In our past research, Co K-edge XAFS analysis of blue decorations of blue-and-white porcelains found that Co^{2+} was the main constituents coordinated with oxygen atoms situated in both octahedral and tetrahedral sites. Considering the ability in absorbing the characteristic frequencies in visible region, it was concluded that $[Co^{2+}O^{4}]$

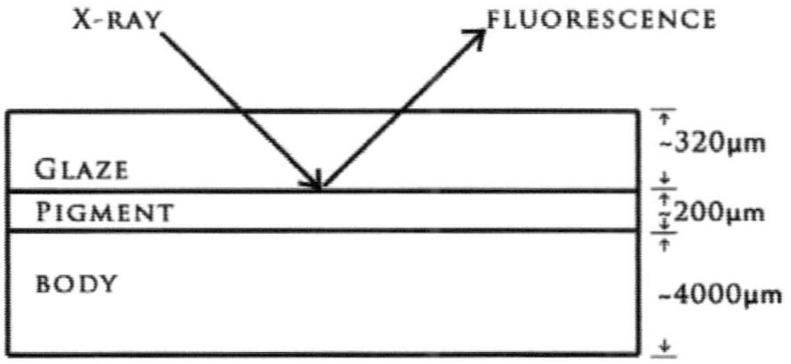

Figure 1: Map of relationship about the glaze, pigment and body across the profile of blue-and-white porcelain

made more contribution to the blue color of blue-and-white porcelain.[3] However, the Co *K*-edge spectra for all samples collected from Yuan dynasty to Ming dynasty with different tones of blue colors were almost identical. Since iron and manganese, which are also co-existed in blue decorations, are usually considered as colorant elements, it is necessary to investigate the oxidation state and local environment of iron and manganese in order to have a thorough understanding of the coloring origin of blue decorations.

However, to investigate the oxidation state of iron and manganese in blue decoration of blue and white porcelain by XAFS technique, there is a problem that should be solved firstly. Because the blue and white porcelain are under-glaze ceramics, the surface of blue pigments were covered a transparence glaze (see Figure 1). Iron and manganese contained in glaze layer will have an impact on the XAFS spectra recorded from the surface of glaze layer directly.

Thus, in this research, Mn and Fe K-edge XANES spectra of blue decoration with glaze layer above or not were obtained and compared. The primary purpose was to investigate the influence of glaze layer during the XAFS measurements nondestructively.

2. Samples and experimental

Samples analyzed in this research included two blue and white porcelains, labeled with HJ-02(Figure 2a) and JS-06 (Figure 2b). both of these were collected from Jingdezhen, Jiangxi province. The experiments were carried out as the following steps: 1) Put samples to holder and focus on one point, then collected the XANES spectra; 2) Cut off the glaze where the above point have been tested, then collected the XANES spectra again.

Mn and Fe K-edge XANES measurements were performed on beamline BL14W at Shanghai Synchrotron Radiation Facility (SSRF). The storage ring energy was operated at 3.5 GeV with a ring current of 130 mA to 210 mA. A Si (111) double-crystal monochromator was used to monochromatize the white beam. And higher harmonics were rejected with the aid of harmonic mirrors of Ni layer. At the site of sample, the flux was about 10^{12} photo/s and the beam size was ca. 1 (H)×1 (V) mm^2. All K-edge XAFS spectra of blue-and-white porcelains were recorded in fluorescence yield mode with silicon drift detector (SDD) at room temperature. The IFEFFIT package was used to analyze the raw spectra using conventional procedures[4].

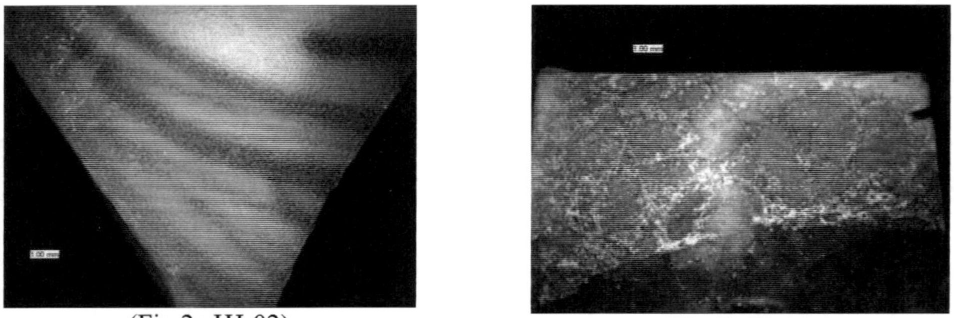

(Fig.2a HJ-02) (Fig.2b JS-06)

Figure 2: Photograph of the samples HJ-02 and JS-06

3. Results and Discussion

Mn *K*-edge XANES spectra of blue decorations with or without glaze layer for HJ and JS were shown in Figure 3. As shown in Figure 3, two findings could be observed. First of all, Mn K-edge XANES spectra of blue color decorations with or without glaze for certain blue-and-white porcelain (sample HJ-02 and JS-06) were almost identical. This finding indicated that the oxidation state of manganese contained in blue decorations was consistent with that of glaze layers. Secondly, no significant shifts of pre-edge (peak A) and white line (peak B) between different samples were observed. Since coloring variations in porcelains are usually changed with the variations of oxidation state of transition element, the almost identical XANES features demonstrated that manganese has no direct effect on the variation of the tone of blue decorations.

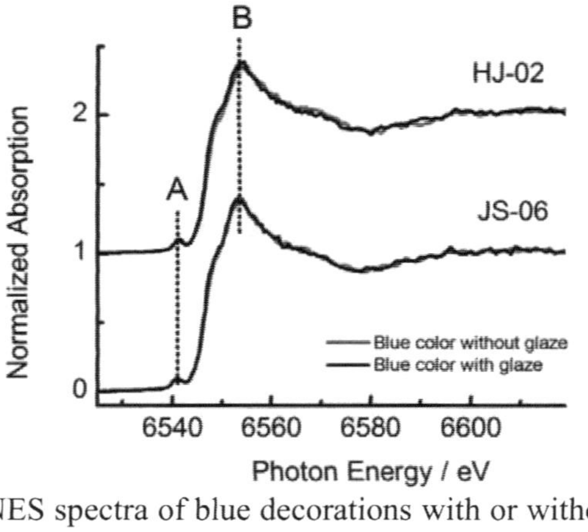

Figure 3: Mn *K*-edge XANES spectra of blue decorations with or without transparent glaze for HJ-02 and JS-06

Fe *K*-edge XANES spectra for different blue-and-white porcelains were shown in Figure 4. Compared with XANES spectra detected directly from the glaze surface and the cut-off glaze result, two findings could be found. For HJ-02 sample, the position of absorption edge of iron in blue decorations without glaze layer shifted to lower energy compared with that of transparent glaze and

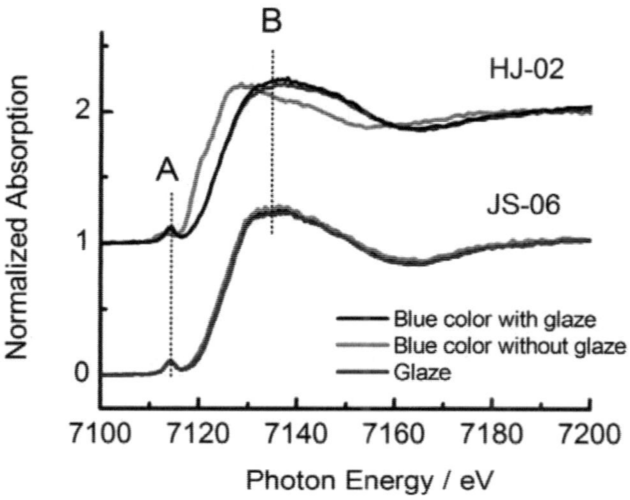

Figure 4: Fe *K*-edge XANES spectra of blue decorations with or without transparent glaze for HJ-02 and JS-06

blue decorations with glaze layer. For JS-06 sample, the position of absorption edge of iron in blue decorations without glaze layer, blue decorations with glaze layer, transparent glaze were almost identical. The above results indicated that the oxidation state of iron in blue decorations and transparent glaze were different. As a result, Fe *K*-edge XANES spectra detected directly from the surface of transparent glaze of blue decorations did not reflect the real information of the oxidation state of iron contained in blue decorations. The inconsistent may be induced by two possibilities. One was influenced by manganese element since iron can be oxidized by manganese. The other possibility was about the firing atmosphere. Because previous studies pointed out that blue and white porcelain were fired at reduction atmosphere[5].

As a kind of under-glaze color, blue-and-white porcelain from the profile view, the location of transparent glaze layer, the pigment layer and the body. During the XANES experiments, the incident X-ray through a transparent glaze layer to reach the pigment layer, the fluorescence signal is also required to pass through the transparent glaze layer after the detector detection. During this process, the transparent glaze layer will affect the XANES measurement of blue decoration in two ways. On one hand, iron and manganese are also existing in transparent glaze (Fe: 0.8-1.92%, Mn 0.1-1.91%). In this research, it is induced that the Mn content exist in the glaze and pigments were consistent by the comparison analysis. However, the oxidation state of Fe demonstrated some shift in different part, indicating that the Fe element in glaze will have infection to the XAFS result. On the other hand, the thickness of transparent glaze layer cannot be ignored too. Under normal circumstance, structural characterization of XAFS experimental spectra contains useful information (the absorption coefficient of the oscillation part) that relative to the absorption edge height of only a few percent. Due to varying of the thickness of the transparent glaze layer in samples (thickness around 300-400μm), when the fluorescence signal of the pigment layer through the transparent glaze, the layer of glaze with what kind of impact for the fluorescence peaks, as well as how this correction is not yet clarity on past and still need more systematic research in future. Thus, results obtained here were not a conclusive identification. If more samples were incorporated, findings we found here could be changed subsequently. Therefore, in future researches, a larger set of samples, as well as reference materials, will be considered in order to get a thorough understanding of the role of iron and manganese on the coloring mechanism of blue decorations of blue-and-white porcelains.

4. Conclusions

In this study, two blue and white porcelains from Jingdezhen, Jiangxi province were selected and the transparent glaze were polished and rubbed off at some areas with blue decorations. Mn and Fe *K*-edge XANES spectra recorded from blue decorations with or without transparent glaze showed that Mn *K*-edge XANES spectra were almost identical while the absorption edge of iron changed. These results directly ruled out the positive influence of oxidation state of Mn in the variation of the tone of blue decorations in blue-and-white porcelains. On the contrary, changes of oxidation state of iron in different samples indicated its positive role in the variation of blue decorations which should be investigated further in the future.

Acknowledgements

This work was supported by the project of National Nature Science Foundation of China (Grant No.10979075). And authors also thank BL14W beamline of Shanghai Synchrotron Radiation Facility for providing valuable time for XANES experiments.

References

[1] J. Li, 1998, *History of Science and Technology in China, Ceramics Volume*. (Science Press, Beijing) chapter 11.
[2] N. Wood, M. S. Tite, C. Doherty and B. Gilmore, *Archaeometry,* 2007, **49** (4), 665–84.
[3] Lihua Wang and Changsui Wang, *J. Anal. At. Spectrom.* 2011, 26, 1796- 01.
[4] B. Ravel, M.Newville, *J. Synchrotron Rad.* 2005. 12, 537-42.
[5] Wu Jun, Li Jiazhi, Guo Jingkun, *Journal of Inorganic Materials,* 1999, 14(1), 143-149.

Journal of Physics: Conference Series **430** (2013) 012067 doi:10.1088/1742-6596/430/1/012067

Local crystal structure of TiNiCu shape memory alloys

A P Menushenkov[1], O V Grishina[1], A A Yaroslavtsev[1], R V Chernikov[2], J Bednarcik[2], A V Shelyakov[1], N N Sitnikov[1], Y V Zubavichus[3] and A A Veligzhanin[3]
[1] National Research Nuclear University "MEPhI", Moscow, 115409, Russia
[2] HASYLAB at DESY, Hamburg, D-22603, Germany
[3] NRC "Kurchatov Institute", Moscow, 123182, Russia

E-mail: menushen@htsc.mephi.ru

Abstract. The features of long-range crystal structure and local environment of Ni and Cu atoms in shape memory alloy $Ti_{50}Ni_{25}Cu_{25}$ are investigated using X-ray diffraction and extended X-ray absorption fine structure spectroscopy (EXAFS) in the temperature range of direct and reverse martensitic transformations. According to XRD study, alloy has the B2 type structure in austenitic phase and B19 type structure (space group *Pmmb*) in martensitic phase. EXAFS demonstrates the difference between Cu-Ti and Ni-Ti bond lengths ~ 0.09 Å not typical for common TiNi lattice. The static disordering is more significant in the local environment of Ni atoms than in local environment of Cu atoms.

1. Introduction

TiNi-based alloys are some of the most widespread shape memory alloys (SMA). While being heated they are able to fully restore their pre-deformation shape as a result of reverse *martensitic transformation* (MT). Under such a behavior these alloys demonstrate considerable mechanical force. Due to their unique properties SMAs are of great interest to find the unusual ways in development of various important technical applications [1,2,3]. The demand of micro- and nanodevices construction drives the production of new types of SMA as well as their further exploration on different scales.
Direct and reverse MTs occur in their own temperature ranges, therefore the material demonstrates hysteretic properties under phase transformation. Characteristic temperatures of MT depend on both the alloy grade and its chemical composition [4,5]. The slight composition changes lead to a shift of these parameters. The transformation properties have the connection with the fine effects taking place in a local crystal structure upon alloys cooling and heating. Therefore, SMA are also important for the understanding of chemical bonding and phase transitions in intermetallics. In present work we investigated the features of crystal structure and local environment in ternary $Ti_{50}Ni_{25}Cu_{25}$ SMA by means of X-ray diffraction (XRD) and extended X-ray absorption fine structure spectroscopy (EXAFS).

2. Experimental

In initial state the samples were thin ribbons obtained by components melt extrusion from the quartz crucible through the thin nozzle on the surface of rotating copper disk, where the sample's cooling took place with the rate of 10^6K/sec. The thickness of the initial amorphous ribbons was 40–45μm. Samples were crystallized by isotropic annealing in the air at 500°C for 4 min (sample No. 3484).
The EXAFS spectra were obtained in transmission mode above *K*-Ni (8333 eV) and *K*-Cu (8979 eV) absorption edges of sample 3484 at STM beamline of Kurchatov synchrotron radiation source

(Moscow, Russia). The EXAFS spectra above *K*-Ti (4966 eV) absorption edge were collected at beamline C of DORIS-III storage ring (HASYLAB, DESY, Hamburg) on the sample No. 967 with the same composition $Ti_{50}Ni_{25}Cu_{25}$ as sample 3484 but with the thickness reduced to 15 μm. XRD patterns of sample 3484 were obtained at beamline BW5 of DORIS-III with the photons of energy 100 keV. All the measurements were carried out upon consistent heating and cooling in the temperature ranges of direct and reverse MTs. The parameters of two samples explored are as follows:

- 3484: $Ti_{50}Ni_{25}Cu_{25}$, A_s=53°C, A_f=65°C, M_s=57°C, M_f=46°C (*K*-Ni, *K*-Cu EXAFS)
- 967: $Ti_{50}Ni_{25}Cu_{25}$, A_s=56°C, A_f=64°C, M_s=54°C, M_f=46°C (*K*-Ti EXAFS, XRD)

The processing of EXAFS spectra was performed using VIPER [6] with standard procedure of EXAFS-oscillations function $\chi(k)$ deconvolution and Fourier-analysis. Magnitudes and phases of back scattering were computed using FEFF-8.20 code [7] within a model of *Pmmb* symmetry group of the martensitic phase and *Pm-3m* symmetry group of the austenitic phase. Wavelet analysis was performed using a set of software modules HAMA [8]. XRD patterns were processed using GSAS [9].

3. Results and discussion

The X-ray diffraction measurements were carried out in the temperature ranges of direct and reverse MT with ~1°C step. The background was approximated by a sum of 8 Chebyshev polynomials. The peaks were assumed to have a symmetric shape and their profiles were fitted by the modified pseudo-Voigt function. The thermal atomic displacement factor was taken as isotropic and equal for nickel and copper atoms. XRD results demonstrate that $Ti_{50}Ni_{25}Cu_{25}$ alloy has the B2 type structure in austenitic phase. For martensitic phase the peaks in X-ray diffraction scans were successfully indexed in terms of the orthorhombic lattice, *Pmmb* space group, B19 type structure. Weak reflections of the residual martensitic phase were detected even at temperatures much higher than A_f, which were taken into account in the further Rietveld refinement. The refined lattice parameters of the orthorhombic structure listed in table 1 are in good agreement with the previous studies of the Ti-Ni-Cu martensitic phase [10]. The lattice parameter of cubic B2 phase is a=3.0413Å. The positions of diffraction peaks vary with the temperature. During the phase transformation the lattice constants change slightly by the value of ~ 0.008 Å. While *a* and *b* unite cell parameters of martensitic B19 phase and the lattice constant *a* of the austenitic cubic B2 phase increase upon heating, the parameter *c* of martensitic phase decreases. The temperature dependences of martensitic phase content changes during the heating and cooling processes were calculated and turned out to be hysteretic. Besides, z-coordinates of Ni/Cu atoms in unit cell of B19 phase at the same temperature differ upon heating and cooling as well.
The major problem in the refinement of polycrystalline TiNi-based alloys is the preferential orientation (texture) of their grains. Refinement of such a multi-texture structure is complicated. The texture was described by the Marsh-Dollase function with one adjustable parameter, which should be less than 1 when preferred orientations exist. The assumption of [110], [001] and [101] texture improve the results significantly.

Table 1. Lattice constants and basis coordinates of $Ti_{50}Ni_{25}Cu_{25}$ martensitic phase (sample 967)

	a=2.906 Å	*b*=4.279 Å	*c*=4.492 Å	a=2.906 Å
Ti	x=0.5	y=0.25	z=0.225	x=0.5
Ni/Cu	x=0	y=0.5	z=0.697	x=0

Qualitative analysis of spectra was performed using wavelet transforms of EXAFS oscillation functions $\chi(k)$. Results of *K*-Ti spectrum wavelet transform for the sample 967 are shown in fig. 1. For the spectrum at 70°C they demonstrate two clear resolved maxima, i.e. two scattering acts are registered on different elements situated in different distances from the absorbing atom. Thus, the closest local environment of Ti atoms consists of two atom types: Ni (or Cu because they have close atomic numbers and consequently close dependences of scattering amplitude vs. the wave vector *k*) and Ti (which atomic number is smaller so the wavelet maximum appears at lower *k*). In martensitic

Journal of Physics: Conference Series **430** (2013) 012067 doi:10.1088/1742-6596/430/1/012067

phase the atoms change their positions and the maxima merge together in one broad peak which corresponds to coordination shells with close radii containing Ni, Cu and Ti atoms. The wavelet transform of the austenitic phase *K*-Ni EXAFS-functions reveal two maxima situated in approximately same distances. Thus, the local environment of Ni and Cu atoms in austenitic phase consists of 2 coordination shells with different atom types (Ni/Cu and Ti).

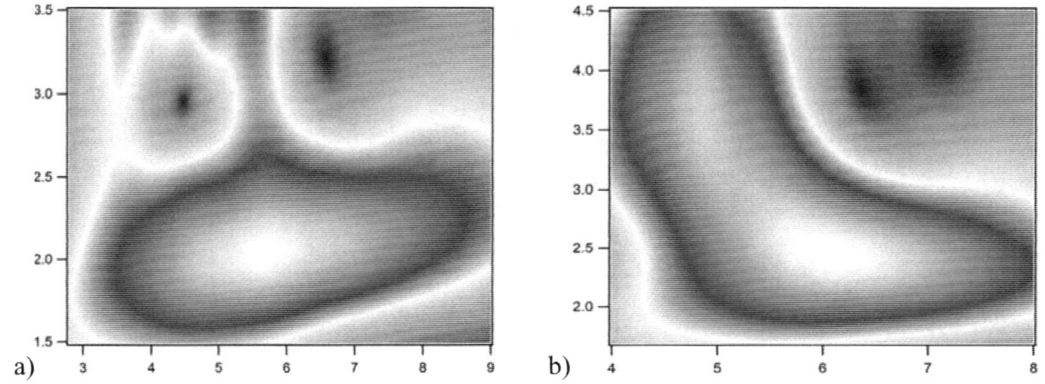

Fig. 1. Wavelet transforms of *K*-Ti EXAFS-functions for the sample 967 at 30°C – martensitic phase (a) and 70°C – austenitic phase (b). Wave vector *k*, $Å^{-1}$ is set along the horizontal axis; radius R, Å is set along the vertical axis

EXAFS-functions above *K*-Ti absorption edge at temperatures 30°C of martensitic phase, 70°C of austenitic phase and at the temperature 55°C in the region of phase transformation collected during the heating and cooling processes are shown in fig. 2a. The shape of $\chi(k)k^2$ function changes with temperature and is not the same at 55°C for the heating and cooling processes, so we can consider the changes of EXAFS line shape hysteretic. Modules of *K*-Ti EXAFS-functions Fourier transforms (FT) are shown in fig. 2b. There are significant changes in the first maxima of FT, e.g. in nearest coordination shells of Ti atoms. With the temperature increase the first peak grows and shifts towards the greater value of interatomic distance.

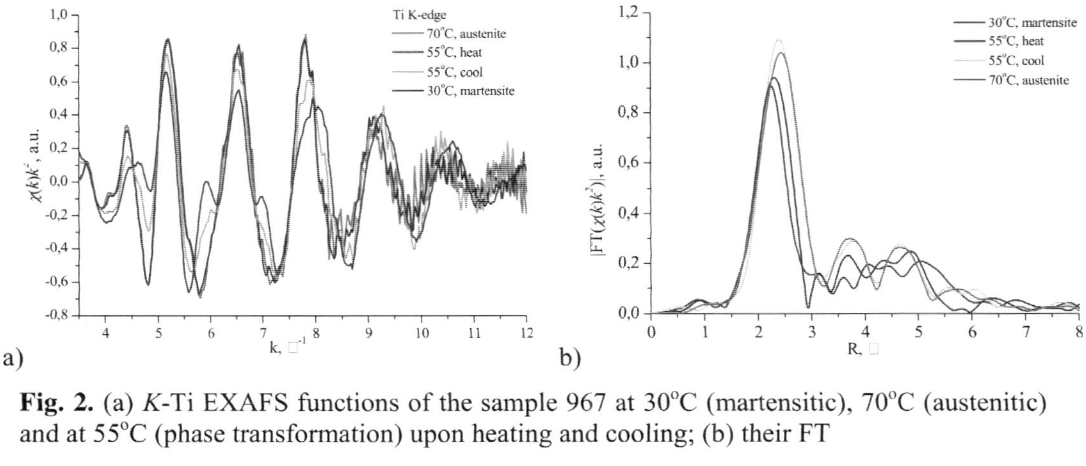

Fig. 2. (a) *K*-Ti EXAFS functions of the sample 967 at 30°C (martensitic), 70°C (austenitic) and at 55°C (phase transformation) upon heating and cooling; (b) their FT

The character of *K*-Ni EXAFS-spectra (sample 3484) line shape change is similar to the case of *K*-Ti EXAFS (sample 967), however, the first large FT maximum of *K*-Ni EXAFS corresponding to the Ti, Ni and Cu shells doesn't change much with temperature, whereas several smaller peaks in the region

3.5–5 Å change significantly. This FT region corresponds to the variety of coordination shells containing Ni and Cu atoms. Therefore, we suppose that the general reversible changes in the local crystalline structure of $Ti_{50}Ni_{25}Cu_{25}$ alloy upon MT occur in the sublattice of Ni and Cu atoms. Note that the spectral compositions of EXAFS-functions are different for Ni and Cu local environment. The first *K*-Ni EXAFS FT maximum is lower than the one for *K*-Cu EXAFS and is shifted towards the smaller R. The results of EXAFS spectra fitting are given in table 2.

Table 2. Results of EXAFS spectra fitting. R error is ±0,003 Å, σ^2 error is ±0.002 Å^2

Sample	Type of bond	Martensite, 30°C			Austenite, 70°C	
		N	R, Å	σ^2, Å^2	R, Å	σ^2, Å^2
967	Ti-Ti	4	2.881	0.012	2.992	0.015
	Ti-Ni	4	2.409	0.029	2.526	0.014
	Ti-Cu	4	2.535	0.012	2.646	0.12
3484	Cu-Ti	8	2.631	0.012	2.632	0.013
	Cu-Cu	2	2.803	0.008	2.821	0.015
	Cu-Ni	2	2.996	0.011	3.026	0.017
	Ni-Ti	8	2.569	0.019	2.571	0.019
	Ni-Ni	2	2.815	0.008	2.834	0.014
	Ni-Cu	2	2.996	0.008	3.026	0.017

Disordering of interatomic distances is characterized by Debye-Waller factor σ^2. In both samples the average static disordering of coordination shell around Cu atom or shells containing Cu atoms is less than such for Ni. The discrepancy between $R_{Ti\text{-}Ni}$ and $R_{Ti\text{-}Cu}$ is 0.06-0.12 Å depending on the temperature.

4. Conclusion

The most significant changes in local crystal structure of $Ti_{50}Ni_{25}Cu_{25}$ alloy under MT occur in the Ni-Cu sublattice. Static disordering of Ni atoms is larger than of Cu atoms in both samples explored. Average Ti-Ni interatomic distance is less than Ti-Cu distance and significantly differs from the value calculated from diffraction data. These discrepancies may drive the local deformation that affects the parameters of the phase transition.

Acknowledgments

This work is partially supported by RFBR (grant No. 11-02-01174-a). Authors thank the HASYLAB program committee for providing the opportunity of XAFS and XRD measurements (projects No. I-20100397 and H-20000015).

References

[1] Yongqing F, Hejun D and Weimin H 2004 *Sensors and Actuators A* **112** 395–408
[2] Van Humbeeck J 1999 *Mat. Sc. & Eng.* A **273-275** 134–48
[3] Otsuka K and Ren X 1999 *Intermetallics* **7** 511–28
[4] Rösner H, Slossmacher P, Shelyakov A V and Glezer A M 2001 *Acta mater.* **49** 1541–1548
[5] Hori K, Namazu T and Inoue S 2010 *Thin Solid Films* **518** S26–S28
[6] Klementev K V 2001 *J.Phys.D:Appl.Phys.* **34** 209
[7] Ankudinov A L, Bouldin C, Rehr J J, Sims J and Hung H 2002 *Phys. Rev. B* **65** 104107-1-11
[8] Funke H, Scheinost A C and Chukalina M 2005 *Phys. Rev. B* **71** 094110
[9] Larson A C, Von Dreele R B 2004 *Los Alamos National Laboratory Report LAUR* 86-748
[10] Potapov P L, Kulkova S E, Shelyakov A V *et al* 2003 *Phys. IV France* **112** 727-730

Effective pair potential for Ca-O bonds in $CaGeO_3$ polymorphs

A Koganemaru[1], A Yoshiasa[2], L Wang[2], T Nakatani[2], A Nakatsuka[3], M Okube[4], H Arima[5] and K Sugiyama[5]

[1] *Faculty of Science, Kumamoto University, Kumamoto 860-8555, Japan*
[2] *Graduate School of Science and Technology, Kumamoto Univ. Kumamoto 860-8555, Japan*
[3] *Graduate School of Science and Engineering, Yamaguchi University, Ube 755-8611, Japan*
[4] *Materials and Structures Lab. Tokyo institute of technology, Yokohama, 226-8502, Japan.*
[5] *Institute for Materials Research, Tohoku University, Sendai 980-8577 ,Japan*

E-mail: ai93kgnmr@gmail.com

Abstract. The $CaGeO_3$ perovskite and garnet were synthesized in a cubic anvil type apparatus under high pressure. The measurements of Ca and Ge K- edge XAFS spectra were carried out in the transmission mode at temperature up to 700 K. The effective pair potentials $V(u)=\alpha u^2/2+\beta u^3/3!$, for Ca-O bond in various phases of $CaGeO_3$ have been investigated by the temperature dependence of EXAFS Debye-Waller factors. The potential coefficient α for the Ca-O bond in perovskite-type $CaGeO_3$ is small, 4.4 eV/Å^2, compared with those in garnet (6.0 eV/Å^2) and wollastonite (6.4 eV/Å^2). The potential for Ca-O bond in perovskite is broader than those in other $CaGeO_3$ polymorphs, which is one reason for the Clausius-Clapeyron's curve for perovskite-garnet phase boundary having a negative slope. The potential coefficients for the Ca-O in perovskite are significantly smaller than those for the longer Ge-Ge distances as the framework vibration though the potential coefficient decreases usually as a result of the larger bond distance.

1. Introduction

High-pressure modification of silicate minerals such as perovskite- and garnet-type $MgSiO_3$ could be the dominant phases in the Earth's mantle, so detailed knowledge of their structures and thermal properties is of great importance for understanding the state of the deep mantle. It is known that phase boundary between perovskite- and garnet-type phases [1] has the negative Clausius-Clapeyron's curve, which has great meaning geophysically (Fig. 1).

Metagermanates are useful as structural analogues of silicate minerals at high pressures because of thermal stability of high pressure phases. $CaGeO_3$ perovskite and garnet structures are isotypic with $MgSiO_3$ perovskite and garnet, respectively. At ambient pressure, metastable high pressure $MgSiO_3$ perovskite and $CaGeO_3$ perovskite phases back-transform to stable phases with heating. The transient $CaGeO_3$ amorphous phase appears during back-transformation, before recrystallization of stable wollastonite-type phase occurs at higher temperatures. The metastable $CaGeO_3$ perovskite undergoes a phase transformation to an amorphous phase at 930 K [2].

In this study, we have determined the precise local structure around Ca in perovskite-, garnet- and wollastonite-type $CaGeO_3$ polymorphs. The analysis of temperature-dependent EXAFS Debye-Waller factor allows us to evaluate the interatomic force constants for neighboring atoms [3,4]. Unique thermal vibration in perovskite-type phase has been introduced.

2. Experimental and analysis

The wollastonite-type $CaGeO_3$ crystals were prepared by solid-state reaction of special grade reagents of CaO and GeO_2. The perovskite- and garnet-type $CaGeO_3$ crystals were synthesized in a cubic anvil type high-pressure apparatus under 10GPa 1250K and 3GPa 1300K, respectively. Single phase specimens of perovskite-, garnet- and wollastonite-type CaGeO3 were identified by X-ray diffraction. Figure 1 shows the synthesized condition for each specimen.

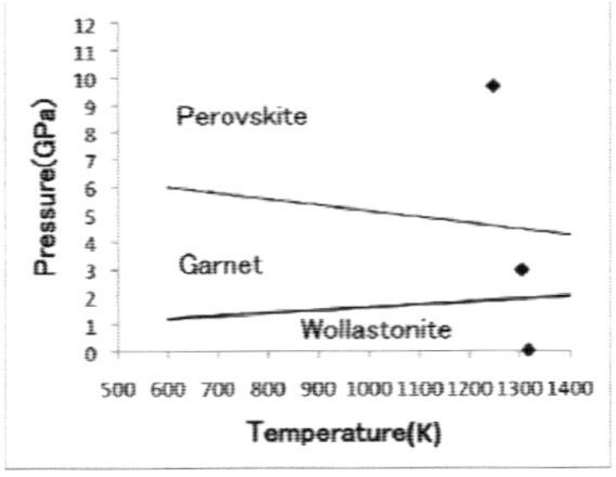

Figure 1. P-T phase diagram of the $CaGeO_3$ system [1] and the synthesized condition for each specimen.

The appropriate amount of fine powder sample and boron nitride powder was mixed and pressed into pellets of <0.15 mm in thickness and 10.0 mm in diameter. All samples had edge-jumps with 0.7 ($\Delta\mu d$), where μ is the linear absorption coefficient and d is the thickness. The measurements of Ca and Ge K-edge XAFS spectra (Fig. 2) were carried out in the transmission mode at beam line BL-9C of the Photon Factory in KEK, Tsukuba (Proposal No. 2010G608). X-ray absorption measurements in the temperature range from 100 to 700 K were made under a helium atmosphere. The synchrotron radiation was monochromatized by a Si(111) double crystal monochromator. Mirrors are used to eliminate higher harmonics. Details of the measurement and analysis were given in reference [5,6].

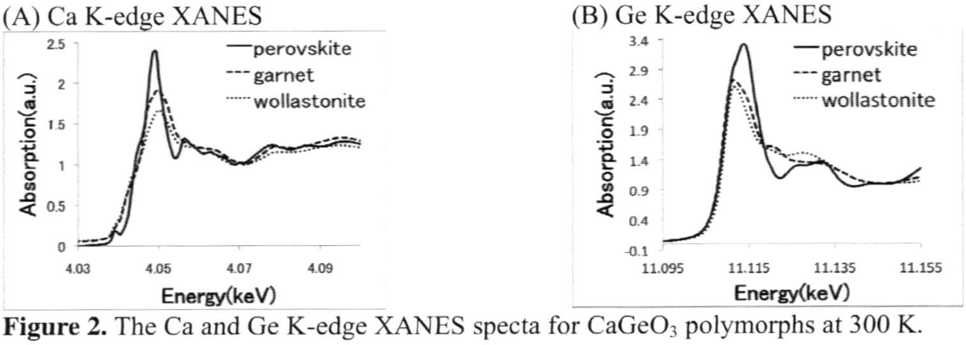

Figure 2. The Ca and Ge K-edge XANES specta for $CaGeO_3$ polymorphs at 300 K.

The EXAFS interference function, $\chi(k)$, was extracted from the measured absorption spectra using the standard procedure [7]. The $\chi(k)$ was normalized using MacMaster coefficients according to the EXAFS workshop report [8]. Figure 3 shows examples of Fourier transforms of the Ca K-edge EXAFS spectra for $CaGeO_3$ polymorphs in the ranges of $2.5<k<10.0$. In quantitative analyses we carried out the Fourier-filtering technique and a nonlinear least-squares fitting method by comparing the observed $\chi(k)_{exp}$ and calculated $\chi(k)_{calc}$. We used the EXAFS formula in the single scattering theory with the cumulant expansion up to the fourth order term. Values of the backscattering amplitude function $|f_B(k;\pi)|$ and the phase shift function $\psi_{AB}(k)$ were calculated using the FEFF3 program [9]. The mean free path λ of the photoelectron is assumed to depend on the wave number, $\lambda(k)=k/\eta$, where η is a constant.

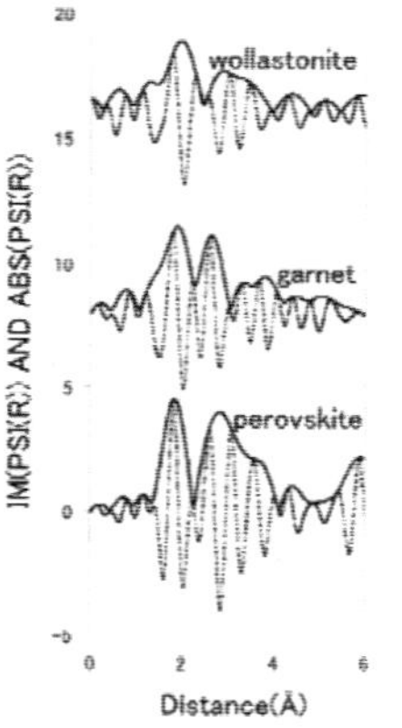

Figure 3. Fourier transforms of the Ca K-edge EXAFS for $CaGeO_3$ polymorphs at 300 K. No phase shift corrections are made.

Table 1. Refined parameters for the Ca-O bonds in perovskite-, garnet- and wollastonite-type $CaGeO_3$ at 700K.

	Perovskite (700K)	Garnet (700K)	Wollastonite (700K)
R(Å)	2.48(2)	2.45(2)	2.37(1)
σ2(Å^2)	0.0295(27)	0.0147(18)	0.0196(20)
σ3 (Å^3)	0.0004(2)	0.0002(2)	0.0002(1)
R-factor (%)	1.4	3.6	0.8

Single-shell fitting was carried out for each nearest-neighbor distance, where the number of neighboring atoms was fixed at the crystallographic value as N_B=8, N_B=7.5 and N_B=6 for the Ca-O bonds in perovskite-, garnet- and wollastonite-type polymorphs, respectively. Because the fourth-order term was negligible, the refinement was performed to the structure parameters R, σ2 and σ3 in each shell by use of the fixed η and ΔE_0 values. The reliability of fit parameters R,between the experimental and calculated EXAFS functions was less than 0.036. Table 1 shows the refined parameters at 700K.

3. Results and Discussion

3.1 Comparison of Ca and Ge K-edge XANES spectra for the polymorphs

XANES spectra are quite sensitive to the three-dimensional local atomic configuration around X-ray absorbing atoms. Each XANES spectrum shows considerably different shape among three phases (Fig. 2). The Ca K-edge XANES specta has a clear pre-edge peak and the highest main peak is sharp in $CaGeO_3$ perovskite. The highest peak top energy of perovskite situates at lower energy side as compared with other phases. On the contrary, the highest main peak of Ge K-edge XANES specta is broad in $CaGeO_3$ perovskite and the peak top of perovskite is located in a high energy position as compared with other phases.

Fourier transforms of the Ca K-edge EXAFS spectra for $CaGeO_3$ polymorphs show also considerably different functions among three phases (Fig. 3). Clear peaks are observed at least 6 Å in perovskite. This should be showing that the distribution correlation function [3,4] in local thermal vibration is large in perovskite.

3.2 Temperature dependence of EXAFS Debye-Waller factors for Ca-O bonds in $CaGeO_3$ polymorphs

Figure 4 shows the temperature dependence of σ2 for Ca-O, bonds in perovskite-, garnet- and wollastonite-type $CaGeO_3$. The Debye-Waller factor σ2 includes the effects of static and dynamic disorders. The static disorder is the configuration disorder, while the dynamic disorder arises from the thermal vibration of atoms. The contribution of the thermal vibration, $\sigma_{thermal}$, can be estimated under the assumption of classical statistical dynamics by the temperature dependence of σ2 [4]. A steep slope in the figures represents a weak bonding. The gradient for the experimental σ2 is equal to k_B/α, if we evaluate the anharmonic effective pair potential $V(u)=\alpha u^2/2+\beta u^3/3!$ from the contribution to the thermal vibration, where k_B is the Boltzmann constant, α and β are the potential coefficients and u is the deviation of the bond distance from the location of the potential minimum. β is calculated from the values of $\sigma 2$ and $\sigma 3$ [10,11]. The potential coefficients α and β for the Ca-O bond in $CaGeO_3$ perovskite are 4.6 eV/Å^2 and -4.4 eV/Å^3, respectively. Those for wollastonite are 6.4 eV/Å^2 and -5.6 eV/Å^3, respectively. The effective pair potential for Ca-O bond in perovskite is broader than those in other $CaGeO_3$ polymorphs. This is one reason for the Clausius-Clapeyron's curve for perovskite-garnet phase boundary having a negative slope. In the expression, $dP/dT=\Delta S/\Delta V$, the dS becomes negative because denser high pressure phase perovskite has larger thermal vibration amplitude.

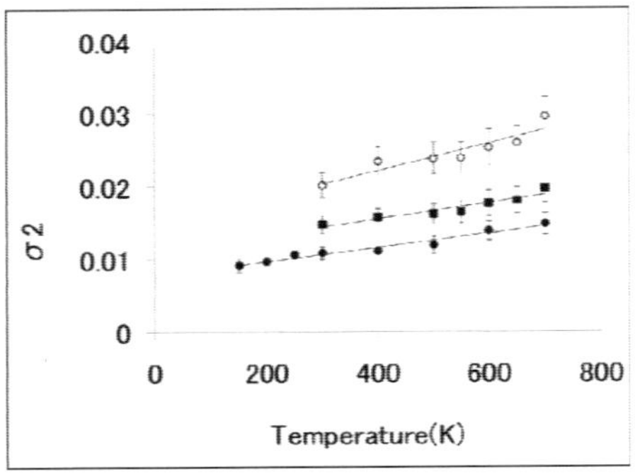

Figure 4. Temperature dependence of $\sigma 2$ for Ca-O bond in $CaGeO_3$ perovskite (°), garnet (•) and wollastonite(▪).

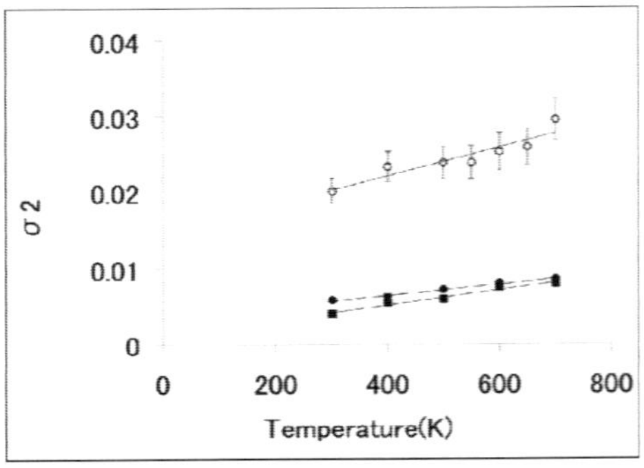

Figure 5. Temperature dependence of $\sigma 2$ for Ca-O bond (°), for Ge-O bond (•) and for Ge-Ge distance (▪) in $CaGeO_3$ perovskite.

3.3 potential coefficients for Ca-O,Ge-O and Ge-Ge distances in perovskite-type $CaGeO_3$

Figure 5 shows the temperature dependence of $\sigma 2$ for the Ca-O (2.48 Å at 700 K), Ge-O (1.90 Å at 700 K) and Ge-Ge (3.73 Å at 700 K) distances in the perovskite-type $CaGeO_3$. The gradient of $\sigma 2$ for the Ca-O bond is larger than those for the Ge-Ge distance as the framework vibration: the potential coefficients for the Ca-O bond is significantly smaller than those for the longer Ge-Ge distances though the potential coefficient decreases usually as a result of the larger bond distance. The potential coefficients α for the Ca-O in perovskite-type $CaGeO_3$ are 4.64 eV/Å^2. The potential coefficients α for the Ge-O and Ge-Ge in perovskite-type $CaGeO_3$ are 9.8 eV/Å^2 and 7.4 eV/Å^2 [12]. The Ca-O bonds in perovskite type compounds have relatively larger mean square amplitudes of vibration than those for Ge-Ge distance as judged by each value of potential coefficients.

Reference

[1] Ross N L, Akaogi M, Navrotsky A, Sasaki J, McMillan P 1986 *J. Geophys Res.* **91** 4685-4696
[2] Andrault D, Itié J P & Farges F 1996 *Amer. Mineral.* **81** 822-832
[3] Yoshiasa A, Nagai T, Ohtaka O, Kamishima O & Shimomura O 1999 *J. Synchrotron Rad.* **6** 43-49
[4] Yoshiasa A & Maeda H 1999 *Solid State Ionics* **121** 175-182
[5] Yoshiasa A, Maeda H, Ishii T & Koto K 1990 *Solid State Ionics* **40/41** 341-344
[6] Yoshiasa A, Tamura T, Kamishima O, Murai K, Ogata K & Mori H 1999 *J. Synchrotron Rad.* **6** 1051-1058
[7] Maeda H 1987 *J. Phys. Soc. Jpn.* **56** 2777-2787
[8] Lytle F W, Sayers D E & Stern E A 1989 *Physica* **B158** 701-722
[9] Rehr J J, Mustre de Leon, J Zabinski, S I & Albers R C 1991 *Am. Chem. Soc.* **113** 5135-5140
[10] Ishii T 1994 *Principle of the theory of EXAFS, Shokabo, Tokyo*
[11] Yoshiasa A, Koto K, Maeda H & Ishii T 1997 *Jpn. J. Appl. Phys.* **36** 781-784
[12] Yoshiasa A, Nakajima K, Murai K & Okube M 2001 *J. Synchrotron Rad.* **8** 940-942

Journal of Physics: Conference Series **430** (2013) 012069 doi:10.1088/1742-6596/430/1/012069

Endohedral fullerenes: a concurrent characterization by means of synchrotron radiation X-ray and IR spectroscopy

Wei Xu[1*], Augusto Marcelli[2,3], Lei Liu[1], Chunru Wang[4] and Ziyu Wu[3*]

[1] Institute of High Energy Physics, Chinese Academy of Sciences, Beijing, 100049, China

[2] Laboratori Nazionali di Frascati - INFN, Frascati, 00044, Italy

[3] NSRL, University of Science and Technology of China, Hefei 230026, China

[4] Institute of Chemistry, Chinese Academy of Sciences, Beijing, 100080, China

Email: xuw@mail.ihep.ac.cn ; marcelli@lnf.infn.it ; wuzy@ustc.edu.cn

Abstract. Endohedral Metal Fullerenes exhibit a great variety of physical and chemical properties depending on the metal inserted into the cage. These systems are molecular conductors, magnets, ferroelectrics and also superconductors representing extremely promising materials for advanced technologies such as nano-medicine. Here we present temperature-dependent XANES and FTIR investigations of two La@C82 EMF isomers. The combinatorial investigation shows that guest ions move inside the cage perturbing the vibrational states of the carbon cage due to the charge transfer dynamics. Moreover, the principal component analysis points out a discrepancy between temperature-dependent FTIR and XANES based on the occurrence of a non-equilibrium process between charge transfer and cage dynamics. We propose to perform simultaneous time-resolved X-ray and infrared spectroscopy studies to resolve the complex interplay among charge, structure and electric properties of these systems.

1. Introduction

The metal endohedral fullerenes (EMFs) are carbon cluster compounds containing one or more metal atoms inside the fullerene cage [1]. The unique structure of EMFs and their great variety of physical and chemical properties due to the metal inserted in the cage is of great interest. Indeed, EMFs can be molecular conductors, magnets or ferroelectric materials, and can be applied in radiotherapy, in tomography and as contrast agents in medicine and in many other advanced technologies. [2,3] For this reason EMFs have been widely investigated [4,5] in particular, to characterize their unique electronic and structural properties. Among the earliest synthesized EMFs [6], La@C82 certainly represents a prototypical species useful to describe their physical-chemical properties. It has nine isomers but only two of them can be easily synthesized in a reasonable amount to be measured. As it is shown in Figure 1, C2v and Cs isomers have a different symmetry or structure and, despite of the many investigations, the interplay between their charge and metal dynamics is still

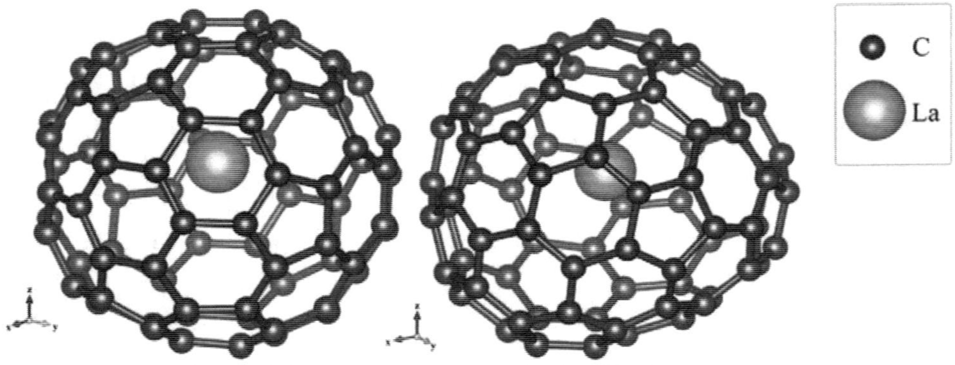

Figure.1 : The schematic view of the C2v and Cs isomers of the metal-endohedral fullerene La@C82.

not known. Usually, lanthanide metals contribute to the carbon cage with three valence electrons [7,8] forming the electrical structure: M3+C823-. Many different physical pictures have been introduced to properly describe the structural model of an EMF system, but none is fully adequate to describe experimental observations.

The X-ray absorption spectroscopy is a local symmetry-sensitive and element-specific probe. However, it is not easy to obtain structural information by looking at the carbon K-edge due to the multiple occupation sites of carbon atoms in the fulleride structure. To obtain structural information is better looking at the absorption edge of the metal guest and, in order to achieve more accurate structural information regarding the relative position of metal ions inside the carbon cage, theoretical simulations can be combined with X-ray absorption spectroscopy [9]. The Fourier Transform infrared spectroscopy (FTIR) is a vibrational technique based on the absorption of infrared radiation by infrared active molecules that absorb energies within the mid-IR region of the electromagnetic spectrum. In this region we mainly probe two types of molecular vibrations: stretching and bending, typically observed in the wavenumber range ~500–4000 cm–1 (i.e., from 2.5 to ~20 μm). They can be used to retrieve dynamic information in molecular systems as well as solid systems [10].

In the last decade, to extract information not achievable by individual techniques performed independently, the combinatorial investigation of complex systems has been continuously expanding. To characterize the complex dynamics of metal endohedral fullerenes as well as to exploit the advantages of the combinatorial approach, we investigated La@C82 isomers performing both temperature-dependent XAS and FTIR experiments.

2. Experiment and data processing

Detailed information about the sample preparation is available elsewhere [9,11]. Temperature dependent X-ray Absorption Spectroscopy (T-XAS) experiments of La@C82 at the La L3-edge for both C2v and Cs isomers, were performed in the transmission mode at the XAFS station of the 1W1B beamline at the Beijing Synchrotron Radiation Facility, using a Si (111) double crystal monochromator. During experimental runs the storage ring operated at the energy of 2.2 GeV, with an electron current decreasing from 120 to 80 mA within approximately 12 hours. The incident and transmitted beam intensities were monitored using two ionization chambers supplied by a continuous flow of nitrogen gas. A partial suppression of high order harmonics has been achieved by detuning the monochromator crystals by 40%. Experiments were performed in vacuum (~3×10-3 torr) in the temperature range 35-300 K. Samples were encapsulated in a hole of diameter 1.5 mm, manufactured at the centre of an aluminium holder.

FTIR experiments were performed using the IR synchrotron radiation source of the DAΦNE storage ring: the SINBAD beam line of the Frascati National Laboratory (LNF) of the INFN. Due to the high circulating electron current (> 2 A) of this low energy (E=0.51 GeV) ring, this beamline represents an almost unique IR source in the mid- and far-IR range. Samples were prepared evaporating a liquid droplet casted on a CVD diamond substrate at room temperature inside a glove box fluxed with high purity N2. Using a BRUKER Equinox 55 interferometer and a Hyperion 3000 microscope, mid-IR spectra of metal-doped fullerenes were collected in the range 600-4000 cm-1 with a resolution of 4 cm-1. Experiments were performed on samples loaded in the ARS Helitran LT-3He flowing cryostat with a KBr beam splitter using a single element MCT IR detector of area 1 mm2.
The difference among temperature dependent XAS spectra at the La L3 edge were obtained using IFEFFIT [12] and SIXPACK [13] packages. All spectra were normalized and background subtracted. The IFEFFIT package was used for conventional data pre-processing while the SIXPACK package for the principal component analysis. IR data were subjected to a polynomial background removal and a peak height normalization using the Bruker-OPUSTM. In order to extract spectral information, difference spectra were obtained by subtracting the corresponding FTIR or XAS spectra at room temperature.

3. Results and discussions

3.1 Temperature dependent X-ray absorption spectroscopy

Due to their different site symmetry and local structure (Figure 1), temperature dependent XAS spectra of the two isomers, shown in Figure 2, are different. The La L3 edge difference of T-XAS of the La@C82 C2v isomer (top panel) exhibits large spectral variations in the region 5475-5525 eV.

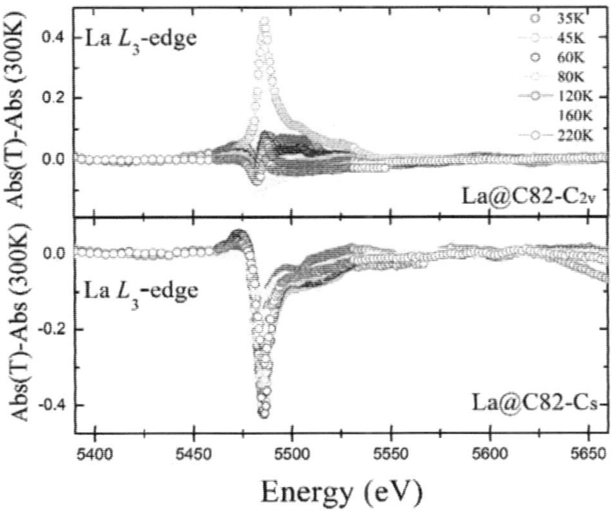

Figure.2: Comparison of La L_3 edge T-XAS spectra vs. temperature of both La@C82 isomers.

This energy range around the La absorption edge at ~5483.5 eV probes the electronic structure (2p-nd transitions) as well as the local geometric configurations around lanthanum metal ions.

Actually, increasing the temperature field both the metal and the carbon atoms of the cage are excited; therefore, the system will experience multiple kinetic states. At low temperature, configurations deviating from the room temperature phase occur around 45 K and 60 K for the C2v and the Cs isomer, respectively. However, for the high symmetry C2v structure, excluding an anomaly at ~45 K the spectral variation is relatively small. On the contrary, the low symmetry Cs structure exhibits large spectral variations at low temperature. As predicted from theory [14], lanthanum atoms move more then 1 Å from its initial position at room temperature. These experiments clearly point out that a large displacement of the lanthanum atom occurs inside the cage also at low temperature and, the motion is really large for the low symmetry isomer. In a previous research, the oscillations of a gadolinium ion inside the carbon cage around the equilibrium position have been described [9]. Actually, with its larger ionic radii, lanthanum ions have a larger amplitude of oscillation then gadolinium

In order to interpret XANES spectra, we performed theoretical calculations for the optimized structures of the two La@C82 isomers. Figure 3 shows the projected density of states as well as the theoretical La L3 edge absorption spectra (top panels). As discussed in Ref. [9], many equilibrium

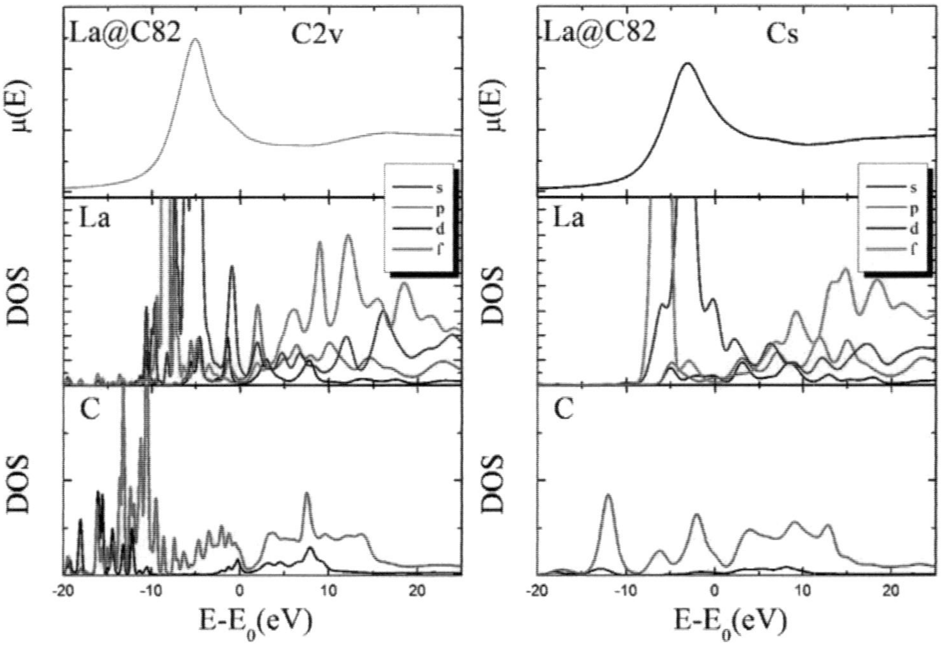

Figure.3: XANESS spectrum (top) and projected density of states (mid and bottom panels) of the two La@C82 isomers.

structures have to be considered and averaged to improve the agreement between theoretical and experimental spectra. To achieve structural information, here we attempted to optimize the metal-fullerene structure neglecting temperature effects but considering the projected density of states. Looking at Figure 3, we may recognize that the local density of La d states of C2v is more structured than Cs. However, the local density of carbon p state is also strongly dependent by the symmetry of the carbon cage. Interestingly, in the XAS spectrum of the C2v isomer the shoulder (at -2 eV) has a La d character while in the Cs isomer the peak at ~5 eV is probably a mixture of La p and d states. The carbon p states distribution is weak and the density of the highest occupied orbital is much larger in the C2v than in the Cs isomer. The local and partial density of states points out that charge transfer effect can be more relevant in the C2v structure.

3.2 Fourier Transform Infrared Spectroscopy

The mid- and far-infrared spectra of EMFs at room temperature have been studied in details by Lebedkin et al. in Ref. 15. His data show that radial and intermediate vibrations of carbons lie within the range 150-600 cm-1 while longitudinal and lateral vibrations of the metal ion inside the cage and external rotational modes of the molecule fall in the range 40-150 cm-1 and below 50 cm-1, in the far-IR and THz ranges, respectively. Our FTIR data refer only to vibrations in the mid-IR range.

In Figure 4, we compare FTIR spectra of the two isomers in the range 600-4000 cm-1. The marked region (1700-4000 cm-1) is not intrinsic to EMF vibrations and due to hydration. We may identify three characteristic vibrations: the CH stretching mode, the C-C stretching mode and the OH stretching mode. If we consider the average phonon energy of EMFs, this is rather large, i.e., up to 0.2 eV [14] and corresponds to ~1600 cm-1 in wavenumber and ~2300 K in temperature. As a consequence, the temperature field can be considered negligible for the vibrational dynamics of these systems.

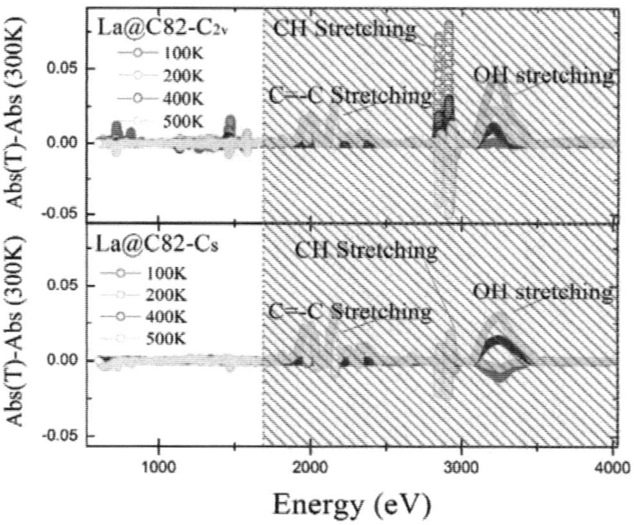

Figure.4: Difference among FTIR spectra of the La@C82 C_{2v} isomer collected at different temperature and the reference spectrum at room temperature (top), the same for the Cs isomer (bottom).

In order to correlate the analysis of XAS and FTIR spectra, we performed a principal component analysis (PCA) on both XAS and FTIR datasets. The principal component analysis [17] is a statistical tool aiming at reducing a large dataset to a minimum size, in order to make possible its interpretation. It is widely used in a variety of fields and this approach has been already used for the analysis of XAS spectra in heterogeneous systems [18] as well as to FTIR data. [19] In Figure 5 we present the PCA of temperature-dependent FTIR and XAS spectra of both C2v and Cs isomers of the La@C82. To evaluate the number of major components, we considered the results of two major tools: the scree plot and the cumulative variance. The first gives the eigenvalue representing the weight of the components of the entire dataset while the cumulative variance measures the deviation of the sum of each weighed components with respect to the average data. [17] Together they give the minimum number of components necessary to reconstruct the entire dataset.

For a very large dataset, an indicator function is designed to show the number of major components. In our case, due to the limited size of the dataset, scree plot and cumulative variance together are sufficient to estimate the major components. In Figure 5 we show how the five T-FTIR spectra of the La@C82:C2v can be reduced to three major components. A similar results is obtained for the T-FTIR dataset of the La@C82:Cs. On the contrary, the analysis of the eight T-XAS spectra of the two isomers returns only two major components. Since FTIR and XAFS datasets were collected

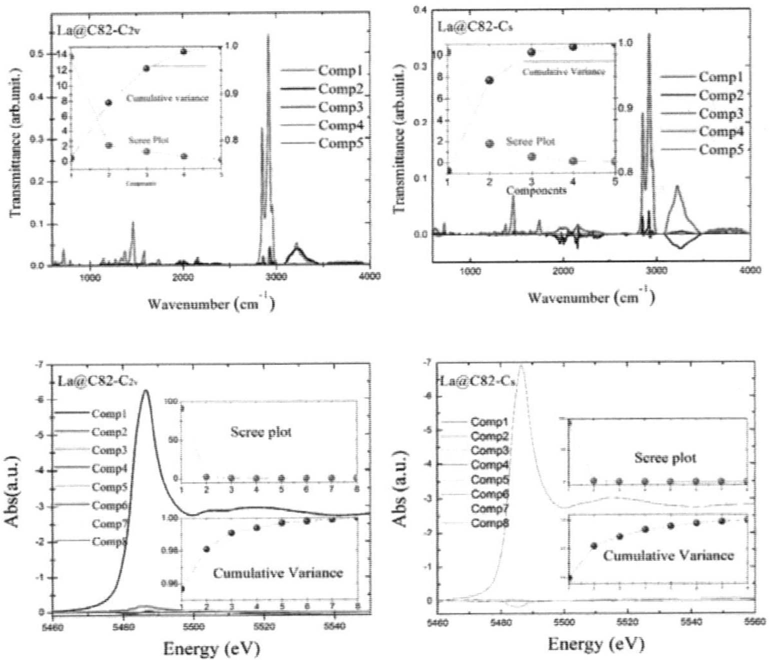

Figure.5: PCA of FTIR (top panels) and XAS (bottom panels) spectra of the two La@C82 isomers.

independently, the discrepancy points out that FTIR and XAFS techniques are not probing exactly the same sample or, in other words, the same structural configurations as a function of the temperature. It is then impossible to correlate the two datasets in a coherent way. To overcome the drawbacks of collection of experimental data in systems out of equilibrium [20,21] with independent techniques we recently proposed a beamline [22] that combines a synchrotron radiation X-ray and an IR beam into a sub-millimetre spot. This apparatus may perform simultaneous sub-sec time-resolved X-ray absorption spectroscopy and FTIR spectroscopy experiments. It can be envisaged that the capabilities provided by a simultaneous beamline may probe the interplay among electronic, vibrational and geometrical structures in metal endohedral fullerenes and in many other complex systems.

4. Conclusion

We presented an investigation of the dynamics of metal endohedral fullerenes by means of temperature-dependent XAS spectroscopy and FTIR spectroscopy. Changes observed in the temperature-dependent XAS spectra of La@C82 isomers can be associated to the dynamics of guest ions inside the carbon cage, in good agreement with previous theoretical reports. [14] Temperature-dependent FTIR spectra also show different temperature behaviours between the two isomers, characterized by different symmetries as well as different charge transfers. We performed an integral analysis on FTIR and XAS datasets by using the Principal Component Analysis. Datasets collected by different spectroscopic methods at different times show different numbers of major components. The discrepancy does not allow resolving the effective dynamics of these systems, while points out the need to perform in these systems simultaneous measurements. In addition to EMFs, many other complex systems and non-equilibrium phenomena can be probed with a simultaneous approach. Moreover, the principal component analysis method can be successfully applied in particular to large datasets collected in a concurrent/simultaneous way on the same sample.

Acknowledgements

This research has been supported by the National Natural Science Foundation of China (Grant No. 11105172).

References:

[1] H. Shinohara, Endohedral metallofullerenes.2000, *Rep. Prog. Phys.* **63**, 843
[2] P. Chawla, V. Chawla, R. Maheshwari, S. A. Saraf, and S. K. Saraf, 2010, *Mini-Rev. Med. Chem.* **10**, 662.
[3] O. Gunnarsson, Alkali-doped fullerides: narrow-band solids with unusual properties, (2004) Singapore: *World Scientific.*
[4] Robert D. Johnson, Mattanjah S. de Vries, Jesse Salem, Donald S. Bethune, and Costantino S. Yannoni,1992, *Nature* **355**, 239.
[5] Koichi Kikuchi, Nobuo Nakahara, Tomonari Wakabayashi, Shinzo Suzuki, Haruo Shiromaru, Yoko Miyake, Kazuya Saito, Isao Ikemoto, Masatsune Kainosho, and Yohji Achiba,1992, *Nature* **357**, 142.
[6] Yan Chai, Ting Guo, Changming Jin, Robert E. Haufler, L. P. Felipe Chibante, Jan Fure, Lihong Wang, J. Michael Alford, and Richard E. Smalley,1991, *J. Phys.Chem.* **95,** 7564.
[7] Shigeru Nagase, Kaoru Kobayashi, Tatsuhisa Kato, and Yohji Achiba,1993, *Chem. Phys. Lett.* **201**, 475.
[8] Kobayashi K and Nagase S 1998 *Chemical Physics Letters* **282,** 325-9
[9] Lei Liu, Bin Gao, Wangsheng Chu, Dongliang Chen, Tiandou Hu, Chunru Wang, Lothar Dunsch, Augusto Marcelli, Yi Luo, and Ziyu Wu, 2008, *Chem.Commun.* (2008) 474.
[10] A. Marcelli, A. Cricenti, W.M. Kwiatek, C. Petibois, *Biotech. Adv.* doi: 10.1016/j.biotechadv.2012.02.012
[11] A. Marcelli, W. Xu, L. Liu, C.R. Wang, W.S. Chu, and Z.Y. Wu,2009, *J. Nanophotonics.* **3**, 031975.
[12] B. Ravel and M. Newville,2005, *J. Synchrotron. Rad.* **12**, 537.
[13] http://sixpack.sams-xrays.com/
[14] P. Jin, C. Hao, S. Li, W. Mi, Z. Sun, J. Zhang, and Q. Hou,2007, *J.Phys.Chem.A.* **79**, 167.
[15] S. Lebedkin, B. Renker, R. Heid, H. Schober, and H. Rietschel,1998, *Appl. Phys. A* **66**,273.
[16] O. Gunnarsson, J. E. Han, E. Koch, and V. H. Crespi, Superconductivity in Complex Systems, ed K A Muller and A BussmannHolder (Berlin: Springer-Verlag Berlin) (2005) 71.
[17] E.R. Malinowski, Factor analysis in chemistry (1991), 2nd edition, (New York: John Wiley & Sons, Inc.)
[18] Thorsten Ressler, Joe Wong, Joseph Roos, and Issac L. Smith,2000, *Energy Environ. Sci.* **34**, 950.
[19] Peiqiang Yu,2005, *J. Agric. Food. Chem.* **53**, 7115.
[20] A. Marcelli, D. Hampai, Wei Xu, L. Malfatti, and P. Innocenzi,2009, *Acta. Phys. Pol. A* **115**,489-500
[21] Wei Xu, Augusto Marcelli, Dariush Hampai, Luca Malfatti, Plinio Innocenzi, Ulrich Schade, and Ziyu Wu,2011, *Rendiconti Lincei*, **22**, S59.
[22] Augusto Marcelli, Wei Xu, Dariush Hampai, Luca Malfatti, Plinio Innocenzi, Ulrich Schade, and Ziyu Wu,2010, *Anal. Bioanal.Chem.* **397**, 2095.

XAFS study on Ca local structure in natural glasses and tektite

T Tobase*[1], L Wang [1], A Yoshiasa[1], M Okube[2], T Nakatani[1], Y Hayasaka[3] and H Isobe[1]

[1] Graduated School of Science and Technology, Kumamoto University, Kumamoto, 860-8555, Japan

[2] Materials and Structures Laboratory, Tokyo institute of technology, Yokohama, 226-8502, Japan

[3] Faculty of Science, Hiroshima University, Higashi, Hiroshima, 739-8511, Japan

E-mail: 097s6018@st.kumamoto-u.ac.jp

Abstract. The local structures of tektite and natural glasses were studied by Ca K-edge X-ray absorption near edge structure (XANES) and extended X-ray absorption fine structure (EXAFS) in order to provide quantitative data on bonding distances and coordination numbers. The pre-edge peak intensities of tektites are 10.7-11.7%, and those of peudotachylite, Kirauea volcanic glass, impactite, pitchstone and perlite are 6.7-10.9%. The main peak shoulder intensities of tektites are 68.3-70.7%, and other natural glasses are 63.0-63.9%. XAFS analysis indicated all tektites possess 7-coordinated Ca, but natural glasses possess 6-, 7- and 8- coordinated Ca. This study indicated that different petrogenesis of natural glasses gives different local structures of calcium.

1. Introduction

In nature, several kinds of glasses are formed by various geological actives. Tektites and impactglasses are produced by impact event (Alvarez *et al.*, 1980); volcanic actives formed volcanic glass;and plate boundary active produced fault rocks. These natural glasses experienced different extreme environments, which should lead changing of local structure (Stebbins & McMillan, 1989; Paris et al., 1994; Mysen & Neuville, 1995; Yarger et al., 1995).

Upon a devastating impact of a giant meteoroid on the Earth, particles of the Earth's surface were melted and catapulted into outer space, where they finally solidified and fell back to the Earth; this series of processes is how tektites were formed (Wang et al., 2011). Impactites are stratigraphically and genetically equivalent to coherent impact melt rocks present in craters developed in crystalline targets (Gordon et al., 2005). Titanium, iron, and aluminum local structures are studied on natural glasses and tektites (Frages, 1997; Wang et al., 2011; Neuville et al., 2004). Only calcium aluminosilicate glasses were investigated (Combes et al., 1991; Poe et al., 2001) and few Ca K-edge XANES structure on natural glasses and tektites have been reported.

2. Experiment

The specimens of tektites are from different strewn fields, and they are tektite (hainanite, indochinite, philippinite, australite, bediasite and moldavite), impact glass (impactite, suevite, and köfelsite), volcanic glass (perlite, obsidian, pitch stone, rhyolitic glass, Kilauea volcanic glasses), and fault rock (pseudotachylite). In order to analysis the local structure of calcium in natural glasses, we used the

XAFS methods. The XAFS measurement of calcium local structure was preformed with a Si (111) double crystal monochromator at the beam line BL-9C of the Photon Factory in National Laboratory for High Energy Physics (KEK), Tsukuba, Japan. Spectra near calcium K-edge were collected in transmission and fluorescence mode at the room temperature. Analyses of XAFS data were performed by using XAFS93 and MBF93 programs (Maeda, 1987; Yoshiasa et al., 1997).

3. Result and discussion

Experimental Ca K-edge XANES spectra of tektites are shown in Fig.1A. All tektites show only one XANES pattern. The first pre-edge transition is observed at 4038.63eV and corresponds to a 1s→3d type transition (Combes et al., 2004). The white line appears at 4049.21eV, corresponding to the main 1s →3p transition. The main white line peak has a shoulder observed near 4045.17eV before peak top. Each main peak white line height is normalized at 4049.21eV as 100%. The intensities of pre-edge and shoulder is 10.7-11.7% and 68.3-70.7%.

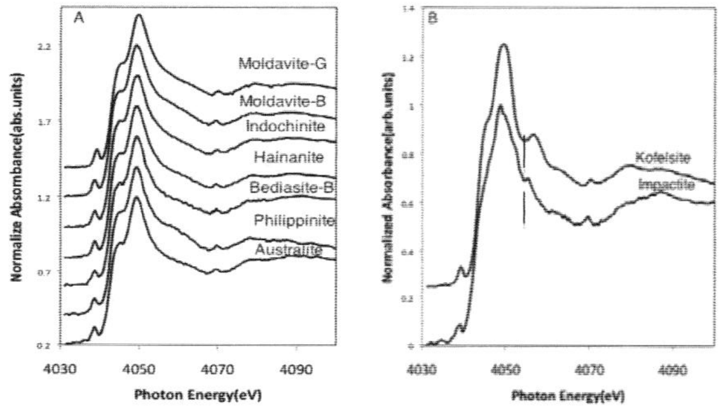

Fig.1. XANES spectra near the Ca K-edge for tektite and impact glasses.

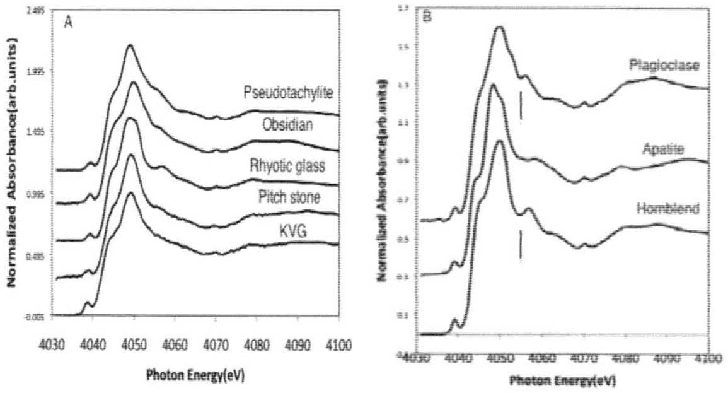

Fig. 2. XANES spectra near the Ca K-edge for natural glasses and minerals

Fig.1B shows XANES spectra of impact glasses. XANES spectra of two impact glass-impactite and köfelsite composed of pre-edge at 4039.132eV with intensity of 8.5%, a shoulder at 4045.17 with intensity of 67.5% and 67.8%, and white line at 4049.208eV. The pre-edge peak height of impact glasses is lower than tektites, this may cause of high distortion around Ca atom in tektites. The main peak shoulder intensity of tektite is a bit higher and broader than impact glasses. All tektites contain 70.53-79.35wt% SiO_2, and impact glasses is 60.16wt% (details are shown in Table 1.). The shoulder intensities are increasing with the content of SiO_2. This result is corresponding to previous study (Neuville et al., 2004). In addition, a slight difference in the trend of post-edge details became apparent in the calcium K-edge XANES spectra of impact glasses, with a possible splitting a contribution.

In Fig.2A, pre-edge and shoulder in natural glass possess more lower intensities than tektites and impact glass, and around 4.0-8.0%, 60.1-68.0% (details are shown in Table 2). But Kilauea volcanic glass (KVG) is an exception that its pre-edge intensity is closely similar with tektites around 11.3%.

Chemistry composition of Kilauea volcanic glasses is distinguishing; it is similar with basalt. However, tektites and natural glasses are similar with rhyolitic glass. Pseudotachylite, Kilauea volcanic glass, and rhyolitic glass possess a same contribution at the post-edge with impact glasses. Compare the tektites and natural glasses with reference minerals: plagioclase, apatite, and hornblende (Fig.2C). Hornblende and plagioclase, in which Ca occupies eight-coordinated site, have a slight shoulder and their post-edge have a contribution. These features are same as impact glasses and natural glasses. The difference in the post-edge features (near 4056.30eV) in impact glasses and natural glasses suggests a modification in calcium environment, being possible that calcium ions partially substitute for other ions in eight-coordinated site. Apatite, which contains six-coordinated calcium, is extremely alike with tektites in post-edge feature. Degree of relaxation of impact glasses and volcanic glasses is heigher than tektites. Different relaxation processes are caused by environment when impact glasses are formed. Temperature and annealing time are most important condition for relaxation.

Table 1. SiO_2, CaO content, pre-edge and shoulder intensities of tektites and impact glasses.

Sample name	SiO_2 (wt%)	CaO (wt%)	Pre-edge intensity (%)	Shoulder intensity (%)
Australite	71.89	3.28	11.7	68.7
Bediasite-Black	74.4	0.57	11.4	70.7
Hainanite	72.96	2.22	11.0	69.1
Moldavite-green	79.35	2.55	11.4	68.3
Moldavite-black	76.67	3.07	11.6	69.8
Indochinite	73.88	1.81	11.0	70.0
Philippinite	70.53	2.96	10.7	69.7
Impactite	-	-	8.5	67.8
Köfelsite	60.16	1.44	8.5	67.5

Table 2. SiO_2, CaO content, pre-edge and shoulder intensities of natural glasses.

Sample name	SiO_2 (wt%)	CaO (wt%)	Pre-edge intensity (%)	Shoulder intensity (%)
Kilauea volcanic glass	50.89	11.19	11.3	65.1
Pitchstone	72.81	0.89	6.1	60.1
Rhyolitic glass	70.93	1.67	6.7	60.3
Obsidian	68.77	2.95	8.0	68.0
Pseudotachylite	61.81	0.43	4.0	62.7

The EXAFS $k^3\chi(k)$ functions were transformed into the radial distribution functions (RDF) for Ca K-edge of five tektites and three impact glasses, as shown in Fig.3. The RDF for Ca atoms in all tektites are very similar, and RDF in natural glasses are slightly shifted in perlite. It indicates that all tektites have a similar local atomic environment around the Ca atoms, but have difference in natural glasses.

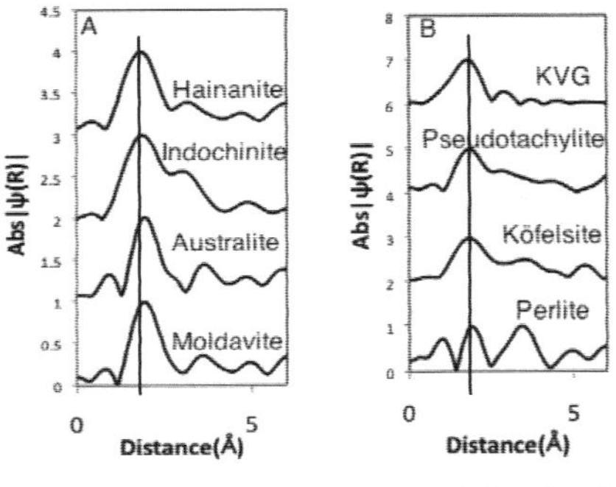

Fig.3. Fourier transforms of the Ca K-edge EXAFS oscillation function $k^3\chi(k)$. No phase shift corrections are made

In order to obtain the further information on structure parameter, we figured out the parameter fitting with analytical EAXFS formula expressed by cumulant expansion up to third order terms in tektites. The obtained structural parameters are summarized in Table 3. Ca-O distances in all tektites are around 2.4 Å. And in natural glasses are from 2.37-2.46 Å, it is more various than tektites. Tektites possess 7-coordinated Ca, but natural glasses possess 6-, 7- and 8- coordinated Ca. Variation in thermodynamic parameter may be an important factor for formation of the local structure in impact glasses and volcanic glasses.

Table 3. The Structure parameters determined by EXAFS for tektites. The expected bonding distances of 6-, 7-and 8- coordinated Ca-O based on the Shannon Ionic Radii: Ca (6-coordinated)-O=2.38Å; Ca (7-coordinated)-O=2.44Å; Ca (8-coordinated)-O=2.50 Å. Uncertainties in the last decimal place are shown in parentheses

Sample name	σ^2(Å)	Coordination number	Ca-O distance(Å)	R-factor (%)
Australite	0.007(2)	7	2.419(23)	1.6
Hainanite	0.005(2)	7	2.400(19)	0.14
Indochinite	0.002(2)	7	2.416(23)	0.14
Moldavite-green	0.006(2)	7	2.406(25)	0.70
Kilauea volcanic glass	0.020(3)	6	2.371(26)	0.47
Pseudotachylite	0.003(2)	6	2.380(23)	0.42
Köfelsite	0.004(1)	7	2.438(19)	6.7
Perlite	0.014(4)	7 and 8	2.464(43)	1.8

4. Acknowledgment

This study was performed under the auspices of the Photon Factory (PAC No. 2011G611). The synchrotron radiation XAFS experimental is performed at the beam line BL9C at the Photon Factory.

References

[1] Alvarez L W, Alvarez W, Asaro F and Michel H V 1980 *Science* **208** 1095-1108
[2] Combes J –M, Brown Jr G E, Waychunas G A, 1991 X-ray Absorption Fine Structure, Proc. Sixth Int. Conf. on XAFS York 312–314
[3] Cormier L, and Neuville D R, 2004 Chem. Geol. **213** 103-113
[4] Frages F, and Brown G E, 1997 *Geochim. Cosmo. Acta.* **61** 1863-1870
[5] Gordon R O, Pascal L, John G S, John P, Darlene S S L, Theodore E B, Charles S C and Brian G (2005) Meteoritics and Planetary Science **40** 1759-1776
[6] Maeda H, 1987 *J. Phy. Soc. Jpn.* **56** 2777-2787
[7] Mysen B, and Neuville D, 1995 *Geochim Cosmochim Acta* **59** 325-342
[8] Neuville D R, Cormier L, Flank A, Briois V, and Massiot D, 2004 *Chem. Geol.* **213** 153-163
[9] Paris E, Dingwell D, Seifert F, Mottana A, and Romano, C, 1994 *Phys. Chem. Miner.* **21** 520-525
[10] Poe B T, Romano C, Zotov N, Cibin G, and Marcelli A, 2001 *Chem. Geol.* **174** 21–31.
[11] Stebbin J F, and McMillan P, 1989 *Am. Mineral* **74** 965-968
[12] Wang L, Yoshiasa A, Okube M, and Takeda T, 2011 *J. Synchrotron Rad.* **18** 885-890
[13] Yarger J L, Diefenbacher J, Smith K H, Poe B, and Yoshiasa A, Koto K, Maeda H, and Ishii T, 1997 *Jpn. J. Appl. Phys.* **36** 781-784

Low Temperature XRF-yield XAFS of low concentration Mo in steel

M Nagoshi[1], T Kawano[2], K Sato[1] and Y Funakawa[1]

[1]Steel Research Laboratory, JFE Steel Corporation, 1, Kawasaki-cho, Chuo-ku, Chiba, 260-0835, Japan
[2]Steel Research Laboratory, JFE Steel Corporation, 1-1, Minamiwatarida-cho, Kawasaki, 210-0085, Japan

E-mail: m-nagoshi@jfe-steel.co.jp

Abstract. Low temperature x-ray fluorescence yield x-ray absorption fine structure measurements have been carried out for 0.2 mass% Mo in steel and carbides extracted from a steel sample. The increase in amplitude of the extended x-ray absorption fine structure, at 8 K, was larger for Mo in bcc-Fe matrix than for Mo in carbides. This is explained by the differences in distortion of the local structure around Mo atoms between the two chemical states. The Ratio method was applied to the EXAFS, measured at 8 K, for the Fe atoms surrounding the Mo in the bcc-Fe matrix. The atomic fraction of the Mo in solution and Mo as carbides in steel was determined by this technique.

1. Introduction

Micro-alloying elements added to steel dramatically change the properties of steel products such as strength and corrosion resistance. It is important to understand the roles of these elements on the properties of the steel for designing micro-alloyed steel products of superior performance.

Nanometer-sized precipitates formed by micro-alloying elements effectively improve the strength of steel. These precipitates have been directly observed by microscopic analytical techniques such as transmission electron microscopy (TEM) [1] and atom probe (AP) microscopy. [2] XAFS is a powerful tool for clarifying the local structures of specific elements forming the nanometer-sized precipitates. [3-5] Especially, x-ray fluorescence (XRF) yield XAFS techniques have the advantages of high detectability and quantitative analysis performance for the chemical states of micro-alloying elements in steel. [6] Therefore XAFS and microscopic techniques are complementary in revealing all the added elements in steel. [3,7]

Low temperature measurements enhance the amplitude of extended x-ray absorption fine structure (EXAFS) due to the reduction of thermal fluctuations of atoms. Although this technique has been used in much basic research [8,9], there are very few studies applying it to micro-alloying elements using XRF-yield XAFS. In this study, low temperature XRF-yield XAFS measurements have been applied to Mo added to steel in order to obtain detailed information about the local structures. XAFS spectra measured for Mo in steel at low temperature and room temperature (RT) are presented and discussed in terms of the local structure of Mo and its analysis technique.

2. Experimental

The XAFS experiment was carried out at beam line BL01B1, SPring-8 (Hyogo, Japan) with a two-crystal Si (111) monochromator. The Mo-K edge XAFS spectra were recorded in the x-ray fluorescence mode with a 19-element Ge solid-state detector. The samples were cooled to 8 K using a He cryostat equipped with the beam line. The analysis of the XAFS spectra was performed with commercial software (Rex 2000, Rigaku Corp.).

Chemical composition of the base steel was 0.04mass%C-1.5mass%Mn. Ti and Mo were added as atomic ratios of Ti/C = 0.5 and Mo/C = 0.6. Two steel samples were prepared. Sample A was obtained by quenching from 900 ℃ to room temperature. This sample was used as a reference for Mo which is substitutional in the bcc-Fe matrix. Sample B was obtained by hot-rolling with a coiling temperature of 650 ℃. The details concerning this steel were reported in ref [10]. Mo-containing carbides, with the diameter of a few nanometers, were observed by TEM for Sample B. Residue extracted from Sample B was obtained by the electrolytic extraction method and measured by XRF-yield XAFS as a reference for the carbides formed in Sample B.

Journal of Physics: Conference Series **430** (2013) 012071 doi:10.1088/1742-6596/430/1/012071

3. Result and discussion

Figure 1 shows Mo-K edge x-ray absorption near edge structure (XANES) spectra measured at RT and 8 K for Sample A, Sample B, and residue extracted from Sample B. The XANES spectra measured at 8 K and at RT are indistinguishable from each other for all samples. These results show that the chemical state of Mo at RT in steel and in carbides hardly changes at 8 K. The XANES spectra for Sample B have the characteristics of those for both Sample A and Residue. This shows that the Mo atoms in Sample B are partially precipitated as carbides and the remainder are retained in the bcc-Fe matrix.

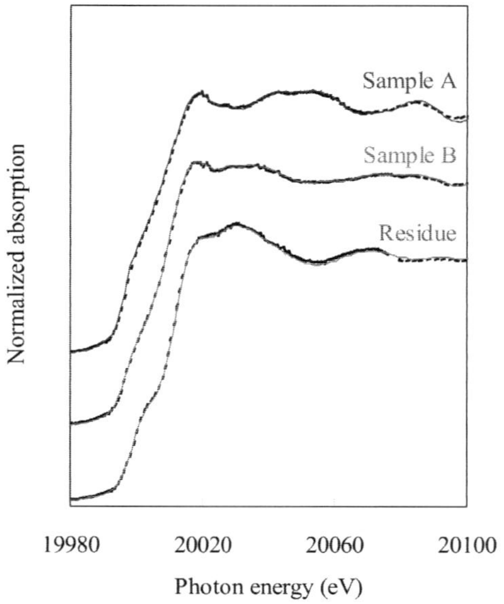

Fig.1 Normalized Mo-K edge x-ray fluorescence XAFS spectra for Sample A, Sample B, and residue extracted from Sample B. The spectra measured at RT and at 8 K are denoted as broken and solid lines, respectively.

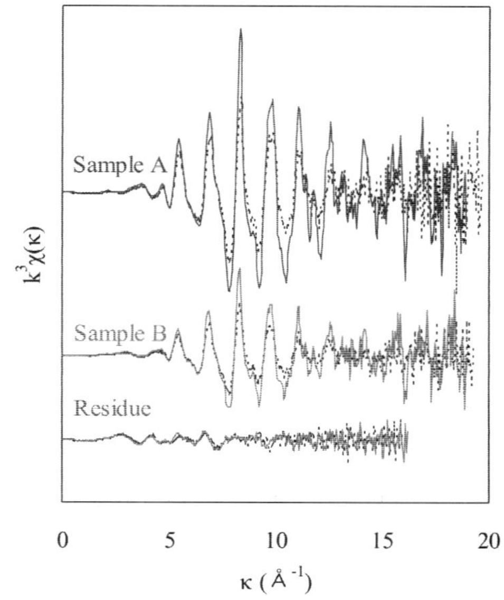

Fig.2 k^3-weighted EXAFS spectra for Sample A, Sample B, and the extracted residue (carbides) from Sample B. The spectra measured at RT and at 8 K are denoted as broken and solid lines, respectively.

Figure 2 shows k^3-weighted EXAFS for Sample A, Sample B, and extracted carbides from Sample B. In the figure, the spectra measured at RT and 8 K are denoted as broken and solid lines, respectively. The EXAFS oscillations for Sample A originated from Fe atoms surrounding Mo atoms in the bcc-iron matrix. The EXAFS for Sample B has almost the same structure as that of Sample A except for the weaker amplitude. This is explained by the reduction of the number of Mo atoms in the bcc-Fe matrix due to the precipitation of Mo as carbides as indicated in Fig.1. The EXAFS oscillation for the carbides is very weak as shown in Fig. 2 and invisible in the EXAFS for Sample B. The difference of the EXAFS amplitude between the steel and carbide samples is caused by the difference of the Deby-Wallar factor (DW factor ; σ) as well as by the differences of the scattering amplitude and the number of the neighboring atoms.

The amplitude of the EXAFS measured at 8 K was higher than that measured at RT and clearer oscillations were observed, especially in the high k region for Samples A and B. The Fourie transforms (FT) derived from the k^3-weighted EXAFS measured at 8 K show discriminating peaks from the 1st to

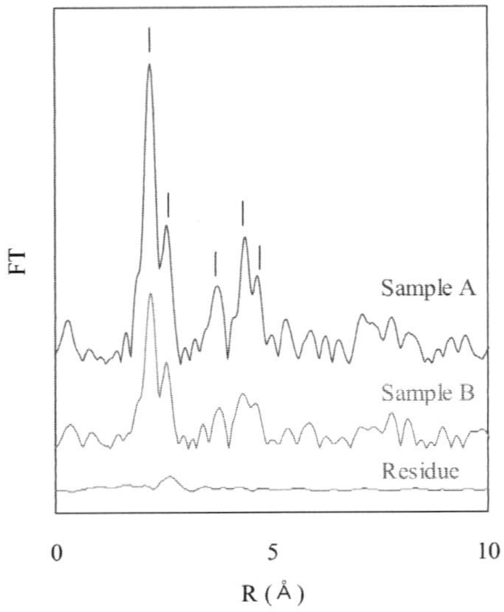

Fig.3 Fourier transforms of k^3-weighted Mo-K edge EXAFS spectra measured at 8 K for Sample A, Sample B, and extracted residue (carbides) from Sample B. The k range for FT is 3 – 19.5 $Å^{-1}$ for the steel samples and 3 – 12 $Å^{-1}$ for the residue sample.

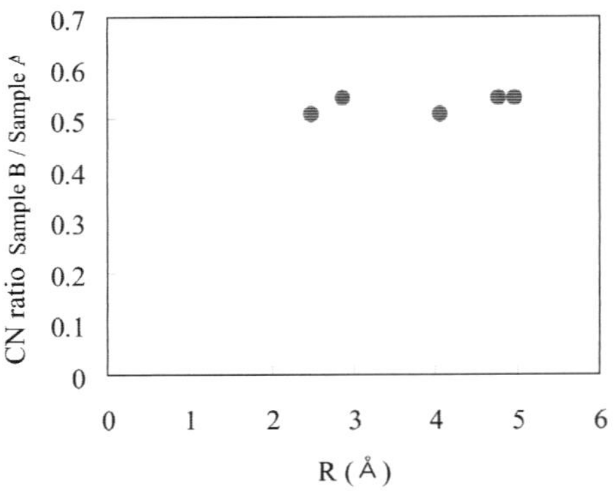

Fig.4 Coordination number ratio of Sample B to Sample A for the 1st to 5th NN Fe atoms surrounding Mo atoms as estimated by the Ratio method.

at least the 5th nearest neighbor (NN) Fe atoms around the Mo in the bcc-Fe matrix, as shown in Fig.3. On the other hand, the increase of EXAFS amplitude at 8K for the carbides is small when compared with those for Samples A and B. (Fig.2)

The DW factor was estimated for the 1st NN Fe atoms around the Mo by the carve-fitting method, assuming that the coordination number (CN) is 8 for Sample A. The DW factor was 0.064 Å at RT and 0.041 Å at 8 K and the difference of σ^2 ($\Delta\sigma^2$[RT – 8 K]) is 0.0024 $Å^2$. The latter value was also confirmed by the Ratio method.[7] The differences of σ^2 were estimated for the first shell as $\Delta\sigma^2$[300 K – 77 K] ~ 0.0047 $Å^2$ for Cu [8] and $\Delta\sigma^2$[300 K – 40 K] ~ 0.005 $Å^2$ for Ni [9] from the previous reports. Our datum for the Mo in the bcc-Fe matrix is about half of the previous results for pure metal. The DW factor can be divided into two parts corresponding to thermal and static fluctuations.[11] The smaller temperature dependence of the DW factor for Sample A suggests a relatively large static fluctuation compared with pure metal. This can be explained by distortion in the bcc- Fe lattice adjacent to the Mo atoms in the bcc-Fe matrix because the atomic radius of bcc- Mo (0.136 nm) is different from bcc- Fe (0.124 nm). The difference in the amplitude of the EXAFS between RT and 8 k for Mo in carbides is remarkably small. We consider that one of the reasons for this result is the large static fluctuation of the carbide structure. The nanometer-sized carbides in this type of steel have been reported as (Ti, Mo)C with a NaCl type structure.[10] The distortion caused by the coexistence of Mo and Ti in the carbide particles and the surface effects resulted from the nanometer size of the particles probably are the reasons for the large static fluctuation around the Mo atoms in the carbides.

At low temperatures, the difference in amplitude of EXAFS oscillation between Mo in the bcc-Fe matrix and in carbides increases, which would improve the quantitative analysis of the atomic fraction of Mo in the bcc-Fe matrix and in the precipitates using the curve-fitting for EXAFS.[6] However, the estimated fraction of the former Mo from the EXAFS measured at 8K, 51 %, was almost the same as that from the EXAFS measured at RT, 50 %, for Sample B.

The Ratio method [8] can be applied to EXAFS measured at 8 K for determining the coordination number (CN) of Fe atoms around Mo atoms in the bcc Fe-matrix because the different NN Fe atoms are clearly distinguished as shown in Fig.3. This method provides relative CNs for each NN Fe atom without fitting techniques. The results are shown in Fig.4 for the 1st to the 5th NN Fe atoms around Mo for Sample B when Sample A is used as reference. The relative CN of Fe atoms for Sample B to Sample A are between 0.51 and 0.54. This is in good agreement with the values estimated by the curve-fitting methods mentioned above.

4. Conclusion

The XRF-mode XAFS measurements were carried out for low concentration Mo in steel at RT and 8 K. The difference in the DW factor between RT and 8 K for Mo in bcc-Fe was lower than those reported for pure metals. Mo in carbides showed very small differences in the EXAFS amplitude between RT and 8 K compared to Mo in the matrix. These results suggest relatively ordered and strongly distorted local structures around Mo atoms in matrix and carbides, respectively. The ratio method was successfully applied for the 1st to the 5th NN Fe atoms surrounding Mo atoms in the bcc-Fe matrix using EXAFS measured at 8K. The low temperature XRF-XAFS is useful for further investigation of the local structures of the micro-alloying elements in steel.

Acknowledgements

The XAFS measurements were carried out at the SPring-8 with the approval of the Japan Synchrotron Radiation Research Institute (Proposal No. 2003A0452-NX-np). The authors thank Dr.T.Uruga for his kind cooperation.

References

[1] Nakamichi H *et al* 2011 *J.Microsc.* **242** Pt 1, 55.
[2] Danoix F *et al* 2006 *Adv. Eng. Mater.* **8** 1202.
[3] Pizzini S *et al* 1990 *Phil.Mag.Lett.* **61** 223.
[4] Watanabe T *et al* 2003 *User Expriment Report Spring-8 (2002A)* **No.9** 177.
[5] Nagoshi M *et al* 2003 User *Expriment Report Spring-8 (2003A)* **No.11** 12.
[6] Nagoshi M *et al* 2005 *Physica Scripta.* **T115**, 480.
[7] Mulholland M D and Seidman D N 2011 *Acta Materialia* **50** 1881.
[8] Stern E A 1975 *Phys.Rev.* B **11** 4836.
[9] Yokoyama T 1998 *Phys.Rev.* B **57** 3423.
[10] Funakawa Y *et al* 2004 *ISIJ int.* **44** 1945.
[11] Bunker G 1983 *Nuclear Instrum. Methods* **207** 437.

Journal of Physics: Conference Series **430** (2013) 012072 doi:10.1088/1742-6596/430/1/012072

X-ray absorption spectroscopy and photoemission study of Bi-doped $LaMnO_3$

H G Zhang[1], Y T Li[1, 2], X G Dong[1], Q T Hou[1], Y C Huang[1], H Liu[1] and Q Li[1]
[1] Physics Department, Southeast University, Nanjing, 211189, P. R. China

[2] College of Science, Nanjing University of Post and Telecommunications, Nanjing, 210046, P. R. China

E-mail: qli@seu.edu.cn

Abstract. The magnetic property of Bi-doped $LaMnO_3$ was studied from perspective of valence state and local distortion given by XPS and XAFS. An isotrivalent Mn^{3+} is proved to be existed in samples by XPS due to trivalent Bi^{3+} ions doped $LaMnO_3$. Compared to the traditional ferromagnetic-paramagnetic phase transition with double exchange interaction between mixed valence states, this phase transition in our samples with isotrivalent Mn ions is understood by superexchange interaction in the way of Mn *K*-edge XAFS. A net ferromagnetic moment exists when the Mn-O-Mn bond angle deviates from 180°. The regular pattern of magnetization with doping concentration is related to the disorder degree of Mn-O and Mn-Mn pairs.

1. Introduction

To pervoskite manganites with colossal magnetoresistance (CMR), the occurrence of ferromagnetic (FM)-paramagnetic (PM) phase transition is generally understood by double exchange (DE) interaction [1], although DE interaction alone can not account for the large magnitude of magnetoresistance [2]. In the DE model, Mn ions should be in mixed-valence states. The ferromagnetic transition with metallic transition can be explained by DE interaction between Mn^{3+} ($t^3{}_ge^1{}_g$) and Mn^{4+} ($t^3{}_{2g}$) via oxygen ions due to the Hund coupling. In fact, the DE interaction is also operative in the electron-doped colossal magnetoresistive manganites with Mn^{2+} ($t^3{}_{2g}e^2{}_g$) /Mn^{3+} ($t^3{}_ge^1{}_g$) mixed-valence states [3].

As is known, $LaMnO_3$ parent compound is an antiferromagnetic (AFM) insulator. Divalent cations or tetravalent ions doped at La-site can induce the mixed-valence states of transition metal Mn ions. However, trivalent ions (like Dy^{3+}, Bi^{3+}) doped $LaMnO_3$ samples also exhibit FM-PM phase transition with negative magnetoresistance [4]. It is known that the chemical valence of Mn ions in trivalent ions doped $LaMnO_3$ samples generally shows consistence with that of their parent compound due to the charge neutrality of the whole system. If so, does the FM-PM phase transition in this system also originate from the mixed-valence states, like in compounds with divalent cations or tetravalent ions doping at La-site? In order to understand the change of magnetic properties in trivalent ions doped $LaMnO_3$, it is essential to investigate the electronic structure of this system, especially the electronic structure of Mn ions and doping ions.

In this paper, the valence states of Mn ions and related magnetic properties in the trivalent Bi^{3+} ions doped $LaMnO_3$ samples are investigated by x-ray photoemission (XPS) and extended x-ray absorption fine structure (EXAFS). An isotrivalent Mn^{3+} is proved to exist in samples. By studying

EXAFS of Mn *K*-edge, traditional FM-PM phase transition in Bi-doped $LaMnO_3$ with isotrivalent Mn ions is understood by superexchange interaction, compared to that with double exchange interaction between mixed valence states. A net ferromagnetic moment exists when the Mn-O-Mn bond angle deviates from 180º. And the regular pattern of magnetization with doping concentration is related to the disorder degree of Mn-O and Mn-Mn pairs.

2. Experimental

The compounds $La_{1-x}Bi_xMnO_3$ (x=0.2, 0.25, 0.3, 0.4) were prepared by solid state method. The high-purity oxides with stoichiometric ratio La_2O_3, Bi_2O_3, MnO_2 were mixed together, and the mixtures are thoroughly ground within petroleum ether. Three times were calcined at 900 ºC, 950 ºC, 1050 ºC in ambient atmosphere for 10 hours, respectively. The structural characterization was carried out by X-ray diffraction (XRD) with Cu-*Kα* at room temperature. XPS spectrum of Mn 2*p* core level with photon energy $h\nu$=1486.7 eV and EXAFS of Mn *K*-edge were performed at National Synchrotron Radiation Laboratory (NSRL), University of Science and Technology of China (USTC). Its binding energy position was calibrated by C 1*s* binding energy of 284.6 eV.

3. Results and Discussion

Fig.1 shows the XRD patterns of $La_{1-x}Bi_xMnO_3$ (x=0.2, 0.25, 0.3, 0.4) samples at room temperature. It is suggested that the well-single phase samples have been synthesized and all the samples can be indexed as rhombohedral structure, which are same as results in literature [4]. A shift of diffraction angle (inset of Fig.1) is observed. Except sample with x=0.4, diffraction angle for another three samples shifts to higher angle with the increasing of doping concentration, which means the contraction of lattice. It is due to the increasing of doping Bi^{3+} (r=0.096nm) with lower ions radius compared to La^{3+} (r=0.106nm) ions. However, when the concentration x increases to 0.4, the unusual behavior of XRD peaks indicates the crystal structure may begin to transform, which can be seen from that the doublet peak (1,0,-1) and (2,0,1) tend to merge to a single peak.

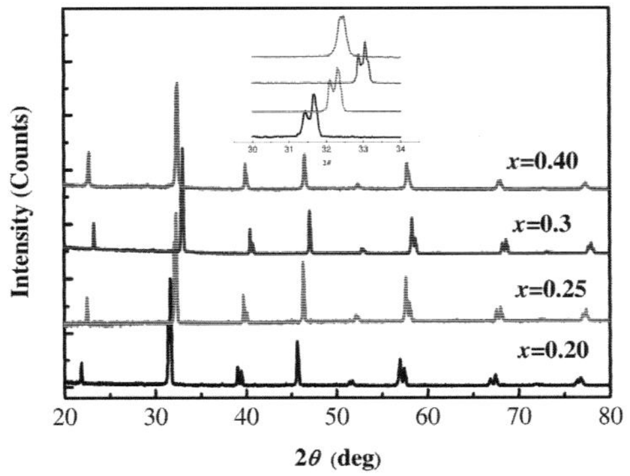

Figure 1. The XRD patterns of $La_{1-x}Bi_xMnO_3$ (x=0.2, 0.25, 0.3, 0.4) samples at room temperature. The inset shows the shift of diffraction angle for the strongest peaks (1,0,-1) and (2,0,1).

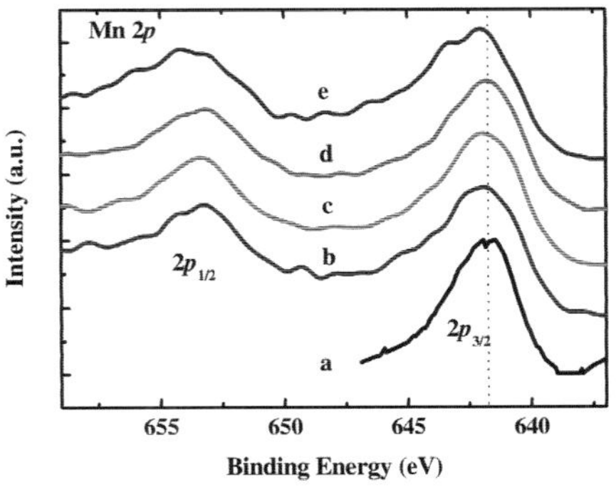

Figure 2. The XPS spectra of Mn 2p core level for $LaMnO_3$ (a) and $La_{1-x}Bi_xMnO_3$ (x=0.2(b), 0.25(c), 0.3(d), 0.4(e)) samples. Data for $LaMnO_3$ (a) comes from Ref [5].

To the valence states of Mn ions, the XPS spectra of Mn 2*p* core level are shown in Fig. 2. Two obviously intense peaks at near 653.2 eV and 641.7 eV are observed, which is the Mn $2p_{1/2}$ and $2p_{3/2}$ states. The 11.5 eV difference between the two states is owing to the spin-orbital splitting. It can be seen from the Figure that there is no obvious change of binding energy for Mn $2p_{3/2}$ and $2p_{1/2}$ levels with doping concentration. The value of binding energy of Mn $2p_{3/2}$ peak for our samples is 641.7 eV, which is consistent with that of parent compound $LaMnO_3$ [5]. It demonstrates that the electronic state of Mn ions is proved to maintain trivalent after trivalent Bi^{3+} ions are doped in $LaMnO_3$.

Now that the valence state of Mn ions with trivalent in prepared samples is determined, it is necessary to check the magnetic property of those sample that whether they show FM-PM phase transition without existence of double exchange interaction between Mn^{3+}-O-Mn^{4+}. The magnetization as a function of temperature (*M-T* curve) presented in Fig. 3 exhibits the traditional transition from FM to PM phase. The transition temperature T_C is obtained from the maximum of d*M*/d*T*. The value of T_C is 139.1, 110.0, 122.3, and 71.1 K for x=0.20, 0.25, 0.30, 0.40 respectively. The magnitude of magnetization at low temperature is 5.47, 6.61, 6.31, 3.24 emu/g with increasing of doping concentration respectively. From the tendency of magnetic variation, the singularity for x=0.25 is displayed.

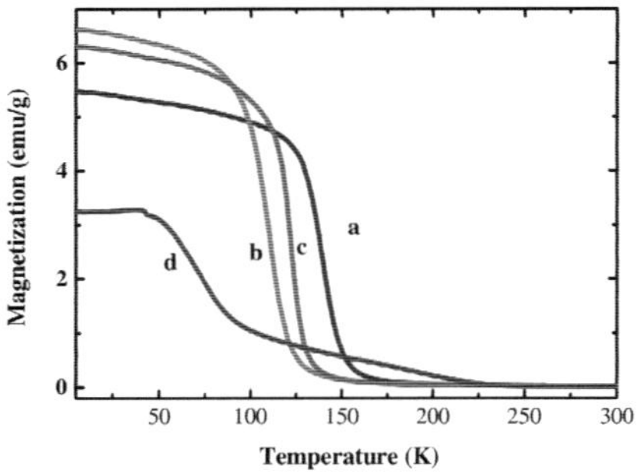

Figure 3. *M-T* curve for $La_{1-x}Bi_xMnO_3$ (x=0.2(a), 0.25(b), 0.3(c), 0.4(d)) samples.

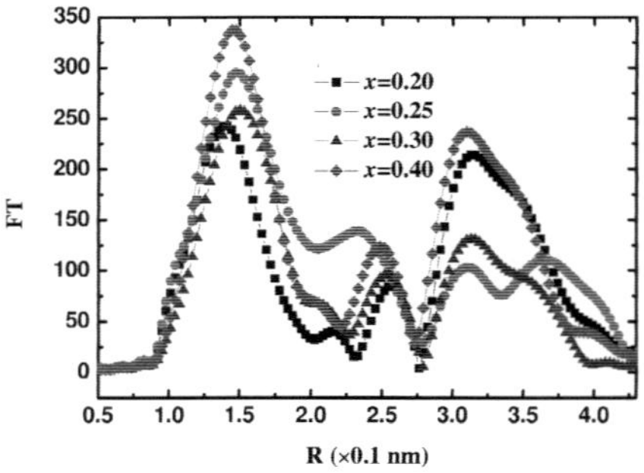

Figure 4. Radial structure distribution function of samples $La_{1-x}Bi_xMnO_3$ (x=0.2, 0.25, 0.3, 0.4)

To the FM to PM phase transition, the DE interaction between Mn^{3+} and Mn^{4+} is usually taken into main consideration. However, under the existence of only isovalent Mn^{3+} ions in our system, the transition to ferromagnetism at low temperature does never be explained by DE again due to the absent of Mn^{4+} ions, but can be understood by super-exchange AFM interaction among Mn^{3+} ions. Although superexchange between Mn^{3+} ions via O^{2-} ions exhibits AFM ordering when Mn^{3+}-O-Mn^{3+} bond angle is 180° based on Goodenough's predictions [6], A FM ordering can be obtained when bond angle of Mn^{3+}-O-Mn^{3+} is bended to lower than 180° with higher lattice distortion. A net FM moment exists when spins antiparallel to each other with a small canting angle. Thus, a larger distortion in the samples would be shown up. The local structural distortion was studied by the following data of EXAFS.

As to understand the tendency of magnetic properties with doping concentration, the local structure around Mn atoms is thus investigated. As literature reported [7-8], high disorder degree of Mn-O distribution (σ^2) is related to the low magnetization. Thus, the radial structure function distribution curve is obtained from extend XAFS after background subtraction, normalization, Fourier transform, which is shown in Fig. 4. The first peak near 0.14 nm is the first shell information manifesting the distribution of Mn-O bond, and the second shell coordination peak near 0.3 nm results mainly from contribution of Mn-Mn, and Mn-La/Bi bonds. The regular pattern of the intensity for the first peak and second peak is plotted in Fig. 5.

It can be seen that the change tendency of the intensity in first shell with doping concentration is basically same as that of magnetization, except for sample x=0.4, while that of the second shell is just completely opposite. It is reported that a low degree disorder of Mn-O bond distribution is favorable for the FM interaction via DE interaction [8], because DE interaction happens through Mn-O bonds. Thus, so does the superexchange interaction between Mn^{3+} ions via O^{2-} ions, like that happened for DE interaction [8]. Thus, sample x=0.2 with a low disorder degree of Mn-O shows a large magnetization. To the singularity of x=0.4, the disorder degree of second shell should be taken into consideration. In fact, the Mn-O-Mn bond angle of near 180° with low degree of disorder is generally to be formed in the perovskite manganite structure. Thus, the degree of disorder in the second shell means the bending degree (deviate from 180°) of Mn-O-Mn bond angle or the amount of such bond bending. It is correspondingly understood that sample x=0.25 displaying higher magnetization is due to the higher bending degree of Mn-O-Mn bond angle or more bending bonds caused by disorder degree of Mn-Mn pairs, which is apt to give rise to ferromagnetic ordering by superexchange interaction. Although the low degree disorder of Mn-O is exhibited for sample x=0.4, the low degree disorder of Mn-Mn pairs as a fundamental to ferromagnetic superexchange determines the low magnetization at low temperature.

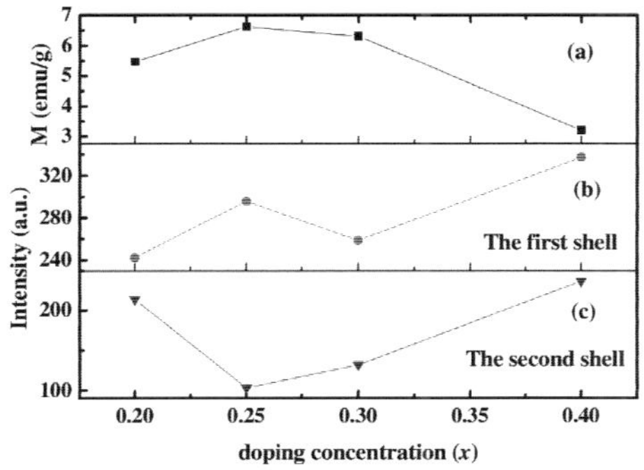

Figure 5. Doping concentration dependence of magnetization at low temperature (a), intensity of the first peak (b) and second peak (c) in radial structure function (from Figure 4).

4. Conclusion

In the paper, an isotrivalent Mn^{3+} in the trivalent Bi^{3+} ions doped $LaMnO_3$ samples is proved to exist by XPS. Compared to the traditional FM-PM phase transition with DE interaction between mixed valence states, this phase transition in Bi-doped $LaMnO_3$ with isotrivalent Mn ions is understood by superexchange interaction in the way of Mn K-edge XAFS. A net FM moment exists when the Mn-O-Mn bond angle deviates from 180°. The regular pattern of magnetization with doping concentration is related to the disorder degree of Mn-O and Mn-Mn pairs.

References

[1] Zener C 1951 *Phys. Rev.* **82** 403
[2] Millis A J *et al* 1995 *Phys. Rev. Lett.* **74** 5144
[3] Han S W *et al* 2004 *Phys. Rev. B* **69** 104406
[4] Zhao Y D *et al* 2004 *J. Magn. Magn. Mater.* **280** 404
[5] Matsuno J *et al* 2002 *Europhys. Lett.* **59(2)** 252
[6] Goodenough J B 1963 *Magnetism and the Chemical Bond* (Inter-science, New York)
[7] Downward L *et al* 2005 *Physica Scripta.* **T115** 629
[8] Jiang Y *et al* 2007 *Phys. Rev. B* **76** 224428

Lattice mismatching effect in composites systems

Xueguang Dong[1], Hongguang Zhang[1], Yongtao Li[1, 2], Hao Liu[1], Kai Chen[1], Qingteng Hou[1], Yongchao Huang[1] and Qi Li[1]

[1] Department of physics, Southeast University, Nanjing, 211189, P. R. China

[2] College of science, Nanjing University of Posts and Telecommunications, Nanjing, 210003, P. R. China.

E-mail: qli@seu.edu.cn

Abstract. The superconductivity can be studied using composites systems. Recently studies are focused on the relation between magnetism and superconductivity by introducing manganese oxide, which exhibits room temperature ferromagnetism, into $La_{2-x}Sr_xCuO_4$ system. Results indicate that the magnetism can have effect on superconductivity properties. However, in this paper it is found that the influence of sintering temperature for superconductor is also very important. The composites systems are synthesized and the experimental results indicate that the local structural distortion can be caused by high sintering temperatures. The distortion leads to changes of energy band around Fermi level and it is commonly recognized that the energy band around Fermi Level is very important for superconducting. While in the low sintering temperatures the local structural distortion is not found so the sintering temperature should be carefully concerned when the composites systems are synthesized.

1. Introduction

High-T_c superconductivity (SC) has been an extensive research topic since it was found in cuprates [1]. Today many issues remain controversial and the basic mechanism has not been found. For example, the relation of pseudogap and SC gap is unclear and the "glue" of cooper pair in high-T_c superconductivity hasn't been found. Further studies are needed. Recently researches are focused on the relation of SC and magnetism which are mutually exclusive in general principles. However the opposite points are proposed. Julie A [2] *et al.* imaged the coexistence of ferromagnetism (FM) and SC. Even amazingly magnetism can be a favorable factor for SC in $CeCu_2Si_2$ [3] and the magnetic induced stripe order in Y123 system, leading to further support that magnetism is an important factor in cuprates superconductors [4].

In order to clarify the relation between SC and antiferromagnetism (AFM), composites systems were proposed. In the composites systems of $(La_{1.85}Sr_{0.15}CuO_4)_x(Lu_2Cu_2O_5)_{1-x}$, it is believed that the enhancement of SC is driven by $Lu_2Cu_2O_5$ [5]. Other works still aim at how the FM can affect SC system using composites systems. In D. Hsu' experiment, the composites of $(La_{1.85}Sr_{0.15}CuO_4)_{1-x}(La_{0.3}Dy_{0.4}Sr_{0.3}MnO_3)_x$ were synthesized, the low ferromagnetic coupling of $La_{0.3}Dy_{0.4}Sr_{0.3}MnO_3$ undergoes a spin glass transition below 70 K. The interaction induced structure distortion leads to the phase separation of LSCO into hole-rich and hole-poor regions [6]. Just recently, in the composites of $(La_{1.85}Sr_{0.15}CuO_4)_{1-x}(La_{2/3}Sr_{1/3}MnO_3)_x$ the electronic phase separation is found which is attributed to the interaction between SC and FM at the grain surface [7]. In the pervious research, magnetism and structural distortion are two important factors in the composites which can give great effect on SC.

In this paper we focus on the lattice mismatching effect and its effect on the electronic structure around Fermi level. The double layered system of $LaSr_2Mn_2O_7$ (LSMO) and $La_{1.85}Sr_{0.15}CuO_4$ (LSCO) are chosen. The two systems both have AFM properties and large lattice mismatch. After mixed the sintering temperatures are carefully chosen by which the interaction can be modulated. The interaction induces local structural distortion in the SC system of LSCO and the interaction effect can be carefully studied by changing sintering temperatures. A series of composite samples are synthesized. In order to carefully investigate the relationship between sintering temperature and local structural distortion, and also to study how the local structural distortion affect the band structure, x-ray photoelectron spectra were measured in National Synchrotron Radiation Laboratory (NSRL) and Beijing Synchrotron Radiation Facility (BSRF). Extended X-ray absorption fine structure (EXAFS) and X-ray absorption near edge structure (XANES) were carried out in NSRL. The structures of the samples are characterized by x-ray diffraction. The two systems are immiscible at the chosen temperatures. The lattice parameters of the samples have almost no change.

2. Samples and experiments

The high quality samples of $La_{1.85}Sr_{0.15}CuO_4$ (LSCO) and $LaSr_2Mn_2O_7$ (LSMO) were synthesized. Polycrystalline samples of nominal composition LSCO are synthesized using high purity powders of La_2O_3, $SrCO_3$, CuO. The mixed powders are heated to 1250°C after grinding and repelletizing. The treatment was repeated three times. Using the same process, the LSMO system was synthesized with the sintering temperature of 1300°C. Then the two systems were mixed with the chemical formula: $(La_{1.85}Sr_{0.15}CuO_4)_{0.85}(LaSr_2Mn_2O_7)_{0.15}$ and sintered at 600°C, 800°C, and 1100°C, respectively. The structures of the samples are characterized by x-ray diffraction. The LSCO and LSMO are in a single phase and the two systems are immiscible after sintered at the above sintering temperatures.

3. Results and discussion

The O1s XPS spectra of the layered LSMO and LSCO are shown in Figure 1(a). The oxygen atoms have different sites which can be confirmed from the peak splitting. The binding energy of O1s is higher in LSCO than in LSMO. In Figure 1(b) the O1s spectra contain the information of two systems. Compared with Figure 1(a), it is believed that the peak A mainly comes from the contribution of LSCO. And the peak B mainly belongs to LSMO system. In the samples sintered at 600°C and 800°C, the binding energy of peak A is almost the same and no changes in shape are found. However, in the sample sintered at 1100°C, the peak A moves toward higher binding energy side and also broadens, which can be explained as the lattice mismatch effect [8, 9]. Our results indicate that in low sintering temperature (600°C, 800°C) the LSCO does not suffer strain from LSMO and no local structure distortion can be found, which will be discussed later. However in the sample sintered at 1100°C, large local lattice distortion happens at the grain boundary of LSCO and LSMO just as shown in Figure 2 (a). In order to clarify the relation between local lattice distortion and electron structure the extended X-ray absorption fine structure (EXAFS) and X-ray absorption near edge structure (XANES) experiments are performed which are shown in Figure 2.

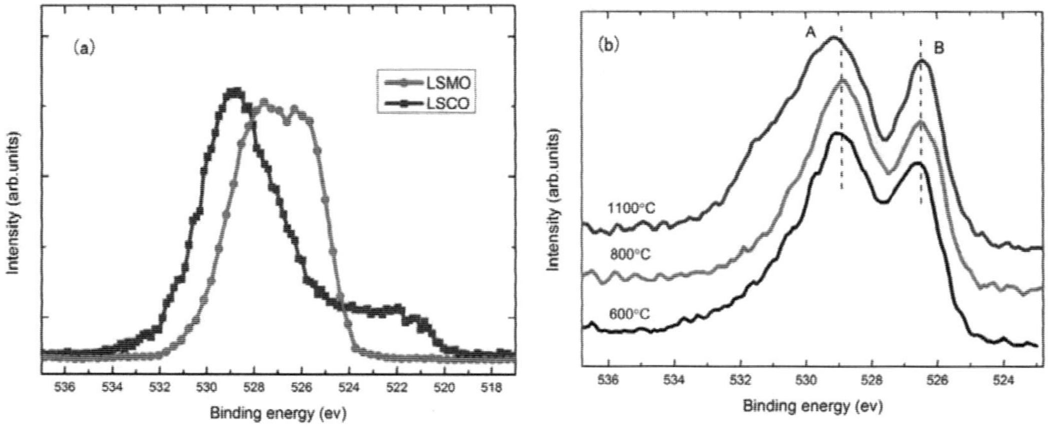

Figure 1. (a): the XPS spectra of O1s in LSCO and LSMO. (b): the XPS spectra of O1s of the composite system with sintering temperatures of 600°C, 800°C, 1100°C.

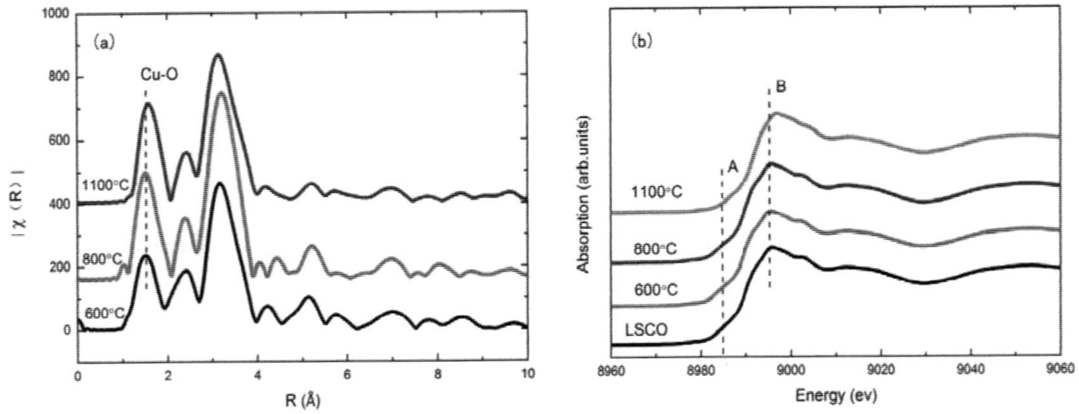

Figure 2. (a) The Fourier transform (FT) curves of the Cu K-edge EXAFS for the samples sintered at 600°C, 800°C, 1100°C. (b) Cu K-edge XANES spectra of the pure LSCO, and samples sintered at 600°C, 800°C, 1100°C.

In Figure 2(a), the bond length $R_{Cu\text{-}O}$ dose not change for samples sintered at 600°C, 800°C, but increases for sample sintered at 1100°C, The results indicate small strain happens in the samples sintered at 600°C, 800°C, but large strain exists for sample sintered at 1100°C. The interaction at the grain surface induces local structural distortions because of the high sintering temperature. In Figure 2(b), a transition of Cu 1s to 3d is generally dipole forbidden. The A feature is due to the Cu 1s to 3d state having a quadruple character [10, 11, 12], and the empty 3d band just above the Fermi level. While B feature arises from Cu 1s to 4p* [13]. In the sample sintered at 1100°C, the features A and B shift to higher energy side comparing with samples sintered at 600°C and 800°C, which means that the electronic structure around Fermi level changes by local lattice distortion and the changes of the electronic structure can probably lead to great effect on SC in LSCO. This result is in agreement with the above mentioned results of O 1s spectra. The experimental results make it explicit the relation between sintering temperature, local structural distortion and the band structure. In the composites the high sintering temperature can affect the band structure around Fermi level by inducing local structural distortion.

4. Conclusion

In our experiments, a series of $(LSCO)_{0.85}(LSMO)_{0.15}$ samples are synthesized by solid state reaction method. The binding energy of O1s is investigated by XPS. EXAFS experiments are carried out in order to identify the local structure. In the samples sintered at 600°C and 800°C, no local distortions are found, but in the sample sintered at 1100°C, large local distortions exist which can be explained as the lattice mismatching effect. The LSMO can give large strain on LSCO and then affect the SC after sintered at high temperature. Suitable sintering temperature should be chosen in order to minimize the local lattice distortion. By XANES experiments the relationship between structural distortion and electronic band is found. The empty 3d band in LSCO can be affected at high sintering temperature when the structural distortion happens.

5. Acknowledgement

The authors are grateful to Fengchun Hu for her helps during the EXAFS experiments and also grateful to the XPS stations of NSRL and BSRF. This work was supported by NSFC in project of 10979016.

References

[1] Bednorz J G, Müller K A 1986 *Phys. B Condens. Matter Quanta* **64** 189
[2] Bert J A, Beena Kalisky *et al* 2011 *Nature Phys* **7** 767
[3] Stockert O, Arndt J, Faulhaber E, Geibel C *et al* 2010 *Nature phys* **7** 119
[4] Wu T, Mayaffre H *et al* 2011 *Nature* **477** 191
[5] Park S, Zhang C L, Lee N, Choi Y J 2011 *Phys. Rev.* B **83** 220509(R)
[6] Daniel H, Geetha K T 2006 *Phys. Rev.* B **74** 214504
[7] Yao X C, Zhang J C *et al* 2011 *Journal of Alloys and Compounds* **509** 5472
[8] Dai H X, Ng C F, and Au C T 2001 *Journal of Catalysis* **197** 251
[9] Kimura T and Tokura Y 2000 *Annu. Rev. Mater. Sci* **30** 451
[10] Hahn J E, Scott R A, Hodgson K O, Chem 1982 *Phys. Lett* **88** 595
[11] Kosugi N, Tokura Y, Takagi H, Uchida S 1990 *Phys. Rev.* B **41** 131
[12] Saini N L, Lanzara A, Bianconi A, Oyanagi H 1998 *Phys. Rev.* B **58** 11768
[13] Zhang C J, Oyanagi H, Kim B H *et al* 2007 *Phys Rev.* B **75** 174504

Journal of Physics: Conference Series **430** (2013) 012074 doi:10.1088/1742-6596/430/1/012074

In situ QXAFS observation of the reduction of Fe_2O_3 and $CaFe_2O_4$

Masao Kimura*[1], Yohei Uemura[2], Toru Takayama[1], Reiko Murao[1], Kiyotaka Asakura[3] and Masaharu Nomura[1]

1 Nippon Steel Corporation, Futtsu, 293-0011, Japan
2 Photon Factory, KEK, Tsukuba, 305-0801, Japan
3 CRC, Hokkaido University, 001-0021, Japan

E-mail: kimura.masao@nsc.co.jp

Abstract *In situ* QXAFS studies of the reduction of α-Fe_2O_3 and $CaFe_2O_4$ were conducted to determine their reduction kinetics and mechanisms. The reduction of α-Fe_2O_3 involved two steps, the first being a very fast process in which Fe^{3+} was reduced to Fe^{2+}, and the second being the reduction of Fe^{2+} to Fe metal over a longer period. In contrast, the reduction of Fe in $CaFe_2O_4$ was a single first-order reaction, although an induction period was clearly observed at the beginning of the reduction process. The reduction processes were successfully studied using a combination of *in situ* QXAFS spectra at the Ca and Fe K-edges.

1. Introduction

Iron is one of the most indispensable metals. It has been used for thousands of years and is used in numerous materials that we use in our day-to-day lives. Demand for iron is still increasing today. To make the best use of iron resources, most pig iron is now produced in blast furnaces by reducing sinters made from fine grains of various iron ore types (α-Fe_2O_3, α-FeOOH, and/or Fe_3O_4) and $CaCO_3$. Small iron ore grains are mixed with $CaCO_3$ and sintered to make relatively large "heterogeneous" sinters (typically a few tens of mm in diameter).

Reduction kinetics, averaged over sinters, have been monitored by quenching sinters during the reduction process and determining the amount of reduced iron by chemical analysis. However, sinter bonding phases, e.g., calcium ferrite and SFCA (silicoferrite of calcium and alumina)」ヤ play important roles in determining the sinter quality parameters, such as reducibility [1-3] and the reduction degradation index (RDI) [4,5]. At high temperatures (e.g., 1273 K), an increase in surface area and the presence of pores less than 1 μm in diameter are important to the reduction properties of the whole sinter [3]. Information on changes in the atomic structures of various Fe-Ca-O phases is, therefore, crucial to our ability to control the reduction properties of the sinter, although these processes have not yet been studied sufficiently. The X-ray absorption fine structure (XAFS) can reveal local structures and the electronic states of target atoms under reaction conditions. In this study, we performed *in situ* QXAFS experiments on the reduction of α-Fe_2O_3 and $CaFe_2O_4$ and investigated their reduction mechanisms.

2. Experimental

In situ QXAFS[6] experiments were conducted at Photon Factory, KEK (KEK-PF), Japan. BL-9A and BL-12C were used for the Fe K-edge XAFS. At BL-9A, high energy resolution and high flux are both achieved; the flux at 6.5 keV was as high as 7×10^{11} photons/s at the maximum stored current of 450 mA. At BL-12C, a fairly high flux, 1×10^{10} photons/s was achieved. A 170-mm-long ion chamber was used for measuring the incident X-ray intensity and a 310-mm-long ion chamber was for measuring the transmitted intensity. Each *in situ* QXAFS spectrum of the Fe K-edge was measured in 20 s with the Si(111) monochromator rotated from 16.90° to 14.20°. BL-9A was used to measure the

Ca K-edge XAFS. Each QXAFS spectrum of the Ca K-edge was measured in 51 s with the Si(111) monochromator rotated from 31.00° to 26.50°. Ni-coated double mirrors were used for eliminating the high harmonic waves. The beam size was cut using slits to 1 × 1 mm^2 at the specimen.

A flow type *in situ* cell was used, details of which have been published previously [7]. The cell was equipped with four rod heaters, and the cell could be heated to above 900 °C. The cell temperature was measured using a K-type thermocouple. A mixture of hydrogen and helium (H_2:He = 20:80 vol. %) was used to reduce the iron oxides. This gas composition was selected for experimental reasons, even though the composition of this reductive gas is different from that of in a real blast furnace (70–90 vol. % CO). The gas flow rate was 100 sccm, and the reduction temperature was 900 °C. A mixture of oxygen and helium (O_2:He = 20:80 vol. %) was introduced into the cell, at 100 sccm, until the reduction temperature was reached.

Normalized XANES data ($\Delta\mu t = 1$) were used to calculate the reduction rate constants. EXAFS data analyses were performed using Athena software [8].

3. Results and Discussion

Figures 1 (a) and (b) shows the evolution of XANES spectra over time around the α-Fe_2O_3 Fe K-edge, obtained by *in situ* QXAFS at time t after the gas was switched from O_2/He to H_2/He, at 900 °C. The XANES spectrum changed dramatically as soon as the H_2–He mixture was introduced, with the white-line intensity of α-Fe_2O_3 (peak at 7129 eV) decreasing and the edge energy shifting to a lower energy. After this, another peak appeared at 7125 eV. The *in situ* XANES features at t = 0.3 and 3.3 min resembled the FeO feature, although the peak intensity at 7125 eV, at t = 0.3 min, was greater than that of FeO. Therefore, it appears that the Fe^{2+} species may be produced with a different local structure from FeO. It is notable that the XANES at 3.3 min after the reduction process began was quite similar to that of FeO. The main peak decreased and the pre-edge region increased as the reaction proceeded, and the shape became similar to that of Fe metal. In summary, the reduction of α-Fe_2O_3 consisted of at least two steps, the first being fast reduction from Fe^{3+} to Fe^{2+}, which finished by 0.3 min. At 0.3 min, the peak intensity of the *in situ* spectrum at 7125 eV was greater than that of FeO, while the overall XANES feature of both spectra resembled each other. This implies that the first step was the reduction of Fe^{3+} to form compounds such as FeO. The second step was a slow reduction process from Fe^{2+} to Fe metal.

Figure 1 (c) shows the dependence of the X-ray absorption coefficients, at specific energies, on time t. Time 0 was when the reduction gas was first introduced. The isosbestic point was found at 7142 eV and another weak isosbestic point was found at 7117 eV. The intensity at 7125 eV (circles) quickly increased and then gradually decreased. The increase was too fast to be analyzed, but the X-ray absorption variation during the decay processes fitted well with first order kinetics, with a rate constant of 0.19(1) min^{-1}.

Figure 2 (a) shows the in situ XANES of $CaFe_2O_4$. An isosbestic point was found at 7142 eV from 3 min after H_2 was introduced. The XANES spectra did not change dramatically immediately after the H_2 mixture was introduced, as it did for α-Fe_2O_3 (see Fig. 2 (b)). Figure 2 (c) shows the X-ray absorption variations at 7117 eV, 7125 eV, and 7142 eV. Absorption at 7125 eV did not change in the period $0 < t < 3$ min, then the intensity gradually decreased. The reduction of $CaFe_2O_4$ appeared to consist of at least two reaction steps. The first step was an induction period in which the local structure and Fe atom electronic states changed. The second step was reduction to Fe, shown by the variation at 7125 eV, which fitted a single exponential function with a 3 min delay. The rate constant for this step was 0.21 min^{-1}.

Journal of Physics: Conference Series **430** (2013) 012074 doi:10.1088/1742-6596/430/1/012074

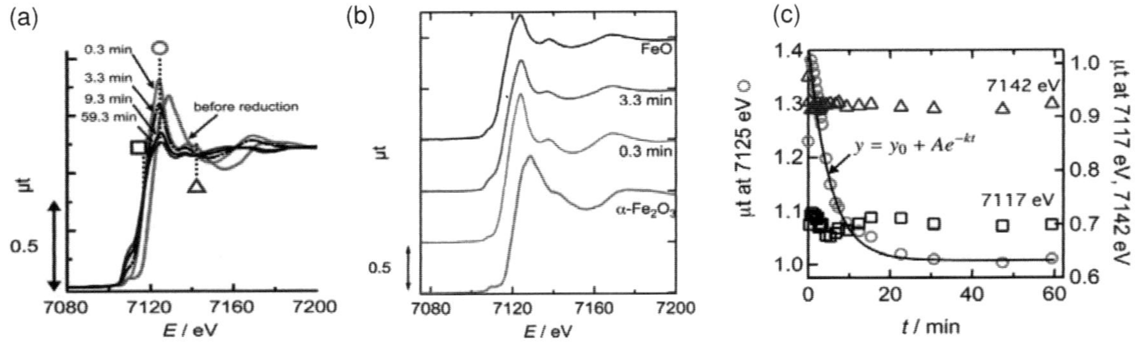

Fig. 1 (a) Time evolution of XANES spectra around the α-Fe_2O_3 Fe K-edge, obtained by *in situ* QXAFS at time *t* after the gas was switched from O_2/He to H_2/He (20% H_2 and 80% He, 100 sccm), at 900 °C, (b) Fe K-edge XANES of α-Fe_2O_3 and during its reduction (at 0.3 min and at 3.3 min) and the Fe K-edge XANES of FeO, (c) variation in X-ray absorption at 7117 eV (squares), 7125 eV (circles), and 7142 eV (triangles).

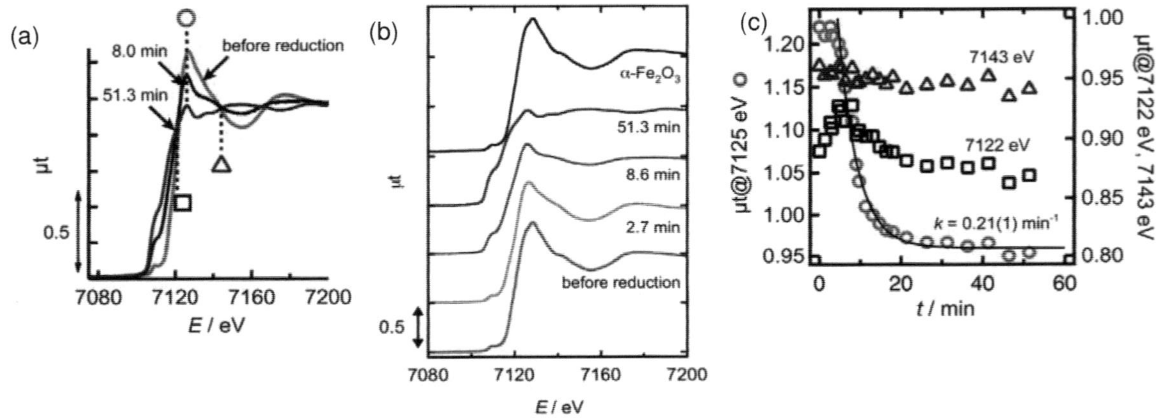

Fig. 2 (a) Time evolution of XANES spectra around the $CaFe_2O_4$ Fe K-edge, obtained by *in situ* QXAFS at time *t* after the gas was switched from O_2/He to H_2/He (20% H_2 and 80% He, 100 sccm), at 900 °C, (b) Fe K-edge XANES of $Ca_2Fe_2O_4$ before and during the reduction process (before reduction, 2.7 min, 8.6 min, and 51.3 min) and the Fe K-edge XANES of α-Fe_2O_3, (c) variation in X-ray absorption at 7122 eV (squares), 7125 eV (circles), and 7143 eV (triangles).

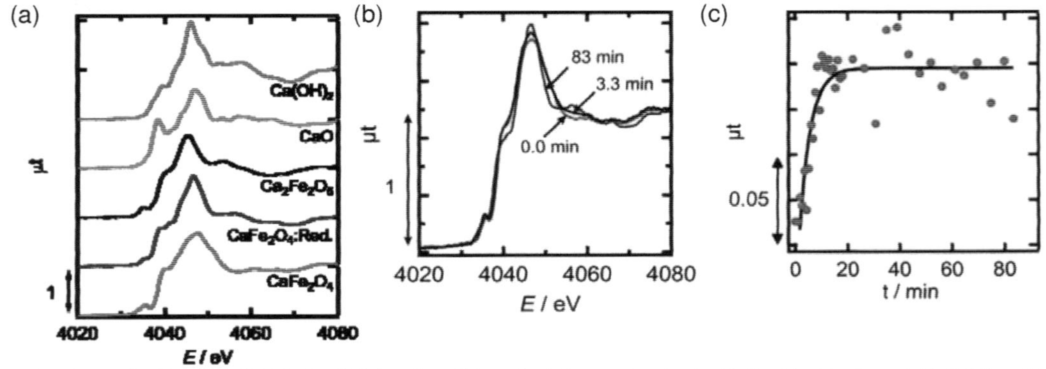

Fig. 3 (a) Ca K-edge XANES of $CaFe_2O_4$, $CaFe_2O_4$ (after reduction), $CaFe_2O_5$, CaO, and $Ca(OH)_2$, (b) time evolution of XANES spectra around the $CaFe_2O_4$ Ca K-edge, obtained by *in situ* QXAFS under reduction gas flow, at 900 °C, (c) the dependence of X-ray absorption at 4047 eV on time *t*

Figure 3 (a) shows Ca K-edge XANES spectra of $CaFe_2O_4$, $CaFe_2O_4$(after reduction), $Ca_2Fe_2O_5$, CaO, and $Ca(OH)_2$ at R.T. Before reduction, the $CaFe_2O_4$ XANES spectrum features were different from those of CaO or $Ca(OH)_2$ because the $CaFe_2O_4$ crystal structure is composed of layers of Fe-O and Ca-O and is different from the CaO and $Ca(OH)_2$ crystal structures. After reduction, the $CaFe_2O_4$ XANES resembled the $Ca(OH)_2$ XANES, indicating that the chemical states of the Ca atoms in $CaFe_2O_4$ changed to $Ca(OH)_2$-like states. Figure 3 (b) shows the evolution of XANES spectra, over time, around the Ca K-edge, obtained by *in situ* QXAFS, with a measuring time of 51 s, under the reducting atmosphere, at 900 °C. The absorption intensity at 4041 eV decreased, and the intensity at 4047 eV increased, after the reduction gas was introduced.

The intensity at 4047 eV was plotted against the reaction time in Fig. 3 (c). It should be noted that an induction period, which was seen at the Fe K-edge XANES in Fig. 2 (b), was not observed. Ca atoms changed once the reaction gas was introduced, and this change started several minutes before the reduction of Fe ions started. In other words, the decomposition of Fe-O and Ca-O layers may precede the reduction of Fe ions. The variation at 4047 eV fitted a single exponential function, and the rate constant was estimated as 0.24 min^{-1}. This value was slightly higher than the rate constant for Fe in $CaFe_2O_4$.

4. Summary

α-Fe_2O_3 and $CaFe_2O_4$ reduction processes were studied using *in situ* QXAFS experiments, and we have suggested the reduction mechanisms involved. The reduction of α-Fe_2O_3 involved two steps. The first step was a very fast process in which α-Fe_2O_3 changed to a FeO-like structure, and the second step involved the reduction of Fe^{2+} to Fe^0 metal over a longer period.

In contrast, the reduction of Fe in $CaFe_2O_4$ was a simple first-order reaction, although an induction period, in which decomposition to $Ca(OH)_2$ and Fe compounds occurred, was clearly observed at the beginning of the reduction process. This was successfully shown by the combination of *in situ* QXAFS at the Ca and Fe K-edges. Although further research is necessary, the existence of Ca atoms appears to have a great effect on the reduction of Fe.

A combination of QXAFS with different edges was used to successfully elucidate the reduction kinetics of Fe-Ca-O phases for specimens in fine powders. The reaction rates are intrinsic to each phase, giving us fundamental information that can be used to distinguish microstructure and porosity factors for sinters from the factors related to the types and amounts of co-existing phases. This will allow us to improve the over-all reduction properties of sinters by controlling each factor.

5. References

[1] N. Sakamoto 1984 *McMaster Symposillm on iron and steelmaking* **12** 137.
[2] K. Kitamura 1985 *AIMElronmaking Proc.* **44** 405.
[3] N. J. Bristow and A. G. Waters 1991 *Transactions of the Institution of Mining and Metallurgy, Section C: Mineral Processing and Extractive Metallurgy* **100** 1.
[4] I. SHIGAKI, M. SAWADA, and N. GENNAI 1986 *Trans. Iron Steel Inst. Jpn* **26** 503.
[5] C. E. Loo, K. T. Wan, and V. R. Howes 1988 *lronmaking Steelmaking* 279.
[6] R. Frahm 1988 *Nucl. Instrum. Methods Phys. Res.* **A270** 578.
[7] M. Kimura, K. Uemura, T. Nagai, Y. Niwa, Y. Inada, and M. Nomura 2009 *J Phys.* Conference Series, **Vol.190** 012163.
[8] B. Ravel and M. Newville *J. Synchrotron Rad.* **12** 537.

Journal of Physics: Conference Series **430** (2013) 012075 doi:10.1088/1742-6596/430/1/012075

Interpretation of EXAFS in scheelite-type AWO_4 (A = Ca, Sr, Ba) compounds using molecular dynamics simulations

Aleksandr Kalinko and Alexei Kuzmin

Institute of Solid State Physics, University of Latvia, Riga, Latvia

E-mail: akalin@latnet.lv

Abstract. In this work we successfully interpret the W L_3-edge EXAFS spectra in scheelite-type AWO_4 (A = Ca, Sr, Ba) compounds using a combination of classical NVT molecular dynamics (MD) and *ab initio* multiple-scattering (MS) theory. The configuration-averaged EXAFS spectra show good agreement with our room temperature experimental data supporting the reliability of the developed force-field models. The contributions from all coordination shells up to 6 Å are elucidated. The contribution of the MS effects into the total EXAFS signal in AWO_4 compounds is small, being around 10%.

1. Introduction

Analysis and interpretation of EXAFS signals in the presence of disorder and multiple-scattering (MS) effects is a challenging task, which can be addressed by the MD-EXAFS method based on molecular dynamics (MD) simulations combined with *ab initio* EXAFS spectra calculation [1]. This method is particularly suitable to access structural information beyond the first coordination shell in crystalline materials [2, 3, 4, 5]. In this work, the MD-EXAFS method is used to clarify the contribution of the outer coordination shells and the importance of the MS effects and thermal disorder in the W L_3-edge EXAFS spectra of AWO_4 (A=Ca, Sr, Ba) tungstates.

Polycrystalline tungstates AWO_4 have tetragonal (space group $I4_1/a$) scheelite-type structure composed of $[WO_4]$ tetrahedra and $[AO_8]$ polyhedra [6, 7]. Recently they gained renewed interest for application as active media in solid state Raman lasers due to the W–O bonds have high internal vibration frequencies [8, 9]. The local atomic structure of AWO_4 (A=Ca, Sr) was studied by the W L_3-edge EXAFS spectroscopy in [10, 11, 12, 13] only within the range of the first coordination shell of tungsten. The analysis of the outer shells and the role of the multiple-scattering (MS) effects in the EXAFS spectra were not addressed. To the best of our knowledge, no EXAFS studies of $BaWO_4$ have been performed until now.

2. Experimental and calculation details

Commercially available powders (99.9% purity) of AWO_4 (A=Ca, Sr, Ba) tungstates were characterized by x-ray powder diffraction and Raman spectroscopy and used in EXAFS studies.

X-ray absorption measurements were performed at the W L_3 (10207 eV) edge in transmission mode at the HASYLAB DESY C1 bending-magnet beamline at 300 K. The x-ray beam intensity

Table 1. Force-field models for $CaWO_4$, $SrWO_4$ and $BaWO_4$ used in the molecular dynamics simulations. Parameters for $CaWO_4$ are from [17], whereas for $SrWO_4$ and $BaWO_4$ they were obtained in the present work.

	$CaWO_4$			$SrWO_4$			$BaWO_4$		
	Buckingham potential, cutoff 14 Å								
	A, eV	ρ, Å	C, eV Å^2	A, eV	ρ, Å	C, eV Å^2	A, eV	ρ, Å	C, eV Å^2
A–O	2312.0	0.2812	0.0	516.15	0.3865	0.0	534.77	0.3959	0.0
O–O	2023.0	0.2674	13.83	4634.08	0.2871	182.81	3475.78	0.2824	108.03
	Morse potential, cutoff 2 Å								
	D_e, eV	a_m Å^{-2}	r_0, Å	D, eV	a_m Å^{-2}	r_0, Å	D, eV	a_m Å^{-2}	r_0, Å
W–O	1.501	2.671	1.8928	0.914	3.109	1.8928	1.240	2.931	1.8928
	Three-body potential, cutoff 2 Å								
	k, eV rad^{-2}	θ_0, deg		k, eV rad^{-2}	θ_0, deg		k, eV rad^{-2}	θ_0, deg	
O–W–O	2.6071	110.524		1.5266	110.524		3.0148	110.524	

was measured by two ionization chambers filled with argon and krypton gases. The higher order harmonics were effectively eliminated by detuning the double-crystal monochromator Si(111) to 60% of the rocking curve maximum, using the beam-stabilization feedback control.

The simulations of the W L_3-edge EXAFS spectra were performed using recently developed method, based on the joined use of classical MD simulations combined with *ab initio* EXAFS spectra calculations [1]. The MD simulations were performed in the NVT ensemble at 300 K using GULP3.1 code [14, 15] for a supercell 3×3×3, containing 324 atoms. A set of 4000 static atomic configurations was obtained during a simulation run of 20 ps with a time step interval of 0.5 fs. These configurations were further used to calculate the W L_3-edge configuration-averaged EXAFS signals using *ab initio* real-space multiple-scattering FEFF8.2 code [16]. The complex exchange-correlation Hedin-Lundqvist potential and default values of muffin-tin radii, as provided within the FEFF8.2 code [16], were employed.

For $CaWO_4$ we have used Senyshyn's force-field model [17] (Table 1), which is able to describe structural, elastic and phonon properties. It consists of the pairwise interatomic potentials and the three body interaction. Note that the three body potential is required to stabilize [WO_4] tetrahedra in the MD simulations. The atomic charges were taken from the DFT calculations and are equal to +2.0 for Ca, +1.68 for W and –0.92 for O atoms. Force-field models for $SrWO_4$ and $BaWO_4$ were constructed based on $CaWO_4$ model (Table 1). The atomic charges were left unchanged, only the short range Buchingham, Morse and three body potential parameters were optimized to reproduce tungstate structure [6] and the highest phonon frequency corresponding to the W–O vibrations [8].

3. Discussion

The configuration-averaged W L_3-edge EXAFS spectra $\chi(k)k^2$ for three tungstates, calculated within the single-scattering (SS) and multiple-scattering (MS) approximations, are compared with the experimental data in Fig. 1. In the analysis of outer coordination shells the influence of the MS effects on the overall EXAFS spectrum should be always considered. However, as one can clearly see, the MS effects play only minor role in the EXAFS spectra for three tungstates, changing their amplitude less than 10%. The phase of the EXAFS spectra is described rather well by our theoretical model in all k-space range, while their amplitude is slightly overestimated, especially, at the end of the spectra. Note that some high-frequency contributions are not present

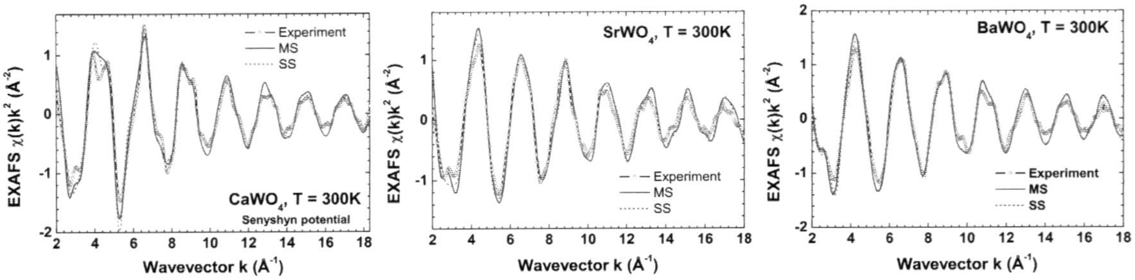

Figure 1. Comparison of the experimental and configuration-averaged W L_3-edge EXAFS spectra $\chi(k)k^2$ at 300 K for $CaWO_4$, $SrWO_4$, and $BaWO_4$. Theoretical EXAFS spectra were calculated within the single-scattering (SS) and multiple-scattering (MS) approximations.

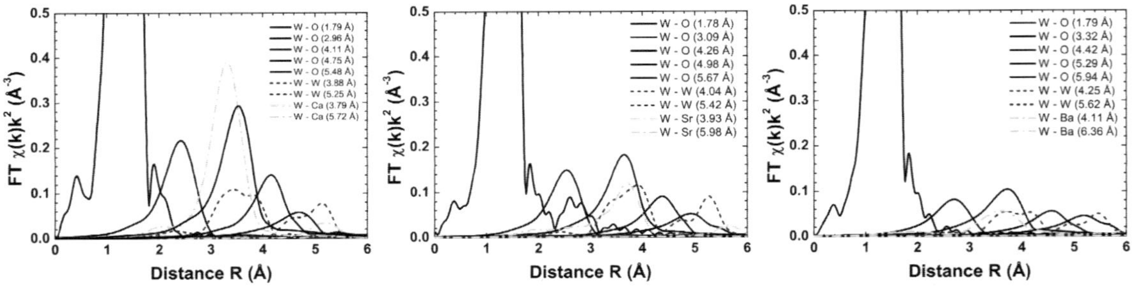

Figure 2. Fourier transforms (FT) of the W L_3-edge EXAFS spectra $\chi(k)k^2$ for separate coordination shells, calculated based on the structural parameters extracted from the results of the MD simulations.

in the configuration-averaged EXAFS spectra due to the limitation induced by the supercell size of 6 Å and resulting in the absence of contributions from more distant (>6 Å) shells.

Because of rather good agreement between calculated and experimental EXAFS spectra, the accurate analysis of the second and further coordination shells can be reliably performed. The origin of peaks in Fourier transforms (FTs) of the EXAFS signals was identified for each shell (Fig. 2). To do that each peak in the radial distribution function (RDF), calculated till 6 Å from the MD simulations, was approximated by the Gaussian function, which depends on three parameters – position, amplitude and width. Such decomposition is possible, because peaks corresponding to the shells either do not overlap or overlap little. Using thus obtained set of structural parameters, the single-scattering EXAFS signals and their FTs for each coordination shell were calculated and are shown in Fig. 2.

As expected, the major contribution in FTs (the first peak at 1.3 Å in Fig. 2) comes from the four nearest oxygens at about 1.78-1.79 Å, forming the first coordination shell of tungsten. The mean square relative displacement values for the W–O bonds are small ($4\text{-}6\times10^{-4}$ Å^2) and are weakly influenced by the A^{2+} ion type. This indicates strong W–O bonding in agreement with high vibrational frequencies observed by Raman spectroscopy [8, 9].

The outer shell contributions are noticeably smaller than from the first shell. The second peak is due to oxygen atoms located in the second coordination shell at about 2.96-3.32 Å. The next three overlapping peaks originate from A (A = Ca, Sr, Ba) atoms at 3.79-4.11 Å, O atoms at 4.11-4.42 Å, and W atoms at 3.88-4.25 Å. It is surprising that the most heavy tungsten atoms give the smallest contribution due to the thermal disorder. Note also a strong decrease of the amplitude for the peak, related to A-atoms, going from Ca to Ba. This effect can be explained by the crystal lattice expansion due to an increase of the divalent cation size ($R(Ca^{2+})$=1.12 Å, $R(Sr^{2+})$=1.26 Å, $R(Ba^{2+})$=1.42 Å). It leads to an increase of the separation between the two groups of the W–A distances: the differences are 0.16 Å for $CaWO_4$, 0.20 Å for $SrWO_4$, and 0.28 Å for $BaWO_4$. As a result, the strong amplitude damping of the corresponding total EXAFS signal occurs, and its FT peak decreases.

The next peak in FT at 4.2-4.5 Å originates from the group of oxygen atoms located at 4.75-5.29 Å. The three most distant peaks are due to the groups of tungsten atoms at 5.25-5.62 Å, oxygen atoms at 5.48-5.94 Å, and divalent atoms at 5.72-6.36 Å.

4. Conclusions

In the present work we performed the W L_3-edge EXAFS study of the sheelite-type compounds ($CaWO_4$, $SrWO_4$, $BaWO_4$) at room temperature. The experimental EXAFS spectra were interpreted using recently developed approach, based on classical molecular dynamics (MD) simulation combined with *ab initio* EXAFS spectra calculation [1].

The classical MD simulations in the NVT ensemble were performed at 300 K using the force-field potential models, which were developed for the $SrWO_4$ and $BaWO_4$ compounds based on the previously published model for $CaWO_4$. The good agreement between the experimental and calculated configuration-averaged EXAFS spectra allowed us to perform detailed analysis of the outer coordination shells contributions, leading to the following conclusions.

It was found that the MS contributions have minor influence (around 10%) on the W L_3-edge EXAFS signal in AWO_4 compounds. The main part of the EXAFS signal comes from oxygen atoms in the first coordination shell. Besides, there are also noticeable contributions from the outer groups of oxygen atoms located up to 6 Å and from A (A = Ca, Sr, Ba) atoms. The latter decreases by going from Ca to Ba due to the crystal lattice expansion inducing the separation between the two A-atoms shells. The contribution from the heavy tungsten atoms is surprisingly small due to the thermal disorder.

Acknowledgments

This work was supported by ESF Project 2009/0202/1DP/1.1.1.2.0/09/APIA/VIAA/141 and Latvian Government Research Grant No. 09.1518. The EXAFS measurements have been financed from the European Community's Seventh Framework Programme under grant agreement No. 226716 (Project I-20100098 EC).

References

[1] Kuzmin A and Evarestov R A 2009 *J. Phys.: Conf. Ser.* **190** 012024
[2] Kalinko A, Evarestov R A, Kuzmin A and Purans J 2009 *J. Phys.: Conf. Ser.* **190** 012080
[3] Anspoks A, Kuzmin A, Kalinko A and Timoshenko J 2010 *Solid State Commun.* **150** 2270
[4] Kuzmin A, Efimov V, Efimova E, Sikolenko V, Pascarelli S and Troyanchuk I O 2011 *Solid State Ionics* **188** 21
[5] Timoshenko J, Kuzmin A and Purans J 2011 *Centr. Eur. J. Phys.* **9** 502
[6] Gurmen E, Daniels E and King J S 1971 *J. Chem. Phys.* **55** 1093
[7] Hazen R M, Finger L W and Mariathasan J W E 1985 *J. Phys. Chem. Solids* **46** 253
[8] Gallucci E, Goutaudier C, Bourgeois F, Boulon G and Cohen-Adad M T 2002 *J. Solid State Chem.* **163** 506
[9] Černý P, Jelínková H, Zverev P G and Basiev T T 2004 *Prog. Quantum Electron.* **28** 113
[10] Balerna A, Bernieri E, Burattini E, Lusis A, Kuzmin A, Purans J and Cikmach P 1991 *Nucl. Instrum. Meth. A* **308** 234
[11] Kuzmin A and Purans J 2001 *Rad. Measurements* **33** 583
[12] Kuzmin A, Kalendarev R, Purans J, Itie J P, Baudelet F, Congeduti A and Munsch P 2005 *Phys. Scripta* **2005** 556
[13] Gracia L, Longo V M, Cavalcante L S, Beltran A, Avansi W, Li M S, Mastelaro V R, Varela J A, Longo E and Andres J 2011 *J. Appl. Phys.* **110** 043501
[14] Gale J D 1996 *Philos. Mag. B* **73** 3
[15] Gale J D and Rohl A L 2003 *Mol. Simulat.* **29** 291
[16] Ankudinov A L, Ravel B, Rehr J J and Conradson S D 1998 *Phys. Rev. B* **58** 7565
[17] Senyshyn A, Kraus H, Mikhailik V B and Yakovyna V 2004 *Phys. Rev. B* **70** 214306

Journal of Physics: Conference Series **430** (2013) 012076 doi:10.1088/1742-6596/430/1/012076

Magnetic Properties of Co doped ZnS Diluted Magnetic Semiconductor

Qingteng Hou[1], Kai Chen[1], Hongguang Zhang[1], Yongtao Li[1, 2], Hao Liu[1], Xueguang Dong [1], Yongchao Huang[1] and Qi Li[1*]

1. Department of physics, Southeast University, Nanjing, 211189, P. R. China

2. College of science, Nanjing University of Posts and Telecommunications, Nanjing, 210003, P. R. China.

*Correspondence author, email: qli@seu.edu.cn

Abstract. The magnetic properties of $Zn_{0.9}Co_{0.1}S$ nanoparticles synthesized by solid state reaction method are presented. The nanoparticles are crystallized with zinc-blende structure. The variation of magnetic properties stems from the changes of the local structure around Co ions and the distribution of Co ions, given by X-ray absorption fine structures.

1. Instruction

Doped transition metal II-VI semiconductors have been investigated extensively due to various optoelectronic applications, such as optical sensors, light emitting displays, electroluminescence devices, etc [1, 2]. Zinc sulfide (ZnS), a II–VI compound semiconductor with wide band gap of 3.67 eV when crystallized with zinc-blende structure and of 3.90 eV for wurtzite structure [3, 4], is an attractive material for current research. Theoretical studies have shown that wide band gap semiconductors may be representative materials for ferromagnetism at room temperature or higher temperature modulated by carriers [5]. It is reasonable to speculate that there most likely exists ferromagnetism in ZnS-doped materials.

As far as the magnetic properties of ZnS doped with cobalt (Co) are concerned, most of the theoretical and experimental studies show that this material has a spin glass or anti-ferromagnetic behavior, paramagnetic and ferromagnetic behaviors [6, 7, 8]. However many experimental results are contradictory [9, 10, 11]. The bound magnetic polaron (BMP) model is based on the fact that when carrier concentration exceeds a certain value an impurity band is formed and the magnetic dopants couple with localized carriers [12, 13]. Lots of observations on magnetic moments in oxide diluted magnetic semiconductor (DMS) materials, such as ZnO–based DMS, support this theory. However, only a few indirect experiments support this model in ZnS-based DMS. In this work, we analyzed the influence of interactions between magnetic ions and defects on the magnetic properties and the influence of annealing temperature on the crystal structure and magnetic properties.

2. Samples and experiments:

In this work, the synthesis of $Zn_{0.9}Co_{0.1}S$ nanoparticles was carried out by solid state reaction. The high-purity reactants with stoichiometric ratio zinc acetate, cobalt acetate and sodium sulfide were mixed together and the mixtures were thoroughly ground with adding sodium chloride and Isopropanol. After the reactions were completed, the $Zn_{0.9}Co_{0.1}S$ nanoparticles were then separated from the reactants by centrifugation, and washed several times with deionized water and ethanol. Finally, the $Zn_{0.9}Co_{0.1}S$ nanoparticles were dried in a desiccator for 10 h at 80 °C. Then the samples were divided into 3 parts and were marked as A, B, C. The sample A was without any treatment, sample B was calcined at 120 °C in vacuum environment for 10 hours and sample C was calcined at 200 °C in vacuum environment for 10 hours.

3. Results and discussions:

The XRD patterns of the samples are shown in Fig. 1. The XRD patterns for $Zn_{0.9}Co_{0.1}S$ samples match well to ZnS with zinc-blende (ZB) structure. This result illustrates that the stable and high quality $Zn_{0.9}Co_{0.1}S$ samples with zinc-blende (ZB) structure were synthesized. The fact that XRD does not show existence of the Co-related phase indicates that Co ions are oxidized in the ZnS matrix and substitute into the ZnS lattice at the Zn^{2+} site. According to Scherer's formula ($D=0.89\lambda/B\cos\theta$), the size of all samples is almost the same, which is about 3 nm. Shiv P. Patel proposes that the ferromagnetism in diluted magnetic semiconductors is related to the surface area of grains [14]. And the ferromagnetism can be enhanced by increasing the area of grain boundary and reducing the volume of grains [15]. All of our samples are basically the same size, so we can rule out the possibility that the variation of magnetism stems from the surface size.

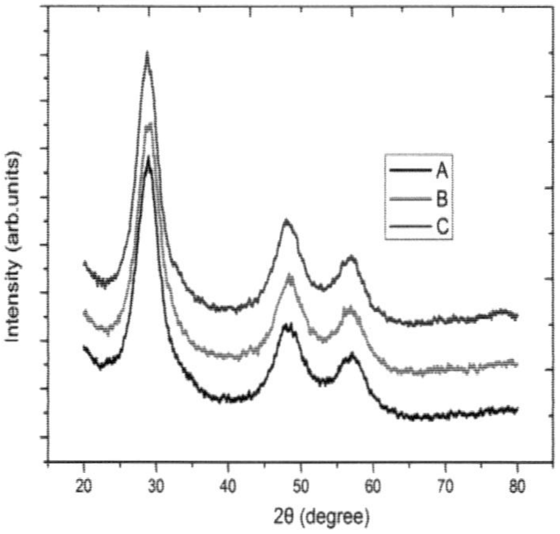

Fig. 1 XRD patterns of $Zn_{0.9}Co_{0.1}S$ samples.

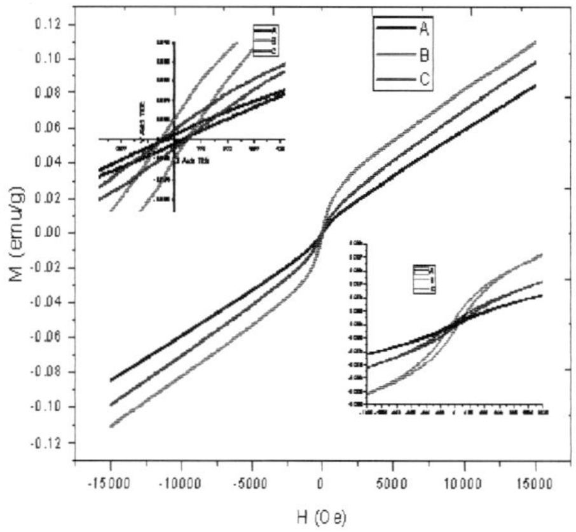

Fig. 2 Magnetic hysteresis loops of $Zn_{0.9}Co_{0.1}S$ samples. Inset shows *M-H* loops at low magnetic field.

The *M-H* loops of all samples measured at 300K are shown in Fig. 2. It appears clearly that the magnetization of all the samples is the superimposition of ferromagnetic and paramagnetic components. Obviously, the magnitudes of paramagnetic components are almost the same which we can obtain from the slope of the paramagnetic component at high magnetic field. The magnitudes of ferromagnetic components change with increasing annealing temperature (saturation magnetizations (M_S) of A, B and C are 0.005emu/g, 0.02emu/g, 0.01emu/g, respectively). The remanent magnetization of the samples A, B, C is 0.0005emu/g, 0.002emu/g and 0.001emu/g respectively.

Co K-edge XANES spectra of $Zn_{0.9}Co_{0.1}S$ nanoparticles are shown in Fig. 3. There are two peaks (M, N) in all samples below the energy of 7730 eV. The peak M is the absorption edge and the peak N is the pre-edge peak, which is ascribed to the transition of the 1s electron of the absorber to the bound state induced by hybridization of the cation and anion states. The M peak of sample C shifts to low energy side by about 4 eV compared with samples A and B, which indicates the valence state variation of Co ions or the changes of local environment around Co ions. The distance between the pre-edge and edge position remains the same in sample A and sample B, while the distance decreases in sample C, which also can be due to the changes of local environment around Co ions.

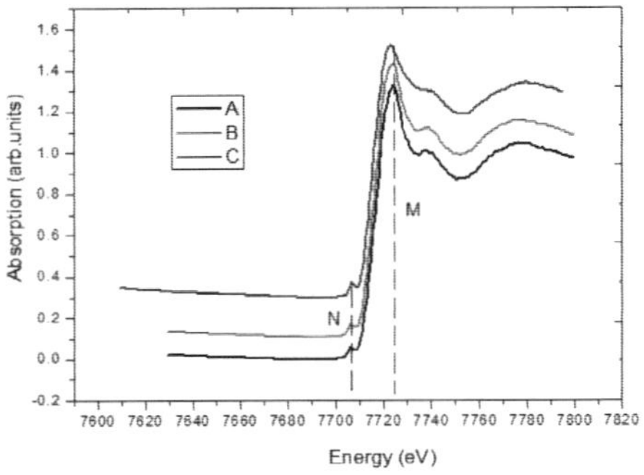

Fig. 3. Co K-edge XANES spectra of $Zn_{0.9}$ $Co_{0.1}S$ nanoparticles.

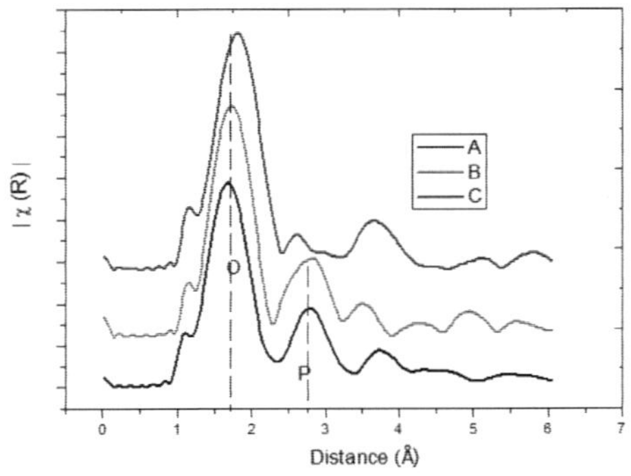

Fig. 4. The Fourier Transform (FT) curves of the Co K-EXAFS of $Zn_{0.9}$ $Co_{0.1}S$ nanoparticles.

Fig. 4 shows the radial structure function for each sample. It can be seen from the above figure that it displays two main peaks below 0.3 nm in sample A and B. The position of the first peak (O) is almost coincident at 0.16nm, while the second peak (P) of sample B shifts to higher distance. This is not surprised if the distribution of Co ions is taken into account. The ionic radii of Co^{2+} and Zn^{2+} are 0.72 Å and 0.74 Å respectively [8]. Sample B was calcined at 120 °C in vacuum environment for 10 hours, so the distribution of Co ions in the lattice becomes more uniform in sample B than that in sample A. The σ (disorder factor) of sample A and B we get by fitting the data is 0.1037 and 0.09 respectively. Therefore, the shift of the second peak is ascribed to redistribution of Co ions with more uniformity after annealing. Besides, as shown in Fig. 4 the sample B has a higher long-range structural order than sample A. When the annealing temperature is raised to 200 °C, the local structure around Co ions has changed. It can be clearly seen that the radial structure function of sample C is largely different from those of sample A and B. The Co K-edge XANES and Co K-EXAFS spectra are compared of possible impurities such as CoO, CoS and Co_2O_3 with those of sample C. We get that a part of Co ions have been oxidized to Co_2O_3 after annealing at 200 °C because of the low degree of vacuum. The experimental data presented here are consistent with the results obtained from Fig. 3.

The PL spectra of all the samples are shown in Fig. 5. The presence of sulfur vacancies is confirmed based on the results of photoluminescence (PL, the excitation wavelength is 325 nm). Results indicate that all samples exhibit broadening and stable blue emission peaks centered at around 400 nm, which can be assigned to the sulfur vacancies [16]. In order to study the influence of Co-doping on luminescence, the PL spectra of ZnS are presented here for comparison. By comparison, we can clearly find that the doping of Co^{2+} into ZnS does not import new energy bands or defect states that can produce new emission peaks. It is certain that anion vacancies produce defect states which are responsible for the observed photoluminescence. With annealing of the sample, these defect states induced by S vacancies decrease in amount and the photoluminescence intensity diminishes.

The concentration of delocalized free carriers is so low that they do not have the capacity to induce room-temperature ferromagnetism by RKKY mechanism. Hence, this ferromagnetism does not arise from the RKKY mechanism, and our results argue for the BMP mechanism, in which S vacancies (V_s) play an important role. Shallow donors such as V_s are energetically aligned with the Co^{2+} leading to effective dopant-defect hybridization as described by perturbation theory. Formation of such a band upon coalescence of bound magnetic polarons in DMSs has been described theoretically [17].

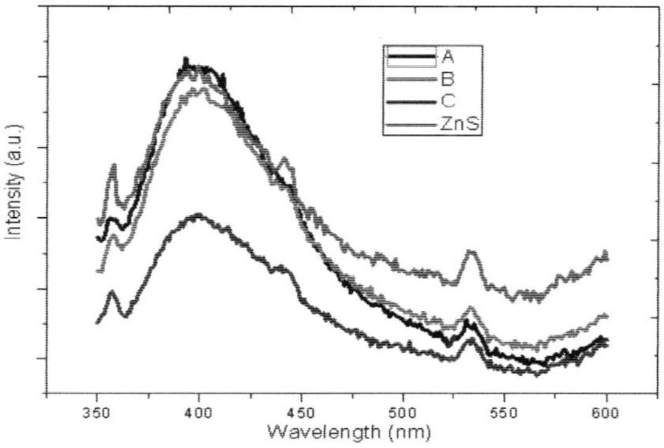

Fig. 5. The PL spectra of $Zn_{0.9}Co_{0.1}S$ and ZnS nano-particles. The excitation for all of these emissions was 325nm.

We consider three possible types of magnetic states for Co in the ZnS matrix: (a) isolated PM Co due to small doping concentrations; (b) antiferromagnetic (AFM) exchange coupling between Co atoms through S; (c) FM exchange coupling (the cobalt ions form bound magnetic polarons). If two cobalt ions are separated by only one S ion, the spin of the two cobalt ions is in antiferromagnetic arrangement by the coupling with the excited states of S ions. According to previous analysis, the distribution of Co ions in sample B is more uniform than that in sample A, leading to the number of the antiferromagnetic Co ions decreases and the number of isolated cobalt ions and the Co ions which form bound magnetic polarons increases. So the M_S of sample B is larger than that of sample A, and the remanent magnetization of the sample B (0.002emu/g) is larger than that of sample A (0.0005emu/g). The local structure around Co ions of the samples has changed and a part of Co ions have been oxidized to Co_2O_3 after annealing at 200 °C, which displays anferromagnetism with the Neel temperature 40K [18]. Therefore, the M_S and remanent magnetization of sample C become smaller than those of sample B, but still larger than those of sample A. With annealing of the sample, defect states induced by S vacancies decrease in content and the photoluminescence intensity diminishes, indicating the number of S vacancies becomes smaller with increasing annealing temperature. However, the number of S vacancies is still large enough to form bound magnetic polarons with Co ions. So the ferromagnetism of our samples mainly depends on the distribution of Co ions, rather than the number of S vacancies.

4. Conclusion:

High quality $Zn_{0.9}Co_{0.1}S$ samples with zinc-blende (ZB) structure were successfully synthesized by solid state reaction at RT. According to Scherer's formula, the size of all samples is almost the same, which is about 3 nm. The ferromagnetism in our samples mainly depends on the distribution of Co ions, rather than the number of S vacancies. When the temperature is raised to 200 °C, the stability of the lattice becomes weaker and a part of Co ions have been oxidized to Co_2O_3.

5. Acknowledgement:

The authors are grateful to Fengchun Hu for her helps during the EXAFS experiments. This work was supported by NSFC in project of 10979016.

References

[1] Pearton S J, Norton D P, et al. 2004 *J. Vac. Sci. Technol.* B 22932-48.
[2] Macdonald A H, Schiffer P and Samarth N 2005 *Nat. Mater.* 4 195-202.

[3] B. Gilbert, B.H. Frazer,et al. 2002 *Phys. Rev.* B. 66 245205.
[4] Shiv P. Patel, J.C. Pivin, et al. 2011 *JMMM.* 323 2734-2740.
[5] Dietl T, Ohno H, Matsukura F, et al. 2000 *Science.* 287 1019.
[6] Sato K, Katayama-Yoshida 2002 *H Phys. Status. Solidi.* B 229 673.
[7] Ren G, Lin Z, Wang C, Liu W, et al. 2007 *Nanotechnology.* 18 035705.
[8] Sambasivama S, Joseph D P,et al. 2009 *J. Solid. State.Chem.*182 2598-2601.
[9] K. Sato, H. Katayama-Yoshida 2002 *Phys. Status. Solidi.* b 229 673.
[10] A. Lewicky, J.K. Furdyna, A.I. Schindler, W. Giriat 1989 *Phys. Rev.* B 40 2379.
[11] T.M. Giebultowicz, P. Klosowski, et al. 1990 *Phys. Rev.* B 41 504.
[12] Dietl T, Ohno H, Matsukura F, et al. 2000 *Science* 287 1019-1022.
[13] S. Riaz.S. Naseem .Y. B. Xu 2011 *J Sol-Gel Sci Technol.* 59 584-59.
[14] Shiv P. Patel, j. c. pivin, et al. 2011 *JMMM*, 323 734-2740.
[15] Boris B. Straumal, Andrei A. Mazilkin, et al. 2009 *Phys. Rev.* B, 79 205206-205212.
[16] C. H. Ye, X. F. Fang, G. H. Li, and L. D. Zhang 2004 *Appl. Phys. Lett.* 85 3035.
[17] J. M. D. COEY, M. VENKATESAN, et al. 2005 *nature materials.* 4 173-179.
[18] Wang Yong qiang, et al. 2007 *Chinese Science Bulletin.* 52(8) 1019-1023.

Journal of Physics: Conference Series **430** (2013) 012077 doi:10.1088/1742-6596/430/1/012077

Structural transformation in Co-doped ZnS Nanoparticles

Kai Chen[1], Qingteng Hou[1], Xueguang Dong[1], Hongguang Zhang[1], Yongtao Li[1, 2], Hao Liu[1], Yongchao Huang[1] and Qi Li[1]

1. Department of physics, Southeast University, Nanjing, 211189, P. R. China

2. College of science, Nanjing University of Posts and Telecommunications, Nanjing, 210003, P. R. China.

Email: qli@seu.edu.cn

Abstract. Co-doped ZnS nanoparticles have been synthesized by co-precipitation method at room temperature and characterized by extended X-ray absorption fine structure (EXAFS) and X-ray absorption near edge structure (XANES). The effect of doping Co on structural transformation is investigated. The results of Co K-edge XANES and Zn K-edge EXAFS indicate that Co ions have substituted Zn in the samples. Zn K-edge XANES measurements reveal that sphalerite and wurtzite structures coexist in the samples. The results of Co K-EXAFS show that the distance between Co and the nearest neighbor S decreases slightly with the content of Co increasing and it is suggested that there is structural transformation around Co.

1. Introduction

As a wide gap semiconductor, pure and doped ZnS have been extensively investigated owing to their wide applications in diluted magnetic semiconductors (DMS), phosphors, photo catalysts, optoelectronic devices [1-4]. However, the structure, optical, electronic and magnetic properties of nano-scale materials may be different from those of their bulk counterparts due to surface adsorption effects and quantum confinement effects. Cubic zinc blende (sphalerite) structure of ZnS, which is a stable phase at room temperature (RT), was reported by a lot of researches [1-4]. However, few researches have reported hexagonal (wurtzite) ZnS as a result of lack of proper synthesis methods [5, 6] at mild temperature, because wurtzite ZnS which can be formed at temperature higher than 1020°C [7] is the high-temperature polymorph of sphalerite.

In order to investigate the structure transformation, we synthesize Co-doped ZnS nanoparticles in which sphalerite and wurtzite coexist and focus on Co role in structural transformation at RT.

2. Experiments

The synthesis of $Zn_{1-x}Co_xS$ (x=0.01, 0.03, 0.05, 0.07) nanoparticles was carried out by co-precipitation method at RT with zinc acetate and cobalt acetate as starting materials. The stoichiometric solution with $NaPO_3$ as stabilizer was mixed with Na_2S solution by continuous stirring until the fine precipitates were formed. After completed reactions, the precipitates were filtered out separately by centrifugation and washed clean with deionized water and ethanol, then dried in a desiccator. The structures of the samples were characterized by x-ray diffraction. There are only sphalerite and wurtzite structures in the samples and no other phases are found. Extended X-ray absorption fine structure (EXAFS) and X-ray absorption near edge structure (XANES) were measured at National Synchrotron Radiation Laboratory (NSRL).

3. Results and Discussion

Co K-edge XANES spectra of $Zn_{1-x}Co_xS$ nanoparticles are shown in Figure 1(a). A comparison with cobalt oxides and metal cobalt indicates a Co^{2+} valency in the samples. Two main features are obvious in the XANES spectra of all samples: a small pre-edge peak and a dominant peak. Generally, the pre-edge peak is explained as a quadrupolar electronic transition from 1s to the 3d final states hybridized with 4p character of the absorber; its intensity strengthens in the local atomic configuration that lacks centrosymmetry [8, 9], since in a tetrahedral local coordination, this peak can be remarkable. The main peak comes from the dominant dipolar transition from 1s to the 4p-related final states. In Figure 1(b), the first peak marked by dashed line represents Zn-S bond and the distance between Zn and the nearest neighbor S does not change in all samples. The results indicate that Co ions have substituted Zn in Co-doped ZnS nanoparticles, corresponding to the analysis of Figure 1(a).

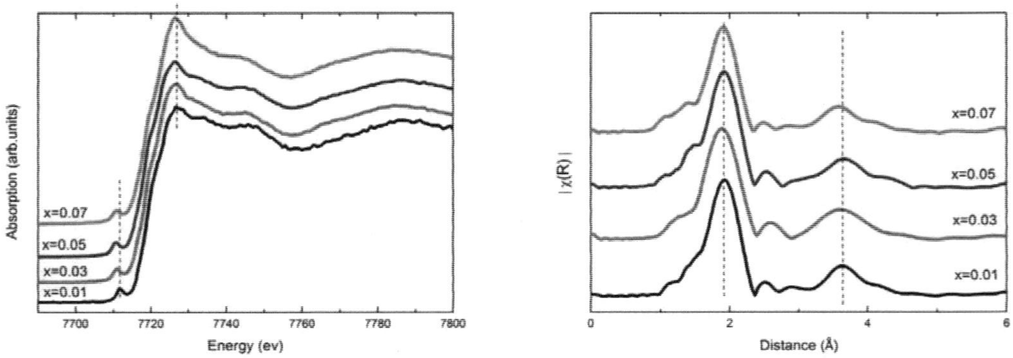

(a) (b)

Figure 1. (a) Co K-edge XANES spectra of $Zn_{1-x}Co_xS$ nanoparticles. (b) The Fourier transform (FT) curves of the Zn K-EXAFS of $Zn_{1-x}Co_xS$ nanoparticles.

In Figure 2(a), there are five peaks marked by A, B, C, D, E in all samples. For the calculated Zn K-edge XANES spectra of wurtzite and sphalerite ZnS based on multiple-scattering theory [10, 11], their shapes are similar to the experimental spectra. However, peak A is much smaller and peak D almost disappear in calculated Zn K-edge XANES of wurtzite ZnS compared with that of sphalerite ZnS. Considering the experimental results of

Figure 2(a), it is concluded that sphalerite and wurtzite structures coexist. In Figure 2(b), the first peak marked by dashed line represents Co-S bond and the distance between Co and the nearest neighbor S decreases slightly with the content of Co increasing. Because the distance between Co and the nearest neighbor S is a little smaller in wurtzite than that in sphalerite, it is suggested that the substitution of Co makes the local structure around Co transform from sphalerite to wurtzite. When the content of doped Co is lower, the second nearest neighbors of Co are almost Zn, and Figure 1(b) shows that the distance of Zn-S does not change, so structure transformation around Co is suppressed. With the content of Co increasing, there are more Co ions at the second nearest site of Co and it makes the structure transformation become easier.

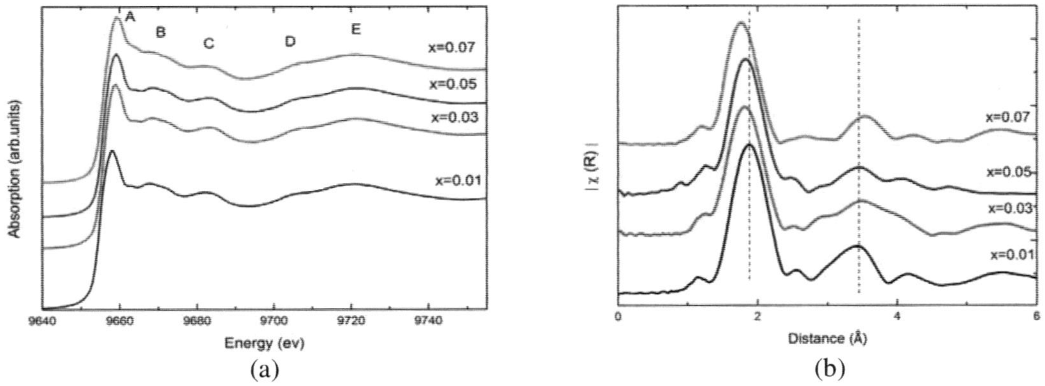

Figure 2. (a) Zn K-edge XANES spectra of $Zn_{1-x}Co_xS$ nanoparticles. (b) The Fourier transformed (FT) curves of the Co K-EXAFS of $Zn_{1-x}Co_xS$ nanoparticles.

In this work, sphalerite and wurtzite structures coexist in the samples. Doped Co ions play an important role in structural transformation from sphalerite to wurtzite. It is possible to control structural transformation by changing the content of Co and find a new way to prepare pure wurtzite structure at RT, which need to be further studied.

4. Acknowledgement

The authors are grateful to Fengchun Hu for her helps during the EXAFS experiments and also grateful to the XPS station of NSRL. This work was supported by NSFC in project of 10979016.

References

[1] Sarkar I, Sanyal M K, Kar S, Mino H, Komori F et al 2007 *Phys. Rev.* B **75** 224409
[2] Stern R A, PerezDieste V et al 2004 *J. Appl. Phys.* **95** 7468
[3] Tablero C 2006 *Phys. Rev.* B **74** 195203
[4] Tsujii N, Kitazawa H, Kido G 2003 *J. Appl. Phys.* **93** 6957
[5] Deng Z X, Wang C, Sun X M, Li Y D 2002 *Inorg. Chem.* **41** 869-873
[6] Yu S H, Yoshimura M 2002 *Adv. Mater.* **14** 296
[7] Birman J L 1959 *Phys. Rev.* **115** 1493
[8] Fernandez-Garcia M, Martinez-Arias A et al 2004 *Chem. Rev.* **104** 4063

[9] Groot F D 2001 *Chem. Rev.* **101** 1779
[10] Sainctavit P, Petiau J, Benfatto M, Natoli C R 1989 *Physica* B **158** 347
[11] Gilbert B, Frazer B H et al 2002 *Phys. Rev.* B **66** 245205

15th International Conference on X-ray Absorption Fine Structure (XAFS15) IOP Publishing
Journal of Physics: Conference Series **430** (2013) 012078 doi:10.1088/1742-6596/430/1/012078

The potential of operando XAFS for determining the role and structure of noble metal additives in metal oxide based gas sensors

Jan-Dierk Grunwaldt[1,*], Michael Hübner[2], Dorota Koziej[3], Nicolae Barsan[2], and Udo Weimar[2]

[1] Institute for Chemical Technology and Polymer Chemistry (ITCP), Karlsruhe Institute of Technology (KIT), 76128 Karlsruhe, Germany
[2] Institute of Physical and Theoretical Chemistry, Tübingen University, 72076-Tübingen, Germany
[3] Department of Materials, ETH Zürich, 8093 Zürich, Switzerland

E-mail: grunwaldt@kit.edu

Abstract. Noble metal additives significantly improve the performance of SnO_2 based sensors. Recently, it has been found that X-ray absorption spectroscopy is an excellent tool to identify their structure under sensing conditions, despite of the low concentrations and the rather thin (50 μm) and highly porous layers. For this purpose a new in situ approach has been established and here we highlight the potential with an overview on the results of Pd-, Pt-, and Au-additives in SnO_2-based sensors at work. Emphasis was laid on recording the structure (by XANES and EXAFS) and performance at the same time. In contrast to earlier studies, Pd- and Pt-additives were observed to be in oxidized and finely dispersed state under sensing conditions excluding a spillover from metallic noble metal particles. However, Au was mainly present as metallic particles in the sensing SnO_2-layer. For the Pt- and Au-doped SnO_2-layers high energy-resolved fluorescence detected X-ray absorption spectra (HERFD-XAS) were recorded not only to minimize the lifetime-broadening but also to eliminate the Au- and Pt-fluorescence effectively and to record range-extended EXAFS.

1. Introduction

The qualitative and quantitative detection of various gases is essential for areas like process monitoring, environmental control, safety, comfort and health. One of the most promising approaches is the use of gas sensors based on semiconducting metal-oxides due to their high sensitivity, long-term stability, low cost in production and relative low power consumption during operation. State-of-the-art semiconducting metal oxide based gas sensors are mostly using porous, 10 – 50 μm thick film SnO_2 layers deposited via screen printing or drop coating onto a transducer [1]. In order to further improve their sensing performance, especially to diminish the cross sensitivity towards water vapor, small quantities of noble metal additives (Pt, Pd and Au) are generally introduced into the sensing layer [2].

Although these promoters are considered to be the most relevant ones since several decades, the understanding of their role is still under strong discussion. This is mainly due to the lack of spectroscopic measurements on real sensors under relevant operation conditions ("operando"). We were recently able to minimize this gap by applying for the first time X-ray absorption spectroscopy under real "operando" conditions and thereby deriving structure-function relationships helping to explain the improving effect of these additives [3-5]. The greatest challenge in these studies is the fact, that the noble metals Pt, Pd and Au are present in very low concentration (0.2 wt. %) in a heavily X-ray absorbing, highly porous SnO_2 matrix (50 μm thick films). In addition, the platinum present in the electrodes and the heater of the conventional sensor devices leads to strong Pt-fluorescence hiding the signals of the gold and platinum additives in SnO_2. In the present study, we give an overview how to tackle the structure of the mentioned noble metal additives in SnO_2-sensors and how the challenges in low concentration, sensor and cell-design can be overcome.

2. Results

In a first step, the structure and role of Pd additives in SnO_2 based gas sensors was elucidated by using X-ray absorption spectroscopy in real sensing operation conditions. For this purpose a new cell was designed (Figure 1, ref. [3]) which allows (i) measuring the variation in resistance (corresponding to the sensor signal) as a function of the atmospheric compositions, (ii) operating the sensing layer at well controlled temperatures and (iii) recording at the same time X-ray absorption spectra in the fluorescence mode at the Pd K-edge. The experiments were performed at the undulator beamline ID26 of the European Synchrotron Radiation Facility (ESRF) using a pair of cryogenically cooled Si(311) crystals. The Pd Kα fluorescence was recorded using a Si photodiode and copper foil as filter. Both XANES and EXAFS analysis showed that Pd at low concentration (< 1wt%) is present on an atomic level in an oxidized state, well dispersed at the surface and in the bulk of the host SnO_2 material [3,6]. Due to the high dispersion 0.2wt%Pd/SnO_2 is only reduced above 200 °C in 1000 ppm H_2/He (Figure 1). A change in the chemical state of the Pd sites during sensing of CO and H_2 could not be observed although tremendous changes in the resistance were observed with 30 ppm H_2 or 50 ppm CO in air. In addition, these Pd species depict a rather dynamic reduction-re-oxidation behavior [3].

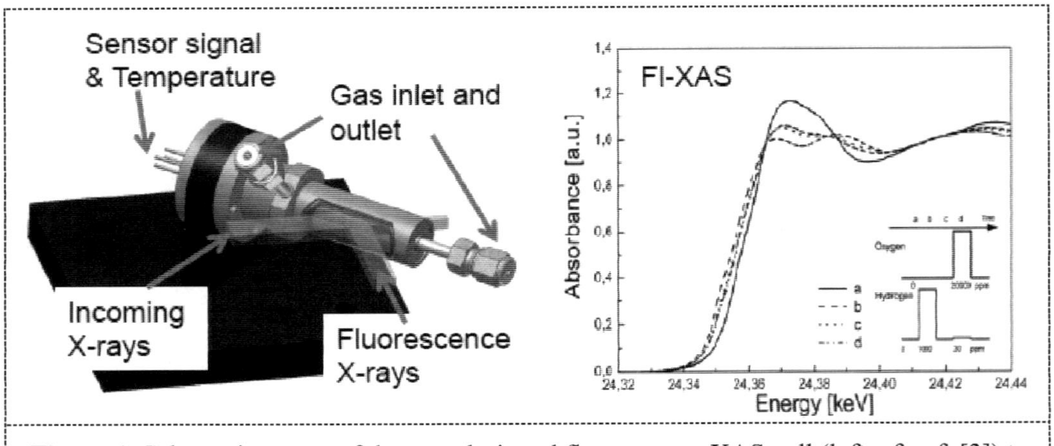

Figure 1. Schematic set-up of the new designed fluorescence XAS-cell (left, cf. ref. [3]) to record the resistance changes simultaneously with the XAS spectra under "operando" conditions given on the right: XANES spectra of 0.2 wt. % Pd:SnO_2 sensor at RT in He (a), in 1000 ppm H_2 in He (b), at 300 °C, partial reoxidation in He (c) under sensing conditions at 300 °C with 30 ppm of H_2 in air after pre-reduction (d); partly reproduced from ref-. [3] with permission from the Royal Society of Chemistry

In case of Pt additives two additional challenges came up. Since Pt was also used for the heater/electrodes in the conventional sensor device, a new sensor substrate - showing similar electrical characteristics - had to be designed. Au was used instead of Pt for the electrodes and the Pt heater was replaced by a Pd/Ag alloy resistance heater on the back (see schematics in Figure 2). Nevertheless, the disturbing gold fluorescence from the new electrode material had to be eliminated (Au $L\alpha_1$ line at 9.442 keV and Pt $L\alpha_2$-line at 9.713 keV). This was achieved by recording high energy resolved fluorescence detection (HERFD) XAS at ID26 at the ESRF with an X-ray emission spectrometer in horizontal plane. Sample, analyzer crystal and photon detector (avalanche photodiode) were arranged in a vertical Rowland geometry [7]. The Pt HERFD-XANES spectra at the L_3-edge were obtained by recording the intensity of the Pt $L\alpha_1$ emission line (9442 eV) as a function of the incident energy (Figure 2). The emission energy was selected using the <660> reflection of four spherically bent Ge crystal analyzers (with R = 1m) aligned at 80° Bragg angle. A combined (incident convoluted with emitted) energy resolution of 1.8 eV was obtained as determined by measuring the elastic peak. This approach succeeded and due to the 0.2 wt. % Pt/SnO_2 layer being on top of the Au electrodes no self-absorption effects were found (Figure 2). This also demonstrates that no remarkable diffusion of gold into the sensing layer has occurred. The experiments further uncovered that under operating conditions – temperature of 300°C in dry air - platinum is in a highly oxidized state and does not change upon CO and H_2 exposure [4]. In addition, also no change in the oxidation state of Pt during similar reduction-reoxidation experiments like in the case of Pd could be observed. Finally, EXAFS analysis uncovered that the highly oxidized Pt is dispersed also on an atomic level in the lattice of SnO_2, on the surface as well as incorporated in the bulk.

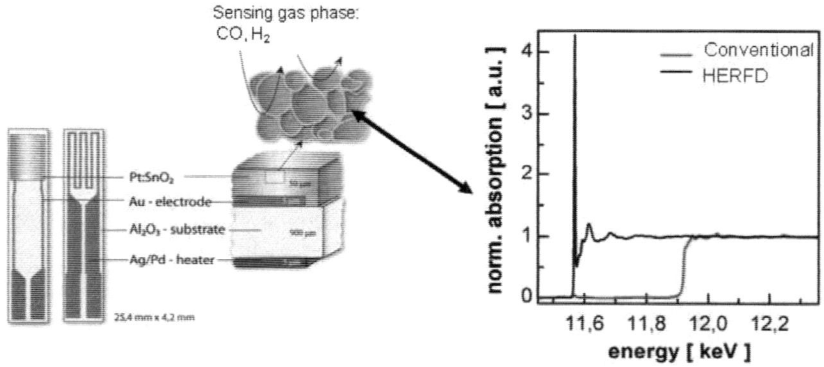

Figure 2. Principal hierarchical structure of the sensor for studying the Pt-doped $Pt:SnO_2$ based sensor with Au-electrode and PdAu-heater (left, cf. Ref. [4]b) and XANES and EXAFS spectra (right, cf. Ref. [4]): Pt L_3 edge EXAFS spectra of the 0.2 wt. % $Pt:SnO_2$ sensor with Au electrodes measured by HERFD-XAS (black) and traditional (red, Pt L_3-edge hardly visible) detection mode.

These results are surprising because in previous studies dealing with model samples (much higher concentrations) or conditions (far away from the operational ones), metallic particles or clusters were found and have been ascribed in some of the models to the improved sensing performance [8,9]. In our samples, Pd and Pt exhibit a quite similar effect, namely on one side a surface effect (surface sensitization) and on the other side a bulk effect (change of the electrical properties of the semiconductor) as illustrated schematically in Figure 3 a and b.

Finally, the focus was on Au promoted SnO_2 based gas sensors, demonstrating an improved sensing and catalytic performance towards CO due to the presence of the additive. Contrary to Pd and Pt - being present in an oxidized state - HERFD-XANES spectra at the Au L_3-edge revealed that Au is present in a bulk metallic form and that the chemical state is not changing during exposure towards CO and H_2 [5]. The sensitization effect of gold additives in SnO_2 based gas sensors is therefore attributed to small gold metallic particles incorporated in the SnO_2 layer. The high chemical activity of these very small Au particles facilitates the dissociation of oxygen as known from CO-oxidation (spill-over effect; see Figure 3 c) and therefore enriches the surface of the metal-oxide with oxygen ions, which are the source of CO and H_2 sensing. A direct electrical connection with the SnO_2 matrix could not be observed.

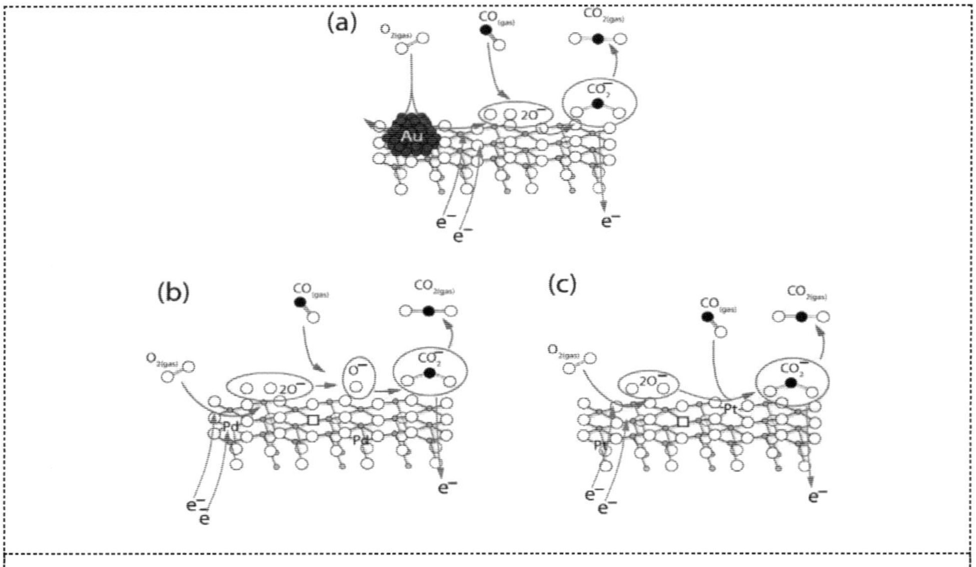

Figure 3. Schematic presentation emphasizing the sensitization effect of (a) Au, (b) Pd (a), and (c) Pt additives in SnO_2 and summarizing the observations in Refs. [3-5].

3. Conclusions and outlook

This overview on differently doped metal oxides underlines the importance of studying the materials as close as possible to the operating conditions, the strength of novel X-ray absorption spectroscopic techniques and the design of in situ cells/devices that allow both monitoring of the structure and sensor performance as known from catalysis. They provided new insight into the sensing mechanism demonstrating a rather similar sensitization effect for Pt and Pd, but a very different mechanism for the Au additive. In future, it will be interesting to extend this to sensors prepared by different preparation techniques and further dopants as well as sensor materials.

4. Acknowledgements

We thank ESRF (Grenoble, F) for beamtime allocation at the insertion device beamline ID26, financial support and Kristina Kvashnina, Marcin Sikora, and Pieter Glatzel for their help and discussion. Moreover, JDG thanks for financial support and original seed funding by the Karl Winnacker grant.

References

[1] Williams D E 1999 Sensors and Actuators B **57** 1
[2] Barsan N, Schweizer-Berberich M and Göpel W 1999 Fresenius' J. Anal. Chem. **365** 287
[3] Koziej D, Huebner M, Barsan B, Weimar U, Sikora M and Grunwaldt J D 2009 Phys. Chem. Chem. Phys. 11 8620
[4] a) Hübner M, Koziej D, Bauer M, Barsan N, Kvashnina K, Rossell M D, Weimar U, Grunwaldt J D 2011 Angew. Chem. Int. Ed. **50** 2841; b) highlighted in ESRF Highlights 2011, p. 17.
[5] Hübner M, Koziej D, Grunwaldt J D, Weimar U and Barsan N 2012 Phys. Chem. Chem. Phys. **14** 13249.
[6] Hübner M, Bârsan N, Weimar U 2012 Sensors and Actuators B **172** 171.
[7] Glatzel P, Bergmann U 2005 Coord. Chem. Rev. **249** 65.
[8] Safonova OV, Neisius T, Ryzhikov A, Chenevier B, Gaskov AM, Labeau M 2005, Chem. Commun. 5202.
[9] Gurlo A, Riedel R 2007 Angew. Chem. Int. Ed. **46** 3826.

Analysis of Cs/GaAs NEA surface by XAFS

K. Tsubota[1], M. Tabuchi[2], T. Nishitani[2], A. Era[1] and Y. Takeda[2]

[1]Graduate school of Engineering, Nagoya University, [2]Synchrotron Radiation Research Center, Nagoya University

E-mail: tsubota@mercury.nagoya-u.ac.jp

Abstract. By exposing a GaAs surface to Cs and O_2, a surface which is called negative electron affinity (NEA) surface can be obtained where the vacuum level shifts lower than the bottom of GaAs conduction band. The Cs/GaAs NEA surface has been attracting great attention in recent years, because of its availability for a highly functional electron source which can be controlled by excitation light. In this work, XAFS and X-ray fluorescence measurements were conducted for Cs/GaAs NEA surface prepared on GaAs (001) and (110) substrates. As the result, As and O were considered as near neighbor atoms of center Cs for both of the samples. Most of Cs formed some kind of Cs-oxide and the Cs atoms at interface between Cs-oxide and GaAs were observed as forming Cs-As bond. Taking account of the ratios of coordination numbers and the fluorescence X-ray yield, the Cs-oxide on GaAs (001) surface forms island like structure in which the body/interface ratio is large, and the Cs-oxide on GaAs (110) surface forms layer like structure in which the body/interface ratio is small.

1. Introduction

Cs adsorbed GaAs surface shows negative electron affinity (NEA) when the surface is properly prepared with O_2. [1,2] At the NEA surface, the vacuum level is lower than the bottom of the conduction band of the base semiconductor, GaAs. Therefore when electron-hole pairs are excited in GaAs by irradiating light, the electrons in the conduction band are easily emitted from the surface. The surface is utilized for highly functional electron source that can be controlled by light, which is so called photo-cathode. The distributions of energy and momentum of electrons emitted from the surface directly reflect the distributions of them in the base semiconductor, which can be controlled by changing the energy of the excitation light. When the energy of the light is tuned in very near to the band gap energy of the base semiconductor, the distributions of energy and momentum emitted from the surface can be so small as those defined by environmental temperature. For example it was reported 8 meV at 90K.[3]

Because of such availability for a highly functional electron source, the Cs/GaAs NEA surface has been attracting great attention in recent years. However, the Cs/GaAs surface has been performed only in an ultra high vacuum environment and the life-time of the surface is reported to be sensitively affected by the background pressure.[4] For the practical use of the Cs/GaAs NEA surface, it is necessary to realize highly durable and long life-time NEA surface by understanding the structure of the NEA surface formed with Cs.

In this work, the Cs/GaAs NEA surface was formed on GaAs (110) substrate to investigate the characteristics of the surface by comparing them with those on GaAs (001) substrate. The local structures around Cs on GaAs (110) and (001) substrates were investigated by using X-ray absorption fine structure (XAFS) measurement at Cs-L_{III} edge, while total amount of Cs on the surface was monitored by fluorescence X-ray yield measurement.

2. Preparation of the NEA surface

An NEA activation chamber specially designed to set on the XAFS measurement line was prepared. The chamber was equipped with ion-pump and NEG (non-evaporative getter) pump to realize ultra-high vacuum (UHV) background up to 10^{-9} Pa order and had 3 Be-windows to introduce X-rays into the UHV environment and to detect fluorescence X-ray emitted from the samples.

After the GaAs (110) or (001) substrate was set on the sample holder in the chamber, the chamber was pumped and baked out to realize the UHV environment. When the UHV environment was realized, the substrate was heated up to more than 580 °C and kept at the temperature for more than 1 hour to obtain a clean surface. Then, the sample was cooled down to R.T. before starting the process to obtain NEA surface. The process, which is so called "Yo-Yo" process [2], was started by the supply of Cs while the sample was biased at -100 V and irradiated by X-ray of 5100eV to monitor photo-emitting electron current and measure the yield of Cs-L_α fluorescence X-ray. When Cs was supplied, the typical vacuum level was about 5.0 x 10^{-7} Pa. Figure 1 schematically shows the change of the photo-current during the Yo-Yo process.

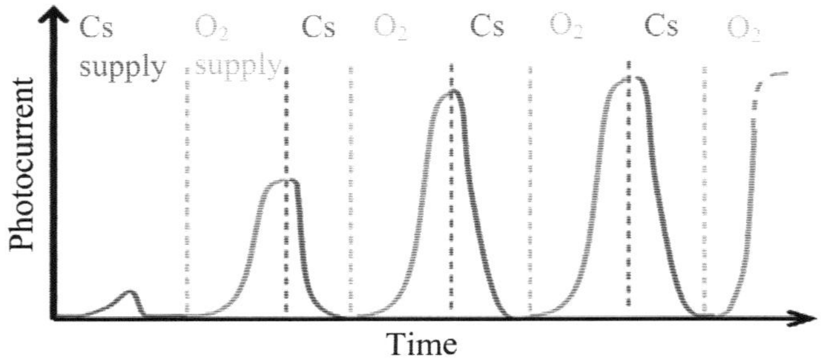

Figure 1. Schematically drawn photo-current change during the Yo-Yo process.

Several minutes after starting Cs supply, a weak photo-current was detected as shown in Figure 1. With the proceeding Cs supply, the photo-current reached to the first small peak and then decreased. When the photo-current became very small to observe, the Cs supply was stopped and O_2 supply was started alternatively. Typically, the timing was 10 to several 10min after the start of Cs supply. When O_2 was supplied, the vacuum level was about 2.0 x 10^{-6} Pa. On the O_2 supply, the photo-current rapidly increased and reached to the second peak which was usually much higher than the first one. Then the O_2 supply was switched to the Cs supply and the photo-current decreased again. When it was observed that the photo-current reached to a valley, the Cs supply was again switched to O_2 supply. After that, O_2 and Cs were alternatively supplied by switching them at the peak and valley of the photo-current. The photo-current peak heights gradually increased with the sequence as shown in Figure 1. Finally, when the increase of the photo-current peaks was saturated, the full process was finished. Since the NEA surface must be prepared with such an artful process, Cs is expected to form some special structure on the surface with oxygen. However the fragility of the surface has prevented us from investigating the structure.

3. In-situ X-ray fluorescence and XAFS measurements

In this work X-ray fluorescence and XAFS measurements were conducted by setting the NEA activation chamber on the XAFS measurement beam line 9A at Photo Factory in KEK in Tsukuba, Japan. By introducing the synchrotron radiation X-ray into the chamber through a Be window, the fluorescence X-ray from the sample was observed through another Be window. The Yo-Yo process

Journal of Physics: Conference Series **430** (2013) 012079 doi:10.1088/1742-6596/430/1/012079

was conducted on line and the amount of Cs on top of the surface was monitored in-situ by measuring the yield of the Cs-L_α fluorescence X-ray.

The XAFS measurements were conducted in fluorescence X-ray detection mode with X-rays polarized normal to the surface for the finally obtained NEA surfaces on GaAs (110) and (001) surfaces. The fluorescence yield measurements were conducted also for the finally obtained NEA surfaces on GaAs (110) and (001) surfaces. During the XAFS measurement, the sample temperature was kept at about -90°C and the measurements spent 16 hours for each sample.

4. Results

Figure 2 shows (a) measured Cs-L_α fluorescence yield and (b) normalized $k^3\chi(k)$ spectra.

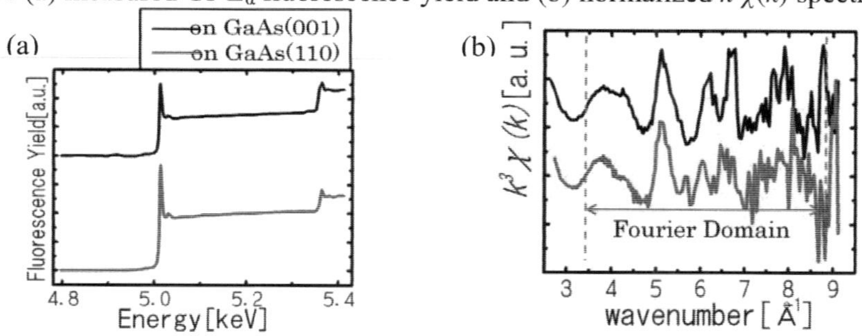

Figure 2. (a) Measured fluorescence X-ray yield and (b) normalized $k^3\chi(k)$spectra

First of all, no change was observed in XANES region in the spectra, which means the chemical status of Cs on GaAs (001) and (110) surface were almost the same. Then we should concentrate on the EXAFS spectra.

As shown in Figure 2, the EXAFS oscillations were observed very weak although the fluorescence yield was not so small. As a result, the S/N rations of the obtained $k^3\chi(k)$ spectra were small. Reported Debye-Waller factors of Cs-related compounds were often very large. [5] It might be the reason that the oscillations were so weak and we need to spend so long time to measure the XAFS spectra.

Figure 3 shows the radial-distribution functions obtained by Fourier transformation of the normalized XAFS spectra.

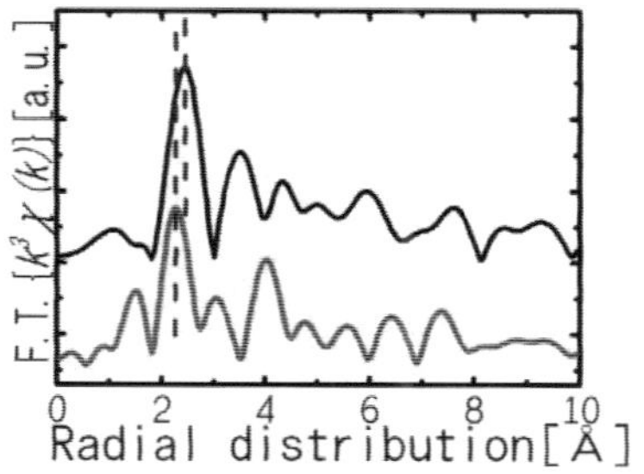

Figure 3. Radial distributions

As shown in Fig. 3, main peaks were observed at around 2.5 Å, and the peaks shifted between the samples on GaAs (110) and (001) surfaces. Curve fittings for the Fourier filtered XAFS spectra for the main peaks were conducted assuming Cs, As and O as near neighbor atoms around center Cs. The best fit curves were obtained when As and O were assumed to be the neighbor atoms for both of the spectra. Obtained best fit parameters were listed in Table I and the best fit curves were shown in

Fig. 4. The error ranges are shown with each parameter in parenthesis in TableI, which were estimated as the ranges where the R-factor increased up to twice of the minimum.

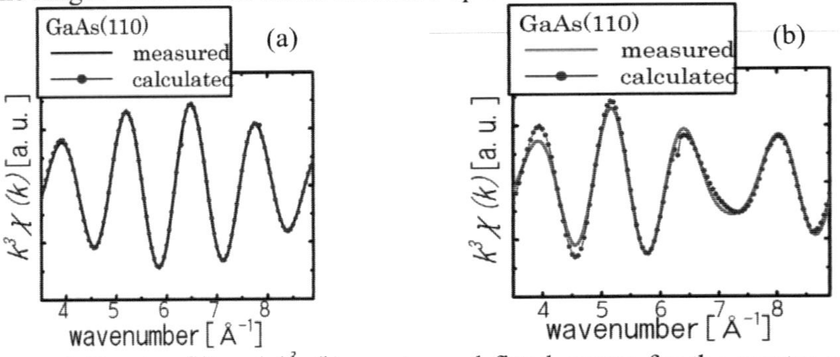

Figure 4. Fourier filtered $k^3\chi(k)$ spectra and fitted curves for the spectra of the samples on (a) GaAs (110) and (b) GaAs(001) substrates.

Table I Fitting parameter				
	GaAs(001)		GaAs(110)	
Assumed atom	Distance	Coordination number	Distance	Coordination number
As	2.83Å (±0.01)	0.04 (±0.01)	3.12Å (+0.02,-0.01)	0.26 (+0.08,-0.07)
O	2.93Å (±0.005)	0.79 (+0.02,-0.03)	2.92Å (±0.01)	0.99 (+0.13,-0.20)
R. Factor	0.02		0.13	

The Cs-As distances were changed between the samples and the change was agreed with the observed peak shift in Fig. 3. The difference in coordination number of As was large while that of O was rather small, although the coordination numbers were not so reliable since the S/N ratio was not so good. Figure 5 shows the intensity of the Cs-L_α fluorescence X-ray from the Cs/GaAs NEA surface that were measured after the full Yo-Yo process on GaAs (001) and (110).

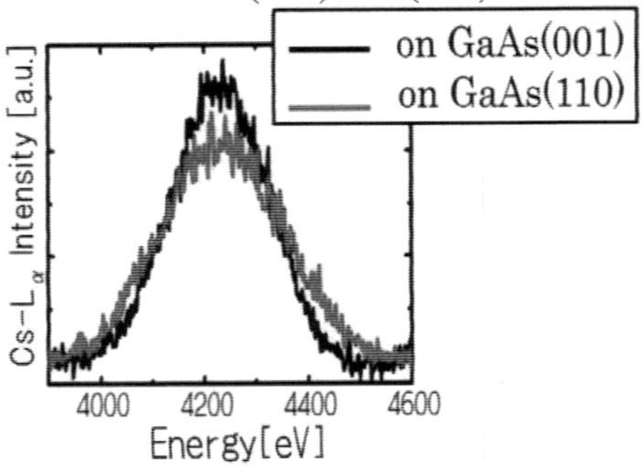

Figure 5. Intensity of the Cs-L_α fluorescence X-ray yield from Cs/GaAs surfaces.

As shown Fig. 5, the peak heights were almost the same, in other words, the amount of Cs on top of the GaAs (001) and (110) surfaces were almost the same.

5. Discussions

The result that the coordination number of O, N_O, is larger than that of As, N_{As}, for both of the samples indicated that most of the Cs atoms on NEA surface bond with O. In other words, most of Cs formed some kind of Cs-oxide. As described before, we had to assume As with also O as neighbours, although the deduced amount of As was small. Since the NEA surface was formed on GaAs substrate, it is naturally speculated that the Cs-oxide was adsorbing on GaAs surface by the Cs-As bonds at the interface.

The difference of the ratios of coordination numbers, N_{As}/N_O, reflects the difference of the ratios of Cs atoms located at the Cs-oxide/GaAs interface and in Cs-oxide body. The ratios N_{As}/N_O were about 0.05 and 0.26 for the samples on GaAs (001) and (110) substrate, respectably. The difference indicated that the ratio (interface area)/(Cs-oxide volume) was smaller for the sample on GaAs (001). On the other hand, the amount of Cs should be almost the same since the Cs-L_α fluorescence X-ray intensity was almost the same for both of the samples. These two results that N_{As}/N_O ratio was different while the amount of Cs was constant might be explained by assuming that the Cs-oxides on GaAs (001) surface forms island like structure in which the interface/body ratio is small and the Cs-oxide on GaAs (110) surface forms layer like structure in which the interface/body ratio is large. These differences of the structure on NEA-Cs/GaAs surface may affect the performance of the NEA surface such as the life-time or the behaviour of photocurrent during Yo
-Yo process.

6. Conclusions

XAFS and X-ray fluorescence measurements were conducted for Cs/GaAs NEA surface prepared on GaAs (001) and (110) substrates. As the result, As and O were considered as near neighbor atoms of center Cs for both of the samples. Most of Cs formed some kind of Cs-oxide and the Cs atoms at interface between Cs-oxide and GaAs were observed as forming Cs-As bond. Taking account of the ratios of coordination numbers and the fluorescence X-ray yield, the Cs atoms on GaAs (001) surface forms island like Cs-oxide structure in which the body/interface ratio is large, and the Cs atoms on GaAs (110) surface forms layer like structure in which the body/interface ratio is small.

Reference

[1] C. Y. Su, W. E. Spicer, and I. Lindau, J.Appl. Phys., 54,pp. 1413-1422 (1983)
[2] K. Togawa et al., Nucl. Instr. And Meth. A, 414, pp.431-445 (1998)
[3] F. Sprenger et al., Nucl. Instr. And Meth., A, 532, pp.298-302(2004)
[4] T. Nakanishi et al., Proceeding of LINAC, pp.201-203, (2004)
[5] L. -M. Peng et al., Acta Cryst., A52, pp.456-470 (1996)

Tracking the Formation of Nano-sized Zinc Oxide from Zinc Peroxide by In Situ XAS and XRD

T Daley [1*], E Raj [2], S Ramos[3], G Cibin[3], A Dent[3], T I Hyde[2] and G Sankar[1*]

[1] Department of Chemistry, University College London (UCL), London, WC1H 0AJ, England

[2] Johnson Matthey Technology Centre, Blounts Court, Sonning Common, Reading RG4 9NH

[3] Diamond Light Source, Harwell Science and Innovation Campus, Didcot, Oxfordshire, OX11 0DE

*Corresponding author, email: t.daley@ucl.ac.uk , g.sankar@ucl.ac.uk

Abstract. In this paper we show how the formation of zinc oxide from zinc peroxide takes place during thermal decomposition, through a combination of X-ray Absorption Spectroscopy (XAS) and X-ray Diffraction (XRD). The time-resolved study allowed us to follow the changes in both the local structure of the material, in particular the conversion of octahedral Zn-O coordination to tetrahedral one through EXAFS and XANES analysis and at the same time the particle growth using both EXAFS and diffraction. Multi-cluster analysis provided information on the phase changes which could be cross-correlated with diffraction results which suggests an abrupt change in the long range coordination followed by a slow growth of zinc oxide NPs. This study allowed us to gain a detailed understanding of zinc peroxide decomposition leading to the formation of zinc oxide.

1. Introduction

Zinc Oxide nanoparticles (NPs) are important for a variety of applications such as pigments, gas sensors, light emitting diodes, photo-detectors, chemical sensors, piezoelectrics and solar cells semi-conductors and as catalysts due to their size-dependant electronic and optical properties[1]. There are a number of well-documented ways of synthesising the specific ZnO NPs, which includes sol-gel processes, precipitation, microemulsion, spray pyrolysis and hydrothermal synthesis.[1a, 2]. More recently it has been shown that zinc oxide NPs can also be formed through the decomposition of zinc peroxide.[3]

Using a combination of X-ray Absorption Spectroscopy (XAS) and X-ray Diffraction (XRD), we show in this study how the formation of zinc oxide from zinc peroxide takes place during the thermal decomposition process, through monitoring the changes in the short and long-range order that occur during the conversion. This time-resolved study facilitated the examination of changes in the local structure of the material, in particular the conversion of the change in Zn-O coordination from octahedral to tetrahedral through the use of EXAFS and XANES analysis. Concurrently, the particle growth was monitored using both EXAFS and diffraction. Multi-cluster analysis enabled us to gain information on the phase changes that occur and this was cross-correlated with the diffraction results obtained and suggested an abrupt decomposition of zinc peroxide followed by a slow conversion to zinc oxide NPs transpired. Overall, this study allowed us to gain a detailed understanding of zinc peroxide decomposition leading to the formation of zinc oxide.

2. Experimental

XAS data were collected on zinc K-edge at the B18 beamline at Diamond Light Source (UK), which operates at 3 GeV with a typical current of *ca* 250 mA. Measurements were performed in transmission mode using ion chambers filled with a mixture of argon and helium. The zinc peroxide sample was made using 20mg of ZnO_2 diluted with fumed silica (130 mg) and the mixture was pressed into a 13mm pellet before loading into a specially designed *in-situ* furnace cell that is capable of measuring combined XRD/XAS data. The cell is a custom built high temperature water cooled cell with gas inlet ports (see Figure 1). For these experiments a flow of air at 10ml/min was used. The zinc peroxide pellets were then heated up to 500°C at a rate of 5°C/min, in air, whilst diffraction and absorption data were recorded sequentially. X-Ray Diffraction data was recorded using a wide angle

Figure 1: Image of the in situ cell utilised in this work showing the important features such as X-ray windows and sample holder.

detector and a wavelength of 1.305Å (corresponding to an energy of 9500eV). Quick-EXAFS data was collected in transmission mode. A zinc reference foil was used to calibrate the energy. XRD data was analysed using Fityk software[4] whilst QuEXAFS data was normalised in Athena and fitted using Artemis software running with IFEFFIT 11[5]. Refinement was carried out using k^3 weighting in the range 2.5-11.5$Å^{-1}$. Throughout the cluster refinement coordination numbers were fixed in order to avoid creating correlations between the co-ordination numbers and the Debye-Waller factor. Fits of the first shell and further analysis of coordination number variation was performed using Excurve (version 9.3).

3. Results and Discussion

Typical XRD, Zn K-edge XANES patterns obtained from the *in situ* measurements are shown in figure 2 It is clear that the initial material is pure crystalline zinc peroxide, which upon heating is converted into zinc oxide [6]. According to the diffraction data this change starts to occur at ca 200°C and the decomposition of ZnO_2 appears to be complete at ca 260°C. However, the ZnO reflections could not be seen during this decomposition period suggesting an amorphous intermediate phase is formed prior to the growth of ZnO phase. However, closer inspection of the peroxide [2 2 1] and oxide [1 0 2] reflections, which do not overlap, reveals the complete disappearance of crystalline peroxide at ca 260°C followed by a gradual increase in the oxide [1 0 2] reflection intensity indicating a slow growth of zinc oxide, which is more apparent above 300°C. Zn K-edge XANES data shows the transition in the local structure from the peroxide to the oxide accentuated by a large decrease in the whiteline intensity (Figure 2) This would correlate to the change from a octahedrally coordinated

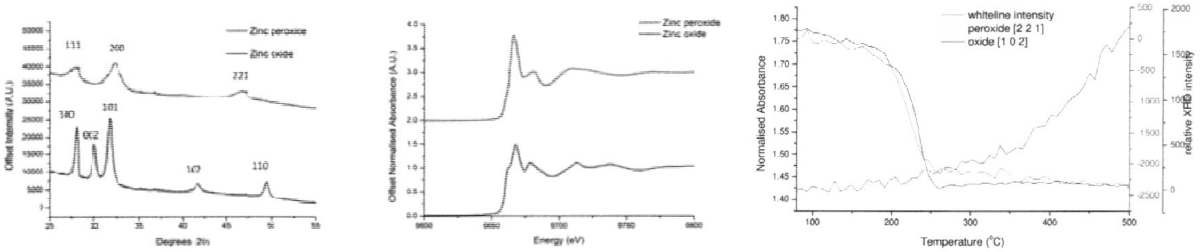

Figure 2: Typical XRD (left) and Zn K-edge XANES (middle) of the starting and final material obtained through the decomposition of ZnO_2. On the right the variation of intensities of (221) and (102) reflections corresponding to ZnO_2 and ZnO along with the changes in the white-line intensity obtained from the XANES are shown.

Zn(II) in the ZnO_2 structure to a tetrahedral one present in ZnO[3]. This transition occurs in the region from 180-250°C and follows the observation seen from XRD data. This suggests the presence of ZnO like phase in the intermediate amorphous material Zn K-edge EXAFS was processed by fitting the starting material to the structure of ZnO_2 and final one with ZnO structure. The initial peroxide model was fitted for the first scan using scattering paths and the refined values for the Zinc peroxide are, Zn-O at ca 2.07Å, Zn-O at ca 2.92Å, and Zn-Zn at ca 3.44Å and they are given in Table 1 and the best fit is shown in Figure 3. A similar procedure was used for fitting the final data using ZnO structure [6b]. These values are in agreement with those reported previously by Sankar et al.[7] By taking the

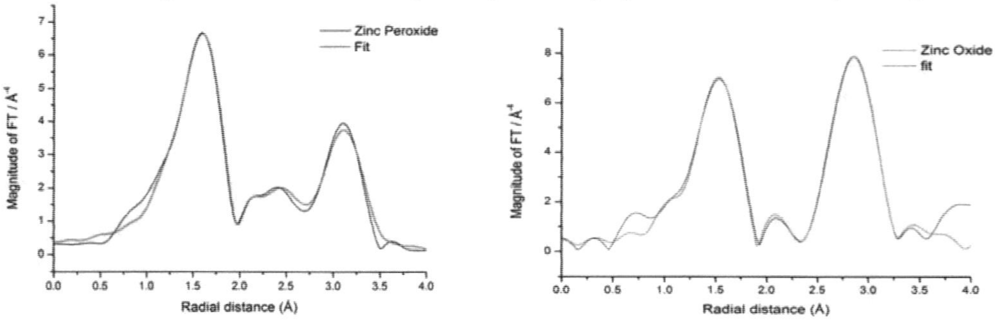

Figure 3: Fourier Transformed $k^3\chi$ data of ZnO_2 at the beginning of the heating process (left) and ZnO at the end of the heating process (right)

parameters from the best fit values from the analysis of the initial and final EXAFS data, a more rigorous analysis of the remaining data sets recorded at various temperatures were performed. This allowed us to estimate the amount of the individual phases present at a given temperature. The analysis revealed the major changes in local structure occurring between 180-250°C, which correlates well with the diffraction data. The results are consistent with other measurements, in particular *In situ* Raman spectroscopic data and TGA studies

Table 1 Structural parameters obtained from the best fit to the Zn K-edge EXAFS of ZnO_2 and ZnO

Phase	scatterer	N	N (error)	R (Å)	Error in R (Å)	σ^2 (Å^2)	error in σ2 (Å2)	R-Factor	$S0_2$
Zinc peroxide	O	6	-	2.07	0.0082	0.011	0.0116	0.00787	0.80
	O	6	-	2.92	0.023	0.0252	0.0185		
	Zn	12	-	3.44	0.0107	0.0171	0.0171		
	O	2	-	3.48	0.023	0.0171	0.0185		
zinc oxide	O	3.8	0.33	1.97	0.0088	0.0015	0.0082	0.00387	0.75
	Zn	6.99	0.40	3.22	0.0083	0.0033	0.023		
	O	1.0717	0.084	3.19	0.0245	0.0011	0.0107		
	Zn	6.99	0.40	3.26	0.0083	0.0033	0.023		

4. Conclusions

In summary, the decomposition of zinc oxide nanoparticles was examined using a combined in situ XAS/XRD experiment. The data indicated changes in the local structure occurring between 180 – 250oC as the material decomposes to form the oxide. Although it appears from diffraction that a very poorly crystalline phase is formed prior to the appearance of reflections of crystalline ZnO phase, both XANES and EXAFS suggest the presence of ZnO like phase in the poorly crystalline, intermediate system. Furthermore, the study revealed that the decomposition temperature may be controlled to tailor the synthesis of zinc oxide NPs from zinc peroxide.

References

[1] a) J. Lee, A. J. Easteal, U. Pal and D. Bhattacharyya, *Current Applied Physics* **2009**, *9*, 792-796; b) M. Jitianu and D. V. Goia, *Journal of Colloid and Interface Science* **2007**, *309*, 78-85

[2] a) S. Baruah and J. Dutta, *Science and Technology of Advanced Materials* **2009**, *10*; b) R. M. Trommer, A. K. Alves and C. P. Bergmann, *Journal of Alloys and Compounds* **2010**, *491*, 296-300

[3] N. Uekawa, N. Mochizuki, J. Kajiwara, F. Mori, Y. J. Wu and K. Kakegawa, *Physical Chemistry Chemical Physics* **2003**, *5*, 1489-1489.

[4] M. Wojdyr, *Journal of Applied Crystallography* **2010**, *43*, 1126-1128

[5] B. Ravel and M. Newville, *Journal of Synchrotron Radiation* **2005**, *12*, 537-541.

[6] a) M. Puselj, Z. Ban and J. Morvaj, *Documenta Chemica Yugoslavica. Croatica Chemica Acta* **1985**, *58*, 283-288; b) D. A. A. Santos, A. D. P. Rocha and M. A. Macedo, *Powder Diffraction* **2008**, *23*, S36-S41

[7] L.-L. Liew, G. Sankar, A. D. Handoko, G. K. L. Goh and S. Kohara, *Journal of Solid State Chemistry* **2012**, *189*, 63-67.

Journal of Physics: Conference Series **430** (2013) 012081 doi:10.1088/1742-6596/430/1/012081

XAFS Study of metal-doped In_2O_3

Steve Heald[1], Marzook S. Alshammari[2,3], S. Alfehaid[2], M. Alotaibi[2], Qi Feng[3] and A. M. H. R. Hakimi[4]

[1] X-ray Science Division, Argonne National Lab, Argonne, IL 60439, USA
[2]The National Centre of Nanotechnology, KACST, P.O Box 6086, Riyadh 11442, Saudi Arabia
[3]Dept. of Physics and Astronomy, Univ. of Sheffield, Sheffield, S3 7RH, UK
[4] Dept. of Materials Science and Metallurgy, Univ. of Cambridge, Cambridge, CB2 3QZ, UK
Email: heald@aps.anl.gov

Abstract. In_2O_3 has attracted interest recently because of its favorable properties for potential applications as diluted magnetic semiconductors. For these applications the material must be doped to produce the desired magnetic properties, and it is essential to determine the fate of the dopants in order to optimize the materials and understand the magnetic measurements. This paper summarizes some recent XAFS studies of a series of In_2O_3 films doped with V, Cr, Co, and Fe. XAFS is shown to be a powerful tool to characterize these materials, providing valence information, and detecting both substitutional doping as well as second phase formation. V and Cr are found to take up well ordered substitutional sites with no evidence for second phase formation. For Co the substitutional sites were much more disordered than for V or Cr. Fe doped samples also displayed a range of results with well-ordered substitutional sites in some samples, while others showed extensive second phase formation including magnetite and Fe metal.

1. Introduction

The search for room temperature diluted magnetic semiconductors has spurred investigation of a variety of metal doped oxides. In_2O_3 is a promising system[1-4]that has seen a lot of previous work in conjunction with applications as transparent conducting films. In that application the primary dopant is Sn, and a number of XAFS studies have been carried out on doped In_2O_3.[5, 6] For magnetic applications the dopants are generally transition metals.

In this paper we present the application of XAFS to a variety of transition metal doped In_2O_3 films. XAFS has been demonstrated to be a powerful method for characterizing doped metal-oxide films[7]. A common problem in such films is the formation of second phases, especially metal phases. For magnetic second phases this can result in a magnetic signal that can confuse the magnetic characterization. These second phases can be difficult to detect since they may be of low concentration, and are often in the form of nanoparticles that are hard to detect by diffraction methods. Both the XANES and EXAFS have been shown to be sensitive to the presence of such nanoparticles. In the XANES the metals show up as a low energy feature in the edge while for the EXAFS the metal-metal bond in the metallic form often shows up as a peak intermediate between the first and second neighbors in metal oxides. For films in which second phase formation is not present the EXAFS can provide valuable information about the substitutional sites, both the position in the lattice and any lattice distortions from the size mismatch of the dopant atoms. The XANES can verify the dopant valence. If it differs from the host In atoms (3+) then charge compensation such as the formation of O vacancies or interstitials must be occurring.

2. Experimental Results

The samples measured include V, Cr, Fe and Co doped In_2O_3 thin films. The films were grown by PLDand sputtering on Al_2O_3 substrates with varying O partial pressures. The samples are summarized in Table 1.

Table 1. Summary of the samples measured by XAFS. All were grown on Al_2O_3 (0001) substrates.

Sample	Content	T (°C)	Pressure	Method
Co-1	In_2O_3/ (5% Co)	330	$6x10^{-4}$Torr	Sputtering
Fe-1	In_2O_3/ (4% Fe)	600	$2x10^{-6}$Torr	PLD
Fe-2	In_2O_3/ (5% Fe) vac. annealed 450°C	450	$2x10^{-6}$Torr	PLD
Fe-3	In_2O_3/ (5% Fe)	450	$2x10^{-6}$Torr	PLD
V-1	In_2O_3/(5%V)	550	$2x10^{-5}$Torr	PLD
Cr-1	In_2O_3/(3%Cr)	600	$2x10^{-6}$Torr	PLD

The samples were measured at beamlines 20-BM-B and 20-ID-B at the Advanced Photon Source. In both cases Si (111) monochromator crystals were used with energy resolution of about $1.4x10^{-4}$. The samples were mounted on a spinner that rotated at a few Hz to minimize distortion from substrate Bragg peaks. In most cases the samples were mounted with an incident angle of a few degrees and the x-ray polarization was perpendicular to the surface. Fluorescence detection was used with either a 4-element Vortex (20-ID) or a 13-element Ge fluorescence detector (20-BM).

Figure 1 shows the EXAFS data and the Fourier transforms for the Cr, Co and V doped samples. The Cr and V samples showed strong XAFS extending to higher k indicating the metal is residing in reasonably well ordered sites. There is a strong similarity indicating the metals are in similar sites, and there is little indication of metal formation in the near edge spectra (not shown). All of this suggests that the Cr and V are entering substitutional sites, and this will be confirmed in the next section. The Co data shows some similarities to the other two, but the XAFS is strongly damped.

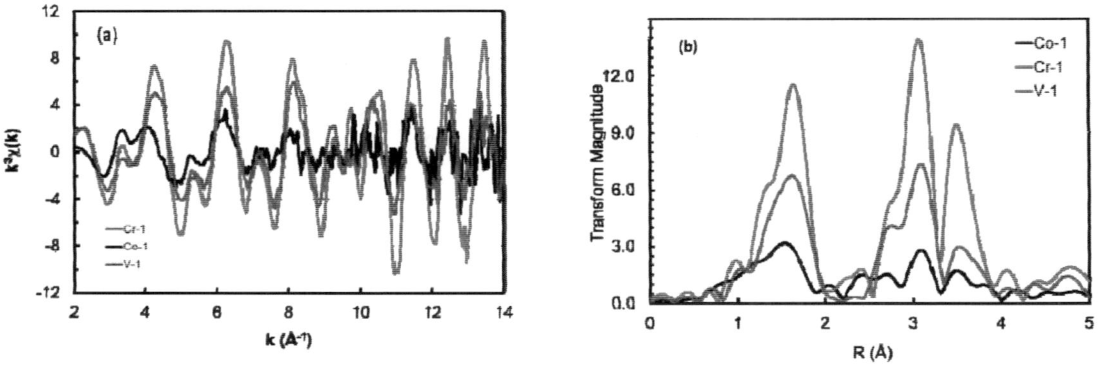

F

Figure 1. Comparison of the (a) EXAFS and (b) Fourier transformed EXAFS for Co, Cr, and V in In_2O_3.

Figure 2a shows the XANES spectra for a set of Fe doped samples. These show a whole range of behavior. For sample Fe-1 the near edge looks very similar to magnetite, and this is confirmed by the EXAFS shown in Fig. 2b. In this sample virtually all of the Fe is in the form of magnetite. Sample Fe-

3 shows a complex structure with an Fe valence ofapproximately 2+. This sample could not be fit with any simple combination of Fe metal, FeO, magnetite, and hematite. Sample Fe-2 had the expected valence similar to Fe 3+, and the EXAFS looks similar to the substitutional spectra in Fig 1.

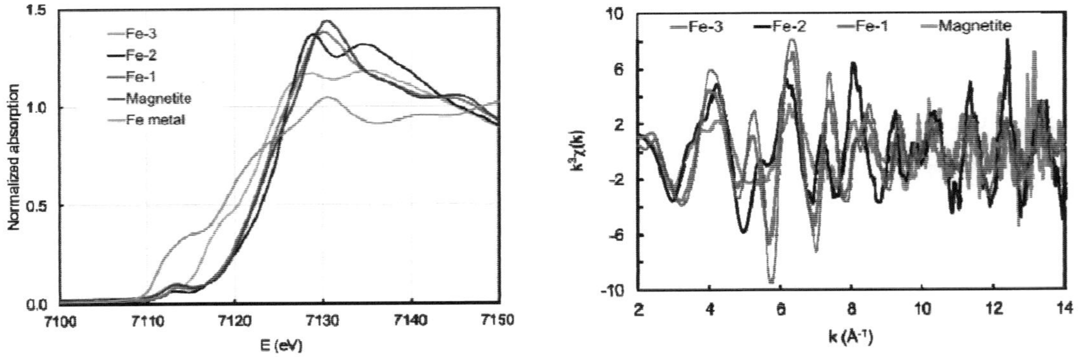

Figure 2. The Fe XANES and EXAFS for some Fe doped In_2O_3 samples.

3. Analysis and Discussion

In_2O_3 has the bixbyite structure[8] that has two In sites with near neighbors as shown in Table 2. Site 1 is three times more numerous than site 2. The main difference is in the first shell. For the more numerous site 1 there is a much larger spread of bonds. This opens the possibility that detailed analysis can distinguish site preference for the dopants. However, in all cases the dopant ionic size is much smaller than the host In atoms, and we expect substantial contraction of the lattice about the dopants. This rearrangement can make it difficult to distinguish between sites.

Since the Cr data had the strongest EXAFS and appears to be substitutional, it was chosen for detailed multi-shell fitting to look for the possibility of distinguishing site preference. Examples of fits for the two sites are shown in Fig. 3. These kept the coordination numbers fixed and varied the r and σ^2 for the four shells. It used a common Δr for the site 1 first neighbors. Both fits are reasonable and the resulting parameters similar. For site 1 the long first shell O distance was a problem. To get a good fit it had to be removed from the fit and the coordination numbers of the two shorter distances increased to 3 each. This indicates that the first shell coordination around the Cr is more symmetric than site 1, and could indicate a preference for site 2. However, in both fits there is a strong contraction to an average near neighbor distance of 2.03 Å. In the course of this contraction the site 1 neighbors could become more symmetrically arranged.

As another example of fitting, Fig. 3c shows a fit to the Fe data from sample 5. This assumes both sites are randomly occupied and used common parameters for the corresponding shells in the two sites. Again it was necessary to move the long first shell O in site 1 to values similar to the other two distances. This fit confirms that in this sample the majority of the Fe is substitutional.

Table 2. Near neighbors for the two In sites in In_2O_3.

	Site 1	Site 2
1st neighbor O	2 @ 2.04 2 @ 2.14 2 @ 2.41	6 @ 2.15
2nd neighbor In	6 @ 3.34[a]	6 @ 3.34
3rd neighbor In	6 @ 3.84[a]	6 @ 3.84
4th neighbor O	6 @ 3.98[a]	6 @ 3.89

[a]Average of very similar distances.

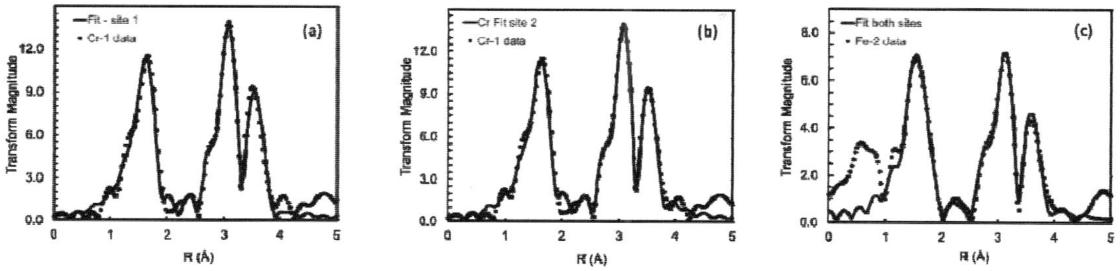

Figure 3. Fitting results for the first four shells for a) the Cr sample using site 1, b) the Cr sample using site 2, and c) the Fe-2 sample using mixed sites 1 and 2.

4. Summary

This paper has shown the potential for XAFS to characterize doped In_2O_3 films with some examples using V, Cr, Fe and Co as dopants. The near edge is sensitive to valence and second phase formation as was shown for some of the Fe doped samples. For the other samples the dopants are found to occupy substitutional sites. These dopants are much smaller than the host In atoms and there is significant contraction of the O near neighbors about the dopant. This complicates the potential for determining any site preference the dopants may have.

Acknowledgements

Use of the Advanced Photon Source, an Office of Science User Facility operated for the U.S. Department of Energy (DOE) Office of Science by Argonne National Laboratory, was supported by the U.S. DOE under Contract No. DE-AC02-06CH11357.This research was also funded by the UK Engineering and Physical Sciences Research Council and also KACST (Saudi Arabia).

References

[1] Gupta, A., et al., *Room temperature ferromagnetism in transition metal (V, Cr, Ti) doped In_2O_3.* Journal of Applied Physics, 2007. **101** 09N513.

[2] Hong, N.H., et al., *Magnetism in transition-metal-doped In_2O_3 thin films.* Journal of Physics: Condensed Matter, 2006. **18** 6897.

[3] Subias, G., et al., *Local structure study of Co-doped indium oxide and indium-tin oxide thin films using x-ray absorption spectroscopy.* Physical Review B, 2009. **79** 094118.

[4] Yu, Z.G., et al., *Origin of ferromagnetism in semiconducting $(In_{1-y}Fe_xCu_y)_2O_3$.* Physical Review B, 2006. **74** 165321.

[5] Nadaud, N., et al., *Structural Studies of Tin-Doped Indium Oxide (ITO) and $In_4Sn_3O_{12}$.* Journal of Solid State Chemistry, 1998. **135** 140-148.

[6] Parent, P., et al., *Structural Study of Tin-Doped Indium Oxide Thin Films Using X-Ray Absorption Spectroscopy and X-Ray Diffraction.* Journal of The Electrochemical Society, 1992. **139** 282-285.

[7] Ney, A., et al., *Advanced spectroscopic synchrotron techniques to unravel the intrinsic properties of dilute magnetic oxides: the case of Co:ZnO.* New Journal of Physics, 2010. **12** 013020.

[8] Morozova, L.V., P.A. Tikhonov, and V.B. Glushkova, *Physicochemical Investigation of the In_2O_3-HfO_2 system in the Indium Oxide-rich region.* Inorg. Mater., 1991. **27** 217-220.

Journal of Physics: Conference Series **430** (2013) 012082 doi:10.1088/1742-6596/430/1/012082

Observation of π backbonding features appearing in Fe 2p X-ray absorption spectra and Fe 1s-4p-1s resonant X-ray emission spectra of $RbMn[Fe(CN)_6]$

Yoshinori Kitajima[1], Yūsuke Nanba[2], Masayuki Tanaka[3], Yuji Koga[3], Akihiro Ueno[3], Kosuke Nakagawa[4], Hiroko Tokoro[4], Shin-ichi Ohkoshi[4], Toshiaki Iwazumi[3], Kozo Okada[2] and Yasuhito Isozumi[5]

[1]Photon Factory, Institute of Materials Structure Science, High Energy Accelerator Research Organization, 1-1 Oho, Tsukuba, Ibaraki 305-0801, Japan
[2]Graduate School of Natural Science and Technology, Okayama University, 3-1-1 Tsushima-naka, Kita-ku, Okayama 700-8530, Japan
[3]Department of Mathematical Sciences, School of Engineering, Osaka Prefecture University, 1-1 Gakuen-cho, Nakaku, Sakai, Osaka 599-8531, Japan
[4]Department of Chemistry, School of Science, The University of Tokyo, 7-3-1 Hongo, Bunkyo-ku, Tokyo 113-0033, Japan
[5]Radioisotope Research Center, Kyoto University, Yoshidakonoe-cho, Sakyo-ku, Kyoto 606-8501, Japan

E-mail: yoshinori.kitajima@kek.jp

Abstract. Satellite features by π backbonding (metal-to-ligand charge transfer) were observed in both Fe 2p ($L_{3,2}$-edge) X-ray absorption spectra and 1s-4p-1s resonant X-ray emission spectra for $RbMn[Fe(CN)_6]$, which can be theoretically explained using common parameters. Fe $L_{3,2}$-edge and Mn $L_{3,2}$-edge X-ray absorption spectra for low temperature phase are also presented.

1. Introduction

Prussian blue analogues are potential candidates for novel magnetic devices. They exhibit the valence transition induced by temperature [1], photo-irradiation [2,3], pressure [3], and electric field [4]. For example, the valence states change from Fe(III)-Mn(II) to Fe(II)-Mn(III) and structural form from the cubic structure to the tetrahedral one accompanying the Jahn-Teller distortion of MnN_6 octahedrons with decreasing temperature in rubidium manganese hexacyanoferrate, $RbMn[Fe(CN)_6]$. That has been clearly shown by Fe K-edge and $L_{3,2}$-edge X-ray absorption spectroscopy (XAS) and X-ray emission spectroscopy (XES) [5,6]. Hexacyano-compounds are composed of a 3-D network structure in which the transition-metal ions are strongly bridged by cyano groups [7]. One of the characteristics of the electronic states of cyanides is π backbonding (metal-to-ligand charge transfer, MLCT). Satellite peaks appearing in Fe $L_{3,2}$-edge XAS have been theoretically reproduced considering both the MLCT and electron correlation [8,9]. The importance of MLCT was confirmed both experimentally and theoretically for simpler model compounds [10,11] but further experiments have been desired for complex systems. To acquire more electronic information, we have adopted resonant X-ray emission spectroscopy (RXES) around Fe K absorption edge for RbMn[Fe(CN)$_6$], together with the model compounds of K_4[Fe(CN)$_6$] and K_3[Fe(CN)$_6$] and also Fe and Mn $L_{3,2}$-edge XAS for low temperature phase.

2. Experimental

Fe $L_{3,2}$-edge XAS experiment was performed at a soft X-ray spectroscopy station BL-11A of the Photon Factory, a synchrotron radiation facility in the Institute of Materials Structure Science of the High Energy Accelerator Research Organization (KEK-PF) [12]. Powder sample of $Rb_{0.94}$Mn[Fe(CN)$_6$]$_{0.98}$ 0.4H_2O was prepared by the method described elsewhere [1] and diffused on a carbon adhesive tape on a copper sample holder. The holder was set on a head of a closed-cycle He refrigerator and a heater was also set on the back side of the sample holder. The sample temperature was measured with a Si-diode thermometer set on the sample holder and controlled with a digital temperature controller. X-ray absorption spectra of model compounds were recorded by total electron yield. While the partial electron yield mode was adopted for temperature-controlled measurements using a microchannel plate (MCP) detector with applying a retarding voltage of -100 V to a mesh in front of the MCP. Although the light source is not an undulator but a bending magnet, sufficient photon flux with a reasonable energy resolution is available thanks to the stable top-up operation at 430 mA beam current of the PF 2.5-GeV storage ring and the MCP detector with a higher efficiency. The energy spread of the incident X-rays is estimated as less than 0.3 eV, which is much smaller than the previously reported spectra [6].

Fe 1s-4p-1s RXES experiment was conducted at BL-15B1 of KEK-PF, where the white synchrotron radiation beam from a bending magnet is monochromatized by a double-crystal Si(111) monochromator. Fe K-edge X-ray absorption spectra were measured by transmission method using powder samples spread over adhesive tapes. The emission spectra were recorded by a curved Ge(440) crystal monochromator installed in a vacuum chamber named "Escargot" spectrometer [13]. The analyzed X-rays were detected by a position-sensitive proportional counter using a charge division method. The overall experimental energy resolution was about 2.4 eV.

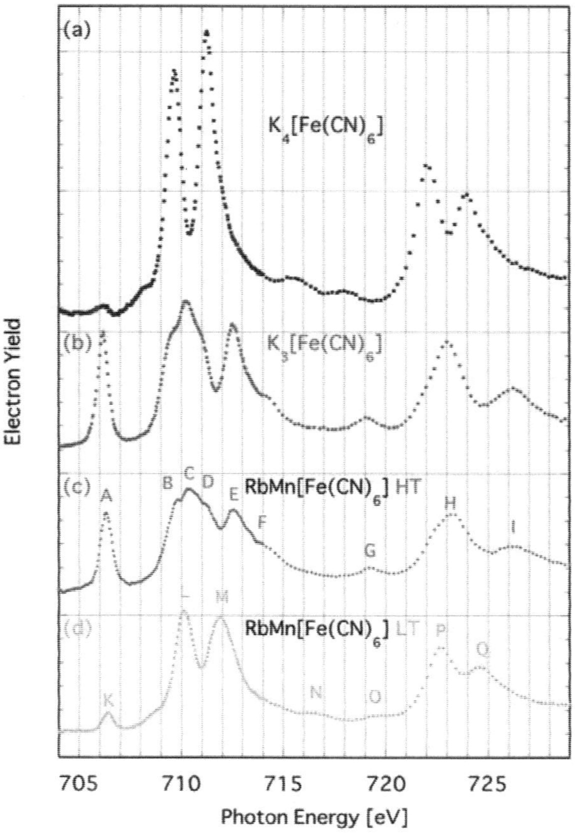

Figure 1. Fe $L_{3,2}$-edge X-ray absorption spectra of (a) K_4[Fe(CN)$_6$], (b) K_3[Fe(CN)$_6$], (c) RbMn[Fe(CN)$_6$] measured at 300 K and (d) at 40 K.

3. Results and Discussion

Figure 1 shows the Fe $L_{3,2}$-edge X-ray absorption spectra measured at 300 K (HT) and 40 K (LT) for RbMn[Fe(CN)$_6$] sample together with standard materials of K_4[Fe(CN)$_6$] 3H_2O and K_3[Fe(CN)$_6$]. HT spectrum is quite similar to that for K_3[Fe(CN)$_6$], where a distinct pre-edge peak A at around 706.3 eV is well known to be assigned to the transition of Fe(III) 2p - t_{2g} [9,11] and is absent in the spectrum of K_4[Fe(CN)$_6$]. This peak was used as a reference for the energy calibration. With a configuration-interaction full-multiplet theory

Journal of Physics: Conference Series **430** (2013) 012082 doi:10.1088/1742-6596/430/1/012082

for a $[Fe(CN)_6]^{3-}$ cluster model, satellite peaks E and I can be associated with the electron configuration of $d^4L(t_{2g})$ (MLCT). Detailed theoretical methods and discussion were described in Ref. 9.

Spectrum for LT phase (Fig. 1(d)) drastically changes from HT and shows some differences in peak position and intensity from $K_4[Fe(CN)_6]$ one, which means that the electronic state of Fe predominantly changes from +3 to +2, according with structural change through temperature-induced phase transition. Possibility of soft X-ray induced phase transition [6] cannot be eliminated completely but we believe that it is negligible because the intensity of the incident beam is much lower than the previous experiments. LT spectrum was also theoretically analyzed taking the crystal field of the tetragonal symmetry into account in detail in Ref. 9. The overall spectral features including satellite peak M were well reproduced by using $[Fe(CN)_6]^{4-}$ cluster model considering MLCT.

In addition, Mn $L_{3,2}$-edge X-ray absorption spectra for HT and LT phases were measured at the same time. HT spectrum (Fig. 2(a)) was again well reproduced theoretically using $[Mn(NC)_6]^{4-}$ cluster model including satellite peak U [9]. Backbonding is suppressed owing to the Hund's coupling in the Mn $L_{3,2}$-edge XAS. Analysis of LT spectrum is now in progress.

Figure 3 shows the results of 1s-4p-1s RXES with Fe K-edge XAS for $RbMn[Fe(CN)_6]$. The excitation energies of RXES shown by thick lines in the right panel are indicated by white arrows in the left one. Three Raman peaks (R1-R3) were observed. Raman peaks also appeared in theoretical calculation with the cluster model using the same parameters as in Fe $L_{3,2}$-edge XAS simulation such as charge transfer energy, Coulomb interaction, exchange interaction, hybridization interaction and crystal field splitting [8,9]. The observed Raman shift energies of 5.1 eV, 9.7 eV and 14.5 eV are close to those calculated for three features associated with $d^6\underline{L}(t_{2g})$, $d^4L(t_{2g})$ and $d^6\underline{L}(e_g)$ CT transitions, respectively. Metal-to-ligand charge transfer (MLCT; $d^4L(t_{2g})$ configuration) is also important for RXES.

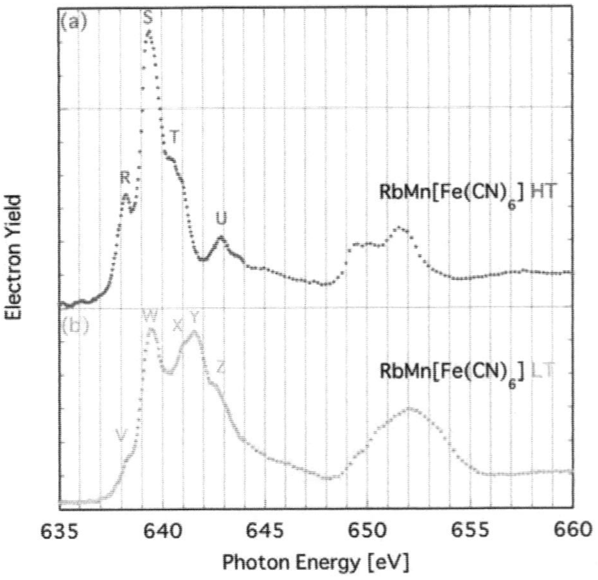

Figure 2. Mn $L_{3,2}$-edge X-ray absorption spectra of $RbMn[Fe(CN)_6]$ measured at (a) 300 K and (b) 40 K.

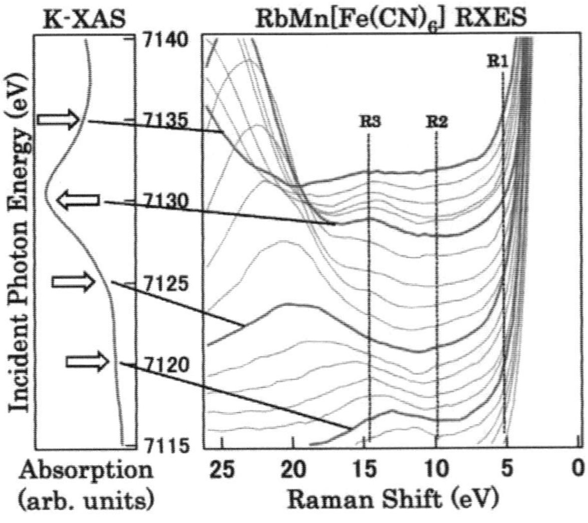

Figure 3. (left) Fe K-edge X-ray absorption spectra and (right) Fe 1s-4p-1s resonant X-ray emission spectra of $RbMn[Fe(CN)_6]$ measured at 297 K. Three Raman peaks were observed at 5.1 eV, 9.7 eV and 14.5 eV.

These three Raman peaks were also observed at similar energies for reference compounds of $K_4[Fe(CN)_6]$ and $K_3[Fe(CN)_6]$ (not shown).

4. Summary

Fe $L_{3,2}$-edge XAS and Fe 1s-4p-1s RXES were conducted for the $Rb_{0.94}Mn[Fe(CN)_6]_{0.98}$ $0.4H_2O$ powder sample. In all the spectra, π backbonding (MLCT) features were observed and theoretically analyzed with a configuration-interaction full-multiplet theory for metal hexacyanide cluster model. Mn $L_{3,2}$-edge X-ray absorption spectra for HT and LT phases were also presented for further theoretical consideration.

This work was performed with the approval of the Photon Factory Program Advisory Committee (Proposal Nos. 2005G244, 2008G093, 2010G159).

References

[1] Ohkoshi S, Tokoro H, Utsunomiya M, Mizuno M, Abe M and Hashimoto K, 2002 *J. Phys. Chem.* B **106** 2423

[2] Tokoro H, Ohkoshi S and Hashimoto K, 2003 *Appl. Phys. Lett.* **82** 1245

[3] Moritomo Y, Hanawa M, Ohishi Y, Kato K, Takata M, Kuriki A, Nishibori E, Sakata M, Ohkoshi S, Tokoro H and Hashimoto K, 2003 *Phys. Rev. B* **68** 144106

[4] Mahfoud T, Molnár G, Bonhommeau S, Cobo S, Salmon L, Demont P, Tokoro H, Ohkoshi S, Boukheddaden K and Bousseksou A, 2009 *J. Am. Chem. Soc.* **131** 15049

[5] Osawa H, Iwazumi T, Tokoro H, Ohkoshi S, Hashimoto K, Shoji H, Hirai E, Nakamura T, Nanao S and Isozumi Y, 2003 *Solid State Comm.* **125** 237

[6] Kitajima Y, Ishiji K, Matsuda T, Tokoro H, Ohkoshi S and Iwazumi T, 2009 *J.Phys.: Conf. Series* **148** 012032

[7] Mallah T, Thiébaut S, Verdaguer M and Veillet P, 1993 *Science* **262** 1554

[8] Nanba Y and Okada K, 2010 *J. Phys. Soc. Jpn.* **79** 114722

[9] Nanba Y and Okada K, 2012 *J. Electron Spectrosc. Relat. Phenom.* **185** 167

[10] Kosugi N, 1998 *J. Electron Spectrosc. Relat. Phenom.* **92** 151

[11] Hocking R K, Wasinger E C, de Groot F M F, Hodgson K O, Hedman B and Solomon E I, *J. Am. Chem. Soc.* **128** 10442

[12] Kitajima Y, Amemiya K, Yonamoto Y, Ohta T, Kikuchi T, Kosuge T, Toyoshima A and Ito K, 1998 *J. Synchrotron Rad.* **5** 729
http://pfwww.kek.jp/sxspec/sx/bl11a_e.html

[13] Iwazumi T, Kobayashi K, Kishimoto S, Nakamura T, Nanao S, Ohsawa D, Katano R and Isozumi Y, 1997 *Phys. Rev. B* **56** R14267

Journal of Physics: Conference Series **430** (2013) 012083 doi:10.1088/1742-6596/430/1/012083

Local and electronic structure around manganese in $Cd_{0.98}Mn_{0.02}Te_{0.97}Se_{0.03}$ studied by XAFS

I Radisavljević[1], N Novaković[1], N Romčević[2] and N Ivanović[1]

[1] Vinča Institute of Nuclear Sciences–University of Belgrade, P.O. Box 522, 11001 Belgrade, Serbia
[2] Institute of Physics–University of Belgrade, Pregrevica 118, 11000 Belgrade, Serbia

E–mail: iva@vin.bg.ac.rs

Abstract. X–ray Absorption Fine Structure (XAFS) technique was employed to study local electronic and structural features of Mn ions incorporated in $Cd_{0.98}Mn_{0.02}Te_{0.97}Se_{0.03}$. XAFS measurements performed at Mn K edge revealed that manganese Mn(II) ions are well incorporated into the host CdTe lattice (cubic zinc–blende structure type) and their immediate surrounding is found to be composed exclusively of Te atoms. The observed preference of Mn ions distribution around Te opposes earlier observations on the similar systems, where preferential Mn–Se over Mn–Te paring was found.

1. Introduction

II–VI semiconductors doped with transition metals, known as diluted magnetic semiconductors (DMS), are of scientific interest because of their wide range of optical, magnetic and electronic transport properties. The ternary nature of DMS allows for tuning the band gap and the lattice constant by varying the composition of the material, while by changing the concentration of the magnetic species the magnetic and magneto–optical properties can be altered. In quaternary systems [1–4] the possibilities to tune the material properties are even more extended owing to the additional degree of freedom an extra ion component contributes to. Another peculiarity of the later systems is that the ion substitution exhibits the nearest–neighbour and site occupation preferences (SOPs) in both cation and the anion sub–lattices [2–4].

This paper focuses on the X–ray Absorption Fine Structure (XAFS) studies of $Cd_{1-x}Mn_xTe_{1-y}Se_y$ (x=0.02, y=0.03). Low concentrations x and y ensure that the trahedrally coordinated zinc–blende (ZB) type structure of the host CdTe is preserved. To gain precise information on local and electronic structure, valence, and short range order of the immediate surrounding of manganese atoms, both X–ray Absorption Near Edge Structure (XANES) and Extended X–ray Absorption Fine Structure (EXAFS) were employed and the results of the preliminary analysis are presented.

2. Experimental

XAFS (XANES/EXAFS) measurements were performed on $Cd_{0.98}Mn_{0.02}Te_{0.97}Se_{0.03}$ single crystals grown by the Bridgman method. Details on the sample's preparation are given in [5]. Manganese K edge data were collected in fluorescence mode at 77 K and 293 K on the HASYLAB A1 Beamline at Deutsches Elektronen–Synchrotron DESY (Hamburg, Germany). The synchrotron radiation source was operating at electron beam energy of 4.45 GeV with the maximum stored current 120 mA. The

A1 station uses double crystal monochromator with two Si–(111) crystal pairs. Samples were oriented at 45° to the incident beam direction and a 7–cell SDD detector was used to collect the spectra. XAFS data processing and analysis were performed using IFEFFIT [6] as implemented in ATHENA and ARTEMIS software packages [7]. The interatomic distances (*r*), mean–square variation of the distances (σ^2) and the edge shift correction (E_0) were treated as free parameters during the fitting procedure. The passive electron reduction factor S_0^2 was determined from the procedure described in [8]. The obtained value S_0^2=0.63 (see Figure 1) was then kept fixed throughout the analysis.

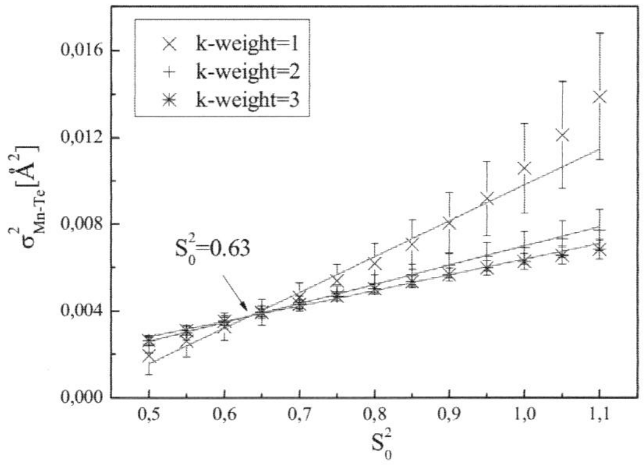

Figure 1. Parameter S_0^2 for the Mn K–edge in $Cd_{0.98}Mn_{0.02}Te_{0.97}Se_{0.03}$, as determined from the first shell fits successively performed with different *k*–weights (1, 2 and 3) and the values of S_0^2 fixed from 0.5 to 1.1, with step 0.05.

3. Results and discussion

The *k*–weighted Mn K edge EXAFS spectra of $Cd_{0.98}Mn_{0.02}Te_{0.97}Se_{0.03}$ taken at T=77 K and T=293 K are presented in Figure 2. The corresponding Fourier transformed spectra with their best fits are presented in Figure 3. The results are summarized in Table 1.

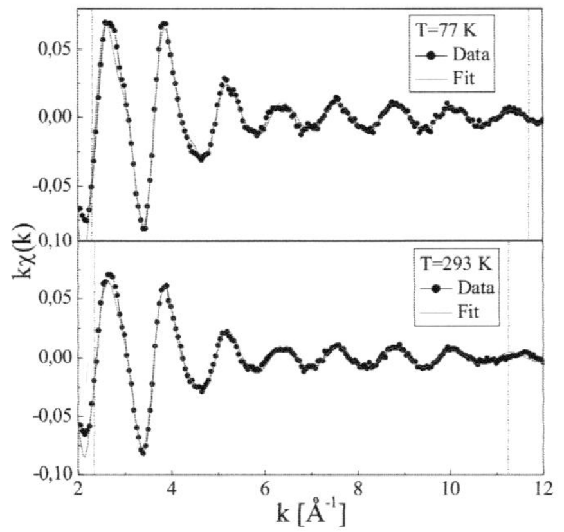

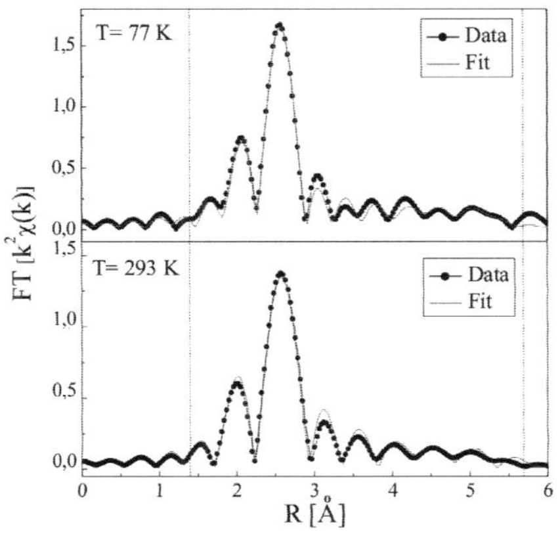

Figure 2. The *k*–weighted EXAFS function $k\chi(k)$ extracted from the Mn K edge absorption data at T=77 K (upper part) and at T=293 K (lower part), with the vertical lines denoting the *k*–ranges over which Fourier transforms (FT) of the data (represented with points) were taken and analyzed. The best fits are represented with lines.

Figure 3. Fourier transforms (FT) of the k^2–weighted EXAFS function $k^2\chi(k)$ extracted from the Mn K edge absorption data at T=77 K (upper part) and at T=293 K (lower part), with the vertical lines denoting the fitting ranges. The experimental data are represented with points and the best fits with lines.

According to these preliminary results, in the investigated sample manganese is surrounded by four equidistant tellurium atoms. Selenium atoms were not found in the immediate environment of Mn, which disagrees with previously observed preferences for Mn–Se paring [2]. The Mn–Te bond length (2.866–2.867 Å, see Table 1) is longer than in ZB form of MnTe (2.75 Å) and approaches the value in NiAs–type MnTe (2.93 Å). This agrees with the theoretical predictions that in CdTe–based alloys the Mn–Te bond length should stay nearly constant around the value in NiAs–type MnTe [9], but opposes the experimental observations, according to which the local Mn–environment stabilizes in ZB–type MnTe [3]. Interatomic distances involving second and third coordinations are within experimental uncertainties the same as in host CdTe (4.58 Å and 5.37 Å, respectively), which confirms the dominance of nearest neighbour interactions in the alloy [3]. In the investigated temperature range (77– 293 K) the obtained interatomic distances of all three coordination shells around Mn are only slightly temperature dependent.

Table 1. Structural parameters as obtained from the Mn K edge EXAFS data analysis: *r*–interatomic distance, σ^2–mean–square variation of the distance between Mn atom and atoms in the surrounding coordination shells, *n*–coordination number, E_0–the edge shift correction and the fit quality factor R.

	T=77 K			T=293 K		
	Mn–Te(1)	Mn–Cd	Mn–Te(2)	Mn–Te(1)	Mn–Cd	Mn–Te(2)
r [Å]	2.866(3)	4.55(4)	5.44(4)	2.867(4)	4.55(6)	5.44(7)
σ^2 [Å^2]	0.0037(2)	0.030(7)	0.024(6)	0.0040(4)	0.037(1)	0.036(2)
n	4	12	12	4	12	12
E_0 [eV]	0.9(4)			0.8(5)		
R–factor	0.020			0.026		

XANES spectrum taken at the Mn K edge at T=77 K is presented in Figure 4 a. The spectrum taken at T=293 K is identical in shape and therefore it was not shown. The pre edge arises from the electronic transitions into the lowest unoccupied states partially formed by empty Mn 3d–states, and contains valuable information about the local symmetry, valence state and degree of hybridisation with Te 5p states. The pre edge (see inset of Figure 4 a) was fitted with a Lorentzian function and an arctangent function is used to approximate the baseline. The normalized pre–edge peaks at the two temperatures are symmetric in shape and their energy position meet at E=6540.5 eV (see Figure 4 b) which corresponds to Mn in 2+ valence state (Mn(II)) [10]. Slightly longer Mn–Te distance at T=293 K yields somewhat lower degree of pd–hybridisation and therefore the intensity of the pre–egde peak is a little lower (see Figure 4 b).

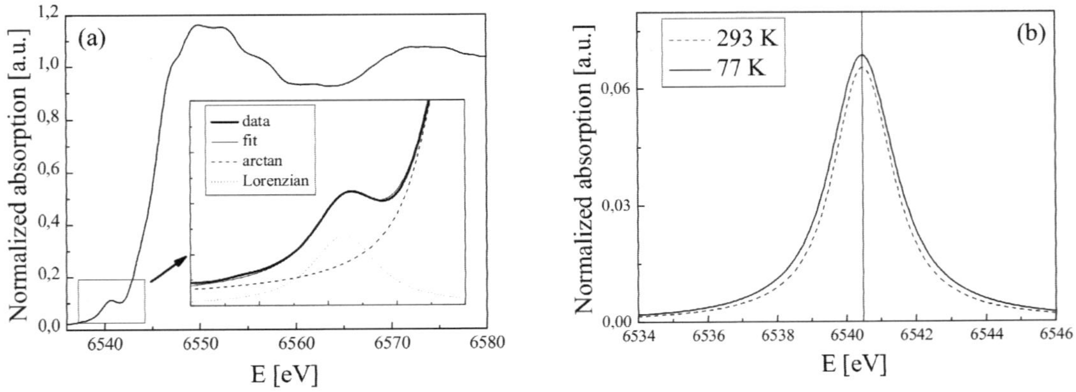

Figure 4. (a) The Mn K–edge XANES spectrum with the normalization of the pre–edge region shown in inset; (b) the pre–edge region normalized.

4. Conclusion
In conclusion, X–ray Absorption Fine Structure (XAFS) technique was employed to study local electronic and structural features of Mn ions incorporated in $Cd_{0.98}Mn_{0.02}Te_{0.97}Se_{0.03}$. Manganese Mn(II) ions are found to be well incorporated into the host CdTe lattice in symmetric local environment composed of four equidistant Te atoms, with the Mn–Te bond length approaching the value in NiAs–type MnTe. The second and third coordination shell atoms seem not to be affected by the Mn presence and stay close to their ideal positions in host CdTe.

Acknowledgements
The authors gratefully acknowledge HASYLAB @ DESY for providing the beamtime and the partial financial support. The A1 beamline scientist Dr. E. Welter is acknowledged for the assistance during the XAFS measurements. This work is supported by Serbian Ministry of Education and Science under the Grant III 45003–4.

References
[1] Chehab R S F and Woolley J C 1987 *Phys. Status Solidi (b)* **139** 213
[2] Pukowska B, Jaglarz J, Such B, Wagner T, Kisiel A and Mycielski A 2002 *J. Alloy. Compd.* **335** 35
[3] Balzarotti A, Motta N, Kisiel A, Zimnal–Starnawska M, Czyżyk M T and Podgórny M 1985 *Phys. Rev.* B **31** 7526
[4] Romčević N, Romčević M, Golubović A, Van Khoi L, Mycielski A, Jovanović Đ, Stojanović D, Nikolić S and Đurić S 2005 *J. Alloy. Compd.* **397** 52
[5] Robouch M B V, Kisiel A and Konior J 2002 *J. Alloy. Compd.* **340** 13
[6] Newville M 2001 *J. Synchrotron Radiat.* **8** 322
[7] Ravel B and Newville M 2005 *J. Synchrotron Radiat.* **12** 537
[8] Ravel B 2001 *EXAFS Analysis with FEFF and FEFFIT* (Part 2: Commentary) p 37
[9] Wei S H and Zunger A 1987 *Phys. Rev.* B **35** 2340
[10] Chalmin E, Farges F and Brown Jr G E 2009 *Contrib. Mineral. Petrol.* **157** 111

Journal of Physics: Conference Series **430** (2013) 012084 doi:10.1088/1742-6596/430/1/012084

XAS spectra of mechanically milled TiO_2

Ikuo Nakai [1], **Akihiro Hattori** [1], **Takahiro Ishii** [1] **and Yingjie Li** [2]

[1] Department of Electrical and Electronic Engineering, Tottori University, Tottori 680-8552, Japan

[2] Inner Mongolia Key Laboratory for Physics and Chemistry of Functional Materials, Physics and Electronic Information College, Inner Mongolia Normal University, Hohhot 010022, China

E-mail: nakai@ele.tottori-u.ac.jp

Abstract. We report the structural phase transformation of a rutile-type TiO_2 during mechanical milling measured by x-ray absorption spectroscopy. The pre-edge peaks near the Ti K edge consists of four components caused by the transitions to the ligand-field-split 3*d* orbitals hybridized with 4*p* states and the 4*p*-4*s* hybridized level. One of the 3*d*-4*p* transitions grows remarkably with increasing the milling time, while the others remain unchanged during the milling. We discuss the features from the structural point of view.

1. Introduction

Titanium dioxides are polymorphic materials with rutile, anatase and brookite structure. They have been investigated experimentally and theoretically because of their interesting chemical and electronic properties [1-10]. X-ray absorption spectroscopy (XAS) such as x-ray absorption near-edge structure (XANES) and extended x-ray absorption fine structure (EXAFS) is one of the powerful techniques for probing these properties and the local structural environment around a specific element. Mechanical milling induces a solid state reaction including the local distortion, defects and interatomic diffusion in materials and is also useful for synthesizing nano-structured materials [11, 12]. In the present work we concentrate on the structural phase transformation of a rutile-type TiO_2 induced by mechanical milling using XAS.

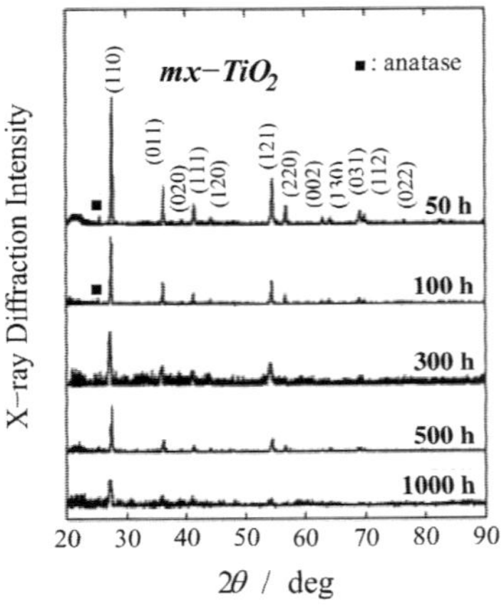

Figure 1. X-ray diffraction patterns of mx-TiO_2 after typical milling time.

2. Experimental details

The starting material was rutile-type TiO_2 powders with purity of more than 99.99%. It was mechanically milled at room temperature in a stainless steel vial of a water-cooled vibratory mill with stainless steel balls under an argon atmosphere up to 1000 h. The frequency of vibration was about 5 Hz. The ratio of ball to powder was 7:1 in weight. Electron probe microanalysis detected no impurity element such as Fe which might have contaminated TiO_2 during milling with the stainless steel vial and balls. Hereafter the milled TiO_2 powders thus obtained will be referred to as mx-TiO_2.

X-ray diffraction patterns were measured using Kα radiation from bright x-ray generator with rotating Cu anticathode. XAS measurements near the Ti K edge were made at 300 K in the transmission mode using a Si (111) double crystal monochromator at the beamline 9A of Photon Factory [13] and at the beamline BL01B1 of SPring-8 [14]. Higher harmonics of x-rays were eliminated by a Rh-coated double crystal mirror with a glancing angle of 7.5 mrad. The signals were accumulated by a couple of ionization chambers filled with mixture of appropriate gases. Especially near the Ti pre-edge peaks, their features were detected at the energy intervals less than 0.1 eV. The energy of x-rays was calibrated using the value of 8979.0 eV at the pre-edge peak for the metallic 5N purity Cu foil 6 μm thick.

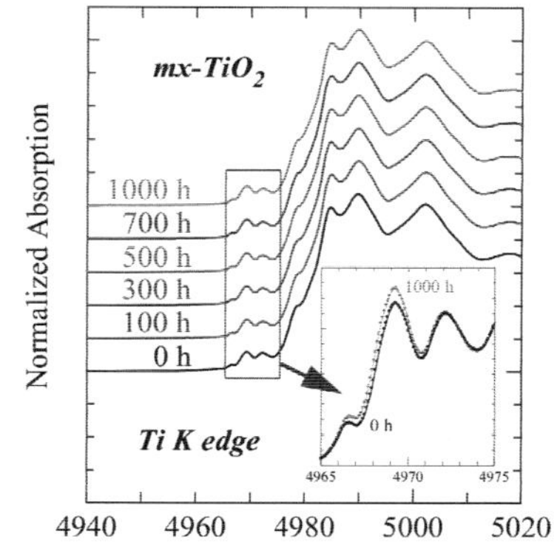

Figure 2. Normalized absorption spectra at Ti K edge of mx-TiO_2 at the typical milling time. The inset shows enlargement near the pre-edge before milling (0 h) and after 1000 h milling.

3. Results and discussion

Figure 1 presents x-ray diffraction patterns of mx-TiO_2 at some typical milling time. The starting material shows the rutile-type structure of which some reflections are labeled by the plane indices; it also includes a small amount of anatase-type structure. The (011) reflection of the anatase-type TiO_2 near 25 degree shown by a filled square rapidly decreases with increasing the milling time and disappears after milling more than 100 h. This means that milling induces the structural phase transformation of the anatase into the rutile [9]. The intensity of the rutile-type structure decreases with milling time and the main reflection around 27 degree shifts to a lower angle, which may suggest that some atomic diffusion and/or defect occur during the milling.

Our aim in the present work is to investigate what happens in mx-TiO_2 by XANES and EXAFS from the viewpoint of the electronic states and the local structure. The normalized absorption spectra of mx-TiO_2 up to the milling time of 1000 h near the Ti K edge are shown as a function of photon energy in Fig. 2 together with the inset enlarged near the pre-edge. The pre-edge configuration appears to be three peaks, of which the central one grows with milling as seen in the inset. To discuss the features of these peaks in detail, we perform the deconvolution process based on the following form with Gaussian functions,

$$\mu(E)=\left\{\frac{1}{2}+\frac{1}{\pi}\arctan\left(\frac{E-E_c}{\Gamma/2}\right)\right\}+\sum_i I_i\exp\left\{-\frac{2(E-E_i)^2}{W_i^2}\right\}. \quad (1)$$

The arctangent function shows the transition from the 1*s* to the continuum states with the energy difference E_c and the core-hole lifetime Γ. The Gaussian functions describe the transition to bound states with the peak energy E_i, the intensity I_i and the width W_i. The deconvolution for mx-TiO_2 before milling and after 1000 h milling is shown in Figs. 3(a) and (b). The upper circle with the thin line represents the observed absorption. The three peaks in the inset of Fig. 2 are decomposed into four Gaussian components as labeled A_1, A_2, A_3 and B. The concave curve is the arctangent function. The lower circle denotes the difference between the sum of these components and the observations. It is noted that

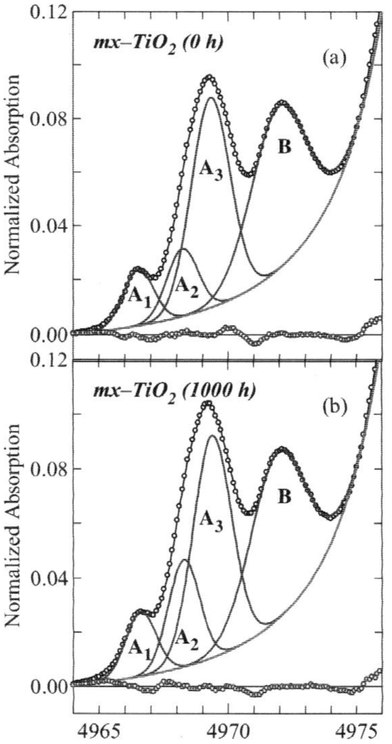

Figure 3. Four components with Gaussian form A_1, A_2, A_3 and B of Ti K pre-edge for mx-TiO_2 (a) before milling and (b) after milling for 1000 h.

enlargement of the central peak in Fig. 2 with milling is mainly attributed to growing the peak A_2. We obtain the lattice constants of the rutile-type TiO_2 before milling to be a=4.5886 Å and c=2.9556 Å from the Rietveld analysis, which means that a Ti atom is surrounded by four O neighbors at the distance of 1.9464 Å and two O atoms located in 1.9779 Å in a slightly distorted octahedral environment. The fivefold Ti 3d orbitals are split into the threefold t_{2g} state with a lower energy and the twofold e_g level with a higher energy due to the ligand field. The peaks A_1, A_2 and A_3 are assigned to the transitions from the 1s to the split 3d orbitals mixed with 4p states. In more detail they are mainly due to the quadrupole transition to the t_{2g} state, the dipole transition to 3d-4p hybridized states and the quadrupole transition to the e_g level. The peak B is mostly ascribed to the dipole transition to 4p-4s hybridized states [4, 6]. The values of E, I and W thus obtained for these peaks are tabulated in Table 1 where the fitting accuracy is about ±0.1 eV for E and W and about ±0.003 for I. There is little change in E and W for all of four peaks during the milling up to 1000 h. On the other hand, a characteristic tendency in I is as follows: it is almost constant for the peaks A_1, A_3 and B within the accuracy of fitting; for the peak A_2 it shows a quite remarkable increase with milling, that is, I at 1000 h is almost twice as large as that of 0 h. To make these features clearer, we plot the intensity ratios of the peaks A_1, A_2 and A_3 due to the split 3d orbitals to the peak B which includes no 3d contribution to the transition in Fig. 4. The ratio I_{A1}/I_B and I_{A3}/I_B are independent of milling time, while I_{A2}/I_B shows an almost linear increase with milling time.

A marked increase in the peak A_2 has been reported in both nanocrystalline anatase and amorphous phases of TiO_2 [5, 7]. It is also observed in a perovskite $PbTiO_3$ that the distortion of Ti atoms from O octahedron increases the central peak [15]. Therefore it appears that the milling induces distorted and/or defective surroundings with low crystallinity in mx-TiO_2.

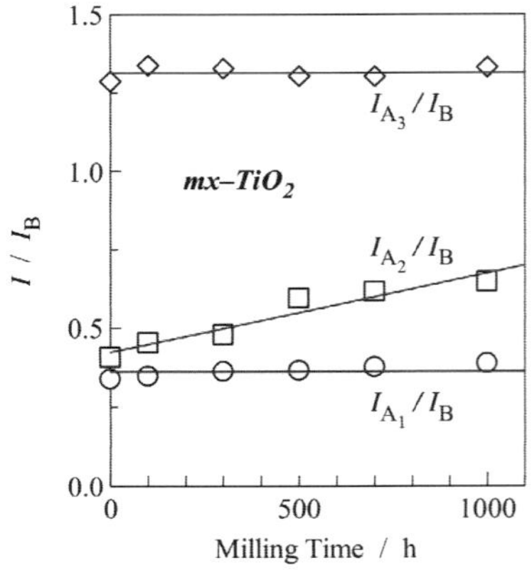

Figure 4. The intensity ratio of the peaks A_1, A_2 and A_3 to the peak B against the milling time.

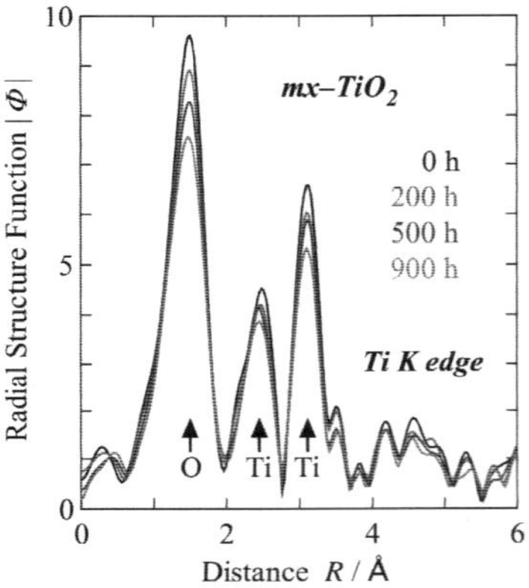

Figure 5. The radial structure functions for mx-TiO_2 at the Ti K edge at the typical milling time.

Figure 5 represents the radial structure function of mx-TiO_2 at typical milling time plotted against distance from an x-ray absorbing Ti atom, which is obtained from Fourier transform of EXAFS spectra near the Ti K edge at 300 K. We confirm that three peaks indicated by arrows reflect positions of nearest neighbor O and Ti atoms and the second Ti neighbors. The magnitude of the radial structure function at the O atoms decreases faster than that of the Ti neighbors during the milling. We here concentrate on the nearest neighboring O atoms because we are interested especially in the distortion

Table 1. The energy E, intensity I and width W in eq. (1) for the peaks A_1, A_2, A_3 and B of mx-TiO_2.

	A_1			A_2			A_3			B		
Milling time / h	E / eV	I	W / eV	E / eV	I	W / eV	E / eV	I	W / eV	E / eV	I	W / eV
0	4966.5	0.021	1.2	4968.2	0.025	1.2	4969.3	0.078	1.5	4972.0	0.061	2.2
100	4966.6	0.021	1.2	4968.2	0.028	1.2	4969.4	0.081	1.5	4972.1	0.061	2.2
300	4966.7	0.022	1.2	4968.2	0.029	1.2	4969.3	0.081	1.5	4971.9	0.061	2.2
500	4966.6	0.023	1.2	4968.3	0.037	1.2	4969.4	0.081	1.5	4972.0	0.062	2.2
700	4966.6	0.023	1.2	4968.3	0.038	1.2	4969.3	0.080	1.5	4971.9	0.061	2.2
1000	4966.6	0.024	1.2	4968.3	0.040	1.2	4969.4	0.082	1.5	4971.9	0.061	2.2

around the Ti site. Although there are two kinds of distance between Ti and O atoms as mentioned above, we assume that they are the same to simplify our analysis. In the fitting the quantities such as the backscattering amplitude, the phase shift and the photoelectron mean free path are estimated from the theoretical standard code EFEE 8.4 [16]. We obtain the interatomic distance and the Debye-Waller factor between Ti and O atoms as shown in Fig. 6. The Ti-O distance appears to be growing rather moderately with increasing the milling time, while the Debye-Waller factor of Ti-O pairs increases clearly at rate proportional to the milling time. The increase in the Debye-Waller factor mainly reflects the decrease in magnitude of the radial structure function shown in Fig. 5, which indicates deviation from the octahedral symmetry on the Ti site surrounded by six O atoms. This distortion of the octahedral symmetry induces change in the ligand field on Ti atoms and therefore the redistribution of the orbital splitting due to $3d$-$4p$ mixing. This is the origin of growing the peak A_2. We can view the rutile-type TiO_2 milled for 1000 h as being at an early stage of the structural transformation to the amorphous phase.

4. Conclusion

We have discussed an early stage of the structural transformation of the rutile-type TiO_2 during the milling from the viewpoint of the electronic states and the local structure using XANES and EXAFS. We conclude that the milling enhances the distortion and therefore redistributes the ligand field on the Ti atom.

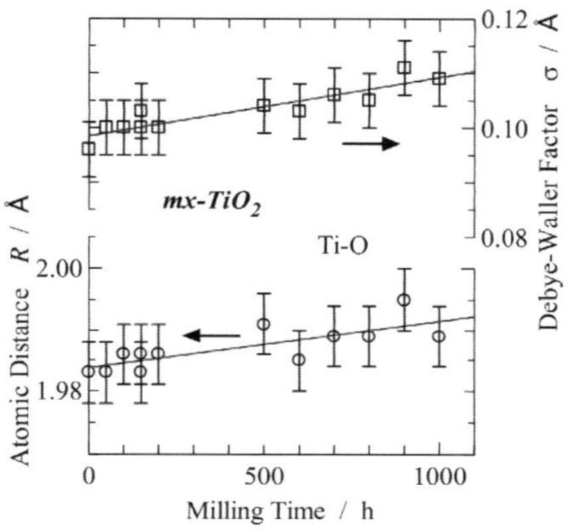

Figure 6. The Debye-Waller factor and the interatomic distance between Ti and O atoms as a function of milling time.

Acknowledgements

The XAFS measurement was performed at Photon Factory under the approval of Photon Factory Program Advisory Committee (proposal Nos. 2007G081 and 2008G606) and at SPring-8 with approval of the Japan Synchrotron Radiation Research Institute (No. 2007B1664).

References

[1] Brydson R, Sauer H, Engel W, Thomas JM, Zeitler E, Kosugi N and Kuroda H 1989 *J. Phys.: Condens. Matter* **1** 797
[2] Wu ZY, Ouvrard G, Gressier P and Natoli CR 1997 *Phys. Rev.* **B55** 10382
[3] Chen LX, Rajh T, Wang Z and Thurnauer C 1997 *J. Phys. Chem.* **B101** 10688
[4] Parlebas JC, Khan MA, Uozumi T, Okada K and Kotani A 1995 *J. Electron Spectrosc. Relat. Phenom.* **71** 117
[5] Luca V, Djajanti S and Howe RF 1998 *J. Phys. Chem.* **102** 10650
[6] Joly Y, Cabaret D, Renevier H and Natoli CR 1999 *Phys. Rev. Lett.* **82** 2398
[7] Matsuo S, Sakaguchi N and Wakita H 2005 *Analytical Sciences* **21** 805
[8] Angelomé PC, Andrini L, Calvo ME, Requejo FG, Bilmes SA and Soler-Illia JAA 2007 *J. Phys. Chem.* **111** 10886
[9] Bégin-Colin S, Girot T, Le Caër G and Mocellin A 2000 *J. Solid State Chem.* **149** 41
[10] Gesenhues U 2005 *J. Appl. Cryst.* **38** 749
[11] Benjamin JS 1970 *Met. Trans.* **1** 2943
[12] Suryanarayana C 2001 *Prog. Mat. Sci.* **46** 1
[13] Nomura M and Koyama A 2001 *Nucl. Instrum. Methods Phys. Res.* **A467-468** 733
[14] Uruga T, Tanida H, Yoneda Y, Takeshita K, Emura S, Takahashi M, Harada M, Nishihata Y, Kubozono Y, Tanaka T, Yamamoto T, Maeda H, Kamishima O, Takabayashi Y, Nakata Y, Kimura H, Goto S and Ishikawa T 1999 *J. Synchrotron Rad.* **6** 143
[15] Sato K, Miyanaga T, Ikeda S and Diop D 2005 *Phys. Scr.* **T115** 359
[16] Ankudinov AL, Nesvizhskii AI and Rehr JJ 2003 *Phys. Rev.* **B67** 115120

X-ray absorption spectroscopy study of Gd^{3+}-loaded ultra-short carbon nanotubes

Q Ma[1], M Jebb[2], M F Tweedle[3] and L J Wilson[2]

[1]DND-CAT, Northwestern Synchrotron Research Center at the Advanced Photon Source, Argonne, IL 60439, USA
[2]Department of Chemistry, the Center for Nanoscale Science and Technology, and the Center for Biological and Environmental Nanotechnology, Rice University, Houston, TX 77251, USA
[3]Radiology Department, The Ohio State College of Medicine, Columbus, OH 43210, USA

E-mail: q-ma@northwestern.edu

Abstract. We present an x-ray absorption spectroscopy study of the local structure around the Gd^{3+}ion loaded in ultra short (20-100 nm) carbon nanotubes (GNTs). X-ray Gd L_3 absorption near edge structure data shows that the 31.2-μM GNT suspension exhibits a clear characteristic of hydration at the [GdO_n] cluster. Extended x-ray absorption fine structure data show that the Gd^{3+} ion is coordinated by about 9 oxygen ions and that this first coordination shell exhibits an asymmetry similar to that found in triclinic Gd-acetate or $Gd[C_2H_3O_2]_3 \cdot 4H_2O$ or GdAc. After correction for the asymmetry using the cumulant of the third order, the Gd-O bond distance is found to be 2.34_5 Å, instead of 2.40_6 Å for a symmetrical (or Gaussian) distribution. It is shorter than that in the Gd-containing MRI contrast agents currently in clinical uses. This may account in part for high proton relaxivity observed for the GNT suspension.

1. Introduction

Contrast agents (CAs) play a prominent role in magnetic resonance imaging in medicine [1]. The lanthanide ion, Gd^{3+}, is usually chosen for MRI CAs because it has a very large magnetic moment ($\mu^2 = 63\ \mu_B^2$) and a symmetric electronic ground state, $^8S_{7/2}$. The aquated Gd^{3+} ion is toxic and hence is sequestered by chelation [2] or encapsulation [3,4] in order to reduce toxicity. As described in [5], ultra-short, single walled carbon nanotubes (US-SWNT) may be suitable for cellular uptake, biocompatibility, and eventual elimination from the body. Therefore, they may be ideal materials to encapsulate atoms, ions, and even small molecules, whose cytotoxicity may be sequestered within the short carbon nanotube (20-100 nm). The Gd-loaded nanotubes, or Gadonanotubes (GNT), have been characterized previously as high-performance MRI CAs [5]. However, the details of the local structure around Gd atoms are yet to be revealed. X-ray absorption spectroscopy techniques are powerful in probing the local structure around a chemical element, and have helped understanding the magnetic property of the nano magnetic spinel ferrites for MRI contrast agent applications [6]. Its ability to probe a length scale up to a few angstroms matches well with the interaction sphere of the solvent and solute. Here we employ these techniques to uncover the electronic and atomic structure of GNT.

2. Experimental Aspects

GNT were prepared according to a previously reported procedure [5]. Briefly, full-length single walled carbon nanotubes (SWNT) prepared by the arc-discharge method were fluorinated and then pyrolyzed to produce US-SWNT as previously reported [7]. The US-SWNT were purified in 12 M hydrochloric acid, debundled by sonication in THF with sodium metal (or in DI water as a debundled control control), and refluxed in 6 M nitric acid for 5 min. The prepared US-SWNT were sonicated for 1 h in 6.7 mM solutions of gadolinium chloride at 1 mg/mL US-SWNT and pH 1.9 and washed with DI water to remove external Gd^{3+} ions. For relaxivity measurements, the resultant GNTs were suspended in 0.17% (w/v) Pluronic® F 108NF Prill Poloxamer 338 (BASF) at about 1 mg/mL, then centrifuged several times at 1100 rcf, collecting supernatant and discarding unsuspended GNTs. Samples were dialyzed in DI water to remove excess surfactant and obtain a stable dilute suspension. The debundled GNTs contained 4.1 wt. % Gd. A 31.2-μM GNT solution was prepared which showed relaxivity of 98 $s^{-1}mM^{-1}$. X-ray absorption was measured both on dry and suspended GNT.

X-ray absorption measurements around the Gd L_3 edge were carried out at the 5BM-D beamline of DND-CAT, at the Advanced Photon Source (Argonne, IL). A double crystal Si(111) monochromator was used for energy selection with $\Delta E/E=1.4\times10^{-4}$. The monochromator is water cooled and driven by a servo motor with an angular resolution of 5×10^{-5} degree. The monochromator position errors are limited to the resolution which is equivalent to < 0.1 eV at the Gd L_3 edge (7243 eV). The incident x-ray intensity was detuned by 65 % of its maximum for harmonic rejection as monitored by an ionization chamber (Oxford Danfysik). The Gd L_3 edge data of dry and suspended GNTs were collected in fluorescence mode. The Gd L_α emission was measured using two 4-element Si-drift vortex detectors (SII NanoTech USA), and x-ray pulses were analyzed using the XMAP processors (XIA). Due to the low concentration of the GNT solution, only the x-ray absorption near edge structure (XANES) was measured, while the extended x-ray absorption fine structure (EXAFS) data were measured on the dry GNT. Commercial Gd-acetate ($Gd[C_2H_3O_2]_3{\cdot}4H_2O$ or GdAc), $Gd(OH)_3$, and $GdCl_3{\cdot}6H_2O$ from Alfa Aesar were used as reference compounds. Fine powder of these reference compounds was spread on the Scotch tape, respectively, which was then folded several times to produce sufficient absorption for x-ray transmission measurements. Extreme care was taken to make sure samples were uniform. A small x-ray beam (0.5×0.5) was used to check the sample uniformity by measuring the transmitted x-ray intensities through the sample. The intensity variation was within ~1% RMS over the measured area of 2×12 mm^2. A 10-mM $GdCl_3$ solution was prepared and measured in fluorescence mode with the solution in a rectangular Plexiglas container with a volume of $10\times15\times2$ mm^3. The data were analyzed using the Athena package [8].

3. Results and Discussion

Figure 1 shows the XANES of 31.2-μM GNT, dry GNT, GdAc, 10-mM $GdCl_3$, and $GdCl_3{\cdot}6H_2O$. The XANES's of the GNT are similar to that of 10-mM $GdCl_3$ in terms of their whiteline intensity which is significantly higher than those of GdAc and $GdCl_3{\cdot}6H_2O$ (by~12%). The whiteline intensity is proportional to the unoccupied density of states (DOS) near the Fermi level. In the case of the L_3 edge (*p*-state ionization) it is the unfilled *5d* state. Despite its complexity, it is affected empirically by the first coordination shell, namely, the coordination number (CN), chemical species, the bond distances (and their spread), and electron localization. The crystalline structure of $GdCl_3{\cdot}6H_2O$ is monoclinic (space group - P2/n) with a negligible spread among three pairs of the Gd-O bonds at 2.412±0.010 Å, and with two Gd-Cl bonds at 2.768 Å [9]. Based on this structure, the amplitude reduction factor, S_0^2, obtained by the FEFF calculation is essentially equal to 1 [10]. With S_0^2 = 1, the 8.6 Gd-O bonds at 2.40_5 Å are found in the first coordination shell in 10-mM $GdCl_3$. This result is very close to the one found in a 6-mM $GdCl_3$ solution (CN=8.9 at 2.412 Å) by Mayanovic et al. [11]. Therefore, this increase in the CN can account for the majority increase (~62%) of the whiteline intensity for 10-mM $GdCl_3$. The rest of the increase is likely due to the replacement of chlorine by oxygen at a shorter distance and with higher electron affinity. In the case of triclinic GdAc (space group – P-1) [12], the

Gd atoms are coordinated by nine oxygen atoms at different distances that spread from 2.368 to 2.571 Å. Assuming S_0^2 = 1 and a single bond distance, only 6.6 oxygen atoms are found at 2.42_1 Å versa 9 oxygen atoms when the bond spread is considered. This show that the large disorder can cause difficulties in determination of the coordination numbers [13]. In addition to the bond spread, the smaller whiteline may also be attributed to electron delocalization over -C-C-H_3 ligands.

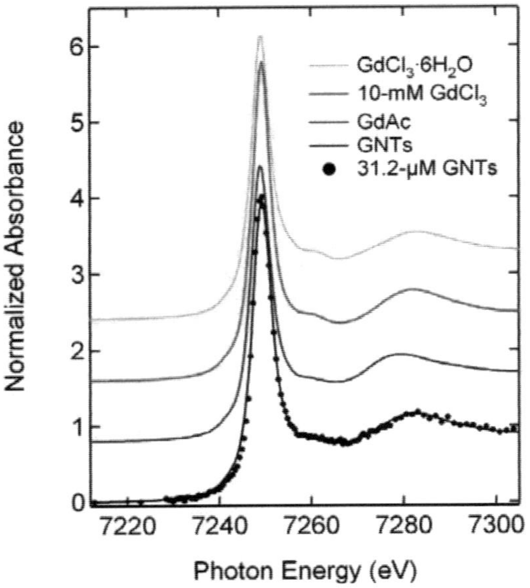

Figure 1. The XANES spectra of $GdCl_3{\cdot}6H_2O$, 10-mM $GdCl_3$, GdAc, dry GNTs and 31.2-μM GNTs.

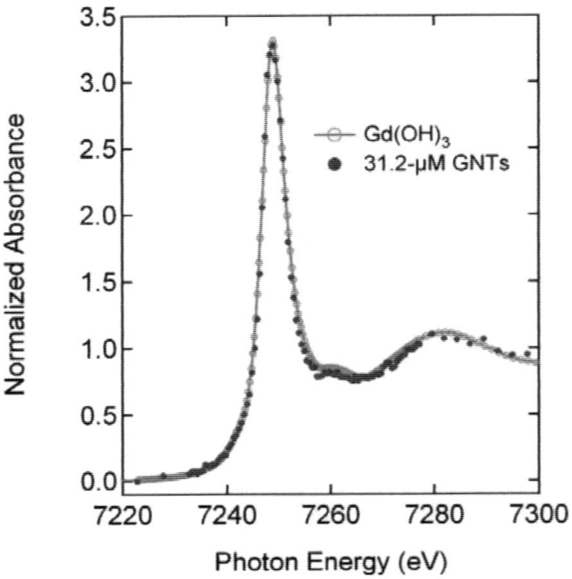

Figure 2. The XANES spectra of $Gd(OH)_3$ and 31.2-μM GNTs. The latter was measured a month after the one (solid dots) shown in Fig. 1.

The peak shape analyses, carried out on the data shown in Figure 1, indicate that they are all approximated quite well by the sum of an arctangent function for transitions to the continuum and a Lorentzian function for the whitelines [14] [15], except for the 31.2-μM GNT whose approximation is somewhat poorer. The whiteline of the GNT solution is narrower and has different symmetry from that of dry GNT (while the threshold is unchanged). A similar peak asymmetry is seen for the whiteline of $Gd(OH)_3$ which is however 18% smaller than those of the GNTs. A month after the data shown in Figure 1 were acquired, a measurement was carried out again on the dry and solution GNTs, respectively. There were no changes for the dry GNT. But the whiteline height of the GNT sulution was lowered. As displayed in Figure 2, its XANES compares nicely with that of $Gd(OH)_3$, except that the peak width is still slightly narrower. The narrow peak may imply a localized d-band. $Gd(OH)_3$ has a hexagonal structure (space group: P63/m) in which Gd is 9-fold coordinated [16]. Therefore, the peak asymmetry of the freshly-prepared suspension is likely caused by hydration at an early stage. The process continued and is likely facilitated by carbon nanotubes which are good transporters of water [17] and protons [18]. Preliminary XANES calculations using the full multiscattering approach support this assignment (not shown). Hydrogen adsorption on small Pt particles (~1.0 nm) on inorganic oxides induced a similar asymmetry on the Pt L_3 whiteline [19].

Figure 3 shows the k^3-weighted EXAFS spectra of the dry GNT, GdAc, and 10-mM $GdCl_3$ together with a short k range of the EXAFS oscillations measured on the 31.2-μM GNT solution. The EXAFS's of GdAc and GNT were similar in that they died off beyond 11 $Å^{-1}$. Despite noise, the EXAFS oscillations of the GNT solution match roughly to those of the dry GNT. It is therefore assumed that the local structures around Gd atoms in these samples are similar. The arrow indicates a sharp multiple electron excitation (ME) peak which is typically seen in the L shell absorption data of rare-earth elements [20]. Removing it or not in the data analysis does not seem to affect the outcomes.

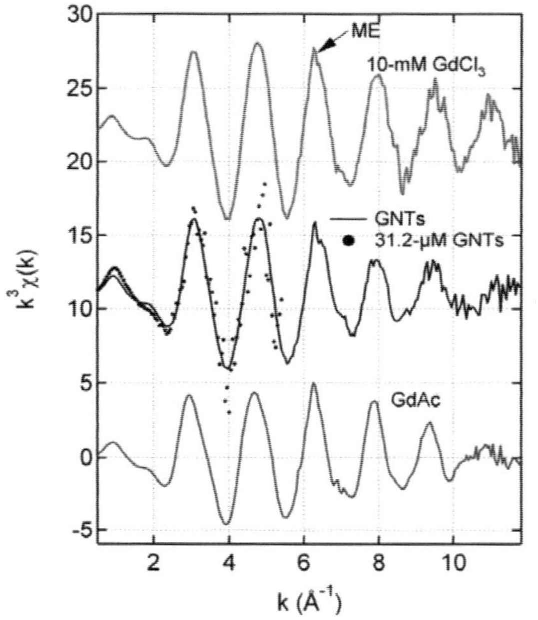

Figure 3. The EXAFS data of GdAc, dry GNTs and 10-mM $GdCl_3$. Solid dots are the data from 31.2-μM GNTs suspension.

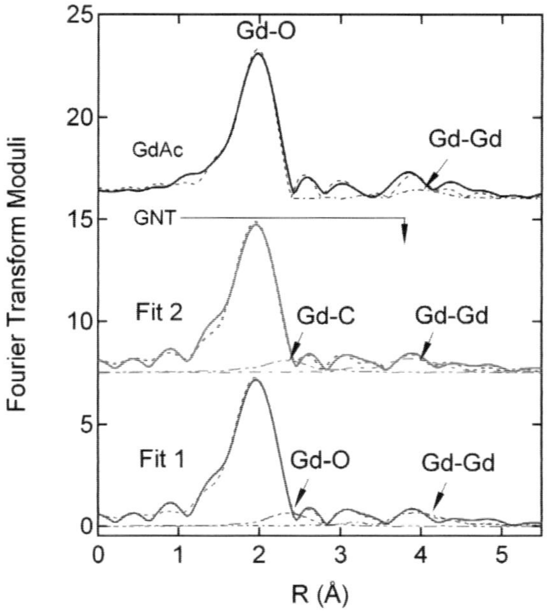

Figure 4. Fourier transforms of the EXAFS data (solid lines) and the fitting results (dashed lines) for GdAc and dry GNTs. See text for details.

Table 1 gives the first coordination structures around Gd for the GNT, GdAc, $Gd(OH)_3$, and 10-mM $GdCl_3$. These results are obtained assuming a single distance by the FEFF fitting for which the phases and amplitudes are obtained using the crystalline structure of $GdCl_3{\cdot}6H_2O$. Unlike the $GdCl_3$ solution, the first coordination shell around Gd in the GNT contains significant asymmetry and is more disordered. Use of a third cumulant term C_3 [10] can greatly reduce the figure of merit. This can be due either to the anharmonic oscillator nature of chemical bonds or to a disorder similar to that discussed above for GdAc. This shell can be fit just as well when the discrete distribution of the Gd-O bonds found in GdAc is considered in the data fitting. A temperature dependence of the EXAFS measurement is in order, to determine the true origin of the asymmetry. Note that the Gd-O bond length of the GNT is shorter than that of $Gd(OH)_3$ or of GdAc. It represents a volume change from ~3% up to ~8% assuming a cube. This is likely a result of confinement in the SWNT.

Figure 4 shows the Fourier transforms of the EXAFS data for GdAc and GNT. Also shown are the fitting results (dash lines) of the data up to 4.5 Å. This is to detect possible Gd-Gd interactions in the GNTs. For GdAc, a full-path fit is presented. It is seen that one Gd-Gd interaction at 4.20 Å only contributes about one third of the apparent peak near 4 Å and that the rest of the peak is due mostly to the multi-scattering (MS) paths. Clearly, there are no directly referable models for the local structure around Gd in a carbon nanotube, but GdAc from which the phases and amplitudes of all possible Gd-C, Gd-O, and Gd-Gd interactions can be produced. Fitting of the GNT data using the paths generated from GdAc produces a reasonably good fit, except for in the region from 2.5 to 3.5 Å where the MS's dominate. This should imply the differences in their structural geometries. Assuming that the MS's are at minimum in the GNT given the disorder and confined geometry, use of the single scattering paths up to 4.5 Å fits the GNT data nicely. Two fitting results presented in Fig. 4 for the GNT were so obtained. Fit 1was obtained with the Gd-O interactions at 2.81Å, 3.54 Å, and 3.72 Å, respectively, and a Gd-Gd interaction existing at 4.13 Å with CN = 1.6 and σ^2 = 0.007 Å^2, in addition to the first coordination shell. Fit 2 was obtained with the Gd-C interactions near 2.81Å, 3.54 Å, and 3.72 Å and the Gd-Gd interaction at 4.14 Å. For the latter, CN = 2.7 and σ^2 = 0.012 Å^2. Clearly, correlation between CN and σ^2 is high (CN/σ^2 = *Constant*). However, due likely to the characteristics of the backscatters that contribute the peak around 4.0 Å, no convergences were achieved if the Gd-Gd interaction was replaced. Therefore, a Gd-Gd interaction is implied with good certainty.

Table 1. First coordination structures around Gd^{3+} ions*

Samples	R(Å)	σ^2(Å^2)	N	$C_3\times10^{-4}$	R-factor
GNTs	2.40_6	0.010	8.4	0	0.014
	2.34_5	0.010		-12	0.004
GdAc	2.42_7	0.007	6.6†	0	0.012
	2.37_2	0.007	(9.0)	-9	0.005
$Gd(OH)_3$	2.43_9	0.010	8.4	0	0.013
	2.41_2	0.010	(9.0)	-6.2	0.010
10-mM $GdCl_3$	2.40_5	0.007	8.6	0	0.007
	2.40_4	0.007		-0.3	0.006

*Data range of k = 2.5 – 10 Å^{-1} are used. A single distance is assumed.
†The likely cause for the small N value is disorder (see the main text).

4. Summary

The electronic and atomic structures around Gd^{3+} ions in the GNT were probed by the Gd L_3 edge x-ray absorption spectroscopy. Hydration of Gd^{3+} ions is clearly indicated in the GNT suspension. It is shown that the Gd-O bond length in the GNT is considerably shorter than the one (=2.463 Å) found in the Gd-containing MRI CAs currently in clinical use [21]. If the protons being relaxed in suspended GNT are experiencing the Gd through these shorter Gd-O bonds, it may account for some of the remarkably high relaxivity observed for GNT (since relaxivity, at least according to prevailing theory [2], is proportional to $(R_{Gd\text{-}H})^6$). Although it is not trivial to uncover possible Gd-C, Gd-O and Gd-Gd interactions at larger distances in GNT, analyses of the dry GNT EXAFS data using the phases and amplitudes extracted from the GdAc structure does give a measure of the likelihood of their existence. Given hydration and localized d-band observed in the GNT, the Gd^{3+} ions likely locate in US-SWNT as small a cluster as $[GdO_9]$. Some may aggregate or form somewhat larger clusters $[GdO_n]_m$, as implied by the likely Gd-Gd interactions.

Acknowledgments
This work was performed at the DuPont-Northwestern-Dow Collaborative Access Team (DND-CAT) located at Sector 5 of the Advanced Photon Source (APS). DND-CAT is supported by E.I. DuPont de Nemours & Co., The Dow Chemical Company and Northwestern University. Use of the APS, an Office of Science User Facility operated for the U.S. Department of Energy (DOE) Office of Science by Argonne National Laboratory, was supported by the U.S. DOE under Contract No. DE-AC02-06CH11357. L. J. Wilson thanks the Welch Foundation (Grant C-0627) for financial support. M. Jebb thanks the Atwell-Welch Graduate Research Fellowship (H-20924) and a National Science Foundation Graduate Research Fellowship (DGE-0940902) for financial support. M. Jebb would like to thank A. Gizzatov for SWNT fluorination, R. Sethi for relaxivity measurements, and S. Phounsavath for AFM assistance.

References

[1] Merbach A E and Toth E (Editors) 2001 *The Chemistry of Contrast Agents in Medical Magnetic Resonance Imaging* (Chichester: John-Wiley and Sons).

[2] Caravan P, Ellison J J, McMurry T. J. and Lauffer R. B 1999 *Chem. Rev.* **99** 2293.
[3] Kato H, Kanazawa Y, Okumura M, Taninaka A, Yokawa T and Shinohara H 2003 *J. Am. Chem. Soc.* **125** 4391.
[4] Bolskar R D, Benedetto A F, Husebo L O, Price R E, Jackson E F, Wallace S, Wilson L J and Alford J M 2004 *J. Am. Chem. Soc.*, **125** 5471.
[5] B. Sitharaman D, Kissell K R, Hartman K B, Tran L A, Baikalov A, Rusakova I, Sun Y, Khant H A, Ludtke S J, Chiu W, Laus S, Tóth É, Helm L, Merbach A E, and Wilson L J 2005 *Chem. Commun.* **31** 3915-3917.
[6] Schultz-Sikma E A, Joshi H M, Ma Q, MacRenaris K W, Eckermann A L, Dravid V P, and Meade T J 2011 *Chem. Mater.* **23** 2657.
[7] Gu Z, Peng H, Hauge R H, Smalley R E, Margrave J L 2002 *Nano Lett.* **2**(9) 1009-1013.
[8] Ravel B and Newville M 2005 *J. Synchrotron Rad.* **12** 537-541.
[9] Marezio M, Plettinger H A, and Zachariasen W H 1961 *Acta Crystallogr.* **14** 234.
[10] Rehr J J and Albers R C 2000 *Rev. Modern Physics* **73** 621.
[11] Mayanovic R A, Anderson A J, Bassett W A, and Chou I M 2007 *Geology* **239** 266–283.
[12] Favas M C, Kepert D L, Skelton B W and White A H 1980 *J. C. S. Dalton* **3** 454.
[13] Stern E A, Ma Y J, Hanske-Petitpierre O and Bouldin C E 1992 *Phys. Rev.* B **45** 687.
[14] Brown M, Peierls R E and Stern E A 1977 *Phys. Rev.* B **15** 738.
[15] Richtmyer F K, Barnes S W and Ramberg E 1934 *Phys. Rev.* **46** 843.
[16] Milligan W O, Mullica D F and Oliver J D 1979 *J. Appl. Cryst.* **12** 411-412.
[17] Hummer G, Rasalah J C and Noworyta J P 2001 *Nature* **414** 188.
[18] Dellago C, Naor M M and Hummer G 2003 *Phys. Rev. Lett.* **1** 90105902.
[19] Asakura K, Kubota T, Chun W J, Iwasava Y, Ohtani K and Fujikawa T 1999 *J. Sync. Rad.* **6** 439.
[20] Solera J A, Garcia J and Proietti M G 1995 *Phys. Rev.* B **51** 2678.
[21] Chang C A, Francesconi L C, Malley M F, Kumar K, Gougoutas J Z, Tweedle M F, Lee D W, Wilson L J 1993 *Inorg. Chem.* 32 3501.

15th International Conference on X-ray Absorption Fine Structure (XAFS15) IOP Publishing
Journal of Physics: Conference Series **430** (2013) 012086 doi:10.1088/1742-6596/430/1/012086

Multiple Scattering Debye-Waller Factors for Arsenate

E. Kim[1], N. Chen[2], Z. Arthur[1], J. Warner[2], G.P. Demopoulos[3], J.W. Rowson[4] and D.T. Jiang[1,*]

[1] Department of Physics, University of Guelph, Guelph, ON, Canada
[2] Canadian Light Source, University of Saskatchewan, Saskatoon, SK, Canada
[3] Department of Materials Engineering, McGill University, Montreal, QC, Canada
[4] Department of Environment Science and Technology, AREVA Resources Canada Inc., Saskatoon, SK Canada

*Corresponding author, email: djiang@uoguelph.ca

Abstract. Debye-Waller factors for the As-O-O triangular multiple scattering paths within the arsenate in $Na_2HAsO_4 \cdot 7H_2O$ are evaluated in terms of magnitude ratio with respect to the Debye-Waller factor of the nearest neighbour As-O shell ($\sigma^2_{As-O-O}/\sigma^2_{As-O}$). The arsenates are studied under two different levels of distortion from an ideal tetrahedron, i.e. a relatively high distortion in the powder form and a nearly ideal tetrahedron in an aqueous solution at pH 14. The Debye-Waller factor ratio $\sigma^2_{As-O-O}/\sigma^2_{As-O}$ is found to be 2.0 and 1.9 for the powder and liquid sample, respectively, appearing to be insensitive to the distortion of the arsenate tetrahedron.

1. Introduction

EXAFS chemical speciation of environmental arsenate-bearing specimens often requires consideration of single scattering (SS) paths beyond the arsenate tetrahedron, where contributions from the multiple scattering (MS) paths within the arsenate could overlap with that from the paths of interest [1-4]. For such systems, experimentally established Debye-Waller (DW) factors for the MS paths are desirable for quantitative structural parameter extraction.

Figure 1 illustrates schematically the most significant MS paths within the arsenate tetrahedron. It has been suggested that the DW of path in Fig. 1c should have a rather simple correlation with the DW of the single scattering (SS) nearest neighbor path in Fig. 1a, as they are fully correlated and therefore the ratio $\frac{\sigma^2_{As-O_1-As-O_1}}{\sigma^2_{As-O}} = 4$ is expected [5]. For the path in Fig. 1d, the DW ratio should be between 2 (fully uncorrelated vibrational motions) to 4 (fully correlated vibrational motions) [5]. What remains completely unclear is the relation between the DW factor of the triangular MS path (Fig. 1b) and that of the SS path (Fig. 1a). Various values of this DW factor ratio have been assumed in the literature, e.g. unity [3] and 1.5 [4], but there has not been an experimental investigation similar to the chromate case [6] and this is what motivates the current study.

2. Experimental

Two sample types were prepared from commercial Na_2HAsO_4•$7H_2O$ powder material (98%, Alfa Aesar): a pellet was made from sodium arsenate ($Na_2HAsO_4 \cdot 7H_2O$) mixed with BN (98%, ~1 μm, Sigma-Aldrich) to reach one absorption length thickness for transmission XAFS and a 250 ppm aqueous solution ($Na_2HAsO_4 \cdot 7H_2O$ + 1M NaOH; pH 14) was made for fluorescence XAFS.

To prepare the solid sodium arsenate and BN mixture pellet sample, the sodium arsenate was first ground with a mortar and pestle and then mixed BN at a 50:50 weight ratio. The resulting powder was then placed in an automatic grinder (Wig-L-bug grinding mill). This mixture powder was then compressed into a pellet. Aqueous samples of sodium arsenate were prepared at arsenic concentrations of 75 mg/L and 250 mg/L. Ground sodium arsenate was added to 1M NaOH (pH 14), and agitated using a vortex generator to ensure dissolution. It was assumed that the added sodium arsenate did not significantly affect the pH of the 1M NaOH. The aqueous solution samples were sealed in plastic sample holders with Kapton tape.

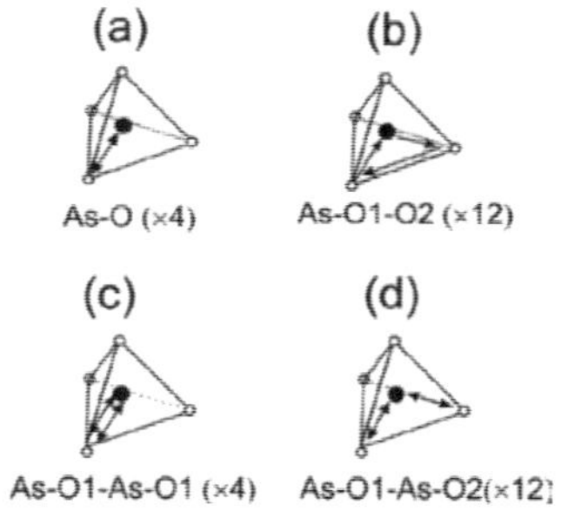

Figure 1. Illustration of the primary multiple scattering paths within arsenate. Filled circles represent arsenic and open circles represent oxygen; numbers in parentheses show path degeneracy.

The XAFS measurements were carried out at the HXMA beamline [7] at the Canadian Light Source (CLS), Saskatoon, Canada. The HXMA wiggler was at 1.9 T; Si(111) monochromator crystals and Rh mirrors were used, and the CLS ring peak current was 250 mA. Transmission XAFS data were collected from the powder pellet sample using standard ion chambers and fluorescence XAFS data from the aqueous solution samples by using a 4-element Vortex detector with filter/solar slits combination.

3. Results

At pH 14, the arsenates in the aqueous solution are fully deprotonated, since arsenate is a triprotic acid with its third pK_a at 11.6 [8, 9]. Thus, using 1M NaOH as the solvent for the solution maximizes the proportion of aqueous AsO_4^{3-}[8] and this solution sample represents the case where the arsenate tetrahedron is virtually free from distortion. In contrast, the powder sample represents a case where the arsenate is subject to the distortion defined by the crystalline structure of the compound; for example, the nearest neighbour As-O distance varies from 1.66 Å to 1.74 Å with a standard deviation of 0.04 Å (in comparison to scorodite, where the standard deviation among the As-O nearest neighbour distances is 0.01 Å).

Figure 2 and 3 show the EXAFS results for the aqueous arsenate solution sample and the powder Na_2HAsO_4•$7H_2O$ sample, respectively; the curve-fitting results are summarized in Table 1. In both solution and powder cases (Fig. 2a and 3a) the multiple scattering (MS) signal could be seen by the presence of feature "A" in the Fourier transform with either k^1 or k^2 weighting. For the k^3 weighting, even though the appearance of the MS is obscured by the emphasis of transform on the shorter interatomic distances, the MS intensity remains strong enough to permit R-space curve fitting analysis (Fig. 2a insert). It is noted that there is also a feature "B" consistently present in the data (Fig. 2a and 3a). For the powder sample, FEFF7 [10] modelling based on the crystallographic data of Na_2HAsO_4•$7H_2O$ [11] suggests that in the region of the feature "B" there is a large group of distant As-O single scattering paths. In the R-space curve-fitting range includes the main As-O shell and the feature "A". For the solution sample, we attribute the feature "B" (Fig. 3a) as the indication of a hydrolysis shell for the aqueous arsenate ion, hence our analysis also involves only the first two peaks (As-O and the MS feature "A"). In the curve-fitting, the DW factor for the MS path As-O_1-As-O_1 is correlated to that of the As-O by a factor of 4 [5], the ratio between the DW of the triangular MS path As-O_1-O_2 and that of the As-O path is left as the free varying parameter to seek its best fitted value;

the DW of MS path As-O_1-As-O_2 is tested for values between factor of 2 to 4 of that of the As-O, i.e. from completely uncorrelated to the fully correlated possibilities, in order to provide an uncertainty estimate on the main result sought, $\sigma^2_{As-O-O}/\sigma^2_{As-O}$.

The difference in the level of the distortion of the arsenate between the powder and pH 14 aqueous solution is reflected in the difference in their As-O shell DW factors (Table 1): significantly smaller DW for the solution indicates that the arsenate tetrahedron in the solution is more symmetrical compared to that in the $Na_2HAsO_4 \cdot 7H_2O$ powder. It is also interesting to notice that the As-O DW of the arsenate in the pH 14 solution is quite close to that in scorodite [4], consistent with the fact that the arsenate in the latter is not significantly distorted. The most significant result of this study is that the ratio $\sigma^2_{As-O-O}/\sigma^2_{As-O}$ was found to be 2.0 and 1.9 for the powder and solution, respectively; this is significantly higher than the values suggested in the previous studies (see Table 2). When taking into account that the uncertainty due to the possible variance of $\sigma^2_{As-O1-As-O2}/\sigma^2_{As-O}$ is about 0.1 for these ratios, it can be concluded that $\sigma^2_{As-O-O}/\sigma^2_{As-O}$ remains virtually the same for these two different distortions of the arsenate tetrahedron (Table 2).

The implication of these new estimates on $\sigma^2_{As-O-O}/\sigma^2_{As-O}$ is that in the previous analyses (Table 2), the contribution of the triangular path MS may have been overestimated as much as by a factor of two. Therefore, the coordination numbers for the shells (e.g. the nearest As-Fe in the poorly-crystalline scorodite [4, 12]) could be re-evaluated more reliably given these new estimates on the MS intensities. These coordination numbers often play a key role in the chemical speciation of ferric arsenate compounds [12].

Table 1. R-space Curve Fitting Results for the Powder and Aqueous Solution (as pH 14) Samples of $Na_2HAsO_4 \cdot 7H_2O$

	Common Parameters		Powder[e]		pH 14 Aqueous Solution[f]	
Paths	R (Å)	N	σ^2 (Å^2)	E_0 (eV)	σ^2 (Å^2)	E_0 (eV)
As-O	1.69	4^b	0.0023	5.8	0.0015	5.0
As-O_1-O_2	*a*	12^b	0.0045	5.8	0.0029	5.0
As-O_1-As-O_1	*a*	4^b	*c*	5.8	*c*	5.0
As-O_1-As-O_2	*a*	12^b	*d*	5.8	*d*	5.0

[a] Correlated to the As-O path length;
[b]Fixed;
[c]Correlated to σ^2_{As-O} by a factor of 4;
[d]Correlated to σ^2_{As-O} by a factor varying from 2 to 4, in order to provide an uncertainty estimate for the main focus, the ratio $\sigma^2_{As-O-O}/\sigma^2_{As-O}$;
[e]WinXAS R-space fit Residual = 8.57; [f]WinXAS R-space fit Residual = 8.71.

Table 2. Summary of the Debye-Waller factor ratios for the triangle As-O_1-O_2 path within arsenate.

DW Factor Ratio	Voegelin et al. [3]	Chen et al. [4]	This study - powder	This study - solution
$\sigma^2_{As-O-O}/\sigma^2_{As-O}$	1	1.5	2.0 ± 0.1[a]	1.9 ± 0.1[a]

[a] Uncertainties estimated by varying $\sigma^2_{As-O_1-As-O_2}/\sigma^2_{As-O}$ between 2.0 to 4.0 during curve-fit.

Journal of Physics: Conference Series **430** (2013) 012086 doi:10.1088/1742-6596/430/1/012086

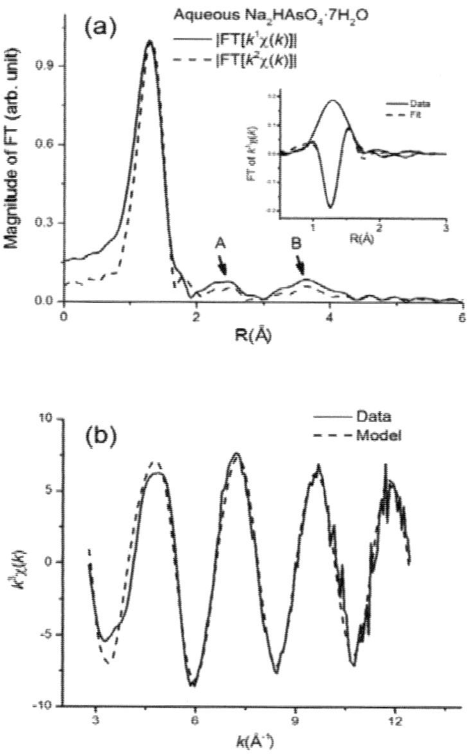

Figure 2. As K-edge EXAFS of arsenate in solution ($Na_2HAsO_4 \cdot 7H_2O$ + 1M NaOH). (a) Magnitudes of the Fourier transform with k^1 (solid) and k^2 (dashed) weightings, respectively. The insert shows the magnitude and the imaginary parts of experimental FT[$k^3\chi(k)$] (solid) and the curve fits (dashed). (b) Raw experimental $k^3\chi(k)$ (solid line) and curve fits (dashed) to the n.n. + MS shells.

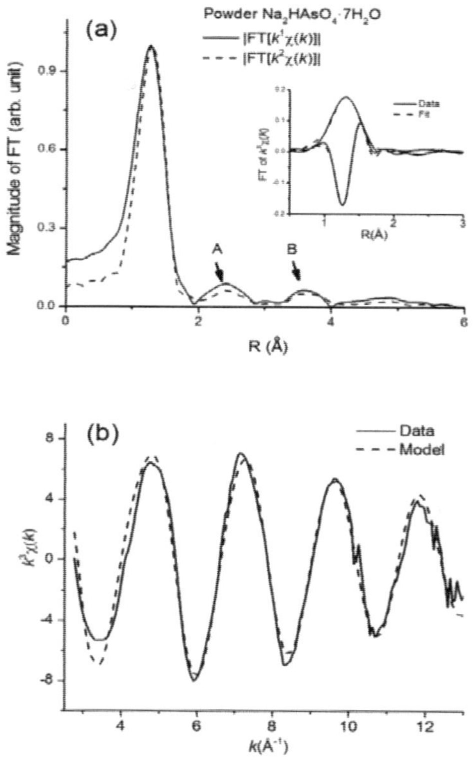

Figure 3. As K-edge EXAFS of powder $Na_2HAsO_4 \cdot 7H_2O$. (a) Magnitudes of the Fourier transform with k^1 (solid) and k^2 (dashed) weightings, respectively. The insert shows the magnitude and the imaginary parts of experimental FT[$k^3\chi(k)$] (solid) and the curve fits (dashed). (b) Raw experimental $k^3\chi(k)$ (solid line) and curve fits (dashed) to the n.n. + MS shells.

Acknowledgements

Research described in this paper was supported by NSERC of Canada through a CRD grant to GPD and DTJ. EK acknowledges the support by a President's Research Assistantship from the University of Guelph. The experiments were performed at the Canadian Light Source, which is supported by the NSERC of Canada, the NRC Canada, the CIHR, the Province of Saskatchewan, Western Economic Diversification Canada, and the University of Saskatchewan.

References

[1] Waychunas G. A., Rea B. A., Fuller C. C. and Davis J. A., 1993, Geochim. Cosmochim. Acta **57**, 2251.
[2] Manceau A., Lanson M and Geoffroy N., 2007, Geochim. Cosmochim. Acta **71**, 95.
[3] Voegelin A., Weber F.-A. and Kretzschmar R., 2007, Geochim. Cosmochim. Acta **71**, 5804.
[4] Chen N., Jiang D.T., Cutler J., Kotzer T., Jia Y.F., Demopoulos G.P. and Rowson J.W., 2009, Geochim. Cosmochim. Acta **73**, 3260.
[5] Hudson E.A., Allen P.G., Terminello L.J., Denecke M.A. and Reich T., 1996, Phys. Rev. B, **54**, 156.
[6] Pandya, K.I., 1994, Phys. Rev. B 50, 15509.
[7] Jiang, D. T., Chen N., Sheng, W., 2007, AIP Proc. **879** (Accelerators and Beams): 800-803.
[8] Sharma, V. and Sohn, M., 2009, Environ. Int. **35**, 743.
[9] Cullen, W. and Reimer, K., 1989, Chem. Rev. **89**, 713.
[10] Rehr, J. J. and Albers, R.C., 2000, Rev. Modern Phys. **72**: 621-654.
[11] Ferraris, G. and Chiari, G., 1970, Acta Cryst. B**26**, 1574.
[12] Jiang D.T., Chen N., Demopoulos G.P. and Rowson J.W., 2010, Geochim. Cosmochim. Acta 74, 4597.

Journal of Physics: Conference Series **430** (2013) 012087 doi:10.1088/1742-6596/430/1/012087

XAFS and XRD Studies of the $Cd_{1-x}Zn_xTe$ Crystal Fine Structure

Feng Tao[1], Zha Gangqiang[1*], Yang Jian[1], Li Jiong[2], Jiang Zheng[2], Xu Lingyan[1], Wang Tao[1] and Jie Wanqi[1]

1. School of Material Science and Engineering, Northwestern Polytechnical University, Xi'an, 710072, China.
2. Shanghai Synchrotron Radiation Facility, Shanghai Institute of Applied Physics, Shanghai, 201204, China.

zha_gq@nwpu.edu.cn

Abstract. $Cd_{1-x}Zn_xTe$ crystal is a new kind of room temperature semiconductor radiation detector material developed in recent years. $Cd_{1-x}Zn_xTe$ is zinc-blende structure, which is similar to CdTe, but the fine structure of $Cd_{1-x}Zn_xTe$ ternary compound semiconductor is different from CdTe and ZnTe binary compound semiconductors. In this contribution, the fine structure of $Cd_{1-x}Zn_xTe$ has been studied by the synchrotron radiation X-ray absorption fine structure (XAFS) technology and X-ray diffraction (XRD) technology. The K-edges XAFS spectra of cadmium, zinc and tellurium in $Cd_{1-x}Zn_xTe$ have been obtained, and the differences between the structures of binary and ternary compound crystals have also been analyzed. The Fourier transform of the k^2-weighted absorption spectra prove that Zn atoms occupy the position of the Cd atoms. The bond lengths of $Cd_{1-x}Zn_xTe$ have also been obtained by extended X-ray absorption fine structure (EXAFS), and the results imply that the local atomic structure of $Cd_{1-x}Zn_xTe$ is distorted. The doping of Zn in the structure contributes to the distortion, which should be responsible for the different properties in $Cd_{1-x}Zn_xTe$. Normalized X-ray absorption near-edge structure (XANES) spectra on Cd K-edge, Zn K-edge and Te K-edge in CdTe, $Cd_{0.96}Zn_{0.04}Te$, $Cd_{0.9}Zn_{0.1}Te$ and ZnTe are also shown.

1. Introduction

Ⅱ - Ⅵ ternary compound semiconductor $Cd_{1-x}Zn_xTe$ is a new kind of room temperature semiconductor radiation detector material developed in recent years. In 1989, R.weil et al. found that $Cd_{1-x}Zn_xTe$ is ferroelectric[1]. This is the first example of ferroelectricity in zinc-blende-based systems. D. J. Fu et al.[2] found the hysteretic behaviour of the electrical polarization and the electrical conductivity of this material. However, ferroelectricity is forbidden in the zinc-blend structure adopted by pure CdTe, and it implies that Zn doping provokes a lattice distortion and reduces the symmetry[3]. Most of the polarization hysteresis measurements reported in the literature have been performed on compositions with low Zn content, such as $Cd_{0.96}Zn_{0.04}Te$, but no structural distortions have been reported for such low Zn concentration. Recent diffraction studies employing synchrotron radiation of $Cd_{0.7}Zn_{0.3}Te$ crystals revealed an undistorted zinc-blende structure[4].

$Cd_{1-x}Zn_xTe$ was thought as a solid solution semiconductor mixed by CdTe and ZnTe, so the lattice constant can be adjusted by the value of x. Because of Zn doping, $Cd_{1-x}Zn_xTe$ has a more covalent bonding and higher stacking fault energy compared to CdTe crystal, which reduces the dislocation density and twin formation[5].

Here, the fine structures of $Cd_{1-x}Zn_xTe$ crystals have been studied by XRD and XAFS techniques. As XAFS signal is sensitive to the local structure and chemical state of X-ray absorbing atom, this experimental technique constitutes a powerful tool for the objective of this study.

2. Experimental

The samples of CdTe, $Cd_{0.96}Zn_{0.04}Te$, $Cd_{0.9}Zn_{0.1}Te$ and ZnTe were grown by the modified vertical Bridgman (MVB) method in our laboratory. Powder samples with a grain size in the micrometer range were obtained by grinding single crystal samples.

XRD studies were made with PANalytical's X'Pert PRO Materials Research Diffractometer using Cu K_α radiation (λ=0.15406 nm).

The X-ray absorption data of the samples were recorded at room temperature at beam line BL14W1 of the Shanghai Synchrotron Radiation Facility (SSRF), China. The Cd K-edge, Te K-edge and Zn K-edge were measured in transmission mode using ion chambers or in the fluorescent mode with silicon drift fluorescence detector. XAFS data were analyzed using the IFEFFIT code[6]. A non-linear best fit of theoretical to experimental spectra has been performed by using FEFF8 code[7] and the graphical interface Artemis[8].

3. Results and discussion

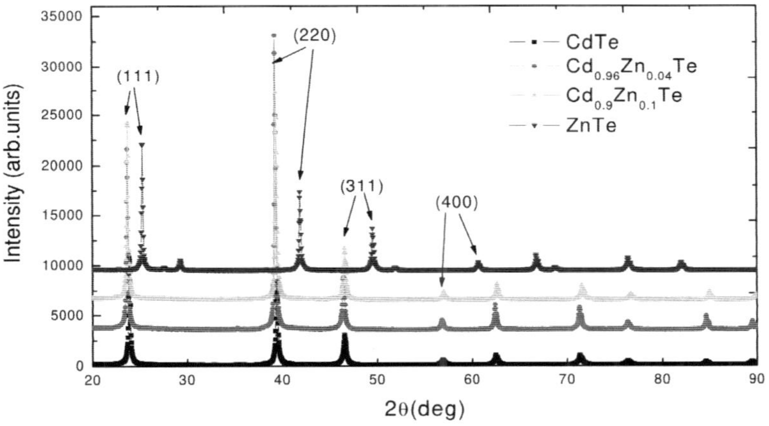

Fig.1. XRD diffraction patterns of CdTe, $Cd_{0.96}Zn_{0.04}Te$, $Cd_{0.9}Zn_{0.1}Te$, and ZnTe powder samples

The XRD spectra of CdTe, $Cd_{0.96}Zn_{0.04}Te$, $Cd_{0.9}Zn_{0.1}Te$, and ZnTe powder samples are shown in Fig.1. From the XRD diffraction pattern, we can obtain the different crystal plane spacing by using the Bragg equation. As we knew the structure of the unit cell, the lattice constant and average bond length of CdTe, $Cd_{0.96}Zn_{0.04}Te$, $Cd_{0.9}Zn_{0.1}Te$ and ZnTe crystals have been calculated with the spacing of (400) crystal plane. The results show in Table.1.

Table 1. The calculated results of plane spacing, lattice constant and average bond lenght

	Spacing of (400)/Å	Lattice constant/ Å	Average bond length/ Å
CdTe	1.6200	6.480	2.8059
$Cd_{0.96}Zn_{0.04}Te$	1.6196	6.478	2.8051
$Cd_{0.9}Zn_{0.1}Te$.	1.6145	6.458	2.7964
ZnTe	1.5252	6.101	2.6418

From Tab.1, we found that the average bond length of $Cd_{1-x}Zn_xTe$ crystal decreased with the increase of x. The results are agreed with the VCA model, Vegard's law[9].

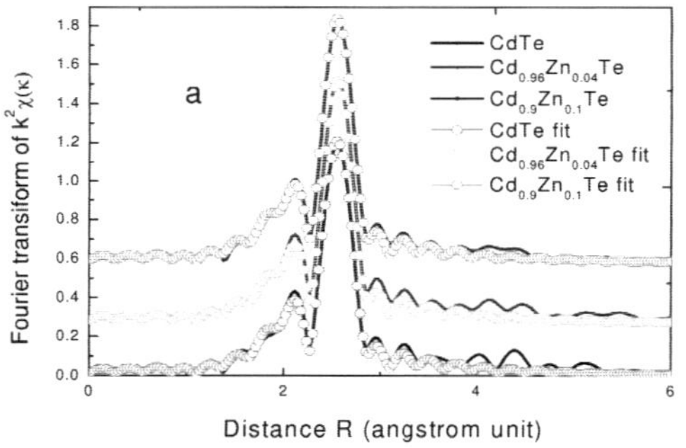

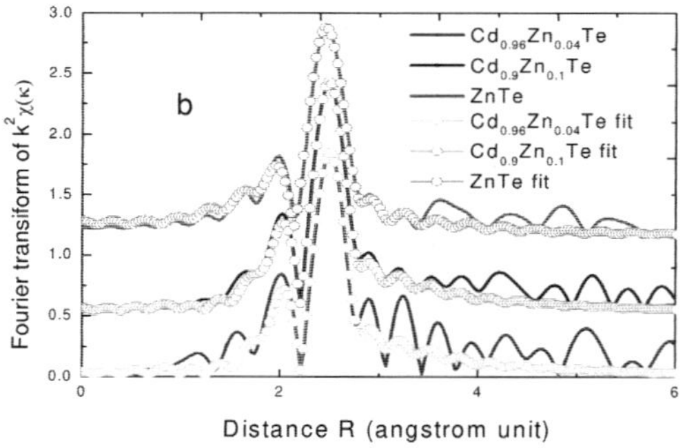

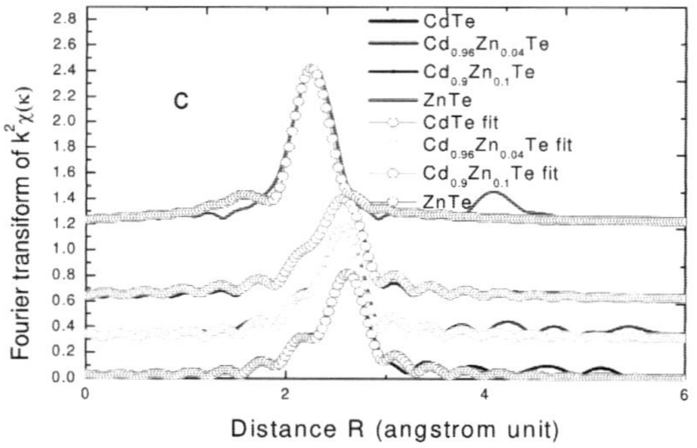

Fig.2. The lines are Fourier transform of the k^2-weighted absorption spectra on Cd (a) in , Zn (b) and Te (c) in CdTe, $Cd_{0.96}Zn_{0.04}Te$, $Cd_{0.9}Zn_{0.1}Te$, and ZnTe, and the circles represent the fitting results.

The local structure of CdTe, $Cd_{0.96}Zn_{0.04}Te$, $Cd_{0.9}Zn_{0.1}Te$, and ZnTe were studied with EXAFS at the Cd K-edge, Zn K-edge and Te K-edge, and the Fourier transform of the k^2-weighted absorption spectra on Cd (Fig.2 a), Zn (Fig.2 b), and Te (Fig.2 c) are presented. The Fourier transformed spectra do not change its character significantly with different Zn contents in our samples. The similarity in shape of the first-neighbor Te peak with the cation Zn or Cd as central atom (with R between 1.5 and 3 Å) can be observed in Fig.2. So the Zn atoms occupy the position of the Cd atoms can be educed.

In Fig.2, simulated EXAFS spectra were also generated using Atermis, based on the known lattice constants calculated by XRD. The bond lengths were fixed according to FEFF8 code. The absorption spectra of Cd and Zn are best fitted with first neighbour Te-shell to the Zn and Cd as central atom in terms of width, amplitude, and position as seen in the Fourier transformed spectra. The Fourier transformed spectra of Te have are best fitted with two distinct bond lengths (Te-Cd and Te-Zn). The results list in Table.2.

Table.2. The results of the EXAFS data fits for first shell around Cd atom and Zn atom

	Bond length/ Å	Coordination number	σ^2	$\triangle E_0$/eV
Cd-Te(CdTe)	2.798	4	0.005890	1.841
Cd-Te($Cd_{0.96}Zn_{0.04}Te$)	2.798	4	0.006309	1.978
Cd-Te($Cd_{0.9}Zn_{0.1}Te$)	2.797	4	0.006019	1.746
Zn-Te(ZnTe)	2.635	4	0.006759	3.325
Zn-Te($Cd_{0.96}Zn_{0.04}Te$)	2.658	4	0.006361	3.286
Zn-Te($Cd_{0.9}Zn_{0.1}Te$)	2.654	4	0.006778	3.161

The results imply that the local atomic structure of ternary compound semiconductor is distorted. It can be observed that the local bond lengths are deviated strongly from Vegard's law as given by XRD data. The results imply the bond lengths retain the values of pure compounds than to the average bond length measured by XRD. As we know above, Zn atoms take place of the Cd atoms in $Cd_{1-x}Zn_xTe$ crystal. So one (or more) of the four chemical bonds connected with Te is short than the others, and this leads to the distortion of local structure. The distortion reduces the symmetry of $Cd_{1-x}Zn_xTe$ crystal. Because of the asymmetry of the $Cd_{1-x}Zn_xTe$ crytal, the centers of positive and negative charges can not coincide, and the crystal has a spontaneous electric polarization. So structural distortion exists in $Cd_{1-x}Zn_xTe$ with low Zn content, and it may be related with the hysteresis of the electric polarization and ferroelectricity.

The Zn-Te bond is stretched with doping of Zn, so it is reasonable to consider the Cd-Te bond is compressed. Cd-Te bond-length of the $Cd_{1-x}Zn_xTe$ is shorter than that of pure CdTe, so the Cd-Te bond in $Cd_{1-x}Zn_xTe$ is stronger, which should be responsible for the low density of dislocation and twin formation in $Cd_{1-x}Zn_xTe$.

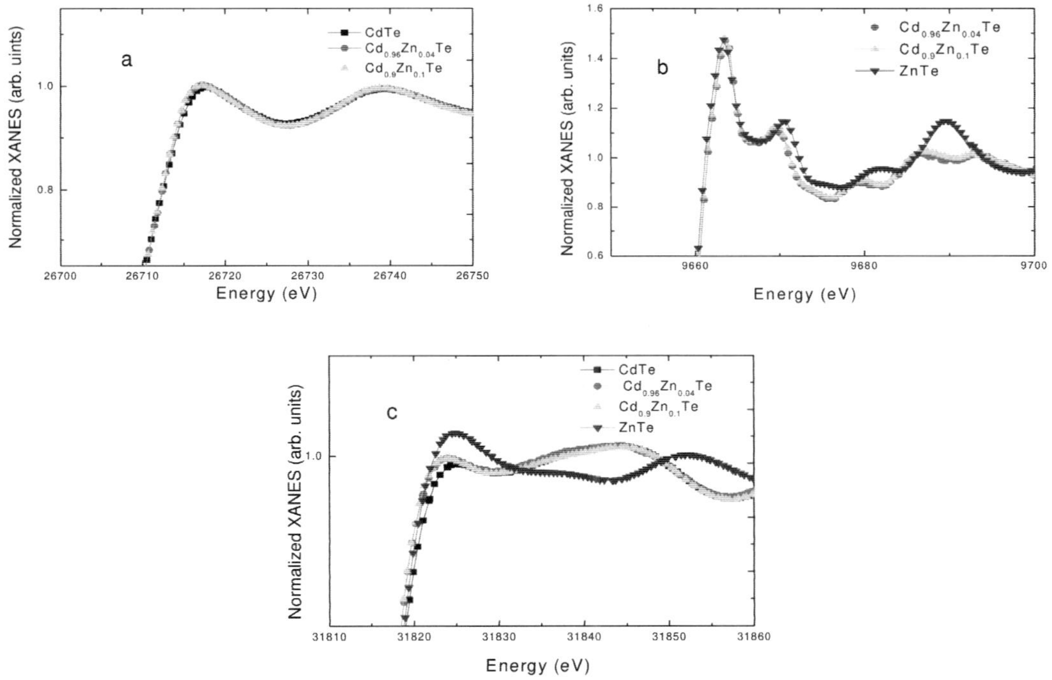

Fig.3. Normalized XANES spectra on Cd K-edge (a), Zn K-edge (b) and Te K-edge (c) in CdTe, $Cd_{0.96}Zn_{0.04}Te$, $Cd_{0.9}Zn_{0.1}Te$ and ZnTe.

Fig.3 shows the comparison of X-ray absorption near-edge structure (XANE) spectra on the Cd K-dege, Zn K-edge and Te K-edge in CdTe, $Cd_{0.96}Zn_{0.04}Te$, $Cd_{0.9}Zn_{0.1}Te$ and ZnTe. The XANES of Cd K-edge (Fig.3 a) are similar, which shows the Cd atoms have the same chemical state in CdTe, $Cd_{0.96}Zn_{0.04}Te$ and $Cd_{0.9}Zn_{0.1}Te$. By comparing the XANES spectra of the Zn K-edge (Fig.3 b), we have found that state of valence electrons in Zn atoms are changed in ZnTe compared with $Cd_{0.96}Zn_{0.04}Te$, $Cd_{0.9}Zn_{0.1}Te$. The XANES of Te K-edge in ZnTe is different from that in CdTe, $Cd_{0.96}Zn_{0.04}Te$ and $Cd_{0.9}Zn_{0.1}Te$ (Fig.3 c). That means the states of the valence electrons of Te are similar in CdTe, $Cd_{0.96}Zn_{0.04}Te$ and $Cd_{0.9}Zn_{0.1}Te$, but change a lot in ZnTe crystal. The spectra show above reveal the original state of Zn-Te bond has changed in $Cd_{1-x}Zn_xTe$. The XANES spectra of CdTe, $Cd_{0.96}Zn_{0.04}Te$, $Cd_{0.9}Zn_{0.1}Te$ and ZnTe need further study to find out the differences.

4. Conclusion

From the XRD and XAFS data analysis, we can conclude that Zn atoms occupy the position of Cd atoms in $Cd_{1-x}Zn_xTe$ crystal. Because of Zn doping, local atomic structure is distorted, and this distortion reduces the symmetry of $Cd_{1-x}Zn_xTe$ crystal. Structural distortion exists in $Cd_{1-x}Zn_xTe$ with low Zn content. Such distortion may be the reason of some special properties in $Cd_{1-x}Zn_xTe$ crystal, such as ferroelectricity, low density of dislocation and twin formation. The XANES spectra show the different chemical states of Zn-Te bond in $Cd_{0.96}Zn_{0.04}Te$, $Cd_{0.9}Zn_{0.1}Te$ and ZnTe.

Acknowledge

The authors thank beamline BL14W1 (Shanghai Synchrotron Radiation Facility) for providing the beam time.

References

[1] R. Weil, R. Nkum, E. Muranevich and L. Benguigui, Phys. Rev. Lett. 1988, 62, 2744.
[2] D. J. Fu, J. C. Lee, S. W. Choi, S. J. Lee, T. W. Kang, M. S. Jang, H. I. Lee and Y. D Woo, Appl. Phys. Lett. 2002, 81, 5207.
[3] I. Stolichnov, E. Colla, N. Setter, T. Wojciechowski, E. Janik and G. Karczewski, Phys. Rev. Lett. 2006, 97, 247601.
[4] Y. Yoneda, N. Matsumoto, K. Suzuya, S. Kohara, and J. Mizuki, Ferroelectrics. 2002, 268, 277.
[5] F. P. Doty, J. F. Schetzina, J. F. Butler, K. A. Bowers, Properties of CdZnTe crystals grown by high pressure Bridgman method, J. Vac. Sci. Technol. 1992, B10,1418.
[6] Newville M, J. Synchrotron Radiat. 2001. 8 314–6.
[7] Rehr J J, Zabinsky S I and Albers R CPhys. Rev. Lett. 1992, 69,3397.
[8] Ravel B and Newville M, J. Synchrotron Radiat. 2005, 12, 537–41.
[9] Motta, N., Balzarotti, A., Letardi, P., Kisiel, A., Czyzyk, M. T., Zimnal-Starnawska, M. and Podgorny, M., J. Crystal Growth. 1985,72,205.

Journal of Physics: Conference Series **430** (2013) 012088 doi:10.1088/1742-6596/430/1/012088

X-ray absorption spectroscopy using BL11 at SAGA-LS, and its applications for materials science

T Okajima[1,2], K Sumitani[1], M Kawamoto[1] and E Kobayashi[1]
[1] *Kyushu Synchrotron Light Research Center*, 8-7 Yayoigaoka, Tosu Saga 841-0005, JAPAN
[2] *Research Center for Synchrotron Light Applications, Kyushu University*, 6-1 Kasugakohen, Kasuga, Fukuoka 816-8505, JAPAN

E-mail: okajima@saga-ls.jp

Abstract. Beamline BL11 at the Saga Light Source (SAGA-LS) is designed for X-ray absorption fine structure (XAFS) experiments in the energy range 2.1–23 keV using a Si(111) double crystal monochrometer. For XAFS experiments, the sample temperature can be controlled in the range of 15–1073 K using a cryostat and furnaces. Besides the transmission measurement mode, we can obtain XAFS spectra in fluorescence measurement mode and conversion electron yield modes. An XAFS measurement system under atmospheric pressure for soft X-rays with energies under 3.5 keV was also installed on this beamline. In addition, we can perform *in situ* XAFS measurements under toxic and harmful gas atmospheres by using a quick XAFS measurement system.

1. Introduction

The Saga Light Source (SAGA-LS) is a compact synchrotron light source with a storage electron energy of 1.4 GeV [1]. The purpose of this facility is to play a pivotal role in the promotion of industrial development through the collaboration between universities and industry in Kyushu. The SAGA-LS has been operated since February 17, 2006. Six prefectural (BL07, BL09A, BL10, BL11, BL12, and BL15) and three other agency beamlines (BL06, BL13, and BL18) are now in operation at SAGA-LS. The prefectural beamlines are open to the public and are available for multiple users such as universities, institutes, and industries. The light sources of these beamlines consist of four bending magnets for BL09A, BL11, BL12 [2], and BL15 [3], an undulator for BL10 [4], and a superconducting wiggler for BL07 [5]. Summaries of these beamlines are listed in Table 1.

The BL11 beamline is mainly designed for X-ray absorption fine structure (XAFS) experiments. This beamline was constructed in the winter of 2010 as the fourth public beamline. In this paper, we introduce beamline BL11 and review some interesting experimental results that have been undertaken with BL11 at SAGA-LS.

Table 1. Beamlines at SAGA-LS

Beamline	[a]Source	Monochrometer	Photon energy	Main experiments	[b]Category
BL06	BM	Double crystal	2.1keV - 23keV	XAFS, SAXS	Kyushu U.
BL07	W	Double crystal	5keV - 35keV	PXD, XAFS, Imaging	Pref.
BL09A	BM	None	White beam	Irradiation, Topography	Pref.
BL10	U	[c]VLS-PGM	40eV - 90eV	PEEM, ARPES	Pref.
BL11	BM	Double crystal	2.1keV - 23keV	XAFS, SAXS	Pref.
BL12	BM	[c]VLS-PGM	40eV - 1500eV	XPS, NEXAFS	Pref.
BL13	U	[c]VLS-PGM	15eV - 600eV	ARPES	Saga U.
BL15	BM	Double crystal	3.5keV - 23keV	XRD,	Pref.
BL18	BM	Multi layered mirror	*ca.* 92eV	Irradiation	Nikon Ltd.

[a] Source : W, wiggler; BM, bending magnet; U, undulator.
[b]Category : Pref., public beamline by Saga Prefecture Government; Kyushu U., Saga U. and Nikon, contract beamline by Kyushu University, Saga University and Nikon Corporation, respectively.
[c]VLS-PGM : Varied-line-spacing plane grating monochrometer.

2. Beamline BL11 at SAGA-LS and the XAFS measurement system

X-rays from a bending magnet with a 1.46-tesla magnetic field and critical energy of 1.9 keV are used. A fixed-exit double-crystal monochrometer with Si(111) crystals was installed and X-rays with photon energies from 2.1 keV to around 23 keV generated by the beamline were used. This beamline has two experimental hutches along its length, as shown in Figure 1, located at 12 m and 20 m from the light source, respectively. After the 1st experimental hutch, a 1 m long rhodium-coated fused silica bent cylindrical mirror is installed. A 0.125 mm thick beryllium window is placed at the end of the front end section. Monochromatic X-rays are guided to the experimental stations through a 0.05 mm thick Kapton® window placed at the end of transport section. The Kapton® window separates the vacuum in the transport channel from the experimental station but lets the X-rays pass. When experiments in the 2nd experimental hutch are performed, the vacuum pass is set in the 1st experimental hutch. Figure 2 shows the observed photon flux at the 2nd experimental hutch with and without the mirror. We can use X-rays of almost 1×10^{10} photons per second at a keV energy range by using a focusing mirror. The photon fluxes are suddenly decreased at the energy range above 10 keV by the cut-off energy determined by the grazing incident angle to the mirror.

The XAFS spectrometer allows X-ray absorption measurements in transmission mode, fluorescence mode, and conversion electron yield mode. Three lengths of the ionization chamber (65, 170, and 310 mm) were prepared for the transmission mode measurements. The current in the ionization chamber was amplified to a voltage, converted to frequency, and then counted. Different gases, namely, helium, nitrogen, argon, and mixtures of these gases, can be injected. For the fluorescence mode measurements, a Lytle-type detector, a germanium 19-element solid-state detector (19-Ge SSD), and a silicon drift-chamber detector (SDD) can be used. The Lytle-type detector contains an aluminium solar slit and a 20-mm long ionization chamber. The elements of the 19-Ge SSD have an effective area of 100 mm^2 × 19 elements and a thickness of 10 mm, enabling detection of more than 95% of the incident X-rays. The element of the SDD has an effective area of 50 mm^2 and a thickness of 0.35 mm. The SDD can be operated without liquid nitrogen. In addition to the detectors, a cryostat and furnaces can be used in the beamline. Using these apparatuses, the sample temperature can be controlled in the range 15–1073 K. The research areas that require this XAFS station include the following examples: the local structure analysis of catalysts for fuel cells and electrode materials for rechargeable lithium-ion batteries, and chemical species analysis of environmental materials.

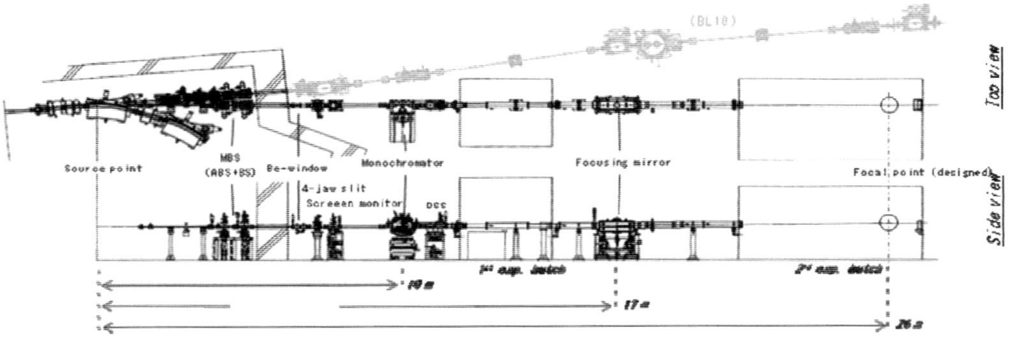

Figure 1. Schematic layout of beamline BL11 at SAGA-LS. MBS is the main beam shutter which contains the photon absorber (ABS) and beam shutter (BS) made of tungsten. DSS is the downstream shutter.

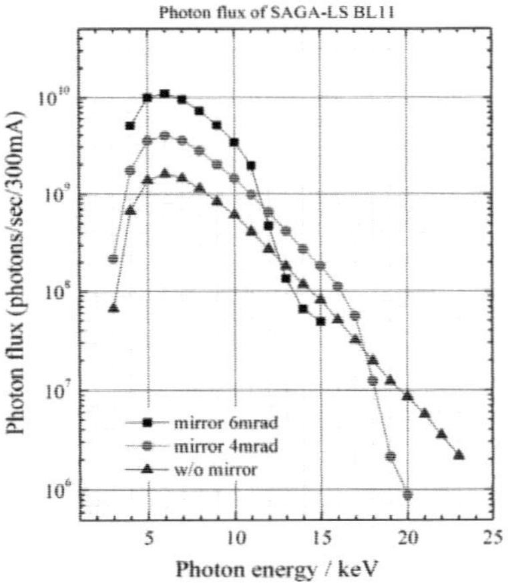

Figure 2. Observed typical photon flux at the XAFS sample position in the 2nd experimental hutch. The photos pass through a slit located just before the sample with a size of 5 mm (horizontal) × 1 mm (vertical). The sample was positioned 22 m from the source point.

3. Experimental results at BL11

3.1 XANES spectroscopy for materials science

With the development of synchrotron light sources, X-ray absorption fine structure (XANES) spectroscopy has become an important branch of X-ray spectroscopy. The XANES spectrum is sensitive to the chemical environment of a specific element, such as chemical bonding, charge state, and magnetic state. XANES is not only an analytical tool for characterization, but a uniquely useful technique for deeply investigating the local environment of selected elements [6–8].

Many investigations have been performed recently using XANES spectroscopy at BL11 [9]. Han *et al.* discovered tetravalent dysprosium (Dy) in the perovskite-type oxide $BaZr_{0.8}Dy_{0.2}O_3$ after heat-treating in pure O_2 gas at 1 atm by using XANES spectroscopy [10]. They prepared two types of $BaZr_{0.8}Dy_{0.2}O_3$ materials, one treated at 1600 °C in pure O_2 gas at 1 atm and the other in H_2 gas at 1 atm. Figure 3 shows the Dy L_3-edge XANES spectra of these materials and the reference samples of Dy_2O_3. The perovskite-type oxide $BaZrO_3$, in which 20% of Zr was replaced by Dy, was prepared by a conventional solid state reaction method. After sintering at 1600 °C in pure O_2 at 1 atm pressure for 24 h, they obtained a black pellet as shown in Figure 3(a). The pellet changed color to white after subsequent heat-treating in H_2 for 120 h at 600 °C. The Dy L_3-edge XANES spectrum of the black pellet shows triple peaks (Figure 3(a)), in contrast to the single peak spectra of the white pellet (Figure 3(b)) and trivalent dysprosium in Dy_2O_3 (Figure 3(c)), which were similar. That is to say, Dy in the black pellet heat-treated at pure O_2 at 1 atm is multivalent and that in the white pellet heat-treated in H_2 is trivalent. These features are also observed for Dy^{4+}-containing fluorides which are reported to exhibit double-peak spectra [11]. In addition to this result, XRD analysis with Rietveld refinement, thermogravimetry experiments, and hydroxide ion measurements at several temperatures led them to conclude that tetravalent dysprosium exists in perovskite-type oxide barium zirconate.

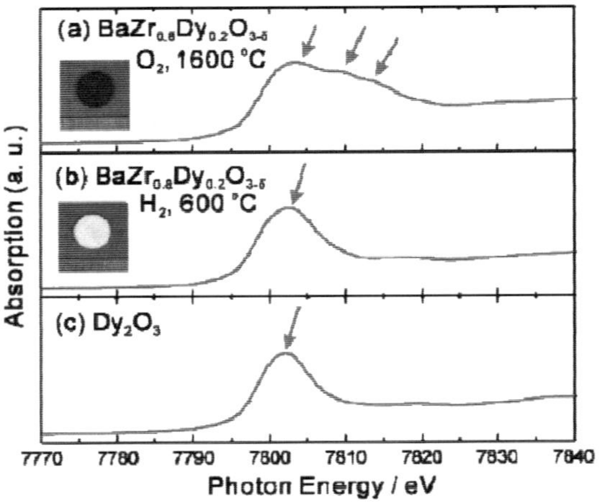

Figure 3. Dy L_3-edge XANES spectra of (a) $BaZr_{0.8}Dy_{0.2}O_{3-\delta}$ heat-treated in O_2 at 1600 °C for 24 h, (b) $BaZr_{0.8}Dy_{0.2}O_{3-\delta}$ subsequently heat-treated in H_2 at 600 °C for 120 h, after the heat-treatment in O_2 at 1600 °C, and (c) Dy_2O_3.

3.2 In situ XAFS measurement of NiO reduction

Quick-XAFS (QXAFS) measurements have been widely used at many synchrotron radiation facilities. In an QXAFS measurement system, XAFS spectra are obtained by moving a monochrometer with a constant velocity. Scanning times of a few minutes can be routinely achieved, with some special systems measuring even below one second.

Metal oxides are used as catalysts for the reforming of hydrocarbons [12, 13]. Among the transition-metal oxides, the reduction of nickel monoxide (NiO) by hydrogen has been the subject of the most extensive fundamental studies, becoming an important reference system [13, 14]. We demonstrated the *in situ* XAFS observation of the reduction of NiO using a QXAFS system. The *in situ* XAFS measurements were carried out in a reaction cell in transmission mode. The cell was made of fused quartz with Kapton® windows. Powders of NiO were mixed with boron nitride powder and packed inside the cell. The specimen was heated to about 1073 K in a 2% H_2/98% N_2 gas with a flow rate of 50 ml/min. The temperature of the specimen was raised continuously at a rate of 10 K/min. The spectra were registered once every minute, and it took about 30 seconds to measure one spectrum. Figure 4 shows the Ni K-edge *in situ* XANES spectra and the Fourier transform of the EXAFS ($k^3\chi(k)$) spectra for NiO powder. The insets of Figure 4(a) show the XANES spectra obtained at temperatures of 310 and 1073 K, and the XANES spectrum of Ni foil at the upper right side. At around 8355 eV, the typical strong white line for NiO in the 310 K spectrum is clearly observed. The intensity of the white line exhibits a monotonic decrease above about 773 K (red solid line) with increasing temperature, while the pre-edge feature at around 8340 eV for metallic Ni appears. A comparison of the spectrum at 1073 K and the spectrum of Ni foil shows that the reductions of Ni in NiO are insufficient. Linear combination analysis of the XANES spectra also showed about 90% of the Ni in NiO was present as Ni in metal after reduction at 1073 K. In Figure 4(b), the peaks at 1.5 Å and 2.5 Å become consistent with the presence of Ni-O (in NiO) and Ni-Ni (in NiO) bonds, respectively. The inset at the upper right side also shows the $k^3\chi(k)$ spectra for Ni metal. This figure indicates that the peak corresponding to the Ni-Ni bond (in Ni metal) appears at 2.0 Å, shown as a dotted line in Figure

4(b). The spectrum also shows the reductions of Ni in NiO are insufficient. These spectra clearly show the reduction from Ni in NiO to Ni metal occurs according to the increase of the temperature in a reduction atmosphere.

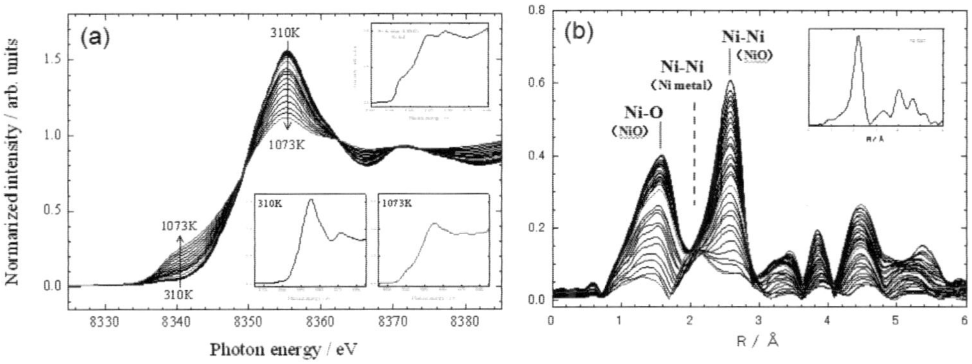

Figure 4. (a) Ni K-edge *in situ* XANES spectra and (b) Fourier transform of EXAFS ($k^3\chi(k)$) spectra for NiO powder.

3.3 *XAFS measurement at sulfur K-edge region*

Materials containing sulfur and phosphor play important roles in various research fields, especially in environmental and biochemical sciences, as well as in many industrial applications in the fields of rubber, cosmetics, pharmaceuticals, pesticides, and fertilizers. XAFS is one of the most suitable methods for observing the many functions and transformations of these species. However, because X-rays with energies between 2 and 4 keV barely penetrate the atmosphere, XAFS spectra are usually measured in vacuum, and for this reason, XAFS measurements of these species have been restricted.

We installed a new XAFS measurement system using helium paths that is easy to operate and allows the collection of XAFS spectra in the soft X-ray region around the sulfur and phosphor K-edges under atmospheric pressure [15]. XAFS spectra can be obtained simultaneously in partial fluorescence yield (PFY) mode using an SDD and in convergent electron yield (CEY) mode. Figure 5(a) shows sulfur K-edge XANES spectra of amino acids obtained by the CEY mode. These acids have sulfur with different chemical bonding states. The XANES structures clearly show the chemical shift according to the chemical bonding states. Figure 5(b) also shows the sulfur K-edge XANES spectrum of residual fuel oil obtained by the PFY mode. The oil specimen was packed in a polyethylene bag and was only attached to the sample holder as shown in the inset. These results clearly show that the newly installed XAFS measurement system using a helium pass is useful for XAFS measurement in the soft X-ray region for the measurement of both solid- and liquid-state specimens.

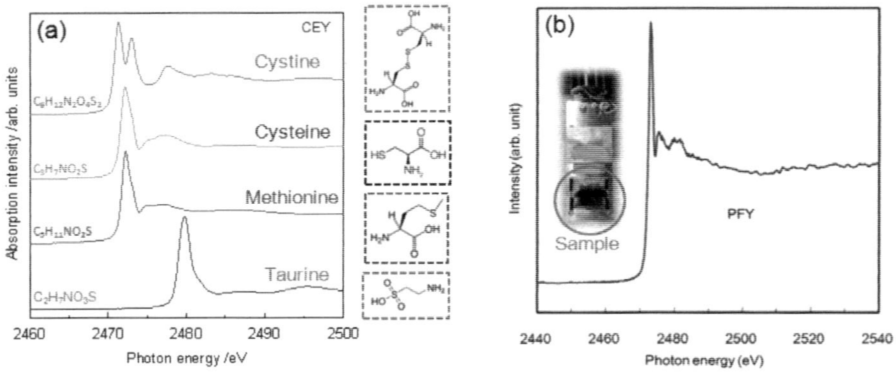

Figure 5. XANES structures showing various chemical states of sulfur: (a) sulfur K-edge XANES spectra of amino acids and (b) sulfur K-edge XANES spectrum of residual fuel oil.

4. Conclusion

In this paper, we introduced the beamline optics and XAFS measurement system in the BL11 beamline. We also presented results of XAFS applications carried out at BL11. These results indicated that BL11 has a performance as good as that of the other XAFS beamlines. An aim of third generation synchrotron radiation facilities is to realize measurements at extreme conditions of time, space, quantity, and so on. The effort to extend the capabilities of XAFS measurement systems is also expanding the range of XAFS applications. As a consequence, stable and versatile XAFS beamlines are becoming more important. It is our aim that the BL11 beamline will become a pioneering facility.

5. References

[1] Yoshida K, Iwasaki Y, Takabayashi Y, Koda S. Okajima T, Setoyama H, Yoshimura D, Tomimasu T and Ohgaki H 2007 AIP Conf. Proc., **879**, 179-183
[2] Kamada M, Kondo Y, Azuma J, Takahashi K, Miyata H, Tsuji J, Hashimoto H, Yamashige H, Matsuo S, Kurisaki S, Wakita H, Furuya K, Harata A, Tochihara H, Watanabe T, Inaba M, Okajima T, Setoyama H, Yoshimura D and Fujimoto S 2007 AIP Conf. Proc., **879**, 623-626
[3] Okajima T, Chikaura Y, Suzuki Y, Tabata M, Soejima Y, Hara K, Haruki R, Nagata K, Hiramatsu N, Khono A, Takumi M, Setoyama H and Yoshimura D 2007 AIP Conf. Proc., **879**, 820-823
[4] Yoshimura D, Setoyama H and Okajima T 2010 AIP Conf. Proc., **1234**, 423-426
[5] Kawamoto M, Sumitani K and Okajima T 2010 AIP Conf. Proc., **1234**, 355-358
[6] Koningsberger DC and Prince R 1998 *X-ray Absorption: Principles, Applications, Techniques of EXAFS, SEXAFS, and XANES in Chemical Analysis* (New York: John Wiley & Sons)
[7] Stöhr J 1996 *NEXAFS Spectroscopy* (Berlin: Springer-Verlag)
[8] Tanaka I, Mizoguchi T and Yamamoto T 2005 J. Am. Ceram. Soc., **88**, 2013
[9] SAGA-LS Users report, http://www.saga-ls.jp/?page=43 (in Japanese)
[10] Han D, Uda T, Nose Y, Okajima T, Murata H, Tanaka I 2012 Adv. Mater., **24**, 2051-2053
[11] Hu Z, Kaindl G, Vandré D, Hoppe R and Wortmann G 1994 J. Alloys Compd., **205**, 263
[12] Thomas J M and Thomas E J 1997 *Principles and Practice of Heterogeneous Catalysis* (New York: VCH)
[13] Kung H H 1989 *Transition Metal Oxides: Surface Chemistry and Catalysis* (New York: Elsevier)
[14] Delmon B 1997 *Handbook of Heterogeneous Catalysts*, ed G Ertl, H Knözinger and J Weikamp (New York: Wiley-VCH) pp 264-277
[15] Okajima T and Ohtani R 2011 Diamond Light Source Proc., **1**, e141

Journal of Physics: Conference Series **430** (2013) 012089 doi:10.1088/1742-6596/430/1/012089

Investigation of Er^{3+} coordination in zinc-lead tellurite bulk glasses and silica-hafnia glass ceramics waveguides.

Raj Kumar Ramamoorthy[1,2], A.K. Bhatnagar[1,3], P.S. Vachhani[2,3], G. Dalba[2], R. Grisenti[2], M. Mattarelli[2], M. Montagna[2], C. Armellini[4] and F. Rocca[4]

[1]School of Engineering Sciences and Technology, University of Hyderabad, Hyderabad-500 046, India

[2]Department of Physics, University of Trento, 38123 Povo (Trento) Italy

[3]School of Physics, University of Hyderabad, Hyderabad-500 046, India

[4]IFN-CNR, Institute for Photonics and Nanotechnologies, Unit "FBK-Photonics" of Trento, 38123 Povo (Trento) Italy

E-mail: rajkr37@gmail.com, rocca@science.unitn.it

Abstract. We present a comparison of the local structure of erbium in zinc-lead tellurite glasses and in silica-hafnia glass ceramics waveguides prepared using the melt quenching or sol-gel and dip coating techniques respectively. The local environment of Er^{3+}ions was probed using the extended x-ray absorption fine structure (EXAFS) at the L_3-edge of erbium at room temperature. Significative differences in the first Er-O coordination shells are evidenced. In tellurite glasses, independently on the ZnO/PbO content, the erbium co-ordination number (N) is constant at about 6.5, the distance (R) remains the same at 2.34 Å and the Debye-Waller (σ^2) is about 0.0150 Å^2. On the other hand, N is lower at about 5.2 for all the silica-hafnia waveguides, whereas the Er-O distance and the DW factors are changed with time and temperature of thermal annealing.

1. Introduction

Erbium doped glasses are highly powerful materials in the field of wavelength division multiplexing (WDM), integrated circuits and fiber amplifier (EDFA) at 1.5 μm, etc. [1, 2, 3]. Even though large pool of materials are available for the photonic applications, the scintillating properties of low loss transmission at 1.5 μm [4] and the recent successes on new materials based on nanocrystals (NC) inside the waveguides [5] make the silicate glasses and films the more suitable for it. On the other hand, tellurite glassy systems urge their importance in similar applications due to high refractive index, large solubility of erbium ions and large flat band width and high gain at 1.5 μm [2]. The above mentioned intriguing properties of these two glassy systems stimulated us to study the reasons behind them. The local crystal field around the Er^{3+} ion substantially plays an important role in determining its spectroscopic properties. The local atomic structure around Er^{3+} ions can be easily probed through the EXAFS spectroscopy [6, 7]; later on, it can be correlated with the optical properties. In this report, we present our

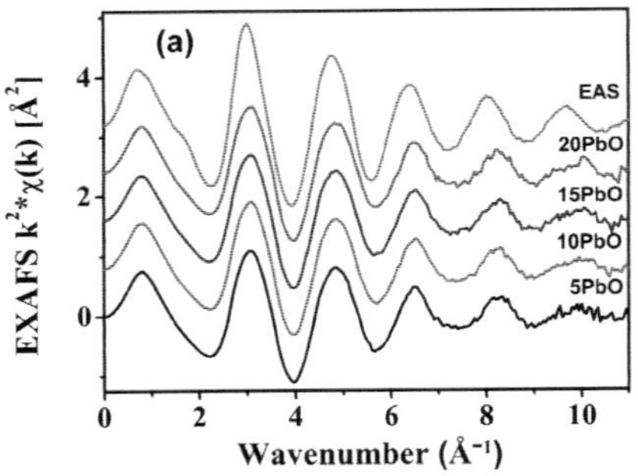

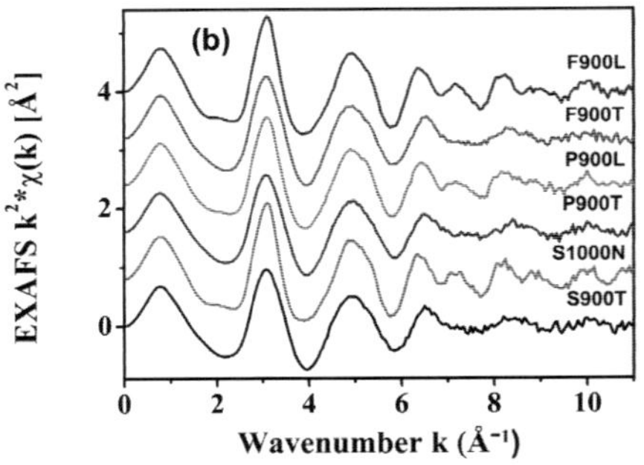

Figure 1. (a) Experimental EXAFS signals $k^2\chi(k)$ at L_3 edge of erbium ions in tellurite glasses and of erbium in aqueous solution used as reference. (b) Experimental EXAFS signals $k^2\chi(k)$ at L_3 edge of erbium ions in selected silica-hafnia waveguides for different heat treatments.

studies on the nearest coordination shells of erbium ions, as determined from the Er^{3+} L_3 edge using EXAFS in both tellurite glasses and silica-hafnia waveguides.

2. Experimental details

Erbium doped Tin-Lead tellurite glasses (TZPE) having composition $(70\ TeO_2-(30-x)\ ZnO - xPbO)_{0.99}-(Er_2O_3)_{0.01}$ were prepared using the melt quenching technique. The tellurite glasses are labeled in accordance with the nominal percentage of PbO x = 5, 10, 15, 20. The 1 mol% Er^{3+} doped [80:20] silica-hafnia waveguides were prepared using the sol-gel route, deposited on the v-SiO_2 substrates by dip-coating technique and finally heat treated at 900 °C for 30 min for the full densification. Hafnia NCs were produced, or inserted inside the waveguides, through two different methods: in the standard one (S) [6], the glassy film was heat treated at 900 °C for 30 min or at 1000 °C for 30 min and 12 hr; in alternative way, pre-grown 2.5 mol% HfO_2 NCs (P) were added together into the standard [80:20] mixture (S) or pre-grown 20 mol% HfO_2 NCs were fully (F) substituted for the HfO_2 in the standard [80:20] mixture, then heat treated at 900 °C for 30 min, 24 hr and 48 hr. The waveguides are labeled as follows: S900T, S1000T, S1000N, P900T, P900M, P900L, F900T, F900M and P900L, where S, P and F indicate the preparation method, the numbers represent the heat treatment temperature and T, N, M and L stand for the heat treatment durations 30 min, 12 hr, 24 hr and 48 hr.
Fluorescence EXAFS measurements were carried out at the L_3 edge of erbium at BM08-GILDA beamline of ESRF. EXTRA [8] and EDAFIT [9] analysis software were used for extracting the signal and for the best-fitting, respectively. EXAFS signals (shown in figure 1(a and b)) were Fourier transformed using Kaiser-Bessel window, then the first shell contributions were singled out and back Fourier transformed for the best-fit, performed in the range 2-10 $\AA^{-1}$ or 2-8 $\AA^{-1}$ for tellurite glasses or silica waveguides, respectively. Bestfit was performed using amplitude and phase shift functions of Er-O in erbium aqueous solution [EAS] calculated using FEFF8 or obtained experimentally from the EAS reference, following a well tested procedure [6, 10].

3. Results and Discussion

3.1. Tellurite glasses

The $k^2\chi(k)$ weighted EXAFS signals of all tellurite glasses have similar oscillation patterns, but the amplitudes are lower than those of the reference EAS. There is no evidence of high frequency oscillations for these glasses: this means that the higher shell contributions are merely smeared out due to the high static disorder. This is well evidenced in the FT moduli shown

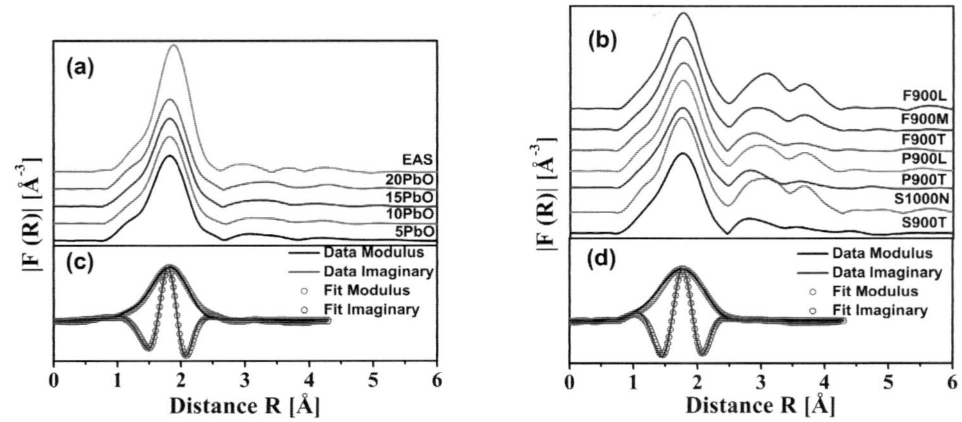

Figure 2. Top: FT of EXAFS signal $k^2\chi(k)$ of tellurite glasses (a) and of selected silica-hafnia waveguides (b). Bottom: experimental and best-fit calculated FT of the first shell of the 5PbO tellurite glass (c) and filtered first shell of F900L silica-hafnia waveguide (d).

in figure 2(a). The experimental and best-fit calculated FT signals are shown in figure 2(c). The obtained structural parameters are reported in table 1. The quantitative analysis gives the N values around 6.5, R constant at about 2.340 $\mathring{A}$ and σ^2 values about 0.0150 $\mathring{A}^2$. The observed N values are similar to those previously obtained by other authors, whereas the R and σ^2 values are larger [7, 11]. The XANES spectra of these four glasses (see Figure 7 in our recent paper [12]) are indistinguishable, even though different from those of Er in aqueous solution, because their white line peak is broader. These data indicate that erbium ions are located in almost the same environment in all tellurite glasses, independently on the compositional change of PbO with ZnO. Erbium ions are probably bonded only to available non-bridging oxygens of the tellurite units of the glasses, not to the modifiers ZnO and PbO. Note that N is larger than in silicate systems [6].

3.2. Silica-Hafnia glass/glass ceramics waveguides

The $k^2\chi(k)$ weighted EXAFS signals of the series of silica-hafnia waveguides, prepared under three different conditions, show the evolution of oscillating signals for increasing time of heat treatment at 900 °C [for inserted NC] and 1000 °C [for standard mixture]. This confirms the formation of outer shells around the erbium ions, as clearly evidenced also in the second and third shells (see figure 2(b)). Since the paper discussion is based on the comparison of the first shell around the Er^{3+} in these two glasses, the detailed analysis of the outer shells is omitted here. In our previous papers [6, 10] on standard waveguides it was clear that the contributions from next nearest shells are mainly related to Er-Hf bonds.

Experimental and best-fit calculated FTs are shown in figure 2(d). This shorter range of best-fit is chosen in order to decrease the importance of outer shells contribution in the singled out first shell analysis. The obtained N value is around 5.2 for the waveguide series, independently on heat treatment; the R value initially increases after heat treatment at 900 °C and 1000 °C and then decreases for the longest heat treatment at the same temperature, while the σ^2 value continuously decreases (see table 1). The obtained first shell structural parameters are in good agreement with our previous literature [6, 10]. The decrease of DW factor values σ^2 indicates the progressive local structural ordering around the erbium ions in the waveguides, occurring mainly because of the presence of erbium ions inside the NCs in all the studied waveguides,

Table 1. First shell (Er-O) structural parameters of erbium ions. Number in parentheses are the numerical uncertainties on the last digit, as obtained from average of bestfits done in different ranges.

	First shell (Er-O)				First shell (Er-O)		
Sample	N	R (Å)	σ^2 (Å^2)	Sample	N	R (Å)	σ^2 (Å^2)
5PbO	6.5(5)	2.340(3)	0.0156(3)	P900T	4.9(5)	2.338(3)	0.0136(3)
10PbO	6.4(5)	2.340(3)	0.0146(3)	P900M	5.1(5)	2.344(2)	0.0125(3)
15PbO	6.6(5)	2.340(3)	0.0150(3)	P900L	5.2(5)	2.340(3)	0.0123(1)
20PbO	6.7(5)	2.340(3)	0.0155(3)	F900T	5.2(5)	2.335(3)	0.0129(3)
S900T	4.9(5)	2.338(3)	0.0134(4)	F900M	5.2(5)	2.342(2)	0.0121(3)
S1000T	5.3(5)	2.342(3)	0.0121(2)	F900L	5.3(5)	2.341(2)	0.0119(2)
S1000N	5.3(5)	2.344(2)	0.0118(1)				

similarly to the situation described in a previous paper [6] for the longest heat treatment at 900 °C and 1000 °C.

4. Conclusion

First shell structural parameters of Er^{3+} ions in tellurite glasses and silica-hafnia waveguides have been compared using EXAFS. In tellurite glasses, EXAFS can detect only the first shell Er-O bonds, with negligible changes when ZnO is substituted by PbO. These results confirm the presence of a large number of non-bridging oxygens from the tellurite units of the glasses, that are at the origin of the enhancement of erbium solubility in these glassy systems. On the other side, EXAFS clearly evidences the nucleation and crystallization of hafnia around erbium atoms in all silica-hafnia waveguides. As already evidenced in previous papers of us [6, 10], the number of available oxygens from HfO_2 nanocrystals is reduced since the first stage of preparation, while the progressive growing of hafnia NCs inside the silica matrix influences the Er-O distances and their static disorder, monitored by the DW factor.

Acknowledgement

The authors acknowledge DST-ITPAR for the financial support and ESRF (Grenoble, France) BM08 beamline for the EXAFS measurement. One of the authors (AKB) acknowledges support from the Indian National Science Academy (New Delhi, India)

References

[1] Miniscalco W J 1991 *J. Lightwave Technol.* **9(2)** 234
[2] Chryssou C E, Pasquale F D and Pitt C W 2000 *IEEE J. Sel. Topics Quantum Electron.* **6** 114
[3] Yamada M, Mori A, Kobayashi K, Ono H, Kanamori T, Oikawa K, Nishida Y and Ohishi Y 1998 *IEEE Photonics Technol. Lett.* **10** 1244
[4] Wang J S, Vogel E M and Snitzer E 2004 *Phil. Mag.* **84** 1659
[5] Lahoz F, Martín I R, Rodríguez-Mendoza U R, Iparraguirre I, Azkargorta J, Mendioroz A, Balda R, Fernández J and Lavín V 2005 *Opt. Mater.* **27** 1762
[6] Afify N D, Dalba G and Rocca F 2009 *J. Phys. D: Appl. Phys.* **42** 115416
[7] Jin Chung W and Yong Gyu Choi 2010 *J. Am. Ceram. Soc.* **93(5)** 1432
[8] Dalba G, Afify N D and Rocca F 2008 *Phys. Chem. Glasses: Eur. J. Glass Sci. Technol. B* **49** 149
[9] Kuzmin A 1995 *Physica B* **208/209** 175–176
[10] Afify N D, Dalba G, Armellini C, Ferrari M, Rocca F and Kuzmin A 2007 *Phys. Rev. B* **76** 024114
[11] Souza Neto N M, Ramos A Y and Barbosa L C 2002 *J. Non-Cryst. Solids* **304** 195
[12] Ramamoorthy R K, Bhatnagar A K, Rocca F, Mattarelli M and Montagna M 2012 *J. Phys.: Condens. Matter* **24** 505101

Journal of Physics: Conference Series **430** (2013) 012090 doi:10.1088/1742-6596/430/1/012090

Local structure of Mo-doped TiO_2 photocatalysts investigated by X-ray absorption fine structure

Y J Feng, C Wang, W R Cheng, J H Huang, T X Zhao, Q H Liu, Z Xie, Z Y Pan* and S Q Wei*
National Synchrotron Radiation Laboratory, University of Science and Technology of China, Hefei, Anhui 230029, P.R. China

*Corresponding authors. E-mail: zhypan@ustc.edu.cn, sqwei@ustc.edu.cn

Abstract. The local structures of Mo-doped TiO_2 photocatalysts are investigated by X-ray absorption fine structure spectroscopy (XAFS) combined with X-ray diffraction (XRD). The XRD patterns show that all the Mo-doped TiO_2 samples maintain the high qualitative anatase TiO_2 structure without any visible diffraction peak related to Mo. The detailed XANES and EXAFS analysis reveal that all of the Mo atoms substitute Ti atoms in the anatase lattice as Mo^{6+} with doping concentration as high as 6 atom %. Combined with theoretical calculation, we suggested that the high electron density of Mo-O causes the strong shrinkage of ~0.1 Å for Mo-O bond length in Mo-doped TiO_2.

1. Introduction

Titanium dioxide has attracted considerable attentions in recent years, due to its strong catalytic activity, high chemical stability, a long lifetime of photo-generated carriers, and water splitting under ultraviolet light as well as potential commercial in environmental cleanup and solar energy utilization. However, the band gap of TiO_2 absorbs only small portion in the ultraviolet region of solar spectrum, which inhibits its photoreaction efficiency under natural sunlight [6, 7, 9]. In order to increase the limited optical absorption of TiO_2 under sunlight, many researchers have studied TiO_2 by adding controlled metal [1, 4, 6, 7, 9] or nonmetal [2, 3, 14]. Several groups have studied the Molybdenum (Mo) ions doped TiO_2, both from experimental and theoretical points of view. For example, Stengl et al. have reported Mo-doped TiO_2 with the Mo content varying from 0 up to 4.7 atom % [12]. Devi et al. have reported Mo doped TiO_2 with the Mo content varying from 0 up to 0.1 atom % [5]. Jeon et al. have studied nanosized Mo doped TiO_2 with the Mo content varying from 0 up to 2.5 atom% [8]. Typically, X-ray diffraction (XRD) was utilized to investigate the structure of Mo doped TiO_2 to confirm the possibility of Mo substitution in TiO_2 lattice [5]. However, those results could hardly

Journal of Physics: Conference Series **430** (2013) 012090 doi:10.1088/1742-6596/430/1/012090

provide effective information on the exact formation of Mo ions occupy sites in the TiO_2 matrix. Theoretically, Yu et al. have performed that the substitutional site Mo ions and the interstitial site Mo ions in the matrix have different impacts on the electronic/optical properties of anatase TiO_2, and substitutional Mo doping is effective in reducing the bandgap of TiO_2 and the redshift is significantly enhanced through increasing Mo concentration [15]. Considering that the optical and electrics of the Mo doped TiO_2 strongly depend on the concentration and the existence sites of the Mo ions in the host structure, it is necessary to investigate the local structural of the Mo atoms in the TiO_2 matrix.

In this work, the photocatalysts Mo doped TiO_2 with high concentration of Mo were synthesized through a co-deposition method. The x-ray absorption fine structure (XAFS) and x-ray diffraction (XRD) techniques were combined to investigate the atomic structure in Mo doped TiO_2 samples. With the advantage of the sensitivity of the XAFS spectroscopy to the local structure around Mo ions, in conjunction with the theoretical calculations, the results reveal that the Mo ions occupied the Ti sites in the TiO_2 lattice and substitutionally doped in TiO_2 matrix.

2. Experimental

The Mo doped TiO_2 samples $Ti_{1-x}Mo_xO_2$ (x= 0.03, 0.06) were prepared via a co-deposition method, briefly, using a solution containing Ti (solution A) and a solution containing Mo (solution B). To prepare solution A, 30 mL tetrabutyl titanate and 5 mL acetylacetone were added into 90 mL ethanol. And a ethanolic solution of nitric acid (10 mL stilled water and 0.55 mL nitric acid in 45 mL ethanol) was dropped in and the solution was stirred for 60 min. Solution B is MoO_3 powder with different amount (0.407 g for 3%, and 0.806 g for 6%) dissolved in 0.55 mL nitric acid, and 25mL aqueous ammonia was added in. Solution B was dropped into solution A during 30 min under stirring and the stirring lasted for another 120 min. After aging of 48 h, the wet gel was dried under vacuum and calcined at 600℃ for 12 h. The final products were gray powder.

The crystal structures of the samples were determined by an X-ray diffractometer operated at 60 kV and 55 mA, using Cu Kα radiation with a θ-2θ scan mode. The Mo K-edge and Ti K -edge XAFS spectra of the prepared samples were collected respectively at beamline BL14W1 of Shanghai Synchrotron Radiation facility (SSRF, China) and beamline U7C of National Synchrotron Radiation Laboratory (NSRL, China). At all the beamlines, a Si (111) double crystal monochromator was used. XAFS data were analyzed by the UWXAFS3.0 software package [13].

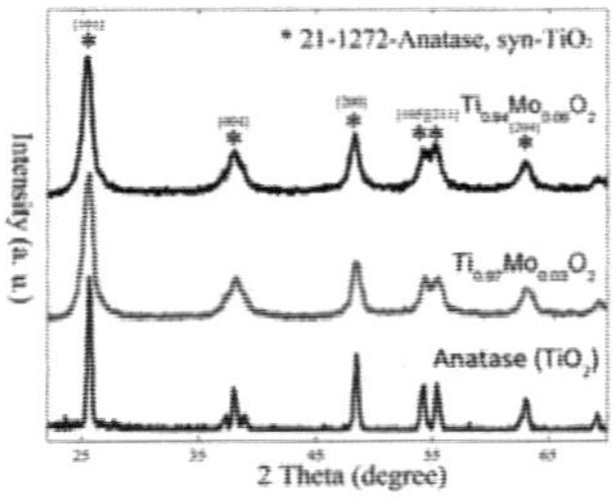

Figure 1. XRD patterns of prepared $Ti_{0.97}Mo_{0.03}O_2$, $Ti_{0.94}Mo_{0.06}O_2$ and anatase TiO_2 samples.

3. Results & Discussion

The XRD patterns of $Ti_{1-x}Mo_xO_2$ (x=0, 0.03, 0.06) are showed in Figure 1. From Figure 1, it can be seen that only six peaks as the characteristic of the anatase TiO_2 ([101], [004], [200], [105], [211] and [204] crystalline faces) can be observed without any visible diffraction peak related to Mo. They point out that all the samples have a good crystallinity with anatase structure, almost consistent with previous studies [5]. In order to determine the detailed structure of Mo atoms in TiO_2, a further study using XAFS is necessary.

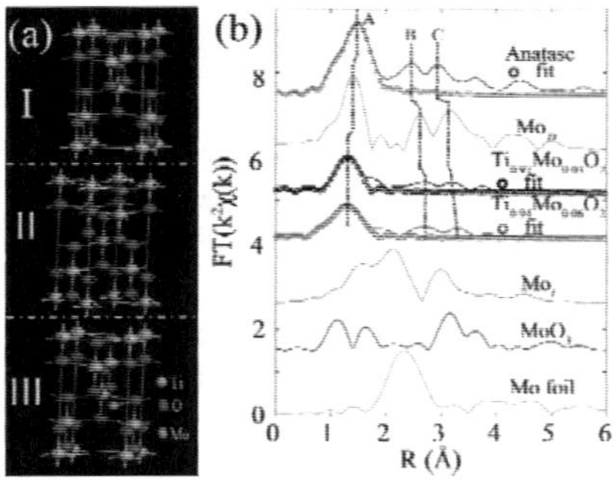

Figure 2. (a) Schematic diagrams of anatase TiO_2 lattice (model I), substitutional Mo (model II, Mo_{Ti}) and interstitial Mo (model III, Mo_i); (b) Mo K-edge EXAFS FTs spectra of $Ti_{0.97}Mo_{0.03}O_2$, $Ti_{0.94}Mo_{0.06}O_2$, the reference materials of crystalline MoO_3 and Mo metal as well as the calculated spectra for two model structures: substitutional Mo (model II, Mo_{Ti}) and interstitial Mo (model III, Mo_i); Ti K-edge EXAFS FTs spectra of anatase TiO_2.

In Figure 2a, we show schematically three representative models built by anatase TiO_2 lattice: anatase TiO_2 lattice (model I), substitutional Mo (model II, Mo_{Ti}) and interstitial Mo (model III, Mo_i) respectively. All the calculations were performed within the supercells constructed from a $2\times 2 \times 2$ standard unit cell of anatase TiO_2 lattice and the optimization of total energy was used to calculate the theoretical Mo K -edge EXAFS spectra of substitutional Mo (model II, Mo_{Ti}) and interstitial Mo (model III, Mo_i).

Figure 2b displays Mo K-edge Fourier transforms (FTs) of EXAFS spectra for $Ti_{0.97}Mo_{0.03}O_2$ and $Ti_{0.94}Mo_{0.06}O_2$ samples. As references, the Mo K-edge functions of Mo foil, MoO_3, Mo_{Ti} model, Mo_i model, and Ti K - edge EXAFS spectrum of anatase TiO_2 are also displayed in Figure 2b. It can be seen that the FT features of $Ti_{0.97}Mo_{0.03}O_2$, $Ti_{0.94}Mo_{0.06}O_2$ samples, and Mo_{Ti} are very close to that of anatase TiO_2 powder, presenting strong peaks at around 1.5 Å corresponding to the nearest Mo-O and two weak peaks B and C from 2.0 Å to 3.5 Å corresponding to the next-nearest Mo-Ti neighbors. These peaks are distinctly different from those of MoO_3 bulk, Mo metal and model Mo_i. Therefore, the existence of MoO_3, Mo metal and Mo occupying the interstitial sites can be easily excluded. This indicates that Mo ions in Mo doped TiO_2 samples occupy the Ti sites in TiO_2 lattice with the doping concentration from 0.03 to 0.06. It is notable that peak A in Mo doped TiO_2 samples presents an evident shift of ~ 0.1 Å toward low R compared with pristine TiO_2. Therefore, it can be deduced that the bond length of Mo-O bond significantly shrinks than that of Ti-O bond.

To further obtain the quantitative structural information, we quantitatively fit the nearest Mo-O coordinations using the basic EXAFS formula [10]. The best fit of the structural parameters obtained for the samples are summarized in the Table I and the results of the fit are shown as empty circles in

Figure 2b. Table I shows that for the Mo atoms in $Ti_{1-x}Mo_xO_2$, there are 4 Mo-O in 1.80 ~ 1.81 Å and 2 Mo-O in 1.85 ~ 1.86 Å, respectively. They are obviously shrink ~ 0.1 Å as to be considered in comparison with the distance of anatase TiO_2 ($R_{Ti\text{-}O}$=1.90, 1.95). The goodness of the fits provides evidence that the real structure of Mo doped TiO_2 can be modeled by the substitutional Mo in TiO_2. However, due to the less thermodynamically stable than the pure phase, Mo doping will enhance structural distortion [12].

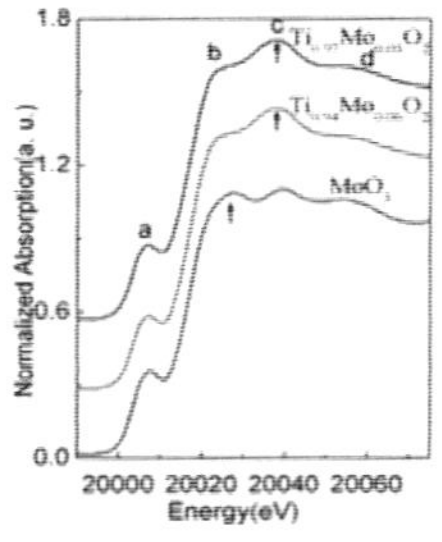

Figure 3. Mo K-edge XNAES spectra of $Ti_{0.97}Mo_{0.03}O_2$, $Ti_{0.94}Mo_{0.06}O_2$, and MoO_3.

The Mo K-edge X-ray absorption near-edge structure spectra (XANES) of $Ti_{0.97}Mo_{0.03}O_2$ and $Ti_{0.94}Mo_{0.06}O_2$, along with the reference materials of crystalline MoO_3 are presented in Figure 3. The absorption edge positions of both Mo doped TiO_2 samples are similar to that of MoO_3, suggesting the existence of Mo^{6+}.

Due to the similarity in the ionic radii of Mo^{6+} (0.062 nm) and Ti^{4+} (0.068 nm), Mo^{6+} can replace Ti^{4+} in the lattice [11]. Figure 4 depicts the electron density distribution of Mo-doped TiO_2 calculated using Density-functional theory (DFT) with generalized gradient approach (GGA). It is clearly that the electron density in Mo-O bonds is higher than that in Ti-O bonds, which may arise from that the Mo^{6+} fetches more electrons from O^{2-} than Ti^{4+} due to its high charge. Therefore, we can induce that the larger intensity of Mo^{6+} attracting O^{2+} than that of Ti^{4+} is the origin of shrinkage in Mo-O band length, which is consistent with the calculated results of model Mo_{Ti}.

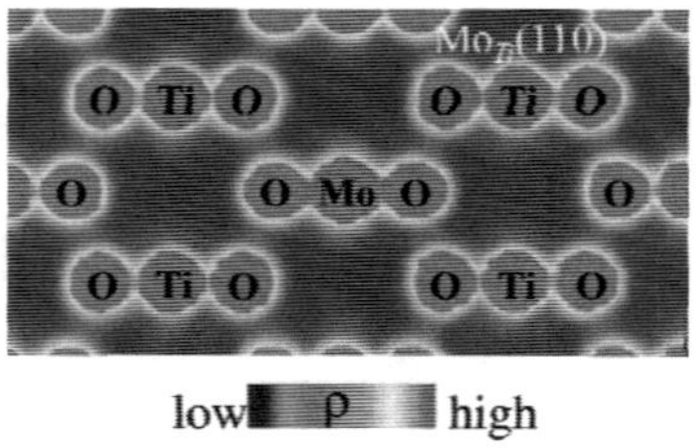

Figure 4. The electron density distribution of substitutional Mo (Mo_{Ti}) calculated using Density-functional theory (DFT) with generalized gradient approach (GGA).

Table I. Structure parameters for Theoretic Anatase TiO_2, Anatase TiO_2, $Ti_{0.97}Mo_{0.03}O_2$, $Ti_{0.94}Mo_{0.06}O_2$ Samples, and calculated substitutional Mo (model II, Mo_{Ti}).

Sample	Coordination	Coordination numbers（N）	R(Å)	σ^2(Å^2)
Theoretic	Ti-O	4	1.92	0
Anatase TiO_2		2	1.97	0
Anatase TiO_2	Ti-O	3.9	1.90±0.01	0.0031±0.0001
		2	1.95±0.01	0.0035±0.0001
$Ti_{0.97}Mo_{0.03}O_2$	Mo-O	3.9±0.1	1.81±0.02	0.0045±0.0001
		1.9±0.1	1.86±0.01	0.0044±0.0002
$Ti_{0.94}Mo_{0.06}O_2$	Mo-O	3.9±0.1	1.80±0.01	0.0050±0.0003
		1.9±0.1	1.85±0.01	0.0050±0.0002
Calculated	Mo-O	4	1.85	0
model Mo_{Ti}		2	1.90	0

4. Conclusions

Using XAFS spectroscopy combined with XRD and theoretical calculation, we studied Mo doped TiO_2 samples with different Mo concentration by co-deposition method. The detailed XANES and EXAFS analysis shows that all of the Mo atoms substitute Ti atoms as Mo^{6+} in the anatase TiO_2 lattice at the higher Mo concentration of 6 atom%. The high electron density of Mo doping causes ~ 0.1 Å shrinkage of Mo-O bond length, which is consistent with the results of calculated model structure of Mo_{Ti}.

Acknowledgments

This work was supported by the National Basic Research Program of China (No. 2012CB825801 and 2012CB825801), the National Natural Science Foundation of China (No. 11135008, 10725522, 10905058, and 11179004). The Authors would like to thank the SSRF and NSRL facilities for the beam time. A special acknowledgement is due to Prof. Yuying Huang, Prof. Zheng Jiang, Dr. Lina Li, and Mr Hongliang Bao for their invaluable support.

References

[1] Anpo M 1997 *Catalysis Surveys from Japan* **1** 169
[2] Asahi R, Morikawa T, Ohwaki T, Aoki K and Taga Y 2001 *Science* **293** 269
[3] Chen X B and Burda C 2008 *J Am Chem Soc* **130** 5018
[4] Choi W Y, Termin A and Hoffmann M R 1994 *J Phys Chem-Us* **98** 13669
[5] Devi L G and Murthy B N 2008 *Catalysis Letters* **125** 320
[6] Fujishima A and Honda K 1972 *Nature* **238** 37
[7] Gratzel M 2001 *Nature* **414** 338
[8] Jeon M S, Yoon W S, Joo H, Lee T K and Lee H 2000 *Appl. Surf. Sci.* **165** 209
[9] Khan S U M, Al-Shahry M and Ingler W B 2002 *Science* **297** 2243
[10] Marczak R, Werner F, Ahmad R, Lobaz V, Guldi D M and Peukert W 2011 *Langmuir* **27** 3920
[11] Schrier J, Demchenko D O and Wang L-W 2007 *Nano Lett* **7** 2377
[12] Stengl V and Bakardjieva S 2010 *J Phys Chem C* **114** 19308
[13] Stern E A, Newville M, Ravel B, Yacoby Y and Haskel D 1995 *Physica B* **208** 117
[14] Wang D H, Jia L, Wu X L, Lu L Q and Xu A W 2012 *Nanoscale* **4** 576
[15] Yu X H, Hou T J, Sun X H and Li Y Y 2012 *Chemphyschem* **13** 1514

Journal of Physics: Conference Series **430** (2013) 012091 doi:10.1088/1742-6596/430/1/012091

Structural and morphological characterization of Mo coatings for high gradient accelerating structures

Y Xu[1,2], B Spataro[1], S Sarti[3], V A Dolgashev[4], S Tantawi[4], A D Yeremian[4], Y Higashi[5], M G Grimaldi[6], L Romano[6], F Ruffino[6], R Parodi[7], C Caliendo[8], A Notargiacomo[9], G Cibin[10] and A Marcelli[1,11]*

[1]*INFN-Laboratori Nazionali di Frascati, Frascati, 00044, Italy*

[2]*Department of Chemical Physics, Anhui Key Laboratory of Biomass Clean Energy, University of Science & Technology of China, Hefei, 230026, China*

[3]*University of Rome Sapienza, Dipartimento di Fisica, 00185 Rome, Italy*

[4]*SLAC National Accelerator Laboratory, Menlo Park, CA 94025, USA*

[5]*KEK 1-1 Oho, Tsukuba, Ibaraki 305, Japan*

[6]*University of Catania, Dipartimento di Fisica e Astronomia & MATIS-IMM-CNR, 95123 Catania, Italy*

[7]*INFN-Genova, 16146 Genova, Italy*

[8]*ISC-CNR, Area della Ricerca di Roma 1, 00040 Monterotondo, Rome, Italy*

[9]*Istituto di Fotonica e Nanotecnologie CNR, Rome, 00156, Italy*

[10]*Diamond Light Source, Chilton, Didcot, Oxon OX110DE, U.K.*

[11]*NSRL, University of Science and Technology of China, Hefei 230026, China*

E-mail: marcelli@lnf.infn.it

Abstract. A series of Mo coatings grown on Al_2O_3 and Cu were prepared via sputtering method. XAS experiments were performed at the Mo K edge to characterize the coating structure and the chemical status of Mo atoms. From the XANES analysis we recognized that all Mo coatings on Al_2O_3 have a slightly disordered structure with a negligible Mo oxide contribution. Moreover, the chemical state is that of the Mo metal with a negligible oxygen contribution. Compared with Cu based materials, molybdenum is an attractive material for accelerator components and a reliable option for RF linear accelerating structures to achieve high accelerating gradients and low breakdown rates, two fundamental parameters for the next generation of linear accelerators.

Journal of Physics: Conference Series **430** (2013) 012091 doi:10.1088/1742-6596/430/1/012091

Introduction

The next generation of linear accelerators is highly demanding in terms of accelerating gradients, one of the main parameter of linear accelerators since it limits the accelerator length and affects the power consumption. The experience shows that obtaining high gradients with a normal conducting structure requires operation at a relatively high frequency. As a consequence to make possible the construction of the future high-energy linear accelerators and X-ray Free Electron Lasers (XFEL) and minimize costs, advanced high frequency accelerating structures are required. Significantly improved manufacture technologies and surface characterization methods are fundamental issues to achieve advanced cavities with performances well beyond the existing copper based device now in operation [1]. Molybdenum is a Group 6 chemical element with the atomic weight of 42 g/mole. Pure molybdenum is characterized by a high melting point, a hot strength and a creep resistance. The pure element has a *Mohs* hardness of 5.5, accounts for the sixth-highest melting point (2623 °C) of any element and easily forms hard, stable carbides. Although it has one of the lowest coefficients of thermal expansion among commercially used metals, molybdenum burns just above 600°C [2]. As a consequence, molybdenum appears as an interesting material for accelerator components and a possible option for RF linear accelerating structures with low breakdowns at high RF power. In particular, if compared with Cu for applications in high gradient accelerating structures, the results of Mo breakdown rate are very promising [1]. A series of Mo coatings grown on Al_2O_3 with different thickness were prepared via the sputtering method at the Laboratori Nazionali di Frascati (LNF) and at ISC-CNR at Monterotondo (Rome). The deposited films are thick metallic disordered layers with different resistivity values. In this contribution, in order to investigate materials for future devices with improved RF performances, we will describe a characterization method of these different molybdenum coatings.

Experiment

Different methods are typically used to characterize the properties and the structure of metallic thin films or coatings. To characterize our Mo coatings we used non-conventional methods. The first is an imaging technique with a high spatial resolution recently introduced, that involves the use of a Focused Ion Beam (FIB) microscope [3] to obtain morphological and dimensional information, the second is a spectroscopic technique: the X-ray Absorption Spectroscopy (XAS), used to investigate the Mo structure at the local atomic level.

Focused Ion Beam (FIB) microscope: A FIB microscope can be usefully considered to characterize metallic coatings. This non-conventional microscopy technique now available has imaging and micromachining capabilities at the nm-μm scale, two characteristics nowadays extremely important and widely used in materials science. Actually, in addition to imaging, a FIB can be also used to prepare extremely thin and oriented crystal sections. Thanks to its versatility and spatial resolution, FIB was employed here to visualize with high spatial resolution the coating morphologies and the cross sections of these films, to evaluate the thickness of the investigated Mo coatings. A typical FIB image of one of the Mo films grown on Al_2O_3 (sample B5) is shown in Figure 1. It refers to a film with a thickness of ~ 615 nm. The thickness of the investigated coatings is reported in Table 1. The FIB characterization has been performed at the LIME laboratory of the *Roma Tre* University.

XAS (X-ray Absorption Spectroscopy): The XAS (X-ray Absorption Spectroscopy) technique is a well-established method capable to perform a quantitative analysis of local structural properties such as geometry and coordination numbers in ordered and disorder systems such as liquids, glasses, coatings or interfaces [4]. To characterize the chemical status of Mo atoms in our samples (see Table 1), we performed X-ray Absorption Spectroscopy experiments at the Mo K edge. XAS experiments were carried out at B18, the Core XAS beamline of the Diamond Light Source (UK). This synchrotron radiation source operates at the electron energy of 3 GeV with ~250 mA current, in the top-up mode. Data were collected using a double crystal monochromator equipped with two Si(111) crystals and coupled to a Pt coated mirror focusing radiation in a spot of ~200x200 μm [2]. Acquisition of Mo K-edge spectra has been performed in the continuous scan mode using a 9-element Ge detector with XSPRESS-II acquisition electronics in the fluorescence detection mode. Experiments have been performed both at normal incidence and at grazing incidence to enhance the signal associated to thin surface layers and in different areas of the Mo coatings to probe the homogeneity of the investigated samples.

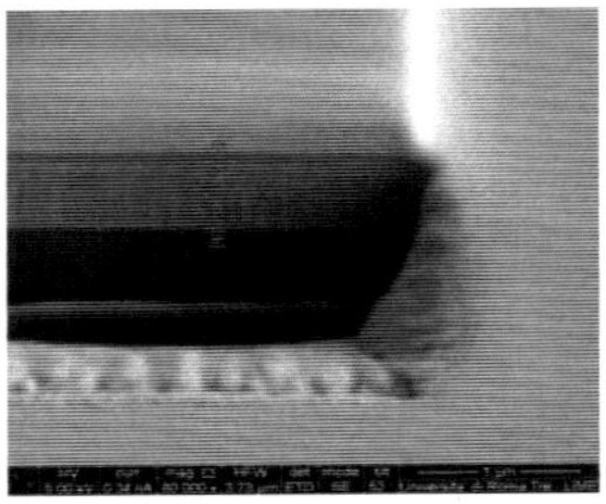

Figure 1 FIB image of the Mo/Al_2O_3 coating (B5).

Table 1 Measured thickness of all investigated Mo coatings.

Sample	Mo coatings	Thickness /nm[a]
A1	Mo/ Cu	630
A2	Mo/Cu	1300
B1	Mo/Al_2O_3	70-75
B2	Mo/Al_2O_3	130-140
B3	Mo/Al_2O_3	205-225
B4	Mo/Al_2O_3	310
B5	Mo/Al_2O_3	615
B6	Mo/Al_2O_3	1030

[a]measured with a Focused Ion Beam (FIB)

Results and discussions

XAS (X-ray Absorption Spectroscopy)

In figure 2a we compared XAS spectra of different Mo coatings with a Mo metallic film. Spectra were normalized by fitting linear polynomials in the pre- and post-edge regions [5]. Looking at the comparison of X-Ray Absorption Near Edge Spectroscopy (XANES) data between Mo coatings and Mo metallic film measured in the transmission geometry we may claim that all coatings have a slightly disordered structure. From the comparison with the metallic Mo film, the observed energy shift indicates that all coatings have a negligible oxygen contribution. Furthermore, spectra are similar except for the sample with a thickness of 1300 nm, which shows a significant Mo oxide contribution, e.g., MoO_2 and/or MoO_3 [6]. To better analyze XANES spectra we plot the edge energy shift relative to the metallic Mo film absorption edge vs. thickness. For all samples the edge has been determined from the normalized absorption at the height corresponding to the value of 0.8. As shown in Figure 2b, for samples grown on Al_2O_3 (B1-B6), the position of the edge shifts linearly towards higher energies. The mechanism is possibly associated to a shift of the Fermi level increasing the film thickness, while EXAFS data (not shown here) point out that the Mo-Mo distance of the main contribution decreases with thickness. Samples have not been annealed and for the thinner samples a clear distortion of the unit cell with the presence of a lower distance component occurs. For samples grown on Cu (A1-A2), XANES spectra are slightly different (in particular for the thicker samples) and the measured energy shift position shows that those samples are well outside the trend associated to the samples grown on Al_2O_3. We explain this behaviour with an expansion of the nest-nearest distance around Mo atoms due to oxygen atoms [6].

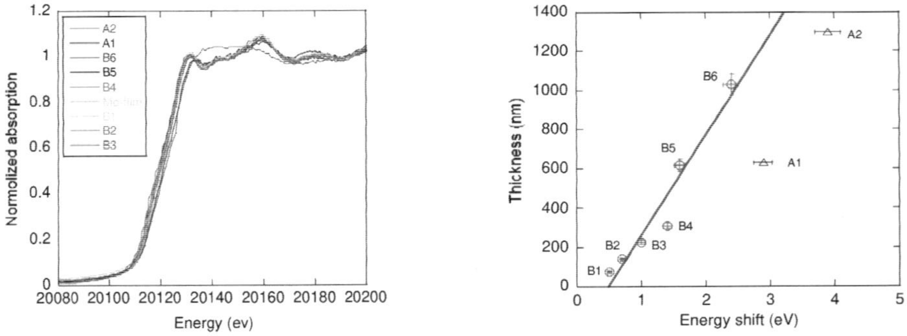

Figure 2 (a) Normalized Mo K-edge XANES spectra of coated samples (Table 1); (b) behaviour of the edge shift vs. thickness of Mo films on Al_2O_3. Cu samples are well outside the linear behaviour.

Conclusions

Large efforts to grow and characterize Mo coatings on Al_2O_3 have been done to optimize performance of coating materials required by the extremely demanding next generation of accelerating structures. In particular, in this contribution we present a XAS characterization of a series of Mo coatings obtained via the sputtering method. As shown by the experimental results we presented, the sputtering method is really a promising approach to obtain homogenous Mo coatings increasing the performances of RF cavities working at high frequencies. Structural properties of these sputtered coatings have been determined by a combined analysis of FIB imaging and XAS spectroscopy. We used a FIB microscope to visualize at high spatial resolution the morphology of these films and to accurately measure their thickness. XAS spectra were used to characterize the chemical state of Mo coatings, their local structure and the possible presence of Mo oxides. Data show that all coatings are made by metallic molybdenum with negligible oxygen contributions. Measurements have shown that modifying the thickness of the films both structural and electronics changes occur. While the Mo bcc structure close to the Mo/substrate boundary is naturally deformed, the Fermi level shifts to higher values. This has direct consequences on the transport properties of the coating material [7]. XAS spectroscopy is really a powerful method for coatings characterization not only because allows to identify the oxidation state but also because it may easily probe order/disorder composition in metallic coatings of different thickness. Because dedicated RF devices with such coatings have to be manufactured and tested at high power, work is still in progress to improve both quality and performances of Mo coatings. A full characterization of conductivity properties and of the behaviour under high fields is under way for these systems and will the subject of a forthcoming publication. Moreover, to optimize coatings performances on film grown on copper, deposition with other metals, different manufacturing and characterization methods are in progress. This extensive characterization is required to improve the RF breakdown performance and a lot of work is really necessary to identify reliable procedures of the next generation of accelerating devices.

Acknowledgments

We thank the Diamond Light Source for access to beamline B18 (Proposal nt1984-1). One of us (X.Y.) gratefully acknowledges the *Joint PhD program* of the *Chinese Scholarship Council*. Thanks are also due to S. Bini and P. Chimenti for technical support.

References

[1] Bini S, Chimenti V, Marcelli A, Palumbo L, Spataro B, Dolgashev V, Tantawi S, Higashi Y, Grimaldi M G, Romano L, Ruffino F, Parodi R and Yeremian A D (2012) Development of X-band accelerating structures for high gradients, Chin. Phys. C **36,** 639-647.

[2] Bini S, Chimenti V, Marcelli A, Palumbo L, Spataro B, Dolgashev V, Tantawi S, Yeremian A D, Higashi Y, Grimaldi M G, Romano L, Ruffino F and Parodi R (2011) Development of X-Band Accelerating Structures for High Gradients, SPARC-RF-11/004

[3] Orloff J, Utlaut M and Swanson L (2003) High Resolution Focused Ion Beams: FIB and its Applications (New York: Kluwer Academic/Plenum Publishers) Chapt. 6 pp 205-279.

[4] Bianconi A, Garcia J, Marcelli A, Benfatto M, Natoli C R and Davoli I, 1985, J. Phys. Coll. **9,** 101-106.

[5] Mottana A, Cibin G and Marcelli A (2004) in: Spectroscopy Methods in Mineralogy (Vienna), eds. Beran A and Libowitzky E (Eötvös University Press, Budapest) pp. 62-95.

[6] Ressler T, Wienold J, Jentoft R E and Neisius T (2002) Bulk structural investigation of the reduction of MoO_3 with propene and the oxidation of MoO_2 with oxygen, J. Catal. **210**, 67-83.

[7] Rocca F, Kuzmin A, Mustarelli P, Tomasi C and Magistris A, 1999, XANES and EXAFS at Mo K-edge in $(AgI)_{1-x}(Ag_2MoO_4)_x$ glasses and crystals, Solid State Ionics **121,** 189-192.

Journal of Physics: Conference Series **430** (2013) 012092 doi:10.1088/1742-6596/430/1/012092

XAFS Study of Arsenical Nickel Hydroxide

N. Chen[1], E. Kim[2], Z. Arthur[2], R. Daenzer[3], J. Warner[1], G.P. Demopoulos[3], Y. Joly[4] and D.T. Jiang[2,*]

[1] Canadian Light Source, University of Saskatchewan, Saskatoon, SK, Canada
[2] Department of Physics, University of Guelph, Guelph, ON, Canada
[3] Department of Materials Engineering, McGill University, Montreal, QC, Canada
[4] Laboratoire de Cristallographie, Centre National de la Recherche Scientifique, Boı^te Postale 166, F-38042 Grenoble Cedex 9, France

*Corresponding author, email: djiang@uoguelph.ca

Abstract To Investigate the role played by nickel co-ions in contributing to the stability of arsenic, fluorescence XAFS measurements at both arsenic K-edge and nickel K-edge, respectively, on amorphous arsenical nickel hydroxide, crystalline arsenical nickel hydroxide, and annabergite reference compounds have been carried out. The XAFS results indicate that the arsenic-bearing nickel hydroxides have a well-defined arsenic local structure with multiple coordination shells, suggesting a compound formation mechanism instead of surface adsorption. The degradation of the arsenic local structure in the crystalline arsenical nickel hydroxide is observed. The XAFS of annabergites are compared to that of the arsenical nickel hydroxide and possible structural models are discussed.

1. Introduction

Coprecipitation of arsenic(V) with ferric iron from hydrometallurgical solutions is widely used for the removal and immobilization of arsenic in uranium mine tailings. Other common co-ions in the industrial solutions such as nickel and aluminum have been found enhancing the arsenic retention under certain coprecipitiaon conditions [1]. In order to understand the process of arsenate coprecipitation with nickel hydrides under the industrially relevant geochemical conditions, we investigate several laboratory prepared reference compounds formed under the similar conditions of the uranium mine tailings neutralization process. The samples include two types: arsenical nickel hydrides and annabergites, and in each group samples of different long range ordering are studied. The analysis of the data are mainly based on the FEFF [2] modeling of EXAFS in the k-space and the FDMNES [3] modeling in the XANES of these samples.

2. Experimental

Five samples were prepared: amorphous arsenical nickel hydroxide (sample-1), crystalline arsenical nickel hydroxide (sample-2); amorphous annabergite, and two crystalline annabergite samples with different crystallinity: sample-4 and sample-5, respectively. Synthesis of the amorphous arsenical nickel hydroxide was performed by adding aqueous arsenate into a nickel sulfate solution, with a final Ni:As molar ratio of 4:1 [4]. Sodium hydroxide was used to maintain a pH of 8 during this process. Crystalline arsenical nickel hydroxide was synthesized using a nearly identical method: in place of nickel sulphate, a commercially-available crystalline nickel hydroxide (theophrastite) was used. Amorphous annabergite was synthesized by separately dissolving nickel sulphate and arsenate, then adding the two solutions together at pH 2 with a Ni:As ratio of 1.5:1 [5]. This combined solution was then fed through a pump into a vessel containing pH 7 de-ionized water; the pH was kept constant by the addition of sodium hydroxide. Crystalline annabergite was produced by first following the method for synthesizing amorphous annabergite. After pumping the Ni-As solution into the pH 7 de-ionized water vessel, the mixture was then filtered, and the filtrate was dried and re-suspended in a reactor with pH 7 de-ionized water. The reactor was then heated to 95 °C for 120 hours.

The Ni and As K-edge XAFS data were collected at the HXMA beamline [6] at the Canadian Light Source. Mono crystals Si(220) and Si(111) were used, respectively, and the beamline Rh mirrors were used for both experiments. Data collection was performed in transmission mode for concentrated samples or in florescence mode by using a 32 element Ge detector. The energy calibration was made by using metal foils downstream to the sample.

3. Results and Discussion

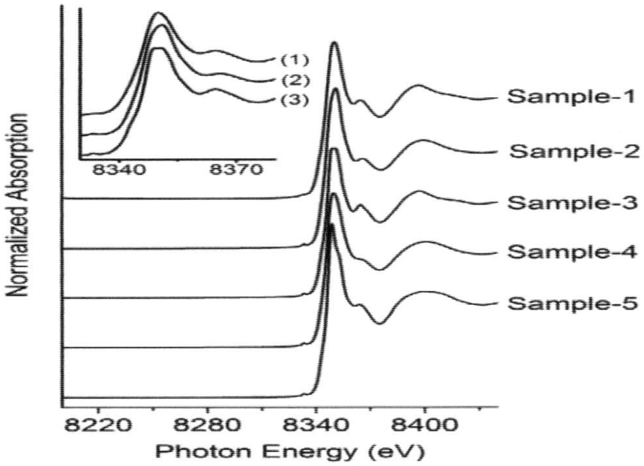

Figure 1. Ni K-edge XANES. (1) crystalline nickel hydroxide; (2) amorphous arsenical nickel hydroxide; (3) crystalline arsenical nickel hydroxide; (4) amorphous annabergite; (5) crystalline annabergite, respectively. The insert highlights the difference in the white line peak shape between sample nickel hydroxide samples with and without arsenic.

Figure 1 shows the Ni K-edge XANES for nickel hydroxide reference data [7] and for sample 2-5 measured in this study. A general similarity exists among these data presumably reflecting the presence of the common NiO_6 octahedron. Close examination reveals the differences between the arsenic-bearing samples (sample 2-5) and the crystalline nickel hydroxide (sample-1). First of all, compared to the pure nickel hydroxide (sample-1) the arsenic-bearing samples all have a significantly more prominent pre-edge feature at 8333.1eV due to the distortion of the NiO_6 octahedron; secondly, they have their whiteline peak either being asymmetric (Fig. 1 insert trace-2) or split into multiple peaks (Fig. 1 insert trace-3). These indicate that the Ni in the arsenical nickel hydroxides have multiple types of Ni local environments in contrast to the single type in the pure nickel hydroxide. Given that the molar ratio of Ni:As is 4:1 in these samples, these observations indicate the sensitivity of the Ni K-edge XANES to the presence of arsenic in the samples, as not all Ni are influenced by the As presence.

The EXAFS at Ni and As K-edges of the samples are shown in Fig. 2. The presence of the low level arsenic (126.22 μmol/g) in the crystalline arsenical nickel hydroxide appears to have a minimal structural impact to the nickel hydroxide matrix: the EXAFS of the crystalline arsenical nickel hydroxide (sample-3) is virtually identical to that of the pure nickel hydroxide in literature [7]. On the other hand, the EXAFS of the amorphous arsenical nickel hydroxide (sample-2) has clear deviations from that of pure nickel hydroxide, e.g. the shifting of feature B (Fig. 2a) and the smearing in the region D (Fig. 2a). These differences are attributed as due to the presence of relatively heavy arsenic loading in this sample (1612.29 μmol/g) and the size of the crystallites in the sample (see below).

Interestingly, these differences also make the amorphous arsenical nickel hydroxide (sample-2) resembling to a certain extent that of the amorphous annabergites (sample-4). FEFF modeling (not shown) based on the nickel hydroxide crystallographic data indicates that when the crystallite size is about 5.7 Ă the calculated EXAFS would qualitatively reproduce all the features of the EXAFS of sample-2 (Fig. 2a), suggesting a morphological interpretation for the XRD "amorphous" nature of the sample-2.

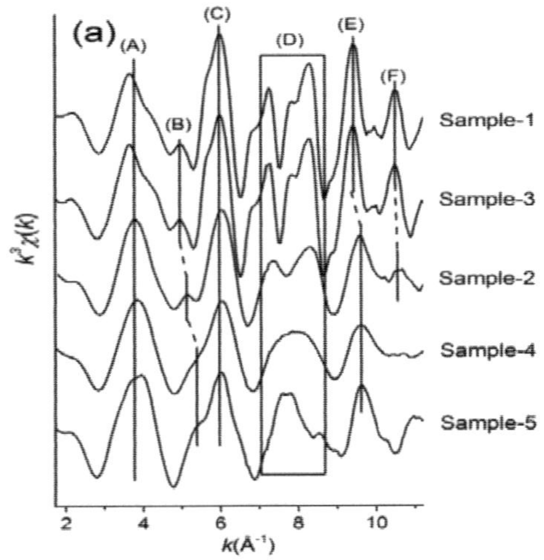

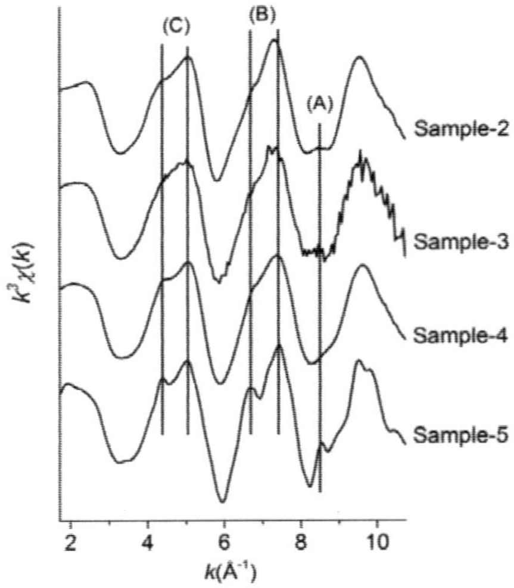

Figure 2. EXAFS $k^3\chi(k)$. (a) Ni K-edge data for sample 1-5; (b) As K-edge data or sample 2-5.

Figure 3b shows the As K-edge $k^3\chi(k)$ of the four arsenic-containing samples. The first noticeable observation is that the arsenical nickel hydroxides (Fig. 3b, sample 2-3) are strikingly similar to those of annabergites (Fig. 3b sample4, 5), regardless of their long range order properties. Secondly, the multiple shell nature of the EXAFS of the arsenical nickel hydroxide (Fig. 3b smaple-2, 3) indicates that the arsenate reacts with the nickel hydroxide leading to multiple coordination shells of arsenic local structure or a compound formation. Combined these observations we conclude that the nickel co-ion seems to lead to formation of an annabergites-like compound in the coprecipitation with arsenic(V) at around pH 8.

To further understand the nature of the arsenical nickel hydroxide we carried out FDMNES [3] simulations of the annabergites experimental XANES data. Figure 4 shows a comparison between the FDMNES results and the As K-edge experimental data. First, from the results in the left and the middle panel of Fig. 4 a clear correlation between the cluster sizes used in the calculation and the XANES features beyond the whiteline. When compared to the experimental data shown in Fig. 4 (right panel) it is apparent that the two crystalline annabergites samples posses both of the feature A and B (Fig.4 middle), this is consistent with their expected large crystallite sizes. On the other hand, for the amorphous annabergite (Fig. 4 right panel: sample-4) the feature B is completely missing; it can be shown that simple broadening of the calculated data for model 4 (Fig. 4 middle, 6.1 Ă cluster size) will not be able to account for the spectral different near feature B. Therefore, these FDMNES calculations appear to be a rather sensitive method in predicting the size of the crystallite in a co-precipitated compound, if a structural model exists.

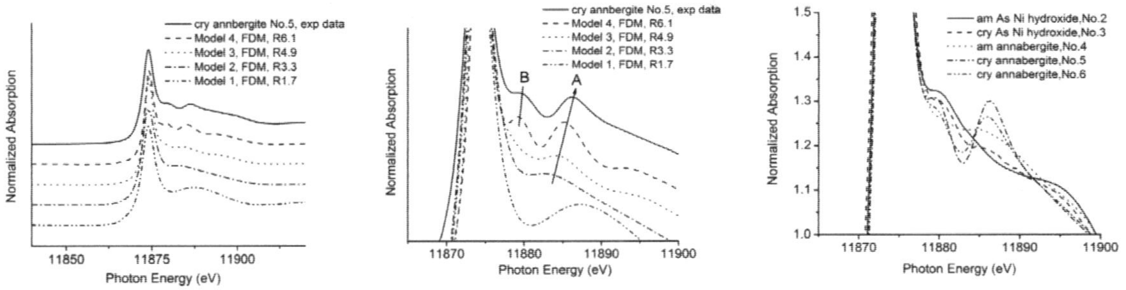

Figure 4. FDMNES calculation for annabergite XANES. *Left panel*: Experimental XANES of crystalline annabergites (solid) and a series of FDMNES calculations (Model 1 to 4) with increasing cluster sizes, 1.7 Ǻ, 3.3 Ǻ, 4.9 Ǻ, and 6.1 Ǻ. *Middle panel*: a close-up view of the same data in the left panel, in the region near the whiteline. Notice that the feature A of the calculated spectra appears first appears when the first As-Ni shell is included in the calculation, then gradually approaches the correct photo energy as the cluster size increases; while the feature B does not appear until the cluster includes the third group of As-Ni interatomic distance at 6.06 Ǻ for the annabergite structure. *Right panel*: Experimental XANES for the samples 2-6 in the same region as that in the middle panel (see text for more discussion).

Finally, for the arsenical nickel hydroxides (Fig. 4 right: sample 2 and 3), two preliminary observations can be made: first, detail XANES comparison in the shown region indicates that the crystalline arsenical nickel hydroxide (Fig.4 right: sample 3) possesses both the A and B features of annabergite but at rather different branching ratio; second, the amorphous arsenical nickel hydroxide (Fig. 4 right: sample 2) does not have the feature A as that in annabergite. These observations reveal the subtle, presumably longer range structural differences exist between the two groups of the compound, even though the EXAFS (Fig. 3b) strongly suggest a similar local atomic structure.

4. Conclusions

Arsenical nickel hydroxide formed under the similar geochemical conditions as that in typical uranium mine tailings neutralization process possesses a well-ordered arsenic local structure which bears strong resemblance to that in the annabergite; FDMNES calculations successfully simulate the experimental XANES of annabergite; the FDMNES results identify XANES features that can be used as indicators for specific atomic cluster sizes at nanoscale.

Acknowledgements

The research described was supported by NSERC of Canada through a CRD grant to GPD and DTJ. The experiments were performed at the Canadian Light Source, which is supported by the NSERC of Canada, the NRC Canada, the CIHR, the Province of Saskatchewan, Western Economic Diversification Canada, and the University of Saskatchewan.

References

[1] De Klerk R.J., et al., 2011, Hydrometallurgy, 111-112, 65.
[2] Rehr, J. J. and Albers, R.C., 2000, Rev. Modern Phys. 72: 621-654.
[3] Joly, Y., 2001, Phys. Rev. B 63, 125120-125129.
[4] Jia Y F, et al., 2006, Environ. Sci. & Technol., 40(10), 3248-3253.
[5] Yuan T.C., et al., 2005, Canadian Metallurgical Quarterly, **44**, pp. 449-456.
[6] Jiang D T, Chen N and Sheng W, 2007, AIP Proc. 879 (Accelerators and Beams), 800-803.
[7] Data from XAFS Data Library: http://cars9.uchicago.edu/cgi-bin/newville/Model-Search.cgi

EXAFS study of PZT ferroelectric thin films of different crystallinities

X P Hu[1], D W Duan[1], K Zhang[1], S Q Chu[2], J Zhang[2], Y N Xie[2], D Guo[3] and J L Cao[1]

[1]Institute for Advanced Materials and Technology, University of Science and Technology Beijing, Beijing 100083, China

[2]Beijing Synchrotron Radiation Facility, Institute of High Energy Physics, Chinese Academy of Sciences, Beijing 100049, China

[3]Institute of Acoustics, Chinese Academy of Sciences, Beijing 100190, China

E-mail: jlcao@mater.ustb.edu.cn

Abstract: Lead zirconate titanate ($PbZr_xTi_{1-x}O_3$, PZT) thin films were deposited on Pt/TiOx/SiO_2/Si substrate by RF-magnetron sputtering with a $Pb_{1.2}Zr_{0.3}Ti_{0.7}O_3$ target at room temperature. The Zr K-edge of PZT thin films of different crystallinities were studied by extended x-ray absorption fine structure (EXAFS) spectrum. Scans were made in fluorescence mode to monitor the structural differences between the amorphous thin films and the crystalline thin films. It was found that the local structure of the amorphous thin films was quite different from that of the crystalline thin films from the analyses of the EXAFS spectra. In the amorphous PZT thin films only one oxygen coordination shell can be found. By contrast, in the crystalline PZT thin films, the EXAFS spectrum was fitted with three shells and Zr-Pb bond was found. The oxygen coordination numbers of the amorphous and crystalline PZT thin films were same. The Zr-O bond distance of the amorphous samples were bigger than the crystalline samples. Thus, the influences of the phase transformation on the ferroelectric properties of PZT thin films were interpreted from the angle of the local chemical structures.

1. Introduction

Lead zirconate titanate ($PbZr_xTi_{1-x}O_3$, PZT) thin films have been of great interest for many years because of the excellent properties and promising applications in non-volatile ferroelectric memories, micro-electro mechanical system(MEMS), and piezoelectric ultrasonic motors [1,2]. Many deposition methods were used to fabricate PZT thin films, such as evaporation [3],RF-magnetron sputtering [4], sol-gel process [5], chemical vapor deposition (CVD) [6], metal organic decomposition (MOD) [7], pulsed laser deposition (PLD) [8] and so on.

Extended X-ray Absorption Fine Structure (EXAFS) spectroscopy is a highly sensitive technique with elemental specificity. As such, EXAFS is an extremely powerful technique to determine the chemical structure of materials, even with a very low abundance or concentration irrespective to the state of the samples [9,10]. By means of theoretical computations, the analysis of the EXAFS spectrum allows the determination of the coordination number, the bond distance, the "Debye-Waller"-like factor, and the nature of neighboring atoms surrounding the excited atoms. So far,

Journal of Physics: Conference Series **430** (2013) 012093 doi:10.1088/1742-6596/430/1/012093

EXAFS has been used in the research fields of environmental chemistry, medical science and so forth [11]. However, the use of EXAFS received little attention for the research of ferroelectric thin films.

In the present study, RF-magnetron sputtering was used to deposit the PZT thin films on $Pt/TiO_x/SiO_2/Si$ substrates [12]. The structures and ferroelectric properties of the PZT thin films of different crystallinities were discussed. Zr K-edge EXAFS spectra of PZT thin films of different crystallinities were collected and the local structure parameters of zirconium atoms were obtained. The relationship between the local structure and ferroelectric properties was discussed.

2. Experimental

The $Pb_{1.2}$ $(Zr_{0.3}Ti_{0.7})$ O_3 target was used because of the lead lose during the sputtering deposition process. PZT thin films of about 200nm were deposited on the $Pt/TiO_x/SiO_2/Si$ substrates. Before each deposition, the chamber was evacuated to a base pressure lower than 8×10^{-4} Pa. The as-deposited PZT films were amorphous while the PZT films become crystallized upon 3min annealing in pure oxygen at 700°C. Pt top electrodes were deposited by sputtering. The crystal structure of the PZT thin films was examined by X-ray diffraction (XRD, MSALXD-3) using Cu Kα radiation and graphite monochromator. The surface structure of thin films was characterized using atomic force microscopy (SPM 5500, Agilent). The hysteresis properties were examined at 5V and 100 kHz by an aixACCT TF analyzer 2000 system.

EXAFS spectra at Zr K-edge of PZT thin films were collected in fluorescence mode at the 1W1B beam line in Beijing Synchrotron Radiation Facility (BSRF). The storage ring was operated at 2.5GeV and 200mA. All the experiments were carried out at room temperature.

Fluorescence EXAFS spectra were summed, calibrated and background subtracted using the Athena program which is implanted in IFEFFIT [13]. Best fits were achieved using Artemis program, also a part of IFEFFIT. The Fourier transformations (FT) of $k^3\chi$ (k) from K space into R space were performed in the 3-9$Å^{-1}$ range using a Hanning window. The R-range used for fitting was between 1.0 and 4.0 Å. The bond distance, coordination numbers, nature of neighboring atoms surrounding the excited atoms were obtained.

3.Results and discussion

3.1. X-Ray Diffraction and AFM

XRD spectra of the amorphous and crystalline PZT thin films are shown in Figure 1. In the figure, PZT (100), (101)/(110), (111) diffraction peaks showed in the crystalline PZT thin films, while no PZT diffraction peak appeared in the amorphous samples. It means that the perovskite structure was formed without any observable secondary phases in the crystalline PZT films which showed a preferential orientation of (111) and are better crystallized than the amorphous films. And the XRD results indicated that the crystalline films exhibit a tetragonal structure.

AFM pictures of the amorphous and crystalline PZT thin films as shown in Figure 2 provide a visualization of the change of surface microstructure. The film morphology changes from smooth with no grain structure (amorphous) to an fine-grained structure (crystalline). The roughness of the annealed sample is 12.9 nm while the amorphous one is 8.47 nm. It is obvious that after crystallized, the roughness of the samples is increased.

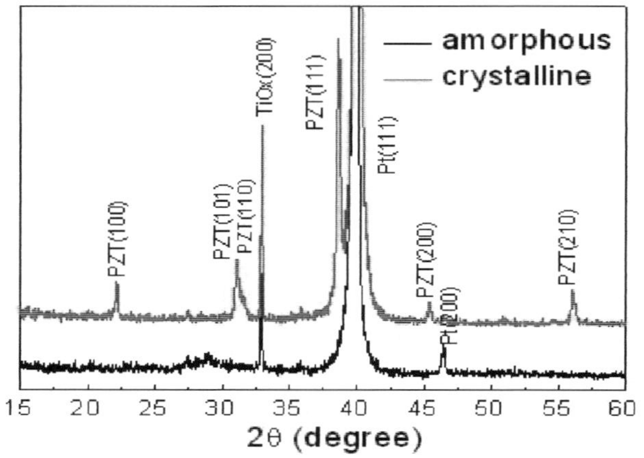

Figure 1 XRD spectra of amorphous and crystalline PZT thin films

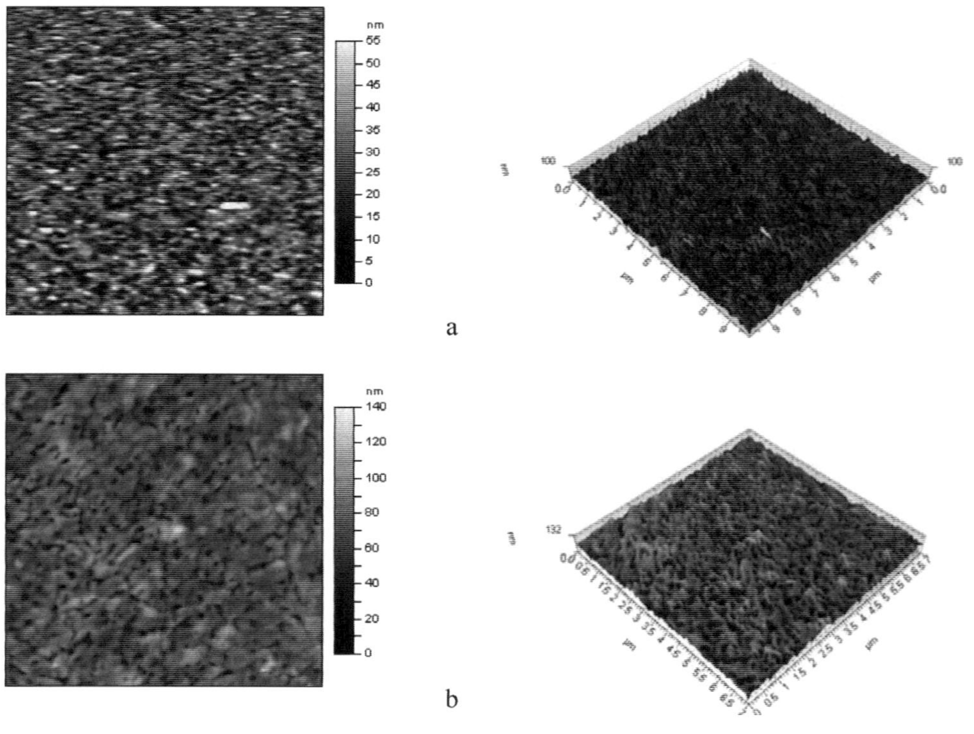

Figure 2 AFM images of the amorphous (a) and crystalline (b) PZT thin films

3.2 P-V properties

The P-V hysteresis loops of the PZT thin films are shown in Figure 3. It can be seen that the typical P-v hysteresis loop of the PZT thin films showed in the crystalline films. The amorphous films did not have ferroelectric properties. The result is consistent with XRD spectra.

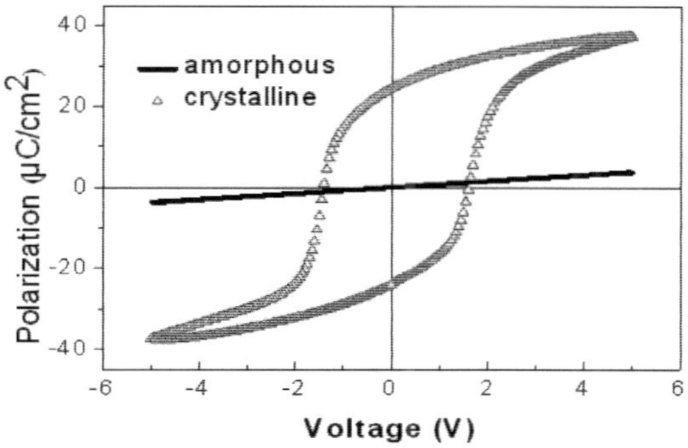

Figure 3 P-V hysteresis loops of the amorphous and crystalline PZT thin films

3.3 EXAFS Analyses

The EXAFS measurements in fluorescence mode of Zr K-edge (17998.0eV) of the PZT thin films samples were performed at the 1W1B beam line in BSRF.

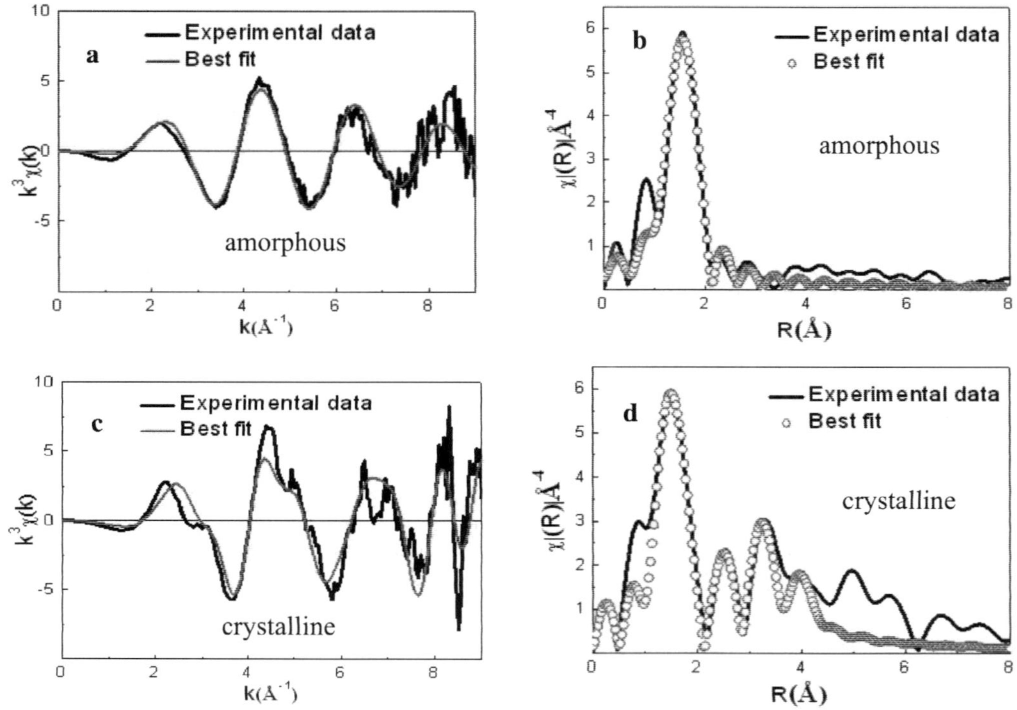

Figure 4 Experimental and calculated $k^3\chi$ (k) functions (a, c) and their Fourier transformation (b, d) of the PZT thin films (Zr K-edge)

Figure 4 shows the experimental and calculated $k^3\chi$ (k) functions and Fourier transform curves obtained from the Zr K-edge EXAFS spectra. As can be seen, the first peaks of the Fourier transform curves of amorphous and crystalline PZT thin films between 1 and 2.2Å present a high intensity and mainly formed by contributions from single path Zr-O. The coordination number N and the nearest Zr-O bond length R were evaluated via best fit of $k^3\chi$ (k), calculated by inverse Fourier transform. According to the calculation, the Zr-O distances of the crystalline samples are shorter than in the amorphous samples. However, the oxygen coordination numbers are nearly the same. Local structure parameters of PZT from Zr K-edge EXAFS spectra are shown in Table 1.

In addition, other intense peaks between 2.2 and 4Å are also observed in the Fourier transform curve of the Zr K-edge EXAFS spectra of the crystalline samples. It does not shown in the Fourier transform curve of the amorphous samples (Fig.4(b) and (d)). In the amorphous PZT thin films only one oxygen coordination shell can be found in the EXAFS spectra at the Zr K-edge. According to calculations, these peaks are mainly formed by contributions from single scattering path Zr-Pb and from some multiple scattering paths. It means that significant differences of the local structure exist between the amorphous and crystalline PZT thin films.

Table 1. Local structure parameters of PZT thin films from Zr K-edge EXAFS spectra.

		R (Å)	N	σ^2 (Å)
crystalline	Zr-O	2.09±0.03	6	0.00686
	Zr-Pb	3.12±0.05	4	0.00625
	Zr-Pb	3.85±0.06	4	0.00286
amorphous	Zr-O	2.14±0.02	6.1±0.2	0.00958

4. Conclusions

We measured Zr K-edge EXAFS spectra for amorphous and crystalline PZT thin films. The structure differences were discussed from the viewpoint of local structure. According to the results , the Zr-O distances of the crystalline samples are shorter than in the amorphous samples whereas the oxygen coordination numbers are nearly the same. Other intense peaks between 2.2 and 4Å are only observed in the Fourier transform curve of the crystalline samples.

According to the comparisons between the EXAFS spectra at the Zr K-edge, the local structure of PZT thin films of different crystallinities are obviously different. The analysis of our EXAFS spectra can be consistently interpreted with the result of XRD, and demonstrated that the ferroelectric property change of PZT films are closely related to the structure changes.

Acknowledgements

The authors acknowledge the support of National Natural Science Foundation of China (10979013), Beijing Natural Science Foundation (2092020), Beijing Nova Program (2007B025) and Fundamental Research Funds for the Central Universities (FRF-SD-12-027A).

References

[1] Mesquita A, Michalowicz A, Roberto V 2009 J . Phys ,**190** 012081
[2] Li J H, Wang C H, Huang X 2006 Chin.J.Semi, **27** (10): 1776-1780

[3] Oikawa M and Toda K 1976 Appl. Phys. Lett., **29** 491.
[4] Castellano R 1980 Ferroelectrics, **28** 387.
[5] Budd K D, Dey S K and Payne D A 1985 Proc. Brit.Ceram. Soc., **36** 107.
[6] Peng C H and Desu S B 1992 AppL Phys. Lett., **61** 1.
[7] Zhu W, Vest R W, Tse M S, Rao M K and Liu Z Q 1994 Materials in Electronics, **5** 173.
[8] Chen X, Kingon A. I, Mantese L, Auciello O.,and Hsieh K Y 1993 Integrated Ferroelectrics, **3** 355.
[9] Sato K, Miyanaga T, Ikeda S and Diop D 2005 Physica Scripta. Vol. T**115** 359–361
[10] Limpijumnong S, Rujirawat S, and Boonchun A 2007 Appl. Phys. Lett., **90** 103113
[11] Gurman S J1982 J. Mater.Sci **17** 1541-1570
[12] Zhang B S, He J, Zhang T J 2006 Electronic components and materials **25** N3
[13] Newville M 2001 J.Synchrotron Rad .**8** 322-324

Journal of Physics: Conference Series **430** (2013) 012094 doi:10.1088/1742-6596/430/1/012094

Organochlorines in surface soil at electronic-waste wire burning sites and metal contribution evaluated using quantitative X-ray speciation

Takashi Fujimori[1,2], Hidetaka Takigami[1] and Masaki Takaoka[2]

[1] Center for Material Cycles and Waste Management Research, National Institute for Environmental Studies (NIES), 16-2 Onogawa, Tsukuba, 305-8506, Ibaraki, Japan
[2] Department of Environmental Engineering, Graduate School of Engineering, Kyoto University, Katsura, Nisikyo-ku, 615-8540, Kyoto, Japan

E-mail: fujimori.takashi.3e@kyoto-u.ac.jp, fujimori.takashi@nies.go.jp

Abstract. Heavy metals and toxic chlorinated aromatic compounds (aromatic-Cls) such as dioxins and polychlorinated biphenyls (PCBs) are found at high concentrations and persist in surface soil at wire burning sites (WBSs) in developing countries in which various wire cables are recycled to yield pure metals. Chlorine K-edge near-edge X-ray absorption fine structure (NEXAFS) is used to detect the specific chemical form of Cl and estimate its amount using a spectrum jump in the solid phase. Quantitative X-ray speciation of Cl was applied to study the mechanisms of aromatic-Cls formation in surface soil at WBSs in Southeast Asia. Relationships between aromatic-Cls and chlorides of heavy metals were evaluated because heavy metals are promoters of the thermochemical solid-phase formation of aromatic-Cls.

1. Introduction

Toxic chlorinated aromatic compounds (aromatic-Cls) such as dioxins, polychlorinated biphenyls (PCBs), and chlorobenzenes are produced by various anthropogenic thermal processes. High concentrations of dioxins and PCBs are found in soil at wire burning sites (WBSs) [1]. Wire burning is performed in developing countries and relatively pure metals (especially copper) have been recovered by direct oil burning of various wires and cables from dismantled electronic-wastes on natural soil surfaces. Figure 1a shows a WBS at Vietnam visited of this study. Because exhausted gas and residual soil at WBSs can increase the health risks of workers and local citizens, knowing the total concentration and thermochemical behavior of aromatic-Cls at WBSs is necessary.

Chlorine K-edge near-edge X-ray absorption fine structure (NEXAFS) can be used to detect a specific chemical form of Cl [2,3] and to estimate Cl concentrations using a spectrum jump [4] in the solid phase. Studies have applied Cl K-edge NEXAFS to natural soils [2] and fly ashes in anthropogenic thermal processes [3]; however, characteristic soils at WBSs have not been studied using this technique. Quantitative X-ray speciation of Cl was applied to evaluate the mechanisms of aromatic-Cls formation in surface soil at WBSs in Southeast Asia. The relationship between aromatic-Cls and chlorides of heavy metals was further evaluated since heavy metals are a well-known promoter of the thermochemical solid-phase formation of aromatic-Cls [5]. Recently, Otsuka et al. analyzed copper chloride as a major chemical form of copper at WBS by using Cu K-edge X-ray absorption spectroscopy [6]. So, Cl K-edge NEXAFS also has the possibility of detection of copper chlorides.

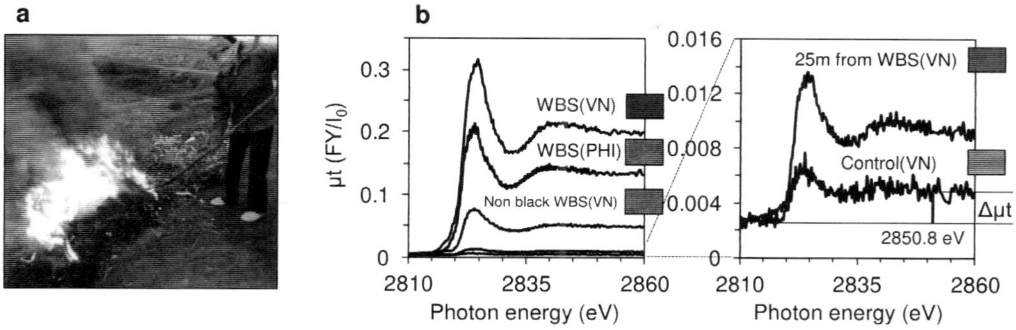

Figure 1 a, Wire burning site (WBS) at Vietnam. **b**, Raw spectra of Cl K-edge NEXAFS. Δμt was measured based on the height at 2850.8 eV from pre-edge baseline. Soil colors are shown in square-box photograph.

2. Materials and methods

Environmental soil samples were collected around Metro Manila, the Philippines (PHI), during August 2010, as well as Hanoi, Vietnam (VN), in January, 2011. Detailed sampling procedures were based on our previous study [7]. Environmental soil samples were selected as blackened soil at WBSs in VN and PHI after direct open-air wire burning, non-black soil at WBS similar to the ash color in VN, surface soil near the WBS (25 m from WBS) in VN, and control surface soil in VN. After drying the soil samples under air atmosphere for approximately 1 week, soil preparations were made. Fig. 1b shows representative soil colors. Powdered soil samples were mounted on carbon tape. Measurements of the Cl K-edge NEXAFS spectrum were taken for each sample between 2810 and 2860 eV detected by X-ray fluorescence using BL-11B at the Photon Factory in Tsukuba, Japan. We used "difference between intensity at 2850.8 and baseline at pre-edge" (denoted by "Δμt") of the raw Cl K-edge NEXAFS spectrum to quantify total chlorine concentration (mg/kg) in a soil sample by referring to a calibration curve (1–1,400 mg/kg) by Leri et al. [4], as shown in Fig. 1b. Furthermore, measurements were taken of a sample with high Cl concentrations (>10,000 mg/kg) using field-portable X-ray fluorescence spectroscopy (FP-XRF, Innov-X alpha; Innov-X Systems, Woburn, MA). High Cl concentrations had similar values between Cl K-edge NEXAFS and FP-XRF. Therefore, Δμt was a useful indicator of Cl quantification below nonmeasurable ranges using the XRF method. Ratios of Cl chemical forms in soil samples were analyzed by linear combination fitting (LCF) using inorganic chlorides (KCl, NaCl, $CaCl_2$, $MgCl_2$), metal chlorides [$Cu_2(OH)_3Cl$, $CuCl_2$, CuCl, $FeCl_3$, $FeCl_2{\cdot}4H_2O$ $PbCl_2$, $ZnCl_2$], and organic chlorides. Organic chlorides showed similar NEXAFS spectra depending on aliphatic and aromatic structures [2,3]. Representative chemicals were used such as 2,3-chlorophenol (aromatic) and polyvinyl chloride (aliphatic). To explore the variability of LCF results, five types of fitting ranges (2810–2830, ~2835, ~2840, ~2845, ~2850 eV) were examined. Next, the average and standard deviation of five types of fitting results were calculated. Fitting example with representative reference spectra is shown in supplementary Fig. S1. By combining the concentrations of total Cl using the Δμt ratio of chlorinated aromatic compounds by LCF, the concentration of aromatic-Cls was estimated. Detailed analytical procedures were described previously [3].

3. Results and discussion

Total Cl concentrations in blackened soils at WBSs showed the highest values of 21,000 mg/kg in at the WBS in VN and 16,000 mg/kg in the WBS in PHI, respectively, based on Δμt analysis (Fig. 1b), which suggests that the interregional similarity was due to informal wire burning activity between VN and PHI. In contrast, control soil (VN) had the lowest concentration of total Cl (120 mg/kg). Therefore, surface soil at the WBS was heavily polluted by chlorine derived from wire and cable insulation-coating materials. Surrounding surface soil at the WBS (25 m from WBS) in VN contained 720 mg/kg Cl. Therefore, contaminated Cl in blackened soil may not spread to the surrounding environment.

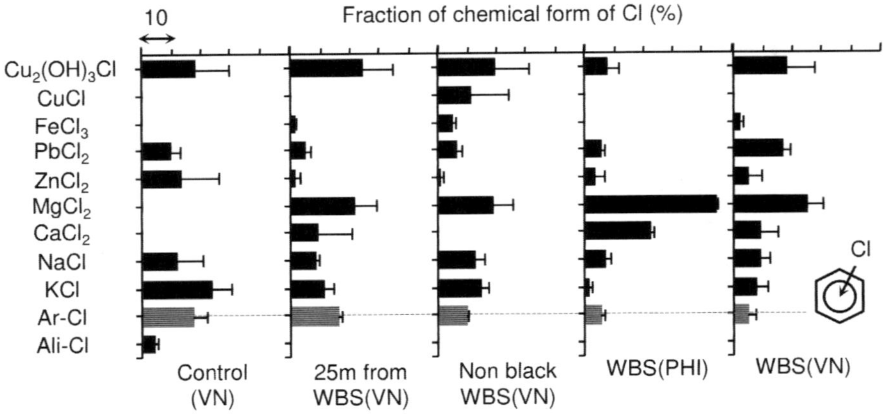

Figure 2 Fraction of Cl chemical forms at various sites. Error bars indicate standard deviation. Ar-Cl and Ali-Cl indicate chlorine connected with aromatic and aliphatic carbon, respectively.

Non-blackened soil at the WBS, such as the ash color in VN, contained high concentrations of total Cl (5,400 mg/kg); however, its concentration was lower than blackened soil at the WBS, partly due to the mixing of blackened soil with natural soil. Blackness of soil at the WBS was thought to be a useful indicator of pollution level of chlorine (and aromatic-Cls described below) as well as heavy metals [6].

Chemical forms of chlorine in soil samples at the WBSs were analyzed using LCF of Cl K-edge NEXAFS spectra. Chlorine was partially connected with aromatic carbon (5.2–18% of total Cl) in all types of surface soils, as shown in Fig. 2. According to the distribution pattern of 11 references, blackened soils at WBSs in PHI and VN showed a relatively similar pattern and were distinguished from other soils, which suggested that wire burning was performed using similar electronic-waste materials in PHI and VN. Control soil contained no $MgCl_2$ or $CaCl_2$, which were detected in blackened soils at the WBS by LCF and showed the highest ratio of Cl connected with aromatic carbon [2]. Non-blackened soil at the WBS in VN was thought to have an average distribution pattern from blackened (VN) and control soil (VN). Therefore, mixing blackened soil with natural soil may contribute to this characteristic distribution pattern. The surrounding surface soil at the WBS (25 m from WBS) in VN showed a high ratio of Cl connected with aromatic carbons (Fig. 2), which reflects natural soil characteristics.

Quantitative speciation of Cl revealed aromatic-Cls concentrations in environmental soil at the WBSs. Figure 3 shows that blackened soil at the WBSs contained high levels of aromatic-Cls. The WBS in VN and the one in PHI were assessed as having 1,000 ± 570 and 1,200 ± 240 mg/kg, respectively, of aromatic-Cls in blackened soils. Non-blackened soil at the WBS in VN showed relatively high aromatic-Cls concentrations (500 ± 47 mg/kg). The lowest and relatively low concentrations of aromatic-Cls were respectively assessed in control soil (18 ± 4.8 mg/kg) and surrounding soil at the WBS in VN (110 ± 13 mg/kg). Compared with global measurements of natural aromatic-Cls concentrations by Asplund and Grimvall [8], blackened and non-blackened soils at the WBSs had concentrations higher than the maximum value (360 mg/kg) measured in natural soils at 26 sampling sites denoted by Natural (World) in Fig. 3. In contrast, aromatic-Cls concentrations in the surrounding soil (25 m from WBS) in VN was at median (75%) levels based on the Natural (World), while control soil (VN) concentrations were under the 25% level. Although some have reported toxic aromatic-Cls concentrated in soil at the WBS, the total concentration of PCDD/Fs and PCBs in WBS soil was only a few mg/kg [1]. Therefore, blackened soil at the WBS contained toxic aromatic-Cls at concentrations of a few hundred mg/kg, which also contained unreported toxins such as chlorobenzenes, chlorinated PAHs, and unknown chlorinated organics, excluding reported toxins (PCDD/Fs and PCBs) [1] and natural background (median: 69 mg/kg, range: 4–360 mg/kg [8]).

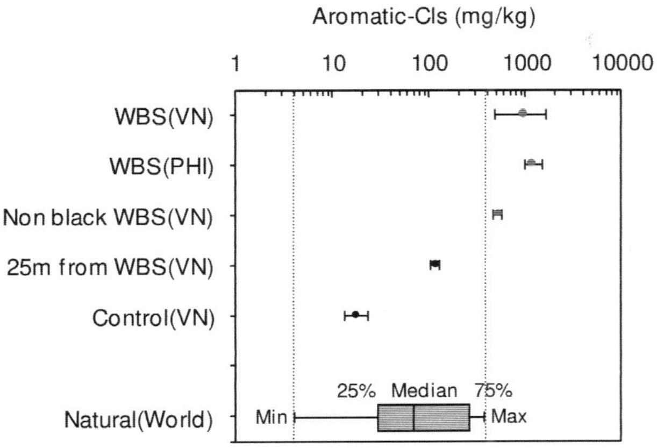

Figure 3 Amount of aromatic-Cls based on quantitative speciation. Error bars of this study reflect the standard deviation of LCF, as shown in Fig. 2. Natural (World), a global study of aromatic-Cls concentrations, was based on a previous study [8].

Quantitative speciation revealed correlations between inorganic chlorides and aromatic-Cls. High positive correlation coefficients were found between aromatic-Cls and $Cu_2(OH)_3Cl$ (R^2 = 0.97), NaCl (0.98), $MgCl_2$ (0.97), $PbCl_2$ (0.92), and KCl (0.85). Previous reports have identified the thermochemical chlorination mechanism of carbon triggered by metal chlorides [5]. Recent copper study did not have the $Cu_2(OH)_3Cl$ in their library [6]. The present study was the first to report detection of $Cu_2(OH)_3Cl$ in soil at the WBSs. $Cu_2(OH)_3Cl$ played a key role as a strong catalytic promoter to generate aromatic-Cls in the solid phase during thermochemical processes [5,9,10]. This suggests that high concentrations of aromatic-Cls in WBSs were derived from thermochemical interactions of carbon matrix in the soil surface or insulation-coating materials with oxychlorination of metal copper in wires and cables, which produced $Cu_2(OH)_3Cl$.

The present studies applied quantitative speciation using Cl K-edge NEXAFS to estimate total Cl content, the chemical form of Cl, aromatic-Cls concentration, and correlations in relation to the formation mechanism of aromatic-Cls in various soils at WBSs. Although the accuracy of this analytical technique compared with other quantitative methods remains unknown, Cl-connected compounds can be quantified at a broad range of concentrations, which can be used to perform quantitative speciation of chlorine in various environmental solid matrices.

Acknowledgements
We thank G. Suzuki (NIES) and S. Sakata (IDE-JETRO) for supporting field sampling; K. Shiota (Kyoto Univ.), Y. Kitajima (KEK-PF, BL-11B), and Y. Inada (KEK-PF, BL-9A) for helping with Cl K-edge NEXAFS measurement at Photon Factory (proposal 2007G069).

References
[1] Wong C S C et al. 2007 *Environ. Pollut.* **145** 434
[2] Myneni S C B 2002 *Science* **295** 1039
[3] Fujimori T et al. 2010 *Anal. Sci.* **26** 1119
[4] Leri A C et al. 2006 *Anal. Chem.* **78** 5711
[5] Fujimori T, Takaoka M and Takeda N 2009 *Environ. Sci. Technol.* **43** 8053
[6] Otsuka M et al. 2012 *Interdisciplinary Studies on Environmental Chemistry* **6** 161
[7] Fujimori T et al. 2012 *J. Hazard. Mater.* **221-222** 139
[8] Asplund G and Grimvall A 1991 *Environ. Sci. Technol.* **25** 1346
[9] Takaoka M et al. 2005 *Environ. Sci. Technol.* **39** 5878
[10] Fujimori T, et al. 2007 *Organohalogen Compds.* **69** 319

Journal of Physics: Conference Series **430** (2013) 012095 doi:10.1088/1742-6596/430/1/012095

Characterization of lead, chromium, and cadmium in dust emitted from municipal solid waste incineration plants

K Shiota, G Imai, K Oshita and M Takaoka
Department of Environmental Engineering, Kyoto University,
Kyotodaigakukatsura4, Nishikyo-ku, Kyoto, 615-8540, Japan

E-mail: kenji.shiota.4x@kyoto-u.ac.jp

Abstract. The dust is emitted from municipal solid waste incinerators (MSWIs). Volatile toxic heavy metals are abundant in smaller dust particles and influence the toxicity of particulate matter such as fine particles <2.5 μm ($PM_{2.5}$). However, little is known about the properties of these metals in fine dust particles. Therefore, X-ray absorption fine structure (XAFS) spectroscopy was used to investigate the chemical states of lead (Pb), chromium (Cr), and cadmium (Cd) in MSWI dust collected for nine particle size fractions at the inlet of the dust collector and the stacks of two MSWI plants. XAFS spectroscopy of the dust in the inlet of the dust collectors showed that finer dust contained predominantly Pb as $PbCl_2$ with some $PbSiO_3$, coarser dust consisted of Cr forms, including more toxic Cr(VI) species, and all dust contained $CdCl_2$. Although the dust collector removed almost all of the Pb, trace amounts of $PbCl_2$ remained in the stack gas after passing through the dust collector.

1. Introduction

In municipal solid waste incineration (MSWI), volatile toxic heavy metals are abundant in smaller dust particles and influence the toxicity of particulate matter (PM) [1] such as fine particles <2.5 μm ($PM_{2.5}$). Although understanding the behavior and chemical states of heavy metals in the fine particles of MSWI dust is important, few data on such properties are yet available [2]. In this study, we examined the heavy metals lead (Pb), chromium (Cr), and cadmium (Cd) because of their toxicity. This study aimed to investigate the chemical states of Pb, Cr, and Cd in MSWI dust collected in particle size fractions at the inlet of the dust collectors and stacks in MSWI plants by using X-ray absorption fine structure (XAFS) spectroscopy.

2. Materials and Methods

Samples of MSWI flue dust were collected in nine particle size fractions (from submicron to about 10 μm) by an Andersen stack sampler (Tokyo Dylec AS-500), which was inserted into the flue gas ducts at the inlets of the dust collectors in two MSWI plants [plants A and B: two different continuous stoker-type MSWIs with bag filter (BF) dust collectors]. In plant B, samples were also collected from the inlet of the stack (figure 1). In these sampling methods, Japanese Industrial Standards (JIS) Z8808 and K0302 were adopted to collect the dust and to calculate the theoretical particle size of each fraction [3]. The dust samples were leached in deionized water. The residues were dissolved with acid.

[1] To whom any correspondence should be addressed.

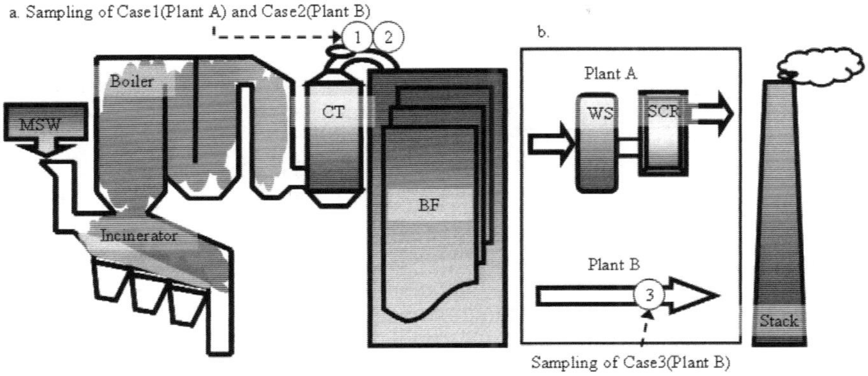

Figure 1 Sampling locations at flue flow in MSWI plants; a. Plant A and plant B have similar flue treatment equipments in flue flow from incinerator to dust collector. Samples were collected at the inlet of BF in Plant A (Case1) and in Plant B (Case2). b. Flue treatment equipments at downstream of BF in plant B is different from that in plant A. Sample was collected at the inlet of stack in Plant B (Case3). CT: Cooling Tower, BF: Bag Filter, WS: Wet Scrubber, SCR: Selective Catalytic Reduction.

Table 1 Property of flue dust samples for Case 1, Case 2, and Case 3; m^3_N: Reduced volume of gas in 0 °C, 1 atm, and dried, Gas phase: amounts in cohesive water and H_2O_2 after removed dust particles.

Characterization of Flue dust fractions							Sampling condition				
				Concentration					Flue gas		
Case	No.	Particle size (µm)	Dust weight (mg)	Pb (µg/g)	Cd (µg/g)	Cr (µg/g)	Plant	Location	Temperature (°C)	moisture (%)	Sampling volume (m^3_N)
1	1-1	11.1<	48.6	360	13	411	A	Inlet of BF	156	20.2	0.069
	1-2	7.2~11.1	31.5	550	18	392					
	1-3	4.9~7.2	14.2	1100	30	555					
	1-4	3.3~4.9	4.1	1800	62	752					
	1-5	2.1~3.3	2.4	5640	139	1230					
	1-6	1.0~2.1	7.7	10400	814	828					
	1-7	0.58~1.0	9.9	11700	595	541					
	1-8	0.42~0.58	6.8	7000	376	247					
	1-9	<0.42	6.9	7200	354	500					
	1-Gas phase (µg/m^3_N)			280	<0.66	<0.024					
2	2-1	0.98~2.0	16.495	7350	620	<17	B	Inlet of BF	173	17.1	0.066
	2-2	<0.40	6.714	12100	750	<7					
	2-Gas phase (µg/m^3_N)			91	8.4	10					
3	3-1	6.0~9.0	1.746	1750	107	-	B	Inlet of stack	163	21.3	40.8
	3-2	1.7~2.7	1.228	6130	210	-					
	3-3	0.33~0.46	0.482	16700	530	-					

These solutions for the chemical compositions of the dust were analyzed with ICP-MS (Yokogawa Electric HP4500). Table 1 shows the properties of the flue dust samples and the sampling conditions.

XAFS spectroscopy was carried out in beamline BL01B1 of Super Photon ring-8GeV (SPring-8), in beamline BL-12C of the Photon Factory (PF), and in beamline NW10A of the Photon Factory–Advanced Ring for pulse X-rays (PF-AR). These beamlines have double crystal monochromators of Si(111) in BL01B1 and BL-12C, and Si(311) in NW10A. The dust fractions were prepared by cutting the dust collection filters and setting them into the synchrotron radiation X-ray. XAFS spectra of the dust fractions were collected for Pb-L_{III}, Cr-K, and Cd-K in the fluorescence mode using a Ge multielement solid-state detector, and those of reference materials were determined in the transmission mode. The software REX2000 ver. 2.5.5 (Rigaku Co. Ltd.) was used for analyzing the chemical forms of each element in the dust by the linear combination fitting technique (LCF) [4][5]. LCF describes the chemical forms of elements in samples based on ratios to reference materials which for LCF were the

Pb compounds $PbCl_2$, $PbSiO_3$, $PbCO_3$, PbO, PbO_2, and $PbSO_4$, the Cd compounds Cd, CdO, and $CdCl_2$, and the Cr compounds Cr metal, $CaCr_2O_4$, $FeCr_2O_4$, Cr_2O_3, $CrCl_3$, and CrO_3.

3. Results and Discussion

Pb-L_{III} XAFS spectra were collected for 11 dust fractions from the inlets of the dust collectors of plants A and B, three fractions from the stack of Plant B [Figure 2(a)], and the reference materials (Figure S1, Supplementary data). The spectra for each fraction had similar shapes. From plant A (1-1–1-9) The LCF result shows that the dust of each fraction predominantly contained Pb as $PbCl_2$ (15–60 mol%) and $PbSiO_3$ (30–60 mol%), with more $PbCl_2$ tending to occur in finer fractions and more $PbSiO_3$ appearing more often in coarse dust fractions. Each fraction contained other low compounds (10–25 mol%) (Figure3 and Table S1, Supplementary data).

The spectra of samples collected from the inlet of the stack at plant B (3-1–3-3) were poor because the amount of Pb in those fractions was very small. LCF result indicates that each fraction at the inlet of the BF (2-1, 2-2) predominantly contained Pb as $PbCl_2$ (29–56 mol%) and $PbSiO_3$ (33–34 mol%), with the finer fraction (2-2) containing more PbO and $PbSO_4$ (10–28 mol%). The spectra of the fractions from the inlet of the stack (3-1–3-3) predominantly contained Pb as $PbCl_2$ (about 50 mol%) as shown by LCF, indicating that some Pb compounds, such as $PbCl_2$, remained in the fractions in the stack although the dust collector removed almost all of the Pb contained in dust.

The results of the dust fractions sampled from the inlets of the BF in plants A and B are discussed first. The results for plants A and B were similar. Takaoka *et al.* [4] and Funatsuki *et al.* [5] reported the primary Pb chemical forms as $PbCl_2$ or $PbSiO_3$ in unclassified, bulk MSWI fly ash using XAFS and LCF. The results of this paper, showing the primary chemical forms of Pb as $PbCl_2$ and $PbSiO_3$, are similar to those reports. $PbCl_2$ may occur as the result of high concentrations of HCl in the flue gas of MSWI [4]. The differences in chemical forms among particle sizes may be explained by melting points or species volatility. MSWIs operate at high temperatures of 800–1000°C, at which $PbCl_2$ is volatile, because its melting and boiling points are 501°C and 950°C, respectively. $PbCl_2$ tended to be more abundant in the finer dust fractions because $PbCl_2$ gas is cooled in the flue gas treatment process and may turn into a fine particle solid through aggregation. In fact, Pb occurred in the flue gas phase, which removed dust particles at the inlet of the BF (Table 1). The amounts of Pb leached to deionized water were about 50wt% in some finer fractions, which supports the idea that $PbCl_2$ is a primary compound because of its high solubility. $PbSiO_3$ is not volatile, having a high melting point, and thus could be carried by the flue gas stream and be abundant in coarse dust fractions. Alternatively, $PbSiO_3$

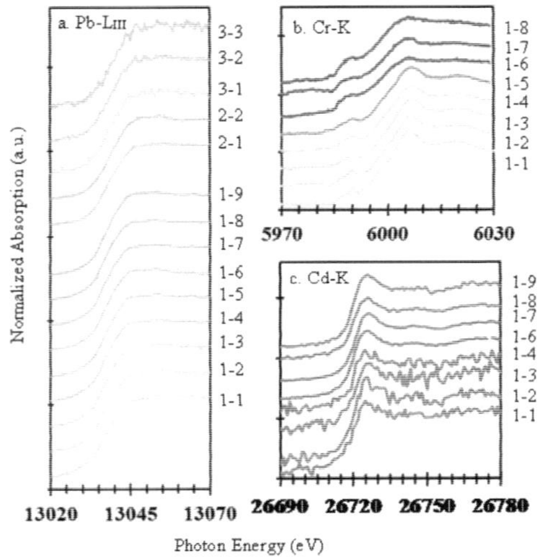

Figure 2 XAFS spectra of Pb-L_{III} (a), Cr-K (b), and Cd-K (c) in flue dust samples; Labels such as 1-1: Fraction number of flue dust samples in Table 1

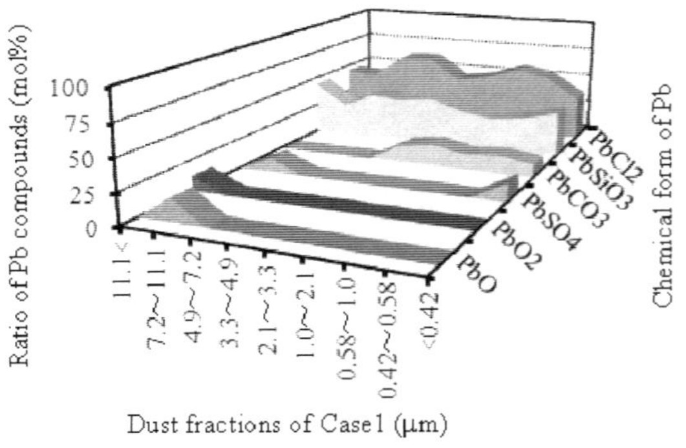

Figure 3 LCF result of Pb compounds in Case 1

might be composed during the vapor phase [6].

For the dust fractions sampled from the inlet of the stack at plant B, an explanation for the existence of Pb at the inlet of the stack is that Pb compounds, such as $PbCl_2$ in the flue gas phase, passed the BF and condensed, although the BF removed almost all of the dust. Further research is needed to better understand the other Pb compounds sampled at the inlet of the stack.

Cr-K XAFS spectra were collected for eight dust fractions from the inlets of dust collectors for only plant A [Figure 2(b)] and the reference materials (Figure S1, Supplementary data). The spectra of coarse and fine fractions showed different shapes on the pre-edge and whiteline. A clear pre-edge peak at about 5993 eV, espesially in the spectra of some coarse fractions (1-1–1-5), lead to the suggestion of the occurrence of more toxic Cr(VI). LCF results showed that coarse fractions predominantly contained Cr as Cr(III) (57–71 mol%) and some Cr(VI) (13–23 mol%). Cr(III) was present as $CaCr_2O_4$ (10–47 mol%), $FeCr_2O_4$ (12–24 mol%), and Cr_2O_3 (17–26 mol%), and Cr(VI) was present as CrO_3. The LCF results also showed that finer fractions contained more Cr metal (13–60 mol%).

Most Cr compounds have high melting points and are not particularly volatile. The particle sizes of the Cr compounds themselves might have caused the differences in the chemical forms observed among dust particle fractions, but this is speculation. Cr compounds at low oxidation states under the high temperatures of the incinerator might have caused some of the Cr(VI) formation.

Cd-K XAFS spectra were collected for eight fractions of dust from the inlets of dust collectors only for plant A [Figure 2(c)] and the reference materials (Figure S1, Supplementary data). Some spectra of coarse fractions (1-1–1-4) were poor because those fractions contained very small amounts of Cd. The spectra of each fraction had similar shapes. The LCF results showed that each fraction predominantly contained Cd as $CdCl_2$ (40–100 mol%), with some Cd metal (0–60 mol%). This result is similar to that of the Funatsuki *et al.* on MSWI fly ash [7]. The ratio of $CdCl_2$ to Cd metal did not show a clear tendency in terms of dust particle size.

Like Pb, $CdCl_2$ may also occur as the result of high concentrations of HCl in the flue gas of MSWIs [4]. $CdCl_2$, with its melting and boiling points of 568°C and 960°C, respectively, is easily volatile under the high temperatures of MSWIs. Pinzani *et al.*, using μ-EXAFS, reported primary chemical forms of $CdCl_2$ and $CdSiO_3$ in particles of MSWI fly ash [2]. The dust fractions examined in the present study were not well fitted using $CdSiO_3$ in LCF, indicating that $CdSiO_3$ was not a primary chemical form in our samples.

4. Conclusion

In summary, XAFS spectroscopy of the dust from the inlets of dust collectors showed that finer dusts predominantly contained Pb as $PbCl_2$ and some $PbSiO_3$, coarser dusts contained Cr forms, including more toxic Cr(VI) species, and all dusts contained $CdCl_2$. Although the dust collector removed almost all of the Pb, some $PbCl_2$ remained in the stack gas after passing through the dust collector.

Acknowledgements

This study was supported by the research grant from Osaka Bay Regional Offshore Environmental Improvement Center and by the aid for presentation at XAFS15 from Kansai Research Foundation for technology promotion. We thank T. Fujimori (Kyoto University) and beamline staffs for help with the XAFS experiments at SPring-8 (Proposal No.2009B1458) and PF (Proposal No.2009G044).

References

[1] Greenberg R R, Zoller W H and Gordon G E 1978 *Environ. Sci. Technol.* **12** 566
[2] Pinzani M C C, Ansell S, Somogyi A, Steenari B and Lindqvist O 2004 *Anal. Chem.* **76** 1596
[3] Shiota K *et al.* 2011 *J. Japan Soc. Atmos. Environ.* **46** 224 (in Japanese)
[4] Takaoka M *et al.* T 2005 *Physica Scripta* **T115** 943
[5] Funatsuki A, Takaoka M, Oshita K and Takeda N 2012 *ANALYTICAL SCIENCE* **28** 481
[6] Biswas P and Zachariah M R. 1997 *Environ. Sci. Technol.* **31** 2455
[7] Funatsuki A *et al.* 2010 *BUNSEKI KAGAKU* **59** 801 (in Japanese)

Journal of Physics: Conference Series **430** (2013) 012096 doi:10.1088/1742-6596/430/1/012096

Analyses on Forms of Sulfurin Chinese Yanzhou Coal and Their Transformation during Pyrolysis by X-ray Absorption Spectroscopy

C H Fu[1], J C Wang[1], Y F Hu[2], M J Wang[1], Y L Zhang[1] and L P Chang[1,*]

1 Key Laboratory of Coal Science and Technology, Taiyuan University of Technology, Ministry of Education and Shanxi Province, Taiyuan, 030024, P.R China
2Canadian Light Source, 101 Perimeter Road, Saskatoon, SK, S7N 0X4, Canada

E-mail:yongfeng.hu@lightsource.ca; lpchang@tyut.edu.cn

Abstract: In the processof pyrolysis, inherent and the added minerals are the main factors to affectthe transformation and distribution of sulfur in coal. In order to cleanly and efficiently use the inferior coal with high-sulfur content, Yanzhou coal containing 5wt % sulfur was chosen as the sample to do pyrolysis experiments in a fixed-bed reactor. The effect of iron additive on the sulfur forms and their transformation during coal pyrolysis wasstudied in this paper. Fe and S forms in raw coal with (FC) and without (RC) iron additive and their chars from different temperature were determined by XANES, and the gaseous products and total sulfur in char were also considered. The results show that addition of Fe in coal can make more sulfur retained in the char. The transformation of sulfur exists in the entire coal pyrolysis process from 200 to 1000 °C, and more inorganic sulfur was produced during the FC pyrolysis at high temperature (700-1000 °C). FeS existing in the FC char from 1000°C occupies about 62% of the total sulfur in char, but there is almost no FeS in the RC char at 1000 °C. The semi-quantitative analysis of XANES data by LCF fitting will also be discussed.

1. Introduction

The energy and environment undergo more and more serious tension with the rapid development of social economy and the mass consumption of coal resource in China. One of the main problems obstructed the effective and extensive utilization of coal is the sulfur existing in coal. Sulfur in coal can be generally divided into two categories, inorganic sulfur and organic sulfur. Understanding the sulfur forms and the transformation during coal pyrolysis plays a role in the theoretical direction for the removal of sulfur.

Pyrolysis, the initial stage of the coal conversion process,is one of main approaches to adjust and control the proportion and composition of coal gas, tar and char. In this process, there exist the release of sulfur from coal and the mutual transformation among different sulfur forms [1,2]. The transformation and distribution of sulfur are determined by coal properties and its pyrolysis conditions. The inherent minerals in coal, such as alkali, alkaline-earth and transition metal, have the ability to change the devolatilization reactions during coal pyrolysis and the transformation of sulfur [3-6]. Iron is the main inherent minerals in coal and the common catalyst for many chemical reactions, which has the efficient, cheap and multi-valent characters [7]. How the inherent and additional iron in coal affecting the transformation of sulfur during pyrolysis are our primary focus.

X-ray absorption spectroscopy (XAS) is a non-destructive technique that can be used to probe the chemical and electronic structure of various materials. S K-edge XAS to qualitatively and quantitatively determine the inorganic sulfur forms has previously applied [8], but the analyses of organic sulfur forms in coal are rarely reported. Based on the above analysis, the present work chose Yanzhou coal with high-sulfur content as the experimental samples, the sample loaded iron was prepared and their pyrolysis experiments were carried out in a fixed-bed reactor. The performance of sulfur transformation during coal pyrolysis and the effect of iron in coal on them were discussed based on the XANES spectra analyses.

2. Experimental

2.1 Samples preparation

Yanzhou raw coal (RC) used as the experiment sample was ground and sieved to the particle sizes of 0.154-0.258 mm. The proximate, ultimate analyses and sulfur forms of RC are given in Table 1. It can be seen that the total sulfur content in RC is 5.0%, which is a typical high-sulfur coal, the pyritic sulfur and organic sulfur have similar proportion and the sulfate sulfur is also significant.

Table 1. The proximate and ultimate analyses and sulfur forms of RC

Proximate analysis /%			Ultimate analysis /% (daf)					Sulfur forms / % (daf)		
M_{ad}	A_d	V_{daf}	C	H	N	S_t	O*	S_s	S_p	S_o*
2.3	13.6	43.4	81.5	5.1	1.3	5.0	7.1	0.9	2.1	2.0

ad: air-dried basis; d: dry basis; daf: dry and ash-free basis. * by difference.
S_t: the total sulfur; S_s: the sulfatesulfur; S_p: the pyritic sulfur; S_o: the organic sulfur.

The coal added iron (FC) with 1% (wt) $FeCl_3$ was prepared by impregnation method according to the following procedures. 0.167 g $FeCl_3 \cdot 6H_2O$ was dissolved in deionized water and then 10 g raw coal was added into this precursor solution, fully blended for 30 min and placed 12 h in the atmosphere, then dried for 12 h at 60 °Cin a vacuum drying oven.

2.2. Pyrolysis

About 0.5 g RC and FC samples are preset on the sintered plate in a fixed-bed quartz reactor (600 mm long and 28 mm ID) under pure argon atmosphere and pyrolyzed by temperature programmed-isothermal thermal evolution course with a gasflow rate of 600 mL/min. Each experiment was carried from room temperature to the preset levels (300, 400, 500, 600, 700, 800, 900 and 1000 °C) at a heating rate of 10 °C/min and then held for 40 minutes. The concentrations of H_2S and COS in outlet gas were determined by gas chromatograph of GC-950 with the FPD detector. The yield of sulfur in gases ($Y_{s\text{-}gas}$) is calculated by the ratio of the sum of H_2S and COS with the total sulfur in raw coal.

The total sulfur content in char from different temperature was measured using a HCS-140 carbon sulfur analyzer. The yield of sulfur retained in char ($Y_{s\text{-}char}$) is expressed according to the ratio of total sulfur in char with total sulfur in raw coal.

2.3. S K-edge XAS spectra

The iron and sulfur forms in coal sample and pyrolyzed chars were detected by the X-ray absorption spectroscopy at Canadian Light Source by using the soft X-ray Micro characterization Beam line (SXRMB). Si (111) double crystal monochromator was used, giving an energy resolving power of 10,000 eV [9]. The experimental sample was placed in a vacuum chamber (base pressure of 1 x 10^{-8}torr), using a double-sided conduct carbon tape. The Fe and S K-edge XAS spectra were recorded in the bulk sensitive fluorescence yield (FLY) with a micro-channel plate detector. The energy scale was calibrated using the Ar K-edge in the ion chamber and a sulfate peak at 2481.6 eV. The accuracy of energies reported here is 0.2 eV.

3. Results and discussion

3.1. The formation of char and transformation of sulfur during coal pyrolysis

The effect of iron added on the formation of H_2S and COS during coal pyrolysis is shown in Figure 1. The addition of iron in coal decreased largely the release of gaseous sulfur H_2S and COS during pyrolysis below 650 °C and shifts the temperature of maximal H_2S and COS release to the low values, due to the possibility that the pyrite may be oxidized to ferrous sulfate and elemental sulfurby $FeCl_3$ solution [7]. However, the release of gaseous sulfur was promoted by iron above 650 °C, which was caused by the decomposition of iron sulfate from the oxidation of pyrite to form iron sulfide and the catalytic influence of the iron on the decomposition of the sulfates. And the restraining effect of iron added to coal during pyrolysis was further provenby the yields of gaseous sulfur in Figure 1.

Figure2shows the yields of char and total sulfur in charsduring coal pyrolysis. The char yields of both RC and FC samples presented a general trend, decreasing with the increasing of temperature. But the addition of iron in RC changed the yield of char atdifferent temperature. Below 450 °C, the char yield of RC was larger than FC, indicating that iron improved the volatilization of coal during low pyrolysis temperatures by catalyzing the breakage of unstable bonds. However, at 450 °C to 900 °C, the case is in reverse,which indicates that the combination of char with the volatile to form new species in char was improved bythe iron. When the pyrolysis temperature is further risen to 1000 °C, their change of char yields performed the same shape, and then tended to be constant, forthe volatiles have been entirely released.

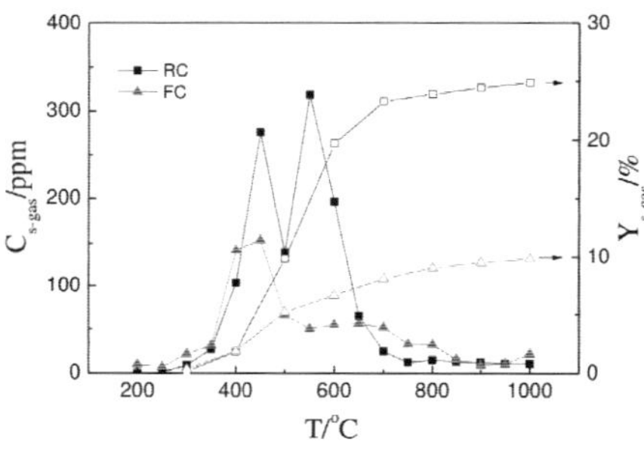

Figure1. Release concentrationsandyieldof H_2S and COS during coal pyrolysis

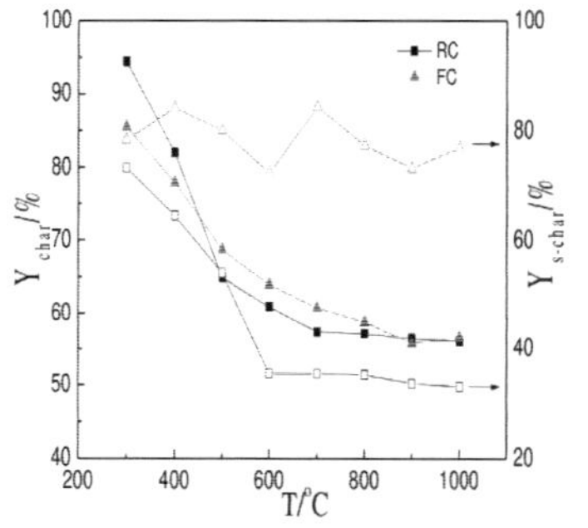

Figure 2. Yields of char and sulfur retained in char during coal pyrolysis

The sulphur retention in char of RC performed a sharp decrease below600 °C, and then kept stable, owing to the mass decomposition of pyrite, unstable organic sulphur and sulphatein coal during pyrolysis. However, above600 °C,the degree of decrease became slightly,which was resulted from the great difficulty of decomposition of the complicated sulfur species. Compared with the change of RC, the sulfur of FC was quite stable and the change of yield of sulphur in charwas not obvious. The difference between RC and FC samples may be owing to the effect of iron added in coal by changing the decomposition behaviour of pyrite and the secondary reactions of sulfur species, such as formation of the complex organic sulphur or stable sulphide from H_2S and COS releasing during pyrolysis.

3.2. The transformation of iron during coal pyrolysis

Figure3shows the XANES spectra of Fe K edge in FC and the corresponding chars at different pyrolysis temperature. Compared with the XANES spectra of the serial Fe model compounds, it can be seen from the XANES spectra that FC and chars samples below 500 °C had obvious edge positions at 7124.0eV, 7135.5 eV, indicating that there existed large amount of ferrous chloride (Fe^{2+})and ferric sulfate and ferrous sulfate during FC coal samples pyrolysis below 500 °C, meanwhile, the high contents of ferrous chloride and ferrous sulfate in the FC were from the oxidation of pyrite by ferric chloride (Fe^{3+}) during the impregnation process of sample. However, the peak position at 7124.0 eV almost disappeared between 600-700 °C, illustrating that $FeCl_2$ decomposed completely and transformed into the other sulfur-containing species.

There appeared obvious edge position edge at around 7131eV in the XANES spectra of the chars of FC above 400 °C, which is related to the characteristic peaks of Fe_2O_3 and FeO, implying that these iron oxides produced from the decomposition of iron sulfate. Subsequently these iron oxides reacted with the H_2S from the pyrolysis volatiles to form FeS retaining in char [10].

3.3. Effect of iron additive on the changes of sulfur forms during coal pyrolysis

Figure4 shows the X-ray fluorescence spectra of FC samples at different pyrolysis temperatures. Based on the experiments of some sulfur-containing model compounds, it can be obtained thepeak positions of these pure substances, such as FeS is at 2470.1 eV, pyrite at 2471.4 eV, CaS at 2477.7 eV, $Fe_2(SO_4)_3$ at 2481.6 eV, DBS sulfone at 2479.3 eV and 2,2'-Bithiophene at 2473.3 eV.

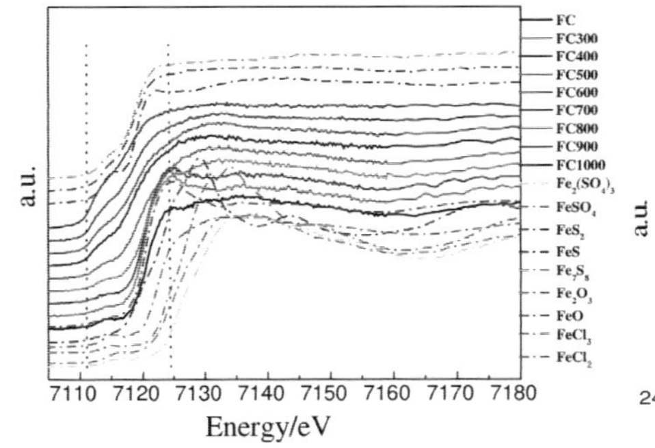

Figure3.XANES results of the iron-containing modal compoundsand samples

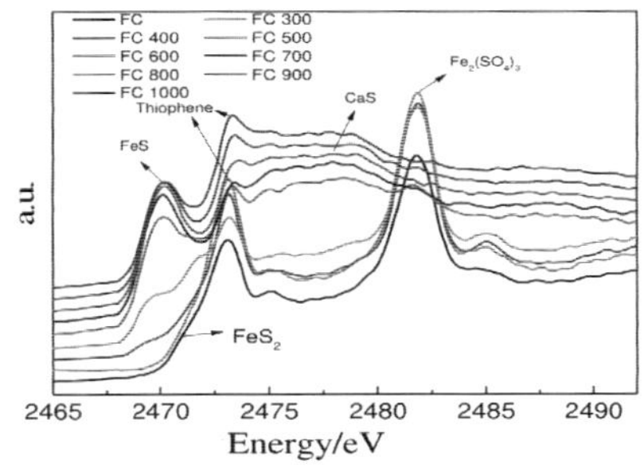

Figure 4. X-ray Fluorescence yields of FC sample and chars from different temperature

It can be seen that the catastrophe point of the sulfur transformation during coal pyrolysis presents in the temperature range of 400-600 °C, from which the decomposition reactions ofunstable sulphur-containing species had been already completed, and new sulphur species began to form. The X-ray Fluorescence spectra of RF indicated the presence of ferric sulphate, thiophene, and pyrite, which was characterized by the edge position at 2481.8 eV, 2473.0 eV, and 2471.3 eV, respectively. Above the temperature of 600 °C, ferric sulphate and pyrite in char decomposed completely to form FeS and other stable sulphide such as CaS (at 2478.0 eV) in char and the sulfur-containing gaseous products. With the increase of pyrolysis temperature, the peak position of thiophene rings began to move to the higher energy position, indicating that condensation reactions of thiophene rings happened to form the complex thiophene rings restraining in char.

The differential percentages of different sulfur species in chars at different pyrolysis temperatures from the results of X-ray Fluorescence spectroscopy are shown in Table 2. It can be seen that the addition of iron played animportant role in the transformation of sulfur in coal during coal pyrolysis. The increase of sulfate on the surface of sample may be due to the oxidation on the surface of sample by air during the coal preparation. In the process of coal pyrolysis, the iron decreased the amount of organic sulfur and sulfate, while increased the inorganic sulfur contents in solid products, indicating that iron catalyzed the breakage of organic sulfur bond to form some sulfur-containing radicals, and then the sulfur-containing radicals pass through the pore channel to release. When the radicals encountered the iron on the surface of pore, reactions of forming inorganic sulfur occurred. From the data of CaS in char, the content of calcium sulfide between 600-800 °C increased by the addition of iron but decreased at above 800°C, indicating that the addition of iron affected the formation of CaS. FeS is the main residue of iron in the char, whose amount is far larger than that of CaS, so sulfur is more susceptible to combine with iron not calcium.

Table2. The differential percentage of different sulfur species between FC and RC samples from the results of X-ray Fluorescence spectroscopy

Sample	FeS	CaS	Pyrite	Sulfide	Thiophene	Sulfoxide	Sulfate

coal	0	0	5.7	2.2	-6.3	0.8	-2.2
Char-300	2	0	11.7	10.8	-34.4	-3.2	15.2
char-400	8	0	11.3	13.7	-45.8	-2.8	15.7
char-500	26.4	0	10	13	-57.2	-2.6	9.6
char-600	37.9	0	0	13.8	-51.1	-4.3	-1.1
char-700	50.2	2.7	0	5.8	-50.8	-2.4	-2.0
char-800	47.7	4.2	0	0	-50.2	0	-1.7
char-900	53.7	-3.9	0	0	-40.5	0	-2.4
char-1000	54.2	-8.1	0	0	-43.7	0	-2.5

4. Conclusion

(1) The pyrolysis temperature of 600°C is a catastrophe point for the sulfur behavior in coal during pyrolysis, where both organic sulfur and inorganic sulfur begins to distribute again.

(2) Iron changed the release of H_2S and COS in the gaseous products dramatically duing pyrolysis. The H_2S from the decomposition of pyrite decreased largely, due to the intermediate sulfur combining with iron rather than the H radicals to form H_2S. Compared with raw coal, the release temperature zone of COS is in advance, for iron added improved the formation of volatile at low temperature.

(3) The addition of iron decreased the formation of organic sulfur and sulfates in char, and increased the formation of inorganic sulfur like FeS, indicating that iron played a catalytic role in breaking the organic sulfur and recombined with free sulfur to reform new sulfur-containing species in char.

Acknowledgements

The authors gratefully thank the financial supports of the National Natural Science Foundation of China (21176165, U1261110) and the Research Fund for Doctoral Program of High Education (20111402110009). We thank the CLS staff for the technical support.

References

[1] Yani S and Zhang D K An experimental study into pyrite transformation during pyrolysis of Australian lignite samples 2010 *Fuel.* **89** 1700
[2] Sugawara K, Tozuka Y, Sugawara T and Nishiyama Y Effect of heating rate and temperature on pyrolysis desulfurization of a bituminous coal 1994 *Fuel Process. Technol.* **37** 73
[3] Telfer M and Zhang D K The influence of water-soluble and acid-soluble inorganic matter on sulphur transformations during pyrolysis of low-rank coals 2001 *Fuel.* **80** 2085
[4] Tyler R J and Schafer H N S Flash pyrolysis of coals: influence of cations on the devolatilization behaviour of brown coals 1980 *Fuel.* **59** 487
[5] Liu Q R, Hu H Q, Zhou Q, Zhu S W and Chen G H Effect of mineral on sulfur behavior during pressurized coal pyrolysis 2004 *Fuel Process. Technol.* **85** 863
[6] Chen H K, Li B Q, Yang J L and Zhang B J Transformation of sulfur during pyrolysis and hydropyrolysis of coal 1998 *Fuel.* **77** 487
[7] Zhao Z B, Li W and Li B Q Catalytic reduction of NO by coal chars loaded with Ca and Fe in various atmospheres 2002 *Fuel.* **81** 1559
[8] Jagtap S B and Wheelock T D Coal desulfurization by ferric chloride 1995 *Fuel Process. Technol.* **43** 227
[9] Braun A, Janousch M, Sfeir J, Kiviaho J, Noponen M, Huggins F E, Smith M J, Wilckens R S, Holtappels P and Graule T Molecular speciation of sulfur in solid oxide fuel cell anodes with X-ray absorption spectroscopy 2008 *J. Power Sources.* **183** 564

Zn distribution and speciation in zinc-containing steelmaking wastes by synchrotron radiation induced μ-XRF and μ-XANES spectroscopy

Lihua Wang[1], Yuying Huang[1] and Xiaoming Lu[2]

[1] Shanghai Synchrotron Radiation Facility, Shanghai Institute of Applied Physics, Chinese Academy of Sciences, Shanghai 201204, P. R. China

[2] Chemical Analysis Laboratory, Baoshan Iron & Steel Co., LTD, Shanghai 201900, P.R.China

Email: huangyuying@sinap.ac.cn

Abstract. Zinc is an important element in steelmaking industry not only for its impact on environment but also for its hazardous influence to steel production. Collecting information about the chemical association of zinc in zinc-containing steelmaking wastes is fundamental for monitoring its behavior during further treatment. In the work presented here, the spatial distribution, and chemical forms of zinc in accumulated particles of zinc-containing steelmaking wastes, including electric arc furnace dust (EAFD) and basic oxygen furnace OG sludge (BOF OG), have been investigated using synchrotron radiation induced μ-XRF and μ-XANES spectroscopy. Results of μ-XRF analysis showed that zinc distributed in two ways. One was shared with iron and its distribution showed a positive correlation with that of iron. The other was accumulated in some well-defined hot spots with high amount and its distribution showed negative correlation with that of iron. For EAFD, results of μ-XANES spectroscopy indicated that zinc was mainly present in the form of $ZnFe_2O_4$ within the whole particles no matter the spots with high or low zinc content. Whereas for BOF OG, $ZnFe_2O_4$ was the main constituent in well-defined hot spots while in other regions zinc was mainly in the form of zinc carbonate. These results indicated that chemical reaction between zinc and other components occurred during the formation of accumulated OG sludge particles. If the above findings could be confirmed by more systematic investigations, it will provide valuable information for treating and utilizing these metallurgical residues.

1. Introduction

In steelmaking industry, zinc is one of those elements which should be paid attention to in particular. This element is usually easy to volatilize during the metallurgical process due to its boiling point below the melting temperature in furnace and subsequently condenses and accumulates to form particles in the form of dust or sludge when passing through the gas-cleaning system. Therefore, a large amount of such kind of residues is generated. Generally these dust and sludge must be disposed of on a hazardous waste disposed site. For this kind of metallurgical residues is often composed of fine particles, it is easy to be discharged into atmosphere and accumulated in soils during the transportation and landfill, which will cause risks on environments. On the other hand, landfill will bring high transportation and disposal cost. These costs could be reduced if metals, such as iron, zinc, and so on, can be recovered or sold. The most common way to utilize these dust and sludge is reusing them as a secondary source of raw material due to the high iron content. However, when they are reused inside enterprises, undesirable consequences will occur. In particular, the presence of zinc will damage the production equipment. Take blast furnace for example. When these dust and sludge are back recycled in blast furnace, zinc will accumulate on the walls, penetrate the lining, take part in the lining's deformation and disintegration and subsequently damage the blast furnace.[1,2] To find an effective and safe method to handle and recycle such kind of steelmaking wastes, a thorough characterization of these particles is a very important step.

Theoretically, knowledge of spatial elemental distribution and speciation in the accumulated particles is necessary for developing new recycling strategies and providing information on potential problems that could be encountered during the processing of such steelmaking wastes. To get these information, synchrotron radiation induced μ-X-ray fluorescence and μ-X-ray absorption fine structure spectroscopy are employed here. The reason why these two techniques are used is that, in previous studies, the power of the combination of synchrotron radiation induced μ-XRF and μ-XANES spectroscopy as micro-analytical tools to investigate the spatial distribution and speciation of heavy metals in small regions of typically a few micrometers in size was demonstrated.[3,4,5] Thus in the present research, elemental maps are created by scanning the particle in a regular grid pattern by the X-ray micro-beam and detecting the induced fluorescence intensities at each position. Depending on these patterns of distribution, μ-XANES spectroscopy were used to determine directly the chemical forms of zinc in micro-sized areas where zinc in high or low content. It is a preliminary research with the purpose to find useful information which might lead to new innovations helpful to create better recycling strategy.

2. Samples and methodologies

2.1 Samples and sample preparing

Zinc-containing steelmaking wastes, including electric arc furnace dust (labeled EAFD) and basic oxygen furnace OG sludge (labeled BOF OG), analyzed in this study were collected from the gas-cleaning system of carbon steelmaking operation in China. In previous studies, chemical analysis showed that the main compositions contained in EAFD were: total Fe 38.0%; ZnO 20.4%; SiO_2 4.1%; MgO 2.5%; CaO 8.3%; Al_2O_3 1.0%; MnO 3.0%; P_2O_5 0.4%. And the main compositions contained in BOF OG were: total Fe 54.1%; ZnO 3.4%; SiO_2 1.2%; MgO 5.7%; CaO 12.7%; Al_2O_3 0.2%; MnO 0.7%; P_2O_5 0.2%. For μ-XRF and μ-XANES analysis, the samples were pressed in an organic resin by cold cure, grounded and polished with silicon carbide paper. During the experiments, the polished cross section was first detected by μ-XRF spectroscopy. And then the areas with different zinc content were analyzed to obtain the chemical information of zinc by means of μ-XANES spectroscopy.

2.2 Methods

Both μ-XRF and μ-XANES measurements were performed at beamline BL15U at Shanghai Synchrotron Radiation Facility (SSRF). SSRF is a third-generation synchrotron operating at a current of between 130

Journal of Physics: Conference Series **430** (2013) 012097 doi:10.1088/1742-6596/430/1/012097

and 210 mA and energy of 3.5 GeV. Beamline BL15U at SSRF is an undulator beamline equipped with liquid-He-cooled Si(111) monochromator that has an energy resolution of 1.5 eV at 10 keV and Si(311) monochromator that has an energy resolution of 0.5 eV at 10 keV. The monochromatorized light was focused by K-B optics system. The sample stage consisted of multiple motor-stages: two for scanning the sample in horizontal and vertical directions and a third one in order to align the sample into the image plane of the focusing device. The minimal step size of the motor-stages was 20 nm.

During the measurements, the intensity of the incoming X-ray was monitored by an ionization chamber, which was used for normalizing the measured X-ray intensity of the sample. For both EAFD and BOF OG, the characteristic X-ray fluorescence of elements contained in particle 1 and 2 were detected by 7-element Si(Li) detector. And the spot size at the sample location were 2×2 μm^2 (H×V). Both the detectors were placed at 90° to the incoming X-ray in order to decrease the intensity of the Compton scattering. All the μ-XRF experiments were performed by using monochromatic radiation at an exciting energy of 10 keV.

Zn K-edge μ-XANES experiments were performed on pure Zn compounds and zinc-containing steelmaking wastes particles as well. The reference compounds used here (Zn, ZnO, $ZnFe_2O_4$, ZnS, $Zn_5(OH)_6(CO_3)_2$, $ZnSiO_3$) were chosen due to their possibility to be present in EAFD and BOF OG sludge. Each reference material was bought from chemical agents in China. And before measurements, each pure compound were mixed with LiF and pressed to form pellets. The reference XANES spectra were recoded in transmission mode. Due to the small dimensions of zinc-containing steelmaking wastes particles, their XANES spectra were recorded in fluorescence mode with the detector of 7-element Si(Li) detector. All the μ-XANES spectra were obtained at room temperature by scanning the energy in the 9610-9740 eV range in 0.5 eV steps with dwelling time of 1s. XANES spectra were analyzed with the aid of WinXAS software. Background was subtracted with Victoreen mode and then normalized in the range of 9680-9730 eV. After that, the normalized XANES spectra could be compared with the XANES spectra of reference materials.

3. Results and discussion

3.1 Zn reference compounds

Zn K-edge μ-XANES spectra of pure Zn compounds, including Zn foil, ZnO, $ZnFe_2O_4$, ZnS, $ZnSiO_3$, $Zn_5(OH)_6(CO_3)_2$ were measured prior to the particle analysis. And normalized XANES spectra were reported in Figure 1. As shown in Fig. 1, each reference material has its characteristics in the absorption area. In brief, for the zinc metal, the white line peak was at around 9669 eV. The XANES spectrum of $ZnFe_2O_4$ showed three resolved peaks at around 9665, 9669, 9673 eV respectively and an additional structure at around 9686.7 eV. For ZnS, the white line peak was at around 9665 eV with the shoulder at around 9662 eV and the additional structure at around 9669 eV. The XANES features of ZnO were the white line peak at around 9669 eV, a shoulder at around 9663 eV, and the additional structure at around 9679 eV. As for $Zn_5(CO_3)_2(OH)_6$, the white line peak was at around 9667 eV with shoulders at around 9665 eV and 9672 eV and the additional structure at around 9684 eV. While Zn K-edge XANES spectrum of $ZnSiO_3$ had two resolved peaks at around 9665 and 9668 eV. Depending on these characteristics, chemical form of Zn distributed in different regions can be investigated.

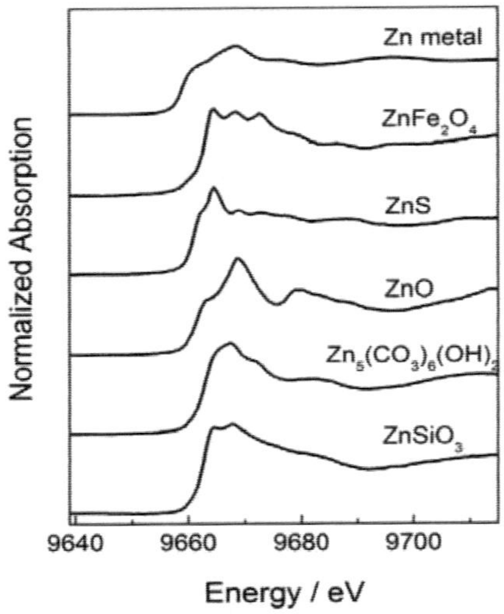

Fig.1. Zn K-edge XANES spectra of the measured Zn reference compounds

3.2 Electric arc furnace dust (EAFD)

Two particles of EAFD were detected by means of μ-XRF and μ-XANES spectroscopy. Figure 2 showed the distribution of elements (Ca, Fe, Mn and Zn) and Zn K-edge XANES spectra taken at interested spots for particle 1 of EAFD (EAFD 1). Here zinc was accumulated in some well-defined hot spots where with high amount and its distribution showed negative correlation with that of iron. And spot A, B, C, D, and E

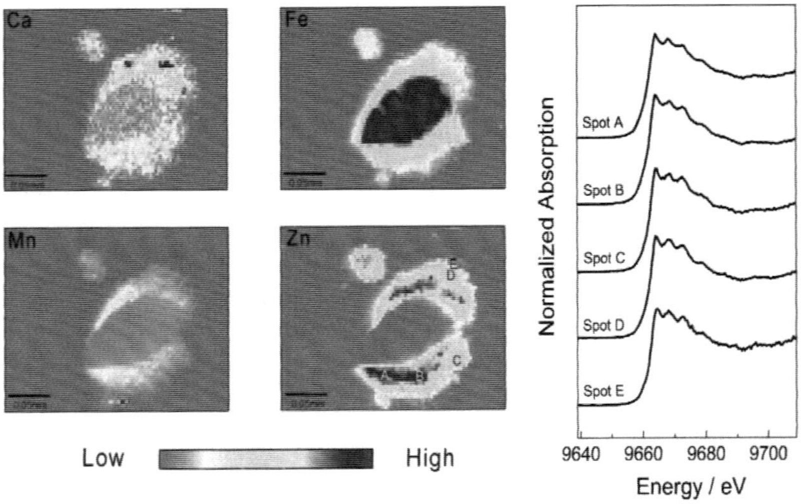

Fig. 2. (Colour online) The distribution of elements (Ca, Fe, Mn and Zn) and Zn K-edge XANES spectra taken at interested spots for particle 1 of EAFD (EAFD 1).

corresponded to the interested spots where zinc content decreasing from high to low. As it can be seen in Figure 2, all the XANES spectra were similar. Comparison with the XANES spectra of Zn pure compounds, it could found that zinc ferrite ($ZnFe_2O_4$) was present as the main constitutes in these analyzed spots within the cross section of EAFD 1.

Figure 3 showed the result of particle 2 of EAFD (EAFD 2). In this case, zinc was shared with iron and its distribution showed a positive correlation with that of iron. Spot A and B corresponded to the interested spots where zinc content were high and low respectively. The XANES spectra of these spots showed the similar results with that of EAFD 1. And zinc ferrite ($ZnFe_2O_4$) was also considered as the main constitute in the analyzed spots within the cross section of EAFD 2.

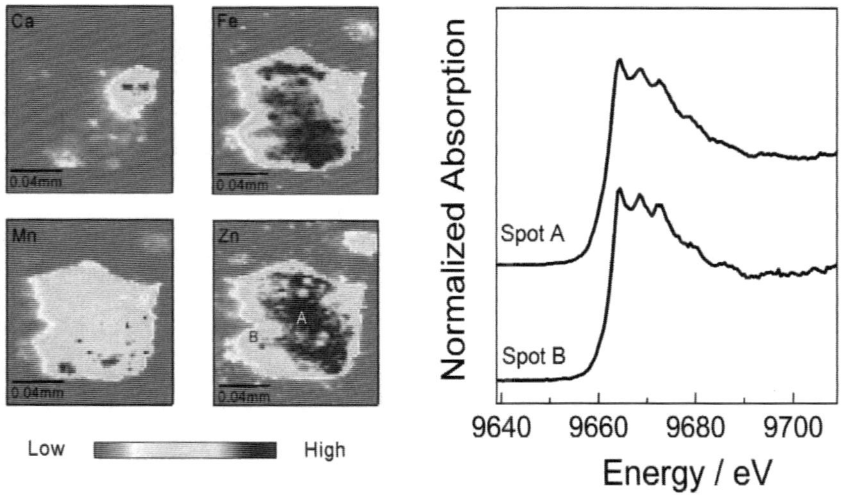

Fig.3. (Colour online) The distribution of elements (Ca, Fe, Mn and Zn) and Zn K-edge XANES spectra taken at interested spots for particle 2 of EAFD (EAFD 2).

3.3 Basic oxygen furnace OG sludge (BOF OG)

There were also two particles of BOF OG which were detected by means of μ-XRF and μ-XANES spectroscopy. Figure 4 showed the distribution of elements (Ca, Fe, Mn and Zn) and Zn K-edge XANES spectra taken at interested spots for particle 1 of BOF OG (BOF OG 1). Here zinc was accumulated in some well-defined hot spots where with high amount. And spot A, B, C, D corresponded to interested spots where with high zinc content, low iron content, low zinc content and high calcium content showed in the distribution maps of Zn, respectively. As it can be seen in Fig. 4, these XANES spectra showed similar features, that were the white line peak at around 9668 eV, shoulders at around 9665 eV and 9671 eV, and an additional structure at around 9685 eV. Comparison with the XANES spectra of zinc pure compounds, the similarity with that of $Zn_5(CO_3)_6(OH)_2$ could be observed, especially for spot D. This finding indicated that zinc carbonate was present as the main constitutes in these analyzed spots within the cross section of BOF OG 1.

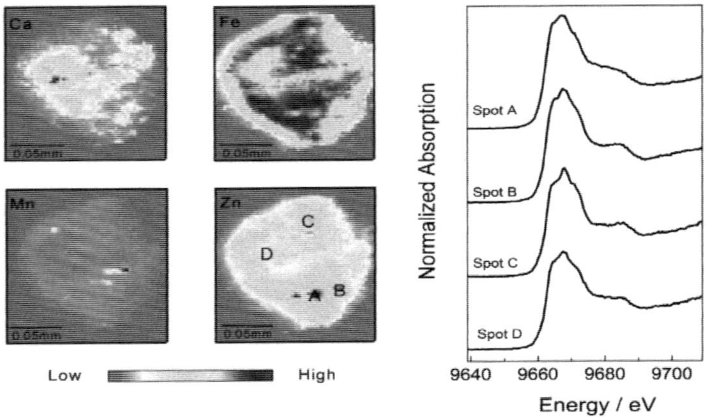

Fig.4. (Colour online)The distribution of elements (Ca, Fe, Mn and Zn) and Zn K-edge XANES spectra taken at interested spots for particle 1 of BOF OG sludge (BOF OG 1)

On the contrary, for particle 2 of BOF OG (BOF OG 2), Zn K-edge XANES spectra showed obvious variations from high Zn content to low Zn content as in Fig. 5. In this case, zinc was also accumulated in some well-defined hot spots with high amount. Spot A were selected for its highest zinc content. And Zn content decreased from high to low as from spot B to spot D. Among these spots, the XANES spectra of spot A showed the triplet structure between 9665 and 9675 eV (see peak a, b, c in Fig. 5). Results obtained by the comparison with these Zn K-edge XANES spectra of Zn pure compounds indicated the significant component of $ZnFe_2O_4$. While in case of spot D, the XANES spectra showed the similar characteristics with that of $Zn_5(CO_3)_6(OH)_2$ which also suggested the significant presence of zinc carbonate as in case of BOF OG 1. And the XANES features of spot B showed the combination of zinc carbonate and $ZnFe_2O_4$ to some extent. For further understand this variation, spot E, which was next to spot A, was chose. And its XANES spectra also showed the combination of $ZnFe_2O_4$ and zinc carbonate.

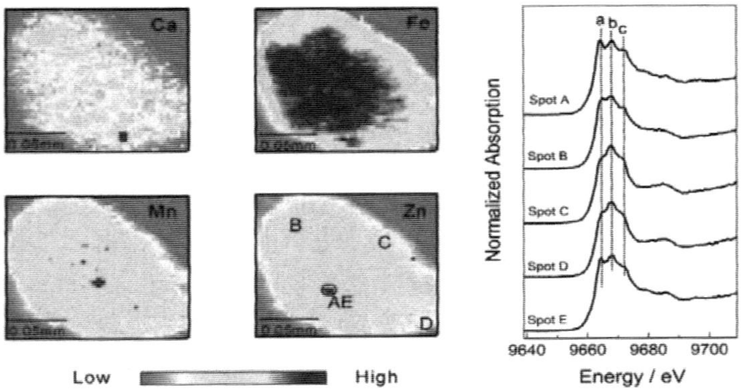

Fig.5. (Colour online)The distribution of elements (Ca, Fe, Mn and Zn) and Zn K-edge XANES spectra taken at interested spots for particle 2 of BOF OG sludge (BOF OG 2).Peak a,b and c refers to the triplet structure between 9665 and 9675 eV.

The variation of zinc species contained in regions with different zinc content indicated that chemical reactions between zinc species and other components occurred during the formation of BOF OG 2. For $ZnFe_2O_4$, it is one of the most common phases that can be found in zinc-containing steelmaking wastes,[6,7] such as electric arc furnace dust, basic oxygen furnace dust, and so on. In general, $ZnFe_2O_4$ has a normal spinel structure which forms at high temperature and in oxidizing conditions. Such spinel structure is very stable and difficult to be destroyed.[8,9] On the contrary, Zn carbonate was a type of very soluble phase. It was seldom mentioned in previous studies.[10] These information pointed out the possibility to separate the fine particles with different zinc species by controlling the relative manufacturing parameters at high temperature. However, how these two phases formed and present together during the formation of BOF OG sludge is difficult to give a clear explanation only by results obtained here. A thorough study is needed in the future.

4. Conclusions

To summarize, μ-XRF analysis showed that zinc distributed in two ways. One was shared with iron and its distribution showed a positive correlation with that of iron. The other was accumulated in some well-defined hot spots with high amount and its distribution showed negative correlation with that of iron. For EAFD, results of μ-XANES spectroscopy indicated that zinc was mainly present in the form of $ZnFe_2O_4$ within the whole particles no matter the spots with high or low zinc content. Whereas for BOF OG, $ZnFe_2O_4$ was the main constituent in well-defined hot spots while in other regions zinc was mainly in the form of zinc carbonate.

Acknowledgements

This research was financially supported by the project of Chinese Academy of Science (No. KJCX2-YW-N43) and National Nature Science Foundation of China (Grant No.11105208). And BL15U beamline at Shanghai Synchrotron Radiation Facility was also appreciated for providing the beamline time to do μ-XRF and μ-XANES experiments.

References

[1] Vymazal J 1985 *Acta Hydrochim. Hydrobiol.* **13** 627
[2] Adamson I Y R *et al* 2000 *Toxicol. Appl. Pharmacol.* **166** 111
[3] Pinzani M C C *et al* 2002 *Environ. Sci. Technol.* **36** 3165
[4] Vincze L *et al* 2002 *Anal. Chem.* **74** 1128
[5] Camerani M C *et al* 2007 *Anal. Chem.* **79** 6496
[6] Oustadakis P *et al* 2010 *J. Hazard.Mater.* **179** 1
[7] Havlík T *et al* 2006 *J.Hazard. Mater.* B **135** 311
[8] Shi L *et al* 2009 *Chinese Resources Comprehensive Utilization* **27**(2) 19
[9] Jha M K *et al* 2001 *Resources, Conservation and Recycling* **33** 1
[10] Sammut M L *et al* 2010 *Chemosphere* **78** 445

Journal of Physics: Conference Series **430** (2013) 012098 doi:10.1088/1742-6596/430/1/012098

Speciation of major heavy metals in ambient fine particles collected in the atmospheres of Xuanwei, Yunan, a higher lung incidence area in China

Senlin Lu[1], Fei Yi [1], Jun Lin[2], Jingjing Ren[1] and Xiaojie Hao[1]

[1]Shanghai Applied Radiation Institute, Shanghai University, Shanghai, 200444, Shanghai

[2]Shanghai Institute of Applied Physics, Chinese Academy of Sciences, Shanghai 201800, China

Email: senlinlv@shu.edu.cn; jlin1978@gmail.com

Abstract. Xuan Wei area, Yunnan province of China, has one of the highest lung cancer mortality rates in China. Heavy metals in the aerosol are potential facts that might attribute to the cause of lung cancer, and speciation of heavy metals could play a key role in toxicity of heavy metals. Therefore, we try to reveal speciation of major heavy metals by using of synchrotron radiation techniques in this study. Mass concentration of chemical elements in ambient particles were analyzed by PIXE (Proton induced X-ray emission), and speciation of Cu was investigated by XAFS (X-ray absorption near-edge structure). Our results demonstrated that mass levels of heavy metals ranked as Fe (791.62 ng/m3) >Zn (479.49 ng/m3) >Pb(154.76 ng/m^3) >Cu(28.51 ng/m^3) >Co (11.79 ng/m^3) >Ni (8.20 ng/m^3). The Cu speciation in the different size fraction was different. Speciation of Cu in fine particles ($PM_{0.56\sim0.32}$) is present as Cu (I), and significant amount of Cu is present as Cu (II) in the ultrafine particles ($PM_{0.1}$).

1. Introduction

Numerous epidemiological and toxicological studies have shown that a statistical association between health effects and ambient particle concentrations, especially the submicron fraction that can be easily penetrated into the alveolar region of the lungs [1-5]. Airborne particulate matter is a very complex mixture that includes particles in a wide range of sizes and with varied composition. These biological effects varied with number, size, and chemical composition of particulate matter [6-8]. Being a rural county with a population of more than 1.4 million in north-east of Yunnan province, China, Xuanwei is famous for its unusually high incidence of lung cancer, which is eight and four times higher than that of the Chinese national average incidence for women and men, respectively[9] . A number of studies of Xuanwei lung cancer has been conducted since 1970s [10,11,12]. However, a recent report by Vermeulen R. [13] et al found that crystalline silica was another contributing factor to the high lung incidence. Most of previous studies focused on PM10 emitted from coal combustion rather than fine or ultrafine particles. This study will focuse on physicochemical characterization of fine and ultrafine particles in the atmosphere of the high lung incidence area and our results could provide new essential data for evaluation of the high lung incidence.

Journal of Physics: Conference Series **430** (2013) 012098 doi:10.1088/1742-6596/430/1/012098

2. Experiments

2.1 Sample collections

The study area located in Laibin town (26。 13.883′ N, 26。 13.883′E), a suburb of Xuanwei, Yunnan Province, China.Size-separated aerosol particles were collected by using a 13-stage MOUDI125B (Micro-Orifice Uniform Deposit Impactor, MSP, Co., Minneapolis, USA) with the flow rate of 10 L/min. The sampling campaigns were performed from April 21 to 25, 2011. A total of three groups (39 samples) of size-segregated ambient particle samples were collected. The sampling period for each group was set at 48 hr. The sampled filters were kept in desiccators for analysis.

2.2 Chemical analysis by PIXE

The concentration of 22 elements (Mg, Al, Si, P, S, Cl, K, Ca, Sc, Ti, V, Cr, Mn, Fe, Co, Ni, Cu, Zn, As, Br, Mo, Pb) in size-resolved particles were determined by PIXE analysis using a 1.7Mev proton beam from a Tandem accelerator in the Department of Science and Technology of Nuclear sciences, Beijing Normal University. The X-rays were detected by means of a Si(Li) detector. The proton beam size is 10 mm × 4 mm.

2.3 X-ray Absorption Fine Stucture analysis

Speciation of toxic chemical copper in the size-resolved particles was investigated by XANES. Cu L-edge XANES spectra were measured at Beamline 4B7B of the Beijing Synchrotron Radiation Facility (BSRF), Beijing, China. The electron energy in the storage ring is ~2.2 Gev with a current of about 100mA. The L-edge absorption of copper was measured with a Monk-Gillieson monochromator. The energy region of X-ray selected is about from 900 to 980 eV. CuO, Cu2O, CuSO4, the possible components of copper in the PM samples, were selected as reference materials. Background removal, normalization and deconvolution of XANES spectra were carried out using IFEFFIT package.

3. Results and discussion

3.1 Mass concentration of the size-resolved particles

Our result showed that mass level of different size particles was different. The highest mass concentration was found in the 1.0-0.56 μm size range (71.78±2.8μg/m3), while the lowest mass level was in the 10-5.6μm size range (9.55±1.2μg/m3). The total mass concentration of the ambient variously size-resolved particles was 316.9 μg/m3.

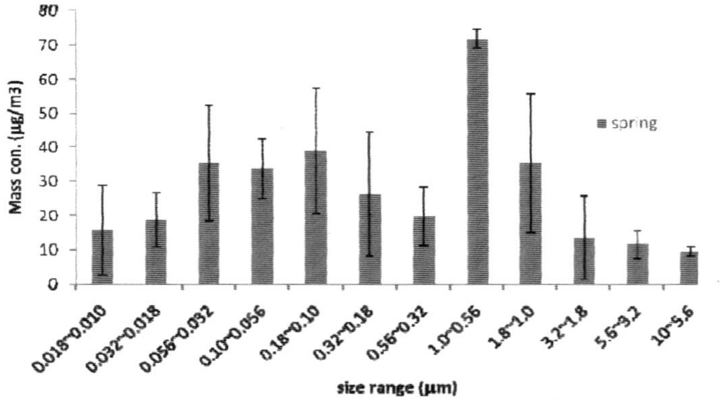

Figure 1: Mass concentrations of size-segregated particles collected in Xuanwei spring atmosphere

3.2 Mass concentration of chemical elements in size-resolved ambient particles

Chemical elemental analysis results showed that the total mass concentration of the 22 chemical elements was 8747.37 ng/m3 (Figure 2). Sulfur was the most abundance element in Laibin ambient particles with mass level 2624.04 ng/m3, and mainly distributed in the fine particles. Being a typical

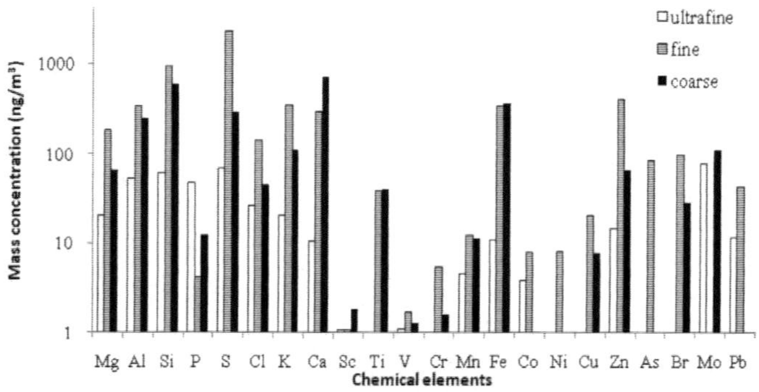

Figure 2: The mass concentration of 22 elements in Xuanwei ambient particles

anthropogenic element, sulfur in ambient particles mainly contributed by coal combustion.Crustal elements, including Ca, Si, Al, Fe (four dominant elements) mainly distributed in the coarse particles. Mass concentrations of Ca in coarse, fine and ultrafine particles was 876.99, 120.6, and 9.81 ng/m3 respectively. Correspondingly, Si was 963.04, 638.44, and 6.33 ng/m3, Al was 386.61, 254.72,5.05 ng/m3, Fe was 521.53, 186.82, 10.07 ng/m3.Anthropogenic elements, such as Zn, Ni, Pb, Cr Co, Cu, were richer in fine particles. The mass levels of Zn and Pb in fine particles was 310.57 and 43.64 ng/m3 respectively. Other metal elements in fine particles ranked in the following order: Cu (28.51 ng/m3) > Mn (28.31 ng/m3) > Ni (8.2 ng/m3) > Cr (7.77 ng/m3), while that for winter was Mn (106.41 ng/m3) > Ni (23.19 ng/m3) > Cr (15.89 ng/m3) > Cu (12.86 ng/m3).

3.3 Speciation of Cu in different ambient size particles

XANES results showed that the spectra of reference materials Cu2O [Cu(I)] and CuO [Cu(II)] is different(Figure 3). The spectrum of Cu+ has a conspicuous preedge peak at approximately 930 eV,

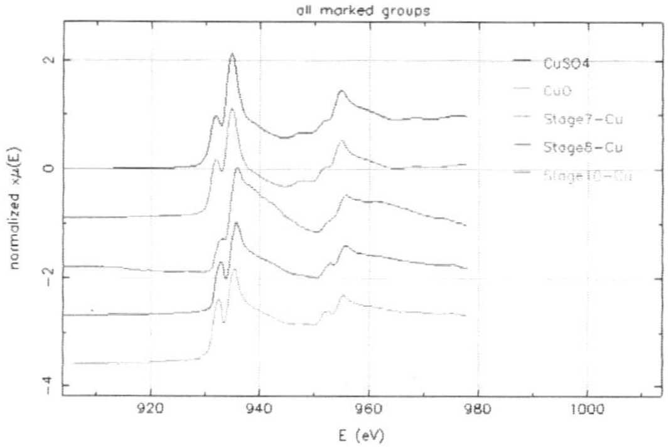

Figure 3: Cu–L edge XANES spectra of the size-fractionated aerosol sample in Xuanwei, reference spectra of Cu(II) sulfate, Cu(II) oxide are plotted for comparison. Particles with diameter 0.56~0.32、0.32~0.18 and 0.1~0.056 μm could be collected through the stage 7、stage 8 and stage 10 respectively.

while Cu2+ exhibits a very weak pre-edge structure. The speciation of Cu in the fine fraction (<2.5 μm) was found the same as the Cu in fine particles (at the two stage, stage7 and stage 8). Cu L-edge absorption spectrum in the ultrafine particles (<0.1 μm) showed differently compared with that in the fine particles. Significant amount of Cu is present as Cu (I) in the ultrafine particles.

4. Conclusions

1. Mass level of the size-resolved ambient particles collected in the atmosphere of Xuanwei high lung cancer area showed differently. The highest mass concentration was found in the 1.0-0.56 μm size range (71.78±2.8μg/m3), while the lowest mass level was in the 10-5.6μm size range (9.55±1.2μg/m3), suggest that fine particles be the main fraction in airborne particles.
2. Mass concentration of chemical elements in the size-resolved particles (aerodynamic diameter, 10 nm ~10μm) in spring varied from 0.24 to 9.92 g/m3, mass levels of heavy metals ranked as Fe (791.62 ng/m3)>Zn (479.49 ng/m3)>Pb(154.76 ng/m3)>Cu(28.51 ng/m3)>Co (11.79 ng/m3)>Ni (8.20 ng/m3).
3. The Cu speciation in the different size fraction was different. Speciation of Cu in fine particles (PM0.56~0.32) is present as Cu (I), and significant amount of Cu is present as Cu (II) in the ultrafine particles (PM 0.1).

Acknowledgements

This research was supported by the NSFC (Grant No. 41273127, 41073073) and by the Shanghai Pujiang Talent Program, Innovation Program of Shanghai Municipal Education Commission (11ZZ80), beamline 4B7B of the Beijing Synchrotron Radiation Facility (BSRF), Beijing, China was greatly appreciated for providing the beamline time to do XANES experiments.

References

[1] Pope III, C.A. 2000. *Aerosol Science & Technology*. **32** (1): p. 4-14..
[2] Selgrade, M.J.K. 2000. *Immunopharmacology* **48** (3): p. 319.
[3] Kampa, M., E. Castanas 2008. *Environmental Pollution* **151** (2): p. 362-367.
[4] Pope III, C.A., R.T. Burnett, D. Krewski, M. Jerrett, Y. Shi, E.E. Calle, M.J. Thun, 2009 . Circulation **120** (11): p. 941-948.
[5] Sacks, J.D., L.W. Stanek, T.J. Luben, D.O. Johns, B.J. Buckley, J.S. Brown, M. Ross, 2011. *Environmental health perspectives* **119** (4): p. 446.
[6] Strak, M., N.A.H. Janssen, K.J. Godri, I. Gosens, I.S. Mudway, F.R. Cassee, E. Lebret, F.J. Kelly, R.M. Harrison, B. Brunekreef 2012.. *Environmental health perspectives,* **120** (8): p.1183–1189.
[7] Dockery, D.W. 2009. *Annals of epidemiology* ,**19** (4): p. 257-263.
[8] Amatullah, H., M.L. North, U.S. Akhtar, N. Rastogi, B. Urch, F.S. Silverman, C.W. Chow, G.J. Evans, and J.A. Scott, 2012. *Inhalation toxicology* **24** (3): 161–171.
[9] Mumford, J., X. He, R. Chapman, S. Cao, D. Harris, X. Li, Y. Xian, W. Jiang, C. Xu, J. Chuang, 1987. *Science*. **235** (4785): p. 217-220.
[10] Mumford, J., R. Chapman, D. Harris, X. He, S. Cao, Y. Xian, X. Li, 1989. *Environment International* **15**(1-6): p. 315-320.
[11] He, X., W. Chen, Z. Liu, R.S. Chapman, 1991. *Environmental health perspectives* **94**: p. 9-13.
[12] Lü, J.G., R.J. Xu, Q. Zhang, J. Liu, C. Liao, F.S. Wei, 2009. *Chinese Science Bulletin*. **54**(19): p. 3528-3535.
[13] Vermeulen, R., N. Rothman, Q. Lan, 2011. *La Medicina del lavoro* **102** (4): p. 362-3.

Journal of Physics: Conference Series **430** (2013) 012099 doi:10.1088/1742-6596/430/1/012099

Speciation of As in calcite by micro-XAFS: Implications for remediation of As contamination in groundwater

Y Yokoyama[1], T Iwatsuki[2], Y Terada[3] and Y Takahashi[1, 4]

[1]Department of Earth and Planetary Systems Science, Graduate School of Science, Hiroshima University, 1-3-1 Kagamiyama,Higashi-Hiroshima, Hiroshima 739-8526, Japan
[2]Mizunami Underground Research Laboratory, Japan Atomic Energy Agency, 1-64, Yamanouchi, Akeyo, Mizunami, Gifu 509-6132, Japan
[3]SPring-8, Japan Synchrotron Radiation Research Institute (JASRI), 1-1-1 Kouto, Sayo-cho, Sayo-gun, Hyogo 679-5158, Japan
[4]Key Laboratory of Petroleum Resources, Chinese Academy of Sciences, Donggang Road, Lanzhou 730000, China

E-mail: yoshiyuka@hiroshima-u.ac.jp (Y Yokoyama)
ytakaha@hiroshima-u.ac.jp (Y Takahashi)

Abstract. To evaluate the role of calcite as a host phase of arsenic (As) in As-contaminated groundwater, distribution behavior of Asbetween natural calcite and groundwater in deep underground was investigated based on As oxidation state. Speciation analyses of As in natural calcite by μ-XRF-XAFS analyses showed (i) preferentialarsenate uptake by calcite, and (ii) promptness of arsenate uptake by minor iron (Fe) carbonate minerals coprecipitated with calcite. These findings suggest that the effect of calcite on As remediation of the As-contamination systems stronglydepends on arsenite to arsenate ratio (i.e., redox condition) in groundwater, and maybe governed bythe amount of Fe coprecipitated with calcite.

1. Introduction

The natural occurrence of arsenic (As)-contamination in groundwater is a serious problem across southern Asia. The mostcommon and widely accepted mechanism of As release togroundwater in alluviumis reductive dissolution of iron (Fe)hydroxides containing As (e.g., [1]). However, in general, the correlation between As and Fe concentrations in the As-contaminated groundwater is poor (e.g., [2]), suggesting that As mobilization/immobilization processes controlling the As concentration in groundwater are also related to other minerals.

Carbonate minerals are ubiquitously found invarious surface environments including areas under reducingconditions.Many As-contaminated groundwaters are slightly saturated with various carbonate minerals [3]. Thus, carbonate minerals have been recentlyof great interest as host phases of Asin groundwater under reducing conditions. Arsenic in natural water is dissolved as trivalent arsenite () and pentavalent arsenate (). Our previous experimental study found that the affinity of arsenate with calcite ($CaCO_3$) is higher than that of arsenite[4]. Siderite ($FeCO_3$) and carbonate greenrust $[Fe^{II}_4Fe^{III}_2(OH)_{12}][4H_2O{\cdot}CO_3]$ were also found to interact more strongly with arsenate than with arsenite [5]. These properties were only observed in laboratory studies, whereas there have beenfew studiesinvestigating Asinteraction with natural carbonate minerals at different oxidation states. In this paper, we report the results of speciation analyses of As both in natural carbonate minerals and in groundwater by micro X-ray fluorescence andX-rayabsorption fine structure (μ-XRF-XAFS) analyses and highperformance liquidchromatography-ICP-MS (HPLC-ICP-MS), respectively, with the aim of estimating the As-scavenging capability of carbonate minerals in groundwater systems.

2. Materials and Method

Natural carbonate minerals, which occur as fracture-filling (FF) secondary minerals at 450 m below ground level (mbgl),were collected from a drilling core(HDB-10) of Neogene marine sedimentary rocks (Wakkanai Formation)in the Horonobe area, northern Hokkaido, Japan.These carbonate minerals originated from carbon dioxide gasproduced by diagenetic processes (e.g., sulfate reduction by sulfate-reducing bacteria and biochemical maturation of organic matter) [6] and saline groundwater (total dissolved solids(TDS)$<$ 22 g/L) which is inferred to originate from fossil seawater [7].Groundwater sample was collectedat 250 mbglwhile maintaining anaerobic conditionsin the shaft ofthe Horonobe Underground ResearchCenter located near the HDB-10 drillingsite.

Concentration of As in the FF carbonate minerals was measured by ICP-MS (Agilent 7700cx, Agilent Technologies) after dissolution in ammonium acetate adjusted to pH 5.0 with acetic acid. Speciation of As in the groundwater sample was performed using HPLC-ICP-MS (pump: Pu-2089 Plus; oven: Co-2065 Plus, JASCO) with anion exchange columns (TSK gel Super IC-AP, Tosoh). A thin section of natural carbonate sampleswas prepared for electronmicroprobemicroanalyzer (EPMA) and μ-XRF-XAFS analyses. Microscopic chemical distribution on the thin section was observed by EPMA (JXA-8200, JEOL). The μ-XRF-XAFS analyses using an X-ray micro beam were performed at BL-4A in KEK-PF (Ibaraki, Japan) and BL37BXUin SPring-8 (Hyogo, Japan).

3. Results and Discussion

Chemical datafrom groundwater in the HDB-10 borehole indicate that groundwater at about 450 mbglis nearly saturated or oversaturated with respect to calcite and sideriteunder areducing condition(pH 6.8; Eh = –234 mV to –148 mV) [8]. Thermodynamic speciation of Asin the groundwater shows that arsenite is dominant dissolved As species, with the ratio of arsenite to total As being above 99.99%. The predominance of arsenite in groundwater was directly observed by the HPLC-ICP-MS analysis of the groundwater samples collected nearthe HDB-10 borehole (Figure 1).

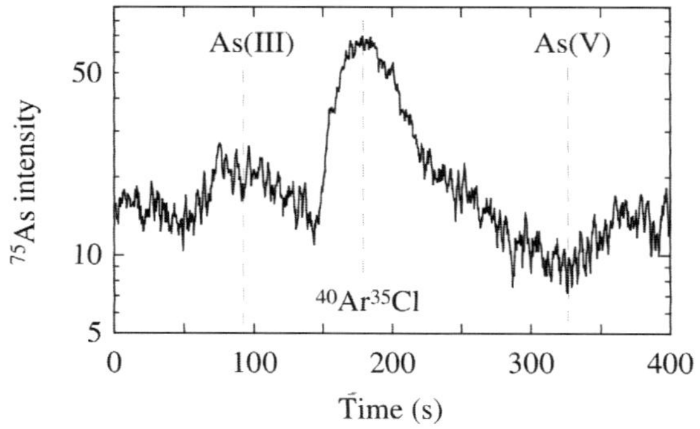

Figure 1. Chromatogram of ^{75}As in the groundwater sample measured by HPLC-ICP-MS.Retention times of arsenite [As(III)], arsenate [As(V)], and $^{40}Ar^{35}Cl$ species are indicated as dashed lines identified by $NaAsO_2$, NaH_2AsO_4, and NaCl solutions, respectively. In this measurement system, $^{40}Ar^{35}Cl$ produced by a large amount of chlorine in groundwater does not interruptidentificationsof As(III) and As(V).

The FF carbonate minerals were identified as calcite with a bulk X-ray diffraction pattern. These minerals also include 1wt.% Fe and 0.7wt.% magnesium (Mg). Bulk Fe K-edge EXAFS spectra for theFF-calciteexhibit a similar oscillatory structure to ankerite [$Ca(Mg,Fe)(CO_3)_2$], indicating that minor ankerite is formed in the FF-calcite (Figure 2).

Two-dimensional μ-XRF maps show that the thin section consists both of the FF-calcite and the rock matrix areas (Figure 3a). Mineral phases of the rock matrix were found to be quartz, silicate, sulfide, and carbonate minerals by the electron microprobe analysis. Distribution of As on the μ-XRF map partially corresponds to that of Fe, and that of pyrite on a backscattered electron(BSE) image. However, the mineralphases where Asaccumulates could not be identified completelyon the BSE image because the probing depth of XRF imaging is deeper than that of electron imaging. Arsenic μ-XANES analysis for the FF-calcite was conducted to investigate As-immobilization to calcite based on its oxidation state. Micro-XANES spectra of As show that As was incorporated into the FF-calcite as arsenate (Figure 3b). In addition, the distribution of Aswithin the FF-calcitephasecorrelates most closely with that of Fe on the XRF maps.These results may suggest that arsenate was distributed in minor Fe carbonate phase of the FF-calcite, as identified by the bulk Fe K-edge EXAFS analysis. However, such Fe carbonate minerals could not be foundmicroscopicallyin the FF-calcite phase.

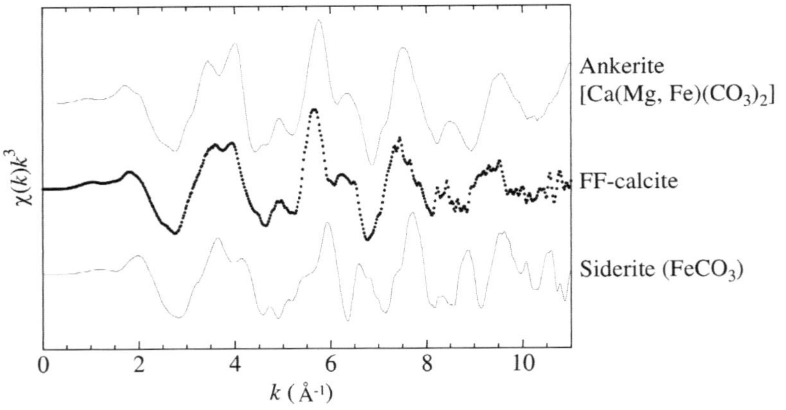

Figure 2. Normalized k^3-weighted EXAFS spectra at the Fe K-edge for Fe in the FF-calcite and the reference compounds (ankerite and siderite).

The FF-calciteincorporatedarsenate despite arsenite is the main species of Asin the groundwater. Thepreferencefor arsenate hasbeen observed experimentallyin previous studies on calcite and Fe carbonate minerals regarding their importance as host phases of As [4, 5]. In the case of calcite, no interaction of arsenite with calcite at circumneutral pH had been observed in the laboratory [9]. These properties indicate that calcite plays a role as a sink of As by incorporating minor arsenate even under reducing conditions, and that its ability depends heavily on arsenite to arsenate ratio in groundwater.

A static groundwater system in the deeper sedimentary rocks at the Horonobearea is expected to result in a nearly equilibrium state between groundwater and surrounding sedimentary rocks (minerals). Thus, based on the concentration of arsenate both in the FF-calcite and groundwater, an apparent distribution coefficient (K) of arsenate could be calculated ($K = [As^V]_{calcite}/[As^V]_{groundwater} =$ 920 L/kg). This value is higher than the K value of arsenate between calcite and water ($K_{arsenate\text{-}calcite} =$ 490 L/kg) determined in our laboratory experiments [9], indicating that the FF-calcite includes a largeramount of arsenate (160 μg/kg) thanthe predicted value for the immobilization of arsenate by calcite. As noted above, the existence of minor Fe carbonate minerals in the FF-calcite may be related to Asincorporation. According to the previously reported K values for arsenate sorption reaction [3], arsenate stronglyinteracts with siderite rather than calcite. Thus, micro-scale Fe carbonate minerals (e.g., siderite or ankerite) in the FF-calcite contribute to the immobilization of arsenate, which consequentlyenhances the capacity of the FF-calcite for As-scavenging. In the rock matrix, authigenic siderite along with pyrite was found, which is also assumed to expand the capability of carbonate minerals as a sink for As in deep underground systems. A similar occurrence of siderite has been reported in sediments of As-contaminated areas [10]. Future work should focus on the evaluation of such siderite based on microscopicspeciation using μ-XAFS technique.

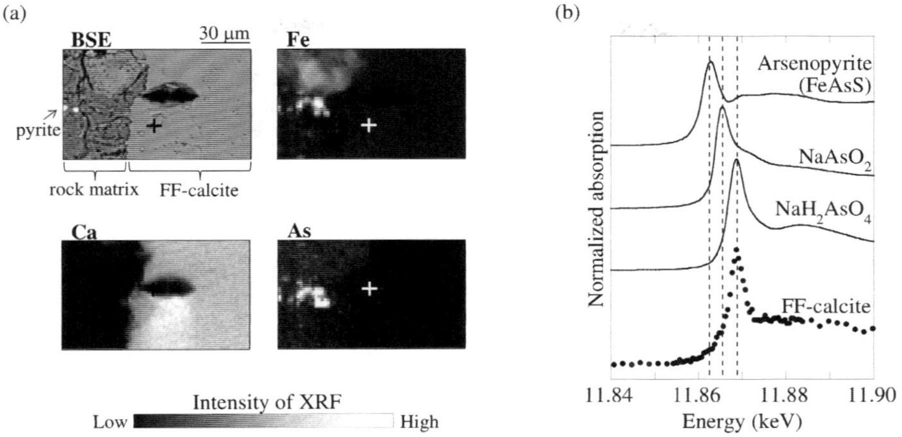

Figure 3.(a) ABSE photograph and XRF mapsfor Ca, Fe, and As in the thin section of the FF-calcite and the rock matrix. A cross in the element distribution maps indicates ananalytical point of As μ-XANES. (b)Micro As K-edge XANES spectra for the point in the FF-calcite and reference compounds (arsenopyrite, $NaAsO_2$, and NaH_2AsO_4).

4. Summary

The results of μ-XRF-XAFS analyses for the FF-calcite formed in deep underground suggested that calcite could be a host phase of As in As-contaminated groundwaterthrough selective incorporation of arsenate. However, the effect of calcite in theremediation ofAs-contamination systemsstronglydepends on arsenite to arsenate ratio (i.e., redox condition) in groundwater. The amount of As incorporated into carbonate minerals can also be enhanced by the amount of Fe carbonate minerals coprecipitated with calcitebecauseminor Fe carbonate minerals in calcite can enhance the capacity of calcite forAs immobilization.

Acknowledgments

We deeply thank Y. Shibata and H. Ishisako for assistance inEMPA analysis and preparation of the thin section, respectively.This study was supported by theJSPS Research Fellowships for Young Scientists. This work hasbeen performed with the approval of KEK-PF(2012G111) and JASRI (2011B1517 and 2012A1240).

References

[1] Nickson R, McArthur J, Burgess W, Ahmed K M, RavenscroftP and Rahman M1998*Nature***395** 338
[2] BGS and DPHE2001*Arsenic Contamination of Groundwater inBangladesh*, ed Kinniburgh D G and SmedleyP L (Keyworth: British Geological Survey)
[3] Smedley P L and Kinniburgh D G 2002 *Appl.Geochem.***17** 517–68
[4] Yokoyama Y, Mitsunobu S, Tanaka K, Itai T and Takahashi Y 2009 *Chem. Lett.***38** 910–1
[5] Jönsson J and Sherman D M 2008*Chem. Geol.***255** 173–81
[6] Iwatsuki T, Mizuno T, Hama K and Kunimaru T 2010 *Groundwater Response to Changing Climate*, ed Taniguchi M and Holman I P (Boca Raton: CRC Press) pp 87–98
[7] Hama K, Kunimaru T, Metcalfe R and Martin A J 2007 *Phys. Chem. Earth***32** 170–80
[8] Kunimaru T, Shibano K, Kurikami H, Tomura G, Hara M and Yamamoto H 2007 *Analysis of Ground Water from Boreholes, River Water and Precipitation in the Underground Research Laboratory Project* (Tokai: JAEA) JAEA-Data/Code 2007-015 (in Japanese)
[9] Yokoyama Y, Tanaka K and Takahashi Y 2012 *Geochim. Cosmochim. Acta***91** 202–19
[10] Pal T and Mukherjee P K 2009 *Environ. Geol.***56** 1211–25

A study on adsorption mechanism of organoarsenic compounds on ferrihydrite by XAFS

M Tanaka,[1]* Y Takahashi,[1] and N Yamaguchi[2]

[1] Graduate School of Science, Hiroshima-University, 1-3-1 Kagamiyama, Higashi-Hiroshima 739-8526, Japan
[2] National Institute for Agro-Environmental Sciences, 3-1-3, Kan-nondai, Tsukuba 305-8505, Japan

E-mail: tanamasa@hiroshima-u.ac.jp

Abstract. Anthropogenic organoarsenic compounds which were used such as agrochemicals, pesticides, and herbicides can have a potential as a source of arsenic pollution in water. In the process, the adsorption of arsenic onto mineral surface in soil may play an important role to affect arsenic distribution in solid-water interface. However, adsorption structures of organoarsenic compounds on the iron-(oxyhydr)oxides are not well known. In this study, extended X-ray absorption fine structure (EXAFS) spectroscopy was employed to know the adsorption structure of methyl- and phenyl-substituted organoarsenic compounds (methylarsonic acid (MMA), dimethylarsinic acid (DMA), phenylarsonic acid (PAA), and diphenylarsinic acid (DPAA) onto ferrihydrite which can be a strong adsorbent of arsenic. EXAFS analysis suggests that the formation of inner-sphere surface complex for all organoarsenic compounds with ferrihydrite regardless of the organic functional groups and the number of substitution. The As-Fe distances are around 3.27 Å, which suggests both mono- and bi-dentate inner-sphere complexes by DFT calculations. The corresponding coordination numbers (CNs) are less than two, suggesting that coexistence of both structures of inner-sphere complexes.

1. Introduction

The water pollution by arsenic has been a serious problem around the world. Its predominant source is inorganic arsenic species in natural environment. In addition, the organoarsenic species such as methyl and phenyl arsenic compounds can also be a pollution source, although organoarsenic compounds are thought to be less toxic than inorganic ones [1,2]. For example, diphenylarsinic acid (DPAA) contamination in well water posed health hazards to the residents at Kamisu Town in Japan [3]. Such organoarsenic compounds are used as agrochemicals, pesticides, herbicides and so on [4,5]. Furthermore, phenylarsenic species were considered as degradation products of chemical warfare agents used during the World Wars I and II [3]. After the World War II, such kind of chemical weapon-related agents were abandoned by Europe, China, Japan and other countries by dumping in the sea or by disposal in geosphere [6], which can migrate afterwards in the environment. Hempel et al. suggested that phenylarsonic acid (PAA) may be transformed microbiologically in anoxic aquifers which indicate a natural attenuation potential [7]. Nakamiya et al. and Harada et al. also reported bio-

Journal of Physics: Conference Series **430** (2013) 012100 doi:10.1088/1742-6596/430/1/012100

degradation of DPAA to PAA in environment [8,9]. These studies indicate that such anthropogenic organoarsenic compounds have a potential to be a secondary source of higher toxic inorganic arsenics.

A number of studies on the migration of inorganic arsenic into groundwater in environment have been done because of the high attention of the arsenic pollution, [10]. However, several problems still remain unclear, especially for organoarsenic species. In its migration process, the strength of adsorption of arsenic onto minerals is an important factor to determine the distribution of arsenic between soil and water. Ferrihydrite is one of the most important iron-(oxyhydr)oxide mineral in natural systems due to its large surface area and characteristic as a strong absorbent of anionic species such as inorganic As in the environment [11]. Thus, ferrihydrite may control the migration of arsenic as its scavenger in soils. Therefore, it is important to extend the adsorption mechanisms of arsenic compounds on iron-(oxyhydr)oxides to organoarsenic species to understand their migration processes in soils. The extended X-ray absorption fine structure (EXAFS) analysis has been employed to obtain the information of the adsorption structures of inorganic arsenic compounds on several adsorbents [12]. However, there have been only a few studies on the adsorption structure of organoarsenic compounds, which are limited to those on nanocrystalline TiO_2 [13], amorphous aluminum oxide [14], and goethite [15], whereas no study has been done on ferrihydrite.

This study focuses on the adsorption mechanism of PAA and DPAA as well as methylarsonic acid (MMA) and dimethylarsinic acid (DMA) on ferrihydrite. In addition, substitution effect on the arsenic-adsorption is also discussed. For the purpose, the EXAFS analysis for these organoarsenic species was conducted.

2. Materials and methods

2.1. Preparation of adsorption samples

Two-line ferrihydrite was synthesized in laboratory as described by Schwertmann and Cornell (2000) [16]. The mineralogy of the 2-line ferrihydrite was confirmed by powder X-ray diffraction. All arsenic solutions samples were prepared with MMA ($CH_3AsO(OH)_2$; Wako), DMA ($(CH_3)_2AsO(OH)$; Tokyo Kasei), PAA ($C_6H_5AsO(OH)_2$; Wako) or DPAA ($(C_6H_5)_2AsO(OH)$; Wako) diluted by MQ water. Adsorption samples for EXAFS were prepared using 5 mL solution and 10 mg ferrihydrite for MMA, DMA, PAA, and DPAA. The pH was adjusted by adding a HNO_3 or NaOH solution to pH 4 and 7, and the ionic strength was fixed at 0.10 M by $NaNO_3$. The samples were shaken by a reciprocation shaker for 24 hours at 25°C and at 130 rpm, and then filtered by a cellulose membrane filter (0.45 μm).

2.2. XAFS measurement of organoarsenic compounds

The K-edge EXAFS spectra of MMA, DMA, PAA, and DPAA adsorbed on ferrihydrite were collected at beamline BL01B1 at SPring-8 (Hyogo, Japan) and BL-12C at Photon Factory (Tsukuba, Japan). A Si(111) double-crystal monochromator was used. Two Rh-coated mirrors placed at above and downstream of the monochromator were used for collimation and focusing of X-ray beam in SPring-8. Harmonic rejection was achieved by adjusting the glancing angle of these two mirrors. The K-edge EXAFS spectra for arsenic adsorbed on ferrihydrite were measured in fluorescence mode using a 19-element Germanium semiconductor detector, and the spectra of reference solution samples were measured in transmission mode. The X-ray energy was calibrated with a single peak at XANES region of $KAsO_2$ at 11865 eV. The measurement was carried out at room temperature under ambient air condition. EXAFS data were analyzed by REX2000 Version 2.5.9 (Rigaku Co. Ltd.). The Fourier transformation of the $k^3\chi(k)$ EXAFS oscillation from k (Å^{-1}) space to R (Å) space was performed in a range of 3.0 – 13.0 Å to obtain a radial structural function (RSF) .

3. Results and discussion

Figures 1a and 1c show k^3-weighted $\chi(k)$ spectra of As K-edge EXAFS for methyl substituted arsenic solutions and those adsorbed on ferrihydrite samples at pH 4 and 7. Deconvoluted spectra of adsorbed

Journal of Physics: Conference Series **430** (2013) 012100 doi:10.1088/1742-6596/430/1/012100

samples at pH 4 and their RSF were shown in Figures 1b and 1d. The structural parameters which were obtained by curve fitting are listed in Table 1. The $k^3\chi(k)$ spectra of the both MMA and DMA adsorption samples show the different shape from that of the solution (Figures 1a and 1c), which clearly indicates that MMA and DMA adsorb on ferrihydrite as inner-sphere complexes at pH 4 and 7.

In the RSF for MMA and DMA (Figures 1b and 1d), the first shell is composed of As-O and As-C shells. The distances of As-O and As-C obtained by the fitting were 1.68 and 1.88 Å for MMA and 1.70 and 1.89 Å for DMA, respectively (Table 1). These results are similar to previous EXAFS study for MMA and DMA adsorbed on TiO_2 by Jing et al. [13] and on goethite by Shimizu et al. [15].

The As-Fe shells appear at around R + ΔR = 3.0 Å for adsorbed samples at pH 4 and 7 (Figure 1b and 1d), and the As-Fe distances and its coordination numbers (CNs) are 3.28 Å and 1.7 for MMA and 3.26 Å and 1.5 for DMA (Table 1), showing that the CN of DMA is smaller than that of MMA. The CN for arsenate onto ferrihydrite in EXAFS analysis was reported as ~ 2.0. This result suggests that arsenate forms bidentate inner-sphere complex with ferrihydrite [12]. Adamescu et al. reported that DFT calculations of DMA-iron-(oxyhydr)oxides model complexes for bidentate and monodentate structures show that the As-Fe distances for these are almost same (3.16-3.34 Å) [17], and around 3.27 Å is in between these distances. It indicates that the adsorption samples can include these both structures. Thus, less than 2.0 of CN may indicate the existence of both bi- and mono-dentate structures. This suggestion is consistent with the results of ATR-FTIR study, which indicated that MMA and DMA were adsorbed on aluminum oxide [14] and iron-(oxyhydr)oxides [17] by both inner-(mono-/bi-dentate) and outer-sphere complexation.

Figures 2a and 2c show k^3-weighted $\chi(k)$ spectra of As K-edge EXAFS for phenyl-substituted arsenic solutions and those adsorbed on ferrihydrite samples at pH 4 and 7 as well as deconvolution of the spectra for adsorbed samples at pH 4. Their RSF are also shown in Figures 2b and 2d. The structural parameters which were obtained by curve fitting are listed in Table 2. The $k^3\chi(k)$ spectra of the PAA and DPAA adsorption samples showed the shape different from that of the solution. This discrepancy clearly indicates that both PAA and DPAA also adsorb on ferrihydrite as inner-sphere complex at pH 4 and 7.

In the RSF for PAA and DPAA, the first shell is composed by oxygen and the nearest carbon atom of phenyl group (C_1) and corresponding As-O and As-C_1 distances obtained by the fitting are ca. 1.68 and 1.84 Å for PAA and 1.68 and 1.90 Å for DPAA, respectively (Table 2). These are similar distances to the case of methyl substituted compounds (Table 1).

The As-Fe shells appear around R + ΔR = 3.0 Å for adsorbed samples at pH 4 and 7. The fitting shows that the As-Fe distances and its CNs were around 3.24 Å and 1.5 for PAA, 3.25 Å and 1.4 for DPAA, respectively (Table 2). The trend of CN is similar to the methyl-substitution case described above. The same factor may govern the adsorption structures of organoarsenic compounds onto ferrihydrite. Thus, the adsorption of these organoarsenic compounds can be dominated by the size of substituent. In other words, the steric hindrance between inter-molecules or molecule and OH functional groups on the surface may inhibit the formation of bidentate surface complex. The CNs of phenyl-substituted compounds are slightly smaller than those of methyl-substituted ones. This result may also support the importance of the steric hindrance, since the size of phenyl group is larger than that of methyl group.

Lafferty et al. reported that the adsorption amount of DMA on ferrihydrite was smaller than that of arsenate and MMA, i.e. the adsorption amount decrease by increasing the number of substitution [18]. It is consistent with the fact that the adsorption amount is smaller for the monodentate structures with smaller stability compared with arsenate, suggesting the relationship between the adsorption structure and adsorption amount. Thus, the results obtained in this study are helpful to understand the migration process of organoarsenic compounds.

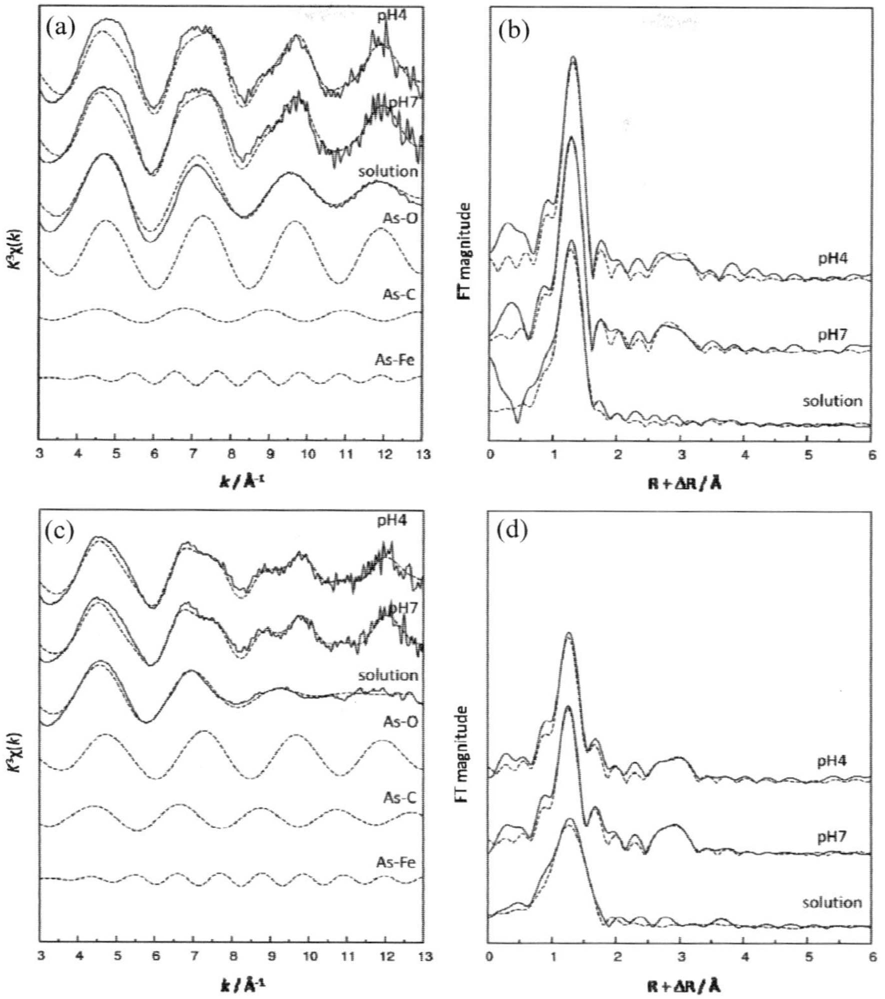

Figure 1. As K-edge EXAFS spectra and corresponding RSFs for MMA (a, b) and DMA (c, d). Solid lines are measured spectra and dashed lines are fitted spectra.

Table 1. Structural parameters obtained from curve fitting of EXAFS spectra for MMA and DMA solution and adsorbed on ferrihydrite at pH 4 and 7

		MMA			DMA		
		solution	pH4	pH7	solution	pH4	pH7
As-O	CN	*3.0*	*3.0*	*3.0*	*2.0*	*2.0*	*2.0*
	R / Å	1.69(1)	1.70(1)	1.69(1)	1.70(1)	1.70(1)	1.70(1)
	σ^2/ Å^2	0.0031(1)	0.002(5)	0.0014(1)	0.0027(1)	0.0008(2)	0.0007(2)
As-C	CN	*1.0*	*1.0*	*1.0*	*2.0*	*2.0*	*2.0*
	R / Å	1.88(1)	1.91(8)	1.90(9)	1.89(1)	1.9 (5)	1.9 (4)
	σ^2/ Å^2	0.001(1)	0.002(2)	0.001(3)	0.0012(1)	0.0013(3)	0.0007(4)
As-Fe	CN		1.7(4)	1.6(4)		1.5(3)	1.6(3)
	R / Å		3.28(2)	3.27(1)		3.26(1)	3.26(1)
	σ^2/ Å^2		*0.0081*	*0.0081*		*0.0081*	*0.0081*
	R / %	1.2	3.0	3.4	2.1	5.8	5.6

Italic: Fixed parameter during fitting

Journal of Physics: Conference Series **430** (2013) 012100 doi:10.1088/1742-6596/430/1/012100

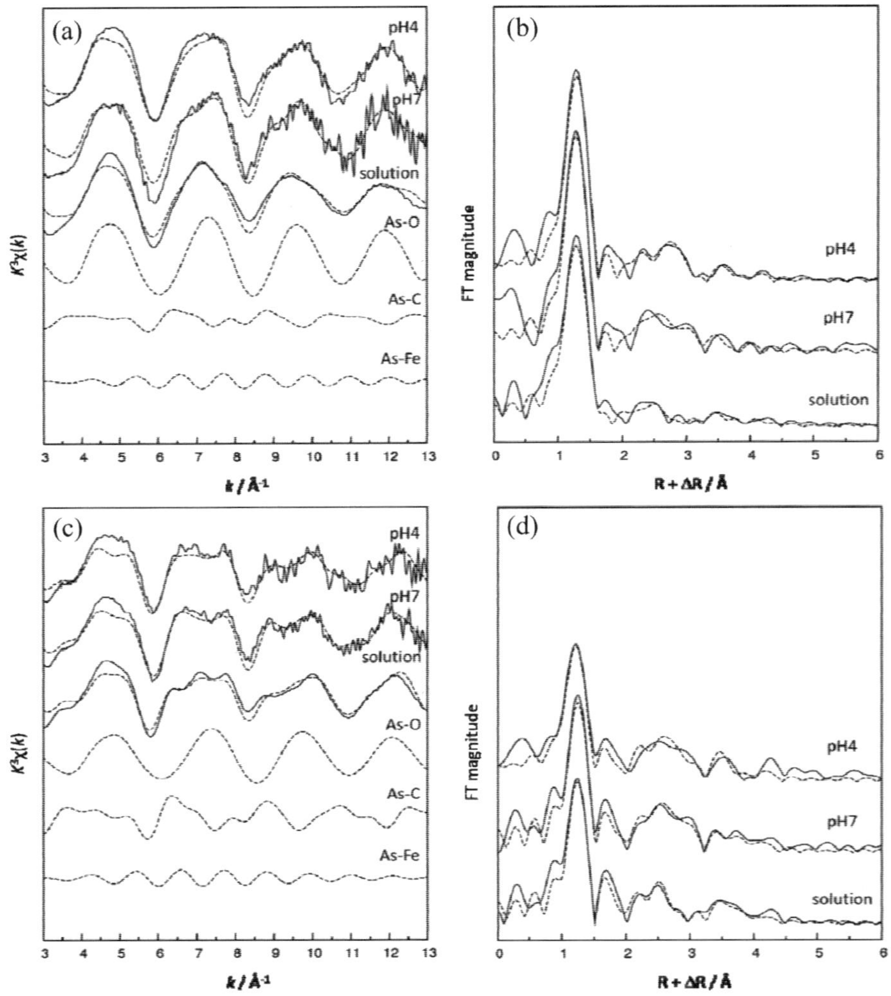

Figure 2. As K-edge EXAFS spectra and corresponding RSFs for PAA (a, b) and DPAA (c, d). Solid lines are measured spectra and dashed lines are fitted spectra.

Table 2. Structural parameters obtained from curve fitting of EXAFS spectra for PAA and DPAA solution and adsorbed on ferrihydrite at pH 4 and 7

		PAA			DPAA		
		Solution	pH4	pH7	solution	pH4	pH7
As-O	CN	*3.0*	*3.0*	*3.0*	*2.0*	*2.0*	*2.0*
	R / Å	1.68(0)	1.69(0)	1.69(0)	1.68(0)	1.68(0)	1.68(0)
	σ^2/ Å^2	0.0037(0)	0.0017(3)	0.0009(1)	0.0014(0)	0.0016(0)	0.0014(1)
As–C_1	CN	*1.0*	*1.0*	*1.0*	*2.0*	*2.0*	*2.0*
	R / Å	1.84(1)	1.88 (2)	1.88 (2)	1.90(1)	1.89(1)	1.88(1)
	σ^2/ Å^2	0.002(1)	0.001(2)	0.001(2)	0.0025(2)	0.0023(2)	0.0025(2)
As-Fe	CN		1.5(4)	1.3(5)		1.4(3)	1.3(2)
	R / Å		3.24(2)	3.27(2)		3.25(1)	3.23(1)
	σ^2/ Å^2		*0.0081*	*0.0081*		*0.0081*	*0.0081*
	R / %	3.2	2.8	4.3	4.1	3.3	3.9

Italic: Fixed parameter during fitting

Journal of Physics: Conference Series **430** (2013) 012100 doi:10.1088/1742-6596/430/1/012100

4. Conclusion

The adsorption structures of methyl- and phenyl-substituted arsenic compounds onto ferrihydrite were investigated by EXAFS analysis. The coordination number of As-Fe shell was smaller than 2, which indicates the existence of both bi- and mono-dentate inner-sphere complexes on ferrihydrite. This result is in contrast to corresponding CN of arsenate on ferrihydrite (= 2). It is possible that this decrease of CN is due to the steric hindrance by the presence of methyl- or phenyl- functional group. This effect may help us to predict the dependence of the mobility of organoarsenic compounds in the soil on the number of organic substituent.

Acknowledgement

This study has been performed with the approval of KEK (2009G655) and SPring-8 (2011B1742 and 2012A1299).

References

[1] Jackson B P and Bertsch P M 2001 *Environ. Sci. Technol.* **35** 4868-73
[2] Huang J-H, Hu K-N and Decker B 2011 *Water Air Soil Pollut.* **219** 401-15
[3] Hanaoka S, Nagasawa E, Nomura K, Yamazawa M and Ishizaki M 2005 *Appl. Organomet. Chem.* **19** 265-75
[4] Bednar A J, Garbarino J R, Ferrer I, Rutherford D W, Wershaw R L, Ranville J F and Wildeman T R 2003 *Sci. Tot. Environ.* **302** 237-45
[5] Dickens R and Hiltbold A E 1967 *Weeds* **15** 299-304
[6] Arao T, Maejima Y and Baba K 2009 *Environ. Sci. Technol.* **43** 1097-101
[7] Hempel M, Daus B, Vogt C and Weiss H 2009 *Environ. Sci. Technol.* **43** 6989-95
[8] Nakamiya K, Nakayama T, Ito H, Edmonds J S, Shibata Y and Morita M 2007 *FEMS Microbiol. Lett.* **274** 184-8
[9] Harada N, Takagi K, Baba K, Fujii K and Iwasaki A 2010 *Biodegradation* **21** 491-9
[10] Mohana D and Pittman C U Jr. 2007 *J. Hazard. Mater.* **142** 1–53
[11] Smedley P L and Kinniburgh D G 2002 *Appl. Geochem.* **17** 517-68
[12] Sherman D M and Randall S R 2003 *Geochim. Cosmochim. Acta* **67** 4223-30
[13] Jing C, Meng X, Liu S, Baidas S, Patraju R, Christodoulatos C and Korfiatis G P 2005 *J. Colloid Interface Sci.* **290** 14-21
[14] Shimizu M, Ginder-Vogel M, Parikh S and Sparks D L 2010 *Environ. Sci. Technol.* **44** 612-7
[15] Shimizu M, Arai Y and Sparks D L 2011 *Environ. Sci. Technol.* **45** 4293-9
[16] Schwertmann U and Cornell R M 2000 *Iron Oxide in the laboratory, 2nd ed.* (Weinheim: WILEY-VCH)
[17] Adamescu A, Mitchell W, Hamilton I P and Al-Abadleh H A 2010 *Environ. Sci. Technol.* **44** 7802-7
[18] Lafferty B J and Loeppert R H 2005 *Environ. Sci. Technol.* **39** 2120-7

Origin of incommensurate satellite reflections on $TbMnO_3$ by resonant x-ray scattering

V Cuartero[1*], J García[2], G Subías[2], J Herrero-Martín[3], J Blasco[2] and C R Natoli[4]

[1]ALBA Synchrotron/CELLS, Cerdanyola del Vallès, 08290, Spain
[2] Instituto de Ciencia de Materiales de Aragón, Departamento de Física de la Materia Condensada,CSIC-Universidad de Zaragoza, Zaragoza, 50009, Spain
[3] Institut de Ciència de Materials de Barcelona, ICMAB-CSIC, Campus Universitari de Bellaterra, E-08193 Bellaterra, Spain.
[4] Laboratori Nazionali di Frascati INFN, 13, I-00044 Frascati, Roma, Italy.
*E-mail: vcuartero@cells.es

Abstract. We have performed a deep spectroscopic study of incommensurate satellite reflections *(h, k, l)* ± *(0, 0.278, 0)* on a $TbMnO_3$ single crystal in order to shed light on its physical origin. Resonant x-ray scattering experiments at the Mn K-edge and Tb L_3-edge have been performed at different temperatures and azimuthal angles, analyzing the outgoing light polarization. Our experimental results agree with a magnetic origin for A and C-type reflections, while F and G-type have the hallmarks of anisotropic tensor susceptibility (ATS) reflections, so that they come purely from the structural modulation.

1. Introduction.

The characterization of the complex magnetoelectric phase diagram of $TbMnO_3$ has become a great challenge in the recent years [1-4]. This multiferroic material shows a collinear antiferromagnetic (AFM) ordering (A-type) of Mn sublattice below T_{N1}~41K [1]. The amplitude of the moments of this magnetic structure is sinusoidally modulated, being the modulation vector q~$0.278b^*$ incommensurate (ICM). Below T_{N2}~27 K Mn moments develop a new component along c resulting in a cycloidal magnetic structure [3]. A net electric polarization along c axis appears at T_{N2} due to the inversion symmetry breaking caused by this non-collinear magnetic structure. In addition, Tb sublattice orders AFM below T_{Tb}~8K.

Under this complex scenario many attempts to characterize the ICM reflections have been performed [2-7], some of them being contradictory. Following Bertaut´s notation [8], the ICM reflections are classified as A-type (*h*+*k*=even, *l*=odd), F-type (*h*+*k*=even, *l*=even), C-type (*h*+*k*=odd, *l*=even) and G-type (*h*+*k*=odd, *l*=odd), independently of its origin: *(i)* structural (ATS, from atomic environment anisotropy) or *(ii)* magnetic (magnetic moment modulations on magnetic ordered phases). Neutron diffraction experiments on single crystals mainly detected A and G-type reflections on the collinear phase [2, 3]. Some authors ascribe the appearance of G-type reflections on the non-collinear (T_{N2}) phase to Tb ordering coupled to Mn [2], while some others also detect C and F-type reflections arising from this coupling as well [3].

Resonant x-ray scattering (RXS) is a very powerful technique to separate Tb and Mn contributions due to its chemical selectivity, but it can also provide information on the nature of the reflections. Magnetic reflections are much weaker than Thomson (standard diffraction) reflections. When tuning the photons out of the absorption edge, both magnetic and Thomson reflections can be detected in the σ–σ´ and π–π´ polarization channels (being σ and π polarizations perpendicular and parallel to the scattering plane, respectively). However, Thomson reflections are not seen in the σ–π´ channel, so reflections that are observed out of resonance in the σ–π´ channel correspond to magnetic reflections [8]. On the other hand, when tuning the photon energy across the absorption edge, both ATS and magnetic reflections can be detected in the σ–π´ channel.

Regarding RXS experiments on $TbMnO_3$ single crystals, some authors claim that only A-type reflections are due to magnetic modulations while F and C reflections do not come from the magnetic modulation [4]. However, other authors claim that F-type reflections are magnetic when appearing in the non-collinear phase [5-7].

We have performed a deep spectroscopic study both at Mn K-edge and Tb L_3-edge of ICM satellite reflections such as $(h, k, l) \pm (0, q, 0)$, being $q=q_{Mn}=0.278$, in order to solve the controversy on the physical origin of these modulations.

2. Experimental section.

$TbMnO_3$ single crystals were grown by the floating zone method at University of Zaragoza. The untwined crystal was cut and polished with faces perpendicular to [100]/[010]/[001] directions. It was characterized by x-ray powder diffraction and magnetic susceptibility measurements, being the lattice parameters a=5.316 Å, b=5.831 Å, c=7.375 Å.

RXS experiments were carried out at ID20 beamline on the ESRF (Grenoble, France). A double crystal Si (111) monochromator was used to select the energy of the beam, which is 99% linearly polarized perpendicular (σ) to the scattering plane. A closed-cycle helium refrigerator was used to study the temperature dependence of the reflections down to 12 K. Azimuthal dependence was also tested by rotating the sample around the scattering vector **Q**. The scattered beam polarization (σ' or π') was analysed putting a Cu (220) crystal between the sample and the detector. All reflections were studied off and on resonance at the Mn K-edge and Tb L_3-edge.

3. Results.

A and F-type reflections are observed at both absorption edges below T_{N1}. The energy dependence of these reflections at the Mn K-edge can be observed at figure 1. Both reflections are resonant in the σ-π' channel, while no intensity was observed in the σ-σ´ one.

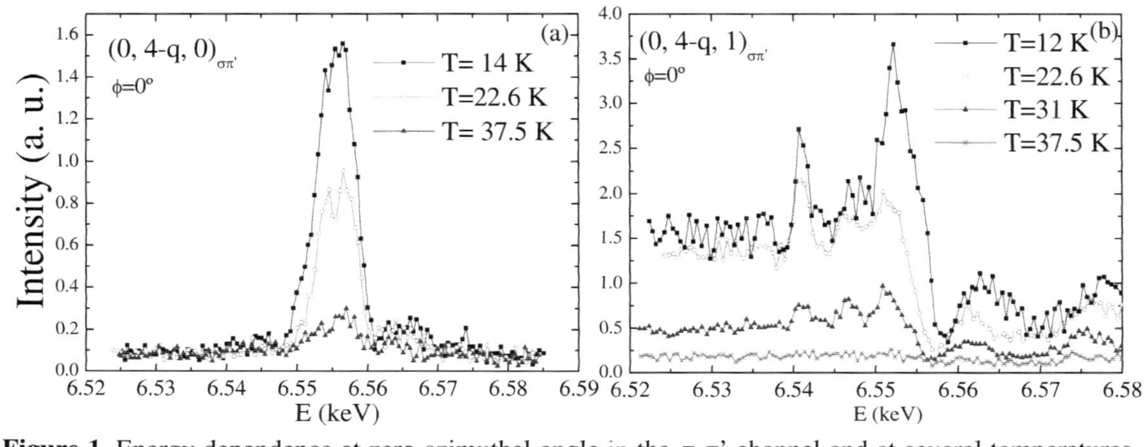

Figure 1. Energy dependence at zero azimuthal angle in the σ-π' channel and at several temperatures below T_{N1} of (a) (0, 4-q, 0) F-type reflection and (b) (0, 4-q, 1) A-type reflection.

The (0, 4-q, 0) F-type reflection shows a main resonance at the Mn K-edge, as it is shown in figure 1(a), but there is no diffracted peak at this reflection out of resonance. The spectral dependence of this reflection is very similar to that observed for the (0, 3, 0) reflection [10], which is forbidden by the group symmetry but seen on resonance due to the ATS contribution [11]. However, A-type reflections show a non-resonant contribution (NRXS) below T_{N1}, which is the main feature of magnetic reflections.

The azimuthal evolutions for A and F type reflections have been checked (not shown here): while the first one does not show any significant change along the ϕ-scan, F-type reflection shows a $\cos^2\theta$ modulation. Azimuthal evolution of A-type reflections is not expected. The scattering factor (proportional to scalar product $\mathbf{k}'\cdot\mathbf{m}$) is almost constant around $\mathbf{Q}$ for this reflection in the non-collinear phase because m_b and m_c show very close values [2].

The same spectroscopic behaviour is found at the Tb L_3-edge for F and A-type reflections, as can be observed on figure 2(a). There is a NRXS contribution for A-type, while F-type reflection can be only detected on resonance in the σ-π' channel. The temperature dependence of both reflections on resonance (Tb L_3-edge) can be seen on figure 2(b). It is very similar to that observed at Mn K-edge, as both reflections appear below T_{N1} and no changes are noticeable across T_{N2}.

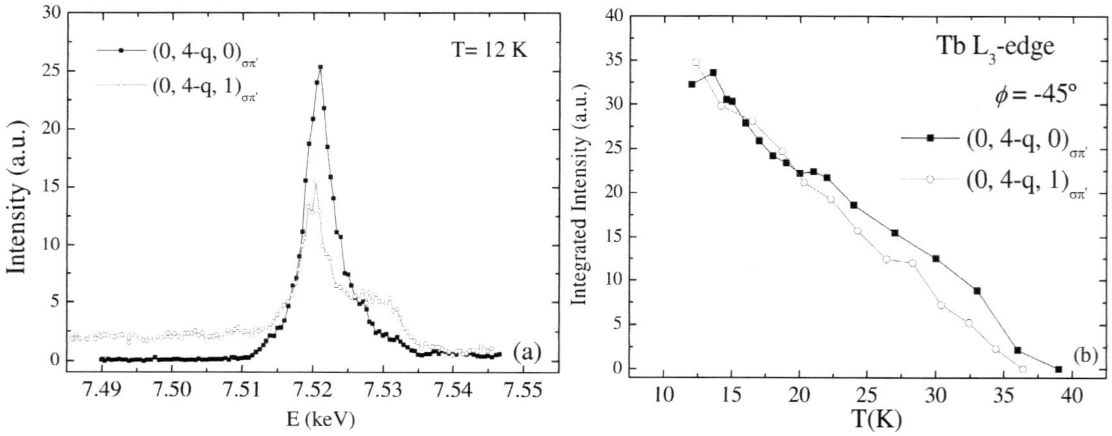

Figure 2. (a) Energy dependence of (0, 4-q, 0) F-type and (0, 4-q, 1) A-type reflections at the Tb L_3-edge in the σ-π' channel at 12 K. **(b)** Temperature evolution of (0, 4-q, 0) F-type and (0, 4-q, 1) A-type at the Tb L_3-edge in the σ-π' channel.

On the other hand, C and G-type reflections appear just below T_{N2} only at Tb L_3-edge in the σ-π' channel. Figure 3(a) shows the resonances at Tb L_3-edge of both reflections. G-type is less intense than C type reflection, so its signal is normalized to the first. It is noteworthy that the maximum of G-type reflection is shifted to high energies with respect to C-type. C-type reflection is also observed out of resonance (magnetic origin) and its temperature dependence is plotted on the inset of figure 3(a) at E=7.5 keV: it disappears above T_{N2}. The temperature dependence of C and G-type on resonance (Tb L_3-edge) is plotted in figure 3(b), both appearing below T_{N2}, in the non-collinear phase.

The azimuthal evolution of (0, 3-*q*, 0) C-type reflection can be observed on the inset of figure 3 (b) and corresponds to the variation of the *c* magnetic moment component around the scattering vector in the non-collinear phase.

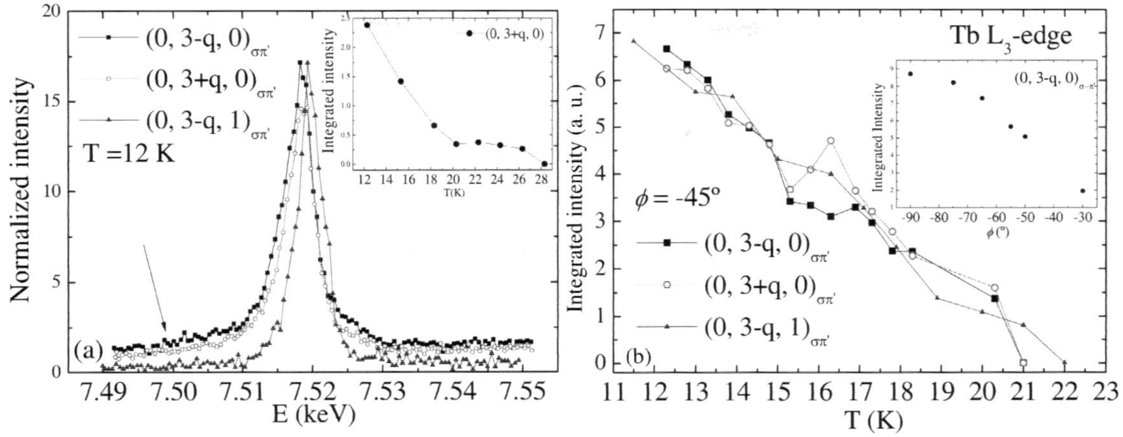

Figure 3. (a) Energy dependence of C (l=0) and G (l=1) type reflections at the Tb L_3-edge and 12 K in the σ-π' channel. Inset: Temperature evolution of the off-resonance intensity of (0, 3+*q*, 0) C-type reflection. **(b)** Temperature evolution of the same reflections on resonance. Inset: Azimuthal evolution of (0, 3-*q*, 0) C-type on resonance.

4. Conclusions

The experimental results above make it possible to assert that A-type reflections, which are observed below T_{N1} at both Mn K-edge and Tb L_3-edge on and off resonance, come from the magnetic modulation. However, F-type reflections are not seen out of resonance, so they cannot be ascribed to the magnetic modulation. On the other hand, its spectral dependence points to an ATS origin, coming from a structural modulation of the out-of-diagonal elements of the Mn and Tb atomic scattering factors, that is, $f_{xy}=f_{xy0}+\delta f_{xy}\cos[2\pi\mathbf{q}(\mathbf{R}_n+\mathbf{r})]$.

On the other hand, C and G-type reflections are only resonant at the Tb L_3-edge and below T_{N2}. C-type reflections are also observed off resonance so that they contribute to non-resonant magnetic scattering. G-type reflections, not detected off-resonance, show the hallmarks of ATS reflections, coming from a structural modulation from the Tb sublattice.

Both Mn and Tb magnetic modulations are coupled with $q=q_{Mn}=0.278$ below T_{N2}, but also the lattice ones, as F and G-type reflections are showing.

Acknowledgments

The authors thank ESRF for granting beam time, especially ID20 staff. Financial support from Spanish MICINN (project FIS08-03951 and MAT2007-61621) and DGA (CAMRADS) is acknowledged.

References.

[1] Kimura T *et al* 2003 *Nature (London)* **426** 55
[2] Kenzelmann M *et al* 2005 *Phys Rev. Lett.* **95** 087206
[3] Aliouane N *et al* 2008 *J. Phys.:Cond. Matt.* **20** 434215
[4] Mannix D *et al* 2007 *Phys. Rev. B* **76** 184420
[5] Prokhnenko O *et al* 2007 *Phys. Rev. Lett.* **99** 177206
[6] Wilkins S B *et al* 2009 *Phys. Rev. Lett.* **103**, 207602
[7] Fabrizi F *et al* 2009 *Phys. Rev. Lett.* **102** 237205
[8] Bertaut E F 1968 *Acta Cryst.* A**24**, 217
[9] Blume M 1985 *J. Appl. Phys.* **57**, 3615
[10] Sánchez M C *et al* 2009 *J. Phys.:Conf. Series* **190** 012087
[11] Subías G *et al* 2007 *Phys. Rev. B* **75**, 235101

Journal of Physics: Conference Series **430** (2013) 012102 doi:10.1088/1742-6596/430/1/012102

Evolution of Mn and Co oxidation state on $TbMn_{1-x}Co_xO_3$ compounds

V. Cuartero [1*], J. Blasco[2], J. García[2], S. Lafuerza[2] and G. Subías[2]

[1]ALBA Synchrotron/CELLS, Cerdanyola del Vallès, 08290, Spain
[2] Instituto de Ciencia de Materiales de Aragón, Departamento de Física de la Materia Condensada,CSIC-Universidad de Zaragoza, Zaragoza, 50009, Spain

*E-mail: vcuartero@cells.es

Abstract. Structural and electronic properties of $TbMn_{1-x}Co_xO_3$ samples have been characterized by means of x-ray and neutron diffraction and x-ray absorption spectroscopic techniques. Crystallographic structure was refined from diffraction data, and samples show the orthorhombic distorted perovskite structure (*Pbnm*), being isostructural to $TbMnO_3$, except for x=0.4, 0.5 and 0.6, where an ordered double perovskite structure ($P2_1/n$) is found. The electronic configuration of Mn and Co ions was obtained along the series from the chemical shift of the edge on x-ray absorption spectra, at the Mn and Co K-edges. Our results agree with a charge redistribution between Mn and Co ions, yielding a mixed valence state Mn^{3+}/Mn^{4+} and Co^{3+}/Co^{2+} for the whole series.

1. Introduction.

The multiferroic state of $TbMnO_3$ is found at low temperatures, below $T_{N2}=T_{FE}\sim27$ K, where a non-collinear magnetic ordering breaks the spatial inversion symmetry allowing a ferroelectric state [1]. The first magnetic transition on this compound takes place at $T_{N1}\sim41$K, when an antiferromagnetic (AFM) collinear ordering arranges. Magnetoelectric properties are highly correlated in this compound, so the main challenge nowadays is to promote the multiferroic state to higher temperatures, keeping magnetoelectric coupling.

In this work, Mn sublattice has been partially substituted by a magnetic ion, such as Co, in order to enhance magnetic interactions. In other rare earth perovskites, Co is normally trivalent. Co^{3+} shows a $3d^6$ electronic configuration on a CoO_6 octahedral environment and can show several spin states depending on the occupation of t_{2g} and e_g orbitals, most of them having a net magnetic moment [2]. There are three configurations that may also change with temperature and composition: low spin (LS), intermediate spin (IS) and high spin (HS) which correspond to S=0, S=1 and S=2 respectively. In addition, Co^{2+} is also a magnetic ion, with S=3/2 on its HS configuration.

Previous works on similar La-based compounds claim the evolution of the oxidation state of Mn and Co from Mn^{3+}/Co^{3+} to Co^{2+}/Mn^{4+} [3]. This would lead to superexchange interactions through oxygen ions. On the other hand, different results are found in the case of $LaMn_{1-x}Ni_xO_3$ series compounds. Mn^{3+}/Mn^{4+} mixed valence state is described for the lower doped compounds, while Ni^{2+}/Ni^{3+} appears at higher Ni concentrations [4]. This mixed valence state can lead to a large variety of competitive magnetic interactions, not found on Co^{2+}/Mn^{4+} valence state, such as AFM superexchange (Ni-O-Ni, Mn^{n+}-O-Mn^{n+}, n=3 or 4), ferromagnetic (FM) superexchange (Ni^{2+}-O-Mn^{4+}) and FM double exchange (Mn^{3+}-O-Mn^{4+}).

Thus, determining the evolution of the oxidation state of Mn and Co is essential to further understand the magnetic properties of $TbMn_{1-x}Co_xO_3$ series.

2. Experimental Section.

All samples were synthesized by solid state chemistry procedures. As a first step, stoichiometric amounts of Tb_4O_7, $MnCO_3$ and Co_3O_4 were mixed, ground and calcined overnight at 1000 ºC in air. After that, they were ground, pressed into pellets and sintered at 1200ºC for 1 day in air. Then, we repeat the last process, but at 1400 ºC for 2 days in Ar atmosphere for x=0.1, 0.2 and air in the case of x=0.3, 0.4, in order to ensure the appropriate oxygen content and the lack of vacancies. O_2 atmosphere was used for higher Co concentrations (x≥0.5). The oxygen content was tested by cerimetric titration and all samples have the correct oxygen content within experimental error (±0.01).

A Rigaku D/max-B diffractometer with a rotating anode was used to preliminarily test the crystallographic structure of the powder samples at room temperature, selecting the Cu K_α wavelength (λ=1.5418 Å). Neutron diffraction powder experiments were performed at D1B and D2B instruments at ILL (Grenoble, France) to properly determine the crystallographic structure and complement absorption experiments.

Finally, x-ray absorption experiments were carried out at BM25-Spline beam line at the ESRF (Grenoble, France), at the Mn and Co K-edge, with a double crystal Si (111) monochromator. The energy resolution of the beam line is $\Delta E/E \sim 10^{-4}$. All spectra were collected at room temperature in transmission mode, using ionization chambers as detectors, so that we could measure up to x=0.6 at the Mn K-edge and down to x=0.3 at the Co K-edge. The powder pellets were mixed with cellulose and we used $TbMnO_3$ (Mn^{3+}) as a reference for Mn K-edge and $TbCoO_3$ (Co^{3+}) for Co K-edge.

3. Results and discussion.

The crystallographic structure has been refined by Rietveld analyses with FullProf package. X-ray diffraction measurements at room temperature have been successfully refined using a simple perovskite model, space group *Pbnm*, for the whole series. The evolution of the lattice parameters has been obtained along the series and can be seen on fig. 1. There is a shrinkage of *a*and *b* axis as Co concentration increases, being stronger the contraction of *b*-axis. In both cases, these axes change their contraction rate from x~0.4, being smaller after this concentration of Co. On the contrary, *c* axis expands up to x~0.4-0.5, when it reaches a maximum and then slightly decreases as Co content increases. The global effect is a contraction of the unit cell volume, with an inflection point for x~0.5. In this central region, the crystallographic structure change and corresponds to a monoclinic cell, as will be explained hereinafter.

When Mn and Co have different oxidation state they have rather different radius. This fact, as will be probed below, can lead to the ordering of the Mn-Co sublattice in a NaCl structure [5]. In this case, the crystallographic structure is a double perovskite $A_2BB'O_6$, with monoclinic space group $P2_1/n$. Neutron diffraction allows distinguishing between Mn and Co, as these ions have very different neutron scattering lengths. The double perovskite structure gives better results for diffraction patterns (not shown here) of intermediate samples (x=0.4, 0.5 and 0.6), with a ~25% of *antisites* defects, while the rest of the samples are consistent with a simple perovskite structure, as can be seen on figure 1.

The values of B-O distances of BO_6 octahedra can be seen in fig. 2 for the whole series. As Co content increases, BO_6 octahedra tend to be more regular. In the composition range 0.4≤x≤0.6, two octahedra can be differentiated corresponding to the two Wyckoff positions *2d* and *2c* of the double perovskite. BO_6 octahedron (*2d*) is composed of ~75% Co atoms and 25% Mn ones. Considering the ionic bond valence model, this corresponds to $Co^{+2.3}$ and $Mn^{+2.8}$. $Co^{+3.0}$ and $Mn^{+3.6}$ is obtained in the case of $B'O_6$, with the opposite atom composition. As a first approximation, B and B' sites favor the existence of Co^{2+} and Mn^{4+} cations respectively, while misplaced atoms might be favored in the trivalent state.

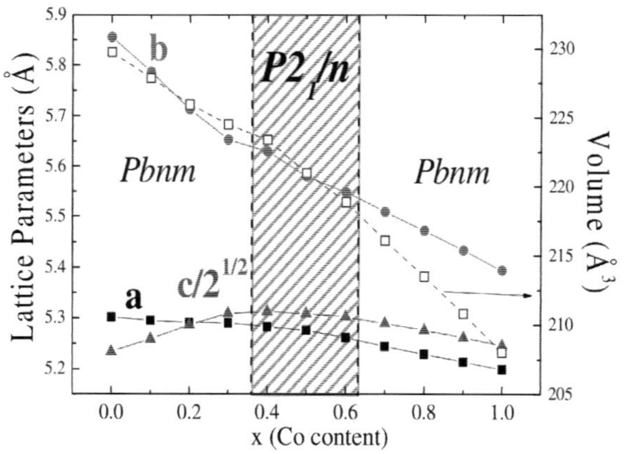

Fig. 1. Evolution of the lattice parameters and the unit cell volume along the series. The crystallographic structure obtained for the different x ranges is indicated. Data for x=0 is taken from ref. [4].

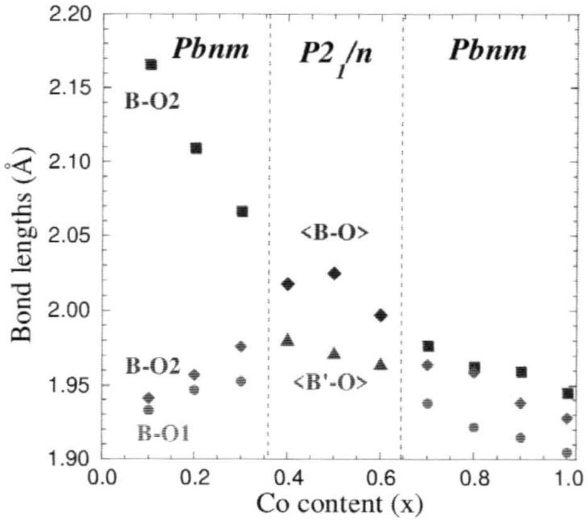

Fig. 2. Interatomic distances in BO_6 octahedra obtained from the refinement of neutron diffraction data from D1B. B-O1=Apical. B-O2=Equatorial. <B-O>= Average.

The oxidation state of Mn and Co has been properly determined by x-ray absorption techniques. Figures 3 and 4 show the normalized x-ray absorption near edge spectra (XANES) measured in transmission at Mn K-edge and Co K-edge respectively. In both cases a shift of the edge at higher energies is observed as Mn is diluted by Co. This chemical shift of the edge is linearly related to the valence state of the absorber atom [6], so that the oxidation state increases when the edge shifts to higher energies, as shown by the arrow on figs. 3 and 4. In addition, the evolution of the pre-peak area at the Co K-edge can be observed in the inset of fig. 4. The integrated absorption of the pre-peak increases as Co content does, so as Co^{2+} tends to Co^{3+} for x>0.5. This is related to the hybridization of *3d* states of Co and *2p* states of O: Co-O covalency increases as the valence state of Co increases.

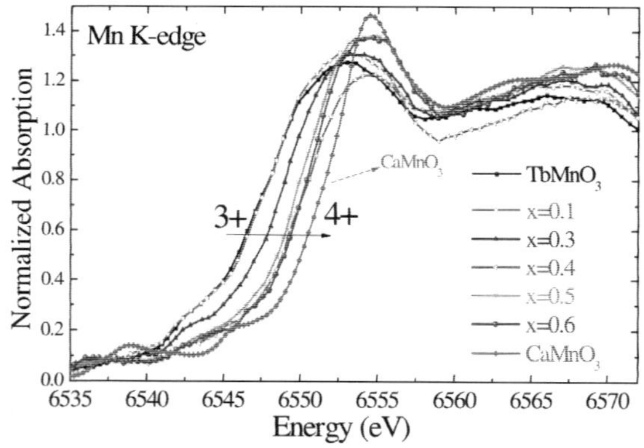

Fig. 3. XANES spectra at the Mn K-edge for 0≤x≤0.6 measured in transmission.

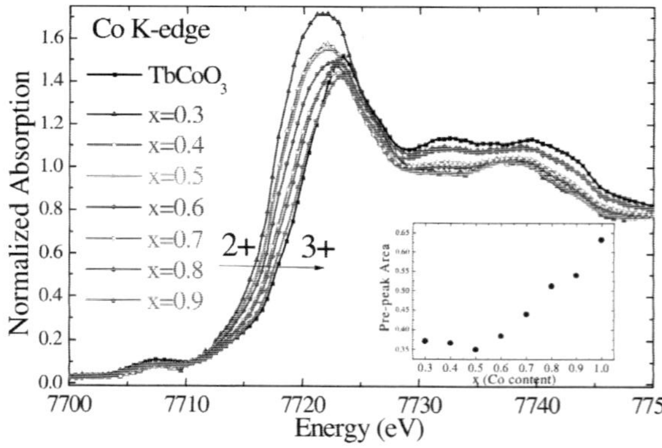

Fig. 4. XANES spectra at the Co K-edge for 0.3≤x≤1 measured in transmission. Inset: Integrated absorption at the pre-edge region.

The valence state has been determined considering the maximum of the derivative of XANES as the energy edge position. In the case of Mn^{3+} and Mn^{4+} ions showing octahedral local structure, the chemical shift is around 4.2 eV [7], while it is 3.5 eV in the case of Co^{2+} and Co^{3+} in an octahedral coordination [8]. Taking this into account, we calculate Mn and Co formal valence and the results are plotted on fig. 5, where it is appreciated that the evolution is not continuous with the composition and a full charge disproportion is not found along the series (dashed lines). The valence of Mn increases when it is diluted with Co from +3 ($TbMnO_3$) up to +3.6 for x=0.5, 0.6, when it tends to saturate. Inversely, Co valence decreases when it is diluted with Mn from +3 ($TbCoO_3$) down to 2.4 (x=0.5, 0.4 and 0.2), also tending to saturation.

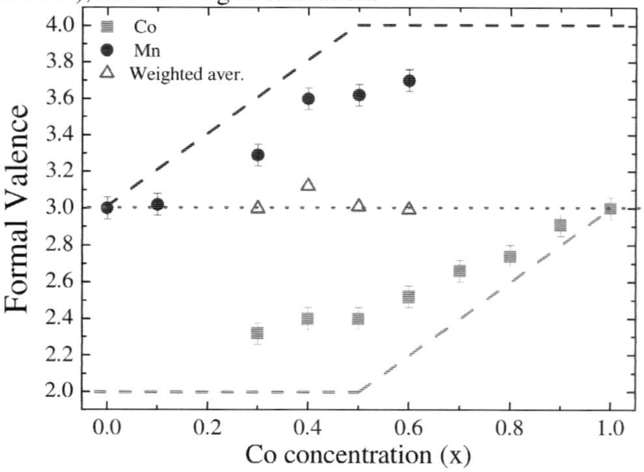

Fig. 5. Evolution of the formal valence of Mn and Co along the series. Green triangles represent the weighted average of Mn and Co valences. Dashed lines represent the tendency of a full charge disproportion. Dotted line indicates the average trivalent state.

We can assert then that there is a charge redistribution along the series, and the formal equilibrium $Mn^{3+}+Co^{3+}\leftrightharpoons Mn^{4+}+Co^{2+}$ is not completely shifted to the right. This yields to a mixed valence state Mn^{3+}/Mn^{4+} and Co^{3+}/Co^{2+}, which is in agreement with the evolution of B-O distances obtained from neutron powder diffraction experiments (table I) and may be due to the existence of *antisite* defects in the intermediate compounds.

4. Conclusions.

$TbMn_{1-x}Co_xO_3$ series has been synthesized by ceramic procedures and all samples are single phase. At intermediate concentrations (x=0.4, 0.5, 0.6) the structure becomes monoclinic $P2_1/n$, and the metal transition sublattice is ordered on a double perovskite structure with *antisites* (~25%). The dilution with Co decreases the distortion on BO_6 octahedra. We have also determined the valence state along the series by x-ray absorption spectroscopy. We found a mixed valence state along the series Mn^{3+}/Mn^{4+} and Co^{3+}/Co^{2+} so that the oxidation of Mn^{3+} is coupled to the reduction of Co^{3+}. The B-O distances obtained by diffraction techniques are in agreement with the electronic information extracted from XANES.

It is noteworthy that several exchange interactions are expected to appear in this series (such as FM superexchange Mn^{4+}-O-Co^{2+}), which makes it very promising to enhance FM correlations.

Acknowledgments.

The authors thank ESRF and ILL for granting beam time, especially the Spanish CRG BM25, D2B and D1B instruments. Financial support from Spanish MICINN (project FIS08-03951 and MAT2007-61621) and DGA (CAMRADS) is acknowledged.

References.

[1] T. Kimura, *et al.* Nature (London) **426**, 55 (2003).
[2] R. Ibarra *et al.* Phys. Rev. B **57** R3217 (1998).
[3] R. I. Dass and J. B. Goodenough, Phys. Rev. B **67** 014401 (2003).
[4] J. Blasco *et al.* Eur. Phys. J. B **30** 469 (2002).
[5] I. O. Troyanchuk *et al.* J. Phys.: Condens. Matter **18** 9541 (1997).
[6] J. García *et al.* J. Synchrotron Rad. **17**, 386 (2010).
[7] G. Subías *et al.* Phys. Rev. B **56**, 8183 (1997).
[8] M. Sikora *et al.* Phys. Rev. B **73**, 094426 (2006).

Journal of Physics: Conference Series **430** (2013) 012103 doi:10.1088/1742-6596/430/1/012103

In situ electronic structure investigation ofMn doped $BiFeO_3$ thin films

Abduleziz Ablat[1,2], Emin Muhemmed[1],Cheng Si[1], Rui Wu[1], Jiaou Wang[1], Haijie Qian[1], Rong Wu[2] and Kurash Ibrahim[1]

[1]Beijing Synchrotron Radiation Facility, Institute of High Energy Physics, CAS, Beijing 100049, China

[2]School of Physical Science and Technology, Xinjiang University, Urumqi 830046, China

Email: abduleziz@ihep.ac.cn, kurash@ihep.ac.cn

Abstract. The electronic structure of $BiFe_{1-x}Mn_xO_3$ ($0 \leq x \leq 0.3$) thin films has been investigated by *in situ* near edge X-ray absorption fine structure (NEXAFS) and photoemission spectroscopy.Fe ions are found to be trivalent for the whole range of x in $BiFe_{1-x}Mn_xO_3$and a hybridization between the Fe 3*d* and O 2*p* states is decreased by increasing Mn concentration. Charge transfer multiplet (CTM) calculation is used to understand the experimental results of the Fe *L*-edge spectra. The experimental results plus the CTM calculation showsthe hybridization degree between the transition metal 3*d* and O 2*p* states areplay important role for the multiferroic properties of underlying systems.

1. Introduction

Multiferroics are multifunctional materials that simultaneously possess ferroelectric, ferromagnetic and ferroelastic ordering. Thiskind of materials has been studied broadly and enthusiasticallyin recent years due to their potential application tospintronics, high temperature sensors, visible light photocatalyst[1] and memory storage system. As ABO_3 perovskite counterparties,rhombohedral $BiFeO_3$(BFO)has attracted much attention due to both magnetic and ferroelectric orderings at ambient conditions (antiferromagnetic below the Neel temperature $T_N \sim 370$ °C , ferroelectric below the Cure temperature $T_C \sim 825$ °C[2]).

Investigations of multiferroic BFO bulk and thin films[3, 4]focus on a high remnant polarization and magnetization behavior.Proper substrates[5], controlling films thickness[6], and A or B site cation substitutions in the lattice[7, 8]were used to improve these properties. However, the origin of enhanced physical properties is still controversial, especiallyenhanced magnetization by cation substitution. In this work, we have investigated electronic structures of $BiFe_{1-x}Mn_xO_3$ ($0 \leq x \leq 0.3$) (BFMO) thin films by employing *in situ*photoemission spectroscopy (PES) and near edge X-ray absorption fine structure (NEXAFS)spectroscopy. Charge transfer multiplet (CTM) calculation[9]was performed to simulate the Fe *L*-edge NEXAFS data.

2. Experiment

The $BiFe_{1-x}Mn_xO_3$ ($0 \leq x \leq 0.3$) films were prepared on the $Pt(111)/TiO_2/SiO_2/Si(100)$ substrates by pulsed laser deposition (PLD)and studied by in situ PES system at 4B9B beam line of the Beijing Synchrotron Radiation Facility (BSRF). The detailed method of sample preparation and investigation of structural, morphological and magnetic properties were reported in elsewhere[10]. The Fe *L*-edgeNEXAFS and PES experiments in the soft X-ray range of the BFMO samples were carried out *in situ* at 4B9B beam line of the BSRF with total electron yield (TEY) detection mode. After the *in situ* measurements, the films were taken out for Fe *K*-edge hard X-ray NEXAFS measurements at 4B9A XRD/SAXS beam line. The Fe *K*-edgeNEXAFS spectra of all samples were made in total fluorescence yield (TFY) mode and the resolution of the monochromator was about 3~4eV in the energy ranges which used for Fe *K*-edge. The simulation of experimentally observed Fe *L*-edge NEXAFS spectra of BFMO films were carried out by charge transfer multiplet for X-ray absorption spectroscopy (CTM4XAS) program that was developed by E. Stavitski and F.M.F.de Groot[11].

3. Result and Discussion

3.1 Photoemission spectra

Figure 1 illustrates wide scan PES spectra of the BFMO films at *hv*=900eV, including the main photoemission peaks of constituent elements in the system. The core-level photoemission peaks and Auger lines of Bi, Fe, Mnand O atoms are dominant. Absence of the C 1*s* signal at ~285eV indicates that the surface is free from contamination, since the measurements are carried out *in situ*in high vacuum. The binding energy in the spectra are calibrated by Pt 4*f* peak from substrate. The corresponding characteristic core-level photoemission peaks and Auger peaks are labeled on the spectra of $BiFe_{1-x}Mn_xO_3$ (x=0.1,0.2,0.3) films. It is obvious that the characteristic peaks of undoped BFO films are the same asthat of doped samples,except the signal ofMn.

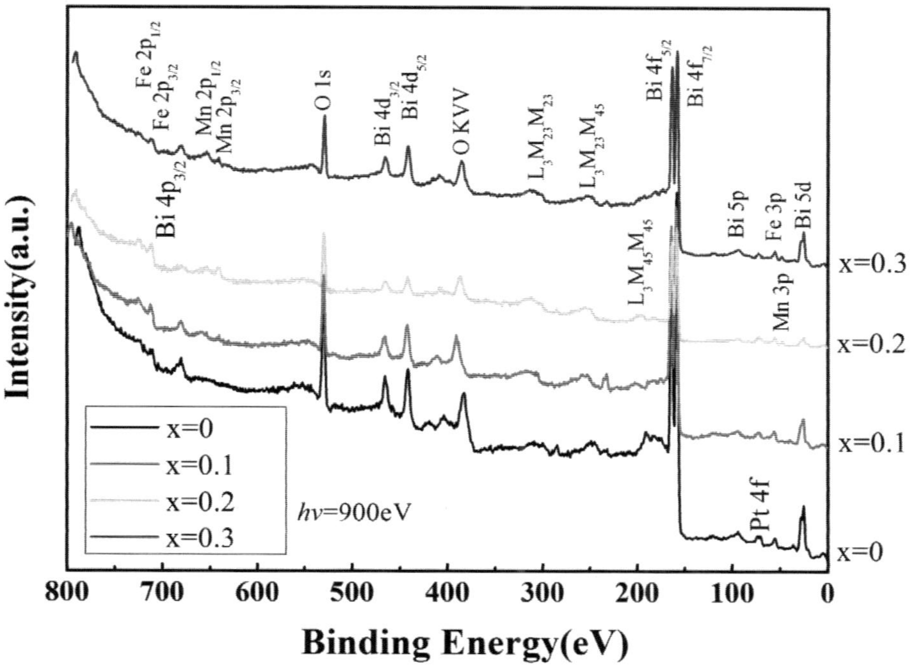

Figure 1. PES survey spectra of $BiFe_{1-x}Mn_xO_3$ ($0 \leq x \leq 0.3$) thin films onPt (111)/TiO_2/SiO_2/Si (100) substrate.

*3.2 Fe K-*edge *NEXAFS spectra*

In order to obtain oxidation state of Fe and influence of Mn doping ininternalregion of the films, we have performed hard X-ray NEXAFS(NEXAFS is also called X-ray Absorption Near Edge Structure, XANES. However, some researcher used the term NEXAFS for soft x-ray absorption spectra and XANES for hard x-ray absorption spectra. Here, we totally used the same term NEXAFS for both Fe *L* and *K*-edge)measurement on BFMO thin films. The probing depth of hard X-ray NEXAFS spectroscopy is much larger than that of soft X-ray PES in section 3.1 and coulddetectthe deeper region of thin films under the surface.

Figure 2 shows the Fe *K*-edge NEXAFS spectra of BFMO films, which are similar tothat of Fe_2O_3[12], except the intensity difference of the three peaks labeled with A, B and C. It implicates thatall the Fe ions in the deeper region of thin films are also in trivalent state.That also provethe valence state of Fe ions inentire region of thin films is not changed by Mndoping [10]. Dongeun et al.[13] reported that peak A corresponds to the 1*s*-3*d* electric quadrupole-forbidden transition, peak B appeared as shoulder peak is caused by the shakedown process in which 2*p* electron in ligand (here the ligand is oxygen) 2*p* band partially transfer to the Fe 3*d*orbitals, so called ligand to metal charge transfer process. Peak C is dipole allowed transition to unoccupied 4*p* orbital. Intensities of peak B and C are more sensitive to Mn doping than peak A and increased byincreasing of Mn doping amount.As is well known,the BFO isa charge transfer type insulator and assume to beFe $3d^5$ + Fe $3d^6\underline{L}$ ($\underline{L}$ represents the ligand hole state O $2p^5\underline{L}$ formed by a charge transfer from O $2p^6$ to Fe $3d^5$) mixture states derived by the hybridization of Fe 3*d*with O 2*p*states.Trend of the peak intensity indicates that the charger transfer from ligand to metal sides is decreased (the Fe $3d^5$ portion increased in Fe $3d^5$ + Fe $3d^6\underline{L}$ mixed states) with Mncontentincreasing and Mn doping has provided more empty 4*p* orbitals for BFO systems. Main causes of the current observation at the Fe *K*-edge, in a self-consistent way with an decreased hybridization strength between Fe 3*d*- O 2*p* states. This is clearly seen affirmatively through the below Fe *L*-edge NEXAFS measurement results and CTM simulation.

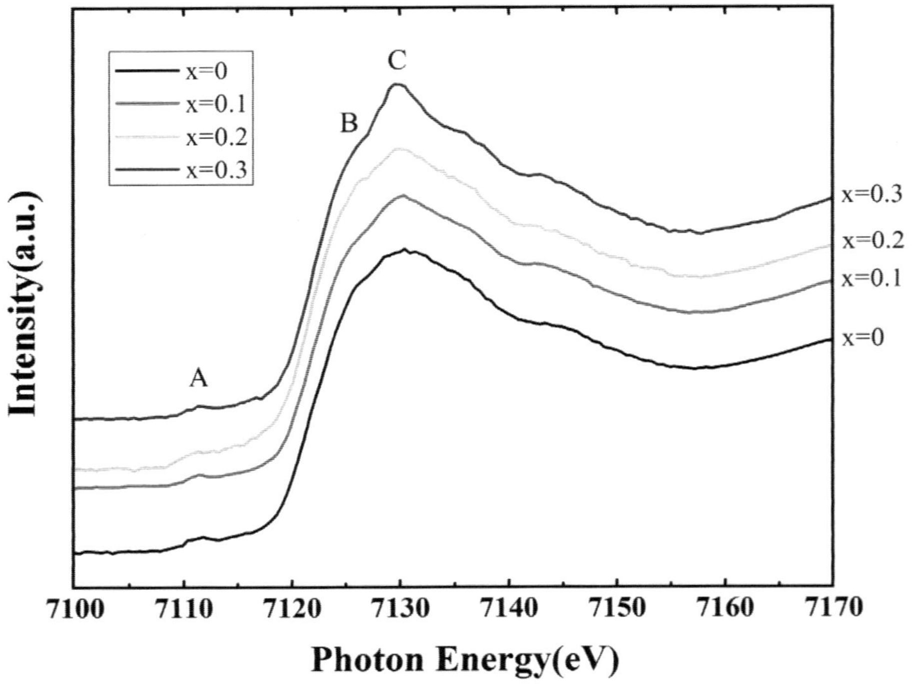

Figure 2. Fe *K*-edge NEXAFS spectra of $BiFe_{1-x}Mn_xO_3$ ($0\leq x\leq 0.3$) thin films as a function of Mn content x.

3.3 Fe L-edge NEXAFS spectra and CTM simulation

To investigate the influence of Mn doping on unoccupied Fe 3*d* states of BFMO systems, we have carried out the Fe *L*-edge NEXAFS spectra of the samples and also simulated the experimental spectra by CTM program. Figure 3(a) shows the normalized NEXAFS spectra of Fe *L*-edge for the BFMO ($0\leq x\leq 0.3$) films. It involves the excitation from the Fe 2*p*state to an unoccupied Fe 3*d* state. There is a strong overlap between the core wave function and the valence wave function. Spin-orbit coupling effects make Fe *L*-edgeabsorption spectrasplit into L_3 and L_2 structures, which further split intolower-lying $t_{2g}\downarrow$and higer-lying $e_g\downarrow$ (where ↓ denotes the minority spin states) sates because of the octahedral crystal ligand field effects. The energy difference between $t_{2g}\downarrow$ and $e_g\downarrow$ states for both L_3 and L_2 levels is about 1.4eV, in corresponding to the crystal field splitting energy, 10Dq. We have reported the influence of Mn doping on the valence states of Fe ions and relative intensityof $t_{2g}\downarrow$ and $e_g\downarrow$states[10]. Here, we will discuss the simulation of spectraby CTM calculation in detail. Figure 3(b) shows the simulation curves for the absorption spectra in Figure 3(a). Atomic multiplet model is chosen, and different crystal field splitting (10Dq) and charge transfer energy (△) are used. 10 Dq varies from 1.0 to 1.8eV to get the best fitting curves(the simulation with the other value of 10Dq is not shown here). When 10Dq is around 1.65eV for all the doped films, the calculation and experiment curves match very well.Especially, the same fit parameter, 10Dq, gives the best fitting results, which indicates that B site doping in ABO_3 perovskite has less influence on the octahedral crystal filed.

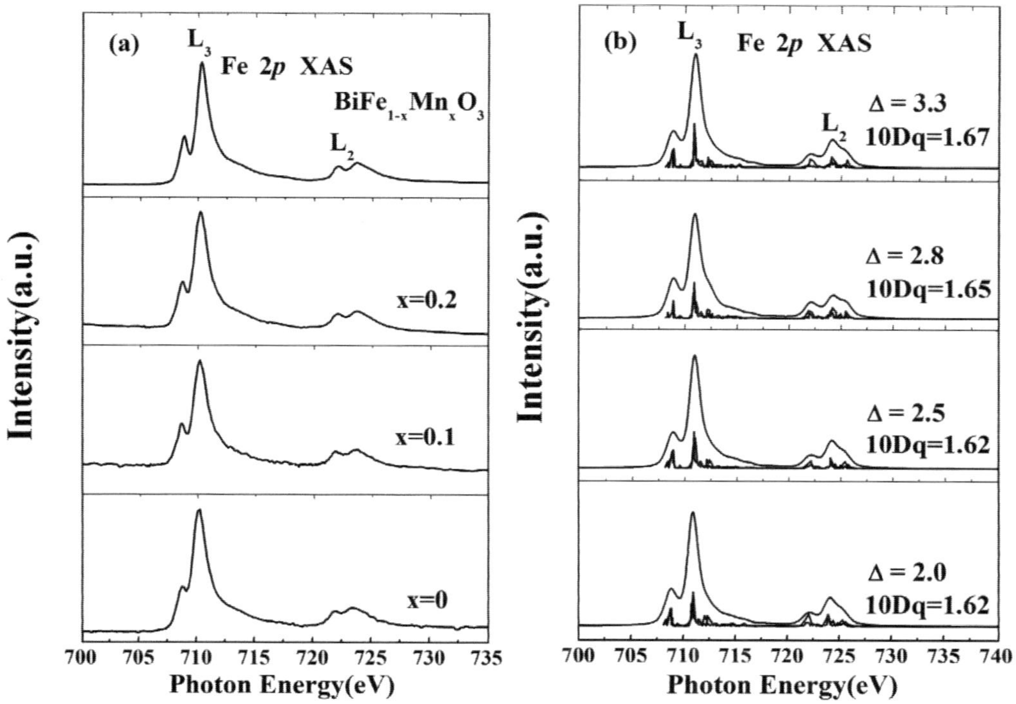

Figure 3. (a) Fe L-edge NEXAFS spectra of BiFe1-xMnxO3 (0⩽x⩽0.3) thin films. (b) calculated spectra of Fe L-edge with various value of charge transfer energy (△).

The other fitting parameter, △,are 2.0, 2.5, 2.8, and 3.3eV for x=0, 0.1, 0.2 and 0.3, respectively. Charge transfer effects refer to charge fluctuations in the initial and final states. In the charge transfer multiplet model[14], the charge transfer energy △ is defined as the energy difference between the lowest states of the 3d5 and Fe 3d6 configuration. It is found that △ decreases with Mn doping increasing, which suggests the portion of Fe 3d6 in the Fe 3d5 + Fe 3d6 mixed state decreasing. It is equal todecreasing of hybridization between Fe 3d and O 2p states.In a word, Mn doping in BFMO systems can modify the charge transfer effect between Fe *3d* orbital and the O *2p* ligand orbitals. The Fe *K*-edge NEXAFS spectra and Fe *L*-edge CTM simulation also prove that Mn doping could induce the variation of hybridization interactions and further change the electron density of state (DOS) distribution between Fe-O and Mn-O pairs.

4. Conclusion

PES and NEXAFS spectroscopy in the soft and hard photon energy range have been used to study

$BiFe_{1-x}Mn_xO_3$ thin films with x=0, 0.1, 0.2 and 0.3. The Fe *K*, *L*-edge NEXAFS spectra reveals that the valence sate of Fe ions remains +3 under all of the doping conditions. Combined with Fe *L*-edge CTM simulation, it can be concluded that Mn doping in BFMO systems modified the charge transfer effect between Fe 3*d* orbital and the O 2*p* ligand orbitals. In conclusion, increasing Mn doping can increase the charge transfer energy, decrease the portion of Fe $3d^6\underline{L}$ in Fe $3d^5$ + Fe $3d^6\underline{L}$ mixed state, and decrease the hybridization between Fe 3*d* and O 2*p* states.

Acknowledgement
The authors are grateful to Giovanni Carabalia for suggestion in CTM simulation. This study was financially supportedby Chinese Academy of Sciences under grant No.H91G750Y21, and National Natural Science Foundation of China (Grant No.11164026).

Reference

[1] Guo RQ, Fang L, Dong W, Zheng FG, Shen MR 2010 *Journal of Physical Chemistry C***114** 21390
[2] Roginska.Ye, Tomashpo.Yy, Venevtse.Yn, Petrov VM, Zhdanov GS 1966 *Soviet Physics Jetp-Ussr***23** 47.
[3] Su WT, Song KX, Huo DX 2011 *J.phys.:Conf.Ser.***263** 012024
[4] Wei J, Zhang C, Xu Z *Materials Research Bulletin*In Press, DOI: 10.1016/j.bbr.2011.03.031
[5] Luo WB, Zhu J, Zeng HZ, Liao XW, Chen H, Zhang WL, et al. 2011 *Journal of Applied Physics***109**104108
[6] Huang FZ, Lu XM, Lin WW, Kan Y, Zhang JT, Chen QD, et al. 2010 *Applied Physics Letters***97**22901
[7] Xu Q, Zhou S, Wen Z, Wu D, Qiu T, Xu M, et al. 2011 *Physics Letters A***375** 1209
[8] Leontyev I, Yuzyuk YI, Janolin P, El-Marssi M, Chernyshov D, Dmitriev V, et al. 2011 *Journal of Physics: Condensed Matter***23** 332201
[9] Stavitski E, De Groot FMF 2010 *Micron***41** 687
[10] Abduleziz Ablat, Emin Muhemmed, Cheng Si, Jiaou Wang, Haijie Qian, Rui Wu, Nian Zhang, Rong Wu, and Kurash Ibrahim 2012 *Journal of nanomaterials* **2012** 3238
[11] *CTM4XAS home pages,*http://www.anorg.chem.uu.nl/CTM4XAS/
[12] Sasaki S 1995 *Review of Scientific Instruments***66** 1573
[13] Lee D, Kim MG, Ryu S, Jang HM, Lee SG 2005 *Applied Physics Letters***86** 22903
[14] de Groot F 2005 *Coordination Chemistry Reviews***249** 31.

15th International Conference on X-ray Absorption Fine Structure (XAFS15) IOP Publishing
Journal of Physics: Conference Series **430** (2013) 012104 doi:10.1088/1742-6596/430/1/012104

XAFS and XRD studies of local structure peculiarities in magnetic $R_2Fe_{17-x}Mn_x$ (R = Ce, Lu) intermetallics

A P Menushenkov[1], A A Yaroslavtsev[1], O V Grishina[1], I A Zaluzhnyy[1], R V Chernikov[2], A G Kuchin[3]

[1] National Research Nuclear University "MEPhI", Kashirskoe sh., 31, Moscow 115409, Russia
[2] HASYLAB at DESY, Notkestrasse, 85, Hamburg D-22607, Germany
[3] Institute of Metal Physics UB RAS, S. Kovalevskoy, 18, Yekaterinburg 620990, Russia

E-mail: menushen@htsc.mephi.ru

Abstract. The rearrangement of cerium and lutetium local environment in $Ce_2Fe_{17-x}Mn_x$ and $Lu_2Fe_{17-x}Mn_x$ intermetallics vs. Mn concentration and temperature was investigated by means of the extended X-ray absorption fine structure (EXAFS) spectroscopy above K-Ce absorption edge and powder X-ray diffraction. The valence state of Ce in $Ce_2Fe_{17-x}Mn_x$ was also studied by X-ray absorption near-edge structure (XANES) spectroscopy above L_3-Ce absorption edge. The correlation between changes in local electronic and crystal structure and the types of magnetic ordering in these compounds is discussed.

1. Introduction

R_2Fe_{17} (R - rare-earth element) intermetallic compounds are extensively studied due to large magnetic moments and multiple magnetic transitions [1, 2]. In compound Ce_2Fe_{17} with the Th_2Zn_{17}-type crystal structure the partial substitution of Mn for Fe causes the origin of unusual magnetic properties: at the low temperature the basic state is ferromagnetic, but in the Mn concentration range $x = 0.5 - 1.0$ it is helical antiferromagnetic [3]. The Lu_2Fe_{17} compound with related Th_2Ni_{17}-type structure exhibit the similar magnetic behavior as Ce_2Fe_{17} being ferromagnetic at low temperature and antiferromagnetic at higher temperature [4]. However, upon Mn-doping the temperature of "ferromagnetic-antiferromagnetic" transition Θ_T in $Lu_2Fe_{17-x}Mn_x$ increases up to Néel temperature and, finally, the compounds in the range $x = 0.7 - 2$ exhibit only ferromagnetic ordering with a near constant $T_C(x)$ value [4].

It is assumed that magnetic ordering in R_2Fe_{17} is dictated generally by $3d$-atoms exchange interaction characterized by the critical distance of sign change (e.g. two Fe atoms in the distance more than 2.45 Å interact ferromagnetically [5]) and, hence, depends on the features of crystal structure. Therefore, the aim of present research was to refine the local crystal structure of $R_2Fe_{17-x}Mn_x$ (R = Ce, Lu) compounds with the methods of extended X-ray absorption fine structure (EXAFS) spectroscopy and X-ray diffraction (XRD), and to check if there are any correlations with magnetic ordering. Besides, the valence state of Ce ion in $Ce_2Fe_{17-x}Mn_x$ was studied by means of X-ray absorption near edge structure (XANES) spectroscopy since it might contribute into non-monotonous behavior of magnetic parameters.

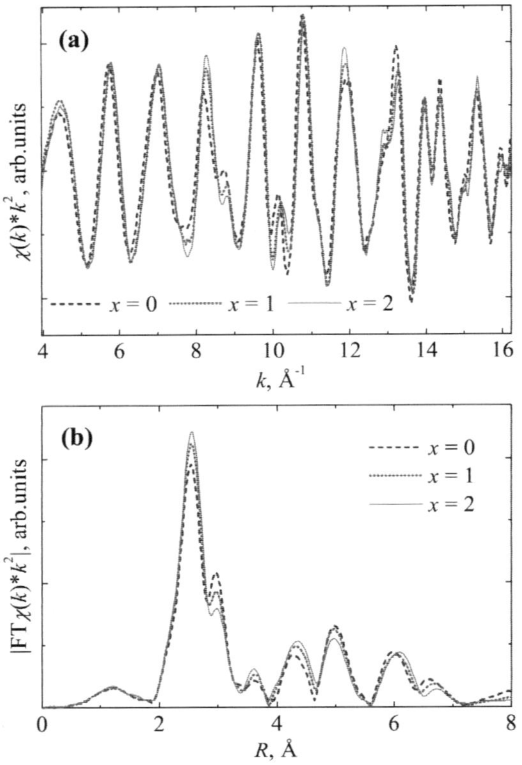

Figure 1. (a) K-Ce EXAFS-functions $\chi(k)k^2$ of $Ce_2Fe_{17-x}Mn_x$ at 8 K; (b) their Fourier-transformation modules

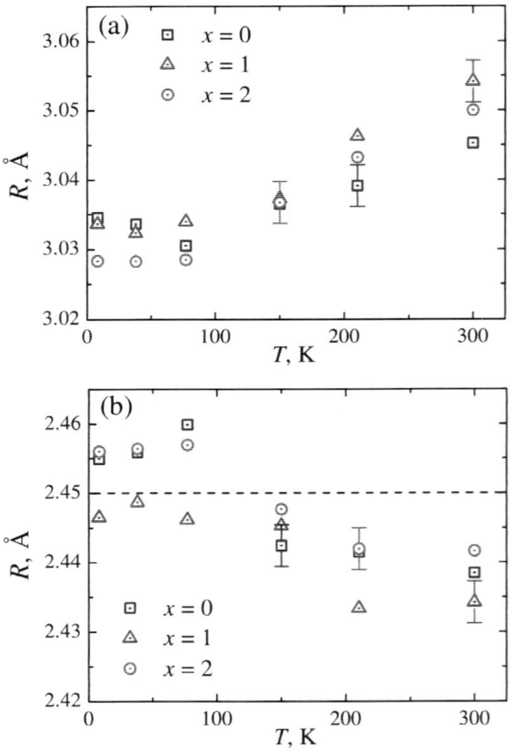

Figure 2. The temperature dependences of (a) $R_{\text{Ce-Fe(Mn)}}(T)$ interatomic distance; (b) distance between $3d$-metal atoms in the position $18f$ of $Ce_2Fe_{17-x}Mn_x$ structure

2. Experimental

The preparation and temperatures of magnetic transitions in $Ce_2Fe_{17-x}Mn_x$ ($x = 0; 1; 2$) and $Lu_2Fe_{17-x}Mn_x$ ($x = 0; 0.2; 0.5; 0.7; 2$) samples are reported in [3, 4, 6]. The X-ray absorption spectra were collected at C beamline of DORIS III storage ring (DESY, Hamburg, Germany) above K-Ce (40443 eV) and L_3-Ce (5723 eV) absorption edges on $Ce_2Fe_{17-x}Mn_x$ and above L_3-Lu (9244 eV) absorption edge on $Lu_2Fe_{17-x}Mn_x$. The energy resolution of double-crystal Si (311) monochromator was about 2.5 eV at 40 keV and 1.2 eV at 6 keV (the Si (111) plate). Low-temperature measurements were carried out using a liquid helium continuous flow cryostat with a temperature control of ±1 K at 300 K and ±0.1 K at 5 K. The XRD patterns of $Lu_2Fe_{17-x}Mn_x$ were collected at B2 beamline of DORIS III using the X-ray wavelength 0.688106 Å and the closed cycle refrigerator for measurements in the 8 – 308 K temperature range.

3. Results and discussion

The EXAFS spectra of $Ce_2Fe_{17-x}Mn_x$ were collected above the high-energy K-Ce (40443 eV) absorption edge so that they are not limited in length by the presence of another edges and contain more information on local structure. Because of the small Mn concentration and the similarity of Fe and Mn photoelectron back scattering phases and amplitudes it was not possible to distinguish between the Fe and Mn contributions in the experimental spectrum. Fig.1 shows

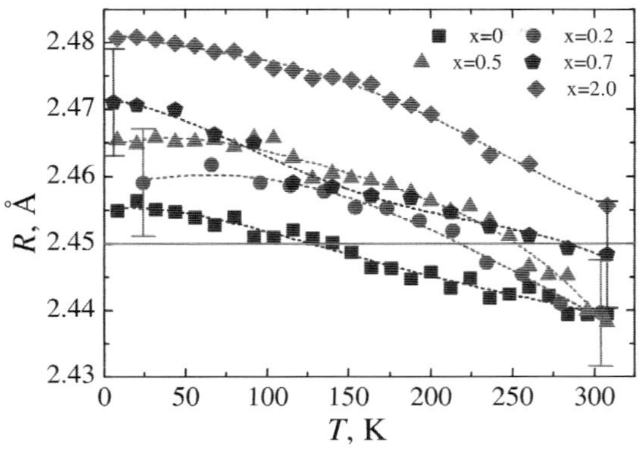

Figure 3. The temperature dependences of the distance between $3d$-metal atoms in position $12j$ of $Lu_2Fe_{17-x}Mn_x$ crystal structure. The critical value 2.45 Å is marked

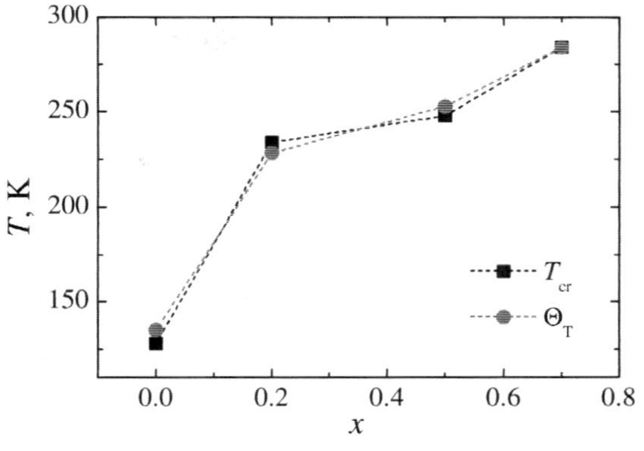

Figure 4. The temperatures T_{cr} obtained from XRD in comparison with the "ferromagnetic-antiferromagnetic" transition temperatures Θ_T in $Lu_2Fe_{17-x}Mn_x$

the EXAFS-functions $\chi(k)k^2$ of $Ce_2Fe_{17-x}Mn_x$ ($x = 0; 1; 2$) samples at 8 K and their Fourier-transformation modules (FT). The first split maximum in the $R \sim 2 - 3.5$ Å region of FT (fig.1(b)) corresponds to 6 coordination shells containing $3d$-metal atoms Fe(Mn). The next small peak at $R \sim 3.7$ Å corresponds to one Ce atom situated along the c axis relative to the absorbing atom. It is clearly seen that the shape of maxima changes with the Mn concentration.

The radius of coordination shell containing 1 cerium atom or the R_{Ce-Ce} distance between two Ce atoms along the c axis monotonously increases from $3.84-3.86$ to $3.88-3.92$ Å in various samples with the temperature decrease from 300 to 8 K. It correlates well with the temperature dependence of cell parameter $c(T)$ [7]. The temperature dependences of the first coordination shell radius $R_{Ce-Fe(Mn)}(T)$ containing $3d$-metal atoms (fig.2(a)) are in qualitative agreement with the cell parameters $a(T)$ [7] for the investigated compounds. In the binary alloy Ce_2Fe_{17} the $R_{Ce-Fe}(T)$ dependence has the minimum at 77 K corresponding to the drastic fall of the $a(T)$ at $T \sim 90$ K [7].

The $R_{Ce-Fe(Mn)}(T)$ and $R_{Ce-Ce}(T)$ dependences obtained and the crystal structure data [7] allowed us to calculate the temperature dependence of distance between the $3d$-metal atoms forming the first coordination shell (crystallographical position $18f$) (fig.2(b)). The figure shows that the critical value of 2.45 Å is exceeded in both Ce_2Fe_{17} and $Ce_2Fe_{15}Mn_2$ at the temperature less than 100 K corresponding to the ferromagnetic ordering region in these alloys. In $Ce_2Fe_{16}Mn$ alloy the distance between the $3d$-metal atoms at the low temperature is close but less than 2.45 Å. It might be that this sample at low temperature is close to the ferromagnetic state which forms in the neighboring alloy $Ce_2Fe_{15.7}Mn_{1.3}$ [6] but does not reach it.

The determination of cerium valence from L_3-Ce XANES spectra of $Ce_2Fe_{17-x}Mn_x$ is reported elsewhere [8, 9]. The main result is that the average oxidation state of cerium does not depend on temperature, but slightly depends on Mn concentration. In all three samples $x = 0; 1; 2$ the valence of Ce is intermediate around $3.30 - 3.33$, however, in ferromagnets $x = 0; 2$ its value is slightly larger, than in antiferromagnet $x = 1$.

In the Th_2Ni_{17}-type crystal structure of Lu_2Fe_{17} intermetallics Lu atoms occupy two different crystallographical positions $2b$ and $2d$, which makes EXAFS analysis of Lu_2Fe_{17}-based compounds extremely complicated. Therefore, we report here only on $Lu_2Fe_{17-x}Mn_x$ study by more self-consistent method of powder X-ray diffraction. The XRD patterns of $Lu_2Fe_{17-x}Mn_x$ were processed by the Rietveld method. The temperature dependences of lattice parameters and atomic coordinates were refined, interatomic distances calculated. The results indicate that the pair of 3*d*-metal atoms in position 12*j* probably gives the main contribution into formation of magnetic properties, since only change of distance between this pair of atoms passes the critical value of 2.45 Å [5] at some temperature T_{cr} in all samples except for $x = 2$ (Fig.3). Within the temperature range 8 − 308 K the distances between other possible pairs of Fe(Mn) atoms are always greater or less than the critical value. The obtained values of T_{cr} increase monotonously with the Mn concentration and are in satisfactory agreement with the temperatures of "ferromagnetic-antiferromagnetic" transition Θ_T (Fig.4) including the $x = 0.7$ sample for which the temperatures of "ferromagnetic-antiferromagnetic" and "antiferromagnetic-paramagnetic" transitions match and Curie temperature T_C appears [4].

In the Th_2Ni_{17}-type crystal structure of $Lu_2Fe_{17-x}Mn_x$ the 12*j* position is analogous to 18*f* position in the Th_2Zn_{17}-type structure of $Ce_2Fe_{17-x}Mn_x$. The temperature and concentration behavior of distance between Fe(Mn) atoms in these positions correlates well with the change of magnetic ordering in both $Ce_2Fe_{17-x}Mn_x$ and $Lu_2Fe_{17-x}Mn_x$. Since the magnetic phase diagrams are quite different for these systems, one can conclude that the exchange interaction between 3*d*-metal atoms in positions 18*f*/12*j* is determinant for the type of magnetic ordering in such compounds. The recent high pressure study of $Lu_2Fe_{17-x}Mn_x$ revealed a probable high impact of Mn atoms on the magnetic transition temperatures in addition to Fe atoms [10].

4. Conclusion

The substitution of Mn for Fe causes the changes in local crystal structure of $R_2Fe_{17-x}Mn_x$ (R = Ce, Lu) compounds. The distances between the Fe atoms in positions 12*j* and 18*f* in $Ce_2Fe_{17-x}Mn_x$ and $Lu_2Fe_{17-x}Mn_x$ alloys respectively exceed the critical value of 2.45 Å [5] exactly in the T and x regions of ferromagnetic ordering. Besides, in all $Ce_2Fe_{17-x}Mn_x$ samples the intermediate valence state of cerium was confirmed, changing with Mn concentration but independent from the temperature. In summary, the results agree with the assumption of magnetic ordering in such systems being governed by exchange interaction between the 3*d*-metal atoms, but leave open the possible impact of an electronic factor.

Acknowledgments

This work is partially supported by RFBR (grant No. 11-02-01174-a) and RAS (project No. 12-T-1012). Authors thank the HASYLAB Program Committee for providing the opportunity of low-temperature XAFS and XRD measurements (projects No. I-20100397 and I-20110292).

References

[1] Buschow K H J 1997 *Handbook of Magnetic Materials* vol 10, ed K H J Buschow (Amsterdam: North-Holland) p 463
[2] Jacobs T H, Buschow K H J, Zhou G F, Liu J P, Li X and de Boer F R 1992 *J. Magn. Magn. Mater.* **104-107** 1275–76
[3] Kuchin A G, Pirogov A N, Khrabrov V I, Teplykh A E, Ermolenko A S and Belozerov E V 2000 *J. Alloys Compds.* **313** 7–12
[4] Kuchin A G, Iwasieczko W, Drulis H and Khrabrov V I 2008 *Solid State Commun.* **146** 446-9
[5] Givord D and Lemaire R 1974 *IEEE Trans. Magn.* **MAG-10** 109–13
[6] Kuchin A G, Mushnikov N V, Bartashevich M I, Prokhnenko O, Khrabrov V I and Lapina T P 2007 *J. Magn. Magn. Mater.* **313** 1–7
[7] Prokhnenko O, Arnold Z, Kamarád J, Ritter C, Isnard O and Kuchin A 2005 *J. Appl. Phys.* **97** 113909-1-8
[8] Menushenkov A P, Yaroslavtsev A A, Grishina O V, Chernikov R V and Kuchin A G 2011 *JETP Letters* **94** 187–191
[9] Yaroslavtsev A A, Menushenkov A P, Grishina O V, Chernikov R V and Kuchin A G 2012 *Solid State Phenom.* **190** 251–254
[10] Arnold Z, Kuchin A and Kamarad J 2012 *Journal of Applied Physics* **111** 07E310-1-3

Journal of Physics: Conference Series **430** (2013) 012105 doi:10.1088/1742-6596/430/1/012105

Local electronic and crystal structure of rare-earth cobalt phosphides RCo_2P_2 (R = La, Ce, Pr, Nd, Eu) studied by XAFS and RIXS

A A Yaroslavtsev[1], A P Menushenkov[1], I A Zaluzhnyy[1], R V Chernikov[2], W Caliebe[2], C M Thompson[3], A A Arico[3], K Kovnir[4] and M Shatruk[3]

[1] National Research Nuclear University "MEPhI", Kashirskoe sh., 31, Moscow 115409, Russia
[2] HASYLAB at DESY, Notkestrasse, 85, Hamburg D-22607, Germany
[3] Department of Chemistry and Biochemistry, Florida State University, Chieftan Way, 95, Tallahassee, FL 32306, USA
[4] Department of Chemistry, University of California at Davis, One Shields Avenue, Davis, CA 95616, USA

E-mail: yalex03@gmail.com

Abstract. A detailed combined study of local electronic and crystal structures was performed for a series of rare-earth cobalt phosphides RCo_2P_2 (R = La, Ce, Pr, Nd, Eu) with peculiar itinerant magnetic ordering schemes using different methods employing the synchrotron radiation: RIXS, XANES and EXAFS. Europium and cerium are shown to be in the intermediate valence state in all compounds. The correlation between rare-earth valence and the observed changes of magnetic ordering in the systems under investigation is discussed.

1. Introduction

The rare-earth cobalt phosphides RCo_2P_2 (R = La, Ce, Pr, Nd, Eu) with the $ThCr_2Si_2$-type crystal structure demonstrate multiple transitions between states with the different ordering of magnetic moments in the sublattices of $4f$ rare-earth elements and $3d$ transition metals. In these materials the magnetic properties are dictated by peculiarities of electronic band structure at the Fermi level, which can be affected by external perturbation, e.g. pressure [1]. The similar modification of electronic structure can be induced by chemical compression via non-isoelectronic substitution in the rare-earth sublattice. While the pure $EuCo_2P_2$ is antiferromagnetic with magnetic Eu($4f$)-sublattice (T_N = 66.5) K, the $Pr_{0.8}Eu_{0.2}Co_2P_2$ compound is unexpectedly ferromagnetic (T_C = 282 K) with magnetic Co($3d$)-sublattice [2]. In the series $La_{1-x}R_xCo_2P_2$ the Ce-substituted phases behave quite differently from the Pr- and Nd-substituted ones [3]. Therefore, the motivation of this work was to use the local-sensitive methods of resonant inelastic X-ray scattering (RIXS) and the X-ray absorption fine structure (XAFS) spectroscopy to explore in detail the features of local electronic and crystal structure of RCo_2P_2, in particular the possible intermediate valence state of Eu and Ce.

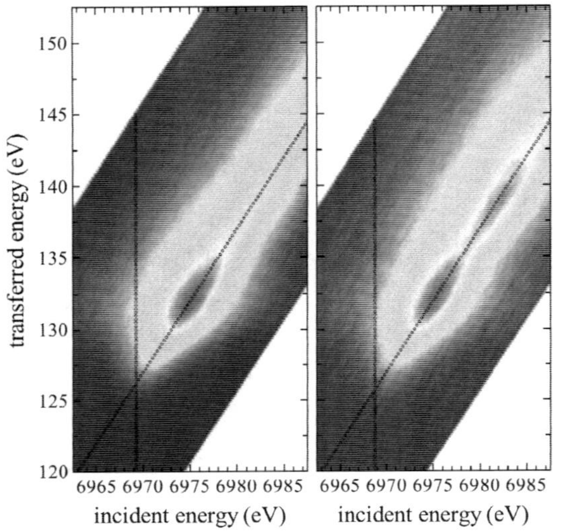

Figure 1. RIXS maps of $EuCo_2P_2$ (left) and $Nd_{0.6}Eu_{0.4}Co_2P_2$ (right). The maximum of Eu $L\beta_{2,15}$ fluorescence intensity is marked with a diagonal line. The emission spectrum at the incident photon energy $h\nu_{in}$ = 6970.7 eV is marked with a vertical line

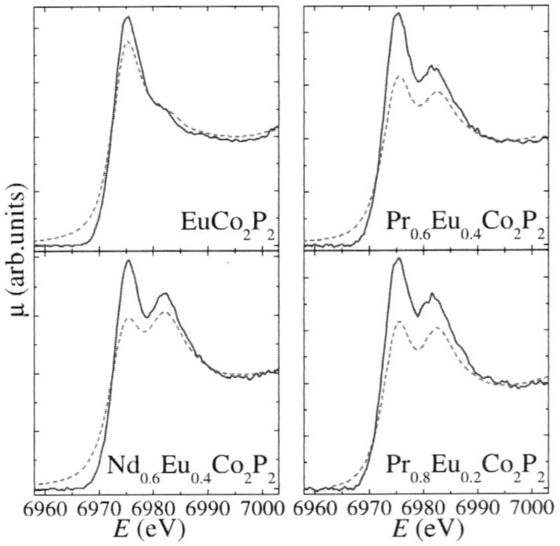

Figure 2. The comparison of high energy resolution fluorescence detected absorption (HERFD) spectra (——) at the L_3-Eu absorption edge with the XANES-spectra (- - - -) for various samples

2. Experimental

The samples $Pr_{1-x}Eu_xCo_2P_2$ (x = 0.1, 0.2, 0.3, 0.4, 1.0), $Nd_{0.6}Eu_{0.4}Co_2P_2$ and $La_{1-x}Ce_xCo_2P_2$ (x = 0.3; 0.6; 0.9) were prepared by standard procedures reported in [2, 3]. The phase purity of bulk products obtained was confrmed by powder X-ray diffraction. The full RIXS maps of $EuCo_2P_2$-based samples were collected at W1 beamline of DORIS-III storage ring (HASYLAB/DESY, Hamburg, Germany) at room temperature using the high-resolution X-ray spectrometer equipped with a spherically bent Si(531) analyzer crystal by measuring the intensity of Eu $L\beta_{2,15}$ emission line with the maximum at 6843.2 eV while scanning the incident photon energy $h\nu_{in}$ with the 0.5 eV step around the L_3-Eu absorption edge (6977 eV). The X-ray absorption spectra of all studied samples were collected at C and A1 beamlines of DORIS-III in the transmission mode above the L_3-Eu (6977 eV), L_3-Ce (5723 eV), L_3-Pr (5964 eV) and K-Co (7709 eV) absorption edges. Energy resolution of the double-crystal Si (111) monochromator at 7 keV was about 1.2 eV. Low-temperature measurements were carried out using a liquid helium continuous flow cryostat with a temperature control of ±1 K at 300 K and ±0.1 K at 5 K.

3. Results and discussion

In curent RIXS study the core-hole decay channel $2p^5 4f^N \epsilon d \rightarrow 4d^9 4f^N \epsilon d$ (Eu $L\beta_{2,15}$) was chosen. Since the $4d$ core-hole has the longer liftime than the $2p$ core-hole, the spectral broadening was reduced from $\Gamma_{2p} \sim 3.9$ eV to $\Gamma_{4d} \sim 2.75$ eV, as shown below. Besides, Eu $L\beta_{2,15}$ emission line is free from any Pr and Nd fluorescense contributions, which improves the precision of Eu valence determination.

The two-dimensional RIXS maps of samples $EuCo_2P_2$ and $Nd_{0.6}Eu_{0.4}Co_2P_2$ are shown in fig. 1. The energy transferred to the sample, i.e. $h\nu_T = h\nu_{in} - h\nu_{out}$, is shown along the vertical axis. In this plot, the emission line at constant photon energy $h\nu_{out}$ is represented by a diagonal line crossing from the lower left to the top right corner of the RIXS map. In

Journal of Physics: Conference Series **430** (2013) 012105 doi:10.1088/1742-6596/430/1/012105

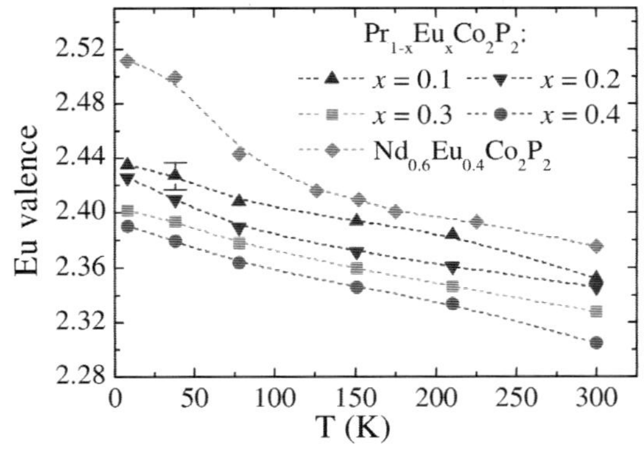

Figure 3. Temperature dependence of Eu valence in $(Pr,Nd)_{1-x}Eu_xCo_2P_2$ samples

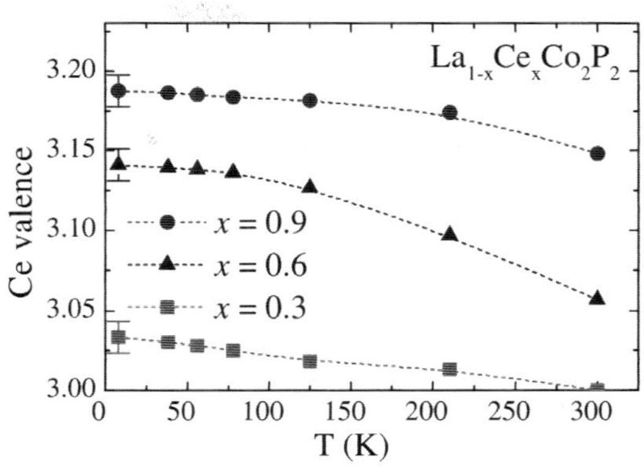

Figure 4. Temperature dependence of Ce valence in $La_{1-x}Ce_xCo_2P_2$ samples

the map of $Nd_{0.6}Eu_{0.4}Co_2P_2$ there are two pronounced resonances at $h\nu_{in,1} \sim 6975.5$ eV and $h\nu_{in,2} \sim 6975.5$ eV, corresponding to the Eu^{2+} and Eu^{3+} configurations. Both resonances are visible at single emission energy $h\nu_{out} = 6843.2$ eV, because the $2p$ and $4d$ levels are localized and equally screened by the $4f$ orbital. In the map of pure $EuCo_2P_2$, only the Eu^{2+} resonance is clearly resolved while the Eu^{3+} resonance is resolved partially. The RIXS spectra of the Pr-doped samples are similar to that of $Nd_{0.6}Eu_{0.4}Co_2P_2$. The presence of two resonances in the RIXS and XANES spectra and a single signal in the ^{151}Eu Mössbauer spectra [1, 2] indicate the intermediate valence state of europium [4]. This is due to the different timescales of Mössbauer spectroscopy (10^{-8} s) and X-ray absorption spectroscopy (10^{-15} s).

Fig. 2 shows the high energy resolution fluorescence detected absorption (HERFD) spectra, taken at a maximum of Eu $L\beta_{2,15}$ fluorescence intensity (shown with a diagonal line in fig. 1), in comparison to the transmission XANES spectra of the samples. Both series of spectra reveal a structure consisting of two maxima at $h\nu_{in,1} \sim 6975.5$ eV and $h\nu_{in,2} \sim 6982.0$ eV that correspond to the Eu^{2+} and Eu^{3+} contributions. The HERFD spectra, however, are broadened much less than the XANES-spectra, which provides an advantage in the accurate determination of the Eu^{2+} to Eu^{3+} resonance amplitude ratio. Besides, the lesser broadening reveals the absence of quadrupolar $2p \rightarrow 4f$ contribution in the pre-edge region.

In order to determine the Eu valence, the split peaks in the HERFD spectra were analyzed the same way as commonly done for XANES spectra. The processing was performed using XANES dactyloscope program [5]. The modeling of experimental spectra and the determination of $4f^7$ (Eu^{2+}) and $4f^6$ (Eu^{3+}) contributions was performed using the conventional fitting with the combination of analytical functions with constrained widths and energy positions [6] and also using the theoretical spectrum of $EuCo_2P_2$ pre-calculated with FDMNES [7] as an integer valence standart. The fitting revealed that the width of individual valence components in the HERFD spectra was $\sim$ 2.75 eV, whereas in the XANES spectra the width varies from 3.63 to 4.08 eV for different samples. The Eu valence in the Nd-containing sample (+2.336) is larger than in the $Pr_{1-x}Eu_xCo_2P_2$ (+2.298 for $x = 0.4$ and +2.341 for $x = 0.2$). This might be explained by the smaller ionic radius of Nd^{3+} as compared to Pr^{3+}. The pure $EuCo_2P_2$ sample also exhibits an intermediate Eu valence of $\sim$ 2.15 [8]. Although the total magnitude of absorption maxima of HERFD spectra is reduced by the self-absorption, the ratio of two valence contributions is affected insignificantly, and the relative error of final valence values is not bigger than 1%.

Since the recording of RIXS maps was very time-demanding (over 6 h for one map), the temperature dependences of Eu valence in $(Pr,Nd)_{1-x}Eu_xCo_2P_2$ and Ce valence in $La_{1-x}Ce_xCo_2P_2$ were obtained only by means of XANES. The Eu valence in $Pr_{1-x}Eu_xCo_2P_2$

increases from +2.30 in $x = 0.4$ sample at 300 K to +2.44 in $x = 0.1$ sample at 8 K (Fig. 3). The Eu valence in Nd-containing sample increases from +2.39 at 300 K to +2.51 at 8 K with the curvature in the direction of valence increase in the low temperature region, which is not observed for the Pr-doped samples. Interestingly, the valence ~ 2.15 of Eu in the pure $EuCo_2P_2$ does not change significantly upon temperature decrease. Thus, in this sample the intermediate valence state of Eu coexists with the antiferromagnetic ordering of Eu sublattice below $T_N = 66.5$ K [1]. The Ce valence in $La_{1-x}Ce_xCo_2P_2$ samples increases from integer +3.00 in $x = 0.3$ sample at 300 K to +3.19 in $x = 0.9$ sample at 8 K (Fig. 4). Notably, the discrepancy between XANES and HERFD results at room temperature is not such a significant: it does not exceed 0.04. This indicates that the common XANES spectroscopy is still precise enough in order of rare-earth intermediate valence determination if the correct fitting parameters are chosen.

The refinement of RCo_2P_2 local crystal structure by EXAFS is reported elsewhere [9]. The most important results are a split of rare-earth coordination shell around Co atom and the correlation between local R-Co and Co-Co distances and Ce valence, which are strongly related to interaction between the rare-earth $4f$ states and Co $3d$ states.

4. Conclusion

The results of RIXS and XANES spectroscopy demonstrate the intermediate valence state of Eu and Ce in all the studied RCo_2P_2 compounds, including pure $EuCo_2P_2$, despite the antiferromagnetic ordering in the Eu sublattice of the latter [1]. The cooling, Pr/Nd substitution for Eu and Ce substitution for La result in a significant increase in the Eu and Ce average oxidation state due to the increased chemical pressure in the lattice. At the same time the Eu valence in $Pr_{0.8}Eu_{0.2}Co_2P_2$ deviates significantly from +3, despite the similarity of structure to $PrCo_2P_2$ with the integer +3 valence of praseodymium and to the high-pressure form of $EuCo_2P_2$, also with the postulated +3 valence of europium [1]. The magnetic properties of $La_{1-x}Ce_xCo_2P_2$ are quite different from the case of structurally related compound $La_{1-x}Pr_xCo_2P_2$ at large Ce concentrations [9], i.e. exactly in samples with the largest deviation of Ce valence from +3. All these facts confirm the assumption [2] that the magnetic ordering in RCo_2P_2 systems is governed not only by the structural factors, but also by the electronic state of the rare-earth ion. In summary, the temperature and concentration dependences of Eu and Ce oxidation states and correlation with local interatomic distances can be formally associated with the hybridization between the partially localized $4f$-orbital of rare-earth and the $3d$-level of Co, which apparently is the reason for the modification of the long-range magnetic order.

Acknowledgments

This work is partially supported by RFBR (grant No. 11-02-01174-a) and NSF (DMR-0955353). Authors thank the HASYLAB program committee for providing the opportunity of RIXS and XAFS measurements (project No. I-20100397).

References

[1] Chefki M, Abd-Elmeguid M M, Micklitz H, Huhnt C, Schlabitz W, Reehuis M and Jeitschko W 1998 *Phys. Rev. Lett.* **80** 802–5
[2] Kovnir K, Reiff W M, Menushenkov A P, Yaroslavtsev A A, Chernikov R V and Shatruk M 2011 *Chem. Mater.* **23** 3021–24
[3] Kovnir K, Thompson C M, Zhou H D, Wiebe C R and Shatruk M 2010 *Chem. Mater.* **22** 1704–13
[4] Khomskii D I 1979 *Physics-Uspekhi* **129** 443–85
[5] Klementiev K V *XANES dactyloscope for Windows* freeware: www.cells.es/Beamlines/CLAESS/software/xanda.html
[6] Röhler J 1975 *J. Magn. Magn. Mater.* **47&48** 175–80
[7] Joly Y 2001 *Phys. Rev.* B **63** 125120-1-10
[8] Yaroslavtsev A A, Menushenkov A P, Chernikov R V *et al* 2012 *JETP Letters* **96** 44–48
[9] Menushenkov A P, Yaroslavtsev A A, Grishina O V *et al* 2012 *Solid State Phenom.* **190** 200–3

XAFS understanding of "repeated" magnetic compensation in $Nd_{0.8}Tb_{0.2}Al_2$

Debdutta Lahiri[1,*], Tomohiro Shibata[2,3], Soma Chattopadhyay[2,3], Nitya Ramanan[1], Swati Pandya[4], P D Kulkarni[4], A Thamizhavel[4], A K Grover[4], S Ramakrishnan[4] and Surinder Sharma[1]

[1] High Pressure and Synchrotron Radiation Physics Division, Bhabha Atomic Research Centre, Mumbai, 400085, India
[2] BCPS Department, Illinois Institute of Technology, Chicago, Illinois 60616, USA
[3]CSRRI-IIT, MRCAT, Advanced Photon Source, Sector 10, Argonne, Illinois 60439, USA
[4]Department of Condensed Matter Physics and Materials Science, Tata Institute of Fundamental Research, Homi Bhabha Road, Colaba, Mumbai 400005, India

* Email: debduttalahiri@yahoo.com; dlahiri@barc.gov.in

Abstract. Magnetic compensation (zero-moment) has important advantage in spintronics as it does not generate any self stray magnetic field. The physics of magnetic compensation (at a well defined temperature T_{comp}) is well understood in terms of competition and compensation between antiferromagnetically coupled magnetic moments of two rare earth ions. The different thermal evolution of the dissimilar rare earth moments can compensate each other at a particular temperature determined by the concentration of the dopant. We have observed, for the first ever time, "repeated" magnetic compensation in both polycrystalline and single crystals of $Nd_{0.8}Tb_{0.2}Al_2$ ($T_{f1} \sim 86$ K and $T_{f2} \sim 34$ K). A possible reason could be the non-random distribution of magnetic rare earth ions on different crystallographic sites, presenting deviation from mean field like situation. The motivation for XAFS derives from detection of these circumstances. XAFS at Nd, Tb L_3 edges reveals that 25% (Nd,Al) anti-site defect exists in $Nd_{0.8}Tb_{0.2}Al_2$. This implies that Nd atoms occupy two different sites. In contrast, Tb preferentially substitutes only one of the Nd sites. The implication of these results upon magnetic compensation is discussed.

1. Introduction

The phenomenon of magnetic compensation (zero-moment) in mixed rare-earth compounds [1] is well known. The concurrent condition of finite spin and zero net-moment is ideal for spintronics [2] as it inhibits the generation of stray magnetic field. The physics of magnetic compensation, at single temperature T_{comp}, is well understood. In an admixed alloy viz. $R_{1-x}R'_x$, with anti-ferromagnetically coupled magnetic moments μ_R and $\mu_{R'}$, magnetic compensation (between R and R') could result if $(1-x)\,\mu_R = x\,\mu_{R'}$. The thermal evolution of μ is monotonic over practical temperature range and solving the equation for particular x, results in single

Journal of Physics: Conference Series **430** (2013) 012106 doi:10.1088/1742-6596/430/1/012106

compensation temperature T_{comp}. We have observed, for the first ever time, "repeated" magnetic compensation ($T_{f1} \sim 86$ K and $T_{f2} \sim 34$ K) in both polycrystalline and single crystals of $Nd_{0.8}Tb_{0.2}Al_2$ [3]. [In this case, R=Nd, R'=Tb]. This is curious since it directly challenges the basis of the above equation viz. monotonic thermal evolution of moments. Additionally, the phenomenon is encouraging since it raises T_{comp} towards practical temperature range (e.g. $T_{f1} > T_{f2}$). Heat capacity, resistivity, ac-susceptibility and XMCD show fingerprints at both these temperatures [3], indicating the involvement of two magnetic sub-lattices. [Fingerprints at T_{f1} are more prominent compared to T_{f2}.] These observations suggest (i) co-existence of different phases; (ii) one phase (corresponding to T_{f1}) more dominant than the other (corresponding to T_{f2}). Although XRD has detected single phase for these compounds, defects on nano-scale (beyond the detection limits of XRD) cannot be ruled out. The motivation for our Nd and Tb L_3-edge XAFS experiments derived from detection of these phases. Since the coupling between Nd and Tb is responsible for compensation, XAFS at Tb L_3 edge is intended to characterize this coupling in terms of Tb-Nd bond-length and coordination.

2. Experimental

Polycrystalline samples of $Nd_{0.8}Tb_{0.2}Al_2$ were prepared by melting together stoichiometric amounts of the end members ($x = 0$ and $x = 1$) in an arc furnace. They were characterized by element-analyzer, XRD and magnetization (SQUID-VSM) prior to XAFS measurements. For XAFS measurements, the polycrystalline samples were further ground to ~5μm particle size and pasted on scotch tape. XAFS data at Nd L3- edge (6.208 keV) and Tb L3-edge (7.514 keV) were collected in transmission-mode at the bending magnet beamline 10-BM of MRCAT (Materials Research Collaborative Access Team) [4], Advanced Photon Source, USA. During XAFS measurement, reference compounds / foils were measured in parallel at respective edges to calibrate the monochromator [Si (111)]. The latter was detuned to remove higher harmonics. For the incident and transmitted ion chambers, [He:N2=1:4] and [N2:Argon=4:1] were used respectively. The data were processed using ATHENA [5] and the structural parameters (bond-length, coordination and DWF) were fit using the programs FEFF8 and FEFFIT [5].

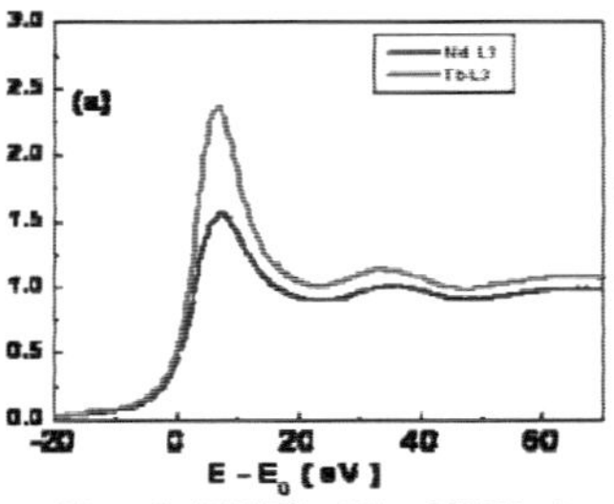

Figure 1a. XANES at Nd and Tb L3 edges for $Nd_{0.8}Tb_{0.2}Al_2$

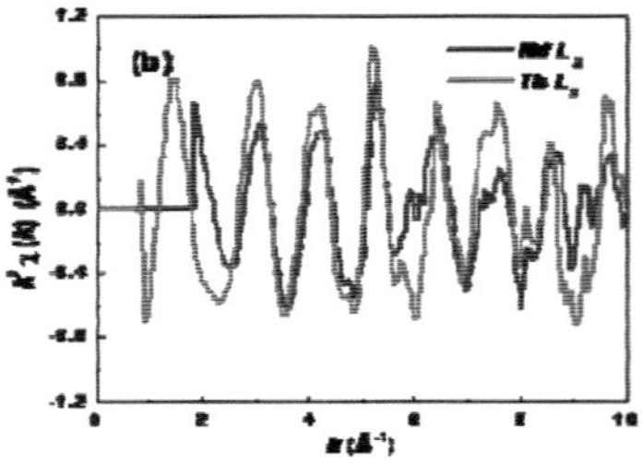

Figure 1b. XAFS at Nd and Tb L3 edges for $Nd_{0.8}Tb_{0.2}Al_2$

3. Results and Discussions

XANES and EXAFS spectra are shown in Figures 1a-b. Nd L3-XANES whiteline is significantly more suppressed (than Tb L3), despite lower DWF *i.e.* better order for Nd bonds (observed from EXAFS analysis). This could be consistent with co-existing different structural phases for Nd site. The oscillations from these phases interfere destructively and reduce Nd L3 XANES whiteline. EXAFS signal shows noise and scan-to-scan irreproducibility beyond 8 $Å^{-1}$ due to large absorption in air (etc.) at these low edges. For fitting, $k^{1,2,3}$-weighted data were transformed over 3-8 Å^{-1}. Due to limited spline range, the Fourier transforms of the data showed relatively large background peak at < 2 Å (Figures 2a-c). However, since the structural peaks of interest are farther away (2.5-4 Å), the contribution of background peak could be ignored during fitting. $NdAl_2$ crystallizes in *f d -3 m* group. The bond lengths for the pure $NdAl_2$ phase were generated using the crystallographic structural parameters in the ATOMS program. Simulation generated for the pure phase, using the NOFIT option in feffit.inp, is compared with the Nd L_3 XAFS data in Figure 2a. S_0^2 and DWF were constrained to be 0.9 [6] and 0.01Å^2 respectively. Figure 2a suggests marked deviation between the simulated and experimental data over 2-2.8 Å. We attempted to correct for this deviation by actually fitting (*i.e.* not using NOFIT) the data. During fitting, coordination was constrained to crystallographic values while bond-lengths and DWF were varied. The fit resulted in corrections to crystallographic $R_{Nd\text{-}Al}$ and $R_{Nd\text{-}Nd}$ by 0.23 Å and 0.14 Å respectively, with DWF ~ 0.02 Å^2 each. [It may be noted that Nd and Tb neighbors are indistinguishable by XAFS]. Even including these corrections, the structure of pure $NdAl_2$ phase failed to fit the experimental data (the peak shape) well, as is clear from Figure 2a and poor fit quality (R-factor = 0.03). Even including coordination vacancy could not improve the fit. These results consolidated our speculation that single structural model may not be sufficient to fit the data and co-existing second structural phase (Phase 2) needs to be considered. "Pure" phase will henceforth be referred to as Phase 1.

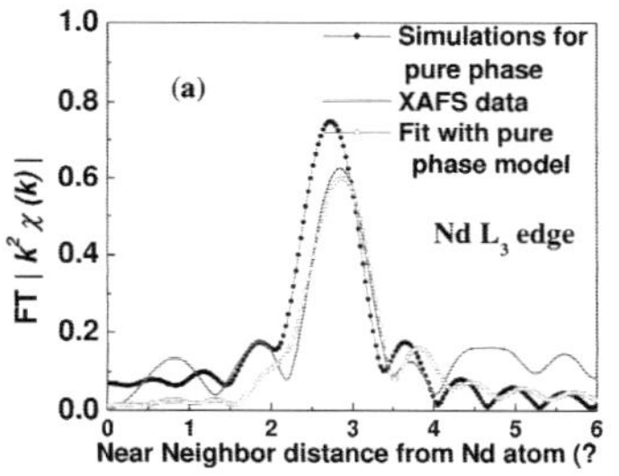

Figure 2a. Comparison of Nd L_3 XAFS data with (i) simulations and (ii) best fit with pure $NdAl_2$ model.

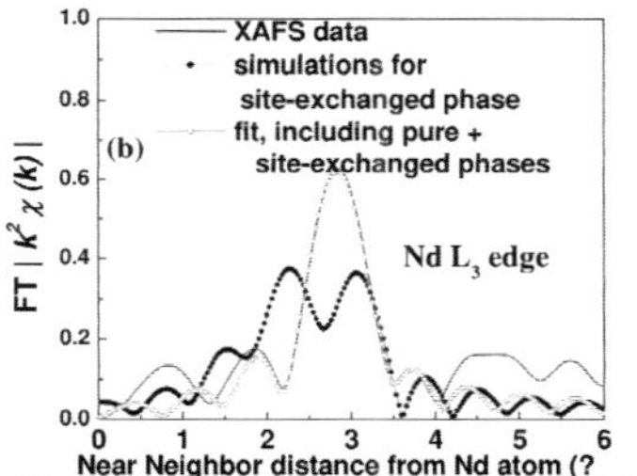

Figure 2b. Comparison of Nd L_3 XAFS data with simulations for site-exchanged phase and fit including both pure and site-exchanged phases.

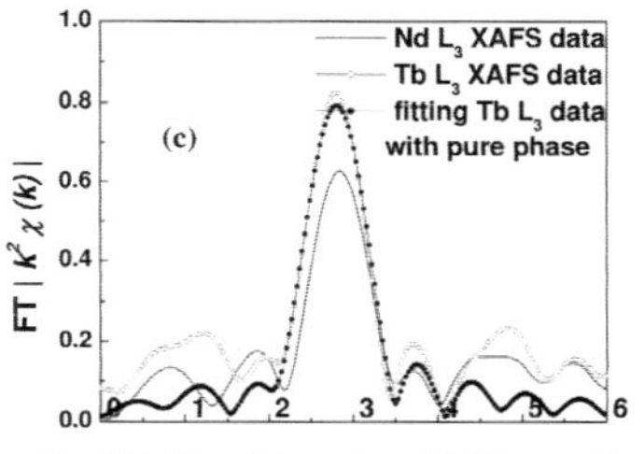

Figure 2c. Comparison of Tb L_3 XAFS with Nd L_3 XAFS and fit with pure phase model

For Phase 2, we considered several models and generated the corresponding transforms. Out of these, Nd-Al anti-site defect appeared to be the most plausible model. Bond lengths, corresponding to the site exchanged phase, were generated by interchanging the *Nd* and *Al* sites in the ATOMS file. XAFS simulation, for site-exchanged phase (Phase 2), is shown in Figure 2b. Subsequent simulations, combining different proportions of pure and site-exchanged phases, show considerable improvement in the peak shape between 2-2.8 Å. To quantify the fraction (x) of anti-site defect, the fit strategy included linear combination of the ordered and site-exchanged phases. Scattering paths, for the two phases, included: (i) $R_{Nd\text{-}Al}$= 3.31Å (N = 12) and $R_{Nd\text{-}Nd}$=3.46 Å (N = 4) for Phase 1 and (ii) $R_{Nd\text{-}Nd}$ = 2.82 Å (N = 6) and $R_{Nd\text{-}Al}$= 3.31Å (N = 6) for Phase 2. The fraction of each phase was duly included in feffit.inp through multiplicative factor "degeneracy": viz. (1-x) for Phase 1 paths and x for Phase 2 paths. Best fit was obtained for $x = 0.25 \pm 0.03$ *i.e.* 75% *Nd* atoms occupy Phase 1 site while 25% Nd atoms occupy Phase 2. The refined bond-lengths are determined to be $R_{Nd\text{-}Al}$=

3.28 Å, $R_{Nd\text{-}Nd}$ = 3.39 Å and $R_{Nd\text{-}Nd}$ = 2.94 Å. Good fit quality is ensured by low R-factor (= 0.007) and demonstrated in Figure 2b.

Tb L_3 dataset was fit, following the same procedure as Nd L_3. Interestingly, the best fit is obtained for Tb *preferentially* substituting only the Nd sites in Phase 1 (with increased DWF: $\sigma^2_{Tb-Al} = 0.01$ Å^2 and $\sigma^2_{Tb-Nd} = 0.03$ Å^2. [Constraining Tb to occupy both the sites degrades fit quality considerably.] This is clear from Figure 2c, where the Tb edge transform, rather than Nd edge, resembles the structure of Phase 1. $R_{\text{Tb-Nd}}$ = 3.46 Å implies that Tb bonds *preferentially* with the Nd atoms in Phase 1 site.

These structural results suggest that Nd-Tb magnetic compensation (at least partial) occurs only at *one* Nd site viz. Phase 1. Nd atoms in Phase 2 remain magnetically uncompensated due to absence of Tb neighbor. Phases 1 and 2 result in different "μ-T" curves. Their weight-averaged μ_{net} -T curve crosses zero magnetization line twice, at T_{f1} and T_{f2}. Thus, repeated compensation is result of structural inhomogeneity, with the latter acting like two magnetic sub-lattices. Detailed theoretical modeling is underway to account for the observed repeated compensation more quantitatively.

4. Conclusion

In summary, we have studied the structure around Nd and Tb sites in $Nd_{0.8}Tb_{0.8}Al_2$, using XAFS. The system is determined to be structurally inhomogeneous, with Nd atoms occupying ideal crystallographic and site-exchanged (with Al) sites in 3:1 ratio. In contrast, Tb dopants preferentially substitute *only* the ideal Nd sites. This causes inhomogeneity in Nd-Tb coupling (or magnetic cancellation) since *only* the Nd atoms in ideal sites find Tb neighbors. Nd atoms, in site-exchanged site, remain magnetically uncompensated due to absence of Tb neighbors. The two structural phases resemble two magnetic sub-lattices, resulting in different magnetic moments (and their thermal evolution) and subsequently, double compensation of the weighted averaged magnetic moment.

5. Acknowledgements

MRCAT operations are supported by the Department of Energy and MRCAT host institutions. Usage of the Advanced Photon Source is supported by the U. S. Department of Energy, Office of Science, Office of Basic Energy Sciences, under Contract No. DE-AC02-06CH11357

6. References

[1] Qiao et al. 2004 *Phys. Rev. B* **70** 134418.
[2] Adachi H and Ino H 1999 *Nature* **401** 148.
[3] Kulkarni P, Sangeetha N S, Ramakrishnan S and Grover A K 2010 *Solid State Physics* (India) **53** 1097; Kulkarni P, Thamizhavel A, Kodama K , Nakamura T, Ramakrishnan S and Grover A K 2011 *AIP.Conf. Proc.* **1349** 1217.
[4] Segre et al 2000, *AIP Conf. Proc.* **521** 419.
[5] Ankudinov et al. 1998 *Phys. Rev. B* **58** 7565; Ravel B and Newville M 2005 *J.Synchrotron Rad.* **12:4** 537.
[6] Mountjoy G et al. 2001 *J. Non-Cryst. Sol.* **279** 20.

Journal of Physics: Conference Series **430** (2013) 012107 doi:10.1088/1742-6596/430/1/012107

Determination of incommensurate lattice modulations in $La_{0.4}Sr_{1.6}MnO_4$ by resonant r-ray scattering

J García [1*], J Herrero Martín[2], G Subías[1], J Blasco[1] and M C Sánchez[1]
[1]Instituto de Ciencia de Materiales de Aragón, Departamento de Física de la Materia Condensada, CSIC-Universidad de Zaragoza, Zaragoza, 50009, Spain
[2]Institut de Ciència de Materials de Barcelona- CSIC, Campus Universitari de Bellaterra, E-08193, Bellaterra, Barcelona, Spain
*E-mail: jgr@unizar.es

Abstract. Superlattice reflections, which appear below the semiconductor-insulator phase transition temperature in $La_{0.4}Sr_{1.6}MnO_4$, have been investigated by Resonant X-ray Scattering. We have observed two types of resonant reflections with modulation wave vectors $\mathbf{q}_{h\pm\varepsilon} = (h\pm\varepsilon, h\pm\varepsilon, 0)$ and $\mathbf{q}_{h\pm2\varepsilon} = (h\pm2\varepsilon, h\pm2\varepsilon, 0)$ (h=integer, ε=0.2) in the *I4/mmm* setting. The energy dependence across the Mn K-edge, together with the azimuthal angle, polarization and temperature evolution have shown that reflections with 2ε periodicity only appear in the σ-σ' channel, involving resonant and non-resonant scattering. On the other hand, reflections with ε periodicity exclusively appear in the σ-π' channel on resonance and the scattered intensity follows a π-periodicity with the azimuthal angle. We demonstrate that this different behaviour discards any possible Mn^{3+}-/Mn^{4+}-like bimodal stripe ordering but it can be easily explained as due to oxygen displacements sinusoidal modulated along and perpendicular to the tetragonal [1, 1, 0] direction.

1. Introduction

Incommensurate structural phases in mixed valence transition metal oxides have been traditionally described as due to the ordering of the extra charges, giving rise to a sequence of stripes with rich and poor electronic density. This scenario has been considered of general relevance in the physics of nickelates, cobaltates and manganites [1-3]. Overdoped manganites show incommensurate structural ordering, where the periodicity of the modulations follows linearly the amount of extra charges [4]. Some authors interpret this fact as evidence for a stripe order with the Mn^{4+} stripes aligned along the diagonal of the perovskite lattice [5], whereas recent electron and x-ray diffraction experiments [6] point to a continuous incommensurate structure. Resonant x-ray scattering (RXS) at the Mn K-edge allows to directly determine the charge and the anisotropic ordering of the Mn atoms. The specific atomic selectivity joined to the energy, azimuth and polarization dependences make this technique ideal to solve this controversy.

We present here a RXS study of the super-lattice reflections ($h\pm2\varepsilon$, $h\pm2\varepsilon$, 0) (so-called CO reflections) and ($h\pm\varepsilon$, $h\pm\varepsilon$, 0) (so-called OO reflections) with ε=0.2 that appear in the insulating phase of the $La_{0.4}Sr_{1.6}MnO_4$ manganite. The different polarization and azimuth behaviour of the two types of resonant reflections has allowed us to discriminate between the stripe and the continuous incommensurate structural models. We demonstrate that both sets of reflections should show the same polarization and azimuthal behaviour in the bimodal stripe model. On the contrary, the experimental results are well described by a sinusoidal modulation of oxygen displacements.

Journal of Physics: Conference Series **430** (2013) 012107 doi:10.1088/1742-6596/430/1/012107

2. Experimental Section

A single crystal of $La_{0:4}Sr_{1.6}MnO_4$ was grown by the floating-zone technique. Powder x-ray diffraction measurements on crushed crystals indicate that it is single phase and crystallizes in a tetragonal structure of space group *I4/mmm* with room-temperature lattice constants a=3.852 Å and c=12.4 Å.

RXS experiments were performed at the ID20 beam line [7] at ESRF. The incident monochromatic beam, energy resolution being 1 eV at the Mn *K* edge, was linearly polarized (99%) with the polarization vector perpendicular to the scattering plane. The $La_{0.4}Sr_{1.6}MnO_4$ single crystal, cut and polished in the (1, 1, 0) plane in the tetragonal setting, was mounted in a four-circle vertical diffractometer equipped with a closed-cycle helium refrigerator. Energy scans of the scattered signal across the Mn K absorption edge and azimuth scans at the resonance energy were measured. The polarization of the scattered light was analysed using a Cu (2,2,0) crystal, which allows the measurement of the σ-σ' and σ-π' channels independently.

3. Results and discussion

The energy dependence of the scattered intensity across the Mn K edge for the (1.6, 1.6, 0) reflection in the σ-σ' channel at different azimuth angles (φ) are shown in figure 1a. We observe scattered intensity at energies below the Mn K absorption edge and a strong enhancement of the intensity at photon energies close to the Mn K edge. The overall intensity at the maxima shows moderate marked azimuthal angle dependence. The scans corresponding to the two polarization channels at resonance (6556 eV) are shown in the inset of figure 1a. It is clear that this reflection is not observed in the σ-π'channel. Figure 1b shows the energy dependence spectra of the (1.8, 1.8, 0) reflection, where a strong resonance is observed in the σ-π' channel at the Mn K edge. The absence of Bragg diffraction in the σ-σ' channel, as shown in the inset of figure 1b on resonance for example, indicates that these reflections are forbidden by symmetry [8]. The shape of the energy spectrum is the same regardless of the azimuth angle but it shows a strong azimuthal angle dependence of π period (figure 1b). The intensity of both type of reflections, (h±0.2, h±0.2, 0) and (h±0.4, h±0.4, 0), decreases with increasing temperature. These reflections vanish at T_{CO}~240 K, which agrees with the change in the electrical resistivity.

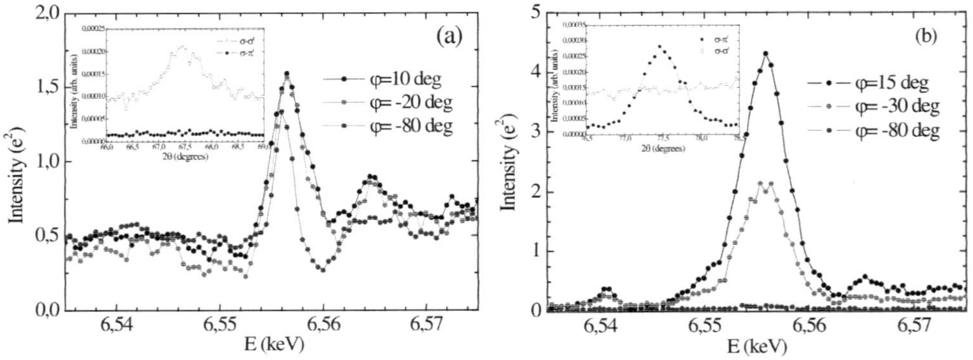

Figure 1. (a) Energy dependence of (1.6, 1.6, 0) reflection at T=80 K close to the Mn K-edge in the σ-σ' channel for three different azimuthal angles. Inset: The signal collected at the Mn K-edge in the σ' and π' detection channels. (b) Energy dependence of (1.8, 1.8, 0) reflection at T=80 K close to the Mn K-edge in the σ-π' channel for three different azimuthal angles. Inset: The signal collected at the Mn K-edge in the σ' and π' detection channels.

We have simulated the intensity of the reported superlattice reflections considering two models, the bimodal stripe pattern and the sinusoidal modulated one. In the first case, we make use of the anomalous atomic scattering tensors (AAST) of anisotropic Mn^{3+} ions ($f^{\alpha\beta}{}_{Mn3+}$) and isotropic Mn^{4+} ions ($f^{\alpha\beta}{}_{Mn4+}$) that are ordered forming stripes. In the sinusoidal case, the terms $f^{\alpha\beta}$ of the AAST will follow a dependence $f^{\alpha\beta}{}_i = f^{\alpha\beta}{}_0 + \Delta f^{\alpha\beta} \cos(2\pi \mathbf{q}\mathbf{r}_i + \gamma)$, where $\Delta f^{\alpha\beta}$ is the amplitude of the sinusoidal modulation, **q** is the modulation vector, $\mathbf{r}_i$ is the Mn position in the supercell and γ is the phase. Figure 2 shows a sketch of the orthorhombic supercell projected in the $\mathbf{a}_o\mathbf{b}_o$ plane. In our case, the incommensurate periodicity matches with the lattice in the [1, 1, 0] direction for five unit cells and we can identify five distinct Mn sites. The two possible stripe arrangements with sequences Mn^{3+}-Mn^{3+}-Mn^{3+}-Mn^{4+}-Mn^{4+} and Mn^{4+}-Mn^{3+}-Mn^{4+}-Mn^{3+}-Mn^{4+}, together with the sinusoidal modulated one are indicated in figure 2.

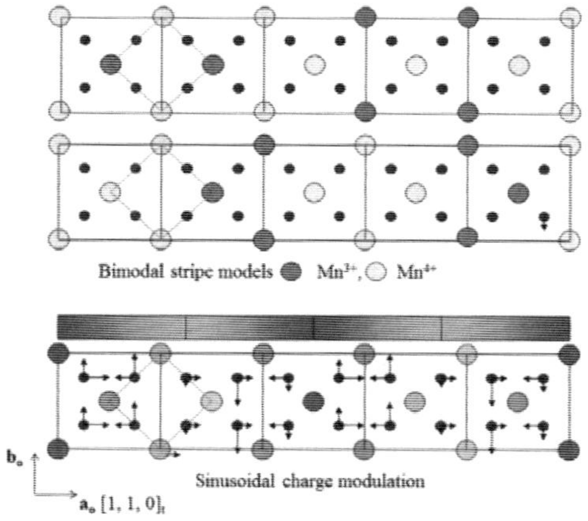

Figure 2. Schematic picture of the proposed superstructural orders in the $\mathbf{a}_o\mathbf{b}_o$ plane showing (lower panel) longitudinal and transverse atomic displacements for the five non-equivalent MnO_6 octahedra in the sinusoidal modulated model and (upper panel) bimodal charge stripe models.

We first discuss the stripe models. The real and imaginary parts of the diagonal terms of the AAST differ on the energy chemical shift, i.e. $f^{\alpha\alpha}{}_{(Mn3+)}(E) = f^{\alpha\alpha}{}_{(Mn4+)}(E\text{-}\Delta E)$, where ΔE is the difference of the absorption edge energy. The anisotropic $f^{\alpha\beta}$ term is zero for the isotropic Mn^{4+} site whereas it is equal to that calculated for the anisotropic Mn^{3+} site in ref. 9. Figure 3 (left panel) shows the simulated energy dependence of the superlattice reflections for the two stripe models compared with the experimental spectra. We obtain a good agreement with the experimental data for the (h±0.4, h±0.4, 0) reflections in the σ-σ´ channel and the (h±0.2, h±0.2, 0) reflections in the σ-π´ channel. However, important σ-σ´and σ-π' contributions are also obtained for the (h±0.2, h±0.2, 0) and (h±0.4, h±0.4, 0) reflections, respectively that are not experimentally observed. Therefore, the bimodal stripe ordering can be completely discarded.

In the following, we show that the sinusoidal modulation of the AAST for the Mn atoms gives the appropriate uncorrelated reflections. Since the Mn AAST mainly depends on the local structure around the scattered atom, a modulation in the atomic position of the nearest neighbour oxygen atoms will induce the same kind of modulation as for the Mn AAST. A uniform compression/expansion of the bond lengths will originate changes in the diagonal AATS terms (monopoles) whereas a deformation maintaining the average bond-distances induces a change in the out of diagonal terms (quadrupoles). We have built a model where oxygen motions either in the $\mathbf{a}_o\mathbf{b}_o$ plane or along the $\mathbf{c}_o$ axis are longitudinal modulated with an amplitude $\Delta x(\Delta z) = \Delta x_0\ (\Delta z_0)\cos\ (2\pi\cdot n\cdot 0.4)$, being n the position of the oxygen atom in the supercell (figure 2). The transverse modulation for the quadrupole mode will follow the expression $\Delta y = \Delta y_0\ [\cos(2\pi\cdot n\cdot\varepsilon) + \mathrm{sen}(2\pi\cdot n\cdot\varepsilon)]$. The same kind of displacements yield a checkerboard sequence of expanded-distorted and compressed-undistorted MnO_6 octahedra for $La_{0.5}Sr_{1.5}MnO_4$ (2ε=0.5). The AAST for each of the five distinct Mn atoms were calculated using the FDMNES code [10]. We have got that the structure factor for the (h±0.4, h±0.4, 0) reflections is $F(h\pm0.4, h\pm0.4, 0) = \Delta F_{yy}\cos^2\varphi + \Delta F_{zz}\sin^2\varphi$ in the σ-σ´ channel, whereas it is given by $F(h\pm0.2, h\pm0.2, 0) = \Delta F_{xy}\cos\theta\cos\varphi$ for the (h±0.2, h±0.2, 0) reflections in the σ-π' channel.

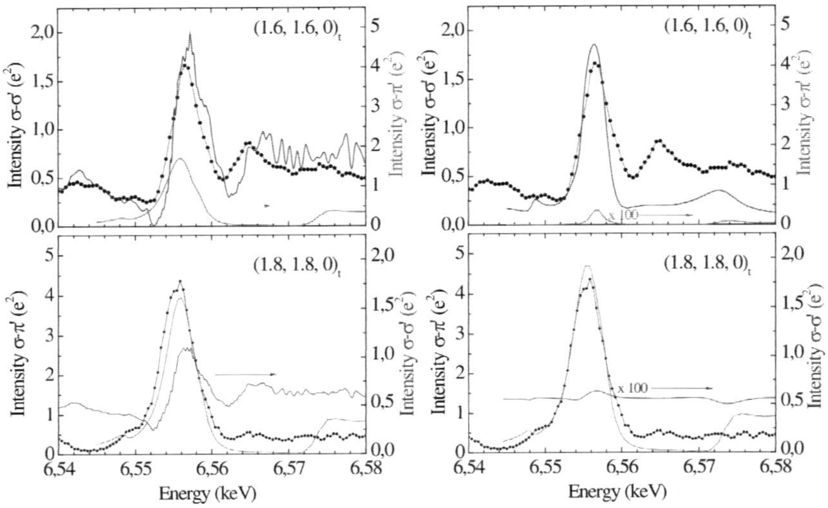

Figure 3. Energy dependence of the intensity for (1.6, 1.6, 0) and (1.8, 1.8, 0) reflections at φ maximum. Line plus circles are the experiment and full lines are the simulations in the σ-σ' (blue) and σ-π' (red) channels for the bimodal charge stripe models (left panel) and the sinusoidal modulations model (right panel).

Similar ΔF_{yy} and ΔF_{zz} amplitudes do barely produce any significant intensity variation of (h±0.4, h±0.4, 0) reflections as a function of φ whereas (h±0.2, h±0.2, 0) reflections show a pronounced $\cos^2\varphi$ dependence as experimentally observed. Moreover, $F(h\pm0.2, h\pm0.2,0)^{\sigma\text{-}\sigma'} = 0$ and $F(h\pm0.4, h+0.4,0)^{\sigma\text{-}\pi'} = 0$. Figure 3 (right panel) compares calculated and experimental spectra for representative reflections of $La_{0.4}Sr_{1.6}MnO_4$. We obtained a satisfactory agreement with Δx_0= 0.0024 Å; Δy_0= 0.014 Å and Δz_0= 0.0012 Å, the values of the refined distortions for the longitudinal, transverse and apical modulations. For the (h±0.4, h±0.4, 0) reflections, ΔF_{yy} and ΔF_{zz} can be related to the chemical shift arising from a charge disproportionation among the different Mn atoms. We obtain a chemical shift of ~0.1 eV, which corresponds to a charge disproportionation of 0.04 electron.

This RXS study demonstrates that the occurrence of (h±0.4, h±0.4, 0) and (h±0.2, h±0.2, 0) reflections in $La_{0.4}Sr_{1.6}MnO_4$ is originated by sinusoidal oxygen motions, longitudinal and transverse to the modulation direction (orthorhombic $\mathbf{a_o}$-vector). Reflections that arise from the transverse modulation are forbidden for Thomson scattering and they are only detected in the σ-π' polarisation channel. Thus, our results confirms that the two modulations with ε=0.2 and 2ε=0.4 have different origin, i.e. the modulation with 2ε=0.4 does not correspond to the second harmonic of the modulation with ε=0.2. In conclusion, we show that a continuous charge-orbital density wave model better describes the low temperature incommensurate-ordered phases in overdoped manganites.

References

[1] Moritomo Y *et al* 1995 *Phys. Rev. B* **51** 3297
[2] Radaelli P G *et al* 1997 *Phys. Rev. B* **55** 3015
[3] Mori S, Chen C H and Cheong S W 1998 *Nature (London)* **392** 473
[4] Larochelle S *et al* 2005 *Phys. Rev. B* **71** 024435
[5] Ulbrich H *et al* 2011 *Phys. Rev. Lett.* **106** 157201
[6] Loudon J C *et al* 2005 *Phys. Rev. Lett.* **94** 097202
[7] Paolasini L *et al* 2007 *J. Synch. Rad.* **14** 301
[8] Dmitrienko V E *et al* 2005 *Acta Crystallogr., Sect. A: Found. Crystallogr.* **61** 481
[9] Herrero-Martín J *et al* 2011 *Phys. Rev. B* **83** 184401
[10] Yoly Y 2001 *Phys. Rev. B* **63** 125120

Journal of Physics: Conference Series **430** (2013) 012108 doi:10.1088/1742-6596/430/1/012108

Structural phase transition and magnetic properties of Er-doped $BiFeO_3$ nanoparticles

Y T Li[1,2], H G Zhang[1], X G Dong[1], Q Li [1,3], W W Mao[2], C L Dong[2], S L Ren[2], X A Li[2] and S Q Wei[3]

[1] Physics Department, Southeast University, Nanjing, 211189, P. R. China

[2] College of Science, Nanjing University of Post and Telecommunications, Nanjing, 210046, P. R. China

[3]National Synchrotron Radiation Laboratory, University of Science and Technology of China, Hefei, 230029, P. R. China

E-mail: qli@seu.edu.cn

Abstract. The structural phase transition and local structural distortion of Er-doped $BiFeO_3$ nanoparticles have been discussed in order to understand the variation of magnetic properties in this system. The X-ray diffraction patterns and X-ray absorption fine structure of these samples demonstrate that there is structural phase transition and no obvious local structural distortion with the increasing of doping concentration. Unfortunately, no ferromagnetic properties have been observed even at a lower temperature. And the X-ray absorption spectra of Fe 2*p* core level of these samples are totally same, especially the energy positions do not shift which means the consistent valence states of Fe ions.

1. Introduction

The compound $BiFeO_3$ with electrical and magnetic ordering is one of the potential materials in novel storage media and spintronics devices [1]. It behaves ferroelectric properties at room temperature since its ferroelectric phase transitional temperature is 836°C, which is testified by the observation of ferroelectric/ferroelastic single domain by polarized light microscopy [2]. Although the ferroelectric nature of $BiFeO_3$, it is actually antiferromagnetic ordering above room temperature. Thus, many researchers are engaged in releasing the ferromagnetism from space-modulated spin structure. As to spin structure of $BiFeO_3$, there are partially filled 3*d* orbits of Fe^{3+} ion which lead to G-type canted antiferromangetic order with an incommensurately space-modulated spin structure along the $(110)_h$ direction on a long wavelength of 62 nm [3, 4]. In order to obtain good ferromagnetic properties at room temperature, many ways have been taken by researchers to destroy the modulated spin structure, such as substitution for A-site by rare-earth element with different ion radius [4] and preparation of thin film deposited on different substrates [5].

In our paper, the samples Er-doped $BiFeO_3$ nanoparticles have been prepared by sol-gel method in order to investigate the influence on magnetism of structural distortion caused by Er^{3+} ions (ion radius 0.088 nm) doping. Moreover, there is little literature until now studying the relationship between magnetism and local structure distortion for Er-doped $BiFeO_3$. Our results demonstrate that no obvious ferromagnetic properties have been observed even at lower temperature, although the structural phase transition and local structural distortion have been caused by Er ions doping.

2. Experiment

Polycrystalline $Bi_{1-x}Er_xFeO_3$ (x = 0, 0.05, 0.1, 0.15) nanoparticles have been prepared by sol-gel method. Highly pure Bi_2O_3, Er_2O_3, $Fe(NO_3)_3{\cdot}9H_2O$ powders were mixed into mixture solution by dilute nitric acid with appropriate composition. Tartaric acid as organic complex was put into the mixture, then preheated at 80 ºC for 4 h until dissolution. The solution was heated at 140 ºC for 5 h. The formed gel precursor was dried at 250 ºC. The target product was obtained by annealing the precursor at 550 ºC for 2 h.

The crystalline qualities of the samples were assessed using X-ray diffractometer (D/Max 2000) with Cu-*K*α radiation at room temperature. The corresponding images of the resulting nanoparticles were obtained via scanning electron microscopy (SEM, HITACHI S4800). The Magnetizations with a function of temperature and magnetic field were performed by PPMS (Quantum Desgin-9). X-ray absorption fine structure (XAFS) spectra of Fe *K*-edge of samples were measured at beamline U7C at National Synchrotron Radiation Laboratory (NSRL), University of Science and Technology of China (USTC). X-ray absorption spectra (XAS) of Fe 2*p* core level of those samples were measured at Beijing Synchrotron Radiation Facility (BSRF).

3. Results and Discussion

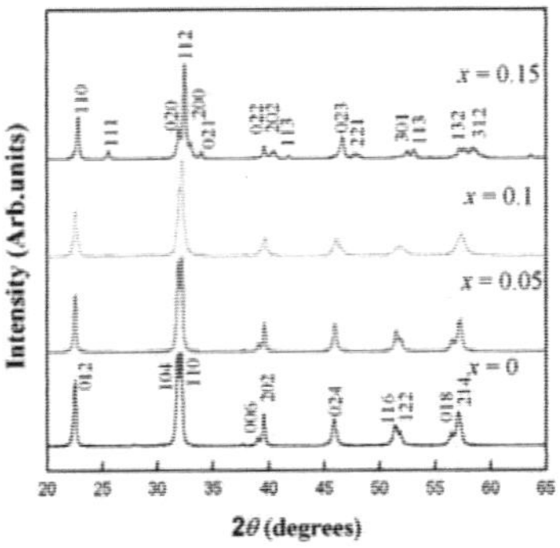

Fig. 1 XRD patterns of $Bi_{1-x}Er_xFeO_3$ (x = 0, 0.05, 0.1, 0.15) nanoparticles.

Fig. 2 SEM images of $Bi_{1-x}Er_xFeO_3$ samples. (a) $x = 0$; (b) $x = 0.05$; (c) $x = 0.1$; (d) $x = 0.15$.

The X-ray diffraction (XRD) patterns are shown in Fig. 1 that there is single-well phase with $BiFeO_3$ perovskite structure and no obvious secondary phase is detected in $Bi_{1-x}Er_xFeO_3$ (x = 0, 0.05, 0.1, 0.15). It also notes that the structural phase transition occurs by increasing doping concentration of Er ions, which is also consistent with the results in literature [4, 6]. For samples x=0, 0.05, they belongs to rhombohedral lattice type with *R3c* space group. The structural transformation begins when $x = 0.1$, a transitional state with triclinic structure is tended to be shown. As $x = 0.15$, an orthorhombic lattice structure with $Pn2_1a$ space group is transformed.

The surface morphologies of the samples are displayed in Fig. 2. The particles are slightly agglomerated into clusters and the grain size is nonuniform which is estimated as about 0.2-0.8 μm. Referring to the doped $Bi_{1-x}Er_xFeO_3$ (x = 0.05-0.15) samples, with the increase of doping concentration, the grain size becomes small, which may be interpreted by the suppression of oxygen vacancy concentration [7].

Fig. 3 presents the radial structure function curves of samples x = 0, 0.05 and 0.1, which is obtained from the normalized XAFS spectra via Fourier transformation using USTCXAFS2.0 code [8]. In this figure, the first peak at near 0.15 nm is the first nearest neighbor coordination of center atom Fe. There is no obvious change as the doping concentration increasing, although the structural phase transition is occurred. It demonstrates that the local structural distortion, especially for the Fe-O-Fe distortion, is not significantly affected by the Er ions doping, although there is unequivocal variation of ions radius between Er ions and Bi ions.

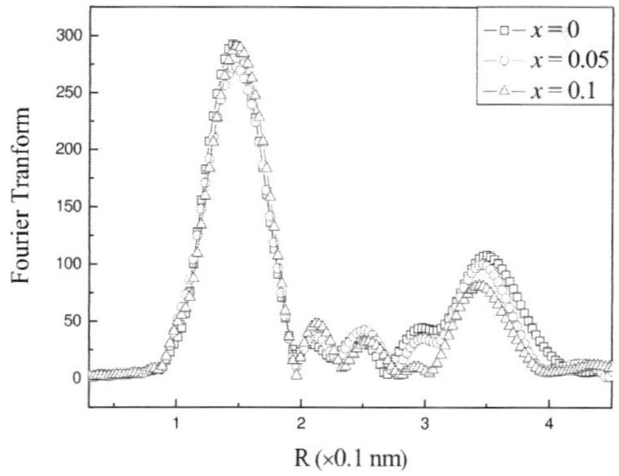

Fig. 3 Fourier transform of Fe *K*-edge XAFS spectra of $Bi_{1-x}Er_xFeO_3$ ($x = 0, 0.05, 0.1$).

Now that there is no obvious local structural distortion of Fe-O-Fe bonds, the space-modulated spin structure would not be destroyed based on the deduction in the part of Introduction. Therefore, the weak ferromagnetism will not be released and no room temperature ferromagnetism will be observed. The temperature dependence of magnetization of samples $x = 0.05$ and 0.1 has been measured which is shown in Fig. 4. The pure $BiFeO_3$ nanoparticle behaves paramagnetism at room temperature which will be published elsewhere [9]. Both the temperature dependence of magnetization with external field $H = 500$ Oe and magnetic field dependence of magnetization at lower temperature $T = 5$K indicate that there is no ferromagnetism observed even at lower temperature and the value of magnetization at lower temperature and higher magnetic filed is quite small, not to mention the magnetic properties at room temperature. These results are inconsistent with literatures reported [4, 6, 10], where structural phase transition occurred by ions doping at Bi-site induces room temperature ferromagnetism. In our case, although structural phase transition occurs, there is no obvious local structure distortion which will not destroy the space-modulated spin structure. From this point of view, non-ferromagnetism of the samples may be easily explained.

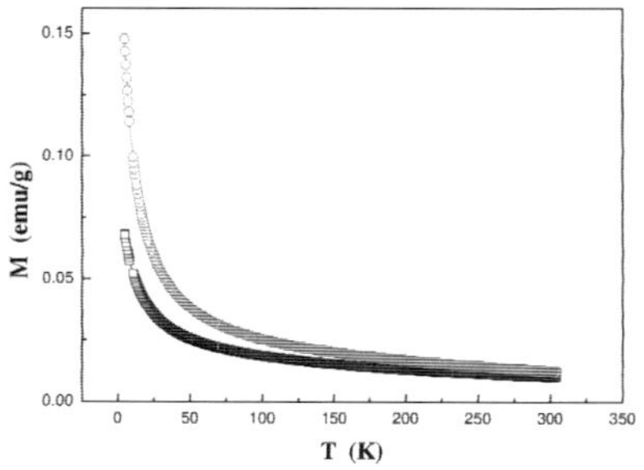

Fig. 4 Temperature dependence of magnetization (M~T) of $Bi_{1-x}Er_xFeO_3$ ($x = 0.05, 0.1$) nanoparticles with an external field $H = 500$ Oe. The inset shows the M~H curves of $Bi_{1-x}Er_xFeO_3$ ($x = 0.05, 0.1$) nanoparticles at temperature $T = 5$ K.

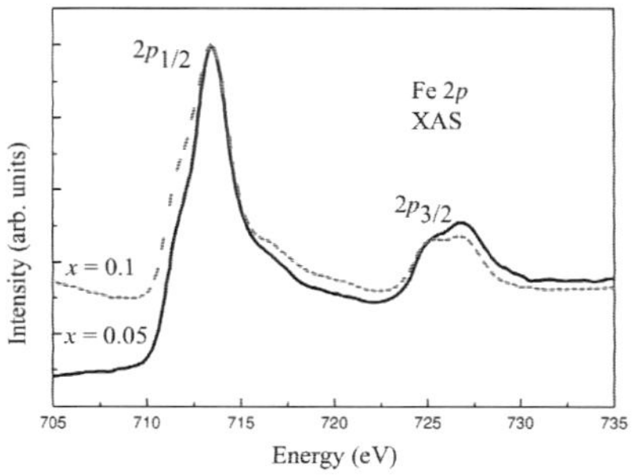

Fig. 5 X-ray absorption spectroscopy of Fe 2*p* core level of $Bi_{1-x}Er_xFeO_3$ ($x = 0.05, 0.1$) nanoparticles at room temperature.

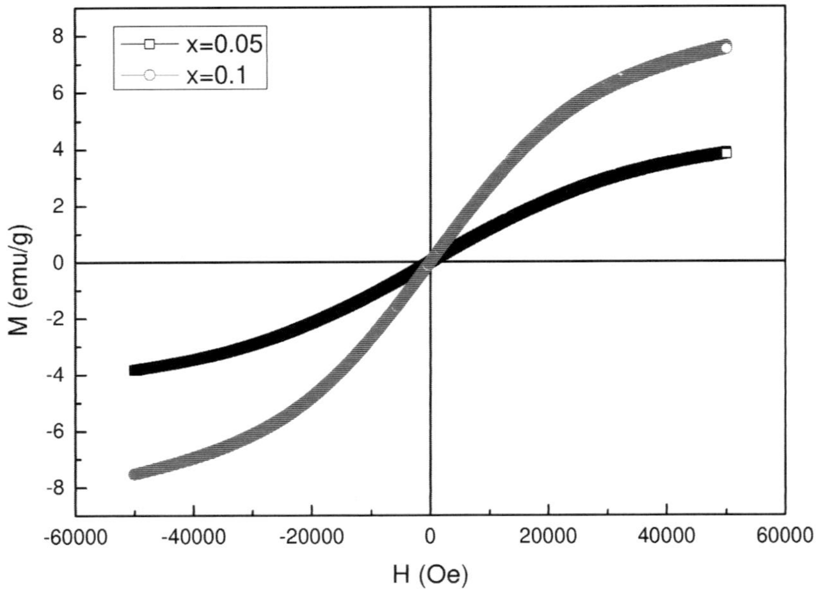

Besides the local structure distortion, variation of valence states of Fe ions is another factor to induce ferromagnetism for doped $BiFeO_3$ samples [11], even though it is not one of the necessary factors [6]. D. Kothari *et al.* [6] have reported the multiferroic properties of Eu doped $BiFeO_3$ which is not due to Fe multiple valence but to the structural distortions. Here, the XAS spectra of Fe $2p$ core level at room temperature are shown in Fig. 5. It can be seen that two main peaks of $2p_{1/2}$ and $2p_{3/2}$ are located at energy position of 726.6eV and 713.4eV, respectively. For the two samples, the energy position of Fe $2p_{1/2}$ is totally same, which means the same valence states of Fe ions and no variation of valence states.

In summary, only structural phase transition can not drive room temperature ferromagnetism without variation of local structural distortion and valence states of Fe ions for Er-doped $BiFeO_3$ samples.

4. Conclusion

In this paper, the structural phase transition and local structural distortion of the Er-doped $BiFeO_3$ nanoparticles have been discussed. The XRD patterns and XAFS of these samples demonstrate that there is structural phase transition but on obvious local structural distortion with the increasing of doping concentration. And the XAS of Fe $2p$ core level of these samples are totally same, which means the consistent valence states of Fe ions. Only the structural phase transition can not drive the magnetism change of Er-doped $BiFeO_3$ nanoparticles. Perhaps there is some kind of correlation between the magnetism of doped $BiFeO_3$ and the local structural distortion and valence states of Fe ions.

5. Acknowledgement

We are indebt to NSFC for its financial support to our research projects No.10979016 and No.51172110 and to NSRL and BSRF for their beamtime. Further more, this work is also supported by Nanjing University of Posts and Telecommunications under research project No. NY211144 and No.NY208027 as well.

References

[1] Wang J *et al* 2003 *Science* **299** 1719

[2] Tabares-Munoz C *et al* 1985 *Jpn. J. Appl. Phys.* **24** 1051
[3] Neaton J B *et al* 2005 *Phys. Rev. B.* **71** 014113.
[4] Yuan G L *et al* 2007 *J. Phys. D: Appl. Phys.* **40** 1196
[5] Li M *et a l* 2007 *J. Phys. D: Appl. Phys.* **40** 1603
[6] Kothari D *et al* 2010 *J. Phys.: Condens. Matter* **22** 356001
[7] Li Y B *et al* 2011 *J Mater Sci: Mater Electron* **22** 323
[8] Wei S Q *et al* 2001 *Phys. Rev. B* **63** 224201
[9] Li Y T *et al* 2012 *Mater. Lett* **87** 117
[10] Zhang S T *et al* 2006 *Appl. Phys. Lett.* **88** 162901
[11] Wang J *et al* 2005 *Science* **307** 1203b

Journal of Physics: Conference Series **430** (2013) 012109 doi:10.1088/1742-6596/430/1/012109

Effect of cobalt doping on the local structure and dynamics of multiferroic $MnWO_4$ and $Mn_{0.7}Co_{0.3}WO_4$

A Kuzmin, A Anspoks, A Kalinko and J Timoshenko
Institute of Solid State Physics, University of Latvia, Kengaraga Street 8, LV-1063 Riga, Latvia
E-mail: a.kuzmin@cfi.lu.lv

Abstract. The local atomic structure and dynamics in multiferroic $MnWO_4$ and $Mn_{0.7}Co_{0.3}WO_4$ have been studied by X-ray absorption spectroscopy at the Co(Mn) K-edge and W L_3-edge. The analysis of the first coordination shell of metal ions using single-shell Gaussian approximation and regularization-like method allowed us to determine a distortion of $Mn(Co)O_6$ and WO_6 octahedra. It was found that the local environment of Co^{2+} ions in $Mn_{0.7}Co_{0.3}WO_4$ is close to that in $CoWO_4$, whereas the presence of cobalt ions reduces the distortion of MnO_6 octahedra in comparison with pure $MnWO_4$.

1. Introduction

Pure manganese tungstate $MnWO_4$ is multiferroic material with spiral long-range magnetic order [1, 2]. It crystallizes in the wolframite-type structure [3] and is isomorphous to other antiferromagnetic transition-metal tungstates [4] such as $CoWO_4$ (T_N=55 K), $NiWO_4$ (T_N=67 K) and $FeWO_4$ (T_N=76 K). $MnWO_4$ contains only one kind of magnetic ion (Mn^{2+}) and undergoes three successive magnetic phase transitions to antiferromagnetically (AF) ordered states at ~13.5 K (T_N), ~12.7 K (T_2) and ~7.6 K (T_1) [3]. The magnetic phase transitions in $MnWO_4$ are responsible for anomalies of the specific heat, the dielectric constant and the magnetic susceptibility [5]. The appearance of simultaneous ferroelectric polarization in the AF2 phase between ~7.6 K and ~12.7 K is explained by the loss of inversion symmetry due to the spiral magnetic order and a strong spin-lattice coupling [1, 6].

The magnetic properties of $MnWO_4$ can be influenced by replacing Mn^{2+} with other magnetic Co^{2+} [7, 8, 9, 10, 11], Fe^{2+} [12], Ni^{2+} [13] and non-magnetic Mg^{2+}, Zn^{2+} [14, 15] ions as well as by substitution of W^{6+} with Mo^{6+} [16] ions. The c-T phase diagram of $Mn_{1-c}Co_cWO_4$ (c<0.3) has been established recently using the temperature-dependent magnetic susceptibility and neutron powder diffraction studies [7]. Upon Co doping the lattice parameters and the unit-cell volume decrease linearly, reflecting smaller radii of Co^{2+} ions. At the same time, the temperature coefficients of the lattice parameters a and b show non-linear behavior [7]. Recent, Raman and infra-red spectroscopy studies of $Mn_{0.85}Co_{0.15}WO_4$ revealed an unusual behaviour for some phonon modes below 100–200 K, which were related to the rigidity of zigzag chains of edge-shared MnO_6 octahedra, directly influencing various exchange interactions [10, 17]. Therefore, the information on the local structure in solid solutions $Mn_{1-c}Co_cWO_4$ is one of the key factors for deeper understanding of their properties.

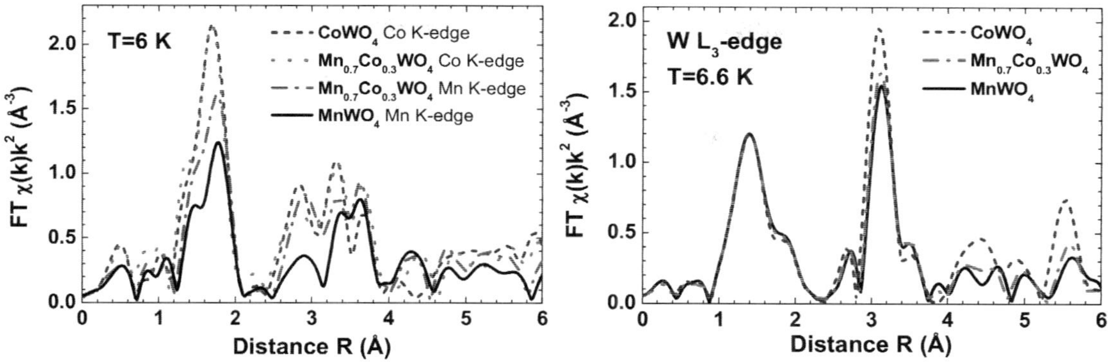

Figure 1. Fourier transforms (FT) of the experimental low-temperature Mn(Co) K-edge and W L3-edge EXAFS spectra $\chi(k)k^2$ in Mn(Co)WO_4 and $Mn_{0.7}Co_{0.3}WO_4$.

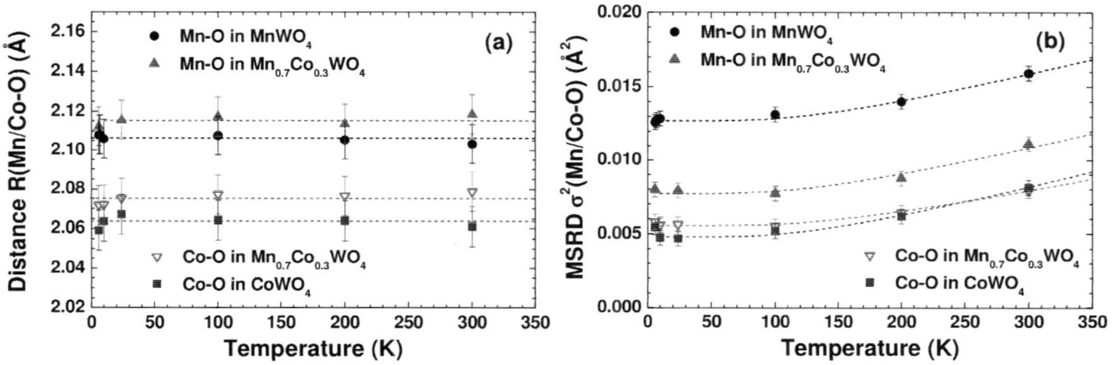

Figure 2. Temperature dependence of (a) the mean interatomic distances R(Mn/Co–O) and (b) the MSRD σ^2 obtained within the single-shell Gaussian approximation for Mn–O and Co–O atom pairs in Mn(Co)WO_4 and $Mn_{0.7}Co_{0.3}WO_4$. Dashed lines in (b) are the Einstein models.

In this work we have performed for the first time temperature dependent (6–300 K) x-ray absorption spectroscopy study of the local structure and dynamics around Mn, Co and W ions in $MnWO_4$, $Mn_{0.7}Co_{0.3}WO_4$ and $CoWO_4$.

2. Experimental and calculation details

$MnWO_4$, $Mn_{0.7}Co_{0.3}WO_4$ and $CoWO_4$ powders were synthesized using co-precipitation technique by a reaction of proper amounts of aqueous solutions of $Co(NO_3)_2{\cdot}6H_2O$ and/or $MnCl_2{\cdot}4H_2O$ salts with that of $Na_2WO_4{\cdot}2H_2O$ at 20°C. The obtained precipitates were washed, dried and finally annealed in air for 8 hours at 800°C. The W L3-edge and Mn(Co) K-edge extended X-ray absorption fine structure (EXAFS) spectra were measured in transmission mode at the HASYLAB/DESY C1 bending-magnet beamline in the temperature range from 6 to 300 K. The x-ray radiation was monochromatized by a detuned Si(111) double-crystal monochromator, and the beam intensity was measured using two ionization chambers filled with argon and krypton gases. EXAFS oscillations $\chi(k)k^2$ were extracted in k-space range up to 18 Å^{-1} following the conventional procedure [18] using the EDA package [19]. Fourier transforms (FTs) of representative low-temperature EXAFS spectra are shown in Fig. 1. The

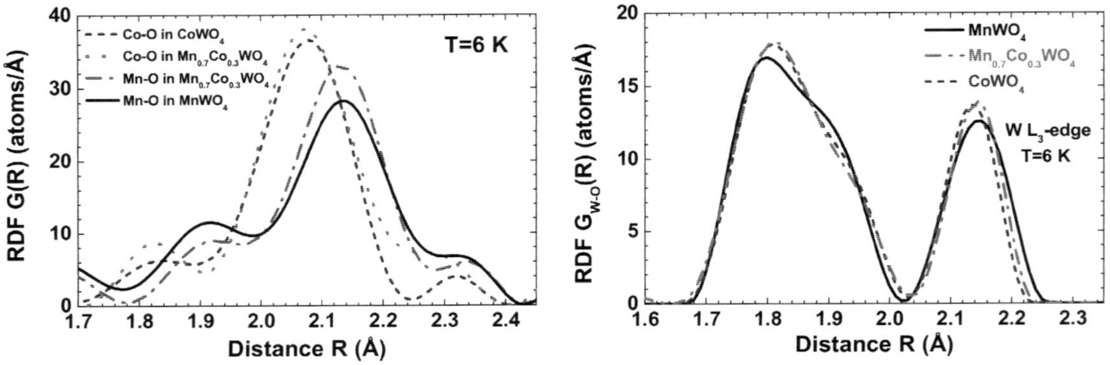

Figure 3. Radial distribution functions (RDFs) for Co–O, Mn–O and W–O atom pairs within the first coordination shell in Mn(Co)WO_4 and $Mn_{0.7}Co_{0.3}WO_4$ reconstructed by regularization-like technique [20].

first shell EXAFS contributions (Fig. 2(a)) were singled out in the R-space range 1.2–2.1 Å and best fitted using single shell Gaussian model to obtain the mean interatomic distances $\langle R \rangle$ and effective mean-square relative displacements (MSRDs) σ^2 for Co–O and Mn–O bonds (Fig. 2(b)). The radial distribution functions (RDFs) within the first coordination shell of metal ions were reconstructed by the regularization-like method [20] (Fig. 3). Theoretical scattering amplitude and phase shift functions, calculated by the *ab initio* FEFF8 code [21, 22], were used in all simulations.

3. Results and discussion

In wolframite-type structure tungsten and $3d$(Mn/Co) ions have distorted octahedral coordination by oxygen atoms, which can be characterized by a distortion parameter $\Delta = (1/6)\sum_{i=1}^{6} |R_i - \langle R \rangle|$. The WO_6 octahedra are strongly distorted with Δ=0.124 Å in $MnWO_4$ and 0.118 Å in $CoWO_4$, so that six oxygens can be divided into two groups of 4 nearest and 2 distant atoms located at 1.8–1.9 Å and 2.1–2.2 Å, respectively [3, 4]. A distortion of the MnO_6 octahedra Δ=0.064 Å is twice smaller, and that of the CoO_6 octahedra Δ=0.048 Å decreases even more. The octahedra distortion is well observed in the RDFs reconstructed from EXAFS spectra (Fig. 3). While RDFs for Co–O and Mn–O pairs are single-peak shaped, the ones for W–O pairs have two maxima in agreement with diffraction data [3, 4]. Note that the W–O RDFs are very close in all three compounds, and the different position of the RDF maxima for Co–O and Mn–O pairs is due to the size difference of Co^{2+} (0.75 Å) and Mn^{2+} (0.83 Å) ions being in the high-spin state [11].

Rather small distortion of the Mn(Co)O_6 octahedra allows one to perform the analysis of the first shell within the single-shell Gaussian approximation. As a result, the mean interatomic distances R(Mn/Co–O) and the MSRDs σ^2 for Co–O and Mn–O bonds were determined (Fig. 2). The Mn(Co)–O distances remain constant upon increasing temperature within ±0.01 Å, and their values in $Mn_{0.7}Co_{0.3}WO_4$ are close to that in pure tungstates.

The analysis of the Mn(Co)–O MSRDs suggests that they contain dynamic, i.e. temperature dependent, and static, caused by octahedra distortion, contributions. The MSRDs temperature dependent part is close in three tungstates and is well approximated by the Einstein model with the characteristic temperature $\theta_E \approx 415$ K.

The main difference in the MSRDs is due to the static octahedra distortion, being the largest in $MnWO_4$ and the smallest in $CoWO_4$. In $Mn_{0.7}Co_{0.3}WO_4$ the MSRD values for Co–O bonds are close to that in $CoWO_4$, suggesting that Co^{2+} ions adopt similar local environment. On the contrary, the MSRD values for Mn–O bonds in $Mn_{0.7}Co_{0.3}WO_4$ are significantly smaller by about 0.005 Å^2 than in $MnWO_4$, but slightly larger by about 0.002 Å^2 than for Co–O bonds. This means that cobalt doping leads to a decrease of the MnO_6 octahedra distortion. Comparison of the Mn(Co) K-edge FTs in Fig. 1 indicates that the behaviour of the outer shells at 2.5–4.0 Å in $Mn_{0.7}Co_{0.3}WO_4$ is also closer to that in $CoWO_4$. These original results confirm the stabilizing role of Co doping in solid solutions $Mn_{1-c}Co_cWO_4$ suggested in [7, 10].

4. Conclusions

Temperature dependent x-ray absorption spectroscopy study of the local structure and dynamics around Mn, Co and W ions in multiferroic $MnWO_4$ and $Mn_{0.7}Co_{0.3}WO_4$ has been performed in comparison with antiferromagnetic $CoWO_4$. The obtained results provide direct evidence of the stabilizing role of cobalt in $Mn_{0.7}Co_{0.3}WO_4$. We have proved that the local environment of Co^{2+} ions in the solid solution is close to that in $CoWO_4$. At the same time, the presence of cobalt ions influences strongly the local environment around Mn ions, which becomes less distorted compared to pure $MnWO_4$. The strongly distorted octahedral environment of tungsten atoms was found to be close in three tungstates.

Acknowledgments

We would like to thank Dr. R. Chernikov for assistance during the experiments at the HASYLAB/DESY beamline. This work was supported by ESF projects 2009/0202/1DP/1.1.1.2.0/09/APIA/VIAA/141, 2009/0216/1DP/1.1.1.2.0/09/APIA/VIAA/044 and Latvian National Research Program IMIS. The EXAFS experiments at HASYLAB/DESY have been supported from the European Community's Seventh Framework Programme (FP7/2007-2013) under grant agreement No. 226716 (project I-20100098 EC).

References

[1] Heyer O, Hollmann N, Klassen I, Jodlauk S, Bohaty L, Becker P, Amydosh J, Lorenz T and Khomskii D 2006 *J. Phys.: Condens. Matter* **18** L471
[2] Taniguchi K, Abe T, Takenobu T, Iwasa Y and Arima T 2006 *Phys. Rev. Lett.* **97** 0972003
[3] Lautenschlager G, Weitzel H, Vogt T, Hock R, Bohm A, Bonnet M and Fuess H 1993 *Phys. Rev.* B **48** 6087
[4] Weitzel H 1970 *Solid State Commun.* **8** 2071
[5] Arkenbout A H, Palstra T T M, Siegrist T and Kimura T 2006 *Phys. Rev.* B **74** 184431
[6] Mostovoy M 2006 *Phys. Rev. Lett.* **96** 067601
[7] Song Y S, Chung J H, Park J and Choi Y N 2009 *Phys. Rev.* B **79** 224415
[8] Song Y S, Yan L, Lee B, Chun S, Kim K, Kim S, Nogami A, Katsufuji T, Schefer J and Chung J 2010 *Phys. Rev.* B **82** 214418
[9] Chaudhury R P, Ye F, Fernandez-Baca J A, Wang Y Q, Sun Y Y, Lorenz B, Mook H A and Chu C W 2010 *Phys. Rev.* B **82** 184422
[10] Maczka M, Ptak M, Hermanowicz K, Majchrowski A, Pikul A and Hanuza J 2011 *Phys. Rev.* B **83** 174439
[11] Urcelay-Olabarria I, Ressouche E, Mukhin A A, Ivanov V Y, Balbashov A M, García-Muñoz J L and Skumryev V 2012 *Phys. Rev.* B **85** 224419
[12] García-Matres E, Stußer N, Hofmann M and Reehuis M 2003 *Eur. Phys. J.* B **32** 35
[13] Ptaka M, Maczka M, Hermanowicz K, Pikul A and Hanuza J 2012 *Spectrochim. Acta* A **86** 85
[14] Meddar L, Josse M, Deniard P, La C, Andre G, Damay F, Petricek V, Jobic S, Whangbo M H, Maglione M and Payen C 2009 *Chem. Mater.* **21** 5203
[15] Chaudhury R P, Ye F, Fernandez-Baca J A, Lorenz B, Wang Y Q, Sun Y Y, Mook H A and Chu C W 2011 *Phys. Rev.* B **83** 014401
[16] Meddar L, Josse M, Maglione M, Guiet A, La C, Deniard P, Decourt R, Lee C, Tian C, Jobic S, Whangbo M H and Payen C 2012 *Chem. Mater.* **24** 353
[17] Tian C, Lee C, Xiang H, Zhang, Payen C, Jobic S and Whangbo M H 2009 *Phys. Rev.* B **80** 104426
[18] Aksenov V L, Kuzmin A Y, Purans J and Tyutyunnikov S I 2001 *Phys. Part. Nucl.* **32** 675
[19] Kuzmin A 1995 *Physica* B **208-209** 175
[20] Kuzmin A and Purans J 2000 *J. Phys.: Condens. Matter* **12** 1959
[21] Ankudinov A L, Ravel B, Rehr J J and Conradson S D 1998 *Phys. Rev.* B **58** 7565
[22] Rehr J J and Albers R C 2000 *Rev. Mod. Phys.* **72** 621

Journal of Physics: Conference Series **430** (2013) 012110 doi:10.1088/1742-6596/430/1/012110

Relation between the atomic distance and the Ce valence of amorphous Ce_xRu_{100-x}

Yingjie Li [1], Ikuo Nakai [2], Yusuke Amakai [3] and Shigeyuki Murayama [3]

[1] Inner Mongolia Key Laboratory for Physics and Chemistry of Functional Materials, Physics and Electronic Information College, Inner Mongolia Normal University, Hohhot 010022, China

[2] Department of Electrical and Electronic Engineering, Tottori University, Tottori 680-8552, Japan

[3] Department of Materials Science and Engineering, Muroran Institute of Technology, Muroran, Hokkaido 050-8585, Japan

E-mail: lyingjie@imnu.edu.cn

Abstract. We report the local structure of amorphous alloys Ce_xRu_{100-x} (x=9, 43 and 80) and a C15 Laves phase compound $CeRu_2$ determined from an extended x-ray absorption fine structure measurements near Ru *K* edge and Ce *K* edge. The structure analyses indicate that the Ce-Ru and Ru-Ru interatomic distances of amorphous Ce_xRu_{100-x} are constant for all of the Ce concentration, while the Ce-Ce distance increases with increasing the Ce content. The interatomic distance of Ru-Ru in the amorphous alloy system is as same as that of the $CeRu_2$ compound. The Ce-Ce interatomic distance of amorphous Ce_xRu_{100-x} is closely connected with the Ce valence.

1. Introduction

The Ce-Ru amorphous alloy system (a-Ce_xRu_{100-x}, where x means the Ce concentration in atomic percent) has a wide variety of electronic properties. It shows the heavy fermion like behavior in the Ce-rich region ($x \geq 67$) and the superconductivity in the Ru-rich concentration ($x \leq 39$) [1-3]. The electronic properties of a-Ce_xRu_{100-x} depend strongly on the Ce 4*f* electrons and therefore the valence of Ce atoms [4]. The C15 cubic Laves phase intermetallic compound $CeRu_2$ (x-$CeRu_2$) shows superconductivity of which the Ce valence is 3.36 [4-6]. We have found the linear relation between the energy of transition to the continuum and the valence of Ce atoms for both a-Ce_xRu_{100-x} and x-$CeRu_2$ and also the Ce valence values of 3.44, 3.29 and 3.13 for a-Ce_xRu_{100-x} of x=9, 43 and 80, respectively. These findings provide evidence that the 4*f* electron is the origin of the heavy fermion like behavior in a-Ce_xRu_{100-x} [4]. In general the electronic properties of alloys and compounds are quite sensitive to their crystallographic structure. Especially the local atomic arrangement around a specific atom influences predominantly to the electronic states of the atom. For example, although both α-cerium and γ-cerium possess a face centered cubic structure, their interatomic distances between Ce atoms are quite different from each other: 3.65 Å for γ-cerium and 3.41 Å for α-cerium [7]. The valence of α-cerium is reported to be tetravalent, while the γ-cerium is trivalent [8]. Also in the case of a-Ce_xRu_{100-x} the local structure is a crucial factor of determining the electronic properties. Because the extended x-ray absorption fine structure (EXAFS) is especially sensitive to the short range order [9], it is one of the best techniques for detecting the local structure of amorphous materials with no long range order. The oscillatory part of EXAFS contains the atomic local structure information such as coordination number and bond length [10, 11]. Here we report on the local structure of the Ce *K* edge and Ru *K* edge in a-Ce_xRu_{100-x}.

2. Experimental Details

Ingots of the Ce_xRu_{100-x} alloys were prepared by arc melting of the appropriate amounts of 99.9% Ce and 99.9% Ru in an argon atmosphere. The amorphous phase of Ce_xRu_{100-x} was fabricated by DC high-rate sputtering. The compound x-$CeRu_2$ was also prepared as a reference material. Another references used were 99.9% cerium trifluoride (x-CeF_3) and 99.9% cerium dioxide (x-CeO_2).

The x-ray diffraction patterns were measured using Cu *K*α radiation in a conventional θ-2θ scan mode. The chemical composition of a-Ce_xRu_{100-x} was fixed to be x=9, 43 and 80 by electron-probe microanalysis.

The EXAFS measurements near Ru and Ce *K* edges were carried out at the beamline NW10A with a Si (311) double crystal monochromator in Photon Factory [11]. The storage ring was operated at 6.5 GeV with an initial current of about 50 mA. All the spectra were measured at 10 K and 300 K in the transmission mode using the ionization chambers filled with a mixed gas of 50% N_2 and 50% Ar in front of the sample and a pure Kr gas behind it for Ru *K* edge; for Ce *K* edge a pure Kr gas was filled in both the ionization chambers. The incident energy of photons was corrected with the energy for a prepeak near the Cu *K* edge.

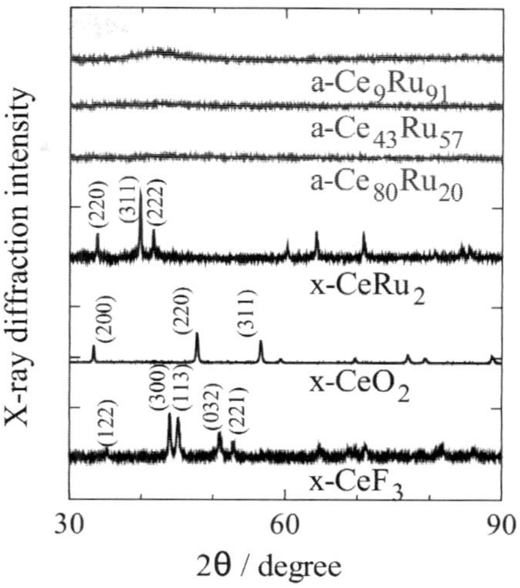

Figure 1. X-ray diffraction patterns of a-Ce_xRu_{100-x} (x=9, 43 and 80) and the compound x-$CeRu_2$, along with cerium dioxide x-CeO_2 and cerium trifluoride x-CeF_3 for comparison. A halo pattern appears around 40 degree for a-Ce_xRu_{100-x}. The typical index of the plane is labelled for the crystalline materials.

3. Results and Discussion

Figure 1 shows the x-ray diffraction patterns of a-Ce_xRu_{100-x} with the references x-$CeRu_2$, x-CeO_2 and x-CeF_3. No reflections except a broad bump near about 40 degree are detected for a-Ce_xRu_{100-x}, which means that it is in the amorphous phase. We obtain the following by the Rietveld refinement of these patterns. The compound x-$CeRu_2$ crystallizes in the C15 Laves phase structure with the lattice constant of 7.5381 Å, in which the Ce atom has twelve nearest neighbors of Ru at 3.1251 Å and four Ce atoms at 3.2641 Å, while Ru is surrounded by six Ru atoms at 2.6651 Å and six Ce atoms at 3.1251 Å. The oxide x-CeO_2 is the CaF_2 structure with the Fm3m space group and lattice constant of 5.4109 Å, where the Ce has twelve neighboring Ce atoms at 3.8262 Å and eight O atoms at 2.3431 Å. The fluoride x-CeF_3 has the P6mmc space group symmetry and lattice constant of 7.0930 Å, in which Ce has four nearest Ce atoms at 4.0570 Å and seven F atoms at 2.4448 Å, 2.4260 Å, 2.3862 Å, 2.4056 Å and 2.3866 Å.

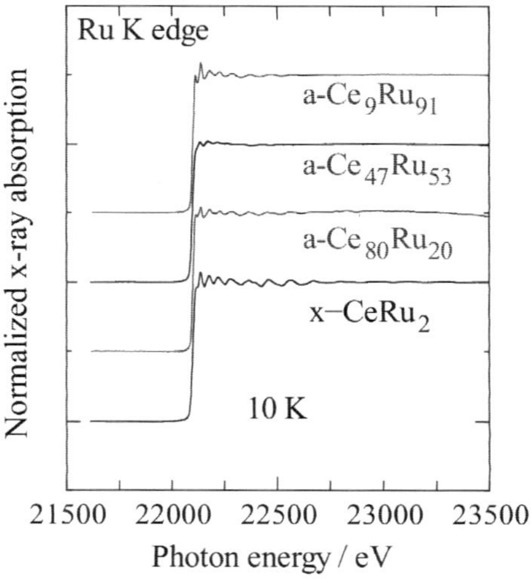

Figure 2. Normalized x-ray absorption near the Ru *K* edge at 10 K for a-Ce_xRu_{100-x} and x-$CeRu_2$ plotted against photon energy.

The EXAFS signals are derived from the routine procedure of background removal, spline smoothing and normalization. Figure 2 represents normalized x-ray absorption spectra near the Ru *K* edge measured at 10 K as a function of photon energy *E* for a-Ce_xRu_{100-x} and x-$CeRu_2$. The radial structure functions $\Phi(R)$ obtained from Fourier transform of $k^3\chi(k)$ are displayed in Figs. 3(a) and (b) for a-Ce_xRu_{100-x} and x-$CeRu_2$. The thick and thin lines denote the magnitude and the imaginary part of $\Phi(R)$. In Fig. 3(a) the peaks at 2.45 Å and 3.45 Å for a-Ce_xRu_{100-x} correspond to the Ru and Ce atoms around the central Ce atom. Surprisingly the peak of the Ce-Ru pair disappears for a-$Ce_{80}Ru_{20}$. We have not yet figured out the puzzle. It is why we have not shown the structural parameter for the Ce-Ru pair in Table 1. In x-$CeRu_2$ the peak at 2.75 Å shows both the Ce and Ru neighbors. In Fig. 3(b) we find one peak at 3.00 Å for a-Ce_xRu_{100-x}, which corresponds to the Ru atom. For x-$CeRu_2$ the peaks at 2.30 Å and 3.00 Å indicate the Ru and Ce atoms. To estimate the structural parameters, we reconvert $\Phi(R)$ into $k^3\chi(k)$ with the Fourier filtering technique for each peaks of $\Phi(R)$. The Fourier filtered $k^3\chi(k)$ spectrum is fitted to the following EXAFS formula by the least squares method [9, 12],

$$\chi(k)=\sum_i N_i\cdot|f_i(k,\pi)|\cdot\exp(-2(\sigma_i^2k^2+R_i/\lambda))/kR_i^2\cdot\sin(2kR_i+\delta_i(k)), \quad (1)$$

where N_i is the coordination number, $f_i(k, \pi)$ is backscattering amplitude, σ_i is Debye-Waller factor, *R* is the interatomic distance between the absorber and the *i*-th neighboring atom, δ_i is the phase shift. The program FEFF8.4 is now widely used for estimate $f_i(k, \pi)$ and δ_i from some model structure [13]. Unfortunately we have no idea about the structure of our amorphous specimens. They are insensitive to the interatomic distance and mainly determined by atomic pair distribution function. Therefore we consider that in the amorphous phase with unknown structure they are similar to those of x-$CeRu_2$. As an example the Fourier filtered spectrum of a-Ce_9Ru_{91} at Ru *K* edge is illustrated in Fig. 4 by the open circle. The solid line shows the best fit to the spectrum. The parameters about the local structure thus determined are listed in Tables 1 and 2. We obtain quite small coordination numbers for a-$Ce_{43}Ru_{57}$. This may be because that the structure function for a-$Ce_{43}Ru_{57}$ in Fig. 3 is weaker than that of the others and that we made a simplified Gaussian analysis for a-$Ce_{43}Ru_{57}$ with higher disorder. The interatomic distance for x-CeF_3, x-CeO_2 and x-$CeRu_2$ are quite close to the corresponding value estimated from the lattice constant using the x-ray diffraction. In a-Ce_xRu_{100-x}

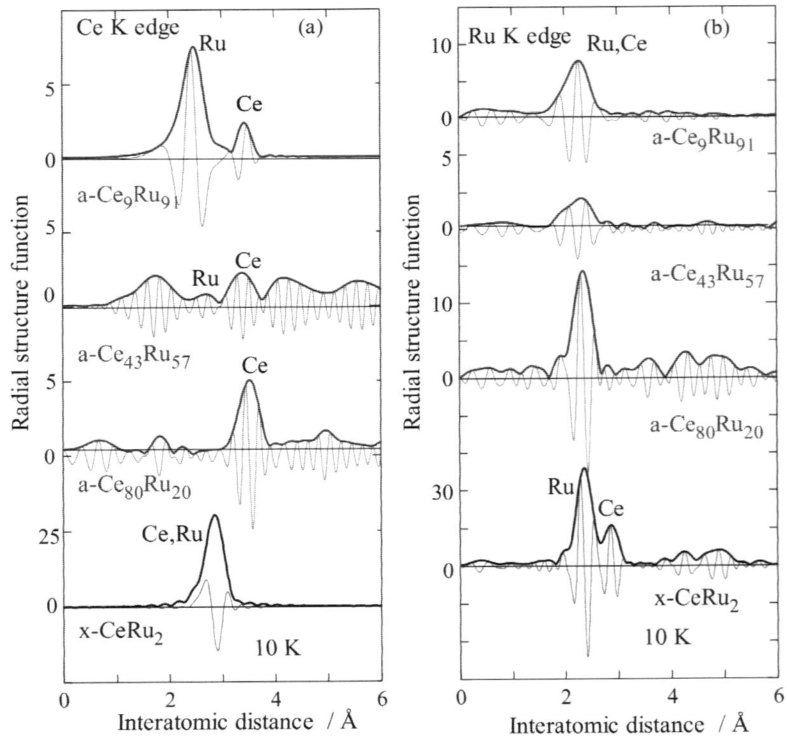

Figure 3. Imaginary part (thin line) and magnitude (thick line) of the radial structure function $\Phi(R)$ for a-Ce_xRu_{100-x} (x=9, 43 and 80) and x-$CeRu_2$ at (a) the Ce *K* edge and (b) the Ru *K* edge. The peaks mainly consist of the labelled atoms.

Table 1. Structural parameters of a-Ce_xRu_{100-x} with x-CeF_3, x-CeO_2 and x-$CeRu_2$ determined from the Ce *K* edge EXAFS where A is an absorbing atom, B is a backscattering atom, *N* is the coordination number, *R* is the interatomic distance, σ is the Debye-Waller factor.

Ce K edge	Temp	A-B	*N*	*R* / Å	σ / Å
x-CeF_3	300 K	Ce-F	7	2.41 (0.01)	0.09 (0.01)
		Ce-Ce	4	4.07 (0.01)	0.09 (0.01)
x-CeO_2	300 K	Ce-O	8	2.33 (0.01)	0.09 (0.01)
		Ce-Ce	12	3.83 (0.01)	0.07 (0.01)
x-$CeRu_2$	300 K	Ce-Ru	12	3.13 (0.01)	0.13 (0.01)
		Ce-Ce	4	3.26 (0.01)	0.10 0.01)
x-$CeRu_2$	10 K	Ce-Ru	12	3.13 (0.01)	0.07 (0.01)
		Ce-Ce	4	3.26 (0.01)	0.06 (0.01)
a-Ce_9Ru_{91}	300 K	Ce-Ru	8.3 (0.4)	2.79 (0.02)	0.11 (0.01)
		Ce-Ce	1.3 (0.1)	3.72 (0.02)	0.08 (0.01)
a-Ce_9Ru_{91}	10 K	Ce-Ru	9.0 (0.4)	2.79 (0.02)	0.10 (0.01)
		Ce-Ce	1.4 (0.1)	3.72 (0.02)	0.07 (0.01)
a-$Ce_{43}Ru_{57}$	300 K	Ce-Ru	0.4 (0.1)	2.79 (0.03)	0.10 (0.01)
		Ce-Ce	0.4 (0.1)	3.75 (0.02)	0.08 (0.01)
a-$Ce_{43}Ru_{57}$	10 K	Ce-Ru	0.9 (0.3)	2.78 (0.03)	0.09 (0.01)
		Ce-Ce	0.9 (0.3)	3.74 (0.02)	0.07 (0.01)
a-$Ce_{80}Ru_{20}$	300 K	Ce-Ru	----	----	----
		Ce-Ce	10.1 (1.1)	3.84 (0.02)	0.10 (0.01)
a-$Ce_{80}Ru_{20}$	10 K	Ce-Ru	----	----	----
		Ce-Ce	10.1 (1.1)	3.84 (0.02)	0.11 (0.01)

Table 2. Structural parameters of a-Ce_xRu_{100-x} and x-$CeRu_2$ determined from the Ru *K* edge EXAFS where A is an absorbing atom, B is a backscattering atom, *N* is the coordination number, *R* is the interatomic distance, σ is the Debye-Waller factor.

Ru K edge	Temp	A-B	*N*	*R* / Å	σ / Å
x-$CeRu_2$	300 K	Ru-Ce	6	3.13 (0.02)	0.13 (0.01)
		Ru-Ru	6	2.67 (0.01)	0.09 (0.01)
x-$CeRu_2$	10 K	Ru-Ce	6	3.13 (0.02)	0.08 (0.01)
		Ru-Ru	6	2.67 (0.01)	0.06 (0.01)
a-Ce_9Ru_{91}	300 K	Ru-Ce	0.8 (0.3)	2.80 (0.02)	0.10 (0.01)
		Ru-Ru	7.8 (1.0)	2.68 (0.02)	0.11 (0.01)
a-Ce_9Ru_{91}	10 K	Ru-Ce	0.8 (0.3)	2.80 (0.02)	0.08 (0.01)
		Ru-Ru	7.8 (1.0)	2.68 (0.02)	0.10 (0.01)
a-$Ce_{43}Ru_{57}$	300 K	Ru-Ce	1.1 (0.2)	2.80 (0.03)	0.12 (0.01)
		Ru-Ru	1.1 (0.2)	2.69 (0.03)	0.11 (0.01)
a-$Ce_{43}Ru_{57}$	10 K	Ru-Ce	1.7 (0.4)	2.80 (0.03)	0.11 (0.01)
		Ru-Ru	1.7 (0.4)	2.69 (0.03)	0.10 (0.01)
a-$Ce_{80}Ru_{20}$	300 K	Ru-Ce	7.7 (1.0)	2.80 (0.03)	0.10 (0.01)
		Ru-Ru	1.9 (0.2)	2.69 (0.02)	0.11 (0.01)
a-$Ce_{80}Ru_{20}$	10 K	Ru-Ce	7.7 (1.0)	2.80 (0.03)	0.09 (0.01)
		Ru-Ru	1.9 (0.2)	2.69 (0.02)	0.10 (0.01)

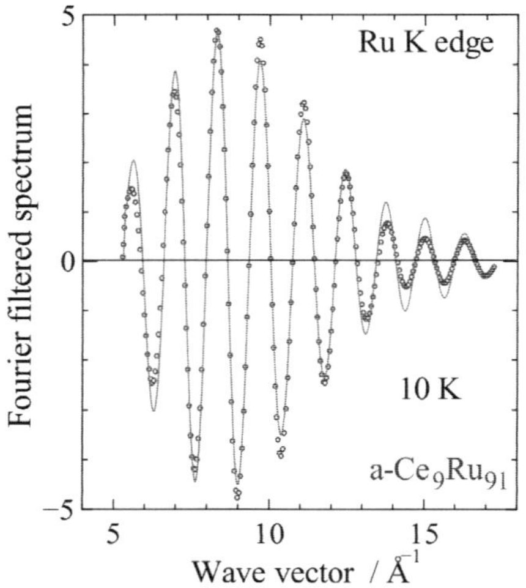

Figure 4. Fourier filtered spectrum $k^3\chi(k)$ (open circle) plotted against wave vector k for a-Ce_9Ru_{91} at Ru K edge. The solid line denotes the best fit.

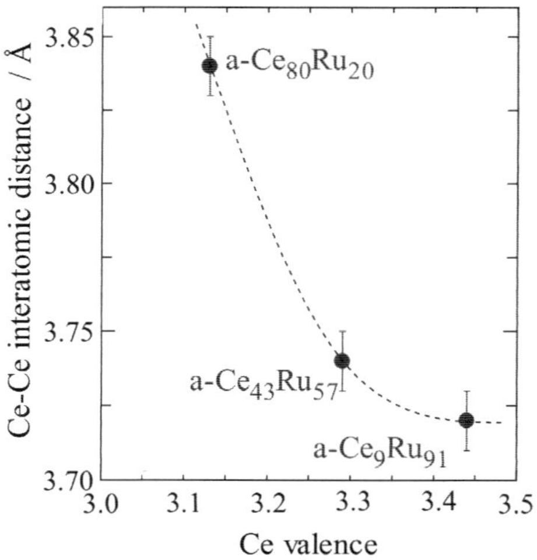

Figure 5. The Ce-Ce interatomic distance as a function of the Ce valence for a-Ce_xRu_{100-x}. The broken line is a guide for the eyes.

we confirm that the Ce-Ru interatomic distance deduced from the Ce K edge is equal to that of the Ru K edge within the accuracy of fitting. In a-Ce_xRu_{100-x} the distance between Ce atoms goes up from 3.72 Å to 3.84 Å with increasing Ce content, which is much longer than that of x-$CeRu_2$. On the other hand the Ce-Ru interatomic distance is rather shorter than that of x-$CeRu_2$ and independent of the Ce fraction. The Ru-Ru distance is also independent of the Ce content and very close to that of x-$CeRu_2$. The Ce-Ru and Ru-Ru distance of a-Ce_xRu_{100-x} does not obey the Vegard's law.

It has been found that a-Ce_xRu_{100-x} shows a rapid increase in the effective magnetic moment per Ce atom derived from the Curie-Weiss law [3] and a linear decrease in the valence of Ce atoms with increase in the Ce concentration [4]. The Ce atom is in the mixed valence state between trivalent $4f^1$ and tetravalent $4f^0$ levels. It can be considered that the Ce valence or the number of $4f$ electrons is a crucial factor in determining the electronic properties of a-Ce_xRu_{100-x}. To check this, we plot the relation of the Ce-Ce interatomic distance against the Ce valence in Fig. 5. The Ce-Ce distance increases rapidly with decreasing the Ce valence or increasing $4f$ electrons. The findings suggest that the volume of the Ce atom should be determined by the number of $4f$ electrons.

In Fig. 6 the Debye-Waller factor is represented for a-Ce_xRu_{100-x} and x-$CeRu_2$. In general it consists of the static disorder due to the structural configuration and the dynamic disorder caused by thermal vibration of atoms [14]. The Debye-Waller factor drawn in Fig. 6 includes the static disorder only because of measuring at 10 K. In a-Ce_xRu_{100-x} it is independent of the Ce content as shown by the broken line as same as the interatomic distance mentioned above and a little larger than that of x-$CeRu_2$. As tabulated in Tables 1 and 2 the Debye-Waller factor at 300 K for a-Ce_xRu_{100-x} is as same as that of 10 K, which means that the

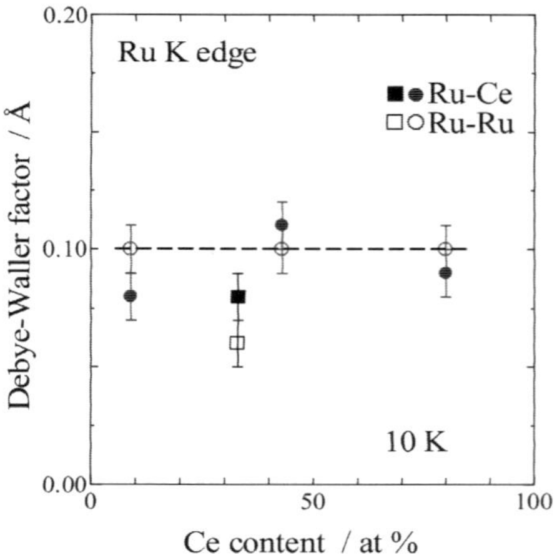

Figure 6. The Debye-Waller factors of the Ru-Ru and Ce-Ru pairs for a-Ce_xRu_{100-x} (circle) and x-$CeRu_2$ (square). The broken line is a guide for the eyes.

disorder almost comes from the static origin. On the other hand for x-$CeRu_2$ it increases with increase in temperature.

4. Conclusion

The structural parameters for a-Ce_xRu_{100-x} such as the coordination number, the interatomic distance and the Debye-Waller factor have been determined from the Ce and Ru *K* edge EXAFS spectra. The nearest neighboring interatomic distance between Ce atoms increases remarkably with increasing the Ce concentration. On the other hand the Ce-Ru and Ru-Ru distances are thoroughly independent of the Ce content, which means that they do no obey the Vegard's law. We have found the strong dependence of the Ce-Ce distance on the Ce valence or the number of 4*f* electrons.

Acknowledgements

The EXAFS measurement was performed under the approval of Photon Factory Program Advisory Committee (proposal Nos. 2007G587 and 2010G099).

References

[1] Obi Y, Murayama S, Amakai Y, Okaba Y and Asano K 2006 *Physica* **B378-380** 857
[2] Amakai Y, Murayama S, Takano H, Mizutani M, Asano K, Obi Y and Takanashi K 2007 *J. Magn. Magn. Mater.* **310** 416
[3] Amakai Y, Murayama S, Obi Y, Takano H, Momono N and Takanashi K 2009 *J. Phys: Conf. Series* **150** 042004
[4] Li Y J, Ohnishi N, Nakai I, Amakai Y and Murayama S 2009 *J. Phys. Soc. Jpn.* **78** 094717
[5] Lawrence JM, Riseborough PS. and Parks RD 1981 *Rep. Prog. Phys.* **44** 1
[6] Chaboy J, Garcia J and Marcelli A 1992 *J. Magn. Magn. Mater.* **104-107** 661
[7] Peterman DJ, Weaver JH, Croft M and Peterson DT 1983 *Phys. Rev.* **B27** 808
[8] Grimberg AJT 1973 *Thesis University of Amsterdam*
[9] Sayers DE, Stem EA and Lytle FW 1971 *Phys. Rev. Lett.* **27** 1204
[10] Stern EA 1974 *Phys. Rev.* **B10** 3027
[11] Nomura M, Koike Y, Sato M, Koyama A, Inada Y and Asakura K 2007 *AIP Conf. Proc.* **882** 896
[12] Grandjean F, Long GJ, Cortes R, Morelli DT and Meisner GP 2000 *Phys. Rev.* **B62** 12569
[13] Ankudinov AL, Nesvizhskii AI and Rehr JJ 2003 *Phys. Rev.* **B67** 115120
[14] Willis BTM and Pryor AW 1975 *Thermal vibrations in crystallography, Cambridge University Press*

Journal of Physics: Conference Series **430** (2013) 012111 doi:10.1088/1742-6596/430/1/012111

W L_{III}-edge XANES and EXAFS studies of $Pb(Fe_{2/3}W_{1/3})O_3$-$PbTiO_3$ multiferroic ceramics

Alexandre Mesquita[1], Bárbara Maraston Fraygola[1], Valmor Roberto Mastelaro[2] and José Antonio Eiras[1]

[1]Departamento de Física, Universidade Federal de São Carlos, São Carlos, SP, Brazil
[2]Instituto de Física de São Carlos, Universidade de São Paulo, São Carlos, SP, Brazil

E-mail: valmor@ifsc.usp.br

Abstract. The local structure of $Pb(Fe_{2/3}W_{1/3})O_3$ (PFW) and $Pb(Fe_{2/3}W_{1/3})O_3$-$PbTiO_3$ (PFW100xPT, x=0.10 and x=0.20) multiferroic ceramics was probed by X-ray absorption spectroscopy at the W L_{III}-edge. The analysis of XANES spectra did not show significative modifications in the intensity or energy of transitions as a function of temperature or $PbTiO_3$ content. The fitting of EXAFS spectra of PFW samples revealed a local rhombohedral symmetry of W atoms around 120 K, which is consistent with the electrical properties of this compound. Concerning the local structure of W atoms in PFW100xPT samples, in all cases, at 290 K, a local tetragonal symmetry was determined although by X-ray diffraction a cubic local symmetry was expected.

1. Introduction

Multiferroic materials have attracted the attention of many researches in the last years due the coexistence of magnetic and electric ordering [1]. The research with these materials is driven by the prospect of controlling charges by applied magnetic fields or spins by applied voltages, and using this ability to construct new forms of multifunctional devices [1]. The magnetoeletric coupling can be observed in multiferroic materials in the regions where magnetic and ferroelectric ordering coexists and in many cases, is mediate via lattice strain [2].

Among these materials, lead iron tungstate $Pb(Fe_{2/3}W_{1/3})O_3$ (PFW) is known as a perovskite-type structure in which Fe^{3+} and W^{6+} are randomly distributed at the octahedral B-site positions and presents ferroelectric and antiferromagnetic ordering. PFW has a frequency dependent permittivity with a broad maximum around 180 K which makes difficult practical applications [2]. Challenges for technological applications require preferentially that the magnetic and ferroelectric ordering coexistence occurs around room temperature and a low electric conductivity, in order to enable the poling of the sample. Thus, the addition of $PbTiO_3$ to PFW structure, forming the $(1\text{-}x)Pb(Fe_{2/3}W_{1/3})O_3$ $-xPbTiO_3$ (PFW–PT) system causes symmetry changes and on the range of temperature where magnetic and ferroelectric ordering coexists, depending on the $PbTiO_3$ contents [3]. Moreover, a transition from relaxor to normal ferroelectric state is induced as the $PbTiO_3$ content increases [2].

Although the long-range structure and the electrical/magnetic properties of these compounds has been characterized in order to better understand the processes of phase transition and electrical/magnetic behaviour of these compounds, as far as authors know, there are no available studies concerning the local structure of PFW-PT compounds probed by X-ray absorption spectroscopy (XAS). Thus, the present paper concerns a study about the local order and electronic structure of PFW and PFW-PT compounds by measuring the W L_{III}-edge EXAFS (extended X-ray absorption fine structure) and XANES (X-ray absorption near edge structure) spectra.

2. Experimental

The two-stage solid-state reaction process first used by Swartz and Shrout to synthesize the perovskite $Pb(Mn_{2/3}Nb_{1/3})O_3$ PMN compound, was used in this study to prepare PFW and PFW-PT samples and experimental details about samples preparation are given elsewhere [2,4]. Ceramic samples of (1 - x)$Pb(Fe_{2/3}W_{1/3})O_3$ –x$PbTiO_3$ (PFW100xPT) compositions were prepared with x = 0.00, 0.10 and 0.20.

EXAFS and XANES spectra at the W L_{III}-edge were collected in transmission mode as a function of the temperature (PFW samples) and sample composition (PFW100xPT samples) using a Si(111) double-crystal monochromator on the D04B-XASF2 beamline at the LNLS (National Synchrotron Light Laboratory) facility, Brazil. EXAFS spectra at the W L_{III}-edge were recorded for each sample between 10100 and 11000 eV using an energy step of 2 eV whereas XANES spectra were recorded between 10130 and 10300 eV with a energy step of 0.3 eV. The extraction and analysis of XANES and EXAFS spectra were performed using the Multi-Platform Applications for X-Ray absorption (MAX) software package [5] whereas theoretical spectra were calculated using FEFF8.2 code [6].

3. Results and discussion

Figure 1 (a) presents the XANES spectra at W L_{III}-edge of PFW sample measured at 120, 224 and 297 K and Figure 1 (b) shows the XANES spectra at W L_{III}-edge of PFW-PT samples as a function of PT content measured at 297 K. The W L_{III}-edge white line mostly derives from electron transitions from the $2p_{3/2}$ state to a vacant 5d state [7]. Figure 1 shows that L_{III} absorption edge presents a clear double peak structure with slightly less intensity of the peak situated at highest energy, which is identified as the signature of the crystal field splitting of the 5d band into t_{2g} and e_g states in a octahedral symmetry [7]. As can be seen in Figure 1 (a) and (b), no changes are observed on the XANES spectra of PFW and PFW-PT samples with the increasing of temperature or the PT content.

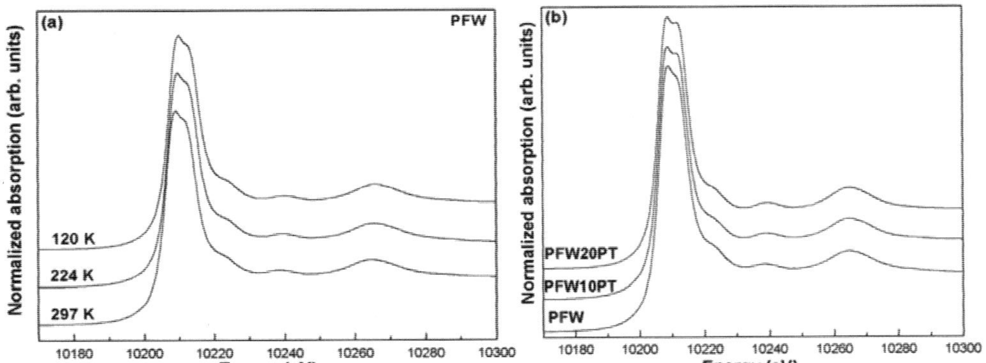

Figure 1. XANES spectra at W L_{III}-edge (a) PFW sample measured at different temperatures and (b) PFW-PT samples measured at 297 K.

In order to identify which atoms or interactions have a more significant contribution on the features observed in the W L_{III}-edge XANES spectra, theoretical spectra were calculated using CRYSTALFFREV (from MAX software package) and FEFF8.2 programs [5, 6] using the crystallographic data of PFW sample obtained from X-ray diffraction pattern collected at 297 K [8]. Figure 2 shows the calculated spectra as a function of the cluster radius around the absorber atom. The cluster radius of 6.0 Å concerns a neighbourhood around W atoms formed by 84 atoms, whereas the cluster radius of 1.98 Å concerns only the first shell of O atoms belonging to the WO_6 octahedra (6 atoms), the cluster radius of 3.45 Å includes up to the first shell of Pb atoms (14 atoms), the cluster radius of 3.98 Å includes up to the first shell of W/Fe atoms (20 atoms) and the cluster radius of 4.45 Å includes up to the second shell of O atoms (44 atoms).

As shown in Figure 2, the spectrum calculated with a cluster radius of 6 Å reproduces quite well the white line of XANES structure and the features up to 60 eV above the edge observed in the XANES experimental spectrum of PFW sample. In the spectrum calculated with a cluster radius of 1.98 Å, it is not observed the transitions located between the edge and 10240 eV, indicating that these transitions would be originated from W-Pb interactions.

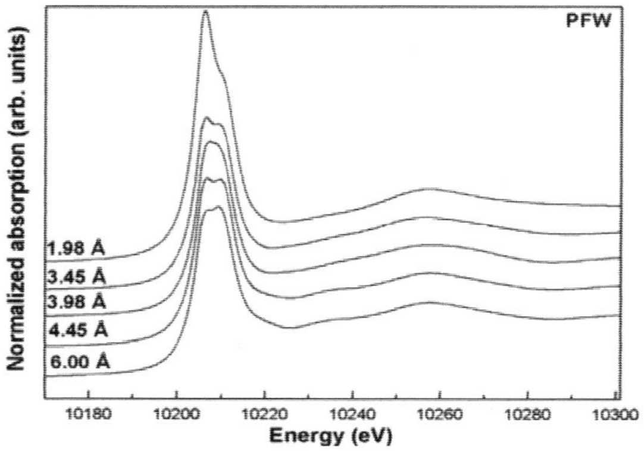

Figure 2. Theoretical XANES spectra at W L_{III}-edge of PFW sample as a function of the cluster radius.

Figure 3 (a) shows the modulus of Fourier transform of PFW sample extracted from W L_{III}-edge EXAFS spectra collected at different temperatures and Figure 3 (b) shows the modulus of Fourier transform of PFW, PFW10PT and PFW20PT samples collected at 297 K.

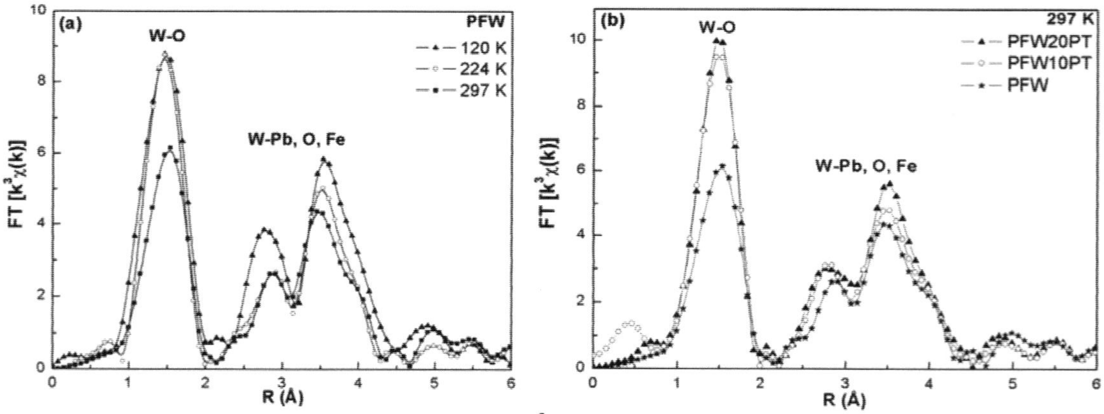

Figure 3. Modulus of Fourier transform of $k^3\chi(k)$ EXAFS spectra (a) PFW sample collected at different temperatures and (b) PFW-PT measured at 297 K.

In order to obtain qualitative structural information from EXAFS spectra, theoretical EXAFS spectra were calculated considering the reported crystalline structure of PFW sample [9]. In a study of PFW compound using neutron diffraction measurements between 10 and 700 K, Ivanov *et al.* showed that the symmetry of PFW remains cubic with a *Pm-3m* space group. Mitoseriu *et al.* also identify a cubic symmetry for PFW at 85 K, but reported a possible separation of diffraction peaks due to a rhombohedral phase with *R3m* space group [8]. However, these authors argue that the difference between these two types of symmetries, cubic and rhombohedral, could not be clearly detected by X-ray diffraction measurements due the experimental accuracy [8]. Thus, theoretical EXAFS spectra were calculated using FEFF8.2 code [6] considering structural models with a cubic (*Pm-3m* space group), rhombohedral (*R*3*m* space group) and tetragonal (*P4mm* space group) [8] symmetries.

According to these structural models, the more intense peak, between 1.0 and 2.0 Å in the Fourier transforms of PFW sample, corresponds to a single scattering interaction between the first six O atoms and the W atoms. The single scattering interactions relative to W-Pb, W-W/Fe and W-O (beyond the first neighbours at WO_6) paths correspond the peaks and shoulders observed between 2.0 and 5.0 Å. This region also includes multiple scattering paths such as W-O-O, W-O-Fe-O, W-Pb-O, W-W-O, W-Fe-O, W-O-W-O, and W-O-O-O interactions.

As can be seen in Figure 3 (a), the position and shape of the peaks do not show significant modifications, indicating that the neighbourhood of W atom does not modify drastically with the temperature. Only the intensity of the peaks in Fourier transform, mainly the intensity relative to first peak, decreases with increasing of the temperature which occurs due to an increasing of the Debye-Waller factor since, according XRD, modifications in the neighbour number are not expected in this range of temperature. In Figure 3 (b), only an intensity increasing on the first peak of Fourier transform is observed as the amount of $PbTiO_3$ increases. This increasing can be associated with the increasing of the temperature of the ferroelectric-paraelectric phase transition [4]. For PFW sample, the temperature of ferroelectric-paraelectric phase transition is around 180 K and thus, the EXAFS spectra collected at 297 K corresponds to a cubic paraelectric phase whereas the temperature of ferroelectric-paraelectric phase transition of PFW20PT sample is around 256 K [4, 8].

In order to obtain quantitative structural results regarding the first W-O coordination shell, the first peak of the Fourier transform curves of PFW and PFW-PT samples (Figure 3) were back Fourier transformed. The experimental and theoretical EXAFS spectra concerning only the first coordination shell are shown on Figure 4. As can be seen on Figure 4(a), the best fit was those using a rhombohedral symmetry at 120 K and a cubic symmetry at 224 K and 297 K. Table I shows the interatomic distance (R), Debye-Waller factor (σ^2) and the quality factor (QF) obtained with the fitting. The fitting results on Table I show that the quality factor of the fit considering a local rhombohedral symmetry for PFW sample measured at 120 K presents a better agreement with experimental data than considering a local cubic symmetry. In the rhombohedral local symmetry, the WO_6 octahedra are formed by three different shells of two O atoms. As the temperature increases above 220 K, the best fit was performed using a local cubic symmetry model, which is in agreement with previous studies [2, 8].

The observation of a local rhombohedral symmetry for W atoms in PFW samples measured at 120 K is in apparent disagreement with previous studies which reported a cubic symmetry for PFW samples from 10 and 700 K [8, 9]. The apparent discrepancy between the XAS and neutron results can be explained in terms of the level of disorder that these techniques can detect [10,11]. As a result, XAS technique allows a better image of the local distortions because, if the symmetry were completely cubic, the structure would correspond to the paraelectric phase instead of a ferroelectric one and therefore, the observation of the ferroelectric hysteresis loops would not be expected. However, our previous study concerning electrical properties of this compound showed ferroelectric hysteresis loops at 80, 100, 160 and 200 K [4].

As illustrated in Figure 4b and Table I, the best fitting of the EXAFS spectra of PFW10PT and PFW20PT samples measured at 297 K gives a local tetragonal symmetry for the first W-O shell formed by two short WO_I and four longer WO_{II} interactions. However, at 297 K, a local cubic symmetry was expected since this temperature is respectively 65 K and 41 K higher than the temperature of the maximum of the dielectric permittivity of PFW10PT and PFW20PT samples which corresponds to the ferroelectric to paraelectric dielectric transition [4].

Thus, EXAFS results of PFW10PT and PFW20PT showed that the local structure around W atoms in these two samples remained tetragonal well above the maximum of their dielectric permittivity curves where the samples were characterized as paraelectric and presenting a cubic long-range order phase according XRD experiments [4]. These findings are similar to those observed in others lead ferroelectric ceramic materials systems [10,11]

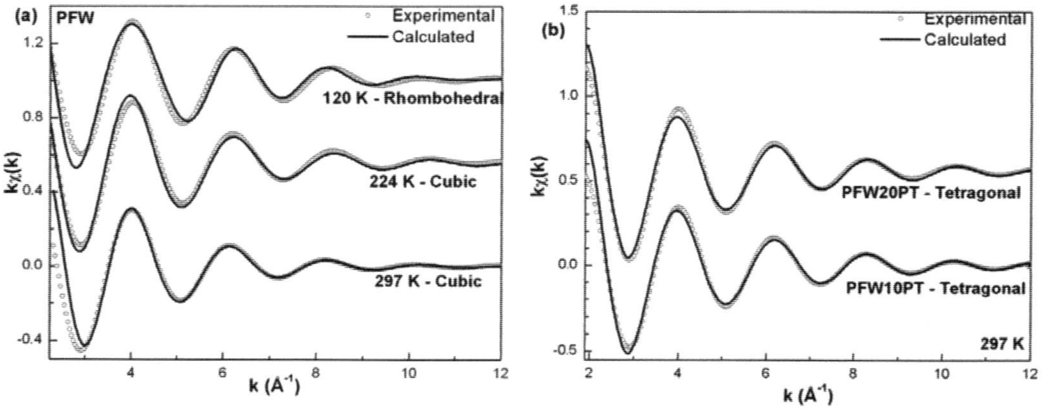

Figure 4. Fitting and back-Fourier-filtered experimental signal of the W-O first shells for (a) PFW sample at 120 K, 224 K and 297 K and (b) PFW-PT samples at room temperature.

Table I – W L_{III}-edge EXAFS fitting structural parameters of PFW as a function of the temperatures and PFW-PT samples as a function of the composition. Different structural models with rhombohedral, tetragonal and cubic symmetries were used during the fitting procedure.

Sample	Temperature	Symmetry	Shell	N	R (Å)	σ^2 (Å^2)	QF
PFW	120 K	Cubic	W-O_I	6	1.86(1)	0.0157(7)	21.71
PFW	120 K	Rhombohedral	W -O_I	2	1.88(2)	0.0085(6)	5.96
			W -O_{II}	2	1.87(2)		
			W -O_{III}	2	2.29(1)		
PFW	224 K	Cubic	W -O_I	6	1.85(1)	0.0069(11)	0.50
PFW	297 K	Cubic	W -O_I	6	1.89(1)	0.0107(7)	1.31
PFW10PT	297 K	Tetragonal	W -O_I	2	1.77(1)	0.0027(6)	0.61
			W -O_{II}	4	1.90(1)		
PFW20PT	297 K	Tetragonal	W -O_I	2	1.78(2)	0.0010(12)	0.56
			W -O_{II}	4	1.90(1)		

4. Conclusions

The local structure of $Pb(Fe_{2/3}W_{1/3})O_3$-$PbTiO_3$ multiferroic ceramics was probed by XAS at W L_{III}-edge . The analysis of XANES spectra did not show significative modifications in intensity or energy of transitions as a function of temperature or $PbTiO_3$ content. Results obtained by the fitting of the EXAFS spectra revealed a local rhombohedral symmetry at 120 K for PFW composition, which is consistent with electrical properties of this compound. On the case of (PFW100xPT) multiferroic ceramics, for all studied compositions, the local structure around W atoms presents a local tetragonal symmetry with two short and four longer W-O bond-lengths well above the maximum of their dielectric temperature curve where a local cubic symmetry was expected.

5. Acknowledgements

Research partially carried out at LNLS National Laboratory of Synchrotron Light, Brazil. This study was supported by FAPESP and CNPq Brazilian financing agencies. The authors wish to thank Mr. Francisco J. Picon and Mrs. Natalia A. Zanardi for the technical assistance during sample preparation.

References

[1] Cheong S-W and Mostovoy M 2007 *Nat. Mater.* **6** 13-20
[2] Eiras J A, Fraygola B M and Garcia D 2010 *Key Eng. Mater.* **434-435** 307-10
[3] Lu C H and Wong Y C 1995 *Ceram. Int.* **21** 413-9
[4] Fraygola B M, Coelho A A, Garcia D and Eiras J A 2012 *Process. Appl. Ceramics* **6** 65-75
[5] Michalowicz A, Moscovici, J., Muller-Bouvet, D., Provost, K. 2009 *J. Phys. Conf. Ser.* **190** 012034
[6] Ankudinov A L, Ravel B, Conradson S D and Rehr J J 1998 *Phys. Rev. B* **58** 7565
[7] Majewski P, Geprags S, Boger A, Opel M, Erb A, Gross R, Vaitheeswaran G, Kanchana V, Delin A, Wilhelm F, Rogalev A and Alff L 2005 *Phy. Rev. B* **72**
[8] Mitoseriu L, Vilarinho P M, Viviani M and Baptista J L 2002 *Mater. Lett.* **57** 609-14
[9] Ivanov S A, Eriksson S G, Tellgren R and Rundlof H 2004 *Mater. Res. Bull.* **39** 2317-28
[10] Sicron N, Ravel B, Yacoby Y, Stern E A, Dogan F and Rehr J J 1994 *Phys. Rev. B* **50** 13168-80
[11] Neves P P, Doriguetto A C, Mastelaro V R, Lopes L P, Mascarenhas Y P, Michalowicz A and Eiras J A 2004 *J. Phys. Chem. B* **108** 14840-9

Local Atomic and Electronic Structure of the Fe dopants in AlN:Fe Nanorods

V L Mazalova[1], Y V Zubavichus[2], D S Chub[1], A A Guda[1], X H Ji[3], S P Lau[4] and A V Soldatov[1]

[1] Department of Physics, Southern Federal University, Zorge 5, Rostov-on-Don, Russia, 344090

[2] NRC "Kurchatov Institute", Moscow, Russia,123182

[3] School of Materials Science and Engineering, South China University of Technology, Guangzhou, P.R. China, 510641

[4] Department of Applied Physics, The Hong Kong Polytechnic University, Hung Hom, Kowloon, Hong Kong, P.R. China

E-mail: mazalova@sfedu.ru

Abstract. Fe-doped AlN nanorods were studied by means of x-ray absorption spectroscopy above the Fe K- and $L_{2,3}$- edges. Theoretical simulations of the x-ray absorption spectra show that Fe atoms mainly substitute Al. A minor fraction of Fe interstitials or Fe-Al-N ternary alloy can be identified as well. Bader's AIM analysis predicts that neutral substitutional Fe_{Al} defect is in 2^+ charge state, though Al in pure AlN is in 3^+ charge state. Fe $L_{2,3}$ absorption spectra and photoluminescence data indicate the coexistence of Fe^{2+}/Fe^{3+} in AlN:Fe nanorods so different charge states of substitutional Fe_{Al} should co-exist.

1. Introduction

Nanostructured AlN doped with iron [1-4] attracts much interest due to its potential applications in spintronic devices, which offer unique opportunity to integrate conventional semiconductor and ferromagnetic technology. The main requirement for the materials in this area is the presence of ferromagnetic properties at room or higher temperatures. However, a full explanation of the magnetic properties of these materials has not been established yet. The first was a model of Zener [5]. It is based on the assumption that the magnetic atoms replace metal atoms in the crystal lattice of the semiconductor. However, in gallium arsenide, doped atoms often occupy interstitial position [6]. The formation of manganese clusters is shown in a number of papers [7]. Other authors, being based on the results of X-ray diffraction, observed a presence of the $GaMn_3N$ phase in GaN:Mn [8]. Other theories such as a model involving magnetic polarons [9] and some other approaches [10-12] have been developed to explain the magnetic properties of diluted magnetic semiconductors.

Nanostructured AlN is a wide band gap semiconductor. Controlling synthesis process of nanostructures from AlN it is possible to check demanded properties. However, understanding of the origin of the ferromagnetism in the Fe-doped AlN presents a challenge in the modern material sciences. There are several reports that Fe-doped AlN nanorods exhibited room temperature ferromagnetism with a saturated magnetic moment of 0.64μB/Fe [13]. It notes that the magnetic moment of TM-doped AlN nanotubes reaches maximum for Fe-doped AlN nanotubes [14]. It is found that the Fe atoms substitute the Al atoms in the lattice when the iron concentration is smaller than 1.2%, while they are embedded into interstitial spaces of the lattice at larger Fe concentrations. A maximum of the saturation magnetization is found to be 2.81 emu/cm^3 for 1.2% of Fe atoms in AlN:Fe film [15].

A careful analysis should be provided for the synthesized samples in order to rule out the presence of secondary phases that induce ferromagnetic properties [16]. This work is devoted to the study of the local atomic structure and charge state of Fe dopants in the AlN:Fe ferromagnetic nanorods. Both near edge and extended regions of x-ray absorption spectra above Fe K-edge are utilized. The multiplet structure of Fe $L_{2,3}$ spectra is used as a fingerprint for the Fe charge state.

2. Methods

AlN:Fe nanorods were prepared in a horizontal tube furnace on Si (100) substrate in the presence of $FeCl_3$ during reaction of heating as iron source [17]. The Fe $L_{2,3}$ XANES spectra were recorded under UHV condition at the Russian-German beam line (RGBL) of BESSYII synchrotron in the total electron yield mode. The Fe K-edge spectrum was measured in the fluorescence mode at the Structural Materials Science beamline [18] of the Kurchatov Synchrotron Radiation Center with a Si avalanche photodiode as a fluorescence detector and an ionization chamber as an incident beam intensity monitor.

X-ray absorption spectra above the Fe K-edge were simulated using self-consistent real space multiple scattering approach, implemented in the Feff9.0 software [19]. Hedin-Lundquist exchange-correlation potential was used in the simulations. Relaxation of atomic positions for model defects in AlN was performed in the full-potential linear augmented plane-wave approximation (LAPW), implemented in the Wien2k program package [20]. The generalized gradient approximation in the scheme of Perdew–Burke–Ernzerhof (GGA-PBE) [21] was used for the treatment of exchange-correlation effects.

3. Results and discussion

The nanorods are 500 nm in length with a sharp tip 15-20 nm in diameter. The spontaneous saturated magnetization and the coercitivity of the AlN:Fe are about 0.64μ_B/Fe and 116 Oe, respectively. Diffraction patterns show that AlN:Fe nanorod is a single wurtzite AlN crystal with the growth direction (0001) with no impurity phases within sensitivity of the method.

The figure 1 shows a magnitude of the Fourier transforms (FT) of $k^3\chi(k)$ data, obtained from EXAFS part of spectra. From the data on the left panel it is clear that no iron clusters or bulk iron phase were created in AlN:Fe nanorods. The best fit of the FT data shown on the right panel was obtained with three contributions, which correspond to Fe-N, Fe-Al, Fe-Al bonds with distances equal to 1.93Å, 2.37Å, 3.01Å, correspondingly. The first and the third contributions can be attributed to the coordination spheres around substitution Fe_{Al} defects, while the second one probably corresponds to the octahedral interstitial position of Fe in AlN lattice.

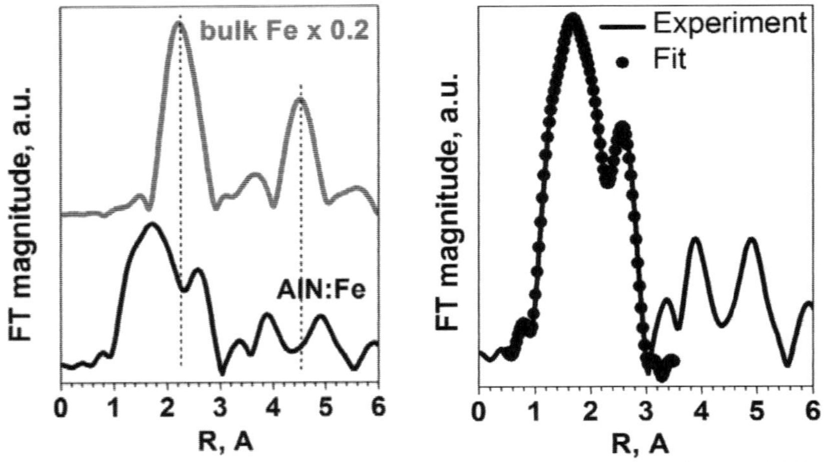

Figure 1. Left panel shows the experimental Fourier transform of Fe K-edge EXAFS spectra for AlN:Fe nanorods and bulk Fe. Right panel shows a simulated and experimental FT with three contributions at different distances Fe-N, Fe-Al and Fe-Al.

Fe K-edge XANES region was used to further verify the structural models proposed by EXAFS. Left panel of the figure 2 shows the results of simulations for the substitutional Fe_{Al} defect and octahedral interstitial. The following AlN lattice parameters were used in the simulations: a=b=3.112, c=4.982, α=β=90, γ=120, space group P6sub3mc, Al (1/3, 2/3, 0), N (1/3, 2/3, 0.382) [22]. The relaxation of AlN lattice around defects was performed within LAPW simulations for a 2×2×2 supercell containing the one point defect.

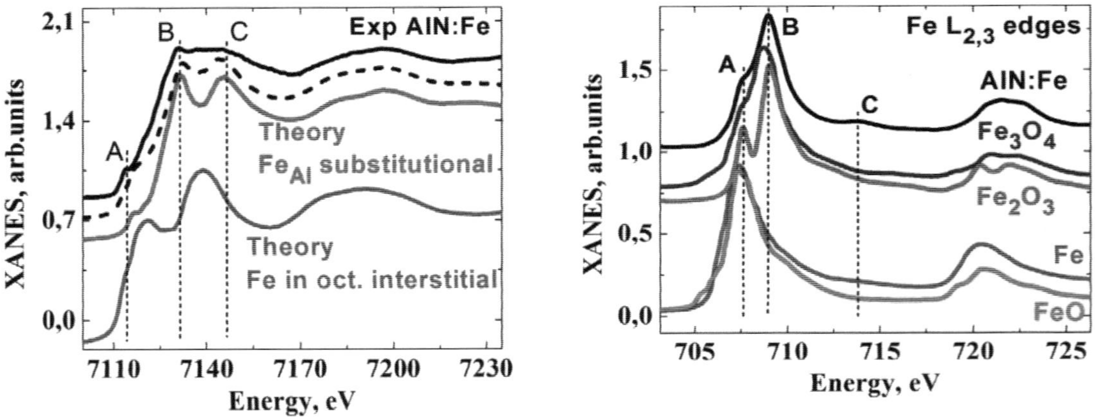

Figure 2. Left panel shows the experimental XANES spectrum for AlN:Fe nanorods above the Fe K-edge and theoretical spectra for the substitutional Fe_{Al} defect and octahedral interstitial Fe_{int} defect. Dashed line is a superposition of 30% interstitial and 70% substitutional positions of Fe atoms. Right panel shows Fe $L_{2,3}$ XANES for AlN:Fe nanorods and series of reference compounds. Data for Fe-oxides are taken from [12].

The simulation for Fe_{Al} substitutional model reproduces well the shape of experimental Fe K-edge spectrum of AlN:Fe nanorods. However, the maxima on the experimental spectrum are broadened especially in the region marked as B and C. This indicates a significant structural disorder in the sample and co-existence of other structural defects in addition to the substitutional ones. A better agreement with the experiment was obtained with a combination of 30% Fe in the octahedral interstitial positions and 70% substitutional defects as shown by the dashed line. However, Fe-Al-N ternary alloys can exist in Fe doped AlN as was shown in the work [23]. Additional experimental analysis should be carried out in order to unambiguously identify the complementary phase to the substitutional Fe_{Al}.

In the table 1 we show the results of Bader's AIM analysis [24] for different compounds. As one can see from the table Al atoms in AlN have a formal charge 3^+. One could expect that Fe in Al site should be in the same charge state. However Bader's AIM analysis performed for the Fe_{Al} defect and reference compounds FeN and FeO shows that without any additional defect charge state of Fe in Al site is closer to 2^+.

Table 1. Bader's AIM analysis of the charge states of atoms in pure AlN, FeN, FeO and AlN:Fe. In the case of substitutional defect Fe_{Al} in AlN the average value for the four nitrogen atoms in the FeN_4 tetrahedron is presented.

	AlN	FeN	FeO	Fe_{Al} in AlN
Al	+2.45	-	-	-
N	-2.45	-1.10	-	-2.19
Fe	-	+1.10	+1.31	+1.40
O	-	-	-1.31	-

The $L_{2,3}$ spectra of transition metals are more sensitive to the charge state rather than to the local atomic structure. Right panel of the Figure 2 shows the Fe $L_{2,3}$ XANES spectra for AlN:Fe nanorods and series of reference compounds. In contrast to the expected charge state Fe^{2+} from the Bader's analysis, Fe in AlN:Fe nanorods shows a multiplet structure which is closed to the Fe^{3+} or a mixture of Fe^{3+}/Fe^{2+} as in Fe_3O_4. Since neutral Fe_{Al} defect results in Fe^{2+} charge state one can conclude that singly charged substitutional Fe_{Al} defect is more energetically favorable. The photoluminescence study of the AlN:Fe nanorods [25] also reveals the fact of the coexistence of Fe^{2+}/Fe^{3+} charge states.

4. Conclusions

Theoretical analysis of X-ray absorption experiments above the Fe K-edges of AlN:Fe nanorods showed that more than 70% of Fe atoms substituted Al ones. Other 30% can be attributed to the octahedral interstitials or Fe-Al-N ternary alloy. The Fe $L_{2,3}$ XANES spectra show that Fe charge state in AlN is a mixture of Fe^{2+}/Fe^{3+}.

Acknowledgements

This work has been supported by the Ministry of Education and Science of Russia (project 2.5440.2011).We would like to thank UGINFO computer center of Southern Federal University for providing the computing time. User Facility Kurchatov Center for Synchrotron Radiation and Nanotechnology is supported via the Russian Federal contract #16.552.11.7003.

References

[1] Zhang Y, Liu W and Niu H 2008 *Solid State Commun* **145** 590
[2] Gao X D, Jiang E Y, Liu H H 2007 *Appl Surf Sci* **253** 5431
[3] Zeng F, Fan B, Yang Y C 2010 *J Vac Sci Technol* B **28** 62
[4] Zhang J-M, Li H-H, Zhang Y, Xu K-W 2011 *Physica* E **43** 1249
[5] Dietl T, Ohno H, Matsukura F, Cibert J, Ferrand D 2000 *Science* **287** 1019
[6] d'Acapito F, Smolentsev G, Boscherini F, Piccin M, Bais G, Rubini S, Martelli F, Franciosi A 2006 *Phys. Rev.* B **73** 035314
[7] Martínez-Criado G, Somogyi A, Ramos S, Campo J, Tucoulou R, Salome M, Susini J, Hermann M, Eickhoff M and Stutzmann M 2005 *Appl. Phys. Lett.* **86** 131927

[8] Granville S *et.al.* 2006 *J. Appl. Phys.* **100** 084310
[9] Chou H, Lin C P, Huang J C A, Hsu H S 2008 *Phys. Rev.* B **77** 245210
[10] Bryksa V, Noltin W 2008 *Phys. Rev.* B **78** 064417
[11] Dionne G F 2007 *J. Appl. Phys.* **101** 09C509
[12] Mounkachi O, Benyoussef A, A. El Kenz, Saidi E H and Hlil E K 2009 *Physica* A **388** 3433
[13] Ji X H, Lau S P, Yu S F, Yang H Y, Herng T S, Sedhain A, Lin J Y, Jiang H X, Teng K S, Chen J S 2007 *Appl Phys Lett* **90** 193118
[14] Zhang J-M, Li H-H, Zhang Y, Xu K-W 2011 *Physica* E **43** 1249
[15] Gao X D, Jiang E Y, Liu H H, Mi W B, Li Z Q, Wu P, Bai H L 2007 *Appl. Surf. Sci.* **253** 5431
[16] Ney A, Opel M, Kaspar T C, Ney V *et.al.* 2010 *New J. Phys.* **12** 013020
[17] Ji X H, Lau S P, Yu S F *et.al.* 2007 *Appl. Phys. Lett.* **90** 193118
[18] Chernyshov A A, Veligzhanin A A, Zubavichus Y V 2009 *Nucl. Instr. Meth. Phys. Res.* A **603** 95
[19] Rehr J J, Kas J J, Vila F D, Prange M P, Jorissen K 2010 *Phys. Chem. Chem. Phys.* **12** 5503
[20] Schwarz K and Blaha P 2003 *Comput. Mater. Sci.* **28** 259
[21] Perdew J P, Burke S, Ernzerhof M 1996 *Phys. Rev. Lett.* **77** 3865
[22] Goldberg Yu 2001 *Aluminum Nitride (AlN)* (in Properties of Advanced SemiconductorMaterials GaN, AlN, InN, BN, SiC, SiGe) eds M E Levinshtein, S L Rumyantsev, M S Shur (New York: John Wiley & Sons Inc) pp 31-47
[23] Regan T J, Ohldag H, Stamm C *et al* 2001 *Phys. Rev.* B **64** 214422
[24] Bader R F W 1991 *Chem. Rev.* **91** 893
[25] Ji X H, Lau S P, Yu S F *et al* 2007 *Appl. Phys. Lett.* **90** 193118

Journal of Physics: Conference Series **430** (2013) 012113 doi:10.1088/1742-6596/430/1/012113

XANES characterization of UO_2/Mo(Pd) thin films as models for ε-particles in spent nuclear fuel

M A Denecke[1], T Petersmann[2], R Marsac[1], K Dardenne[1], T Vitova[1], T Prüßmann[1], M Borchert[3], U Bösenberg[3], G Falkenberg[3] and G Wellenreuther[3]
[1]Karlsruhe Institute of Technology, Institut für Nukleare Entsorgung, Postfach 3640, D-76021 Karlsruhe, Germany
[2]European Joint Research Center, Institute for Transuranium Elements (ITU), D-76344 Eggenstein-Leopoldshafen, Germany
[3]Deutsches Elektronen-Synchrotron (DESY), Notkestr. 85, D-22607 Hamburg, Germany

E-mail: melissa.denecke@kit.edu

Abstract. X-ray absorption near edge structure (XANES) is recorded for nano- and micro-particles formed in Mo doped UO_2 thin films (TFs) prepared by in situ sputter co-deposition and 1000°C tempering. These UO_2/Mo TFs are intended to serve as synthetic models for ε-particles in spent nuclear fuel. We find that when Si is used as substrate, nano-sized zero-valent Mo particles form as desired. However, these are embedded in USi_3, which forms at the high temper temperatures. Micron-sized Mo-particles are formed when SiO_2 is used as substrate. Using focussed X-ray beams of varying size (500µm, 25µm and 5µm), these particles are characterized to be predominantly hexavalent Mo oxides, potentially with tetrahedral coordinated $[Mo(VI)O_4]^{2-}$ at the surface and a MoO_3-like phase in the bulk. These TFs are poor synthetic models for spent fuel ε-particles but do offer the opportunity to study changes in surface structures in response to stress/charge as a function of particle size.

1. Introduction

Spent nuclear fuel (SNF) consists of 95% UO_2 and 5% mixture of radionuclides (RNs) and fission products, heterogeneously distributed throughout the fuel matrix in different phases, e.g., gases (Xe, Kr, I), oxides (including transuranic elements Pu, Np, etc.) and metallic precipitates or so-called ε-particles (Pd, Mo, Rh, Ru, etc.)[1,2]. Systematic investigations of SNF matrix corrosion as an important source term for the mobilization of RNs are essential for evaluating SNF disposal safety [3,4]. We prepare UO_2/Mo thin films (TFs) for potential use as models to study the influence of ε-particles on the surface corrosion of SNF. The TFs provide a less complex system and lower radiation fields than actual SNF samples, which allow systematic mechanistic investigations even in non-radioactive laboratories.

We determine how well these TF models mimic the SNF surface by thoroughly characterizing their composition, structure and morphology using numerous methods. Studies of UO_2/Mo TFs having varying Mo content prepared via *in situ* sputter co-deposition on different substrates (Si, SiO_2 or $LaAlO_3$) by means of 2-dimensional (2D) and 3-dimensional synchrotron-based imaging methods (scanning nano-/micro-XRF, scanning nano-XRD and XRF, holographic and ptychographic tomography) reveal that nano- and micro-sized Mo-particles form in TFs prepared on Si and SiO_2

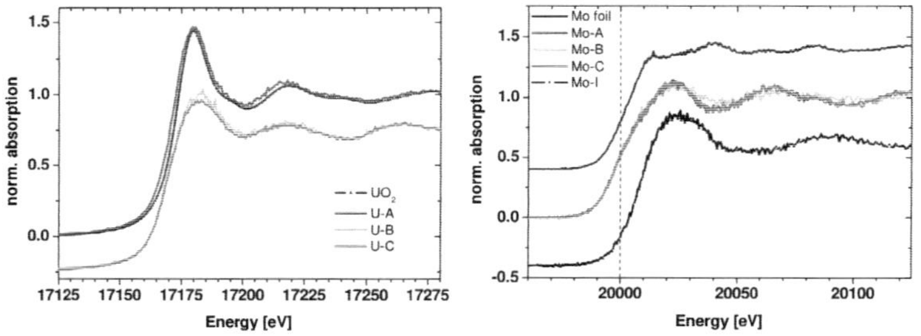

Figure 1: (left) U L3 XANES of UO_2 and sample H measured at different positions. (right) Mo K- XANES of a Mo^0-foil (edge energy indicated as vertical dashed line), recorded at different positions of sample H and of a Mo hot spot in sample I (Mo-I). Designations A,B,C refer to the sample H phase at the position where the spectrum was measured: A= UO_2, B= USi_3, C= mixed phase region.

substrates and tempered at 1000°C (600 °C was not sufficient); when $LaAlO_3$ is used as substrate and tempered at 1000°C, Mo does not aggregate but is found concentrated at the substrate surface [5]. In this report we characterize the Mo-particles observed to form in these TFs using X-ray absorption near-edge structure (XANES) at the Mo K-edge and the L3-edges of both Mo and U.

2. Experimental

Mo L3 edge spectra are recorded at the INE-Beamline for actinide research [6] at the ANKA synchrotron in fluorescence mode, using the low energy setup available there and a Si(111) double crystal monochromator (DCM) calibrated against the first inflection point in the Mo L3-XANES of a Mo^0-foil (2.52 keV). Both scanning XRF measurements and XANES measurements using a focused X-ray beam are performed at the hard X-ray Micro/Nanoprobe beamline P06 at PETRA III [7] and at HASYLAB beamline-L experimental station (DORIS at DESY). The X-ray beam is focused at P06 using a Kirkpatrick-Baez (KB) system equipped with Rh-coated silica mirrors, which in our experiment delivered an enlarged focal spot size of approximately 2μm (H) × 7μm (V) due to monochromator instabilities. At beamline-L a polycapillary half-lens is used to focus the beam to 35 μm (H) × 22 μm (V) at 23 keV. At both beamlines a Si(111) DCM is used and the beam energy calibrated against the first inflection point in the Mo K-XANES of a Mo^0-foil reference, defined as 20.0 keV.

Two UO_2 TFs doped with 8 and 20 wt% Mo and around 500 nm thick are prepared via co-deposition of U and Mo onto SiO_2 and Si substrates and tempering, designated as sample I and H, respectively (sample preparation is described in [5]). Samples are mounted at a 45° angle to the incident beam on a positioning stage for scanning in the y and z directions (2D) and positioning the x-axis into the beam focus. A Vortex-90EX silicon drift detector (SII Nanotechnology USA) is used at the INE-Beamline and beamline-L and a Vortex-EM at P06 for collecting X-ray fluorescence radiation. Mo K (U L3) edge XANES are recorded from Mo (U)-rich areas identified in distribution maps reconstructed from the scanning μ-XRF data.

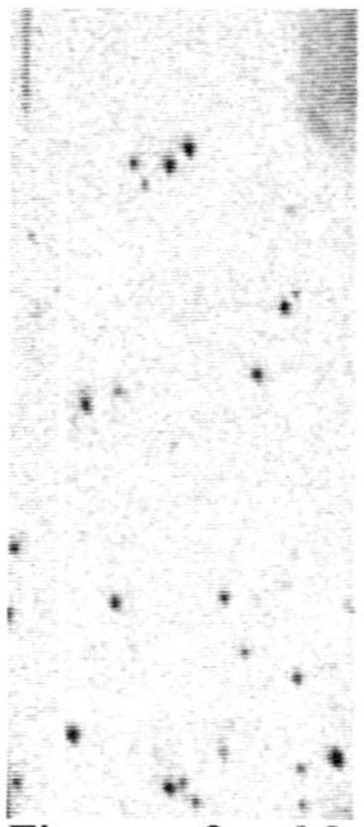

Figure 2: Mo distribution in a 5000 x 2100 μm^2 area of sample I, 25 μm step size and 0.5 s counting time.

3. Results and discussion

In previous μXRF/μ-XRD studies performed at the ESRF ID22Ni using a 100 nm focussed beam, we found that for TFs prepared on Si substrates and tempered at 1000°C (sample H) the UO_2 partially reacts with the substrate, thereby forming USi_3 with Mo localized as 100-400 nm nanoparticles in USi_3 grain boundaries [5]. Uranium L3 XANES recorded at beamline-L using a ~25 μm focussed-beam at different sample areas identified to be composed of UO_2, USi_3 and a mixture of these two phases are depicted in Figure 1 (left). The XANES from the UO_2 area (U-A) is similar to a bulk UO_2 reference spectrum but with a broadened white line at ~17180 eV, as well as indication of a lengthening of the average U-O distance, evident as a shift of the first EXAFS resonance at ~17216 eV to lower energy (due to the relation $\Delta E \propto 1/R(U\text{-}O)^2$ [8]). The U L3 XANES spectra measured at the sample area comprised the mixed UO_2/USi_3 phases (U-C) are more similar to the XANES recorded in an area composed solely of USi_3 (U-B) suggesting that U is mostly present as USi_3 in the mixed phase.

All Mo K XANES recorded at various areas randomly distributed across sample H lie within 1.2 eV to one another and although their edge energies are near that for Mo(0), their XANES fingerprint does not resemble Mo^0-foil at all. The Mo K-XANES recorded at exactly the same sample positions as the U L3 spectra are shown in Figure 1 (right). The edge energies are near that for Mo(0) and well below energies expected for Mo(IV) and Mo(VI) (around 20003.5 and 20007.5 eV [9]). In contrast to the U L3 results above, Mo K XANES recorded in the mixed phase area of sample H (Mo-C) is essentially identical to that from the UO_2 area (Mo-A) and less similar to the USi_3 sample area XANES (Mo-B). The bulk Mo L3 edge data measured at the INE-Beamline at ANKA for sample H at different positions on the sample is shown in Figure 3 (lower curves, top). The L3 edge spectra line up with that for the Mo(0) L3 XANES and lie below the $Mo(IV)O_2$ and $Mo(VI)O_3$ reference spectra, indicating that the mean valence of Mo nanoparticles observed to form in the tempered UO_2/Mo TFs on Si is zero. Similar to the K edge data, the sample H XANES fingerprint does not resemble that for Mo^0-foil; the white lines are at least twice as intense and the resonance above the white line visible in the Mo(0) spectrum is absent. Presumably these differences result from the small size of the particles (narrowing of final state 4d band and loss of extended coordination structure).

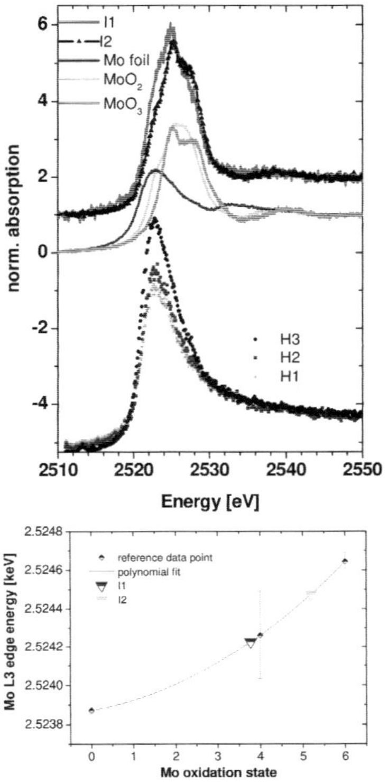

Figure 3: (top) Mo L_3- XANES recorded at different positions on sample H (lower curves) and for sample I (upper curves), compared to Mo(0), Mo(IV), and Mo(VI) reference spectra (middle curves). (bottom) Calibration curve of L3-edge energies versus Mo oxidation states. See text for details.

In Mo distributions reconstructed from scans of small areas of sample I using a nano-sized beam at ESRF ID22-NI, no Mo hot spots were found. In large area XRF scans of sample I (Figure 2) measured at beamline-L with a µ-sized beam, 25-150 µm sized Mo hot spots are revealed, which are distributed randomly and widely spaced in the UO_2 TF. A Mo K-XANES recorded from one of the larger hot spots is shown in Figure 1 (right, bottom curve). The spectrum show a single broad main absorption and its edge energy lies approximately at 20007.3 eV, which is close to the value expected for Mo(VI) [9]. Bulk Mo L3 edge data (Figure 3, top) for sample I recorded at two different positions on the sample show well-resolved features, as the life time broadening due to the L3 core hole is considerably reduced. The white line is split into t_{2g} and e_g states for the (distorted) O_h symmetry of the reference oxides; the t_{2g} being partially occupied in Mo(IV) (d^2). The sample I spectra qualitatively more resemble the Mo(VI) L3 signal with a strong t_{2g} intensity. The L3 XANES energy position is evaluated from the value of the arctan step function obtained in XANES fits, compared to the reference spectra. The result of this analysis is shown in Figure 3 (bottom). The relationship between reference oxidation state and edge energy was not strictly linear so we modelled the calibration curve with a polynomial function. Using this curve, the Mo oxidation state in sample I determined for the two L3 XANES spectra is 3.8 (I1) and 5.2 (I2). For comparison, fits to the spectra as linear combination of the MoO_2 and MoO_3 reference spectra gave slightly different mean Mo valence values in I1 and I2, 4.7 and 5.4, respectively. The quality of least square fit results, however, indicates that these two reference compounds are not sufficient to model sample I L3 spectra. This is likely due to the presence of tetrahedral coordinated Mo(VI) in this sample (see below).

In order to explore the reason for measuring different mean Mo oxidation states in one and the same sample I, we use a ~5 µm beam at the P06 microprobe experimental station to measure Mo K XANES of individual Mo particles <5µm in size and compared these to those recorded at different positions of a particle at least 10 times larger. The results are shown in Figure 4. Particles studied having dimensions of the beam size or smaller exhibit a strong pre-peak at around 20005 eV, which is a signature for tetrahedral coordinated $[Mo(VI)O_4]^{2-}$ (cf. the XANES for $CaMoO_4$). This feature is absent or reduced to a shoulder (similar to MoO_3) on the rising absorption edge in most pixels studied (spectra 1-3), but clearly (and reproducibly) visible in the XANES measured on the outer rim of the particle (#4). The Mo K energy position at half the edge jump of the XANES without a pre-peak all lie about 8 eV above the Mo(0) XANES, indicative of hexavalent molybdenum. We find no evidence for Mo(IV). Summarizing results for this particular particle: we observe a mixture of a MoO_3-like phase and also tetrahedral coordinated $[Mo(VI)O_4]^{2-}$, which is particularly evident at the pixel located at the particle rim. The average XANES from spectra 1 through 4 is shown in Figure 4 (#5), where the main absorption peak is broadened and the contribution of tetrahedral coordinated Mo(VI) reduced to a mere shoulder, which is imaginably a trend towards the sample I spectrum in Figure 1 measured with a ~25 µm beam. We interpret the two observations that 1) both the smallest particles and the rim of a large particle exhibit the largest tetrahedral coordinated Mo(VI) signal and that 2) the particles studied have Mo in its hexavalent state as suggesting charge compensation at the Mo-particles surface occurs through reduction of the number of O atoms coordinated to Mo (tetrahedral surface versus octahedral bulk coordination).

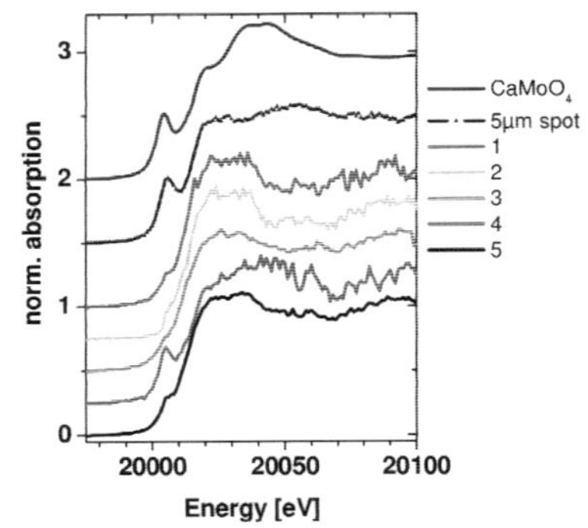

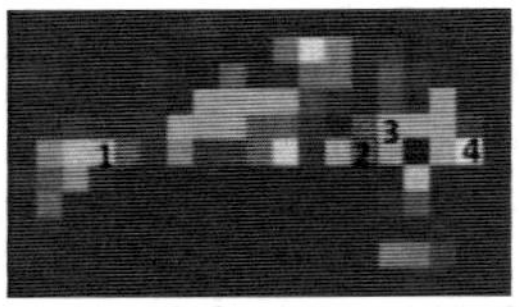

Figure 4: (left) Mo K-XANES measured in sample I for a 5 µm pixel Mo hot spot and a ~60 µm wide Mo particle at the various positions (1-4) indicated in the Mo distribution map (above). Each pixel is 5 x 5 µm^2. XANES #5 is the average of spectra 1-4.

4. Conclusions

These results show that none of the substrates or temper programs used produces satisfactory SNF ε-particle models. While nano- and micron-sized Mo particles are formed in the UO_2/Mo TF systems studied, these particles are either not embedded in the desired UO_2 substrate or are present in oxide form, instead of the desired metallic state. The oxide particles, however, do offer the opportunity to study changes in surface structures in response to stress/charge as a function of particle size. It may be possible to prepare Mo(0)-particles embedded in UO_2 on Si substrates through tempering at temperatures below 1000°C, but above 600 °C.

Acknowledgments
We thank PETRAIII, HASYLAB and ANKA for the granted beamtime.

References

[1] Kleykamp, H 1985 *J Nucl Materials* **131**(2-3) 221
[2] Bruno J, Ewing R 2006 *Elements* **2** 34.
[3] Johnson LH, et al. Spent Fuel. In Radioactive waste forms for the future, Lutze, W (Ed) 1988, pp 635-698
[4] Shoesmith DW 2000 *J Nucl Materials* **282** 1
[5] Denecke MA, et al. 2012 *MRS Spring Meeting Proceedings* **1444**, mrss12-1444-y01-05doi:10.1557/opl.2012.1159
[6] Rothe J, et al. 2012 *Rev Sci Instrum* **83** 043105
[7] Schroer CG, et al. 2010 *Nucl Instrum Methods* **A616** 93
[8] Denecke MA *Proceedings of the OECD-NEA Workshop on Speciation, Techniques and Facilities for Radioactive Materials at Synchrotron Light Sources*, Grenoble, France, 4-6 October 1999, pp. 135-41.
[9] Ressler T, Jentoft RE, Weinhold J, Günter MM, Timpe O 2000 *J Phys Chem* **B104**, 6360

Journal of Physics: Conference Series **430** (2013) 012114 doi:10.1088/1742-6596/430/1/012114

Characterization of U(VI)-phases in corroded cement products by micro(μ)-spectroscopic methods

J Rothe[1], B Brendebach[2], C Bube[1], K Dardenne[1], MA Denecke[1], B Kienzler[1], V Metz[1], T Prüßmann[1], K Rickers-Appel[3], D Schild[1], E Soballa[1] and T Vitova[1]
[1] Karlsruhe Institute of Technology, Institute for Nuclear Waste Disposal (KIT-INE), P.O. Box 3640, D-76021 Karlsruhe, Germany
[2] Gesellschaft für Anlagen und Reaktorsicherheit (GRS) mbH, Schwertnergasse 1, D-50667Köln, Germany
[3]HASYLAB at DESY, Notkestr. 85, D-22607 Hamburg, Germany

E-mail: joerg.rothe@kit.edu

Abstract. Cementation is an industrial scale conditioning method applied to fix and solidify liquid low and intermediate level radioactive wastes (LLW/ILW) prior to underground disposal in geological formations.To assist prognosis of the long-term safety of cemented waste, alteration of uranium doped cement productswas studied in chloride-rich solutions relevant for final LLW/ILW disposal in rock salt. After long-time exposure of the full-scale LLW/ILW simulates to concentrated NaCl and $MgCl_2$ brines, solid samples were retrieved for chemical and mineralogical analysis with an emphasis on uranium speciation in the corroded cement matrix.Bulk and recent spatially resolved micro(μ) U L3-XAFS measurements point to the occurrence of a diuranate type U(VI) phase forming throughout the corroded cement monoliths. U-enriched hot spots with dimensions up to several tens of μm turn out to be generally X-ray amorphous.

1. Introduction

Cementation of waste solutions containing low to intermediate level concentrations of radionuclidesis considered as an appropriate conditioning method to fix and solidify radioactive wastes with negligible heat production(low-level and intermediate-level waste, LLW and ILW, respectively) prior to disposal. Cementation of LLW/ILW isused in many countries on an industrial scale for treatment of waste streams stemming from basic research, engineering applications, medical treatment or reprocessing activities. Cemented LLW/ILW containing in total some 100 tons of uranium was disposed of in the AsseII[1] and ERAM salt mines in Northern Germany from 1967 to 1978. For safety assessment and evaluation of possible closure concepts, it is mandatory to understand the release and retention reactions of uranium and other radionuclides in case of water intrusion into the disposal caverns. Alteration of radionuclide bearing cement products in dilute aqueous solutions has been studied quite extensively. Yet, in chloride-rich solutions, which are relevant for final disposal of cemented LLW/ILW in rock salt, thermodynamic data and understanding of both the cement corrosion process and radionuclide leaching behaviour are still lacking.Hardened cement paste or concrete consists of a multitude of phases whichcomplicate the investigation of radionuclide leaching andretention mechanisms. As laboratory batch experiments conducted with hardened cement paste

alone may not provide sufficient insight into the complexity of long-term cement chemistry, full-scale leaching and corrosionexperimentswith simulated waste products were performed with U(VI) doped cement monolithsin the Asse II mine. Cement blocks produced by an industrial process simulating relevant LLW/ILW compositions were exposed to NaCl and $MgCl_2$ brines for more than two decades.Solid samples for mineralogical laboratory analysis and synchrotron radiation (SR) based speciation techniques wereobtained after recovery of corroded monoliths from the mine.
Initial bulk powder U L3-XANES, laser spectroscopy and XRD measurements were interpreted as indicating the presence of a Ca-U(VI)-silicate phase, e.g., uranophane, in the corroded waste products [2]. U L3-XANES white line (WL) and multiple scattering (MS, cf. spectra in figure 1) features of the corroded cement samples appeared significantly broadened compared to corresponding features in crystalline reference compound spectra. Coexistence of different U(VI) coordination environments associated with high structural disorder or highly distorted U(VI) coordination environments in the corroded cement were assumed to explain the difference between the corroded cement sample and reference XANES. However, recent μ-Raman spectroscopy data indicate that U-rich aggregates found in the corroded cement matrix with dimensions up to several tens of μm are more likely diuranate phases [3]. The present investigation extends initial SR based investigations to spatially resolved techniques - including μ-XAFS (μ-XANES/μ-EXAFS), μ-XRFand μ-XRD-to better characterize U-rich volumes of the generally heterogeneoussalt brine corroded cement matrices.

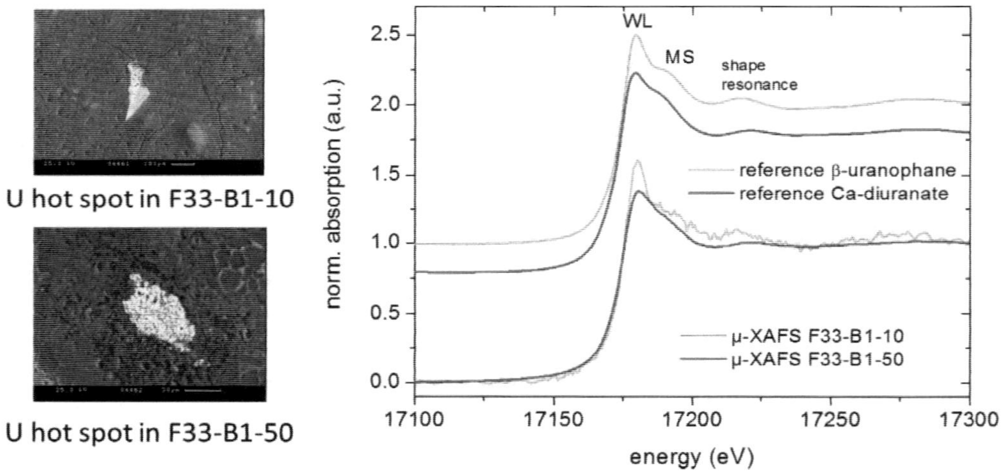

Figure 1. left – SEM images of U-rich hot spots in samples F33-B1-10 ($MgCl_2$ system, top) and F33-B1-50 (bottom, cf. text for nomenclature); right – U L3 μ-XANES spectra recorded at these spots compared to bulk XANES of crystalline U(VI) references Ca-diuranate and β-uranophane(spectra vertically shifted for clarity).

2. Experimental

2.1. Full-scale corrosion experiments

Cemented full-scale LLW/ILW simulateswere fabricated in the 1980sbased onliquid waste compositions simulating thoseresulting from spent nuclear fuel reprocessing in the PUREX process.For monoliths doped with natural uranium, aqueous solutions containing 1.02 kg of U(VI) in the form of $(NH_4)_2U_2O_7$ were added to Ordinary Portland Cement, the water/cement ratios adjusted to 0.5 and the mixtures filled in 200 l steel drums.Details on monolith production and compositioncan be found elsewhere([2] and references therein). After complete hydration, these monoliths (total mass 336 kg) were removed from the drums, immersed in 400 l steel canisters filled with saturated NaCl or $MgCl_2$ brines and stored in the Asse II salt mine at the 490 m level. Two uranium doped monoliths-'F31' corroded in NaCl and 'F33' corroded in $MgCl_2$- were recovered from the leaching vessels after

18 and 17 years, respectively.Several vertical bore holes were cored from the top parallel to the central monolith cylinder axis to probe the degree of corrosive cement degradation and U speciationat varying positions along the cylinder length and surface distances. Samples are identifiedaccordinglyusing the nomenclature 'brine-distance-vertical position', e.g., F33-B1-10 denotes a sample extracted from bore core B1(cored at 21 cm distance from the cylinder surface) in the upper 0-10 cm of the monolith corroded in $MgCl_2$ (cf. details in [2]).

2.2 SR based analysis methods
Thin sectionscut from bore cores F31-B1-10, F31-B1-50, F33-B1-10 and F33-B1-50 mounted on glass slides are used for SR-based characterization.Regions of interest are preselected fromScanning Electron Microscopy (SEM) backscattering images, showing high contrast for U rich aggregates (hot spots) embedded in the corroded cement matrix (figure 1 - left), and identified with a visible light microscope (VLM) at the SR experimental stations.Two beamlines are used for spatially resolved measurements - HASYLAB BL-L at DESY (Hamburg, Germany) and the INE-BL at ANKA located at the KIT North Campus (Karlsruhe, Germany) [4]. At HASYLAB, a small wavelength bandpass from a multilayer monochromator (max. output tuned to 0.70246 Å/17.650 keV) is focused into a spot of <20 μm on the sample surface by a single bounce capillary mounted on a hexapod positioning unit.Sample slidesaremounted on a three-axis positioning stage with the thin section samplesurface at a45° angle to theincident beam. The y and z stage axes (the x-axis determines the focus) are used to scan selected areas and precisely position the samples in the beam path. A silicon drift detector (SDD, SIINT Vortex-90EX)is used for collecting X-ray fluorescence radiation.U hot spots identified in element distribution maps reconstructed from the scanning μ-XRF data are selected for μ-XRD analysis. Diffraction patterns are collected in Laue transmission mode by a CCD camera (MAR CCD, MARUSA Co.) mounted perpendicular to the beam ~15.5 cmdownstream from the sample. XRD geometry parameters are calibrated against a LaB_6 reference sample. At ANKA, a similar setup and single bounce capillary is used for μ-XRD measurements, but diffraction patterns are registered on erasable X-ray sensitive films (PerkinElmer). Apolycapillary optic is used to focus monochromatic radiation delivered by a double crystal monochromator (DCM) equipped with a pair of Ge<422> crystals and reflected on a toroidal focussingmirror to a beam spot-size of 25-30 μm at the sample surface for μ-XRF and μ-XAFS measurements. A SDD (Vortex-60EX) is used here, too. U L3 μ-XAFS spectra, recorded by registering windowed U Lαfluorescence intensity, are normalized to the total flux impinging on the sample measured by a pin-diode mounted behind the polycapillary. The DCM energy is calibrated relative to the first inflection point in the K-XANES of a Y foil defined as 17.038 keV.

3. Results and Discussion

3.1 U L3 μ-XAFS/μ-XRF
The uranium speciation in numerous U hot spots and in the surrounding corroded cement matrix of the four thin section samples are investigated. U-rich hot spots identified in μ-XRF element distribution maps generally exhibit U L3-XANES energy positions characteristic for U(VI) (WL peak position at ~17.180 keV), thus clearly confirming preservation of the hexavalent uranium state in the cement after long-time exposure to NaCl and $MgCl_2$ brines. This finding is also in accordance with previous U speciation results based on bulk XAFS spectra obtained for F31 and F33 monolith drill dust samples [2]. μ-XANES features obtained for most U hot spots and in more diffuse U distributions located nearby are nearly indistinguishable from each other and from previous bulk spectra. These spectra obviously represent a prevalent U(VI) species formedthroughout the corroded cement matrices. The XANES for the F33-B1-50 U hot spot is representative of this major species. The spectra exhibit features significantly different from those typical for U(VI)dioxo-moieties (e.g., theU(VI) mineralβ-uranophane$Ca(UO_2)_2(SiO_3OH)_2 \cdot 5(H_2O)$; cf. figure 1). Both μ-XANES WL features and the post-edge 'shape resonance' are moresimilar to a diuranate (CaU_2O_7/$Na_2U_2O_7 \cdot xH_2O$) type phase (figure 1 - right, red curves).The U coordination in diuranates is characterized by an elongated axial oxygen distance (1.92 Å vs. 1.79Å for uranophane) and a reduced average equatorial oxygen distance (2.31 Å vs.

2.36Å for uranophane). This coordination results in WLdampening and a decrease in the energy difference between the WL and axial oxygenMS feature observed for diuranates (cf., e.g., [5]). These effects are easily modelled by feff8.4 [6]U L3-XANES calculations. Correspondingly, in bulk U L3-EXAFS measurements of pure Ca/Na-diuranate reference samples the Fourier transformation (FT) magnitude peaks associated with axial and equatorial oxygen neighbours merge into a single broad first shell peak [2]. This same broad FT peak is observed in a μ-EXAFS spectrum recorded at a U hot spot insample F33-B1-50 (not shown). Nevertheless, at least one U-rich area in sample F33-B1-10 (corroded in $MgCl_2$ brine - thin section taken near the F33 monolith surface) is found, which exhibits distinct features similar to a uranophane-like U(VI) phase (figure 1 - right, green curves).Thermodynamic solubility calculations performed for the U-doped cement/salt brine systems suggest both diuranates and uranophane as possible solubility limiting phases in the cement/$MgCl_2$ system, while ruling out the formation of schoepite, another mineral with the dioxo moiety U corrdination structure.In contrast, measured U(VI) concentrations in the cement/NaClbrineare only in agreement with a diuranate solubility limiting phase [7]. No XANES spectra similar to the uranophane reference were found in the F31 samples investigated. Note, however, that the total sample volume studied in these spatially resolved experiments is necessarily small and, hence, any exclusion of the presence of such phases cannot be made with great certainty.
In addition to U-rich hot spots anduranium ubiquitously present in diffuse distributions across the thin section samples, a number of other elementsassociated with the cement and its additives are identified in μ-XRF distribution maps. As a general trend, Fe showsa clear anti-correlation with U enriched areas, whereas K, Sr and Rb distributions are correlatedto U in most of these hot spots. Other transition metals (Ti, Mn, Cu, Zn) show no or only weak U correlations (figure 2).

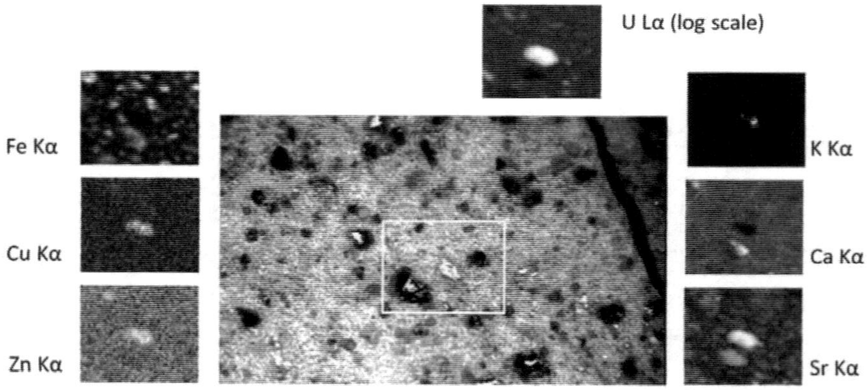

Figure 2. VLM image of area around U hot spot in F33-B1-50 (same hot spot as depicted in SEM image, figure 1 bottom left) and μ-XRF element distribution maps recorded for the marked area (scan area 600μm×500μm, 10μm×10μm steps, beam spot size ~25 μm).

3.2 μ-XRD
μ-XRD measurements of numerous U-rich hot spots,including those depicted on the left hand side of figure 1, performed at both SR sources show these to be generally X-ray amorphous;no evidence is obtained for the presence of any crystalline uranium phase. Weak diffraction patterns identified for some of these hot spots can be attributed to cement corrosion products.

4. Conclusions
It is well-established from multiple scattering XAFS theory that relatively small clusters of some tens of atoms surrounding a chosen absorber are generally sufficient to produce XANES spectral features. However, such nm-sized atomic clusters may be too small – and/or too distorted – to sufficiently

diffract impinging X-rays into detectable XRD patterns. This might explain the absence of any μ-XRD signal from uranium hot spots in the present study. The U-rich areas in the corroded cement matrix, which are clearly discernible in both SEM images and in μ-XRF maps, are X-ray amorphous and therefore not crystalline phases. More likely, they are mere smallatomic clusters or assemblies of small clusters (crystallites) forming under the chemical conditions of the long-time cement corrosion experiments.

Acknowledgement
HASYLAB at DESY and ANKA at KIT are acknowledged for providing beamtime.

References

[1]Kappei G 2007*Proc. Int. Conf. on Radioactive Waste Disposal in Geological Formations (Braunschweig)* p 396
[2] Kienzler B, Metz V, Brendebach B, Finck N, Plaschke M, Rabung T, Rothe J and Schild D 2010 *Radiochim. Acta***98** 675
[3] Bube C, Metz V, Schild D, Lagos M, Bohnert E, Garbev K, Altmaier M, Kienzler B 2011 *Proc. 1st Int. Symp. on Cement-based Materials for Nuclear Wastes (Avignon)*on CD-ROM O319
[4] Rothe J, Butorin S, Dardenne K, Denecke M A, Kienzler B, Löble M, Metz V,Seibert A, Steppert M, Vitova T, Walther C and Geckeis H 2012 *Rev. Sci. Instrum.***83** 043105
[5] Allen P G, Shuh D K, Bucher J J, Edelstein N M, Palmer C E A, Silva R J, Nguyen S N, Marquez L N and Hudson E A 1996 *Radiochim. Acta***75** 47
[6] Ankudinov A L, Nesvizhskii A I and Rehr J J 2003*Phys. Rev.* B **67** 115120
[7] Bube C, Metz V, Rothe J, Dardenne K, Schild D, Lagos M, Finck N, Bohnert E, Kienzler B 2012 (in prep.)

Journal of Physics: Conference Series **430** (2013) 012115 doi:10.1088/1742-6596/430/1/012115

Comparative investigation of N donor ligand-lanthanide complexes from the metal and ligand point of view

T Prüßmann[1], MA Denecke[1], A Geist[1], J Rothe[1], P Lindqvist-Reis[1], M Löble[1], F Breher[1], DR Batchelor[1], C Apostolidis[5], O Walter[5], W Caliebe[2], K Kvashnina[3], K Jorissen[4], JJ Kas[4], JJ Rehr[4], T Vitova[1]

[1] Karlsruhe Institute of Technology (KIT), P.O. Box 3640, D-76021 Karlsruhe, Germany
[2] HASYLAB at DESY, Notkestr. 85, 22603 Hamburg, Germany
[3] European Synchrotron Radiation Facility, BP 220, F-38043 Grenoble Cedex, France
[4] Department of Physics, University of Washington, Seattle, WA 98195-1560, USA
[5] Institute for Transuranium Elements, European Commission, Joint Research Center, D-76125 Karlsruhe, Germany

E-mail: Tim.Pruessmann@kit.edu

Abstract. N-donor ligands such as *n*-Pr-BTP (2,6-bis(5,6-dipropyl-1,2,4-triazin-3-yl)pyridine) studied here preferentially bind *An*(III) over *Ln*(III) in liquid-liquid separation of trivalent actinides from spent nuclear fuel. The chemical and physical processes responsible for this selectivity are not yet well understood. We present systematic comparative near-edge X-ray absorption structure (XANES) spectroscopy investigations at the Gd L_3 edge of $[GdBTP_3](NO_3)_3$, $[Gd(BTP)_3](OTf)_3$, $Gd(NO_3)_3$, $Gd(OTf)_3$ and N K edge of $[Gd(BTP)_3](NO_3)_3$, $Gd(NO_3)_3$ complexes. The pre-edge absorption resonance in Gd L_3 edge high-energy resolution X-ray absorption near edge structure spectra (HR-XANES) is explained as arising from $2p_{3/2} \rightarrow 4f/5d$ electronic transitions by calculations with the *FEFF9.5* code. Experimental evidence is found for higher electronic density on Gd in $[Gd(BTP)_3](NO_3)_3$ and $[Gd(BTP)_3](OTf)_3$ compared to Gd in $Gd(NO_3)_3$ and $Gd(OTf)_3$, and on N in $[Gd(BTP)_3](NO_3)_3$ compared to *n*-Pr-BTP. The origin of the pre-edge structure in the N K edge XANES is explained by density functional theory (DFT) with the *ORCA* code. Results at the N K edge suggest a change in ligand orbital occupancies and mixing upon complexation but further work is necessary to interpret observed spectral variations.

1. Introduction

One of the major steps in the partitioning and transmutation (P&T) strategy for reduction of the longterm radiotoxicity of spent nuclear fuel is the separation of $5f$ elements from their chemically similar $4f$ counterparts. This separation is necessary, as the lanthanide (Ln(III)) fission products have large neutron absorption cross sections and thereby compromise transmutation efficiency in the nuclear fission process. Selective liquid-liquid extraction of actinides (An(III)) from Ln(III) has been demonstrated using soft donor extracting agents such as heterocyclic N-donor ligands, e.g., bistriazinylpyridines (BTP) [1] and bistriazinylbipyridines (BTBP) [2] with separation factors (SF) for Am(III) over Eu(III) greater than 100 (SF = distribution ratio D_{Am}/D_{Eu}; $D_M = [M]_{org}/[M]_{aq}$). Optimization of the partitioning ligands is a topic of global interest, as

Journal of Physics: Conference Series **430** (2013) 012115 doi:10.1088/1742-6596/430/1/012115

the present ligands do not yet fulfill all criteria for application in an industrial process. Such optimization implies basic understanding of the chemical and physical processes responsible for ligand selectivity for An(III) over Ln(III). It was shown that BTPs act as tridentate ligands, forming 1:3 complexes with An(III) and Ln(III) in the solid state and in organic solution [3, 4]. Spectroscopic data available demonstrate thermodynamically favored An(III) extraction from nitric acid solution into the organic phase over Ln(III). However, we found with extended X-ray absorption fine structure (EXAFS) spectroscopy only subtle structural differences between the $[An(III)(BTP)_3](NO_3)_3$ and $[Ln(III)(BTP)_3](NO_3)_3$ complexes in organic solution, which may not fully explain An(III)/Ln(III) selectivity [4, 5]. Additionally, the Ln elements exhibit a trend in selectivity as a function of their ionic radii [6]. Systematic comparative X-ray absorption near edge structure (XANES) spectroscopy investigations at the Ln L_3 edge and N K edge can help differentiate bonding mechanisms between An and Ln bound to different ligands and counter ions, e.g., molecular orbital energy differences and relative electronic populations. We present comparative Gd L_3 high-energy resolution (HR-) XANES and conventional N K edge XANES spectroscopy combined with quantum chemical calculations to investigate the electronic structure of Gd(III) complexes from both the metal and the ligand point of view.

2. Experimental

Samples were prepared by drying iso-propanol solutions of $Gd(NO_3)_3$ (2 mmol/L) and $[Gd(BTP)_3](NO_3)_3$ (2 mmol/L $Gd(NO_3)_3$, 10 mmol/L BTP (n-Pr-BTP (2,6-bis(5,6-dipropyl-1,2,4- triazin-3-yl)pyridine) unless noted otherwise) on appropriate substrates for measurement. This leads to a 40% excess of BTP in the complex solution to ensure the formation of 1:3 complexes. We dried 2 μl (N K-edge) and 10 μl (Gd L_3 edge) on pieces of aluminium foil (5x5 mm^2) and on plastic holders for the N K edge and Gd L_3 edge experiments, respectively. Powder samples of $Gd(CF_3SO_3)_3$ (CF_3SO_3=OTf) and $[Gd(BTP)_3](OTf)_3$ were measured as pellets by mixing and pressing 2 mg compound with 20 mg cellulose. HR-XANES spectra at the Gd L_3 edge (7243 eV) were collected at the W1 beamline (HASYLAB) and ID26 beamline (ESRF). Incident X-rays were tuned to energies from 40 eV below to 70 eV above the Gd L_3 edge. The X-rays emitted from the sample were energy analyzed by a Johann spectrometer in dispersive geometry (W1) [7] or scanning geometry (ID26) [8] and were detected by a CCD camera (W1) or avalanche photodiode (ID26). The $L\alpha_1$ emission line (6053.4 eV) was diffracted and focused by one spherically bent Si(333) (W1) or Ge(333) (ID26) crystal with a 1 m radius of curvature at a Bragg angle $\theta = 78.4°$ (W1) or $\theta = 70.15°$ (ID26). N K edge spectra were measured at the UE52-PGM beamline (BESSY) using partial electron yield detection. A PGM (Plane Grating Monochromator) with a 1200 l/mm grating was used to tune the incident X-Rays from 380-420eV. The $[Gd(BTP)_3](OTf)_3$ spectrum and the f and d angular momentum projected density of states (f-, d-DOS) were calculated using the ab initio multiple scattering theory based *FEFF9.5* code [9]. The selfconsistent field (SCF card) and full multiple scattering (FMS card) calculations were performed on a cluster of 152 atoms corresponding to one $[Gd(BTP)_3](OTf)_3$ molecule. A Hedin-Lundqvist type exchange potential was used. The f-electron DOS was calculated self-consistently by using the UNFREEZEF card. The Fermi level was set to 3 eV below the calculated value of -9.6 eV in order to reproduce the pre-edge structure in the spectrum. To reach convergence, the core-hole type was set to random phase approximation (RPA card) and the core-hole potential was calculated for a cluster of 47 atoms (SCREEN card). The N K and Gd L_3 edge spectra were modeled with an arctangent function for the edge step and PseudoVoigt (PV) functions for all other features using the WINXAS software [10]. The shape parameter α of the PseudoVoigt functions was set to 0.5 to account for both experimental and core-hole lifetime broadening effects. Time dependent density functional theory (TD-DFT) calculations of the N K edge of H-BTP were performed using the *ORCA* package [11] with a def2-TZVP basis set, B3LYP functional, RIJCOSX SCF approximation and ZORA as scalar relativistic approximation.

3. Results and discussion

3.1. Gd L_3 edge

Gd L_3 edge HR-XANES spectra of $[Gd(BTP)_3](NO_3)_3$ and $Gd(NO_3)_3$ are shown in figure 1 a. The $[Gd(BTP)_3](NO_3)_3$ spectrum is shifted 0.4 eV to lower energies compared to the $Gd(NO_3)_3$ spectrum indicating greater cation charge in the $[Gd(BTP)_3](NO_3)_3$ complex. This trend is preserved if NO_3^- is exchanged with an OTf^- anion. The $[Gd(BTP)_3](OTf)_3$ and $Gd(OTf)_3$ spectra are plotted in figure 1 b showing a 0.5 eV energy shift of $[Gd(BTP)_3](OTf)_3$ to lower energies. The additional pre-edge intensity at about 7241 eV, more noticeable for $[Gd(BTP)_3](NO_3)_3$ compared to $Gd(NO_3)_3$, can be assigned to $2p_{3/2}$ photoelectron transitions to $4f$ and $5d$ final states. The good experimental energy resolution of the ID26 data allowed us to model the WL and pre-edge areas (not shown here) and quantify the energy difference between these two resonances to be 0.4 eV smaller in the $[Gd(BTP)_3](OTf)_3$ than in the $Gd(OTf)_3$ spectrum. The origin of the pre-edge feature is revealed by calculations with the *FEFF9.5* code. The calculated $[Gd(BTP)_3](OTf)_3$ spectrum and the f-, d-DOS are plotted in figure 1 c. The f-DOS has high intensity at the energy position of the pre-edge feature suggesting that this feature arises from a mixture of $2p \rightarrow 4f$ and $2p \rightarrow 5d$ electronic transitions; however, the $4f$ contribution dominates. Even though the direct bonding partners of Gd change from N in $[Gd(BTP)_3](OTf)_3$ to O in $Gd(OTf)_3$, hardly any difference between the areas of the pre-edges and WLs is detectable, indicating that the relative electronic populations of the Gd $4f$ and $5d$ states are not significantly influenced by bonding with the n-Pr-BTP molecule. Nevertheless, the -0.4 eV relative energy shift and the 0.4 eV decrease in the energy difference between pre-edge and WL for $[Gd(BTP)_3](OTf)_3$ over $Gd(OTf)_3$ are clear indications for better screening of the $2p$ core-hole due to higher charge density on Gd in $[Gd(BTP)_3](OTf)_3$ than in $Gd(OTf)_3$. Future studies will investigate additional $Ln(BTP)_3$ and $An(BTP)_3$ compounds to obtain relative energy positions and oscillatory strengths of $2p_{3/2} \rightarrow 4f/5f$ and $2p_{3/2} \rightarrow 5d/6d$ electronic transitions, which might be correlated with BTP extraction properties.

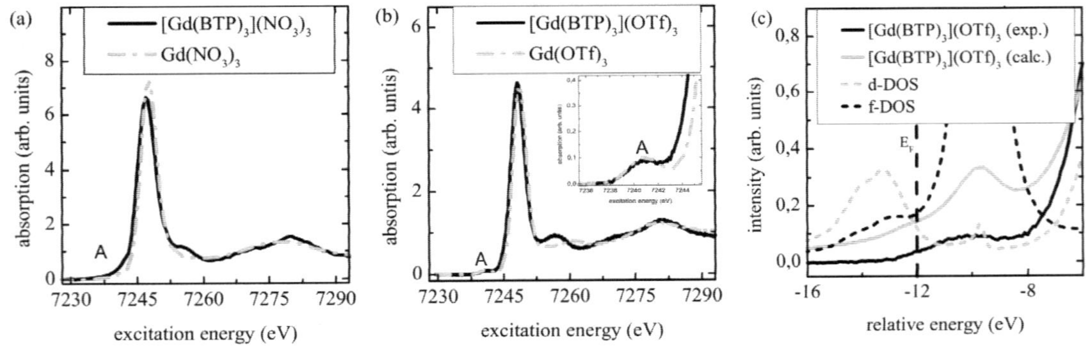

Figure 1. Gd L_3 edge HR-XANES spectra of (a) $[Gd(BTP)_3](NO_3)_3$ & $Gd(NO_3)_3$, (b) $[Gd(BTP)_3](OTf)_3$ & $Gd(OTf)_3$. Pre-edge calculated with *FEFF9.5* for (c) $[Gd(BTP)_3](OTf)_3$, including Gd f-, d-DOS.

3.2. N K edge

The N K edge XANES spectra a show a distinct pre-edge at about 399 eV (figure 2) arising from excitations of the N $1s$ electron to antibonding π^* orbitals. This feature can be successfully modeled with three PV functions. The $[Gd(BTP)_3](NO_3)_3$ pre-edge feature is shifted 0.1 eV to lower energies compared to the BTP, suggesting higher charge density on N in the complex (better screening of antibonding π^*; see figure 2 b). The energy difference between features 1 and 2 is 0.52 eV and 0.46 eV in the $[Gd(BTP)_3](NO_3)_3$ and the BTP spectra, respectively, and feature 1 has 25% lower intensity (area) compared to feature 2 in the $[Gd(BTP)_3](NO_3)_3$ compared to the BTP spectrum, which indicates variations in orbital mixing and occupancies. In figure 2 b, feature 3 is better resolved in the BTP spectrum. However, we noticed a dependence on exposure time for this feature indicating that this part of the spectrum is susceptible to radiation damage. The theoretical N 1s spectrum of H-BTP calculated with TD-DFT (see figure 2 c) is shifted about 13 eV to lower energies, due to the omission of corehole and relativistic effects and to approximations used in the functional [12]. The calculated excitation bands are grouped around three energy positions representing the three features observed in the measurement. Each band can be described as a mixture of excitations into the LUMO (lowest unoccupied molecular orbital) and LUMO+1 and +2. These orbitals are localized on all N atoms of the ligand, which makes it difficult to distinguish between the different Gd-N metal-to-ligand bonds and to make any conclusions about their properties (e.g., orbital mixing and charge transfer). In future studies, we hope to overcome these difficulties by investigating several structural motifs of the BTP architecture (e.g., the pyridyl or triazine moiety [13, 14]) and f-element complexes thereof.

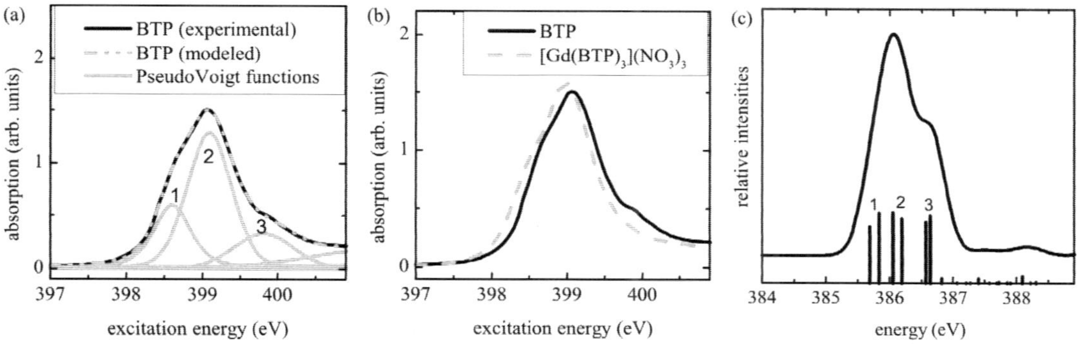

Figure 2. (a) an example of the pre-edge of N K edge XANES pre-peak modeled with PV; (b) N K XANES pre-peak of BTP and $[Gd(BTP)_3](NO_3)_3$; (c) TD-DFT calculation of N K H-BTP spectrum.

4. Conclusions

By performing quantum chemical calculations with the *FEFF9.5* code, we showed that the *Ln* L_3 HRXANES pre-edge feature in the $[Gd(BTP)_3](OTf)_3$ spectrum originates from a mixture of $2prightarrow4f$ and $2p \rightarrow 5d$ electronic transitions. Additionally, the 0.4 eV energy shift and the reduction in energy difference between pre-edge and WL by 0.4 eV in the $[Gd(BTP)_3](OTf)_3$ spectrum (see figure 1 b) indicated higher electronic density on the Gd in $[Gd(BTP)_3](OTf)_3$ compared to $Gd(OTf)_3$. No significant differences in electronic population of $4f$ and $5d$ states were detected for these materials. TD-DFT calculations of the N K pre-edge showed that the three absorption resonances associated with the pre-peak near 399 eV are due to 1s electron excitations to the LUMOs localized on all N atoms and that specialized investigations on single structural and chemical motifs are necessary to interpret variations in orbital mixing and charge transfer resulting from complexation of the ligand by an f-element.

Acknowledgments
We gratefully acknowledge KIT and Helmholtz Association of German Research Centers for financial support (VH-NG-734). This work is partially funded by the German Federal Ministry of Education and Research (BMBF) under contracts 02NUK020A, 02NUK012A and 02NUK012D. We thank HASYLAB, BESSY and ANKA for the granted beamtime.

References
[1] Kolarik Z, Mullich U and Gassner F 1999 *Solvent Extraction and Ion Exchange* **17** 1155–1170
[2] Geist A, Hill C, Modolo G, Foreman M R S J, Weigl M, Gompper K and Hudson M J 2006 *Solvent Extraction and Ion Exchange* **24** 463–483
[3] Drew M, Guillaneux D, Hudson M, Iveson P, Russell M and Madic C 2001 *Inorganic Chemnistry Communications* **4** 12–15
[4] Denecke M, Rossberg A, Panak P, Weigl M, Schimmelpfennig B and Geist A 2005 *Inorganic Chemnistry* **44** 8418–8425
[5] Banik N L, Schimmelpfennig B, Marquardt C M, Brendebach B, Geist A and Denecke M A 2010 *Dalton Transactions* **39**(21) 5117–5122
[6] Steppert M, Walther C, Geist A and Fanghanel T 2009 *New Journal of Chemistry* **33**(12) 2437–2442
[7] Welter E, Machek P, Drager G, Bruggmann U and Froba M 2005 *Journal Of Synchrotron Radiation* **12** 448–454
[8] Glatzel P, Sikora M, Smolentsev G and Fernández-García M 2009 *Catalysis Today* **145** 294 – 299
[9] Rehr J J, Kas J J, Prange M P, Sorini A P, Takimoto Y and Vila F 2009 *Comptes Rendus Physique* **10** 548 – 559
[10] Ressler T 1998 *Journal Of Synchrotron Radiation* **5** 118–122
[11] Becker U, Ganyushin D, Hansen A, Liakos D, Kollmar C, Kossmann S, Petrenko T, Reimann C, Riplinger C, Sivalingam K, Wezisla B and Wennmohs F 2010 *ORCA - An Ab Initio, DFT and Semiempirical electronic structure package* v2.8 ed
[12] Martin R L and Shirley D A 1977 *Many-electron theory of photoemission - Electron Spectroscopy, Theory, Techniques and Applications* vol 1 (Academic Press)
[13] Girnt D, Roesky P W, Geist A, Ruff C M, Panak P J and Denecke M A 2010 *Inorganic Chemnistry* **49** 9627–9635
[14] Bremer A, Geist A and Panak P J 2012 *Dalton Transactions* **41** 7582–7589

Identification of hexanuclear Actinide(IV) carboxylates with Thorium, Uranium and Neptunium by EXAFS spectroscopy

Christoph Hennig[1,*], Shinobu Takao[2], Koichiro Takao[3], Stephan Weiss[1], Werner Kraus[4], Franziska Emmerling[4], Michel Meyer[5] and Andreas C Scheinost[1]

[1] Helmholtz-Zentrum Dresden-Rossendorf, Institute of Resource Ecology, Bautzner Landstr. 400, D-01314 Dresden, Germany

[2] Department of Engineering Science, University of Electro-Communications, Chofu, Tokyo 182-8585, Japan

[3] Department of Materials and Life Science, Seikei University, 3-3-1, Kichijoji-kitamachi, Musashino-shi, Tokyo 180-8633, Japan

[4] BAM Federal Institute for Materials Research and Testing, Richard-Willstätter-Str. 11, D-12489 Berlin, Germany

[5] Institut de Chimie Moléculaire de l'Université de Bourgogne, UMR 6302 du CNRS, 9, avenue A. Savary, BP 47870, 21078 Dijon, France

E-mail: hennig@esrf.fr

Abstract. Hydrated actinide(IV) ions undergo hydrolysis and further polymerization and precipitation with increasing pH. The resulting amorphous and partly crystalline oxydydroxides $AnO_n(OH)_{4-2n}{\cdot}xH_2O$ can usually be observed as colloids above the An(IV) solubility limit. The aging process of such colloids results in crystalline AnO_2. The presence of carboxylates in the solution prevents the occurrence of such colloids by formation of polynuclear complexes through a competing reaction between hydrolysis and ligation. The majority of recently described carboxylates reveals a hexanuclear core of $[An_6(\mu_3\text{-}O)_4(\mu_3\text{-}OH)_4]^{12+}$ terminated by 12 carboxylate ligands. We found that the An(IV) carboxylate solution species remain often preserved in crystalline state. The An(IV) carboxylates show An–An distances which are ~ 0.03 Å shorter than the An–An distances in AnO_2 like colloids. The difference in the distances could be used to identify such species in solution.

1. Introduction

Carboxylates comprise a large group of organic ligands, including peptides and proteins as well as natural organic material like humic and fulvic acids. The carboxylates form stable complexes with various metal ions. These carboxylato complexes are often highly soluble in aqueous solution and have to be considered with respect to metal migration in the natural environment.

Previous investigations by potentiometric titration on tetravalent actinides with several carboxylates assumed exclusively mononuclear solution species [1]. The recent discovery of polynuclear An(IV) carboxylate complexes in aqueous solution [2–5] makes it likely that the thermodynamic data have to be reassessed at least to some extend. The discovery of such polynuclear complexes in solution

* To whom any correspondence should be addressed.

indicated that the species appear already in the aqueous phase and remain preserved in the crystalline state, which is in contrast to other solid structures with infinite $-(RCOO-M^{4+})_n-$ chains that appear only during the crystallization process.
Tetravalent actinides as U(IV), Th(IV), Np(IV) and Pu(IV) are strikingly alike with respect to their strong tendency to hydrolyze in aqueous solution. The hydrolysis products undergo polymerization, resulting in the formation of amorphous, nanometer sized $AnO_n(OH)_{4-2n}{\cdot}xH_2O$ particles which usually afford stable aqueous colloidal suspensions [6]. The complexation of An(IV) with carboxylates results in a competing reaction between hydrolysis and ligation [4]. The recently reported structures of polynuclear An(IV) carboxylates in aqueous solution were determined with EXAFS spectroscopy. The question arises if this technique is able to differentiate polymeric hydrolysis products, whose structure can be considered as precursors of AnO_2, from polynuclear An(IV) carboxylates. At the first glance it occurs not to be difficult to differentiate these structures, as each absorbing center is surrounded by 12 next An(IV) neighbors in the AnO_2 structures, while there are only 4 neighbors in the hexanuclear An(IV) carboxylate clusters. However, these coordination numbers can strongly deviate from the limiting values either through structural disorder [7, 8] or through pH dependent species fraction [4–6]. Therefore, an alternative way to distinguish these species might be to compare the interatomic metal-metal distances. The subsequent discussion is focused on that issue.

2. Results and Discussion

The hydrolyzed An(IV) ions undergo polymerization and condensation through different mechanisms. As such, olation and oxolation processes were suggested by Henry et al. [9].

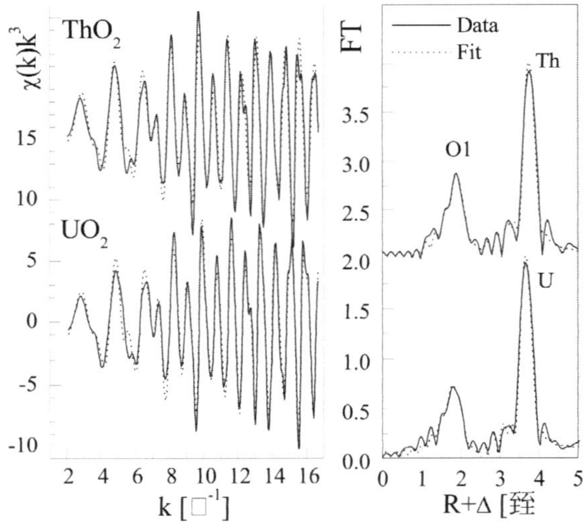

Figure 1. L_3-edge k^3-weighted EXAFS data (left) and corresponding Fourier Transforms (right) of ThO_2 and UO_2.

Olation results in a hydroxo bond between the metal ions. It originates from a nucleophilic attack of a hydroxylated metal ion onto a hydrated metal cation by replacement of the water molecule by a hydroxo bond between both metal centers ($M{-}OH + M{-}OH_2 \rightarrow M{-}OH{-}M + H_2O$). Oxolation appears if hydroxo groups of two hydroxylated metal ions react together by releasing a water molecule and formation of an oxo bond between the metal ions ($M{-}OH + OH{-}M \rightarrow M{-}O{-}M + H_2O$). Further polymerization results in formation of amorphous An(IV) oxyhydroxides $AnO_n(OH)_{4-2n}{\cdot}xH_2O$(am). With increasing pH there is a preferred tendency toward oxolation [9]. It has been observed that the amorphous structure undergoes an aging process that transforms the amorphous and disordered

$AnO_n(OH)_{4-2n}{\cdot}xH_2O$(am) structure into ordered and crystalline AnO_2(cr) [10]. In the AnO_2 structures each An^{IV} is coordinated by 8 oxygen atoms in the first coordination shell and surrounded by 12 An^{IV} atoms in the second coordination shell. The crystal structure can be described by a juxtaposition of octahedral polyhedra (Figure 3a) which are structurally similar to the octahedral core found in the hexanuclear An(IV) carboxylate clusters (Figure 3b, top). The structural features of selected AnO_2 samples were characterized by An L_3-edge EXAFS spectroscopy. Figure 1 displays the corresponding ThO_2 and UO_2 spectra. The metal-metal interaction under discussion is the peak at $R + \Delta = 3.6$ Å. The EXAFS data fit reveals a Th–Th distance of 3.957(2) Å for ThO_2, which is in close agreement with the reported crystallographic values, e.g. 3.957 Å [11], 3.958 Å [12] and 3.960 Å [13]. Similarly, the U–U distance deduced from the EXAFS data fit of a UO_2 sample, 3.867(2) Å, correspond well with that found by X-ray diffraction measurements, e.g. 3.868 Å [14] and 3.869 Å [13].

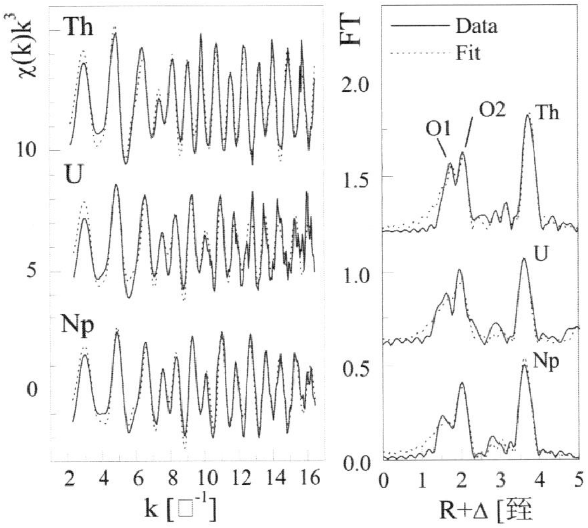

Figure 2. L_3-edge k^3-weighted EXAFS data (left) and corresponding Fourier Transforms (right) of An(IV) glycinates.

The presence of carboxylates prevents the formation of An(IV) oxyhydroxides. Indeed, solution studies of An(IV) carboxylates revealed several polynuclear species [2–5], most of them possessing a hexanuclear $[An_6(\mu_3\text{-}O)_4(\mu_3\text{-}OH)_4]^{12+}$ core terminated by 12 carboxylate ligands. Besides the above mentioned solution studies of An(IV) carboxylates, several solid-state structures of An(IV) carboxylates have been described in the literature [4, 15–17]. We report herein as examples for the hexanuclear An(IV) carboxylates three aqueous samples of An(IV) glycinates with An = Th(IV), U(IV) and Np(IV), investigated by EXAFS spectroscopy. The An L_3-edge EXAFS spectra of these glycinates are shown in Figure 2. Because the EXAFS data alone are not sufficient to determine the complete complex structure, a solid sample is used here as reference. The species $[Th_6(\mu_3\text{-}O)_4(\mu_3\text{-}OH)_4(H_2O)_6(Gly)_6(HGly)_6]^{6+}$ was preserved in a crystal [5]. The crystal structure of this compound (Figure 3b, bottom) consists of a hexanuclear $[Th_6(\mu_3\text{-}O)_4(\mu_3\text{-}OH)_4]^{12+}$ core with six Th(IV) atoms arranged at the corners of a nearly regular octahedron. The Th–Th edge distances are deviating from each other by less than 0.01 Å. Each face of the octahedron is bridged alternatively by either a μ_3-O or a μ_3-OH group. The twelve edges of the octahedron are bridged by chelating carboxylic groups of glycine ligands. Each Th atom shows a coordination number of 9 and is bound to two μ_3-O, two μ_3-OH, four oxygen atoms from the carboxyl groups of glycine, and a terminal water molecule. The different metal-oxygen distances involving either oxo, hydroxo, or carboxylate oxygen atoms give rise in the EXAFS spectra to a splitting of the oxygen shell at $R + \Delta = 1.9$ Å. Each corner atom of the hexanuclear cluster has 4 next An neighbors, as evidenced by the peak at $R + \Delta = 3.6$ Å. The extracted An–An distances for the various glycinate complexes are: 3.921(2), 3.829(2), and 3.816(2) Å for Th(IV), U(IV), and Np(IV) respectively. Most remarkably, these An–An distances are significantly different from the corresponding values found for AnO_2 samples.

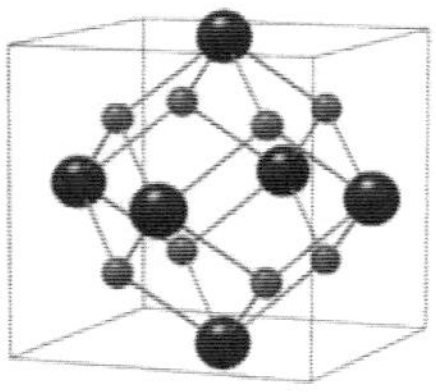

Figure 3a. Unit cell of AnO_2 with the section of the face-centered octatedral polyhedron.

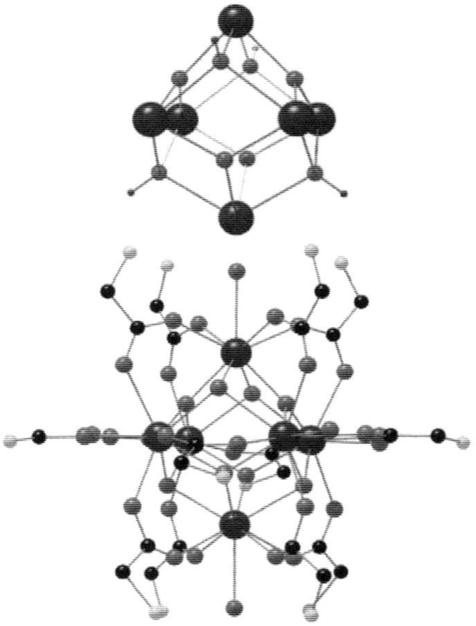

Figure 3b. Hexanuclear complex unit $[Th_6(\mu_3\text{-}O)_4(\mu_3\text{-}OH)_4(H_2O)_6(Gly)_6(HGly)_6]^{6+}$ (bottom) and the simplified $[An_6(\mu_3\text{-}O)_4(\mu_3\text{-}OH)_4(H_2O)_6]^{12+}$ core (top) from Ref. [6]

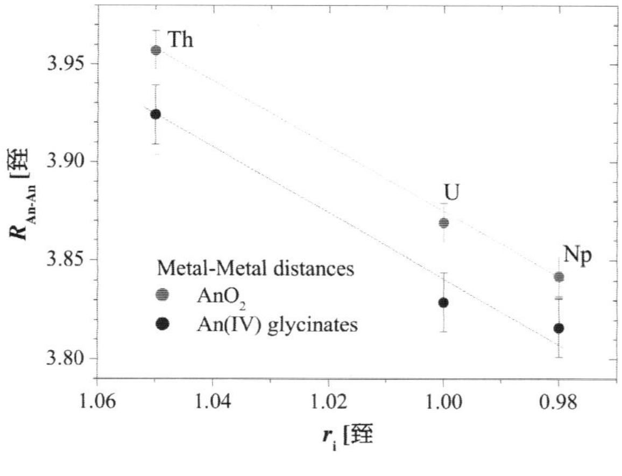

Figure 4. Correlation of the An–An distances in AnO_2 and hexanuclear An(IV) glycinates with the Shannon ionic radius r_i [18].

Figure 4 compares the An–An distances of AnO_2 (An = Th, U, and Np) with those found for the An(IV) glycinate complexes. The latter are ~ 0.03 Å shorter than the former, suggesting a tighter packing of the metal centers in the carboxylic-bridged structures. The same tendency can be observed for other An(IV) carboxylates [4, 5].

3. Conclusion

The presence of carboxylates results in a competing reaction between hydrolysis and ligation affording hexanuclear complexes. These An(IV) carboxylate clusters prevail in solution at pH values where the olation and oxolation reaction products as $AnO_n(OH)_{4-2n}{\cdot}xH_2O$(am) and AnO_2(cr) would appear in the absence of carboxylic or amino acids. Characteristic features for $[An_6(\mu_3\text{-}O)_4(\mu_3\text{-}OH)_4(H_2O)_6(L)_{12}]$ structural units are: a specific splitting of the An–O shell, an An^{4+} coordination number of 9, and An–An distances which are ~ 0.03 Å shorter than in AnO_2.

4. Acknowledgements

We thank Philippe Marin (CEA Cadarache) for providing the UO_2 sample.

5. References

[1] Casellato U and Vigato P A 1978 *Chem. Rev.* **26** 85
[2] Toraishi T, Farkas I, Szabo Z and Grenthe I 2002 *J. Chem. Soc., Dalton Trans.* 3805
[3] Takao S, Takao K, Kraus W, Emmerling F, Scheinost A C, Bernhard G and Hennig C 2009 *Eur. J. Inorg. Chem.* 4771
[4] Takao K, Takao S, Scheinost A C, Bernhard G and Hennig C 2012 *Inorg. Chem.* **51** 1336
[5] Hennig C, Takao S, Takao K, Weiss S, Kraus W, Emmerling F, Scheinost A C 2012 *J. Chem. Soc., Dalton Trans.* **41** 12818
[6] Rothe J, Denecke M A, Neck V, Müller R and Kim J I 2002 *Inorg. Chem.* **41** 249
[7] Rothe J, Walther C, Brendebach B, Büchner S, Fuss M, Denecke M A and Geckeis H 2009 *J. Phys.: Conf. Ser.* **190** 012188;
[8] Hennig C, Weiss S, Banerjee D, Brendler E, Honkimäki Cuello G, Ikeda-Ohno A, Scheinost A C, Zänker H (2012) *Geochim. Cosmochim. Acta* DOI 10.1016/j.gca.2012.10.051
[9] Henry M, Jolivet J P and Livage J 1992 *Structure and Bonding* **77** 153
[10] Schofield E J, Veeramani H, Sharp J O, Suvorova E, Bernier-Latmani R, Mehta A, Stahlman J, Webb S M, Clark D L, Conradson S D, Ilton E S and Bargar J R 2008 *Environ. Sci. Technol.* **42** 7898
[11] Leigh H D and McCartney E R 1974 *J. Am. Ceram. Soc.* **57** 192
[12] Taylor D 1984 *Transactions and Journal of the British Ceramic Society* **83** 32
[13] Idiri M, Le Bihan T, Heathman S and Rebizant J 2004 *Phys. Rev. B* **70** 014113
[14] Lander G H and Mueller M H 1974 *Phys. Rev. B* **10** 1994
[15] Mougel V, Biswas B, Pécaut J and Mazzanti M 2010 *Chem. Commun.* **46** 8648
[16] Ok K M and O'Hare D 2008 *Dalton Trans.* 5560
[17] Knope K E, Wilson, R E, Vasiliu M, Dixon D A and Soderholm L 2011 *Inorg. Chem.* **50** 9696
[18] Shannon R D 1976 *Acta Cryst.* **A32** 751

Journal of Physics: Conference Series **430** (2013) 012117 doi:10.1088/1742-6596/430/1/012117

Actinide and lanthanide speciation with high-energy resolution X-ray techniques

T Vitova[1], M A Denecke[1], J Göttlicher[2], K Jorissen[3], J J Kas[3], K Kvashnina[4], T Prüßmann[1], J J Rehr[3] and J Rothe[1]
[1]Karlsruhe Institute of Technology, Institute for Nuclear Waste Disposal, P.O. Box 3640, D-76021 Karlsruhe, Germany
[2]Karlsruhe Institute of Technology, Institute for Synchrotron Radiation, P.O. Box 3640, D-76021 Karlsruhe, Germany
[3]Department of Physics, University of Washington, Seattle, WA 98195, USA
[4]European Synchrotron Radiation Facility (ESRF), 6 Rue Jules Horowitz, BP 220, 38043 Grenoble Cedex 9, France

E-mail: Tonya.Vitova@kit.edu

Abstract. High-energy resolution X-ray absorption spectroscopy (HR-XAS) and Resonant inelastic X-ray scattering (RIXS) combined with quantum theoretical tools are gaining importance for understanding electronic and coordination structures of actinide (An) and lanthanide (Ln) materials. HR-XAS is successfully used to remove lifetime broadening by registering the partial fluorescence yield emitted by the sample, thereby yielding highly resolved X-ray absorption near edge structure spectra (HR-XANES), which often display resonant features not observed in conventional XANES. We demonstrate the structural characterization capabilities of these novel techniques by comparative discussion of U M_4/L_3-HR-XANES and L_3-valence band RIXS (L_3-VB-RIXS) spectra of two model U(VI) minerals. We show that the *ab initio* multiple scattering theory based code *FEFF9.5* is an effective tool for calculations of An and Ln L_3-HR-XANES and L_3-RIXS spectra as it successfully reproduces dipole and quadrupole transitions in the same spectrum.

1. Introduction

High-energy resolution X-ray absorption near edge structure spectroscopy (HR-XANES) at the actinide (An), lanthanide (Ln) L_3 edge and An $M_{4,5}$ as well as An/Ln L_3 valence-band and core-to-core resonant inelastic X-ray scattering (VB-RIXS, CC-RIXS) are currently developing techniques providing new insights into the An and Ln electronic structure [1,2]. The highly efficient X-ray emission spectrometers developed in the last years minimise the experimental spectral broadening to values significantly smaller than the core-hole lifetime broadening and thereby approach the energy resolution limit for HR-XANES spectra [3]. These instruments are based mostly on an array of bent analyser crystals in a Rowland circle geometry diffracting and focusing the fluorescence onto a detector and, for example, resolve pre-edge structures in Ln and U L_3 edge HR-XANES (L_3-HR-XANES) spectra [1,2]. Only few HR-XANES for U materials have been reported so far and the U pre-edge is clearly visible in U(VI) spectra [1]. This spectral resonance is interpreted as resulting from $2p_{3/2}\rightarrow 5f$ electronic transitions, as corroborated by calculations with the *FDMNES* code. However, the calculated energy difference between the pre-edge and most intense resonance (white line, WL) is smaller (4.5 eV) than the experimental one (7 eV) [1]. We present calculations with the *ab initio FEFF9.5* code [4], which agree well with the experimental spectrum. Additionally, we demonstrate that *FEFF9.5* is able to calculate the pre-edge ($2p_{3/2} \rightarrow 4f/5f$) and the WL ($2p_{3/2} \rightarrow 5d/6d$) in both the An/Ln L_3-HR-XANES and Ln L_3-CC-RIXS spectra, whereas other theoretical approaches calculate these absorption resonances separately [2]. The capability of the HR-XANES technique to measure the relative energy difference between An/Ln *f* and *d* states in a system of materials with small structural variations is extremely valuable for tuning theoretical approaches. It will also shed light onto the degree of stabilisation of the An *5f* valence orbitals across the actinide series and onto the amount of mixing in molecular orbitals between An/Ln valence with ligand orbitals. We also compare and discuss the sensitivity of the U L_3/M_4-HR-XANES and L_3-VB-RIXS to geometric structural changes for the same U(VI) oxidation state.

2. Experimental part and calculation details

The U L_3/M_4-HR-XANES, L_3-VB-RIXS of $Ca(UO_2)_2(PO_4)2{\cdot}10\text{-}12(H_2O)$ (meta-autunite), $NaCa_3(UO_2)(CO_3)_3(SO_4)F{\cdot}10(H_2O)$ (schroeckingerite) and Yb L_3-HR-XANES, L_3-CC-RIXS of Yb_2O_3 were recorded at the ID26 beamline, ESRF, Grenoble [3]. The synchrotron radiation was monochromatized by a Si(311) (U L_3 17166 eV, Yb L_3 8944 eV)/Si(111) (U M_4 3728 eV) double crystal monochromator (DCM). The experimental energy resolution was about 1.8 eV (U L_3), 0.8 eV (U M_4) or 1.2 eV (Yb L_3), measured as the width of the quasi-elastic peak. For each excitation energy, the emitted photons from the sample were diffracted by spherically bent Ge(777) (U $L\alpha_1$ 13614 eV), Si(220) (U M_β 3339.8 eV) or Si(620) (Yb $L\alpha_1$ 7416 eV) analyzer crystals and focused onto a X-ray silicon drift detector (SDD KETEK). The sample, crystal and detector were positioned in Rowland circle geometry with 1m diameter, equal to the bending radius of the crystals. The conventional fluorescence mode U M_4-XANES of $(UO_2)_8O_2(OH)_{12}{\cdot}12H_2O$ (schoepite) was recorded at the INE-Beamline, ANKA, Karlsruhe [5] using a pair of Si(111) crystals in the DCM and detecting U M_β fluorescence with a photo diode. The U/Yb L_3-HR-XANES and Yb L_3-CC-RIXS calculations were performed with the *ab initio* multiple-scattering (MS) theory based *FEFF9.5* code [4]. The atomic potentials were calculated self consistently on a cluster of 260 (U)/100 (Yb) atoms in the presence of a core-hole (U)/RPA (Yb) [4]. The exchange correlation Hedin-Lundquist potential was employed, reducing the $2p_{3/2}$ core-hole lifetime broadening (U: 8.1 eV, Yb: 4.3 eV) to 2.8 eV (U)/2 eV (Yb) and correcting the Fermi energy by -1 eV (U)/-8 eV(Yb). Full MS calculations were done on a cluster of 250 (U)/100 (Yb) atoms. Dipole ($2p_{3/2} \rightarrow 5d/6d$) and quadrupole transitions ($2p_{3/2} \rightarrow 4f/5f$) were taken into account. The Yb L_3-CC-RIXS calculations were performed by using the RIXS card (see [6]).

3. Results and discussion

The calculated spectra of Yb L_3-HR-XANES, L_3-CC-RIXS of Yb_2O_3 and U L_3-HR-XANES, U unoccupied *f* and *d* angular momentum projected density of states (*d* and *f*-DOS) of schroeckingerite are plotted in figure 1 a, b and figure 2 d. The *FEFF9.5* code successfully reproduces all spectral features at correct energy positions for U and at positions differing within few eVs from the experiment for Yb; some intensity differences are also observed. Yb_2O_3 characterises with Yb1 and Yb2 sites in octahedral and distorted octahedral local coordination environment, respectively (ICSD 163730). The weighted Yb1 and Yb2 calculated spectra including only dipole or quadruple transitions are shown in figure 1 a. The Yb L_3-HR-XANES spectrum is a linear combination of the two calculated spectra (1/4 Yb1 and 3/4 Yb2). The A region from the pre-edge structure originates mainly from $2p_{3/2} \rightarrow 4f$ electronic transitions but $2p_{3/2} \rightarrow 5d$ transitions are also well distinguishable. The energy difference between features A and B2 (13.4 eV) is equal, whereas between B1 and B2 (2.8 eV) and B2 and C (12.1 eV) is smaller than in the experimental spectrum (B2-A=13.4 eV, B2–B1=3.3 eV, C-B2=16.7 eV). The Yb $L\alpha_1$ on the energy transfer scale is plotted as a function of the excitation energy in figure 1 b (Yb L_3-CC-RIXS). The Yb_2O_3 L_3-CC-RIXS has intense pre-edge structure at about 8935 eV and a double structure in the WL region. The left RIXS plot shows the *FEFF9.5* calculation result using only dipole transitions ($2p_{3/2} \rightarrow 5d$), the middle plot the result obtained including quadrupole

Journal of Physics: Conference Series **430** (2013) 012117 doi:10.1088/1742-6596/430/1/012117

transitions ($2p_{3/2}\rightarrow 4f$). The pre-edge absorption resonance is weak in the left plot (not visible on this scale) but has higher intensity and position comparable to the experimental spectrum (right) in the middle plot suggesting $2p_{3/2}\rightarrow 4f$ dominates but mixes with $2p_{3/2}\rightarrow 5d$ transitions in this part of the spectrum. This result partially confirms previous atomic multiplet calculations attributing the pre-edge to $2p_{3/2}\rightarrow 4f$ electronic transitions, which however calculated only the pre-edge structure [2]. In case of uranium, $2p_{3/2}$ electronic transitions to *5f* valence states contribute to the pre-edge region at about -9 eV, as at this energy the U *f*-DOS has large intensity (see figure 2 d). The WL and post-edge regions of the U L_3-HR-XANES spectrum is well described by dipole $2p_{3/2}\rightarrow 6d$ transitions and resemble the U *d*-DOS as expected.

We use two U(VI) minerals, meta-autunite and schroeckingerite, to compare the structural sensitivity of U M_4/L_3-HR-XANES and U L_3-VB-RIXS spectra. Meta-autunite and schroeckingerite possess tetragonal (P4/nmm) (ICSD 20378) and triclinic (P -1) (AMCSD 0015705) crystal structures, respectively. In both mineral phases $U(VI)O_2^{2+}$ dioxo cations are coordinated solely by O atoms in the equatorial plane (figure. 2c,g). Second coordination neighbours are P/O atoms (meta-autunite) or C atoms (schroeckingerite). In figure 2a, the meta-autunite and schroeckingerite U L_3-HR-XANES spectra (2a top) are compared to their conventional fluorescence XANES spectra (2a bottom). All U L_3-HR-XANES features are sharper than in the conventional XANES and an additional pre-edge feature A is resolved. The U M_4-HR-XANES of these minerals also exhibit significantly improved energy resolution (figure 2f) which is shown here for schoepite where the distinct fine structure in the M_4-HR-XANES, is not resolved in the conventional fluorescence XANES (figure 2e). Only subtle differences between the mineral phases are observed in their U M_4-HR-XANES spectra such as the energy shift of feature D (figure. 2f inset). In contrast, the different U coordination environments in these mineral structures are clearly reflected in distinct variations of their U L_3-HR-XANES features. The An $M_{4,5}$-edges are of special interest as they directly probe unoccupied An *5f* electronic state densities ($3d\rightarrow 5f$ transitions) and potentially can reveal subtle bonding differences such as metal-to-ligand charge donation. In addition, An M_4-XANES is expected to be especially useful for An speciation, since the reduced core-hole life-time broadening at the M_4 edge compared to the L_3 edge theoretically sharpens spectral features, thereby facilitating detection of minority valence species in oxidation state mixtures. However, our results on these two minerals indicate that the An L_3 edge might be more sensitive to changes in coordination environment for the same An oxidation state. For example, feature C in the U L_3-HR-XANES spectrum is an indicator for the U-O axial distance and shifts to lower energy upon lengthening of one or both U-O axial bonds [1]. The longest U-O axial distance in meta-autunite is about 0.2 Å longer than in schroeckingerite; however, feature C does not shift detectable to lower energies (figure. 2a). The U L_3-HR-XANES spectrum of schroeckingerite is calculated by elongation of the U-O1 bond from 1.8 to 2.5 Å compared to the spectrum of the unmodified structure in figure 2 h. Moreover, the pre-edge feature A has higher intensity, whereas the intensity of the WL (feature B) decreases. This trend is also visible in the experimental spectrum of meta-autunite compared to schroeckingerite. Notice that the reduction of WL intensity is stronger in the experiment than in the calculation, suggesting influence of additional structural effects. Nevertheless, this simulation implies that features A, B and C in L_3-HR-XANES can assist characterizing U-O bond distances more effectively than conventional XANES or M_4-HR studies.

Figure 1 b shows the U L_3-VB-RIXS spectra on the energy transfer scale of meta-autunite and schroeckingerite measured at an excitation energy at the maximum of the U L_3-HR-XANES WL (17176 eV). The intense peak at 0 eV is due to the elastically scattered incident photons. In the final state, the excited U atom has an electron promoted to an empty *6d* states and a core-hole in the filled part of the VB; this mimics a direct excitation from the VB [7]. Therefore, the U L_3-VB-RIXS spectrum, with maximum at about 12.7 eV, describes unoccupied *6d* states. The energy difference between the elastic peak and the VB-RIXS spectrum measures the energy gap between the highest occupied state and the lowest unoccupied valence state [7]. The U L_3-VB-RIXS of meta-autunite exhibits a broader VB feature compared to schroeckingerite, suggesting a larger energy spread of *6d* states and maybe less mixing with ligand valence orbitals. Comparison of the U L_3-HR-XANES WL areas, which is proportional to the oscillatory strength of $2p_{2/3}$ electronic transitions to U *6d* states, corroborates this interpretation; the U *d* states in the U L_3-HR-XANES of meta-autunite have larger energy spread than schroeckingerite. Attempts to quantify this difference by fitting the U L_3-HR-XANES remain unsuccessful, as no unique model using Pseudo Voigt functions was found.

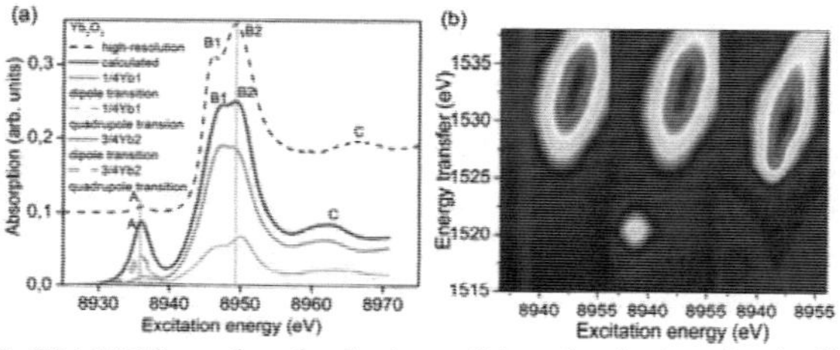

Figure 1. (a) Yb L_3-HR-XANES and calculation, (b) calculation including dipole (left) or dipole and quadrupole (middle) transitions, Yb L_3-CC-RIXS (right) of Yb_2O_3.

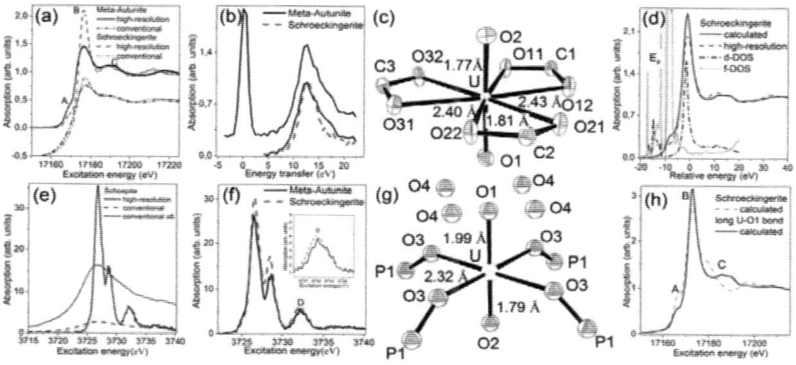

Figure 2. (a) U L_3-HR and -XANES, (b) U L_3-VB-RIXS, 3.6 Å atomic cluster for schroeckingerite (c) and meta-autunite (g), (d) U L_3-HR-XANES and (h) calculation for schroeckingerite, (e) U M_4-HR and -XANES for schoepite, (f) U M_4-HR-XANES of meta-autunite and schroeckingerite.

4. Conclusions

The *FEFF9.5* code emerges as a very useful tool for calculations of An/Ln L_3-HR-XANES and RIXS spectra, as it successfully reproduces all spectral features in both. The U(VI) L_3-HR-XANES spectra tend to be more sensitive to changes in geometric structure compared to U M_4-HR-XANES and VB-RIXS for the same U oxidation state. However, the techniques are complimentary and coupled to *ab initio* quantum chemical calculations provide detailed geometric and electronic structural information, which, e.g., can help to explain reduction mechanisms of Ln(III) to Ln(II) and can also be valuable for understanding the selectivity of N-donor ligands to complex An(III) over Ln(III) ions relevant for the separation and transmutation strategy for reduction the radiotoxicity of spent nuclear fuel.

Acknowledgments: We gratefully acknowledge KIT and the Helmholtz Association of German Research Centers for the financial support (VH-NG-734). We thank ESRF and ANKA for the granted beamtime.

References

[1] Vitova T *et al.* 2010 *PRB* **82** 235118
[2] Kvashnina K O, Butorin S M and Glatzel P 2011 *J. Anal. At. Spectrom.* **26** 1265
[3] Glatzel P and Bergmann U 2005 *Coord. Chem. Rev.* **249** 65
[4] Rehr J J, Kas J J, Prange M P, Sorini A P, Takimoto Y and Vila F 2009 *C. R. Physique* **10** 548 59
[5] Rothe J *et al.* 2012 *Rev. Sci. Instrum.* **83** 043105
[6] Kas J J, Rehr J J, Soininen J A, and Glatzel P 2011 *PRB* **83** 235114
[7] Glatzel P, Singh J, Kvashnina K O and van Bokhoven J A 2012 *JACS* **132** 2557

Journal of Physics: Conference Series **430** (2013) 012118 doi:10.1088/1742-6596/430/1/012118

Interatomic distances for overlapping shells in disordered systems: model-less approach

Yuri Babanov, Dmitry Ponomarev and Yuri Salamatov

Institute of Metal Physics, Russian Academy of Sciences, 18 S.Kovalevskaya str., 620990 Ekaterinburg, Russia

E-mail: babanov@imp.uran.ru

Abstract. The regularization method of solving ill-posed problem is used to determine five partial interatomic distances from combined two EXAFS spectra without building any model of atomic structure. The mathematical procedure and the experimental results of the EXAFS analysis for $Ni_cZn_{1-c}O$ (c = 0.0, 0.3, 0.5, 0.7, 1.0) solid solutions with the rock salt (rs) crystal structure synthesized from the binary oxide powders are discussed. The measurements were performed using synchrotron facilities (Russian Research Centre "Kurchatov Institute", Moscow). The Ni and Zn K absorption spectra were recorded in transmission mode at room temperature. It was shown the ideal rock salt lattice is distorted and long-range order exists only at the average (Vegard law).

1. Introduction

A series of the $Ni_cZn_{1-c}O$ solid solutions with the rock salt (rs) crystal structure belongs to the family of wide-band-gap semiconductors with strong ionic character of chemical bonds. They are of great interest both from a fundamental and an applied point of view. These solutions has been used in wide range of applications such as p-type transparent conducting film, chemical sensors, electro-chromic and high-speed devices. Recent EXAFS studies show that local structure in disordered systems does not coincide with information obtained from X-ray diffraction [1-3]. It has been shown that Vegard's law is locally not obeyed.

In the case of the systems with a disorder, it has been known for some time [1-4] that treating EXAFS data using a standard analysis may cause large errors. There is a serious problem for the systems like $Ni_cZn_{1-c}O$ solid solutions which is characterized by the coordination spheres, overlapping in r-space. In contrast to the conventional Fourier transformation and fitting we propose a new procedure of obtaining local information for three-component solid oxide solutions. For the first time, the method of combining two EXAFS spectra for two-component system (Ge-Si disordered alloys) was presented in [1]. Later this method was applied for $ZnBr_2$ aqueous solution [4].

This paper reports the results of EXAFS investigation of the local structure for rs $Ni_cZn_{1-c}O$ solid solutions. There is some disagreement in the literature concerning the first shell structure of rs $Ni_cZn_{1-c}O$ solutions; in particular whether or not there is a measurable variation of the nearest-neighbor interatomic distances as a function of alloy composition.

2. Mathematical algorithm

Let us consider a solid solution based on the oxides NiO / ZnO. These solid solutions were prepared by mixing the oxide powders in appropriate proportions: $(NiO)_c(ZnO)_{1-c} \equiv Ni_cZn_{1-c}O$.

Let Ni - 1, Zn - 2, O - 3. The local atomic structure of a homogeneous three-component material can be described by five partial pair correlation functions: g_{11} (r), g_{12} (r) = g_{21} (r), g_{13} (r) and g_{21} (r) = g_{12} (r), g_{22} (r), g_{23} (r). The function $g_{ij}(r)$ is the probability to find atom j at the distance r from atom i, by definition. The integral equations for two EXAFS spectra at Ni and Zn K absorption edges: $\chi_1(k) \equiv \chi_{Ni}(k)$ and $\chi_2(k) \equiv \chi_{Zn}(k)$ - can be written as follows

$$k\chi_1(k) = 4\pi\rho_0 \left\{ \begin{array}{l} \int_0^\infty c g_{11}(r) f_1(k,r) \exp[-2r/\lambda_1(k)] \sin[2kr + \psi_{11}(k,r)] dr + \\ (1-c)\int_0^\infty g_{12}(r) f_2(k,r) \exp[-2r/\lambda_1(k)] \sin[2kr + \psi_{12}(k,r)] dr + \\ \int_0^\infty g_{13}(r) f_3(k,r) \exp[-2r/\lambda_1(k)] \sin[2kr + \psi_{13}(k,r)] dr \end{array} \right\}$$

$$k\chi_2(k) = 4\pi\rho_0 \left\{ \begin{array}{l} c\int_0^\infty g_{21}(r) f_1(k,r) \exp[-2r/\lambda_2(k)] \sin[2kr + \psi_{21}(k,r)] dr + \\ (1-c)\int_0^\infty g_{22}(r) f_2(k,r) \exp[-2r/\lambda_2(k)] \sin[2kr + \psi_{22}(k,r)] dr + \\ \int_0^\infty g_{23}(r) f_3(k,r) \exp[-2r/\lambda_2(k)] \sin[2kr + \psi_{23}(k,r)] dr \end{array} \right\} \quad (1)$$

Here c - the atomic concentration of molecules such as NiO, f_1(k, r) - modulus of the amplitude of backscattering, $\psi_{12}(k, r) = 2\delta_1(k) + \varphi_2(k, r)$ - the total phase shift for the ion pair 1-2 which is a peculiar "mark" for a specific pair of atoms, $\lambda_1(k)$ - the mean free path of the photoelectron ejected from the ion of type 1. All atomic scattering characteristics (amplitude functions, phase functions and mean free paths) were calculated by FEFF-8.1 code [5].

The integral operator A_{ij} generates a contribution of a specific atomic pair ij to the normalized oscillating part, as it acts on the partial PCF g_{ij}. We propose to use the dependence of the integral operator A_{ij} on the atomic scattering characteristics $\delta_{ii}(k)$ and $\phi_j(k,r)$ for determining partial PCFs in multicomponent systems from EXAFS data.

Let us assume for simplicity: $g_1(r) = g_{13}(r) = g_{NiO}(r)$, $g_2(r) = g_{11}(r) = g_{NiNi}(r)$, $g_3(r) = g_{12}(r) = g_{NiZn}(r)$, $g_4(r) = g_{22}(r) = g_{ZnZn}(r)$, $g_5(r) = g_{23}(r) = g_{ZnO}(r)$. In operator notation Eq.1 may be rewritten as

$$\begin{aligned} u_1 &= (A_{11}g_1 + A_{12}g_2 + A_{13}g_3) \\ u_2 &= (A_{23}g_3 + A_{24}g_4 + A_{25}g_5) \end{aligned} \quad (2)$$

Here vectors u_1, u_2 – input EXAFS data for both edges.

For the numerical solution of integral equations system (2) can be written as

$$\begin{bmatrix} u_1 \\ u_2 \end{bmatrix} = \begin{bmatrix} A_{11} & A_{12} & A_{13} & 0 & 0 \\ 0 & 0 & A_{23} & A_{24} & A_{25} \end{bmatrix} \times \begin{bmatrix} g_1 \\ g_2 \\ g_3 \\ g_4 \\ g_5 \end{bmatrix} \quad \text{or } u = Ag. \tag{3}$$

where A is an integral operator, g is an unknown function describing 5 functions g_{ij}, u is an input (experimental) data for two EXAFS spectra. The numerical solution of Eq. 3 is an ill-posed problem. In this paper, the Tikhonov's method of regularization is used [6]. A full description of the mathematical scheme for one EXAFS spectrum and 3 unknown functions can be found in Ref. [7].

3. Experimental

A series of metastable Ni_cZn_{1-c} O solid solutions with the rock salt crystal structure have been synthesized from the binary oxide powders NiO and ZnO by quenching from 7.7 GPa and 1450-1650 K. Phase composition of the samples, as well as structural properties and stoichiometry of synthesized solid solutions have been studied by X-ray powder diffraction, both conventional and with synchrotron radiation [8]. The widest (c = 0.0, 0.3, 0.5, 0.7, 1.0) composition range of the existence of individual rock salt solid solution has been established for the NiO-ZnO system. All samples were prepared at Moscow State University [8].

The measurements were performed using synchrotron facilities (The Structural Materials Science end-station at the Kurchatov Centre for Synchrotron Radiation and Nanotechnology (KCRNT), Moscow, Russia).

The Ni and Zn K edge absorption spectra were recorded in transmission mode at room temperature for rs $Ni_cZn_{1-c}O$ (c = 0.0, 0.3, 0.5, 0.7, 1.0) solid solutions.

4. Results and discussion

Fig. 1 shows EXAFS spectra for two edges: Ni and Zn. These data were used for determination partial pair correlation functions of ternary oxide solutions as a result of solution for inverse ill-posed problem.

The Tikhonov's solutions for rs $Ni_{0.3}Zn_{0.7}O$ solid solution are depicted on Fig. 2. The maximal values of these curves correspond to interatomic distances. The numerical results obtained for all samples are given in Table 1. It is obvious, that we have a high resolution for the overlapping coordination spheres. The second feature - interatomic distances for pairs NiO and ZnO are almost independent of solution composition of our alloys and are close to the crystallographic distances for NiO (2.088 Å) and ZnO (2.140 Å). This suggests that for the rs NiO/ZnO oxide solid solutions it may be valid Pauling model [9] (see Table 2):

$$a_c = 2\{cr_O + [cr_{Ni} + (1-c)r_{Zn}]\}. \tag{4}$$

Here r_O, r_{Ni} and r_{Zn} are ionic radii of O^{2-}, Ni^{2+} and Zn^{2+}.

In order to obtain the lattice parameters for rs-NiO/ZnO solid solutions from EXAFS data as a function of concentration c it is necessary to submit the values of bond lengths R_{NiO} and R_{ZnO} from Tabl.1 according to formula:

$$a_c = 2\{cR_{NiO} + (1-c)R_{ZnO}\}. \tag{5}$$

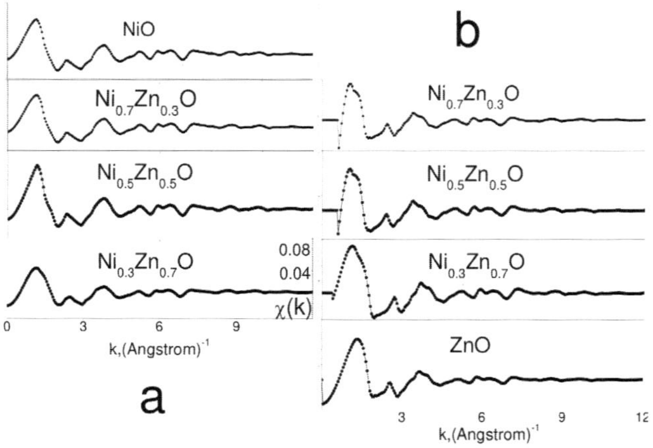

Figure 1. Experimetal EXAFS $\chi(k)$ specta in rs $Ni_c,Zn_{1-c}O$: a) Ni K-edge, b) Zn K-edge.

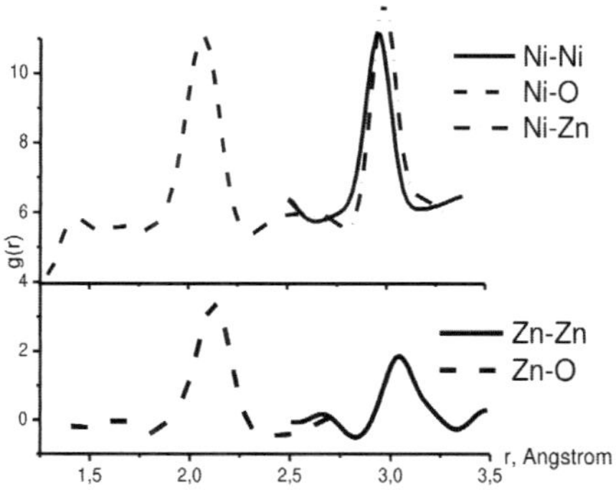

Figure 2. Partial PCFs (Tikhonov's solution) for rs-$Ni_{0.3}Zn_{0.7}O$

Table 1. Bond lengths for rs-NiO/ZnO solid solutions.

Sample	Ni-O	Ni-Ni	Ni-Zn	Zn-Zn	Zn-O
NiO	2,079	2,952	-	-	-
Ni0,7Zn0,3O	2,078	2,947	3,014	3,050	2,106
Ni0,5Zn0,5O	2,071	2,952	2,983	3,041	2,119
Ni0,3Zn0,7O	2,083	2,966	2,989	3,025	2,126
ZnO	-	-	-	3,017	2,135

Table 2. Lattice parameter for rs-NiO/ZnO solid solutions.

Sample	Diffraction	Pauling	EXAFS
NiO	4,176	4,18	4,16
Ni0,7Zn0,3O	4,2088	4,21	4,1847
Ni0,5Zn0,5O	4,22887	4,23	4,1904
Ni0,3Zn0,7O	4,24902	4,25	4,2104
ZnO	4,28	4,28	4,26

- A method for determining partial interatomic distances from combined two EXAFS spectra for three-component solid oxide solutions has been presented.
- We have demonstrated a high resolution for bond lengths of close spacing atoms (Ni, Zn) in the Periodical Table.
- These results were obtained without any assumptions concerning interatomic distances for multicomponent systems

Acknowledgments

The authors would like to thank Andrey Baranov and Petr Sokolov (Moscow State University) for a preparation of the samples and also Yan Zubavichus (RRC "Kurchatov Institute") for an assistance in performing the experiment. This work was partially supported by grant № 12-П-2-1032 (the program of the Presidium of the RAS).

References

[1] Babanov Y, Deev A, Ruts Yu *Nuclear Instruments and methods* A 2000 **448** 368-371.
[2] Kuzmin A *et al J. Phys.: Condens. Matter* 1995 **7** 9357-9368.
[3] Lebedev A, Sluchinskaya I and Munro I *J. Synchrotron Rad.* 2001 **8** 800–802.
[4] Babanov Y, Kamensky I , Hazemann J, Calzavarac Y, Raoux D *NIM* A 2007 **575** 155–158.
[5] Zabinsky S, Rehr J, Ankudinov A, Albers R, Eller M, *Phys. Rev.* B 1995 **52** 2995.
[6] Tikhonov A and Arsenin V 1981 *Solution of Ill-Posed Problems* (NY John Willey & Sons, Inc.).
[7] Babanov Y, Salamatov Y *et al J. Electron Spectroscopy and Rel. Phenom.* 2009 **175** 27–30
[8] Baranov A *et al High pressure Research* 2008 **28** 515-519.
[9] Pauling L 1953 *The Nature of the Chemical Bond* (Ithaca, NY: Cornell University Press).

Journal of Physics: Conference Series **430** (2013) 012119 doi:10.1088/1742-6596/430/1/012119

An in-situ heater for the XAS beamline (12-ID) in Australia

B Johannessen[1], Z S Hussain[2], D R East[2] and M A Gibson[2]

[1] Australian Synchrotron, 800 Blackburn Rd, Clayton VIC 3168, Australia
[2] CSIRO, Division of Process Science and Engineering, Normanby Rd, Clayton VIC 3168, Australia

E-mail: Bernt.Johannessen@synchrotron.org.au

Abstract. To accommodate for a growing number of requests by our user community an in-situ heater has been commissioned for the X-ray absorption spectroscopy (XAS) beamline 12-ID at the Australian Synchrotron. Here, we present an in-situ method for calibrating the temperature of the heating stage based on an anharmonic, correlated Einstein model. Specifically, we show that a temperature-dependant study of a bulk metallic foil (7.5 μm Cu) can be used to accurately calibrate the temperature of the heater. We also present the temperature-dependant coordination number, bond length, Debye-Waller factor, and third order cumulant to the bond length distribution function of the material from 18K to 1074K. At the higher temperatures we find that the atomic structure is comparable to that of an amorphous or liquid material indicating a gradual transition from crystalline to disordered atomic structure.

1. Introduction

Temperature-dependant X-ray absorption spectroscopy (XAS) is now a sought after technique in a range of fields, including solid-state physics to investigate issues such as vibrational properties of Co nanoparticles [1], chemical disorder in bulk amorphous GaSb [2], and the bonding environment in InN epilayers [3]. Many XAS beamlines around the world now have in-situ heaters enabling measurements to be performed at above room temperature, including beamline BM23 at the European Synchrotron Radiation Facility (*controlled atmosphere oven*, 300 – 3000K [4]), beamline 5-BM-D at the Advanced Photon Source (*displex cryostat*, 15 – 600K [5]), and beamline BL01B1 at Spring-8 (*muffle furnace*, 205 – 1870K [6]).

In response to a growing number of requests by our expanding user community, the XAS beamline at the Australian Synchrotron has recently commissioned an in-situ heater for high temperature XAS studies. A water-cooled, controlled-atmosphere Linkam TS1500 heating stage has been integrated into the second beamline experimental end station (Hutch C), see figure 1.

Certainly, one of the most critical parameters associated with any heating device is its temperature and the calibration of the said parameter. Commonly, a thermocouple is used to report the temperature of the sample; however, it is necessarily an inaccurate method limited by imperfect thermal equilibrium between the sample and the thermocouple, which is affected by the distance between them and the source of heat. Here, we present a method for accurately determining the sample temperature based on an anharmonic, correlated Einstein model.

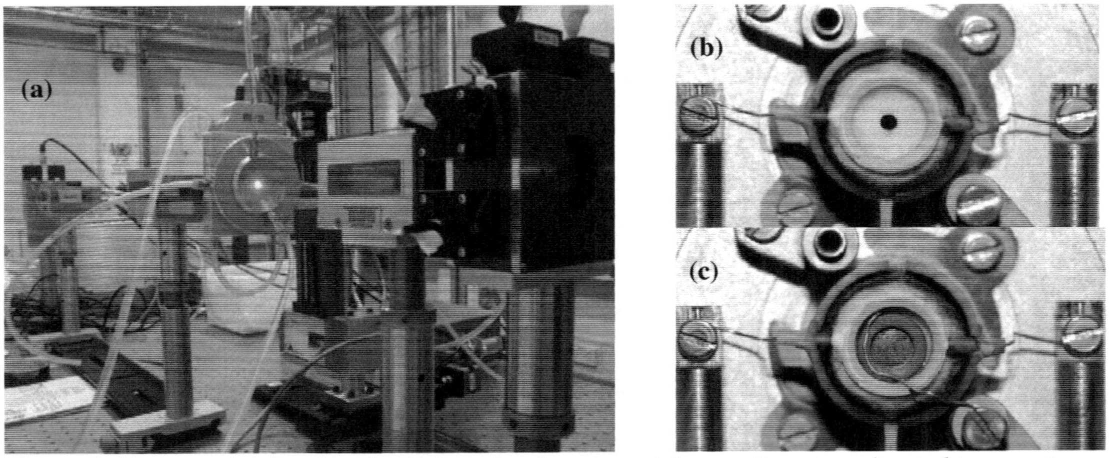

Figure 1. Photographs showing the Linkam heating stage; (a) operating at an elevated temperature and situated between two ion chambers on the optical table, (b) ceramic sample cup, (c) ceramic sample cup with Cu foil held in place by Ir wire and stainless washer.

2. Methods

A bulk Cu foil (7.5 μm) was measured in transmission mode at beamline 12-ID at the Australian Synchrotron as a function of temperature. The extended X-ray absorption fine structure (EXAFS) was collected at the Cu K edge (8979 eV) whilst controlling the temperature first using a Gifford-McMahon type cryostat from 18K to 280K and then using the Linkam heater from 300K to 1200K (as reported by an in-built thermocouple). The heater was used under a flowing Ar atmosphere with water-cooling and with the Cu foil held in place underneath a stainless washer with an Ir wire (see figure 1).

The spectra were analyzed using the IFEFFIT program package [7], and with a forward Fourier Transform (FT) from 3.0 to 10.7 $Å^{-1}$ (see figures 2 (a) and (c)) and backward FT from 1.2 to 2.8 Å (see figures 2 (b) and (d)). We observe in figure 2 that the EXAFS and FT amplitudes decrease gradually as the measurement temperature increases. This is indicative of a reduction in coordination number and / or increase in atomic disorder. The amplitude reduction factor and threshold energy were determined at 18K and kept constant when fitting spectra at all other temperatures. The photoelectron scattering-path amplitudes and phases were calculated ab initio using FEFF8 [8]. From the first nearest neighbouring shell of atoms structural parameters were extracted, specifically the coordination number, bond length, Debye-Waller factor and third order cumulant to the bond length distribution function [9].

The Debye-Waller factor (σ^2), or mean square relative displacement of the bond length distribution, consists of a static and a dynamic component. The former originates from inherent crystallographic disorder, while the latter describes thermal vibrations of the atoms around their equilibrium position. In order to quantify the thermal component an anharmonic, correlated Einstein model is often used. The Debye-Waller factor can then be written as [10]:

$$\sigma^2 = \frac{\hbar}{2\omega\mu}\,\frac{1+\exp\left(-\theta_E/T\right)}{1-\exp\left(-\theta_E/T\right)} + \sigma_s^2 \qquad (1)$$

where μ is the reduced mass, θ_E the Einstein temperature, ω the Einstein frequency, and σ_s is the static component of the Debye-Waller factor.

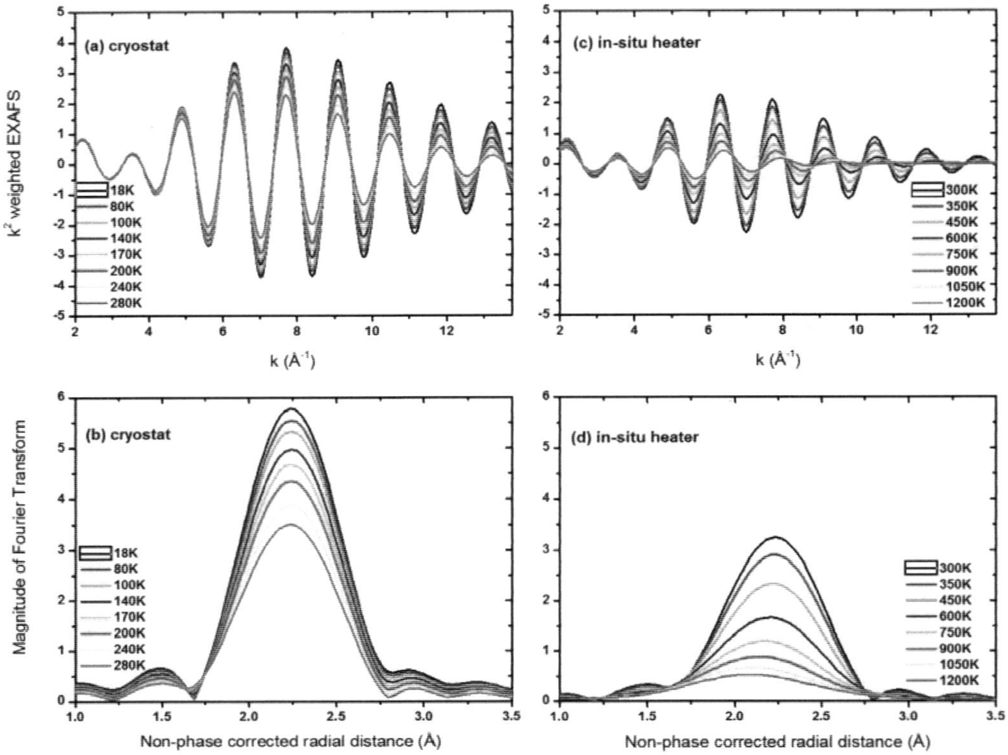

Figure 2. EXAFS data of a bulk Cu foil, (a) and (b) using a cryostat, and (c) and (d) using an in-situ heating stage. The EXAFS and Fourier Transform amplitudes decrease gradually with increasing measurement temperature.

3. Results and Discussion

Figure 3 shows the Debye-Waller factor as a function of temperature. Clearly, as the temperature increases the Debye-Waller factor increases, consistent with increasing thermal lattice vibrations. For the cryostat measurements the data were fitted using equation (1). From the fit the Einstein temperature was determined to be (244.3 ± 1.4) K. This is in excellent agreement with literature [11, 12] and as such one can then justify extrapolating the fit to beyond 280K (the highest measurement temperature using the cryostat).

It is evident that the extrapolated fit does not match with the measured data points using the in-situ heater and the reported temperature. Specifically, the reported temperature ($T_{Reported}$) is greater than that for the extrapolated fit, which is a more precise measure for the real sample temperature (T_{Real}). This is readily understood when inspecting the design of the heater and noting that the heater element and the thermocouple are both separated from the sample by a ~ 2 mm thick ceramic wall.

The discrepancy between the reported and real temperatures gives rise to the following temperature calibration equation:

$$T_{Real} = 68 + 0.8 T_{Reported} \ [K] \quad (2)$$

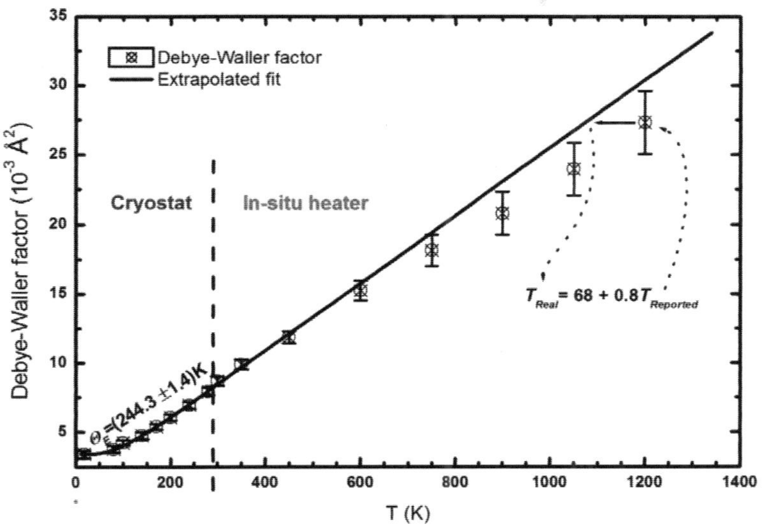

Figure 3. Debye-Waller factor as a function of (reported) temperature. Measurements were performed both in the cryostat and the heater. The solid line indicates an extrapolated Einstein temperature fit to the cryostat data.

Figure 4 shows the temperature-dependant evolution of the four structural parameters that were extracted from the system after applying the calibration to the reported temperature (equation (2)). Specifically, we find that as the temperature increases the coordination number (eventually) decreases, while the bond length, Debye-Waller factor and third order cumulant all increase. This finding is equivalent to a broadening of the bond length distribution function related to a low degree of correlations between atoms. Again, this is in excellent agreement with previous studies up to ~ 500K [11]. At higher temperatures the structure of the Cu foil is comparable to that of an amorphous or liquid material (though we note that the bulk Cu melting point was not reached) [13, 14].

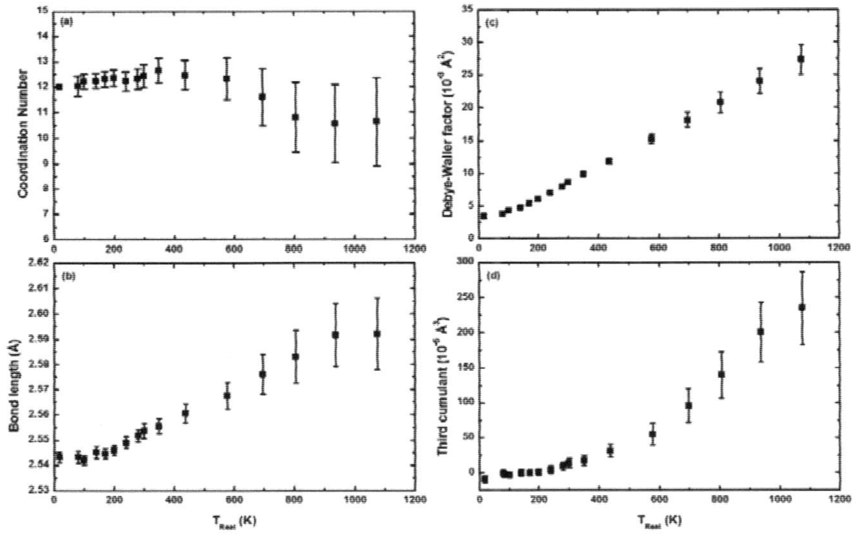

Figure 4. Cu foil structural parameters as a function of sample temperature; (a) coordination number, (b) bond length, (c) Debye-Waller factor, and (d) third order cumulant of the bond length distribution.

Journal of Physics: Conference Series **430** (2013) 012119 doi:10.1088/1742-6596/430/1/012119

4. Conclusion

We have shown that the temperature of an in-situ heating stage can be accurately calibrated using XAS irrespective of where the in-built thermocouple is located relative to the sample. For a bulk Cu foil we have extracted the coordination number, bond length, Debye-Waller factor, and third order cumulant to the bond length distribution function. We found that while the coordination number decreased above ~ 600K, all the other parameters increased gradually and this is consistent with a highly disordered structure typically observed for amorphous or liquid materials, though we note that the maximum Cu foil temperature was below the bulk melting point. This indicates, as far as atomic structure is concerned, that there is a gradual transition from the highly ordered crystalline structure to the highly disordered and thermally excited structure.

Acknowledgement

This research was undertaken at the XAS beamline at the Australian Synchrotron, Victoria, Australia. ZSH acknowledges the support of the CSIRO Office of the Chief Executive Science Program through the provision of a Postdoctoral Fellowship.

References

[1] Sprouster D J, Giulian R, Araujo L L, Kluth P, Johannessen B, Cookson D J, Foran G J and Ridgway M C 2010 *J. Appl. Phys.* **107** 014313
[2] Sapelkin A V, Bayliss S C, Lyapin A G, Brazhkin V V and Dent A J 1997 *Phys. Rev. Lett.* **56** 11531
[3] Katsikini M, Pinakidou F, Paloura E C, Komninou Ph, Georgakilas A and Welter E 2008 *Phys. Stat. Sol.* A **205** 2611
[4] www.esrf.eu
[5] www.aps.anl.gov
[6] www.spring8.or.jp
[7] Newville M 2001 *J. Synchrotron Radiat.* **8** 322
[8] Ankudinov A L, Ravel B, Rehr J J and Conradson S D 1998 *Phys. Rev.* B **58** 7565
[9] Stern E A 1974 *Phys. Rev.* B **10** 3027
[10] Frenkel A I, Hills C W and Nuzzo R G 2001 *J. Phys. Chem.* B **105** 12689
[11] Fornasini P, Dalba G, Grisenti R, Purans J, Sanson A, Vaccari M and Rocca F 2004 *Phys. Stat. Sol.* C **1** 3085
[12] Dalba G and Fornasini P 1997 *J. Synchrotron Radiat.* **4** 243
[13] Johannessen B, Kluth P, Llewellyn D J, Foran G J, Cookson D J and Ridgway M C 2007 *Phys. Rev.* B **76** 184203
[14] Sprouster D J, Giulian R, Araujo L L, Kluth P, Johannessen B, Kirby N, Nordlund K and Ridgway M C 2010 *Phys. Rev.* B **81** 155414

High-pressure X-ray absorption fine structure in the diamond anvil cell and its applications in geological materials

Xinguo Hong[1,*], Matthew Newville[2] and Thomas S. Duffy[3]

[1] Mineral Physics Institute, State University of New York at Stony Brook, Stony Brook, NY 11794, USA
[2] Consortium for Advanced Radiation Sources, University of Chicago, Chicago, Illinois 60637, USA
[3] Department of Geosciences, Princeton University, Princeton, New Jersey 08544, USA

E-mail: xhong@bnl.gov; xinguo.hong@gmail.com

Abstract. Although coupling the X-ray absorption fine structure (XAFS) technique to the diamond anvil cell (DAC) has long been recognized as potentially important for understanding the behavior and evolution of the local and electronic structure of matter under extreme conditions, the DAC has been regarded as poorly suited for XAFS due to the DAC imposed glitches. Recently, an iterative method was proposed to distinguish and eliminate the serious interference of the diamond Bragg peaks, and succeeded in acquiring high quality 'reflection-free' XAFS spectra at high pressure under the DAC environment. In this paper, we use this method and present demonstration XAFS spectra for GeO_2 glass, which is the archetype of network forming glasses with important geophysical implications, at pressures to 64 GPa. This is far above the change in coordination from tetrahedra to octahedra occurring at 6-13 GPa. The results provide important insight into the structural polymorphism of GeO_2 glass at high pressure.

1. Introduction

Diamond anvil cells (DAC) are commonly used to investigate the high-pressure phase behavior of condensed matter. It is well known that XAFS including the near edge (XANES) and extended (EXAFS) spectra can offer critical information on the local electronic structure and atomic arrangement of absorber atoms in materials at high pressure or ambient conditions [1-4]. The coupling of the diamond anvil cell technique to the XAFS technique is of general importance in physics, chemistry, materials science and earth sciences, and has long been regarded as potentially important for understanding the behavior and evolution of the local and electronic structure of matter under extreme conditions [5-7]. However, the use of diamond anvil cells results in distortion of XAFS spectra due to the Bragg reflections from the diamond anvils, restricted sample thickness, and reduced signal quality due to absorption by diamond at low energy. The Bragg reflection of the diamond anvils can suddenly reduce the transmitted intensity and makes a strong contribution to the measured attenuation of x-ray photons, i.e. glitches, at certain x-ray energies and orientations of the diamonds.

Because of the DAC Bragg glitches, which are often more intense than EXAFS and XANES oscillations and spoil the high-quality XAFS data necessary for precise structural information, diamond anvils has long been recognized as not suited for the conventional energy scan XAFS measurements in high-pressure experiments [5]. In recent work, it was found that by using nano-polycrystalline diamond instead of single crystal anvils, the influence of diamond diffraction could be significantly mitigated [8, 9]. However, nano-polycrystalline diamond is not widely available and small DAC-induced ripples are still visible in the resulting EXAFS data [8].

Another method to acquire high quality X-ray absorption spectra (XAFS) over a long energy range at high pressure using the diamond anvil cell has been recently developed [10]. It was found that the DAC imposed glitches, whose energy positions are very sensitive to the relative orientation between the DAC and incident X-ray beam, can be effectively eliminated using an iterative algorithm based on redundant measurements over a small angle range of DAC orientations, e.g. within ± 3° around the X-ray beam [10]. It has been demonstrated that XAFS spectra obtained for crystalline GeO_2 recorded by traditional Scotch Magic tape and the DAC method, respectively, show nearly identical good quality to 440 eV above the absorption edge. High pressure 'reflection-free' XAFS data were obtained for GeO_2 glass up to 55 GPa.

The polyamorphism of classic network-forming glasses such as GeO_2 or SiO_2 is of great interest due to their Earth science implications [11] and importance in glass theory [12]. These network-forming glasses have been intensively studied for several decades [11]. Due to the recent findings of pressure-induced intermediate states in GeO_2 glass [10, 13-15], considerable renewed investigations have been performed both experimentally and theoretically during the last few years [16-22].

It is desirable to make an exhaustive check on the reliability of the emerging method of Ref. 8. It is also important to apply the proposed method to probe the possible new polyamorphsim in GeO_2 glass at pressures much higher than the completion pressure of the change from tetrahedral to octahedral coordination at 13-15 GPa. To this end, we have carried out high-pressure Ge K-edge XAFS experiments. High quality XAFS spectra of GeO_2 glass have been obtained at pressures up to 64 GPa. It should be mentioned that the developed method is not limited by the applied pressure of the DAC, and should be applicable for in situ XAFS measurements using DAC to ultra-high pressures in the Mbar region.

2. Experimental

XAFS experiments were carried out on the Ge K-edge (11.103 eV) of GeO_2 glass in transmission mode at the GeoSoilEnviroCARS bending magnet beamline 13-BM-D, Advanced Photon Source (APS), Argonne National Laboratory. The X-rays were monochromatized using a Si(111) double-crystal monochromator. Higher harmonics were rejected by detuning the second crystal of the monochromator to reduce the total intensity to ~50% of the intensity at full tune. A 1000-mm Pt-coated mirror was used to focus the beam in the vertical to ~50 μm. The horizontal X-ray beam was then focused down to a diameter of approximately 15 μm at the sample position using a 200-mm long Kirkpatrick-Baez (KB) mirror. Details of the equipment and X-ray optics of this beamline are described elsewhere [23, 24].

GeO_2 glass samples with good quality as determined by X-ray examination at ambient condition were made by quenching GeO_2 melt annealed at 1600°C for six hours, and generously donated by Liping Huang from Rensselaer Polytechnic Institute, New York. Details of the GeO_2 glass preparation can be found elsewhere [13]. Princeton-manufactured standard and large-opening symmetrical DACs with 300-μm culets and 2.1 mm thickness for each anvil, were employed for XAFS measurements. Rhenium gaskets (250-μm thick) were pre-indented to about 40-μm thickness, and a hole of approximately 100-μm in diameter in the gasket center was made using a laser cutting system installed at GSECARS. This served as the sample chamber for loading of amorphous GeO_2 powder. Pressures were determined using the pressure-dependent fluorescence of small ruby balls which were scattered oppositely at the corner of sample chamber. The DAC position was placed on the rotation center of the diffractometer prior to spectral collection so as guarantee that the X-rays pass through the same sample area at each diamond cell orientation with respect to the incident X-ray beam. This is necessary to avoid sample inhomogeneities and thickness effects due to lack of parallelism or deformation of diamond anvils at high pressure [14] [25].

XAFS spectra were collected by scanning the monochromator energy from 10,953 to 11,747 eV at 5 eV steps before the main edge, 0.5 eV steps from -10 eV to 25 eV across the main edge (11,103 eV), and 0.05 $Å^{-1}$ steps in photoelectron wave number above the main edge to 13 $Å^{-1}$ in k-space. Each XAFS spectrum consists of 310 data points, and the signals from the ion chambers were typically recorded for 1 s before and across the main edge while a k-weighted collecting time was applied above the main edge at each energy point. It took about 7 minutes to collect an XAFS scan at each specific DAC orientation relative to the incident X-ray beam covering an angular range of ±3°. Seven scans, which were collected at an angle step of 1°, have 2,170 data points. In total, it took about one hour to complete all 7 spectra at each pressure step. Other details regarding sample preparation and XAFS experiments can be found in Ref. [10], and will not be duplicated here.

To identify and remove the DAC-imposed glitches in the raw absorption spectra, an iterative method, which is described in detail in Ref. [10], was employed. Further XAFS data processing and analysis were then performed with the ATHENA and ARTEMIS programs [26] of the IFEFFIT package [27]. Theoretical models for the XAFS were constructed with FEFF (Refs.[28] and [29]) by using crystallographic atomic positions of GeO_2. The models were fitted to the data by using ARTEMIS [26] which was also used for error analysis and calculated goodness-of-fit parameters.

3. Results and discussion

Figure 1 shows the multiple datasets of transmission Ge K-edge XAFS for GeO_2 glass at 64 GPa with the diamond anvil cell at different angle settings of 0°, ±1°, ±2°, ±3° with respect to the incident beam. All these spectra were overlapped and aligned using the iterative algorithm [10] for the elimination of glitches. The overlap is excellent for all of the spectra with glitches removed, and after three iterations, the standard deviation of seven independent spectra to the average one is as low as 0.1% in the whole spectra.

It has been pointed out that high data redundancy of DAC orientations to the X-ray beam would significantly improve the data quality of the final composite XAFS spectra [10]. Usually a measurement of seven DAC orientations respective to the incident X-ray beam should be sufficient to get a smooth composite spectrum (bottom curve, Figure 1) suitable for further data analysis and modeling. It should be mentioned that each composite point is typically based on at least three independent XAFS scans, as shown in Figure 1.

It can be noted that the composite data shows a clear drop starting at 11,562 eV, which coincides with the L3-edge energy of platinum (11,564 eV). This is due to the platinum layer employed to coat the upstream vertical focusing mirror at sector13-BM-D of the APS. An alternative coating material on the mirror would eliminate such an artefact in Ge k-edge XAFS. The coating material of focusing mirrors should be considered for a dedicated high-pressure XAFS beamline for geosciences in the future. To allow reliable data extraction of μ(E), the extent of the energy range is then limited below 11,560 eV.

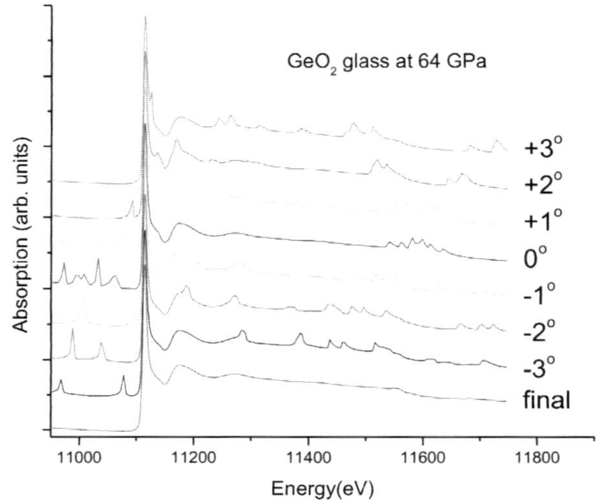

Figure 1. Ge K-edge XAFS of GeO_2 glass at 64 GPa obtained by classic energy-scan transmission mode across the diamond anvils (300 μm culet) at different angle settings. The X-ray absorption spectrum shows extra peaks due to Bragg diffraction by the single-crystal diamond anvils. The bottom curve shows the spectrum obtained after application of the glitch removal algorithm.

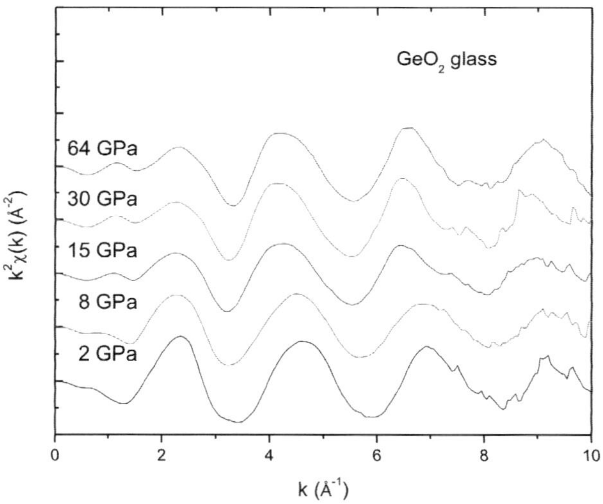

Figure 2. The k^2-weighted $\chi(k)$ plots of GeO_2 glass under pressure. Due to the L_3 edge of Pt (11,564 eV) coating layer in the focusing mirror, all data are limited to 11,560 eV.

The k^2-weighted $\chi(k)$ plots of GeO_2 glass under high pressures are shown in Figure 2. As pressure increases, the EXAFS oscillations shift to lower k values, which mean some elongation of local bond lengths as a result of coordination change in the glass. However, it appears that these oscillations shift to higher k values at pressures above 15 GPa. In other words, the material exhibits the normal reduction of bond length with compression above 15 GPa. It can be noticed that the most remarkable change in EXAFS oscillations occurs in the k range between 6-8.5 $Å^{-1.}$ Considering the margin effect of the forward Fourier transformation window, the extent of EXAFS oscillations should be above 9.5 $Å^{-1}$, otherwise the limited data range will impair the reliability of further structure modeling. The independent points are 3, 4 and 6, for the k maximum values of 7, 8, and 10 $Å^{-1}$, respectively, suggesting the importance of wide k space coverage for a reliable structure modeling even for a simple first shell analysis. The good data quality shown in Figure 2 illustrates the validity of the iterative algorithm [10].

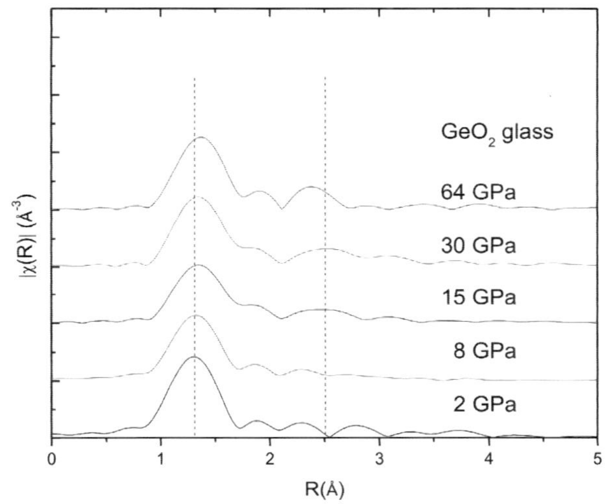

Figure 3. Fourier transformation modulus, $|\chi(R)|$, of k^2 $\chi(k)$ for GeO_2 glass under high pressures. Dashed lines as guide for eye.

Figure 3 shows the Fourier transform modulus, $|\chi(R)|$, of the k^2 $\chi(k)$ plots that are exhibited in Figure 2. While a slightly progressive elongation in the Ge-O bond length at 1.3 Å is observed, a more evident change occurs in the range of 2.3-3.0 Å, where the intermediate range order (IRO) or interchain bonds are located, as observed by a recent EXAFS study of GeO_2 glass to 44 GPa [20]. Of interest is that the overall spectrum at 30 GPa is quite similar to that at 15 GPa, suggesting that the transition from tetrahedral to octahedral coordination of Ge atoms should be essentially completed before 15 GPa, rather than the previous estimation of as high as 30 GPa [20]. The smaller k range (<8 $Å^{-1}$) in the previous work [20] might be the source of the difference [29]. Nevertheless, the IRO region of 64 GPa data shows a considerable enhancement in comparison with those of 30 and 15 GPa, suggesting a sign of possible new polyamorphsim.

To crosscheck the possible amorphous-amorphous transition at pressures above 15 GPa in GeO_2 glass, it is necessary to inspect if there exists any significant evidence in the XANES spectra, as XANES is sensitive to the valence state and local symmetries around the Ge atom. Figure 4(a) shows the Ge K-edge XANES spectra of GeO_2 glass at pressures of 15, 20, 30, and 64 GPa, and Figure 4b is the difference spectra compared to the spectrum at 15 GPa. The XANES spectra of GeO_2 glass at 20 and 30 GPa looks quite similar except for the height of the white line. Little change above the edge is observed. As shown in Figure 4b, a strong peak appears at E= 11,116 eV, and there is a dramatic change in oscillation for the spectrum of 64 GPa, confirming the EXAFS observation as discussed above, i.e. the possible existence of a new polyamorphsim, which was first reported in a previous paper from an independent high-pressure XAFS measurement [10]. This provides clear evidence that at pressure above 15 GPa the octahedral form in GeO_2 glass would change to a more compact arrangement of atoms, i.e. the so-called "post-octahedral transition" as noted in X-ray diffraction experiments [13]. Detailed structural modeling will be published in a separate paper.

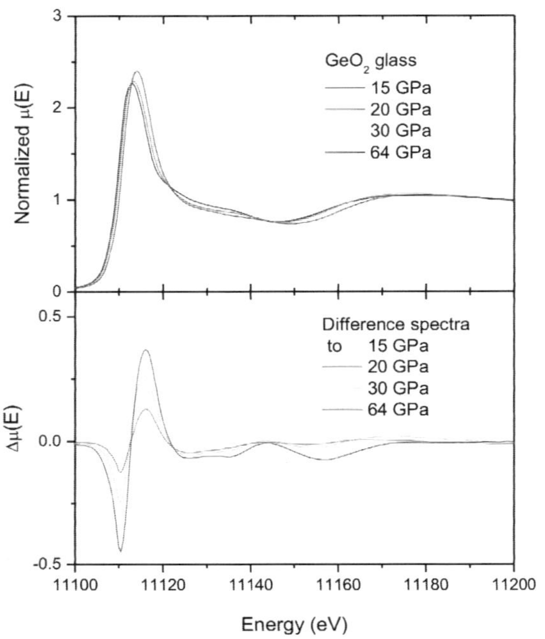

Figure 4. (a) Ge K-edge XANES spectra of GeO_2 glass obtained at pressures of 15, 20, 30, and 64 GPa; (b) the difference spectra compared to the spectrum at 15 GPa.

In summary, we have shown that the serious problem caused by the interference of diamond Bragg peaks can be effectively addressed based on repeated measurements over a small angular range of DAC orientation relative to the x-ray beam. Using the developed iterative method [10], we can eliminate these DAC imposed glitches effectively. We have demonstrated that 'reflection-free' high-quality XAFS spectra can be achieved at high pressure up to 64 GPa using the diamond anvil cell for GeO_2 glass. The obtained results provide insight into the structural polymorphism of GeO_2 glass at high pressure. With the advent of sub-micron focused x-ray beams at synchrotron facilities, there would be no upper limit in pressure for the proposed iterative method in XAFS data reduction.

4. Acknowledgment

We would like to acknowledge S. Sutton, M. Rivers, G. Shen and V. Prakapenka and for their help and useful discussions. We also thank L.P. Huang for her generosity in providing glassy GeO_2 samples. Assistance from X.M. Yu and S. Lin are also acknowledged. This work is partially supported by COMPRES under NSF grant of EAR 11-57758. The GSECARS sector is supported by the NSF (Earth Sciences Instrumentation and Facilities Program) and DOE (Geoscience Program).

5. References:

1. Sayers, D.E., E.A. Stern, and F.W. Lytle, *New Technique for Investigating Noncrystalline Structures: Fourier Analysis of the Extended X-Ray—Absorption Fine Structure.* Physical Review Letters, 1971. **27**(18): p. 1204-1207.
2. Ingalls, R., G.A. Garcia, and E.A. Stern, *X-Ray Absorption at High Pressure.* Physical Review Letters, 1978. **40**(5): p. 334-336.
3. Rehr, J.J. and R.C. Albers, *Theoretical approaches to x-ray absorption fine structure.* Reviews of Modern Physics, 2000. **72**(3): p. 621-654.
4. Filipponi, A., *EXAFS for liquids.* Journal of Physics: Condensed Matter, 2001. **13**(7): p. R23-R60.
5. Ingalls, R., et al., *Extended x-ray absorption fine structure of NaBr and Ge at high pressure.* Journal of Applied Physics, 1980. **51**(6): p. 3158-3163.
6. Ohsumi, K., et al., *Exafs Measurement under High-Pressure by Diamond Anvil Cell.* Journal De Physique, 1986. **47**(C-8): p. 189-192.
7. Tranquada, J.M. and R. Ingalls, *X-ray-absorption study of CuBr at high pressure.* Physical Review B, 1986. **34**(6): p. 4267-4277.
8. Baldini, M., et al., *High-pressure EXAFS measurements of crystalline Ge using nanocrystalline diamond anvils.* Physical Review B, 2011. **84**(1): p. 014111.
9. Ishimatsu, N., et al., *Glitch-free X-ray absorption spectrum under high pressure obtained using nano-polycrystalline diamond anvils.* Journal of Synchrotron Radiation, 2012. **19**(5): p. 768-772.
10. Hong, X., et al., *High quality x-ray absorption spectroscopy measurements with long energy range at high pressure using diamond anvil cell.* Review of Scientific Instruments, 2009. **80**(7): p. 073908-10.
11. Heaney, P.J., *Silica, Physical Behavior, Geochemistry and Materials Applications*, ed. C.T.P.G.V.G. P. J. Heaney. 1994, Washington, DC: Mineralogical Society of America. 1-39.
12. Wright, A.C., *Neutron scattering from vitreous silica. V. The structure of vitreous silica: What have we learned from 60 years of diffraction studies?* Journal of Non-Crystalline Solids, 1994. **179**: p. 84-115.
13. Hong, X., et al., *Intermediate states of GeO2 glass under pressures up to 35 GPa.* Physical Review B (Condensed Matter and Materials Physics), 2007. **75**(10): p. 104201-5.
14. Hong, X., et al., *Density measurements of noncrystalline materials at high pressure with diamond anvil cell.* Review of Scientific Instruments, 2007. **78**(10): p. 103905-6.
15. Shen, G., et al., *Distinct thermal behavior of GeO2 glass in tetrahedral, intermediate, and octahedral forms.* Proceedings of the National Academy of Sciences, 2007. **104**(37): p. 14576-14579.
16. Vaccari, M., et al., *Structural changes in amorphous GeS2 at high pressure.* Physical Review B, 2010. **81**(1): p. 014205.
17. Drewitt, J.W.E., et al., *Structure of GeO2 glass at pressures up to 8.6 GPa.* Physical Review B, 2010. **81**(1): p. 014202.
18. Mei, Q., et al., *High-pressure x-ray diffraction measurements on vitreous GeO2 under hydrostatic conditions.* Physical Review B, 2010. **81**(17): p. 174113.
19. Dario, M. and et al., *High-pressure behaviour of GeO2 : a simulation study.* Journal of Physics: Condensed Matter, 2010. **22**(15): p. 152102.
20. Baldini, M., et al., *High-pressure EXAFS study of vitreous GeO2 up to 44 GPa.* Physical Review B, 2010. **81**(2): p. 024201.

21. Funamori, N. and T. Sato, *Density contrast between silicate melts and crystals in the deep mantle: An integrated view based on static-compression data.* Earth and Planetary Science Letters, 2010. **295**(3-4): p. 435-440.
22. Sato, T. and N. Funamori, *Sixfold-Coordinated Amorphous Polymorph of SiO2 under High Pressure.* Physical Review Letters, 2008. **101**(25): p. 255502.
23. Newville, M., et al., *Micro-beam X-ray absorption and fluorescence spectroscopies at GSECARS: APS beamline 13ID.* Journal of Synchrotron Radiation, 1999. **6**(3): p. 353-355.
24. Shen, G., et al., *Facilities for high-pressure research with the diamond anvil cell at GSECARS.* Journal of Synchrotron Radiation, 2005. **12**(5): p. 642-649.
25. Hemley, R.J., et al., *X-ray Imaging of Stress and Strain of Diamond, Iron, and Tungsten at Megabar Pressures.* Science, 1997. **276**(5316): p. 1242-1245.
26. Ravel, B. and M. Newville, *ATHENA, ARTEMIS, HEPHAESTUS: data analysis for X-ray absorption spectroscopy using IFEFFIT.* Journal of Synchrotron Radiation, 2005. **12**(4): p. 537-541.
27. Newville, M., *EXAFS analysis using FEFF and FEFFIT.* Journal of Synchrotron Radiation, 2001. **8**(2): p. 96-100.
28. Zabinsky, S.I., et al., *Multiple-scattering calculations of x-ray-absorption spectra.* Physical Review B, 1995. **52**(4): p. 2995.
29. Ankudinov, A.L., et al., *Real-space multiple-scattering calculation and interpretation of x-ray-absorption near-edge structure.* Physical Review B, 1998. **58**(12): p. 7565.

Journal of Physics: Conference Series **430** (2013) 012121 doi:10.1088/1742-6596/430/1/012121

Local structure of Titanium in natural glasses probed by X-ray absorption fine structure

L Wang[1], A Yoshiasa[1], M Okube[2], T Nakatani[1], Y Hayasaka [3] and H Isobe[1]

[1] Graduated School of Science and Technology, Kumamoto University, 860-8555, Japan

[2] Materials and Structures Laboratory, Tokyo institute of technology, Yokohama, 226-8502, Japan

[3] Faculty of Science, Hiroshima University, Higashi, Hiroshima, 739-8511, Japan

E-mail:wangling.1984@live.cn

Abstract. Synchrotron radiation has been used to collect titanium K-edge absorption spectra of a suite of natural glasses (tektites, impact glasses, fault rocks and volcanic glasses). XANES and XAFS analysis provided the qualitative and quantitative information of Ti oxidation state, Ti-O distance and site geometry. Tektites possess four-, five-, six-coordinated Ti, whereas fault rock-pseudotachylite, volcanic glasses and impact glass only presented five- and six-coordinated Ti. This study indicated that different petrogenesis of natural glasses has different local structures of titanium.

1. Introduction

In natural, several kinds of glasses are formed by various geological actives. Tektites and impact glasses are produced by impact event (Alvarez *et al.*, 1980); volcanic actives formed volcanic glass; and plate boundary active produced fault rocks. These natural glasses experienced different extreme environments, which should lead changing of local structure (Stebbins & McMillan, 1989; Paris et al., 1994; Mysen & Neuville, 1995; Yarger et al., 1995). We studied Ti local structure of tektite (2011). This present study completed natural glasses results.

2. Experiment

The specimens are tektites: hainanite, indochinite, philippinite, australite, bediasite and moldavite; impact glass-impactite, suevite and köfelsite; volcanic glasses- Darwin glass, perlite, obsidian, Kilauea volcanic glasses; and fault rock- pseudotachylite. In order to analysis the local structure of Ti in natural glasses, we used the XAFS methods. The XAFS measurement of Ti local structure was preformed with a Si (111) double crystal monochromator at the beam line BL-9C of the Photon Factory in National Laboratory for High Energy Physics (KEK), Tsukuba, Japan. Spectra near Ti K-edge were collected in transmission and fluorescence mode at the room temperature. Analysis of XAFS data were performed by XAFS93 and MBF93 programs. (Maeda, 1987; Yoshiasa et al., 1997)

3. Result and discussion

Experimental Ti-K edge XANES spectra of typical tektites are shown in Fig.1A. The highest peak position of moldavite-green at 4984.7eV is shown by dotted line for comparison. The pre-edge peak heights at 4967.3eV change with coordinated environments. The detailed XANES spectra of tektites from three strawnfield were explained by Wang et al., 2011. These spectra are divided into three types according to pre-edge and XANES shape: Type I, indochinite; Type II, hainanite-core, hainanite-rim, philippinite and Type III, moldavite-green. The normalized intensity at 4984.7eV is regard as 100% in moldavite-green. The pre-edge peak heights at 4967.3eV for Type I, II and III are 59-49%, 49-47%, 14%, respectively. The white line peak and shoulder heights at 4984.7eV for Type I, II and III are around 86%, 90% and 100%, respectively. Meanwhile, the crest energy at 4984.7eV in moldavite-green is more intense than other tektites.

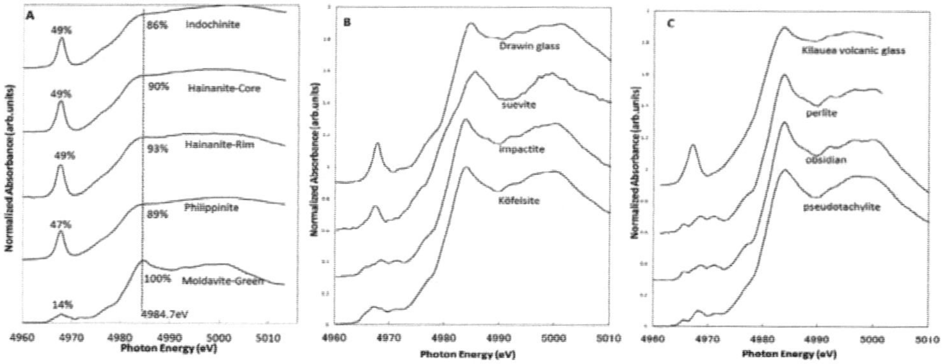

Figure 1. XANE spectra near the Ti K-edge for typical tektites (A) (Wang 2011), impact glasses (B) and volcanic glasses and fault rock-pseudotachylite (C).

Fig.1B shows XANES spectra of impact glasses. Darwin glass, suevite, and köfelsite have a large pre-edge peak with shoulder. Pre-edge intensities of these three samples are 25.7%, 15.7% and 11.4%. This feature is comparable with tektite Type II and III. Impactite possesses pre-edge, which contains a main peak and two subordinate peaks. The back of XANES regions of these glasses is similar with moldavite-green, which has intense crest edge around 4984eV. The variation of XANES feature in impact glass suggested the relaxation is diverse. The various relaxations are caused by the environment when impact glasses are formed. Temperature and annealing time are most important conditions for relaxation.

Fig.1C shows XANES spectra of volcanic glasses and fault rock-pseudotachylite. Kilauea volcanic glass is an exception of this group, and only possesses a large pre-edge peak with shoulder. In addition, it has a wider crest (4984eV) than other volcanic glasses. Chemistry composition of Kilauea volcanic glasses, which contains 50.89wt% SiO_2 and 2.55wt% TiO_2, is distinguishing; it is similar with basalt. But other samples are similar with rhyolite ($SiO_2 \geq 70$wt %). Their different forming processes are suggested that the Kilauea volcanic glass is distinguishable from other volcanic glasses by XANES pattern. Fault rock-pseudotachylite has a similar XANES spectrum with normal volcanic glasses but possesses a bit broader crest.

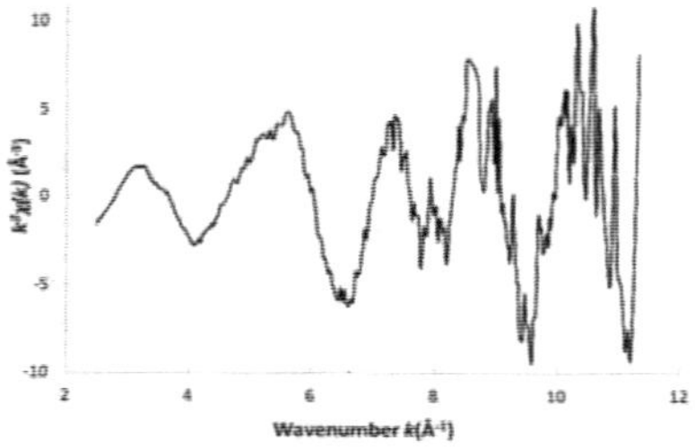

Figure 2. Observed Ti K-edge EXAFS $k^3\chi(k)$ oscillation of a tektite

The EXAFS $k^3\chi(k)$ functions (Fig.2) were transformed into the radial distribution functions (RDF) for Ti K-edge , as shown in Fig.3. In Fig.3A, the RDF of Ti atom in Type I- indochinite, Type II-hainanite-rim, hainanite-core, and philippinite are similar and Type III- moldavite-green is discriminated from others. In Fig.3B, Suevite shows the same feature with Type II tektites; impactite, darwin glass, and köfelsite also show the same feature with Type III tektite. The mainly peak located around 1.5 Å and it is the same position with moldavite-green. It indicates that impact glasses have two kinds of local atomic environment around the Ti atom. In Fig.3C, Kilauea volcanic glass and perlite show the same feature with Type II tektites; obsidian and pseudotachylite show the same feature with Type III tektite. The impact glasses, volcanic glasses, and fault rock are almost having the same structure with Type II and Type III tektites.

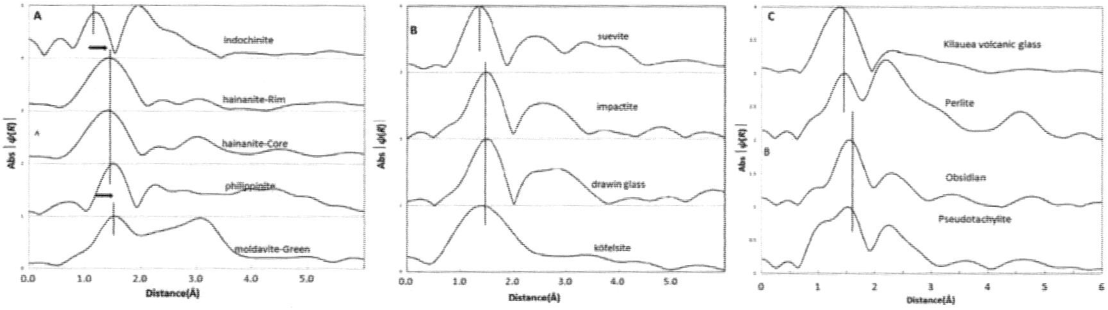

Figure 3. Fourier transforms of the Ti K-edge EXAFS oscillation function $k^3\chi(k)$. No phase shift corrections are made. In A (Wang 2011), The first nearest peaks corresponding Ti-O bonds increase with increasing coordination number of Ti in tektite (arrows). Three types of tektites can be recognized. Impact glass has the main peak position, which between Type II and Type III tektites. In B and C, all non impact related glasses possess the mainly peak that located around 1.5 Å.

In order to obtain the further information on structure parameter, we figured out the parameter fitting with analytical EAXFS formula expressed by cumulant expansion up to third order terms in tektites and natural glasses. The obtained structural parameters are summarized in Table 1.

Table 1. The structure parameters determined by EXAFS and TiO_2 composition for natural glasses. The expected bonding distances of 4-, 5-and 6- coordinated Ti-O based on the Shannon Ionic Radii: Ti(4-coordinated)-O=1.80 Å; Ti (5-coordinated)-O=1.89 Å; Ti (6-coordinated)-O=1.99 Å. Uncertainties in the last decimal place are shown in parentheses.

Sample name	σ^2(Å)	Coordination number	Ti-O distance(Å)	R-factor (%)	TiO_2(wt%)
indochinite	0.009(1)	4	1.812(5)	4.3	0.8
hainanite-core	0.005(1)	4 and 5	1.868(4)	0.3	0.87
hainanite-rim	0.006(1)	5	1.888(6)	0.1	0.87
philippinite	0.003(1)	5 and 6	1.921(5)	2.7	0.81
moldavite-green	0.003(1)	6	2.001(4)	3.5	0.11
suevite	0.001(0)	5	1.887(2)	2.5	0.64
Kilauea volcanic glass	0.002(2)	5	1.900(1)	0.330	2.55
perlite	0.000(0)	5	1.902(1)	0.46	0.9
pseudotachylite	0.005(0)	5 and 6	1.950(2)	2.3	0.78
impactite	0.002(0)	5 and 6	1.952(2)	0.22	-
Darwin glass	0.005(0)	5 and 6	1.963(4)	0.17	0.63
obsidian	0.000(0)	5 and 6	1.969(2)	0.94	0.76
köfelsite	0.006(0)	6	1.983(3)	2.1	0.66

Based on the local Ti-O distances, coordination numbers, and radial distribution function determined by EXAFS analysis, we can classify the tektites into three types. Indochinite is classified in Type I, where Ti occupy 4-coordinated (tetrahedral site) and Ti-O distances are 1.812-1.835 Å. Hainanite-rim, hainanite-core, and philippinite are in Type II, where Ti occupy 5-coordinated site (trigonal bi-pyramidal or tetragonal pyramidal site) and Ti-O distances are 1.868-1.921 Å. Moldavite-green is in Type III, where Ti occupy the 6-coordinated site (octahedral site) and Ti-O distance is 2.00-1.96 Å. Impact glasses, fault rocks and volcanic glasses have fewer variety structure than tektites.

Only 5 - and 6- coordinated titanium existed with Ti-O distance of 1.89-1.90 Å, 1.95-1.98 Å. Titanium in natural glasses are from 0.11-2.55 wt%, the composition does not influence the local structure.

Farges, F. and Brown, G.E. (1997) discussed the pre-edge intensity gradually decreased by 4-.5- and 6- coordination Ti. Fig.4 shows the relationship between Ti-O distance, pre-edge, and crest intensity in all tektites and natural glasses. The pre-edge feature corresponded to the previous study. In addition, the crest intensity trends to higher with increasing of Ti-O distance. It is very easy to identify the tektites and natural glasses.

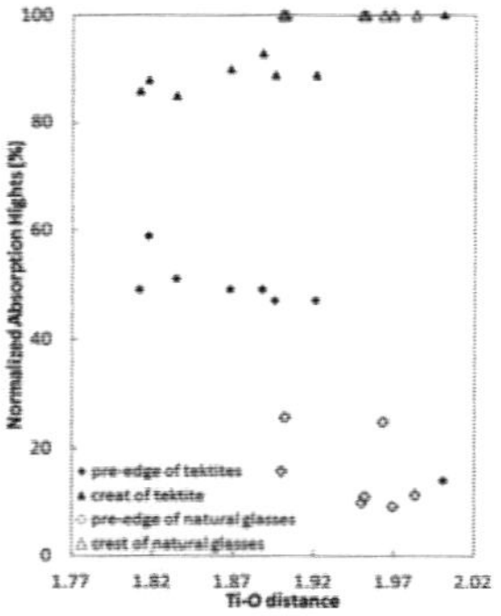

Figure 4. relationship between Ti-O distance and special absorption energies (pre-edge, and crest) in all tektites and natural glasses

4. Summary

Analysis of local structure is helping to compare formation conditions of natural glasses and distinguish them. Tektite possessed 4-, 5-, 6-coordinated Ti. All of them had a single pre-edge peak with different normalized heights. But fault rock-pseudotachylite, volcanic glasses and impact glasses, such as impactite possessed fewer variety local structures and tektites, only presented 5- and 6-coordinated Ti with Ti-O distances of 1.89-1.90 Å, 1.95-1.98 Å, and their XANES pre-edge features are similar with TiO_2 anatase, which has a main peak and two subordinate peaks. The variation of XANES feature in impact glass suggested the relaxation is diverse. The various relaxations are caused by the environment when impact glasses are formed. Temperature and annealing time are most important conditions for relaxation. This study indicated that different petrogenesis of natural glasses has different local structures of titanium. Tektite flied from atmosphere to outer space and once more returned to the earth, in this process, tektite rapidly experienced changes of annealing time, temperature and pressure. But fault rock- pseudotachylite and volcanic glasses experienced only a single tremendous change of temperature or pressure. Impact glasses are close to impact crater, so changes of temperature and pressure are limited and slowly. This reason leads the local structure of tektite is more various than other natural glasses and the four-coordinated Ti are unique in natural glasses.

5. Acknowledgement

This study was performed under the auspices of the Photon Factory (PAC No. 2010G512).

References

[1] Alvarez L W, Alvarez W, Asaro F and Michel H V 1980 *Science.* **208** 1095
[2] Farges F and Brown G E 1997 *Geochim. Cosmo. Acta.* **61** 1863
[3] Maeda H 1987 *J. Phy. Soc. Jpn.* **56** 2777
[4] Mysen B and Neuville D 1995 *Geochim. Cosmochim. Acta.* **59** 325
[5] Paris E,Dingwell D, Seifert F, Mottana A and Romano C 1994 *Phys. Chem. Miner.* **21** 520-525
[6] Stebbin J F and McMillan P 1989 *Am.Mineral.* **74** 965
[7] Wang L, Yoshiasa A, Okube M and Takeda T 2011 *J. Synchrotron Rad.* **18** 885
[8] Yarger J L, Diefenbacher J, Smith K H, Poe B and McMillan P F 1995 *Science.* **270** 1964
[9] Yoshiasa A, Koto K, Maeda H and Ishii T 1997 *Jpn. J. Appl. Phys.* **36** 781

Journal of Physics: Conference Series **430** (2013) 012122 doi:10.1088/1742-6596/430/1/012122

Insights from X-ray absorption/fluorescence spectroscopy and ab-initio molecular dynamics on concentration and complexation of Zr and Hf in aqueous fluids at high pressure and temperature

M Wilke[1]*, S Jahn[1], C Schmidt[1], J Dubrail[1], K Appel[2], M Borchert[2], K Kvashnina[3], S Pascarelli[3] and CE Manning[4]

[1] Deutsches GeoForschungsZentrum GFZ, Telegrafenberg, 14473 Potsdam, Germany;
[2] Deutsches Elektronensynchrotron DESY, Notkestr. 85, 22607 Hamburg, Germany
[3] European Synchrotron Radiation Facility, 6 Rue Jules Horowitz, 38043 Grenoble, France
[4] Department of Earth and Space Sciences, University of California, Los Angeles, CA 90095-1567, USA

max@gfz-potsdam.de

Abstract. High-field-strength elements such as Zr and Hf are important geochemical indicators for processes in the deep Earth. Aqueous fluids play a significant role in the transport of heat and matter in these systems. However, concentrations and complexation of Zr and Hf in these fluids at high pressure (P) and temperature (T) conditions are largely unknown. Zr/Hf contents were determined in-situ at P and T in aqueous fluids containing NaOH, HCl or Na-Al silicate components in equilibrium with zircon or baddeleyite. Concentrations are strongly enhanced in comparison to pure H_2O. The variation of Zr/Hf contents with fluid composition and their respective variation with P and T point to differences in the Zr/Hf complexation. High-resolution Zr K-edge XANES spectra were collected in-situ at P and T as well as Hf L_3-edge XANES spectra. Analysis of spectra evidences considerable differences in the coordination of Zr and Hf as a function of fluid composition. $^{[7]}$Zr is found in NaOH solution. $^{[7]}$Zr is also found in HCl solution, however, coordinated by both chlorine and oxygen. In fluids with Na-silicate components, Zr and Hf are 6-fold coordinated in zircono-silicate complexes, likely complexed with Na and Si. Further insight into the speciation was obtained by ab-initio molecular dynamics simulations of Zr monomers in H_2O-NaOH and H_2O-HCl. In basic solutions, Zr is five-fold coordinated by oxygen, whereas in acidic solutions, mixed oxychloride complexes are formed. The oxychloride complexes also seem to be present in experimental fluids, whereas the Zr coordination in basic fluids is underestimated by the simulations. The most likely explanation for this discrepancy is the formation of Zr oligomers in the fluid.

1. Introduction

High-field-strength elements (e.g. Ti, Zr, Hf, Nb, Ta) in natural rocks are important petrogenetic indicators for the geochemical evolution of the Earth's crust and upper mantle. Zr and Hf are predominantly hosted by the accessory mineral zircon $(Zr,Hf)SiO_4$, which occurs in nature as a Zr-rich solid solution between the end-members zircon and hafnon. Dissolution and re-crystallization of zircon therefore strongly controls the mobility of Zr and Hf in processes of the deep Earth (e.g., [1]). However, very little is known about the solubility of zircon in aqueous solutions at high pressures and temperatures as well as the incorporation mechansm of Zr and Hf in aqueous fluids.

In this study, we used a hydrothermal diamond cell (HDAC) to determine Zr and Hf contents in aqueous fluids in-situ at high P and T by X-ray fluorescence (XRF) spectroscopy. We particularly focused on the influence of Na-Al silicate components dissolved in the aqueous solution. With the same apparatus, the complexation of Zr and Hf in fluids was constrained by in-situ XANES measurements at the Zr K and Hf L_3-edges, performed on the fluids in equilibrium with zircon or hafnon. Finally, the complexation of these two elements was investigated using ab-initio molecular dynamics (MD) simulations.

2. Experimental

2.1. Diamond anvil cell and sample composition

In-situ high T and high P experiments were performed using a modified Bassett-type HDAC ([2]). This cell has a recess with a diameter of 200 μm and a depth of up to 80 μm in the culet face of the lower anvil. This modification allows acquisition of the fluorescence signal from the fluid in the recess at 90° to the incident beam. Synthetic zircon or hafnon crystals, a silicate glass chip and distilled water were loaded into the HDAC. In case of HCl or NaOH-solutions, only a crystal and the solution were loaded. The glass compositions ranged from $Na_2Si_3O_7$, $Na_2Si_3O_7$ + 5 wt% Al_2O_3, to $NaAlSi_3O_8$. At run conditions the glass chip had dissolved in the aqueous fluid to form a homogeneous fluid. The pressure at the desired experimental T was estimated from the equation of state of water [3] and the vapor-liquid homogenization temperature.

2.2. X-ray fluorescence and absorption spectroscopy

For the zircon solubility study, XRF spectra were collected at beamline L, HASYLAB, using a high bandwidth multilayer (Ni/C) monochromator, and an excitation energy of 21 keV. A single-bounce capillary was used to focus the beam to a spot of about 11 μm in diameter [2]. In addition, a polycapillary was placed in front of the Si drift-chamber solid-state detector. This permitted confocal measurement of the XRF, which substantially improves the fluorescence to background ratio by reducing the signal from inelastic scattering (Fig. 1, [4]).

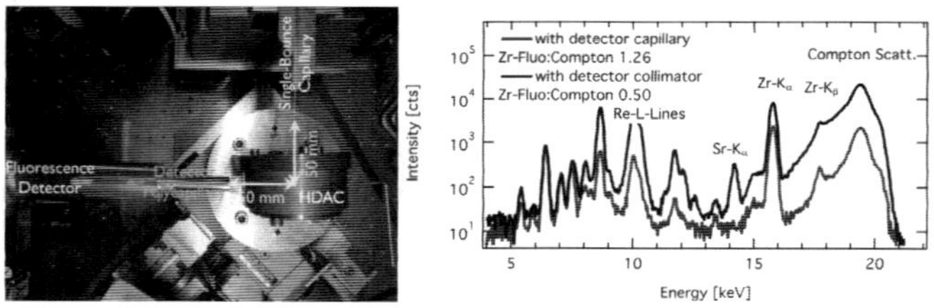

Figure 1: (left) Photograph showing a top view of the confocal arrangement at beamline L, HASYLAB. (right) Comparison of XRF spectra obtained with and without detector polycapillary.

Hafnium L_3-edge XANES spectra were collected at beamline ID24, ESRF. Details on the dispersive optical scheme are described in ref. [5]. The optics is based on a vertically and a horizontally focusing mirror in Kirkpatrick-Baez geometry and a horizontally bent Si (111) Laue polychromator crystal. The focal spot size achieved with this setup was < 5 μm FWHM. XANES spectra in fluorescence mode were recorded by a stepwise scan of a 100 μm slit, positioned just downstream of the polychromator. A Si-drift detector together with a polycapillary half-lens was used to record the fluorescence signal in a confocal arrangement (see above). XRF spectra were also used to determine the solubility of hafnon.

High-resolution XANES spectra at the Zr-K-edge were acquired at beamline ID26 using a high-resolution wavelength dispersive spectrometer in Rowland geometry [6] for fluorescence detection and a Si (111) double-crystal monochromator. The spot size was 80 x 120 μm^2. High-resolution X-ray emission detection was achieved with 4 spherically bent Ge crystals (bending radius 1004 mm) in (888) orientation and a Si drift detector. During XANES scans the emission spectrometer was tuned to the $K_{\alpha 1}$ emission line of Zr (15775.1 eV). The convoluted resolution of the monochromator and the enission spectrometer was 2.8 eV.

2.3. Ab-initio Molecular dynamics simulation and XANES calculation

Ab initio molecular dynamical simulations were performed using the Car–Parrinello molecular dynamics code CPMD [7] which implements the density functional theory (DFT) using a plane-wave basis set and pseudo potentials for the core electrons. The exchange-correlation energy was calculated using the Perdew-Burke-Ernzerhof modification to the Generalized Gradient Approximation and Trouillier-Martins-type pseudopotentials were used. Gas phase optimized clusters of $[Zr(OH)_n]^{4-n}$ and $[Hf(OH)_n]^{4-n}$ with n=1 to 8 were immersed in simulation boxes containing 32 water molecules at densities corresponding to a pressure range of 0.4 GPa to 2 GPa at 1000 K. XANES spectra were calculated using Feff 9.03 [8]. Input structures were either estimated from crystal structures or snapshots taking from the MD trajectories were used.

3. Results and Discussion

3.1. Zirconium and Hafnium concentrations in aqueous fluids

Zircon solubility is strongly enhanced by addition of $Na_2Si_3O_7$ (NS3) to H_2O (Fig. 2). The degree of enhancement increases with $Na_2Si_3O_7$ concentration. The Zr content of fluids containing 10 wt% $Na_2Si_3O_7$ reached up to 86±2 ppm Zr at 550°C and 400 MPa. At 30 wt% $Na_2Si_3O_7$, the maximum Zr concentration was 997±6 ppm at 600°C and 440 MPa. Zircon solubility in these fluids decreases considerably with pressure and increases slightly with temperature. Addition of Al_2O_3 decreases the zircon solubility. In H_2O-$NaAlSi_3O_8$ fluids, Zr concentrations are in the sub-ppm to ppm range. Zr concentrations in NaOH solutions in equilibrium with baddeleyite reached up to 390±2 ppm at 600°C and 930 MPa and increase with pressure and temperature (see also [9]). The solubility of hafnon shows quite similar systematics as zircon. However, the maximum absolute Hf concentrations in NS3-bearing fluids are much higher, by a factor of 1.5 on molar basis. The different dependence of the Zr and Hf contents on pressure and temperature in silicate-H_2O fluids and in NaOH-solutions points to different complexation of Zr and Hf. For Na-Al silicate-bearing fluids, Zr and Hf likely complex with silicate components in the fluid.

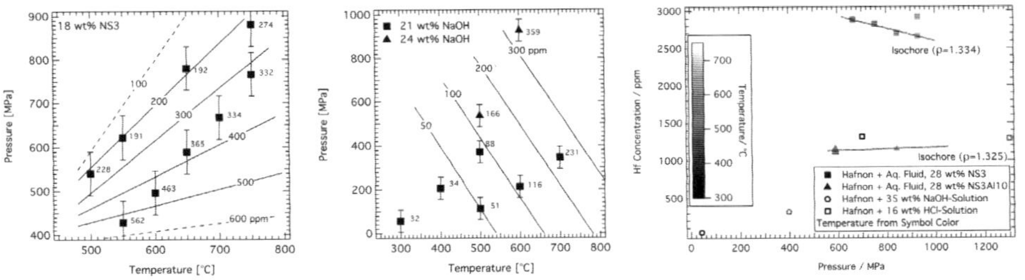

Figure 2: (left) solubility of zircon in aqueous fluid containing 18 wt% $Na_2Si_3O_7$ (NS3) as a function of pressure and temperature. Tags indicate Zr concentrations in the fluid and lines indicate isopleths of Zr content. (middle) solubility of monoclinic ZrO2 in NaOH-solution. Tags and lines as before. (right) solubility of hafnon in aqueous fluids with dissolved NS3, NaOH, and HCl-solutions as indicated.

3.2. XANES at Zr K-edge and Hf L_3-edge – constraints on complexation in fluid

In-situ XANES spectra collected on Zr in H_2O-$Na_2Si_3O_7$, H_2O-$Na_2Si_3O_7$+Al_2O_3, H_2O-NaOH, and H_2O-HCl fluids in equilibrium with zircon provide evidence for strong differences in the Zr complexation between these fluids (Fig. 3). Comparison of XANES spectra to those of model compounds and ab-initio simulation of XANES spectra revealed [8]Zr for the standard solution and [7]Zr for the NaOH solution. In the HCl-solution, Zr is coordinated by both O and Cl. [6]Zr and [6]Hf are found for the Na-Al-silicate-bearing solutions and formation of alkali zircono-silicate complexes is indicated by the similarity of XANES spectra to spectra simulated based on the local structure around Zr in the alkali zircono-silicates vlasovite and catapleite [8]. Alkali zircono-silicate complexes are responsible for the enhancement of Zr concentrations in Na-Al-silicate-bearing solutions.

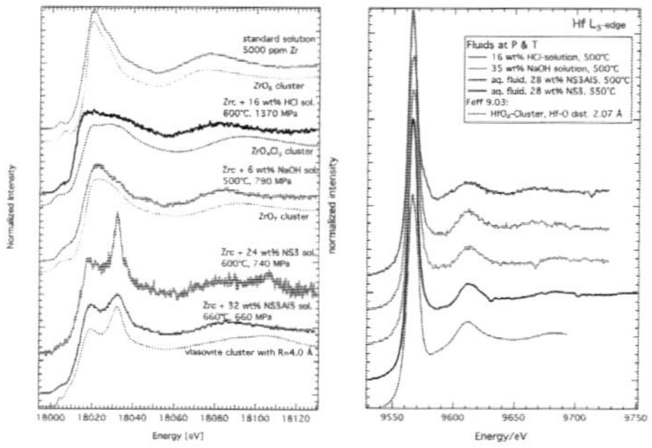

Figure 3: (left) Experimental and calculated XANES spectra at the Zr K-edge of aqueous fluids at P and T for various compositions as indicated. (right) Experimental and calculated XANES spectra at the Hf L_3-edge of aqueous fluids at P and T for fluid compositions as indicated.

3.3. MD simulation of Zr-monomers in aqueous solution

For solutions with $[Zr(OH)_n]^{4-n}$ and with n=1 to 8, Zr and Hf is five to six-fold coordinated by oxygens (Fig. 4). XANES spectra calculated based on monomers of the MD results do not reproduce the experimental spectrum of Zr in NaOH solution. The experimental spectrum indicates further contributions by a higher coordination shell indicating polymerization of Zr with other components. In HCl-solutions mixed oxychloride complexes are formed in MD runs, which is consistent with the experimental results (Fig. 3).

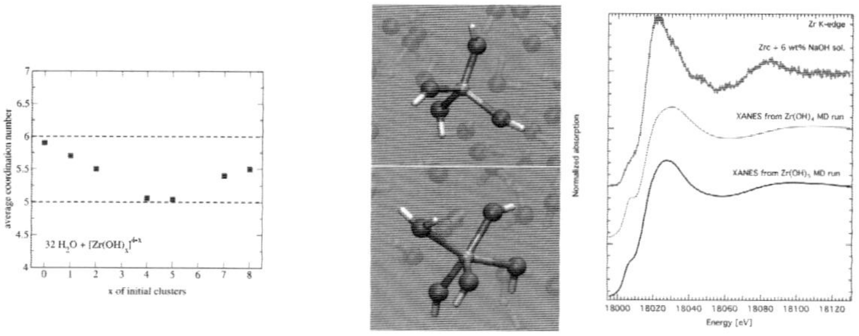

Figure 4: (left) Zr coordination in fluids with $[Zr(OH)_n]^{4-n}$ with n=1 to 8. (middle) snapshot of MD runs for 4-fold and 5-fold coordinated Zr monomers. (right) XANES spectra calculated with Feff based on indicated MD simulations in comparison to Zr measured in NaOH solution.

References

[1] Hoskin PWO, Schaltegger U 2003 *Rev. Min. Geochem.* **53** 27
[2] Schmidt C, Rickers K, Wirth R, Nasdala L, Hanchar J 2006 *Am. Mineral.* **91** 1211
[3] Wagner W, Pruss A 2002 *J. Phys. Chem. Ref. Data* **31** 387
[4] Wilke M, Appel K, Vincze L, Schmidt C, Borchert M, Pascarelli S 2010 J. Synchr. Rad. **17** 669
[5] Pascarelli S, Mathon O, Munoz M, Mairs T, Susini J 2006 *J. Synchrotron Rad.* **13** 351
[6] Glatzel P, Bergmann U 2005 *Coord. Chem. Rev.* **249** 65
[7] www.cpmd.org
[8] Rehr JJ, Kas JJ, Prange MP, Sorini AP, Takimoto Y, Vila F 2008 *C. R. Phys.* **10** 548
[9] Wilke M, Schmidt C, Dubrail J, Appel K, Borchert M, Kvashnina K, Manning CE 2012 *Earth. Planet. Sci. Let.* **349-350** 15

Journal of Physics: Conference Series **430** (2013) 012123 doi:10.1088/1742-6596/430/1/012123

Time-resolved combined XAS and UV-Vis applied to the study of the cerium catalysed BZ reaction

M Hagelstein[1], T Liu[1], S Mangold[1] and M Bauer[2]

[1]Karlsruhe Institute of Technology, Karlsruhe, Germany
[2]University of Kaiserslautern, Kaiserslautern, Germany

michael.hagelstein@kit.edu

Abstract. X-ray absorption spectroscopy has been applied to the study of the Belousov-Zhabotinsky reaction. Ultraviolet-visible spectroscopy has been used as the reference spectroscopic technique. The extent of the reaction has been followed continuously by acquiring the X-ray absorption at specific photon energies during the process. Oxidation states as well as local structural data of the cerium catalyst and the bromine reactant species during self sustained chemical oscillations have been determined.

1. Introduction

The Belousov-Zhabotinsky (BZ) reaction [1, 2] is the most prominent model reaction for self-sustained chemical oscillations. The basis for the description of the reaction pathways and kinetics has been described [3].

X-ray absorption spectroscopy (XAS) provides invaluable insight into the electronic structure and short range geometrical order of condensed matter. Not only to well ordered crystalline materials, it has been also applied to the study of disordered or nanostructured systems, i.e. catalysts, amorphous, glasses or liquids. For many years this method has been combined with other experimental techniques (microscopy, ultraviolet-visible (UV-Vis), infrared (IR) or Raman spectroscopy) to obtain a full picture of the structure on different length scales and resolved in time [4].

After the initiation of the BZ reaction, the concentration of constituents oscillates between different states, either in terms of oxidation or molecular states. The analysis of spectra as well as of electrical potential during the reaction has allowed a detailed analysis of the reaction mechanism. The overall chemistry can be described by process T without regard for the detailed organic chemistry [5].

$$BrO_3^- + 3RH + H^+ \rightarrow 2ROH + RBr + H_2O \qquad (T)$$

The organic moiety R is oxidized. Cerium ions are used to catalyse the reaction. The process T describes the overall chemistry but not the detailed evolution of the reaction. Today many species, reactions and overall reaction models are reported [6]. The reaction remains after initiation far from chemical equilibrium for many hours. After an induction period, self-sustained chemical oscillations are observed. The oscillation period of the BZ reaction is typically in the seconds to minutes range.

2. Experimental

The experiment reported here has been performed at the ANKA-XAS beamline at the absorption edges Ce L_3 and Br K. XAS and UV-Vis spectroscopy have been applied using a specifically designed in-situ cell. To monitor the compositional and structural changes in the reaction, quick scan mode XAS was employed together with a time scan mode at fixed monochromatic photon energy.

2.1. Sample and sample environment

The chemicals $Ce(NO_3)_3 \cdot 6H_2O$, $C_3H_4O_4$ (malonic acid), $KBrO_3$ were used as received and dissolved in 1 M acidic solution (H_2SO_4) using deionised water. The sample solution was kept in a stirred batch reactor. The chemical solution was pumped into the sample cell for combined XAS and UV-Vis experiments using a gear pump and thin tubes. The small inner aperture of the tubes and high pumping speed assured the concurrence of chemical reaction states in both vessels.

Glass fibre optics was applied connecting the sample cell to the grating/diode array based Ocean Optics UV-Vis spectrometer. The X-ray beam is transmitted through kapton windows and a nominal solution thickness of 5 mm.

The liquid acidic solutions have been mixed and filled into the batch reactor and into the programmable syringe pump. During the experiment S1 stored in the batch reactor was continuously pumped through the sample cell. In order to initiate the reaction, a defined quantity S2 was added to S1 in the batch reactor using the syringe pump. The stirred batch reactor was purged with helium.

Table 1: The composition of the solutions S1 and S2 in the initial state after mixing is summarized. The chemicals have been dissolved in 1M H_2SO_4 aqueous solution.

Experiment	Solution S1	Solution S2
Ce L_3 edge	0.05 mol/l $Ce(NO_3)_3 \cdot 6H_2O$ and 0.4 mol/l $C_3H_4O_4$	0.06 mol/l $KBrO_3$
Br K edge	0.06 mol/l $KBrO_3$	0.002 mol/l $Ce(NO_3)_3 \cdot 6\ H_2O$ and 0.04 mol/l $C_3H_4O_4$

2.2. XAS set-up

X-ray absorption spectroscopy data (in transmission and fluorescence mode) were taken at the XAS beamline at ANKA. The synchrotron is operated at 2.5 GeV energy and typical electron currents of 100 to 200 mA. The bending magnet beamline is equipped with a Si(111) double crystal monochromator together with ion chambers and a Ge multi-element fluorescence detector. Typically time scan mode at fixed photon energy with a time resolution of 0.2 s has been applied, interrupted by QEXAFS scans (typically 30 s scan time) during reaction and step by step EXAFS scans of stable states. EXAFS spectra of a set of reference compounds in solid and liquid state were acquired for the Ce L_3 edge as well as the Br K edge.

2.3. UV-Vis set-up

UV-Vis spectra were measured using an Ocean optics USB200 device. The spectral range comprises 178 to 874 nm wavelength. The data were corrected for background and dark signal. A single spectrum is taken by averaging five spectra which are acquired in 1.2 s. During the BZ reaction changes are observed in the spectral range of 250 to 450 nm. The UV absorption in cerium oxides is generated by a charge-transfer transition between O2*p* and Ce 4*f* bands [7]. The increasing absorption is attributed to the oxidation of Ce^{3+} to Ce^{4+}.

3. Results and discussion

The time dependent variation of the catalyst species has been measured at the Ce L_3 edge and the absorption of the reactant species at the Br K edge.

3.1. Time resolved experiment at the CeL$_3$ edge

The Ce L_3 edge XAS of Ce^{3+} ($4f^1$) in hydrated cerium samples shows a strong resonance absorption line at the photon energy of 5.728 keV. The near edge structure of Ce^{4+} ($4f^0$) of solid CeO_2 powder is considerably different with two less pronounced maxima at 5.732 and 5.739 keV. The resonance absorption features have been studied in detail in order to deduce valence states of rare earth elements dissolved in liquids [8] as well as in solids [9, 10]. The observed XANES of the BZ solution is characteristic for Ce^{3+}. The small change of the white line intensity during the reaction (as seen in the inlet plot) indicates that only a small amount of the trivalent ions is oxidized to Ce^{4+}. It has been estimated that the molar ratio Ce^{3+}/Ce^{4+} during all phases of the BZ reaction should be larger than 0.99.

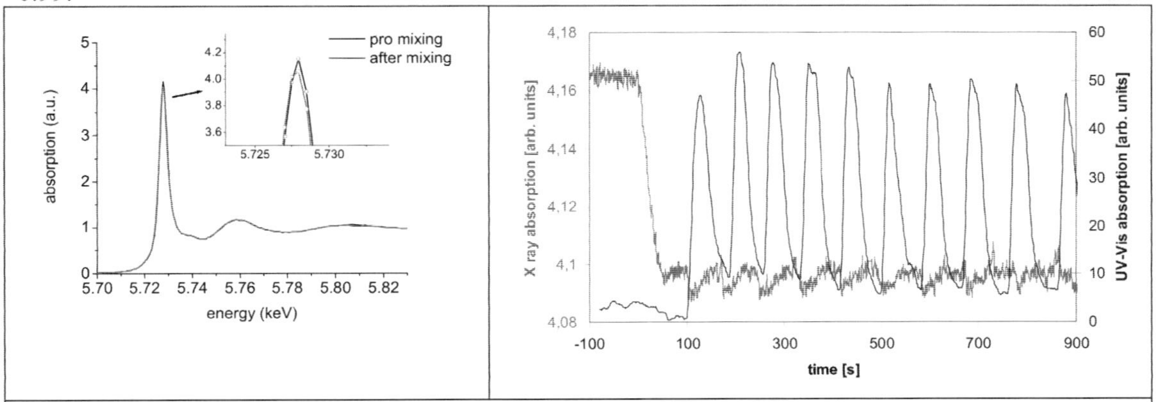

Fig. 1: Normalised X-ray absorption spectra at the Ce L_3 edge for the solution before and after the initiation of the reaction (left), the prominent white line was zoomed in and displayed in the inlet. The white line at 5.728 keV has been used for single photon energy time scans. The two spectra have been acquired before (black) and after initiation of the BZ reaction (red). Temporal variation of the X-ray fluorescence signal I_f/I_0 measured in monochromatic mode at the photon energy of 5.728 keV (left scale, red line) and the UV-Vis absorption integrated over the wavelength range 400 to 440 nm (right scale, blue line).

3.2. Time resolved experiment at the Br K edge

The spectra at the K absorption edge of bromine vary strongly as a function of Br oxidation state. The spectra of solid $KBrO_3$ and of this substance dissolved in water are characterized by a white line.

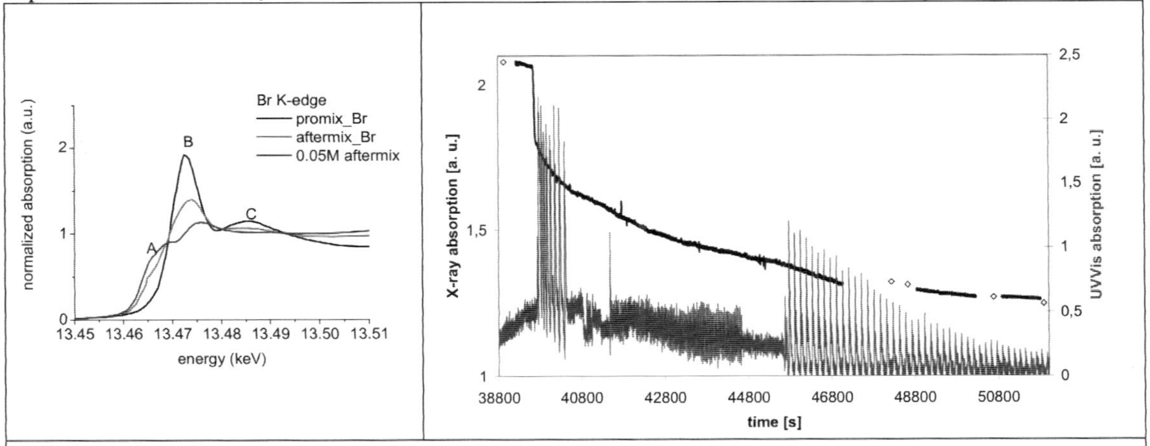

Fig. 2 left: Br K edge XANES of Br-containing solutions prior to mixing and at the two different stages of the reaction. The strong white line typical for BrO_3^- ions (black line, feature B) before initiation of the reaction is steadily damped during the course of the reaction. The final state spectrum (blue line) is characteristic for bromine bound to an organic moiety. Right: Temporal variation of the white line (blue line and blue diamonds) at the photon energy of 13.4725 keV (feature B) displayed together with the temporal variation of the UV-Vis absorption (dark red line) during the course of the BZ reaction. The reaction has been initiated at t_i = 39645 s. During the gaps of the time scan mode the QEXAFS spectra have been acquired.

The strong white line typical for bromate BrO_3^- ions corresponding to the reference spectrum of $KBrO_3$ in solution characterises the initial spectrum (fig. 2, see also [11]). Upon reaction the white line (feature B) is damped and shifts to higher photon energy, two shoulders prior to the white line and at the post edge develop (feature A and C, fig. 2).

The white line at the photon energy 13.473 keV is used for the acquisition of time scans. During the course of the reaction the intensity of the white line steadily decreases (fig. 2 right). The final state spectrum (fig. 2 left, blue spectrum) is characteristic for bromine bound to an organic moiety. The spectrum resembles a spectrum of $CHBr_3$ [12]. The quick consumption of bromate ions is also supported by the EXAFS analysis. The analysis of the intermediate state sample reveals only a weak coordination peak at the Br-O position, implying that a mixture of multiple Br species may coexist. The final product of Br shows two weak but distinct peaks with the first peak very different from Br-O coordination. The two peaks can be assigned to Br-C and Br-Br coordination shells. The result implies the reaction with malonic acid to form brom-malonic acid, indicating that the reaction of brome with malonic acid leads to a product that is a multiply halogenated carbohydrate. A Br-Br shell at a distance of 3.22 Å with a coordination number 1.8 is determined. The formation of multiply halogenated compounds during the BZ reaction has been reported [13].

4. Conclusions

A detailed analysis revealed that fast temporal variations observed in the UV-Vis signal correspond to the oscillatory variations in the X-ray absorption. The absorption determined with the two experimental techniques during the course of the reaction shows a well defined frequency as well as phase relationship. With the oxidation of Ce^{3+} a reduction of bromine ions should take place at the same time. The oxidation of cerium ions lead to an increased UV-Vis absorption, as does the reduction of bromine ions leads to an increase of the X-ray absorption signal of the white line at the Br K edge. The acquisition of the X-ray absorption signal interlaced with QEXAFS spectra allowed the determination of kinetic as well as structural data of the catalyst and one of the reactant species.

Acknowledgements

The authors wish to thank Wolfgang Mexner for software development required for the analysis of time resolved data. The provision of synchrotron radiation by the synchrotron radiation facility ANKA at KIT (reference MR-142) and ESRF (reference CH200) is kindly acknowledged.

References

[1] Belousov B P 1958 Sbornik Referatov po Radiotsionnoi Meditsine, Moscow Medgiz pp 145-147
[2] Zhabotinsky A M 1964 Biofizika **9** pp 306-311
[3] Field R J, Körös E and Noyes R M 1972 J. Am. Chem. Soc. **94** pp 8649-8664
[4] Bauer M and Bertagnolli H 2009 Chem. Phys. Chem. **10** pp 2197-2200
[5] Noyes R M 1990 J. Phys. Chem. **94** pp 4404-4412
[6] Hegedüs L, Wittmann M, Noszticzius Z, Yan S, Sirimugkala A, Försterling H-D and Field R J 2001 Faraday Discuss. **120** pp 21-38
[7] Tsunekawa S, Fukuda T and Kasuya A 2000 J. Appl. Phys. **87** pp 1318-1321
[8] Sham T K 1989 Phys. Rev. **B 40** 9 p 6045-6051
[9] Röhler J 1985 J. Magnetism and Magnetic Materials **47&48** pp 175-180
[10] Sham T K, Gordon R A and Heald S M 2005 Phys. Rev. **B 72** p 035113
[11] Levason W, Ogden J S, Spicer M D and Young N A 1990 Journal of the Chemical Society Dalton Transactions: Inorganic Chemistry (1) pp 349-53
[12] Feiters M, Küpper F C and Meyer-Klaucke W 2005 J. Synchr. Radiation **12** pp 85-93
[13] Degn H 1966 Nature **213** pp 589-590

Quick-Scanning QEXAFS in grazing incidence: Surface science in sub-seconds

D Lützenkirchen-Hecht, J Stötzel, O Müller, J Just and R Frahm*

Bergische Universität Wuppertal, Gaußstr. 20, 42097 Wuppertal, Germany

*E-mail: frahm@uni-wuppertal.de

Abstract. Quick scanning X-ray absorption spectroscopy was combined with the grazing incidence geometry for the time resolved in situ investigation of surfaces and thin films. The results demonstrate that a time resolution of about 50 ms for a single spectrum is feasible using a dedicated monochromator. In-situ investigations performed during the sputter deposition of thin gold films demonstrate the capabilities of this new approach.

1. Introduction

Many important physico-chemical phenomena such as e.g. catalysis, the oxidation or corrosion of materials by gases and liquids and coating procedures are taking place at the surfaces of materials. All of those processes are strongly time dependent, so that fast surface sensitive in-situ probes are necessary to study those phenomena. The X-ray absorption fine structure (EXAFS) spectroscopy is a well established tool that can also be used for studies of non-crystalline solids or liquids, and it provides accurate short range structure information such as bond distances, coordination numbers and the local atomic disorder [1]. EXAFS can be made surface sensitive by using the grazing incidence geometry: For incidence angles below the critical angle of total reflection the penetration depth of the X-rays amounts to only few nanometers, and thus the reflected X-ray beam only contains information about the near surface region of the studied samples [2-6]. Here we have combined the grazing incidence geometry with the Quick-scanning EXAFS (QEXAFS) method [7]. Recent experimental developments of QEXAFS instrumentation have substantially improved the time resolution reliably to about 50 ms for a single spectrum [8-13]. Here we will show for the first time that thin film growth by sputtering can be studied in situ on a sub-second timescale by combining the Quick-scanning EXAFS technique and the grazing incidence geometry.

2. Experimental details

The reflection mode QEXAFS spectra were recorded using the QEXAFS monochromator installed at the SuperXAS beamline at the Swiss Light Source (Villigen, Switzerland) [10]. The basic setup and the monochromator system are described in detail elsewhere [8]. The storage ring was operated in the top-up mode with an electron beam of 2.4 GeV at about 400 mA. The X-ray beam from the superbend magnet was vertically collimated on the QEXAFS channel-cut monochromator crystal to improve the energy resolution and to increase the available flux, while a bent cylindrical mirror behind the monochromator focused the beam onto the sample, resulting in a spot size of 100 μm x 100 μm and a flux of about 10^{12} photons/s at the position of the sample [10]. The beam incident on the samples was

Journal of Physics: Conference Series **430** (2013) 012124 doi:10.1088/1742-6596/430/1/012124

further defined by a slit system. All QEXAFS data presented here were taken using a Si(111) channel-cut crystal and a 0.30° excenter disc (see [9] for details), resulting in scan ranges of about 750 eV at the Au L_3-edge (11919 eV). The Bragg angle of the monochromator crystal was measured simultaneously with the acquired spectra using a specially developed fast angular encoder system [9]. N_2- and Ar-filled ionization chambers were used as detectors for the incident and reflected X-rays. A gold metal foil placed between the second and a third ionization chamber was measured simultaneously to the actual sample as a reference. DC sputter deposition was performed in a miniaturized sputter chamber equipped with a magnetron source (55 mm diameter) and a resistively heated sample holder [14]. The cell has two large area Kapton windows for the incident and reflected X-rays, electrical feedthroughs for the resistive sample heater and the temperature measurements, and a gas inlet that is regulated by a gas-flow controller. The cell is evacuated by a turbomolecular pump (Pfeiffer TMU 071) that enables a base pressure of typically 10^{-5} mbar, while the Ar-gas pressure was adjusted in the range of $2x10^{-2}$ mbar during the thin film depositions on float glass substrates.

The continuously collected spectra were separated, averaged, calibrated and normalized by in-house software tools [11]. The further reflection mode XAFS data analysis included a background subtraction and the extraction of the reflectivity fine structure $\Delta R(E)$ by using the Athena software which is part of the IFEFFIT EXAFS data analysis software [15]. In contrast to conventional transmission or fluorescence mode EXAFS, both the real $\delta(E)$ and the imaginary part $\beta(E)$ of the complex refractive index $n(E)=1-\delta(E)-i^*\beta\,(E)$ contribute to the reflectivity fine structure depending on the grazing angle and the photon energy E. Furthermore, for a thin film on a substrate, the presence of surface and interface roughness may change the distribution of the X-ray absorbing species within the penetration depth of the X-rays. Thus the respective contributions are affected in the detected reflectivity EXAFS signal, so that reflectivity EXAFS data cannot simply be analyzed using standard routines (see e.g. [1, 6]). If the X-ray penetration depth is larger than the film thickness, the absorption can not be directly extracted from the measured reflectivity EXAFS spectra. For the extraction of short range order structural information, the measured data have to modeled on an ab-initio basis [4, 6, 16]. Here we have calculated the reflectivity using the Fresnel theory and the Distorted Wave Born Approximation (DWBA) including surface and interface roughness [6]. The input parameters of these calculations, especially the film thickness and the roughness parameters, were refined until the results matched closely with the experimental data.

3. Results and discussion

In Fig. 1, the reflection mode QEXAFS spectra collected during the sputter deposition of a thin gold film on a glass substrate are presented for an incidence angle of $\Theta=0.23°$. Due to the fact that the critical angle of total reflection for glass is only about 0.2°, the measured reflectivities are low at the start of the experiment. However, the reflectivity continuously increases during gold deposition. The measurements were performed with 10 Hz oscillation frequency of the channel-cut monochromator crystal, i.e. one single spectrum was measured within 50 ms. The displayed data were obtained by averaging over 10 subsequent spectra measured going up in Bragg angle, so that each spectrum contains the information measured during 1 s. It should be stressed that the measurements are technically very demanding: Due to the low reflectivity of the glass substrate, which amounts to less than 0.03, the intensity in the second and third ionization chambers are very small at the beginning of the experiment, i.e. prior to the gold deposition. Thus a high amplification in the used current amplifiers would be desirable in order to obtain a high quality signal with low noise. However, in the course of the deposition process, the reflectivity of the gold layer increases by more than a factor of 30 to values of about 0.9 in the pre-edge region of the Au L_3-edge. Thus, at the end of the film growth, significantly smaller current amplification is required in order not to exceed the range of the current amplifiers. Since the current amplification can not be changed in the course of the experiment, its value has to be selected on the basis of the maximum signals that are measured during the entire experiment.

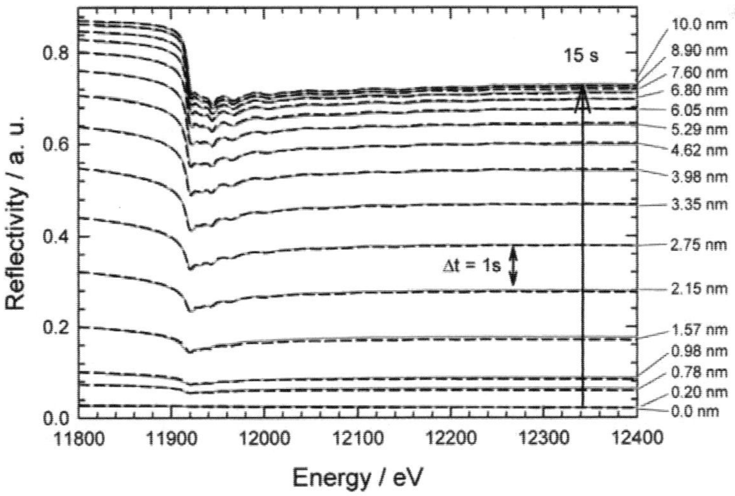

Figure 1. Reflection mode ($\Theta = 0.23°$) QEXAFS data at the Au L_3-edge collected during the sputter deposition of a thin gold film on a float glass substrate at room temperature. The spectra (solid grey lines) were measured with 10 Hz oscillation frequency, 10 subsequent up spectra were averaged and displayed. Fits of the experimental data using the distorted wave Born approximation (DWBA) are shown as dotted black lines. The thickness of the gold layer of each simulation is indicated.

Distinct signatures of the Au L_3 absorption edge are already visible after only 2 s of sputtering, and well developed fine structure oscillations are detectable after a few seconds of deposition. Here the XANES features typical for metallic gold are clearly visible after about 5 s, corresponding to a film thickness of about 2.75 nm. With increasing film thickness, all the features in the spectra become more pronounced.

For a more detailed analysis of the experimental data, we have fitted the spectra assuming a polycrystalline thin gold film on the glass substrate. Using the $\delta(E)$ and $\beta(E)$ values extracted from a transmission mode EXAFS spectrum of a gold foil and tabulated data for the glass substrate [17], we have calculated the energy dependent reflectivities of the thin films by applying the DWBA. By a variation of the film thickness t, the vacuum side surface roughness σ_1 and the substrate side interface roughness σ_2, the experimental data can be modelled, and calculated reflectivity data are included in Fig. 1. In Fig. 2, the influence of the different parameters on the calculated spectra in comparison to the experimental data measured 9 s after the start of the experiment is shown. Single spectra measured in 50 ms are plotted, which hardly can be distinguished from the best fit results. As can be deduced from Fig. 2(a), a small variation of the film thickness (t_f) from 4.59 nm to 4.65 nm causes significant changes in the reflectivity spectra. For constant vacuum and substrate side roughness parameters, an increase of t_f results in an overall increase of the reflectivity in the entire energy range, and thus, the best thickness value can be determined to be 4.62 nm in the present case. The Fourier-transformed data shown in the insert of Fig. 2(a) only show minor changes, namely a small increase of all peaks with increasing thickness. In Fig. 2(b) the influence of the vacuum side surface roughness on the calculated spectra is shown for a film thickness $t_f = 4.62$ nm. Obviously, the roughness mainly influences the post-edge region of the spectra, where the films with smaller roughness exhibit a larger reflectivity, and thus the influence of film thickness and roughness can be discriminated.

The increased edge jump also leads to slightly higher peaks in the Fourier-transformed data in the insert of Fig. 2(b) (see also [6]). The substrate side interface roughness only plays a minor role in the present case as the incidence angle Θ is significantly larger than the critical angle of the substrate, so only a small fraction of the impinging X-rays are scattered back to the surface and contribute to the signal measured in the detector [6]. As can be concluded from these plots, the calculated reflection mode EXAFS spectra are sensitively depending especially on the thickness t_f and the surface roughness σ of the deposited gold thin films, and thus, the determination of those parameters from the X-ray data measured in-situ during thin film growth is possible. This is also illustrated in Fig. 1, where the calculated reflectivity spectra are also shown, and the determined film thicknesses are indicated as a function of time. As can be deduced from this figure, the correct description of the X-ray data is possible from the initial stages of material deposition to the film growth on scales of several nm. This is a unique capability – of course the film thickness and roughness may be obtained by other techniques. However there are only very few techniques that may be applied in-situ during sputtering, but non of those may also yield the atomic structure of the deposited films on a sub-second time scale.

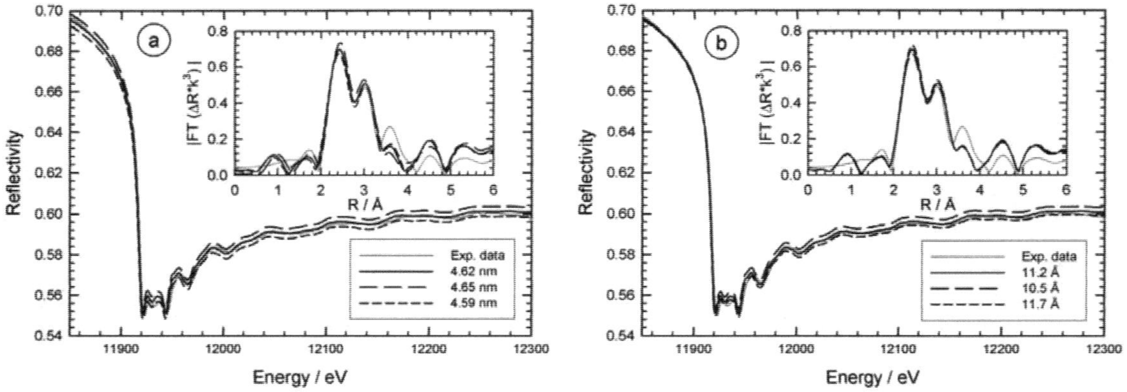

Figure 2. Experimental reflectivity EXAFS (Θ=0.23°) data measured after 9 s of gold sputter deposition (full grey line) and simulated spectra with varying thickness and surface roughness parameters: **(a)** Variation of the thickness around t_f=4.62 nm. Vacuum side surface roughness σ_1=11.2 Å, substrate side roughness σ_2=7.0 Å. **(b)** Variation of σ_1 around 11.2 Å for t_f=4.62 nm and σ_2=7.0 Å. In the inserts, the Fourier-transformed data are shown (data are not phase-shift corrected; k-range for the FT: 2 $Å^{-1}$ < k < 9.8 $Å^{-1}$).

4. Conclusions

The results have shown that a detailed investigation of thin film growth processes is feasible using a combination of quick-scanning EXAFS and the grazing incidence geometry. The measured data can be modeled quantitatively as a function of time using the Distorted wave Born approximation, yielding precise values e.g. for the thickness, roughness and the roughness scaling behaviour of the growing films. The determined values will be given in a detailed publication, which furthermore will also include the oxidation of thin films.

5. References

[1] Koningsberger D C, Prins R 1988 *X-ray absorption: Principles, Applications, Techniques of EXAFS, SEXAFS and XANES*, (New York: Wiley)
[2] Parratt L G 1954 *Phys. Rev.* **95** 359
[3] Martens G and Rabe P 1981 *J. Phys. C: Solid State Phys.* **14** 1523
[4] Borthen P and Strehblow H-H 1995 *Phys. Rev.* B **52** 3017
[5] Benzi F, Davoli I, Rovezzi M and d'Acapito F 2008 *Rev. Sci. Instrum.* **79** 103902
[6] Keil P and Lützenkirchen-Hecht D 2009 *J. Synchrotron Rad.* **16** 443
[7] Frahm R 1989 *Rev. Sci. Instrum.* **60** 2515
[8] Frahm, R, Griesebock B, Richwin R, et al. D 2004 *AIP Conf. Proc.* **705** 1411
[9] Stötzel J, Lützenkirchen-Hecht D, Fonda E, et al. 2008 *Rev. Sci. Instr.* **79** 083107
[10] Frahm R, Nachtegaal M, Stötzel J, et al. 2010 *AIP Conf. Proc.* **1234** 251.
[11] Stötzel J, Lützenkirchen-Hecht D and Frahm R 2012 *J. Synchrotron Rad.* (in print)
[12] Baudelet F, Belkhou R, Briois V, et. al. 2005 *Oil & Gas Sci. Technol.* **60** 849
[13] Uruga T, Tanida H, Inoue K, Yamazaki H and Irie T 2007 *AIP Conf. Proc.* **882** 914
[14] Lützenkirchen-Hecht D, Bruder K, Haake U, et. al. 2005 *Rev. Sci. Instrum.* **76** 073905.
[15] Ravel B and Newville N 2005 *J. Synchrotron Rad.* **12** 537
[16] Souza-Neto N M, Ramos A Y, Tolentino H C N, et al. 2009 *J. Appl. Cryst.* **42** 1158
[17] Henke B L, Gullikson E M and Davis J C 1993 *Atom. Data Nucl. Data Tables* **54** 181

The Use of Differential EXAFS Analysis for the determination of Small Structural Differences between two closely-related Ruthenium Complexes

D. Gianolio[1]*, E. Borfecchia[2], C. Garino[2], T. Ruiu[2], C. Lamberti[2] and L. Salassa[3]

[1]Diamond Light Source Ltd., Science and Technology Facility Council, Harwell Campus, Didcot OX11 0DE (UK)
[2]Department of Chemistry and NIS Centre of Excellence, University of Turin, via P. Giuria 7, 10125 Turin (Italy)
[3]CIC biomaGUNE Paseo Miramón 182, 20009 Donostia–San Sebastián (Spain)

E-mail: diego.gianolio@diamond.ac.uk

Abstract. X-ray Absorption Spectroscopy (XAS) is a sensitive and powerful technique in revealing the structure of a material, providing as well high accuracy on interatomic distances. Nevertheless, when dealing with systems that differ only by small structural features, a standard data analysis might be unable of discriminating between such differences. A differential approach was proposed by Bressler, Chergui, and co-workers [2003, *Phys. Rev. Lett.* **90**, 047403][2006, *J. Phys. Chem B*, **110**, 14035][2009 *Angew. Chem., Int. Ed.* **48** 2711] to solve this problem and differentiate between excited and unexcited state structures during pump-and-probe transient XAS experiments. In this contribution, we apply the differential data analysis procedure to the study of two closely-related molecular complexes, namely *cis*-$[Ru(bpy)_2(py)_2]^{2+}$ and *cis*-$[Ru(bpy)_2(py)(H_2O)]^{2+}$, characterized under static conditions. It is herein demonstrated that the method, based on a direct fit of differential curves, is able to reveal the small differences present between the two structures which, conversely, could not be resolved by standard EXAFS fitting of full spectra.

1. Introduction

A major challenge for X-ray spectroscopy is to improve sensitivity and accuracy in order to increase the capability to discriminate minimal differences between similar structures. In the case of EXAFS spectroscopy, the quality of the experimental data reached nowadays allows to obtain structural parameters with accuracy up to 0.01 Å (if the experimental energy range, number of points and acquisition time are properly selected to obtain the best signal to noise ratio). Nevertheless, when dealing with complex structures (where the distribution of atoms around the absorbing atom is very heterogeneous, both for the variety of elements and their bond lengths), the accuracy that can be reached is limited also by the fitting procedure. In fact, the mathematical correlations between fit variables can increase the errors on the optimized parameters. Research for the highest accuracy is not crucial when the aim of an EXAFS study is to reveal the structure of an unknown system, but it becomes essential if the goal is comparing similar systems. When structural differences are close to the technique limits, there is a high risk of losing such information. This problem is common in pump-and-probe experiments, where the aim is to appreciate small time-dependent changes between the ground state geometry and the structural features of photo-generated transient species and

Journal of Physics: Conference Series **430** (2013) 012125 doi:10.1088/1742-6596/430/1/012125

photoproducts. For this kind of time resolved studies, Bressler, Chergui and co-workers[1-2] specifically developed the so-called X-ray transient absorption spectroscopy (XTA) which is based on a differential approach for both data acquisition and analysis. The technique directly records differential spectra, measuring for each point of the spectrum the value of excited and unexcited system. The differential approach was adopted successfully for light-responsive molecules but also in case of systems sensitive to magnetic fields[3] and changes of temperature[4-5], pressure[6-7], and pH[8-10]. A recently proposed strategy to obtain quantitative structural refinement from XTA data is based on a direct fitting of the differential signal, in energy or wavevector (k) space, by means of systematic variations of structural parameters. It was shown that this method, besides emphasizing the variations between the spectra, can increase the sensitivity of the experiment and reduce possible systematic errors during the fitting procedure[11-16].

In our contribution we present an alternative application for this differential data analysis approach by studying two distinct systems measured under static conditions and highlighting their small structural differences.

In particular, we performed a study of the two closely-related molecular complexes *cis*-$[Ru(bpy)_2(py)_2]^{2+}$ (**1**) and *cis*-$[Ru(bpy)_2(py)(H_2O)]^{2+}$ (**2**) (where bpy = bipyridine and py = pyridine).

Complex **1** upon UV and visible light excitation in water solution releases one of the py ligand with high yield (ca. 20%)[17]; the subsequent coordination of a water molecule results in the stable aqua photoproduct **2**. Such kind of photoactivatable complexes has shown unique properties that have been successfully exploited for several applications. They are promising cage compounds for their ability to release bioactive ligands[18-20], and the high reactivity of the aqua photoproducts formed under light irradiation can be exploited to design photocontrollable anticancer agents [21-24]. Aiming at a thorough structural characterization of these systems and understanding their photoreactivity is fundamental for further developments. We recently have undertaken a comprehensive investigation on **1** and **2** and their light-driven dynamics by means of a multi-technique characterization, based on TR-WAXS (Wide Angle X-ray Scattering)[25], XTA (X-ray Transient Absorption) and OTA (Optical Transient Absorption) experiments supported by DFT (density functional theory) calculations[26]. The results presented in this work set the bases for a full comprehension of the system in static conditions by revealing small structural differences otherwise not appreciable. Moreover the data analysis here described gives an important contribution to the correct interpretation of the time-resolved EXAFS studies.

2. Experimental Section

Sample Preparation

The complexes *cis*-$[Ru(bpy)_2(py)_2](ClO_4)_2$[27] and *cis*-$[Ru(bpy)_2(py)(H_2O)](ClO_4)_2$[28] were synthesized as described in the literature and characterized accordingly. To increase the water solubility of *cis*-$[Ru(bpy)_2(py)_2]^{2+}$, the chloride derivative *cis*-$[Ru(bpy)_2(py)_2]Cl_2$ was prepared by anion exchange of a methanol solution of *cis*-$[Ru(bpy)_2(py)_2](ClO_4)_2$ on a Dowex weakly basic anion exchanger resin (purchased from Aldrich).

EXAFS data acquisition

The EXAFS spectrum for **1** was measured during an XTA experiment performed at the 11-ID-D beamline (APS, Chicago). The data were acquired in fluorescence mode; the sample was a 1 mM water solution flowing in a properly designed cell[29-30]. The EXAFS spectrum of **2** was collected in transmission mode at BM26A (ESRF, Grenoble). The sample (aqueous solution) was measured in a specific liquid cell with a thickness optimized to obtain an edge jump close to unit. Both the experiments were performed at the Ru K-edge (22117 eV) and the spectra were calibrated in energy thanks to the simultaneous collection of a reference metal foil.

Journal of Physics: Conference Series **430** (2013) 012125 doi:10.1088/1742-6596/430/1/012125

Computational details

The geometry optimization of **1** and **2** was performed with the DFT method, employing the Gaussian 09 (G09) program package[31]. The geometries were fully optimized with the PBE0[32-33] functional at the LanL2DZ/6-311G** level[34-35], using the CPCM method[36-38] to include the solvent effect (water).

3. Results and Discussions

3.1. Comparison of X-ray absorption spectra and results of the standard EXAFS fitting procedure

Observing the data reported in Figure 1a, it is evident that the spectra of **1** and **2** are almost identical both in the XANES and EXAFS regions. This is not surprising since, from and EXAFS perspective, the two structures are very similar. The major difference between **1** and **2** is the change of one of the ligands from a py ring to a water molecule (see inset in Figure 1a). In terms of EXAFS signal contributions, the first shell signal is expected to be only minimally perturbed since the N atom of the py and the O of the water molecule are almost iso-electronic. The higher shells suffer of signal loss due to lacking scattering paths related to the py ring, however the cumulative contribution of such paths can be estimated to be only 1/6 of the global signal (loss of 1 over 6 rings, 2 py rings and 4 rings associated in 2 bpy units).

Despite these considerations, we performed a test to check if it was possible or not to discriminate the two systems by a standard EXAFS analysis. Initially we tried to obtain a high quality fit on the spectrum of **1**. It was found that the best reproduction of the experimental curve, considering both fit goodness and physical meaningfulness of the parameters, was obtained fitting in R-space (range ΔR = 1.0 – 5.0 Å) the k-weighted FT function, transformed in the 2.5 – 10.8 $Å^{-1}$ k-range ($2\Delta k\Delta R/\pi \sim 21$). Phase and amplitude functions of each path were calculated by the FEFF code[39] using the DFT optimized structure as input. All the EXAFS paths up to R = 5.0 Å were included in the fitting model. To limit the number of optimized variables, a common amplitude factor (S_0^2) and an energy shift (ΔE) were adopted. Both py and bpy ligands were considered as rigid objects, thereby all path lengths were calculated according to geometrical constraints as a consequence of a radial translation along the corresponding Ru–N axis. In this way only other four parameters were optimized during the fit: $R_{Ru-N(py)}$ and $R_{Ru-N(bpy)}$ and their associated Debye-Waller (DW) factors $\sigma^2_{N(py)}$ and $\sigma^2_{N(bpy)}$. Once performed the fit, we obtained an outstanding agreement between experimental and simulated curve as visible in Figure 1b and confirmed by the R-factor close to 1%. The results (Table 1) show a slight general contraction with respect to the distances predicted by DFT but mainly validated the calculated structure, in agreement with what obtained from previous studies on other similar complexes.[20,24-25,40]

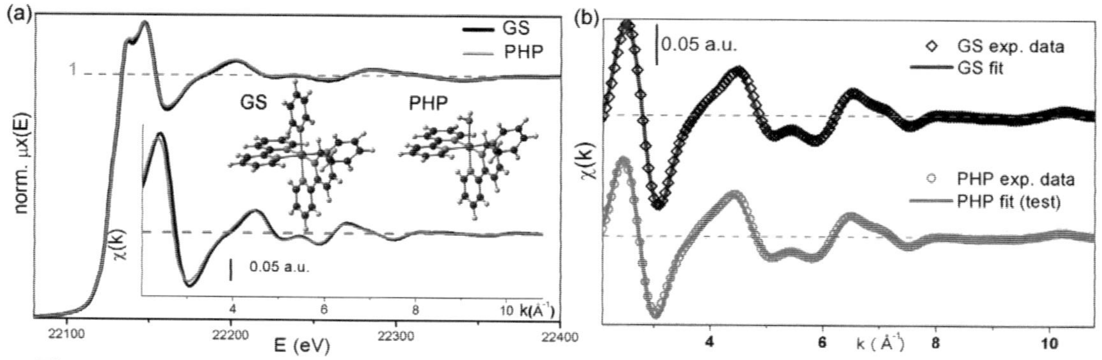

Figure 1. (a) Normalized $\mu x(E)$ spectra for GS (black line) and for PHP (orange line), in the inset $\chi(k)$ EXAFS curves for GS and PHP (same colors). Schemes of GS and PHP structures are also reported (Ru purple, N blue, C grey, O red). (b) Upper part: experimental $\chi(k)$ spectrum (black diamond) and its best fit (dark gray line) for GS; lower part: experimental $\chi(k)$ spectrum (orange circles) of PHP and its simulated curve (red line) obtained by a test fit using the same conditions and the same theoretical structure adopted for GS.

For the fit of **2,** close attention was paid to maintain exactly the same conditions adopted in the case of **1**, in terms of fitting-space, number of parameters and geometrical constraints. In order to obtain the best fit of experimental data, the normal procedure would have required calculating the EXAFS paths from the structure of **2** to generate the simulated curve. However, our aim was firstly to test if it was possible to discriminate the two systems, so we adopted the same paths previously calculated from the structure of **1.** In this way it was possible to establish if the differences between the two structures were negligible controlling if they could be compensated on a pure mathematical plane (small variations of the parameter values leading to an overall good fit). The simulated curve ensued from this test fit showed an outstanding agreement with experimental spectrum of **2** (Figure 1b) and the R-factor obtained was of 1%, as in the case of the best fit for the spectrum of **1**. Moreover, all the refined parameters values were comparable with the ones previously found for the fit of **1** within their experimental error. The loss of a py unit in **2** was well compensated by a slight increase of the associated DW factor σ^2_{py}, from (0.0032 ± 0.0008) $Å^2$ to (0.0041 ± 0.0009) $Å^2$. In view of these results, it was demonstrated that the sensitivity of the standard EXAFS analysis was not sufficient to disclose the differences between the two structures.

Parameters	**1** standard fit [DFT]	**2** test fit (with **1** structure)
R-factor	0.011	0.010
S_0^2	0.95 ± 0.06	0.94 ± 0.06
ΔE (eV)	0.8 ± 0.5	–0.4 ± 0.6
$R_{N(bpy)}$ (Å)	2.05 ± 0.02 [2.08]	2.04 ± 0.01
$\sigma^2_{N(bpy)}$ ($Å^2$)	0.0026 ± 0.0008	0.0023 ± 0.0008
$R_{N(py)}$ (Å)	2.09 ± 0.03 [2.14]	2.12 ± 0.03
$\sigma^2_{N(py)}$ ($Å^2$)	0.0032 ± 0.0008	0.0041 ± 0.0009

Table 1. Results of standard EXAFS fit on GS spectrum and of a test fit on PHP spectrum using GS model. $R_{N(L)}$ bond distances in the GS structure optimized by DFT calculation are reported for comparison, separately averaged for each type of ligand L.

3.2. The differential analysis: method and results

Once established that the standard EXAFS analysis method is not suitable for our systems, we tried to apply the differential analysis. First of all, we calculated the experimental differential curve (Figure 2a) by the subtraction of the two spectra $\chi(k)_2 - \chi(k)_1$, were $\chi(k)_2$ and $\chi(k)_1$ indicate the EXAFS spectra of **1** and **2**. Afterward, for a direct fitting of this curve, it was necessary to generate simulated differential signals starting from hypothetic structures of **2**. Hence, the following procedure was adopted. Most of the paths previously included in the simulation of the EXAFS spectrum of **1** were kept, and only the ones involving C atoms of the py ring missing in **2** were removed. The paths involving the N atom of the py unit were maintained, since this atom is substituted by the almost isoelectronic oxygen of the water molecule; however they were described with two additional parameters, namely $R_{Ru-O(H2O)}$ and σ^2_O. Fixing the amplitude S_0^2 at the value obtained during the fit of **1,** only 7 parameters remained to be refined: the energy shift (ΔE), 3 bond length variations (ΔR_{bpy}, ΔR_{py}, ΔR_O) referred to the DFT optimized structure of **1** and their corresponding 3 DWs (σ^2_{bpy}, σ^2_{py}, σ^2_O). All these parameters were initially set at values obtained during the fit of **1**, or derived by the structural changes predicted by DFT calculations (the new ΔR_O, σ^2_O refer to the values of the N_{py}). Then, a systematical variation was planned as explained in more details hereinafter. For each combination of parameter values (correlated to a structural variation), the sum of all EXAFS contributions was calculated, obtaining a series of simulated curve for **2**. After subtracting the simulated curve for **1** (best fit), a series of simulated differential curves was created with the purpose to be compared with the differential experimental data.

The simultaneous exploration of the whole 7-dimensional parameter space, adopting a sampling step fine enough to appreciate small structural differences, would be highly time-consuming. For this reason, a properly selected consecutive series of 2D or 3D sub-spaces was initially explored with a gross sampling step trying to approach the global minimum. During all these partial minimization cycles, it was observed that variations in the DW factors were negligible in terms of influence on the differential curves with respect to the variation of the other parameters. Therefore, for the subsequent steps of the analysis, σ^2_{bpy} and σ^2_{py} were fixed to the values of the best fit for **1**, while σ^2_O was set to 0.005 Å^2, a local minimum of one minimization cycle. In this way the working dimensions were reduced to 4, including the energy shift and the three distortions along the Ru–N(bpy), Ru–N(py) and Ru–O(H_2O) bonds axes. Relying on the structural variations predicted by DFT and on the minima found in the sub-spaces previously explored, a refined analysis with restricted ranges and finer steps for each parameter was performed next. Among the grid of hundreds of simulated and tested curves, the lower R-factor value (0.147) was obtained in correspondence of the following combination of parameter values: {$\Delta E = -1.0$ eV; $\Delta R_{N(bpy)} = -0.03$ Å; $\Delta R_{N(py)} = -0.07$ Å; $\Delta R_{O(H2O)} = -0.02$ Å} which represent the results of the fit. Starting from the DFT structure of **1** and applying the bond length variations (ΔR_L) obtained it is then possible to calculate the structure of **2** optimized by means of the differential fit (Table 2).

Parameters	DFT		EXAFS	
	1	**2**	**1** standard fit	**2** differential fit
R-factor			0.011	0.147 #
S_0^2			0.95 ± 0.06	0.95
ΔE (eV)			0.8 ± 0.5	–1.0 ± 0.5
$R_{N(bpy)}$ (Å)	2.08	2.07	2.05 ± 0.02	2.05 ± 0.01
$dR_{N(bpy)}$ (Å)*		–0.01		0.00
$\sigma^2_{N(bpy)}$ (Å^2)			0.0026 ± 0.0008	0.0026
$R_{N(py)}$ (Å)	2.14	2.13	2.09 ± 0.03	2.07 ± 0.01
$dR_{N(py)}$ (Å)*		–0.01		–0.02
$\sigma^2_{N(py)}$ (Å^2)			0.0032 ± 0.0008	0.0032
$R_{O(H2O)}$ (Å)	---	2.21	---	2.12 ± 0.01
$dR_{O(H2O)}$ (Å)*		+0.07		+0.03
$\sigma^2_{O(H2O)}$ (Å^2)			---	0.005 ± 0.001

Table 2. Results from standard EXAFS analysis of **1** and from differential analysis of **2**. The results are compared to DFT predicted distances averaged per type of ligand. The errors on the refined parameters for **2** correspond to the step adopted for the differential parameter variation. *$dR_L = R_{L(2)} - R_{L(1)}$ reported in the table do not correspond with ΔR_L, the fit parameters. #The two R-factors are not directly comparable because calculated differently.

The simulated differential curve, corresponding to the minimum found, correctly reproduces all the features of the experimental differential curve (see Figure 2a), providing a further validation of the robustness of the method and of the results obtained. Observing the differences $dR_L = R_{L(2)} - R_{L(1)}$ between the refined structures of **2** and **1** resulted by the fits, it is worth noting that the differential analysis could detect the expected structural changes. The refined Ru–O bond length for the water molecule in **2** was found to be noticeably higher (+0.03 Å) than the corresponding Ru–N(py) bond length for py ring in **1**. Also the corresponding DWs were relevantly different: the value of 0.005 Å^2 relative to the H_2O ligand is higher with respect to 0.0032 Å^2 found for the py, as expected substituting a pyridine ring with a smaller water molecule. A remarkable achievement obtained by means of the differential analysis is the possibility to capture the small rearrangements in the other ligands around the Ru centre induced by the presence of either py or water. In particular, the bpy average distances remained unvaried, while the bond length for the remaining py ring appeared to be contracted of –0.02 Å in **2**.

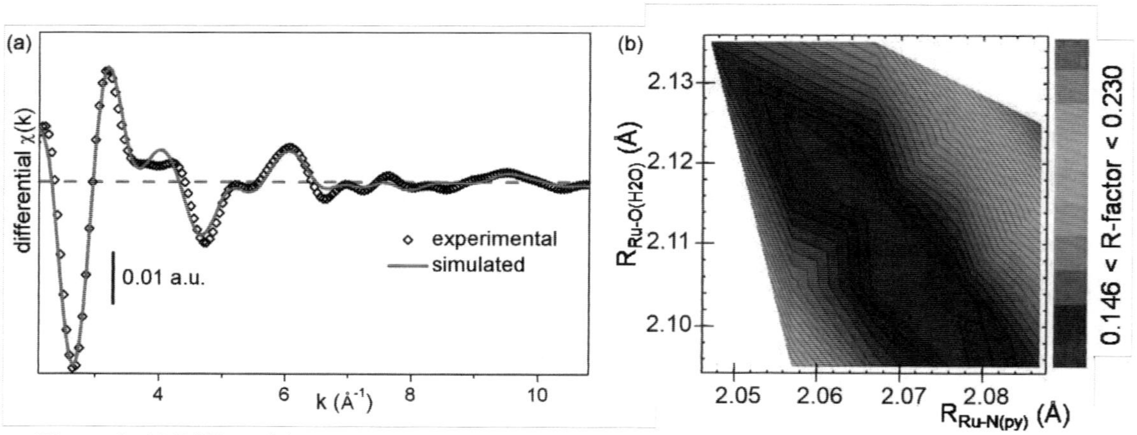

Figure 2. (a) Differential EXAFS curve calculated by subtraction of **2** – **1** spectra. Experimental data (black diamonds) and best fit (red line). (b) 2D cut of the 4D R-factor surface, obtained by fixing the other parameters to the values found in correspondence of the minimum.

Furthermore, the results of the differential refinement can be compared with the dR_L calculated from the DFT optimized structures (see Table2). The fit results and the theoretical predictions agree in indicating a longer Ru–O(H_2O) bond in **2** with respect to the Ru–N(py) bond in **1**. It is difficult to comment upon the values related to bpy and py because the parameterization adopted for the data analysis still contains many approximations necessary to reduce the number of variables. In this regard, it is worth noting that EXAFS analysis could afford only information averaged on all atoms of the same element belonging to the same unit (bpy or py). On the contrary, DFT calculations provided the single atomic positions for every atom of the complex. Examining the single Ru–N distances optimized by DFT for complexes **1** and **2**, four of them are predicted to be mainly unchanged, while the Ru–N(bpy) of the bpy ring in *trans* to the Ru–O bond is expected to be shortened of 0.04 Å.

Hence, it is possible that the differential analysis was able to reveal this shortening predicted by the DFT. However, such contraction was incorrectly related to a variation of the Ru–N(py) bond. Many causes could have contributed to this misinterpretation. Firstly it is not straightforward, in terms of EXAFS signal, to distinguish between a py ring or one of the rings of a bpy. On top of that, the rigidity imposed during the parameterization to the bpy units would not allow the fit to move a single Ru-N(bpy) bond (EXAFS paths 4-times degenerated) while the Ru–N(py) bond length was the only parameter which could be changed independently. For these reasons, it was hypothesized that the detected shortening of a Ru–N bond should be assigned to the Ru–N(bpy) in *trans* positions with respect to the water molecule. This hypothesis was also supported by the strong anti-correlation found between ΔR_{py} and ΔR_O variables. This anti-correlation appears as a diagonal line of minima in a 2D cut of the 4D R-factor surface (Figure 2b), where the other two parameters have been fixed to the optimized values of the best fit.

4. Conclusions

Differential EXAFS analysis is a powerful method with applicability that ranges from the study of spectra measured in static conditions to excited-state systems. In particular, the application of this method can be really useful when studying systems characterized by small differences, close to the state-of-art sensitivity limit of the XAS technique, as in the case of *cis*-$[Ru(bpy)_2(py)_2]^{2+}$ and *cis*-$[Ru(bpy)_2(py)(H_2O)]^{2+}$. It is worth pointing out that such approach could require a remarkably higher amount of time compared to a standard analysis. However, the direct fitting on differential curves provides a superior accuracy for the derived structural parameters when compared to conventional EXAFS fits of the individual (not differential) spectra, where the absolute measure of the distances is affected by errors which do not allow distinguishing slight structural modifications. In order to adopt such a complex analysis for static EXAFS spectra, it is necessary to deal with experimental data correctly calibrated in energy and normalized so that they could be safely compared and subtracted.

With respect to the specific case proposed, the differential fit allowed the discrimination between the longer Ru–O bond and higher DW associated to the water molecule in *cis*-$[Ru(bpy)_2(py)(H_2O)]^{2+}$ with respect to Ru–N and DW related to the py ring in *cis*-$[Ru(bpy)_2(py)_2]^{2+}$. In addition, it was possible to appreciate also a small structural rearrangement of one of the other ligands around the Ru atom; the measure of a shorter Ru-N bond in *cis*-$[Ru(bpy)_2(py)(H_2O)]^{2+}$ was finally assigned to the nitrogen of a bpy unit in trans position with respect to the water molecule with the help of considerations based on the results of DFT optimized structures. DFT calculations played a key role when the complexity of the case of study required approximations in the parameterization. Moreover, the computational study assisted the selection of guessed values and their variation ranges during the exploration of the n-dimensional space of variables, when looking for the best fit. DFT calculations were relevant both in providing structures used as input model for the simulation of EXAFS contributions, and in testing the reliability of the data interpretation by means of a synergic cross-comparison.

References

[1] Bressler C 2004 *Chem. Rev.* **104** 1781-1812
[2] Bressler C, Chergui M 2010 *Annu. Rev. Phys. Chem.* **61** 263-282
[3] Pettifer R F, Mathon O, Pascarelli S, Cooke M D and Gibbs M R J 2005 *Nature (London)* **435** 78-81
[4] Yokoyama T, Murakami Y, Kiguchi M, Komatsu T and Kojima N 1998 *Phys. Rev. B* **58** 14238-14244
[5] Boca R, Vrbova M, Werner R and Haase W 2000 *Chem. Phys. Lett.* **328** 188-196
[6] Baudelet F, Pascarelli S, Mathon O, Itie J P, Polian A, d'Astuto M and Chervin J C 2005 *J. Phys.: Condens. Matter* **17** S957-S966
[7] Itie J P, Baudelet F, Congeduti A, Couzinet B, Farges F and Polian A 2005 *J. Phys.: Condens. Matter* **17** S883-S888
[8] Frank P, Carlson R M K, Carlson E J and Hodgson K O 2003 *Coord. Chem. Rev.* **237** 31-39
[9] Flogeac K, Guillon E and Aplincourt M 2005 *Eur. J. Inorg. Chem.* 1552-1558
[10] Jezowska-Bojczuk M, Szczepanik W, Mangani S, Gaggelli E, Gaggelli N and Valensin G 2005 *Eur. J. Inorg. Chem.* 3063-3071
[11] Van der Veen R M *et al* 2009 *Angew. Chem., Int. Ed.* **48** 2711-2714
[12] Benfatto M, Della Longa S, Hatada K, Hayakawa K, Gawelda W, Bressler C and Chergui M 2006 *J. Phys. Chem. B* **110** (29) 14035-14039
[13] Saes M, Bressler C, Abela R, Grolimund D, Johnson S L, Heimann P A and Chergui M 2003 *Phys. Rev. Lett.* **90** 047403-047401
[14] Saes M *et al* 2004 *Rev. Sci. Instrum.* **75** 24-30
[15] Saes M, Gawelda W, Kaiser M, Tarnovsky A, Bressler C, Chergui M, Johnson S L, Grolimund D and Abela R 2003 *Synchrotron Radiat. News* **16** 12
[16] Gawelda W, Bressler C, Saes M, Kaiser M, Tarnovsky A,Grolimund D, Johnson S L, Abela R and Chergui M 2005 *Phys. Scr.* **T115** 102
[17] Durham B, Caspar J V, Nagle J K and Meyer T J 1982 *J. Am. Chem. Soc.* **104** 4803-4810
[18] Respondek T, Garner R N, Herroon M K, Podgorski I, Turro C and Kodanko J J 2011 *J. Am. Chem. Soc.* **133** 17164-17167
[19] Zayat L, Calero C, Albores P, Baraldo L and Etchenique R 2003 *J. Am. Chem. Soc.* **125** 882-883
[20] Salassa L, Garino C, Salassa G, Nervi C, Gobetto R, Lamberti C, Gianolio D, Bizzarri R and Sadler P J 2009 *Inorg. Chem.* **48** 1469-1481
[21] Barragan F, Lopez-Senin P, Salassa L, Betanzos-Lara S, Habtemariam A, Moreno V, Sadler P J and Marchan V 2011 *J. Am. Chem. Soc.* **133** 14098-14108

[22] Betanzos-Lara S, Salassa L, Habtemariam A, Novakova O, Pizarro A M, Clarkson G J, Liskova B, Brabec V and Sadler P J 2012 *Organometallics* **31** 3466-3479
[23] Betanzos-Lara S, Salassa L, Habtemariam A and Sadler P J 2009 *Chem. Commun.* 6622-24
[24] Salassa L *et al* 2010 *Organometallics* **29** 6703-6710
[25] Salassa L *et al* 2010 *Inorg. Chem.* **49** 11240-11248
[26] Borfecchia E *et al* submitted to *Dalton Trans.*
[27] Dwyer F P, Goodwin H A and Gyarfas E C 1963 *Aust. J. Chem.* **16** 544-548
[28] Hua X and Lappin A G 1995 *Inorg. Chem.* **34** 992-994
[29] Chen L X 2004 *Angew. Chem., Int. Ed.,* **43** 2886–2905
[30] Chen L X 2005 *Annu. Rev. Phys. Chem.* **56** 221–254
[31] Frisch M J *et al* 2009 *Gaussian, Inc.*, Wallingford, CT, Revision A. 1st edn.
[32] Adamo C and Barone V 1999 *J. Chem. Phys.* **110** 6158-6170
[33] Adamo C, Scuseria G E and Barone V 1999 *J. Chem. Phys.* **111** 2889-2899
[34] Hay P J and Wadt W R 1985 *J. Chem. Phys.* **82** 270-283
[35] McLean A D and Chandler G S 1980 *J. Chem. Phys.* **72** 5639-5648
[36] Barone V and Cossi M 1998 *J. Phys. Chem. A* **102** 1995-2001
[37] Cossi M and Barone V 2001 *J. Chem. Phys.* **115** 4708-4717
[38] Cossi M, Rega N, Scalmani G and Barone V 2003 *J. Comput. Chem.* **24** 669-681
[39] Ravel B and Newville M 2005 *J. Synchrot. Radiat.* **12** 537-541
[40] Salassa L *et al* 2009 *J. Phys.: Conf. Series* **190** Art. n. 012141

Improving the sensitivity of QEXAFS using modulation excitation spectroscopy

J Stötzel[1], D Lützenkirchen-Hecht[1], R Frahm[1] and J-D Grunwaldt[2]

[1] Fachbereich C – Physik, Bergische Universität Wuppertal, Gaußstr. 20, 42097 Wuppertal, Germany
[2] Institute for Chemical Technology and Polymer Chemistry, Karlsruhe Institute of Technology (KIT), 76131 Karlsruhe, Germany

E-mail: stoetzel@uni-wuppertal.de, grunwaldt@kit.edu

Abstract. Modulation excitation spectroscopy (MES) is a promising tool to analyze time-resolved EXAFS data during reproducible catalytic processes and was applied in two different experiments to evaluate the benefits compared to conventional analysis strategies. Thereby, phase sensitive detection (PSD) was applied directly to the x-ray absorption spectra as well as to the Fourier transformed extracted EXAFS. In both cases the results were discussed with respect to the achievable level of noise reduction as well as the possibilities to better understand the underlying mechanisms of the reactions, more precisely the dynamics and possible intermediate species. The results clearly demonstrate serious limitations for the application of MES in time-resolved EXAFS studies as soon as more sophisticated steps than averaging over several modulation cycles are performed

1. Introduction

In catalysis and materials science often only small but rapid structural changes occur under varying reaction conditions. In this case modulation excitation spectroscopy (MES) has received much attention which makes use of monitoring the structural changes by periodically changing the conditions [1,2]. In case of reversible structural changes the alternating part of the signal, which is the frequency response of the active species to the applied excitation, can be separated from the often dominant non-varying part. Furthermore, by considering only frequencies that can be assigned to the excitation, it is possible to efficiently suppress random noise. For catalytic applications, the technique has been widely used in infrared spectroscopy [2]. In addition, it has been found useful when studying materials by Raman spectroscopy and XRD [3].

In a first step we straight forwardly improved the signal-to-noise ratio by averaging over all cycles [4] which is possible due to the exactly known cycle duration T and the fact that T is an integer multiple of the measurement time for one spectrum. With this approach the original time resolution of the experiment is maintained while photon statistics are significantly improved. Afterwards, the more advanced phase sensitive detection (PSD) approach can be applied, where the spectra of the cycle are studied with the modulation frequencies only and the measured response is resolved dependent on the phase shift to the applied modulation [5]. An alternative way to increase sensitivity is low pass filtering followed by conventional XAS data analysis. Presently, it is strongly discussed which of these approaches is more useful or how they complement each other. Hence, we present here a comparison of the two approaches by analysing QEXAFS data in two model reactions [6].

2. Results

A Pt-Rh/Al_2O_3 catalyst was investigated during catalytic partial oxidation of methane in 6%CH_4/3%O_2/He and in 5%H_2/He by periodically switching the gas atmospheres every 60 s and collecting two absorption spectra each second during three full cycles with T = 120 s. Reproducing the collected absorption data with Fourier series (FS) or applying phase-sensitive detection (PSD), both with the first order frequency $\omega = 2\pi/T$ only, yield the same results either as function of time or phase as shown in Figure 1b. The non-varying part of the signal, which is the spectrum resulting from averaging over all spectra, can be added to proceed with standard XAFS analysis tools. Alternatively the results can be compared directly to difference spectra of reference compounds. A stong smoothing effect can be observed by comparing the results to the difference spectra of the raw data shown in Figure 1a. However, here evidently the significant changes in the spectra occur around 90 s (switch to methane), as clearly recognizable with the minimum whiteline feature at about 11570 eV in Figure 1a. With first order FS/PSD all changes are forced to follow a sinusoidal function with $\omega = 2\pi/T$. Thus there is a symmetry leading to an artificial maximum in whiteline intensity at 30 s. Applying low pass filtering with a cut-off slightly above ω yields the difference spectra in Figure 1c. The quality of noise suppression is similar to the one achieved with the FS/PSD approach, but it is still possible to localize the major changes in the spectra at 90 s. Considering these results low pass filtering seems to be the preferred choice and it cannot be confirmed that FS/PSD is an outstanding tool to for example increase surface sensitivity in catalytic experiments.

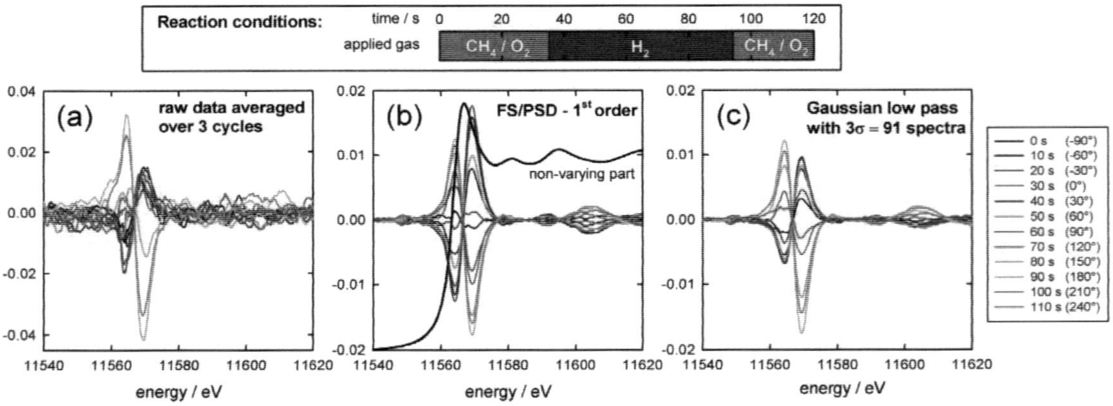

Figure 1. Pt-Rh catalyst measured at Pt L_3-edge: (a) difference spectra, (b) application of FS/PSD, (c) application of low pass filtering. The phases in the legend are valid for (b) only, where also the vertically scaled and shifted non-varying part is displayed.

As a second example a Cu/Al_2O_3 catalyst at 322 °C was investigated during periodic reduction and reoxidation induced by switching between 21%O_2/He and 5%H_2/He gas streams every 60 s (see also [7]). A combination of XANES linear combination analysis (LCA) and EXAFS fitting was applied to the low pass filtered 1200 spectra of one cycle. The results are shown in Figure 2. An intermediate Cu(I) state is vsible during both transitions and detailed information about the sample composition during the entire experiment is available. It was already shown that the application of low pass filtering on absorption spectra is superior to FS/PSD approaches. Thus, in Figure 3 results are presented where first order PSD was applied to the Fourier transformed EXAFS data. The application of PSD is more promising compared to the absorption spectra because distinct shells as function of the atomic distance are obtained. Due to the high level of smoothing it is possible to immediately recognize the typical patterns of oxidized and reduced copper for different phases. However, it is much more difficult to find evidences for the intermediate Cu(I) state via the PSD approach in this case as the first shells of CuO and Cu_2O show strong overlapping. Moreover, following the intensity of the Cu-O and Cu-Cu peaks shown in the inset of Figure 3 once more demonstrates the problem to correctly reproduce the dynamics of the reaction with this first order approach, especially compared to the LCA results in Figure 2, where it is clear that the Cu(0) contribution is not well represented by a sinusoidal function.

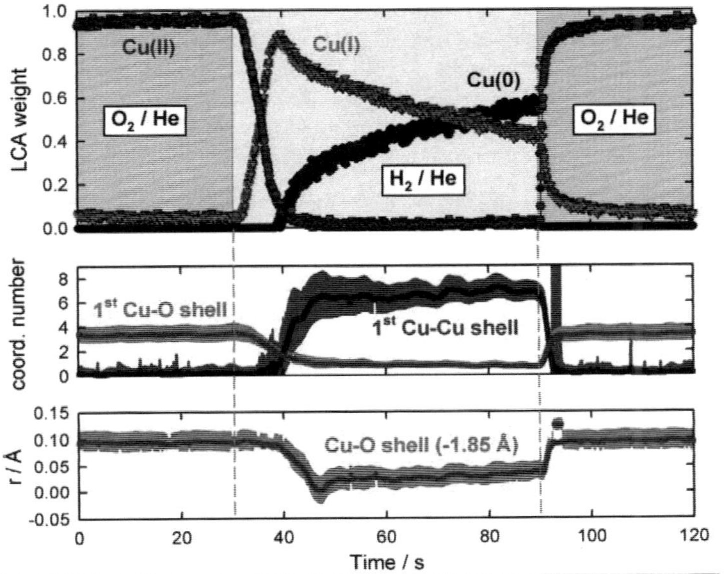

Figure 2. Linear combination analysis (top) and EXAFS analysis (coordination number and distance of Cu-O shell) results of a Cu/Al_2O_3 catalyst at 322°C during reduction/re-oxidation cycles.

In addition to the shown investigations an approach was developed using the calculated phases in R-space to more directly access the phase information as function of the local distance R [6]. Theoretically it should be possible to obtain three different phases here according to the three different shell configurations of the contributing species. However, due to the close relationship of CuO and Cu_2O (overlapping peaks in radial distribution function) in this example as well as the typically high disorder contributions in catalysis in general (small particles, high temperatures), the usefulness of this approach still has to be evaluated in more detail with more experimental data.

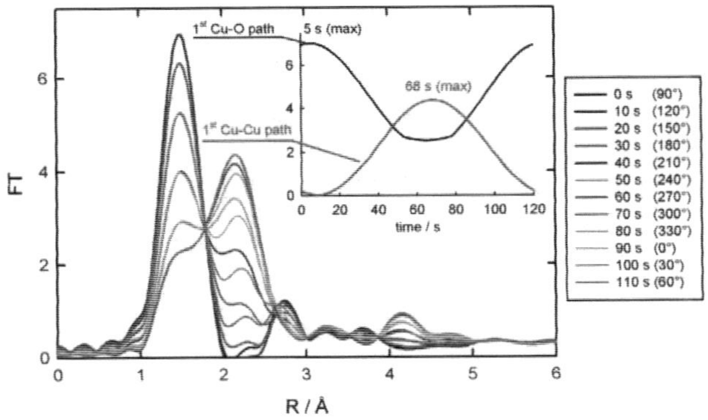

Figure 3. FS/PSD applied to the 1200 Fourier transformed spectra of the Cu/Al_2O_3 catalyst (322°C) during one reduction/reoxidation cycle. The non-varying part is added to the FS/PSD spectra here and the inset shows the development of the maxima at about 1.4 and 2.2 Å representing the first Cu-O and Cu-Cu backscattering paths.

3. Discussion and conclusion

The results show that modulation excitation spectroscopy in combination with PSD can be used to significantly reduce noise in X-ray absorption spectra. Thereby, a low pass character is achieved which can be controlled by the number of contributing orders, while the dynamics of the resulting

Journal of Physics: Conference Series **430** (2013) 012126 doi:10.1088/1742-6596/430/1/012126

spectra as function of time only consist of frequencies directly linked to the modulation. However, low pass filtering with similar cut-off frequencies yields spectra with about the same signal-to-noise ratios while no artefacts are introduced as it is coercively the case for low order PSD. In contrast to other techniques like infrared spectroscopy, where PSD is successfully applied, the EXAFS spectra do not provide well separated features, also not in R-space.

We can conclude that modulation exitation spectroscopy combined with low pass filtering and linear combination analysis is therefore a much better tool to improve data quality in time resolved X-ray absorption spectroscopy than PSD which deterioates time resolution and only results in difference spectra. The presented results further suggest that averaging over several modulation cycles and applying controlled filtering (enhanced sensitivity) followed by conventional XANES/EXAFS data analysis are the best approaches to analyze time-resolved XAS data (see also ref. [6]).

4. Acknowledgement

We thank Maarten Nachtegaal of the SuperXAS beamline (SLS, Switzerland) for his support before and during the beamtime. Matthias Beier (Technical University of Denmark) and Bertram Kimmerle are acknowledged for their help with the experiments. SLS is acknowledged for providing beamtime and financial support.

References

[1] Baurecht D and Fringeli U P 2001 *Rev. Sci. Instrum.* **72** 3782

[2] (a) Bürgi T and Baiker A 2002 *J. Phys. Chem. B* **106** 10649 (b) Urakawa A, Bürgi T and Baiker A 2006 *Chem. Phys.* **324** 653 (c) Urakawa A, Bürgi T and Baiker A 2008 *Chem. Eng. Sci.* **63** 4902

[3] Urakawa A, van Beek W., Monrabal-Capilla M, Galán-Mascarós J R, Palin L and Milanesio M. 2011 *J. Phys. Chem. C* **115** 1323

[4] Stötzel J, Lützenkirchen-Hecht D, Frahm R, Kimmerle B, Baiker A, Nachtegaal M, Beier M J and Grunwaldt J-D 2009 *J. Phys.: Conf. Ser.* **190** 012162

[5] (a) Ferri D, Kumar M S, Wirz R, Eyssler A, Korsak O, Hug P, Weidenkaff A and Newton M A 2010 *Phys. Chem. Chem. Phys.* **12** 5634 (b) Ferri D, Newton M A and Nachtegaal M 2011 *Top. Catal.* **54** 1070

[6] Stötzel J, Lützenkirchen-Hecht D, Frahm R and Grunwaldt J-D *submitted for publication*

[7] Stötzel J, Lützenkirchen-Hecht D, Frahm R, Kimmerle B, Baiker A, Nachtegaal M, Beier M J and Grunwaldt J-D 2009 *J. Phys.: Conf. Ser.* **190** 012153

Journal of Physics: Conference Series **430** (2013) 012127 doi:10.1088/1742-6596/430/1/012127

Applications of soft x-ray magnetic dichroism

G van der Laan

Diamond Light Source, Chilton, Didcot, Oxfordshire OX11 0DE, United Kingdom

E-mail: `gerrit.vanderlaan@diamond.ac.uk`

Abstract. Applications of x-ray magnetic circular and linear dichroism (XMCD and XMLD) are reviewed in the soft x-ray region, covering the photon energy range 0.4-2 keV, which includes important absorption edges such as the $3d$ transition metal $L_{2,3}$ and rare earth $M_{4,5}$. These techniques enable a broad range of novel and exciting studies such as on the electronic properties and magnetic ordering of novel nanostructured systems. XMCD has a sensitivity better than 0.01 monolayer (at the surface) and due to simple detection methods, such as electron yield and fluorescence yield, it has become a workhorse technique in physics and materials science. It is the only element-specific technique able to distinguish between the spin and orbital parts of the magnetic moments. The applications are vast, e.g., in x-ray holographic imaging, XMCD gives a spatial resolution of tens of nm. While many studies in the past were centered on physics, more recently new applications have emerged in areas such as chemistry, biology and earth and environmental sciences. For instance, XMCD allows the determination of the cation occupations in spinels and other ternary oxides. In scanning transmission x-ray microscopy (STXM), XMCD enables us to map biogenic magnetite redox changes resulting in a surprising degree of variation on the nanoscale. Another recent development is ferromagnetic resonance (FMR) detected by time-resolved XMCD which opens the door to element-, site- and layer-specific dynamical measurements. By exploiting the time structure of the pulsed synchrotron radiation from the storage ring the relative phase of precession in the individual magnetic layers of a multilayer stack can be determined.

1. Introduction

The effect of magnetic x-ray dichroism was discovered in the *Laboratoire pour l'Utilisation du Rayonnement Electromagnetique* (LURE) at Université Paris-Sud in France in 1986 [1], only one year after its prediction [2]. A sum rule, derived in 1992, relates the integrated intensity of the x-ray magnetic circular dichroism (XMCD) spectrum over the x-ray absorption edge to the ground-state orbital magnetic moment [3]. This has transformed XMCD into a powerful standard technique to determine spin and orbital magnetic moments, providing insight in the element-specific microscopic origin of anisotropic magnetic properties, such as the magnetocrystalline effect, easy direction of magnetization, magnetostriction, coercivity and magnetic hysteresis. X-ray magnetic linear dichroism (XMLD) makes it possible to measure antiferromagnets, which have no net total magnetic moment.

The soft x-ray region covers the important edges for transitions from core levels into empty magnetic states: the $3d$ transition metal $L_{2,3}$ ($2p \rightarrow 3d$, at 0.5-1 keV), rare earth $M_{4,5}$ ($3d \rightarrow 4f$, at 0.8-1.6 keV) and actinide $N_{4,5}$ ($4d \rightarrow 5f$, at 0.7-1.5 keV). Soft x-ray excitations have considerable advantages over hard x-rays. They give huge resonances corresponding to dipole-allowed transitions directly into the magnetic valence state. Polarization dependent selection rules result in a multiplet structure that serves as a fingerprint for the ground state electronic and magnetic structure.

The XMCD is defined as the difference between the x-ray absorption spectra (XAS) for antiparallel and parallel configuration of the light helicity and magnetic field direction. In a simple picture, the XMCD process can be visualized as a two-step model. The $2p$ core level is split in a $j = 3/2$ state (L_3 edge) and $j = 1/2$ state (L_2 edge), where spin and orbit are coupled parallel and antiparallel, respectively. In the first step, the emission with the light helicity vector parallel (antiparallel) to the $2p$ orbital moment results in excited electrons of preferred spin up (down) direction. In the second step these excited electrons have to find a place in the $3d$ valence band, and if there are less spin-up than spin-down *holes* available, the XMCD spectrum displays net negative L_3 and positive L_2 peaks [cf. Fig. 1(b)].

Interesting applications arise when the XMCD or XMLD effect is used in magnetic x-ray holography, x-ray detected ferromagnetic resonance (XFMR) or x-ray photoemission electron microscopy (PEEM). Magnetic dichroism in magnetic x-ray resonant scattering enables the investigation of (anti)ferromagnetic lattices, magnetic roughness versus chemical roughness, relative spin orientations, non-collinear moments and magnetic domain structures [4, 5, 6].

2. Advantages of soft x-ray spectroscopy

It is useful to emphasize the main advantages of soft x-ray spectroscopy [7].

(i) *Element-specificity.* Sensitivity to a chemical element is obtained by tuning the x-rays to the resonance energy of the core to valence transition. When various layers contain different elements, this turns into a layer specific probe.

(ii) *Orbital sensitivity.* The light interacts only with the orbital part of the wave function, so that spin and orbital properties can be separated. This allows to study specific orbital effects, such as orbital ordering and Jahn-Teller effect.

(iii) *Selection rules.* Electric-dipole transitions from the ground state can reach only a limited subset of final states, thereby providing a fingerprint for the specific ground state. This gives a high sensitivity to crystal field, spin-orbit interaction, site symmetry and spin configuration [8].

(iv) *Magnetic sensitivity.* The light polarization dependence makes the absorption sensitive to the ground state magnetic polarization. The magnetic dichroism is strong because it arises from the electric-dipole and not the magnetic-dipole transitions [8].

(v) *High energy resolution.* The core hole life time broadening is of similar magnitude as the instrumental resolution. This narrow intrinsic broadening, in the order of a few hundred meV, makes it possible to resolve the detailed multiplet structure and charge-transfer satellites.

(vi) *Well-defined core-level states.* X-ray transitions involving a core state are much more straightforward to calculate than optical transitions. The wave function of the core state is well defined and mixing (hybridization) between core and valence states can be neglected. In localized materials the presence of spin-orbit and electrostatic interactions leads to a unique multiplet structure. The core-level state is therefore an ideal probe to study the local electronic configuration of the valence states.

(vii) *Fundamental spectra.* The angular dependent part of the spectra can be separated from the physical (dynamical) part. This provides the powerful concept of fundamental spectra [9, 10, 11].

(viii) *Sum rules.* Summation over a core-level state with a good quantum number (such as total angular momentum, j) fulfils the completeness relation. This provides the theoretical basis of the sum rules.

(ix) *Femtosecond time scale.* In localized systems, the x-ray transition is much faster than the rearrangement of the valence electrons. This means that calculations can be done in the limit of the sudden approximation. Compared to Mössbauer spectroscopy, XAS is much faster, perceiving in magnetite the Fe^{2+} and Fe^{3+} cations on the octahedral sites as 'frozen' [12, 13]. In

Journal of Physics: Conference Series **430** (2013) 012127 doi:10.1088/1742-6596/430/1/012127

itinerant systems such as metals, the femtosecond time scale of the core-hole clock can be used to monitor the electronic screening processes [14, 15].

3. X-ray absorption—the basics

3.1. The magneto-optical effect

It is illustrative to compare the XMCD effect with the magneto-optical Faraday or Kerr effect. The optical spectra are due to excitations from occupied valence band to unoccupied conduction band states, as schematically shown in Fig. 1(a). Selection rules ($\Delta m = -1$ and $+1$) for the orbital component result in a different transition probability for left- and right-circularly polarized light into the unoccupied polarized valence band. Since the light does not interact directly with the spin, but only indirectly via the spin-orbit interaction, the dichroism increases in first instance with the spin-orbit interaction, which in the optical region is relatively weak (~50-100 meV for $3d$ atoms). However, a very strong dichroism is obtained in the case of XMCD, since the core-level spin-orbit interaction is much larger (tens of eV), resulting in pure $j = \ell \pm s$ levels, as depicted in Fig. 1(b).

Magnetic dichroism in photoemission can be described similarly as in x-ray absorption, but there are some differences. In absorption the dichroism is a consequence of the Pauli principle. The polarized light creates polarized electrons which can not go freely into the magnetic $3d$ shell. In contrast, in x-ray photoemission spectroscopy (XPS) the core electron is ejected into a continuum shell which has no interaction, and the dichroism arises from the electrostatic interaction between the core hole and the polarized $3d$ electrons [9]. Magnetic dichroism furthermore plays an important role in second-order coherent processes, such as resonant diffraction, resonant inelastic x-ray scattering (RIXS), resonant photoemission (RESPES), etc., as has been described elsewhere [16].

3.2. Calculational aspects

For localized many-electron systems, transitions are of the type $d^n \rightarrow p^5 d^{n+1}$ (or $f^n \rightarrow d^9 f^{n+1}$ in case of rare earths and actinides). The polarized XAS spectra can be calculated using multiplet theory [17], which puts the spin-orbit and electrostatic interaction on an equal footing [18]. The electrostatic interaction includes the *p*-*d* and *d*-*d* Coulomb and exchange interactions. The wave functions of the initial- and final-state configurations are calculated in intermediate coupling using Cowan's atomic Hartree-Fock (HF) code with relativistic correction [19, 20, 8, 17]. The HF values of the Slater parameters are normally reduced to 70-80% to account for configuration interaction and screening effects. Such a reduction was found to give the optimal scaling for the rare-earth $M_{4,5}$ XAS [21, 17]. The spin-orbit parameters are usually kept within a few percent of their atomic values.

Crystal- and ligand-field interactions are taken into account using Butler's chain of groups method [22, 23]. Hybridization (mixing) can also be taken into account, resulting in charge-transfer satellites [24]. The calculations show that the spectral shape of XAS, XMCD and XMLD is strongly sensitive to the character of the $3d^n$ ground state [8, 25].

At finite temperatures a Boltzmann distribution is taken over the energy levels of the initial state [2]. This is especially important when the ground-state degeneracy is lifted by the $3d$ spin-orbit interaction, giving rise to a bunch of low energy levels, such as in the case of CoO [26].

3.3. Sum rules

Sum rules provide a theory-independent method to obtain ground-state properties. They are valid for localized as well as itinerant electron systems, and in fact theory independent. Sum rules for the isotropic XAS relate the total integrated intensity to the number of valence holes, n_h and relate the core-level spin-orbit split branching ratio to the angular part of the spin-orbit

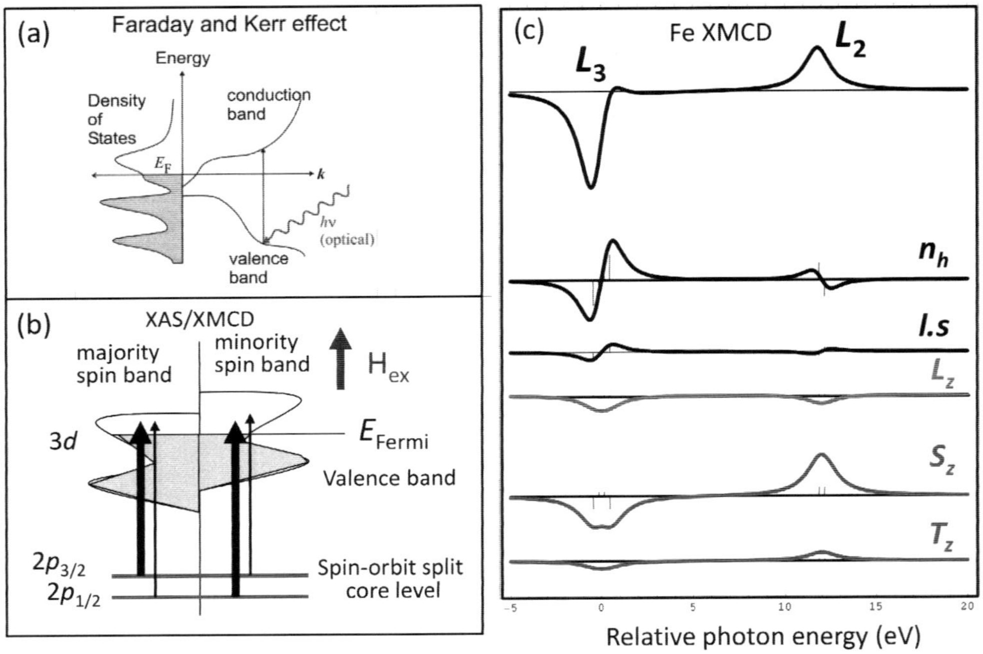

Figure 1. (a) Magneto-optical Faraday and Kerr effect. (b) XAS/XMCD effect in core-level spectroscopy. (c) Relative contributions to the energy dependence of the Fe $2p$ XMCD spectrum arising from the different ground state moments for iron metal. n_h = number of holes, $\ell \cdot s$ = scalar part of spin-orbit interaction, L_z = orbital moment, S_z = spin moment and T_z = magnetic dipole term. The red and blue spectra relate to the sum rules for the orbital and spin magnetic moments, respectively. The upper spectrum shows the sum of all contributions, which resembles the experimental Fe $L_{2,3}$ XMCD spectrum. For further details see Ref. [27].

interaction, $\ell \cdot s$ [28, 29, 30, 31]. This sum rule for the isotropic spectrum has recently been generalized to electric-multipole transitions, which are important in nonresonant inelastic x-ray scattering (NIXS) [32].

Sum rules for XMCD enable us to separate the orbital moment from the spin moment [3, 33, 34]. Although in $3d$ transition metals the orbital moment is usually small (~10% of the total magnetic moment), it represents that part of the magnetization which is coupled to the lattice and, therefore, determines magnetic properties such as the easy-direction of magnetization, the magneto-crystalline anisotropy and the magnetostriction.

It is useful to put the sum rules in a more general perspective. X-ray absorption [35, 34] and photoemission [36] provide a whole set of sum rules, relating integrated spectral intensities to ground state properties . The reason that excited states (typically at 1 keV) still give information about ground state properties is that by integrating over all possible states of the core hole its influence is averaged out. A necessary condition for this is that the core level is characterized by a good quantum number, such as the total angular momentum j. In other words, the jj mixing between the L_3 and L_2 edges must be small, which is fulfilled when the core spin-orbit splitting is much larger than the core-valence electrostatic interaction [37].

The moment transferred by the light determines which combinations of ground state operators are accessed. The integral over the isotropic intensity reveals the number of valence holes. The integral over the circular dichroism is related to the orbital magnetic moment, and the linear dichroism is related to the charge quadrupole moment. The light does not interact directly with the electron spin, but spin dependent operators can be obtained when the integrated intensities are weighted appropriately over the two spin-orbit split core levels. Using these weighted integrals the ground state spin-orbit coupling is obtained from the isotropic spectrum; the spin moment and magnetic dipole term are obtained from the circular dichroism; and the spin-orbit-induced quadrupole moment is obtained from the linear dichroism [38].

3.4. Ground state moment contributions

We will now illustrate the sum rules from the perspective of the different ground-state moments. In Fig. 1(c) the Fe $L_{2,3}$ XMCD spectrum is decomposed into its contributions from the separate ground-state moments [27, 39]. It can be seen that only for the orbital moment, L_z, the XMCD has a non-zero intensity integrated over both edges. For the spin moment, S_z, and magnetic dipole term, T_z, the XMCD does not vanish over each separate edge but integrated over both edges the intensity is zero. For the number of valence holes, n_h, and the spin-orbit interaction, $\ell \cdot s$, the integrated intensity of of XMCD vanishes for each edge, but they still give an energy dependence across each edge. This is also the case for the higher moments, such as the quadrupole moments, as demonstrated elsewhere [27]. Hence, this not only illustrates that the integrated intensities over the L_3 and L_2 edges are proportional to the orbital and spin moments, but also that they do not depend on any of the other ground state moments. Since all moments contribute to the energy dependence, the sum rules in their differential formulation [40] are not exact.

3.5. Influence of the 3d spin polarization

The main contributions in the XMCD normally come from S_z and n_h, with the other ground-state moments usually much smaller. From Fig. 1(c) it is clear that n_h gives a strong energy dependence across each absorption edge. To simplify the analysis, we will condense S_z and n_h into a single parameter, i.e., the spin polarization P_S which is the difference between the amount of spin-up, $n_\uparrow$, and spin-down, $n_\downarrow$, *holes* (where $n_\uparrow \geq n_\downarrow$), divided by the total amount of holes,

$$P_S \equiv \frac{n_\uparrow - n_\downarrow}{n_\uparrow + n_\downarrow} = -\frac{S_z}{n_h}. \tag{1}$$

Fig. 2(a) shows the XMCD with the spin polarization P_S stepwise varying from 0 (where $n_\uparrow = n_\downarrow$) to 1 (where $n_\downarrow = 0$). The L_3 line shape of the XMCD changes from very symmetric for $P_S = 1$ to very asymmetric for $P_S = 0$ [39]. This trend of increasing asymmetry is confirmed experimentally as seen in Fig. 2(b), which shows the XMCD of the pure metals along the $3d$ transition-metal series [41], in which P_S reduces going from Ni towards V metal. Note that the experimental spectra for Ni and V metal also reveal some peculiarities. Nickel metal shows a pronounced satellite structure due to correlation effects in the ground state [24]. The spectrum of vanadium metal has a complicated structure with overlap between the L_3 and L_2 features [42].

The peak asymmetry of the L_3 XMCD is inversely proportional to the spin polarization [39]. The positive lobe at the high energy side of the L_3 in the XMCD can be taken as a measure for an unoccupied spin-down band. It is absent in hard ferromagnets where the spin-down band is full and the holes are in the spin-up band. The experimental spectra in Fig. 2(b) show indeed that the positive lobe is absent in Ni and Co metal and starts to appear in Fe metal, in good agreement with the calculated majority and minority $3d$ spin density of states [43]. XMCD measurements from galfenol (Ga_xFe_{1-x} alloy) thin films reveal a gradual change in magnitude of the positive lobe as a function of composition, which can be related to a change in the electronic structure of this material that exhibits extraordinary magnetostrictive properties

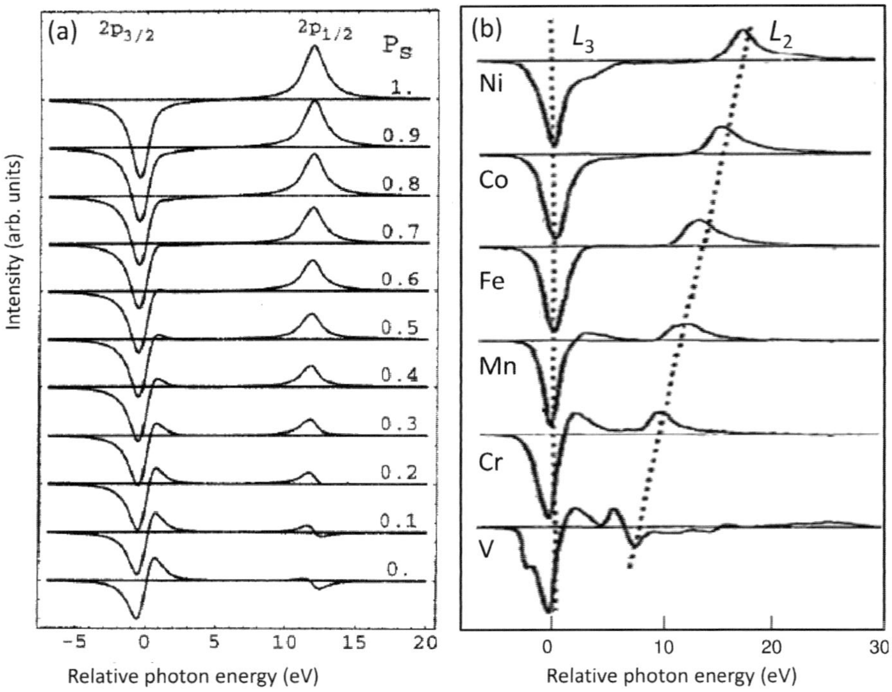

Figure 2. (a) Calculated $3d$ metal $L_{2,3}$ XMCD for varying $3d$ spin polarization $P_S = -S_z/n_h$, with $\ell \cdot s$, L_z, T_z and $2p$ spin-orbit splitting kept at constant values [39]. (b) Experimental $L_{2,3}$ XMCD spectra of the $3d$ transition metals [41]. The $2p$ spin-orbit splitting gradually reduces for lighter elements, as shown by the dotted lines.

[44]. In magnetostrictive thin films the adaption to new dimensions is inhibited by bonding to the substrate; and applying a magnetic field in the film plane does not result in an expansion or contraction along the field direction. Instead a magnetoelastic stress is created that leads to a change in the magnetocrystalline anisotropy.

4. Detection and surface sensitivity

In the soft x-ray region the absorption at resonance is very strong and proportional to the number of $3d$ or $4f$ holes. Henke's data tables [45], which are valid for off-resonance excitation, provide no assistance here. For instance, the x-ray attenuation length for pure Fe just below the L_3 edge is 591 nm, but at the maximum of the L_3 and L_2 edge it is 16.7 and 32 nm, respectively [28]. At the rare earth M_5 edge the x-ray attenuation length is even shorter, i.e., 6.2 nm for Gd and and 12.1 nm for Dy. It was at the rare earth $M_{4,5}$ edges that saturation effects in total electron yield (TEY) detection were first quantified [21, 46].

TEY is by far the most practiced detection method in the soft x-ray region. It gives a high surface sensitivity with a sampling depths of $\sim$3 nm [47, 48]. As long as the sampling depth remains much shorter than the x-ray attenuation length, the TEY signal represents the true (photon energy dependent) absorption coefficient very well. Due to its high surface sensitivity, TEY is ideal for monolayer coverages and magnetic adatoms, capable of measuring the XMCD of 0.005 of a monolayer coverage [49].

In transmission measurements, samples have to be very thin because of the strong absorption. Samples can be produced as thin films and multilayers, although one should take care to avoid saturation effects [50]. Thin films grown on MgO or sapphire can be measured using the luminescence from the substrate, which is excited by the transmitted x-rays not absorbed in the sample.

Fluorescence yield (FY) probes deeper than TEY, but the signal is low in the soft x-ray region. Also the spectrum obtained by FY can have a different shape as the x-ray absorption, due to the specific excitation de-excitation process [51]. This is no real problem for measurements done at fixed photon energy, such as in the case of XMCD hysteresis loops [52] and XFMR [53].

5. Spintronics applications

5.1. Device development

Introduction of controlled thin film deposition and lithographic patterning methods has brought about a renaissance in magnetism research. The possibility to engineer multilayers of magnetic and non-magnetic metals on the nanoscale led to the discovery in the late eighties of interlayer exchange coupling and giant magnetoresistance (GMR). Confining the dimensions of a magnetic structure to the magnetic exchange length scale or the domain wall width results in intriguing phenomena, unknown to bulk materials. Apart from their scientific interest, the technological potential of GMR devices and nanostructured materials is an important driving force for the study of magnetic thin films and nanostructures. GMR magnetic field sensors were introduced within 10 years of the discovery of the GMR effect and are nowadays indispensable for automotive applications and data storage. Tunnel magneto-resistant junctions form the core of non-volatile magnetic random access memories (MRAM).

Many methods are used to characterize the magnetism of thin films and confined magnetic structures. Among these, magneto-optical methods offer the advantage not to be hampered by magnetic and electric fields. However, magneto-optics in the visible range is not element-specific and moreover the spatial resolution is limited by the wavelength. Element-specific XMCD on magnetic elements has opened up new possibilities [1, 2, 54], which strongly benefit from the advent of undulator devices with variable linear and circular polarization.

Organic magnetic materials possess a number of chemical properties that may provide new processing strategies and new applications, such as electro-optic and opto-magnetic materials.

Scattering experiments are mostly performed using hard x-rays, which offer high spatial resolution and large penetration, hence a narrow Darwin width of the Bragg peaks. However, in the soft x-ray range the resonant magnetic scattering cross-sections are orders of magnitude larger, and if the soft x-ray wavelengths are not too large for determination of the unit cell structure, they are ideally suited to resolve the micromagnetic structure of domains and artificially structured devices [5, 55].

5.2. Time resolved spectroscopy and imaging

Knowledge of the time domain structure is critically important due to the need for increased read-write speed of magnetic information. Interesting magnetic phenomena are displayed in a wide range of exploratory materials, such as biomagnets, molecular magnets, oxide magnets and soft magnets. Fundamental questions exist about the nature of domain wall motion and precessional reorientation in confined magnetic structures [56].

Time resolved x-ray measurements complement optical pump-probe studies that aim to determine the most effective means of reversing the magnetization of a thin film or nanostructure in the shortest possible time. So far the role of interfaces and edges upon the reversal process is poorly understood. The local nature of time-resolved XMCD measurements may provide new insight into this problem, which is of vital importance to the future of the data storage industry.

The pulsed nature of synchrotron radiation provides an inherent capability for stroboscopic measurements of picosecond magnetic processes. The timing pulse generated by the synchrotron can be used to trigger a pulsed magnetic field in a stripline or micro-coil structure. Time resolved

Journal of Physics: Conference Series **430** (2013) 012127 doi:10.1088/1742-6596/430/1/012127

XMCD has been used to resolve the nanosecond dynamics of individual layers within spin and tunnel valve structures [57], while time resolved spin polarized photoemission was used to isolate the signal from the surface region [58]. Time resolved magnetic images with spatial resolution of down to 20 nm were obtained with x-ray photoelectron emission microscopy (XPEEM) [59, 60]. Other synchrotron-based studies include full-field magnetic transmission microscope images of magnetization dynamics.

Increased data rates are required in magnetic recording technology in order to exploit recent gains in areal density and storage capacity. High-speed magnetic reorientation is a generic problem underlying the operation of all parts of hard disk systems, MRAM, and to some extent magneto-optical recording. The existing technology is at a watershed where thermal activation and domain wall motion are being supplanted by precessional mechanisms. Different material properties, for example anisotropy and damping, are required in these high frequency applications, and it is necessary to characterize the materials at the frequencies at which they are used.

5.3. Diluted magnetic semiconductors

Diluted magnetic semiconductors are expected to play an important role in interdisciplinary materials science and future spintronics because the integration of charge and spin degrees of freedom will deliver new types of devices for information storage and processing. Magnetic semiconducting materials such as EuO have been known for many decades, but have remained a scientific curiosity due to the fact that the ferromagnetic phase was always only found at low temperatures requiring helium cryogenics.

In the past decade or so, the concept of a diluted magnetic semiconductor, where a magnetic species is doped into a conventional semiconducting material has received large attention, as it potentially offers the promise of room temperature ferromagnetism in a semiconducting material [61]. Probably the best studied system to date is $Mn_xGa_{1-x}As$, which is widely accepted as a truly ferromagnetic material, with a Curie temperature below room temperature [62]. Fig. 3(a) shows the ferromagnetic coupling between two Mn atoms in the zinc blende crystal structure of GaAs. The experimental Mn $L_{2,3}$ XMCD of $(Ga_{0.93}Mn_{0.07})As$ thin film is reproduced in Fig. 3(b) and agrees very well with multiplet calculations [63]. The orbital and spin magnetic moments are 0.15 μ_B and 4.3 μ_B, respectively, as derived from the sum rules for the integrated $L_{2,3}$ intensities.

A large number of other systems has been investigated both theoretically and experimentally, with GaN and the semiconducting metal oxides such as ZnO, TiO_2 and SnO_2 receiving large interest due to the theoretical results predicting room temperature ferromagnetism for the correct level of dopant Mn, Fe or Co atoms [61]. Experimental results so far are mixed, as all these materials rely on levels of magnetic species far above the solubility limit, so that non-equilibrium growth techniques are needed to prepare them successfully. Failure will result in phase segregation leading to clusters of the magnetic atoms, which will exhibit a ferromagnetic signal to bulk probes such as conventional magnetometers. XMCD is precisely the technique whereby such cluster dominated systems and secondary phases can be distinguished from those that are truly dilute [64].

6. Environmental sciences

6.1. Cation occupancies

In nature, spinel ferrites are common constituents of igneous rocks, often in concentrations of economic importance and are the main contributors to rock magnetism [65]. Complex chemistries and cationic distributions result in a range of electric and magnetic properties that have motivated extensive mineralogical and technological investigations.

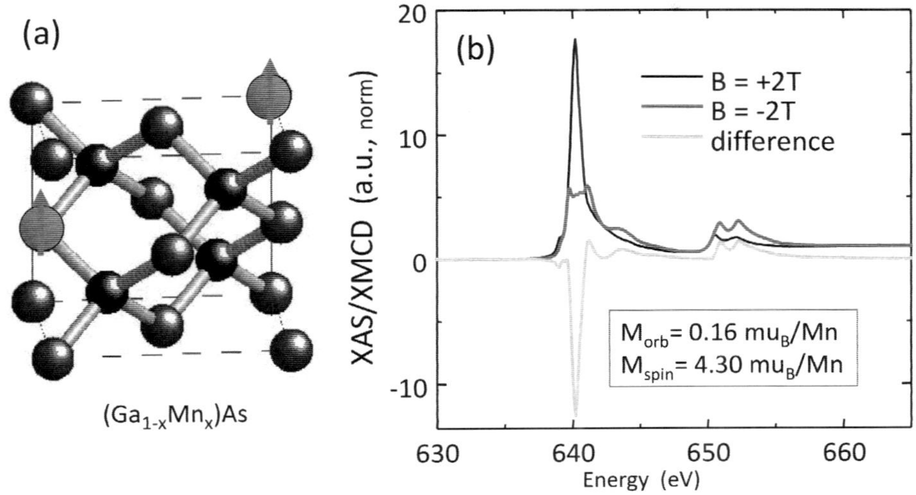

Figure 3. (a) Zinc blende crystal structure of GaAs with ferromagnetic coupling between two substitutional Mn atoms on Ga sites (Ga = blue spheres, Mn = green spheres and As = black spheres). (b) Experimental Mn $L_{2,3}$ XAS for positive (black line) and negative (red line) applied magnetic field and resulting XMCD (green line) of 50 nm ($Ga_{0.93}Mn_{0.07}$)As film [63].

Magnetite has a cubic spinel crystal structure with a unit cell containing 32 oxygen ions, eight Fe^{3+} ions at tetrahedral ( ) sites, eight Fe^{2+} and eight Fe^{3+} ions at octahedral [ ] sites, i.e., $(Fe^{3+})[Fe^{2+}Fe^{3+}]O_4$. Its crystal structure is shown in Fig. 4(a). The magnetism exhibited by magnetite is a result of the super-exchange interaction between the tetrahedral and octahedral sites, which yields magnetic moments that in both sublattices are antiparallel to each other. The distribution of iron within the spinel lattice is such that the magnetic moments of Fe^{3+} ($\sim$5 μ_B per atom) on both lattice sites cancel each other, resulting in a net magnetization due to Fe^{2+} ($\sim$4 μ_B per atom). The substitution of Fe^{2+} or Fe^{3+} cations in magnetite with transition metal dopants, such as cobalt, zinc, nickel or chromium, serves as a method in which to change the magnetic properties of the magnetic nanoparticles [66].

For an unambiguous determination of the relative site occupations in spinels and other ternary oxides, XMCD is the only available technique [12, 13]. The XMCD is dependent upon the magnetic moments of the Fe cation sublattice spinels in the magnetite structure (i.e., spin up and spin down), valence state (i.e., d count) and site symmetry (crystal field). The distribution of Fe cations within the magnetite structure can be determined by comparison of the XMCD to atomic multiplet calculations [8, 25]. As shown in Fig. 4(b), each peak in the L_3 XMCD of Fe_3O_4 corresponds primarily to a different site in the magnetite, the first negative peak corresponds to octahedral Fe^{2+}, the second positive peak to tetrahedral Fe^{3+} and the last negative peak to octahedral Fe^{3+} [13]. In titanomagnetite (magnetite-ulvöspinel solid solutions) the XMCD contains furthermore an extra positive peak at the low energy side of the L_3 XMCD, which is due to tetrahedral Fe^{2+} [67].

6.2. Bionanomagnetism

Bacterial surfaces can be used to pattern, template and scaffold metal nanocluster deposition, superseding chemical manufacturing for technologically important metal-based materials that are limited in several ways. Bacteria can also scavenge metals from liquid waste enzymatically, and perform nanofabrication of catalytic nanoparticles from precious metal

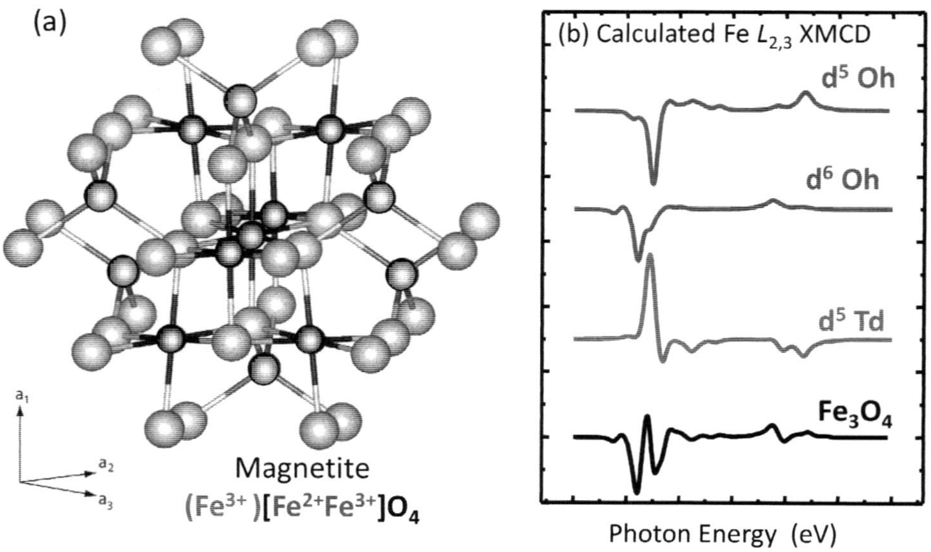

Figure 4. (a) Crystal structure of magnetite with octahedral (blue spheres) and tetrahedral (red spheres) sublattices and oxygen atoms (grey spheres). (b) Calculated multiplet structure of the XMCD for Fe d^6 O_h, Fe d_5 T_d and Fe d^5 O_h, and the XMCD summed over the three sites for the stoichiometric composition $(Fe^{3+})[Fe^{2+}Fe^{3+}]O_4$ [13].

wastes. The mechanisms of microbial metal reduction have been addressed with emphasis on the environmental impact and biotechnical application of Fe^{3+}-reducing bacteria [68, 69, 70]. A diversity of unrelated Fe^{3+}-reducing bacteria can convert ferrihydrite into the magnetic spinel magnetite. One of the long-term goals is to realize the biotechnological potential of anaerobic Fe^{3+}-reducing bacteria, by understanding and controlling their ability to produce nanocrystalline magnetic particles suitable for use with information management technologies. Successful development of these biomineralizing bacteria can revolutionize the manufacture of nanocrystals and provide low energy, environmentally friendly processing methodologies. In addition, this novel application of biotechnology has the potential to convert natural and waste Fe^{3+} oxides that are bulky and difficult to handle, to a high value product that is easy to process.

The bio-produced spinels are ideally studied by XMCD, where the relative peak intensities of the three sharp peaks in the Fe L_3 XMCD with signature $(-, +, -)$ are proportional to the site occupancies of the Fe^{2+} (d^6 O_h), Fe^{3+} (d^5 T_d) and Fe^{3+} (d^5 O_h) cations, respectively [13]. Using XMCD the cation site occupancy of biogenic magnetite has been compared to polygenic ferrite spinels [71, 72]. XMCD has been successfully used to study the incorporation of vanadium [73], chromium [74, 75, 76, 77], manganese [78], cobalt [78, 79, 66], nickel [78], zinc [80], arsenic [81], technetium [75] and palladium [82] into the biomagnetite.

6.3. Reaction mapping

XMCD can be combined with scanning transmission x-ray microscopy (STXM) [83]. The combination of STXM and XMCD has been used to spatially resolve the distribution of different carbon and iron species associated with *Shewanella oneidensis* cells [84]. The bacteria couple the reduction of Fe^{3+}-oxyhydroxides to the oxidation of organic matter in order to conserve energy for growth. Findings suggest a dominant role for surface contact-mediated electron transfer in this study and also the inhomogeneity of magnetite species on the submicron scale present

Journal of Physics: Conference Series **430** (2013) 012127 doi:10.1088/1742-6596/430/1/012127

in microbial reactions. This also illustrates the applicability of this new synchrotron-based technique for high-resolution characterization of the microbe–mineral interface, which is pivotal in controlling the chemistry of the Earth's critical zone.

7. Low-temperature physics

A very exciting and new area is low temperature physics with polarized soft x-rays. Magnetic, orbital and charge ordering phenomena are induced by rather weak interactions and usually only manifest at temperatures of a few Kelvin. This is currently the realm of neutron scattering. However, x-ray spectroscopies offer many advantages and the community now relying on neutron scattering will be able to take advantage of this new possibility to study, e.g., heavy-fermions, Kondo systems, frustrated magnets, and superconductors. The combination of low temperature and high magnetic field also allows the study of paramagnets and bioinorganic molecules of interest in the life sciences. The possibility to study magnetism under these extreme conditions using XMCD and XMLD will open up many interesting and unpredictable new avenues of research.

8. Magnetic imaging with coherent x-rays

8.1. Speckle patterns

Synchrotron radiation, like any other finite source, can be made coherent at the expense of flux, which means that a third-generation synchrotron source is able to produce a sufficient coherent flux [85]. The conditions for both longitudinal and transverse coherence have to be fulfilled. The longitudinal (or temporal) coherence length is determined by the resolving power of the monochromator at given wavelength. This condition is normally fulfilled because of the high energy resolution of the monochromator. The transverse (or spatial) coherence is obtained from the diffraction limit and depends on the distance and size of the source. The required transverse coherence can be achieved by spatial filtering, i.e., by inserting a pinhole with a diameter of typically a few tens of microns at close range in front of the sample. Finally, a charge-coupled device (CCD) camera is an essential requisite to record speckle patterns of the scattered radiation [86].

The speckle pattern is caused by the random path differences between radiation scattered from different locations in the sample irradiated with the coherent light, which interfere at the detector in the far field to produce an interference pattern [89]. The speckle pattern is sensitive to the particular configuration of the sample, whereas incoherent light scattering is only sensitive to statistical averages of the structure. To reconstruct the local configuration of the illuminated objects from the speckle pattern would require the measurement of the total signal (real and imaginary part), while only the intensity (i.e., the squared amplitude) can be detected and the phase is lost. So far only under very favorable circumstances, and for a small number of objects, the phase problem has been solved and the illuminated sample has been reconstructed directly from its coherent scattering pattern. However, progress has been made in this area by developing efficient algorithms and large computing power.

The presence of magnetic speckle, displaying a very high magnetic contrast, has opened up the possibility to use soft x-ray magnetic scattering for the study of static and dynamic magnetic disorder on length scales relevant for nanomagnetism [6, 90]. As an example, Fig. 5 shows the magnetic speckle from a FePd thin film with a perpendicular magnetic domain structure [91].

Combined with the time structure of synchrotron radiation, speckle measurements enable us to study rapid fluctuations in the local magnetic order, giving access to dynamical processes, and ultimately leading to improvements in the read-write speed of magnetic storage devices. This photon in—photon out technique allows us to study reversal processes as a function of magnetic field [91, 88, 92]. It enables to determine the degree to which the microscopic domain pattern is retained after magnetization cycling [93]. Since soft x-ray absorption spectra show a huge

Journal of Physics: Conference Series **430** (2013) 012127 doi:10.1088/1742-6596/430/1/012127

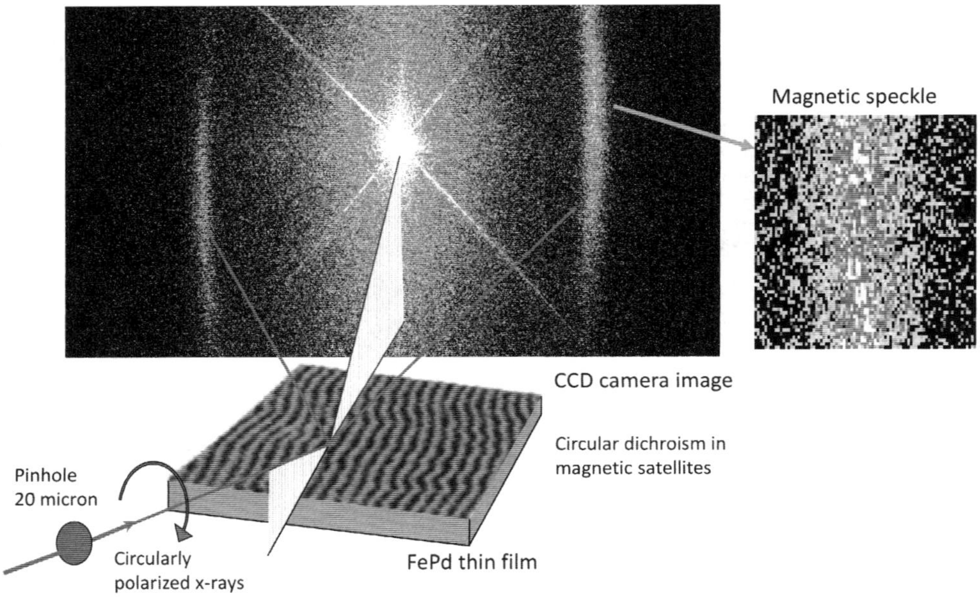

Figure 5. Schematic layout of coherent x-ray scattering in reflection geometry using a pinhole to give high coherence and a CCD camera to capture the scattering pattern. The resonant magnetic x-ray scattering from a FePd thin film with a perpendicular magnetic domain structure gives magnetic satellites along the direction of the magnetic periodicity [87]. Using coherent radiation the satellites display a speckle pattern which contains information about the local magnetic order of the domain structure [88].

magnetic dichroism compared to their hard x-ray counterparts, the magnetic peak intensities are more straightforward to analyse. With the advent of x-ray free electron lasers this new technique clearly holds great promises for the future of magnetic scattering, with imaging of dynamic disorder among the possibilities.

8.2. Magnetic imaging by x-ray holography
Fourier transform holography (FTH) is a well established lensless technique for imaging the perpendicular component of magnetic domains [94]. Holographic imaging, like x-ray scattering, is suitable not only for the remanent state [as typical for PEEM or magnetic force microscopy (MFM)] but can also be used under applied electric or magnetic fields and shows great potential for studying the dynamics of multiferroic thin films. A recent development in FTH has reduced the restriction on the reference size to allow a much wider range of possibilities. The technique, known as 'holography with extended reference by autocorrelation linear differential operator' (HERALDO) [95] permits the use of larger objects as references without compromising the spatial resolution. The HERALDO experiment is schematically displayed in Fig. 6 where it is used to image a Co/Pt multilayer with perpendicular magnetic domain structure [96].

9. Ferromagnetic resonance detected by XMCD
Ferromagnetic resonance (FMR) can be combined with XMCD to a technique called x-ray detected FMR (XFMR) [53, 98, 99, 100, 101]. XFMR offers unique element-specificity to probe the properties of multilayered films. Standard FMR cannot be used to study multilayer dynamics directly, since the signal is an average over the whole system. The precession frequency of the

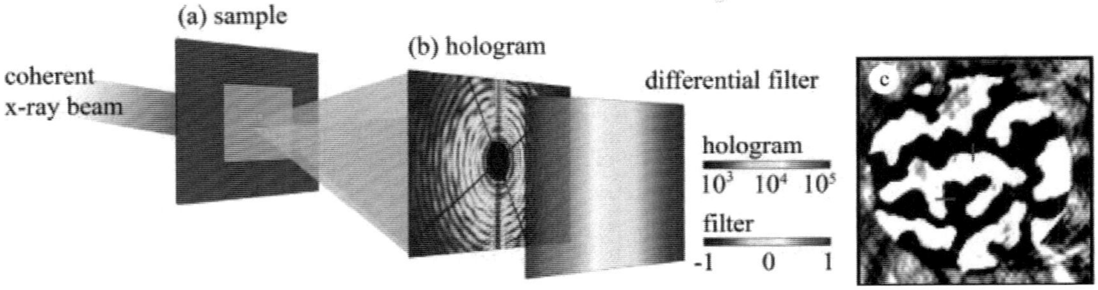

Figure 6. Holography with extended reference by autocorrelation linear differential operator (HERALDO). (a) The object and reference slit are illuminated by a coherent x-ray beam. (b) The hologram formed by the interference of the scattered x-rays from the object and reference slit is recorded on the CCD camera in the far field. A linear differential filter, defined by the derivative of the slit direction, is multiplied by the hologram prior to performing a Fourier transform to retrieve the reconstructed image. (c) After Fourier transformation the image shows the out-of-plane magnetic domain structure of the object, in this case a [Co(5 Å)/Pt(10 Å)]×30 multilayer. Field-of-view = 1.5 μm and spatial resolution = 30 nm [96].

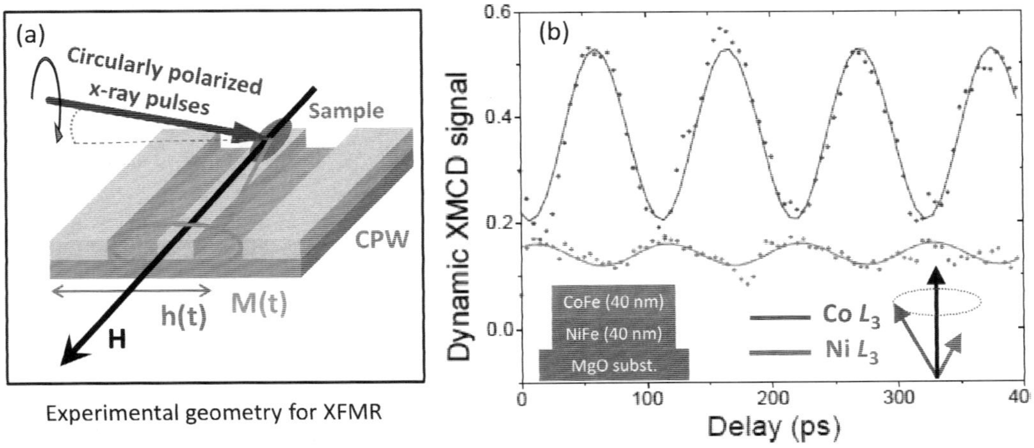

Figure 7. (a) Experimental geometry of x-ray ferromagnetic resonance (XFMR) experiment. The sample (pictured as the red disc) is mounted on the central strip of the coplanar waveguide (CPW). The magnetization $\mathbf{M}(t)$ is forced into a state of steady precession about the applied bias field $\mathbf{H}$ by an in-plane continuous microwave field $h(t)$ in the CPW. The circularly polarized x-ray pulses from the synchrotron are incident at grazing angle onto the sample in transverse geometry in order to detect stroboscopically the in-plane oscillatory component of the magnetization $\mathbf{M}(t)$ at variable phase delay. (b) The dynamic XMCD signals at the Co and Ni L_3 edges from a MgO/NiFe(40 nm)/CoFe(40 nm) bilayer stack as a function of the delay time between the x-ray pulse and the phase of the microwave field. Shown is the high energy mode (the 'optical resonance') of the 9.5 GHz precession at $\mathbf{H}$ = 65 Oe, which shows that the Co and Ni magnetizations are precessing in anti-phase [97].

magnetization obtained by time-resolved XMCD measurements can be used to determine the magnetic parameters of magnetic thin films and nanostructures. These parameters include the internal dipolar field strength, the magnetocrystalline anisotropy constants, the exchange bias field, interlayer exchange coupling constants, the gyromagnetic ratio and the phenomenological damping constant.

The damping of the magnetization precession in thin films has been subject of extensive and vigorous debate with conflicting views about the correct functional form of the damping torque. The phenomenological damping is thought to arise from a combination of the following effects: intrinsic damping due to the spin-orbit interaction; eddy current damping; two magnon scattering from defects and impurities; dephasing of spin wave modes in confined geometries; and higher order spin wave processes. Transition metals can be doped with rare earths and different elements placed at the film interfaces to modify the damping.

The experimental set-up for XFMR is schematically illustrated in Fig. 7(a). The magnetization dynamics is driven by the microwave field of a coplanar waveguide (CPW). The oscillating current in the CPW generates a time-varying magnetic field that can be phase-locked to the x-ray pulses of the storage ring. A magnetic sample will absorb the microwaves of given frequency at the ferromagnetic resonance condition, resulting in a steady precession about the applied bias field. High sensitivity in FMR can be achieved using a vector network analyser (VNA), which can perform fast frequency sweeps. By sweeping both the microwave frequency and the applied field the resonances can be mapped. XMCD is sensitive to the magnetization component along the direction of the x-rays. Element specificity is obtained by tuning the photon energy to the absorption edge of the element of interest. Phase-resolved measurements of XFMR are made by varying the delay between x-ray pulse arrival and phase of excitation.

Applied to magnetic bilayers, multiple modes can be found describing the dynamics of the coupled layers, with a low and high energy mode corresponding to in-phase and anti-phase oscillation, respectively, between the layer magnetizations. XFMR in transverse geometry reveals for each mode the amplitudes and relative phase of the precessing magnetization in the adjacent layers. The phase delay between the microwave phase and the x-ray pulse for the high energy mode from a NiFe(40 nm)/CoFe(40 nm) bilayer is shown in Fig. 7(b).

10. Angular dependent measurements

10.1. Angular dependence in XMCD

For an isotropic system, which is magnetically saturated by an external field, the XMCD signal scales with the angle θ between the field and x-ray beam as $\cos\theta$ [34]. Consequently, in transverse geometry, i.e., perpendicular orientation of x rays and field ($\theta = 90°$), the XMCD signal vanishes. In systems with cubic magnetic anisotropy, the XMCD spectrum is slightly different along $\langle 001\rangle$ and $\langle 111\rangle$ directions, due to the different symmetry of these axes (C_4 and C_3, respectively) [102]. The angular dependence of the intensity of the XMCD spectral features can be well described by the lowest order term for the cubic anisotropy [102].

The geometry of the XMCD experiment is described by three vectors, i.e., the light helicity, $\hat{\mathbf{P}}$, the magnetization direction, $\hat{\mathbf{M}}$, and the crystal frame orientation, $\hat{\mathbf{Z}}$. For uniaxially distorted symmetry the XMCD signal can be decomposed as,

$$I_{\mathrm{XMCD}} = I_0 \hat{\mathbf{P}} \cdot \hat{\mathbf{M}} + I_2 \left[\frac{3}{2} (\hat{\mathbf{P}} \cdot \hat{\mathbf{Z}})(\hat{\mathbf{Z}} \cdot \hat{\mathbf{M}}) - \frac{1}{2} \hat{\mathbf{P}} \cdot \hat{\mathbf{M}} \right], \tag{2}$$

where

$$I_0 = \frac{1}{3}(I_x + I_y + I_z), \tag{3}$$

$$I_2 = \frac{2}{3} I_z - \frac{1}{3}(I_x + I_y), \tag{4}$$

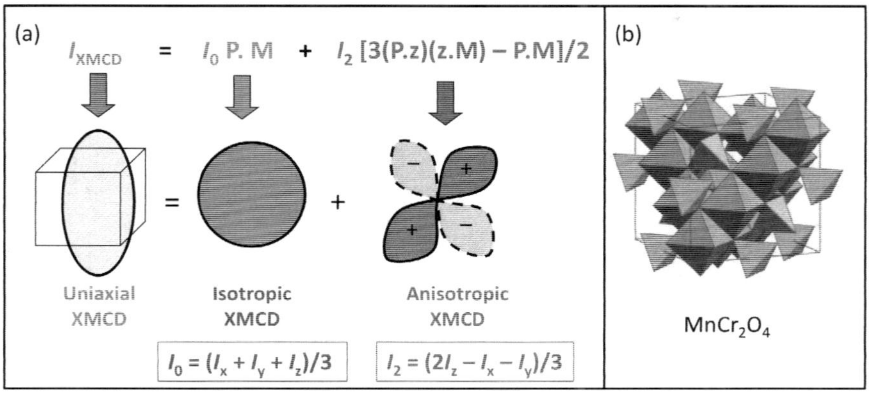

Figure 8. (a) Decomposition of the XMCD of an axially symmetric system into its isotropic and anisotropic contributions I_0 and I_2, respectively, where z is along the uniaxial direction. For the anisotropic XMCD the angular dependence is shown for $\hat{\mathbf{P}} \perp \hat{\mathbf{M}}$ [cf. Eq. (5)]. (b) Crystal structure of $MnCr_2O_4$.

are the isotropic and anisotropic XMCD contributions, respectively [34]. These have a simple graphical interpretation, as illustrated in Fig. 8(a).

In transverse geometry, where $\hat{\mathbf{P}} \cdot \hat{\mathbf{M}} = 0$, we obtain from Eq. (2) the angular dependence

$$I_{\mathrm{XMCD}}(\hat{\mathbf{M}}) = \frac{3}{4} I_2 \sin 2\nu = \frac{1}{2}(I_z - I_x) \sin 2\nu, \tag{5}$$

where $\nu = \angle(\hat{\mathbf{Z}}, \hat{\mathbf{M}})$ and taking $I_y = I_x$. The angular dependence of the anisotropic XMCD for $\hat{\mathbf{P}} \perp \hat{\mathbf{M}}$ is shown in Fig. 8. Therefore, in systems with axial magnetic anisotropy, such as uniaxial, tetragonal or trigonal symmetry of the lattice, a non-vanishing integral of the XMCD signal measured in transverse geometry corresponds to a non-zero component of the orbital magnetic moment perpendicular to the spin moment, a situation encountered in thin films with strong uniaxial magnetic anisotropy [103, 104].

The decomposition into isotropic and anisotropic parts can be applied—not only to the integrated $L_{2,3}$ intensities, i.e., sum rules—but also to the XMCD spectrum itself. XMCD measurements on $MnCr_2O_4/SrTiO_3$, with its crystal structure shown in Fig. 8(b), have revealed that the angular-dependent XMCD is strongly sensitive to strain-induced electronic structure variations [105]. This is a general phenomenon that provides an invaluable tool for studying and tailoring the physical characteristics of transition metal oxide nanoarchitectures through strain at interfaces of layered and nanocomposite heterostructures.

10.2. Angular Dependence in XMLD

In cubic crystals the XMLD depends strongly on the directions of the linear polarization, $\hat{\mathbf{E}}$, and spin axis, $\hat{\mathbf{M}}$, with respect to the crystallographic directions. A large anisotropy in the XMLD was first observed experimentally for the Mn^{2+} $L_{2,3}$ edge in $(Ga_{1-x}Mn_x)As$, which was shown to be in agreement with atomic multiplet calculations [106]. This effect is also present in metallic systems, as was predicted by *ab initio* band structure calculations [107] and more recently observed in, e.g., Co-based Heusler alloys [108] and Ga_xFe_{1-x} [44].

The most intensively studied case is that of the antiferromagnetic NiO. The angular dependence of the XMLD at the Ni^{2+} $L_{2,3}$ edges in a lattice with cubic symmetry can only be described as a linear combination of two fundamental spectra with distinctly different spectral

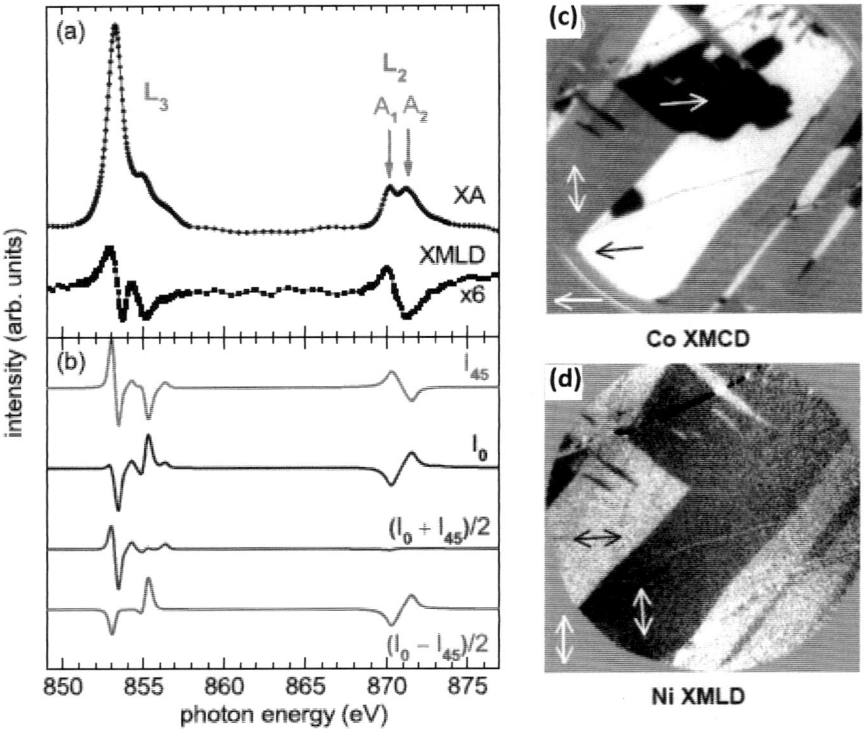

Figure 9. (a) Ni $L_{2,3}$ XAS and XMLD measured as the difference between the two local XAS spectra with in-plane linear x-ray polarization $\hat{\mathbf{E}} \parallel [110]$. The two peaks of the L_2 edge, indicated by A_1 and A_2, have different magnetic contrast. (b) Multiplet calculations for the XMLD spectra I_{45} and I_0 and the isotropic and anisotropic parts $\frac{1}{2}(I_0 + I_{45})$ and $\frac{1}{2}(I_0 - I_{45})$, respectively. The calculation shows that the L_2 XMLD contains only an anisotropic part. Right-hand side of the figure shows PEEM images of 2.5 nm Co/NiO(001) with field-of-view = 15 μm. (c) Ferromagnetic Co domains measured with circular polarization and (d) antiferromagnetic NiO domains measured with linear polarization. The polarization direction is indicated by the arrow in the bottom left corner of each panel. The other arrows indicate the deduced spin directions or axes of the different domains. In the Co XMCD PEEM image (c) the light and dark areas have the spin parallel and antiparallel to the light helicity vector, respectively. The grey areas in the Co image have the spin perpendicular to the light helicity. In the Ni XMLD PEEM image (d) the light and dark areas have their spin axis along $[1\bar{1}0]$ and $[110]$, respectively. This means that the ferromagnetic Co and antiferromagnetic NiO layers are coupled perpendicular.

features [109, 26, 110, 111, 112]—and not just by a single spectrum as previously thought. Previous studies, such as by Alders *et al.* [113], Stöhr *et al.* [114] on NiO and Scholl *et al.* [115] on $LaFeO_3$, identified only a single XMLD spectrum.

The in-plane (001) XMLD spectra, defined as,

$$I_{\mathrm{XMLD}} = I_{\mathrm{XAS}}(\mathbf{E} \parallel \mathbf{M}) - I_{\mathrm{XAS}}(\mathbf{E} \perp \mathbf{M}), \tag{6}$$

for $\mathbf{E} \parallel [100]$ and $\mathbf{E} \parallel [110]$ are indicated by I_0 and I_{45}, respectively. The subscript value denotes the angle between the linear polarization vector $\mathbf{E}$ with the [100] direction. The two fundamental spectra for the Ni^{2+} $L_{2,3}$ XMLD in cubic symmetry are shown in Fig. 9(b). Comparison with the calculated I_0 and I_{45} spectra shows that at the L_2 edge, $I_{45} \approx -I_0$, whereas the L_3 structure is completely different. Therefore, to determine unambiguously the spin direction both the L_2 and L_3 structure need to be considered.

The I_0 and I_{45} contributions have their maxima along $\langle 100 \rangle$ and $\langle 110 \rangle$, respectively [112]. Alternatively, one can decompose the XMLD into an 'isotropic' term, $\frac{1}{2}(I_0+I_{45})$, which is rigidly coupled to **H** and an 'anisotropic' term, $\frac{1}{2}(I_0 - I_{45})$, which is coupled to the crystal field frame [112].

XMLD-PEEM allows to determine the antiferromagnetic spin spin axes of magnetic domains. The spin coupling in Co/NiO(001) can be studied by comparing the PEEM images of ferromagnetic (FM) Co measured by XMCD [Fig. 9(c)] and antiferromagnetic (AFM) NiO measured by XMLD [Fig. 9(d)] [110, 111, 112]. Comparison of the spin directions unambiguously shows a perpendicular coupling between the FM Co and AFM NiO moments. Such a perpendicular coupling, which appears for certain Co thickness and interface roughness, was theoretically predicted by Koon [116] and later by Schulthess and Butler [117].

11. Conclusions

Different applications of x-ray magnetic dichroism have been summarized. The possibilities of magnetic dichroism in soft x-ray spectroscopy are vast and almost endless. XMCD and XMLD offer substantial advantages, such as element-specificity, spin and orbital moment analysis, magnetic anisotropy determination, magnetic imaging, surface sensitivity and time domain resolution. The technique provides a unique probe able to solve a large variety of different problems, such as the valency determination of the constituents in ternary oxides, the magnetic origin in dilute ferromagnetic semiconductors and the effects of electron correlation on the electronic structure.

Acknowledgments

The author is grateful to all his collaborators who made contributions to the research presented in this article.

References

[1] van der Laan G, Thole B T, Sawatzky G A, Goedkoop J B, Fuggle J C, Esteva J M, Karnatak R C, Remeika J P and Dabkowska H A 1986 *Phys. Rev. B* **34** 6529
[2] Thole B T, van der Laan G and Sawatzky G A 1985 *Phys. Rev. Lett.* **55** 2086
[3] Thole B T, Carra P, Sette F and van der Laan G 1992 *Phys. Rev. Lett.* **68** 1943
[4] Hase T P A, Pape I, Tanner B K, Dürr H A, Dudzik E, van der Laan G, Marrows C H and Hickey B J 2000 *Phys. Rev. B* **61** R3792
[5] van der Laan G 2006 *Curr. Opin. Solid State Mater. Sci.* **10** 120
[6] van der Laan G 2008 *C. R. Physique* **9** 570
[7] van der Laan G 2001 *Lect. Notes Phys.* **565** Foreword, pp. V–VII
[8] van der Laan G and Thole B T 1991 *Phys. Rev. B* **43** 13401
[9] Thole B T and van der Laan G 1991 *Phys. Rev. B* **44** 12424
[10] van der Laan G and Thole B T 1993 *Phys. Rev. B* **48** 210
[11] Thole B T and van der Laan G 1994 *Phys. Rev. B* **49** 9613
[12] Morall P, Schedin F, Case G S, Thomas M F, Dudzik E, van der Laan G and Thornton G 2003 *Phys. Rev. B* **67** 214408
[13] Pattrick R A D, van der Laan G, Henderson C M B, Kuiper P, Dudzik E and Vaughan D J 2002 *Eur. J. Mineral.* **14** 1095
[14] Braicovich L, Ghiringhelli G, Tagliaferri A, van der Laan G, Annese E and Brookes N B 2005 *Phys. Rev. Lett.* **95** 267402
[15] Braicovich L and van der Laan G 2008 *Phys. Rev. B* **78** 174421
[16] van der Laan G and Thole B T 1995 *J. Phys.: Condens. Matter* **7** 9947
[17] van der Laan G 2006 *Lect. Notes Phys.* **697** 143
[18] Moore K T and van der Laan G 2009 *Rev. Mod. Phys.* **81** 235

[19] Cowan R D 1968 *J. Opt. Soc. Am.* **58** 808
[20] Cowan R D 1981 *The Theory of Atomic Structure and Spectra* (Berkeley, CA: University of California Press)
[21] Thole B T, van der Laan G, Fuggle J C, Sawatzky G A, Karnatak R C and Esteva J M 1985 *Phys. Rev. B* **32** 5107
[22] Butler P H 1981 *Point Group Symmetry, Applications, Methods and Tables* (Plenum Press, N.Y.)
[23] Thole B T, van der Laan G and Butler P H 1988 *Chem. Phys. Lett.* **149** 295
[24] van der Laan G and Thole B T 1992 *J. Phys.: Condens. Matter* **4** 4181
[25] van der Laan G and Kirkman I W 1992 *J. Phys.: Condens. Matter* **4** 4189
[26] van der Laan G, Arenholz E, Chopdekar R V and Suzuki Y 2008 *Phys. Rev. B* **77** 064407
[27] van der Laan G 1997 *Phys. Rev. B* **55** 8086
[28] Thole B T and van der Laan G 1988 *Phys. Rev. B* **38** 3158
[29] van der Laan G and Thole B T 1988 *Phys. Rev. Lett.* **60** 1977
[30] Thole B T and van der Laan G 1988 *Phys. Rev. A* **38** 1943
[31] van der Laan G and Thole B T 1996 *Phys. Rev. B* **53** 14458
[32] van der Laan G 2012 *Phys. Rev. Lett.* **108** 077401
[33] Carra P, Thole B T, Altarelli M and Wang X 1993 *Phys. Rev. Lett.* **69** 2307
[34] van der Laan G 1998 *Phys. Rev. B* **57** 112
[35] Carra P, König H, Thole B T and Altarelli M 1993 *Physica* **192B** 182
[36] Thole B T and van der Laan G 1993 *Phys. Rev. Lett.* **70** 2499
[37] van der Laan G, Moore K T, Tobin J G, Chung B W, Wall M A and Schwartz A J 2004 *Phys. Rev. Lett.* **93** 097401
[38] van der Laan G 1999 *Phys. Rev. Lett.* **82** 640
[39] van der Laan G 1997 *J. Phys.: Condens. Matter* **9** L259
[40] Strange P 1994 *J. Phys.: Condens. Matter* **6** L491
[41] O'Brien W L, Tonner B P, Harp G R and Parkin S S P 1994 *J. Appl. Phys.* **76** 6462
[42] Huttel Y, van der Laan G, Johal T K, Telling N D and Bencok P 2003 *Phys. Rev. B* **68** 174405
[43] Chassé A, Dürr H A, van der Laan G, Kucherenko Y and Yaresko A N 2003 *Phys. Rev. B* **68** 214402
[44] Arenholz E, van der Laan G, McClure A and Idzerda Y 2010 *Phys. Rev. B* **82** 180405R
[45] Henke B L, Lee P, Tanaka T J, Shimabukuro R L and Fujikawa B K 1982 *At. Data Nucl. Data Tables* **27** 1
[46] van der Laan G and Thole B T 1988 *J. Electron Spectrosc. Relat. Phenomen.* **46** 123
[47] Lee J, Lauhoff G, Fermon C, Hope S, Bland J A C, Schillé J P and van der Laan G 1997 *J. Phys.: Condens. Matter* **9** L137
[48] Nakajima R, Stöhr J and Idzerda Y U 1999 *Phys. Rev. B* **59** 6421
[49] Shelford L R, Hesjedal T, Collins-McIntyre L, Dhesi S S, Maccherozzi F and van der Laan G 2012 *Phys. Rev. B* **86** 081304(R)
[50] Chen C T, Idzerda Y U, Lin H J, Smith N V, Meigs G, Chaban E, Ho G H, Pellegrin E and Sette F 1995 *Phys. Rev. Lett.* **75** 152
[51] Goedkoop J B, Brookes N B, van Veenendaal M and Thole B T 1997 *J. Electron Spectrosc. Relat. Phenomen.* **86** 143
[52] Stenning G B G, Bowden G J, Gregory S A, Beaujour J M L, de Groot P A J, van der Laan G, Shelford L R, Bencok P, Steadman P, Drobrynin A N and Hesjedal T 2012 *Appl. Phys. Lett.* **101** 072412
[53] Marcham M K, Keatley P S, Neudert A, Hicken R J, Cavill S A, Shelford L R, van der Laan G, Telling N D, Childress J R, Katine J A, Shafer P and Arenholz E 2011 *J. Appl. Phys.* **109** 113920
[54] Schütz G, Wagner W, Wilhelm W, Kienle P, Zeller R, Frahm R and Materlik G 1987 *Phys. Rev. Lett.* **58** 737
[55] Beale T A W, Hase T P A, Iida T, Endo K, Steadman P, Marshall A R, Dhesi S S, van der Laan G and Hatton P D 2010 *Rev. Sci. Instrum.* **81** 073904
[56] Hillebrands B and Ounedjela K (eds) 2001 and 2003 *Spin Dynamics in Confined Magnetic Structures* (*Topics in Applied Physics* vol I and II) (Springer-Verlag)
[57] Bonfim M, Ghiringhelli G, Montaigne F, Pizzini S, Brookes N B, Petroff F, Vogel J, Camarero J and Fontaine A 2001 *Phys. Rev. Lett.* **86** 3646
[58] Sirotti F, Girlando S, Prieto P, Floreano L, Panaccione G and Rossi G 2000 *Phys. Rev. B* **61** R9221
[59] Vogel J, Kuch W, Bonfim M, Camarero J, Pennec Y, Offi F, Fukumoto K, Kirschner J, Pizzini S and Fontaine A 2003 *Appl. Phys. Lett.* **82** 2299
[60] Krasyuk A, Oelsner A, Nepijko S A, Kuksov A, Schneider C M and Schonhense G 2003 *Appl. Phys. A* **76** 863
[61] Dietl T, Ohno H, Matsukura F, Cibert J and Ferrand D 2000 *Science* **287** 1019
[62] Wang M, Campion R P, Rushforth A W, Edmonds K W, Foxon C T and Gallagher B L 2008 *Appl. Phys. Lett.* **93** 132103

[63] Edmonds K W, Farley N R S, Johal T K, van der Laan G, Campion R P, Gallagher B L and Foxon C T 2005 *Phys. Rev. B* **71** 064418
[64] Edmonds K W, Freeman A A, Farley N R S, Wang K Y, Campion R P, Gallagher B L, Foxon C T, van der Laan G and Arenholz E 2007 *J. Appl. Phys.* **102** 023902
[65] Dunlop D J and Özdemir Ö 1997 *Rock Magnetism. Fundamental and frontiers* Cambridge Studies in Magnetism (Cambridge University Press)
[66] Coker V S, Telling N D, van der Laan G, Pattrick R A D, Pearce C I, Arenholz E, Tuna F, Winpenny R and Lloyd J R 2009 *ACS Nano* **3** 1922
[67] Pearce C I, Henderson C M B, Telling N D, Pattrick R A D, Charnock J M, Arenholz E, Tuna F, Coker V S and van der Laan G 2010 *Am. Mineral.* **95** 425
[68] Coker V S, Pattrick R A D, van der Laan G and Lloyd J R 2006 *Magnetoreception and Magnetosomes in Bacteria* Microbiology Monograph ed Schüler D (Springer, Heidelberg) p 275
[69] Lloyd J R, Pearce C I, Coker V S, Pattrick R A D, van der Laan G, Cutting R, Vaughan D J, Paterson-Beedle M, Mikheenko I P, Yong P and Macaskie L E 2008 *Geobiology* **6** 285
[70] Macaskie L E, Mikheenko I P, Yong P, Deplanche K, Murray A J, Paterson-Beedle M, Coker V S, Pearce C I, Pattrick R A D, Vaughan D, van der Laan G and Lloyd J R 2010 *Hydrometallurgy* **104** 483
[71] Coker V S, Pearce C I, Lang C, van der Laan G, Pattrick R A D, Telling N D, Schüler D, Arenholz E and Lloyd J R 2007 *Eur. J. Mineral.* **19** 707
[72] Staniland S, Ward B, Harrison A, van der Laan G and Telling N D 2007 *P. Natl. Acad. Sci. USA* **104** 19524
[73] Coker V S 2012 to be published
[74] Telling N D, Coker V S, Cutting R S, van der Laan G, Pearce C I, Pattrick R A D, Arenholz E and Lloyd J R 2009 *Appl. Phys. Lett.* **95** 163701
[75] Cutting R S, Coker V S, Telling N D, Kimber R L, Pearce C I, Ellis B L, Lawson R S, van der Laan G, Pattrick R, Vaughan D J, Arenholz E and Lloyd J R 2010 *Environ. Sci. Technol.* **44** 2577
[76] Byrne J M, Telling N D, Coker V S, Pattrick R A D, van der Laan G, Arenholz E, Tuna F and Lloyd J R 2011 *Nanotechnology* **22** 455709
[77] Crean D E, Coker V S, van der Laan G and Lloyd J R 2012 *Environ. Sci. Technol.* **46** 3352
[78] Coker V S, Pearce C I, Pattrick R A D, van der Laan G, Telling N D, Charnock J, Arenholz E and Lloyd J R 2008 *Am. Mineral.* **93** 1119
[79] Staniland S, Williams W, Telling N D, van der Laan G, Harrison A and Ward B 2008 *Nat. Nanotechnol.* **3** 158
[80] Byrne J M, Telling N D, Coker V S, Wincott P L, Vaughan D J, Pattrick R A D, van der Laan G, Arenholz E, Tuna F and Lloyd J R 2012 *Adv. Funct. Mater.* in print
[81] Coker V S, Gault A G, Pearce C I, van der Laan G, Telling N D, Charnock J M, Polya D A and Lloyd J R 2006 *Environ. Sci. Technol.* **40** 7745
[82] Coker V S, Bennett J A, Telling N D, Henkel T, Charnock J M, van der Laan G, Pattrick R A D, Pearce C I, Cutting R S, Shannon I J, Wood J, Arenholz E, Lyon I C and Lloyd J R 2010 *ACS-Nano* **4** 2577
[83] Lam K P, Hitchcock A P, Obst M, Lawrence J R, Swerhone G D W, Leppard G G, Tyliszczak T, Karunakaran C, Wang J, Kaznatcheev K, Bazylinski D A and Lins U 2010 *Chemical Geology* **270** 110
[84] Coker V S, Byrne J M, Telling N D, van der Laan G, Lloyd J R, Hitchcock A P, Wang J and Pattrick R A D 2012 *Geobiology* **10** 347
[85] Beutier G, Marty A, Livet F, van der Laan G, Stanescu S and Bencok P 2007 *Rev. Sci. Instrum.* **78** 093901
[86] Beutier G, van der Laan G, Marty A and Livet F 2008 *Eur. Phys. J. Appl. Phys.* **42** 161
[87] Dürr H A, Dudzik E, Dhesi S S, Goedkoop J B, van der Laan G, Belakhovsky M, Mocuta C, Marty A and Sampson Y 1999 *Science* **284** 2166
[88] van der Laan G, Chesnel K, Belakhovsky M, Marty A, Livet F, Collins S P, Dudzik E, Haznar A and Attané J P 2003 *Superlattice Microst.* **34** 107
[89] van der Laan G 2004 *Physica B* **345** 137
[90] Beutier G, van der Laan G, Chesnel K, Marty A, Belakhovsky M, Collins S P, Dudzik E, Toussaint J C and Gilles B 2005 *Phys. Rev. B* **71** 184436
[91] Chesnel K, Belakhovsky M, Livet F, Collins S P, van der Laan G, Dhesi S S, Attané J P and Marty A 2002 *Phys. Rev. B* **66** 172404
[92] Chesnel K, Belakhovsky M, van der Laan G, Livet F, A Marty G Beutier S P C and Haznar A 2004 *Phys. Rev. B* **70** 180402
[93] Pierce M S, Moore R G, Sorensen L B, Kevan S D, Hellwig O, Fullerton E E and Kortright J B 2003 *Phys. Rev. Lett.* **90** 175502
[94] Eisebitt S, Lüning J, Schlotter W F, Lörgen M, Hellwig O, Eberhardt W and Stöhr J 2004 *Nature* **432** 885
[95] Guizar-Sicairos M and Fienup J R 2008 *Opt. Lett.* **33** 2668
[96] Duckworth T, Ogrin F, Dhesi S, Langridge S, Whiteside A, Moore T, Beutier G and van der Laan G 2011

Opt. Express **19** 16223
[97] Shelford L R 2012 to be published
[98] Martin T, Woltersdorf G, Stamm C, Dürr H A, Maththeis R, Back C H and Bayreuther G 2009 *J. Appl. Phys.* **105**
[99] Arena D A, Vescovo E, Kao C C, Guan Y and Bailey W E 2006 *Phys. Rev. B* **74** 064409
[100] Goulon J, Rogalev A, Wilhelm F, Jaouen N, Goulon-Ginet C, Goujon G, Youssef J B and Indendom M V 2005 *JETP Lett.* **82** 696
[101] Boero G, Rusponi S, Bencok P, Popovic R S, Brune H and Gambardella P 2005 *Appl. Phys. Lett.* **87** 152503
[102] Edmonds K W, van der Laan G, Freeman A A, Farley N R S, Johal T K, Campion R P, Foxon C T, Gallagher B L and Arenholz E 2006 *Phys. Rev. B* **96** 117207
[103] Dürr H A and van der Laan G 1996 *Phys. Rev. B* **54** R760
[104] Dürr H A, van der Laan G and Thole B T 1996 *Phys. Rev. Lett.* **76** 3464
[105] van der Laan G, Chopdekar R V, Suzuki Y and Arenholz E 2010 *Phys. Rev. Lett.* **105** 067405
[106] Freeman A A, Edmonds K W, van der Laan G, Farley N R S, Johal T K, Arenholz E, Campion R P, Foxon C T and Gallagher B L 2006 *Phys. Rev. B* **73** 233303
[107] Kunes J and Oppeneer P M 2003 *Phys. Rev. B* **67** 024431
[108] Telling N D, Keatley P S, van der Laan G, Hicken R J, Arenholz E, Sakuraba Y, Oogane M, Ando Y, Takanashi K, Sakuma A and Miyazaki T 2008 *Phys. Rev. B* **78** 184438
[109] Arenholz E, van der Laan G, Chopdekar R V and Suzuki Y 2007 *Phys. Rev. Lett.* **98** 197201
[110] Arenholz E, van der Laan G and Nolting F 2008 *Appl. Phys. Lett.* **93** 162506
[111] Ohldag H, van der Laan G and Arenholz E 2009 *Phys. Rev. B* **79** 052403
[112] van der Laan G, Telling N D, Potenza A, Dhesi S S and Arenholz E 2011 *Phys. Rev. B* **83** 064409
[113] Alders D, Tjeng L H, Voogt F C, Hibma T, Sawatzky G A, Chen C T, Vogel J, Sacchi M and Iacobucci S 1998 *Phys. Rev. B* **57** 11623
[114] Stöhr J, Scholl A, Regan T J, Anders S, Lüning J, Scheinfein M R, Padmore H A and White R L 1999 *Phys. Rev. Lett.* **83** 1862
[115] Scholl A, Stöhr J, Lüning J, Seo J W, Fompeyrine J, Siegwart H, Locquet J P, Nolting F, Anders S and Fullerton E E 2000 *Science* **287** 1014
[116] Koon N C 1997 *Phys. Rev. Lett.* **78** 4865
[117] Schulthess T C and Butler W H 1998 *Phys. Rev. Lett.* **81** 4516

Journal of Physics: Conference Series **430** (2013) 012128 doi:10.1088/1742-6596/430/1/012128

Local structure and magnetism of Cu-doped ZnO via Cu K-edge XAS and XMCD: theory and experiment

O Šipr[1], J Vackář[1], P S Vachhani[2,3], R K Ramamoorthy[2,4], G Dalba[2], A K Bhatnagar[3,4] and F Rocca[5]

[1] Institute of Physics ASCR, v. v. i., CZ-162 53 Prague, Czech Republic
[2] Department of Physics, University of Trento, 38123 Povo (Trento), Italy
[3] School of Physics, University of Hyderabad, Hyderabad-500 046, India
[4] School of Engineering Sciences & Technology, University of Hyderabad, Hyderabad 500 046, India
[5] IFN-CNR, Institute for Photonics and Nanotechnologies, Unit "FBK-Photonics", 38123 Povo (Trento), Italy

E-mail: sipr@fzu.cz

Abstract. We investigate theoretically the sensitivity of Cu K-edge XANES and XMCD of Cu-doped ZnO (Cu:ZnO) to the local structure. We find that the sensitivity is high enough to enable to distinguish between geometries conceivable for Cu:ZnO. However, one cannot distinguish between Cu and Zn in further coordination shells. Likewise, presence of oxygen vacancies in the ab plane does not alter XANES or x-ray linear dichroism (XLD). By analysing experimental XANES and EXAFS signals of Cu:ZnO pellets (which are paramagnetic) and Cu:ZnO films (which are ferromagnetic) we find that for the pellets, Cu is in Zn-substitutional sites while for the films, the coordination of Cu is as in CuO. Cu K-edge XMCD measurements were performed for both types of samples but detectable signal was obtained only for the pellets. The theory confirms that this XMCD signal originates from magnetic moments localised at Cu atoms located in Zn-substitutional sites.

1. Introduction

Magnetism of wurtzite ZnO doped by transition metal atoms still evades a consistent explanation [1]. Numerous results on structure and magnetism of pure and doped ZnO were obtained but a coherent picture is difficult to obtain. One of the problems is that it is hard to study magnetism and structure simultaneously "on the same footing". X-ray absorption (XAS) and x-ray magnetic circular dichroism (XMCD) spectroscopy is a very useful tool in this respect.

In this paper we investigate the potential of XAS and XMCD to investigate Cu-doped ZnO (Cu:ZnO). Our main focus is on theory but to put this work in context, we also present some new experimental data which will be analysed in a greater detail elsewhere.

2. Computational framework

Electronic structure was calculated within a fully relativistic KKR Green's-function formalism, using the SPRKKR code [2]. The Cu impurities were treated within the embedded Green's function technique, meaning that — unlike in the super-cell approach — there is no interaction between the impurities. The Cu impurity is located either at the Zn-substitutional (Cu_{Zn}) or

Journal of Physics: Conference Series **430** (2013) 012128 doi:10.1088/1742-6596/430/1/012128

at the interstitial site (Cu_i) of the underlying wurtzite structure. To simulate vacancies, which have attracted a lot of attention lately [3], we substituted O atoms by empty sites (V_O) in some cases. Structural relaxation was not attempted (substituting Cu_{Zn} for Zn changes the distances by less than 0.01 Å [4]).

We imposed the atomic sphere approximation (ASA) to the potential, with additional empty spheres inserted at the interstitial sites. Most of the spectra were calculated for the ground state potential. Namely, simple ways of incorporating the core hole used in XAS such as the the final state approximation or the transition state approximation seem to be unsuitable for XMCD [5, 6]. However, to assess the core hole effect at least approximatively, we did some calculations also using the transition state potential.

3. Experiment

We investigated spectra of Cu:ZnO pellets and of oriented Cu:ZnO thin films. Polycrystalline targets of $Zn_{1-x}Cu_xO$ were synthesized by mixing ZnO and CuO powders and calcinating the mixture at 400°C. To get polycrystalline pellets, the powders were ground, pelletized and sintered at 1000°C. To get oriented $Zn_{1-x}Cu_xO$ thin films of thickness of about 300 nm, we used pulsed laser deposition (PLD) to deposit the targets material onto *c*-cut sapphire substrate. The Cu concentration x was 2%, 4%, 7% or 10%.

XAS was measured at the ESRF at the BM08-Gilda beamline. The x-ray linear dichroic (XLD) spectra were recorded as a difference between XAS for two perpendicular linear polarisations at the ID12 beamline. The XMCD spectra were recorded at 7 K at the ID12 beamline, in an external magnetic field of ±3 T.

More detailed description of the sample preparation and a thorough analysis of our measurements will be published elsewhere.

4. Results and discussion

4.1. Magnetic properties

Magnetic measurements were carried out using SQUID in the temperature range from 5 to 350 K. The polycrystalline pellets samples were paramagnetic, with signatures of antiferromagnetic interaction between local moments (obtained by fit to the Curie-Weiss law). The oriented thin films were all ferromagnetic at room temperature [7, 8].

4.2. Sensitivity of XANES to structure type

To be able to analyse XAS properly, it is desirable to know the structural sensitivity of the spectra for our specific situation. Therefore, we calculated Cu *K*-edge x-ray absorption near-edge structure (XANES) for Cu in a number of structures: Cu_{Zn} and Cu_i in ZnO, CuO, and also hypothetic Cu metal with structure of ZnO. Apart from orientationally-averaged XANES for polycrystals, we calculated also the XLD as the difference between the spectra of monocrystals for two perpendicular orientations of the x-rays polarisation vector. The results are shown in figure 1. There are clear differences between XAS and especially XLD for Cu in different environments, one can thus distinguish between Cu_{Zn} and Cu_i and also between CuO or Cu clustering. It does not matter in this respect whether the core hole is included in the calculations or not.

Our calculations further show that Cu and Zn are practically interchangeable as concerns their photoelectron scattering properties. That means that one cannot distinguish by XAS whether Cu is surrounded by Zn or Cu in more distant shells. We calculated also Zn *K*-edge XAS and XLD and found that there is practically no difference whether Zn is in pure ZnO or whether it is close to a Cu impurity. And finally we found that if there is V_O in the *ab* plane next to the Cu impurity, both XAS and XLD look very similar to spectra without vacancies

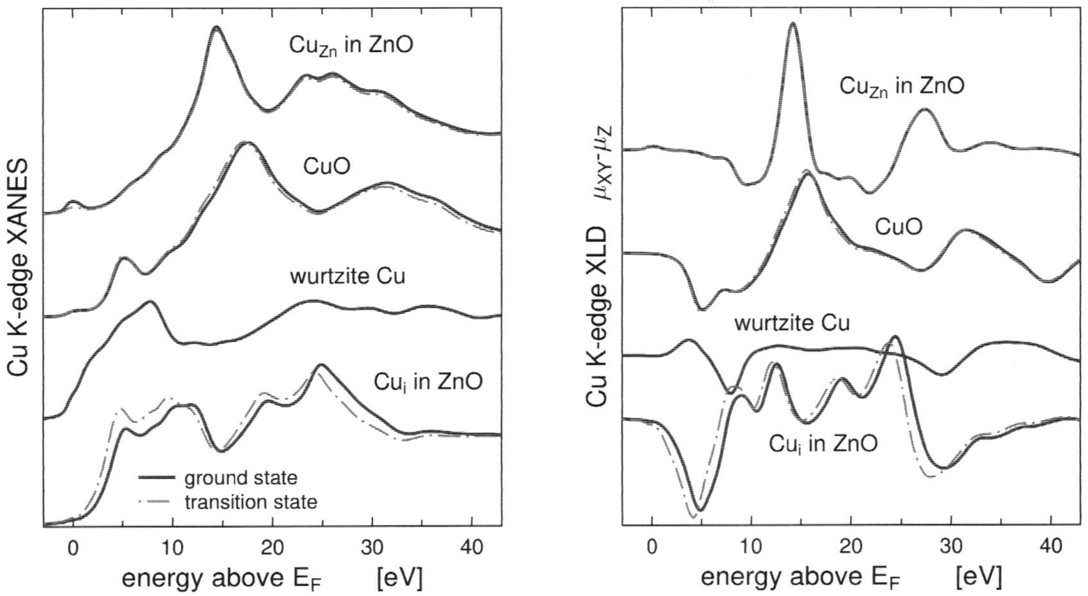

Figure 1. Cu K-edge XAS (left) and XLD (right) for Cu_{Zn} in ZnO, CuO, hypothetic Cu with structure of ZnO and for Cu_i in ZnO. Full grey lines correspond to calculations for ground state potential, dash-dotted lines lines to transition state potential.

(similar conclusions were reached by Ciatto *et al.* [3]). We do not show the spectra here for brevity.

4.3. Structure of Cu:ZnO pellets and films

Experimental Cu K-edge XANES of Cu:ZnO pellets and films differ quite a lot (figure 2 left): spectrum of the pellets generally resembles spectrum of ZnO while spectrum of the films resembles spectrum of CuO. Moreover, a pronounced pre-peak for the pellets looks like a signature of Cu in a tetrahedral environment [9] while a faint pre-peak at a bit higher energy for the films reminds the quadrupolar pre-peak observed in CuO [10]. This suggests that Cu is incorporated as Cu_{Zn} in the wurtzite lattice in the pellets while it segregates as CuO in the films.

We measured also the Cu K-edge EXAFS, in order to have a more accurate and reliable information about the structure. The Fourier-transforms of k^2-weighted EXAFS signals $k^2\chi(k)$ for the x=7% samples are shown in the right panel of figure 2. For other concentrations, the situation is similar. Together with a quantitative analysis that will be published elsewhere, this confirms that Cu occupies the Zn-substitutional site in Cu:ZnO pellets while it is in a CuO-like environment in Cu:ZnO films.

4.4. Cu K-edge XMCD spectra

By calculating electronic structure of Cu impurities in ZnO, we found that Cu_i is nonmagnetic, while Cu_{Zn} is magnetic. We calculated also the magnetic structure for Cu_{Zn} with vacancies, with V_O either in the first or in the second shell of O atoms in the ab plane of the wurtzite structure (the Cu–V_O distance is 1.98 Å and 3.80 Å, respectively). Table 1 summarises the results for spin magnetic moments μ_{spin} and orbital magnetic moments μ_{orb} evaluated within the atomic

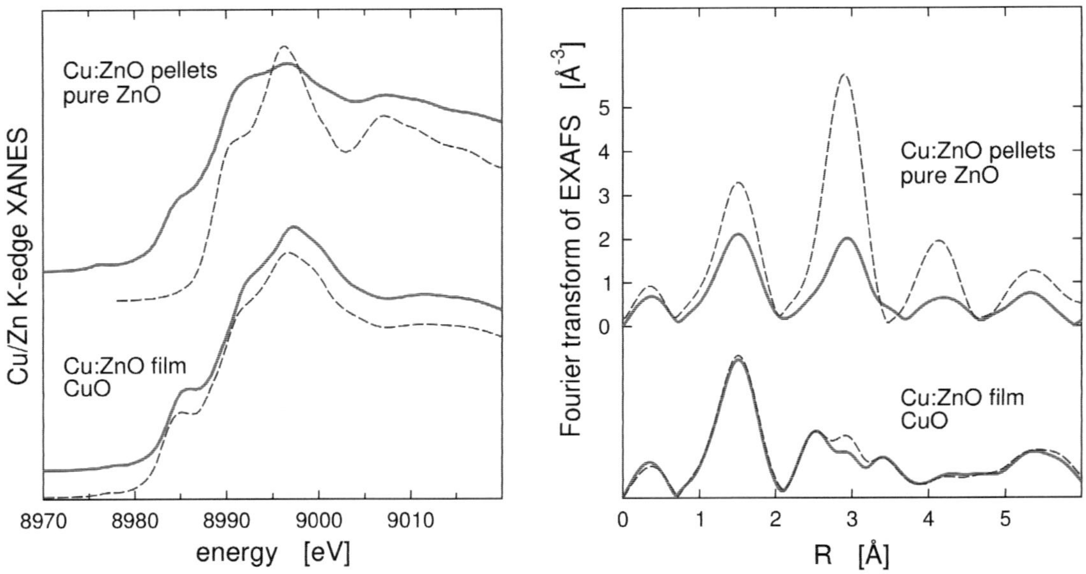

Figure 2. Experimental Cu K-edge XANES (left) and Fourier-transformed EXAFS (right) of $Zn_{0.93}Cu_{0.07}O$ films and pellets (solid grey lines) compared to experimental data for Cu K-edge of CuO and to Zn K-edge of ZnO, respectively (dashed blue lines). The Zn K-edge XANES of ZnO was shifted in energy to be aligned with the Cu K-edge spectra. Vertical shift was added between XANES of Cu:ZnO and reference materials for clarity.

Table 1. Magnetic moments and numbers of holes for Cu_{Zn} in ZnO with no vacancies and with V_O in the ab plane, either in the 1st or in the 2nd shell of O atoms. The unit is μ_B.

	μ_{spin}(tot)	μ_{spin}(Cu)	μ_{orb}(Cu)	$\mu_{\text{orb}}^{(p)}$(Cu)	$n_h^{(p)}$(Cu)	$n_h^{(d)}$(Cu)
no vacancy	1.70	0.59	0.14	−0.0016	5.52	0.98
vacancy in 1st shell	0.54	0.18	0.06	−0.0005	5.60	0.88
vacancy in 2nd shell	1.18	0.52	0.30	−0.0096	5.52	0.95

sphere around Cu (its radius is 1.30 Å). We show also the total spin magnetic moment in the whole system, which demonstrates that the polarisation cloud extends quite far away from the Cu atom. Magnetic moments for the nearest neighbours of Cu_{Zn} are 0.19 μ_B for the three O atoms in the ab plane and 0.09 μ_B for the apical O atom. The Cu K-edge XMCD probes the p component of μ_{orb}, therefore we show it in table 1 as well. Finally, we present also the numbers of holes in the p and d band of Cu_{Zn} as these quantities are of interest when utilising the XMCD sum rules [11].

Theoretical Cu K-edge XMCD of Cu_{Zn} evaluated either for the ground state or for the transition state potential is shown in the left panel of figure 3. XANES is shown as well so that peak positions in XAS and in XMCD can be compared (note that the XMCD signal was multiplied by 100). A prominent XMCD peak appears at the pre-edge. Its intensity strongly depends on the orientation of the magnetic field **M** with respect to the wurtzite lattice. This anisotropy as well as the average XMCD peak intensity depends on how the core hole is treated.

To investigate the effect of vacancies, we calculated XMCD also for Cu_{Zn} with V_O in the first

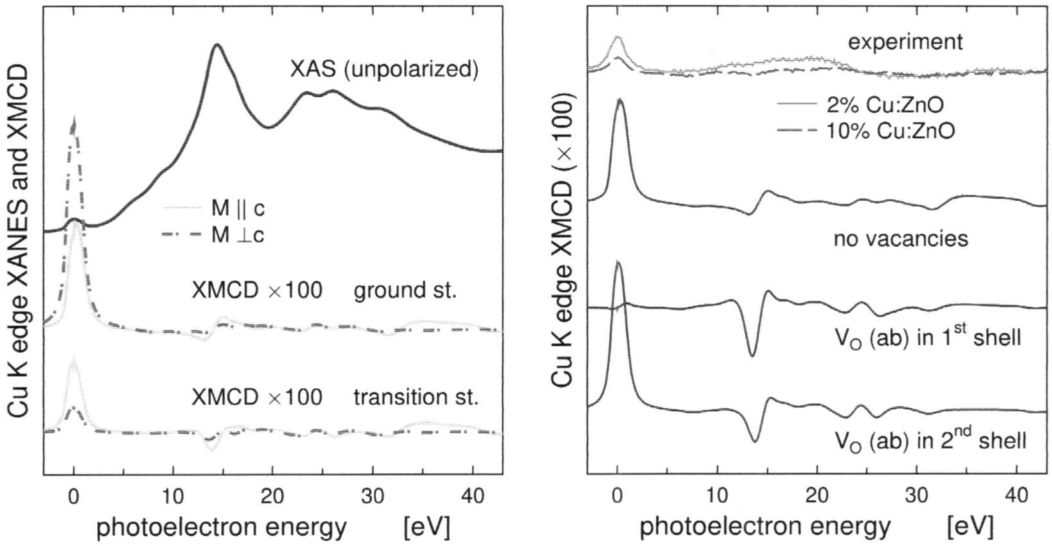

Figure 3. Left: Calculated Cu K-edge XMCD for Cu_{Zn} in single-crystal ZnO, with **M** parallel or perpendicular to the c axis, for the ground state and the transition state potential. XANES for Cu_{Zn} in polycrystalline ZnO for the ground state potential is also shown. Right: Calculated Cu K-edge XMCD for Cu_{Zn} in ZnO without or with O vacancies in the ab plane, for ground state potential, with $\mathbf{M}\|c$ (three lower curves) together with experimental Cu K-edge XMCD for $Zn_{0.98}Cu_{0.02}O$ and $Zn_{0.90}Cu_{0.10}O$ pellets.

or in the second shell of oxygens in the ab plane. The results are shown in the right panel of figure 3 ($\mathbf{M}\|c$, ground state potential). Presence of a vacancy next to Cu_{Zn} nearly suppresses the sharp XMCD pre-edge feature while a more distant vacancy affects the “fine XMCD structure” in the energy region corresponding the the main XANES peak.

Experimental Cu K-edge XMCD was measured both for paramagnetic Cu:ZnO pellets and ferromagnetic Cu:ZnO films. Detectable signal was found only for the pellets and it is shown in the top of the right panel of figure 3 (scaled so that experimental and theoretical XANES intensities match). We investigated also the magnetic-field-strength dependence of the XMCD intensity and found that it exhibits a paramagnetic pattern. For the Cu:ZnO films, the Cu K-edge XMCD is zero (within the limit of the noise).

5. Discussion

By measuring XAS and XMCD at the same edge we can unambiguously connect local geometric and magnetic structure. Use of the K edge makes it possible to exploit the EXAFS data to find that Cu:ZnO materials can indeed have quite different local structures; this may be the reason why divergent results on properties of Cu:ZnO were obtained in the past [12, 13, 14, 15].

Our XMCD measurements show that, surprisingly, Cu atoms are not magnetic in ferromagnetic Cu:ZnO films. This is consistent with the emerging picture that ferromagnetism of doped ZnO is not directly connected to the magnetism of dopant atoms [16]. Paramagnetic character of Cu atoms in Cu:ZnO pellets, where Cu is substitutional, reminds similar situation found earlier for Co-doped ZnO [17]. Another point to mention is that our calculations show that oxygen vacancies do not suppress magnetism of substitutional Cu, even though they affect it significantly.

Our embedded impurity calculations give larger total magnetic moments around Cu_{Zn} than supercell calculations [4, 18, 19] — this indicates that in supercell calculations the impurities still interact. Relatively large theoretical μ_{orb}/μ_{spin} ratio obtained for Cu_{Zn} is consistent with analogous situations for ad-atoms and impurities [20]. Finally, our calculations suggest that the core hole may affect the XMCD spectra significantly, stressing the need to have a practically manageable formalism for including the core hole also in XMCD calculations, where the final state or transition state approximations are inappropriate [5, 6].

6. Conclusions

Cu K-edge XANES and XLD of Cu in different environments is sufficiently different to distinguish between geometries conceivable for Cu:ZnO. However, because of similar scattering properties of Cu and Zn, XAS is not able to tell these two atoms one from another if the structure is otherwise the same.

Local environments of Cu atoms in paramagnetic Cu:ZnO pellets and in ferromagnetic Cu:ZnO films are quite different: for the pellets, Cu is in Zn-substitutional sites while for the films, the coordination of Cu is as in CuO. Copper atoms in paramagnetic pellets are magnetic. For ferromagnetic films, copper atoms are non-magnetic. This suggests that ferromagnetism of transition metal doped ZnO is not directly linked to magnetism of the dopant but rather is connected with structural changes induced by these dopants.

Acknowledgments

This work was supported by project 108/11/0853 of the GA ČR and by common project of AS CR and CNR "Interplay between structural and electronic properties of nanostructures". The experimental work has been carried out under the India-Trento program for advanced research (ITPAR) of the DST, New Delhi and University of Trento. We would also like to thank A. Rogalev and F. Wilhelm for their help in carrying out and discussing the the XANES and XMCD measurements at ID12 of ESRF and to A. Kuzmin for his cooperation in this project.

References

[1] Zunger A, Lany S and Raebiger H 2010 *Physics* **3** 53
[2] Ebert H, Ködderitzsch D and Minár J 2011 *Rep. Prog. Phys.* **74** 096501
[3] Ciatto G, Di Trolio A, Fonda E, Alippi P, Testa A M and Amore Bonapasta A 2011 *Phys. Rev. Lett.* **107** 127206
[4] Ye L H, Freeman A J and Delley B 2006 *Phys. Rev.* B **73** 033203
[5] Meinert M, Schmalhorst J, Wulfmeier H, Reiss G, Arenholz E, Graf T and Felser C 2011 *Phys. Rev.* B **83** 064412
[6] Šipr O, Minár J, Scherz A, Wende H and Ebert H 2011 *Phys. Rev.* B **84** 115102
[7] Vachhani P S, Dalba G, Ramamoorthy R K, Rocca F, Šipr O and Bhatnagar A K 2012 *J. Phys.: Condens. Matter* to be published
[8] Vachhani P S 2012 private communication
[9] Yamamoto T 2008 *X-Ray Spectrom.* **37** 572
[10] Bocharov S, Kirchner T, Dräger G, Šipr O and Šimůnek A 2001 *Phys. Rev.* B **63** 045104
[11] Stöhr J and Siegmann H C 2006 *Magnetism: From Fundamentals to Nanoscale Dynamics* (Berlin: Springer)
[12] Keavney D J, Buchholz D B, Ma Q and Chang R P H 2007 *Appl. Physics Lett.* **92** 012501
[13] Ma Q, Buchholz D B and Chang R P H 2008 *Phys. Rev.* B **78** 214429
[14] Thakur P, Bisogni V, Cezar J C, Brookes N B, Ghiringhelli G, Gautam S, Chae K H, Subramanian M, Jayavel R and Asokan K 2010 *J. Appl. Phys.* **107** 103915
[15] Herng T S, Qi D, Berlijn T, Yi J B, Yang K S, Dai Y, Feng Y P, Santoso I, Sánchez-Hanke C, Gao X Y, Wee A T S, Ku W, Ding J and Rusydi A 2010 *Phys. Rev. Lett.* **105** 207201
[16] Ogale S 2010 *Adv. Mater.* **22** 3125
[17] Ney A, Ollefs K, Ye S, Kammermeier T, Ney V, Kaspar T C, Chambers S A, Wilhelm F and Rogalev A 2008 *Phys. Rev. Lett.* **100** 157201
[18] Park M S and Min B I 2003 *Phys. Rev.* B **68** 224436
[19] Huang L M, Rosa A L and Ahuja R 2006 *Phys. Rev.* B **74** 075206
[20] Šipr O, Bornemann S, Minár J, Polesya S, Popescu V, Šimůnek A and Ebert H 2007 *J. Phys.: Condens. Matter* **19** 096203

Journal of Physics: Conference Series **430** (2013) 012129 doi:10.1088/1742-6596/430/1/012129

Passivating effect of Si(111)-(√3×√3)Ag and Si_3N_4/Si(111)-(8×8) buffer layers

K. Eguchi[1], Y. Takagi[1,2], T. Nakagawa[1,2] and T. Yokoyama[1,2]

[1] The Graduate University for Advanced Studies (SOKENDAI), Myodaiji-cho, Okazaki 444-8585, Japan

[2] Department of Molecular Structure, Institute for Molecular Science, Myodaiji-cho, Okazaki 444-8585, Japan

E-mail: keguchi@ims.ac.jp

Abstract. We have investigated the magnetic properties of iron deposited on Si(111)-(7×7), Si(111)-(√3×√3)Ag and Si_3N_4/Si(111)-(8×8) by x-ray absorption spectroscopy and x-ray magnetic circular dichroism measurements. By choosing the surface, it is possible to control the interaction of Fe atoms with Si atoms and to govern the spin magnetic moment of iron. The magnetic moment of 1.6 ML-Fe is almost lost on the clean surface, while that of 1.6ML-Fe on the Si_3N_4 surface remains completely. About a half of the magnetic moment of bcc-Fe is obtained by depositing Fe on the Ag modified Si(111) surface. In addition, it is found that the magnetic moment of Fe on the Ag modified Si(111) surface is lost easily by annealing at 500 K. On the other hand, that of Fe on the Si_3N_4 surface is sufficiently kept even when the sample is annealed at 500 K.

1. Introduction

It is important to control the magnetic properties of ferromagnetic transition metals on Si substrates because of the technological applications such as magnetic memories and spintronics. However, when Fe is deposited on a clean Si surface at room temperature (RT), Fe atoms form FeSi and/or $FeSi_2$ at low coverage due to strong interaction with Si atoms. Berling *et al.* reported that the magnetic moment of iron atoms in iron silicides depends on the chemical composition of iron silicides [1]. The Fe atoms in Fe-rich silicides carry the magnetic moment, while the Fe magnetic moment is zero in FeSi. Therefore, it is difficult to keep nonzero magnetic moments of Fe atoms at the interface between deposited Fe and the clean Si surface. In order to form ferromagnetic bare iron and/or magnetic iron silicides on the Si substrate, it is required to suppress the chemical interaction of Fe atoms with the clean Si surface.

One way to govern the magnetic moment of Fe on Si(111) is modifying the Si surface [2-9]. It has been attempted to suppress the formation of the iron silicide by inserting metals [3-8], semiconductors [4] and insulators [2, 9]. For instance, Eguchi *et al.* succeeded in effective suppression of the formation of iron silicides by nitridation of the clean Si(111) surface [9]. In this case, the Fe magnetic moments were measured by *in situ* x-ray magnetic circular dichroism (XMCD) and the magnetic moment of Fe atoms on the Si_3N_4 surface is fully maintained. In other cases that Fe deposited on metal coated Si surfaces, the iron silicidation is also suppressed somewhat [3, 6]. Zavaliche *et al.* have studied the suppression effect of the Au modified Si(001) surface and they observed the ferromagnetic behavior of 2.3 ML-Fe film on the Au modified Si surface by the magneto-optical Kerr effect [3].

Takagi *et al.* have reported that the magnetic property of Fe deposited on a Si(111)-(√3×√3)Ag by *ex situ* superconducting quantum interference device (SQUID) measurements. In their report, ferromagnetic iron silicide dots are obtained when Fe atoms deposited on the (√3×√3)Ag surface annealed at 500 K. Additionally, they concluded the chemical composition of the dots is close to Fe_3Si because the magnetic moment of dots is the same value as that of Fe_3Si [6]. Thus, it is clear that metal coated Si surface partially prevent silicidation reaction of Fe atoms. However, the magnetic moment of Fe has not quantitatively evaluated by an *in situ* measurement.

In this work, we investigated the magnetic property of Fe deposited on Si(111)-(√3×√3)Ag by *in situ* x-ray absorption spectroscopy (XAS) and XMCD measurements. Compared with the results of Fe deposited on clean Si(111) and Si_3N_4 surfaces, we have confirmed the passivation effect of the (√3×√3)Ag buffer layer. In addition, we investigated a thermal effect for the magnetic properties of Fe deposited on the Si_3N_4 surface and on the (√3×√3)Ag surface.

2. Experimental

The sample preparation and the XAS and XMCD measurements were carried out in an ultrahigh vacuum (UHV) chamber, of which the base pressure was ~10^{-8} Pa. The sample was degassed in a UHV chamber by direct heating at ~800 K for 5-12 hours. A reconstructed Si(111)-(7×7) surface was obtained by repeated flashing at 1200 K. Thermal nitridation was conducted at 1130 K with ammonia gas (1×10^{-4} Pa, 45 sec). Then, the substrate was subsequently cooled down slowly to room temperature, which leads to the formation of an Si_3N_4/Si(111)-(8×8) surface [9, 10]. Silver of 1 monolayer (ML) was deposited on the clean Si(111) at 800 K, which leads to the formation of a √3×√3 reconstructed structure [11, 12]. The ordering and the cleanness of the Si_3N_4 and (√3×√3)Ag surfaces were verified by low energy electron diffraction and Auger electron spectroscopy. Fe was deposited on Si_3N_4/Si(111)-(8×8), Si(111)-(√3×√3)Ag and Si(111)-(7×7) substrates by using a commercial electron bombardment evaporator and the thickness was estimated by quartz crystal oscillator. In the present work, we define a unity monolayer for Fe as an atomic concentration of Fe of 1.217×10^{15} atoms/cm^2, which corresponds to the atomic density of bcc Fe(001) plane.

The *in situ* XAS and XMCD measurements were performed in the total electron yield mode by using the bending magnet Beamline 4B of UVSOR-II in Institute for Molecular Science (IMS). The details of the XMCD measurement system have been reported in the literature [13, 14]. The circular polarization factors were estimated to be P_c = 0.55 ~ 0.71 from the storage ring parameters, which were verified by the XMCD measurements of a standard sample. The XMCD spectra were taken by switching the magnetic field of ±5 T applied to the sample in parallel and antiparallel geometries with respect to the x-ray beam, leaving the x-ray helicity unchanged. The x-ray incidence angles with respect to the surface normal were 0° (normal x-ray incidence) and 55° (grazing incidence).

3. Results and discussions

Figure 1(a) shows the circularly polarized Fe $L_{2,3}$-edge XAS spectra of Fe deposited on Si(111)-(7×7), Si(111)-(√3×√3)Ag and Si_3N_4/Si(111)-(8×8) at RT at the incidence angle of 55°. The spectra were normalized by the edge jumps. The previous study on iron silicide has shown that additional peaks are observed at a higher photon energy of the main peaks [15, 16]. The spectral feature can be used as a fingerprint to confirm whether Fe atoms are bonded to Si atoms. In the present measurement, the shoulder structure does not appear in the spectrum of Fe on the Si_3N_4/Si(111)-(8×8), indicating that the Fe atom on the Si_3N_4 surface does not interact strongly with Si atoms. In contrast, that of Fe on Si(111)-(7×7) has an additional peak at a higher photon energy of the main L_3 peak. Namely, the iron silicide is formed on the clean Si(111) surface. For Fe on Si(111)-(√3×√3)Ag, although no additional satellite peak is observed in the spectrum, the peak width of the spectrum of Fe L_3 edge is broad in comparison with that of Fe on the Si_3N_4 surface. The full width of half minimum (FWHM) of the μ^- spectrum of Fe on the (√3×√3)Ag surface, actually, is 4.9 eV. This value is an intermediate one between 3.2 eV in the μ^- spectrum of Fe on the Si_3N_4 surface and 6.8 eV in that of Fe on the clean surface. These features imply that the Fe atoms deposited on the (√3×√3)Ag surface chemically react with the Si atoms and partially become an iron silicide.

The magnetization (*M-H*) curves were also obtained by recording the electron yield with the photon energy fixed at the Fe L_3 peak top for each sample (Fig. 1(b)). The amplitude corresponds to the difference in the L_3 peak intensity of the normalized XAS spectra at ±5 T. The data in the vicinity of $H = 0$ are omitted due to strong suppression of the electron yield by the Lorentz force. Fe atoms are clearly magnetized on the Si_3N_4 surface due to the effective suppression of reaction between Fe atoms and Si atoms. On the other hand, it is found that Fe atoms on the clean surface are not magnetized at all. In the case of Fe deposited on the (√3×√3)Ag surface, the saturation magnetization is somewhat smaller than that on the Si_3N_4 surface. The magnetic moment of the iron atom decreases with increasing the composition ratio of silicon in iron silicides, and that is almost vanished at the Si:Fe=1:1 [1]. Therefore, the results indicate that Si-rich iron silicide is formed on the clean surface while Fe-rich iron silicide is formed on the (√3×√3)Ag surface.

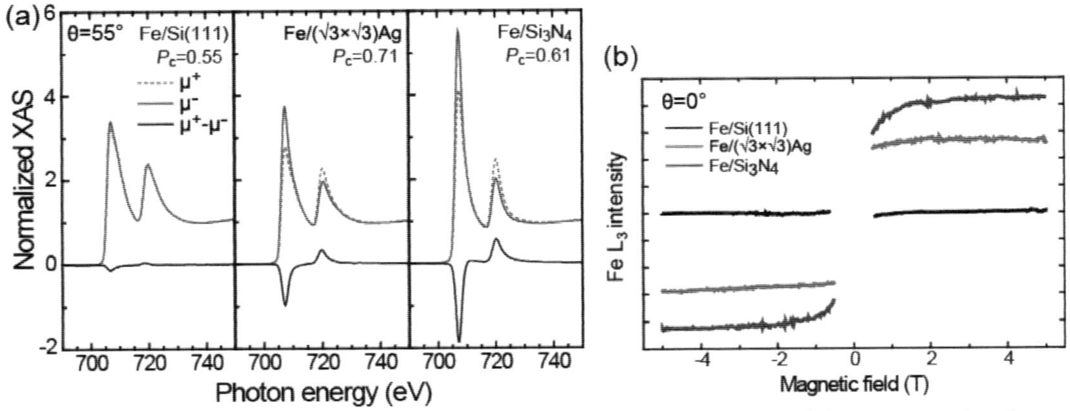

Figure 1. (Color online) (a) Fe $L_{2,3}$-edge circularly polarized XAS spectra of the Fe on each substrate. The Fe thickness on clean, (√3×√3)Ag and Si_3N_4 surfaces are 1.6, 2.0 and 1.6 ML, respectively. The spectra are taken at $T = 5$ K, $\mu_0 H = \pm 5$ T, $\theta = 55°$. μ^+ (red dashed line) and μ^- (blue solid line) spectra correspond to the ones with the electron spin parallel and anti-parallel to the photon angular momentum, respectively. (b) Magnetization curves of Fe on each surface at $T = 5$ K, obtained by recording the electron yield with the photon energy fixed at the Fe L_3 peak top of 707 eV. This data are taken at normal x-ray incidence ($\theta = 0°$).

Figure 2 shows the Fe $L_{2,3}$-edge XMCD spectra of Fe deposited on Si(111)-(7×7), Si(111)-(√3×√3)Ag and Si_3N_4/Si(111)-(8×8), taken at incident x-ray angles $\theta = 0°$ and 55°, a magnetic field of ±5 T, and a temperature of 5 K. The XMCD signal of Fe become larger in order of Fe/Si(111), Fe/(√3×√3)Ag and Fe/Si_3N_4. A sum-rule analysis was performed using the formula of the orbital magnetic moment (m_{orb}) and the effective spin magnetic moment (m_{eff}) [17, 18]. For the quantitative evaluation of the spin and orbital magnetic moments, the *d*-hole number n_d is experimentally estimated in a similar manner to the previous studies [13, 14]. In the present analysis, a XMCD spectrum (not shown) of bulk bcc Fe (n_d is assumed to be 3.40) was used as the standard. Then, the angle-dependent XMCD analysis at saturation was performed to eliminate the contribution of the spin dipole moment (m_T) and obtained the pure spin magnetic moment (m_{spin}) [19].

The results of the XMCD analysis are tabulated in Table 1. The spin magnetic moment is almost lost for Fe/Si(111). When small amount of Fe is deposited on the clean Si(111) surface, it is transformed into a nonmagnetic FeSi amorphous layer even at RT [20]. Additionally, the thickness of a magnetically dead layer is estimated as ~2 ML from photoelectron spectroscopic results [21]. The present XMCD result is consistent with the previous reports on the Fe/Si(111) system. On the other hand, the spin magnetic moment of Fe/Si_3N_4/Si(111) is 2.62 μ_B, which is larger than the bulk value of

Table 1. Results of the Fe $L_{2,3}$-edge XMCD analysis for Fe on each substrate with/without annealing. For the orbital magnetic moment m_{orb}, surface parallel (//) and normal (⊥) components are given separately.

	m_{spin} (μ_B)	m_{orb}(//) (μ_B)	m_{orb}(⊥) (μ_B)
clean	0.17	0.15	0.00
(√3×√3)Ag	1.25	0.17	0.11
Si_3N_4	2.62	0.15	0.18
annealed (√3×√3)Ag	0.32	0.03	0.01
annealed Si_3N_4	2.19	0.11	0.13

2.2 μ_B. It is found by the scanning tunneling microscopy (STM) measurements that the uniform nanoparticles of Fe grow on the Si_3N_4 surface at the initial stage [9]. Thus, the large magnetic moment can be explained by the less chemical interaction of Fe atoms with silicon atoms and the enhancement of the magnetic moment due to the size effect.

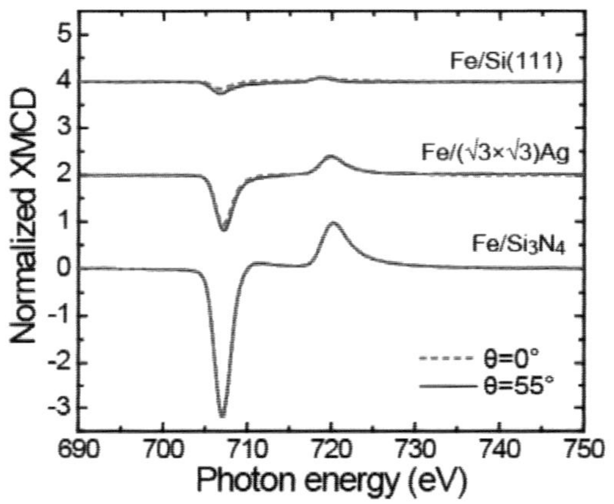

Figure 2. (Color online) Normalized Fe $L_{2,3}$-edge XMCD spectra of Fe on each substrate, taken at T = 5 K and$\mu_0 H = \pm 5$ T. The Fe thickness is the same as that in figure 1.

The spin magnetic moment of Fe deposited on the (√3×√3)Ag is 1.25 μ_B, which indicates that the magnetic Fe-rich iron silicide is formed on the (√3×√3)Ag surface. Compared to the Fe/Si(111) result, it is clear that the (√3×√3)Ag surface suppress a silicidation reaction between deposited Fe atoms and the silicon substrate. However, the magnetic moment is smaller than that of Fe_3Si (1.6 μ_B) [22] and is almost equal to that of Fe_2Si [1]. Therefore, the chemical composition of iron silicide grown on the (√3×√3)Ag surface is close to Fe_2Si. This mean that the passivation of (√3×√3)Ag is not strong enough to prevent the silicidation completely.

We also investigated the annealing effect. The XAS spectra of Fe deposited on the (√3×√3)Ag and Si_3N_4 surfaces annealed at 500 K are shown in Figure 3(a). The peak profile of Fe on the Si_3N_4 surface is similar to that of Fe deposited on the surface at RT and the FWHM of the μ^- spectrum of annealed Fe is 0.2 eV wider than that of the μ^- spectrum of the Fe without annealing. On the other hand, a shoulder structure is clearly seen at a higher photon energy side of the main peak in the spectra of Fe on the (√3×√3)Ag surface. The silicidation of the Fe atoms on the (√3×√3)Ag surface proceeds by annealing, which means that thermal resistance of the (√3×√3)Ag surface is not strong enough to prevent the silicidation reaction at high temperature.

Figure 3(b) shows the normalized XMCD spectra for the Fe atoms deposited on the both substrates and annealed at 500 K, and the results of the XMCD analysis are also tabulated in Table 1. The XMCD intensity of Fe on the Si_3N_4 surface becomes slightly weaker and the spin magnetic moment of the Fe atom decreases from 2.62 μ_B to 2.19 μ_B. The value is, however, still similar to that of the bulk

Fe. Two possible reasons for the decrease in the magnetic moment can be proposed. One reason is that the iron silicide is formed at defects of the Si_3N_4 buffer layer, e.g. steps and pits. Another one is that

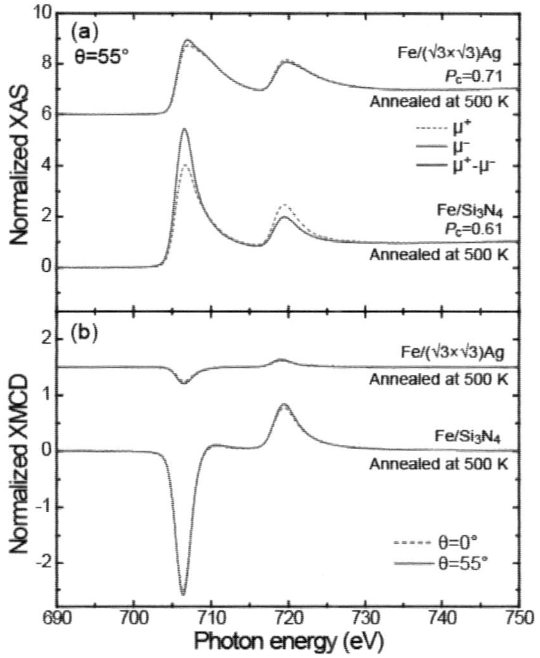

Figure 3. (Color online) (a) Fe $L_{2,3}$-edge circularly polarized XAS spectra of the Fe on (√3×√3)Ag and Si_3N_4 surfaces annealed at 500 K. (b) Normalized Fe $L_{2,3}$-edge XMCD spectra of Fe on each substrate annealed at 500 K. The XAS and XMCD spectra are taken at $T = 5$ K and $\mu_0 H = \pm 5$ T.

Fe nanoparticles are coalesced with each other and make larger nanoparticles by annealing to suppress the size effect. We will here emphasize that the passivation of the Si_3N_4 buffer layer is efficient to avoid the silicidation reaction between the Fe atoms and the Si atoms at least 500 K.

In contrasted to the magnetic moment of Fe on the Si_3N_4 buffer layer, that of Fe on the (√3×√3)Ag surface after annealing decreases significantly in comparison with that before annealing. It is found that the passivation effect of the (√3×√3)Ag surface is rather sensitive to the heating than that of the Si_3N_4 surface. However, it is noted that Fe atoms on the annealed (√3×√3)Ag surface carry the nonzero magnetic moment, although it is completely vanished for the Fe atoms deposited on the clean Si(111) surface annealed at 500 K (not shown). In addition, we also note that the small magnetic moment (0.32 μ_B) is inconsistent with previous study on Fe/Si(111)-(√3×√3)Ag at annealed 500 K, which reported that the nano dots whose composition ratio is Fe_3Si (1.6 μ_B) grow on the (√3×√3)Ag surface [6]. This discrepancy is explained by the difference of the measurement method. The magnetic moment of Fe atoms estimated by SQUID was calculated by dividing the whole magnetic moment with the volume of the dots estimated by STM. In this method, they only took account of the Fe atoms above the surface. Consequently, the average magnetic moment is estimated highly in comparison with the XMCD measurement because the value excludes the contribution from nonmagnetic iron atoms which diffuse into the Si substrate.

The difference of thermal resistance between Si_3N_4 and (√3×√3)Ag can be interpreted as the difference of the decomposition temperature. Si_3N_4 is thermally stable even above 2000 K, whereas the Si(111)-(√3×√3)Ag structure breaks at 900 K [23]. This means that the silicidation reaction is enhanced on the (√3×√3)Ag surface than on the Si_3N_4 surface. The Si_3N_4 surface is superior to the (√3×√3)Ag surface as a buffer layer to suppress the iron silicide formation.

Journal of Physics: Conference Series **430** (2013) 012129 doi:10.1088/1742-6596/430/1/012129

4. Summary

We investigated the magnetic properties of Fe grown on the Si_3N_4/Si(111)-(8×8) and Si(111)-(√3×√3)Ag substrates with and without annealing at 500 K by XAS and XMCD measurements. The spin magnetic moment of Fe can be controlled by modifying the Si surface; those of Fe atoms without annealing are 0.17, 1.25 and 2.62 μ_B on the clean, (√3×√3)Ag, and Si_3N_4 surfaces, respectively. Moreover, the annealing effect was studied for Fe on Si_3N_4 and (√3×√3)Ag surfaces. The iron silicidation is effectively suppressed on the Si_3N_4 surface even when the sample is annealed at 500 K. On the (√3×√3)Ag surface, the iron silicide are formed easily by annealing at 500 K and the spin magnetic moment of Fe atom is almost lost on the (√3×√3)Ag surface.

References

[1] Berling D, Gewinner G, Hanf M, Hricovini K, Hong S, Loegel B, Mehdaoui A, Pirri C, Tuilier M, and Wetzel P 1999 *Journal of Magnetism and Magnetic Materials* **191**, 331.

[2] Heim K R, Hembree G G, and Scheinfein M R 1994 *Journal of Applied Physics* **76**, 8105.

[3] Zavaliche F, Wulfhekel W, Xu H, and Kirschner J 2000 *Journal of Applied Physics* **88**, 5289.

[4] Eddrief M, Wang Y, Etgens V, Mosca D, Maurice J L, George J, Fert A, and Bourgognon C 2001 *Physical Review* B **63**, 1.

[5] Castrucci P 2001 *Surface Science* **476**, 43.

[6] Takagi Y, Nishimura A, Nagashima A, and Yoshino J 2002 *Surface Science* **514**,167.

[7] Paredis K, Smeets D, and Vantomme A 2009 *New Journal of Physics* **11**, 093019.

[8] Paredis K, Smeets D, and Vantomme A 2009 *Nanoscale research letters* **4**, 1447.

[9] Eguchi K, Takagi Y, Nakagawa T, and Yokoyama T 2012 *Physical Review* B **85**, 174415.

[10] Kim J W and Yeom H W 2003 *Physical Review* B **67**, 035304.

[11] Katayama M, Williams R, Kato M, Nomura E, and Aono M 1991 *Physical Review Letters* **66**, 2762.

[12] Takahashi T, Nakatani S, Okamoto N, Ishikawa T, and Kikuta S 1988 *Japanese Journal of Applied Physics* **27**, L753.

[13] Yokoyama T, Nakagawa T, and Takagi Y 2008 *International Reviews in Physical Chemistry* **27**, 449.

[14] Nakagawa T, Takagi Y, Matsumoto Y, and Yokoyama T 2008 *Japanese Journal of Applied Physics* **47**, 2132.

[15] Krumme B, Weis C, Herper HC, Stromberg F, Antoniak C, Warland A, Schuster E, Srivastava P, Walterfang M, Fauth K, Minár J, Ebert H, Entel P, Keune W, and Wende H 2009 *Physical Review* B **80**, 144403.

[16] Sirotti F, De Santis M and Rossi G 1993 *Physical Review* B **48**, 8299.

[17] Thole B, Carra P, Sette F, and van~der Laan G 1992 *Physical Review Letters* **68**, 1943.

[18] Carra P, Thole B, Altarelli M, and Wang X 1993 *Physical Review Letters* **70**, 694.

[19] Stöhr J and König H 1995 *Physical Review Letters* **75**, 3748.

[20] Alvarez J, Vázquez de Parga A, Hinarejos J, de la Figuera J, Michel E, Ocal C, and Miranda R 1993 *Physical Review* B **47**, 16048.

[21] Gomoyunova M V, Malygin D E, andPronin I I 2008 *Physics of the Solid State* **50**, 1579.

[22] Niculescu V, Burch T, and Budnick J 1983 *Journal of Magnetism and Magnetic Materials* **39**, 223.

[23] Gotoh Y and Ino S 1978 *Japanese Journal of Applied Physics* **17**, 2097.

Journal of Physics: Conference Series **430** (2013) 012130 doi:10.1088/1742-6596/430/1/012130

Full field spectroscopic imaging at the ANKA-XAS- and -SUL-X-Beamlines

Stefan Mangold[1]*, Ralph Steininger[1], Tomy dos Santos Rolo[1] and Jörg Göttlicher[1]

[1] Karlsruhe Institute of Technology, Institute for Synchrotron Radiation (KIT-ANKA), P.O. Box 3640, D-76021 Karlsruhe, Germany

stefan.mangold@kit.edu

Abstract. Full field transmission spectroscopic imaging is a powerful method for screening complex and large systems to map the distribution of valence state and electronic configuration of elements. Here, we describe an optimized hardware set-up and data evaluation tool chain for spectroscopic X-ray imaging. We tested this system successful on a ferruginous quartz (SiO_2) sample.

1. Introduction

Full field transmission spectroscopy can be applied in radiography and tomography mode and is started to be used to gain insight into complex and large sample systems in catalysis, environmental science and biology [1]. Its usefulness can be attributed to the fact that spectral information is collected in parallel, thus avoiding the time consuming scanning of the sample through a micro-focussed X-ray beam. To demonstrate the full potential of this technique, we optimized measurement and data evaluation procedures and tested it at the ANKA-XAS beamline on a ferruginous quartz sample containing iron (oxyhydr)oxides of up to three different structural environments. Measurements were performed in radiography mode, collecting 440.000 spectra simultaneously. Similar data evaluation procedures have been reported from other groups [2].

2. Experimental

2.1. Hardware

As shown in figure 1, the sample was placed between an ionization chamber as primary flux monitor and the detector system. Due to the absence of any focusing mirrors the ANKA-XAS beamline delivers a beam of very uniform intensity across 9 x 1 mm^2 and is therefore well suited for radiography of laterally extended samples. The SUL-X beamline with a 27 pole wiggler has higher flux that is needed for tomography, but at a small beam size of about 0.5 x 0.3 mm^2. The smaller homogeneous beam size is not a limitation, since the transmitted signal through the cylindrical sample needs to be on the order of 20% for a good signal to noise ratio in the case of tomography.

We used a high-resolution X-ray imaging detector system, where the visible light image created by the single crystal scintillating screen is relayed by a diffraction limited microscope objective onto the camera sensor. The detector system consisted of a YAG fluorescence screen, a vertically reflecting mirror for the visible light, a Mitutoyu long working distance microscope objective and a PCO camera (type 'edge' and '4000') [3]. With a 2 by 2 binning, a physical pixel size of 9 μm and an optical

Journal of Physics: Conference Series **430** (2013) 012130 doi:10.1088/1742-6596/430/1/012130

magnification of 3.6 X, the effective pixel size on the sample was 5 by 5 μm^2. This detector configuration gave rise to 440.000 X-ray Absorption Near Edge Structure (XANES) data sets in each radiogram. Normalization images were acquired for each energy step by moving the sample out of the beam. Using these additional images, the influence of the beam profile, the detector system and its energy dependence were eliminated.

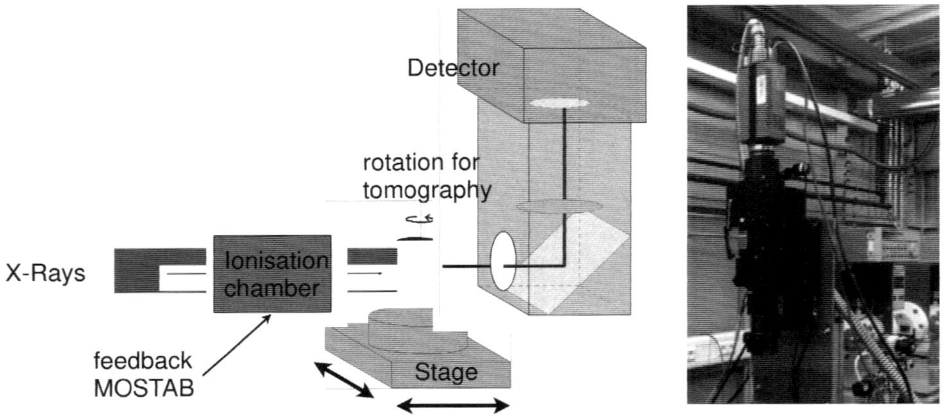

Figure 1. Sketch and photograph of the imaging set-up

2.2. Software

To retrieve the maximum information from the raw data, a prototype of an advanced data evaluation tool chain has been developed in the language IgorPro [4]. The steps performed to calculate the spectra in each pixel are described in table 1.

Table 1: Data evaluation steps

Data evaluation step	Description	Time needed per core (standard intel CPU – not optimized)
Calculation of spectra	Calculation of spectra by ln(measurement without sample/ measurement with sample) for each detector pixel (ln = natural logarithm)	< 5 min per 2Gb
Calculation of the vertical energy shift	Used for correction of the vertical energy drift over the image	9h (440000 spectra); only once per edge
Correction of the vertical energy shift	Used to correct the beamline dependent vertical energy shift	< 1 min (440000 spectra)
Edge filtering	Filter out all pixels, which contains no reasonable spectrum	< 2min
PCA analysis	Detection of components (PCA = principal component analysis)	~15 min

Due to the illumination of the flat monochromator crystal with a divergent beam, a small energy drift in the vertical direction on the detector is expected and can be seen in figure 2 and 3. The drift is corrected by a reference measurement with the metal foil of the element of interest. The shift per spectrum was determined by aligning all spectra to a spectrum from the center of the image. The remaining shift was reduced from 2 eV to below 0.2 eV between all XANES data sets of the radiogram (figure 4). While the determination of the energy drift over the metal foils up to now takes a few hours per core, the correction of the energy drift needs less than a minute for a complete radiogram. To exploit the full potential of the principal component analysis (PCA) and to speed up the data correction steps, XANES spectra containing no important physical information were removed from the radiographs. A simple but very stable edge-screening algorithm accomplished this by selecting only spectra with an edge jump of sufficient height. Additionally, an early prototype of a highly reproducible automatic background correction is in use.

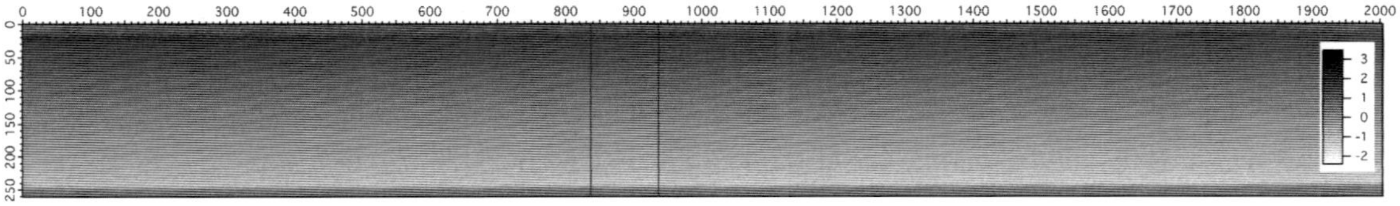

Figure 2. Relative energy distribution across the beam profile at ANKA-XAS

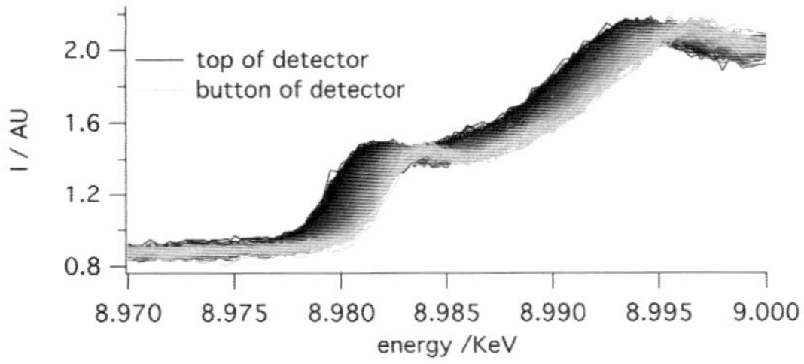

Figure 3. Edge region of the raw Cu spectra across the vertical beam profile.

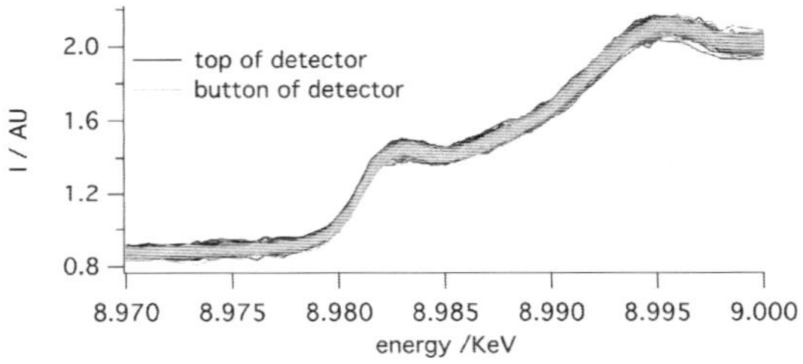

Figure 4. Cu spectra (edge region) of vertical beam profile, energy calibrated to the central spectrum.

3. Results and discussion

Prior to the full filed data evaluation, goethite (FeOOH) and hematite (Fe_2O_3) have been determined in this study by µ-XANES spectroscopy in the ferruginous quartz test sample at the SUL-X beamline to derive reference spectra (figure 5). A third spectrum belongs to a not yet identified black phase of most probably mixed Fe(II,III) valence. In the full field measurements, irrelevant spectra were removed by using the robust edge filter prior to the processing with the PCA. For the studied sample this reduced the number of spectra to process by a factor of ten (figure 6).

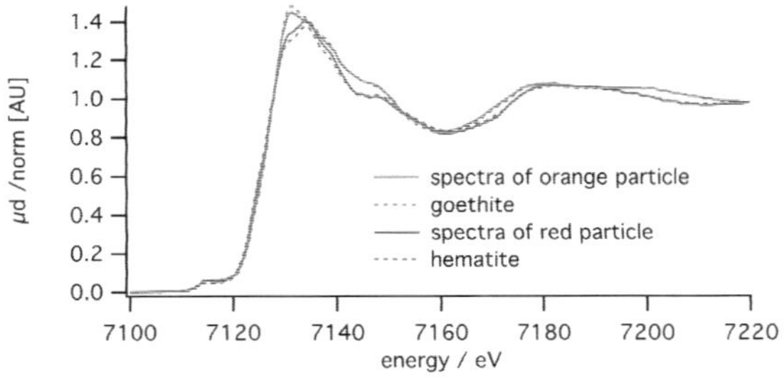

Figure 5. Fe K-edge µ-XANES spectra of orange and red particles with goethite and hematite reference spectra (measured at the SUL-X beamline; Si(111) monochromator crystal pair, beam size at sample position ~ 50 x 50 µm2)

Journal of Physics: Conference Series **430** (2013) 012130 doi:10.1088/1742-6596/430/1/012130

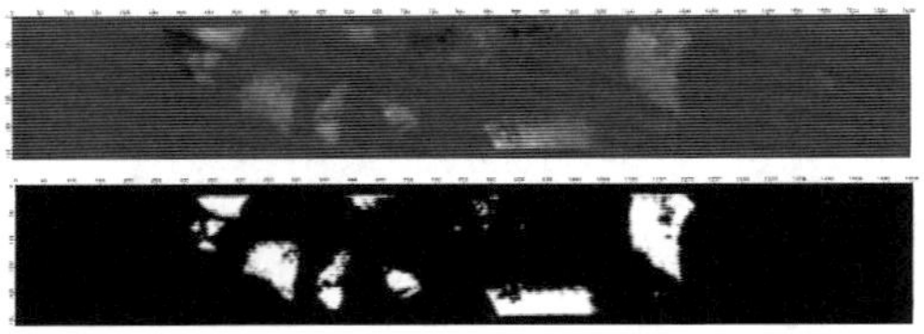

Figure 6. Top: transmission image of the sample; bottom: white areas are used for further processing.

We were able to extract three important components from the PCA (figure 7). They are sufficient to reconstruct all spectral types extracted from the full field measurements.

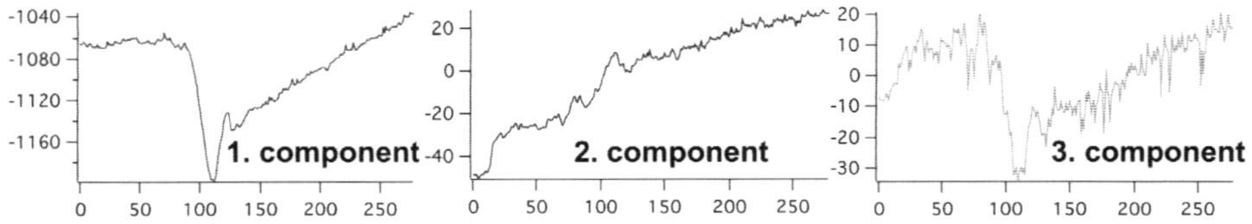

Figure 7. Graphs show the first 3 components of the PCA analysis.

Figure 8 shows the distribution of the two unambiguously identified spectral components across the sample area (goethite (FeOOH) green, hematite (Fe_2O_3) blue). The particle shown in the optical microscope image mainly consists of goethite FeOOH (orange) and hematite Fe_2O_3 (red). The lower part of the particle shows only a very low amount of iron and is therefore not suited for transmission measurements.

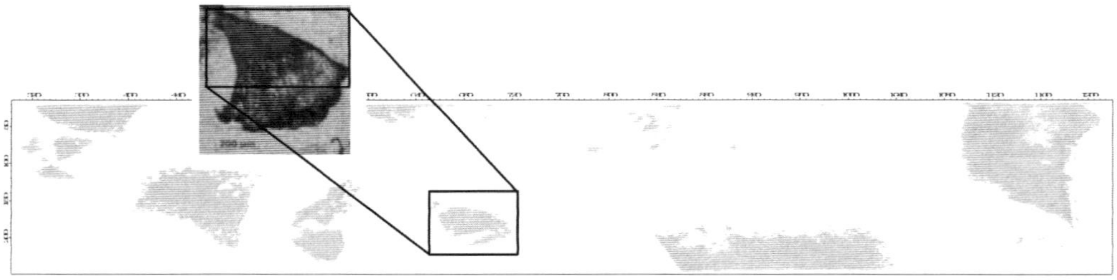

Figure 8. Comparison of light microscope image and the results of spectral analysis (goethite FeOOH, green; hematite Fe_2O_3, blue):

4. Conclusion and outlook

The test measurements illustrate that full field spectroscopic radiography is a powerful tool for screening large samples. Nevertheless, the raw data has to be carefully corrected to eliminate undesired effects introduced by the beam line and detector system. To further improve the method, a background correction on the complete data set is required. We are developing such a background correction method based on a robust algorithm with a combined pre- and post-edge fit. Each step of the procedure is described in table 2 and is based on simple, but very fast fits (time of all steps 0.1 s per core).

Journal of Physics: Conference Series **430** (2013) 012130 doi:10.1088/1742-6596/430/1/012130

Table 2: Steps of the background fit

Fitting step	Description
Pre-edge background correction	Options: line fit and polynomial (second degree), edge position is automatically detected
Sigmoidal fit	Needed to correct the pre- and post-edge in one step afterwards
Combined pre- and post-edge background	Fits the rest of the spectra with a polynomial fit (edge region is omitted)
Adding sigmoidal fit and interpolate data	Adding the data of the sigmoidal fit produces to the spectrum again; the data will be interpolated subsequently

Optimized detector systems are currently in the design phase, which will decrease the required integration time by one order of magnitude, enabling us to perform in-situ studies of catalysts and other fast processes, as well as to increase the sample throughput significantly.

References

[1] Meirer, F.; Cabana, J.; Liu, Y.; Mehta, A.; Andrews, J.C.; Pianettac, P., 2011, J. Synchrotron Rad. 18, 773–781

[2] Liu, Y.; F. Meirer, F.; Williams P.A.; Wang J.; Andrews, J.C.; Pianetta, P., 2012, J. Synchrotron Rad. (2012). 19, 281-287

[3] Cecilia, A.; Rack, A.; Douissard, P.-A.; Martin, T.; dos Santos Rolo, T.; Vagovič, P.; Hamann, E.; van de Kamp, T.; Riedel, A.; Fiederle, M. & Baumbach, T., LPE grown LSO:Tb scintillator films for high-resolution X-ray imaging applications at synchrotron light sources, Nuclear Instruments and Methods in Physics Research Section A: Accelerators, Spectrometers, Detectors and Associated Equipment, 2011, 648, Supplement 1, S321 - S323

[4] IgorPro - http://www.wavemetrics.com/

Journal of Physics: Conference Series **430** (2013) 012131 doi:10.1088/1742-6596/430/1/012131

On the relation between X-ray Photoelectron Spectroscopy and XAFS

N Mårtensson[1], J Söderström[1], S Svensson[1], O Travnikova[2], M Patanen[2], C Miron[2], L J Sæthre[3], K J Børve[3], T D Thomas[4], J J Kas[5], F D Vila[5] and J J Rehr[5]

[1] Department of Physics and Astronomy, Uppsala University, Box 516, 751 20 Uppsala, Sweden
[2] Synchrotron SOLEIL, l'Orme des Merisiers, Saint-Aubin, BP 48, 91192 Gif-sur-Yvette Cedex, France
[3] Department of Chemistry, University of Bergen, Allegaten 41, NO-5007 Bergen, Norway
[4] Department of Chemistry, Oregon State University, Corvallis, OR 97331, USA
[5] Department of Physics, Box 351560,University of Washington, Seattle, WA 98195-1560, USA

E-mail: `nils.martensson@fysik.uu.se`

Abstract. XAFS and X-ray Photoelectron Spectroscopy (XPS) are element specific techniques used in a great variety of research fields. The near edge regime of XAFS provides information on the unoccupied electronic states of a system. For the detailed interpretation of the XAFS results, input from XPS is crucial. The combination of the two techniques is also the basis for the so called core-hole clock technique. One of the important aspects of photoelectron spectroscopy is its chemical sensitivity and that one can obtain detailed information about the composition of a sample. We have for a series of carbon based model molecules carefully investigated the relationship between core level photoelectron intensities and stoichiometry. We find strong EXAFS-like modulations of the core ionization cross sections as function of photon energy and that the intensities at high photon energies converge towards values that do not correspond to the stoichiometric ratios. The photoelectron intensities are dependent on the local molecular structure around the ionized atoms. These effects are well described by molecular calculations using multiple scattering theory and by considering the effects due to monopole shake-up and shake-off as well as to intramolecular inelastic scattering processes.

1. Introduction

X-ray Absorption Spectroscopy (XAS) or X-ray Absorption Fine Structure (XAFS) measurements and core-level X-ray Photoelectron Spectroscopy (XPS) or Electron Spectroscopy for Chemical Analysis (ESCA) are two powerful spectroscopic techniques. Both techniques involve the excitation of a core electron. This makes the techniques atom specific and furthermore they provide information about the local structure and electronic structure around individual atomic sites in a system. In XAFS the absorption of a monochromatic X-ray beam is probed as function of the X-ray energy. This is done either by recording the degree of attenuation of the X-ray beam or by monitoring a secondary signal due to the decay of the excited state, such as electrons or fluorescent light, which is considered to be proportional to the absorption. In XPS one is analyzing the kinetic energy of the photoelectrons that are emitted due to the excitation by X-rays of well-defined energy.

Journal of Physics: Conference Series **430** (2013) 012131 doi:10.1088/1742-6596/430/1/012131

XAS or XAFS provides unique information on the structure as well as on the electronic structure of a system [1, 2, 3]. The near-edge regime can be used to study the unoccupied electronic states of a system. The technique can also be used to study the the orientation of various molecular species, e.g. adsorbed molecules. Due to the dipole selection rules the XAS intensities will depend strongly on the orientation of the molecule relative to the E-vector of the incident X-rays. In the case of Extended XAFS (EXAFS) the absorption is recorded over an extended energy range above a core level edge. EXAFS gives detailed information on the local arrangement of atoms in terms of bond distances and coordination numbers around the absorbing atomic species. By recording the EXAFS spectra using electron yield measurements, the technique can also be made very surface sensitive. In that case the technique is called Surface EXAFS or SEXAFS.

XPS has unique capabilities for studying the properties of solids and surfaces [4]. Through the chemical shift one can distinguish different chemical species of the same element. The chemical shift also provides information on properties such as charge distribution, oxidation state, ionicity, etc. Much XPS work today is performed using synchrotron radiation. The high brilliance of the synchrotron radiation allows much improved monochromatization of the beam and much higher spectral resolution can be achieved than with a conventional Al K_α X-ray source. The tunability of the excitation source furthermore opens up many more possibilities. By varying the photon energy the cross section of different processes can be modified relative to each other. In certain cases specific transitions can be much enhanced due to resonant photoemission. The tunability also makes it possible to vary the information depth of the probe. The mean free path of the photoelectrons is strongly energy dependent with a minimum, typically at kinetic energies of about 50 eV, corresponding to only a couple of atomic layers [4].

The basis for much of the quantitative use of XPS is that the cross section is considered to be a purely atomic property, whereby the core line intensities can be used as a direct measure of the relative amounts of different atomic species. However, we have recently shown that there are large variations in the cross sections that may extend several hundred eV above the ionization thresholds [5]. It is also well-known that the intensities are affected by elastic scattering and diffraction. This provides a method for obtaining detailed structural information [6, 7] also from photoelectron spectroscopy measurements. This requires that the spectra are recorded with angular resolution. One interesting aspect of this structural technique is that one can separately investigate the diffraction for electrons originating from different chemically shifted species in the system. For electron kinetic energies above several hundred eV the results can often be treated considering only single scattering. In the low kinetic energy regime the scattering angles are larger and multiple scattering becomes important.

In this contribution we discuss some important links between XAFS and XPS. At first we discuss the importance of having access to XPS data for the detailed study of unoccupied electronic states using XAFS. After that we briefly discuss the core-hole clock technique where one can study the time evolution of different core excited states. A combination of XAFS and XPS is often needed in these studies. We then focus on our recent results where we have found strongly non-stoichometric intensity ratios for the core ionization cross sections in a set of molecular model systems [5]. The connections to EXAFS as well as the implications for the quantitative use of XPS are discussed.

2. X-ray Photoelectron Spectroscopy and the near-edge regime of XAFS

One of the important aspects of the near-edge regime in XAFS is that it probes the unoccupied valence electron states in a system. A core electron at a specific site is excited to an initially unoccupied valence level. Usually the transitions can be accurately interpreted in terms of dipole selection rules. This implies that one obtains a symmetry projected view of the unoccupied electronic structure. For an s core level one reaches p-symmetry states, for a p core level one

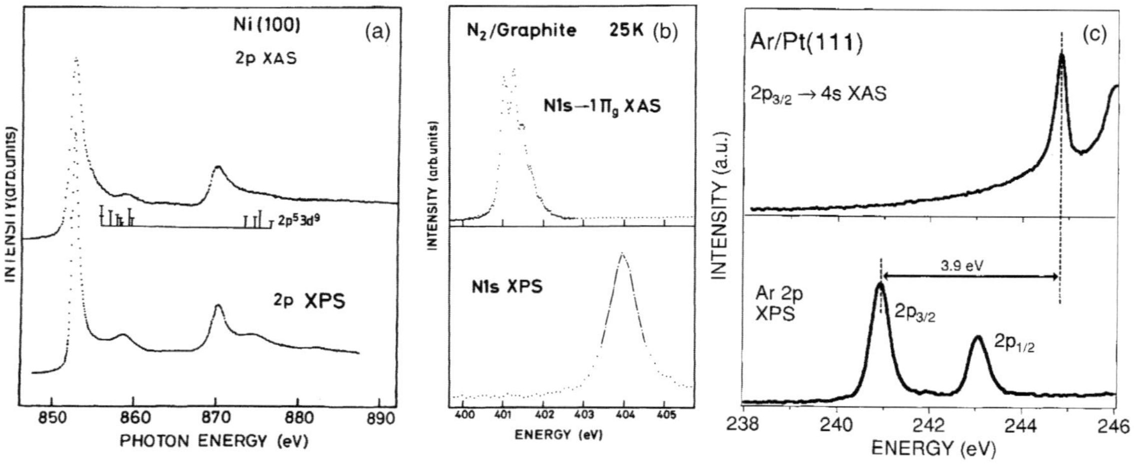

Figure 1. X-ray Absorption Spectra (XAS) and X-ray Photoelectron Spectra (XPS) plotted on common energy scales for three different situations. In the XA spectra the photon energy is given while the XP spectra are plotted on a binding energy scale relative to the Fermi level. (a): The 2p to 3d XA and the 2p XP spectra from Ni metal [8]. (b): The N1s to $1\pi_g$ XA and N1s XP spectra for one monolayer of N_2 physisorbed on graphite [9]. (c): The Ar 2p to 4s XA and 2p XP spectra of Ar physisorbed on Pt(111) [10].

becomes sensitive to the s- and d-symmetry states of that atomic species, etc. By combining results from different core levels one can in this way get very detailed information on the unoccupied density of states of a system.

When interpreting the XAFS results in terms of unoccupied valence electron states it is necessary to account for the effects of the core hole. This requires that the results are combined with information obtained from XPS core-level spectra. First of all one needs to partition the XA energy into two parts, one related to the excitation of a core electron to the Fermi level and one related to the excitation of an electron from the Fermi level to the initially unoccupied valence state. Furthermore, in many cases the core hole may influence the unoccupied states by the additional attractive potential. In order to identify these effects and to be able to account for them it is usually necessary to have access to information from XPS.

In Fig. 1, X-ray Absorption Spectra (XAS) are compared to X-ray Photoelectron Spectra (XPS) for three drastically different situations. Ni(100) [8] represents a metallic system. Nitrogen on graphite [9] and argon on Pt(111) [10] are both weakly adsorbed (physisorbed) systems. In all cases the comparison is made by plotting the spectra on common energy scales. For the XA spectra the photon energy scale is used. For the XP spectra the binding energy is plotted relative to the Fermi level. We immediately note that there are pronounced differences in terms of energy alignment of the XAS and XPS features for the different systems.

In the case of the Ni 2p spectra the two spectra are essentially identical [8]. Ni has an unfilled but nearly filled 3d band. The 2p X-ray absorption process creates a state with a fully occupied 3d-band for this particular system. This implies that the energy of the final state is well-defined which is manifested by a very narrow XA spectrum. The fact that the main peak in XPS occurs at the same energy implies that the same state is created through the screening process. Through the complete screening of the core hole state, the XPS final state corresponds to the lowest core excited state and thus it provides the energy onset for the XA spectrum. This is generally the case for metallic systems. For the empty valence electron states the XPS energy can thus be viewed as providing the Fermi level position.

Turning to N_2 physisorbed on graphite we note that the XAS absorption feature lies 3 eV below the XPS peak position [9]. The XAS final state is a neutral core excited state. The higher energy of the XPS peak implies that the neutral final state is not reached through screening, although this state is energetically more favorable. The required charge transfer from the substrate is too slow to occur with any significant probability. XPS can therefore not be used to locate the Fermi level in this situation. This energy is instead given by the onset of the XA spectrum.

For Ar physisorbed on Pt(111) we observe the reverse situation. The XAS final state occurs at 3.9 eV higher energy than the XPS feature [10]. The energy difference is again a signature of a very weak coupling between the adsorbate and the substrate. However, in this case the XPS line position corresponds to the lowest core excited state and can therefore be used to locate the position of the Fermi level for the valence electron states as seen in XAS. This energy onset cannot be accurately identified from the XA spectrum. Even if there is some intensity at lower energies than the main resonance, there is no sign of a sharp edge. One interesting aspect of the XAS results for adsorbed Ar is that they can be used to obtain information on the properties of adsorbed potassium atoms. The reason is that a core ionized argon atom is isoelectronic to a potassium atom (equivalent core approximation). From experimental Ar XAS data, combined with calculated corrections for the equivalent core error, one can derive unique information relevant to the properties of potassium adsorbed on a number of different surfaces [11, 12].

3. The core-hole clock technique

One interesting consequence of the fact that different states are reached in the XAS and XPS processes, as shown in Figs. 1b and c, is that characteristically different core-hole decay spectra are obtained for the two processes. In certain cases this provides a way to study the time evolution of the different excited states. Even if the screening for the weakly bonded systems in Figs. 1b and c is slow it occurs on a time-scale, comparable to the time scale of the core-hole decay. These processes will lead to characteristic features in the decay spectra that can be studied in different situations. This is the basis for the so called core-hole clock technique [9, 13, 14].

In the case of N_2 physisorbed on graphite, the Auger decay spectra after an N1s core ionization contains additional features characteristic of the decay from a neutral excited state. As seen from Fig. 1b, XPS produces only the ionic state. This implies that the neutralized state has been reached through a charge transfer process from the substrate to the molecule that occurs during the life-time of the core-hole. By measuring the intensity ratio between the two types of decay features, originating from neutral or ionic initial states, respectively, and by knowing the core hole decay rate, the charge transfer rate can be determined. For Ar/Pt(111), as seen in Fig. 1c, a related behavior is observed. In this case the neutral $2p^{-1}4s$ core excited state may become ionized through electron hopping to the substrate before the Auger decay process. Again, signatures characteristic of neutral and ionic adsorbate states are observed in parallel and the hopping rates from the adsorbate to the substrate can be determined from the observed intensity ratios between the two types of decay spectra.

The core-hole clock technique can provide information on different processes, which are difficult to study in other ways. The lifetimes of many frequently used core levels are in the low fs time regime, which opens up the technique for various ultrafast phenomena. It has for instance been used to determine the charge transfer rate for one of the fundamental subprocesses in a model system for a dye sensitized solar cell [15]. By using more short-lived core hole states, the technique has also been used to study processes in the attosecond regime [16].

Journal of Physics: Conference Series **430** (2013) 012131 doi:10.1088/1742-6596/430/1/012131

4. Electron scattering effects in Photoelectron Spectroscopy

XAFS and XPS are related also in another important way. In both techniques the elastic scattering of the excited electrons plays an important role. In EXAFS the X-ray absorption is measured as function of photon energy. The absorption occurs at a specific site where a core electron is excited to the continuum. The photoelectron wave may then be backscattered against the neighboring sites in the lattice. This leads to interference effects which manifest themselves as modulations of the absorption cross section as the photon energy is varied. This provides important information about the local structure around individual sites in the system. One of the important aspects of this technique is that it does not require any type of long range order of the system.

In the case of Photoelectron Diffraction (PhD) one is analyzing the emitted photoelectrons in an angular resolved way [6, 7]. The interference between the directly emitted and the various scattered waves creates a characteristic angular pattern from which structural information can be deduced. One advantage with this technique is that one can separately obtain structural data from chemically shifted components for the same element. PhD measurements can be performed in two different ways. One possibility is to measure the photoemission intensity as function of emission angle at a fixed energy. Another possibility is to perform the measurements at a fixed angle but by varying the photon energy. The photoelectron diffraction technique requires short-range order, although not necessarily long-range order. For disordered systems it is generally assumed that all diffraction effects cancel and that they do not need to be considered. This assumption lies behind much of the use of photoelectron spectroscopy for detailed quantitative analysis.

However, when measuring angle integrated photoelectron spectra from solids, Rothberg et al. observed pronounced photon energy dependent intensity modulations [17]. The results were compared to EXAFS data for the same system and it was concluded that the photon energy dependent variations could be interpreted in the same way as EXAFS. This was proposed to provide an alternative way to record EXAFS spectra and the acronym Photoelectron EXAFS or PEXAFS was introduced. This technique has then the additional advantage that one can use the chemical sensitivity provided by photoelectron spectroscopy to separately study different chemical species of the same element. PEXAFS has been used in a number of structural studies, e.g. for semiconductor surfaces [18, 19, 20] where the results have been interpreted using EXAFS codes.

The proposal that the PEXAFS measurements create results identical to EXAFS has, however, been questioned. First of all, from theoretical considerations it has been concluded that even a 2π integration of the photoemission signal is insufficient in order to yield the EXAFS spectrum [21, 22]. The foundations for PEXAFS has also been investigated experimentally. Scanned energy PhD spectra were recorded by Toomes et al. for a large range of angles for a disordered surface of oxygen adsorbed on Cu(111) [23]. They found that very similar spectra were obtained at all investigated angles. From this they concluded that these spectra also represent the angle integrated results. These spectra were then compared to results obtained by SEXAFS for the same surface. They found first of all that the intensity modulations in PEXAFS were much more pronounced than in SEXAFS. Furthermore, when analyzing the PEXAFS results using EXAFS theory, quite different results were obtained in terms of bond distances and coordination numbers as compared to SEXAFS. The conclusion from this investigation was therefore that the angle integrated photoelectron spectroscopy results obtained in this way are still much affected by PhD effects.

In order to investigate the photon energy dependence of the photoemission intensity in more detail we have performed measurements for a series of molecular model systems in the gas phase. This choice of systems has several advantages. The gas phase molecules are by nature randomly oriented. In this way we can perform measurements that correspond to a complete

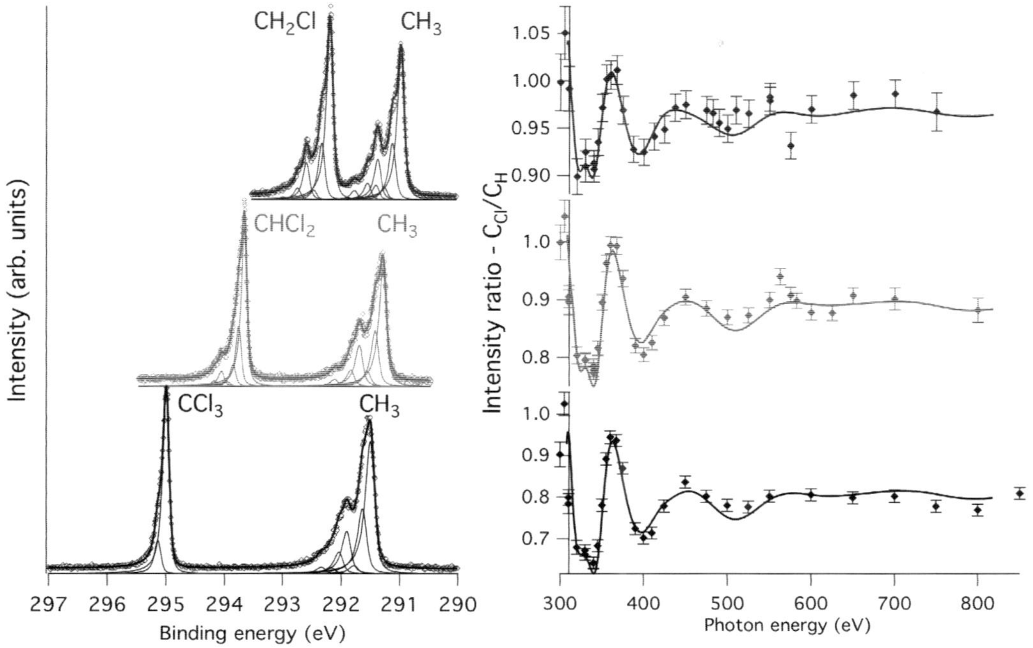

Figure 2. Experimental spectra of CH_3CCl_3, CH_3CHCl_2 and CH_3CH_2Cl. All spectra used to determine intensity ratios in the present investigation have been recorded at the magic angle. The spectrum of CH_3CH_2Cl shown in this figure, however, was recorded at an angle of 0° between the polarization vector and the direction of the escaping photoelectrons in order to maximize intensity and reduce statistical errors. The right-hand panels show for each molecule the measured C1s intensity ratio between the Cl substituted and non-substituted carbon atoms as function of photon energy. The solid lines are the calculated ratios using the FEFF code, see text.

4π integration of the photoemission intensities. We have furthermore chosen molecules that contain the same element in different chemical environments. In this way one can measure intensity ratios between chemically shifted peaks in the same spectra instead of monitoring absolute intensities originating from individual atomic species, whereby all normalization issues are avoided. Furthermore, the concentration ratios are exactly known. All measurements used in the cross section determinations have been performed at the magic angle of 54.7° between the polarization direction of the ionizing x-rays and the propagation direction of the escaping photoelectron to avoid any angular dependent effects.

In one series of measurements we have recorded the C1s core level intensities for CH_3CCl_3, CH_3CHCl_2, and CH_3CH_2Cl over a wide photon energy range [5]. Examples of high resolution C1s photoelectron spectra for the three molecules are shown in the left hand panels of Fig. 2. The spectra in Fig. 2 have been fitted using a number of vibrational components. The parameters derived from these fits have then been used to fit all the spectra recorded at different photon energies in order to derive accurate intensity ratios.

The experimental intensity ratios for the three molecules are shown in the right hand panels of Fig. 2. In all cases we note large oscillations of the intensity ratios as function of photon energy. The oscillations extend several hundred eV above the C1s ionization threshold. In Fig. 3 we compare the experimental intensity ratios for the three molecules. We immediately

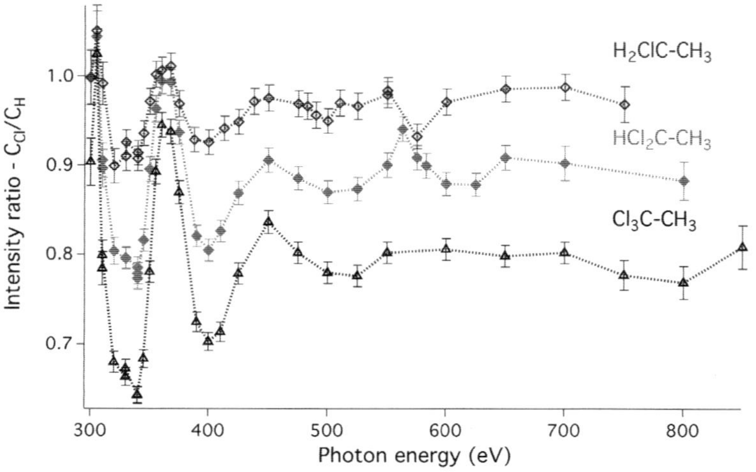

Figure 3. Comparison of the photon energy dependent C1s intensity ratio between the Cl substituted and non-substituded carbon atoms for CH_3CCl_3, CH_3CHCl_2 and CH_3CH_2Cl. The experimental lines are connected by dotted lines in order to help separating the three curves

see that the oscillatory pattern is very similar for the three molecules and that the amplitude of the oscillations increases with the number of Cl atoms in the molecule. The oscillations are damped with energy and the intensity ratios approach asymptotic values at high photon energies, signficantly different from the stoichiometric ratio of one. The asymptotic intensities are lower for the chlorine substituted carbon atoms and they decrease as the number of chlorine substituents increases.

We interpret the oscillations as due to an EXAFS-like modulation of the photoionization cross section, dominated by backscattering from the chlorine atoms. In order to test this assumption we have performed theoretical multiple-scattering calculations using the latest FEFF codes [3, 24]. The calculations include Debye-Waller factors, chemical shifts, and estimates of inelastic losses. The calculated intensity ratios are compared to the experimental results in Fig. 2. We have in this figure normalized the FEFF calculated ratios to match the experimental values at our highest photon energies. It is immediately clear that the theoretical (FEFF) results describe the observed intensity variations very well. We therefore conclude that the observed intensity oscillations are entirely due to the variations in the absorption cross section. Furthermore, we find that they can be handled to a good approximation by present theoretical tools. These findings are consistent with the predicted equivalence between angular integrated photoemission spectra and x-ray absorption spectra (XAS), provided that a full 4π integration can be performed [22].

We have also performed measurements for a number of other molecules that support the present picture. We have for instance investigated molecules where the different carbon atoms have very different types of atomic coordination, such as ethyl trifluoroacetate (CF_3-CO-O-C_2H_5), a molecule that has been frequently used as a showcase of chemical effects in photoelectron spectroscopy (see e.g. Ref.[25]). This leads to characteristically different oscillatory patterns for the different carbon atoms. Again the results are well reproduced by the FEFF calculations. These results will be presented in forthcoming publications.

We also need to consider the fact that the measured intensity ratios approach values that are different from the stoichiometric ones at higher photon energies. This is due in part to shake-up and shake-off processes, which transfer intensity from the main line to a satellite spectrum. This has also to be accounted for when photoelectron spectroscopy is used quantitatively. According to the Manne - Åberg theorem [26] the shake probability is linked to the relaxation energy and hence to the core level shifts. The chemical shifts are due to the combined effect of ground state effects, which can often be discussed in terms of charge transfer, and relaxation effects. From Fig. 2 we see that each electronegative chlorine substituent causes a C1s shift to higher binding energies. This is consistent with the expectations from a simple ground state model since the chlorine substituents make the carbon atoms more positive. However, a detailed treatment of core-level shifts requires much more elaborate considerations. Especially one needs to take the relaxation effects into account. From our calculations we see that the relaxation energy increases with the number of polarizable chlorine substituents. According to the Manne - Åberg theorem [26] this is consistent with a larger shake probability and hence to a reduced main line intensity for the chlorine substituted carbon atoms, as seen in Fig. 2.

The sum of the intensity of the monopole shake-up and shake-off processes can be calculated from the square of the overlap integral, S_0^2, between the initial-state wave function with the active core electron annihilated and the relaxed final state wave function. This value is typically around 0.75 for the present C1s ionizations. However, the detailed value depends on the atomic coordination and it gets consistently lower as the number of chlorine substituents increases. We have also considered one additional process, which is usually not accounted for in free molecules. We denote this process intramolecular inelastic scattering. This is a loss due to the propagation by the photoelectron through the ionized molecule. This effect can be estimated semiclassically [27], in terms of the density-dependent mean-free path of the photoelectron, and it yields further energy dependent reductions. The calculated ratios originating from these two processes range from 0.85 - 0.91 for CH_3CCl_3, 0.90 - 0.95 for CH_3CHCl_2 and 0.94 - 0.97 for CH_3CH_2Cl, over the photon energy range from 400 - 800 eV. These values vary roughly linearly with the number of Cl atoms in agreement with experiment and they are comparable to, although somewhat higher than the experimental values of 0.79, 0.88, and 0.96. The discrepancies are likely due to the approximations used in the calculations. Correlation effects are for instance not included in the calculation of the factors S_0^2. Furthermore, we do not know if the experimental ratios have fully converged at the highest photon energies used in the present experiment.

Our results are important in connection with the use of photoelectron spectroscopy for detailed analytical purposes. This technique is normally considered to be quantitative in a very accurate way. Most often it is assumed that the cross section is an atomic property, largely independent of the surroundings. It is furthermore assumed that elastic scattering effects can be neglected for disordered systems. It is, however, clear that one has to go beyond this approximation. The core ionization cross section is a property of the absorbing atom as well as of its surrounding atomic shells. This is in particular the case for relatively low photoelectron kinetic energies. This energy regime is often of special interest since it optimizes the surface sensitivity of the probe. When performing surface sensitive measurements it may also be of interest to vary the information depth. Within certain limits this can be made in two different ways. One possibility is to change the effective information depth by varying the angle of emission for the photoelectrons. This requires a relatively well-defined surface. For complex samples, for instance liquid samples, depth-profiling must be made in a different way. In that case one can achieve different information depths by recording spectra at different photon energies, leading to different kinetic energies of the photoelectrons. This, however, requires that one can distinguish the intensity variations caused by different probing depths and different relative abundancies of the probed species from those variations that are due to photon energy dependent cross section variations [5, 17, 23, 28, 29].

For the present molecular system, studied in the gas phase, the necessary corrections can be accurately obtained by applying the FEFF code to account for the scattering effects and by calculating the S_0^2 factors and the effects of intramolecular inelastic scattering in order to estimate the intensity reduction due to shake-up and shake-off. For condensed systems the same type of calculations can provide a first approximation for the necessary cross section corrections. However, more experimental and theoretical work is needed in order to determine what additional corrections have to be made when studying solid samples, considering the fact that a 4π integration is no longer possible to achieve. The results by Toomes et al. [23] indicate that also for a disordered solid system there remain strong PhD effects. However, the adsorbate system that was used in their study cannot be considered to be a completely random system and further studies are needed in order to develop a strategy for handling the necessary corrections for condensed samples. Studies of a liquid system could provide additional information on the cross section variations for a truly disordered system.

5. Summary

There are important links between XAFS and XPS. When using XAFS to study the unoccupied electronic structure, the XPS core level spectra are needed for instance in order to determine the Fermi level position for the unoccupied electronic states. By comparing the results from the two techniques one can also derive important information on the character of the chemical bonds in a system. The comparison between XAFS and XPS is also important when investigation the dynamics of various excited states and for the use of the so-called core-hole clock technique. For a molecular model system we have furthermore shown that the core level photoionization cross sections show strong modulations as function of photon energy, which for these molecular systems can be directly related to EXAFS. The observed intensities are also affected by shake-up and shake-off processes, the magnitude of which are depending on the chemical surroundings of the ionized atoms. These effects lead to strongly non-stoichometric core level intensities. This has important consequences when using XPS for quantitative purposes. We show that the present molecular results can be accurately reproduced by FEFF calculations for the intensity modulations, by calculating the monopole shake-up and shake-off intensities by standard quantum chemical methods and by considering losses due to intramolecular inelastic scattering in terms of the density-dependent mean-free path of the photoelectrons.

Acknowledgments

We are grateful to E. Robert and the SOLEIL staff for excellent technical support. This work has been supported by the European Union Seventh Framework Programme FP7/2007-2013 under Grant Agreement No. 252781 (O. T.) and I3 program (LS, SS), by the Swedish Research Council (SS, NM), by the Norwegian Research Council (KB, LS), by Knut and Alice Wallenbergs Foundation (JS), from U.S. Department of Energy Basic Energy Sciences Grant No. DE-FG03-97ER45623 (JJR, JJK, and FDV) and by Triangle de la Physique, France under Contract No. 2007-010T (SS).

Appendix A. Experiment

The measurements were performed at the PLEIADES beamline at the SOLEIL national synchrotron radiation facility, France. The beamline uses two quasi-periodic undulators with 256 mm and 80 mm period to cover the 7-400 eV and 35-1000 eV energy ranges, respectively, with any kind of polarization available from 55 eV. The plane grating beamline operates without entrance slit and it is equipped with four varied line spacing gratings. The gratings have varied grove depths in the direction perpendicular to the beam propagation axis allowing optimization of the grating efficiencies throughout the whole energy range. An ultimate resolving power of approximately 10^5 is achievable at 50 eV. In this study we especially made use of the high photon flux available at relatively high photon energies [30]. Some of the characteristics of the beamline can e.g. be found in Refs. [31, 32, 33].

The measurements were performed using a wide-angle lens VG-Scienta R4000 electron spectrometer installed at a fixed position, with the electron detection axis perpendicular to the storage-ring plane. The polarization vector of the X-ray light has been set at the magic angle of 54.7 with respect to the electron detection axis. The analyzer was also mounted perpendicular to the propagation direction of the ionizing radiation i.e. in the so-called dipole plane in order to minimize non-dipole effects. The degree of linear polarization of the photon beam was better than 98%.

The samples were used to fill a differentially pumped gas cell equipped with a series of electrodes allowing to compensate the local potentials (so called plasma potentials) associated with the ion density gradient created along the beam propagation axis within the gas cell.

The total instrumental broadening was less than 250 meV for all the spectra. The spectra were fitted with PCI profiles [34] with calculated asymmetry parameters using IgorPro with the SPANCF curve fitting package [35]. The fits were used to derive accurate total peak intensities. For some photon energies additional peaks, such as chlorine derived Auger transitions had to be added. The Gaussian line width parameters were constrained to be the same for all of the components in a spectrum, and the Lorentzian line width parameters had to be the same for all of the components within a given peak.

Appendix B. Theory

The XAS calculations were carried out using the real-space multiple-scattering code FEFF9 [24]. Muffin-tin scattering potentials and phase shifts were calculated self-consistently for configurations including a core-hole. The spectral intensities were calculated as $\mu_i = Z_i \mu_{0i}(1 + X_i)$, where μ_{0i} is the atomic background cross-section in the one-electron approximation, X_i is the extended fine structure for the carbon atom C_i, and $Z_i = S_{0i}^2 A_i$ is an amplitude reduction factor that accounts for intrinsic and intramolecular scattering losses. The quantity S_{0i}^2 corresponds to the fraction of photoelectrons emitted without shake-up or shake-off. The X-ray absorption was calculated for each carbon site using the multiple-scattering path expansion in FEFF from 300 - 1000 eV, with about 300 paths. Ab initio Debye-Waller factors at 300K were included, obtained using the Lanczos algorithm in Ref. [36] with a dynamical matrix calculated using Gaussian03. The atomic background absorption μ_{0i} was calculated in the dipole approximation. Since the embedded atomic potentials are nearly identical for the substituted (C_{Cl}) and non-substituted (C_H) carbon atoms, these factors cancel in the calculated ratios $R = \mu_{C_{Cl}}/\mu_{C_H}$. The observed oscillatory structure is approximately given by the difference $X_R \approx X_{C_{Cl}} - X_{C_H}$. This fine structure is dominated by the scattering from the Cl atoms in the C-Cl bonds, since the C-C scattering cancels to leading order in the ratios, and C-H-scattering is negligible.

Estimates of S_{0i}^2 were calculated from $S^2 = |\langle 0|0'\rangle|$ overlap integrals between single determinant, restricted Hartree-Fock wave functions from the Gaussian03 code. The results show that the overlap integrals are dominated by atomic relaxation and that they are quite similar for both C atoms. For CH_3CCl_3, for example, we obtain 0.708 for the CCl hole and 0.755 for the CH hole, yielding an intrinsic ratio of 0.937. Moreover there are substantial contributions from intramolecular inelastic scattering. These have been estimated semi-classically in terms of the energy dependent mean free path of the photoelectron, averaged over solid angle. This quantity is strongly energy dependent, and yields ratios varying from about 0.90 to 0.96 between photoelectron energies 100 and 400 eV above threshold (photon energies between about 400 to 700 eV). Consequently the net ratio is a combination of monopole shake-up and shake-off and intramolecular scattering losses.

References

[1] Stern E A and Heald S M 1983 *Basic Principles and Applications of EXAFS* (*Handbook of Synchrotron Radiation* vol 1-B) (North-Holland) chap 10, pp 995–1014
[2] Stohr J 1992 *NEXAFS Spectroscopy* (*Springer Series in Surface Sciences* vol 25) (Springer-Verlag: Berlin)
[3] Rehr J and Albers R 2000 *Reviews of Modern Physics* **72** 621–654 ISSN 0034-6861
[4] Hufner S 2003 *Photoelectron Spectroscopy: Principles and Applications* (Springer-Verlag Berlin Heidelberg New York)
[5] Soderstrom J, Martensson N, Travnikova O, Patanen M, Miron C, Saethre L J, Borve K J, Rehr J J, Kas J J, Vila F D, Thomas T D and Svensson S 2012 *Physical Review Letters* **108**(19) 193005 (4 pp.) ISSN 0031-9007
[6] Fadley C 1992 *Synchrotron Radiation Research: Advances in Surface and Interface Science* (Plenum Press, New York,) p 421
[7] Woodruff D P 2007 *Surface Science Reports* **62** 1–38 ISSN 0167-5729
[8] Zdansky E O F, Nilsson A, Tillborg H, Bjorneholm O, Martensson N, Andersen J N and Nyholm R 1993 *Physical Review B* **48** 2632–2641 ISSN 1098-0121
[9] Bjorneholm O, Nilsson A, Sandell A, Hernnas B and Martensson N 1992 *Physical Review Letters* **68** 1892–1895 ISSN 0031-9007
[10] Karis O, Nilsson A, Weinelt M, Wiell T, Puglia C, Wassdahl N, Martensson N, Samant M and Stohr J 1996 *Physical Review Letters* **76** 1380–1383 ISSN 0031-9007
[11] Sandell A, Hjortstam O, Nilsson A, Bruhwiler P, Eriksson O, Bennich P, Rudolf P, Wills J, Johansson B and Martensson N 1997 *Physical Review Letters* **78** 4994–4997 ISSN 0031-9007
[12] Sandell A, Bruhwiler P A, Nilsson A, Bennich P, Rudolf P and Martensson N 1999 *Surface Science* **429** 309–319 ISSN 0039-6028
[13] Bruhwiler P A, Karis O and Martensson N 2002 *Reviews of Modern Physics* **74** 703–740 ISSN 0034-6861
[14] Wurth W and Menzel D 2000 *Chemical Physics* **251** 141–149 ISSN 0301-0104
[15] Schnadt J, Bruhwiler P A, Patthey L, O'Shea J N, Sodergren S, Odelius M, Ahuja R, Karis O, Bassler M, Persson P, Siegbahn H, Lunell S and Martensson N 2002 *Nature* **418** 620–623 ISSN 0028-0836
[16] Fohlisch A, Feulner P, Hennies F, Fink A, Menzel D, Sanchez-Portal D, Echenique P M and Wurth W 2005 *Nature* **436** 373–376 ISSN 0028-0836
[17] Rothberg G M, Choudhary K M, den Boer M L, Williams G P, Hecht M H and Lindau I 1984 *Physical Review Letters* **53** 1183–1186 ISSN 0031-9007
[18] Kim S T, Choudhary K M, Shah S N, Lee J H, Rothberg G M, den Boer M L and Williams G P 1987 *Journal of Vacuum Science & Technology A-Vacuum Surfaces and Films* **5** 623–626 ISSN 0734-2101
[19] Mangat P S, Soukiassian P, Schirm K M, Spiess L, Tang S P, Freeman A J, Hurych Z and Delley B 1993 *Physical Review B* **47** 16311–16321 ISSN 0163-1829
[20] Mangat P S and Soukiassian P 1995 *Physical Review B* **52** 12020–12025 ISSN 1098-0121
[21] Lee P A 1976 *Physical Review B* **13** 5261–5270 ISSN 0163-1829
[22] de Leon J M, Rehr J J, Natoli C R, Fadley C S and Osterwalder J 1989 *Physical Review B* **39** 5632–5639 ISSN 0163-1829
[23] Toomes R L, Woodruff D P, Polcik M, Bao S, Hofmann P and Schindler K M 2000 *Surface Science* **445** 300–308 ISSN 0039-6028
[24] Rehr J J, Kas J J, Prange M P, Sorini A P, Takimoto Y and Vila F 2009 *Comptes Rendus Physique* **10** 548–559 ISSN 1631-0705
[25] Travnikova O, Børve K J, Patanen M, Soderström J, Miron C, Sæthre L J, Martensson N and Svensson S 2012 *Journal of Electron Spectroscopy and Related Phenomena* – ISSN 0368-2048 URL `http://www.sciencedirect.com/science/article/pii/S0368204812000552`
[26] Manne R and Aberg T 1970 *Chemical Physics Letters* **7** 282–284 ISSN 0009-2614
[27] Aryasetiawan F and Gunnarsson O 1998 *Reports on Progress in Physics* **61** 237–312 ISSN 0034-4885
[28] Plesiat E, Argenti L, Kukk E, Miron C, Ueda K, Decleva P and Martin F 2012 *Physical Review A* **85** ISSN 1050-2947
[29] Di Tommaso D and Decleva P 2005 *Journal of Chemical Physics* **123** 064311 ISSN 0021-9606
[30] Travnikova O, Liu J C, Lindblad A, Nicolas C, Soderstrom J, Kimberg V, Gel'mukhanov F and Miron C 2010 *Physical Review Letters* **105** ISSN 0031-9007
[31] Thomas T D, Kukk E, Ueda K, Ouchi T, Sakai K, Carroll T X, Nicolas C, Travnikova O and Miron C 2011 *Physical Review Letters* **106** ISSN 0031-9007
[32] Miron C, Nicolas C, Travnikova O, Morin P, Sun Y, Gel'mukhanov F, Kosugi N and Kimberg V 2012 *Nature Physics* **8** 135–138 ISSN 1745-2473
[33] Soderstrom J, Lindblad A, Grum-Grzhimailo A N, Travnikova O, Nicolas C, Svensson S and Miron C 2011 *New Journal of Physics* **13** ISSN 1367-2630

[34] van der Straten P, Morgensterns R and Niehaus A 1988 *Zeitschrift fur Physik D-Atoms Molecules and Clusters* **8** 35–45 ISSN 0178-7683

[35] Kukk E Spectral analysis by curve fitting macro package/spancf/2000 URL `http://www.physics.utu.fi/en/research/material science/Fitting.html`

[36] Vila F D, Rehr J J, Rossner H H and Krappe H J 2007 *Physical Review B* **76** ISSN 1098-0121

Journal of Physics: Conference Series **430** (2013) 012132 doi:10.1088/1742-6596/430/1/012132

The combined study of Dy_2O_3-HfO_2 nanocrystallites formation by means of EXAFS, PDF, XRD and SAXS

A A Yaroslavtsev[1], A P Menushenkov[1], D S Leshchev[1], J Bednarcik[2], R V Chernikov[2], Y V Zubavichus[3], V V Popov[1], V F Petrunin[1] and S A Korovin[1]

[1] National Research Nuclear University "MEPhI", Kashirskoe sh., 31, Moscow 115409, Russia
[2] HASYLAB at DESY, Notkestrasse, 85, Hamburg D-22607, Germany
[3] NRC "Kurchatov Institute", Akademika Kurchatova sq., 1 , Moscow 123182, Russia

E-mail: yalex03@gmail.com

Abstract. The process of double oxide Dy_2O_3-HfO_2 nanocrystallites formation upon annealing to 1600°C was investigated with the combination of few X-ray and synchrotron methods. XRD and PDF show the mean crystal structure to be fluorite typical for the compounds of Ln_2O_3-MeO_2 type. The ordering of structure and the growth of nanocrystallites upon annealing was estimated independently from XRD patterns, PDF functions and SAXS scattering curves. Neither transition to pyrochlore phase, nor signs of recently shown in related Gd_2O_3-HfO_2 compounds pyrochlore-type superstructure is observed. However, EXAFS and the short range order of PDF clearly indicate the discrepancy between Dy and Hf local environment. Such a disorder of local crystal structure can not be described by the fluorite model.

1. Introduction

The binary compounds $Ln_{2+x}Me_{2-x}O_{7-x/2}$ forming in the two-component Ln_2O_3-MeO_2 systems (Ln - rare-earth cation, Me - IVB subgroup cation: Ti, Zr, Hf) are attractive from the both scientific (fluorite-pyrochlore phase transition [1, 2, 3]; spin-glass, spin-ice and spin-fluid behavior [4, 5]) and practical point of view (thermal barrier coatings [6], neutron-absorbing materials [7], oxygen-ionic conductors [3], nuclear waste forms [8]). Regarding the the cation ordering the most interesting are compounds with the ionic radii ratio $R_{Ln^{3+}}/R_{Me^{4+}} \sim 1.45$ in which the fluorite ($Fm\bar{3}m$ space group) and pyrochlore ($Fd\bar{3}m$ space group) crystal structures might coexist at room temperature according to the thermodynamic estimation [2, 9].

Previously we reported on the pyrochlore-type superstructure with the doubled lattice parameter compared to fluorite observed in the nanocrystalline powders $Gd_2Hf_2O_7$ and $Gd_2Zr_2O_7$ synthesized at 1200°C/6 h [10, 11]. The pyrochlore structure was assumed theoretically in Dy_2O_3-HfO_2 [2], and the probable formation of defect pyrochlore structure was experimentally shown in $Dy_2Hf_2O_7$ [12]. Besides, the modulated pyrochlore-like structure is possible in Dy_2HfO_5 powders prepared at 1600°C/3 h [10]. Therefore, the aim of present work was the detailed investigation of binary nanocrystalline compounds formation in the Dy_2O_3-HfO_2 system with molar component ratio 1 : 1 by the combination of modern experimental techniques using the X-ray and synchrotron radiation.

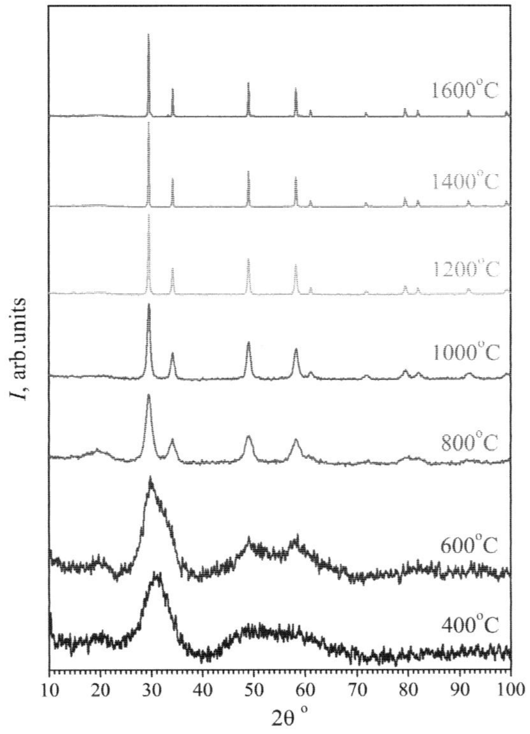

Figure 1. The powder XRD patterns of Dy_2HfO_5 samples prepared by isothermal annealing of precursor at different temperatures

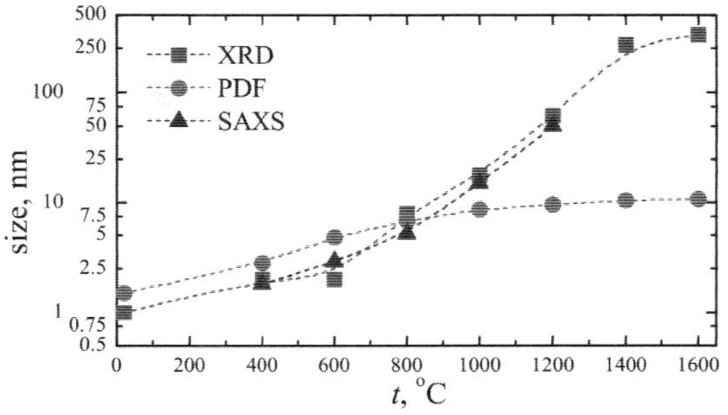

Figure 2. The effect of annealing temperature on the size of nanocrystalline particles

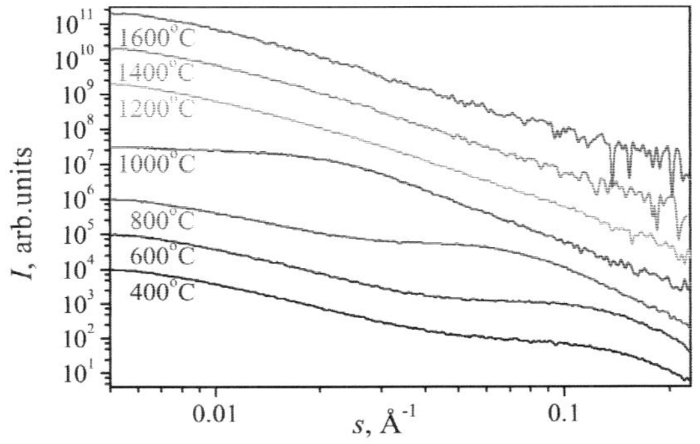

Figure 3. The SAXS curves of Dy_2HfO_5 samples annealed at different temperatures

2. Experimental

The Dy_2O_3-HfO_2 samples were prepared by coprecipitation from a solution of $Dy(NO_3)_3$ and $HfOCl_2$ salts and annealing of the resulting mixed hydroxides (precursors) in the 400 – 1600°C/3 h temperature range as reported in [10, 11]. The powder XRD patterns were measured at "DRON-UN1" diffractometer using the CuK_α radiation, 0.05° step and 10 s exposure, and processed by Rietveld method.

The powder XRD patterns for the pair distribution function (PDF) study were collected at BW5 beamline of DORIS-III positron storage ring (DESY, Hamburg, Germany) in transmission geometry using the photon energy of 100 keV ($\lambda = 0.1239$ Å). The full structure factors and pair distribution functions were calculated from experimental XRD patterns with PDFgetX2 [13]. PDFgui [14] was used to fit the experimental data. The X-ray absorption spectra were measured at C beamline of DORIS-III in transmission geometry above the L_3-Dy (7790 eV) and L_3-Hf (9561 eV) absorption edges at temperature $\sim$ 110 K. The extraction and fitting of EXAFS-functions $\chi(k)$ was performed using VIPER software pack [15]. Amplitudes and phases of photoelectron back scattering were calculated with FEFF code [16] using the parameters of fluorite structure and diffraction data.

The anomalous XRD measurements were performed at STM beamline of Kurchatov synchrotron radiation source (Moscow, Russia) in transmission geometry using the imaging plate detector. The two wavelengths were used: $\lambda_1 = 1.3183$ Å ($E_1 = 9405$ eV) and $\lambda_2 = 1.2976$ Å ($E_2 = 9555$ eV), the latter is close to the L_3-Hf (9561 eV) absorption edge. The SAXS curves were measured also at STM beamline using the X-ray photons of wavelength $\lambda = 1.033$ Å.

Journal of Physics: Conference Series **430** (2013) 012132 doi:10.1088/1742-6596/430/1/012132

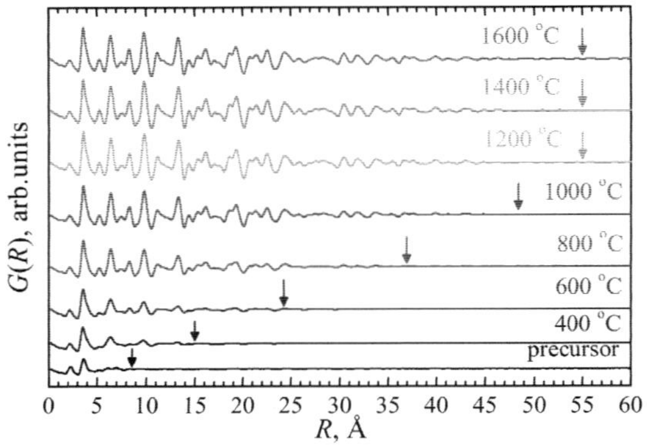

Figure 4. The full pair distribution functions $G(R)$ of Dy_2HfO_5 samples annealed at different temperatures

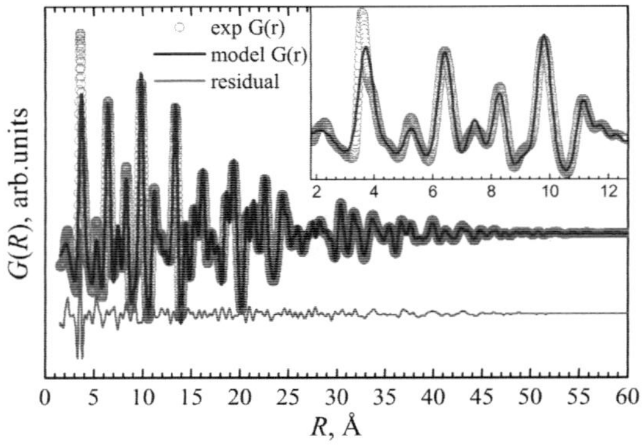

Figure 5. The comparison of experimental and model PDF for the sample annealed at 1600°C

3. Results and discussion

The powder X-ray diffraction study shows that the composition of amorphous precursor is $Dy_2O_3{\cdot}HfO_2 \cdot (200-250)H_2O$. However, the presence of two broad maxima in XRD patterns at $2\Theta \sim 30-31°$ and $2\Theta \sim 50-52°$ as well as two "metal-oxygen" and "metal-metal" coordination shell contributions in EXAFS spectra indicate the crystal structure of precursor particles not to be completely disordered and tend to form the fluorite structure. Upon isothermal annealing at the temperatures up to 600° the samples preserve the amorphous structure (fig.1), but the degree of order increases and partially crystallized clusters of $\sim 2\,\mathrm{nm}$ size with the fluorite-like structure are observed in samples annealed at $400-600°$ [10]. The most significant increase of the coherent-scattering region is observed in the $800-1400°$ annealing temperature range (fig.2). All XRD patterns are described quite well with the fluorite structure model as shown by Rietveld refinement.

The more precise values of nanocrystallite size vs. different heat treatment were estimated from the small angle scattering curves (fig.3). Depending on the annealing temperature the three types of scattering curves are observed (fig.3). In the low-temperature region $400-600°C$ the dominating size of scattering centers is clearly visible as the bending of the curves, which is related to the switching from one asymptotic scattering mode (the Porod mode, the slope about -4 in double logarithmic scale) to another (the Guinier mode, the almost horizontal plateau). The position of bending correlates with the dominating size of particles. The SAXS curves shape in the region of small scattering vectors $s < 0.03\,\text{Å}^{-1}$ indicates the formation of aggregates confirmed by additional LTNA study of surface areas. In the transition region $600-1200°C$ the curves shape undergoes the maximum change, which corresponds to the transfer from the nanocrystalline to well crystalline state. In the high-temperature region $\geq 1200°C$ the SAXS curves are almost identical for all the samples. The linear shape of curves (the Porod mode) corresponds to scattering on the large centers with the size much bigger than 50 nm. The sizes of SAXS scattering centers are in quite good agreement with the XRD coherent-scattering regions for the samples annealed at relatively low temperature $\leq 1000°C$ (fig.2).

Atomic pair distribution functions (PDF) $G(r)$ study demonstrates the formation of cubic fluorite structure upon annealing of amorphous precursor. Fig.4 shows the gradual formation of clear $G(r)$ maxima corresponding to separate coordination shells in the local environment of atoms. Besides, the annealing temperature increase causes the rise of structure order resulting in increase of the interatomic distances range, in which the pair correlation function $G(r)$ maxima are still distinguishable (fig.4). For the low annealing temperature samples ($\leq 800°C$) the

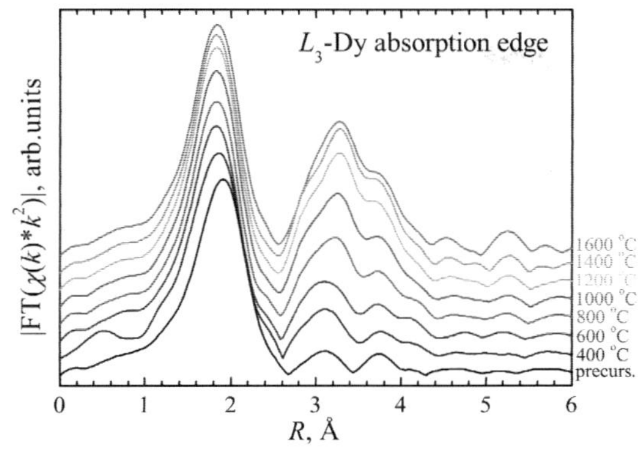

Figure 6. The modules of EXAFS function Fourier transforms for Dy_2HfO_5 samples annealed at different temperatures

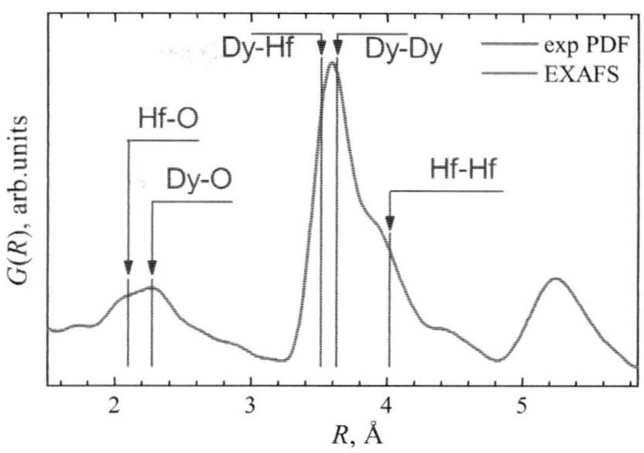

Figure 7. The comparison of PDF for the sample annealed at 1600°C and the interatomic distances obtained from EXAFS

obtained particle sizes are in qualitative correlation with the XRD coherent-scattered regions and the grain sizes estimated by SAXS. In case of higher annealing temperature > 1000°C the error of crystalline size estimation by PDF increases, which might also be due to a lack of angular resolution of imaging plate detector [17]. While, at low annealing temperature the accuracy of PDF in estimating the particle size is quite high (fig.2).

The pair distribution function $G(r)$ of sample annealed at 1600°C was fitted with theoretical PDF based on the model of HfO_2 oxide fluorite structure (fig.5). As result, the lattice parameter ($a = 5.248$ Å) and the anisotropic displacement parameters (U_{ii}(Hf)=0.0291 Å^2 and U_{ii}(O)=0.1227^2) were refined. The fig.5 shows that the fluorite model describes well the long-range order of crystal structure but doesn't adequately describe the local structure up to 8 Å because doesn't take into account the presence of two different scattering centers Dy and Hf.

The study of EXAFS spectra revealed the significant difference of Dy and Hf local environments for the whole temperature series of Dy_2HfO_5 samples, e.g. the discrepancy between the first coordination shell radii or interatomic Dy-O and Hf-O distances exceeds 0.2 Å. The significant variation is obtained for the second bunch of interatomic "metal-metal" distances also. It means that Dy^{3+} and Hf^{4+} cations with the different ionic radii distort the local environment in different way while formally remaining in the similar crystallographic position of the fluorite structure. Upon the annealing temperature increase the first maximum of EXAFS spectra corresponding to the coordination shell with 8 oxygen atoms shifts slightly to smaller interatomic distances, while its intensity decreases (fig.6). At the same time the second maximum appears, which corresponds to the coordination shells with 12 atoms of Dy and Hf.

The combined EXAFS and PDF study demonstrates that upon the annealing temperature increase to 1600°C the processes of ordering and Dy/Hf local environment differentiation come to saturation. The detailed investigation of PDF variation vs. annealing temperature confirms the discrepancy between Dy-O / Hf-O and Dy-Dy / Dy-Hf / Hf-Hf local interatomic distances revealed by EXAFS. The first two peaks of $G(r)$ function which correspond to first two coordination shells around metal ion have a very asymmetrical shape, and the positions of different contributions to these peaks coincide with the interatomic distances determined from EXAFS spectra (fig.7). Thus, the data independently obtained from EXAFS and PDF are in good agreement with each other and point out the difference of the samples local crystal structure from the fluorite.

There is some inconsistency between the results of local sensitive EXAFS and PDF methods and XRD results, which point out the fluorite-type structure without any ordering of Dy and

Hf cation positions. Note that Dy and Hf atoms are hardly distinguishable by the conventional XRD due to a small difference in their atomic scattering factors. For more detailed study of long-range order role in the local distortions of fluorite structure reflected in the discrepancy between Dy and Hf local environments, the measurements of anomalous synchrotron radiation diffraction were performed, which allow to enhance the contrast between the scattering power of atoms. The results indicate that even for well crystallized Dy_2HfO_5 sample (1600°C/3 h) the resonant diffraction pattern is absent from the superstructure reflexes corresponding to pyrochlore-type unit cell with of the doubled spacing comparing to the fluorite unit cell. Such a sharp rise of superstructure reflexes (in particular with 331 index of pyrochlore unit cell) was shown in our recent study of related $Gd_2Hf_2O_7$ samples annealed at 1200°C/6 h [10, 11].

4. Conclusion

The evolution of Dy_2O_3-HfO_2 (1 : 1) crystal structure parameters upon the isothermal annealing of initial mixed amorphous hydroxide was studied by the complex of synchrotron methods. In the low annealing temperature region (400 − 600°C) the gradual ordering of precursor amorphous structure occurs and the nanoclusters of < 2 nm size form. The transition region (600 − 1200°C) corresponds to the maximum change of atomic and crystal structure parameters, such as XRD coherent scattering region, interatomic distances etc. In the high annealing temperature region ($\geq$ 1200°C) the crystallization of samples saturates and only the insignificant changes of atomic and crystal structure parameters are observed. The resulting nanocrystalline oxide compounds have the fluorite-type crystal structure with the pronounced discrepancy between Dy and Hf local environments, but without the superstructural ordering of cation positions.

Acknowledgments

This work is partially supported by RFBR (grant No. 11-02-01174-a). Authors thank the HASYLAB Program Committee for providing the opportunity of XAFS and PDF measurements (projects No. I-20100397 and H-20000015).

References

[1] Subramanian M A, Aravamudan G and Subba Rao G V 1983 *Prog. Solid State Chem.* **15** 55–143
[2] Stanek C R 2003 *Atomic scale disorder in fluorite and fluorite related oxides* (Ph.D. Thesis. London. UK) 195p
[3] Shlyakhtina A V and Shcherbakova L G 2011 *Solid State Ionics* **192** 200–4
[4] Sosin S S, Prozorova L A and Smirnov A I 2005 *Phys. Usp.* **48** 83-90
[5] Gardner J S, Gingras M J P and Greedan J E 2010 *Rev. Modern Phys.* **82** 53–107
[6] Xu Q., Pan W., Wang J., Wan C, Qi L, Miao H, Mori K and Torigoe T 2006 *J. Amer. Ceram. Soc.* **89** 340–2
[7] Risovany V D, Zakharov A V, Muraleva E M, Kosenkov V M and Latypov R N 2006 *J. Nucl. Mater.* **355** 163–70
[8] Sickafus K E, Minervini L, Grimes R W, Valdez J A, Ishimaru M, Li F, McClellan K J and Hartmann T 2000 *Science* **289** 748–51
[9] Ushakov S V, Navrotsky A, Tangeman J A and Helean K B 2007 *J. Amer. Ceram. Soc.* **90** 1171–6
[10] Popov V V, Zubavichus Ya V, Petrunin V F, Menushenkov A P, Kashurnikova O V, Korovin S A, Chernikov R V and Yaroslavtsev A A 2011 *Glass Physics and Chemistry* **37** 511–9
[11] Popov V V, Petrunin V F, Korovin S A, Menushenkov A P, Kashurnikova O V, Chernikov R V, Yaroslavtsev A A and Zubavichus Ya V 2011 *Russian Journal of Inorganic Chemistry* **56** 1536–42
[12] Mandal B P, Garg N, Sharma S M and Tyagi A K 2006 *J. Solid State Chem.* **179** 1990–4
[13] Qiu X, Thompson J W and Billinge S J L 2004 *J. Appl. Crystal.* **37** 678
[14] Farrow C L, Juhas P, Liu J W, Bryndin D, Božin E S, Bloch J, Proffen Th and Billinge S J L 2007 *J. Phys.: Condens. Matter.* **19** 335219
[15] Klementev K V 2001 *J. Phys. D: Appl. Phys.* **34** 209
[16] Rehr J J and Albers R C 2000 *Rev. Mod. Phys.* B **72** 621–54
[17] Proffen Th, Billinge S J L, Egami T and Louca D 2003 *Z. Kristallogr.* **218** 132–143

Journal of Physics: Conference Series **430** (2013) 012133 doi:10.1088/1742-6596/430/1/012133

Characterization of CuHal-intercalated carbon nanotubes with x-ray absorption spectroscopy combined with x-ray photoelectron and resonant photoemission spectroscopies

M Brzhezinskaya[1], A Generalov[2,3], A Vinogdradov[3] and A Eliseev[4]

[1]Institut für Nanometeroptik und Technologie, Helmholtz-Zentrum Berlin für Materialien und Energie GmbH, Albert-Einstein-Str. 15, 12489 Berlin, Germany

[2]St. Petersburg State University, Ulyanovskaya Str. 1, 198504 St. Petersburg, Russia

[3]Technical University Dresden, Mommsenstrasse 9, 01062 Dresden, Germany

[4]Moscow State University, Leninskii Gori 1, 119991 Moscow, Russia

E-mail: maria.brzhezinskaya@helmholtz-berlin.de

Abstract. Encapsulated single-walled carbon nanotubes (SWCNTs) with inner channels filled by different compounds present the new class of composite materials. Such CNTs give opportunity to form 1*D* nanocrystals as well as quantum nanowires with new physical and chemical properties inside the tubes. The present study is aimed to characterize the possible chemical interaction between CuHal (Hal=I, Cl, Br) and SWCNTs in CuHal@SWCNTs and electronic structure of the latter using high-resolution near edge X-ray absorption fine structure (NEXAFS) spectroscopy combined with high-resolution X-ray photoelectron spectroscopy and resonant photoemission spectroscopy. The present study has shown that there is a chemical interaction between the filler and π-electron subsystem of CNTs which is accompanied by changes of the atomic and electronic structure of the filler during the encapsulating it inside CNTs.

1. Introduction

Owing to unique structural and electronic properties, single-walled carbon nanotubes (SWCNTs) could cover the full range of properties important for technology [1]. These potential applications of SWCNTs essentially depend upon the ability to modify their intrinsic properties by manipulating their structure and electronic properties. A promising route for modification of SWCNTs is a chemical modification via substitution of carbon atoms or the addition of atoms or polyatomic groups (so-called functionalization) [2]. An alternative procedure is to add electron acceptors or donors in a controlled manner (intercalation) by encapsulating corresponding atoms (compounds) inside SWCNTs. It generates unique one-dimensional (1*D*) nanostructures that have a large potential [3].

The present study is aimed to characterize the possible chemical interaction between CuHal (Hal=I, Cl, Br) and SWCNTs in CuHal@SWCNTs and electronic structure of the latter using high-resolution near edge X-ray absorption fine structure (NEXAFS) spectroscopy combined with high-resolution X-ray photoelectron spectroscopy and resonant photoemission spectroscopy.

Journal of Physics: Conference Series **430** (2013) 012133 doi:10.1088/1742-6596/430/1/012133

2. Methods

SWNTs were obtained by catalytic arc-discharge method. The pristine SWCNTs have diameters 1.4-1.6 nm. Pre-opened SWNTs (0.025 g) were grinded in agate mortar with CuCl, CuBr or CuI (Aldrich, 99 wt.%) in molar ratio 1:2, 1:1,5 or 1:1, accordingly, then evacuated in quartz ampoules at 1 Pa for 1 hour and sealed. The ampoules were heated at a rate of 1K/min up to the temperature 100 K above the melting point of CuHal. This temperature was maintained for 6 hours, whereupon the samples were slowly cooled at a rate around 0.02 K/min. Tthe CuHal@SWNT were directly observed using TEM. Microphotographs confirm continuous intercalation of CuHal into SWNT channels. The loading factors amount to 20% for CuCl@SWNT, 50% for CuBr@SWNT and ~90% for CuI@SWNT [4].

The x-ray absorption, x-ray photoelectron and photoemission spectra of SWCNTs, CuHal@SWCNTs, CuI and CuO were measured at the RG beamline at BESSY II (Berlin).

The NEXAFS spectra were obtained by recording total electron yield (TEY) of x-ray photoemission in the mode of measurement of the drain current of the sample. The samples were positioned at an angle of ~45° with respect to the incident beam. The energy resolutions ΔE of the monochromator in the range of the Cu $2p$ x-ray absorption edge ($h\nu$~935 eV) and C $1s$ x-ray absorption edge ($h\nu$~285 eV) were equal to ~250 and 70 meV, respectively.

Photoelectron spectra were measured in the normal photoemission registration mode by using the Phoibos 150 spherical analyzer from Specs. The analyzer energy calibration was performed based on photoelectron spectra of Au atom $4f_{7/2,5/2}$ electrons. The monochromator was calibrated by recording the main line of C $1s$ photoelectron spectra exited by radiation reflected from the diffraction grating in the 1st and 2nd diffraction orders.

3. Results and discussion

Fig. 1 presents C $K(1s)$ absorption spectra of initial SWCNTs and CuHal@SWCNTs measured under the same experimental conditions. The A and B-C peaks in the SWCNT spectrum are associated with dipole-allowed electron transitions from $1s$ carbon states to the unoccupied π and σ symmetry states in the conduction band that are formed from carbon $\pi 2p_z$ and $\sigma 2p_{x,y}$ states oriented perpendicular to and parallel to the carbon layer plane (graphene), respectively.

π-resonance A (285.4 eV) of the pristine SWCNTs contents in its fine structure four shoulders at 284.8, 285.0, 285.4 and 285.8 eV (Fig. 1b). Features S_1*, S_2* and S_3* are van Hove singularities (vHs's) of semiconductor SWNTs, while M_1* is vHs of metallic SWNTs [5]. Our pristine SWCNTs contain 70% of semiconductor SWCNTs and 30% of metallic ones. These features occur in the CNT x-ray absorption spectrum due to their quasi-1D character [6]. vHs's are associated with transitions of C1s electrons to unoccupied levels S_1*, S_2* and S_3* in semiconductor SWNTs and M_1* in metallic SWCNTs.

Fig. 1 also illustrates the changes of the x-ray absorption spectra after incorporation of CuHal into the SWNT channels. It is seen that C $1s$ absorption spectra for CuHal@SWCNTs (Fig. 1a) are practically identical to the one for pristine SWCNTs except for an additional small shoulder-like peak A_1 at the low-energy side of the π-resonance A. It is obvious that such a new peak reflects the changes in the spectrum of empty electron states which are take place as a result of chemical interaction.

At the same time, no pronounced variations in the spectra σ-part are observed in the C1s absorption spectra. In the photon energy range of 303 to 310 eV, specific structure that is characteristic of the sp^3-hybridized carbon atoms is not observed. This means that filling of the tubes with CuHal does not cause corrugation or deformation of the CNT wall graphene mesh. The mesh may be corrugated or deformed only if the carbon atom hybridization character changes from sp^2 to sp^3. Therefore, the absence of corrugation evidences that carbon atoms remain in the state of the sp^2 hybridization.

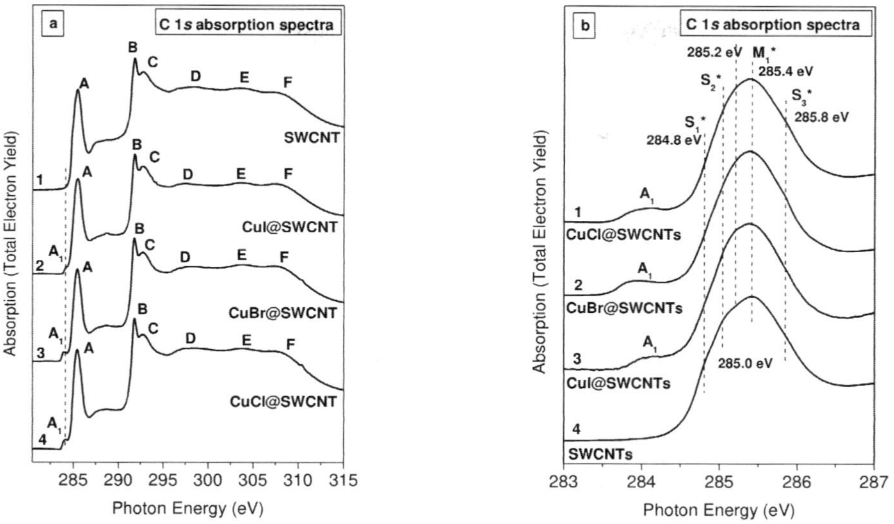

Figure 1. C1*s* absorption spectra of the SWNTs and CuHal@SWCNTs.

Cu 2*p* spectra for CuHal@SWCNTs, and reference samples CuI and CuO are presented in Fig. 2 (bottom panel). If the dipole rules for 2*p* electron transitions are taken into account, so these spectra reflect local density of empty electron states that have mainly Cu 4*s*- and Cu 3*d*-character. It can be seen from the Fig. 2 that the Cu $2p_{3/2}$ NEXAFS spectra of CuI@SWCNT and CuI show appreciable differences between their absorption structures. The main distinction is an appearance of narrow (full width at half maximum FWHM=1.1 eV) low-energy peak A_1 in the CuI@SWCNT spectrum. With consideration for the valence electron configuration of the Cu^+ ion ($3d^{10}4s^0$) in CuI, the appearance of narrow low-energy peak A_1 in the Cu $2p_{3/2}$ spectrum of the composite is reasonable to associate with Cu $2p_{3/2}$ electron transitions to empty 3*d* electron states that are lacking in pristine CuI and appear in CuI@SWCNT as a result of changes in electronic structure of CuI owing to its encapsulation into SWCNTs or to chemical interaction between the filler and the CNT in CuI@SWCNT. This is accompanied by the change in the valence electron configuration of the Cu^+ ion from $3d^{10}4s^0$ for CuI to $3d^{10\text{-}x}$ for CuI@SWCNT. The direct comparison of Cu $2p_{3/2}$ spectra for the composite and CuO confirms this interpretation. The spectrum for CuO with the valence electron configuration of Cu^{2+} ion, $3d^9$, has the similar narrow (FWHM=1.0 eV) low-energy peak A_1 as well. It is well known that this peak is due to the Cu $2p_{3/2}$ - 3*d* electron transition. Comparison of relative intensities of peaks A_1 from the spectra for CuO and CuI@SWCNT, that were normalized equally, gives the magnitude of x= 0.06e for the decrease of the Cu 3*d* charge in the composite. The general resemblance of absorption structures for spectra of CuI and CuI@SWCNT, except for the peak *A*, is probably evidence for conservation of a tetrahedron coordination of absorbing Cu ion as well as relatively small magnitude of effects under consideration. In this case the broadening and smearing of some absorption structures can be related to a distortion of the tetrahedron environment of Cu ions by iodine atoms.

The peak A^* intensity increases gradually in transitions from CuI@SWCNT to CuBr@SWCNT and then to CuCl@SWCNT thus evidencing sequential decrease in the number of 3*d*-electrons in copper atoms in this series. This shall result in the increase in the value of electron transfer from Cu atoms. Indeed, the value of effective electron transfer from Cu atoms increases in the CuI@SWCNT-CuBr@SWCNT-CuCl@SWCNT series. It is 0.6e for CuBr@SWCNT and 1.0e for CuCl@SWCNT. There are two ways of charge transfer from copper atoms to CuHal@SWCNTs: either to carbon atoms, or to haloid atom. Thereby, the copper atom configuration is $[Ar]3d^{9.94}4s^0$ in CuI@SWCNTs, $[Ar]3d^{9.4}4s^0$ in CuBr@SWCNTs, and $[Ar]3d^94s^0$ in CuCl@SWCNTs.

Cu $2p_{3/2}$ and C $1s$ spectra for CuI@SWCNTs, have been aligned in energy by using the binding energy separation of 648.0 eV between the Cu $2p_{3/2}$ and C $1s$ core levels (Fig. 2). The positions of the peaks A_1 in both spectra are nearly coincide within 1 eV. In the framework of quasimolecular approach, this coincidence of the absorption peaks A_1 in both the energetically aligned spectra can be regarded as a result of the transitions of Cu $2p_{3/2}$ and C $1s$ electrons to the same empty state of CuI@SWCNTs which has hybridized Cu $3d$-C$2p_z$ character. In this case, slight difference in positions of peak A_1 in C $1s$ and Cu $2p_{3/2}$ spectra can be associated with a different influence of C $1s^{-1}$ and Cu $2p_{3/2}^{-1}$ holes on the core-excited state under consideration.

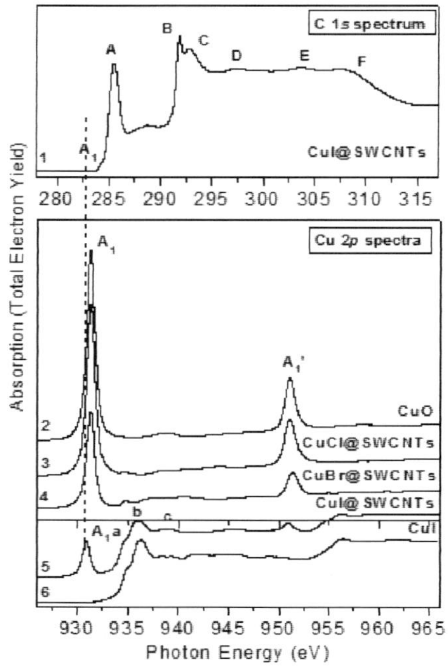

Figure 2. Cu *L*-edge spectra of CuHal@SWNTs, CuI, CuO and *CK*-edge spectrum of CuI@SWNTs.

4. Conclusions

Electronic structure of single-walled carbon nanotubes intercalated by copper halogenides (CuI@SWCNT, CuBr@SWCNT and CuCl@SWCNT) obtained by capillary technique was study. High-resolution X-ray C $1s$ and Cu $2p$ absorption and photoelectron spectra of CuHal@SWCNTs were measured for the first time. The present study has shown that encapsulation of CuHal (CuI, CuBr and CuCl) into SWCNTs is accompanied by the changes in electronic structure of CuHal. These changes are caused by the modifications in atomic structure of CuHal in the new nanomaterials CuHal@SWCNT and the chemical interaction between the filler and carbon nanotubes. At the same time, this interaction does not change the hybridization of C atoms from sp^2 to sp^3. It was found decreasing of electron charge of Cu atoms in the row CuI-CuI@SWCNT-CuBr@SWCNT-CuCl@SWCNT-CuO as a result of electron transfer.

Acknowledgements

This work was financially supported by the Russian Foundation for Basic Research (project no. 09-02-01278), the bilateral Program "RGL at BESSY" and a Marie Curie FP7-Reintegration-Grants within the 7th European Community Framework Program (Project No. PCIG10-GA-2011-297905).

References

[1] Jorio A, Dresselhaus G, Dresselhaus M S (Eds.) 2008 *Carbon Nanotubes: Introduction to the Important and Exciting Aspects of Carbon-Nanotube Science and Technology* (Berlin Heidelberg: Springer-Verlag)
[2] Burghard M 2005 *Surf. Sci. Reports* **58** 1
[3] Sloan J, Kirkland A I, Hutchison J L, Green M L H 2002 *Chem. Commun.* 1319
[4] Eliseev A A, Yashina L V, Verbitsky N I, Brzhezinskaya M M, Kharlamova M V, Chernysheva M V, Lukashin A V, Kiselev N A, Kumskov A S, Freitag B et al. 2012 *Carbon* **50** 4021
[5] Kramberger C, Rauf H, Shiozawa H, Knupfer M, Buchner B, Pichler T, Batchelor D, Kataura H 2007 *Phys. Rev. B* **75** 235437
[6] Geim A K, Novoselov K S 2007 *Nature Mater.* **6** 183

Combined study of structural properties on metal-organic frameworks with same topology but different linkers or metal

D. Gianolio,[1,*], J.G. Vitillo[2] B. Civalleri,[2] S. Bordiga,[2] U. Olsbye,[3] K. P. Lillerud,[3] L. Valenzano,[4] and C. Lamberti,[2]

[1] Diamond Light Source Ltd., Harwell Science and Innovation Campus, OX11 0DE Didcot, United Kingdom
[2] Department of Chemistry and NIS Centre of Excellence, and INSTM Reference Center, University of Turin, Via P. Giuria 7, I-10125 Turin, Italy
[3] inGAP centre of Research-based Innovation, Department of Chemistry, University of Oslo, Sem Saerlandsvei 26, N-0315 Oslo, Norway
[4] Department of Physics, Michigan Technological University, 1400 Townsend Dr, Houghton, MI 49931-1295, USA

Corresponding author's e-mail: diego.gianolio@diamond.ac.uk

Abstract. The recently discovered UiO-66/67/68 class of isostructural metallorganic frameworks (MOFs) [2008 *J. Am. Chem. Soc.* **130** 13850] has attracted great interest because of its remarkable stability at high temperatures, high pressures and in presence of different solvents acids and bases [2011 *Chem. Mater.* **23** 1700]. UiO-66 is obtained connecting $Zr_6O_4(OH)_4$ inorganic cornerstones with 1,4-benzene-dicarboxylate (BDC) linkers, while for the isostructural UiO-67, the longer 4,4′ biphenyl-dicarboxylate (BPDC) is used as linker [2012 *Phys. Chem. Chem. Phys.* **14** 1614]. Hf-UiO-66 is built with the same linker of UiO-66 but, in this case, $Hf_6O_4(OH)_4$ blocks substitute the $Zr_6O_4(OH)_4$ ones [2012, *Phys. Rev. B* **86** 125429]. In all cases isostructural cubic MOFs are obtained. Here we report a systematic EXAFS investigation of UiO-66 and UiO-67 (at Zr K-edge) and of Hf-UiO-66 (at Hf L_3-edge) before and after the removal of solvent in their pores, needed to make the channel volume available. The study reveals that, for the three systems, the perfect $M_6O_4(OH)_4$ (M = Zr or Hf) octahedron forming the cornerstones of the as prepared material undergoes a remarkable local rearrangement into a distorted M_6O_6 octahedron, with the loss of two water molecules under the desolvation process. Notwithstanding the high crystallinity of the materials, this rearrangement, moving M atoms from 8-fold to 7-fold coordination and distorting the M-M distance of the octahedron side by more than 0.2 Å, escape XRPD detection (because not ordered on the long range scale) but is clearly observed by EXAFS owing to its local sensitivity.

1. Introduction

Metal-Organic Frameworks (MOFs) have received much attention in the past decade and a huge number of new materials with a wide range of different compositions and properties have been reported.[1, 2] The most sought after MOF materials are those with internal open-pore systems exhibiting some of the main properties of the traditional porous materials, zeotype and mesoporous materials. The highest specific surface area materials are now found within MOFs[3] that, being constructed directly from simple organic molecules (acting as linkers or spacers) and metal centers (acting as connecting corner units), allow to tailor their chemical composition and physico-chemical properties to a large extent towards any given application. Although the industrial application of MOFs is still limited to few cases,[4] these materials being three-dimensional polymers of alternating metal/metal cluster and organic units have already shown potentialities in: (i) gas storage;[5, 6] (ii) gas separation and purification;[7] liquid phase separation;[8] (iii) drug delivery;[9] (iv) catalysis;[10] (v) molecular sensors;[11] (vi) contrast agents for tomography imaging inside living tissues ;[12] and (vii) interim radioactive waste scavenging.[13]

In the past we have widely documented the relevance in combining XRPD and EXAFS to disclose the structure of such complex materials,[14-23] exhibiting unit cells as large as several thousand of $Å^3$. Here we present a compared EXAFS study the metal center edges on three isostructural MOFs: UiO-66,[24] UiO-67[25] and Hf-UiO-66.[13]

2. Experiment and Methods

2.1. Materials

MOFs materials have been synthesized at the inGAP centre of the Oslo University. For details on the synthesis of the different materials the reader should refer to the following papers: UiO-66,[24] UiO-67,[25] and Hf-UiO-66.[13] Desolvation of UiO-66 and UiO-67 has been done inside a cell suitable for in situ EXAFS data collection,[26] heating the sample at 300 °C (5 °C/min) and keeping the sample at 300 °C for 30 min; data collection has then been done at room temperature. Hf-UiO-66 was desolvated inside a 1 mm capillary heated in situ under inert flow directly on the X-ray beam with a heater gun up to 300 °C; data collection has been done at 300 °C.

2.2. XAFS data collection and analysis.

In all cases XAFS data collection has been performed in transmission mode on ESRF beamlines. For UiO-66 Zr K-edge data collection has been performed at BM29 beamline, while for UiO-67 and Hf-UiO-66 we have used BM01B beamline at Zr K- and Hf L_3-edges, respectively.

The extraction of the $\chi(k)$ function was performed using the Athena code.[27] For each sample, three consecutive EXAFS spectra were collected and corresponding $\chi(k)$ functions were averaged before data analysis.[28] EXAFS data analysis was performed using the Arthemis software.[27] Phase and amplitudes for all single scattering (SS) and multiple scattering (MS) paths were calculated by FEFF6 code,[29] according to the strategy outlined elsewhere.[30]

2.3. Theoretical methods.

The theoretical investigation of UiO-66 and UiO-67 MOF (in its hydroxylated and dehydroxylated structures) was carried out with periodic density functional theory (DFT) calculations employing the hybrid B3LYP functional as implemented in the CRYSTAL program[31] as described elsewhere.[25, 30]

3. Results and Discussion

The recently discovered UiO-66/67/68 class of isostructural MOFs[24] has attracted great interest because of its remarkable stability at high temperatures, high pressures and in presence of different solvents acids and bases.[30] UiO-66 is built up connecting $Zr_6O_4(OH)_4$ inorganic cornerstones with 1,4-benzene-dicarboxylate (BDC) linker; while for the isostructural UiO-67 is formed using as linker the longer 4,4′ biphenyl-dicarboxylate (BPDC)[25] (Figure 1a). Hf-UiO-66 is obtained keeping the UiO-66 linker (BDC) and substituting the $Zr_6O_4(OH)_4$ blocks with $Hf_6O_4(OH)_4$ cornerstones (inset in Figure 1e). XRPD data testify the quality of the synthesis, see Figure 1b,e. Due to the rigidity of the framework several isostructural UiO MOFs have been prepared and tested for stability and gas adsorption capacity. Kandiah et al.[32] studied the thermal and chemical stabilities of isostructural UiO-66-X (X= NH_2, Br and NO_2) and observed the lower stability of this analogue with respect to parent UiO-66. Conversely, as documented by the thermogravimetry studies (reported in Figure 1c,f), UiO-67 and Hf-UiO-66 show thermal and chemical stability similar to the one of UiO-66 and they exhibit the expected surface area, as determined by low temperature volumetric N_2 adsorption isotherms (Figure 1d,g). Such a high stability is related to the fact that each Zr- (Hf-) octahedron is 12-fold connected to adjacent octahedra. This connectivity is very common for metals, resulting in the highly packed *fcc* structure, but it is still almost unique in MOF topologies.

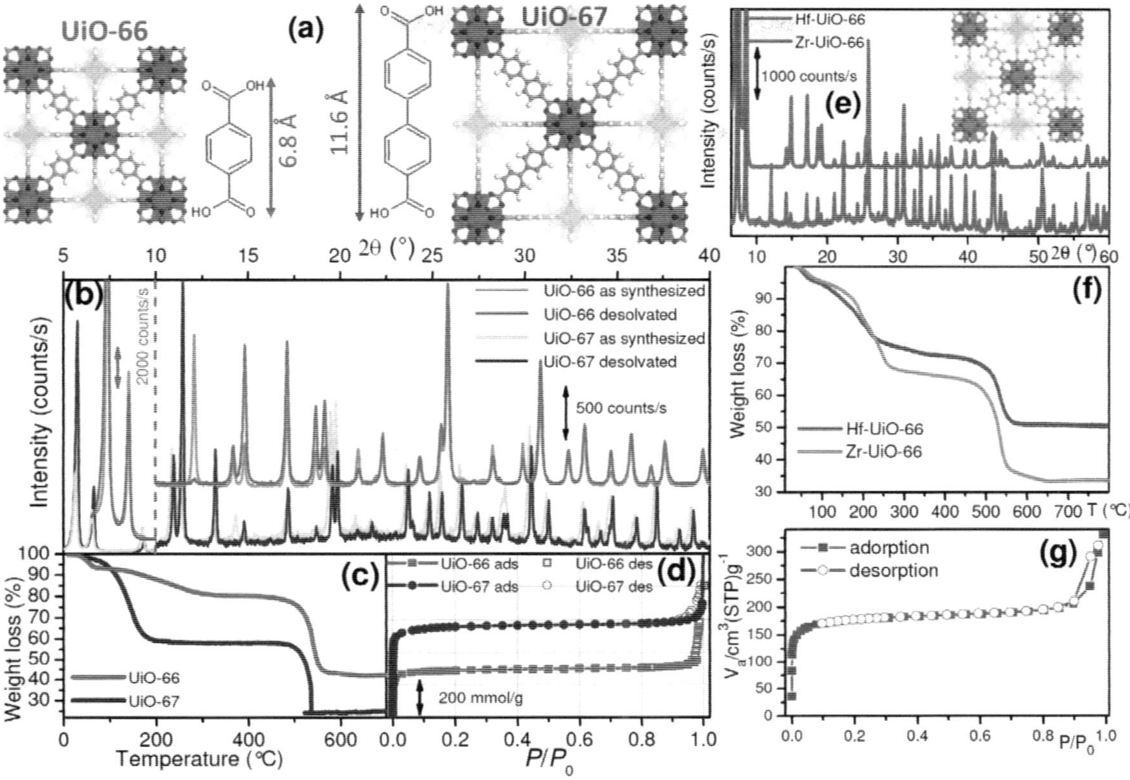

Figure 1. Part (a): Comparison of the dimension of linker and overall structure, for the isostructural UiO-66 and UiO-67 MOFs. Part (b): Comparison of the XRPD patterns (λ=1.540 Å) for UiO-66 and UiO-67 materials as prepared (orange and cyan lines) and activated at 300 °C (red and blue lines). The patterns in the 10-40° 2θ region are amplified by a factor 4. Patterns related to UiO-66 are vertically translated for clarity. Part (c): TGA curves of UiO-66 and UiO-67 samples, red and blue lines respectively. In both cases, the heating ramp was of 5°C/min in N_2 flow (100ml/min). Part (d): volumetric N_2 adsorption isotherms recorded at 77 K on UiO-66 (red squares) and UiO-67 (blue circles). Empty and filled scatters refer to the adsorption and desorption branches. Part (e): XRPD pattern (λ=1.540 Å) of Hf-UiO-66 (green) and Zr-UiO-66 (red) in their solvated forms (as prepared). The inset reports the Hf-UiO-66 structure. Part (f) TGA curve of Hf-UiO-66, it shows the weight loss relative to the initial mass (green curve). The weight loss for Zr-UiO-66 (orange curve) is also reported as a reference. Since Hf is 41% heavier than Zr, the Zr-UiO-66 shows both higher initial and breakdown losses. Part (g): volumetric N_2 adsorption/desorption isotherm for Hf-UiO-66 recorded at 77 K.

All MOFs were synthesized in presence of a solvent that has to be removed to make the large pore volume available for any practical applications. The desolvation process left almost unchanged the XRPD patterns of such materials (Figure 1b), besides a gain of intensity of the basal reflections (due to the removal of the electron density inside the pores[17, 33, 34]), all peaks remains in almost the same 2θ position with small intensity changes. Conversely, a huge modification of the EXAFS spectra is obtained in all cases, see Figure 2.

In the three hydroxylated materials, the structure determined from the Rietveld refinement of the XRPD corresponding patterns resulted in a straightforward interpretation of the complex EXAFS signals, see first three columns in Table 1. The dramatic modification undergone by the EXAFS spectrum upon dehydroxylation (see Figure 2) makes the data analysis not so straightforward. In the case of UiO-66 (see Figure 2b, but similar effects are observed also in the two other cases) the changes are basically explained in terms of three main effects: (i) small contraction of the first M-O shell accompanied by a small decrease in coordination (erosion of the shoulder around 1.9 Å); (ii) relevant distortion of the second shell contribution showing a maximum that moves from 3.17 Å to 2.91 Å, with a shoulder at 3.41 Å, thus reflecting an important splitting of the R_{M1} distances of the octahedron sides; (iii) the almost complete disappearance of the weak contribution around 4.7 Å, due to the M-M SS signal of the octahedron diagonal (R_{M2}). For the three systems, differently to the hydroxylated cases, the 3D model obtained from the Rietveld refinement of XRPD data in the highly symmetric *Fm-3m* space group was inadequate to simulate the experimental datum. The origin of this failure was, obviously due to the inability of the model to account for two different R_{M1} and R_{M2} distances. For both UiO-66[30] and UiO-67[25] cases, the failure of the XRPD model was overcome by using the optimized geometry obtained by ab initio periodic calculations.

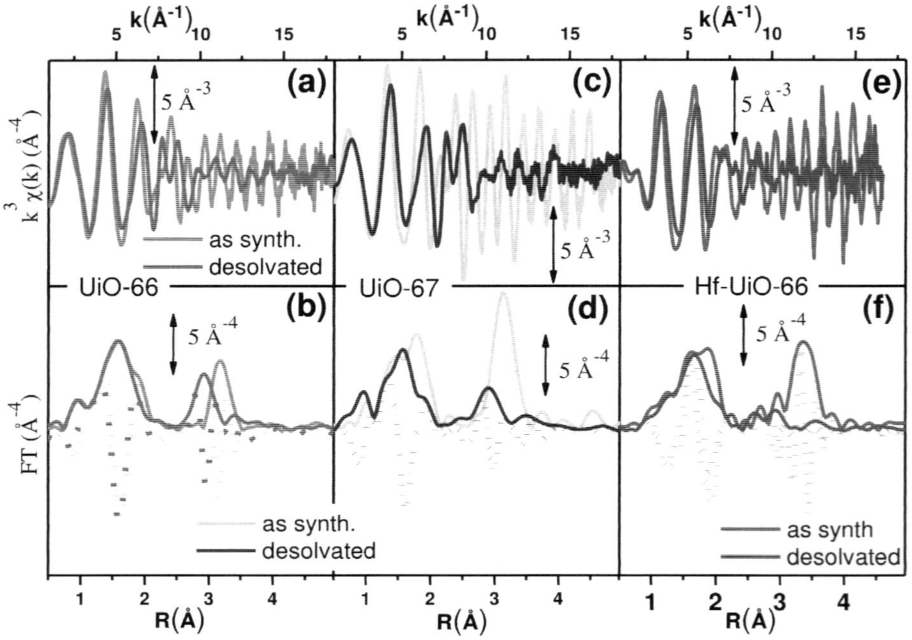

Figure 2. k- (top panels) and R-space (bottom panels) EXAFS data collected on UiO-66, UiO-67 and Hf-UiO-66, parts (a,b), (c,d) and (e,f), respectively. Both as prepared (hydroxylated) and desolvated (dehydroxylated) forms of the three different isostructural MOFs have been measured. With the exception of desolvated Hf-UiO-66 sample (collected at 573 K), all the remaining spectra were collected at 300 K.

The inorganic cornerstones of the as synthesized materials are perfect $M_6(OH)_4O_4$ octahedron (see model in Figure 3b), with 6 equivalent M at the vertex, 12 equivalent M-M1 sides and 3 equivalent and M-M2 diagonals. Upon desolvation 2 structural water molecules are lost per cornerstone unit (Figure 3a), that evolves from $M_6(OH)_4O_4$ to M_6O_6.[13, 25, 30] The new M_6O_6 octahedron is compressed (2 opposite vertexes approaching, see model in Figure 3c) resulting in the shortening of 8 of the 12 edges, and the elongation of the other 4 edges. To take into account this variation we simulated the EXAFS contribution with two independently parameterized paths fixing for the degeneration a ratio of 1/3 and 2/3 with respect to the case of the single contribution. For the three systems, this combined XRPF, EXADS and DFT approach allowed a full interpretation of the EXAFS data in both hydroxylated and dehydroxylated forms.[13, 25, 30]

4. Conclusions

EXAFS spectroscopy allows to detect the evolution from $M_6(OH)_4O_4$ to M_6O_6 (M = Zr or Hf) of the inorganic cornerstones of UiO-66, UiO-67 and Hf-UiO-66 MOFs occurring in the desolvation process, that escaped XRPD detection. On Zr-UiO-66 and Zr-UiO-67, periodic calculations performed with CRYSTAL code[31] at DFT level of theory support EXAFS data.

Journal of Physics: Conference Series **430** (2013) 012134 doi:10.1088/1742-6596/430/1/012134

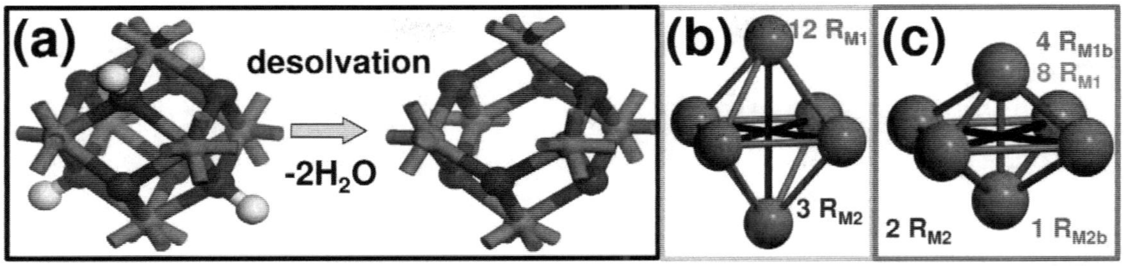

Figure 3. Part (a): Stick and ball representation of the dehydroxylation undergone by the inorganic M6O4(OH)4 cornerstone upon thermal treatment at 300 °C in vacuo resulting in a distorted M6O6 cluster (M =Zr or Hf). Red, blue and cyan colors refer to M, O and H atoms, respectively. Part (b): Stick and ball representation of the perfect M6 octahedron, showing 12 equivalent RM1 sides and 3 equivalent RM2 = √2 RM1 diagonals. Part (c): Stick and ball representation of a compressed M6 octahedron. The 12 sides are now split into 4 in-plane long MM1b sides and 8 prismatic short MM1a sides, while the 3 diagonals evolve into 2 in-plane long MM2b and 1 orthogonal short MM2a diagonals. For clarity, O atoms are omitted in parts (b) and (c).

Table 1. Summary of the EXAFS refinement obtained on the hydroxylated and dehydroxylated forms of UiO-66, UiO-67 and Hf-UiO-66. Parameters without error bars were not optimized. The EXAFS refinement of the hydroxylated materials was obtained using as input model the optimized structure from Rietveld refinement of the corresponding XRPD patterns. The EXAFS refinement of the dehydroxylated materials was obtained using as input a model optimized by *ab initio* calculations. With this approach the coordination number (N) of each contribution is fixed by the model stoichiometry. Refinement of the experimental amplitude is done by optimizing the overall amplitude factor S_0^2 only. The fitting of the higher shells was possible only adopting the axial compressed model of the M_6O_6 octahedron represented in Figure 3c where the twelve sides R_{M1} are split into eight short prismatic distances (R_{M1}, N=8/12) and four long planar ones (R_{M1b}, N=4/12) and where the three diagonals R_{M2} are split into a short axial diagonal and (R_{M2a}, involving two M atoms out of six; N=1/3) and two long planar diagonals (R_{M2b}, involving four M atoms out of six; N=2/3): Figure 3c.

	UiO-66	Hydro UiO-67	Hf-UiO-66	UiO-66	UiO-67	Hf-UiO-66
		hydroxylated			dehydroxylated	
T (K)	300	300	300	300	300	573
R-factor	0.01	0.04	0.02	0.02	0.04	0.05
Δk ($Å^{-1}$)	2.0-18.0	2.0-16.2	2.0-16.0	2.0-15.0	2.0-15.0	2.0-15.0
ΔR (Å)	1.0-5.3	1.0-5.3	1.0-5.5	1.0-5.3	1.0-5.3	1.0-3.9
Indip. points	43	38	40	35	35	25
N. variables	14	13	10	15	15	9
ΔE_0 (eV)	5 ± 1	1 ± 1	2.2 ± 0.6	5	1	2.4 ± 0.7
S_0^2	1.17 ± 0.08	1.17	0.91 ± 0.06	1.17	1.17	0.91
$R_{\mu3\text{-}O}$ (Å)	2.087 ± 0.008	2.12 ± 0.02	2.12 ± 0.01	2.06 ± 0.01	2.096 ± 0.007	2.06 ± 0.01
$\sigma^2(\mu_3\text{-}O)$ ($Å^2$)	0.0036±0.0009	0.005 ± 0.002	0.005 ±0.002	0.008 ± 0.003	0.006 ± 0.001	0.009 ± 0.002
R_{O1} (Å)	2.235 ± 0.008	2.26 ± 0.01	2.25 ± 0.01	2.221 ± 0.007	2.249 ± 0.007	2.19 ± 0.01
σ^2(O1) ($Å^2$)	0.0074±0.0008	0.006 ± 0.001	0.005	0.007 ± 0.002	0.004 ± 0.001	0.009
R_C (Å)	3.19 ± 0.02	3.40 ± 0.06	3.23 ± 0.06	3.17 ± 0.04	3.15 ± 0.04)	3.22 ± 0.05
σ^2(C) ($Å^2$)	0.004 ± 0.002	0.012 ± 0.002	0.014 ± 0.011	0.009 ± 0.009	0.004 ± 0.003	0.016 ± 0.013
R_{M1} (Å)	3.511 ± 0.007	3.512 ± 0.006	3.510 ± 0.005	3.35 ± 0.01	3.365 ± 0.015	3.31 ± 0.03
σ^2(M1) ($Å^2$)	0.007 ± 0.001	0.004 ± 0.001	0.0042 ± 0.0004	0.009 ± 0.001	0.009 ± 0.002	0.009 ± 0.004
R_{M1b} (Å)	-	-	-	3.74 ± 0.02	3.80 ± 0.03	3.45 ± 0.06
σ^2(M1b) ($Å^2$)	-	-	-	0.009 ± 0.002	0.008 ± 0.003	0.009
R_{M2} (Å)	4.99 ± 0.04	4.95 ± 0.03	4.964	4.14 ± 0.07)	4.15 ± 0.07	-
σ^2(M2) ($Å^2$)	0.010 ± 0.006	0.004 ± 0.002	0.008 ± 0.002	0.008 ± 0.006	0.006 ± 0.004	-
R_{M2b} (Å)	-	-	-	5.30 ± 0.04	5.46 ± 0.05	-
σ^2(M2b) ($Å^2$)	-	-	-	0.008	0.006	-

5. Acknowledgements

We are indebted with the staffs of BM29 (particularly C. Prestipino) and BM01 (particularly H. Emerich) beamlines at the ESRF for their competent assistance during XAFS data collection. M. H. Nilsen and S. Jakobsen are acknowledged for sample synthesis and fruitful discussions.

References

[1] Ferey G and Serre C 2009 *Chem. Soc. Rev.* **38** 1280
[2] Long J R and Yaghi O M 2009 *Chem. Soc. Rev.* **38** 1213
[3] Furukawa H, Ko N, Go Y B, Aratani N, Choi S B, Choi E, Yazaydin A O, Snurr R Q, O'Keeffe M, Kim J and Yaghi O M 2010 *Science* **329** 424
[4] Czaja A, Trukhan T and Müller U 2009 *Chem. Soc. Rev.* **38** 1284
[5] Murray L J, Dinca M and Long J R 2009 *Chem. Soc. Rev.* **38** 1294
[6] Sumida K, Rogow D L, Mason J A, McDonald T M, Bloch E D, Herm Z R, Bae T H and Long J R 2012 *Chem. Rev.* **112** 724
[7] Li J R, Sculley J and Zhou H C 2012 *Chem. Rev.* **112** 869
[8] Cychosz K A, Ahmad R and Matzger A J 2010 *Chem. Sci.* **1** 293
[9] Horcajada P, Gref R, Baati T, Allan P K, Maurin G, Couvreur P, Ferey G, Morris R E and Serre C 2012 *Chem. Rev.* **112** 1232
[10] Corma A, Garcia H and Xamena F X L 2010 *Chem. Rev.* **110** 4606
[11] Kreno L E, Leong K, Farha O K, Allendorf M, Van Duyne R P and Hupp J T 2012 *Chem. Rev.* **112** 1105
[12] Lin W, deKrafft K, Boyle W, Burk L and Zhou O 2012 *J. Mater. Chem.* **22** 8139
[13] Jakobsen J, Gianolio D, Wragg D, Nilsen M H, Emerich H, Bordiga S, Lamberti C, Olsbye U, Tilset M and Lillerud K P 2012 *Phys. Rev. B* **86** art. n. 125429
[14] Prestipino C, Regli L, Vitillo J G, Bonino F, Damin A, Lamberti C, Zecchina A, Solari P L, Kongshaug K O and Bordiga S 2006 *Chem. Mater.* **18** 1337
[15] Szeto K C, Lillerud K P, Tilset M, Bjorgen M, Prestipino C, Zecchina A, Lamberti C and Bordiga S 2006 *J. Phys. Chem. B* **110** 21509
[16] Szeto K C, Prestipino C, Lamberti C, Zecchina A, Bordiga S, Bjorgen M, Tilset M and Lillerud K P 2007 *Chem. Mater.* **19** 211
[17] Hafizovic J, Bjorgen M, Olsbye U, Dietzel P D C, Bordiga S, Prestipino C, Lamberti C and Lillerud K P 2007 *J. Am. Chem. Soc.* **129** 3612
[18] Bonino F, Chavan S, Vitillo J G, Groppo E, Agostini G, Lamberti C, Dietzel P D C, Prestipino C and Bordiga S 2008 *Chem. Mater.* **20** 4957
[19] Masciocchi N, Galli S, Colombo V, Maspero A, Palmisano G, Seyyedi B, Lamberti C and Bordiga S 2010 *J. Am. Chem. Soc.* **132** 7902
[20] Bordiga S, Bonino F, Lillerud K P and Lamberti C 2010 *Chem. Soc. Rev.* **39** 4885
[21] Rossin A, Di Credico B, Giambastiani G, Peruzzini M, Pescitelli G, Reginato G, Borfecchia E, Gianolio D, Lamberti C and Bordiga S 2012 *J. Mater. Chem.* **22** 10335
[22] Borfecchia E, Maurelli S, Gianolio D, Groppo E, Chiesa M, Bonino F and Lamberti C 2012 *J. Phys. Chem. C* **116** 19839
[23] Valenzano L, Vitillo J G, Chavan S, Civalleri B, Bonino F, Bordiga S and Lamberti C 2012 *Catal. Today* **182** 67
[24] Cavka J H, Jakobsen S, Olsbye U, Guillou N, Lamberti C, Bordiga S and Lillerud K P 2008 *J. Am. Chem. Soc.* **130** 13850
[25] Chavan S, Vitillo J G, Gianolio D, Zavorotynska O, Civalleri B, Jakobsen S, Nilsen M H, Valenzano L, Lamberti C, Lillerud K P and Bordiga S 2012 *Phys. Chem. Chem. Phys.* **14** 1614
[26] Lamberti C, Prestipino C, Bordiga S, Berlier G, Spoto G, Zecchina A, Laloni A, La Manna F, D'Anca F, Felici R, D'Acapito F and Roy P 2003 *Nucl. Instrum. Meth. B* **200** 196
[27] Ravel B and Newville M 2005 *J. Synchrot. Radiat.* **12** 537
[28] Lamberti C, Bordiga S, Arduino D, Zecchina A, Geobaldo F, Spanò G, Genoni F, Petrini G, Carati A, Villain F and Vlaic G J 1998 *Phys. Chem. B* **102** 6382
[29] Zabinsky S I, Rehr J J, Ankudinov A, Albers R C and Eller M J 1995 *Phys. Rev. B* **52** 2995
[30] Valenzano L, Civalleri B, Bordiga S, Nilsen M H, Jakobsen S, Lillerud K-P and Lamberti C 2011 *Chem. Mater.* **23** 1700
[31] Dovesi R, Orlando R, Civalleri B, Roetti C, Saunders V R and Zicovich-Wilson C M 2005 *Z. Kristallogr.* **220** 571
[32] Kandiah M, Nilsen M H, Usseglio S, Jakobsen S, Olsbye U, Tilset M, Larabi C, Quadrelli E A, Bonino F and Lillerud K P 2010 *Chem. Mater.* **22** 6632
[33] Milanesio M, Artioli G, Gualtieri A F, Palin L and Lamberti C 2003 *J. Am. Chem. Soc.* **125** 14549
[34] Agostini G, Lamberti C, Palin L, Milanesio M, Danilina N, Xu B, Janousch M and van Bokhoven J A 2010 *J. Am. Chem. Soc.* **132** 667

Journal of Physics: Conference Series **430** (2013) 012135 doi:10.1088/1742-6596/430/1/012135

XANES study of the role of Sr doping in $LaCoO_3$

Hao Liu[1], Xueguang Dong[1], Yongtao Li[1, 2], Hongguang Zhang[1], Kai Chen[1], Qingteng Hou[1], Yongchao Huang[1] and Qi Li[1*]

[1] Department of physics, Southeast University, Nanjing, 211189, P. R. China

[2] College of science, Nanjing University of Posts and Telecommunications, Nanjing, 210003, P. R. China.

*Corresponding author, email: qli@seu.edu.cn

Abstract: With the comparisons of experimental XAFS data of Sr doped LaCoO3 samples taken at room temperature and LDA+U (with U = 0,4eV and 8eV, respectively) model calculations of $LaCoO_3$, it is found that the experimental XAFS data of Sr doped samples can be well reconstructed if the samples are phase separated. When U takes 4eV or 8eV, the ground state of $LaCoO_3$ is Mott insulator and the curve shape of calculated XAFS data is even the same as that of the experimental data; when U=0 the HOMO-LUMO band gap vanishes and the calculated XAFS data gets an extra peak. The curve can be well fitted with experimental data of Sr doped $LaCoO_3$ sample. It shows that the doped $LaCoO_3$ sample is not traditional Mott insulator but with some properties of charge transfer type, that is, when Sr ions are doped in $LaCoO_3$, the trivalent Co ions become tetravalent and leave an extra hole in each Co e_g orbital. The charge carriers can hop back and forth between trivalent Co ions and tetravalent ones in which way close the band gap.

1. Introduction:

There is a long argument about the spin state of Co ions (1) and the corresponding local structure of CoO_6 octahedral (2) in $LaCoO_3$, i.e., it is thought that intermediate spin state of Co ions is Jahn-Teller active while high-spin state or low-spin state of Co ions is not. These issues are determined by the local electronic structure of the materials, e.g., the competition of Hund energy and crystal field energy of CoO_6 octahedron can dominate the electron occupancy of sub-orbitals of Co 3d orbital (3). Near edge X-ray absorption fine structure (NEXAFS) is very sensitive to local structure and can be properly used in the determination of local electronic structure of CoO_6 octahedron in $LaCoO_3$.

Previous experiments with extended X-ray absorption fine structure (EXAFS) have obtained many interesting results, like in reference (2) the authors have obtained that few Jahn-Teller active Co ions exist in $La_{1-x}Sr_xCoO_3$ for $x < 0.35$. It has been interpreted that this is mainly because the charge carriers hop in these samples much faster than the phonon time scale, and the wave function of the conducting charges may spread over several sites to form metallic clusters. In reference (4) the authors have found that there was no evidences of the variations of edge positions of Co K edge in $LaCoO_3$ although different amount of Sr ions were doped. It is further argued that the holes are in fact doped on O sites rather than on Co sites. For transition metal oxides, pure ionic bonding is rare but with some degree of covalent bonding, thus the doped holes, even if on O sites, would lead to some variation on Co sites, especially the outer shells of Co ions, i.e., e_g orbits (considering the octahedral crystal field). An important parameter is the electron correlation strength U that we consider in this report.

Previous ab-initio calculations considering nonzero U values have already got many important results, i.e., in reference (5), spin-orbit coupling has been considered to take an important role in determining the electronic structure of Co ions in this series of materials. There have been also some disagreements between model calculations (1, 5) and experimental results (6).

Journal of Physics: Conference Series **430** (2013) 012135 doi:10.1088/1742-6596/430/1/012135

The properties of doped $LaCoO_3$ are different from those of pure $LaCoO_3$, i.e., when a certain amount of charge carriers are introduced by doping Sr ions at La sites, the samples will become ferromagnetically metallic (7). It is also thought that the doped $LaCoO_3$ is phase separated (8), but what is the role of Sr ions in $La_{1-x}Sr_xCoO_3$ is still unclear. In the present work, with the combination of NEXAFS experiment and ab-initio calculation, we would like to give a localized picture of the influences of Sr doping on LaCoO3.

2. Experiments and Calculations:

$La_{1-x}Sr_xCoO_3$ (x = 0.0, 0.1, 0.2 and 0.3) samples were synthesized by traditional solid-state reaction method. Moderate amount of La_2O_3, $SrCO_3$, and Co_2O_3 were properly mixed and ground, followed by calcining for several days between 1000°C ~1100°C, then the mixture was pressed into pellets and annealed at 1200°C. The crystal structure of the samples was characterized by x-ray diffraction measurements. The local structure was measured by x-ray absorption fine structure measurement (XAFS) at Co K edge at National Synchrotron Radiation Laboratory (NSRL), University of Science and Technology of China.

The ab-initio calculations were carried out with the QUANTUM ESPRESSO (9) software packages. As there were arguments about the crystal structure of $LaCoO_3$ series which was firmly relative to the spin state ground state of Co ions, the I2/a spacegroup was chosen since there are un-equivalent Co-O bond lengths in this structure model (10). In fact, we also carried out the calculations with the R-3C space-group (which has equivalent Co-O bond length) to cross-check the results, and the R-3C spacegroup structure model did not give out obvious difference from the I2/a model calculations.

3. Results and Discussions:

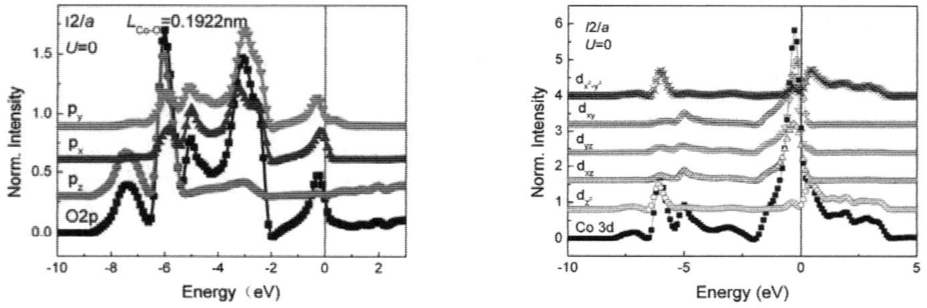

Figure 1. The calculated band structure of $LaCO_3$ in I2/a structure with U=0 shows that the ground state is metallic.

Fig. 1 shows the calculated results of the state density of O 2p orbitals in $LaCoO_3$ with U=0. As shown in the figure, the O (we simply show the ones with the Co-O bond length of 0.1922 nm, the results from the O ions with Co-O bond length of 0.198 nm is even the same as the results from 0.1922 nm-long Co-O bonds) p_x and p_y orbitals have some intensities around the Fermi level. At the same time, there is a prominent amount of intensity around Fermi level of $LaCoO_3$ that comes from Co t_{2g} (d_{xy}, d_{yz}, d_{zx}) orbitals.When U takes 4 eV or 8 eV, as shown in Fig. 2, the e_g orbitals lie above the Fermi level, and the t_{2g} orbitals lie below the Fermi level. The band gap is open and there is little intensity around the Fermi level, thus the ground state of the samples is insulator. The NEXAFS of the samples have also been calculated with the xspectra program as shown in the upper panel of Fig. 3. The curves' shape of the U=4 eV and U=8 eV are even the same. The calculated NEXAFS data of the U=0 model has an extra peak compared with the nonzero-U model calculations.

The experimental NEXAFS data of both doped and un-doped samples is shown in the lower panel of Fig. 3. The undoped samples have only a broad peak at the absorption edge. When Sr concentration increases, another peak appears at energy lower than the absorption edge of undoped $LaCoO_3$. The intensity of this new peak increases with the increase of Sr concentration. The experimental XANES curves of the samples can be well fitted with the calculated ones as shown in Fig. 4. The main peak of the experimental data of the undoped samples is corresponding to the calculated edges with U=4 eV or 8 eV.

The experimental data of the doped samples can be well fitted with the combination of calculated curves with U =0 and the nonzero-U models.

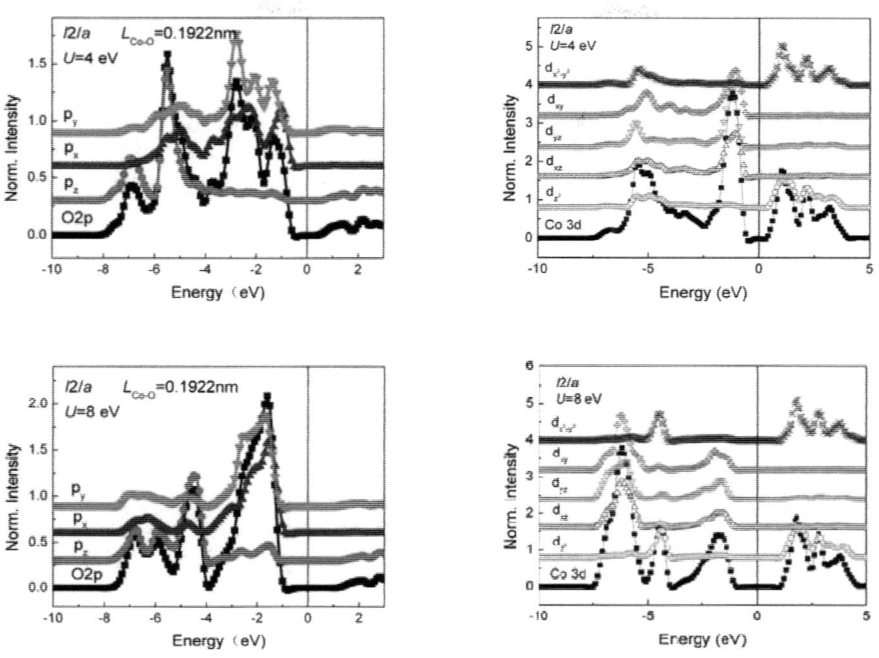

Figure 2. The open of the band gap when *U* takes 4 eV or 8 eV.

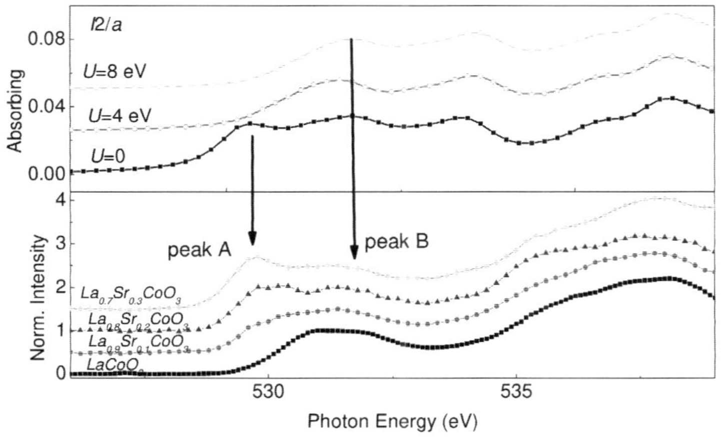

Figure 3. The comparison of calculated XAFS with experimental data.

Previous works about the Sr ions doping in $LaCoO_3$ have the results that when the Sr concentration of $La_{1-x}Sr_xCO_3$ reaches a considerable value, the ground state of the samples become metallic. The possible explanation is that the doped Sr ions will introduce some charge carriers (holes) around Co ions, the charge carriers can hop back and forth to make the samples metallic. With the comparison of experimental NEXAFS data of Sr doped $LaCoO_3$ samples taken at room temperature and LDA+U (with U = 0, 4 eV and 8 eV, respectively) model calculations of $LaCoO_3$, it is found that the experimental NEXAFS data of Sr doped samples can be well reconstructed if the samples are phase separated (11), i.e., at the Co sites around the Sr ions, the experimental NEXAFS data of Co K edge can be well fitted with the calculated curves of U=0. While at sites far from Sr ions, the experimental curves are most like the

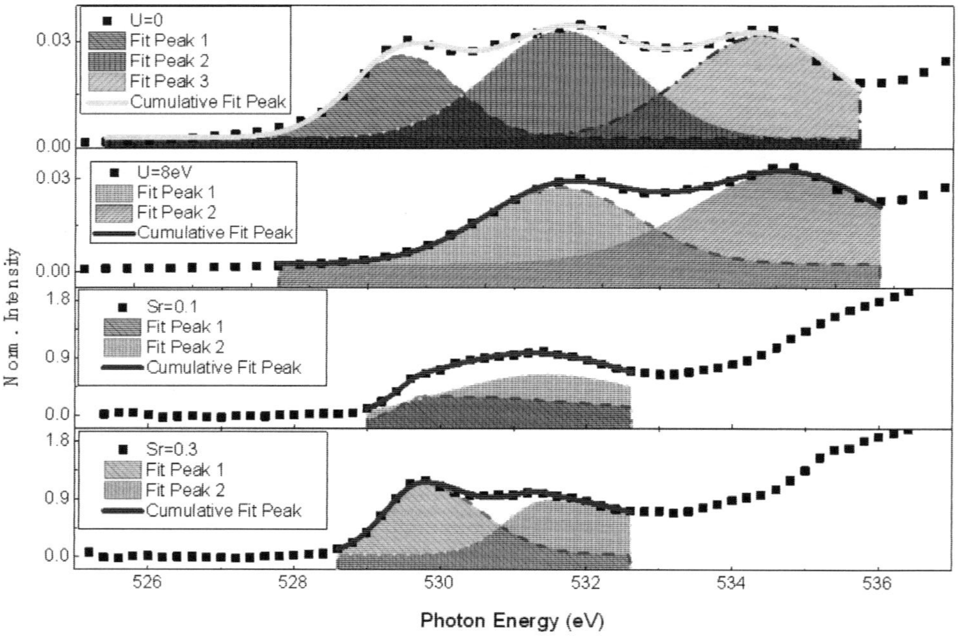

Figure 4. The fit results of calculated XAFS (for U=0 and U=8eV) with experimental data(for Sr=0.1 and 0.3).

calculated curves with nonzero U model. That is also the reason why the relative intensity of peak A (the absorption peak located at energy lower than the one of un-doped $LaCoO_3$, see Fig.3) in the experimental NEXAFS data of doped $LaCoO_3$ increases with the increase of Sr concentration.

When U is zero the HOMO-LUMO band gap vanishes and the Co t_{2g} orbits and O p_x, p_y will form the "π" bonds. When U takes 4 eV or 8 eV, the bands split into two parts as shown in Fig. 2. The energy split of the U=8 eV case is larger than the U=4 eV case, thus there is little density of state around the Fermi level. Comparing with the experimental data, the differences between the undoped $LaCoO_3$ and Sr doped $LaCoO_3$ are mainly that when Sr ions are doped, the electrons of the outer shells (e_g orbitals) of Co^{3+} ions which are very near to Sr^{2+} ions get lost, thus the interaction between t_{2g} electrons and e_g electrons vanishes, then the screening effect disappears, leading to the bonding of Co t_{2g} electrons with O p orbits. As the electrons can hop between O ions and Co ions, metallic clusters will form.

4. Conclusions:

From these results, we conclude that the doped samples are not traditional Mott insulator but with some properties of charge transfer type, that is, when Sr ions are doped in $LaCoO_3$, the nearby trivalent Co ion becomes tetravalent and leaves an extra hole in e_g orbital. The charge carriers can hop back and forth between trivalent Co ions and tetravalent ones through the O ions bridge, in which way the band gap is closed (12).

References:

1. A. Podlesnyak *et al.*, Spin-State Transition in $LaCoO_3$: Direct Neutron Spectroscopic Evidence of Excited Magnetic States. *Phys. Rev. Lett.* **97**, 247208 (2006).
2. N. Sundaram *et al.*, Local Structure of $La_{1-x}Sr_xCoO_3$ Determined from EXAFS and Neutron Pair Distribution Function Studies. *Phys. Rev. Lett.* **102**, 026401 (2009).

Journal of Physics: Conference Series **430** (2013) 012135 doi:10.1088/1742-6596/430/1/012135

3. L. Siurakshina, B. Paulus, V. Yushankhai, E. Sivachenko, Quantum chemical study of Co^{3+} spin states in $LaCoO_3$. *Eur. Phys. J. B* **74**, 53 (2010).
4. Y. Jiang *et al.*, Study of the local distortions of the perovskite system $La_{1-x}Sr_xCoO_3$ (0 <= x <= 0.35) using the extended x-ray absorption fine structure technique. *Phys. Rev. B* **80**, 144423 (2009).
5. S. K. Pandey *et al.*, Investigation of the spin state of Co in $LaCoO_3$ at room temperature: Ab initio calculations and high-resolution photoemission spectroscopy of single crystals. *Phys. Rev. B* **77**, 045123 (2008).
6. R. F. Klie *et al.*, Direct Measurement of the Low-Temperature Spin-State Transition in $LaCoO_3$. *Phys. Rev. Lett.* **99**, 047203 (2007).
7. D. Louca, J. L. Sarrao, Dynamical Disorder of Spin-Induced Jahn-Teller Orbitals with the Insulator-Metal Transition in Cobaltites. *Phys. Rev. Lett.* **91**, 155501 (2003).
8. J. Wu, C. Leighton, Glassy ferromagnetism and magnetic phase separation in $La_{1-x}Sr_xCoO_3$. *Phys. Rev. B* **67**, 174408 (2003).
9. G. Paolo *et al.*, QUANTUM ESPRESSO: a modular and open-source software project for quantum simulations of materials. *J. Phys.: Condens. Matter* **21**, 395502 (2009).
10. N. Sundaram *et al.*, Local Structure of La1-xSrxCoO3 Determined from EXAFS and Neutron Pair Distribution Function Studies. *Phys. Rev. Lett.* **102**, 026401 (2009).
11. B. Raveau, The crucial role of mixed valence in the magnetoresistance properties of manganites and cobaltites. *Phil. Trans. R. Soc. A* **366**, 83 (January 13, 2008, 2008).
12. Y. Tokura *et al.*, Thermally induced insulator-metal transition in $LaCoO_3$:A view based on the Mott transition. *Phys. Rev. B* **58**, R1699 (1998).

Journal of Physics: Conference Series **430** (2013) 012136 doi:10.1088/1742-6596/430/1/012136

Study on Coloration Mechanism of Chinese Ancient Ceramics by X-ray Absorption Near-edge Structure

Y H Peng†, Z Xie†, J F He, Q H Liu, Z Y Pan, W R Cheng and S Q Wei*

National Synchrotron Radiation Lab, University of Science and Technology of China, Hefei 230029
*Corresponding authors. E-mail: sqwei@ustc.edu.cn
†These authors contributed equally to this work.

Abstract. The Fe *K*-edge X-ray absorption near-edge structure (XANES) spectra of a series of ceramic shards were measured by fluorescence mode to reveal the color-generating techniques of Chinese porcelain. The analysis disclosed relationships among the chemical form of the iron, the firing conditions and the colors of the ceramics. The results indicate that the coloration for different ceramics depend on the valence states of iron as the main color element in glaze and the proportion of Fe^{2+} and Fe^{3+} was attributed to the baking technology. The findings provide important information for archaeologist on the coloration researches.

1. Introduction

The invention of porcelain was China's great contribution to the world civilization [1]. With the increase of production, the prosperity of the government porcelain kilns greatly promoted the porcelain baking technology. To reveal the information about the production technology, numerous researches have been focused on the color-generating elements in ancient ceramics glaze among which chemical structure analyses of the transition elements is most popular [2-5]. Therefore, amounts of advanced scientific techniques are used to investigate the archaeological materials.

Proton induced x-ray emission (PIXE) technique was performed to revealed that the glaze of ancient Ru porcelain has very high iron content [6-7]. X-ray photoelectron spectroscopy (XPS) was applied to the moon-white glaze on ancient Ru ceramics in order to investigate the origins of the colour [8]. Mössbauer spectroscopy and Neutron activation analysis (NAA) reveal that the different coloration on the glaze is originated from the conditions of production processes [9-10]. Also, synchrotron radiation technologies, such as SRXRF and XANES, have been applied in porcelain structure analysis [5,11-12]. However, the production of porcelain involved roughly the raw materials, a delicate refinement procedure, and a developed firing technology, so there is no uniform interpretation on the color-generating mechanism for different typical porcelain. It is a debate problem in the archaeology and ancient ceramics research. To clarify the principle of the coloration for Fe as the color element, four different ancient ceramics were selected as prototypes.

In the present paper, fluorescence x-ray absorption fine structure was used to investigate the Fe chemical form and atomic structure in glaze for four ancient ceramics. With the advantage of the sensitivity of the XANES spectroscopy to the local structure around Fe atoms, in conjunction with linear combination fitting (LCF) method, we aim to reveal the production techniques of different ancient ceramics. Our findings indicate that the coloration mechanism depends on the firing conditions.

2. Experimental

Four classical ancient porcelain shards were studied by XANES. Ru and Cyan porcelain were from Ru kilns in Henan province and Jingdezhen in Jiangxi province, respectively. Xin'an ceramic was produced in Anhui Province and Qionglai sample was from Sichuan Province. They were marked by RC-1, RC-2, RC-3, CC, XA and QL, respectively. Reference samples used as standards for iron in various oxidations states included iron metal, FeO, Fe_2O_3 and Fe_3O_4.

The Fe *K*-edge XANES spectra were recorded in the fluorescence mode at the U7C beamline in the National Synchrotron Radiation Laboratory (NSRL) and at the BL14W1 beamline in the Shanghai Synchrotron Radiation Facility (SSRF). The energy of storage ring of NSRL is 800MeV and the average current is 200 mA, and that of SSRF is 3.5 GeV and 300 mA, respectively. The resolution of double crystal monochromater Si (111) is about 1~2 eV at Fe *K*-edge. The seven-element Ge solid state detector (SSD) calibration and the solid angle determination were performed using a set of appropriate approach. Through all experiments, no damage caused by x-ray could be observed in the samples.

3. Results and discussion

According to the XRF results, the content of iron is the most condensed of transition element among Mn, Cu, Ni and Zn, with the ratio of more than two magnitudes. It is a well known fact that iron is one of most important elements for coloration and the colors of porcelain are closely related to the chemical states of iron [6]. To obtain basic information on various oxidation states of iron, the reference compounds were measured. Figure 1(a) displayed the Fe *K*-edge XANES spectra of metallic iron, Fe_3O_4, Fe_2O_3 and FeO. It is understood that the absorption edge shifts to a higher energy as the valence state of the absorber atom in sample increases [11]. It was found that the absorption edge energy of metallic iron is located at the lowest energy level among samples. The pre-edge peak and the absorption edge energy of magnetite (Fe^{2+}, Fe^{3+}) are located at lower position than those in compounds containing trivalent iron. Compared with metallic iron, iron oxides exhibit a small pre-edge peak on the lower-energy side of the absorption edge, corresponding to transitions from 1s to 3d levels. The pre-edge peak also shifts to a higher energy with an increase in the valence state of the absorber atom [6,11]. Thus, it is found that each iron compound has its own characteristic XANES spectrum.

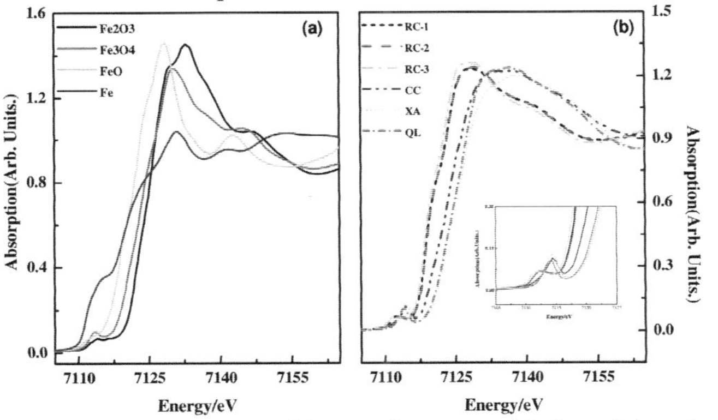

Figure 1 Fe *K*-edge XANES spectra of iron reference oxidations (a) and ancient ceramics samples (b) and the pre-edge peak of reference and samples (inset).

The valence states of transition elements in ancient ceramics maybe influenced by many factors, such as the geological processes, the decoration procedure and the production courses. In order to exclude the interference of a number of factors, we firstly chose the series of RC ceramics to study since they used the same clay and the same decoration techniques. Fe *K*-edge of XANES spectra of three ancient RC ceramics shards were shown in Figure 1(b). It was found that the spectra of ancient RC ceramics are characterized by fine structures with a pre-edge peak at 7112 eV and the absorption at 7121 eV. The oxidation states of iron in these ceramics were determined to be divalent and trivalent, since the energies of the pre-edge peak and the absorption edge are close to magnetite. It is also observed that the intensities of the pre-edge peaks of ancient RC ceramics are weak (see Fig 1(b). inset), which also reflected the low chemical states of iron in ancient RC samples. Therefore, it is expected that the main chemical form of iron is Fe^{2+} and Fe^{3+} in ancient RC ceramics, which is consistent with the previous report [8,12].

It was found that the differences between the XANES spectra measured on three ancient RC samples could not be characterized by their pre-edge peak and absorption edge. But the glaze colors of them are distinct with each other, which are moon-white (light yellow), gray-green and bean-green for RC-1, RC-2 and RC-3, respectively. To reveal the origins of the coloration mechanism, the XANES spectra of three samples were studied by linear combination fitting (LCF) method with fitting results shown in Figure 2. As seen from Fig. 2(a-c), Fe_3O_4 is the main component for all RC samples, in good agreement with the results of the previous report [8]. According to the results, the relationships of the proportions of Fe^{2+} and Fe^{3+} and the main wavelength were summarized in Table I. It was seen that the proportion of Fe^{2+}/Fe^{3+} is greater than 1.0 in all RC samples glaze which reflected more divalent iron in chemical form for ancient RC ceramics. As is well known, the color-generating mechanism is closely related to the valence states of iron. The proportion of divalent and trivalent irons reflected the firing conditions during the production processes. The greater Fe^{2+}/Fe^{3+} revealed that the ancient RC ceramics were produced under reducing conditions. The larger proportion of Fe^{2+}/Fe^{3+} in glaze is, the closer green colors of the ancient RC ceramics are. The smaller proportion of Fe^{2+}/Fe^{3+} in glaze is, the closer yellow colors of RC porcelain are.

Table I Colors, proportion of Fe^{2+} and Fe^{3+} and main wavelengths for the ancient ceramics.

Samples	Color	Proportion Fe^{2+}/Fe^{3+}	Main wavelength
RC-1	Gray - green	1.39	535
RC-2	Bean-green	1.22	560
RC-3	Moon-yellow	1.16	585
CC	Light-Cyan	0.90	598
XA	Dark-Gray	0.42	633
QL	Dark-Brown	0.27	645

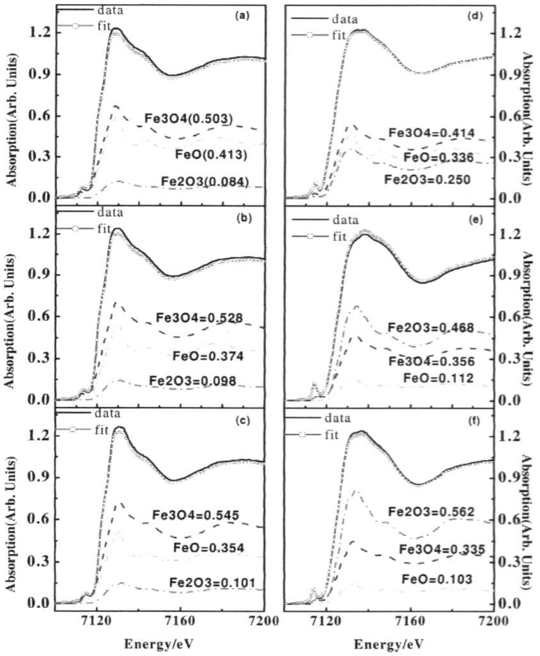

Figure 2 Fe *K*-edge XANES fitting results of ancient samples by linear combination fitting method.

Excepted for the ancient Ru kiln, there are many other kilns using Fe as the coloring agent. To confirm the universality of color-generating mechanism for all porcelain that the Fe is the main color element, other three typical samples were subsumed. Fe *K*-edge XANES spectra of three samples were also shown in Figure 1(b). It was found that the differences between the XANES spectra measured on each ceramics shards could be characterized easily by their absorption edge energy and the pre-edge peak. The absorption edge of CC from Jingdezhen is located at 7124 eV and the pre-edge is at 7114 eV. The pre-edge peak and the absorption edge of ancient XA and QL located at 7114 eV and 7126 eV, respectively. The oxidation states of iron in these ceramics were determined to be entirely trivalent, since the energies of the pre-edge peak and the absorption edge are close to hematite. It is also observed that the intensities of the pre-edge peaks of ancient XA, QL, and CC ceramics are stronger than those of RC samples (see Fig 1b. inset). As a conclusion, the observed increase in the intensity of the pre-edge peak for XA, QL and CC porcelains suggests that these ceramics had been fired under the oxidation conditions since the chemical form of iron is mainly trivalent iron in their glaze.

Ancient CC, XA and QL porcelains are also fitted by the linear combination fitting method and the results were displayed in Fig. 2 (d-f) and Table I. It is instructive to clarify that ancient CC, XA and QL ceramics are also the same coloration principle for different kilns. The main form of iron for CC is magnetite and the proportion of Fe^{2+} and Fe^{3+} close to 1.0. The color of glaze was shown light cyan due to the interaction of divalent and trivalent iron under firing processes. For ancient XA and QL samples, the color element is also iron with the form of Fe_2O_3, so these samples show dark brown. According to discussion above, we can conclude that the color-generating element of ancient RC, CC, XA and QL ceramics is iron and the color-generating mechanism depends on the valence states of transition element which were caused by the firing conditions. However, the relationship of the transition elements and coloration techniques is very complex. Whether the coloration mechanism is related with interaction of all transition elements, there is not a clear interpretation. To reveal the origins of the questions, further research and precise approach are needed.

4. Conclusions

In this report, four typical ancient famous ceramics were studied by using fluorescence XANES. The results revealed that the color-generating mechanism of ancient ceramics is depended on the chemical states of the transition elements (the main color element) in the glaze. It is suggested that the colors of ancient RC samples vary with the proportion of Fe^{2+} and Fe^{3+} that change from cyan through bean green to yellow with the increase of trivalent oxidation state of iron. The ancient XA and QL ceramics show dark brown because of the chemical form of trivalent iron in glaze. The Cyan ceramics is light green due to the interaction of the divalent and trivalent iron.

Acknowledgments

This work was supported by the National Natural Science Foundation of China (Grants No. 11135008, 11105151, 10979044, and 10979041) and the Chinese Academy of Sciences Special Grant for Postgraduate Research, Innovation and Practice. The authors would like to thank NSRL and SSRF for the synchrotron radiation beamtime.

References

[1] Li J Z *History of Science and Technology of China:Ceramics Volume.SciencePress* 1998, *Beijing*
[2] Yu K N and Miao J M 1999 *Appl. Radiat. Isot.* ***51*** 279
[3] Wu J, Leung P L, Li J Z, Stokes M J and Li M T W 2000 *X-Ray Spectrom.* ***29*** 239
[4] Matsunaga M and Nakai I 2004 *Archaeometry.* ***46*** 103
[5] Padovani S, Borgia I, Brunetti B, Sgamellotti A, Giulivi A, D'Acapito F, Mazzoldi P, Sada C and Battaglin G 2004 *Appl. Phys. A* ***79*** 229
[6] Matsunaga M and Nakai I 2004 *Archaeometry.* ***46*** 103
[7] Zhu D, Cheng H S and Lin J W 2006 *Nucl. Instrum. Methods Phys. Res., Sect. B* ***249*** 633
[8] Wang J, Z H, Li W and Li G 1995 *Chin. Ceramic.* ***31*** 27
[9] Zhang B, Z W, Li G, Xie J and Gao Z 1999 *J. Zhengzhou Univer.* ***31*** 1
[10] Chevalier R, Coey J M D and Bouchez R 1976 *J. Phys. (Paris), Colloques* ***37*** 861
[11] Calas G, Manceau A and Petiau J 1998 *in Synchrotron radiation applications in mineralogy and petrology (ed. A. Barto-Kyriakidis)Theophrastus Publications, Athens, Greece. 77-96*
[12] Zhang M, Wang C, Jin P, Wei S, Xu W, Chen D and Wu Z 2008 *Nucl. Tech.* ***31*** 648

AUTHOR INDEX

AUTHOR INDEX

AUTHOR INDEX

AUTHOR INDEX

AUTHOR INDEX

AUTHOR INDEX